PLASTICITY

A Treatise on Finite Deformation of Heterogeneous Inelastic Materials

Providing a basic foundation for advanced graduate study and research in the mechanics of solids, this treatise contains a systematic development of the fundamentals of finite inelastic deformations of heterogeneous materials. The book combines the mathematical rigor of solid mechanics with the physics-based micro-structural understanding of the materials science, to present a coherent picture of finite inelastic deformation of single and polycrystalline metals, over broad ranges of strain rates and temperatures. It also includes a similarly rigorous and experimentally based development of the quasi-static deformation of cohesionless granular materials that support the applied loads through contact friction. Every effort has been made to provide a thorough treatment of the subject, rendering the book accessible to students in solid mechanics and in the mechanics of materials. This is the only book that integrates rigorous mathematical description of finite deformations seamlessly with mechanisms based on micromechanics in order to produce useful results with relevance to practical problems.

OTHER TITLES IN THIS SERIES

All the titles listed below can be obtained from good booksellers or from Cambridge University Press. For a complete series listing visit

http://publishing.cambridge.org/stm/mathematics/cmma

Topographic Effects in Stratified Flows
PETER BAINES

Inverse Methods in Physical Oceanography
ANDREW F. BENNETT

Wave Interactions and Fluid Flows
A.D.D. CRAIK

Theory and Computation in Hydrodynamic Stability
W. CRIMINALE, T.L. JACKSON AND R.D. JOSLIN

Theory of Solidification
STEPHEN H. DAVIS

Dynamic Fracture Mechanics
L.B. FREUND

Finite Plastic Deformation of Crystalline Solids
K.S. HAVNER

Turbulence, Coherent Structures, Dynamical Systems and Symmetry
PHILIP HOLMES, J.L. LUMLEY AND GAL BERKOOZ

Acoustics of Fluid-Structure Interactions
M.S. HOWE

The Dynamics of Fluidized Particles
ROY JACKSON

Benard Cells and Taylor Vortices
E.L. KOSCHMIEDER

Ocean Acoustic Tomography
WALTER MUNK, PETER WORCESTER, AND CARL WUNSCH

Turbulent Combustion
NORBERT PETERS

Colloidal Dispersions
W.B. RUSSELL, D.A. SAVILLE, AND W.R. SCHOWALTER

Vortex Dynamics
P.G. SAFFMAN

The Structure of Turbulent Shear Flow
A.A. TOWNSEND

Buoyancy Effects in Fluids
J.S. TURNER

Reciprocity in Elastodynamics
J.D. ACHENBACH

Theory and Coputation of Hydrodynamic Stability
W.O. CRIMINALE, T.L. JACKSON, AND R.D. JOSLIN

PLASTICITY

A Treatise on Finite Deformation of Heterogeneous Inelastic Materials

S. NEMAT-NASSER
Center of Excellence for Advanced Materials
Department of Mechanical and Aerospace Engineering
University of California
San Diego, USA

CAMBRIDGE
UNIVERSITY PRESS

CAMBRIDGE UNIVERSITY PRESS
Cambridge, New York, Melbourne, Madrid, Cape Town, Singapore, São Paulo, Delhi

Cambridge University Press
The Edinburgh Building, Cambridge CB2 8RU, UK

Published in the United States of America by Cambridge University Press, New York

www.cambridge.org
Information on this title: www.cambridge.org/9780521108065

First published 2004
This digitally printed version 2009

A catalogue record for this publication is available from the British Library

ISBN 978-0-521-83979-2 hardback
ISBN 978-0-521-10806-5 paperback

CONTENTS

PREFACE

This book is a treatise on the mathematical and experimental foundations of the finite deformation of crystalline metals and granular materials. It combines the mathematical rigor of solid mechanics with the physics-based microstructural understanding of the materials science, to present a coherent picture of finite inelastic deformation of single- and poly-crystalline metals, over broad ranges of strain rates and temperatures. It also includes a similarly rigorous and experimentally-based development of the quasi-static deformation of cohesionless granular materials that support the applied loads through contact friction.

The book has grown out of my lecture notes for graduate-level courses on the fundamentals of continuum plasticity and the mechanics of large deformations. These courses were initiated at the University of California, San Diego (UCSD), in the late 60's, and were then continued at Northwestern University in the 70's and early 80's. During this period at Northwestern University, it became clear to me that true advances in the basic understanding of the mechanics of materials, and particularly the inelastic deformation of metals and geomaterials, can be achieved only by moving beyond the traditional phenomenological approach to plasticity models that are based on the classical concepts of rate-independent yielding and empirical data fitting, and to exploring the basic micromechanisms of the phenomena through direct observations, coordinated with indirect systematic experimentation and micromechanical modeling. This then necessitates integrating the traditional mathematical and computational modeling of the solid mechanics community, with extensive macroscopic and microstructural experimental characterization of real materials, in the spirit of the materials science community, demanding novel approaches to both teaching and research in this area, and requiring, at a minimum, new experimental techniques and facilities to investigate the underpinning micromechanisms of deformation.

Hence, in the mid 80's, I spearheaded the creation of a unified mechanics and materials program at UCSD, and sought to develop there a state-of-the-art materials characterization laboratory that now includes many novel testing facilities and techniques. At the same time, an integrated materials science program was launched at UCSD, that built on the contributions of researchers from the physical, natural, and basic engineering and oceanographic sciences, who had interest in materials science and technology. Serving as the founding director of this program, I taught several courses that combined mechanics and materials, and directed a Ph.D. research program that included many outstanding graduate students eager to learn the fundamentals of both solid mechanics and materials, and who were willing to undertake both experimental and theoretical research in the mechanics of materials. It is in this integrated environment that the present version of this book has now emerged.

In addition to my own work, the book liberally draws (with commensurate citation) from my collaborative research with former graduate students, post-doctoral associates, visiting scientists, and other coworkers. Many of these students started their graduate education knowing little about mechanics and materials, but ended their study making significant new experimental and theoretical/computational contributions that have helped to improve my knowledge of the subject. They all have studied at least some sections of one of the several early versions of this book, and have made contributions that have served to improve the work. I am most grateful to them.

There are several former students and coworkers whose work and contributions have been pivotal to a number of topics covered in this book. The idea of writing a book on the micromechanics of large deformations started with my collaboration with Dr. Monte Mehrabadi in the late 70's at Northwestern University, where a draft of a chapter on kinematics was started but was not fully completed. This work was later published,[1] and its results together with further extensions, are included in Section 1.5 of Chapter 1 and in various sections of Chapter 2. In the late 70's, I was most fortunate to work with Professor Jes Christoffersen and Dr. Mehrabadi to seek to relate the continuum concepts of stress and strain to the micromechanics of contact forces and sliding and rolling of the grains within a granular mass. This effort was indeed successful, leading to significant results[2] that are summarized in Section 7.2.

Regarding the content of Chapter 7, it was Professor M. Oda who worked with Mr. John Schmidt (the technician working in my laboratory) and me to produce the biaxial testing system shown in Figure 7.4.2, page 547, and together with Dr. Konishi and Mr. Zong-Lian Qiu, to perform and analyze many of the tests on biaxial deformation of photoelastic granules. These and related contributions are liberally used and referenced in Chapter 7, including some theoretical work with both Mehrabadi and Oda.[3]

Another significant contributor to my work, was my former student and post-doctoral fellow, Dr. B. Balendran, who worked on both computational algorithms and micromechanics of granular materials, and helped with many aspects of the related materials in Chapters 5 and 7, as well as some of the basic mathematical results presented in Chapters 1 and 2.[4]

---

[1] Mehrabadi, M. M. and Nemat-Nasser, S. (1987), Some basic kinematical relations for finite deformations of continua, *Mech. Mat.*, Vol. 6, 127-138.

[2] Christoffersen, J., Mehrabadi, M. M., and Nemat-Nasser, S. (1981), A micromechanical description of granular material behavior, *J. Appl. Mech.*, Vol. 48, 339-344. [This work was completed in the fall of 1979, and was subsequently distributed as a report dated January, 1980. I presented the results at the 1980 Congress of Theoretical and Applied Mechanics, in Toronto, Canada.]

[3] Mehrabadi, M. M., Nemat-Nasser, S., and Oda, M. (1982), On Statistical description of stress and fabric in granular materials, *Int. J. Num. Anal. Meth. Geomech.*, Vol. 6, 95-108.

Oda, M., Nemat-Nasser, S., and Mehrabadi, M. M. (1982), A statistical study of fabric in a random assembly of spherical granules, *Int. J. Num. Anal. Methods Geomech.*, Vol. 6, 77-94.

Oda, M., Konishi, J., and Nemat-Nasser, S. (1982), Experimental micromechanical evaluation of strength of granular materials: effects of particle rolling, *Mech. Mat.*, Vol. 1, No. 4, 269-283.

[4] Balendran, B. and Nemat-Nasser, S. (1993), Double sliding model of cyclic deformation of granular materials, including dilatancy effects, *J. Mech. Phys. Solids*, Vol. 41, No. 3, 573-612.

Others with intellectual contributions that have affected the contents of this book, include my former graduate students Professors T. Iwakuma, M. Obata, G. Subhash, M. Zikry, M. Rashid, and Y-F. Li, and Drs. A. Thakur, N. Okada, T. Okinaka, and R. Kapoor. I am also grateful to Professor Benjamin Loret for proofreading Chapters 4 and 7, and to Mr. L. Ni and Dr. J. Zhang who have read and commented on the contents of Chapter 5. Mr. Ni's collaborative work with me on constitutive algorithms, is included in this chapter.[5] I am especially thankful to Dr. M. Scheidler of Army Research Laboratory, who read and commented on the contents of Chapters 1, 2, 3, and 4,[6] which resulted in much improvement.

The manuscript was proofread in its various versions, several times, by my wife Éva, resulting in important consistency and grammatical improvements. I, however, take full responsibility for any errors that remain. In addition, my daughter Shiba and my assistant Lauri Jacobs helped with various aspects of the word processing, and Dr. Masoud Beizaie assisted in preparing some of the figures and helped with several other tasks. I am most grateful for all these contributions. Except for Chapter 9, I have formatted the book using *ditroff*. Most of the figures and the graphs are constructed by *pic* and *grap*. In this connection, I wish to thank Frank Dwyer who helped to move old ditroff editing and formatting tools from an old Sun system to a new one, which allowed me to continue to use many macros that were constructed particularly to tailor the figures and equations to my taste. Chapter 9 was formatted using *grof*, and I wish to thank my graduate student Mr. Sai Sarva who helped to solve a number of formatting problems, thereby rendering this chapter consistent with the rest of the book.

Sia Nemat-Nasser, La Jolla, California

June, 2004

Balendran, B. and Nemat-Nasser, S. (1995), Integration of inelastic constitutive equations for constant velocity gradient with large rotation, *Appl. Math. Comput.*, Vol. 67, No. 1-3, 161-195.

Balendran, B. and Nemat-Nasser, S. (1996), Derivative of a function of a nonsymmetric second-order tensor, *Q. Appl. Math.*, Vol. LIV, No. 3, 583-600.

[5] Nemat-Nasser, S. and Ni, L. (1994), Effective constitutive algorithms in elastoplasticity and elastoviscoplasticity, *European J. Appl. Math.*, Vol. 5, Part 3, 313-336.

[6] Dr. Scheidler's comments on the material contained in the original version of Section 4.9, led to a complete rewriting of this section which now includes new results on elastic anisotropy in continuum metal plasticity, and clarification of the notion of plastic spin.

PRÉCIS

This book contains a systematic development of the fundamentals of finite inelastic deformations of heterogeneous materials. It is a treatise aimed to provide a basic foundation for advanced graduate study and research in the mechanics of materials, including single crystal and polycrystal metals, and granular materials. A reader with minimal exposure to continuum mechanics and some vector and tensor calculus should be able to master the mathematical necessities that are covered in the first two chapters. An effort is made to provide sufficient detail in order to render the book accessible to students in analytical, computational, and experimental mechanics and the mechanics of materials.

To guide the reader, each chapter is preceded by a brief description of its contents. There are nine chapters, each divided into several sections and subsections. Many sections also begin with a brief description of their contents. Each chapter ends with a list of cited references.

As a foundation, the geometrical, kinematical, and dynamical ingredients are treated in Chapters 1, 2, and 3. For the most part, coordinate-independent vector and tensor notation is used. This however, is augmented by frequent component representation of various expressions in indicial notation, rendering the book accessible to a broader audience. The continuum theories of the rate-independent and rate-dependent deformation of metals and geomaterials (granular materials and rocks) are developed in Chapter 4, where, based on a general framework, many specific cases are detailed and explicit equations useful for computational simulations are given. Included in this chapter (Section 4.8) are detailed presentations of dislocation-based rate- and temperature-dependent models of metals, together with experimentally-obtained values of the corresponding constitutive parameters. Chapter 5 is devoted to techniques for the integration of continuum constitutive equations. Both rate-independent and rate-dependent deformations, including the effects of thermal softening, friction, and dilatancy, are considered. Included are forward-gradient integration techniques, as well as a more efficient technique based on a plastic-predictor/elastic-corrector method, recently developed by the author and his coworkers. Each computational method is described, computational steps are listed, and illustrative examples are provided, leading to a comparative evaluation of various methods. Chapter 6 contains the fundamentals of finite elastoplastic deformation of single crystals, from a micromechanical point of view, starting with a review of the crystal systems and certain elementary topics in the theory of dislocations. Physically-based constitutive relations for single crystals are formulated in this chapter on the basis of slip-induced plastic deformation and the accompanying elastic lattice distortion. The notions of self- and latent-hardening are critically examined, and slip models which directly account for both the temperature- and strain-rate effects, are presented and applied to predict the polycrystal flow stress of both bcc (commercially pure tantalum) and fcc

(OFHC copper) metals over a broad range of strains, strain rates, and temperatures. Chapter 7 covers the micromechanics of finite elastoplastic deformation of densely-packed granular materials. Here, physically-based constitutive relations are developed for particulate materials that carry the applied loads through frictional contacts, based on the slip- and rolling-induced (anisotropic) inelastic deformation, accompanied by shear-induced and pressure-dependent inelastic volumetric changes. In Chapter 8 the mathematical foundation of the transition from micro to macro variables is laid out. Exact results on averaging techniques, valid at finite deformations and rotations, are developed, giving explicit equations for the calculation of the generalized Eshelby tensor and its conjugate, within a nonlinear finite-deformation setting. Aggregate properties and averaging *models* are presented in this chapter, including the Taylor, the self-consistent, and the double-inclusion models. Special advanced experimental methods are reviewed in Chapter 9, and some typical experimental results on large strain, high strain-rate deformation of several metals are given. Included also are experimental results on the deformation and shearbanding of cohesionless frictional granular materials. What follows is a more detailed description of the contents of each chapter.

Chapter 1 includes a treatment of second-order tensors, tensor equations, and a class of isotropic functions of second-order tensors. Both spectral (Sections 1.2 and 1.3) and coordinate-independent (Section 1.5) representations are given. A number of important identities (Section 1.4) are developed, which are then used throughout the book. Tensors and a class of tensor-valued functions with distinct real or complex eigenvalues, as well as cases with repeated eigenvalues, are examined in detail, providing explicit equations which are then used throughout the book. The time derivative of tensor-valued isotropic functions of a time-dependent, second-order tensor, is considered in Section 1.6. Again, cases with distinct and repeated eigenvalues are studied in detail. These results provide powerful tools for computational algorithms, and this is discussed and illustrated, in Chapter 5.

Chapter 2 is devoted to the kinematics of finite deformations and rotations. After a brief review of the basic elements of the kinematics of deformation and its description in Lagrangian and Eulerian triads (Section 2.1), and polar decomposition (Section 2.2), a comprehensive account of various strain measures (Section 2.3), their rates and the associated spin tensors (Sections 2.4 and 2.5), is provided. Various measures of the material spin are examined in Section 2.6. In Sections 2.7 and 2.8, the relations between the deformation-rate and the stretch tensors are examined, and in Section 2.9, relations among various spin tensors are outlined. Section 2.10 is devoted to an examination of various strain rates and the manner by which they relate to one another. Both spectral representation and coordinate-independent expressions are given. Relations among strain measures, strain-rate measures, and spin tensors are detailed.

A discussion of the stress and stress-rate measures, and the balance relations are contained in Chapter 3. Various stress measures, conjugate to the strain measures developed in Chapter 2, are worked out in detail and the relations among them are established (Sections 3.1 and 3.2). Then, stress-rate measures are developed and general relations connecting the objective stress rates are presented (Sections 3.4 and 3.5). Balance relations and boundary-value problems are briefly formulated (Section 3.6), together with the principle of virtual work (Section 3.7), relevant to nonlinear finite-deformation problems. This chapter also includes a weak form of the equations of motion, for application to a finite-element formulation of finite-deformation problems (Section 3.8). Explicit finite-element equations are developed and several important issues are clarified. Other topics discussed in this chapter are the kinematics, dynamics, and balance relations at surfaces of discontinuity (Section 3.9). Finally, the chapter includes a comprehensive account of general variational principles for finite deformation of hyperelastic materials, as well as the corresponding incremental and linearized formulations (Section 3.10).

The continuum theories of elastoplasticity are developed in Chapter 4, beginning with a discussion of the thermodynamics of inelasticity (Section 4.1), elastic and inelastic potentials, and the normality rule. Then, rate-independent theories of plasticity, with both smooth yield surfaces (Section 4.2) and yield surfaces with corners (vertex models, Section 4.4) are presented in a general setting, in terms of an arbitrary strain measure and its conjugate stress measure. Based on this, several commonly used plasticity models with isotropic, kinematic, and combined isotropic-kinematic hardening, are developed, and explicit equations useful for computational implementation, are given in various subsections of Section 4.2. The questions of dilatancy, pressure-sensitivity, friction, and their constitutive modeling are examined in some detail (Section 4.3), providing specific illustrative examples and comparisons with experimental results. The often-used deformation theories of plasticity are also examined, starting from nonlinear elasticity and systematically leading to the vertex model of elastoplasticity (Section 4.5). Another topic covered in this chapter is the physical basis of the noncoaxiality of the plastic strain rate and the deviatoric stress (Section 4.6). As is shown, this property is an integral part of the response of all frictional materials (Section 4.7). In this sense, the non-Schmid effects for crystal plasticity and pressure-sensitivity and frictional effects in the deformation of granular materials, are formulated in a unified manner, seeking to make the vast literature available in these two, seemingly unrelated, scientific inquiries, equally accessible to students and researchers interested in metal plasticity and/or the inelastic deformation of geomaterials. The chapter includes both rate-dependent and rate-independent theories. The rate- and temperature-dependent models are examined in some detail in Section 4.8. First, several empirical models are discussed. Then, based on the mechanisms of plastic deformation by the motion of dislocations, physically-based models for the inelastic deformation of bcc and fcc metals are developed and used to predict the flow stress of commercially pure tantalum, molybdenum, niobium, vanadium, several steels and titanium alloys, and OFHC copper, over a broad range of strains, strain rates, and temperatures, arriving at excellent correlation with experimental results with few free constitutive parameters. These parameters are then tabulated. This chapter also includes a detailed development (Section 4.9) of general continuum anisotropic elastoplasticity in terms of the decomposition of the deformation gradient into an elastic and a nonelastic contribution. It is shown that such a decomposition can always be reformulated and reduced to a (unique) purely inelastic and a (unique) purely elastic constituent that involves no rigid-body rotation, and an accompanying (unique) rigid-body rotation. Objective spin tensors associated with elastic and inelastic strain rates are identified and formulae are given which explicitly express these spins as linear and homogeneous functions of the corresponding deformation rates. Included in this section is the examination of the requirements of the material frame indifference, the general elastic response in terms of an elastic potential, and a detailed description of how various rate quantities can be calculated explicitly, once the elastic potential and the constitutive relations for the plastic strain-rate tensor are given. Several, commonly misunderstood issues in finite inelastic deformations of continua, are critically examined and correct results are given, emphasizing some of the inherent difficulties with the treatment of anisotropic elastoplasticity in terms of the decomposition of the total deformation gradient into elastic and nonelastic parts. Small elastic deformations and very small elastic strains are also considered as special cases. Also discussed are the notion of the backstress and its objective time rate of change, as well as an alternative additive decomposition of the total deformation gradient.

Chapter 5 addresses the question of the integration of continuum constitutive equations, which plays a central role in computational modeling and numerical solutions of finite-deformation problems. The aim is to provide a solid footing for the development of constitutive algorithms which have the necessary accuracy, stability, and efficiency, for implementation into computational codes. After some brief historical comments in Section 5.1, the consequences of kinematical assumptions, such as the incrementally constant velocity gradient and the unidirectional stretch, are examined in detail and the exact coordinate-independent relation between the velocity gradient and the incremental deformation gradient is formulated (Section 5.2). For illustration of various constitutive algorithms, the J_2-flow theories of Chapter 4 are used, including both the rate-dependent and rate-independent cases, as well as the effects of thermal softening (Section 5.3). In Section 5.4, the mathematical basis of the recently proposed plastic-predictor/elastic-corrector method is outlined, and error estimates are given. Section 5.5 is devoted to a detailed account of the application of the singular perturbation method to solve the stiff constitutive equations which govern the variation of the stress magnitude. This section is closed by presenting a modified outer solution which is explicit, simple, accurate, and can be used directly in large-scale computations. Section 5.6 provides a detailed account of computational algorithms for proportional loading and unloading. In the algorithm, the deformation-rate tensor is assumed to remain codirectional with the stress-difference tensor which is defined as the deviatoric part of the Kirchhoff stress less the (deviatoric) backstress. Various integration methods are discussed, illustrated, and compared, for both the rate-dependent and rate-independent flow stress, considering loading as well as unloading. The computational steps are outlined in tables, through specific numerical examples. Computational algorithms based on the assumption of unidirectional stretch are examined in Section 5.7, and those based on the assumption of constant velocity gradient are detailed in Section 5.8. The results are compared for both rate-dependent and rate-independent cases, and their relative merits are discussed. The relation with the classical radial-return method is studied and generalized radial return techniques for application to the cases with large rotations, kinematic hardening, and noncoaxiality of the strain rate and stress, as well as cases with dilatancy and pressure-sensitivity, are formulated. For large deformations and the elastic-perfectly-plastic model, exact integration formulae are presented and used to develop accurate and efficient algorithms, when the model includes workhardening and thermal softening, as well as noncoaxiality and other effects. Several useful tensorial identities are worked out in Appendices 5.A and 5.B, providing explicit exact results that can be directly implemented in large-scale numerical codes.

The elastoplastic deformation of single crystals is examined in Chapter 6. The chapter includes (Section 6.1) brief discussions of crystal structure, crystal plasticity based on dislocations and slip, and various topics in the dislocation theory, necessary for a fundamental understanding of the micromechanics of crystal deformation, including the characterization of dislocations, their elastic interactions which underlie certain aspects of strain hardening in metals, and other relevant issues. Section 6.1 is concluded by developing an explicit expression for a slip rate in terms of the density of the associated mobile dislocations, their average velocity, and the magnitude of the Burgers vector. The kinematics of the finite deformation of single crystals is addressed in Section 6.2. Various decompositions of the deformation gradient and its rate are discussed, providing detailed accounts of alternative representations of the deformation and spin tensors corresponding to various possible reference states. The elasticity of crystals is examined in Section 6.3, where explicit general rate-constitutive relations are presented, again using possible alternative reference states. Various objective stress rates are considered, the associated constitutive relations are produced, and their equivalence is established. The notions of self- and latent-hardening are critically examined in Section 6.4. The slip models which directly account for both the temperature- and strain-rate effects, are presented in Section 6.5. The short- and long-range barriers that the dislocations must overcome in their motion, are identified. The hardening issue of the rate-independent models in light of the short- and the long-range resistance to the motion of the dislocations is reexamined. Explicit results for bcc and fcc crystals are produced, taking into account the temperature, strain rate, and the long-range hardening effects, and the results are illustrated using the experimental data of commercially pure tantalum and OFHC copper. For these crystal structures, general constitutive algorithms for crystal plasticity calculations are also presented, and illustrative examples are given. As in Chapter 4, Section 4.8, dislocation-based crystal plasticity naturally involves several length scales relating to the dislocation densities and various microstructural characteristics of the material. The developed constitutive relations for crystalline slip, explicitly include three such length scales, directly related to the metal's microstructure. These results are new, and unrelated to the continuum gradient plasticity theories that are also briefly examined at the end of Section 6.5, together with couple-stress theories, size- versus length-scale effects, and the classical continuously distributed dislocation theory. Several issues associated with the microscopic (*i.e.*, at the dislocation and grain lengths) and continuum strain gradients are critically reviewed, and an attractive dislocation-based gradient model is produced that is valid at large deformations and reflects the spatial variations in the dislocation activities corresponding to the crystal slip. In this section, it is emphasized that for a sample with only a few interacting crystals, geometric and textural incompatibilities must be explicitly addressed, as they may profoundly affect the overall sample response.

The fundamentals of finite elastoplastic deformation of densely-packed granular materials, are discussed in Chapter 7 from a microscopic point of view. In Section 7.1, an overview of the micromechanics of the flow of frictional granules is given, illustrating the controlling influence of the induced anisotropy or fabric on the accompanying dilatancy or densification. Stress measures are formulated in terms of the contact forces and the granular microstructure in Section 7.2, where the basic concepts associated with fabric anisotropy and fabric measures, relations between the stress and fabric measures, and, finally, the connection between the back stress and the induced fabric anisotropy are discussed in some detail. A number of fabric measures are then presented for densely-packed rigid spheres in Section 7.3. Experimental results on rod-like photoelastic granules, subjected to cyclic shearing, as well as to monotonically applied biaxial compression, are then examined in Section 7.4, and various fabric tensors are correlated with the corresponding stress and strain. The remaining part of this chapter is devoted to the kinematics and constitutive modeling of frictional granules. A unified constitutive formulation at the meso-scale is presented in Section 7.5, and some of its implications are discussed, including the close relation between the double-slip theory of single crystals and that of granular materials, and the generalization to three-dimensional constitutive relations. The resulting constitutive relations account for pressure sensitivity, friction, dilatancy (densification), and, most importantly, the fabric (anisotropy) and its evolution in the course of deformation. The presented theory fully integrates the micromechanics of frictional granular assemblies at the micro- (grains), meso- (large collections of grains associated with sliding planes), and macro- (continuum) scales. Illustrative examples are given with comparison with experimental results.

Chapter 8 is devoted to the development of the exact mathematical tools necessary for obtaining average quantities and the homogenization of heterogeneous elastoplastic materials at finite strains and rotations. The choice of deformation and stress measures and their rates is examined and it is shown that the deformation gradient and the nominal stress, as well as their rates, are particularly useful quantities, with a number of consequential averaging properties. A set of general identities valid at arbitrary strains and rotations, is provided in Section 8.1, and the resulting implications are discussed for uniform boundary tractions and traction rates, and linear boundary displacements and displacement rates. These identities are valid for both rate-independent and rate-dependent materials, whereas only the rate-independent materials are considered in Sections 8.2 to 8.6 and most of Section 8.7. In Section 8.2, it is shown how a homogenization formalism of linear elasticity, in terms of Eshelby's tensor and its conjugate, can be modified and applied to general finite-strain and finite-rotation problems, with exact, calculable results, provided that the rate of the deformation gradient and the nominal stress rate are used as the kinematical and dynamical variables. Homogenization with the aid of an eigenvelocity gradient and an eigenstress rate, is formulated, the consistency conditions are established, and concentration tensors for ellipsoidal inhomogeneities are given explicitly. Green's functions for the rate quantities are discussed in Section 8.3, and formulae for the calculation of the generalized Eshelby tensor and its conjugate are presented in a fully nonlinear finite-deformation setting, based on suitable rate kinematical and dynamical variables. The double- and multiple-inclusion problems are discussed in Section 8.4. The connections among strong ellipticity, inception of localized deformations, and Green's function are discussed in Section 8.5. Appendix A of this chapter gives explicit formulae for the calculation of the Eshelby tensor in two dimensions. Selected homogenization models are discussed in Section 8.6 for general composites, and in Section 8.7 for polycrystals, where the application of the exact results to the problem of estimating the overall mechanical response of a finitely deformed heterogeneous representative volume element (RVE) is outlined and the overall effective pseudo-modulus tensor (or pseudo-compliance tensor) of the RVE is calculated for rate-independent elastoplastic materials. Included in this section are comments on the Sack's and Taylor's models, as well as the self-consistent and double-inclusion models.

A discussion of two classes of special experimental methods and some related experimental results are included in Chapter 9. The first class is for the characterization of the response of metals over a broad range of strains, strain rates, and temperatures, and the second is for the characterization of quasi-static deformation of frictional granules, supplementing the experimental techniques and results on the biaxial loading and simple shearing of photoelastic granules, presented in Section 7.4. After a brief historical account of the origin of the Hopkinson technique, the classical Kolsky bar method is examined in Section 9.1. Then, in Sections 9.2 and 9.3, some recent novel Hopkinson techniques that allow for recovery experiments at various strain rates and temperatures, are discussed in detail. By these techniques it is possible to obtain the isothermal, as well as adiabatic, flow stress of many metals at high strains over a broad range of temperatures and strain rates. This is discussed and illustrated in terms of the experimental data that are given in Sections 4.8 and 6.5 of Chapters 4 and 6, for copper, tantalum, molybdenum, niobium, vanadium, titanium, and steels. Included in Section 9.3 are techniques for direct and indirect measurement of the fraction of work that is used to generate heat within the material, at high strain rates and for large plastic deformations. Section 9.4 details a new triaxial Hopkinson technique that allows simultaneously subjecting a sample to axial and lateral pressures at various strain rates. Finally, the chapter includes a detailed discussion of the triaxial deformation of cohesionless saturated granular materials, and X-ray techniques recently developed to monitor the deformation and the resulting shearbands in cyclic shearing under confining pressure, for this class of materials. Experimental results on densification of drained samples and liquefaction of saturated undrained samples are also presented in Section 9.5, and in Section 9.6, the X-ray techniques are discussed and illustrative results are given.

1

GEOMETRY

This chapter includes some basic topics in three-dimensional tensors and tensor-valued functions. Second-order tensors and certain arbitrary functions of second-order tensors are considered. These are basic in the description of the kinematics of finite deformation. In addition, time-derivatives of time-dependent second-order tensors and their tensor-valued functions are developed in coordinate-independent form. For clarity in presentation, components with respect to a fixed rectangular Cartesian coordinate system are also considered. While the derivation of the results utilizes eigenvectors and their reciprocals, the final expressions are given in direct coordinate-independent form.

More specifically, the notation is introduced in Section 1.1, and the spectral representation of second-order tensors is reviewed in Section 1.2, where formulae for calculating the eigenvalues of second-order tensors are summarized. In Section 1.3, oblique coordinates are considered, reciprocal base vectors are introduced, and spectral representations of second-order tensors and their integral powers are summarized. In these sections, cases with distinct and repeated eigenvalues are examined separately. Both symmetric and nonsymmetric real-valued tensors are examined, including cases with complex-valued eigenvalues and eigenvectors. Section 1.4 provides a set of identities for second-order tensors and gives the solutions to some useful linear tensor equations. These results provide tools for later application throughout the book. Certain isotropic tensor-valued functions of a second-order tensor are examined in Section 1.5, including a detailed study of cases where the tensor-valued argument may admit distinct or repeated eigenvalues. When the tensor argument depends on a scalar variable, say, time, its derivative and the time derivative of its isotropic functions are considered in Section 1.6, providing explicit coordinate-independent expressions for these derivatives. The material in this chapter serves as a background for the topics that are covered later on.

A reader new to the field of continuum mechanics may wish to read Sections 1.1 and 1.2 with care and then move to Chapter 2, using the remaining sections of this chapter as reference material.

1.1. NOTATION

For simplicity, a background fixed rectangular Cartesian coordinate system is used; see Figure 1.1.1. The unit coordinate base vectors are denoted by $\mathbf{e}_1$, $\mathbf{e}_2$, and $\mathbf{e}_3$. Both subscript and direct notation are used throughout. Depending on the occasion, the unit coordinate triad, $(\mathbf{e}_1, \mathbf{e}_2, \mathbf{e}_3)$, is collectively denoted by $\mathbf{e}_i$ or $\mathbf{e}_A$ or $\mathbf{e}_a$, i, A, a = 1, 2, 3. Vectors are generally designated by bold-face letters, such as **a**, **b**, which have the representation $\mathbf{a} = a_i\,\mathbf{e}_i$, $\mathbf{b} = b_j\,\mathbf{e}_j$ with respect to the $\mathbf{e}_i$-triad, where $a_i = \mathbf{a} \cdot \mathbf{e}_i$ and $b_j = \mathbf{b} \cdot \mathbf{e}_j$. Here and throughout, the scalar product of two vectors is denoted by a dot between them, and the summation convention on repeated indices is used. For example, the scalar product of real-valued vectors **a** and **b** is[1]

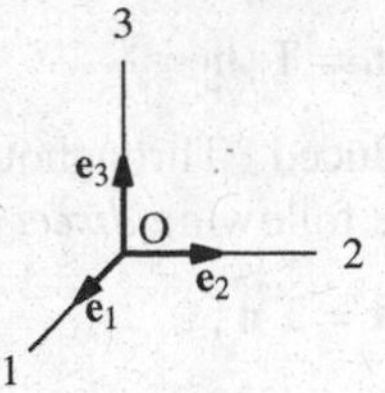

Figure 1.1.1
Rectangular Cartesian coordinates with corresponding unit base vectors

$$\mathbf{a} \cdot \mathbf{b} = (a_i\,\mathbf{e}_i) \cdot (b_j\,\mathbf{e}_j) = a_i\,b_j\,\mathbf{e}_i \cdot \mathbf{e}_j = a_i\,b_j\,\delta_{ij} = a_i\,b_i$$
$$= a_1\,b_1 + a_2\,b_2 + a_3\,b_3\,, \tag{1.1.1a}$$

where the *Kronecker delta* is defined by

$$\mathbf{e}_i \cdot \mathbf{e}_j = \delta_{ij} = \begin{cases} 1 \ \text{when} \ i = j \\ 0 \ \text{when} \ i \neq j \end{cases}. \tag{1.1.1b}$$

If the magnitude of **a** is a, and that of **b** is b, then $\mathbf{a} \cdot \mathbf{b} = a_i\,b_i = a\,b\cos\theta$, where θ is the angle between the two vectors.

The cross product of **a** and **b** is a vector **c** normal to their plane, forming a right-handed triad with them, and having magnitude $a\,b\sin\theta$. The cross product of the base vectors $\mathbf{e}_i$ and $\mathbf{e}_j$ is

$$\mathbf{e}_i \times \mathbf{e}_j = e_{ijk}\,\mathbf{e}_k\,, \quad i, j, k = 1, 2, 3\,, \tag{1.1.2a}$$

where the permutation symbol e_{ijk} equals (+ 1, − 1, 0) depending on whether ijk forms (even, odd, no) permutation of 1 2 3. Then

$$\mathbf{a} \times \mathbf{b} = a_i\,b_j\,e_{ijk}\,\mathbf{e}_k\,. \tag{1.1.2b}$$

An ordered pair of vectors, written as $\mathbf{a} \otimes \mathbf{b}$, forms a second-order tensor with the following coordinate representation:

$$\mathbf{a} \otimes \mathbf{b} = a_i\,b_j\,\mathbf{e}_i \otimes \mathbf{e}_j\,. \tag{1.1.3}$$

---

[1] Mostly, vectorial and tensorial quantities are real-valued. However, complex-valued quantities are also considered when necessary, *e.g.*, complex-valued eigenvectors of real-valued second-order tensors, as in Subsection 1.3.7.

1.1.1. Contraction

When a general second-order tensor $\mathbf{T}$ with components T_{ij} *operates* on a vector $\mathbf{n}$ with components n_i, a vector, say, $\mathbf{t}$, with components

$$t_i = T_{ij}\, n_j \tag{1.1.4a}$$

is produced. Throughout, *matrix operation rules are used,* so that (1.1.4a) also has the following *direct* (coordinate-independent) representation:

$$\mathbf{t} = \mathbf{T}\,\mathbf{n}\,; \tag{1.1.4b}$$

a second-order tensor followed by a vector implies *contraction*. The components of $\mathbf{T}$ are $T_{ij} = \mathbf{e}_i \bullet \mathbf{T}\,\mathbf{e}_j$. The notation $\mathbf{n}\,\mathbf{T}$ is also used for $\mathbf{n} \bullet \mathbf{T}$ which is a vector with components $n_i\, T_{ij}$. If $\mathbf{T} = \mathbf{T}(\mathbf{x})$ is a tensor-valued, differentiable function, its divergence is written as

$$\boldsymbol{\nabla}\mathbf{T} \equiv \boldsymbol{\nabla} \bullet \mathbf{T} = \left[\mathbf{e}_i \frac{\partial}{\partial x_i}\right] \bullet (T_{jk}\, \mathbf{e}_j \otimes \mathbf{e}_k) = (\partial T_{ij}/\partial x_i)\, \mathbf{e}_j\,. \tag{1.1.5a}$$

The tensorial operation of $\boldsymbol{\nabla}$ on $\mathbf{T}$ is represented by

$$\boldsymbol{\nabla} \otimes \mathbf{T} \equiv (\partial T_{ij}/\partial x_k)\, \mathbf{e}_k \otimes \mathbf{e}_i \otimes \mathbf{e}_j\,. \tag{1.1.5b}$$

If $\mathbf{S}$ and $\mathbf{E}$ are two second-order tensors with respective components S_{ij} and E_{ij}, then the single and double contraction of their products, respectively, is a second-order tensor with components $S_{ij}\, E_{jk}$ and a scalar $S_{ij}\, E_{ji}$, as follows:

$$\begin{aligned} \mathbf{S}\,\mathbf{E} &\equiv \mathbf{S} \bullet \mathbf{E} = S_{ij}\, E_{jk}\, \mathbf{e}_i \otimes \mathbf{e}_k \\ \mathrm{tr}\,(\mathbf{S}\,\mathbf{E}) &\equiv \mathbf{S} : \mathbf{E} = S_{ij}\, E_{ji}\,. \end{aligned} \tag{1.1.6a,b}$$

Similarly, if $\boldsymbol{\mathcal{L}}$ is a fourth-order tensor with components $\mathcal{L}_{ijkl}$, then

$$\boldsymbol{\mathcal{L}}\,\mathbf{E} \equiv \boldsymbol{\mathcal{L}} : \mathbf{E} = \mathcal{L}_{ijlk}\, E_{kl}\, \mathbf{e}_i \otimes \mathbf{e}_j\,. \tag{1.1.7}$$

The reader should note the order of contractions in these expressions, since alternative conventions are used elsewhere in the literature.[2] The transpose of a tensor is designated by superscript T. For example, $\mathbf{E}^T = E_{ji}\, \mathbf{e}_i \otimes \mathbf{e}_j$ has components E_{ji}, when the components of $\mathbf{E} = E_{ij}\, \mathbf{e}_i \otimes \mathbf{e}_j$ are E_{ij}.

1.1.2. Scalar and Dyadic Products of Complex-valued Vectors

Consider two complex-valued vectors $\mathbf{z}_1 = \mathbf{x}_1 + i\,\mathbf{y}_1$ and $\mathbf{z}_2 = \mathbf{x}_2 + i\,\mathbf{y}_2$, with $\mathbf{x}_1$, $\mathbf{y}_1$, $\mathbf{x}_2$, and $\mathbf{y}_2$ being real-valued vectors, and $\sqrt{-1} = i$. First, introduce the notion of the *simple product* of $\mathbf{z}_1$ and $\mathbf{z}_2$, as follows:

$$\begin{aligned} \mathbf{z}_1\, \mathbf{z}_2 &\equiv (\mathbf{x}_1 + i\,\mathbf{y}_1) \bullet (\mathbf{x}_2 + i\,\mathbf{y}_2) \\ &= (\mathbf{x}_1 \bullet \mathbf{x}_2 - \mathbf{y}_1 \bullet \mathbf{y}_2) + i\,(\mathbf{x}_1 \bullet \mathbf{y}_2 + \mathbf{y}_1 \bullet \mathbf{x}_2)\,. \end{aligned} \tag{1.1.8a}$$

[2] Nemat-Nasser and Hori (1993, 1999), for example, use the convention $\boldsymbol{\mathcal{L}} : \mathbf{E} = \mathcal{L}_{ijkl}\, E_{kl}\, \mathbf{e}_i \otimes \mathbf{e}_j$.

Then, define the *scalar product* of these complex vectors, $\mathbf{z}_1$ and $\mathbf{z}_2$, by[3]

$$\mathbf{z}_1 \bullet \mathbf{z}_2 \equiv \mathbf{z}_1 \overline{\mathbf{z}}_2 = (\mathbf{x}_1 \bullet \mathbf{x}_2 + \mathbf{y}_1 \bullet \mathbf{y}_2) - i\,(\mathbf{x}_1 \bullet \mathbf{y}_2 - \mathbf{y}_1 \bullet \mathbf{x}_2), \tag{1.1.8b}$$

where superimposed bar stands for complex conjugate. Hence, for any complex number λ, obtain

$$(\lambda \mathbf{z}_1) \bullet \mathbf{z}_2 = \lambda\,(\mathbf{z}_1 \bullet \mathbf{z}_2), \quad \mathbf{z}_1 \bullet (\lambda \mathbf{z}_2) = \overline{\lambda}\,(\mathbf{z}_1 \bullet \mathbf{z}_2). \tag{1.1.9a,b}$$

Consider another complex-valued vector $\mathbf{z}_3$, and note that

$$(\mathbf{z}_1 + \mathbf{z}_2) \bullet \mathbf{z}_3 = \mathbf{z}_1 \bullet \mathbf{z}_3 + \mathbf{z}_2 \bullet \mathbf{z}_3,$$

$$\mathbf{z}_1 \bullet (\mathbf{z}_2 + \mathbf{z}_3) = \mathbf{z}_1 \bullet \mathbf{z}_2 + \mathbf{z}_1 \bullet \mathbf{z}_3. \tag{1.1.10a,b}$$

Define the *operation* of a second-order complex-valued tensor, $\mathbf{Z} = \mathbf{X} + i\,\mathbf{Y}$, on a complex-valued vector, $\mathbf{z} = \mathbf{x} + i\,\mathbf{y}$, by

$$\mathbf{Z}\,\mathbf{z} \equiv (\mathbf{X} + i\,\mathbf{Y})(\mathbf{x} + i\,\mathbf{y}) = (\mathbf{X}\,\mathbf{x} - \mathbf{Y}\,\mathbf{y}) + i\,(\mathbf{X}\,\mathbf{y} + \mathbf{Y}\,\mathbf{x}), \tag{1.1.11a}$$

where $\mathbf{X}$ and $\mathbf{Y}$ are real-valued second-order tensors. Then, for a complex number λ, it follows that

$$\mathbf{Z}\,(\lambda \mathbf{z}) = (\lambda \mathbf{Z})\,\mathbf{z} = \lambda\,\mathbf{Z}\,\mathbf{z}. \tag{1.1.12}$$

Now, define the operation $\mathbf{z} \bullet \mathbf{Z}$, as follows:

$$\mathbf{z} \bullet \mathbf{Z} = \overline{\mathbf{Z}}^{\mathrm{T}}\,\mathbf{z}. \tag{1.1.11b}$$

Then, it follows from (1.1.11b) that for any complex number λ,

$$(\lambda \mathbf{z}) \bullet \mathbf{Z} = \lambda\,(\mathbf{z} \bullet \mathbf{Z}), \quad \mathbf{z} \bullet (\lambda \mathbf{Z}) = \overline{\lambda}\,\mathbf{z} \bullet \mathbf{Z}. \tag{1.1.13a,b}$$

Further, it follows from (1.1.8b) and (1.1.11a,b) that

$$\mathbf{z}_1 \bullet \mathbf{Z}\,\mathbf{z}_2 = (\mathbf{z}_1 \bullet \mathbf{Z}) \bullet \mathbf{z}_2 = \mathbf{z}_1 \bullet (\mathbf{Z}\,\mathbf{z}_2). \tag{1.1.14}$$

In line with the definition of a simple product, (1.1.8a), introduce the *dyadic product* of two complex-valued vectors, $\mathbf{z}_1$ and $\mathbf{z}_2$, as

$$\mathbf{z}_1 \otimes \mathbf{z}_2 = (\mathbf{x}_1 \otimes \mathbf{x}_2 - \mathbf{y}_1 \otimes \mathbf{y}_2) + i\,(\mathbf{x}_1 \otimes \mathbf{y}_2 + \mathbf{y}_1 \otimes \mathbf{x}_2). \tag{1.1.15a}$$

It now follows that

$$(\mathbf{z}_1 \otimes \mathbf{z}_2)^{\mathrm{T}} = \mathbf{z}_2 \otimes \mathbf{z}_1. \tag{1.1.15b}$$

In view of (1.1.15), for any complex number λ, obtain

$$(\lambda \mathbf{z}_1) \otimes \mathbf{z}_2 = \mathbf{z}_1 \otimes (\lambda \mathbf{z}_2) = \lambda\,\mathbf{z}_1 \otimes \mathbf{z}_2. \tag{1.1.16}$$

Further, for any complex-valued vector $\mathbf{z}_3$, it follows from (1.1.11a,b) that

$$(\mathbf{z}_1 \otimes \mathbf{z}_2)\,\mathbf{z}_3 = (\mathbf{z}_2 \bullet \overline{\mathbf{z}}_3)\,\mathbf{z}_1, \quad \mathbf{z}_3 \bullet (\mathbf{z}_1 \otimes \mathbf{z}_2) = (\mathbf{z}_3 \bullet \mathbf{z}_1)\,\overline{\mathbf{z}}_2. \tag{1.1.17a,b}$$

---

[3] See, *e.g.*, Gantmacher (1960), Volume I, Chapter 9. This definition ensures nonzero length for nonzero vectors that are considered in the present work.

Finally, introduce the *inner product* of two complex-valued tensors, $\mathbf{Z}_1 = \mathbf{X}_1 + i\,\mathbf{Y}_1$ and $\mathbf{Z}_2 = \mathbf{X}_2 + i\,\mathbf{Y}_2$, by

$$\mathbf{Z}_1\,\mathbf{Z}_2 = (\mathbf{X}_1\,\mathbf{X}_2 - \mathbf{Y}_1\,\mathbf{Y}_2) + i\,(\mathbf{X}_1\,\mathbf{Y}_2 + \mathbf{Y}_1\,\mathbf{X}_2)\,, \tag{1.1.18}$$

where $\mathbf{X}_1$, $\mathbf{X}_2$, $\mathbf{Y}_1$, and $\mathbf{Y}_2$ are real-valued second-order tensors. Hence, the inner product of two dyadic products is given by

$$(\mathbf{z}_1 \otimes \mathbf{z}_2)\,(\mathbf{z}_3 \otimes \mathbf{z}_4) = (\mathbf{z}_2 \cdot \bar{\mathbf{z}}_3)\,\mathbf{z}_1 \otimes \mathbf{z}_4\,. \tag{1.1.19}$$

1.2. SPECTRAL REPRESENTATION OF SECOND-ORDER TENSORS

A brief account of coordinate transformation is given first. Then *eigenvalues*, *eigenvectors*, and the corresponding spectral representations of real-valued, second-order tensors are discussed. Finally, explicit expressions for *eigenvalues* of both symmetric and nonsymmetric, real-valued, second-order tensors are presented, including cases with repeated *eigenvalues*.

1.2.1. Coordinate Transformation

A second-order tensor $\mathbf{A}$ is expressed in terms of an orthogonal unit coordinate triad $\mathbf{e}_i$, i = 1, 2, 3, as

$$\mathbf{A} = \sum_{i,j=1}^{3} A_{ij}\,\mathbf{e}_i \otimes \mathbf{e}_j\,, \tag{1.2.1a}$$

where A_{ij}, i, j = 1, 2, 3, are the rectangular Cartesian components of $\mathbf{A}$, defined by

$$A_{ij} = \mathbf{e}_i \cdot \mathbf{A}\,\mathbf{e}_j\,. \tag{1.2.1b}$$

A second-order tensor provides a *linear transformation* which maps a *vector* $\mathbf{X}$ into another *vector* $\mathbf{x} = \mathbf{A}\,\mathbf{X}$. Let $\tilde{\mathbf{e}}^i$ be the result of the transformation of the rectangular unit coordinate triad by the second-order tensor $\mathbf{A}$,

$$\tilde{\mathbf{e}}^i = \mathbf{A}\,\mathbf{e}_i\,, \quad i = 1, 2, 3\,. \tag{1.2.2}$$

In general, $\tilde{\mathbf{e}}^i$, i = 1, 2, 3, is not an *orthonormal* vector set, *i.e.,* it is not a set of mutually orthogonal vectors of unit length. Denote by $\mathbf{R}$ the *rotation* tensor which transforms (rotates) the rectangular unit coordinate triad $\mathbf{e}_i$, i = 1, 2, 3, into another rectangular unit coordinate triad $\hat{\mathbf{e}}_i$, i = 1, 2, 3, as follows:

$$\hat{\mathbf{e}}_i = \mathbf{R}\,\mathbf{e}_i\,, \quad \mathbf{e}_i = \mathbf{R}^{-1}\,\hat{\mathbf{e}}_i\,, \quad i = 1, 2, 3\,. \tag{1.2.3a,b}$$

Rotation tensors $\mathbf{R}$ and $\mathbf{R}^{-1}$ can be expressed as

$$\mathbf{R} = \sum_{i=1}^{3} \hat{\mathbf{e}}_i \otimes \mathbf{e}_i\,, \quad \mathbf{R}^{-1} = \sum_{i=1}^{3} \mathbf{e}_i \otimes \hat{\mathbf{e}}_i = \mathbf{R}^{\mathrm{T}}\,. \tag{1.2.4a,b}$$

The components of the rotation tensor **R**, with respect to the $\mathbf{e}_i$- and $\hat{\mathbf{e}}_i$-triad, respectively, are

$$R_{ij} = \mathbf{e}_i \cdot \mathbf{R}\, \mathbf{e}_j = \mathbf{e}_i \cdot \hat{\mathbf{e}}_j\,, \quad \hat{R}_{ij} = \hat{\mathbf{e}}_i \cdot \mathbf{R}\, \hat{\mathbf{e}}_j = \mathbf{e}_i \cdot \hat{\mathbf{e}}_j = R_{ij}\,. \tag{1.2.4c,d}$$

The components of a second-order tensor **A** with respect to two different coordinate triads are related as follows:

$$\hat{A}_{ij} = \hat{\mathbf{e}}_i \cdot \mathbf{A}\, \hat{\mathbf{e}}_j = (\hat{\mathbf{e}}_i \cdot \mathbf{e}_k)\, A_{kl}\, (\mathbf{e}_l \cdot \hat{\mathbf{e}}_j), \quad (\text{k, l summed})\,. \tag{1.2.5}$$

1.2.2. Eigenvalues and Eigenvectors

Now seek to establish whether there exists a unit vector $\hat{\mathbf{e}}$ such that its transformation by **A** involves no rotation, leaving $\hat{\mathbf{e}}$ parallel to itself, *i.e.*,

$$\mathbf{A}\, \hat{\mathbf{e}} = \lambda\, \hat{\mathbf{e}}\,, \tag{1.2.6a}$$

where λ is a scalar. Any such vector is called an *eigenvector* (principal vector) and the corresponding scalar λ is called an *eigenvalue* (principal value) of that tensor. Keeping in mind that λ and hence $\hat{\mathbf{e}}$ may be complex-valued, note that,

$$\mathbf{A}\, \hat{\mathbf{e}} = \sum_{i,\, j=1}^{3} A_{ij}\, (\mathbf{e}_j \cdot \bar{\hat{\mathbf{e}}})\, \mathbf{e}_i\,, \quad \lambda\, \hat{\mathbf{e}} = \sum_{i,\, j=1}^{3} \lambda\, \delta_{ij}\, (\mathbf{e}_j \cdot \bar{\hat{\mathbf{e}}})\, \mathbf{e}_i\,. \tag{1.2.6b,c}$$

Therefore,

$$(A_{ij} - \lambda\, \delta_{ij})\, (\mathbf{e}_j \cdot \bar{\hat{\mathbf{e}}}) = 0\,, \quad i = 1, 2, 3, \quad (\text{j summed})\,. \tag{1.2.6d}$$

For nontrivial solutions, the determinant of the coefficients in the above three linear and homogeneous equations for three unknowns, $\mathbf{e}_j \cdot \bar{\hat{\mathbf{e}}}$, $j = 1, 2, 3$, must vanish, leading to the following characteristic equation whose roots are the three principal values λ_k, $k = 1, 2, 3$, of **A**:

$$\lambda^3 - I_{\mathbf{A}}\, \lambda^2 + II_{\mathbf{A}}\, \lambda - III_{\mathbf{A}} = 0\,, \tag{1.2.7a}$$

where $I_{\mathbf{A}}$, $II_{\mathbf{A}}$, and $III_{\mathbf{A}}$ are the *basic invariants* of **A**,

$$I_{\mathbf{A}} = \mathrm{tr}(\mathbf{A}) = \lambda_1 + \lambda_2 + \lambda_3\,,$$

$$II_{\mathbf{A}} = \frac{1}{2}\, \{I_{\mathbf{A}}^2 - \mathrm{tr}(\mathbf{A}^2)\} = \lambda_1 \lambda_2 + \lambda_2 \lambda_3 + \lambda_3 \lambda_1\,,$$

$$III_{\mathbf{A}} = \frac{1}{3}\, \{\mathrm{tr}(\mathbf{A}^3) - I_{\mathbf{A}}^3 + 3\, I_{\mathbf{A}}\, II_{\mathbf{A}}\} = \det(\mathbf{A}) = \lambda_1 \lambda_2 \lambda_3\,, \tag{1.2.7b-d}$$

and where tr stands for the trace and det for the determinant of the corresponding argument. This is verified by direct calculation.

1.2.3. Symmetric Tensors

For a general three-dimensional, real-valued, second-order tensor **A**, the roots of (1.2.7a) may be complex. In such a case, at least one real-valued vector is transformed parallel to itself by tensor **A**.

The principal values are always real when $\mathbf{A}$ is real-valued and symmetric, as may easily be verified. Denote the principal unit vector associated with the principal value λ_i by $\hat{\mathbf{e}}_i$,

$$\mathbf{A}\,\hat{\mathbf{e}}_i = \lambda_i\,\hat{\mathbf{e}}_i\,,\quad i = 1, 2, 3\,,\quad (\text{i not summed})\,. \tag{1.2.8a}$$

Since $\hat{\mathbf{e}}_j \cdot \mathbf{A}\,\hat{\mathbf{e}}_i = \hat{\mathbf{e}}_i \cdot \mathbf{A}\,\hat{\mathbf{e}}_j$ for the symmetric tensor $\mathbf{A}$,

$$(\lambda_i - \lambda_j)\,(\hat{\mathbf{e}}_i \cdot \hat{\mathbf{e}}_j) = 0\,,\quad (\text{i and j not summed})\,. \tag{1.2.8b}$$

Hence, the eigenvectors associated with distinct eigenvalues of a symmetric real-valued tensor are mutually orthogonal. If all three eigenvalues are equal, then every vector is an eigenvector. If two eigenvalues are equal, say $\lambda_1 \neq \lambda_2 = \lambda_3$, then the eigenvector associated with λ_1 is unique,[4] and any unit vector orthogonal to this eigenvector is an eigenvector. Therefore, it is always possible to choose an orthonormal set of eigenvectors as the coordinate triad. The components of any real-valued symmetric tensor $\mathbf{A}$ with respect to its *principal coordinate triad* then are

$$\hat{A}_{ij} = \hat{\mathbf{e}}_i \cdot \mathbf{A}\,\hat{\mathbf{e}}_j = \lambda_i\,\delta_{ij},\quad (\text{i not summed})\,. \tag{1.2.9a}$$

The *spectral* representation of $\mathbf{A}$ hence is

$$\mathbf{A} = \sum_{i=1}^{3} \lambda_i\,\hat{\mathbf{e}}_i \otimes \hat{\mathbf{e}}_i\,. \tag{1.2.9b}$$

Then, $\mathbf{A}^2$ and $\mathbf{A}^n$ have spectral representations,

$$\mathbf{A}^2 = \sum_{i=1}^{3} \lambda_i^2\,\hat{\mathbf{e}}_i \otimes \hat{\mathbf{e}}_i\,,\quad \mathbf{A}^n = \sum_{i=1}^{3} \lambda_i^n\,\hat{\mathbf{e}}_i \otimes \hat{\mathbf{e}}_i\,, \tag{1.2.10a,b}$$

where n is an integer. Therefore, $\mathbf{A}$ satisfies the following characteristic equation, known as the *Hamilton-Cayley theorem:*

$$\mathbf{A}^3 - I_{\mathbf{A}}\,\mathbf{A}^2 + II_{\mathbf{A}}\,\mathbf{A} - III_{\mathbf{A}}\,\mathbf{1} = \mathbf{0}\,, \tag{1.2.11a}$$

where $\mathbf{1}$ is the three-by-three identity tensor with components δ_{ij}. In component form, and with repeated indices summed, this becomes

$$A_{ij}\,A_{jk}\,A_{kl} - I_{\mathbf{A}}\,A_{ij}\,A_{jl} + II_{\mathbf{A}}\,A_{il} - III_{\mathbf{A}}\,\delta_{il} = 0\,. \tag{1.2.11b}$$

1.2.4. Nonsymmetric Tensors

Examine now the *eigenvalues* of a general (not necessarily symmetric) real-valued tensor $\mathbf{A}$. Denote the *deviatoric* part of tensor $\mathbf{A}$ by $\mathbf{A}'$

$$\mathbf{A}' = \mathbf{A} - \frac{1}{3}\,I_{\mathbf{A}}\,\mathbf{1}\,. \tag{1.2.12a}$$

[4] It is seen from (1.2.8a) that the sense and the magnitude of an eigenvector may be chosen arbitrarily. The magnitude is fixed here by choosing a unit vector. Also, the sense of the eigenvectors will be chosen such that they form a *right-handed* unit triad.

Then, from (1.2.8a),

$$\mathbf{A}'\hat{\mathbf{e}}_i = (\lambda_i - I_{\mathbf{A}}/3)\,\hat{\mathbf{e}}_i\,. \tag{1.2.12b}$$

This shows that the *eigenvectors* of **A** and **A′** are the same, and their *eigenvalues* are related by

$$\lambda_i' = \lambda_i - I_{\mathbf{A}}/3\,, \quad i = 1, 2, 3, \tag{1.2.12c}$$

where λ_i', i = 1, 2, 3, are the *eigenvalues* of **A′**. These are the roots of the characteristic equation

$$\lambda'^3 + II_{\mathbf{A}'}\lambda' - III_{\mathbf{A}'} = 0\,. \tag{1.2.13}$$

Let

$$\lambda' = x + y\,, \quad xy = a \equiv -II_{\mathbf{A}'}/3\,. \tag{1.2.14a,b}$$

Substitution of (1.2.14a,b) into (1.2.13) yields

$$x^6 - 2\,b\,x^3 + a^3 = 0\,, \quad b \equiv III_{\mathbf{A}'}/2\,. \tag{1.2.14c,d}$$

This can be solved for x^3,

$$x^3 = b \pm \sqrt{b^2 - a^3}\,. \tag{1.2.15}$$

It has three roots for x, one of which is always real. The corresponding y is given by (1.2.14b). Then the *eigenvalues* of **A** satisfy

$$\lambda = I_{\mathbf{A}}/3 + x + y\,. \tag{1.2.16}$$

Three cases are considered, depending on the values of b^2 and a^3, as follows:

(i) $b^2 > a^3$

In this case, from (1.2.15) and (1.2.14b), x and y are given by

$$x = (b \pm \sqrt{b^2 - a^3})^{1/3}\,\{\cos(2\,n\,\pi/3) + i\,\sin(2\,n\,\pi/3)\}\,,$$
$$y = (b \mp \sqrt{b^2 - a^3})^{1/3}\,\{\cos(2\,n\,\pi/3) - i\,\sin(2\,n\,\pi/3)\}\,,$$
$$n = 0, 1, 2, \quad i = \sqrt{-1}\,. \tag{1.2.17a,b}$$

Hence, the *eigenvalues* of **A** can be expressed as,

$$\lambda_{n+1} = \frac{1}{3}\,I_{\mathbf{A}} + \{(b + \sqrt{b^2 - a^3})^{1/3} + (b - \sqrt{b^2 - a^3})^{1/3}\}\cos(2\,n\,\frac{\pi}{3})$$
$$+\,i\,\{(b + \sqrt{b^2 - a^3})^{1/3} - (b - \sqrt{b^2 - a^3})^{1/3}\}\sin(2\,n\,\frac{\pi}{3})\,, \quad n = 0, 1, 2\,. \tag{1.2.17c}$$

Note that, in this case, only one *eigenvalue*, say λ_1, is real for the real-valued tensor **A**, where

$$\lambda_1 = I_{\mathbf{A}}/3 + (b + \sqrt{b^2 - a^3})^{1/3} + (b - \sqrt{b^2 - a^3})^{1/3}\,. \tag{1.2.18a}$$

The other two (complex) *eigenvalues* then are,

$$\lambda_2 = \alpha + i\,\beta, \quad \lambda_3 = \alpha - i\,\beta,$$

$$\alpha = (I_{\mathbf{A}} - \lambda_1)/2, \quad \beta = \sqrt{II_{\mathbf{A}} - \alpha^2 - 2\,\alpha\,\lambda_1}\,. \tag{1.2.18b-e}$$

(ii) $b^2 < a^3$

In this case, from (1.2.15) and (1.2.14b), x and y are given by

$$x = \sqrt{a}\,\{\cos(\frac{2\,n\,\pi + \phi}{3}) \pm i\,\sin(\frac{2\,n\,\pi + \phi}{3})\},$$

$$y = \sqrt{a}\,\{\cos(\frac{2\,n\,\pi + \phi}{3}) \mp i\,\sin(\frac{2\,n\,\pi + \phi}{3})\}, \quad \phi = \cos^{-1}(\frac{b}{a^{3/2}})\,. \tag{1.2.19a-c}$$

Then the *eigenvalues* of **A** are given by

$$\lambda_{n+1} = \frac{I_{\mathbf{A}}}{3} + 2\,\sqrt{a}\cos(\frac{2\,n\,\pi + \phi}{3}), \quad n = 0, 1, 2\,. \tag{1.2.20}$$

(iii) $b^2 = a^3$

In this case, (1.2.15) and (1.2.14b) yield

$$x = \sqrt{a}\,\{\cos(\frac{2\,n\,\pi}{3}) + i\,\sin(\frac{2\,n\,\pi}{3})\},$$

$$y = \sqrt{a}\,\{\cos(\frac{2\,n\,\pi}{3}) - i\,\sin(\frac{2\,n\,\pi}{3})\}, \quad n = 0, 1, 2\,. \tag{1.2.21a,b}$$

Then the *eigenvalues* of **A** are,

$$\lambda_1 = \frac{I_{\mathbf{A}}}{3} + 2\,\sqrt{a}, \quad \lambda_2 = \lambda_3 = \frac{I_{\mathbf{A}}}{3} - \sqrt{a}\,. \tag{1.2.22a,b}$$

1.2.5. Complex-valued Eigenvectors

In the case of $b^2 < a^3$, the *eigenvalues* λ_2 and λ_3 are complex-valued. These are then each other's complex conjugate. The corresponding *eigenvectors* $\hat{\mathbf{e}}_2$, and $\hat{\mathbf{e}}_3$ are also complex-valued. Since $\lambda_3 = \bar{\lambda}_2$, it follows from

$$\mathbf{A}\,\hat{\mathbf{e}}_2 = \lambda_2\,\hat{\mathbf{e}}_2\,, \quad \mathbf{A}\,\hat{\mathbf{e}}_3 = \lambda_3\,\hat{\mathbf{e}}_3\,, \tag{1.2.23a,b}$$

that $\hat{\mathbf{e}}_3 = \bar{\hat{\mathbf{e}}}_2$, where superimposed bar denotes complex conjugate.

1.3. REPRESENTATION OF TENSORS IN OBLIQUE COORDINATE TRIADS

The spectral representation of second-order nonsymmetric real-valued tensors is given in this section. Cases with distinct real, distinct complex, and repeated eigenvalues are examined in some detail. Real-valued eigenvalues and eigenvectors are examined first in Subsections 1.3.2 to 1.3.6. Then complex-valued eigenvalues and eigenvectors are considered in Subsection 1.3.7.

1.3.1. Reciprocal Base Vectors

Two sets of three noncoplanar vectors, $\{\mathbf{e}_i,\ i = 1,2,3\}$ and $\{\mathbf{e}^i,\ i = 1,2,3\}$, are said to be *reciprocal* if they satisfy

$$\mathbf{e}_i \cdot \mathbf{e}^j = \delta_i^j, \tag{1.3.1a}$$

where δ_i^j is the *Kronecker delta*. In general, $\mathbf{e}_i$ and $\mathbf{e}^i$, $i = 1, 2, 3$, are not unit vectors. If $\mathbf{e}_1$, $\mathbf{e}_2$, and $\mathbf{e}_3$ are mutually orthogonal unit vectors, then

$$\mathbf{e}_i = \mathbf{e}^i, \quad \mathbf{e}_i \cdot \mathbf{e}_j = \delta_{ij}. \tag{1.3.1b,c}$$

A general nonsymmetric tensor **A** can be expressed in terms of an *oblique* coordinate triad and its *reciprocal*,

$$\mathbf{A} = \sum_{i,j=1}^{3} A_j^i\, \mathbf{e}_i \otimes \mathbf{e}^j, \quad A_j^i = \mathbf{e}^i \cdot \mathbf{A}\, \mathbf{e}_j, \tag{1.3.2a,b}$$

where A_j^i are the components of **A** with respect to the *orthonormal bases*[5] $\mathbf{e}_i \otimes \mathbf{e}^j$, $i, j = 1, 2, 3$. The transpose of **A** can be expressed with respect to the orthonormal bases $\mathbf{e}^i \otimes \mathbf{e}_j$, $i, j = 1, 2, 3$, as

$$\mathbf{A}^T = \sum_{i,j=1}^{3} A_i^j\, \mathbf{e}^i \otimes \mathbf{e}_j. \tag{1.3.2c}$$

The product of two general nonsymmetric tensors, **A** and **B**, can be expressed as

$$\mathbf{C} = \mathbf{A}\,\mathbf{B}, \quad \mathbf{C} = \sum_{i,j,k=1}^{3} A_k^i\, B_j^k\, \mathbf{e}_i \otimes \mathbf{e}^j. \tag{1.3.3a,b}$$

Since $\mathrm{tr}(\mathbf{e}_i \otimes \mathbf{e}^j) = \delta_i^j$, it follows that

$$\mathrm{tr}(\mathbf{A}) = A_i^i, \quad \mathrm{tr}(\mathbf{A}\,\mathbf{B}) = A_j^i\, B_i^j,$$

$$\mathrm{tr}(\mathbf{A}^2) = A_j^i\, A_i^j, \quad \mathrm{tr}(\mathbf{A}^3) = A_j^i\, A_k^j\, A_i^k, \quad (i, j, k, \text{ summed}). \tag{1.3.3c-f}$$

Any vector **a** can be expressed with respect to the oblique coordinate triads $\{\mathbf{e}_i,\ i = 1, 2, 3\}$ and $\{\mathbf{e}^i,\ i = 1, 2, 3\}$, as

$$\mathbf{a} = \sum_{i=1}^{3} (\mathbf{a} \cdot \mathbf{e}^i)\, \mathbf{e}_i,$$

$$\mathbf{a} = \sum_{i=1}^{3} (\mathbf{a} \cdot \mathbf{e}_i)\, \mathbf{e}^i. \tag{1.3.4a,b}$$

[5] Note that, in view of (1.3.1a), $\mathbf{e}^k \cdot (\mathbf{e}_i \otimes \mathbf{e}^j)\, \mathbf{e}_l = \delta_i^k\, \delta_l^j$.

1.3.2. Real Eigenvalues and Eigenvectors

If the real-valued **a** is an *eigenvector* of **A** corresponding to the real-valued *eigenvalue* λ, then

$$\mathbf{A}\,\mathbf{a} = \lambda\,\mathbf{a}, \quad (A^i_j - \lambda\,\delta^i_j)(\mathbf{a}\cdot\mathbf{e}^j) = 0, \quad (\text{j summed}). \tag{1.3.5a,b}$$

For nontrivial solutions $\mathbf{a}\cdot\mathbf{e}^j$, $j = 1, 2, 3$, the determinant of the coefficient matrix must vanish,

$$\det(A^i_j - \lambda\,\delta^i_j) = 0. \tag{1.3.5c}$$

In view of (1.3.3), this leads to the characteristic equation (1.2.7a). Similarly, if **b** is the *eigenvector* of $\mathbf{A}^T$ corresponding to the *eigenvalue* μ (both real-valued), then

$$\mathbf{A}^T\mathbf{b} = \mu\,\mathbf{b}, \quad (A^i_j - \mu\,\delta^i_j)(\mathbf{b}\cdot\mathbf{e}_i) = 0, \quad (\text{i summed}). \tag{1.3.6a,b}$$

For nontrivial solutions $\mathbf{b}\cdot\mathbf{e}_i$, the same characteristic equation (1.2.7a) results. Therefore, the *eigenvalues* of **A** and $\mathbf{A}^T$ are the same while the *eigenvectors* may not be.

Subtract the scalar product of (1.3.6a) with **a**, from the scalar product of (1.3.5a) with **b**, to obtain

$$(\lambda - \mu)(\mathbf{a}\cdot\mathbf{b}) = \mathbf{b}\cdot\mathbf{A}\,\mathbf{a} - \mathbf{a}\cdot\mathbf{A}^T\mathbf{b} = 0. \tag{1.3.7a}$$

This shows that

$$\mathbf{a}\cdot\mathbf{b} = 0, \quad \text{for } \lambda \neq \mu. \tag{1.3.7b}$$

Therefore, the *eigenvectors* of **A** and $\mathbf{A}^T$, corresponding to two different *eigenvalues*, are orthogonal.

Let the real-valued *eigenvalues* of **A** be λ_i, i = 1, 2, 3. Denote its *eigenvector* corresponding to the *eigenvalue* λ_1 by $\hat{\mathbf{e}}_1$. Let $\tilde{\mathbf{e}}^2$ and $\tilde{\mathbf{e}}^3$ be two nonparallel *arbitrary* vectors on the plane P normal to $\hat{\mathbf{e}}_1$. Then

$$\hat{\mathbf{e}}_1\cdot\mathbf{A}^T\tilde{\mathbf{e}}^2 = \tilde{\mathbf{e}}^2\cdot\mathbf{A}\,\hat{\mathbf{e}}_1 = \lambda_1(\tilde{\mathbf{e}}^2\cdot\hat{\mathbf{e}}_1) = 0,$$
$$\hat{\mathbf{e}}_1\cdot\mathbf{A}^T\tilde{\mathbf{e}}^3 = \tilde{\mathbf{e}}^3\cdot\mathbf{A}\,\hat{\mathbf{e}}_1 = \lambda_1(\tilde{\mathbf{e}}^3\cdot\hat{\mathbf{e}}_1) = 0. \tag{1.3.8a,b}$$

This shows that both $\mathbf{A}^T\tilde{\mathbf{e}}^2$ and $\mathbf{A}^T\tilde{\mathbf{e}}^3$ are orthogonal to $\hat{\mathbf{e}}_1$. Therefore, they can be expressed as (Figure 1.3.1)

$$\mathbf{A}^T\tilde{\mathbf{e}}^2 = \tilde{A}^2_2\,\tilde{\mathbf{e}}^2 + \tilde{A}^2_3\,\tilde{\mathbf{e}}^3, \quad \mathbf{A}^T\tilde{\mathbf{e}}^3 = \tilde{A}^3_2\,\tilde{\mathbf{e}}^2 + \tilde{A}^3_3\,\tilde{\mathbf{e}}^3, \tag{1.3.9a,b}$$

where the coefficients of $\tilde{\mathbf{e}}^2$ and $\tilde{\mathbf{e}}^3$ on the right-hand side are the corresponding components of the resulting vector in the $\tilde{\mathbf{e}}^2$, $\tilde{\mathbf{e}}^3$-coordinate system. Hence, $(\tilde{A}^3_2\,\tilde{\mathbf{e}}^2 + \alpha\,\tilde{\mathbf{e}}^3)$ which lies in the plane normal to $\hat{\mathbf{e}}_1$, is an *eigenvector* of $\mathbf{A}^T$ for some non-zero α, corresponding to the eigenvalue λ if[6]

---

[6] If $\alpha = 0$, then $\tilde{\mathbf{e}}^2$ is an eigenvector. Also if $\tilde{A}^3_2 = 0$, then $\tilde{\mathbf{e}}^3$ is an eigenvector. Hence, assume neither quantity is zero.

$$\lambda = \tilde{A}_2^{\cdot 2} + \alpha = \frac{1}{\alpha}(\tilde{A}_2^{\cdot 3}\,\tilde{A}_3^{\cdot 2} + \alpha\,\tilde{A}_3^{\cdot 3}). \tag{1.3.10a}$$

This yields a quadratic equation for α,

$$\alpha^2 + (\tilde{A}_2^{\cdot 2} - \tilde{A}_3^{\cdot 3})\,\alpha - \tilde{A}_2^{\cdot 3}\,\tilde{A}_3^{\cdot 2} = 0. \tag{1.3.10b}$$

Therefore, *at least one eigenvector* of $\mathbf{A}^T$ lies in the plane normal to $\hat{\mathbf{e}}_1$. The roots of (1.3.10b) are

$$\alpha_1 = \frac{1}{2}\,\{\tilde{A}_3^{\cdot 3} - \tilde{A}_2^{\cdot 2} + \sqrt{(\tilde{A}_2^{\cdot 2} - \tilde{A}_3^{\cdot 3})^2 + 4\,\tilde{A}_3^{\cdot 2}\,\tilde{A}_2^{\cdot 3}}\},$$

$$\alpha_2 = \frac{1}{2}\,\{\tilde{A}_3^{\cdot 3} - \tilde{A}_2^{\cdot 2} - \sqrt{(\tilde{A}_2^{\cdot 2} - \tilde{A}_3^{\cdot 3})^2 + 4\,\tilde{A}_3^{\cdot 2}\,\tilde{A}_2^{\cdot 3}}\}. \tag{1.3.10c,d}$$

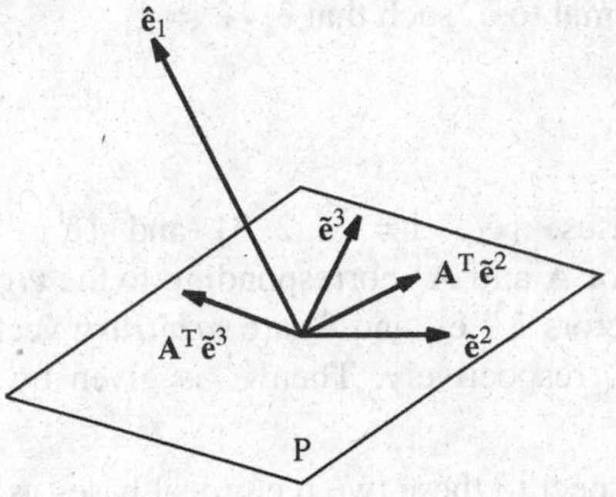

Figure 1.3.1

Plane P, formed by two arbitrary nonparallel vectors $\tilde{\mathbf{e}}^2$ and $\tilde{\mathbf{e}}^3$ normal to $\hat{\mathbf{e}}_1$, contains $\mathbf{A}^T\tilde{\mathbf{e}}^2$ and $\mathbf{A}^T\tilde{\mathbf{e}}^3$, as well as two *eigenvectors* of $\mathbf{A}^T$

If α_1 and α_2 are distinct, then there exist two distinct *eigenvalues* λ_2 and λ_3 for $\mathbf{A}^T$, with respective *eigenvectors* $\hat{\mathbf{e}}^2$ and $\hat{\mathbf{e}}^3$, in the plane normal to $\hat{\mathbf{e}}_1$, where

$$\lambda_2 = \frac{1}{2}\,\{\tilde{A}_2^{\cdot 2} + \tilde{A}_3^{\cdot 3} + \sqrt{(\tilde{A}_2^{\cdot 2} - \tilde{A}_3^{\cdot 3})^2 + 4\,\tilde{A}_3^{\cdot 2}\,\tilde{A}_2^{\cdot 3}}\},$$

$$\lambda_3 = \frac{1}{2}\,\{\tilde{A}_2^{\cdot 2} + \tilde{A}_3^{\cdot 3} - \sqrt{(\tilde{A}_2^{\cdot 2} - \tilde{A}_3^{\cdot 3})^2 + 4\,\tilde{A}_3^{\cdot 2}\,\tilde{A}_2^{\cdot 3}}\},$$

$$\hat{\mathbf{e}}^2 = \tilde{A}_2^{\cdot 3}\,\tilde{\mathbf{e}}^2 + \alpha_1\,\tilde{\mathbf{e}}^3, \qquad \hat{\mathbf{e}}^3 = \tilde{A}_2^{\cdot 3}\,\tilde{\mathbf{e}}^2 + \alpha_2\,\tilde{\mathbf{e}}^3. \tag{1.3.11a-d}$$

When α_1 and α_2 are identical, then, from (1.3.10), it follows that the quantity under the square root in (1.3.11) must vanish,

$$(\tilde{A}_2^{\cdot 2} - \tilde{A}_3^{\cdot 3})^2 + 4\,\tilde{A}_3^{\cdot 2}\,\tilde{A}_2^{\cdot 3} = 0. \tag{1.3.12a}$$

Then, the corresponding *eigenvalues* are identical,

$$\lambda_2 = \lambda_3 = (\tilde{A}_2^{\cdot 2} + \tilde{A}_3^{\cdot 3})/2, \qquad \text{for } \alpha_1 = \alpha_2 = (\tilde{A}_3^{\cdot 3} - \tilde{A}_2^{\cdot 2})/2. \tag{1.3.12b}$$

Therefore, in general,[7] when α_1 and α_2 are identical, there exists only one *eigenvector*, say $\hat{\mathbf{e}}^2$, of $\mathbf{A}^T$ on the plane normal to $\hat{\mathbf{e}}_1$,

---

[7] For the special case where $\tilde{A}_2^{\cdot 3} = \tilde{A}_3^{\cdot 2} = 0$ and $\tilde{A}_2^{\cdot 2} = \tilde{A}_3^{\cdot 3}$, it follows from (1.3.9) that, both $\tilde{\mathbf{e}}^2$ and $\tilde{\mathbf{e}}^3$ are *eigenvectors* corresponding to the *eigenvalue* $\lambda_2 = \lambda_3 = \tilde{A}_2^{\cdot 2} = \tilde{A}_3^{\cdot 3}$. Therefore, in this case, any vector on the plane normal to $\hat{\mathbf{e}}_1$ is an *eigenvector* corresponding to the *eigenvalue* $\lambda_2 = \lambda_3$, *i.e.*, $\mathbf{A}^T(\tilde{\mathbf{e}}^2 + \alpha\,\tilde{\mathbf{e}}^3) = \lambda_2(\tilde{\mathbf{e}}^2 + \alpha\,\tilde{\mathbf{e}}^3)$, for any α.

$$\hat{\mathbf{e}}^2 = \tilde{A}_2^3\,\tilde{\mathbf{e}}^2 + \frac{1}{2}\,(\tilde{A}_3^3 - \tilde{A}_2^2)\,\tilde{\mathbf{e}}^3\,. \tag{1.3.13a}$$

In this case (1.3.9) is expressed as

$$\mathbf{A}^T\,\tilde{\mathbf{e}}^2 = \lambda_2\,\tilde{\mathbf{e}}^2 + \frac{\tilde{A}_2^2 - \tilde{A}_3^3}{2\,\tilde{A}_2^3}\,\hat{\mathbf{e}}^2\,,$$

$$\mathbf{A}^T\,\tilde{\mathbf{e}}^3 = \lambda_2\,\tilde{\mathbf{e}}^3 + \hat{\mathbf{e}}^2\,. \tag{1.3.13b,c}$$

Hence, if $\lambda_2 = \lambda_3$, then for an arbitrary vector $\hat{\mathbf{e}}^3$ on the plane normal to $\hat{\mathbf{e}}_1$,

$$\mathbf{A}^T\,\hat{\mathbf{e}}^3 = \lambda_2\,\hat{\mathbf{e}}^3 + \hat{A}_2^3\,\hat{\mathbf{e}}^2\,, \qquad \hat{A}_2^3 = \hat{\mathbf{e}}_2 \cdot \mathbf{A}^T\,\hat{\mathbf{e}}^3\,, \tag{1.13.d,e}$$

where $\hat{\mathbf{e}}^2$ is an *eigenvector* of $\mathbf{A}^T$, corresponding to the *eigenvalue* $\lambda_2 = \lambda_3$, and $\hat{\mathbf{e}}_2$ is an arbitrary vector on the plane normal to $\hat{\mathbf{e}}^3$ such that $\hat{\mathbf{e}}_2 \cdot \hat{\mathbf{e}}^2 = 1$.

1.3.3. Spectral Representation of $\mathbf{A}^m$

Consider the two reciprocal bases, $\{\hat{\mathbf{e}}_i,\ i = 1, 2, 3\}$ and $\{\hat{\mathbf{e}}^i,\ i = 1, 2, 3\}$. Let $\hat{\mathbf{e}}_1$ and $\hat{\mathbf{e}}^2$ be *eigenvectors* of $\mathbf{A}$ and $\mathbf{A}^T$ corresponding to the *eigenvalues* λ_1 and λ_2, respectively. The vectors $\hat{\mathbf{e}}^3$, $\hat{\mathbf{e}}_3$, and $\hat{\mathbf{e}}_2$ are *arbitrary* vectors on the planes normal to $\hat{\mathbf{e}}_1$, $\hat{\mathbf{e}}^2$, and $\hat{\mathbf{e}}^3$, respectively. Then $\hat{\mathbf{e}}^1$ is given by the reciprocal condition (1.3.1a).

The tensor $\mathbf{A}$ is expressed with respect to these two reciprocal bases as

$$\mathbf{A} = \sum_{i,j=1}^{3} \hat{A}_j^i\,\hat{\mathbf{e}}_i \otimes \hat{\mathbf{e}}^j\,,$$

$$\hat{A}_j^i = \hat{\mathbf{e}}^i \cdot \mathbf{A}\,\hat{\mathbf{e}}_j\,. \tag{1.3.14a,b}$$

Since $\hat{\mathbf{e}}_1$ and $\hat{\mathbf{e}}^2$ are *eigenvectors* of $\mathbf{A}$ and $\mathbf{A}^T$, respectively,

$$\hat{A}_1^i = \lambda_1\,\delta_1^i\,, \qquad \hat{A}_i^2 = \lambda_2\,\delta_i^2\,. \tag{1.3.15a,b}$$

Substitution of (1.3.15) into (1.3.14) results in

$$\mathbf{A} = \lambda_1\,\hat{\mathbf{e}}_1 \otimes \hat{\mathbf{e}}^1 + \lambda_2\,\hat{\mathbf{e}}_2 \otimes \hat{\mathbf{e}}^2 + \hat{A}_3^3\,\hat{\mathbf{e}}_3 \otimes \hat{\mathbf{e}}^3$$
$$+ \hat{A}_2^1\,\hat{\mathbf{e}}_1 \otimes \hat{\mathbf{e}}^2 + \hat{A}_3^1\,\hat{\mathbf{e}}_1 \otimes \hat{\mathbf{e}}^3 + \hat{A}_2^3\,\hat{\mathbf{e}}_3 \otimes \hat{\mathbf{e}}^2\,. \tag{1.3.16a}$$

Then, for any integer m,

$$\mathbf{A}^m = (\lambda_1)^m\,\hat{\mathbf{e}}_1 \otimes \hat{\mathbf{e}}^1 + (\lambda_2)^m\,\hat{\mathbf{e}}_2 \otimes \hat{\mathbf{e}}^2 + (\hat{A}_3^3)^m\,\hat{\mathbf{e}}_3 \otimes \hat{\mathbf{e}}^3$$
$$+ \sum_{i=0}^{m-1} (\lambda_1)^i\,(\lambda_2)^{m-1-i}\,\hat{A}_2^1\,\hat{\mathbf{e}}_1 \otimes \hat{\mathbf{e}}^2$$
$$+ \sum_{i=0}^{m-1} (\lambda_1)^i\,(\hat{A}_3^3)^{m-1-i}\,\hat{A}_3^1\,\hat{\mathbf{e}}_1 \otimes \hat{\mathbf{e}}^3$$
$$+ \sum_{i=0}^{m-1} (\lambda_2)^i\,(\hat{A}_3^3)^{m-1-i}\,\hat{A}_2^3\,\hat{\mathbf{e}}_3 \otimes \hat{\mathbf{e}}^2$$
$$+ \sum_{i,j=0}^{m-2} (\lambda_1)^i\,(\lambda_2)^j\,(\hat{A}_3^3)^{m-2-i-j}\,\hat{A}_2^3\,\hat{A}_3^1\,\hat{\mathbf{e}}_1 \otimes \hat{\mathbf{e}}^2\,. \tag{1.3.16b}$$

1.3.4. Distinct Real Eigenvalues

Consider the case where all three *eigenvalues* are distinct. The *eigenvectors* of **A** and $\mathbf{A}^T$ corresponding to the *eigenvalue* λ_3 can be chosen as $\hat{\mathbf{e}}_3$ and $\hat{\mathbf{e}}^3$, respectively. When $\hat{\mathbf{e}}^1$ and $\hat{\mathbf{e}}_2$ are also *eigenvectors* of $\mathbf{A}^T$ and **A**, respectively, then

$$\hat{A}_3^3 = \lambda_3, \quad \hat{A}_2^1 = \hat{A}_3^1 = \hat{A}_2^3 = 0. \tag{1.3.17a,b}$$

Hence, (1.3.16) reduces to

$$\mathbf{A} = \sum_{i=1}^{3} \lambda_i \mathbf{E}_i, \quad \mathbf{A}^m = \sum_{i=1}^{3} (\lambda_i)^m \mathbf{E}_i, \quad \mathbf{E}_i \equiv \hat{\mathbf{e}}_i \otimes \hat{\mathbf{e}}^i. \tag{1.3.17c-e}$$

Now, seek to obtain a coordinate-independent expression for $\mathbf{E}_i$, i = 1, 2, 3. Consider the following representation of $\mathbf{E}_i$:

$$\mathbf{E}_i = a_0^i \mathbf{1} + a_1^i \mathbf{A} + a_2^i \mathbf{A}^2. \tag{1.3.18a}$$

The scalar product of (1.3.18a) with $\mathbf{E}_j$ yields

$$a_0^i + a_1^i \lambda_j + a_2^i (\lambda_j)^2 = \delta_j^i. \tag{1.3.18b}$$

The solution of this system of equations produces the required coefficients in the right-hand side of (1.3.18a), leading to

$$\mathbf{E}_i = \frac{1}{(\lambda_i - \lambda_j)(\lambda_i - \lambda_k)} \{\mathbf{A}^2 - (\lambda_j + \lambda_k)\mathbf{A} + \lambda_j \lambda_k \mathbf{1}\}, \tag{1.3.18c}$$

where i, j, k are the permutations of 1, 2, 3.

1.3.5. Two Repeated Real Eigenvalues

In this case, there may be *only one eigenvector* for **A** and $\mathbf{A}^T$ corresponding to the *eigenvalue* $\lambda_2 = \lambda_3$. Denote the unique *eigenvector* of $\mathbf{A}^T$ corresponding to the *eigenvalue* $\lambda_2 = \lambda_3$ by $\hat{\mathbf{e}}^2$. Since $\lambda_1 \neq \lambda_3$, it follows from (1.3.10) that **A** has two *distinct eigenvectors* corresponding to the *eigenvalues* λ_1 and $\lambda_3 = \lambda_2$, on the plane normal to $\hat{\mathbf{e}}^2$. Therefore, if each of the two tensors, **A** and $\mathbf{A}^T$, has a unique *eigenvector* corresponding to the *eigenvalue* $\lambda_2 = \lambda_3$, then these *eigenvectors* are orthogonal. Hence, $\hat{\mathbf{e}}_3$ which is orthogonal to $\hat{\mathbf{e}}^2$ in (1.3.16), can be chosen as the *eigenvector* of **A** corresponding to the *eigenvalue* $\lambda_2 = \lambda_3$. Then, $\hat{\mathbf{e}}^1$ which is orthogonal to $\hat{\mathbf{e}}_3$ in (1.3.16), can be chosen as the *eigenvector* of $\mathbf{A}^T$ corresponding to the *eigenvalue* λ_1. Therefore,

$$\hat{A}_3^3 = \lambda_2 = \lambda_3, \quad \hat{A}_3^1 = \hat{A}_2^1 = 0. \tag{1.3.19a,b}$$

In general, $\hat{\mathbf{e}}^2$ may be the only *eigenvector* of $\mathbf{A}^T$ corresponding to the *eigenvalue* $\lambda_2 = \lambda_3$. Hence, in general, $\hat{\mathbf{e}}_2$ and $\hat{\mathbf{e}}^3$ may not be the *eigenvectors* of **A** and $\mathbf{A}^T$, respectively. In such a case, (1.3.16) reduces to

$$\mathbf{A} = \sum_{i=1}^{3} \lambda_i \hat{\mathbf{e}}_i \otimes \hat{\mathbf{e}}^i + \hat{A}_2^3 \hat{\mathbf{e}}_3 \otimes \hat{\mathbf{e}}^2,$$

$$\mathbf{A}^m = \sum_{i=1}^{3} (\lambda_i)^m \hat{\mathbf{e}}_i \otimes \hat{\mathbf{e}}^i + m(\lambda_2)^{m-1} \hat{A}_2^3 \hat{\mathbf{e}}_3 \otimes \hat{\mathbf{e}}^2 . \tag{1.3.19c,d}$$

Since λ_1 and λ_2 are the roots of

$$\lambda^2 - (\lambda_1 + \lambda_2)\lambda + \lambda_1 \lambda_2 = 0,$$

it follows from (1.3.19c) that

$$\mathbf{A}^2 - (\lambda_1 + \lambda_2)\mathbf{A} + \lambda_1 \lambda_2 \mathbf{1} = (\lambda_2 - \lambda_1)\mathbf{L}, \quad \mathbf{L} = \hat{A}_2^3 \hat{\mathbf{e}}_3 \otimes \hat{\mathbf{e}}^2 . \tag{1.3.20a,b}$$

Note that,

$$\mathbf{A}\mathbf{L} = \mathbf{L}\mathbf{A} = \lambda_2 \mathbf{L}, \quad \mathbf{L}^2 = \mathbf{0}. \tag{1.3.20c,d}$$

Since,

$$\sum_{i=1}^{3} \hat{\mathbf{e}}_i \otimes \hat{\mathbf{e}}^i = \mathbf{1}, \tag{1.3.21}$$

A can be expressed as

$$\mathbf{A} = \lambda_1 \mathbf{E}_1 + \lambda_2 (\mathbf{1} - \mathbf{E}_1) + \mathbf{L}, \quad \mathbf{E}_1 = \hat{\mathbf{e}}_1 \otimes \hat{\mathbf{e}}^1 . \tag{1.3.22a,b}$$

This yields a coordinate-independent expression for $\mathbf{E}_1$,

$$\mathbf{E}_1 = \frac{1}{\lambda_1 - \lambda_2} (\mathbf{A} - \mathbf{L} - \lambda_2 \mathbf{1})$$
$$= \frac{1}{(\lambda_1 - \lambda_2)^2} \{\mathbf{A}^2 - 2\lambda_2 \mathbf{A} + (\lambda_2)^2 \mathbf{1}\}. \tag{1.3.22c}$$

Substitution of (1.3.20b) and (1.3.21) into (1.3.19d) results in a coordinate-independent expression for $\mathbf{A}^m$,

$$\mathbf{A}^m = (\lambda_1)^m \mathbf{E}_1 + (\lambda_2)^m (\mathbf{1} - \mathbf{E}_1) + m(\lambda_2)^{m-1} \mathbf{L},$$
$$\mathbf{L} = \frac{-1}{\lambda_1 - \lambda_2} \{\mathbf{A}^2 - (\lambda_1 + \lambda_2)\mathbf{A} + \lambda_1 \lambda_2 \mathbf{1}\}. \tag{1.3.22d,e}$$

1.3.6. Three Repeated Real Eigenvalues

When all three *eigenvalues* are equal, say, to λ, there may be only one *eigenvector* for **A** and $\mathbf{A}^T$. If **A** has *only one eigenvector*, then, from (1.3.5b) and (1.3.6a), $\mathbf{A}^T$ also has *only one eigenvector*. It also follows from (1.3.8) that, if $\hat{\mathbf{e}}_1$ is an *eigenvector* of **A**, then there exists *at least one eigenvector* of $\mathbf{A}^T$ which is orthogonal to $\hat{\mathbf{e}}_1$. Therefore, if each of the two tensors, **A** and $\mathbf{A}^T$, has a unique *eigenvector*, then these *eigenvectors* are orthogonal. Indeed, even if **A** and $\mathbf{A}^T$ have *unique eigenvectors*, two reciprocal bases, $\{\hat{\mathbf{e}}_i, i = 1, 2, 3\}$ and $\{\hat{\mathbf{e}}^i, i = 1, 2, 3\}$, can always be chosen such that $\hat{\mathbf{e}}_1$ and $\hat{\mathbf{e}}^2$ are *eigenvectors* of **A** and $\mathbf{A}^T$, respectively.

It follows from (1.3.13) that, for $\lambda_2 = \lambda_3 = \lambda$,

$$\hat{A}_1^1 = \hat{A}_2^2 = \hat{A}_3^3 = \lambda . \tag{1.3.23a}$$

In this case, for the reciprocal bases, choose the orthogonal vectors $\hat{\mathbf{e}}_1$ and $\hat{\mathbf{e}}^2$ which are the *eigenvectors* of **A** and $\mathbf{A}^T$, respectively, while the other base vectors are arbitrary nonparallel vectors satisfying the orthogonality condition (1.3.1a). Then, (1.3.16) reduces to

$$\mathbf{A} = \lambda\,\mathbf{1} + + \hat{A}_2^1\,\hat{\mathbf{e}}_1 \otimes \hat{\mathbf{e}}^2 + \hat{A}_3^1\,\hat{\mathbf{e}}_1 \otimes \hat{\mathbf{e}}^3 + \hat{A}_2^3\,\hat{\mathbf{e}}_3 \otimes \hat{\mathbf{e}}^2\,,$$

$$\mathbf{A}^m = \lambda^m\,\mathbf{1} + + m\,\lambda^{m-1}\,\{\hat{A}_2^1\,\hat{\mathbf{e}}_1 \otimes \hat{\mathbf{e}}^2 + \hat{A}_3^1\,\hat{\mathbf{e}}_1 \otimes \hat{\mathbf{e}}^3 + \hat{A}_2^3\,\hat{\mathbf{e}}_3 \otimes \hat{\mathbf{e}}^2\}$$

$$+ \frac{1}{2}\,m\,(m-1)\,\lambda^{m-2}\,\hat{A}_2^3\,\hat{A}_3^1\,\hat{\mathbf{e}}_1 \otimes \hat{\mathbf{e}}^2\,. \qquad (1.3.23b,c)$$

These expressions show that

$$\mathbf{A} = \lambda\,\mathbf{1} + \mathbf{L}\,, \qquad \mathbf{L} = \hat{A}_2^1\,\hat{\mathbf{e}}_1 \otimes \hat{\mathbf{e}}^2 + \hat{A}_3^1\,\hat{\mathbf{e}}_1 \otimes \hat{\mathbf{e}}^3 + \hat{A}_2^3\,\hat{\mathbf{e}}_3 \otimes \hat{\mathbf{e}}^2\,,$$

$$\mathbf{L}^2 = \hat{A}_2^3\,\hat{A}_3^1\,\hat{\mathbf{e}}_1 \otimes \hat{\mathbf{e}}^2\,, \qquad \mathbf{A}\,\mathbf{L} = \mathbf{L}\,\mathbf{A} = \lambda\,\mathbf{L} + \mathbf{L}^2\,. \qquad (1.3.24a\text{-}d)$$

Hence, $\mathbf{A}^m$ can be expressed in coordinate-independent form as follows:

$$\mathbf{A}^m = \lambda^m\,\mathbf{1} + m\,\lambda^{m-1}\,\mathbf{L} + \frac{1}{2}\,m(m-1)\,\lambda^{m-2}\,\mathbf{L}^2\,, \qquad \mathbf{L} = \mathbf{A} - \lambda\,\mathbf{1}\,. \qquad (1.3.25a,b)$$

1.3.7. Complex-valued Eigenvalues and Eigenvectors

Now, consider the case when the real-valued nonsymmetric tensor **A** has real and complex *eigenvalues*. Let **a** be the *eigenvector* of **A** corresponding to an *eigenvalue* (real or complex) λ. Then[8]

$$\mathbf{A}\,\mathbf{a} = \lambda\,\mathbf{a}\,. \qquad (1.3.26a)$$

Similarly, if **b** is the *eigenvector* of $\mathbf{A}^T$ corresponding to an *eigenvalue* (real or complex) μ, then

$$\mathbf{b} \bullet \mathbf{A} = \mathbf{A}^T\mathbf{b} = \mu\,\mathbf{b}\,. \qquad (1.3.26b)$$

Subtract the scalar product of (1.3.26b) with **a** from the scalar product of **b** with (1.3.26a) to obtain

$$(\bar{\lambda} - \mu)\,\mathbf{b} \bullet \mathbf{a} = \mathbf{b} \bullet \mathbf{A}\,\mathbf{a} - (\mathbf{A}^T\mathbf{b}) \bullet \mathbf{a} = 0\,. \qquad (1.3.26c)$$

This shows that

$$\mathbf{b} \bullet \mathbf{a} = 0\,, \qquad \text{for } \bar{\lambda} \neq \mu\,. \qquad (1.3.27a)$$

Therefore, the *eigenvectors* of **A** and $\mathbf{A}^T$ are *orthogonal*, if the corresponding *eigenvalues* are not each other's complex conjugate. For real-valued tensors, if λ is an *eigenvalue* with *eigenvector* **a** then $\bar{\lambda}$ is also an *eigenvalue* with *eigenvector* $\bar{\mathbf{a}}$. Hence, it follows from (1.3.27a) that

---

[8] See Subsection 1.1.2, especially expressions (1.1.11) to (1.1.18), for the definition of complex-valued vector products used here and elsewhere in this book.

$$\mathbf{b} \cdot \bar{\mathbf{a}} = \mathbf{a} \cdot \bar{\mathbf{b}} = \bar{\mathbf{a}} \cdot \mathbf{b} = 0, \quad \text{for } \lambda \neq \mu. \tag{1.3.27b}$$

Now, consider the case when the three-dimensional, real-valued nonsymmetric tensor **A** has a real *eigenvalue* λ_1 and two complex *eigenvalues* λ_2 and $\lambda_3 = \bar{\lambda}_2$. Denote the *eigenvectors* of **A** and $\mathbf{A}^T$ corresponding to the *eigenvalues* λ_i by $\hat{\mathbf{e}}_i$ and $\hat{\mathbf{e}}^i$, i = 1, 2, 3, respectively. Since, $\lambda_3 = \bar{\lambda}_2$, it follows that

$$\lambda_1 \neq \lambda_2 \neq \lambda_3 \neq \lambda_1, \tag{1.3.28}$$

and from (1.3.27b), that

$$\hat{\mathbf{e}}^j \cdot \bar{\hat{\mathbf{e}}}_i = \hat{\mathbf{e}}_j \cdot \bar{\hat{\mathbf{e}}}^i = 0, \quad \text{for } i \neq j. \tag{1.3.29}$$

Therefore, the vector sets $\{\hat{\mathbf{e}}_i,\ i = 1, 2, 3\}$ and $\{\bar{\hat{\mathbf{e}}}^i,\ i = 1, 2, 3\}$ are reciprocal if they are normalized such that

$$\hat{\mathbf{e}}_j \cdot \bar{\hat{\mathbf{e}}}^i = \delta_j^i. \tag{1.3.30a}$$

Similarly, the vector sets $\{\bar{\hat{\mathbf{e}}}_i,\ i = 1, 2, 3\}$ and $\{\hat{\mathbf{e}}^i,\ i = 1, 2, 3\}$ are reciprocal if they are normalized such that

$$\hat{\mathbf{e}}^j \cdot \bar{\hat{\mathbf{e}}}_i = \delta_i^j. \tag{1.3.30b}$$

For each eigenvalue λ_i,

$$\mathbf{A}\,\hat{\mathbf{e}}_i = \lambda_i\,\hat{\mathbf{e}}_i, \quad \mathbf{A}^T\hat{\mathbf{e}}^i = \hat{\mathbf{e}}^i \cdot \mathbf{A} = \lambda_i\,\hat{\mathbf{e}}^i, \quad (\text{i not summed}). \tag{1.3.31a,b}$$

From (1.3.31a) and (1.3.31b) respectively, obtain

$$\mathbf{A}\,\hat{\mathbf{e}}_i \cdot \bar{\hat{\mathbf{e}}}_j = \lambda_i\,\delta_i^j, \quad \hat{\mathbf{e}}^i \cdot \mathbf{A}\,\hat{\mathbf{e}}_j = \lambda_i\,\delta_j^i, \quad (\text{i not summed}). \tag{1.3.31c,d}$$

Moreover, any real-valued second-order tensor, say, **B**, can be resolved with respect to the eigenvectors of **A** and $\mathbf{A}^T$, as follows:

$$\mathbf{B} = \sum_{i,j=1}^{3} B_j^i\,\hat{\mathbf{e}}_i \otimes \hat{\mathbf{e}}^j \quad B_j^i = (\mathbf{B}\,\hat{\mathbf{e}}_j) \cdot \bar{\hat{\mathbf{e}}}^i = \hat{\mathbf{e}}^i \cdot \mathbf{B}\,\bar{\hat{\mathbf{e}}}_j. \tag{1.3.31e,f}$$

Now, in view of (1.3.31c,d), tensors **A** and $\mathbf{A}^T$ have the following spectral representation:

$$\mathbf{A} = \sum_{i=1}^{3} \lambda_i\,\hat{\mathbf{e}}_i \otimes \hat{\mathbf{e}}^i, \quad \mathbf{A}^T = \sum_{i=1}^{3} \lambda_i\,\hat{\mathbf{e}}^i \otimes \hat{\mathbf{e}}_i. \tag{1.3.32a,b}$$

Then, for any integer m,

$$\mathbf{A}^m = \sum_{i=1}^{3} \lambda_i^m\,\hat{\mathbf{e}}_i \otimes \hat{\mathbf{e}}^i. \tag{1.3.32c}$$

Now, consider the tensor

$$\mathbf{E}_0 = \sum_{i=1}^{3} \hat{\mathbf{e}}_i \otimes \hat{\mathbf{e}}^i. \tag{1.3.33a}$$

It follows that

$$\mathbf{E}_0\,\hat{\mathbf{e}}_i = \hat{\mathbf{e}}_i, \quad i = 1, 2, 3. \tag{1.3.33b}$$

Since any real vector **a** can be expressed as a linear combination of $\hat{\mathbf{e}}_i$'s, i = 1, 2, 3, it follows that

$$\mathbf{E}_0\,\mathbf{a} = \mathbf{a}\,. \quad \mathbf{1} = \sum_{i=1}^{3} \hat{\mathbf{e}}_i \otimes \hat{\mathbf{e}}^i\,. \tag{1.3.33c,d}$$

Following Subsection 1.3.4, denote $\hat{\mathbf{e}}_i \otimes \mathbf{e}^i$ (i not summed) by $\mathbf{E}_i$ and consider the following representation of $\mathbf{E}_i$:

$$\mathbf{E}_i = a_0^i\,\mathbf{1} + a_1^i\,\mathbf{A} + a_2^i\,\mathbf{A}^2\,. \tag{1.3.34a}$$

Substitute (1.3.32a,c) and (1.3.33d) into (1.3.34a) and equate the coefficients of $\mathbf{E}_j$, j = 1, 2, 3, to obtain

$$a_0^i + a_1^i\,\lambda_j + a_2^i\,\lambda_j^2 = \delta_j^i\,, \quad j = 1,\ 2,\ 3\,. \tag{1.3.34b}$$

The solution to this system of equations leads to

$$\mathbf{E}_i = \frac{1}{(\lambda_i - \lambda_j)\,(\lambda_i - \lambda_k)}\,\{\mathbf{A}^2 - (\lambda_j + \lambda_k)\,\mathbf{A} + \lambda_j\,\lambda_k\,\mathbf{1}\}\,, \tag{1.3.35}$$

where i, j, k are the permutations of 1, 2, 3.

1.4. IDENTITIES FOR SECOND-ORDER TENSORS

Various kinematical quantities are related to each other through identities involving second-order tensors and their principal invariants. The Hamilton-Cayley theorem is an identity which involves only one tensor and its invariants. As an extension to this, Rivlin (1955) develops an identity which involves three second-order tensors and their principal invariants. Rivlin uses this identity to obtain three identities relating two second-order tensors. These identities yield the Hamilton-Cayley theorem as a special case. These and related issues are examined in this section, including explicit solutions to certain tensor equations which are important for the study of the kinematics of deformation.

1.4.1. Rivlin's Identities

Rivlin's results are based on the following identity:

$$\det \begin{bmatrix} \delta_{ij} & \delta_{il} & \delta_{in} & \delta_{iq} \\ \delta_{kj} & \delta_{kl} & \delta_{kn} & \delta_{kq} \\ \delta_{mj} & \delta_{ml} & \delta_{mn} & \delta_{mq} \\ \delta_{pj} & \delta_{pl} & \delta_{pn} & \delta_{pq} \end{bmatrix} = 0\,, \quad i, j, k, l, m, n, p, q = 1, 2, 3\,. \tag{1.4.1}$$

Multiply (1.4.1) by $A_{kl}\,B_{mn}\,C_{pq}$, and sum on repeated indices to obtain a second-order tensor equation as follows:

$$\begin{aligned}
&\mathbf{A\,B\,C + B\,C\,A + C\,A\,B + C\,B\,A + A\,C\,B + B\,A\,C}\\
&\quad = \{\mathrm{tr}(\mathbf{A})\,\mathrm{tr}(\mathbf{B})\,\mathrm{tr}(\mathbf{C}) - \mathrm{tr}(\mathbf{A})\,\mathrm{tr}(\mathbf{B\,C}) - \mathrm{tr}(\mathbf{B})\,\mathrm{tr}(\mathbf{C\,A}) - \mathrm{tr}(\mathbf{C})\,\mathrm{tr}(\mathbf{A\,B})\\
&\quad + \mathrm{tr}(\mathbf{A\,B\,C}) + \mathrm{tr}(\mathbf{A\,C\,B})\}\,\mathbf{1}\\
&\quad - \{\mathrm{tr}(\mathbf{B})\,\mathrm{tr}(\mathbf{C}) - \mathrm{tr}(\mathbf{B\,C})\}\,\mathbf{A} - \{\mathrm{tr}(\mathbf{C})\,\mathrm{tr}(\mathbf{A}) - \mathrm{tr}(\mathbf{C\,A})\}\,\mathbf{B}\\
&\quad - \{\mathrm{tr}(\mathbf{A})\,\mathrm{tr}(\mathbf{B}) - \mathrm{tr}(\mathbf{A\,B})\}\,\mathbf{C}\\
&\quad + 2\,\mathrm{tr}(\mathbf{A})\,\mathbf{B\,C} + 2\,\mathrm{tr}(\mathbf{B})\,\mathbf{C\,A} + 2\,\mathrm{tr}(\mathbf{C})\,\mathbf{A\,B}\,. \qquad (1.4.2)
\end{aligned}$$

Substitution of $\mathbf{B} = \mathbf{C} = \mathbf{A}$ into (1.4.2) yields the Hamilton-Cayley theorem, (1.2.11). In this theorem, *tensor* **A** *need not be symmetric.*

Substituting $\mathbf{C} = \mathbf{A}$ into (1.4.2), obtain another identity relating two second-order arbitrary tensors **A** and **B**,

$$\begin{aligned}
&\mathbf{A}\,(\mathbf{A\,B + B\,A}) + (\mathbf{A\,B + B\,A})\,\mathbf{A} - \mathrm{I}_{\mathbf{A}}\,(\mathbf{A\,B + B\,A}) + \mathrm{II}_{\mathbf{A}}\,\mathbf{B} - \mathbf{A\,B\,A}\\
&\quad = \mathrm{tr}(\mathbf{B})\,\mathbf{A}^2 + \{\mathrm{tr}(\mathbf{A\,B}) - \mathrm{I}_{\mathbf{A}}\,\mathrm{tr}(\mathbf{B})\}\,\mathbf{A}\\
&\quad + \{\mathrm{tr}(\mathbf{A}^2\,\mathbf{B}) - \mathrm{I}_{\mathbf{A}}\,\mathrm{tr}(\mathbf{A\,B}) + \mathrm{II}_{\mathbf{A}}\,\mathrm{tr}(\mathbf{B})\}\,\mathbf{1}\,. \qquad (1.4.3)
\end{aligned}$$

Replace **B** in (1.4.3) by $\mathbf{A\,B + B\,A}$, and use the Hamilton-Cayley theorem to arrive at another identity,

$$\begin{aligned}
&\mathbf{A}\,(\mathbf{A\,B + B\,A})\,\mathbf{A} + \mathrm{III}_{\mathbf{A}}\,\mathbf{B} - \mathrm{I}_{\mathbf{A}}\,\mathbf{A\,B\,A}\\
&\quad = \mathrm{tr}(\mathbf{A\,B})\,\mathbf{A}^2 + \{\mathrm{tr}(\mathbf{A}^2\,\mathbf{B}) - \mathrm{I}_{\mathbf{A}}\,\mathrm{tr}(\mathbf{A\,B})\}\,\mathbf{A} + \mathrm{III}_{\mathbf{A}}\,\mathrm{tr}(\mathbf{B})\,\mathbf{1}\,. \qquad (1.4.4)
\end{aligned}$$

Similarly, replace **B** in (1.4.3) by $\mathbf{A\,B\,A}$ to obtain

$$\begin{aligned}
&\mathbf{A}^2\,\mathbf{B}\,\mathbf{A}^2 + \mathrm{III}_{\mathbf{A}}\,(\mathbf{A\,B + B\,A}) - \mathrm{II}_{\mathbf{A}}\,\mathbf{A\,B\,A}\\
&\quad = \mathrm{tr}(\mathbf{A}^2\,\mathbf{B})\,\mathbf{A}^2 + \{\mathrm{III}_{\mathbf{A}}\,\mathrm{tr}(\mathbf{B}) - \mathrm{II}_{\mathbf{A}}\,\mathrm{tr}(\mathbf{A\,B})\}\,\mathbf{A} + \mathrm{III}_{\mathbf{A}}\,\mathrm{tr}(\mathbf{A\,B})\,\mathbf{1}\,. \qquad (1.4.5)
\end{aligned}$$

Identities (1.4.2) to (1.4.5) are referred to as Rivlin's identities.

1.4.2. Other Related Identities

Eliminate the term $\mathbf{A\,B\,A}$ from (1.4.3) and (1.4.4) to arrive at

$$\begin{aligned}
(\mathrm{I}_{\mathbf{A}}\,\mathrm{II}_{\mathbf{A}} - \mathrm{III}_{\mathbf{A}})\,\mathbf{B} &= \mathbf{A}\,(\mathbf{A\,B + B\,A})\,\mathbf{A} + \mathrm{I}_{\mathbf{A}}^2\,(\mathbf{A\,B + B\,A})\\
&\quad - \mathrm{I}_{\mathbf{A}}\,\{\mathbf{A}\,(\mathbf{A\,B + B\,A}) + (\mathbf{A\,B + B\,A})\,\mathbf{A}\} + \{\mathrm{I}_{\mathbf{A}}\,\mathrm{tr}(\mathbf{B}) - \mathrm{tr}(\mathbf{A\,B})\}\,\mathbf{A}^2\\
&\quad - \{\mathrm{I}_{\mathbf{A}}^2\,\mathrm{tr}(\mathbf{B}) - 2\,\mathrm{I}_{\mathbf{A}}\,\mathrm{tr}(\mathbf{A\,B}) + \mathrm{tr}(\mathbf{A}^2\,\mathbf{B})\}\,\mathbf{A}\\
&\quad + \mathrm{III}_{\mathbf{A}}\,\{\mathrm{I}_{\mathbf{A}}\,\mathrm{tr}(\mathbf{A}^{-1}\,\mathbf{B}) - \mathrm{tr}(\mathbf{B})\}\,\mathbf{1}\,. \qquad (1.4.6)
\end{aligned}$$

Following Scheidler (1994), express identity (1.4.6) as

$$\begin{aligned}
\mathrm{III}_{\tilde{\mathbf{A}}}\,\mathbf{B} &= \tilde{\mathbf{A}}\,(\mathbf{A\,B + B\,A})\,\tilde{\mathbf{A}} + \mathrm{tr}(\tilde{\mathbf{A}}\,\mathbf{B})\,\mathbf{A}^2 - \mathrm{tr}(\tilde{\mathbf{A}}^2\,\mathbf{B})\,\mathbf{A}\\
&\quad + \mathrm{III}_{\mathbf{A}}\,\mathrm{tr}(\mathbf{A}^{-1}\,\tilde{\mathbf{A}}\,\mathbf{B})\,\mathbf{1}\,, \qquad (1.4.7a)
\end{aligned}$$

with

$$\tilde{\mathbf{A}} = \mathrm{I}_{\mathbf{A}}\,\mathbf{1} - \mathbf{A} \tag{1.4.7b}$$

defined such that

$$\mathrm{III}_{\tilde{\mathbf{A}}} = \mathrm{I}_{\mathbf{A}}\,\mathrm{II}_{\mathbf{A}} - \mathrm{III}_{\mathbf{A}} = (\lambda_2+\lambda_3)(\lambda_3+\lambda_1)(\lambda_1+\lambda_2)\,, \tag{1.4.7c}$$

where λ_i, i = 1, 2, 3, are the eigenvalues of **A**.

No constraints are placed on second-order tensors **A** and **B** in deriving identities (1.4.3) to (1.4.7). Denote by $F(\mathbf{A})$ the collection of all three-dimensional second-order tensors **H** which satisfy $\mathrm{tr}(\mathbf{H}) = \mathrm{tr}(\mathbf{A}\,\mathbf{H}) = \mathrm{tr}(\mathbf{A}^2\,\mathbf{H}) = 0$, for a given second-order tensor **A**; *i.e.*, set

$$F(\mathbf{A}) \equiv \{\mathbf{H} \mid \mathrm{tr}(\mathbf{H}) = \mathrm{tr}(\mathbf{AH}) = \mathrm{tr}(\mathbf{A}^2\mathbf{H}) = 0;\ \mathbf{A}\ \text{given}\}\,. \tag{1.4.8}$$

For example, if **A** is symmetric, then all skewsymmetric tensors belong to $F(\mathbf{A})$. However, not every $\mathbf{H} \in F(\mathbf{A})$ is skewsymmetric, even if **A** is symmetric.

Substituting **H** for **B** in (1.4.3) to (1.4.7), obtain the following identities for any $\mathbf{H} \in F(\mathbf{A})$:

$$\mathbf{A}(\mathbf{A}\mathbf{H}+\mathbf{H}\mathbf{A}) + (\mathbf{A}\mathbf{H}+\mathbf{H}\mathbf{A})\mathbf{A} - \mathrm{I}_{\mathbf{A}}(\mathbf{A}\mathbf{H}+\mathbf{H}\mathbf{A}) + \mathrm{II}_{\mathbf{A}}\mathbf{H} - \mathbf{A}\mathbf{H}\mathbf{A} = \mathbf{0}\,,$$

$$\mathbf{A}(\mathbf{A}\mathbf{H}+\mathbf{H}\mathbf{A})\mathbf{A} + \mathrm{III}_{\mathbf{A}}\mathbf{H} - \mathrm{I}_{\mathbf{A}}\mathbf{A}\mathbf{H}\mathbf{A} = \mathbf{0}\,,$$

$$\mathbf{A}^2\mathbf{H}\mathbf{A}^2 + \mathrm{III}_{\mathbf{A}}(\mathbf{A}\mathbf{H}+\mathbf{H}\mathbf{A}) - \mathrm{II}_{\mathbf{A}}\mathbf{A}\mathbf{H}\mathbf{A} = \mathbf{0}\,,$$

$$(\mathrm{I}_{\mathbf{A}}\mathrm{II}_{\mathbf{A}} - \mathrm{III}_{\mathbf{A}})\mathbf{H} = \mathbf{A}(\mathbf{A}\mathbf{H}+\mathbf{H}\mathbf{A})\mathbf{A} + \mathrm{I}_{\mathbf{A}}^2(\mathbf{A}\mathbf{H}+\mathbf{H}\mathbf{A}) - \mathrm{I}_{\mathbf{A}}\{\mathbf{A}(\mathbf{A}\mathbf{H}+\mathbf{H}\mathbf{A}) + (\mathbf{A}\mathbf{H}+\mathbf{H}\mathbf{A})\mathbf{A}\}\,,$$

$$\mathrm{III}_{\tilde{\mathbf{A}}}\mathbf{H} = \tilde{\mathbf{A}}(\mathbf{A}\mathbf{H}+\mathbf{H}\mathbf{A})\tilde{\mathbf{A}}\,, \tag{1.4.9a-e}$$

where $\tilde{\mathbf{A}}$ is defined by (1.4.7b).

1.4.3. Solution of AX+XA = B, Given A and B

Consider the linear tensor equation

$$\mathbf{A}\mathbf{X} + \mathbf{X}\mathbf{A} = \mathbf{B}\,, \quad A_{ik}X_{kj} + X_{ik}A_{kj} = B_{ij}\,, \tag{1.4.10a,b}$$

where **A** and **B** are two arbitrary second-order tensors. Seek to solve this for **X** when **A** and **B** are given. From the Hamilton-Cayley theorem and (1.4.10),

$$\mathrm{tr}(\mathbf{X}) = \frac{1}{2\,\mathrm{III}_{\mathbf{A}}}\{\mathrm{tr}(\mathbf{A}^2\mathbf{B}) - \mathrm{I}_{\mathbf{A}}\,\mathrm{tr}(\mathbf{A}\mathbf{B}) + \mathrm{II}_{\mathbf{A}}\,\mathrm{tr}(\mathbf{B})\}\,,$$

$$\mathrm{tr}(\mathbf{A}\mathbf{X}) = \mathrm{tr}(\mathbf{B})/2\,, \quad \mathrm{tr}(\mathbf{A}^2\mathbf{X}) = \mathrm{tr}(\mathbf{A}\mathbf{B})/2\,. \tag{1.4.11a-c}$$

Therefore, $\mathbf{X} \in F(\mathbf{A})$ when $\mathbf{B} \in F(\mathbf{A})$ for nonsingular $\mathbf{A}$ (*i.e.*, for $\mathrm{III}_{\mathbf{A}} \neq 0$); Scheidler (1994). It follows that, for a nonsingular $\mathbf{A}$, when $\mathbf{B} \in F(\mathbf{A})$, the solution of (1.4.10) is readily obtained from (1.4.9e), as

$$\mathrm{III}_{\tilde{\mathbf{A}}}\,\mathbf{X} = \tilde{\mathbf{A}}\,\mathbf{B}\,\tilde{\mathbf{A}} \quad \text{for } \mathbf{B} \in F(\mathbf{A}), \tag{1.4.12a}$$

where $\tilde{\mathbf{A}}$ is given in terms of $\mathbf{A}$ by (1.4.7b). A similar result is obtained by Mehrabadi and Nemat-Nasser (1987) for symmetric $\mathbf{A}$ and skewsymmetric $\mathbf{B}$ and $\mathbf{X}$.

For an arbitrary tensor $\mathbf{B}$ which may not belong to $F(\mathbf{A})$, solution (1.4.12a) does not generally hold. However, from (1.4.7b) and (1.4.10),

$$\begin{aligned} &2\,\mathrm{tr}(\tilde{\mathbf{A}}\,\mathbf{X}) = \mathrm{tr}(\tilde{\mathbf{A}}\,\mathbf{A}^{-1}\,\mathbf{B}), \quad 2\,\mathrm{tr}(\tilde{\mathbf{A}}^2\,\mathbf{X}) = \mathrm{tr}(\tilde{\mathbf{A}}^2\,\mathbf{A}^{-1}\,\mathbf{B}), \\ &2\,\mathrm{tr}(\tilde{\mathbf{A}}\,\mathbf{A}^{-1}\,\mathbf{X}) = \mathrm{tr}(\tilde{\mathbf{A}}\,\mathbf{A}^{-2}\,\mathbf{B}). \end{aligned} \tag{1.4.13a-c}$$

Therefore, the general solution to tensor equation (1.4.10) is obtained from identity (1.4.7a) by replacing $\mathbf{B}$ with $\mathbf{X}$ and using (1.4.10) and (1.4.13). This solution is

$$\begin{aligned} \mathrm{III}_{\tilde{\mathbf{A}}}\,\mathbf{X} &= \tilde{\mathbf{A}}\,\mathbf{B}\,\tilde{\mathbf{A}} \\ &+ \frac{1}{2}\,\{\mathrm{tr}(\tilde{\mathbf{A}}\,\mathbf{A}^{-1}\,\mathbf{B})\,\mathbf{A}^2 - \mathrm{tr}(\tilde{\mathbf{A}}^2\,\mathbf{A}^{-1}\,\mathbf{B})\,\mathbf{A} + \mathrm{III}_{\mathbf{A}}\,\mathrm{tr}(\tilde{\mathbf{A}}\,\mathbf{A}^{-2}\,\mathbf{B})\,\mathbf{1}\}. \end{aligned} \tag{1.4.12b}$$

This general solution of (1.4.12b) readily reduces to (1.4.12a) for $\mathbf{B} \in F(\mathbf{A})$. Hoger and Carlson (1984) use a different approach to obtain the following form for the general solution of (1.4.10):

$$\begin{aligned} 2\,\mathrm{III}_{\mathbf{A}}\,(\mathrm{I}_{\mathbf{A}}\,\mathrm{II}_{\mathbf{A}} - \mathrm{III}_{\mathbf{A}})\,\mathbf{X} &= \mathrm{I}_{\mathbf{A}}\,\mathbf{A}^2\,\mathbf{B}\,\mathbf{A}^2 - \mathrm{I}_{\mathbf{A}}^2\,(\mathbf{A}^2\mathbf{B}\,\mathbf{A} + \mathbf{A}\,\mathbf{B}\,\mathbf{A}^2) \\ &+ (\mathrm{I}_{\mathbf{A}}\,\mathrm{II}_{\mathbf{A}} - \mathrm{III}_{\mathbf{A}})\,(\mathbf{A}^2\,\mathbf{B} + \mathbf{B}\,\mathbf{A}^2) \\ &+ (\mathrm{I}_{\mathbf{A}}^3 + \mathrm{III}_{\mathbf{A}})\,\mathbf{A}\,\mathbf{B}\,\mathbf{A} - \mathrm{I}_{\mathbf{A}}^2\,\mathrm{II}_{\mathbf{A}}\,(\mathbf{A}\,\mathbf{B} + \mathbf{B}\,\mathbf{A}) \\ &+ (\mathrm{I}_{\mathbf{A}}^2\,\mathrm{III}_{\mathbf{A}} + \mathrm{I}_{\mathbf{A}}\,\mathrm{II}_{\mathbf{A}}^2 - \mathrm{II}_{\mathbf{A}}\,\mathrm{III}_{\mathbf{A}})\,\mathbf{B}. \end{aligned} \tag{1.4.12c}$$

This reduces to (1.4.12b) when identities (1.4.3) to (1.4.5) are used. Scheidler (1994) proposes two other methods to solve (1.4.10) for an arbitrary $\mathbf{B}$, but the resulting solutions do not readily reduce to (1.4.12a) for $\mathbf{B} \in F(\mathbf{A})$; the reduction, however, can be effected by the use of identities (1.2.11) and (1.4.3) to (1.4.5). The following results are based on Scheidler (1994).

Pre- and post-multiplication of (1.4.10) by $\mathbf{A}$ results in

$$\begin{aligned} &\mathbf{A}\,(\mathbf{A}\,\mathbf{X}) + (\mathbf{A}\,\mathbf{X})\,\mathbf{A} = \mathbf{A}\,\mathbf{B} = \frac{1}{2}\,(\mathbf{A}\,\mathbf{B} + \mathbf{B}\,\mathbf{A}) + \frac{1}{2}(\mathbf{A}\,\mathbf{B} - \mathbf{B}\,\mathbf{A}), \\ &\mathbf{A}\,(\mathbf{X}\,\mathbf{A}) + (\mathbf{X}\,\mathbf{A})\,\mathbf{A} = \mathbf{B}\,\mathbf{A} = \frac{1}{2}\,(\mathbf{A}\,\mathbf{B} + \mathbf{B}\,\mathbf{A}) - \frac{1}{2}(\mathbf{A}\,\mathbf{B} - \mathbf{B}\,\mathbf{A}). \end{aligned} \tag{1.4.14a,b}$$

Then, for nonsingular $\mathbf{A}$, by moving $\frac{1}{2}(\mathbf{A}\,\mathbf{B} + \mathbf{B}\,\mathbf{A})$ to the left-hand side of (1.4.14a,b), it follows that

$$\mathbf{X} = \frac{1}{2}\,\mathbf{A}^{-1}\,\{\mathbf{B} + \mathbf{M}_{\mathbf{A}}(\mathbf{B})\} = \frac{1}{2}\,\{\mathbf{B} - \mathbf{M}_{\mathbf{A}}(\mathbf{B})\}\,\mathbf{A}^{-1}, \tag{1.4.12d}$$

where $\mathbf{M_A(B)}$ is the solution of

$$\mathbf{A\,M_A(B) + M_A(B)\,A = A\,B - B\,A}\,, \tag{1.4.14c}$$

obtained from (1.4.9e) as

$$\mathbf{M_A(B)} = \frac{1}{\mathrm{III}_{\tilde{\mathbf{A}}}}\,\tilde{\mathbf{A}}\,(\mathbf{A\,B - B\,A})\,\tilde{\mathbf{A}}\,. \tag{1.4.14d}$$

Alternatively, pre-multiplication followed by post-multiplication of expression (1.4.10) by $\mathbf{A}$ yields

$$\begin{aligned}\mathbf{A\,(A\,X\,A) + (A\,X\,A)\,A = A\,B\,A} \\ = \frac{1}{4}\,\{\mathbf{A\,(A\,B + B\,A) + (A\,B + B\,A)\,A}\} \\ - \frac{1}{4}\,\{\mathbf{A\,(A\,B - B\,A) - (A\,B - B\,A)\,A}\}\,.\end{aligned} \tag{1.4.15}$$

Hence, for nonsingular $\mathbf{A}$,

$$\mathbf{X} = \frac{1}{4}\,\mathbf{A}^{-1}\,\{\mathbf{A\,B + B\,A - N_A(B)}\}\,\mathbf{A}^{-1}\,, \tag{1.4.12e}$$

where $\mathbf{N_A(B)}$ is the solution of

$$\mathbf{A\,N_A(B) + N_A(B)\,A = A\,(A\,B - B\,A) - (A\,B - B\,A)\,A}\,, \tag{1.4.12f}$$

obtained from (1.4.9e) as

$$\mathbf{N_A(B)} = \frac{1}{\mathrm{III}_{\tilde{\mathbf{A}}}}\,\tilde{\mathbf{A}}\,(\mathbf{A}^2\,\mathbf{B} + \mathbf{B}\,\mathbf{A}^2 - 2\,\mathbf{A\,B\,A})\,\tilde{\mathbf{A}}\,. \tag{1.4.16}$$

Substituting (1.4.16) into (1.4.12e), now obtain

$$2\,\mathrm{III}_{\tilde{\mathbf{A}}}\,\mathbf{X} = \tilde{\mathbf{A}}\,\mathbf{B}\,\tilde{\mathbf{A}} + \mathrm{II}_{\mathbf{A}}\,\mathbf{B} + \mathrm{I}_{\mathbf{A}}\,\mathrm{III}_{\mathbf{A}}\,\mathbf{A}^{-1}\,\mathbf{B}\,\mathbf{A}^{-1} - \mathrm{III}_{\mathbf{A}}\,(\mathbf{A}^{-1}\,\mathbf{B} + \mathbf{B}\,\mathbf{A}^{-1})\,. \tag{1.4.12g}$$

1.4.4. Solution of $\mathbf{A\,X - X\,A = B}$, Given $\mathbf{A}$ and $\mathbf{B}$

Consider the linear tensor equation

$$\mathbf{A\,X - X\,A = B}\,, \tag{1.4.17}$$

where $\mathbf{A}$ is a general second-order (symmetric or nonsymmetric) tensor and $\mathbf{B} \in F(\mathbf{A})$. Examine separately the case where $\mathbf{A}$ has distinct *eigenvalues* and the case where it has repeated *eigenvalues*.

A with Distinct Eigenvalues: First consider the case where all three *eigenvalues* of $\mathbf{A}$ are distinct. Then, the general solution $\mathbf{X}$ can be split into a part $\overset{*}{\mathbf{X}}$, coaxial with $\mathbf{A}$, and a (normal) remaining part $\mathbf{H_A(X)}$ which belongs to $F(\mathbf{A})$,[9]

$$\mathbf{X} = \overset{*}{\mathbf{X}} + \mathbf{H_A(X)}\,, \quad \overset{*}{\mathbf{X}} = \sum_{i=1}^{3}\,(\hat{\mathbf{e}}^i \cdot \mathbf{X}\,\hat{\mathbf{e}}_i)\,\hat{\mathbf{e}}_i \otimes \hat{\mathbf{e}}^i\,,$$

---

[9] When $\mathbf{A}$ has two complex eigenvalues, then $\overset{*}{\mathbf{X}} = \sum_{i=1}^{3}\,(\hat{\mathbf{e}}^i \cdot \mathbf{X}\,\bar{\hat{\mathbf{e}}}_i)\,\hat{\mathbf{e}}_i \otimes \hat{\mathbf{e}}^i$, where superbar denotes

$$\mathbf{H}_\mathbf{A}(\mathbf{X}) \in F(\mathbf{A}), \tag{1.4.18a-c}$$

where $\hat{\mathbf{e}}_i$ and $\hat{\mathbf{e}}^i$, $i = 1, 2, 3$, are the *eigenvectors* of $\mathbf{A}$ and $\mathbf{A}^T$, respectively. The coaxial part $\overset{*}{\mathbf{X}}$ has the form

$$\overset{*}{\mathbf{X}} = a_0\,\mathbf{1} + a_1\,\mathbf{A} + a_2\,\mathbf{A}^2, \tag{1.4.18d}$$

where a_0, a_1, and a_2 are constants.

Now, consider only the normal part $\mathbf{H}_\mathbf{A}(\mathbf{X})$ and seek to obtain its coordinate-independent expression. Pre- and post-multiply (1.4.17) by $\mathbf{A}$, and subtract to obtain

$$\mathbf{A}^2\,\mathbf{H}_\mathbf{A}(\mathbf{X}) + \mathbf{H}_\mathbf{A}(\mathbf{X})\,\mathbf{A}^2 - 2\,\mathbf{A}\,\mathbf{H}_\mathbf{A}(\mathbf{X})\,\mathbf{A} = \mathbf{A}\,\mathbf{B} - \mathbf{B}\,\mathbf{A}. \tag{1.4.19}$$

Use identity (1.4.9a) in this expression to arrive at

$$\mathbf{C}\,\mathbf{H}_\mathbf{A}(\mathbf{X}) + \mathbf{H}_\mathbf{A}(\mathbf{X})\,\mathbf{C} = \mathbf{A}\,\mathbf{B} - \mathbf{B}\,\mathbf{A}, \tag{1.4.20a}$$

where $\mathbf{C}$ is given by

$$\mathbf{C} = 3\,\mathbf{A}^2 - 2\,\mathrm{I}_\mathbf{A}\,\mathbf{A} + \mathrm{II}_\mathbf{A}\,\mathbf{1}. \tag{1.4.20b}$$

Then, with $\mathrm{III}_{\tilde{\mathbf{C}}} \neq 0$, it follows from (1.4.9e) that

$$\mathbf{H}_\mathbf{A}(\mathbf{X}) = \frac{1}{\mathrm{III}_{\tilde{\mathbf{C}}}}\,\tilde{\mathbf{C}}\,(\mathbf{A}\,\mathbf{B} - \mathbf{B}\,\mathbf{A})\,\tilde{\mathbf{C}}, \qquad \tilde{\mathbf{C}} = \mathrm{I}_\mathbf{C}\,\mathbf{1} - \mathbf{C}. \tag{1.4.21a,b}$$

This can be rewritten as

$$(4\,\mathrm{II}^3_{\mathbf{A}'} + 27\,\mathrm{III}^2_{\mathbf{A}'})\,\mathbf{H}_\mathbf{A}(\mathbf{X}) = 4\,\mathrm{II}^2_{\mathbf{A}'}(\mathbf{B}\,\mathbf{A}' - \mathbf{A}'\,\mathbf{B}) - 3\,\mathrm{II}_{\mathbf{A}'}\,\mathbf{A}'\,(\mathbf{B}\,\mathbf{A}' - \mathbf{A}'\,\mathbf{B})\,\mathbf{A}'$$
$$- 9\,\mathrm{III}_{\mathbf{A}'}(\mathbf{B}\,\mathbf{A}'^2 - \mathbf{A}'^2\,\mathbf{B}), \tag{1.4.21c}$$

where

$$\mathbf{A}' = \mathbf{A} - \frac{1}{3}\,\mathrm{I}_\mathbf{A}\,\mathbf{1}. \tag{1.4.21d}$$

The same result is obtained by Mehrabadi and Nemat-Nasser (1987) and Guo *et al.* (1992), assuming $\mathbf{A}$ is symmetric and $\mathbf{X}$ is skewsymmetric. A similar result is obtained by Balendran and Nemat-Nasser (1995) for skewsymmetric $\mathbf{A}$ and symmetric $\mathbf{X}$.

A with Repeated Eigenvalues: Now, consider the case where two of the three *eigenvalues* of $\mathbf{A}$ are identical, say, $\lambda_2 = \lambda_3$. Then $\mathbf{A}$ can be split into two parts, $\bar{\mathbf{A}}$ and $\mathbf{L}$, such that[10]

$$\mathbf{A} = \bar{\mathbf{A}} + \mathbf{L}, \qquad \mathbf{L} = \frac{1}{\lambda_2 - \lambda_1}\,\{\mathbf{A}^2 - (\lambda_1 + \lambda_2)\,\mathbf{A} + \lambda_1\,\lambda_2\,\mathbf{1}\},$$

---

the complex conjugate.

[10] Note that $\mathbf{A}$ is real-valued. Thus $\bar{\mathbf{A}}$ is just another tensor that satisfies the conditions defined by (1.4.22). In what follows, it is assumed that the eigenvalues and eigenvectors of $\mathbf{A}$ are all real-valued. When there are complex eigenvalues and eigenvectors, then the corresponding expressions, *e.g.*, (1.4.24), must be modified in accord with (1.3.31e,f).

$$\bar{\mathbf{A}}^2 - (\lambda_1 + \lambda_2)\,\bar{\mathbf{A}} + \lambda_1 \lambda_2\, \mathbf{1} = \mathbf{0}, \quad \bar{\mathbf{A}}\,\mathbf{L} = \mathbf{L}\,\bar{\mathbf{A}} = \lambda_2\,\mathbf{L}, \quad \mathbf{L}^2 = \mathbf{0}. \tag{1.4.22a-e}$$

Denote the *eigenvectors* of $\mathbf{A}$ and $\mathbf{A}^T$, corresponding to the *eigenvalue* λ_1, by $\hat{\mathbf{e}}_1$ and $\hat{\mathbf{e}}^1$, respectively. Let $\hat{\mathbf{e}}_2$ and $\hat{\mathbf{e}}_3$ be two nonparallel arbitrary vectors on the plane normal to $\hat{\mathbf{e}}^1$. Then $\bar{\mathbf{A}}$ can be expressed as

$$\bar{\mathbf{A}} = \sum_{i=1}^{3} \lambda_i\, \hat{\mathbf{e}}_i \otimes \hat{\mathbf{e}}^i. \tag{1.4.22f}$$

Split the general solution $\mathbf{X}$ of (1.4.17) into a part $\mathbf{H}_\mathbf{A}(\mathbf{X}) \in F(\mathbf{A})$, and a remaining part $\overset{*}{\mathbf{X}}$, as follows:

$$\mathbf{X} = \overset{*}{\mathbf{X}} + \mathbf{H}_\mathbf{A}(\mathbf{X}), \quad \overset{*}{\mathbf{X}} = (\hat{\mathbf{e}}^1 \cdot \mathbf{X}\,\hat{\mathbf{e}}_1)\,\hat{\mathbf{e}}_1 \otimes \hat{\mathbf{e}}^1 + \sum_{i,j=2}^{3} (\hat{\mathbf{e}}^i \cdot \mathbf{X}\,\hat{\mathbf{e}}_j)\,\hat{\mathbf{e}}_i \otimes \hat{\mathbf{e}}^j,$$

$$\mathbf{H}_\mathbf{A}(\mathbf{X}) = \sum_{i=2}^{3} \{(\hat{\mathbf{e}}^1 \cdot \mathbf{X}\,\hat{\mathbf{e}}_i)\,\hat{\mathbf{e}}_1 \otimes \hat{\mathbf{e}}^i + (\hat{\mathbf{e}}^i \cdot \mathbf{X}\,\hat{\mathbf{e}}_1)\,\hat{\mathbf{e}}_i \otimes \hat{\mathbf{e}}^1\}. \tag{1.4.23a-c}$$

Note that $\hat{\mathbf{e}}_1$ is an *eigenvector* of $\overset{*}{\mathbf{X}}$. However, the other two *eigenvectors* of $\mathbf{X}$, in general, do not coincide with those of $\mathbf{A}$. Therefore, in general, $\overset{*}{\mathbf{X}}$ cannot be expressed as a linear combination of $\mathbf{1}$, $\mathbf{A}$, and $\mathbf{A}^2$.

Now, consider only the normal part $\mathbf{H}_\mathbf{A}(\mathbf{X})$ and seek to obtain its coordinate-independent expression. For $\mathbf{H}_\mathbf{A}(\mathbf{X})$, (1.4.17) can be rewritten as

$$\mathbf{A}\,\mathbf{H}_\mathbf{A}(\mathbf{X}) - \mathbf{H}_\mathbf{A}(\mathbf{X})\,\mathbf{A} = \mathbf{B} - \sum_{i,j=2}^{3} \{\hat{\mathbf{e}}^i \cdot \mathbf{B}\,\hat{\mathbf{e}}_j\}\,\hat{\mathbf{e}}_i \otimes \hat{\mathbf{e}}^j. \tag{1.4.24}$$

Pre- and post-multiply (1.4.24) by $\bar{\mathbf{A}}$ and subtract to obtain

$$\bar{\mathbf{A}}\,\{\mathbf{A}\,\mathbf{H}_\mathbf{A}(\mathbf{X}) - \mathbf{H}_\mathbf{A}(\mathbf{X})\,\mathbf{A}\} - \{\mathbf{A}\,\mathbf{H}_\mathbf{A}(\mathbf{X}) - \mathbf{H}_\mathbf{A}(\mathbf{X})\,\mathbf{A}\}\,\bar{\mathbf{A}} = \bar{\mathbf{A}}\,\mathbf{B} - \mathbf{B}\,\bar{\mathbf{A}} \tag{1.4.25a}$$

which can be rearranged into

$$\mathbf{A}\,\{\bar{\mathbf{A}}\,\mathbf{H}_\mathbf{A}(\mathbf{X}) - \mathbf{H}_\mathbf{A}(\mathbf{X})\,\bar{\mathbf{A}}\} - \{\bar{\mathbf{A}}\,\mathbf{H}_\mathbf{A}(\mathbf{X}) - \mathbf{H}_\mathbf{A}(\mathbf{X})\,\bar{\mathbf{A}}\}\,\mathbf{A} = \bar{\mathbf{A}}\,\mathbf{B} - \mathbf{B}\,\bar{\mathbf{A}}. \tag{1.4.25b}$$

From (1.4.22f) and (1.4.23c),

$$\bar{\mathbf{A}}\,\mathbf{H}_\mathbf{A}(\mathbf{X}) + \mathbf{H}_\mathbf{A}(\mathbf{X})\,\bar{\mathbf{A}} = (\lambda_1 + \lambda_2)\,\mathbf{H}_\mathbf{A}(\mathbf{X}), \quad \mathbf{L}\,\mathbf{H}_\mathbf{A}(\mathbf{X})\,\mathbf{L} = \mathbf{0}, \tag{1.4.26a,b}$$

and, in view of (1.4.22d),

$$\mathbf{L}\,\{\bar{\mathbf{A}}\,\mathbf{H}_\mathbf{A}(\mathbf{X}) - \mathbf{H}_\mathbf{A}(\mathbf{X})\,\bar{\mathbf{A}}\} - \{\bar{\mathbf{A}}\,\mathbf{H}_\mathbf{A}(\mathbf{X}) - \mathbf{H}_\mathbf{A}(\mathbf{X})\,\bar{\mathbf{A}}\}\,\mathbf{L}$$
$$= (\lambda_2 - \lambda_1)\,\{\mathbf{L}\,\mathbf{H}_\mathbf{A}(\mathbf{X}) + \mathbf{H}_\mathbf{A}(\mathbf{X})\,\mathbf{L}\}. \tag{1.4.26c}$$

Substitute (1.4.26c) into (1.4.26a) to obtain

$$\bar{\mathbf{A}}^2\,\mathbf{H}_\mathbf{A}(\mathbf{X}) + \mathbf{H}_\mathbf{A}(\mathbf{X})\,\bar{\mathbf{A}}^2 - 2\,\bar{\mathbf{A}}\,\mathbf{H}_\mathbf{A}(\mathbf{X})\,\bar{\mathbf{A}}$$
$$= \bar{\mathbf{A}}\,\mathbf{B} - \mathbf{B}\,\bar{\mathbf{A}} + (\lambda_1 - \lambda_2)\,\{\mathbf{L}\,\mathbf{H}_\mathbf{A}(\mathbf{X}) + \mathbf{H}_\mathbf{A}(\mathbf{X})\,\mathbf{L}\}. \tag{1.4.27a}$$

In view of (1.4.22c), identity (1.4.9a) becomes

$$\bar{\mathbf{A}}\,\mathbf{H}_{\mathbf{A}}(\mathbf{X})\,\bar{\mathbf{A}} = \lambda_2\,\{\bar{\mathbf{A}}\,\mathbf{H}_{\mathbf{A}}(\mathbf{X}) + \mathbf{H}_{\mathbf{A}}(\mathbf{X})\,\bar{\mathbf{A}}\} - \lambda_2^2\,\mathbf{H}_{\mathbf{A}}(\mathbf{X})\,. \tag{1.4.27b}$$

Substitute (1.4.22c), (1.4.26a), and (1.4.27b) into (1.4.27a) to obtain

$$\mathbf{H}_{\mathbf{A}}(\mathbf{X}) = \frac{1}{(\lambda_1-\lambda_2)^2}\,(\bar{\mathbf{A}}\,\mathbf{B} - \mathbf{B}\,\bar{\mathbf{A}}) + \frac{1}{\lambda_1-\lambda_2}\,\{\mathbf{L}\,\mathbf{H}_{\mathbf{A}}(\mathbf{X}) + \mathbf{H}_{\mathbf{A}}(\mathbf{X})\,\mathbf{L}\}\,. \tag{1.4.28a}$$

Then, it follows from (1.4.22d) and (1.4.26b) that

$$\mathbf{L}\,\mathbf{H}_{\mathbf{A}}(\mathbf{X}) + \mathbf{H}_{\mathbf{A}}(\mathbf{X})\,\mathbf{L} = \frac{1}{(\lambda_1-\lambda_2)^2}\,\{\mathbf{L}\,(\bar{\mathbf{A}}\,\mathbf{B} - \mathbf{B}\,\bar{\mathbf{A}}) + (\bar{\mathbf{A}}\,\mathbf{B} - \mathbf{B}\,\bar{\mathbf{A}})\,\mathbf{L}\}\,. \tag{1.4.28b}$$

Substitution of (1.4.28b) into (1.4.28a) results in the final desired expression,

$$\begin{aligned}\mathbf{H}_{\mathbf{A}}(\mathbf{X}) = {} & \frac{1}{(\lambda_1-\lambda_2)^2}\,(\bar{\mathbf{A}}\,\mathbf{B} - \mathbf{B}\,\bar{\mathbf{A}}) \\ & + \frac{1}{(\lambda_1-\lambda_2)^3}\,\{\mathbf{L}\,(\bar{\mathbf{A}}\,\mathbf{B} - \mathbf{B}\,\bar{\mathbf{A}}) + (\bar{\mathbf{A}}\,\mathbf{B} - \mathbf{B}\,\bar{\mathbf{A}})\,\mathbf{L}\}\,.\end{aligned} \tag{1.4.28c}$$

1.5. ISOTROPIC TENSOR-VALUED FUNCTIONS

Isotropic tensor-valued functions of symmetric second-order tensors are used to express various strain measures in the kinematics of finite deformation. These strain measures are often expressed in spectral form with respect to the principal triad of the *right stretch tensor* (*left stretch tensor*, if the strain measure is *Eulerian*); Hill (1968). Using the Hamilton-Cayley theorem, a strain measure of this kind can be represented in the coordinate-independent form. Ting (1985) obtains closed-form expressions for the general isotropic functions of a symmetric second-order tensor. Among the various strain measures, logarithmic strain is considered to have certain advantages in constitutive modeling; Hill (1970). However, the relation between the rate of stretch (or deformation rate) and the material time derivative of the logarithmic strain is complicated. Hill (1970) obtains this relation in component form with respect to the principal triad of the right stretch tensor. Later, Hill (1978) obtains a relation between the deformation rate and the material time derivative of a general strain measure. This too is in component form with respect to the principal triad of the right stretch tensor. Gurtin and Spear (1983) obtain a relation for the Jaumann derivative[11] of the logarithmic strain in terms of the deformation rate and the spin of the principal stretches. Carlson and Hoger (1986) obtain a coordinate-independent expression for the derivative of the isotropic tensor-valued functions of a symmetric second-order tensor. Hoger (1986) uses this expression to obtain a coordinate-independent expression for the material time derivative of the logarithmic strain. Scheidler (1991a) provides an alternative proof for Hill's (1978) formula when the principal stretches are repeated. Scheidler (1991b) then gives approximate coordinate-independent formulae for the time derivatives of the generalized strain tensors.

Thus, the isotropic tensor-valued functions of symmetric second-order tensors and their time derivatives, have received considerable attention in the kinematics of finite deformation. Those of nonsymmetric tensors seem to have been addressed only recently. In this section, exact explicit coordinate-independent expressions are given for a class of isotropic tensor-valued functions of nonsymmetric second-order tensors, following Balendran and Nemat-Nasser (1996).

1.5.1. A Class of Isotropic Tensor-valued Functions of Real-valued Second-order Tensors

Let f(x) be a suitably differentiable function, admitting a uniformly convergent Taylor series expansion. Then, to any desired degree of accuracy, f(x) can be represented by

$$f(x) = \sum_{m=0}^{n} \alpha_m x^m, \tag{1.5.1a}$$

where α_i's are constants, and n is a suitable large integer. Consider a class of tensor-valued functions $\mathbf{f}(\mathbf{A})$ that can be expressed as follows:

$$\mathbf{f}(\mathbf{A}) = \sum_{m=0}^{n} \alpha_m \mathbf{A}^m. \tag{1.5.1b}$$

This tensor-valued function is isotropic. Hence,

$$\mathbf{f}(\mathbf{R}^T \mathbf{A}\, \mathbf{R}) = \mathbf{R}^T \mathbf{f}(\mathbf{A})\, \mathbf{R},$$

for any proper orthogonal (rotation) tensor **R**. At focus here is the class (1.5.1b) of isotropic tensor-valued functions of any real-valued second-order tensor with distinct, repeated, or complex-valued eigenvalues.

Substituting the coordinate-independent expressions for $\mathbf{A}^m$, m = 1, 2, ..., n, from (1.3.17d), (1.3.22d), and (1.3.25a) into (1.5.1b), obtain

$$\mathbf{f}(\mathbf{A}) = \mathbf{f}(\bar{\mathbf{A}}) + f'(\lambda)\,\mathbf{L} + \frac{1}{2} f''(\lambda)\,\mathbf{L}^2, \tag{1.5.2a}$$

where

$$\mathbf{f}(\bar{\mathbf{A}}) = \sum_{m=0}^{n} \alpha_m \bar{\mathbf{A}}^m, \quad \bar{\mathbf{A}} = \mathbf{A} - \mathbf{L}, \tag{1.5.2b,c}$$

$$\mathbf{L} = \mathbf{0}, \quad \text{for } \lambda_1 \neq \lambda_2 \neq \lambda_3 \neq \lambda_1,$$

[11] See Subsection 2.10.2, page 84.

$$\mathbf{L} = \frac{1}{\lambda_2-\lambda_1}\{\mathbf{A}^2-(\lambda_1+\lambda_2)\mathbf{A}+\lambda_1\lambda_2\mathbf{1}\}, \quad \text{for } \lambda_1 \neq \lambda_2 = \lambda_3 \equiv \lambda,$$

$$\mathbf{L} = \mathbf{A}-\lambda\mathbf{1}, \quad \text{for } \lambda_1 = \lambda_2 = \lambda_3 \equiv \lambda. \tag{1.5.3a-c}$$

The coordinate-independent expression for $\mathbf{f}(\bar{\mathbf{A}})$ is given by

$$\mathbf{f}(\bar{\mathbf{A}}) = \sum_{i=1}^{3} f(\lambda_i)\mathbf{E}_i, \quad \text{for } \lambda_1 \neq \lambda_2 \neq \lambda_3 \neq \lambda_1,$$

$$\mathbf{f}(\bar{\mathbf{A}}) = f(\lambda_1)\mathbf{E}_1 + f(\lambda_2)(\mathbf{1}-\mathbf{E}_1), \quad \text{for } \lambda_1 \neq \lambda_2 = \lambda_3 \equiv \lambda,$$

$$\mathbf{f}(\bar{\mathbf{A}}) = f(\lambda)\mathbf{1}, \quad \text{for } \lambda_1 = \lambda_2 = \lambda_3 \equiv \lambda, \tag{1.5.4a-c}$$

where

$$\mathbf{E}_i = \frac{1}{(\lambda_i-\lambda_j)(\lambda_i-\lambda_k)}\{\mathbf{A}^2-(\lambda_j+\lambda_k)\mathbf{A}+\lambda_j\lambda_k\mathbf{1}\}, \quad (\text{j, k not summed}). \tag{1.5.4d}$$

Substituting (1.5.3) and (1.5.4) into (1.5.2), obtain

$$\mathbf{f}(\mathbf{A}) = a_0\mathbf{1} + a_1\mathbf{A} + a_2\mathbf{A}^2, \tag{1.5.5}$$

where a_0, a_1, and a_2 are functions of the *eigenvalues* of **A**. These coefficients are given explicitly in terms of the *eigenvalues* of **A**, as discussed below.

1.5.2. A with Distinct Real Eigenvalues

When **A** admits three distinct real *eigenvalues*, substitution of (1.5.3a) and (1.5.4a,d) into (1.5.2a) results in

$$a_0 = \sum_{i=1}^{3} \frac{\lambda_j\lambda_k f(\lambda_i)}{(\lambda_i-\lambda_j)(\lambda_i-\lambda_k)},$$

$$a_1 = -\sum_{i=1}^{3} \frac{(\lambda_j+\lambda_k) f(\lambda_i)}{(\lambda_i-\lambda_j)(\lambda_i-\lambda_k)},$$

$$a_2 = \sum_{i=1}^{3} \frac{f(\lambda_i)}{(\lambda_i-\lambda_j)(\lambda_i-\lambda_k)}, \quad (\text{j, k not summed}). \tag{1.5.6a-c}$$

These can be rearranged into

$$\begin{bmatrix} a_0 \\ -a_1 \\ a_2 \end{bmatrix} = \frac{1}{\Delta}\begin{bmatrix} \lambda_2\lambda_3 & \lambda_3\lambda_1 & \lambda_1\lambda_2 \\ \lambda_2+\lambda_3 & \lambda_3+\lambda_1 & \lambda_1+\lambda_2 \\ 1 & 1 & 1 \end{bmatrix}\begin{bmatrix} f(\lambda_1)(\lambda_2-\lambda_3) \\ f(\lambda_2)(\lambda_3-\lambda_1) \\ f(\lambda_3)(\lambda_1-\lambda_2) \end{bmatrix}, \tag{1.5.6d}$$

where

$$\Delta = -(\lambda_1-\lambda_2)(\lambda_2-\lambda_3)(\lambda_3-\lambda_1). \tag{1.5.6e}$$

Alternatively, the scalar product of (1.5.5) with $\mathbf{E}_i$ yields three linearly independent linear equations for a_0, a_1, and a_2,

$$f(\lambda_i) = a_0 + a_1\lambda_i + a_2\lambda_i^2, \quad i = 1, 2, 3. \tag{1.5.6f}$$

The solution of these linear equations is the same as (1.5.6d,e).

1.5.3. A with Distinct Complex Eigenvalues

Denote the real *eigenvalue* of **A** by λ_1. For a real-valued tensor **A**, the other two *eigenvalues*, say λ_2 and λ_3, are each other's complex conjugate. Denote them by $\lambda_2 = \alpha + i\,\beta$ and $\lambda_3 = \alpha - i\,\beta$. Then, from (1.5.4d), $\mathbf{E}_i$, $i = 1, 2, 3$, can be expressed in terms of λ_1, α, and β as follows:

$$\mathbf{E}_1 = \frac{1}{\Delta}\{\mathbf{A}^2 - 2\alpha\mathbf{A} + (\alpha^2+\beta^2)\,\mathbf{1}\},$$
$$\mathbf{E}_2 = (\mathbf{E}^R + i\,\mathbf{E}^I)/2, \quad \mathbf{E}_3 = (\mathbf{E}^R - i\,\mathbf{E}^I)/2,$$
$$\mathbf{E}^R = \frac{-1}{\Delta}\{\mathbf{A}^2 - 2\alpha\mathbf{A} - \lambda_1(\lambda_1 - 2\alpha)\,\mathbf{1}\},$$
$$\mathbf{E}^I = \frac{1}{\beta\Delta}\{(\lambda_1-\alpha)\,\mathbf{A}^2 + (\alpha^2-\beta^2-\lambda_1^2)\,\mathbf{A} - \lambda_1(\alpha^2-\beta^2-\lambda_1\alpha)\,\mathbf{1}\},$$
$$\Delta = (\lambda_1-\alpha)^2 + \beta^2. \tag{1.5.7a-f}$$

Then, from (1.5.2a), (1.5.3a), and (1.5.4a), **f**(**A**) can be expressed as

$$\mathbf{f}(\mathbf{A}) = f(\lambda_1)\,\mathbf{E}_1 - f^R\,\mathbf{E}^R + f^I\,\mathbf{E}^I, \tag{1.5.8a}$$

where

$$f^R = \mathrm{Re}\{f(\alpha + i\,\beta)\}, \quad f^I = \mathrm{Im}\{f(\alpha + i\,\beta)\}. \tag{1.5.9a,b}$$

For example, if $f(\lambda) = ln\,(\lambda)$, then f^R and f^I are given by

$$f^R = \frac{1}{2}\,ln\,(\alpha^2+\beta^2), \quad f^I = \tan^{-1}(\beta/\alpha). \tag{1.5.9c,d}$$

Similarly, if $f(\lambda) = e^\lambda$, then

$$f^R = e^\alpha \cos\beta, \quad f^I = e^\alpha \sin\beta. \tag{1.5.9e,f}$$

Note that, when β goes to zero, f^I/β goes to $f'(\alpha)$.

Substitution of (1.5.7) into (1.5.8a) results in (1.5.5) with

$$\begin{bmatrix} a_0 \\ -a_1 \\ a_2 \end{bmatrix} = \frac{1}{\Delta}\begin{bmatrix} \alpha^2+\beta^2 & \lambda_1(\lambda_1-2\alpha) & \lambda_1(\alpha^2-\beta^2-\alpha\lambda_1) \\ 2\alpha & -2\alpha & \alpha^2-\beta^2-\lambda_1^2 \\ 1 & -1 & \alpha-\lambda_1 \end{bmatrix}\begin{bmatrix} f(\lambda_1) \\ f^R \\ f^I/\beta \end{bmatrix}. \tag{1.5.8b}$$

The same expressions are obtained for the coefficients a_0, a_1, and a_2, by substituting $\lambda_2 = \alpha + i\,\beta$ and $\lambda_3 = \alpha - i\,\beta$ into (1.5.6d).

1.5.4. A with Two Repeated Eigenvalues

In this case, it follows from (1.5.4b,d) that

$$\mathbf{E}_1 = \frac{1}{(\lambda_1-\lambda_2)^2}\{\mathbf{A}^2 - 2\lambda_2\mathbf{A} + \lambda_2^2\,\mathbf{1}\},$$

$$\mathbf{f}(\overline{\mathbf{A}}) = \{f(\lambda_1) - f(\lambda_2)\}\,\mathbf{E}_1 + f(\lambda_2)\,\mathbf{1}\,. \tag{1.5.10a,b}$$

Then, substitution of (1.5.10) and (1.5.3b) into (1.5.2a) results in (1.5.5) with

$$\begin{aligned}
a_0 &= \frac{\lambda_1 f(\lambda_2) - \lambda_2 f(\lambda_1)}{\lambda_1 - \lambda_2} + a_2\,\lambda_1\,\lambda_2\,,\\
a_1 &= \frac{f(\lambda_1) - f(\lambda_2)}{\lambda_1 - \lambda_2} - a_2\,(\lambda_1 + \lambda_2)\,,\\
a_2 &= \frac{f(\lambda_1) - f(\lambda_2) - (\lambda_1 - \lambda_2)\,f'(\lambda_2)}{(\lambda_1 - \lambda_2)^2}\,.
\end{aligned} \tag{1.5.11a-c}$$

Alternatively, note that, from (1.3.22d), $\mathbf{A}$ and $\mathbf{A}^2$ can be expressed as

$$\begin{aligned}
\mathbf{A} &= \lambda_1\,\mathbf{E}_1 + \lambda_2\,(\mathbf{1} - \mathbf{E}_1) + \mathbf{L}\,,\\
\mathbf{A}^2 &= \lambda_1^2\,\mathbf{E}_1 + \lambda_2^2\,(\mathbf{1} - \mathbf{E}_1) + 2\,\lambda_2\,\mathbf{L}\,.
\end{aligned} \tag{1.5.12a,b}$$

Substitution of (1.5.2a), (1.5.4b), and (1.5.12) into (1.5.5) results in

$$\begin{aligned}
\mathbf{f}(\mathbf{A}) &= f(\lambda_1)\,\mathbf{E}_1 + f(\lambda_2)\,(\mathbf{1} - \mathbf{E}_1) + f'(\lambda_2)\,\mathbf{L}\\
&= (a_0 + a_1\,\lambda_1 + a_2\,\lambda_1^2)\,\mathbf{E}_1\\
&\quad + (a_0 + a_1\,\lambda_2 + a_2\,\lambda_2^2)\,(\mathbf{1} - \mathbf{E}_1)\\
&\quad + (a_1 + 2\,a_2\,\lambda_2)\,\mathbf{L}\,.
\end{aligned} \tag{1.5.13a}$$

This yields three linearly independent linear equations for the coefficients a_0, a_1, and a_2,

$$\begin{aligned}
f(\lambda_i) &= a_0 + a_1\,\lambda_i + a_2\,\lambda_i^2\,, \quad i = 1,\ 2\,,\\
f'(\lambda_2) &= a_1 + 2\,a_2\,\lambda_2\,.
\end{aligned} \tag{1.5.13b,c}$$

The solution of these three equations is the same as (1.5.11a-c).

Note that, when $\mathbf{A}$ is symmetric and $\lambda_2 = \lambda_3$, from (1.3.20a),

$$\mathbf{A}^2 - (\lambda_1 + \lambda_2)\,\mathbf{A} + \lambda_1\,\lambda_2\,\mathbf{1} = \mathbf{0}\,. \tag{1.5.14}$$

In this case, substitute (1.5.11) into (1.5.5) and in view of (1.5.14), obtain

$$\mathbf{f}(\mathbf{A}) = \frac{\lambda_1 f(\lambda_2) - \lambda_2 f(\lambda_1)}{\lambda_1 - \lambda_2}\,\mathbf{1} + \frac{f(\lambda_1) - f(\lambda_2)}{\lambda_1 - \lambda_2}\,\mathbf{A}\,. \tag{1.5.15}$$

1.5.5. A with Three Repeated Eigenvalues

In this case, substitution of (1.5.3c) and (1.5.4c) into (1.5.2a) results in (1.5.5) with

$$\begin{aligned}
a_0 &= f(\lambda) - \lambda\,f'(\lambda) + \frac{1}{2}\,\lambda^2\,f''(\lambda)\,,\\
a_1 &= f'(\lambda) - \lambda\,f''(\lambda)\,, \quad a_2 = \frac{1}{2}\,f''(\lambda)\,.
\end{aligned} \tag{1.5.16a-c}$$

Alternatively, note that, from (1.3.25a), $\mathbf{A}$ and $\mathbf{A}^2$ can be expressed as

$$\mathbf{A} = \lambda\,\mathbf{1} + \mathbf{L}, \quad \mathbf{A}^2 = \lambda^2\,\mathbf{1} + 2\lambda\,\mathbf{L} + \mathbf{L}^2. \tag{1.5.17a,b}$$

Then, substitution of (1.5.2a), (1.5.4c), and (1.5.17) into (1.5.5) results in

$$\begin{aligned}\mathbf{f}(\mathbf{A}) &= f(\lambda)\,\mathbf{1} + f'(\lambda)\,\mathbf{L} + \frac{1}{2}\,f''(\lambda)\,\mathbf{L}^2 \\ &= (a_0 + a_1\,\lambda + a_2\,\lambda^2)\,\mathbf{1} + (a_1 + 2\,a_2\,\lambda)\,\mathbf{L} + a_2\,\mathbf{L}^2.\end{aligned} \tag{1.5.18a}$$

This yields three linearly independent linear equations for a_0, a_1, and a_2,

$$\begin{aligned}&f(\lambda) = a_0 + a_1\,\lambda + a_2\,\lambda^2, \\ &f'(\lambda) = a_1 + 2\,a_2\,\lambda, \quad f''(\lambda) = a_2.\end{aligned} \tag{1.5.18b-d}$$

The solution of these equations is also the same as (1.5.16).

Note that, if $\mathbf{A}$ is symmetric and $\lambda_1 = \lambda_2 = \lambda_3 \equiv \lambda$, then $\mathbf{A} = \lambda\,\mathbf{1}$. In this case, substitution of (1.5.16) into (1.5.5) results in

$$\mathbf{f}(\mathbf{A}) = f(\lambda)\,\mathbf{1}. \tag{1.5.19}$$

1.6. DERIVATIVE OF ISOTROPIC TENSOR-VALUED FUNCTION

This section deals with derivatives of second-order tensor-valued functions of a scalar variable, say, time t, as well as the derivatives of a class of functions of such second-order tensors. First, explicit formulae are given for the time derivative of a general second-order tensor, say, $\mathbf{A} = \mathbf{A}(t)$, examining cases with distinct and repeated eigenvalues separately. Then, a class of isotropic functions of $\mathbf{A}$ is considered and explicit expressions for the time derivative of such a function are obtained, again treating cases where $\mathbf{A}$ has distinct or repeated eigenvalues separately.

Consider the isotropic tensor-valued function $\mathbf{f}(\mathbf{A})$, where $\mathbf{A}$ (not necessarily symmetric) is a differentiable function of t. Denote the derivatives of $\mathbf{A}$ and $\mathbf{f}(\mathbf{A})$ with respect to t by $\dot{\mathbf{A}}$ and $\dot{\mathbf{f}}(\mathbf{A})$, respectively. Here, following Balendran and Nemat-Nasser (1996), exact explicit coordinate-independent expressions are obtained for the derivative of a class of tensor-valued functions of a nonsymmetric tensor.[12]

[12] For a symmetric tensor $\mathbf{A}$, the components of $\dot{\mathbf{f}}(\mathbf{A})$ with respect to the principal coordinates of $\mathbf{A}$ are reported by Truesdell (1961); see also Truesdell and Noll (1965, page 145), and Wang and Truesdell (1973, page 443).

1.6.1. Derivative of a Second-order Tensor

Following the results of Subsection 1.3.7, any real-valued second-order tensor $\dot{\mathbf{A}}$ can be expressed as

$$\dot{\mathbf{A}} = \sum_{i,j=1}^{3} (\hat{\mathbf{e}}^i \cdot \dot{\mathbf{A}}\,\bar{\hat{\mathbf{e}}}_j)\,\hat{\mathbf{e}}_i \otimes \hat{\mathbf{e}}^j\,. \tag{1.6.1}$$

This is split into a part $\overset{*}{\mathbf{A}}$ which is coaxial with $\mathbf{A}$, and a remaining part[13] $\mathbf{H}_\mathbf{A}(\dot{\mathbf{A}}) \in F(\mathbf{A})$, as follows:

$$\dot{\mathbf{A}} = \overset{*}{\mathbf{A}} + \mathbf{H}_\mathbf{A}(\dot{\mathbf{A}})\,. \tag{1.6.2}$$

When $\lambda_1 \neq \lambda_2 \neq \lambda_3 \neq \lambda_1$, then

$$\overset{*}{\mathbf{A}} = \sum_{i=1}^{3} (\hat{\mathbf{e}}^i \cdot \dot{\mathbf{A}}\,\bar{\hat{\mathbf{e}}}_i)\,\hat{\mathbf{e}}_i \otimes \hat{\mathbf{e}}^i\,, \tag{1.6.3a}$$

$$\mathbf{H}_\mathbf{A}(\dot{\mathbf{A}}) = \mathbf{A}\mathbf{X} - \mathbf{X}\mathbf{A}\,, \qquad \mathbf{X} = \sum_{i,j=1}^{3} \frac{1-\delta_{ij}}{\lambda_i - \lambda_j} (\hat{\mathbf{e}}^i \cdot \dot{\mathbf{A}}\,\bar{\hat{\mathbf{e}}}_j)\,\hat{\mathbf{e}}_i \otimes \hat{\mathbf{e}}^j\,, \tag{1.6.3b,c}$$

where $\hat{\mathbf{e}}_i$ and $\hat{\mathbf{e}}^i$, i = 1, 2, 3, are the *eigenvectors* of $\mathbf{A}$ and $\mathbf{A}^T$, respectively, and superimposed bar denotes complex conjugate. Note that, since the *eigenvectors* of $\overset{*}{\mathbf{A}}$ and $\mathbf{A}$ are the same, $\overset{*}{\mathbf{A}}$ is coaxial with $\mathbf{A}$ and can be expressed as a linear combination of $\mathbf{1}$, $\mathbf{A}$, and $\mathbf{A}^2$.

When $\lambda_1 \neq \lambda_2 = \lambda_3$, and $\hat{\mathbf{e}}_1$ is an *eigenvector* of $\mathbf{A}$, then

$$\overset{*}{\mathbf{A}} = (\hat{\mathbf{e}}^1 \cdot \dot{\mathbf{A}} \cdot \hat{\mathbf{e}}_1)\,\hat{\mathbf{e}}_1 \otimes \hat{\mathbf{e}}^1 + \sum_{i,j=2}^{3} (\hat{\mathbf{e}}^i \cdot \dot{\mathbf{A}} \cdot \hat{\mathbf{e}}_j)\,\hat{\mathbf{e}}_i \otimes \hat{\mathbf{e}}^j\,,$$

$$\mathbf{H}_\mathbf{A}(\dot{\mathbf{A}}) = \sum_{i=2}^{3} \{(\hat{\mathbf{e}}^1 \cdot \dot{\mathbf{A}} \cdot \hat{\mathbf{e}}_i)\,\hat{\mathbf{e}}_1 \otimes \hat{\mathbf{e}}^i + (\hat{\mathbf{e}}^i \cdot \dot{\mathbf{A}} \cdot \hat{\mathbf{e}}_1)\,\hat{\mathbf{e}}_i \otimes \hat{\mathbf{e}}^1\} = \bar{\mathbf{A}}\mathbf{X} - \mathbf{X}\bar{\mathbf{A}}\,,$$

$$\mathbf{X} = \frac{1}{\lambda_1 - \lambda_2} \sum_{i=2}^{3} \{(\hat{\mathbf{e}}^1 \cdot \dot{\mathbf{A}} \cdot \hat{\mathbf{e}}_i)\,\hat{\mathbf{e}}_1 \otimes \hat{\mathbf{e}}^i - (\hat{\mathbf{e}}^i \cdot \dot{\mathbf{A}} \cdot \hat{\mathbf{e}}_1)\,\hat{\mathbf{e}}_i \otimes \hat{\mathbf{e}}^1\}\,, \tag{1.6.4a-c}$$

where $\bar{\mathbf{A}} = \mathbf{A} - \mathbf{L}$; see (1.3.20). Note that $\hat{\mathbf{e}}_1$ is an *eigenvector* of both $\mathbf{A}$ and $\overset{*}{\mathbf{A}}$. However, in general, other *eigenvectors* (if any) of $\mathbf{A}$ and $\overset{*}{\mathbf{A}}$ may not be the same. Therefore, $\overset{*}{\mathbf{A}}$ cannot be expressed completely as a linear combination of $\mathbf{1}$, $\mathbf{A}$, and $\mathbf{A}^2$. In view of this, split $\overset{*}{\mathbf{A}}$ into two parts, as follows:

$$\overset{*}{\mathbf{A}} = \overset{\circ}{\mathbf{A}} + \overset{*}{\mathbf{L}}\,,$$

$$\overset{\circ}{\mathbf{A}} \equiv \dot{A}_1^1\,\mathbf{E}_1 + \frac{1}{2}(\dot{A}_2^2 + \dot{A}_3^3)(\mathbf{1} - \mathbf{E}_1) + \dot{A}_2^3\,\hat{\mathbf{e}}_3 \otimes \hat{\mathbf{e}}^2\,, \qquad \mathbf{E}_1 = \hat{\mathbf{e}}_1 \otimes \hat{\mathbf{e}}^1\,,$$

$$\overset{*}{\mathbf{L}} \equiv \frac{1}{2}(\dot{A}_2^2 - \dot{A}_3^3)(\hat{\mathbf{e}}_2 \otimes \hat{\mathbf{e}}^2 - \hat{\mathbf{e}}_3 \otimes \hat{\mathbf{e}}^3) + \dot{A}_3^2\,\hat{\mathbf{e}}_2 \otimes \hat{\mathbf{e}}^3\,. \tag{1.6.5a-d}$$

Since the *eigenvectors* of $\mathbf{A}$ and $\overset{\circ}{\mathbf{A}}$ are the same, $\overset{\circ}{\mathbf{A}}$ is coaxial with $\mathbf{A}$ and can be completely expressed as a linear combination of $\mathbf{1}$, $\mathbf{A}$, and $\mathbf{A}^2$.

---

13 See definition (1.4.8), page 29.

When $\lambda_1 = \lambda_2 = \lambda_3 \equiv \lambda$, then

$$\overset{*}{\dot{\mathbf{A}}} = \dot{\mathbf{A}}, \quad \mathbf{H}_\mathbf{A}(\dot{\mathbf{A}}) = \mathbf{0}. \tag{1.6.6a,b}$$

1.6.2. Coordinate-independent Representation

The coordinate-independent expression for $\mathbf{H}_\mathbf{A}(\dot{\mathbf{A}})$ is obtained by solving (see Section 1.4.4) the following equations, depending on the nature of the *eigenvalues* of $\mathbf{A}$:

$$\mathbf{A}\,\mathbf{H}_\mathbf{A}(\dot{\mathbf{A}}) - \mathbf{H}_\mathbf{A}(\dot{\mathbf{A}})\,\mathbf{A} = \mathbf{A}\,\dot{\mathbf{A}} - \dot{\mathbf{A}}\,\mathbf{A}, \quad \text{for } \lambda_1 \neq \lambda_2 \neq \lambda_3 \neq \lambda_1,$$

$$\overline{\mathbf{A}}\,\mathbf{H}_\mathbf{A}(\dot{\mathbf{A}}) - \mathbf{H}_\mathbf{A}(\dot{\mathbf{A}})\,\overline{\mathbf{A}} = \overline{\mathbf{A}}\,\dot{\mathbf{A}} - \dot{\mathbf{A}}\,\overline{\mathbf{A}}, \quad \text{for } \lambda_1 \neq \lambda_2 = \lambda_3 \equiv \lambda, \tag{1.6.7a,b}$$

where $\overline{\mathbf{A}}$ is defined by (1.5.2c). This leads to

$$\begin{aligned}\mathbf{H}_\mathbf{A}(\dot{\mathbf{A}}) = \frac{1}{4\,\mathrm{II}_{\mathbf{A}'}^3 + 27\,\mathrm{III}_{\mathbf{A}'}^2}\{&18\,\mathrm{III}_{\mathbf{A}'}^2\,\dot{\mathbf{A}} - 6\,\mathrm{II}_{\mathbf{A}'}\,\mathrm{III}_{\mathbf{A}'}\,(\mathbf{A}'\,\dot{\mathbf{A}} + \dot{\mathbf{A}}\,\mathbf{A}') \\ &+ 2\,\mathrm{II}_{\mathbf{A}'}^2\,\mathbf{A}'\,\dot{\mathbf{A}}\,\mathbf{A}' - 4\,\mathrm{II}_{\mathbf{A}'}^2\,(\mathbf{A}'^2\,\dot{\mathbf{A}} + \dot{\mathbf{A}}\,\mathbf{A}'^2) \\ &- 9\,\mathrm{III}_{\mathbf{A}'}\,\mathbf{A}'(\mathbf{A}'\,\dot{\mathbf{A}} + \dot{\mathbf{A}}\,\mathbf{A}')\,\mathbf{A}' - 6\,\mathrm{II}_{\mathbf{A}'}\,\mathbf{A}'^2\,\dot{\mathbf{A}}\,\mathbf{A}'^2\},\end{aligned}$$

$$\text{for } \lambda_1 \neq \lambda_2 \neq \lambda_3 \neq \lambda_1, \tag{1.6.8a}$$

and

$$\mathbf{H}_\mathbf{A}(\dot{\mathbf{A}}) = \frac{1}{(\lambda_1 - \lambda_2)^2}\{(\lambda_1 + \lambda_2)(\overline{\mathbf{A}}\,\dot{\mathbf{A}} + \dot{\mathbf{A}}\,\overline{\mathbf{A}}) - 2\lambda_1\lambda_2\,\dot{\mathbf{A}} - 2\,\overline{\mathbf{A}}\,\dot{\mathbf{A}}\,\overline{\mathbf{A}}\},$$

$$\text{for } \lambda_1 \neq \lambda_2 = \lambda_3 \equiv \lambda. \tag{1.6.8b}$$

1.6.3. Representation of $\dot{\mathbf{f}}(\mathbf{A})$

Differentiation of (1.5.1) with respect to t leads to

$$\dot{\mathbf{f}}(\mathbf{A}) \equiv \mathbf{g}(\mathbf{A}, \dot{\mathbf{A}}) = \sum_{m=0}^{n-1} \alpha_{m+1} \sum_{k=0}^{m} \mathbf{A}^k\,\dot{\mathbf{A}}\,\mathbf{A}^{m-k}. \tag{1.6.9}$$

Substitution of (1.6.2) into (1.6.9) results in

$$\dot{\mathbf{f}}(\mathbf{A}) = \mathbf{g}(\mathbf{A}, \overset{*}{\dot{\mathbf{A}}}) + \mathbf{g}\{\mathbf{A}, \mathbf{H}_\mathbf{A}(\dot{\mathbf{A}})\},$$

$$\mathbf{g}(\mathbf{A}, \overset{*}{\dot{\mathbf{A}}}) = \sum_{m=0}^{n-1} \alpha_{m+1} \sum_{k=0}^{m} \mathbf{A}^k\,\overset{*}{\dot{\mathbf{A}}}\,\mathbf{A}^{m-k},$$

$$\mathbf{g}\{\mathbf{A}, \mathbf{H}_\mathbf{A}(\dot{\mathbf{A}})\} = \sum_{m=0}^{n-1} \alpha_{m+1} \sum_{k=0}^{m} \mathbf{A}^k\,\mathbf{H}_\mathbf{A}(\dot{\mathbf{A}})\,\mathbf{A}^{m-k}. \tag{1.6.10a-c}$$

Now, consider the term $\mathbf{A}^{k}\,\overset{*}{\mathbf{A}}\,\mathbf{A}^{m-k}$ in the double series for $\mathbf{g}(\mathbf{A}\,,\,\overset{*}{\mathbf{A}})$.

Distinct Eigenvalues: When the *eigenvalues* of $\mathbf{A}$ are distinct,

$$\mathbf{A}^{k}\,\overset{*}{\mathbf{A}}\,\mathbf{A}^{m-k} = \sum_{i=1}^{3} (\lambda_i)^m\,(\hat{\mathbf{e}}^i \cdot \dot{\mathbf{A}}\,\bar{\hat{\mathbf{e}}}_i)\,\hat{\mathbf{e}}_i \otimes \hat{\mathbf{e}}^i = \mathbf{A}^m\,\overset{*}{\mathbf{A}} = \overset{*}{\mathbf{A}}\,\mathbf{A}^m\,. \qquad (1.6.11)$$

Substitution of (1.6.11) into (1.6.10b) results in

$$\mathbf{g}(\mathbf{A}\,,\,\overset{*}{\mathbf{A}}) = \mathbf{f}'(\mathbf{A})\,\overset{*}{\mathbf{A}} = \overset{*}{\mathbf{A}}\,\mathbf{f}'(\mathbf{A})\,,$$

$$\mathbf{f}'(\mathbf{A}) = \sum_{m=0}^{n} (m+1)\,\alpha_{m+1}\,\mathbf{A}^m\,. \qquad (1.6.12a,b)$$

Then

$$\dot{\mathbf{f}}(\mathbf{A}) = \mathbf{f}'(\mathbf{A})\,\overset{*}{\mathbf{A}} + \mathbf{g}\{\mathbf{A}\,,\,\mathbf{H}_A(\dot{\mathbf{A}})\}\,, \qquad \text{for } \lambda_1 \neq \lambda_2 \neq \lambda_3 \neq \lambda_1\,. \qquad (1.6.13)$$

Two Repeated Eigenvalues: When the *eigenvalues* of $\mathbf{A}$ are repeated, say $\lambda_1 \neq \lambda_2 = \lambda_3$, it follows from (1.3.22d) that

$$\mathbf{A}^{k}\,\overset{*}{\mathbf{A}}\,\mathbf{A}^{m-k} = \{(\lambda_1)^k\,\mathbf{E}_1 + (\lambda_2)^k\,(\mathbf{1}-\mathbf{E}_1) + k\,(\lambda_2)^{k-1}\,\mathbf{L}\}\,\overset{*}{\mathbf{A}}\,\{(\lambda_1)^{m-k}\,\mathbf{E}_1$$
$$+ (\lambda_2)^{m-k}\,(\mathbf{1}-\mathbf{E}_1) + (m-k)\,(\lambda_2)^{m-k-1}\,\mathbf{L}\}\,,$$

$$\mathbf{E}_1 = \hat{\mathbf{e}}_1 \otimes \hat{\mathbf{e}}^1\,, \qquad \mathbf{L} = \hat{A}_2^3\,\hat{\mathbf{e}}_3 \otimes \hat{\mathbf{e}}^2\,. \qquad (1.6.14a\text{-}c)$$

Note the following relation from (1.6.4a) and (1.6.14):

$$\overset{*}{\mathbf{A}}\,\mathbf{E}_1 = \mathbf{E}_1\,\overset{*}{\mathbf{A}} = \dot{A}_1^1\,\mathbf{E}_1\,,$$

$$\mathbf{E}_1\,(\mathbf{1}-\mathbf{E}_1) = (\mathbf{1}-\mathbf{E}_1)\,\mathbf{E}_1 = \mathbf{E}_1\,\mathbf{L} = \mathbf{L}\,\mathbf{E}_1 = \mathbf{0}\,. \qquad (1.6.15a,b)$$

Substitution of (1.6.15) into (1.6.14a) results in

$$\mathbf{A}^{k}\,\overset{*}{\mathbf{A}}\,\mathbf{A}^{m-k} = \dot{A}_1^1\,(\lambda_1)^m\,\mathbf{E}_1 + (\lambda_2)^m\,(\mathbf{1}-\mathbf{E}_1)\,\overset{*}{\mathbf{A}}\,(\mathbf{1}-\mathbf{E}_1)$$
$$+ (m-k)\,(\lambda_2)^{m-1}\,\overset{*}{\mathbf{A}}\,\mathbf{L} + k\,(\lambda_2)^{m-1}\,\mathbf{L}\,\overset{*}{\mathbf{A}}$$
$$+ k\,(m-k)\,(\lambda_2)^{m-2}\,\mathbf{L}\,\overset{*}{\mathbf{A}}\,\mathbf{L}\,, \qquad \text{for } m \neq 0\,,$$

$$\mathbf{A}^{k}\,\overset{*}{\mathbf{A}}\,\mathbf{A}^{m-k} = \overset{*}{\mathbf{A}} = \mathring{\mathbf{A}} + \overset{*}{\mathbf{L}}\,, \qquad \text{for } m = k = 0\,. \qquad (1.6.16a,b)$$

Note that, from (1.6.5) and (1.6.14),

$$\overset{*}{\mathbf{A}}\,(\mathbf{1}-\mathbf{E}_1) = (\mathbf{1}-\mathbf{E}_1)\,\overset{*}{\mathbf{A}} = (\mathbf{1}-\mathbf{E}_1)\,\overset{*}{\mathbf{A}}\,(\mathbf{1}-\mathbf{E}_1)$$
$$= \frac{1}{2}\,(\dot{A}_2^2 + \dot{A}_3^3)\,(\mathbf{1}-\mathbf{E}_1) + \dot{A}_2^3\,\hat{\mathbf{e}}_3 \otimes \hat{\mathbf{e}}^2 + \overset{*}{\mathbf{L}}\,,$$

$$\mathbf{L}\,\overset{*}{\mathbf{A}} = \mathbf{L}\,\mathring{\mathbf{A}} + \frac{1}{2}\,(\dot{A}_2^2 - \dot{A}_3^3)\,\hat{A}_2^3\,\hat{\mathbf{e}}_3 \otimes \hat{\mathbf{e}}^2 + \hat{A}_2^3\,\dot{A}_3^2\,\hat{\mathbf{e}}_3 \otimes \hat{\mathbf{e}}^3$$
$$= \mathbf{L}\,\mathring{\mathbf{A}} + \frac{1}{2}\,(\dot{A}_2^2 - \dot{A}_3^3)\,\mathbf{L} + \mathrm{tr}(\dot{\mathbf{A}}\,\mathbf{L})\,\hat{\mathbf{e}}_3 \otimes \hat{\mathbf{e}}^3\,,$$

$$\overset{*}{\mathbf{A}}\,\mathbf{L} = \mathring{\mathbf{A}}\,\mathbf{L} - \frac{1}{2}\,(\dot{A}_2^2 - \dot{A}_3^3)\,\hat{A}_2^3\,\hat{\mathbf{e}}_3 \otimes \hat{\mathbf{e}}^2 + \hat{A}_2^3\,\dot{A}_3^2\,\hat{\mathbf{e}}_2 \otimes \hat{\mathbf{e}}^2$$
$$= \mathring{\mathbf{A}}\,\mathbf{L} - \frac{1}{2}\,(\dot{A}_2^2 - \dot{A}_3^3)\,\mathbf{L} + \mathrm{tr}(\dot{\mathbf{A}}\,\mathbf{L})\,\hat{\mathbf{e}}_2 \otimes \hat{\mathbf{e}}^2\,,$$

$$\mathbf{L}\overset{*}{\dot{\mathbf{A}}}\mathbf{L} = \hat{A}_2^3\dot{A}_3^2\hat{A}_2^3\,\hat{\mathbf{e}}_3\otimes\hat{\mathbf{e}}^3 = \mathrm{tr}(\dot{\mathbf{A}}\,\mathbf{L})\,\mathbf{L}\,. \tag{1.6.17a-d}$$

Then, from (1.6.16) and (1.6.17),

$$\begin{aligned}\sum_{k=0}^{m}\mathbf{A}^k\overset{*}{\dot{\mathbf{A}}}\mathbf{A}^{m-k} &= \dot{A}_1^1\,(m+1)\,(\lambda_1)^m\,\mathbf{E}_1 + \frac{1}{2}\,(\dot{A}_2^2+\dot{A}_3^3)\,(m+1)\,(\lambda_2)^m\,(\mathbf{1}-\mathbf{E}_1)\\ &\quad + (m+1)\,(\lambda_2)^m\,\{\dot{A}_2^3\,\hat{\mathbf{e}}_3\otimes\hat{\mathbf{e}}^2 + \overset{*}{\mathbf{L}}\}\\ &\quad + \frac{1}{2!}\,m\,(m+1)\,(\lambda_2)^{m-1}\,\{\mathring{\mathbf{A}}\mathbf{L}+\mathbf{L}\,\mathring{\mathbf{A}}+\mathrm{tr}(\dot{\mathbf{A}}\,\mathbf{L})\,(\mathbf{1}-\mathbf{E}_1)\}\\ &\quad + \frac{1}{3!}\,m\,(m+1)\,(m-1)\,(\lambda_2)^{m-2}\,\mathrm{tr}(\dot{\mathbf{A}}\,\mathbf{L})\,\mathbf{L}\\ &= (m+1)\,\mathbf{A}^m\,\mathring{\mathbf{A}} + (m+1)\,(\lambda_2)^m\,\overset{*}{\mathbf{L}}\\ &\quad + \frac{1}{2!}\,m\,(m+1)\,(\lambda_2)^{m-1}\,\mathrm{tr}(\dot{\mathbf{A}}\,\mathbf{L})\,(\mathbf{1}-\mathbf{E}_1)\\ &\quad + \frac{1}{3!}\,m\,(m+1)\,(m-1)\,(\lambda_2)^{m-2}\,\mathrm{tr}(\dot{\mathbf{A}}\,\mathbf{L})\,\mathbf{L}\,.\end{aligned} \tag{1.6.18}$$

Now, substitution of (1.6.18) into (1.6.10b), in view of (1.5.1), leads to

$$\begin{aligned}\mathbf{g}(\mathbf{A}\,,\,\overset{*}{\dot{\mathbf{A}}}) &= \mathbf{f}'(\mathbf{A})\,\mathring{\mathbf{A}} + f'(\lambda_2)\,\overset{*}{\mathbf{L}}\\ &\quad + \frac{1}{2!}\,f''(\lambda_2)\,\mathrm{tr}(\dot{\mathbf{A}}\,\mathbf{L})\,(\mathbf{1}-\mathbf{E}_1) + \frac{1}{3!}\,f'''(\lambda_2)\,\mathrm{tr}(\dot{\mathbf{A}}\,\mathbf{L})\,\mathbf{L}\,.\end{aligned} \tag{1.6.19a}$$

Therefore, from (1.6.10a),

$$\begin{aligned}\dot{\mathbf{f}}(\mathbf{A}) &= \mathbf{f}'(\mathbf{A})\,\mathring{\mathbf{A}} + \frac{1}{2!}\,f''(\lambda_2)\,\mathrm{tr}(\dot{\mathbf{A}}\,\mathbf{L})\,(\mathbf{1}-\mathbf{E}_1)\\ &\quad + \frac{1}{3!}\,f'''(\lambda_2)\,\mathrm{tr}(\dot{\mathbf{A}}\,\mathbf{L})\,\mathbf{L} + f'(\lambda_2)\,\overset{*}{\mathbf{L}} + \mathbf{g}\{\mathbf{A}\,,\,\mathbf{H}_A(\dot{\mathbf{A}})\}\,,\\ &\qquad \text{for } \lambda_1 \neq \lambda_2 = \lambda_3\,.\end{aligned} \tag{1.6.19b}$$

Three Repeated Eigenvalues: In this case, it follows from (1.3.25) that

$$\begin{aligned}\mathbf{A}^k\dot{\mathbf{A}}\mathbf{A}^{m-k} &= \{\lambda^k\,\mathbf{1} + k\,\lambda^{k-1}\,\mathbf{L} + \frac{1}{2}\,k(k-1)\,\lambda^{k-2}\,\mathbf{L}^2\}\,\dot{\mathbf{A}}\,\{\lambda^{m-k}\,\mathbf{1}\\ &\quad + (m-k)\,\lambda^{m-k-1}\,\mathbf{L} + \frac{1}{2}\,(m-k)\,(m-k-1)\,\lambda^{m-k-2}\,\mathbf{L}^2\}\\ &= \lambda^m\,\dot{\mathbf{A}} + k\,\lambda^{m-1}\,\mathbf{L}\,\dot{\mathbf{A}} + \frac{1}{2}\,k(k-1)\,\lambda^{m-2}\,\mathbf{L}^2\,\dot{\mathbf{A}}\\ &\quad + (m-k)\,\lambda^{m-1}\,\dot{\mathbf{A}}\,\mathbf{L} + (m-k)\,k\,\lambda^{m-2}\,\mathbf{L}\,\dot{\mathbf{A}}\,\mathbf{L}\\ &\quad + \frac{1}{2}\,(m-k)\,k\,(k-1)\,\lambda^{m-3}\,\mathbf{L}^2\,\dot{\mathbf{A}}\,\mathbf{L}\\ &\quad + \frac{1}{2}\,(m-k)\,(m-k-1)\,\lambda^{m-2}\,\dot{\mathbf{A}}\,\mathbf{L}^2\\ &\quad + \frac{1}{2}\,k(m-k)\,(m-k-1)\,\lambda^{m-3}\,\mathbf{L}\,\dot{\mathbf{A}}\,\mathbf{L}^2\end{aligned}$$

$$+\frac{1}{4}(m-k)(m-k-1)k(k-1)\lambda^{m-4}\mathbf{L}^2\dot{\mathbf{A}}\mathbf{L}^2. \qquad (1.6.20a)$$

Then

$$\begin{aligned}\sum_{k=0}^{m}\mathbf{A}^k\dot{\mathbf{A}}\mathbf{A}^{m-k} &= (m+1)\lambda^m\dot{\mathbf{A}}+\frac{1}{2!}m(m+1)\lambda^{m-1}(\mathbf{L}\dot{\mathbf{A}}+\dot{\mathbf{A}}\mathbf{L})\\ &+\frac{1}{3!}(m-1)m(m+1)\lambda^{m-2}(\mathbf{L}^2\dot{\mathbf{A}}+\dot{\mathbf{A}}\mathbf{L}^2+\mathbf{L}\dot{\mathbf{A}}\mathbf{L})\\ &+\frac{1}{4!}(m-2)(m-1)m(m+1)\lambda^{m-3}(\mathbf{L}\dot{\mathbf{A}}\mathbf{L}^2+\mathbf{L}^2\dot{\mathbf{A}}\mathbf{L})\\ &+\frac{1}{5!}(m-3)(m-2)(m-1)m(m+1)\lambda^{m-4}\mathbf{L}^2\dot{\mathbf{A}}\mathbf{L}^2.\end{aligned} \qquad (1.6.20b)$$

Substitution of (1.6.20b) into (1.6.9), in view of (1.5.1), results in

$$\begin{aligned}\dot{\mathbf{f}}(\mathbf{A}) &= f'(\lambda)\dot{\mathbf{A}}+\frac{1}{2!}f''(\lambda)(\mathbf{L}\dot{\mathbf{A}}+\dot{\mathbf{A}}\mathbf{L})\\ &+\frac{1}{3!}f'''(\lambda)(\mathbf{L}^2\dot{\mathbf{A}}+\dot{\mathbf{A}}\mathbf{L}^2+\mathbf{L}\dot{\mathbf{A}}\mathbf{L})\\ &+\frac{1}{4!}f''''(\lambda)(\mathbf{L}\dot{\mathbf{A}}\mathbf{L}^2+\mathbf{L}^2\dot{\mathbf{A}}\mathbf{L})\\ &+\frac{1}{5!}f'''''(\lambda)\mathbf{L}^2\dot{\mathbf{A}}\mathbf{L}^2, \quad \text{for } \lambda_1=\lambda_2=\lambda_3\equiv\lambda.\end{aligned} \qquad (1.6.20c)$$

1.6.4. Expression for $\mathbf{g}\{\mathbf{A}, \mathbf{H}_\mathbf{A}(\dot{\mathbf{A}})\}$

Now, consider the term $\mathbf{g}\{\mathbf{A}, \mathbf{H}_\mathbf{A}(\dot{\mathbf{A}})\}$ of $\dot{\mathbf{f}}(\mathbf{A})$ in (1.6.10c). Since $\mathbf{A}\bar{\mathbf{A}}=\bar{\mathbf{A}}\mathbf{A}$, where $\bar{\mathbf{A}}$ is defined by (1.5.2b), substitution of (1.6.3b) and (1.6.4b) into (1.6.10c) results in

$$\begin{aligned}\mathbf{g}\{\mathbf{A}, \mathbf{H}_\mathbf{A}(\dot{\mathbf{A}})\} &= \mathbf{A}\,\mathbf{g}(\mathbf{A}, \mathbf{X})-\mathbf{g}(\mathbf{A}, \mathbf{X})\,\mathbf{A}, \quad \text{for } \lambda_1\neq\lambda_2\neq\lambda_3\neq\lambda_1,\\ \mathbf{g}\{\mathbf{A}, \mathbf{H}_\mathbf{A}(\dot{\mathbf{A}})\} &= \bar{\mathbf{A}}\,\mathbf{g}(\mathbf{A}, \mathbf{X})-\mathbf{g}(\mathbf{A}, \mathbf{X})\,\bar{\mathbf{A}}. \quad \text{for } \lambda_1\neq\lambda_2=\lambda_3.\end{aligned} \qquad (1.6.21a,b)$$

Therefore, $\mathbf{g}\{\mathbf{A}, \mathbf{H}_\mathbf{A}(\dot{\mathbf{A}})\}$ is normal to $\mathbf{A}$. A closed-form coordinate-independent expression for $\mathbf{g}\{\mathbf{A}, \mathbf{H}_\mathbf{A}(\dot{\mathbf{A}})\}$ cannot be easily obtained through differentiation of (1.5.1). Hence, instead, consider (1.5.5).

Differentiation of (1.5.5) results in

$$\dot{\mathbf{f}}(\mathbf{A}) = a_1\dot{\mathbf{A}}+a_2(\mathbf{A}\dot{\mathbf{A}}+\dot{\mathbf{A}}\mathbf{A})+c_0\mathbf{1}+c_1\mathbf{A}+c_2\mathbf{A}^2, \qquad (1.6.22)$$

where the unknown coefficients c_0, c_1, and c_2 are functions of $\mathbf{A}$ and $\dot{\mathbf{A}}$. Substitute (1.6.2) and (1.6.5) into (1.6.22), to obtain

$$\begin{aligned}\dot{\mathbf{f}}(\mathbf{A}) &= a_1\mathbf{H}_\mathbf{A}(\dot{\mathbf{A}})+a_2\{\mathbf{A}\mathbf{H}_\mathbf{A}(\dot{\mathbf{A}})+\mathbf{H}_\mathbf{A}(\dot{\mathbf{A}})\mathbf{A}\}\\ &+a_1\overset{*}{\mathbf{A}}+a_2(\mathbf{A}\overset{*}{\mathbf{A}}+\overset{*}{\mathbf{A}}\mathbf{A})+c_0\mathbf{1}+c_1\mathbf{A}+c_2\mathbf{A}^2,\\ &\qquad \text{for } \lambda_1\neq\lambda_2\neq\lambda_3\neq\lambda_1,\end{aligned}$$

$$\dot{\mathbf{f}}(\mathbf{A}) = a_1\,\mathbf{H_A}(\dot{\mathbf{A}}) + a_2\,\{\mathbf{A}\,\mathbf{H_A}(\dot{\mathbf{A}}) + \mathbf{H_A}(\dot{\mathbf{A}})\,\mathbf{A}\} + (a_1 + 2\,a_2\,\lambda_2)\,\overset{*}{\mathbf{L}}$$
$$+ a_2\,\mathrm{tr}(\dot{\mathbf{A}}\,\mathbf{L})\,(\mathbf{1} - \mathbf{E}_1) + a_1\,\overset{\circ}{\mathbf{A}} + a_2\,(\mathbf{A}\,\overset{\circ}{\mathbf{A}} + \overset{\circ}{\mathbf{A}}\mathbf{A}) + c_0\,\mathbf{1} + c_1\,\mathbf{A} + c_2\,\mathbf{A}^2,$$
$$\text{for } \lambda_1 \neq \lambda_2 = \lambda_3. \qquad (1.6.23\text{a,b})$$

Comparing the terms which are normal to **A** in (1.6.13), (1.6.19b), and (1.6.23), obtain the coordinate-independent expression for the normal part of $\dot{\mathbf{f}}(\mathbf{A})$, as

$$\mathbf{g}\{\mathbf{A}, \mathbf{H_A}(\dot{\mathbf{A}})\} = a_1\,(\dot{\mathbf{A}} - \overset{*}{\mathbf{A}}) + a_2\,(\mathbf{A}\,\dot{\mathbf{A}} + \dot{\mathbf{A}}\,\mathbf{A} - 2\,\mathbf{A}\,\overset{*}{\mathbf{A}}),$$
$$\text{for } \lambda_1 \neq \lambda_2 \neq \lambda_3 \neq \lambda_1,$$
$$\mathbf{g}\{\mathbf{A}, \mathbf{H_A}(\dot{\mathbf{A}})\} + f'(\lambda_2)\,\overset{*}{\mathbf{L}} = a_1\,(\dot{\mathbf{A}} - \overset{*}{\mathbf{A}}) + a_2\,(\mathbf{A}\,\dot{\mathbf{A}} + \dot{\mathbf{A}}\,\mathbf{A} - 2\,\mathbf{A}\,\overset{*}{\mathbf{A}})$$
$$+ a_2\,\mathrm{tr}(\dot{\mathbf{A}}\,\mathbf{L})\,(\mathbf{1} - \mathbf{E}_1), \quad \text{for } \lambda_1 \neq \lambda_2 = \lambda_3. \qquad (1.6.24\text{a,b})$$

Substitution of (1.6.24) into (1.6.13) and (1.6.19b) results in

$$\dot{\mathbf{f}}(\mathbf{A}) = a_1\,\dot{\mathbf{A}} + a_2\,(\mathbf{A}\,\dot{\mathbf{A}} + \dot{\mathbf{A}}\,\mathbf{A}) + \{\mathbf{f}'(\mathbf{A}) - a_1\,\mathbf{1} - 2\,a_2\,\mathbf{A}\}\,\overset{*}{\mathbf{A}},$$
$$\text{for } \lambda_1 \neq \lambda_2 \neq \lambda_3 \neq \lambda_1, \qquad (1.6.25\text{a})$$
$$\dot{\mathbf{f}}(\mathbf{A}) = a_1\,\dot{\mathbf{A}} + a_2\,(\mathbf{A}\,\dot{\mathbf{A}} + \dot{\mathbf{A}}\,\mathbf{A}) + \{\mathbf{f}'(\mathbf{A}) - a_1\,\mathbf{1} - 2\,a_2\,\mathbf{A}\}\,\overset{\circ}{\mathbf{A}}$$
$$+ \frac{1}{2!}\,\{f''(\lambda_2) - 2\,a_2\}\,\mathrm{tr}(\dot{\mathbf{A}}\,\mathbf{L})\,(\mathbf{1} - \mathbf{E}_1)$$
$$+ \frac{1}{3!}\,f'''(\lambda_2)\,\mathrm{tr}(\dot{\mathbf{A}}\,\mathbf{L})\,\mathbf{L}, \qquad \text{for } \lambda_1 \neq \lambda_2 = \lambda_3. \qquad (1.6.25\text{b})$$

Therefore, to complete the derivation of the time derivative of $\mathbf{f}(\mathbf{A})$, it is only necessary to find coordinate-independent expressions for $\mathbf{f}'(\mathbf{A})$, $\overset{*}{\mathbf{A}}$, and $\overset{\circ}{\mathbf{A}}$ in terms of **A** and $\dot{\mathbf{A}}$.

When all three *eigenvalues* are distinct, a simple method is to note that $\mathbf{f}'(\mathbf{A})$ and $\overset{*}{\mathbf{A}}$ are coaxial with **A**, and write

$$\overset{*}{\mathbf{A}} = \sum_{i=1}^{3} \mathrm{tr}(\dot{\mathbf{A}}\,\mathbf{E}_i)\,\mathbf{E}_i, \quad \mathbf{f}'(\mathbf{A}) = \sum_{i=1}^{3} f'(\lambda_i)\,\mathbf{E}_i$$
$$\text{for } \lambda_1 \neq \lambda_2 \neq \lambda_3 \neq \lambda_1, \qquad (1.6.26\text{a,b})$$

where $\mathbf{E}_i$, $i = 1, 2, 3$, is given by (1.5.4d). Substitution of (1.6.26a,b) into (1.6.25a) results in

$$\dot{\mathbf{f}}(\mathbf{A}) = a_1\,\dot{\mathbf{A}} + a_2\,(\mathbf{A}\,\dot{\mathbf{A}} + \dot{\mathbf{A}}\,\mathbf{A}) + \sum_{i=1}^{3} \{f'(\lambda_i) - a_1 - 2\,a_2\,\lambda_i\}\,\mathrm{tr}(\dot{\mathbf{A}}\,\mathbf{E}_i)\,\mathbf{E}_i,$$
$$\text{for } \lambda_1 \neq \lambda_2 \neq \lambda_3 \neq \lambda_1. \qquad (1.6.27\text{a})$$

When any two of the three *eigenvalues* are equal, say, $\lambda_1 \neq \lambda_2 = \lambda_3$, the tensors coaxial with **A** can be expressed as follows:

$$\mathbf{f}'(\mathbf{A}) = f'(\lambda_1)\,\mathbf{E}_1 + f'(\lambda_2)\,(\mathbf{1} - \mathbf{E}_1) + f''(\lambda_2)\,\mathbf{L},$$
$$\overset{\circ}{\mathbf{A}} = \mathrm{tr}(\dot{\mathbf{A}}\,\mathbf{E}_1)\,\mathbf{E}_1 + \frac{1}{2}\,\{\mathrm{tr}(\dot{\mathbf{A}}) - \mathrm{tr}(\dot{\mathbf{A}}\,\mathbf{E}_1)\}\,(\mathbf{1} - \mathbf{E}_1) + \dot{A}_2^3\,\hat{\mathbf{e}}_3 \otimes \hat{\mathbf{e}}^2, \qquad (1.6.28\text{a,b})$$

where the coefficient $\dot{A}_2^3$ is still unknown. However, it is not necessary to evaluate $\dot{A}_2^3$, in order to obtain an explicit expression for $\dot{\mathbf{f}}(\mathbf{A})$. Substitution of (1.6.28) into (1.6.25b), in view of (1.5.10), results in

$$\begin{aligned}\dot{\mathbf{f}}(\mathbf{A}) = a_1 \dot{\mathbf{A}} + a_2 (\mathbf{A}\dot{\mathbf{A}} + \dot{\mathbf{A}}\mathbf{A}) + \{f'(\lambda_1) - a_1 - 2a_2\lambda_1\} \operatorname{tr}(\dot{\mathbf{A}}\mathbf{E}_1)\mathbf{E}_1 \\ + \frac{1}{2!}\{f''(\lambda_2) - 2a_2\}\{\operatorname{tr}(\dot{\mathbf{A}}) - \operatorname{tr}(\dot{\mathbf{A}}\mathbf{E}_1)\}\mathbf{L} \\ + \frac{1}{2!}\{f''(\lambda_2) - 2a_2\}\operatorname{tr}(\dot{\mathbf{A}}\mathbf{L})(\mathbf{1} - \mathbf{E}_1) \\ + \frac{1}{3!} f'''(\lambda_2)\operatorname{tr}(\dot{\mathbf{A}}\mathbf{L})\mathbf{L}, \quad \text{for } \lambda_1 \neq \lambda_2 = \lambda_3. \qquad (1.6.27b)\end{aligned}$$

The same result is obtained by taking limits as λ_3 goes to λ_2, in (1.6.27a). Note that, when λ_3 goes to λ_2,

$$(\mathbf{E}_2 - \mathbf{E}_3)/2 \to (\lambda_2 - \lambda_3)\mathbf{L}, \quad (\mathbf{E}_2 + \mathbf{E}_3) \to (\mathbf{1} - \mathbf{E}_1), \quad \text{as } \lambda_3 \to \lambda_2. \qquad (1.6.29a,b)$$

Using Rivlin's identities (1.4.3)-(1.4.5) in (1.6.20c) and noting that all three invariants of $\mathbf{L}$ are zero, arrive at

$$\begin{aligned}\dot{\mathbf{f}}(\mathbf{A}) = f'(\lambda)\dot{\mathbf{A}} + \frac{1}{2!} f''(\lambda)(\mathbf{L}\dot{\mathbf{A}} + \dot{\mathbf{A}}\mathbf{L}) \\ + \frac{1}{3!} f'''(\lambda)\{\operatorname{tr}(\dot{\mathbf{A}}\mathbf{L}^2)\mathbf{1} + \operatorname{tr}(\dot{\mathbf{A}}\mathbf{L})\mathbf{L} + \operatorname{tr}(\dot{\mathbf{A}})\mathbf{L}^2\} \\ + \frac{1}{4!} f''''(\lambda)\{\operatorname{tr}(\dot{\mathbf{A}}\mathbf{L}^2)\mathbf{L} + \operatorname{tr}(\dot{\mathbf{A}}\mathbf{L})\mathbf{L}^2\} \\ + \frac{1}{5!} f'''''(\lambda)\operatorname{tr}(\dot{\mathbf{A}}\mathbf{L}^2)\mathbf{L}^2, \quad \text{for } \lambda_1 = \lambda_2 = \lambda_3 = \lambda. \qquad (1.6.27c)\end{aligned}$$

The same result is obtained by taking limits as λ_2 goes to $\lambda_1 = \lambda$ in (1.6.27b). Note the following limits in (1.6.27b) when $\delta\lambda \equiv \lambda_1 - \lambda_2$ goes to zero:

$$\begin{aligned}\delta\lambda\,\mathbf{E}_1 + \mathbf{L} \to \mathbf{A} - \lambda\mathbf{1}, \quad \delta\lambda\,\mathbf{E}_1 - \mathbf{L} \to \frac{2}{\delta\lambda}(\mathbf{A} - \lambda\mathbf{1})^2, \\ \frac{1}{\delta\lambda^2}\{f'(\lambda_1) - a_1 - 2a_2\lambda_1\} \to \frac{1}{6} f'''(\lambda) + \frac{\delta\lambda}{12} f''''(\lambda) + \frac{\delta\lambda^2}{40} f'''''(\lambda), \\ \frac{1}{\delta\lambda} f''(\lambda_2) - 2a_2 \to -\frac{1}{3} f'''(\lambda) - \frac{\delta\lambda}{12} f''''(\lambda) - \frac{\delta\lambda^2}{60} f'''''(\lambda), \quad \text{as } \delta\lambda \to 0. \qquad (1.6.30a\text{-}d)\end{aligned}$$

When the *eigenvalues* λ_2 and λ_3 are complex, say, $\lambda_2 = \alpha + i\beta$ and $\lambda_3 = \alpha - i\beta$, then (1.6.27a) can be rewritten as follows:

$$\begin{aligned}\dot{\mathbf{f}}(\mathbf{A}) = a_1 \dot{\mathbf{A}} + a_2 (\mathbf{A}\dot{\mathbf{A}} + \dot{\mathbf{A}}\mathbf{A}) + \{f'(\lambda_1) - a_1 - 2a_2\lambda_1\} \operatorname{tr}(\dot{\mathbf{A}}\mathbf{E}_1)\mathbf{E}_1 \\ + \frac{1}{2}(f'^{R} - a_1 - 2a_2\alpha)\{\operatorname{tr}(\dot{\mathbf{A}}\mathbf{E}^{R})\mathbf{E}^{R} - \operatorname{tr}(\dot{\mathbf{A}}\mathbf{E}^{I})\mathbf{E}^{I}\} \\ - \frac{1}{2}(f'^{I} - 2a_2\beta)\{\operatorname{tr}(\dot{\mathbf{A}}\mathbf{E}^{I})\mathbf{E}^{R} + \operatorname{tr}(\dot{\mathbf{A}}\mathbf{E}^{R})\mathbf{E}^{I}\}, \qquad (1.6.27d)\end{aligned}$$

where

$$\mathbf{E}_1 = \frac{1}{\Delta}\{\mathbf{A}^2 - 2\alpha\mathbf{A} + (\alpha^2+\beta^2)\,\mathbf{1}\},$$
$$\mathbf{E}^{\mathrm{R}} = \frac{-1}{\Delta}\{\mathbf{A}^2 - 2\alpha\mathbf{A} - \lambda_1(\lambda_1 - 2\alpha)\,\mathbf{1}\},$$
$$\mathbf{E}^{\mathrm{I}} = \frac{1}{\beta\Delta}\{(\lambda_1-\alpha)\mathbf{A}^2 + (\alpha^2-\beta^2-\lambda_1^2)\mathbf{A} - \lambda_1(\alpha^2-\beta^2-\alpha\lambda_1)\,\mathbf{1}\},$$
$$\mathrm{f}^{\mathrm{R}} = \mathrm{Re}\{\mathrm{f}(\alpha+i\,\beta)\}, \quad \mathrm{f}^{\mathrm{I}} = \mathrm{Im}\{\mathrm{f}(\alpha+i\,\beta)\},$$
$$\Delta = (\lambda_1-\alpha)^2 + \beta^2. \tag{1.6.31a-f}$$

Observe the following limits when β goes to zero:

$$\mathbf{E}_1 \to \frac{1}{\lambda_1-\alpha}(\bar{\mathbf{A}} - \alpha\,\mathbf{1}), \quad \mathbf{E}^{\mathrm{R}} \to \mathbf{1} - \mathbf{E}_1, \quad \mathbf{E}^{\mathrm{I}} \to \frac{-1}{\beta}\mathbf{L},$$
$$\mathrm{f}^{\mathrm{I}} \to \beta\,\mathrm{f}''(\alpha), \quad \mathrm{f}^{\mathrm{R}} - a_1 - 2a_2\alpha \to \frac{-\beta^2}{3}\mathrm{f}'''(\alpha), \quad \text{as } \beta \to 0, \tag{1.6.32a-e}$$

where

$$\bar{\mathbf{A}} = \mathbf{A} - \mathbf{L}, \quad \mathbf{L} = \frac{1}{\alpha-\lambda_1}\{\mathbf{A}^2 - (\lambda_1+\alpha)\,\mathbf{A} + \alpha\lambda_1\,\mathbf{1}\}. \tag{1.6.32f,g}$$

This shows that (1.6.27d) reduces to (1.6.27b) when the *eigenvalues* λ_2 and λ_3 are real and equal.

1.7. REFERENCES

Balendran, B. and Nemat-Nasser, S. (1995), Integration of inelastic constitutive equations for constant velocity gradient with large rotation, *Appl. Math. Comput.*, Vol. 67, 161-195.

Balendran, B. and Nemat-Nasser, S. (1996), Derivative of a function of a non-symmetric second-order tensor, *Q. Appl. Math.*, Vol. 54, 583-600.

Carlson, D. E. and Hoger, A. (1986), The derivative of a tensor-valued function of a tensor, *Q. Appl. Math.*, Vol. 44, 409-423.

Gantmacher, F. R. (1960), *The theory of matrices*, Chelsea, New York.

Guo, Z.-H., Lehmann, T., Liang, H., and Man, C.-S. (1992), Twirl tensors and the tensor equation $\mathbf{AX} - \mathbf{XA} = \mathbf{C}$, *J. Elasticity*, Vol. 27, 227-245.

Gurtin, M. E. and Spear, K. (1983), On the relationship between the logarithmic strain rate and the stretching tensor, *Int. J. Solids Struct.*, Vol. 19, 437-444.

Hill, R. (1968), On constitutive inequalities for simple materials-I, *J. Mech. Phys. Solids*, Vol. 16, 229-242.

Hill, R. (1970), Constitutive inequalities for isotropic elastic solids under finite strain, *Proc. R. Soc. Lond. A*, Vol. 314, 457-472.

Hill, R. (1978), Aspects of invariance in solid mechanics, *Adv. Appl. Mech.*, Vol. 18, 1-75.

Hoger, A. (1986), The material time derivative of logarithmic strain, *Int. J. Solids Struct.*, Vol. 22, 1019-1032.

Hoger, A. and Carlson, D. E. (1984), Determination of the stretch and rotation in the polar decomposition of the deformation gradient, *Q. Appl. Math.*, Vol. 42, 113-117.

Mehrabadi, M. M. and Nemat-Nasser, S. (1987), Some basic kinematical relations for finite deformations of continua, *Mech. Mat.*, Vol. 6, 127-138.

Nemat-Nasser, S. and Hori, M. (1993), *Micromechanics: overall properties of heterogeneous solids,* Elsevier, Amsterdam; second revised edition (1999), Elsevier, Amsterdam.

Rivlin, R. S. (1955), Further remarks on the stress-deformation relations for isotropic materials, *J. Rat. Mech. Anal.*, Vol. 4, 681-701.

Scheidler, M. (1991a), Time rates of generalized strain tensors, part I: component formulas, *Mech. Mat.*, Vol. 11, 199-210.

Scheidler, M. (1991b), Time rates of generalized strain tensors, part II: approximate basis-free formulas, *Mech. Mat.*, Vol. 11, 211-219.

Scheidler, M. (1994), The tensor equation $\mathbf{A}\,\mathbf{X}+\mathbf{X}\,\mathbf{A}=\mathbf{\Phi}(\mathbf{A}\,,\,\mathbf{H})$, with applications to kinematics of continua, *J. Elasticity*, Vol. 36, 117-153.

Ting, T. C. T. (1985), Determination of $\mathbf{C}^{1/2}$, $\mathbf{C}^{-1/2}$ and more general isotropic tensor functions of $\mathbf{C}$, *J. Elasticity*, Vol. 15, 319-323.

Truesdell, C. (1961), General and exact theory of waves in finite elastic strain, *Arch. Rat. Mech. Anal.*, Vol. 8, 263-296.

Truesdell, C. and Noll, W. (1965), *The nonlinear field theories of mechanics*, Springer-Verlag, Berlin.

Wang, C.-C. and Truesdell, C. (1973), *Introduction to rational elasticity*, Noordhoff, Leiden, The Netherlands.

2

KINEMATICS

The basic kinematical concepts necessary for a systematic study of the finite deformation of continua are presented in this chapter. Most of the results are standard and can be found in advanced treatises on continuum mechanics, but a number of important nonstandard results are also given. These include various strain, strain rate, and spin measures, relations among them, and their spectral representation, as well as their general coordinate-independent description. The coordinate-independent relations among various spin tensors presented in Sections 2.6-2.9, play a major role in identifying relations among various objective stress fluxes used in the constitutive modeling. Exact explicit coordinate-invariant expressions for the material time derivative of any Lagrangian strain measure and the Jaumann rate of change of any Eulerian strain measure are given in terms of the deformation rate tensor.

More specifically, various measures necessary to describe the finite deformation and rotation of a typical material element, are presented in Section 2.1, in terms of the deformation gradient and the right Cauchy-Green tensor, their principal values, and principal directions. Polar decomposition of the deformation gradient is given in Section 2.2, together with the spectral representation of the right and left Cauchy-Green tensors, right and left stretch tensors, and the rotation tensor. In this section, explicit formulae are presented for the direct evaluation of the right and left stretch tensors and their invariants, and the rotation tensor, involved in polar decomposition, without the need for eigenvector calculations. The relation between the Lagrangian and Eulerian triads is discussed, and the transformation of the volume and surface elements is formulated. Strain measures are examined in Section 2.3, beginning with the fundamental work by Hill (1968). Then explicit coordinate-independent expressions for both Lagrangian and Eulerian strain measures are developed, in terms of the stretch tensors. Expressions for the velocity and acceleration fields are summarized in Section 2.4, in terms of both Lagrangian and Eulerian variables, and their relation to one another is given. The deformation-rate and spin tensors are introduced in Section 2.5, and in Sections 2.6, 2.7, and 2.8, various rate measures are discussed and their explicit coordinate-independent expressions are developed. Section 2.9 provides relations among various spin tensors, and Section 2.10 gives coordinate-independent expressions for various strain-rate measures.

2.1. MOTION AND DEFORMATION OF A CONTINUUM

Let a deformable body B be in configuration C_0 at the initial time t_0; see Figure 2.1.1. Denote by C its configuration at the current instant $t > t_0$. The mapping from C_0 to C is denoted by[1]

$$\begin{aligned} x_i &= x_i(X_1, X_2, X_3, t) \\ &= x_i(X_A, t), \qquad i, A = 1, 2, 3, \end{aligned} \tag{2.1.1a}$$

or

$$\mathbf{x} = \mathbf{x}(\mathbf{X}, t) \tag{2.1.1b}$$

which defines particle positions $\mathbf{x}$ of components x_i in C in terms of their initial positions $\mathbf{X}$ of components X_A in C_0, both taken with respect to a fixed background rectangular Cartesian coordinate system, defined by the fixed unit triad $\mathbf{e}_i$, $i = 1, 2, 3$; Figure 2.1.1.

Assume that mapping (2.1.1) is one-to-one and as many times differentiable as required. Hence, the Jacobian of the transformation,

$$J = \det\left|\frac{\partial x_i}{\partial X_A}\right|, \tag{2.1.2a}$$

is finite and positive, where det denotes determinant.

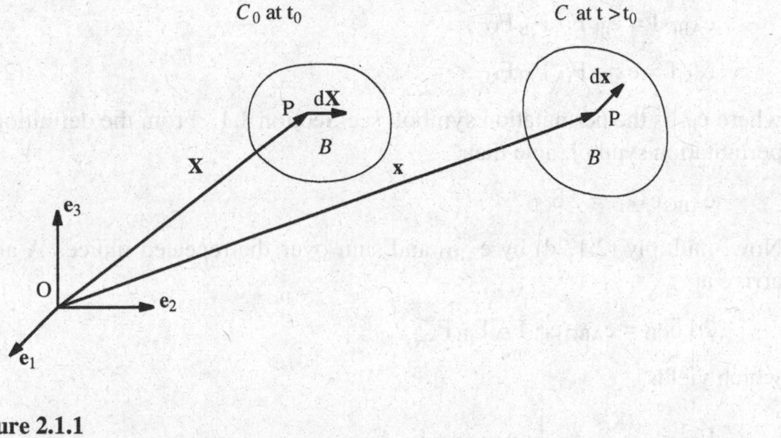

Figure 2.1.1

A deformable body B is in configuration C_0 at the initial time t_0, and in configuration C at the current time t; $\mathbf{X}$ and $\mathbf{x}$ are the corresponding particle positions, and $\mathbf{e}_i$ is a fixed unit triad

---

[1] As is common in most of the continuum mechanics literature, the *function* and its *value* are denoted by the same symbol, *e.g.*, $\mathbf{x}(\mathbf{X}, t)$ and $\mathbf{x}$. This notation is used throughout the present book.

2.1.1. Deformation Gradient

Let $d\mathbf{X} = dX_A\,\mathbf{e}_A$ be an elementary *material* line element emanating from a typical particle P in the reference configuration C_0; see Figure 2.1.1. Mapping (2.1.1) deforms this line element into $d\mathbf{x} = dx_i\,\mathbf{e}_i$, given by

$$dx_i = \frac{\partial x_i}{\partial X_A}\,dX_A = x_{i,A}\,dX_A \tag{2.1.3a}$$

or, when written in direct notation, by

$$d\mathbf{x} = \mathbf{F}\,d\mathbf{X}, \quad \mathbf{F} \equiv x_{i,A}\,\mathbf{e}_i \otimes \mathbf{e}_A, \tag{2.1.3b,c}$$

where a comma followed by an index denotes partial differentiation with respect to the corresponding coordinate, repeated indices are summed, and $\mathbf{F} = (\partial\mathbf{x}/\partial\mathbf{X})^T$ is the *deformation gradient*,[2]

$$F_{iA} \equiv x_{i,A}, \quad \mathbf{F} = F_{iA}\,\mathbf{e}_i \otimes \mathbf{e}_A. \tag{2.1.4a,b}$$

The Jacobian of the transformation (2.1.2a) then is

$$J = \det\mathbf{F}. \tag{2.1.2b}$$

The determinant of the 3×3 matrix $[F_{iA}]$ is given by

$$\begin{aligned} J &= e_{ABC}\,F_{1A}\,F_{2B}\,F_{3C} \\ &= e_{ijk}\,F_{i1}\,F_{j2}\,F_{k3} \end{aligned} \tag{2.1.2c}$$

which can be rewritten as

$$\begin{aligned} e_{ABC}\,J &= e_{ijk}\,F_{iA}\,F_{jB}\,F_{kC}, \\ e_{ijk}\,J &= e_{ABC}\,F_{iA}\,F_{jB}\,F_{kC}, \end{aligned} \tag{2.1.2d,e}$$

where e_{ijk} is the permutation symbol; see Section 1.1. From the definition of the permutation symbol, note that,

$$e_{ABC}\,e_{ABD} = 2\,\delta_{CD}.$$

Now, multiply (2.1.2d) by e_{ABD} and sum over the repeated indices A and B to arrive at

$$2\,J\,\delta_{CD} = e_{ABD}\,e_{ijk}\,F_{iA}\,F_{jB}\,F_{kC} \tag{2.1.2f}$$

which yields

$$\begin{aligned} F^{-1}_{Ck} &= \frac{\partial X_C}{\partial x_k} = \frac{1}{2J}\,e_{ABC}\,e_{ijk}\,F_{iA}\,F_{jB}, \\ \mathbf{F}^{-1} &= X_{A,i}\,\mathbf{e}_A \otimes \mathbf{e}_i = F^{-1}_{Ai}\,\mathbf{e}_A \otimes \mathbf{e}_i, \end{aligned} \tag{2.1.4c,d}$$

$$J = \frac{1}{6}\,e_{ABC}\,e_{ijk}\,F_{iA}\,F_{jB}\,F_{kC}. \tag{2.1.2g}$$

---

[2] Note that $\partial\mathbf{x}/\partial\mathbf{X} \equiv \boldsymbol{\nabla}^0 \otimes \mathbf{x} = x_{i,A}\,\mathbf{e}_A \otimes \mathbf{e}_i$. Throughout the book, a gradient with respect to $\mathbf{X}$ is denoted by $\boldsymbol{\nabla}^0$, while that with respect to $\mathbf{x}$ is denoted by $\boldsymbol{\nabla}$, unless the meaning is obvious from the context, in which case the latter notation is used.

2.1.2. Principal Stretches

From (2.1.3b), the square of the current length ds of the material line element $d\mathbf{X}$ of initial length dS is calculated as follows:

$$(ds)^2 = d\mathbf{x} \cdot d\mathbf{x} = (\mathbf{F}\, d\mathbf{X}) \cdot (\mathbf{F}\, d\mathbf{X}) = d\mathbf{X} \cdot (\mathbf{F}^T \mathbf{F}\, d\mathbf{X})$$
$$= F_{iA} F_{iB}\, dX_A\, dX_B\,. \tag{2.1.5a}$$

Introduce the symmetric positive-definite tensor,

$$\mathbf{C} = \mathbf{F}^T \mathbf{F}\,, \quad C_{AB} = F_{iA} F_{iB}\,, \tag{2.1.6a,b}$$

called the *right Cauchy-Green tensor*, and observe from (2.1.5a) that

$$(ds)^2 = C_{AB}\, dX_A\, dX_B\,, \quad (ds)^2 = d\mathbf{X} \cdot \mathbf{C}\, d\mathbf{X}\,, \tag{2.1.5b,c}$$

defines the positive-definite quadratic form characterizing an *ellipsoid* (known as the *Lagrangian ellipsoid*) in the reference configuration C_0; see Figure 2.1.2. The deformation measure $\mathbf{C}$, therefore, has positive eigenvalues with the corresponding directions defined by the orientation of the principal axes of the *Lagrangian ellipsoid*, (2.1.5b).

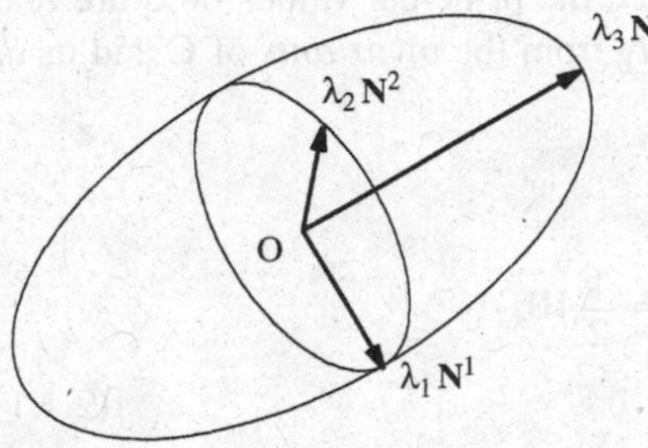

Figure 2.1.2
Lagrangian ellipsoid with principal axes $\lambda_a \mathbf{N}^a$, a = 1, 2, 3

The ratio of the current length ds to the initial length dS of a material line element is its *stretch*,

$$\lambda = \frac{ds}{dS}\,. \tag{2.1.7a}$$

If $\mathbf{N}$ is a unit vector defining the initial orientation of the line element,

$$d\mathbf{X} = dS\, \mathbf{N}\,, \tag{2.1.8}$$

then, from (2.1.5b), the squared stretch of this element is given by the *normal component* of $\mathbf{C}$ in the direction of $\mathbf{N}$,

$$\lambda^2 = C_{AB} N_A N_B = \mathbf{N} \cdot \mathbf{C} \mathbf{N}\,. \tag{2.1.7b}$$

To obtain the squared principal stretches (which equal the eigenvalues of $\mathbf{C}$), seek orientations corresponding to extreme values of the squared stretch. These are the directions of the principal axes of the *Lagrangian ellipsoid*. Equivalently, seek directions of elements in the reference configuration, say, $\mathbf{N}$, which the transformation **CN** does not rotate. Therefore, $\mathbf{C}\mathbf{N} = \mu\mathbf{N}$ for some positive μ. With either approach, the following eigenvalue problem results:

$$(\mathbf{C} - \mu\,\mathbf{1})\mathbf{N} = \mathbf{0}, \qquad (C_{AB} - \mu\,\delta_{AB})\,N_B = 0, \tag{2.1.9a,b}$$

where **1** is the identity tensor with components δ_{AB}, the *Kronecker delta*. For nontrivial solutions, the determinant of the coefficients in the above three homogeneous equations for three unknowns, N_B, B = 1, 2, 3, must vanish, leading to the following characteristic equation whose roots are the squared principal stretches:

$$\mu^3 - I_C\,\mu^2 + II_C\,\mu - III_C = 0\,. \tag{2.1.9c}$$

Here I_C, II_C, and III_C are the basic invariants of **C**; replace **A** by **C** in (1.2.7b-d). The principal (or eigen-) values of **C** are positive and real. The corresponding principal directions are mutually orthogonal.

Let λ_a^2 and $\mathbf{N}^a$, a = 1, 2, 3, denote the principal values and principal directions of **C**. Then, it follows that

$$\mathbf{C} = \sum_{a=1}^{3} \lambda_a^2\,\mathbf{N}^a \otimes \mathbf{N}^a, \qquad C_{AB} = \sum_{a=1}^{3} \lambda_a^2\,N_A^a\,N_B^a \tag{2.1.10a,b}$$

is the *spectral* representation of **C**. Since the principal values of **C** are real and positive, they can be calculated *explicitly* from the *invariants* of **C** and its *deviatoric* part, $\mathbf{C}' = \mathbf{C} - I_C\,\mathbf{1}/3$. First set

$$a = \frac{1}{9}\,(I_C^2 - 3\,II_C) = -\frac{1}{3}\,II_{C'},$$

$$b = \frac{1}{54}\,(2\,I_C^3 - 9\,I_C\,II_C + 27\,III_C) = \frac{1}{2}\,III_{C'},$$

$$\phi = \cos^{-1}(b/a^{3/2}), \tag{2.1.11a-c}$$

where $II_{C'}$ and $III_{C'}$ are the basic invariants of $\mathbf{C}'$. Then obtain

$$\lambda_1^2 = 2\sqrt{a}\cos(\frac{\phi}{3}) + \frac{1}{3}\,I_C, \quad \lambda_2^2 = 2\sqrt{a}\cos(\frac{2\pi+\phi}{3}) + \frac{1}{3}\,I_C,$$

$$\lambda_3^2 = 2\sqrt{a}\cos(\frac{4\pi+\phi}{3}) + \frac{1}{3}\,I_C\,. \tag{2.1.12a-c}$$

Note that if $a^3 < b^2$, then two of the three roots of the cubic equation (2.1.9c) are complex. However, all three roots of (2.1.9c) must be real. Therefore, for a symmetric positive-definite tensor **C**,

$$a^3 \geq b^2\,. \tag{2.1.11d}$$

For $a^3 > b^2$, the eigenvalues (principal stretches) are distinct. If $a^3 = b^2$, then at least two of the principal stretches are equal. In such a case, it follows that

$$\lambda_1^2 = 2\sqrt{a} + \frac{1}{3}\,I_C,$$

$$\lambda_2^2 = \lambda_3^2 = -\sqrt{a} + \frac{1}{3}\,I_C\,. \tag{2.1.12d,e}$$

If a = b = 0, then all three eigenvalues are equal, leading to

$$\lambda_1^2 = \lambda_2^2 = \lambda_3^2 = \frac{1}{3}\,I_C\,. \tag{2.1.12f}$$

2.2. POLAR DECOMPOSITION

Transformation (2.1.3) in general rotates and distorts material line elements. To uncouple rigid rotation from pure distortion, the *polar decomposition* of **F** is used,

$$\mathbf{F} = \mathbf{R}\,\mathbf{U} = \mathbf{V}\,\mathbf{R}\,, \qquad F_{iA} = R_{iB}\,U_{BA} = V_{ij}\,R_{jA}\,, \tag{2.2.1a,b}$$

where the symmetric positive-definite tensors **U** and **V** are called the *right-* and *left-stretch tensors*, and **R** is a *proper orthogonal matrix*,

$$\mathbf{R}^{-1} = \mathbf{R}^{T}\,, \qquad \det \mathbf{R} = +1\,. \tag{2.2.2a,b}$$

From (2.1.6) it follows that

$$\begin{aligned}\mathbf{C} &= \mathbf{F}^{T}\,\mathbf{F} = (\mathbf{R}\,\mathbf{U})^{T}\,(\mathbf{R}\,\mathbf{U}) = \mathbf{U}^{2} \\ &= (\mathbf{V}\,\mathbf{R})^{T}\,(\mathbf{V}\,\mathbf{R}) = \mathbf{R}^{T}\,\mathbf{V}^{2}\,\mathbf{R} = \mathbf{R}^{T}\,\mathbf{B}\,\mathbf{R}\,,\end{aligned} \tag{2.2.3}$$

where

$$\mathbf{B} = \mathbf{V}^{2} = \mathbf{F}\,\mathbf{F}^{T} = \mathbf{R}\,\mathbf{C}\,\mathbf{R}^{T}\,. \tag{2.2.4}$$

The positive-definite, symmetric tensor **B** is called the *left Cauchy-Green tensor* (recall that **C** is the *right Cauchy-Green tensor*).

In Subsection 2.2.2, the tensors **U**, **V**, and **R** are evaluated directly, explicitly, and exactly from **C**, **B**, and their basic invariants, without *eigenvector* calculations.

2.2.1. Spectral Representation

The principal values of **C** and **B** are the same, but their principal directions are not. In fact, from (2.2.4) it follows that **R** rotates the principal directions $\mathbf{N}^a$ of **C** into the principal directions $\mathbf{n}^a$ of **B**,

$$\mathbf{n}^{a} = \mathbf{R}\,\mathbf{N}^{a}\,. \tag{2.2.5}$$

Furthermore, (2.2.3) states that **U** and **C** are coaxial in the sense that they have the same principal directions. Similarly, **V** and **B** are coaxial. The principal values of **C** or **B** are the squares of those of **U** or **V**.

These results are transparently displayed through the spectral (*i.e.*, in the principal coordinates) representation of the involved tensorial quantities, as follows:

$$\mathbf{C} = \sum_{a=1}^{3} \lambda_a^2 \mathbf{N}^a \otimes \mathbf{N}^a, \quad \mathbf{U} = \sum_{a=1}^{3} \lambda_a \mathbf{N}^a \otimes \mathbf{N}^a,$$

$$\mathbf{B} = \sum_{a=1}^{3} \lambda_a^2 \mathbf{n}^a \otimes \mathbf{n}^a, \quad \mathbf{V} = \sum_{a=1}^{3} \lambda_a \mathbf{n}^a \otimes \mathbf{n}^a. \tag{2.2.6a-d}$$

Furthermore, the rotation is

$$\mathbf{R} = \sum_{a=1}^{3} \mathbf{n}^a \otimes \mathbf{N}^a, \quad R_{iA} = \sum_{a=1}^{3} n_i^a N_A^a. \tag{2.2.7a,b}$$

Finally, the spectral representation of the deformation gradient is

$$\mathbf{F} = \sum_{a=1}^{3} \lambda_a \mathbf{n}^a \otimes \mathbf{N}^a, \quad F_{iA} = \sum_{a=1}^{3} \lambda_a n_i^a N_A^a, \tag{2.2.8a,b}$$

which displays the polar decomposition, and from which all the other above results stem directly.[3]

It has been tacitly implied above that the principal stretches λ_a are all distinct. In the special case, when two of these are equal, only one distinct principal direction results. Normal to this direction, any two mutually orthogonal directions are principal. When all three stretches are equal, any three mutually orthogonal directions are principal, and the deformation is locally isotropic, *i.e.*, every direction undergoes the same expansion or contraction. The ellipsoids corresponding to **C** and **B** then are spheres.

2.2.2. Basic Invariants and Direct Evaluation of U, R, and V

As in (1.2.7b-d), the basic invariants of a second-order tensor are denoted by Roman numerals carrying subscripts designating that tensor. Since the right and left Cauchy-Green tensors, **C** and **B**, have equal principal values, their basic invariants coincide,

$$I_{\mathbf{C}} = I_{\mathbf{B}} = \lambda_1^2 + \lambda_2^2 + \lambda_3^2,$$

$$II_{\mathbf{C}} = II_{\mathbf{B}} = \lambda_1^2\lambda_2^2 + \lambda_2^2\lambda_3^2 + \lambda_3^2\lambda_1^2,$$

$$III_{\mathbf{C}} = III_{\mathbf{B}} = \lambda_1^2\lambda_2^2\lambda_3^2. \tag{2.2.9a-c}$$

Similarly, the basic invariants of the right- and left-stretch tensors, **U** and **V**, are

$$I_{\mathbf{U}} = I_{\mathbf{V}} = \lambda_1 + \lambda_2 + \lambda_3,$$

$$II_{\mathbf{U}} = II_{\mathbf{V}} = \lambda_1\lambda_2 + \lambda_2\lambda_3 + \lambda_3\lambda_1,$$

$$III_{\mathbf{U}} = III_{\mathbf{V}} = \lambda_1\lambda_2\lambda_3 = J. \tag{2.2.10a-c}$$

The basic invariants of the Cauchy-Green tensors are related to those of the stretch tensors as follows:

---

[3] See Truesdell and Toupin (1960), Truesdell and Noll (1965), and Chadwick (1976).

$$I_C = I_U^2 - 2\,II_U\,, \qquad II_C = II_U^2 - 2\,I_U\,III_U\,,$$
$$III_C = III_U^2\,. \tag{2.2.11a-c}$$

The basic invariants of the stretch tensors, **U** and **V**, can be written in terms of the basic invariants of the Cauchy-Green tensors, **C** and **B**. These are given by (2.2.14), (2.2.19), and (2.2.20), as is discussed in the sequel.

From the Hamilton-Cayley theorem, (1.2.11), express the stretch tensors in terms of the Cauchy-Green tensors, as follows[4] (Ting, 1985):

$$\mathbf{U} = \frac{1}{III_U - I_U\,II_U}\,\{\mathbf{C}^2 - (I_U^2 - II_U)\,\mathbf{C} - I_U\,III_U\,\mathbf{1}\}\,,$$
$$\mathbf{V} = \frac{1}{III_V - I_V\,II_V}\,\{\mathbf{B}^2 - (I_V^2 - II_V)\,\mathbf{B} - I_V\,III_V\,\mathbf{1}\}\,. \tag{2.2.12a,b}$$

The denominator on the right side of these equations is never zero. To show this, consider the consequence of $III_U = I_U\,II_U$. Then the characteristic equation of **U** becomes

$$(\lambda - I_U)\,(\lambda^2 + II_U) = 0\,.$$

Three real eigenvalues (stretches) result only if $II_U \leq 0$. Then one of the principal stretches would be negative. This is impossible since all the principal stretches are positive; see (2.2.10b). Therefore,

$$III_U - I_U\,II_U = III_V - I_V\,II_V \neq 0\,. \tag{2.2.12c}$$

Once **U** and **V** are calculated from (2.2.12a,b), the rotation tensor **R** is directly obtained without the calculation of the eigenvectors of **U** or **V**, as follows:

$$\mathbf{R} = \mathbf{F}\,\mathbf{U}^{-1} = \mathbf{V}^{-1}\,\mathbf{F}\,,$$
$$\mathbf{U}^{-1} = \frac{1}{III_U}\,(\mathbf{C} - I_U\,\mathbf{U} + II_U\,\mathbf{1})\,,$$
$$\mathbf{V}^{-1} = \frac{1}{III_V}\,(\mathbf{B} - I_V\,\mathbf{V} + II_V\,\mathbf{1})\,. \tag{2.2.13a-c}$$

Since all three principal stretches are positive and nonzero, $III_U = III_V > 0$.

Equations (2.2.12a,b) and (2.2.13a-c) can be employed to explicitly calculate the rotation and stretch tensors, **R** and **U** or **V**, directly from the deformation gradient **F**, without any approximation or a need for eigenvector calculations. Alternatives are: (1) to calculate the *eigenvalues* of **C** from (2.1.12a-c) and (2.1.11a-c); and (2) to calculate the basic invariants of **U** directly from those of **C**.

---

[4] Use the Hamilton-Cayley theorem for **U**, multiply the resulting equation by **U**, and use the theorem again to substitute for $\mathbf{U}^3$, leading to (2.2.12a).

Since the eigenvalues of **U** are real and positive, they are obtained from (2.1.12a-c) and (2.1.11a-c) by direct evaluation of standard expressions. Then the basic invariants of **U** or **V** are calculated from these eigenvalues. Finally, **U** is obtained from (2.2.12a) and **R** from (2.2.13a,b).

Alternatively, observe that

$$\mathrm{III_U} = \sqrt{\mathrm{III_C}}, \quad \mathrm{I_U^2} = \mathrm{I_C} + 2\,\mathrm{II_U},$$
$$\mathrm{II_U^2} = \mathrm{II_C} + 2\,\mathrm{III_U}\,\mathrm{I_U}, \tag{2.2.14a-c}$$

from which it follows that

$$\mathrm{I_U^4} - 2\,\mathrm{I_C}\,\mathrm{I_U^2} - 8\,\mathrm{III_U}\,\mathrm{I_U} + (\mathrm{I_C^2} - 4\,\mathrm{II_C}) = 0. \tag{2.2.15}$$

The four roots of this quartic equation for $\mathrm{I_U}$ are

$$\mathrm{I_U} = \lambda_1 + \lambda_2 + \lambda_3, \quad \mathrm{I_1} \equiv \lambda_1 - \lambda_2 - \lambda_3,$$
$$\mathrm{I_2} \equiv -\lambda_1 + \lambda_2 - \lambda_3, \quad \mathrm{I_3} \equiv -\lambda_1 - \lambda_2 + \lambda_3. \tag{2.2.16a-d}$$

Therefore, $\mathrm{I_U}$ is the greatest of all four roots.

In general, $\mathrm{I_U}$ is not the only positive root of (2.2.15). Since the sum of any other two roots is negative, at most only two roots are positive. Therefore, if $(\mathrm{I_C^2} - 4\,\mathrm{II_C})$ is nonpositive, the quartic equation (2.2.15) has a unique positive root which is $\mathrm{I_U}$. However, if $(\mathrm{I_C^2} - 4\mathrm{II_C})$ is positive, there are two positive roots[5] with $\mathrm{I_U}$ being the greater. Furthermore, only the root $\mathrm{I_U}$ is greater than $\sqrt{\mathrm{I_C}}$.

To show that only the root $\mathrm{I_U}$ exceeds $\sqrt{\mathrm{I_C}}$, first note the relations (2.2.16) among the principal stretches. Then, for $\mathrm{I_1} \geq 0$,

$$\lambda_1 \geq \lambda_2 + \lambda_3, \quad \lambda_1(\lambda_2 + \lambda_3) \geq (\lambda_2 + \lambda_3)^2. \tag{2.2.17a,b}$$

It follows from (2.2.16b) and (2.2.17a,b) that,

$$\begin{aligned} \mathrm{I_1^2} &= (\lambda_1 - \lambda_2 - \lambda_3)^2 \\ &= \lambda_1^2 - 2\lambda_1(\lambda_2 + \lambda_3) + (\lambda_2 + \lambda_3)^2 \\ &\leq \lambda_1^2 - (\lambda_2 + \lambda_3)^2 \\ &= \mathrm{I_C} - 2(\lambda_2^2 + \lambda_3^2 + \lambda_2\lambda_3) \leq \mathrm{I_C}. \end{aligned} \tag{2.2.17c}$$

This shows that $\mathrm{I_1} \leq \sqrt{\mathrm{I_C}}$. Similarly, it can be shown that $\mathrm{I_2}$ and $\mathrm{I_3}$ are also less than or equal to $\sqrt{\mathrm{I_C}}$. Since all three principal stretches are positive and nonzero, it follows from (2.2.14b) that

$$\mathrm{I_U} > \sqrt{\mathrm{I_C}}. \tag{2.2.17d}$$

Therefore, $\mathrm{I_U}$ is the only root of the quartic equation (2.2.15) which is greater

---

[5] Hoger and Carlson (1984) solve quartic equation (2.2.15) for $\mathrm{I_U}$, under the assumption that a unique positive $\mathrm{I_U}$ exists for any symmetric positive-definite tensor **C**. Through a counterexample, Sawyers (1986) shows that $\mathrm{I_U}$ is not a unique positive root of (2.2.15). Stickforth (1987) comments that $\mathrm{I_U}$ is that unique solution of (2.2.15) which is greater than $\sqrt{\mathrm{I_C}}$.

than $\sqrt{I_C}$.

Instead of solving the quartic equation (2.2.15) for I_U, consider

$$(\lambda_2+\lambda_3)^2 = \lambda_2^2+\lambda_3^2+2\,\lambda_2\lambda_3$$

$$= I_C-\lambda_1^2+\frac{2}{\lambda_1}\,III_U\,. \tag{2.2.18}$$

Thus, the first invariant of the stretch tensor can be expressed in terms of the invariants of the Cauchy-Green tensors and one of their *eigenvalues* as follows:[6]

$$I_U = \lambda_1+\sqrt{I_C-\lambda_1^2+2\sqrt{III_C}/\lambda_1}\,. \tag{2.2.19}$$

An eigenvalue of **C** is given by

$$\lambda_1^2 = \frac{1}{3}\,I_C+2\sqrt{a}\cos(\phi/3)\,, \tag{2.2.20a}$$

where

$$\phi \equiv \cos^{-1}(b/a\sqrt{a})\,,\qquad b = III_{C'}/2\,,\qquad a = -\frac{1}{3}\,II_{C'}\,, \tag{2.2.20b-d}$$

and $\mathbf{C}'$ is defined by

$$\mathbf{C}' = \mathbf{C}-\frac{1}{3}\,I_C\,\mathbf{1}\,. \tag{2.2.20e}$$

Once I_U is known, II_U is obtained from (2.2.14b).

The calculation alternatives presented above involve evaluation of simple standard expressions. It appears that the calculation based on the eigenvalues of **C** is no more involved than the direct evaluation of the basic invariants. This alternative also yields the principal stretches directly. In summary, the *polar decomposition of* **F** *can be performed exactly and directly by evaluating simple standard expressions, without a need for eigenvector calculations, as outlined above.* This is of importance for large-scale numerical computations of finite-deformation and rotation problems.

2.2.3. Lagrangian and Eulerian Triads

At each material point, the mapping $d\mathbf{x} = \mathbf{F}\,d\mathbf{X}$, characterized by the deformation gradient **F**, identifies two unit orthogonal triads. One consists of the unit vectors $\mathbf{N}^a$ which are the principal directions of the right-stretch tensor **U**, called the *Lagrangian triad*, and the other consists of the unit vectors $\mathbf{n}^a$ which are the principal directions of the left-stretch tensor **V**, called the *Eulerian triad*. The rotation **R** maps the Lagrangian triad into the Eulerian triad; see

[6] Equation (2.2.19) reduces to the solution obtained by Hoger and Carlson (1984) when λ_1 is replaced by $\sqrt{(\zeta+2I_C)/2}$. The expression for ζ, however, involves the square root of b^2-a^3 which is negative for the symmetric positive-definite tensor **C**, rendering such a solution complex-valued and hence unacceptable.

Figure 2.2.1. The deformation gradient **F** is decomposed into pure stretch **U** along the directions of the Lagrangian triad, $\mathbf{N}^a$, followed by the rotation of this triad into the Eulerian triad, $\mathbf{n}^a$, or it is decomposed into the rotation of the $\mathbf{N}^a$-triad into the $\mathbf{n}^a$-triad first, and then followed by pure stretch **V** along the $\mathbf{n}^a$-directions.

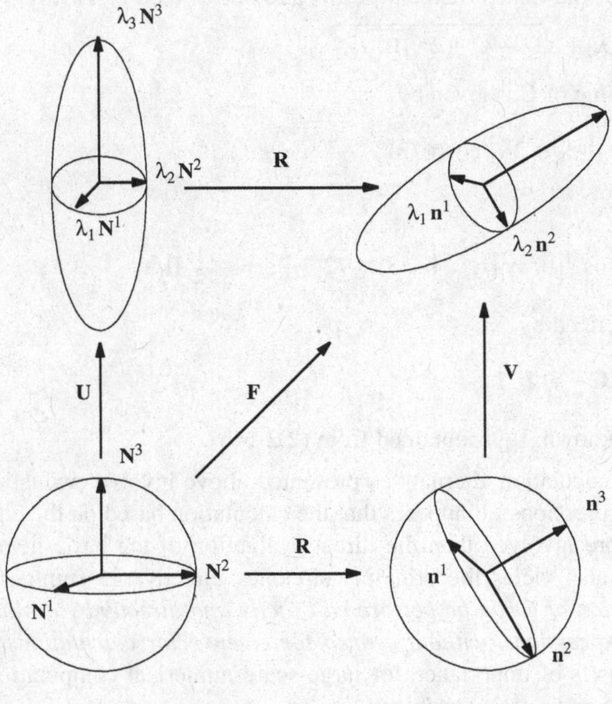

Figure 2.2.1

Polar decomposition of **F**: the rotation **R** maps the Lagrangian triad $\mathbf{N}^a$ into the Eulerian triad $\mathbf{n}^a$; **U** and **V** are pure stretches along the corresponding principal directions

It is convenient to view the Lagrangian triad $\mathbf{N}^a$ as the result of the rotation of the reference coordinate triad $\mathbf{e}_a$, a = 1, 2, 3, induced by the Lagrangian rotation tensor $\mathbf{R}^L$,

$$\mathbf{R}^L = \sum_{a=1}^{3} \mathbf{N}^a \otimes \mathbf{e}_a, \tag{2.2.21a}$$

so that

$$\mathbf{N}^a = \mathbf{R}^L \mathbf{e}_a\,. \tag{2.2.21b}$$

In a similar way, the Eulerian triad $\mathbf{n}^a$ is obtained by rotating the reference triad $\mathbf{e}_a$ through the Eulerian rotation $\mathbf{R}^E$, given by

$$\mathbf{R}^E = \sum_{a=1}^{3} \mathbf{n}^a \otimes \mathbf{e}_a\,, \tag{2.2.22a}$$

so that

$$\mathbf{n}^a = \mathbf{R}^E \mathbf{e}_a\,. \tag{2.2.22b}$$

The rotation of the $\mathbf{e}_a$-triad into the Eulerian $\mathbf{n}^a$-triad, $\mathbf{R}^E$, then may be viewed as the rotation of the $\mathbf{e}_a$-triad into the Lagrangian $\mathbf{N}^a$-triad, $\mathbf{R}^L$, followed by the rotation of the $\mathbf{N}^a$-triad into the Eulerian $\mathbf{n}^a$-triad, $\mathbf{R}$; see Figure 2.2.2. Hence,

$$\mathbf{R}^E = \mathbf{R}\,\mathbf{R}^L\,. \tag{2.2.23}$$

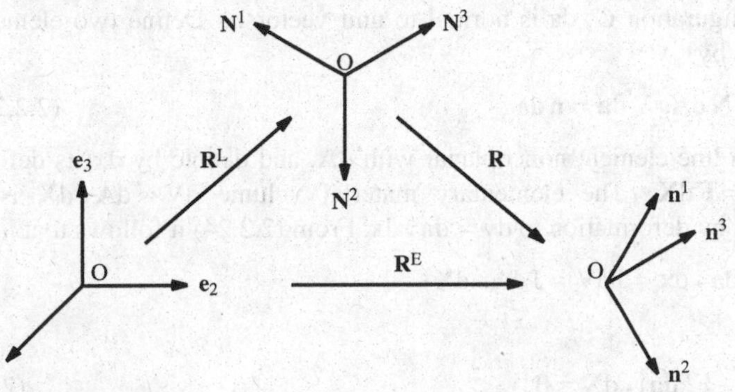

Figure 2.2.2

The rotation of the $\mathbf{e}_a$-triad into the Eulerian $\mathbf{n}^a$-triad, $\mathbf{R}^E$, consists of the rotation of the $\mathbf{e}_a$-triad into the Lagrangian $\mathbf{N}^a$-triad, $\mathbf{R}^L$, followed by the rotation of the $\mathbf{N}^a$-triad into the Eulerian $\mathbf{n}^a$-triad, $\mathbf{R}$

2.2.4. Transformation of Volume Elements

From (2.2.10c), it follows that a material element of volume dV in initial configuration C_0 is mapped by deformation into dv such that

$$\frac{dv}{dV} = \frac{ds_1\, ds_2\, ds_3}{dS_1\, dS_2\, dS_3} = \lambda_1 \lambda_2 \lambda_3 = J\,, \tag{2.2.24}$$

where dS_a and ds_a, a = 1, 2, 3, are the lengths of elements along the Lagrangian and Eulerian triads, respectively.

A deformation which preserves the volume is called *isochoric*. If the right- and left-stretch tensors of *any* deformation are scaled by the cube root of J, the resulting stretch tensors correspond to isochoric deformations (they distort material neighborhoods but do not expand or contract them):

$$\bar{\mathbf{U}} = \mathrm{J}^{-1/3}\,\mathbf{U}, \quad \bar{\mathbf{V}} = \mathrm{J}^{-1/3}\,\mathbf{V}. \tag{2.2.25a,b}$$

It is clear that, since

$$\mathrm{III}_{\bar{\mathrm{U}}} = \mathrm{III}_{\bar{\mathrm{V}}} = \det\bar{\mathbf{U}} = \det\bar{\mathbf{V}} = 1, \tag{2.2.26}$$

$\bar{\mathbf{U}}$ and $\bar{\mathbf{V}}$ induce pure distortion, whereas $\mathrm{J}^{1/3}\,\mathbf{1}$ produces pure volume expansion or contraction.

2.2.5. Transformation of Surface Elements

Let dA be a material surface element in the initial configuration C_0 normal to unit vector **N**, and let it be transformed by the deformation to da in current configuration C; da is normal to unit vector **n**. Define two elementary area vectors by

$$\mathrm{d}\mathbf{A} = \mathbf{N}\,\mathrm{dA}, \quad \mathrm{d}\mathbf{a} = \mathbf{n}\,\mathrm{da}. \tag{2.2.27a,b}$$

Let d**X** be a line element noncoplanar with dA, and denote by d**x** its deformed image, $\mathrm{d}\mathbf{x} = \mathbf{F}\,\mathrm{d}\mathbf{X}$. The elementary material volume $\mathrm{dV} = \mathrm{d}\mathbf{A} \cdot \mathrm{d}\mathbf{X}$ is then transformed by deformation to $\mathrm{dv} = \mathrm{d}\mathbf{a} \cdot \mathrm{d}\mathbf{x}$. From (2.2.24) it follows that

$$\mathrm{dv} = \mathrm{d}\mathbf{a} \cdot \mathrm{d}\mathbf{x} = \mathrm{J}\,\mathrm{dV} = \mathrm{J}\,\mathrm{d}\mathbf{A} \cdot \mathrm{d}\mathbf{X}$$

or

$$(\mathrm{J}\,\mathrm{d}\mathbf{A} - \mathbf{F}^{\mathrm{T}}\,\mathrm{d}\mathbf{a}) \cdot \mathrm{d}\mathbf{X} = 0. \tag{2.2.28}$$

For arbitrary d**X**, it follows that

$$\mathrm{J}\,\mathrm{d}\mathbf{A} = \mathbf{F}^{\mathrm{T}}\,\mathrm{d}\mathbf{a} \tag{2.2.29a}$$

or

$$\mathrm{J}\,\mathrm{dA}_{\mathrm{B}} - \mathrm{da}_{\mathrm{i}}\,\mathrm{F}_{\mathrm{iB}} = 0. \tag{2.2.29b}$$

Equations (2.2.29a,b) also result from direct evaluation of the cross product of two distinct material line elements emanating from a typical particle. Expressions (2.2.29a,b) may be rewritten as

$$\mathrm{d}\mathbf{a} = \mathrm{J}\,\mathbf{F}^{-\mathrm{T}}\,\mathrm{d}\mathbf{A} \tag{2.2.30a}$$

or

$$\mathrm{da}_{\mathrm{i}} = \mathrm{J}\,\mathrm{X}_{\mathrm{B,i}}\,\mathrm{dA}_{\mathrm{B}}, \tag{2.2.30b}$$

where $\mathbf{F}^{-\mathrm{T}} \equiv (\mathbf{F}^{-1})^{\mathrm{T}}$, and the components of $\mathbf{F}^{-1}$ are $\mathrm{X}_{\mathrm{A,i}}$.

2.3. STRAIN MEASURES

For finite deformation, various strain measures coaxial with the Lagrangian or Eulerian triads have been introduced in the literature. Those coaxial with the Lagrangian triad are called *material strain measures* by Hill (1968). This class of strain measures has the general spectral representation

$$\mathbf{E} = \sum_{a=1}^{3} E_a \mathbf{N}^a \otimes \mathbf{N}^a \tag{2.3.1a}$$

or, when components with respect to the fixed background rectangular Cartesian coordinate system (*i.e.*, the $\mathbf{e}_A$-triad) are considered,

$$E_{AB} = \sum_{a=1}^{3} E_a N_A^a N_B^a , \tag{2.3.1b}$$

where

$$E_a = f(\lambda_a) . \tag{2.3.1c}$$

Denote the corresponding *Eulerian strain measure* by $\boldsymbol{\varepsilon}$ which has the general spectral representation

$$\boldsymbol{\varepsilon} = \sum_{a=1}^{3} f(\lambda_a)\, \mathbf{n}^a \otimes \mathbf{n}^a . \tag{2.3.2a}$$

With respect to the fixed rectangular Cartesian coordinate system, $\boldsymbol{\varepsilon}$ has the components

$$\varepsilon_{ij} = \sum_{a=1}^{3} f(\lambda_a)\, n_i^a n_j^a . \tag{2.3.2b}$$

2.3.1. Explicit Representation of General Strain Measures in Terms of U and V

In Section 1.3, for an arbitrary real-valued second-order tensor, $\mathbf{A}$, explicit expressions are given for the corresponding orthonormal bases, $\mathbf{e}_i \otimes \mathbf{e}^i$, in terms of $\mathbf{A}$, where $\mathbf{e}_i$, and $\mathbf{e}^i$ are *eigenvectors* of $\mathbf{A}$ and $\mathbf{A}^T$, respectively, corresponding to the same *eigenvalue*. These are used in what follows, to obtain coordinate-independent explicit representations of the Lagrangian and Eulerian strain measures, $\mathbf{E}$ and $\boldsymbol{\varepsilon}$, in terms of $\mathbf{U}$ and $\mathbf{V}$, respectively, for *any* given function $f(\lambda)$. Since strain measures are considered, all eigenvalues and eigenvectors are real.

Denote the orthonormal bases $\mathbf{N}^a \otimes \mathbf{N}^a$ and $\mathbf{n}^a \otimes \mathbf{n}^a$ by $\hat{\mathbf{E}}_a$ and $\mathbf{E}_a$, respectively. Then, (2.3.1) and (2.3.2) can be rewritten as follows:

$$\mathbf{E} = \sum_{a=1}^{3} f(\lambda_a)\, \hat{\mathbf{E}}_a , \quad \boldsymbol{\varepsilon} = \sum_{a=1}^{3} f(\lambda_a)\, \mathbf{E}_a , \quad \text{for } \lambda_1 \neq \lambda_2 \neq \lambda_3 \neq \lambda_1 ,$$

$$\mathbf{E} = f(\lambda_1)\, \hat{\mathbf{E}}_1 + f(\lambda_2)\, (\mathbf{1} - \hat{\mathbf{E}}_1) ,$$

$$\boldsymbol{\varepsilon} = f(\lambda_1)\, \mathbf{E}_1 + f(\lambda_2)\, (\mathbf{1} - \mathbf{E}_1) , \quad \text{for } \lambda_1 \neq \lambda_2 = \lambda_3 ,$$

$$\mathbf{E} = f(\lambda)\,\mathbf{1}\,, \quad \boldsymbol{\varepsilon} = f(\lambda)\,\mathbf{1}\,, \quad \text{for } \lambda_1 = \lambda_2 = \lambda_3 \equiv \lambda\,. \tag{2.3.3a-f}$$

Distinct Principal Stretches: When the principal stretches are distinct, from (1.3.18c), $\hat{\mathbf{E}}_a$ and $\mathbf{E}_a$ are expressed as

$$\hat{\mathbf{E}}_a = \frac{1}{(\lambda_a - \lambda_b)(\lambda_a - \lambda_c)} \{\mathbf{U}^2 - (\lambda_b + \lambda_c)\,\mathbf{U} + \lambda_b \lambda_c\,\mathbf{1}\}\,,$$

$$\mathbf{E}_a = \frac{1}{(\lambda_a - \lambda_b)(\lambda_a - \lambda_c)} \{\mathbf{V}^2 - (\lambda_b + \lambda_c)\,\mathbf{V} + \lambda_b \lambda_c\,\mathbf{1}\}\,; \tag{2.3.4a,b}$$

a, b, and c are cyclic permutations of 1, 2, and 3 (no sum).

Once the expressions for $\hat{\mathbf{E}}_a$ and $\mathbf{E}_a$ from (2.3.4) are introduced into (2.3.3a,b), the strain measures $\mathbf{E}$ and $\boldsymbol{\varepsilon}$ take on the following explicit forms:

$$\mathbf{E} = a_0\,\mathbf{1} + a_1\,\mathbf{U} + a_2\,\mathbf{U}^2\,, \quad \boldsymbol{\varepsilon} = a_0\,\mathbf{1} + a_1\,\mathbf{V} + a_2\,\mathbf{V}^2\,, \tag{2.3.5a,b}$$

where a_0, a_1, and a_2 are explicit functions of the principal stretches, given by

$$\begin{bmatrix} a_0 \\ -a_1 \\ a_2 \end{bmatrix} = \frac{1}{\Delta} \begin{bmatrix} \lambda_2 \lambda_3 & \lambda_3 \lambda_1 & \lambda_1 \lambda_2 \\ (\lambda_2 + \lambda_3) & (\lambda_3 + \lambda_1) & (\lambda_1 + \lambda_2) \\ 1 & 1 & 1 \end{bmatrix} \begin{bmatrix} f(\lambda_1)\,(\lambda_2 - \lambda_3) \\ f(\lambda_2)\,(\lambda_3 - \lambda_1) \\ f(\lambda_3)\,(\lambda_1 - \lambda_2) \end{bmatrix}, \tag{2.3.5c}$$

and

$$\Delta = -(\lambda_1 - \lambda_2)(\lambda_2 - \lambda_3)(\lambda_3 - \lambda_1)\,. \tag{2.3.5d}$$

Two Repeated Principal Stretches: When the principal stretches are repeated, say $\lambda_2 = \lambda_3$, from (1.3.22c), $\hat{\mathbf{E}}_1$ and $\mathbf{E}_1$ are given by

$$\hat{\mathbf{E}}_1 = \frac{1}{(\lambda_1 - \lambda_2)^2} \{\mathbf{U}^2 - 2\lambda_2\,\mathbf{U} + \lambda_2^2\,\mathbf{1}\}\,,$$

$$\mathbf{E}_1 = \frac{1}{(\lambda_1 - \lambda_2)^2} \{\mathbf{V}^2 - 2\lambda_2\,\mathbf{V} + \lambda_2^2\,\mathbf{1}\}\,. \tag{2.3.6a,b}$$

Since λ_1 and λ_2 are the roots of

$$\lambda^2 - (\lambda_1 + \lambda_2)\,\lambda + \lambda_1 \lambda_2 = 0\,, \tag{2.3.7a}$$

it follows from (2.2.6b,d) that

$$\mathbf{U}^2 - (\lambda_1 + \lambda_2)\,\mathbf{U} + \lambda_1 \lambda_2\,\mathbf{1} = \mathbf{0}\,,$$

$$\mathbf{V}^2 - (\lambda_1 + \lambda_2)\,\mathbf{V} + \lambda_1 \lambda_2\,\mathbf{1} = \mathbf{0}\,. \tag{2.3.7b,c}$$

Substituting (2.3.7b,c) into (2.3.6a,b), obtain

$$\hat{\mathbf{E}}_1 = \frac{1}{\lambda_1 - \lambda_2} (\mathbf{U} - \lambda_2\,\mathbf{1})\,, \quad \mathbf{E}_1 = \frac{1}{\lambda_1 - \lambda_2} (\mathbf{V} - \lambda_2\,\mathbf{1})\,. \tag{2.3.8a,b}$$

Then, explicit coordinate-independent expressions are obtained for the strain measures $\mathbf{E}$ and $\boldsymbol{\varepsilon}$, by substituting (2.3.8) into (2.3.3c,d), arriving at

$$\mathbf{E} = \frac{f(\lambda_1) - f(\lambda_2)}{\lambda_1 - \lambda_2}\,\mathbf{U} + \frac{\lambda_1 f(\lambda_2) - \lambda_2 f(\lambda_1)}{\lambda_1 - \lambda_2}\,\mathbf{1}\,,$$

$$\boldsymbol{\varepsilon} = \frac{f(\lambda_1) - f(\lambda_2)}{\lambda_1 - \lambda_2}\,\mathbf{V} + \frac{\lambda_1 f(\lambda_2) - \lambda_2 f(\lambda_1)}{\lambda_1 - \lambda_2}\,\mathbf{1}\,. \qquad (2.3.9\text{a,b})$$

Now, consider the explicit expressions (2.3.5) for the strain measures. Substituting (2.3.7b,c) into (2.3.5a,b), obtain

$$\mathbf{E} = (a_0 - a_2\,\lambda_1\,\lambda_2)\,\mathbf{1} + \{a_1 + a_2\,(\lambda_1 + \lambda_2)\}\,\mathbf{U}\,,$$

$$\boldsymbol{\varepsilon} = (a_0 - a_2\,\lambda_1\,\lambda_2)\,\mathbf{1} + \{a_1 + a_2\,(\lambda_1 + \lambda_2)\}\,\mathbf{V}\,. \qquad (2.3.10\text{a,b})$$

Then, from (2.3.9) and (2.3.10), it follows that

$$a_0 = \frac{\lambda_1 f(\lambda_2) - \lambda_2 f(\lambda_1)}{\lambda_1 - \lambda_2} + a_2\,\lambda_1\,\lambda_2\,,$$

$$a_1 = \frac{f(\lambda_1) - f(\lambda_2)}{\lambda_1 - \lambda_2} - a_2\,(\lambda_1 + \lambda_2)\,. \qquad (2.3.11\text{a,b})$$

Therefore, when the principal stretches are repeated, the coefficients a_0, a_1, and a_2 are not unique but are related by (2.3.11a,b). The limiting values of these coefficients as λ_2 goes to λ_3 in (2.3.5), are also related by (2.3.11a,b), with

$$a_2 = \frac{f(\lambda_1) - f(\lambda_2) - (\lambda_1 - \lambda_2)\,f'(\lambda_2)}{(\lambda_1 - \lambda_2)^2}\,. \qquad (2.3.11\text{c})$$

Substitution of (2.3.11) into (2.3.5), in view of (2.3.7), also results in (2.3.9).

Three Repeated Principal Stretches: When all three principal stretches are identical, it follows from (2.2.6b,d) that

$$\mathbf{U} = \lambda\,\mathbf{1}\,, \quad \mathbf{V} = \lambda\,\mathbf{1}\,. \qquad (2.3.12\text{a,b})$$

Then the explicit expressions (2.3.5a,b) for the strain measures, become

$$\mathbf{E} = \boldsymbol{\varepsilon} = (a_0 + a_1\,\lambda + a_2\,\lambda^2)\,\mathbf{1}\,. \qquad (2.3.13)$$

It follows from (2.3.3e,f) and (2.3.13) that

$$a_0 + a_1\,\lambda + a_2\,\lambda^2 = f(\lambda)\,. \qquad (2.3.14\text{a})$$

Therefore, when all three principal stretches are identical, the coefficients a_0, a_1, and a_2 are not unique but are related by (2.3.14a). The limiting values of the coefficients as λ_2 goes to $\lambda_1 = \lambda$ in (2.3.11), are also related by (2.3.14a), with

$$a_1 = f'(\lambda) - \lambda\,f''(\lambda)\,, \quad a_2 = \frac{1}{2}\,f''(\lambda)\,. \qquad (2.3.14\text{b,c})$$

Substitution of (2.3.14) into (2.3.5), in view of (2.3.12), also results in (2.3.3e,f).

In summary, *any material and Eulerian strain measure can be evaluated explicitly without a need for eigenvector calculations*. In particular, the general strain measures can be expressed explicitly in the form (2.3.5a,b), with the coefficients given by (2.3.5c,d). These expressions continue to hold as limiting values, even when two or all three principal stretches are equal.

2.3.2. Required Properties of Strain Measures

It is reasonable to require that the strain should vanish at no deformation, should be positive and increasing when the corresponding fibers are monotonically extending, and should reduce to the usual linear measure, when linearized. These are satisfied when the function $f(\lambda)$ in (2.3.1) is chosen such that

$$f(1) = 0, \quad f'(1) = 1, \quad \text{and } f'(\lambda) > 0, \tag{2.3.15a-c}$$

where prime denotes differentiation. A subclass of strain measures which comply with these requirements is (Seth, 1964; Hill, 1968)

$$E_a = \frac{1}{2m}(\lambda_a^{2m} - 1), \tag{2.3.16}$$

where m is real. Various commonly used strain measures result for special values of m. Some of these are listed below.

2.3.3. Lagrangian and Eulerian Strain Tensors

For m = 1, the so-called Lagrangian and Eulerian strain tensors result,

$$\mathbf{E}^L = \sum_{a=1}^{3} \frac{1}{2}(\lambda_a^2 - 1)\,\mathbf{N}^a \otimes \mathbf{N}^a = \frac{1}{2}(\mathbf{C} - \mathbf{1}),$$

$$\boldsymbol{\varepsilon}^E = \sum_{a=1}^{3} \frac{1}{2}(\lambda_a^2 - 1)\,\mathbf{n}^a \otimes \mathbf{n}^a = \frac{1}{2}(\mathbf{B} - \mathbf{1}). \tag{2.3.17a,b}$$

2.3.4. Biot's Strain Measure

For m = 1/2, it follows that

$$\mathbf{E}^B = \sum_{a=1}^{3} (\lambda_a - 1)\,\mathbf{N}^a \otimes \mathbf{N}^a = \mathbf{U} - \mathbf{1} \tag{2.3.18a}$$

which is introduced by Biot (1965). The corresponding Eulerian measure is

$$\boldsymbol{\varepsilon}^B = \sum_{a=1}^{3} (\lambda_a - 1)\,\mathbf{n}^a \otimes \mathbf{n}^a = \mathbf{V} - \mathbf{1}. \tag{2.3.18b}$$

2.3.5. Logarithmic Strain Measure

With a suitable limiting process, the usual logarithmic strain measure is obtained, as m is taken to zero,

$$\mathbf{E}^l = \sum_{a=1}^{3} \ln\lambda_a\,\mathbf{N}^a \otimes \mathbf{N}^a = \ln\mathbf{U},$$

$$\boldsymbol{\varepsilon}^l = \sum_{a=1}^{3} \ln\lambda_a\,\mathbf{n}^a \otimes \mathbf{n}^a = \ln\mathbf{V}. \tag{2.3.19a,b}$$

The logarithmic strain measures $\mathbf{E}^l$ and $\boldsymbol{\varepsilon}^l$ are easily expressed explicitly as quadratic functions of **U** and **V**, as discussed in Subsection 2.3.1.

When interpreting $ln\,\mathbf{U}$ (or $ln\,\mathbf{V}$) as a series expansion of **U** (or **V**), different series result for $\lambda < 2$ and $\lambda > 2$. Therefore, it may be more convenient to interpret (2.3.19) in terms of exponential functions,

$$\mathbf{U} = \exp\{\mathbf{E}^l\} = \sum_{\alpha=1}^{\infty} \frac{1}{\alpha!}(\mathbf{E}^l)^{\alpha},$$

$$\mathbf{V} = \exp\{\boldsymbol{\varepsilon}^l\} = \sum_{\alpha=1}^{\infty} \frac{1}{\alpha!}(\boldsymbol{\varepsilon}^l)^{\alpha}. \qquad (2.3.20a,b)$$

Repeated use of the Hamilton-Cayley theorem produces

$$\mathbf{U} = \phi_0\,\mathbf{1} + \phi_1\,\mathbf{E}^l + \phi_2(\mathbf{E}^l)^2, \qquad (2.3.21a)$$

$$\mathbf{V} = \phi_0\,\mathbf{1} + \phi_1\,\boldsymbol{\varepsilon}^l + \phi_2(\boldsymbol{\varepsilon}^l)^2, \qquad (2.3.21b)$$

where ϕ_0, ϕ_1, and ϕ_2 are functions of the basic invariants of $\mathbf{E}^l$, which can be expressed in terms of the logarithms of the principal stretches. In general, the transcendental (implicit) representation of $\mathbf{E}^l$ and $\boldsymbol{\varepsilon}^l$, given by (2.3.20a,b), may not be very useful for specific applications. Indeed, in light of the results given in Subsection 2.3.1, explicit expressions for $ln\,\mathbf{U}$ and $ln\,\mathbf{V}$ are obtained from (2.3.3) to (2.3.14) by simply identifying f(x) with $ln\,(\mathrm{x})$.

2.3.6. Other Strain Measures

Other strain measures have been introduced in the literature. Two additional measures are[7]

$$\mathbf{E}(\mathrm{m} = -1/2) = \sum_{\mathrm{a}=1}^{3} (1 - \lambda_{\mathrm{a}}^{-1})\,\mathbf{N}^{\mathrm{a}} \otimes \mathbf{N}^{\mathrm{a}} = \mathbf{1} - \mathbf{U}^{-1},$$

$$\boldsymbol{\varepsilon}(\mathrm{m} = -1/2) = \sum_{\mathrm{a}=1}^{3} (1 - \lambda_{\mathrm{a}}^{-1})\,\mathbf{n}^{\mathrm{a}} \otimes \mathbf{n}^{\mathrm{a}} = \mathbf{1} - \mathbf{V}^{-1}, \qquad (2.3.22a,b)$$

and

$$\mathbf{E}(\mathrm{m} = -1) = \sum_{\mathrm{a}=1}^{3} \frac{1}{2}(1 - \lambda_{\mathrm{a}}^{-2})\,\mathbf{N}^{\mathrm{a}} \otimes \mathbf{N}^{\mathrm{a}} = \frac{1}{2}(\mathbf{1} - \mathbf{C}^{-1}),$$

$$\boldsymbol{\varepsilon}(\mathrm{m} = -1) = \sum_{\mathrm{a}=1}^{3} \frac{1}{2}(1 - \lambda_{\mathrm{a}}^{-2})\,\mathbf{n}^{\mathrm{a}} \otimes \mathbf{n}^{\mathrm{a}} = \frac{1}{2}(\mathbf{1} - \mathbf{B}^{-1}). \qquad (2.3.23a,b)$$

The latter is called the Almansi strain by Hill (1978), while Wang and Truesdell (1973) refer to $\mathbf{E}(\mathrm{m} = -1)$ as the Almansi strain. According to Truesdell and Toupin (1960) and Prager (1961), the Almansi strain is $\boldsymbol{\varepsilon}(\mathrm{m} = -1)$. From (2.2.13a-c), strain measures (2.3.22) are explicitly written in terms of **U** and **V** and their common basic invariants. Similar comments apply to (2.3.23); *e.g.*, use the Hamilton-Cayley theorem to calculate $\mathbf{C}^{-1}$ as a quadratic function of **C**,

[7] For an account, see Truesdell and Toupin (1960).

$$\mathbf{C}^{-1} = \frac{1}{\mathrm{III_C}} (\mathbf{C}^2 - \mathrm{I_C}\,\mathbf{C} + \mathrm{II_C}\,\mathbf{1})\,, \tag{2.3.24}$$

with a similar expression for $\mathbf{B}^{-1}$.

2.4. VELOCITY AND ACCELERATION

2.4.1. Velocity

Partial time differentiation of (2.1.1) with **X** held fixed, yields the velocity of the particle initially at **X**,

$$\dot{\mathbf{x}} = \frac{\partial \mathbf{x}(\mathbf{X}, t)}{\partial t}\,, \qquad \dot{x}_i = \frac{\partial x_i(\mathbf{X}, t)}{\partial t}\,. \tag{2.4.1a,b}$$

Since the mapping (2.1.1) is one-to-one and invertible, at least in principle, the inverse mapping may be written as

$$\mathbf{X} = \mathbf{X}(\mathbf{x}, t)\,, \qquad X_A = X_A(x_1, x_2, x_3, t) \tag{2.4.2a,b}$$

which identifies the initial position **X** of the particle that at time t occupies position **x**. Substitution from (2.4.2) into (2.4.1) yields the *Eulerian representation* (*i.e.*, in terms of current particle position) of the velocity field,

$$\mathbf{v} = \mathbf{v}(\mathbf{x}, t) \equiv \dot{\mathbf{x}}(\mathbf{X}(\mathbf{x}, t), t)\,,$$

$$v_i = v_i(\mathbf{x}, t) \equiv \dot{x}_i(\mathbf{X}(\mathbf{x}, t), t)\,; \tag{2.4.1c,d}$$

this is the velocity of the particle which at time t is at place **x**.

2.4.2. Acceleration

Time differentiation of (2.4.1a,b) with **X** held fixed, yields the acceleration of particle **X**,

$$\ddot{\mathbf{x}} = \frac{\partial^2 \mathbf{x}(\mathbf{X}, t)}{\partial t^2}\,, \qquad \ddot{x}_i = \frac{\partial^2 x_i(\mathbf{X}, t)}{\partial t^2}\,. \tag{2.4.3a,b}$$

Either by direct substitution for **X** from (2.4.2) or by *material time differentiation* (*i.e.*, time differentiation, keeping particle **X** fixed) of (2.4.1c,d), the Eulerian description of the acceleration field results,

$$\mathbf{a} = \mathbf{a}(\mathbf{x}, t) = \frac{\partial \mathbf{v}}{\partial t} + \frac{\partial \mathbf{v}}{\partial x_j}\, v_j\,,$$

$$a_i = a_i(\mathbf{x}, t) = \frac{\partial v_i}{\partial t} + \frac{\partial v_i}{\partial x_j}\, v_j\,. \tag{2.4.3c,d}$$

The last term, often called convected, emerges as a result of the rate of change of the particle position.

2.5. DEFORMATION-RATE AND SPIN TENSORS

Information on the rate of deformation of two neighboring particles, **X** and **X** + d**X**, results from the examination of the material derivative of (2.1.3),

$$d\dot{\mathbf{x}} = \dot{\mathbf{F}}\, d\mathbf{X},$$

$$d\dot{x}_i = \dot{F}_{iA}\, dX_A = \frac{\partial \dot{x}_i}{\partial X_A}\, dX_A\,. \qquad (2.5.1a,b)$$

On the other hand, at a given instant (*i.e.*, t is fixed), the instantaneous rate of deformation of a given neighborhood may be described in terms of relative velocities of two neighboring particles, one at **x**, and the other at **x** + d**x**; *i.e.*,

$$\mathbf{v}(\mathbf{x}+d\mathbf{x},\, t) - \mathbf{v}(\mathbf{x},\, t) = \frac{\partial \mathbf{v}}{\partial x_i}\, dx_i + \dots \qquad (2.5.2)$$

which, to the first order, is

$$d\mathbf{v} = \frac{\partial \mathbf{v}}{\partial x_i}\, dx_i = \mathbf{L}\, d\mathbf{x}\,,$$

$$dv_i = L_{ij}\, dx_j\,, \qquad (2.5.3a\text{-}c)$$

where

$$\frac{\partial v_i}{\partial x_j} = L_{ij} \qquad (2.5.4a)$$

is the *velocity gradient*. From the chain rule of differentiation it follows that

$$\frac{\partial \dot{x}_i}{\partial X_A} = \frac{\partial v_i}{\partial x_j}\frac{\partial x_j}{\partial X_A} = L_{ij}\, F_{jA}\,, \qquad (2.5.5)$$

so that

$$\mathbf{L} = \dot{\mathbf{F}}\,\mathbf{F}^{-1}\,. \qquad (2.5.4b)$$

2.5.1. Deformation-rate and Spin Tensors

The velocity gradient is decomposed into the symmetric part, **D**, called the *deformation-rate tensor*, and the skewsymmetric part, **W**, called the *spin tensor*,

$$\mathbf{L} = \mathbf{D} + \mathbf{W}\,, \qquad L_{ij} = D_{ij} + W_{ij}\,, \qquad (2.5.6a,b)$$

$$\mathbf{D} = \frac{1}{2}(\mathbf{L} + \mathbf{L}^T), \quad D_{ij} = \frac{1}{2}(v_{i,j} + v_{j,i})\,, \qquad (2.5.7a,b)$$

$$\mathbf{W} = \frac{1}{2}(\mathbf{L} - \mathbf{L}^T), \quad W_{ij} = \frac{1}{2}(v_{i,j} - v_{j,i})\,. \qquad (2.5.8a,b)$$

The deformation rate, **D**, characterizes the instantaneous rate of distortion. The spin, **W**, the instantaneous rate of rigid-body rotation of the material neighborhood. Equation (2.5.3a) now becomes

$$\mathbf{dv} = \mathbf{D}\,\mathbf{dx} + \mathbf{W}\,\mathbf{dx}$$

$$= \mathbf{D}\,\mathbf{dx} + \boldsymbol{\omega} \times \mathbf{dx}\,, \tag{2.5.9}$$

where $\boldsymbol{\omega}$, the axial vector of the skewsymmetric tensor $\mathbf{W}$, is the instantaneous angular velocity vector; it relates to the spin tensor by

$$\omega_i = -\frac{1}{2}\,e_{ijk}\,W_{jk}\,, \tag{2.5.10}$$

and is called the *vorticity* of the velocity field.

2.5.2. Stretching, Shearing, and Volumetric Deformation Rate

Let ds be the length of $\mathbf{dx}$ along the unit vector $\mathbf{n}$,

$$\mathbf{dx} = ds\,\mathbf{n}\,,$$

$$(ds)^2 = \mathbf{dx} \cdot \mathbf{dx}\,. \tag{2.5.11a,b}$$

Then, material differentiation of (2.5.11b) yields

$$\frac{(ds)^{\cdot}}{ds} = (\mathbf{D}\,\mathbf{n}) \cdot \mathbf{n}$$

$$= D_{ij}\,n_i\,n_j\,. \tag{2.5.12}$$

Hence, the *stretching* in the $\mathbf{n}$-direction, *i.e.*, the time rate of change of length per unit current length of an element with current orientation $\mathbf{n}$, is equal to the normal component of the deformation-rate tensor, $\mathbf{D}$, in the $\mathbf{n}$-direction.

Consider two currently orthogonal line elements, one with orientation $\mathbf{n}$ and the other with orientation $\mathbf{m}$. Denote by $\dot{\theta}$ the rate of decrease in the angle between these elements, Then, half of this rate of decrease is

$$-\frac{1}{2}\,\dot{\theta} = D_{ij}\,n_i\,m_j \tag{2.5.13}$$

which is the corresponding rate of *shear* or *shearing*. In particular, D_{12}, D_{23}, and D_{31} are the rates of shear of orthogonal elements currently in the corresponding coordinate directions. Similarly, from (2.5.12), D_{11}, D_{22}, and D_{33} are the rates of stretch of elements in the corresponding coordinate directions.

Since tr $\mathbf{W} = 0$, take the trace of (2.5.4b) and using (2.5.6), (2.1.4c), and (2.1.2g), obtain

$$\mathrm{tr}\,\mathbf{D} = \dot{J}/J = \frac{(dv)^{\cdot}}{dv} \tag{2.5.14a,b}$$

which is the rate of change of volume per unit current volume. The *deviatoric* part of $\mathbf{D}$ is defined by

$$\mathbf{D}' = \mathbf{D} - \frac{1}{3}\,(\mathrm{tr}\,\mathbf{D})\,\mathbf{1}\,. \tag{2.5.15}$$

It represents the instantaneous rate of pure distortion, since its trace is zero.

2.6. OTHER SPIN TENSORS

In the course of deformation, both the Lagrangian and the Eulerian triads, $\mathbf{N}^a$ and $\mathbf{n}^a$, a = 1, 2, 3, continually change. Let $\mathbf{\Omega}^L$ and $\mathbf{\Omega}^E$ be the corresponding *spin* tensors such that

$$\dot{\mathbf{N}}^a = \mathbf{\Omega}^L \mathbf{N}^a, \quad \dot{\mathbf{n}}^a = \mathbf{\Omega}^E \mathbf{n}^a. \tag{2.6.1a,b}$$

Direct time differentiation of the Lagrangian and the Eulerian rotations, (2.2.21b) and (2.2.22b), yields

$$\dot{\mathbf{R}}^L = \mathbf{\Omega}^L \mathbf{R}^L, \quad \dot{\mathbf{R}}^E = \mathbf{\Omega}^E \mathbf{R}^E, \tag{2.6.2a,b}$$

or, equivalently,

$$\mathbf{\Omega}^L = \dot{\mathbf{R}}^L (\mathbf{R}^L)^T, \quad \mathbf{\Omega}^E = \dot{\mathbf{R}}^E (\mathbf{R}^E)^T. \tag{2.6.2c,d}$$

Let the spin of the Lagrangian triad relative to the Eulerian triad be denoted by $\mathbf{\Omega}^R$, so that

$$\dot{\mathbf{R}} = \mathbf{\Omega}^R \mathbf{R}, \quad \mathbf{\Omega}^R = \dot{\mathbf{R}} \mathbf{R}^T, \tag{2.6.3a,b}$$

where $\mathbf{R}$, given by (2.2.7), rotates the Lagrangian triad into the Eulerian triad. Direct differentiation of (2.2.23) results in

$$\mathbf{\Omega}^E = \mathbf{\Omega}^R + \mathbf{R}\,\mathbf{\Omega}^L \mathbf{R}^T \quad \text{or} \quad \mathbf{\Omega}^L = \mathbf{R}^T (\mathbf{\Omega}^E - \mathbf{\Omega}^R)\,\mathbf{R}. \tag{2.6.4a,b}$$

Since $\mathbf{\Omega}^E$ and $\mathbf{\Omega}^L$ can be viewed as the spin of the background $\mathbf{e}_i$-triad relative to the Eulerian and Lagrangian triads, respectively, (2.6.4) states that the spin of the $\mathbf{e}_i$-triad relative to the Eulerian triad is equal to the spin of the Lagrangian triad relative to the Eulerian triad plus the spin of the $\mathbf{e}_i$-triad relative to the Lagrangian triad transformed to the Eulerian triad.

Let Ω^E_{ab} and Ω^R_{ab} be the components of $\mathbf{\Omega}^E$ and $\mathbf{\Omega}^R$ with respect to the *Eulerian triad*,

$$\mathbf{\Omega}^E = \sum_{a,b=1}^{3} \Omega^E_{ab}\, \mathbf{n}^a \otimes \mathbf{n}^b, \tag{2.6.5a}$$

$$\mathbf{\Omega}^R = \sum_{a,b=1}^{3} \Omega^R_{ab}\, \mathbf{n}^a \otimes \mathbf{n}^b, \tag{2.6.6a}$$

and let Ω^L_{ab} be the components of $\mathbf{\Omega}^L$ with respect to the *Lagrangian triad*,

$$\mathbf{\Omega}^L = \sum_{a,b=1}^{3} \Omega^L_{ab}\, \mathbf{N}^a \otimes \mathbf{N}^b. \tag{2.6.7}$$

Then, from (2.6.4a,b), it follows that

$$\Omega^E_{ab} = \Omega^L_{ab} + \Omega^R_{ab}. \tag{2.6.4c}$$

It is emphasized that this relation among the components of the three spin tensors is valid only if $\mathbf{\Omega}^L$ is expressed in the Lagrangian triad, as in (2.6.7), and the other two spin tensors are expressed in the Eulerian triad, as in (2.6.5a) and (2.6.6a). It is convenient to introduce

$$\hat{\mathbf{\Omega}}^{\mathrm{E}} = \mathbf{R}^{\mathrm{T}}\mathbf{\Omega}^{\mathrm{E}}\,\mathbf{R} = \sum_{a,\,b=1}^{3} \Omega^{\mathrm{E}}_{ab}\,\mathbf{N}^{a} \otimes \mathbf{N}^{b}, \tag{2.6.5b}$$

$$\hat{\mathbf{\Omega}}^{\mathrm{R}} = \mathbf{R}^{\mathrm{T}}\mathbf{\Omega}^{\mathrm{R}}\,\mathbf{R} = \sum_{a,\,b=1}^{3} \Omega^{\mathrm{R}}_{ab}\,\mathbf{N}^{a} \otimes \mathbf{N}^{b}, \tag{2.6.6b}$$

and observe from (2.6.4) that

$$\mathbf{\Omega}^{\mathrm{L}} = \hat{\mathbf{\Omega}}^{\mathrm{E}} - \hat{\mathbf{\Omega}}^{\mathrm{R}}. \tag{2.6.4d}$$

2.7. RELATIONS BETWEEN RIGHT STRETCH RATE AND DEFORMATION RATE

Differentiate (2.2.1a) with respect to time, substitute into (2.5.4b), and use (2.5.6)-(2.5.8) and (2.6.3), to arrive at

$$\mathbf{R}^{\mathrm{T}}(\mathbf{W} - \mathbf{\Omega}^{\mathrm{R}})\,\mathbf{R} = \frac{1}{2}\,(\dot{\mathbf{U}}\,\mathbf{U}^{-1} - \mathbf{U}^{-1}\,\dot{\mathbf{U}}), \tag{2.7.1a}$$

$$\mathbf{R}^{\mathrm{T}}\mathbf{D}\,\mathbf{R} = \frac{1}{2}\,(\dot{\mathbf{U}}\,\mathbf{U}^{-1} + \mathbf{U}^{-1}\,\dot{\mathbf{U}}), \tag{2.7.2a}$$

where $\dot{\mathbf{U}}$ is the material time derivative of the right-stretch tensor $\mathbf{U}$.

Let D_{ab} and W_{ab} be the components of the deformation rate tensor $\mathbf{D}$ and spin tensor $\mathbf{W}$, with respect to the *Eulerian triad*,

$$\mathbf{D} = \sum_{a,\,b=1}^{3} D_{ab}\,\mathbf{n}^{a} \otimes \mathbf{n}^{b}, \tag{2.7.3a}$$

$$\mathbf{W} = \sum_{a,\,b=1}^{3} W_{ab}\,\mathbf{n}^{a} \otimes \mathbf{n}^{b}, \tag{2.7.4a}$$

and set

$$\hat{\mathbf{D}} = \mathbf{R}^{\mathrm{T}}\mathbf{D}\,\mathbf{R} = \sum_{a,\,b=1}^{3} D_{ab}\,\mathbf{N}^{a} \otimes \mathbf{N}^{a}, \tag{2.7.3b,c}$$

$$\hat{\mathbf{W}} = \mathbf{R}^{\mathrm{T}}\mathbf{W}\,\mathbf{R} = \sum_{a,\,b=1}^{3} W_{ab}\,\mathbf{N}^{a} \otimes \mathbf{N}^{b}. \tag{2.7.4b,c}$$

Equations (2.7.1a) and (2.7.2a) then become

$$\hat{\mathbf{W}} - \hat{\mathbf{\Omega}}^{\mathrm{R}} = \frac{1}{2}\,(\dot{\mathbf{U}}\,\mathbf{U}^{-1} - \mathbf{U}^{-1}\,\dot{\mathbf{U}}),$$

$$\mathbf{U}\,\dot{\mathbf{U}} - \dot{\mathbf{U}}\,\mathbf{U} = 2\,\mathbf{U}\,(\hat{\mathbf{W}} - \hat{\mathbf{\Omega}}^{\mathrm{R}})\,\mathbf{U}, \tag{2.7.1b,c}$$

$$\hat{\mathbf{D}} = \frac{1}{2}\,(\dot{\mathbf{U}}\,\mathbf{U}^{-1} + \mathbf{U}^{-1}\,\dot{\mathbf{U}}),$$

$$\mathbf{U}\,\dot{\mathbf{U}} + \dot{\mathbf{U}}\,\mathbf{U} = 2\,\mathbf{U}\,\hat{\mathbf{D}}\,\mathbf{U}, \tag{2.7.2b,c}$$

where $\hat{\mathbf{\Omega}}^{\mathrm{R}}$ is given by (2.6.6b). Note that, if in the above equations the right-stretch tensor $\mathbf{U}$ is replaced by its distortional part $\overline{\mathbf{U}}$ defined by (2.2.25a), then

(2.7.1b) remains unchanged, while in (2.7.2b) the left-hand side changes to $\hat{\mathbf{D}}' = \mathbf{R}^T \mathbf{D}' \mathbf{R}$ which relates to the deviatoric part $\mathbf{D}'$ of the deformation-rate tensor. This follows from the fact that

$$\mathrm{tr}\, \bar{\mathbf{U}}^{-1} \dot{\bar{\mathbf{U}}} = 0 \tag{2.7.5}$$

which is verified by time differentiation of $\mathrm{III}_{\bar{U}} = 1$.

With respect to the Lagrangian triad, (2.7.1b) and (2.7.2b) have the following component representation (Hill, 1968, 1978):

$$\mathrm{W}_{ab} - \Omega^R_{ab} = \frac{\lambda_a - \lambda_b}{2\lambda_a \lambda_b} \dot{\mathrm{U}}_{ab} \quad \text{(no sum)}, \tag{2.7.1d}$$

$$\mathrm{D}_{ab} = \frac{\lambda_a + \lambda_b}{2\lambda_a \lambda_b} \dot{\mathrm{U}}_{ab} \quad \text{(no sum)}, \tag{2.7.2d}$$

where λ_a, a = 1, 2, 3, are the principal stretches.

From (2.7.1b) and (2.7.2b) it follows that

$$\dot{\mathbf{U}} = \hat{\mathbf{D}}\, \mathbf{U} + (\hat{\mathbf{W}} - \hat{\mathbf{\Omega}}^R)\, \mathbf{U}, \tag{2.7.6a}$$

and since the left-hand side is symmetric, it follows that

$$\mathbf{U}\, (\hat{\mathbf{W}} - \hat{\mathbf{\Omega}}^R) + (\hat{\mathbf{W}} - \hat{\mathbf{\Omega}}^R)\, \mathbf{U} = \mathbf{U}\, \hat{\mathbf{D}} - \hat{\mathbf{D}}\, \mathbf{U}. \tag{2.7.7a}$$

In terms of components with respect to the Lagrangian triad, this becomes (Hill, 1978)

$$\mathrm{W}_{ab} - \Omega^R_{ab} = \frac{\lambda_a - \lambda_b}{\lambda_a + \lambda_b} \mathrm{D}_{ab}, \quad \text{(no sum)}. \tag{2.7.7b}$$

Use identity (1.4.9d) to solve (2.7.7a) directly for $\hat{\mathbf{W}} - \hat{\mathbf{\Omega}}^R$, and obtain (Mehrabadi and Nemat-Nasser, 1987),

$$(\mathrm{I_U\, II_U} - \mathrm{III_U})\, (\hat{\mathbf{W}} - \hat{\mathbf{\Omega}}^R) = \mathrm{I}_U^2\, (\mathbf{U}\, \hat{\mathbf{D}} - \hat{\mathbf{D}}\, \mathbf{U}) - \mathrm{I_U}\, (\mathbf{U}^2\, \hat{\mathbf{D}} - \hat{\mathbf{D}}\, \mathbf{U}^2)$$
$$+ \mathbf{U}^2\, \hat{\mathbf{D}}\, \mathbf{U} - \mathbf{U}\, \hat{\mathbf{D}} \mathbf{U}^2. \tag{2.7.7c}$$

As pointed out by Mehrabadi and Nemat-Nasser (1987), (2.7.6a) can be written as

$$\dot{\mathbf{U}} = \frac{1}{2}\, (\mathbf{U}\, \hat{\mathbf{D}} + \hat{\mathbf{D}}\, \mathbf{U}) + \frac{1}{2}\, (\hat{\mathbf{W}} - \hat{\mathbf{\Omega}}^R)\, \mathbf{U} - \frac{1}{2}\, \mathbf{U}\, (\hat{\mathbf{W}} - \hat{\mathbf{\Omega}}^R). \tag{2.7.6b}$$

Thus, $(\mathbf{U}\, \hat{\mathbf{D}} + \hat{\mathbf{D}} \mathbf{U})/2$ is the rate of change of $\mathbf{U}$ as observed in a frame with an instantaneous spin of $(\hat{\mathbf{W}} - \hat{\mathbf{\Omega}}^R)/2$.

Substitute from (2.7.7c) into (2.7.6b) and using the Hamilton-Cayley theorem, obtain the following general result (Mehrabadi and Nemat-Nasser, 1987):

$$
\begin{aligned}
(\mathrm{I_U II_U - III_U})\,\dot{\mathbf{U}} = {} & (\mathrm{I_U^2 + II_U})\,\mathbf{U}\,\hat{\mathbf{D}}\,\mathbf{U} + \mathbf{U}^2\,\hat{\mathbf{D}}\,\mathbf{U}^2 \\
& - \mathrm{I_U}\,(\mathbf{U}^2\,\hat{\mathbf{D}}\,\mathbf{U} + \mathbf{U}\,\hat{\mathbf{D}}\,\mathbf{U}^2) \\
& - \mathrm{III_U}\,(\mathbf{U}\,\hat{\mathbf{D}} + \hat{\mathbf{D}}\,\mathbf{U}) + \mathrm{I_U III_U}\,\hat{\mathbf{D}} \qquad (2.7.8a)
\end{aligned}
$$

which, after some manipulation using identities (1.4.3) to (1.4.5), can be rearranged into the following general form:

$$
\begin{aligned}
& (\mathrm{I_U II_U - III_U})\,\dot{\mathbf{U}} \\
& = \{(2\,\mathrm{II_U^2} - \mathrm{I_U III_U})\,\mathrm{tr}(\hat{\mathbf{D}}) + (\mathrm{III_U} - 2\,\mathrm{I_U II_U})\,\mathrm{tr}(\mathbf{U}\,\hat{\mathbf{D}}) + 2\,\mathrm{II_U}\,\mathrm{tr}(\mathbf{U}^2\,\hat{\mathbf{D}})\}\,\mathbf{1} \\
& + \{(\mathrm{III_U} - 2\,\mathrm{I_U II_U})\,\mathrm{tr}(\hat{\mathbf{D}}) + (\mathrm{I_U^2} + \mathrm{II_U})\,\mathrm{tr}(\mathbf{U}\,\hat{\mathbf{D}}) - \mathrm{I_U}\,\mathrm{tr}(\mathbf{U}^2\,\hat{\mathbf{D}})\}\,\mathbf{U} \\
& + \{2\,\mathrm{II_U}\,\mathrm{tr}(\hat{\mathbf{D}}) - \mathrm{I_U}\,\mathrm{tr}(\mathbf{U}\hat{\mathbf{D}}) + \mathrm{tr}(\mathbf{U}^2\hat{\mathbf{D}})\}\,\mathbf{U}^2 \\
& + 2\,(\mathrm{I_U III_U} - \mathrm{II_U^2})\,\hat{\mathbf{D}} + 2\,(\mathrm{I_U II_U} - \mathrm{III_U})\,(\mathbf{U}\,\hat{\mathbf{D}} + \hat{\mathbf{D}}\,\mathbf{U}) \\
& - 2\mathrm{II_U}\,(\mathbf{U}^2\,\hat{\mathbf{D}} + \hat{\mathbf{D}}\,\mathbf{U}^2)\,. \qquad (2.7.8b)
\end{aligned}
$$

Alternatively, to obtain (2.7.8a) directly, use (1.4.10) and (1.4.12f) in (2.7.2c) (Scheidler, 1994).

2.7.1. Alternative Coordinate-independent Expressions

Now, seek to obtain an alternative coordinate–independent expression for $\dot{\mathbf{U}}$ in terms of $\hat{\mathbf{D}}$. This may involve the *eigenvalues* of $\mathbf{U}$, but requires no *eigenvector* calculations. Split the deformation rate $\hat{\mathbf{D}}$ into a part $\hat{\overset{\Delta}{\mathbf{D}}}$ coaxial with $\mathbf{U}$ and a remainder part $\mathbf{H_U}(\hat{\mathbf{D}})$, normal to $\mathbf{U}$, as follows:

$$
\begin{aligned}
& \hat{\mathbf{D}} = \overset{\hat{\Delta}}{\mathbf{D}} + \mathbf{H_U}(\hat{\mathbf{D}})\,, \\
& \overset{\hat{\Delta}}{\mathbf{D}} = \sum_{a=1}^{3} \mathrm{tr}(\hat{\mathbf{D}}\,\hat{\mathbf{E}}_a)\,\hat{\mathbf{E}}_a\,, \quad \text{for } \lambda_1 \neq \lambda_2 \neq \lambda_3 \neq \lambda_1\,, \\
& \overset{\hat{\Delta}}{\mathbf{D}} = \mathrm{tr}(\hat{\mathbf{D}}\,\hat{\mathbf{E}}_1)\,\hat{\mathbf{E}}_1 + \frac{1}{2}\,\{\mathrm{tr}(\hat{\mathbf{D}}) - \mathrm{tr}(\hat{\mathbf{D}}\,\hat{\mathbf{E}}_1)\}(\mathbf{1} - \hat{\mathbf{E}}_1)\,, \\
& \qquad \text{for } \lambda_1 \neq \lambda_2 = \lambda_3\,, \\
& \overset{\hat{\Delta}}{\mathbf{D}} = \mathrm{tr}(\hat{\mathbf{D}})\,\mathbf{1}\,, \quad \text{for } \lambda_1 = \lambda_2 = \lambda_3 \equiv \lambda\,, \qquad (2.7.9a\text{-}d)
\end{aligned}
$$

where, as given by (2.3.4a) and (2.3.8a),

$$
\begin{aligned}
& \hat{\mathbf{E}}_a = \frac{1}{(\lambda_a - \lambda_b)\,(\lambda_a - \lambda_c)}\,\{\mathbf{U}^2 - (\lambda_b + \lambda_c)\,\mathbf{U} + \lambda_b\,\lambda_c\,\mathbf{1}\}\,, \\
& \qquad \text{for } \lambda_1 \neq \lambda_2 \neq \lambda_3 \neq \lambda_1\,, \\
& \hat{\mathbf{E}}_1 = \frac{1}{\lambda_1 - \lambda_2}\,(\mathbf{U} - \lambda_2\,\mathbf{1})\,, \quad \text{for } \lambda_1 \neq \lambda_2 = \lambda_3\,. \qquad (2.7.10a,b)
\end{aligned}
$$

From (2.7.2c), furthermore, it follows that

$$\mathbf{U}\,\mathbf{H}_U(\dot{\mathbf{U}})+\mathbf{H}_U(\dot{\mathbf{U}})\,\mathbf{U}=2\,\mathbf{U}\,\mathbf{H}_U(\hat{\mathbf{D}})\,\mathbf{U}\,,$$

$$\mathbf{H}_U(\dot{\mathbf{U}})=\dot{\mathbf{U}}-\overset{\Delta}{\hat{\mathbf{U}}}\,,\quad \overset{\Delta}{\hat{\mathbf{U}}}=\mathbf{U}\,\overset{\Delta}{\hat{\mathbf{D}}}=\overset{\Delta}{\hat{\mathbf{D}}}\,\mathbf{U} \tag{2.7.11a-c}$$

which, using identity (1.4.9b-d), can be solved for $\mathbf{H}_U(\dot{\mathbf{U}})$ to yield

$$\mathbf{H}_U(\dot{\mathbf{U}})=\frac{2}{I_U\,II_U-III_U}\{I_U\,III_U\,\mathbf{H}_U(\hat{\mathbf{D}})+II_U\,\mathbf{U}\,\mathbf{H}_U(\hat{\mathbf{D}})\,\mathbf{U}$$

$$-\,III_U\,[\mathbf{U}\,\mathbf{H}_U(\hat{\mathbf{D}})+\mathbf{H}_U(\hat{\mathbf{D}})\,\mathbf{U}]\}\,. \tag{2.7.11d}$$

Substitution of (2.7.9) into (2.7.11) results in

$$\dot{\mathbf{U}}=\mathbf{K}^U+\sum_{a=1}^{3}\{\lambda_a\,\mathrm{tr}(\hat{\mathbf{D}}\,\hat{\mathbf{E}}_a)-\mathrm{tr}(\mathbf{K}^U\,\hat{\mathbf{E}}_a)\}\,\hat{\mathbf{E}}_a\,,$$

$$\text{for }\ \lambda_1\neq\lambda_2\neq\lambda_3\neq\lambda_1\,,$$

$$\dot{\mathbf{U}}=\mathbf{K}^U+\{\lambda_1\,\mathrm{tr}(\hat{\mathbf{D}}\,\hat{\mathbf{E}}_1)-\mathrm{tr}(\mathbf{K}^U\,\hat{\mathbf{E}}_1)\}\,\hat{\mathbf{E}}_1$$

$$+\,\{\lambda_2\,\mathrm{tr}[\hat{\mathbf{D}}\,(\mathbf{1}-\hat{\mathbf{E}}_1)]-\mathrm{tr}[\mathbf{K}^U\,(\mathbf{1}-\hat{\mathbf{E}}_1)]\}\,(\mathbf{1}-\hat{\mathbf{E}}_1)\,,$$

$$\text{for }\ \lambda_1\neq\lambda_2=\lambda_3\,,$$

$$\dot{\mathbf{U}}=\hat{\mathbf{D}}\,,\quad \text{for }\ \lambda_1=\lambda_2=\lambda_3=\lambda\,, \tag{2.7.12a-c}$$

where

$$\mathbf{K}^U=\frac{2}{I_U\,II_U-III_U}\{I_U\,III_U\,\hat{\mathbf{D}}+II_U\,\mathbf{U}\,\hat{\mathbf{D}}\,\mathbf{U}-III_U\,(\mathbf{U}\,\hat{\mathbf{D}}+\hat{\mathbf{D}}\,\mathbf{U})\}\,. \tag{2.7.12d}$$

Note that, for $\lambda_2=\lambda_3$,

$$\mathbf{U}\,(\mathbf{1}-\hat{\mathbf{E}}_1)=(\mathbf{1}-\hat{\mathbf{E}}_1)\,\mathbf{U}=\lambda_2\,(\mathbf{1}-\hat{\mathbf{E}}_1)\,. \tag{2.7.13a}$$

Hence, it follows from (2.7.12d) that

$$\mathrm{tr}[\mathbf{K}^U\,(\mathbf{1}-\hat{\mathbf{E}}_1)]=\frac{2}{I_U\,II_U-III_U}\,(I_U\,III_U+II_U\,\lambda_2^2-2\,III_U\,\lambda_2)\,\mathrm{tr}[\hat{\mathbf{D}}\,(\mathbf{1}-\hat{\mathbf{E}}_1)]$$

$$=\lambda_2\,\mathrm{tr}[\hat{\mathbf{D}}\,(\mathbf{1}-\hat{\mathbf{E}}_1)]\,. \tag{2.7.13b}$$

Therefore, (2.7.12b) reduces to

$$\dot{\mathbf{U}}=\mathbf{K}^U+\{\lambda_1\,\mathrm{tr}(\hat{\mathbf{D}}\,\hat{\mathbf{E}}_1)-\mathrm{tr}(\mathbf{K}^U\,\hat{\mathbf{E}}_1)\}\,\hat{\mathbf{E}}_1\,. \tag{2.7.12e}$$

Finally, note that all the above relations hold also if $\mathbf{U}$ is replaced by $\overline{\mathbf{U}}$, the distortional part of $\mathbf{U}$ (see (2.2.25)), provided that $\hat{\mathbf{D}}$ is replaced by $\hat{\mathbf{D}}'$, the deviatoric part of $\hat{\mathbf{D}}$. For example, (2.7.8b) becomes

$$(I_{\overline{U}}\,II_{\overline{U}}-1)\,\dot{\overline{\mathbf{U}}}=$$

$$\{\mathrm{tr}(\overline{\mathbf{U}}\,\hat{\mathbf{D}}')+II_{\overline{U}}\,\mathrm{tr}(\overline{\mathbf{U}}^{-1}\,\hat{\mathbf{D}}')\}\,\mathbf{1}+II_{\overline{U}}\,\mathrm{tr}(\mathbf{U}\hat{\mathbf{D}}')\,\overline{\mathbf{U}}$$

$$+\,\mathrm{tr}(\overline{\mathbf{U}}^{-1}\,\hat{\mathbf{D}}')\,\overline{\mathbf{U}}^{-1}+2\,(I_{\overline{U}}+III_{\overline{U}}^{2})\,\hat{\mathbf{D}}'$$

$$-2(\overline{\mathbf{U}}\,\hat{\mathbf{D}}'+\hat{\mathbf{D}}'\,\overline{\mathbf{U}})-2\mathrm{II}_{\overline{\mathbf{U}}}(\overline{\mathbf{U}}^{-1}\hat{\mathbf{D}}'+\hat{\mathbf{D}}'\,\overline{\mathbf{U}}^{-1}). \tag{2.7.14}$$

2.8. RELATIONS BETWEEN LEFT STRETCH RATE AND DEFORMATION RATE

Differentiate $\mathbf{F}=\mathbf{V}\,\mathbf{R}$ in (2.2.1) with respect to time, substitute into (2.5.4b), and use (2.6.3) to obtain

$$\dot{\mathbf{V}}=\mathbf{L}\,\mathbf{V}-\mathbf{V}\,\boldsymbol{\Omega}^{\mathrm{R}}. \tag{2.8.1a}$$

This yields

$$\mathbf{V}\,\dot{\mathbf{V}}+\dot{\mathbf{V}}\,\mathbf{V}=\mathbf{L}\,\mathbf{V}^2+\mathbf{V}^2\,\mathbf{L}^{\mathrm{T}}=\mathbf{V}^2\,\mathbf{D}+\mathbf{D}\,\mathbf{V}^2-\mathbf{V}^2\,\mathbf{W}+\mathbf{W}\,\mathbf{V}^2. \tag{2.8.1b}$$

Denote by $\overset{\circ}{\mathbf{V}}$ the rate of change (*Jaumann rate*) of $\mathbf{V}$ as observed in a frame which rotates with the material spin $\mathbf{W}$,

$$\overset{\circ}{\mathbf{V}}\equiv\dot{\mathbf{V}}-\mathbf{W}\,\mathbf{V}+\mathbf{V}\,\mathbf{W}. \tag{2.8.2a}$$

Substitute (2.8.2a) into (2.8.1b) to arrive at

$$\begin{aligned}\mathbf{V}\,\overset{\circ}{\mathbf{V}}+\overset{\circ}{\mathbf{V}}\,\mathbf{V}&=\mathbf{V}^2\,\mathbf{D}+\mathbf{D}\,\mathbf{V}^2\\&=\mathbf{V}\,(\mathbf{V}\,\mathbf{D}+\mathbf{D}\,\mathbf{V})+(\mathbf{V}\,\mathbf{D}+\mathbf{D}\,\mathbf{V})\,\mathbf{V}-2\,\mathbf{V}\,\mathbf{D}\,\mathbf{V}.\end{aligned} \tag{2.8.2b}$$

This can be rewritten as

$$\mathbf{V}(\mathbf{V}\,\mathbf{D}+\mathbf{D}\,\mathbf{V}-\overset{\circ}{\mathbf{V}})+(\mathbf{V}\,\mathbf{D}+\mathbf{D}\,\mathbf{V}-\overset{\circ}{\mathbf{V}})\mathbf{V}=2\,\mathbf{V}\,\mathbf{D}\,\mathbf{V}. \tag{2.8.2c}$$

Comparison of (2.8.2c) with (2.7.2c) shows that an expression for $\mathbf{V}\,\mathbf{D}+\mathbf{D}\,\mathbf{V}-\overset{\circ}{\mathbf{V}}$ can be obtained by replacing $\mathbf{U}$ and $\hat{\mathbf{D}}$ by $\mathbf{V}$ and $\mathbf{D}$, respectively, in the expression for $\dot{\mathbf{U}}$ given by (2.7.8). In this manner, $\overset{\circ}{\mathbf{V}}$ is obtained as follows:

$$\begin{aligned}-(\mathrm{I_V}\,\mathrm{II_V}-\mathrm{III_V})\,\overset{\circ}{\mathbf{V}}\\&=(\mathrm{I_V^2}+\mathrm{II_V})\,\mathbf{V}\,\mathbf{D}\,\mathbf{V}+\mathbf{V}^2\,\mathbf{D}\,\mathbf{V}^2-\mathrm{I_V}\,(\mathbf{V}^2\,\mathbf{D}\,\mathbf{V}+\mathbf{V}\,\mathbf{D}\,\mathbf{V}^2)\\&\quad-\mathrm{I_V}\,\mathrm{II_V}\,(\mathbf{V}\,\mathbf{D}+\mathbf{D}\,\mathbf{V})+\mathrm{I_V}\,\mathrm{III_V}\,\mathbf{D}\\&=\{(2\,\mathrm{II_V^2}-\mathrm{I_V}\,\mathrm{III_V})\,\mathrm{tr}(\mathbf{D})+(\mathrm{III_V}-2\,\mathrm{I_V}\,\mathrm{II_V})\,\mathrm{tr}(\mathbf{V}\,\mathbf{D})+2\,\mathrm{II_V}\,\mathrm{tr}(\mathbf{V}^2\,\mathbf{D})\}\,\mathbf{1}\\&\quad+\{(\mathrm{III_V}-2\,\mathrm{I_V}\,\mathrm{II_V})\,\mathrm{tr}(\mathbf{D})+(\mathrm{I_V^2}+\mathrm{II_V})\,\mathrm{tr}(\mathbf{V}\,\mathbf{D})-\mathrm{I_V}\,\mathrm{tr}(\mathbf{V}^2\,\mathbf{D})\}\,\mathbf{V}\\&\quad+\{2\,\mathrm{II_V}\,\mathrm{tr}(\mathbf{D})-\mathrm{I_V}\,\mathrm{tr}(\mathbf{V}\mathbf{D})+\mathrm{tr}(\mathbf{V}^2\mathbf{D})\}\,\mathbf{V}^2\\&\quad+2\,(\mathrm{I_V}\,\mathrm{III_V}-\mathrm{II_V^2})\,\mathbf{D}+(\mathrm{I_V}\,\mathrm{II_V}-\mathrm{III_V})\,(\mathbf{V}\,\mathbf{D}+\mathbf{D}\,\mathbf{V})\\&\quad-2\,\mathrm{II_V}\,(\mathbf{V}^2\,\mathbf{D}+\mathbf{D}\,\mathbf{V}^2).\end{aligned} \tag{2.8.3}$$

With respect to the Eulerian triad, (2.8.2b) has the component representation,

$$D_{ab} = \frac{\lambda_a + \lambda_b}{\lambda_a^2 + \lambda_b^2} \overset{\circ}{V}_{ab} . \tag{2.8.2d}$$

Since $\dot{\mathbf{V}}$ is symmetric, it follows from (2.8.1a) that

$$\mathbf{V}\,(\mathbf{\Omega}^R - \mathbf{W}) + (\mathbf{\Omega}^R - \mathbf{W})\,\mathbf{V} = \mathbf{D}\,\mathbf{V} - \mathbf{V}\,\mathbf{D} \tag{2.8.4a}$$

which has the following representation with respect to the Eulerian triad:

$$W_{ab} - \Omega^R_{ab} = \frac{\lambda_a - \lambda_b}{\lambda_a + \lambda_b} D_{ab} = \frac{\lambda_a - \lambda_b}{\lambda_a^2 + \lambda_b^2} \overset{\circ}{V}_{ab} . \tag{2.8.4b,c}$$

Use identity (1.4.9d) in (2.8.4a) to deduce that

$$(I_V\, II_V - III_V)\,(\mathbf{W} - \mathbf{\Omega}^R) = I_V^2\,(\mathbf{V}\,\mathbf{D} - \mathbf{D}\,\mathbf{V}) - I_V\,(\mathbf{V}^2\,\mathbf{D} - \mathbf{D}\,\mathbf{V}^2)$$
$$+ (\mathbf{V}^2\,\mathbf{D}\,\mathbf{V} - \mathbf{V}\,\mathbf{D}\,\mathbf{V}^2) . \tag{2.8.5a}$$

Substitution of $\mathbf{U} = \mathbf{R}^T\mathbf{V}\,\mathbf{R}$ into (2.7.7c) also results in (2.8.5a). This is a useful equation, particularly when relations among various stress-rate measures are of interest. It shows that the spin $\mathbf{W}$ equals the spin $\mathbf{\Omega}^R$ plus terms which are independent of the spin (*i.e.*, they are *objective*) and only linearly dependent on the deformation rate $\mathbf{D}$,

$$\mathbf{W} = \mathbf{\Omega}^R + (I_V\, II_V - III_V)^{-1}\{I_V^2\,(\mathbf{V}\,\mathbf{D} - \mathbf{D}\,\mathbf{V})$$
$$- I_V\,(\mathbf{V}^2\,\mathbf{D} - \mathbf{D}\,\mathbf{V}^2) + (\mathbf{V}^2\,\mathbf{D}\,\mathbf{V} - \mathbf{V}\,\mathbf{D}\,\mathbf{V}^2)\} . \tag{2.8.5b}$$

In certain computer codes, the kinematical quantities $\mathbf{U}$, $\mathbf{V}$, and $\mathbf{R}$ are automatically computed incrementally without the eigenvalue calculation necessary to evaluate (2.7.7b). The general form (2.8.5b) provides a direct relation between results based on the Jaumann stress rate and those based on the stress rate associated with the spin tensor $\mathbf{\Omega}^R$. This is further discussed in Section 3.4. For this purpose, it is convenient to write (2.8.5b) as

$$\mathbf{\Omega}^R = \mathbf{W} + \mathbf{\Delta} , \tag{2.8.6a}$$

where

$$\mathbf{\Delta} = (I_V\, II_V - III_V)^{-1}\,\{I_V^2\,(\mathbf{D}\,\mathbf{V} - \mathbf{V}\,\mathbf{D})$$
$$- I_V\,(\mathbf{D}\,\mathbf{V}^2 - \mathbf{V}^2\,\mathbf{D}) + (\mathbf{V}\,\mathbf{D}\,\mathbf{V}^2 - \mathbf{V}^2\,\mathbf{D}\,\mathbf{V})\,\} \tag{2.8.6b}$$

is an objective quantity which vanishes when the deformation-rate tensor is zero.

2.8.1. Alternative Coordinate-independent Expressions

Now, seek to obtain an alternative coordinate–independent expression for the Jaumann rate of $\mathbf{V}$ in terms of the deformation-rate tensor $\mathbf{D}$. Split the deformation rate $\mathbf{D}$ into a part $\overset{\Delta}{\mathbf{D}}$ coaxial with $\mathbf{V}$, and a remainder $\mathbf{H}_V(\mathbf{D})$, normal to $\mathbf{V}$, as follows:

$$\mathbf{D} = \overset{\Delta}{\mathbf{D}} + \mathbf{H}_V(\mathbf{D}) ,$$

$$\overset{\Delta}{\mathbf{D}} = \sum_{a=1}^{3} \mathrm{tr}(\mathbf{D}\,\mathbf{E}_a)\,\mathbf{E}_a\,, \quad \text{for } \lambda_1 \neq \lambda_2 \neq \lambda_3 \neq \lambda_1\,,$$

$$\overset{\Delta}{\mathbf{D}} = \mathrm{tr}(\mathbf{D}\,\mathbf{E}_1)\,\mathbf{E}_1 + \frac{1}{2}\{\mathrm{tr}(\mathbf{D}) - \mathrm{tr}(\mathbf{D}\,\mathbf{E}_1)\}(\mathbf{1} - \mathbf{E}_1)\,,$$

$$\text{for } \lambda_1 \neq \lambda_2 = \lambda_3\,,$$

$$\overset{\Delta}{\mathbf{D}} = \mathrm{tr}(\mathbf{D})\,\mathbf{1}\,, \quad \text{for } \lambda_1 = \lambda_2 = \lambda_3 \equiv \lambda\,, \tag{2.8.7a-d}$$

where, as given by (2.3.5b) and (2.3.8d),

$$\mathbf{E}_a = \frac{1}{(\lambda_a - \lambda_b)(\lambda_a - \lambda_c)}\{\mathbf{V}^2 - (\lambda_b + \lambda_c)\,\mathbf{V} + \lambda_b\,\lambda_c\,\mathbf{1}\}\,,$$

$$\text{for } \lambda_1 \neq \lambda_2 \neq \lambda_3 \neq \lambda_1\,,$$

$$\mathbf{E}_1 = \frac{1}{\lambda_1 - \lambda_2}(\mathbf{V} - \lambda_2\,\mathbf{1})\,, \quad \text{for } \lambda_1 \neq \lambda_2 = \lambda_3\,. \tag{2.8.8a,b}$$

Then, (2.8.2b) can be rewritten as

$$\mathbf{V}\,\mathbf{H}_\mathbf{V}(\dot{\mathbf{V}}) + \mathbf{H}_\mathbf{V}(\dot{\mathbf{V}})\,\mathbf{V} = \mathbf{V}^2\,\mathbf{H}_\mathbf{V}(\mathbf{D}) + \mathbf{H}_\mathbf{V}(\mathbf{D})\,\mathbf{V}^2\,,$$

$$\mathbf{H}_\mathbf{V}(\dot{\mathbf{V}}) = \overset{\circ}{\mathbf{V}} - \overset{\Delta}{\mathbf{V}}\,, \quad \overset{\Delta}{\mathbf{V}} = \mathbf{V}\,\overset{\Delta}{\mathbf{D}} = \overset{\Delta}{\mathbf{D}}\,\mathbf{V}\,. \tag{2.8.9a-c}$$

This can be solved for $\mathbf{H}_\mathbf{V}(\dot{\mathbf{V}})$, using the identity (1.4.9d),

$$\mathbf{H}_\mathbf{V}(\dot{\mathbf{V}}) = \frac{1}{\mathrm{I}_\mathbf{V}\,\mathrm{II}_\mathbf{V} - \mathrm{III}_\mathbf{V}}\{2\,(\mathrm{II}_\mathbf{V}^2 - \mathrm{I}_\mathbf{V}\,\mathrm{III}_\mathbf{V})\,\mathbf{H}_\mathbf{V}(\mathbf{D})$$

$$- (\mathrm{I}_\mathbf{V}\,\mathrm{II}_\mathbf{V} - \mathrm{III}_\mathbf{V})\,[\mathbf{V}\,\mathbf{H}_\mathbf{V}(\mathbf{D}) + \mathbf{H}_\mathbf{V}(\mathbf{D})\,\mathbf{V}]$$

$$+ 2\,\mathrm{II}_\mathbf{V}\,[\mathbf{V}^2\,\mathbf{H}_\mathbf{V}(\mathbf{D}) + \mathbf{H}_\mathbf{V}(\mathbf{D})\,\mathbf{V}^2]\}\,. \tag{2.8.9d}$$

Then, substitution of (2.8.7) into (2.8.9) results in

$$\overset{\circ}{\mathbf{V}} = \mathbf{K}^\mathrm{V} + \sum_{a=1}^{3}\{\lambda_a\,\mathrm{tr}(\mathbf{D}\,\mathbf{E}_a) - \mathrm{tr}(\mathbf{K}^\mathrm{V}\,\mathbf{E}_a)\}\,\mathbf{E}_a\,,$$

$$\text{for } \lambda_1 \neq \lambda_2 \neq \lambda_3 \neq \lambda_1\,,$$

$$\overset{\circ}{\mathbf{V}} = \mathbf{K}^\mathrm{V} + \{\lambda_1\,\mathrm{tr}(\mathbf{D}\,\mathbf{E}_1) - \mathrm{tr}(\mathbf{K}^\mathrm{V}\,\mathbf{E}_1)\}\,\mathbf{E}_1\,,$$

$$\text{for } \lambda_1 \neq \lambda_2 = \lambda_3\,,$$

$$\overset{\circ}{\mathbf{V}} = \mathbf{D}\,, \quad \text{for } \lambda_1 = \lambda_2 = \lambda_3 = \lambda\,, \tag{2.8.10a-c}$$

where

$$\mathbf{K}^\mathrm{V} = \frac{1}{\mathrm{I}_\mathbf{V}\,\mathrm{II}_\mathbf{V} - \mathrm{III}_\mathbf{V}}\{2\,(\mathrm{II}_\mathbf{V}^2 - \mathrm{I}_\mathbf{V}\,\mathrm{III}_\mathbf{V})\,\mathbf{D} + 2\,\mathrm{II}_\mathbf{V}\,(\mathbf{V}^2\,\mathbf{D} + \mathbf{D}\,\mathbf{V}^2)$$

$$- (\mathrm{I}_\mathbf{V}\,\mathrm{II}_\mathbf{V} - \mathrm{III}_\mathbf{V})\,(\mathbf{V}\,\mathbf{D} + \mathbf{D}\,\mathbf{V})\}\,. \tag{2.8.10d}$$

Note that for $\lambda_1 \neq \lambda_2 = \lambda_3$,

$$\mathbf{V}\,(\mathbf{1}-\mathbf{E}_1) = (\mathbf{1}-\mathbf{E}_1)\,\mathbf{V} = \lambda_2\,(\mathbf{1}-\mathbf{E}_1)\,. \tag{2.8.11a}$$

Hence, it follows from (2.8.10d) that

$$\mathrm{tr}\{\mathbf{K}^{V}\,(\mathbf{1}-\mathbf{E}_1)\} = \lambda_2\,\mathrm{tr}\{\mathbf{D}\,(\mathbf{1}-\mathbf{E}_1)\}\,. \tag{2.8.11b}$$

Equation (2.8.11b) has been used in deriving (2.8.10b).

2.9. RELATIONS AMONG VARIOUS SPIN TENSORS

Recall (2.6.1a), $\dot{\mathbf{N}}^{a} = \boldsymbol{\Omega}^{L}\,\mathbf{N}^{a}$, and differentiate the expression

$$\mathbf{U} = \sum_{a=1}^{3} \lambda_a\,\mathbf{N}^{a}\otimes\mathbf{N}^{a},$$

to obtain (Mehrabadi and Nemat-Nasser, 1987)

$$\begin{aligned}\dot{\mathbf{U}} &= \sum_{a,\,b=1}^{3} \{\dot{\lambda}_a\,\delta_{ab} + \Omega^{L}_{ab}\,(\lambda_b-\lambda_a)\}\mathbf{N}^{a}\otimes\mathbf{N}^{b}\\ &= \overset{*}{\mathbf{U}} + \boldsymbol{\Omega}^{L}\,\mathbf{U} - \mathbf{U}\,\boldsymbol{\Omega}^{L}\\ &= \sum_{a,\,b=1}^{3} \dot{U}_{ab}\,\mathbf{N}^{a}\otimes\mathbf{N}^{b}\,,\end{aligned} \tag{2.9.1a}$$

where

$$\overset{*}{\mathbf{U}} = \sum_{a=1}^{3} \dot{\lambda}_a\,\mathbf{N}^{a}\otimes\mathbf{N}^{a}. \tag{2.9.1b}$$

From (2.9.1a) it follows that

$$\dot{U}_{ab} = \begin{cases} \dot{\lambda}_a & \text{when } a=b\\ (\lambda_b-\lambda_a)\,\Omega^{L}_{ab} & \text{when } a\neq b \quad (\text{no sum})\end{cases} \tag{2.9.1c}$$

are the components of $\dot{\mathbf{U}}$ in the Lagrangian triad. Note that, from (2.9.1a),

$$\mathrm{tr}(\dot{\mathbf{U}}) = \sum_{a=1}^{3} \dot{\lambda}_a\,. \tag{2.9.1d}$$

Therefore, in view of (2.7.9) and (2.7.11c), (2.9.1b) can be rewritten as

$$\overset{*}{\mathbf{U}} = \sum_{a=1}^{3} \dot{\lambda}_a\,\hat{\mathbf{E}}_a = \overset{\Delta}{\mathbf{U}}\,, \quad \text{for } \lambda_1 \neq \lambda_2 \neq \lambda_3 \neq \lambda_1\,,$$

$$\begin{aligned}\overset{*}{\mathbf{U}} &= \dot{\lambda}_1\,\hat{\mathbf{E}}_1 + \frac{1}{2}\,(\dot{\lambda}_2+\dot{\lambda}_3)\,(\hat{\mathbf{E}}_2+\hat{\mathbf{E}}_3) + \frac{1}{2}(\dot{\lambda}_2-\dot{\lambda}_3)\,(\hat{\mathbf{E}}_2-\hat{\mathbf{E}}_3)\\ &= \overset{\Delta}{\mathbf{U}} + \frac{1}{2}(\dot{\lambda}_2-\dot{\lambda}_3)\,(\hat{\mathbf{E}}_2-\hat{\mathbf{E}}_3)\,, \quad \text{for } \lambda_1 \neq \lambda_2 = \lambda_3\,,\end{aligned}$$

$$\overset{*}{\dot{\mathbf{U}}} = \frac{1}{3}\,\mathrm{tr}(\dot{\mathbf{U}})\,\mathbf{1} + \sum_{a=1}^{3} \{\dot{\lambda}_a - \mathrm{tr}(\dot{\mathbf{U}})/3\}\,\hat{\mathbf{E}}_a$$

$$= \overset{\Delta}{\mathbf{U}} + \sum_{a=1}^{3} \{\dot{\lambda}_a - \mathrm{tr}(\dot{\mathbf{U}})/3\}\,\hat{\mathbf{E}}_a, \quad \text{for } \lambda_1 = \lambda_2 = \lambda_3, \tag{2.9.2a-c}$$

where

$$\hat{\mathbf{E}}_a = \mathbf{N}^a \otimes \mathbf{N}^a,$$

and $\overset{\Delta}{\mathbf{U}}$ is defined by (2.7.11c). Therefore, in general, $\overset{*}{\dot{\mathbf{U}}}$ is not equal to $\overset{\Delta}{\mathbf{U}}$. Indeed $\overset{\Delta}{\mathbf{U}}$ can be expressed as a linear combination of $\mathbf{1}$, $\mathbf{U}$, and $\mathbf{U}^2$. This is not the case for $\overset{*}{\mathbf{U}}$, if the *eigenvalues* are repeated. However, from (2.9.2a-c),

$$\mathbf{U}\overset{*}{\dot{\mathbf{U}}} - \overset{*}{\dot{\mathbf{U}}}\mathbf{U} = \mathbf{0} \tag{2.9.2d}$$

which, in view of (2.9.1a), yields

$$\mathbf{U}\dot{\mathbf{U}} - \dot{\mathbf{U}}\mathbf{U} = 2\,\mathbf{U}\,\mathbf{\Omega}^L\,\mathbf{U} - \mathbf{U}^2\,\mathbf{\Omega}^L - \mathbf{\Omega}^L\,\mathbf{U}^2. \tag{2.9.2e}$$

With reference to (2.7.1b) and (2.9.2e), $\hat{\mathbf{W}} - \hat{\mathbf{\Omega}}^R$ is related to $\mathbf{\Omega}^L$ by

$$2\,\mathbf{U}(\hat{\mathbf{W}} - \hat{\mathbf{\Omega}}^R)\,\mathbf{U} = 2\,\mathbf{U}\,\mathbf{\Omega}^L\,\mathbf{U} - \mathbf{U}^2\,\mathbf{\Omega}^L - \mathbf{\Omega}^L\,\mathbf{U}^2. \tag{2.9.3a}$$

With the aid of identities (1.4.9b,c) and (1.2.11), this reduces to

$$2\,\mathrm{III}_U(\hat{\mathbf{W}} - \hat{\mathbf{\Omega}}^R) = \mathrm{I}_U\,\mathbf{U}\,\mathbf{\Omega}^L\,\mathbf{U} - \mathrm{II}_U(\mathbf{U}\,\mathbf{\Omega}^L + \mathbf{\Omega}^L\,\mathbf{U}) + 3\mathrm{III}_U\,\mathbf{\Omega}^L. \tag{2.9.3b}$$

In the Lagrangian triad, (2.9.3a) becomes

$$W_{ab} - \Omega^R_{ab} = \frac{(\lambda_a - \lambda_b)^2}{2\,\lambda_a\,\lambda_b}\,\Omega^L_{ba} \quad (\text{no sum}). \tag{2.9.3c}$$

Furthermore, from (2.6.4d) and (2.9.3a,b), it follows that

$$2\,\mathbf{U}(\hat{\mathbf{\Omega}}^E - \hat{\mathbf{W}})\,\mathbf{U} = \mathbf{U}^2\,\mathbf{\Omega}^L + \mathbf{\Omega}^L\,\mathbf{U}^2,$$

$$2\,\mathrm{III}_U(\hat{\mathbf{W}} - \hat{\mathbf{\Omega}}^E) = \mathrm{I}_U\,\mathbf{U}\,\mathbf{\Omega}^L\,\mathbf{U} - \mathrm{II}_U(\mathbf{U}\,\mathbf{\Omega}^L + \mathbf{\Omega}^L\,\mathbf{U}) + \mathrm{III}_U\,\mathbf{\Omega}^L \tag{2.9.4a,b}$$

which has the following component representation in the Lagrangian triad (Hill, 1978):

$$\Omega^E_{ab} - W_{ab} = \frac{\lambda_a^2 + \lambda_b^2}{2\,\lambda_a\,\lambda_b}\,\Omega^L_{ab} \quad (\text{no sum}). \tag{2.9.4c}$$

To find an expression for $\mathbf{\Omega}^L$ in terms of $\mathbf{U}$ and its rate, use (2.9.1a), (2.9.2a), and the solution (1.4.21c) of the tensor equation (1.4.17), to arrive at

$$(4\,\mathrm{II}^3_{U'} + 27\,\mathrm{III}^2_{U'})\,\mathbf{\Omega}^L = 4\,\mathrm{II}^2_{U'}(\mathbf{U}'\dot{\mathbf{U}} - \dot{\mathbf{U}}\mathbf{U}') - 9\,\mathrm{III}_{U'}(\mathbf{U}'^2\dot{\mathbf{U}} - \dot{\mathbf{U}}\mathbf{U}'^2)$$

$$- 3\,\mathrm{II}_{U'}(\mathbf{U}'^2\dot{\mathbf{U}}\mathbf{U}' - \mathbf{U}'\dot{\mathbf{U}}\mathbf{U}'^2), \tag{2.9.5a}$$

where

$$\mathbf{U}' = \mathbf{U} - \frac{1}{3}\,\mathrm{I}_U\,\mathbf{1}. \tag{2.9.5b}$$

Observe that (2.9.5a) holds in general between the spin of the principal axes of any tensor $\mathbf{U}$ and its rate. If $\dot{\mathbf{U}}$ is replaced by another rate, then $\mathbf{\Omega}^L$ must be replaced by the spin of the principal axes of $\mathbf{U}$ relative to the observer who measures the rate of change. For example, the spin $\mathbf{\Omega}^E$ is related to $\mathbf{V}$ and its rate by

$$(4\,\mathrm{II}^3_{\mathbf{V}'} + 27\,\mathrm{III}^2_{\mathbf{V}'})\,\mathbf{\Omega}^E = 4\,\mathrm{II}^2_{\mathbf{V}'}(\mathbf{V}'\dot{\mathbf{V}} - \dot{\mathbf{V}}\mathbf{V}') - 9\,\mathrm{III}_{\mathbf{V}'}(\mathbf{V}'^2\dot{\mathbf{V}} - \dot{\mathbf{V}}\mathbf{V}'^2)$$
$$- 3\,\mathrm{II}_{\mathbf{V}'}(\mathbf{V}'^2\dot{\mathbf{V}}\mathbf{V}' - \mathbf{V}'\dot{\mathbf{V}}\mathbf{V}'^2)\,, \qquad (2.9.6)$$

where the definition of $\mathbf{V}'$ is similar to (2.9.5b).

Substitution from (2.7.8) into (2.9.5a) yields an expression for $\mathbf{\Omega}^L$ in terms of $\mathbf{U}$ and $\mathbf{D}$; Mehrabadi and Nemat-Nasser (1987). Alternatively, observe from (2.7.2b) and (2.9.1), that

$$2\,\hat{\mathbf{D}} = \overset{*}{\mathbf{U}}\mathbf{U}^{-1} + \mathbf{U}^{-1}\overset{*}{\mathbf{U}} - \mathbf{C}\mathbf{U}^{-1}\mathbf{\Omega}^L\mathbf{U}^{-1} + \mathbf{U}^{-1}\mathbf{\Omega}^L\mathbf{U}^{-1}\mathbf{C}\,. \qquad (2.9.7a)$$

Solve (2.9.7a) for $\mathbf{U}^{-1}\mathbf{\Omega}^L\mathbf{U}^{-1}$, using (1.4.17) and (1.4.21c) to arrive at

$$(4\,\mathrm{II}^3_{\mathbf{C}'} + 27\,\mathrm{III}^2_{\mathbf{C}'})\,\mathbf{U}^{-1}\mathbf{\Omega}^L\mathbf{U}^{-1} =$$
$$4\,\mathrm{II}^2_{\mathbf{C}'}(\mathbf{C}'\hat{\mathbf{D}} - \hat{\mathbf{D}}\mathbf{C}') - 9\mathrm{III}_{\mathbf{C}'}(\mathbf{C}'^2\hat{\mathbf{D}} - \hat{\mathbf{D}}\mathbf{C}'^2)$$
$$- 3\,\mathrm{II}_{\mathbf{C}'}(\mathbf{C}'^2\hat{\mathbf{D}}\mathbf{C}' - \mathbf{C}'\hat{\mathbf{D}}\mathbf{C}'^2)\,. \qquad (2.9.7b)$$

A simpler equation results if (2.7.6a) is employed in conjunction with (2.9.3a),

$$2\,(\mathbf{U}\hat{\mathbf{D}} - \hat{\mathbf{D}}\mathbf{U}) = \mathbf{U}\mathbf{\Omega}^L + \mathbf{\Omega}^L\mathbf{U} - (\mathbf{U}^2\mathbf{\Omega}^L\mathbf{U}^{-1} + \mathbf{U}^{-1}\mathbf{\Omega}^L\mathbf{U}^2) \qquad (2.9.8a)$$

which, with respect to the Lagrangian triad, has the following component representation (Hill, 1978):

$$\Omega^L_{ab} = \frac{2\,\lambda_a\lambda_b}{\lambda_b^2 - \lambda_a^2}\,\mathrm{D}_{ab}\,, \qquad \lambda_a \neq \lambda_b\,, \quad (\text{no sum})\,. \qquad (2.9.8b)$$

Furthermore, from this and (2.9.4c), obtain

$$\Omega^E_{ab} - \mathrm{W}_{ab} = \frac{\lambda_b^2 + \lambda_a^2}{\lambda_b^2 - \lambda_a^2}\,\mathrm{D}_{ab}, \quad (\text{no sum})\,. \qquad (2.9.4d)$$

To express the spin of the Eulerian triad, consider the left Cauchy-Green tensor,

$$\mathbf{B} = \sum_{a=1}^{3} \lambda_a^2\,\mathbf{n}^a \otimes \mathbf{n}^a\,,$$

differentiate, and use (2.6.1b) to obtain

$$\dot{\mathbf{B}} = \mathbf{V}\dot{\mathbf{V}} + \dot{\mathbf{V}}\mathbf{V} = \overset{*}{\mathbf{B}} + \mathbf{\Omega}^E\mathbf{B} - \mathbf{B}\mathbf{\Omega}^E\,. \qquad (2.9.9a,b)$$

Substitute (2.8.1b) into (2.9.9a,b) to arrive at

$$\mathbf{B}\mathbf{D} + \mathbf{D}\mathbf{B} - \overset{*}{\mathbf{B}} = (\mathbf{\Omega}^E - \mathbf{W})\,\mathbf{B} - \mathbf{B}\,(\mathbf{\Omega}^E - \mathbf{W})\,. \qquad (2.9.9c)$$

This can be solved using the solution (1.4.21c) of (1.4.17), to express $(\mathbf{\Omega}^E - \mathbf{W})$ in terms of $\mathbf{B}$ and $\mathbf{D}$, as follows:

$$(4\,\mathrm{II}_{\mathbf{B}'}^3 + 27\,\mathrm{III}_{\mathbf{B}'}^2)\,(\boldsymbol{\Omega}^{\mathrm{E}} - \mathbf{W}) = \frac{4}{3}\,\mathrm{II}_{\mathbf{B}'}\,(2\,\mathrm{I}_{\mathbf{B}}\,\mathrm{II}_{\mathbf{B}'} + 9\,\mathrm{III}_{\mathbf{B}'})\,(\mathbf{B}'\mathbf{D} - \mathbf{D}\,\mathbf{B}')$$
$$+ (4\,\mathrm{II}_{\mathbf{B}'}^2 - 6\,\mathrm{I}_{\mathbf{B}}\,\mathrm{III}_{\mathbf{B}'})\,(\mathbf{B}'^2\,\mathbf{D} - \mathbf{D}\,\mathbf{B}'^2)$$
$$- (9\,\mathrm{III}_{\mathbf{B}'} + 2\,\mathrm{I}_{\mathbf{B}}\,\mathrm{II}_{\mathbf{B}'})\,(\mathbf{B}'^2\,\mathbf{D}\,\mathbf{B}' - \mathbf{B}'\,\mathbf{D}\,\mathbf{B}'^2)\,, \tag{2.9.10}$$

where $\mathbf{B}'$ is defined similarly to $\mathbf{C}'$ in (2.2.20e). Since $\overset{*}{\mathbf{B}}\,\mathbf{B}^{-1} - \mathbf{B}^{-1}\,\overset{*}{\mathbf{B}} = \mathbf{0}$, write (2.9.9c) as

$$\mathbf{B}\,\mathbf{D}\,\mathbf{B}^{-1} - \mathbf{B}^{-1}\,\mathbf{D}\,\mathbf{B} = 2\,(\boldsymbol{\Omega}^{\mathrm{E}} - \mathbf{W}) - \mathbf{B}\,(\boldsymbol{\Omega}^{\mathrm{E}} - \mathbf{W})\,\mathbf{B}^{-1} - \mathbf{B}^{-1}\,(\boldsymbol{\Omega}^{\mathrm{E}} - \mathbf{W})\,\mathbf{B}\,, \tag{2.9.11a}$$

and, in view of $\mathbf{C} = \mathbf{R}^{\mathrm{T}}\,\mathbf{B}\,\mathbf{R}$, obtain

$$\mathbf{C}\,\hat{\mathbf{D}}\,\mathbf{C}^{-1} - \mathbf{C}^{-1}\,\hat{\mathbf{D}}\,\mathbf{C} = 2\,(\hat{\boldsymbol{\Omega}}^{\mathrm{E}} - \hat{\mathbf{W}}) - \mathbf{C}\,(\hat{\boldsymbol{\Omega}}^{\mathrm{E}} - \hat{\mathbf{W}})\,\mathbf{C}^{-1} - \mathbf{C}^{-1}\,(\hat{\boldsymbol{\Omega}}^{\mathrm{E}} - \hat{\mathbf{W}})\,\mathbf{C}\,. \tag{2.9.11b}$$

•

2.10. STRAIN RATES

First consider the description of the strain rates in the Lagrangian and Eulerian triads. Then examine the corresponding coordinate–independent expressions.

2.10.1. Strain Rates in Lagrangian Triad

Consider the general material strain measure introduced in Section 2.3,

$$\mathbf{E} = \sum_{a=1}^{3} \mathrm{E}_a\,\mathbf{N}^a \otimes \mathbf{N}^a\,, \qquad \mathrm{E}_a = \mathrm{f}(\lambda_a)\,,$$

and define its time rate of change by

$$\dot{\mathbf{E}} \equiv \sum_{a,\,b=1}^{3} \dot{\mathrm{E}}_{ab}\,\mathbf{N}^a \otimes \mathbf{N}^b\,, \tag{2.10.1a}$$

where $\dot{\mathrm{E}}_{ab}$ are the components of $\dot{\mathbf{E}}$ in the Lagrangian triad. On the other hand, direct time differentiation of $\mathbf{E}$ yields

$$\dot{\mathbf{E}} = \sum_{a,\,b=1}^{3} \{\dot{\mathrm{E}}_a\,\delta_{ab} + (\mathrm{E}_a - \mathrm{E}_b)\,\Omega^{\mathrm{L}}_{ba}\}\,\mathbf{N}^a \otimes \mathbf{N}^b$$
$$= \sum_{a,\,b=1}^{3} \{\mathrm{f}'(\lambda_a)\,\dot{\lambda}_a\,\delta_{ab} + \{\mathrm{f}(\lambda_a) - \mathrm{f}(\lambda_b)\}\,\Omega^{\mathrm{L}}_{ba}\}\,\mathbf{N}^a \otimes \mathbf{N}^b\,. \tag{2.10.1b}$$

Substitution of (2.9.1c) and (2.7.2d) into this expression results in an equation relating $\dot{\mathrm{E}}_{ab}$ and D_{ab} (Hill, 1978). Since (2.9.1c) is not directly invertable when two principal stretches are equal, the resulting equation does not hold in such a case. To avoid this difficulty, follow Scheidler (1991) and obtain,

$$\dot{\mathbf{E}} = \hat{\mathcal{G}}\,\dot{\mathbf{U}}\,, \tag{2.10.2a}$$

where $\hat{\mathcal{G}}$ is a fourth-order tensor, having the following components[8] with respect to both the Lagrangian and Eulerian triads (no sum on repeated indices):

$$\mathcal{G}_{aaaa} = f'(\lambda_a)\,,$$

$$\mathcal{G}_{abab} = \mathcal{G}_{abba} = \begin{cases} \dfrac{1}{2} f'(\lambda_a)\,, & \text{for } \lambda_a = \lambda_b\,,\ \ a \neq b\,, \\[2ex] \dfrac{f(\lambda_a) - f(\lambda_b)}{2\,(\lambda_a - \lambda_b)}\,, & \text{for } \lambda_a \neq \lambda_b \end{cases}$$

$$\mathcal{G}_{abcd} = 0\,, \quad \text{for a, b} \neq \text{c or d}\,. \tag{2.10.3a-c}$$

In the Lagrangian triad, (2.10.2a) becomes

$$\dot{E}_{ab} = \mathcal{G}_{abcd}\,\dot{U}_{cd}\,. \tag{2.10.2b}$$

Substitute (2.7.2d) into (2.10.2b), and in view of (2.10.3), obtain (no sum)

$$\dot{E}_{ab} = \begin{cases} \lambda_a f'(\lambda_a)\, D_{ab} & \text{for } \lambda_a = \lambda_b, \\[2ex] (E_a - E_b)\, \dfrac{2\,\lambda_a \lambda_b}{\lambda_a^2 - \lambda_b^2}\, D_{ab} & \text{for } \lambda_a \neq \lambda_b\,. \end{cases} \tag{2.10.4}$$

From this expression, the strain rates associated with various strain measures introduced in Section 2.3 can be obtained. For example, for the logarithmic strain, $\mathbf{E}^l = ln\,\mathbf{U}$, $E_a^l = ln\,\lambda_a$, and $f'(\lambda_a) = \lambda_a^{-1}$, and therefore (2.10.4) yields (no sum)

$$\dot{E}_{ab}^l = \begin{cases} D_{aa} & \text{for } \lambda_a = \lambda_b \\[2ex] \dfrac{E_a^l - E_b^l}{\sinh(E_a^l - E_b^l)}\, D_{ab} & \text{for } \lambda_a \neq \lambda_b\,. \end{cases} \tag{2.10.5a}$$

Since $\lim_{x \to 0} \dfrac{x}{\sinh x} = 1$, the rate of change of the logarithmic strain is given by

$$\dot{\mathbf{E}}^l = \sum_{a,\,b\,=\,1}^{3} \left\{ \frac{E_a^l - E_b^l}{\sinh(E_a^l - E_b^l)}\, D_{ab} \right\} \mathbf{N}^a \otimes \mathbf{N}^b\,. \tag{2.10.5b}$$

Moreover, the following identities can be easily verified with the aid of the results presented in Section 2.9 (no sum):

$$W_{ab} - \Omega_{ab}^R = D_{ab} \tanh \frac{1}{2}\,(E_a^l - E_b^l)\,; \tag{2.10.6}$$

---

[8] This component formula was reported by Truesdell (1961). See also Truesdell and Noll (1965, page 145), Bowen and Wang (1970), Chadwick and Ogden (1971), and Wang and Truesdell (1973, page 443).

$$\Omega^{E}_{ab} - W_{ab} = \Omega^{L}_{ab} \cosh(E^{l}_{a} - E^{l}_{b})$$

$$= D_{ab} \coth(E^{l}_{a} - E^{l}_{b})\,; \tag{2.10.7}$$

$$\Omega^{L}_{ab} = D_{ab} \,/\, \sinh(E^{l}_{a} - E^{l}_{b})\,. \tag{2.10.8}$$

2.10.2. Strain Rates in Eulerian Triad

Similarly, since Eulerian strain measures are tensor-valued isotropic functions of the left stretch tensor **V**, their time derivatives can be expressed as

$$\dot{\boldsymbol{\varepsilon}} = \boldsymbol{\mathcal{G}}\,\dot{\mathbf{V}}, \tag{2.10.9a}$$

where the fourth-order tensor $\boldsymbol{\mathcal{G}}$ has components $\mathcal{G}_{abcd}$, given by (2.10.3) with respect to the Eulerian triad. Since $\boldsymbol{\mathcal{G}}$ is independent of the spin of the principal axes of **V**, (2.10.9a) holds also for

$$\overset{\circ}{\boldsymbol{\varepsilon}} = \boldsymbol{\mathcal{G}}\,\overset{\circ}{\mathbf{V}}, \tag{2.10.9b}$$

where

$$\overset{\circ}{\boldsymbol{\varepsilon}} = \dot{\boldsymbol{\varepsilon}} - \mathbf{W}\boldsymbol{\varepsilon} + \mathbf{W}\boldsymbol{\varepsilon}\,. \tag{2.10.10}$$

With respect to the Eulerian triad, (2.10.9b) has the following representation:

$$\overset{\circ}{\varepsilon}_{ab} = \mathcal{G}_{abcd}\,\overset{\circ}{V}_{cd}\,. \tag{2.10.9c}$$

Substitute (2.8.2c) into (2.10.9c), and in view of (2.10.3), arrive at (no sum)

$$\overset{\circ}{\varepsilon}_{ab} = \begin{cases} \lambda_a f'(\lambda_a)\, D_{ab} & \text{for } \lambda_a = \lambda_b \\ \{f(\lambda)_a - f(\lambda)_b\} \dfrac{\lambda_a^2 + \lambda_b^2}{\lambda_a^2 - \lambda_b^2}\, D_{ab} & \text{for } \lambda_a \neq \lambda_b\,. \end{cases} \tag{2.10.11}$$

For example, set $f(\lambda) = \ln(\lambda)$, and obtain the rate of change of the Eulerian logarithmic strain as observed in a frame which rotates with spin **W**,

$$\overset{\circ}{\boldsymbol{\varepsilon}}{}^{l} = \sum_{a,\,b=1}^{3} \left\{ (\varepsilon^{l}_{a} - \varepsilon^{l}_{b}) \coth(\varepsilon^{l}_{a} - \varepsilon^{l}_{b})\, D_{ab} \right\} \mathbf{n}^{a} \otimes \mathbf{n}^{b}\,. \tag{2.10.12}$$

2.10.3. Coordinate-independent Expressions

To arrive at a general, coordinate-independent expression for the time rate of change of the strain measures, following Balendran and Nemat-Nasser (1996), take the time derivative of expression (2.3.5a),

$$\dot{\mathbf{E}} = a_1\,\dot{\mathbf{U}} + a_2\,(\mathbf{U}\dot{\mathbf{U}} + \dot{\mathbf{U}}\mathbf{U}) + c_0\,\mathbf{1} + c_1\,\mathbf{U} + c_2\,\mathbf{U}^2\,, \tag{2.10.13}$$

where the unknown coefficients c_0, c_1, and c_2 are scalar functions of **U** and $\dot{\mathbf{U}}$. Substitute (2.7.11b) into (2.10.13), to obtain

$$\dot{\mathbf{E}} = \overset{\Delta}{\mathbf{E}} + a_1\,(\dot{\mathbf{U}} - \overset{\Delta}{\mathbf{U}}) + a_2\,(\mathbf{U}\,\dot{\mathbf{U}} + \dot{\mathbf{U}}\,\mathbf{U} - 2\,\mathbf{U}\,\overset{\Delta}{\mathbf{U}}),$$
$$\overset{\Delta}{\mathbf{E}} \equiv a_1\,\overset{\Delta}{\mathbf{U}} + a_2\,(\mathbf{U}\,\overset{\Delta}{\mathbf{U}} + \overset{\Delta}{\mathbf{U}}\,\mathbf{U}) + c_0\,\mathbf{1} + c_1\,\mathbf{U} + c_2\,\mathbf{U}^2\,. \qquad (2.10.14\text{a,b})$$

Since $\mathbf{U}$ and $\overset{\Delta}{\mathbf{U}}$ are coaxial (*i.e*, have the same *eigenvectors*), $\overset{\Delta}{\mathbf{E}}$ is also coaxial with $\mathbf{U}$. From (2.10.1b), the part of $\dot{\mathbf{E}}$ which is coaxial with $\mathbf{U}$ is

$$\overset{\Delta}{\mathbf{E}} = \sum_{a=1}^{3} f'(\lambda_a)\,\mathrm{tr}(\dot{\mathbf{U}}\,\hat{\mathbf{E}}_a)\,\hat{\mathbf{E}}_a, \quad \text{for } \lambda_1 \neq \lambda_2 \neq \lambda_3 \neq \lambda_1,$$
$$\overset{\Delta}{\mathbf{E}} = f'(\lambda_1)\,\mathrm{tr}(\dot{\mathbf{U}}\,\hat{\mathbf{E}}_1)\,\hat{\mathbf{E}}_1 + f'(\lambda_2)\,\mathrm{tr}(\dot{\mathbf{U}} - \dot{\mathbf{U}}\,\hat{\mathbf{E}}_1)\,(\mathbf{1} - \hat{\mathbf{E}}_1),$$
$$\text{for } \lambda_1 \neq \lambda_2 = \lambda_3,$$
$$\overset{\Delta}{\mathbf{E}} = f'(\lambda)\,\mathrm{tr}(\dot{\mathbf{U}})\,\mathbf{1}, \quad \text{for } \lambda_1 = \lambda_2 = \lambda_3 = \lambda. \qquad (2.10.15\text{a-c})$$

With (2.10.15) in (2.10.14a), arrive at

$$\dot{\mathbf{E}} = a_1\,\dot{\mathbf{U}} + a_2\,(\mathbf{U}\,\dot{\mathbf{U}} + \dot{\mathbf{U}}\,\mathbf{U}) + \{\mathbf{f}'(\mathbf{U}) - a_1\,\mathbf{1} - 2\,a_2\,\mathbf{U}\}\,\overset{\Delta}{\mathbf{U}}. \qquad (2.10.16)$$

The coordinate-independent expressions for $\overset{\Delta}{\mathbf{U}}$ and $\mathbf{f}'(\mathbf{U})$ now are

$$\overset{\Delta}{\mathbf{U}} = \sum_{a=1}^{3} \mathrm{tr}(\dot{\mathbf{U}}\,\mathbf{E}_a)\,\hat{\mathbf{E}}_a, \quad \mathbf{f}'(\mathbf{U}) = \sum_{a=1}^{3} f'(\lambda_a)\,\hat{\mathbf{E}}_a,$$
$$\text{for } \lambda_1 \neq \lambda_2 \neq \lambda_3 \neq \lambda_1, \qquad (2.10.17\text{a,b})$$

$$\overset{\Delta}{\mathbf{U}} = \mathrm{tr}(\dot{\mathbf{U}}\,\hat{\mathbf{E}}_1)\,\hat{\mathbf{E}}_1 + \mathrm{tr}(\dot{\mathbf{U}} - \dot{\mathbf{U}}\,\hat{\mathbf{E}}_1)\,(\mathbf{1} - \hat{\mathbf{E}}_1),$$
$$\mathbf{f}'(\mathbf{U}) = f'(\lambda_1)\,\hat{\mathbf{E}}_1 + f'(\lambda_2)\,(\mathbf{1} - \hat{\mathbf{E}}_1), \quad \text{for } \lambda_1 \neq \lambda_2 = \lambda_3,$$
$$\overset{\Delta}{\mathbf{U}} = \mathrm{tr}(\dot{\mathbf{U}})\,\mathbf{1}, \quad \mathbf{f}'(\mathbf{U}) = f'(\lambda)\,\mathbf{1}, \quad \text{for } \lambda_1 = \lambda_2 = \lambda_3 = \lambda, \qquad (2.10.17\text{c-f})$$

where, as given in (2.3.5) and (2.3.8),

$$\hat{\mathbf{E}}_a = \frac{1}{(\lambda_a - \lambda_b)\,(\lambda_a - \lambda_c)}\,\{\mathbf{U}^2 - (\lambda_b + \lambda_c)\,\mathbf{U} + \lambda_b\,\lambda_c\,\mathbf{1}\}, \qquad (2.10.18\text{a})$$
$$\text{for } \lambda_1 \neq \lambda_2 \neq \lambda_3 \neq \lambda_1,$$

$$\hat{\mathbf{E}}_1 = \frac{1}{\lambda_1 - \lambda_2}\,(\mathbf{U} - \lambda_2\,\mathbf{1}), \text{ for } \lambda_1 \neq \lambda_2 = \lambda_3. \qquad (2.10.18\text{b})$$

Substitution of (2.10.17) into (2.10.16) results in

$$\dot{\mathbf{E}} = a_1\,\dot{\mathbf{U}} + a_2\,(\mathbf{U}\,\dot{\mathbf{U}} + \dot{\mathbf{U}}\,\mathbf{U}) + \sum_{a=1}^{3} \{f'(\lambda_a) - a_1 - 2\,a_2\,\lambda_a\}\,\mathrm{tr}(\dot{\mathbf{U}}\,\hat{\mathbf{E}}_a)\,\hat{\mathbf{E}}_a,$$
$$\text{for } \lambda_1 \neq \lambda_2 \neq \lambda_3 \neq \lambda_1,$$

$$\dot{\mathbf{E}} = a_1\,\dot{\mathbf{U}} + a_2\,(\mathbf{U}\,\dot{\mathbf{U}} + \dot{\mathbf{U}}\,\mathbf{U}) + \{f'(\lambda_1) - a_1 - 2\,a_2\,\lambda_a\}\,\mathrm{tr}(\dot{\mathbf{U}}\,\hat{\mathbf{E}}_1)\,\hat{\mathbf{E}}_1,$$
$$\text{for } \lambda_1 \neq \lambda_2 = \lambda_3,$$

$$\dot{\mathbf{E}} = f'(\lambda)\,\dot{\mathbf{U}}, \quad \text{for } \lambda_1 = \lambda_2 = \lambda_3 = \lambda, \qquad (2.10.19\text{a-c})$$

where a_1 and a_2 are given by (2.3.6) and (2.3.11).

Similarly, coordinate–independent expressions for the Jaumann rate of the Eulerian strain measures are given by

$$\overset{\circ}{\boldsymbol{\varepsilon}} = a_1 \overset{\circ}{\mathbf{V}} + a_2 (\mathbf{V}\overset{\circ}{\mathbf{V}} + \overset{\circ}{\mathbf{V}}\mathbf{V}) + \sum_{a=1}^{3} \{f'(\lambda_a) - a_1 - 2a_2\lambda_a\} \operatorname{tr}(\overset{\circ}{\mathbf{V}}\mathbf{E}_a)\,\mathbf{E}_a, \quad \text{for } \lambda_1 \neq \lambda_2 \neq \lambda_3 \neq \lambda_1,$$

$$\overset{\circ}{\boldsymbol{\varepsilon}} = a_1 \overset{\circ}{\mathbf{V}} + a_2 (\mathbf{V}\overset{\circ}{\mathbf{V}} + \overset{\circ}{\mathbf{V}}\mathbf{V}) + \{f'(\lambda_1) - a_1 - 2a_2\lambda_a\} \operatorname{tr}(\overset{\circ}{\mathbf{V}}\mathbf{E}_1)\,\mathbf{E}_1, \quad \text{for } \lambda_1 \neq \lambda_2 = \lambda_3,$$

$$\overset{\circ}{\boldsymbol{\varepsilon}} = f'(\lambda)\,\overset{\circ}{\mathbf{V}}, \quad \text{for } \lambda_1 = \lambda_2 = \lambda_3 = \lambda. \tag{2.10.20a-c}$$

2.10.4. Summary of Coordinate-independent Expressions for General Strain Rates

Substituting (2.7.12) and (2.8.10) into (2.10.19) and (2.10.20), now obtain explicit coordinate-independent expressions for the material time derivative of any Lagrangian strain measure and the Jaumann rate of change of any Eulerian strain measure, in terms of the deformation rate tensor, as follows:

I. Distinct Eigenvalues

$$\dot{\mathbf{E}} = \mathbf{K}^E + \sum_{a=1}^{3} \{\lambda_a f'(\lambda_a) \operatorname{tr}(\hat{\mathbf{D}}\hat{\mathbf{E}}_a) - \operatorname{tr}(\mathbf{K}^E\hat{\mathbf{E}}_a)\}\,\hat{\mathbf{E}}_a,$$

$$\overset{\circ}{\boldsymbol{\varepsilon}} = \mathbf{K}^\varepsilon + \sum_{a=1}^{3} \{\lambda_a f'(\lambda_a) \operatorname{tr}(\mathbf{D}\mathbf{E}_a) - \operatorname{tr}(\mathbf{K}^\varepsilon\mathbf{E}_a)\}\,\mathbf{E}_a, \tag{2.10.21a,b}$$

$$\mathbf{K}^E \equiv \frac{2}{I_U II_U - III_U} \{a_1 I_U III_U \hat{\mathbf{D}} + a_3 \mathbf{U}\hat{\mathbf{D}}\mathbf{U} - a_1 III_U (\mathbf{U}\hat{\mathbf{D}} + \hat{\mathbf{D}}\mathbf{U})\},$$

$$\mathbf{K}^\varepsilon = \frac{1}{I_V II_V - III_V} \{2a_1 (II_V^2 - I_V III_V)\mathbf{D} + a_4 II_V (\mathbf{V}^2\mathbf{D} + \mathbf{D}\mathbf{V}^2) - a_1 (I_V II_V - III_V)(\mathbf{V}\mathbf{D} + \mathbf{D}\mathbf{V})\}, \tag{2.10.22a,b}$$

$$\hat{\mathbf{E}}_a = \frac{1}{(\lambda_a - \lambda_b)(\lambda_a - \lambda_c)} \{\mathbf{U}^2 - (\lambda_b + \lambda_c)\mathbf{U} + \lambda_b\lambda_c \mathbf{1}\},$$

$$\mathbf{E}_a = \frac{1}{(\lambda_a - \lambda_b)(\lambda_a - \lambda_c)} \{\mathbf{V}^2 - (\lambda_b + \lambda_c)\mathbf{V} + \lambda_b\lambda_c \mathbf{1}\}, \tag{2.10.23a,b}$$

$$a_1 = -\sum_{a=1}^{3} \frac{(\lambda_b + \lambda_c) f(\lambda_a)}{(\lambda_a - \lambda_b)(\lambda_a - \lambda_c)}, \quad a_2 = \sum_{a=1}^{3} \frac{f(\lambda_a)}{(\lambda_a - \lambda_b)(\lambda_a - \lambda_c)},$$

$$a_3 = a_1 II_U + a_2 (I_U II_U - III_U), \quad a_4 = 2a_1 II_U + a_2 (I_U II_U - III_U), \tag{2.10.24a-d}$$

where a, b, c are cyclic permutations of 1, 2, and 3.

II. Two Repeated Eigenvalues

$$\dot{\mathbf{E}} = \mathbf{K}^E + \{\lambda_1 f'(\lambda_1)\,\mathrm{tr}(\hat{\mathbf{D}}\,\hat{\mathbf{E}}_1) - \mathrm{tr}(\mathbf{K}^E\,\hat{\mathbf{E}}_1)\}\,\hat{\mathbf{E}}_1\,,$$

$$\overset{\circ}{\boldsymbol{\varepsilon}} = \mathbf{K}^\varepsilon + \{\lambda_1 f'(\lambda_1)\,\mathrm{tr}(\hat{\mathbf{D}}\,\mathbf{E}_1) - \mathrm{tr}(\mathbf{K}^E\,\mathbf{E}_1)\}\,\mathbf{E}_1\,, \qquad (2.10.21c,d)$$

$$\mathbf{K}^E \equiv \frac{2}{I_U II_U - III_U}\,\{a_1 I_U III_U \hat{\mathbf{D}} + a_3 \mathbf{U}\hat{\mathbf{D}}\mathbf{U} - a_1 III_U(\mathbf{U}\hat{\mathbf{D}} + \hat{\mathbf{D}}\mathbf{U})\},$$

$$\mathbf{K}^\varepsilon \equiv \frac{1}{I_V II_V - III_V}\,\{2a_1(II_V^2 - I_V III_V)\,\mathbf{D} + a_4 II_V(\mathbf{V}^2\mathbf{D} + \mathbf{D}\mathbf{V}^2)$$

$$- a_1(I_V II_V - III_V)(\mathbf{V}\mathbf{D} + \mathbf{D}\mathbf{V})\}, \qquad (2.10.22c,d)$$

$$\hat{\mathbf{E}}_1 = \frac{1}{\lambda_1 - \lambda_2}(\mathbf{U} - \lambda_2 \mathbf{1}), \qquad \mathbf{E}_1 = \frac{1}{\lambda_1 - \lambda_2}(\mathbf{V} - \lambda_2 \mathbf{1}),$$

$$(2.10.23c,d)$$

$$a_1 = -2\lambda_2 \frac{f(\lambda_1) - f(\lambda_2)}{(\lambda_1 - \lambda_2)^2} + \frac{\lambda_1 + \lambda_2}{\lambda_1 - \lambda_2} f'(\lambda_2),$$

$$a_2 = \frac{f(\lambda_1) - f(\lambda_2) - (\lambda_1 - \lambda_2) f'(\lambda_2)}{(\lambda_1 - \lambda_2)^2},$$

$$a_3 = a_1 II_U + a_2(I_U II_U - III_U), \qquad a_4 = 2a_1 II_U + a_2(I_U II_U - III_U).$$

$$(2.10.24e\text{-}h)$$

III. Three Repeated Eigenvalues

$$\dot{\mathbf{E}} = \lambda f'(\lambda)\,\hat{\mathbf{D}}, \qquad \overset{\circ}{\boldsymbol{\varepsilon}} = \lambda f'(\lambda)\,\mathbf{D}. \qquad (2.10.21e,f)$$

2.11. REFERENCES

Balendran, B. and Nemat-Nasser, S. (1996), Derivative of a function of a non-symmetric second-order tensor, *Q. Appl. Math.*, Vol. 54, 583-600.

Biot, M. A. (1965), *Mechanics of incremental deformations*, John Wiley & Sons, New York.

Bowen, R. M. and Wang, C.-C. (1970), Acceleration waves in inhomogeneous isotropic elastic bodies, *Arch. Rat. Mech. Anal.*, Vol. 38, 13-45.

Chadwick, P. (1976), *Continuum mechanics*, Allen and Unwin, London.

Chadwick, P. and Ogden, R. (1971), A theorem of tensor calculus and its application to isotropic elasticity, *Arch. Rat. Mech. Anal.*, 54-68.

Hill, R. (1968), On constitutive inequalities for simple materials - I, *J. Mech. Phys. Solids*, Vol. 16, 229-242.

Hill, R. (1978), Aspects of invariance in solid mechanics, *Adv. Appl. Mech.*,

Vol. 18, 1-75.

Hoger, A. and Carlson, D. E. (1984), Determination of the stretch and rotation in the polar decomposition of the deformation gradient, *Q. Appl. Math.*, Vol. 42, 113-117.

Mehrabadi, M. M. and Nemat-Nasser, S. (1987), Some basic kinematical relations for finite deformations of continua, *Mech. Mat.*, Vol. 6, 127-138.

Prager, W. (1961), *Introduction to mechanics of continua*, Ginn, Boston.

Sawyers, K. (1986), Comments on the paper determination of the stretch and rotation in the polar decomposition of the deformation gradient by A. Hoger and D. E. Carlson, *Q. Appl. Math.*, Vol. 44, 309-311.

Scheidler, M. (1991), Time rates of generalized strain tensors - I, component formulas, *Mech. Mat.*, Vol. 11, 199-210.

Scheidler, M. (1994), The tensor equation $\mathbf{AX} + \mathbf{XA} = \mathbf{\Phi}(\mathbf{A}, \mathbf{H})$, with applications to kinematics of continua, *J. Elasticity*, Vol. 36, 117-153.

Seth, B. R. (1964), Generalized strain measure with applications to physical problems, in *Second-order effects in elasticity, plasticity, and fluid dynamics*, Pergamon, Oxford.

Stickforth, J. (1987), The square root of a three-dimensional positive tensor, *Acta. Mech.*, Vol. 67, 233-235.

Ting, T. C. T. (1985), Determination of $C^{1/2}$, $C^{-1/2}$ and more general isotropic tensor functions of C, *J. Elasticity*, Vol. 15, 319-323.

Truesdell, C. (1961), General and exact theory of waves in finite elastic strain, *Arch. Rat. Mech. Anal.*, Vol. 8, 263-296.

Truesdell, C. and Noll, W. (1965), *The nonlinear field theories of mechanics*, Springer-Verlag, Berlin.

Truesdell, C. and Toupin, R. (1960), The classical field theories, in *Principles of classical mechanics and field theory*, Springer-Verlag, Berlin, 226-793.

Wang, C.-C. and Truesdell, C. (1973), *Introduction to rational elasticity*, Noordhoff, Leiden, The Netherlands.

3

STRESS AND STRESS-RATE MEASURES, AND BALANCE RELATIONS

Certain commonly used stress and stress-rate measures are discussed in this chapter, together with the balance relations, including conditions at surfaces of discontinuity, boundary-value problems and their finite-element formulation, and general variational principles.

More specifically, the stress measures, conjugate to the strain measures developed in Chapter 2, are worked out in detail and the relations among them are established (Sections 3.1 and 3.2). A coordinate-independent expression for the stress measure conjugate to the general material strain measure which has been introduced in Chapter 2, is presented. Objective stress-rate measures are introduced in Section 3.3, and general relations connecting a broad class of objective stress rates are presented in Sections 3.4 and 3.5. Balance relations and boundary-value problems are briefly formulated (Section 3.6), together with the principle of virtual work (Section 3.7), relevant to nonlinear finite-deformation problems. This chapter also includes a weak form of the equations of motion, for application to a finite-element formulation of finite-deformation problems. Explicit finite-element equations are developed in Section 3.8, and a number of important issues pertaining to large deformations, are considered. The kinematics, dynamics, and balance relations at surfaces of discontinuity are discussed in Section 3.9. Finally, the chapter includes a comprehensive account of general variational principles for finite deformation of hyperelastic materials, as well as the corresponding incremental and linearized formulations (Section 3.10). In particular, for finite-deformation hyperelasticity, a variational theorem is given in terms of a general strain measure, $\mathbf{E}$*, its conjugate stress,* $\mathbf{S}$*, and the displacement field,* $\mathbf{u}$ *(three independent fields), encompassing all the field equations, boundary conditions, and any possible jump conditions. The notation follows that defined in Section 1.1 of Chapter 1.*

3.1. STRESS MEASURES

The most commonly used stress measure is the *Cauchy* or *true stress*, here denoted by $\boldsymbol{\sigma} = \sigma_{ij}\,\mathbf{e}_i \otimes \mathbf{e}_j$. It is symmetric, $\boldsymbol{\sigma} = \boldsymbol{\sigma}^T$. In terms of $\boldsymbol{\sigma}$, the tractions, $\mathbf{t}^{(n)}$, transmitted at a point across an elementary area with unit normal vector $\mathbf{n}$ are

$$\mathbf{t}^{(n)} = \boldsymbol{\sigma}^T \mathbf{n}$$

$$= \boldsymbol{\sigma}\,\mathbf{n}. \tag{3.1.1a}$$

Hence, σ_{ij} for fixed i and j is the ith component of the tractions across an element which is currently normal to the x_j-coordinate direction. The total force $d\mathbf{f}^{(n)}$ on the elementary area da then is

$$d\mathbf{f}^{(n)} = \mathbf{t}^{(n)}\,da$$

$$= \boldsymbol{\sigma}\,\mathbf{da}, \tag{3.1.2a}$$

where notation (2.2.27b) is employed, *i.e.*, $\mathbf{da} = \mathbf{n}\,da$. In component form, these become

$$t_i^{(n)} = \sigma_{ij}\,n_j = \sigma_{ji}\,n_j, \tag{3.1.1b}$$

$$df_i^{(n)} = \sigma_{ij}\,da_j. \tag{3.1.2b}$$

3.1.1. Balance of Angular and Linear Momentum

The balance of angular momentum for any elementary volume is guaranteed by the symmetry of the Cauchy stress. If $\mathbf{b}$ is the body force per unit mass and $\mathbf{a}$ is the acceleration, (2.4.3), then the balance of linear momentum for an elementary current volume v with boundary ∂v requires (see Figure 3.1.1)

$$\int_{\partial v} \mathbf{t}^{(n)}\,da = \int_{v} \rho\,(\mathbf{a} - \mathbf{b})\,dv. \tag{3.1.3}$$

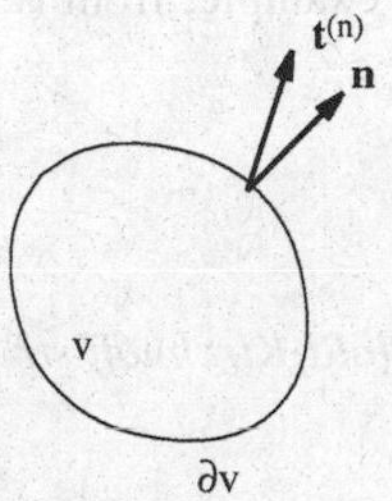

Figure 3.1.1

A deformed body with volume v bounded by ∂v, and surface tractions $\mathbf{t}^{(n)}$ on an elementary area with unit normal $\mathbf{n}$

In (3.1.3), $\rho = \rho(\mathbf{x}, t)$ is the current mass density. From (3.1.1), (3.1.3), and the Gauss theorem obtain, for an *arbitrary* v,

$$\nabla \cdot \boldsymbol{\sigma} + \rho\, \mathbf{b} = \rho\, \mathbf{a} = \rho\, \dot{\mathbf{v}} \tag{3.1.4a}$$

or, in component form,

$$\sigma_{ij,i} + \rho\, b_j = \rho\, a_j = \rho\, \{ \frac{\partial v_j}{\partial t} + \frac{\partial v_j}{\partial x_k} v_k \}, \quad \text{(i and k summed)}, \tag{3.1.4b}$$

where $\nabla \equiv \mathbf{e}_i \dfrac{\partial}{\partial x_i}$.

The quantity $\mathrm{tr}(\boldsymbol{\sigma}\mathbf{D})/\rho$ is the rate of stress-work per unit mass. It is invariant with respect to the change of the strain measure and the reference state. It can be used as a basis for identifying stress measures conjugate to given strain measures; Hill (1968, 1972).

The *Kirchhoff stress*, defined by

$$\boldsymbol{\tau} = (\rho_0/\rho)\, \boldsymbol{\sigma} = J\, \boldsymbol{\sigma}, \tag{3.1.5}$$

is also conjugate to **D**, the deformation rate, in the sense that

$$\frac{1}{\rho_0} \mathrm{tr}(\boldsymbol{\tau}\mathbf{D}) = \frac{1}{\rho} \mathrm{tr}(\boldsymbol{\sigma}\mathbf{D}),$$

$$\frac{1}{\rho_0} \tau_{ij}\, D_{ji} = \frac{1}{\rho} \sigma_{ij}\, D_{ji}. \tag{3.1.6a,b}$$

Hence, $\mathrm{tr}(\boldsymbol{\sigma}\mathbf{D})$ is the rate of stress-work per unit current (with mass density ρ) volume, whereas $\mathrm{tr}(\boldsymbol{\tau}\mathbf{D})$ is the rate of stress-work per unit reference (with mass density ρ_0) volume.

3.1.2. Second Piola-Kirchhoff Stress

To obtain the stress measure **S** conjugate to any member of the material strain measure **E** of class (2.3.1), set

$$\frac{1}{\rho_0} \mathrm{tr}(\mathbf{S}\dot{\mathbf{E}}) = \frac{1}{\rho_0} \mathrm{tr}(\boldsymbol{\tau}\mathbf{D}). \tag{3.1.6c}$$

For a given strain measure **E**, compute $\dot{\mathbf{E}}$ as a linear and homogeneous function of **D**, and obtain the corresponding stress measure from (3.1.6c). For certain strain measures, this may be an easy task. For example, from (2.3.17a), for the Lagrangian strain $\mathbf{E}^L = (\mathbf{C} - \mathbf{1})/2$,

$$\dot{\mathbf{E}}^L = \mathbf{F}^T \mathbf{D}\, \mathbf{F},$$

$$\dot{E}^L_{AB} = D_{ij}\, F_{iA}\, F_{jB}. \tag{3.1.7a,b}$$

Hence, its conjugate stress, called the *second Piola-Kirchhoff stress* and denoted by $\mathbf{S}^P$ here, becomes

$$\mathbf{S}^P = \mathbf{F}^{-1} \boldsymbol{\tau}\, \mathbf{F}^{-T} = J\, \mathbf{F}^{-1} \boldsymbol{\sigma}\, \mathbf{F}^{-T},$$

$$S^P_{AB} = J\, \sigma_{ij}\, F^{-1}_{Ai}\, F^{-1}_{Bj}, \tag{3.1.8a,b}$$

where $J = \rho_0/\rho \equiv \det \mathbf{F}$, and $F^{-1}_{Ai} \equiv \partial X_A/\partial x_i = X_{A,i}$; see (2.1.2) and (2.1.4).

3.1.3. General Relations in Principal Coordinates

To obtain a general relation from (3.1.6c), express $\mathbf{S}$ in the Lagrangian triad and $\boldsymbol{\tau}$ in the Eulerian triad as

$$\mathbf{S} = \sum_{a,\,b=1}^{3} S_{ab}\,\mathbf{N}^a \otimes \mathbf{N}^b\,,$$

$$\boldsymbol{\tau} = \sum_{a,\,b=1}^{3} \tau_{ab}\,\mathbf{n}^a \otimes \mathbf{n}^b\,. \tag{3.1.9a,b}$$

Then, using (2.10.1a) and (2.10.4), observe from (3.1.6b) that

$$\sum_{a,\,b=1}^{3} S_{ab}\Big\{\lambda_a f'(\lambda_a)\, D_{ab}\,\delta_{ab} + (E_a - E_b)\,\frac{2\lambda_a\lambda_b}{\lambda_a^2-\lambda_b^2}\,D_{ab}\Big\}$$

$$= \sum_{a,\,b=1}^{3} \tau_{ab}\,D_{ab}\,. \tag{3.1.10}$$

Compare the coefficients of D_{ab} on both sides to obtain (Hill, 1978),

$$S_{ab} = \begin{cases} \dfrac{\lambda_a^2-\lambda_b^2}{E_a-E_b}\,\dfrac{\tau_{ab}}{2\lambda_a\lambda_b}\,, & \lambda_a \neq \lambda_b\,,\\[2ex] \dfrac{\tau_{ab}}{\lambda_a f'(\lambda_a)}\,, & \lambda_a = \lambda_b\,, \end{cases} \tag{3.1.11}$$

where repeated indices are *not* summed.

3.1.4. General Coordinate-independent Relations

To arrive at a general coordinate-independent expression for the stress measure $\mathbf{S}$ conjugate to *any* general material strain measure $\mathbf{E}$ of Chapter 2, first observe from (3.1.6c) that

$$\mathrm{tr}(\mathbf{S}\,\dot{\mathbf{E}}) = \mathrm{tr}(\boldsymbol{\tau}\,\mathbf{D}) = \mathrm{tr}(\mathbf{S}^P\,\dot{\mathbf{E}}^L)\,. \tag{3.1.12}$$

Then note that the material time derivatives of $\mathbf{E}$ and $\mathbf{E}^L$ are related by

$$\dot{\mathbf{E}}^L = \frac{\partial \mathbf{E}^L}{\partial \mathbf{E}} : \dot{\mathbf{E}}\,,$$

$$\dot{E}^L_{ij} = \frac{\partial E^L_{ij}}{\partial E_{kl}}\,\dot{E}_{kl}\,. \tag{3.1.13a,b}$$

Substitution of (3.1.13) into (3.1.12) results in

$$S_{ij} = S^P_{kl}\,\frac{\partial E^L_{kl}}{\partial E_{ij}}\,. \tag{3.1.14a}$$

Denote the principal values (*eigenvalues*) of the strain measures $\mathbf{E}$ and $\mathbf{E}^L$ by E_a and E^L_a, $a = 1, 2, 3$, respectively. From (2.3.1c),

$$E_a = f(\lambda_a)\,, \qquad a = 1,\ 2,\ 3\,, \tag{3.1.15a}$$

and the Lagrangian strain can be expressed as

$$E^L_a \equiv (\lambda_a^2 - 1)/2 = g(E_a)\,. \tag{3.1.15b}$$

For example,

$$f(\lambda) = ln\,(\lambda) \quad \rightarrow \quad g(E) = \frac{1}{2}\,\{\exp(2\,E) - 1\}\,,$$

$$f(\lambda) = \lambda - 1 \quad \rightarrow \quad g(E) = \frac{1}{2}\,(E^2 + 2\,E)\,,$$

$$f(\lambda) = 1 - 1/\lambda \quad \rightarrow \quad g(E) = \frac{1}{2}\left\{\frac{1}{(1-E)^2} - 1\right\},$$

$$f(\lambda) = 1 - 1/\lambda^2 \quad \rightarrow \quad g(E) = \frac{1}{2}\left\{\frac{1}{1-E} - 1\right\}. \tag{3.1.16a-d}$$

Therefore, from (3.1.15b) and (2.3.1a), the *Lagrangian strain measure*, $\mathbf{E}^L$, can be expressed in terms of any general material strain measure, $\mathbf{E}$, as

$$\mathbf{E}^L = \sum_{a=1}^{3} g(E_a)\,\mathbf{N}^a \otimes \mathbf{N}^b\,. \tag{3.1.17a}$$

For a suitably smooth function g(E), the Taylor series expansion yields

$$g(E) = \sum_{m=0}^{N} \beta_m E^m\,, \tag{3.1.17b}$$

where β_m's are constants and N is a suitable large integer. Hence,

$$\mathbf{E}^L = \sum_{m=0}^{N} \beta_m \mathbf{E}^m\,. \tag{3.1.17c}$$

Differentiation of this with respect to $\mathbf{E}$ results in

$$\frac{\partial E^L_{ij}}{\partial E_{kl}} = \sum_{m=0}^{N-1} \beta_{m+1}\left\{\sum_{n=0}^{m} (\mathbf{E}^n)_{ik}\,(\mathbf{E}^{m-n})_{jl}\right\}. \tag{3.1.17d}$$

Since $\mathbf{E}$ is symmetric, from (3.1.17d) it follows that

$$\frac{\partial E^L_{ij}}{\partial E_{kl}} = \frac{\partial E^L_{kl}}{\partial E_{ij}}\,. \tag{3.1.18}$$

Then, substitution of (3.1.18) into (3.1.14a) leads to

$$S_{ij} = \frac{\partial E^L_{ij}}{\partial E_{kl}}\,S^P_{kl}\,. \tag{3.1.14b}$$

Note from (3.1.17b) that the material time derivative of $\mathbf{E}^L$ can be expressed as a function of $\mathbf{E}$ and $\dot{\mathbf{E}}$,

$$\dot{\mathbf{E}}^L = \mathbf{h}(\mathbf{E},\,\dot{\mathbf{E}}) \tag{3.1.19a}$$

which is linear and homogeneous in $\dot{\mathbf{E}}$. It thus follows from (3.1.13b), (3.1.14b), and (3.1.15), that

$$\mathbf{S} = \mathbf{h}(\mathbf{E},\,\mathbf{S}^P) \tag{3.1.19b}$$

which is also linear and homogeneous in $\mathbf{S}^P$. Note that, $\mathbf{S}$ in (3.1.19b) is obtained by simple substitution of $\mathbf{S}^P$ for $\dot{\mathbf{E}}$ in the right-hand side of (3.1.19a).

Since $\mathbf{E}$ and $\mathbf{E}^L$ are coaxial, by repeated application of the Hamilton-Cayley theorem in (3.1.17b), obtain

$$\mathbf{E}^L = b_0\,\mathbf{1} + b_1\,\mathbf{E} + b_2\,\mathbf{E}^2\,, \tag{3.1.20}$$

where the coefficients b_0, b_1, and b_2 are functions of the principal invariants of $\mathbf{E}$. For a general strain measure $\mathbf{E}$, it is not feasible to express these coefficients explicitly in terms of the invariants of $\mathbf{E}$. However, they can be expressed explicitly as functions of the principal strains E_a, $a = 1, 2, 3$, following the procedure outlined in Section 2.3; see (2.3.6), (2.3.11), and (2.3.14). Note that $\dot{\mathbf{E}}$ is defined through (2.10.19). Indeed, (3.1.19a) can be rendered explicit by simply replacing in (2.10.19), $\mathbf{U}$ by $\mathbf{E}$, $\dot{\mathbf{U}}$ by $\dot{\mathbf{E}}$, $f(\lambda_a)$ by $g(E_a)$, and λ_a by E_a, respectively, and then changing the coefficients a_0, a_1, a_2 respectively to b_0, b_1, b_2. In this manner, the following results are obtained from (2.10.19).

(i) Distinct Principal Strains

$$\mathbf{S} = b_1\,\mathbf{S}^P + b_2\,(\mathbf{E}\,\mathbf{S}^P + \mathbf{S}^P\,\mathbf{E}) + \sum_{a=1}^{3}\{g'(E_a) - b_1 - 2\,b_2\,E_a\}\,\mathrm{tr}(\mathbf{S}^P\,\hat{\mathbf{E}}_a)\,\hat{\mathbf{E}}_a\,,$$

$$\hat{\mathbf{E}}_a = \frac{1}{(E_a - E_b)\,(E_a - E_c)}\,\{\mathbf{E}^2 - (E_b + E_c)\,\mathbf{E} + E_b\,E_c\,\mathbf{1}\}\,,$$

$$b_1 = -\sum_{a=1}^{3}\frac{(E_b + E_c)\,g(E_a)}{(E_a - E_b)\,(E_a - E_c)}\,,$$

$$b_2 = \sum_{a=1}^{3}\frac{g(E_a)}{(E_a - E_b)\,(E_a - E_c)}\,. \tag{3.1.21a-d}$$

(ii) Two Repeated Principal Strains

$$\mathbf{S} = b_1\,\mathbf{S}^P + b_2\,(\mathbf{E}\,\mathbf{S}^P + \mathbf{S}^P\,\mathbf{E}) + \{g'(E_1) - b_1 - 2\,b_2\,E_1\}\,\mathrm{tr}(\mathbf{S}^P\,\hat{\mathbf{E}}_1)\,\hat{\mathbf{E}}_1\,,$$

$$\hat{\mathbf{E}}_1 = \frac{1}{E_1 - E_2}\,(\mathbf{E} - E_2\,\mathbf{1})\,,$$

$$b_1 = -\,2\,E_2\,\frac{g(E_1) - g(E_2)}{(E_1 - E_2)^2} + \frac{E_1 + E_2}{E_1 - E_2}\,g'(E_2)\,,$$

$$b_2 = \frac{g(E_1) - g(E_2) - (E_1 - E_2)\,g'(E_2)}{(E_1 - E_2)^2}\,. \tag{3.1.22a-d}$$

(iii) Three Repeated Principal Strains

$$\mathbf{S} = g'(E)\,\mathbf{S}^P\,. \tag{3.1.23}$$

For the special cases, where the coefficients b_0, b_1, and b_2 can be expressed explicitly in terms of the principal invariants of $\mathbf{E}$, substitution of (3.1.20) into (3.1.14b) results in

$$\mathbf{S} = \{\frac{\partial b_0}{\partial I_{\mathbf{E}}}\,\mathrm{tr}(\mathbf{S}^P) + \frac{\partial b_0}{\partial II_{\mathbf{E}}}\,\mathrm{tr}(I_{\mathbf{E}}\,\mathbf{S}^P - \mathbf{E}\,\mathbf{S}^P) + III_{\mathbf{E}}\,\frac{\partial b_0}{\partial III_{\mathbf{E}}}\,\mathrm{tr}(\mathbf{E}^{-1}\,\mathbf{S}^P)\}\,\mathbf{1}$$
$$+\{\frac{\partial b_1}{\partial I_{\mathbf{E}}}\,\mathrm{tr}(\mathbf{S}^P) + \frac{\partial b_1}{\partial II_{\mathbf{E}}}\,\mathrm{tr}(I_{\mathbf{E}}\,\mathbf{S}^P - \mathbf{E}\,\mathbf{S}^P) + III_{\mathbf{E}}\,\frac{\partial b_1}{\partial III_{\mathbf{E}}}\,\mathrm{tr}(\mathbf{E}^{-1}\,\mathbf{S}^P)\}\,\mathbf{E}$$
$$+\{\frac{\partial b_2}{\partial I_{\mathbf{E}}}\,\mathrm{tr}(\mathbf{S}^P) + \frac{\partial b_2}{\partial II_{\mathbf{E}}}\,\mathrm{tr}(I_{\mathbf{E}}\,\mathbf{S}^P - \mathbf{E}\,\mathbf{S}^P) + III_{\mathbf{E}}\,\frac{\partial b_2}{\partial III_{\mathbf{E}}}\,\mathrm{tr}(\mathbf{E}^{-1}\,\mathbf{S}^P)\}\,\mathbf{E}^2$$
$$+\,b_1\,\mathbf{S}^P + b_2\,(\mathbf{E}\,\mathbf{S}^P + \mathbf{S}^P\,\mathbf{E})\,. \tag{3.1.24}$$

In deriving (3.1.24), the following general relations are used:

$$\frac{\partial I_{\mathbf{E}}}{\partial \mathbf{E}} = \mathbf{1}\,, \qquad \frac{\partial II_{\mathbf{E}}}{\partial \mathbf{E}} = I_{\mathbf{E}}\,\mathbf{1} - \mathbf{E}\,,$$
$$\frac{\partial III_{\mathbf{E}}}{\partial \mathbf{E}} = III_{\mathbf{E}}\,\mathbf{E}^{-1}\,, \tag{3.1.25a-c}$$

where

$$\mathbf{E}^{-1} = \frac{1}{III_{\mathbf{E}}}\,\{\mathbf{E}^2 - I_{\mathbf{E}}\,\mathbf{E} + II_{\mathbf{E}}\,\mathbf{1}\}\,. \tag{3.1.25d}$$

3.1.5. First-order Accurate Expressions

To obtain an accurate to the first order expression for the stress measure conjugate to the material strain measure **E**, consider the Taylor series expansion of the function g(E) about E = 0,

$$E^L = g(E) = g(0) + g'(0)\,E + \frac{1}{2}\,g''(0)\,E^2 + O(E^3)\,. \tag{3.1.26}$$

Furthermore, from (3.1.15) obtain

$$g'(E) = \frac{\partial E^L}{\partial \lambda}\,\frac{\partial \lambda}{\partial E} = \frac{\lambda}{f'(\lambda)}\,,$$
$$g''(E) = \left\{\frac{1}{f'(\lambda)} - \frac{\lambda\,f''(\lambda)}{\{f'(\lambda)\}^2}\right\}\frac{1}{f'(\lambda)}\,. \tag{3.1.27a,b}$$

Since, from (2.3.15),

$$\lambda = 1\,, \quad f'(1) = 1\,, \quad \text{when } E = 0\,, \tag{3.1.28a-c}$$

in view of (3.1.15) and (3.1.28), arrive at

$$g(0) = 0\,, \quad g'(0) = 1\,, \quad g''(0) = 1 - f''(1)\,. \tag{3.1.27c-e}$$

Substitute (3.1.26) in the spectral representation of $\mathbf{E}^L$, and, in view of (3.1.27), obtain

$$\mathbf{E}^L = \mathbf{E} + \frac{1}{2}\,\{1 - f''(1)\}\,\mathbf{E}^2 + O(\mathbf{E}^3)\,. \tag{3.1.29a}$$

This and (3.1.14b) then yield (Hill, 1968)

$$\mathbf{S} = \mathbf{S}^{\mathrm{P}} + \frac{1}{2}\{1 - f''(1)\}(\mathbf{E}\,\mathbf{S}^{\mathrm{P}} + \mathbf{S}^{\mathrm{P}}\,\mathbf{E}) + O(\mathbf{E}^2). \tag{3.1.29b}$$

The stress measure conjugate to a strain measure in the class (2.3.1) can be obtained independently of any coordinate system by substituting the appropriate expression for g(E) into (3.1.21)-(3.1.23) or for b_0, b_1, and b_2 in terms of the principal invariants of **E** into (3.1.24). In many cases, this leads to an exact expression for the conjugate stress measure. For example, the stress measures corresponding to the strains in Section 2.3 may be identified as given below.

3.1.6. Stress Conjugate to Lagrangian Strain Measure

For the Lagrangian strain, $E_a = E_a^L = \frac{1}{2}(\lambda_a^2 - 1)$. The conjugate stress, from (3.1.11), is

$$S_{ab}^{P} = \frac{\tau_{ab}}{\lambda_a \lambda_b} \tag{3.1.30a}$$

which is the component representation of

$$\mathbf{S}^{\mathrm{P}} = \mathbf{F}^{-1}\,\boldsymbol{\tau}\,\mathbf{F}^{-\mathrm{T}} = \mathbf{U}^{-1}\,\hat{\boldsymbol{\tau}}\,\mathbf{U}^{-1},$$

$$\hat{\boldsymbol{\tau}} = \mathbf{R}^{\mathrm{T}}\,\boldsymbol{\tau}\,\mathbf{R}, \tag{3.1.30b,c}$$

with respect to the Lagrangian triad, where $\mathbf{U}^{-1}$ is given explicitly in terms of **U** by (2.2.13b). As pointed out before, the stress measure conjugate to the Lagrangian strain is called the second Piola-Kirchhoff stress.

3.1.7. Stress Conjugate to Biot's Strain Measure

For Biot's strain, $E_a = E_a^B = \lambda_a - 1$. From (3.1.11), it follows that

$$S_{ab}^{B} = \frac{1}{2}(\lambda_a^{-1} + \lambda_b^{-1})\,\tau_{ab} \tag{3.1.31a}$$

which is the component representation of

$$\mathbf{S}^{\mathrm{B}} = \frac{1}{2}(\mathbf{U}^{-1}\,\hat{\boldsymbol{\tau}} + \hat{\boldsymbol{\tau}}\,\mathbf{U}^{-1}), \tag{3.1.31b}$$

with respect to the Lagrangian triad. Substitution of $b_0 = b_2 = 1/2$, $b_1 = 1$, and (3.1.30b) into (3.1.24) also yields the same tensor form for $\mathbf{S}^{\mathrm{B}}$.

3.1.8. Stress Conjugate to Logarithmic Strain Measure

Here, $E_a^l = \ln \lambda_a$. Then

$$S_{ab}^{l} = \frac{\sinh(E_a^l - E_b^l)}{E_a^l - E_b^l}\,\tau_{ab}. \tag{3.1.32a}$$

Substitution of

$$\mathbf{E} = \mathbf{E}^l\,, \quad \mathrm{g(E)} = \frac{1}{2}\{\exp(2\,\mathrm{E}) - 1\}\,,$$

into (3.1.21)-(3.1.23) yields a coordinate-independent expression for $\mathbf{S}^l$. For an approximate expression, substitute $f(\lambda) = ln\,(\lambda)$ into (3.1.29), to obtain

$$\mathbf{S}^l = \mathbf{S}^\mathrm{P} + \mathbf{E}^l\,\mathbf{S}^\mathrm{P} + \mathbf{S}^\mathrm{P}\,\mathbf{E}^l + \mathrm{O}\{(\mathbf{E}^l)^2\}\,. \tag{3.1.32b}$$

3.1.9. Stresses Conjugate to Other Strain Measures

For the Almansi strain measure, $E_a^A = (1 - \lambda_a^{-2})/2$. Then

$$S_{ab}^C = \lambda_a\,\lambda_b\,\tau_{ab} \tag{3.1.33a}$$

which is the component representation of

$$\mathbf{S}^\mathrm{C} = \mathbf{U}\,\hat{\boldsymbol{\tau}}\,\mathbf{U} = \mathbf{F}^\mathrm{T}\boldsymbol{\tau}\,\mathbf{F}\,, \tag{3.1.33b}$$

with respect to the Lagrangian triad. The stress tensor $\mathbf{S}^\mathrm{C} = \mathbf{F}^\mathrm{T}\boldsymbol{\tau}\,\mathbf{F}$, conjugate to the Almansi strain measure $\mathbf{E}^\mathrm{A}$, is called the *convected stress* (Truesdell and Toupin, 1960).

3.1.10. Nominal Stress

The stress tensor conjugate to the deformation gradient $\mathbf{F}$ is called the *nominal stress* (Hill, 1968). Since $\mathbf{L} = \dot{\mathbf{F}}\,\mathbf{F}^{-1}$, it follows that

$$\begin{aligned}\mathrm{tr}\,(\boldsymbol{\tau}\,\mathbf{D}) = \mathrm{tr}\,(\boldsymbol{\tau}\,\mathbf{L}) &= \mathrm{tr}\,(\mathrm{J}\,\mathbf{F}^{-1}\,\boldsymbol{\sigma}\,\dot{\mathbf{F}}) \\ &\equiv \mathrm{tr}\,(\mathbf{S}^\mathrm{N}\,\dot{\mathbf{F}})\,.\end{aligned} \tag{3.1.34a,b}$$

Hence, the nominal stress, $\mathbf{S}^\mathrm{N}$, is given by

$$\begin{aligned}&\mathbf{S}^\mathrm{N} = \mathbf{F}^{-1}\,\boldsymbol{\tau} = \mathrm{J}\,\mathbf{F}^{-1}\,\boldsymbol{\sigma}\,, \\ &S_{Ai}^N = F_{Aj}^{-1}\,\tau_{ji} = J\,F_{Aj}^{-1}\,\sigma_{ji}\,.\end{aligned} \tag{3.1.35a,b}$$

Let da be an elementary area with unit normal $\mathbf{n}$. Let dA with unit normal $\mathbf{N}$ be the initial configuration of this elementary area. Define by $\mathbf{df}^{(n)}$ the total force acting on the elementary area da,

$$\mathbf{df}^{(n)} \equiv \mathbf{t}^{(n)}\,\mathrm{da} = (\mathbf{n}\,\boldsymbol{\sigma})\,\mathrm{da}\,, \tag{3.1.36}$$

where $\mathbf{t}^{(n)}$ is the traction vector (measured per unit area) acting on an elementary area with unit normal $\mathbf{n}$. Substitute (2.2.30) and (3.1.35) into (3.1.36) to arrive at

$$\mathbf{df}^{(n)} = (\mathbf{N}\,\mathbf{S}^\mathrm{N})\,\mathrm{dA} = \mathrm{d}\mathbf{A}\,\mathbf{S}^\mathrm{N}\,, \tag{3.1.37}$$

where $\mathrm{d}\mathbf{A} = \mathbf{N}\,\mathrm{dA}$. Hence, $\mathbf{N}\,\mathbf{S}^\mathrm{N}$ measures tractions per unit initial area of an element with the initial unit normal $\mathbf{N}$. The transpose of $\mathbf{S}^\mathrm{N}$ is called the *first*

Piola-Kirchhoff stress (Truesdell and Toupin, 1960),

$$\mathbf{T}^R = (\mathbf{S}^N)^T,$$

$$T^R_{iA} = J\sigma_{ij}\, F^{-1}_{Aj}\,. \qquad (3.1.38a,b)$$

3.1.11. Summary of Relations Among Various Stress Measures

Table 3.1.1 gives the expressions for various commonly used stress tensors introduced in this section, and relates these expressions to each other.

Table 3.1.1

Relations among various stress measures

Name, Notation	$\boldsymbol{\sigma}$	$\boldsymbol{\tau}$	$\mathbf{S}^N$	$\mathbf{T}^R$	$\mathbf{S}^P$
Cauchy $\boldsymbol{\sigma}$	$\boldsymbol{\sigma}$	$\frac{1}{J}\boldsymbol{\tau}$	$\frac{1}{J}\mathbf{F}\mathbf{S}^N$	$\frac{1}{J}\mathbf{T}^R\mathbf{F}^T$	$\frac{1}{J}\mathbf{F}\mathbf{S}^P\mathbf{F}^T$
Kirchhoff $\boldsymbol{\tau}$	$J\boldsymbol{\sigma}$	$\boldsymbol{\tau}$	$\mathbf{F}\mathbf{S}^N$	$\mathbf{T}^R\mathbf{F}^T$	$\mathbf{F}\mathbf{S}^P\mathbf{F}^T$
Nominal $\mathbf{S}^N$	$J\mathbf{F}^{-1}\boldsymbol{\sigma}$	$\mathbf{F}^{-1}\boldsymbol{\tau}$	$\mathbf{S}^N$	$(\mathbf{T}^R)^T$	$\mathbf{S}^P\mathbf{F}^T$
1st Piola-Kirchhoff $\mathbf{T}^R$	$J\boldsymbol{\sigma}\mathbf{F}^{-T}$	$\boldsymbol{\tau}\mathbf{F}^{-T}$	$(\mathbf{S}^N)^T$	$\mathbf{T}^R$	$\mathbf{F}\mathbf{S}^P$
2nd Piola-Kirchhoff $\mathbf{S}^P$	$J\mathbf{F}^{-1}\boldsymbol{\sigma}\mathbf{F}^{-T}$	$\mathbf{F}^{-1}\boldsymbol{\tau}\mathbf{F}^{-T}$	$\mathbf{S}^N\mathbf{F}^{-T}$	$\mathbf{F}^{-1}\mathbf{T}^R$	$\mathbf{S}^P$

3.2. PHYSICAL RELATIONS AMONG COMMONLY USED STRESS MEASURES

To examine the physical relations among the commonly used stress measures, consider a convected system of coordinates which is embedded in, and deforms with the material. Consider a special case[1] when the particles are marked by their initial rectangular Cartesian coordinates. A material line element initially parallel to a given coordinate axis then deforms into a *curvilinear coordinate*. This is exemplified in Figure 3.2.1. From this figure define the *covariant base vectors* of the curvilinear coordinates by

$$\mathbf{g}_A = \mathbf{F}\,\mathbf{e}_A = F_{iA}\,\mathbf{e}_i = x_{i,A}\,\mathbf{e}_i\,, \qquad A, i = 1, 2, 3\,. \qquad (3.2.1a\text{-}c)$$

The corresponding *metric tensor* is (see (2.2.3)),

[1] The results obtained in this section can be extended to cases where the embedded coordinates are not initially rectangular and Cartesian; see Truesdell and Toupin (1960).

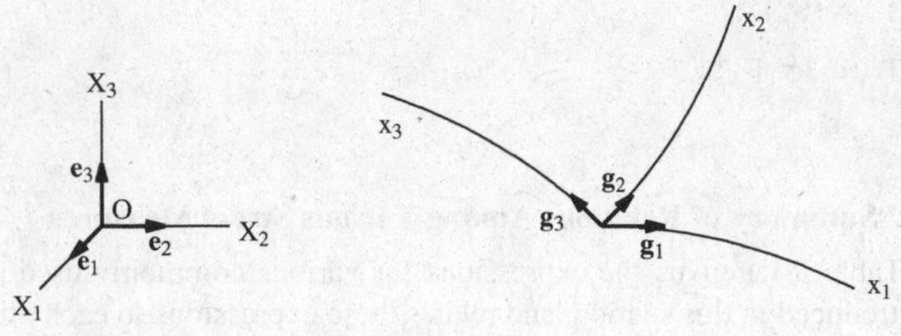

Figure 3.2.1

Embedded initially rectangular Cartesian coordinate system $\mathbf{X}_A = X_A \mathbf{e}_A$ deforms with the material into curvilinear coordinates with covariant base vectors $\mathbf{g}_A$, A = 1, 2, 3

$$\mathbf{g}_A \cdot \mathbf{g}_B = x_{i,A}\, x_{i,B} \equiv C_{AB}\,, \tag{3.2.2}$$

i.e., the components of the right Cauchy-Green tensor.

Denote the corresponding *contravariant base vectors* which are reciprocal to the covariant base vectors, by $\mathbf{g}^A$, A = 1, 2, 3. These are defined such that

$$\mathbf{g}^A \cdot \mathbf{g}_B = \delta^A_B\,, \tag{3.2.3}$$

where δ^A_B is the Kronecker delta. Substitution of (3.2.1) into (3.2.3) results in

$$(\mathbf{g}^A \cdot \mathbf{e}_i)\, F_{iB} = \delta^A_B \tag{3.2.4a}$$

which yields

$$\mathbf{g}^A \cdot \mathbf{e}_i = F^{-1}_{Ai}\,, \quad \mathbf{g}^A = F^{-1}_{Ai}\, \mathbf{e}_i\,. \tag{3.2.4b,c}$$

Consider an infinitesimal rectangular parallelepiped at a typical material point P. In the undeformed configuration, the edges of this parallelepiped are parallel to the rectangular Cartesian coordinates. The homogeneous deformation, defined by $\mathbf{F}$(P), maps this parallelepiped into an infinitesimal oblique parallelepiped whose edges are along the covariant base vectors at P. The faces of this parallelepiped are normal to the contravariant base vectors; see Figure 3.2.2. For example, the face formed by $\mathbf{g}_1$ and $\mathbf{g}_2$ is normal to $\mathbf{g}^3$, with other relations following a cyclic permutation of 1, 2, 3.

Denote the initial area of the face of the parallelepiped which is initially normal to $\mathbf{e}_A$, by $d\mathbf{A}_A = dA_A\, \mathbf{e}_A$ (subscript A not summed). Then, in the deformed configuration, it becomes

$$d\mathbf{a}_A = \frac{1}{2}\, dA_A\, e_{ABC}\, \mathbf{g}_B \times \mathbf{g}_C\,, \quad \text{(A not summed)}\,, \tag{3.2.5}$$

where e_{ABC} is the permutation symbol. Substitute (3.2.1) into (3.2.5) and use

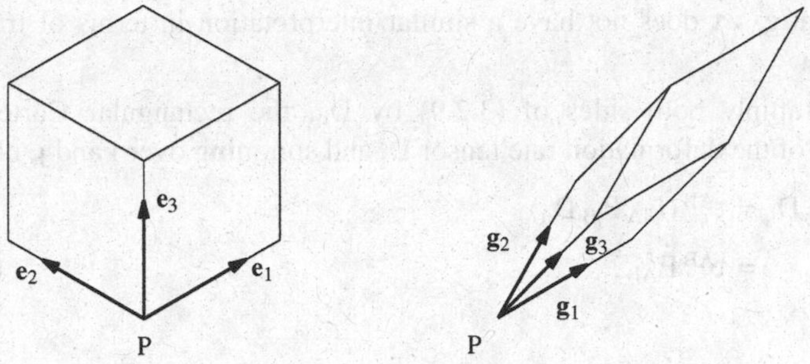

Figure 3.2.2

A rectangular parallelepiped deforms into an oblique parallelepiped

(2.1.4b) and (3.2.4c) to obtain

$$d\mathbf{a}_A = J\,dA_A\,F_{Ai}^{-1}\,\mathbf{e}_i = J\,dA_A\,\mathbf{g}^A, \quad \text{(A not summed)}. \tag{3.2.6}$$

This shows that the vector $J\,\mathbf{g}^A$ defines the current area of a unit area normal to $\mathbf{e}_A$ in the initial configuration.

Denote by $\mathbf{T}^A$ the tractions per unit reference area which is initially normal to $\mathbf{e}_A$. Then

$$\mathbf{T}^A\,dA_A = d\mathbf{a}_A \cdot \boldsymbol{\sigma} = J\,dA_A\,\mathbf{g}^A \cdot \boldsymbol{\sigma}, \quad \text{(A not summed)},$$

which yields

$$\mathbf{T}^A = \mathbf{g}^A \cdot \boldsymbol{\tau}. \tag{3.2.7}$$

3.2.1. Contravariant Components of Kirchhoff Stress

Denote the components of the Kirchhoff stress $\boldsymbol{\tau}$, with respect to the convected coordinates with covariant base vectors $\mathbf{g}_A$, by τ^{AB},

$$\boldsymbol{\tau} = \tau^{AB}\,\mathbf{g}_A \otimes \mathbf{g}_B, \quad \tau^{AB} = \mathbf{g}^A \cdot \boldsymbol{\tau}\,\mathbf{g}^B = \mathbf{T}^A \cdot \mathbf{g}^B = \mathbf{g}^A \cdot \mathbf{T}^B. \tag{3.2.8a,b}$$

Now, in view of (3.2.1),

$$\tau_{ij} = J\,\sigma_{ij} = F_{iA}\,F_{jB}\,\tau^{AB}. \tag{3.2.9}$$

Comparison of this result with (3.1.30a) shows that the matrix of the contravariant components τ^{AB} of the Kirchhoff stress $\boldsymbol{\tau}$, *i.e.* $[\tau^{AB}]$, is equal to the matrix of the rectangular Cartesian components S^P_{AB} of the second Piola-Kirchhoff stress tensor, *i.e.*, $[\tau^{AB}] = [S^P_{AB}]$. Note, however, that the two stress *tensors* are *not* equal, since they do not have the same components in the *same* coordinate system, *i.e.*, $\boldsymbol{\tau} \neq \mathbf{S}^P$, as can be seen from (3.1.8). Thus, it follows from (3.2.8b) that $S^P_{AB} = \tau^{AB}$, B = 1, 2, 3, are the *contravariant components* of the tractions per unit reference area normal to $\mathbf{e}_A$,

$$\mathbf{T}^A = \tau^{AB}\,\mathbf{g}_B = S^P_{AB}\,\mathbf{g}_B\,. \tag{3.2.10}$$

Note that $\mathbf{g}_A \cdot \boldsymbol{\tau}$ does not have a similar interpretation in terms of tractions per unit area.

Multiply both sides of (3.2.9) by D_{ij}, the rectangular Cartesian components of the deformation rate tensor $\mathbf{D}$, and summing over i and j, obtain

$$\tau_{ij}\,D_{ij} = \tau^{AB}\,(F_{iA}\,F_{jB}\,D_{ij}) = \tau^{AB}\,\dot{E}^L_{AB}\,, \tag{3.2.11a,b}$$

where

$$\dot{E}^L_{AB} = F_{iA}\,F_{jB}\,D_{ij} \tag{3.2.12a}$$

are the rectangular Cartesian components of the Lagrangian strain rate (see (3.1.7)),

$$\dot{\mathbf{E}}^L = \mathbf{F}^T\,\mathbf{D}\,\mathbf{F}\,. \tag{3.2.12b}$$

From (3.2.11) and (3.1.8), it now follows that

$$\mathrm{tr}(\boldsymbol{\tau}\,\mathbf{D}) = \tau^{AB}\,\dot{E}^L_{AB} = S^P_{AB}\,\dot{E}^L_{AB} = \mathrm{tr}(\mathbf{S}^P\,\dot{\mathbf{E}}^L)\,. \tag{3.2.13a,b}$$

3.2.2. Cartesian-contravariant Components of Kirchhoff Stress

Now, consider the mixed coordinate bases $\mathbf{e}_i \otimes \mathbf{g}_A$, i = 1, 2, 3. Denote the components of the Kirchhoff stress with respect to these coordinate bases by $\tau_i^{\,A}$, *i.e.*,

$$\boldsymbol{\tau} = \tau_i^{\,A}\,\mathbf{e}_i \otimes \mathbf{g}^A = \tau_i^{\,A}\,\mathbf{g}_A \otimes \mathbf{e}_i\,,$$

$$\tau_i^{\,A} = \mathbf{e}_i \cdot \boldsymbol{\tau}\,\mathbf{g}^A = \mathbf{g}^A \cdot \boldsymbol{\tau}\,\mathbf{e}_i = \mathbf{T}^A \cdot \mathbf{e}_i\,. \tag{3.2.14a,b}$$

Then, in view of (3.2.1),

$$\boldsymbol{\tau} = \tau_{ij}\,\mathbf{e}_i \otimes \mathbf{e}_j\,, \quad \tau_{ij} = \tau_{ji} = \tau_i^{\,A}\,F_{jA}\,. \tag{3.2.14c,d}$$

Comparison of this result with (3.1.35) and (3.1.38) yields

$$S^N_{Ai} = T^R_{iA} = \tau_i^{\,A}\,, \tag{3.2.15}$$

where S^N_{Ai} and T^R_{iA} are the rectangular Cartesian components of the *nominal stress*, $\mathbf{S}^N$, and the *first Piola-Kirchhoff stress*, $\mathbf{T}^R$, respectively. It follows from (3.2.14) and (3.2.15) that $S^N_{Ai} = T^R_{iA}$ are the rectangular Cartesian components of the tractions per unit reference area normal to $\mathbf{e}_A$,

$$\mathbf{T}^A = S^N_{Ai}\,\mathbf{e}_i = T^R_{iA}\,\mathbf{e}_i\,. \tag{3.2.16}$$

Multiply (3.2.14c) by L_{ij} and use the symmetry condition $\tau_{ij} = \tau_{ji}$, to obtain

$$\tau_{ij} D_{ij} = \tau_i^A \dot{F}_{iA}$$
$$= S^N_{Ai} \dot{F}_{iA} = \mathrm{tr}(\mathbf{S}^N \dot{\mathbf{F}}), \qquad (3.2.17a,b)$$

where $\dot{F}_{iA}$ are the rectangular Cartesian components of $\dot{\mathbf{F}}$.

3.2.3. Covariant-contravariant Components of Kirchhoff Stress

Now, consider the components $\tilde{\tau}^A_B$ of the Kirchhoff stress $\boldsymbol{\tau}$ with respect to the reciprocal bases $\mathbf{g}_A \otimes \mathbf{g}^B$, A, B = 1, 2, 3, as follows:

$$\boldsymbol{\tau} = \tilde{\tau}^A_B\, \mathbf{g}_A \otimes \mathbf{g}^B, \quad \tilde{\tau}^A_B = \mathbf{g}^A \cdot \boldsymbol{\tau}\, \mathbf{g}_B = \mathbf{T}^A \cdot \mathbf{g}_B. \qquad (3.2.18a,b)$$

Then, in view of (3.2.1) and (3.2.4c), it follows that $\tilde{\tau}^A_B$ are the rectangular Cartesian components of the stress tensor $\mathbf{S}^F$, defined by

$$\mathbf{S}^F \equiv \mathbf{F}^{-1} \boldsymbol{\tau} \mathbf{F}, \quad S^F_{AB} = F^{-1}_{Ai} \tau_{ij} F_{jB}. \qquad (3.2.19a,b)$$

It is seen from (3.2.18) and (3.2.19) that S^F_{AB}, B = 1, 2, 3, are the *covariant components* of the tractions per unit reference area normal to $\mathbf{e}_A$.

The strain rate $\dot{\mathbf{E}}^F$ conjugate to the stress $\mathbf{S}^F$ is obtained from

$$\mathrm{tr}(\mathbf{S}^F \dot{\mathbf{E}}^F) = \mathrm{tr}(\boldsymbol{\tau} \mathbf{L}) = \mathrm{tr}(\mathbf{L} \boldsymbol{\tau}) = \mathrm{tr}(\dot{\mathbf{F}} \mathbf{S}^F \mathbf{F}^{-1})$$

which yields

$$\dot{\mathbf{E}}^F = \mathbf{F}^{-1} \dot{\mathbf{F}}. \qquad (3.2.20)$$

The strain measure $\mathbf{E}^F$ is an isotropic function of the deformation gradient $\mathbf{F}$. However, a closed-form expression for $\mathbf{E}^F$ in terms of $\mathbf{F}$ does not seem readily obtainable.

3.2.4. Covariant Components of Kirchhoff Stress

Denote the components (*covariant*) of the Kirchhoff stress $\boldsymbol{\tau}$ with respect to the *contravariant base vectors* $\mathbf{g}^A$, by $\tilde{\tau}_{AB}$,

$$\boldsymbol{\tau} = \tilde{\tau}_{AB}\, \mathbf{g}^A \otimes \mathbf{g}^B, \quad \tilde{\tau}_{AB} = \mathbf{g}_A \cdot \boldsymbol{\tau}\, \mathbf{g}_B. \qquad (3.2.21a,b)$$

Note that, $\mathbf{g}_A \cdot \boldsymbol{\tau}$ is not related to the force per unit area.

In view of (3.2.4c),

$$\boldsymbol{\tau} = \tau_{ij}\, \mathbf{e}_i \otimes \mathbf{e}_j, \quad \tau_{ij} = F^{-T}_{iA} \tilde{\tau}_{AB} F^{-1}_{Bj}. \qquad (3.2.21c,d)$$

Comparison of this result with (3.1.33) shows that the matrix of the covariant components $\tilde{\tau}_{AB}$ of the Kirchhoff stress $\boldsymbol{\tau}$ is equal to the matrix of the rectangular Cartesian components S^C_{AB} of the convected stress measure $\mathbf{S}^C$, *i.e.*,

$$[\tilde{\tau}_{AB}] = [S^C_{AB}]. \qquad (3.2.21e)$$

Multiply both sides of (3.2.21d) by D_{ij}, sum on i and j, and obtain

$$\tau_{ij} D_{ij} = \tilde{\tau}_{AB} (F^{-1}_{Ai} D_{ij} F^{-T}_{jB})$$

$$= S^{C}_{AB}\,\dot{E}^{A}_{AB}\,, \qquad (3.2.22a,b)$$

where

$$\dot{E}^{A}_{AB} = F^{-1}_{Ai}\,D_{ij}\,F^{-T}_{jB} \qquad (3.2.22c)$$

are the rectangular Cartesian components of the Almansi strain rate.

3.2.5. Cartesian-covariant Components of Kirchhoff Stress

Denote the components of the Kirchhoff stress with respect to the mixed coordinate bases $\mathbf{e}_i \otimes \mathbf{g}^A$ by $\tilde{\tau}^i_A$,

$$\boldsymbol{\tau} = \tilde{\tau}^i_A\,\mathbf{e}_i \otimes \mathbf{g}^A\,,$$

$$\tilde{\tau}^i_A = \mathbf{e}_i \cdot \boldsymbol{\tau}\,\mathbf{g}_A\,. \qquad (3.2.23a,b)$$

Then, in view of (3.2.4c),

$$\boldsymbol{\tau} = \tau_{ij}\,\mathbf{e}_i \otimes \mathbf{e}_j\,,$$

$$\tau_{ij} = \tau_{ji} = \tilde{\tau}^i_A\,F^{-1}_{Aj}\,. \qquad (3.2.23c,d)$$

Therefore, the components $\tilde{\tau}^i_A$ are equal to the corresponding rectangular Cartesian components of the tensors $\mathbf{T}^n$ and $\mathbf{S}^n$, defined by,[2]

$$\mathbf{T}^n \equiv \boldsymbol{\tau}\,\mathbf{F}\,,$$

$$\mathbf{S}^n \equiv \mathbf{F}^T\,\boldsymbol{\tau}\,, \qquad (3.2.24a,b)$$

respectively, *i.e.*,

$$\tilde{\tau}^i_A = T^n_{iA} = S^n_{Ai}\,. \qquad (3.2.24c)$$

Multiply (3.2.23d) by L_{ij}, use $\tau_{ij} = \tau_{ji}$, sum on i and j, and obtain

$$\tau_{ij}\,D_{ij} = \tilde{\tau}^i_A\,F^{-1}_{Aj}\,L_{ij}$$

$$= \mathrm{tr}(\mathbf{S}^n\,\dot{\mathbf{E}}^n)\,, \qquad (3.2.25a,b)$$

where $\dot{\mathbf{E}}^n$ is the material time derivative of the *nonsymmetric* strain measure $\mathbf{E}^n$, conjugate to the *convected nominal stress* $\mathbf{S}^n$,

$$\dot{\mathbf{E}}^n = \mathbf{F}^{-1}\,\dot{\mathbf{F}}\,\mathbf{F}^{-1}\,, \qquad \mathbf{E}^n \equiv \mathbf{1} - \mathbf{F}^{-1}\,. \qquad (3.2.26a,b)$$

---

[2] Note that the *nominal* and the *first Piola-Kirchhoff* stress tensors, $\mathbf{S}^N$ and $\mathbf{T}^R$, relate to $\mathbf{S}^n$ and $\mathbf{T}^n$ by $\mathbf{S}^n = \mathbf{F}^T\mathbf{F}\,\mathbf{S}^N = \mathbf{C}\,\mathbf{S}^N$, and $\mathbf{T}^n = \mathbf{T}^R\,\mathbf{F}^T\mathbf{F} = \mathbf{T}^R\,\mathbf{C}$; see Table 3.1.1.

3.3. STRESS-RATE MEASURES

In this section, several commonly used stress-rate measures are discussed. In Section 3.4, relations among various objective stress-rate measures are examined. Stress rates suitable for the description of constitutive relations for materials must be invariant under the rigid spin of the coordinates. A stress rate which satisfies this requirement is called *objective*.

3.3.1. Jaumann Stress Rate

The most commonly used objective stress rate is the *Jaumann* stress rate which measures the stress rate in a coordinate system corotational with the material element, *i.e.*, the spin of the coordinate system is the spin $\mathbf{W}$ of the material neighborhood, where $\mathbf{W}$ is defined by (2.5.8). Consider, for example, the Kirchhoff stress $\boldsymbol{\tau}$ with components τ_{ij} in the fixed rectangular Cartesian coordinate system with coordinate base vectors $\mathbf{e}_i$. At a point in a material neighborhood, let $\mathbf{e}_i^*$ be another set of rectangular Cartesian base vectors, currently coincident with the fixed base vectors $\mathbf{e}_i$, but spinning with the material neighborhood. Therefore,

$$\dot{\mathbf{e}}_i^* = \mathbf{W}\,\mathbf{e}_i^* . \tag{3.3.1}$$

Since

$$\boldsymbol{\tau} = \tau_{ij}\,\mathbf{e}_i \otimes \mathbf{e}_j = \tau_{ij}^*\,\mathbf{e}_i^* \otimes \mathbf{e}_j^* , \tag{3.3.2}$$

the use of (3.3.1) in the material time derivative of $\boldsymbol{\tau}$ results in

$$\dot{\boldsymbol{\tau}} = \overset{\circ}{\boldsymbol{\tau}} + \mathbf{W}\,\boldsymbol{\tau} + \boldsymbol{\tau}\,\mathbf{W}^{\mathrm{T}} , \tag{3.3.3}$$

where

$$\overset{\circ}{\boldsymbol{\tau}} \equiv \dot{\tau}_{ij}^*\,\mathbf{e}_i^* \otimes \mathbf{e}_j^* \tag{3.3.4}$$

is the Jaumann rate of the Kirchhoff stress. Therefore,

$$\overset{\circ}{\boldsymbol{\tau}} = \dot{\boldsymbol{\tau}} - \mathbf{W}\,\boldsymbol{\tau} - \boldsymbol{\tau}\,\mathbf{W}^{\mathrm{T}} \quad \text{or} \quad \overset{\circ}{\tau}_{ij} = \dot{\tau}_{ij} - W_{ik}\,\tau_{kj} - W_{jk}\,\tau_{ki} . \tag{3.3.5a,b}$$

3.3.2. Convected Stress Rate

Other objective stress-rate measures can be constructed in a manner similar to that discussed above. Two particularly interesting objective stress rates are the *convected stress rates* associated with the rate of change of the covariant and contravariant components of the stress tensor in a coordinate system which is instantaneously spinning *and deforming* with the material neighborhood. Consider again the Kirchhoff stress, as in (3.3.2), but now let the starred coordinate system spin *and deform* with the material, such that

$$\dot{\mathbf{e}}_i^* = \mathbf{L}\,\mathbf{e}_i^* , \tag{3.3.6}$$

where $\mathbf{L}$ is the velocity gradient. Denote the Kirchhoff stress rate in the starred coordinate system by

$$\overset{\nabla}{\boldsymbol{\tau}} \equiv \dot{\tau}^*_{ij}\,\mathbf{e}_i^* \otimes \mathbf{e}_j^*\,, \tag{3.3.7}$$

and, following a procedure similar to that of Subsection 3.3.1, obtain

$$\overset{\nabla}{\boldsymbol{\tau}} = \dot{\boldsymbol{\tau}} - \mathbf{L}\,\boldsymbol{\tau} - \boldsymbol{\tau}\,\mathbf{L}^{\mathrm{T}}. \tag{3.3.8}$$

The material time derivative of (3.1.30b) leads to

$$\dot{\mathbf{S}}^{\mathrm{P}} = \mathbf{F}^{-1}\{\dot{\boldsymbol{\tau}} - \mathbf{L}\,\boldsymbol{\tau} - \boldsymbol{\tau}\,\mathbf{L}^{\mathrm{T}}\}\mathbf{F}^{-\mathrm{T}}. \tag{3.3.9a}$$

Substitute (3.3.9a) into (3.3.8) to obtain

$$\overset{\nabla}{\boldsymbol{\tau}} = \mathbf{F}\,\dot{\mathbf{S}}^{\mathrm{P}}\,\mathbf{F}^{\mathrm{T}} = \mathbf{F}\,(\mathbf{F}^{-1}\,\boldsymbol{\tau}\,\mathbf{F}^{-\mathrm{T}})^{\boldsymbol{\cdot}}\,\mathbf{F}^{\mathrm{T}}. \tag{3.3.9b}$$

Since the rectangular Cartesian components of the *second Piola-Kirchhoff* stress are the contravariant components of the Kirchhoff stress, the stress rate $\overset{\nabla}{\boldsymbol{\tau}}$ may be called *the contravariant Kirchhoff stress rate*.

Another objective stress rate is obtained by considering the rate of change of the stress in a convected coordinate system which is instantaneously coincident with the background fixed rectangular Cartesian coordinate system but *deforms* such that

$$\dot{\mathbf{e}}_i^* = -\mathbf{L}^{\mathrm{T}}\mathbf{e}_i^*\,. \tag{3.3.10}$$

Denote the corresponding objective Kirchhoff stress rate by

$$\overset{\nabla}{\boldsymbol{\tau}}{}^* \equiv \dot{\tau}^*_{ij}\,\mathbf{e}_i^* \otimes \mathbf{e}_j^*\,, \tag{3.3.11a}$$

and obtain

$$\overset{\nabla}{\boldsymbol{\tau}}{}^* = \dot{\boldsymbol{\tau}} + \mathbf{L}^{\mathrm{T}}\,\boldsymbol{\tau} + \boldsymbol{\tau}\,\mathbf{L}\,. \tag{3.3.11b}$$

The material time derivative of (3.1.33b) yields

$$\dot{\mathbf{S}}^{\mathrm{C}} = \mathbf{F}^{\mathrm{T}}\{\dot{\boldsymbol{\tau}} + \mathbf{L}^{\mathrm{T}}\,\boldsymbol{\tau} + \boldsymbol{\tau}\,\mathbf{L}\}\,\mathbf{F}\,. \tag{3.3.12a}$$

Substitute (3.3.12a) into (3.3.11b) to obtain

$$\begin{aligned} \overset{\nabla}{\boldsymbol{\tau}}{}^* &= \mathbf{F}^{-\mathrm{T}}\,\dot{\mathbf{S}}^{\mathrm{C}}\,\mathbf{F}^{-1} \\ &= \mathbf{F}^{-\mathrm{T}}\,(\mathbf{F}^{\mathrm{T}}\,\boldsymbol{\tau}\,\mathbf{F})^{\boldsymbol{\cdot}}\,\mathbf{F}^{-\mathrm{T}}. \end{aligned} \tag{3.3.12b,c}$$

Since the rectangular Cartesian components of the *convected stress* $\mathbf{S}^{\mathrm{C}}$ are the covariant components of the Kirchhoff stress, this stress rate may be called *the covariant Kirchhoff stress rate.*

It is interesting to note that the Jaumann rate of the Kirchhoff stress is the average of the covariant and contravariant Kirchhoff stress rates, *i.e.*,

$$\overset{\circ}{\boldsymbol{\tau}} = \frac{1}{2}\,(\overset{\nabla}{\boldsymbol{\tau}} + \overset{\nabla}{\boldsymbol{\tau}}{}^*)\,. \tag{3.3.13}$$

Note that all the above results also apply to the Cauchy stress, $\boldsymbol{\sigma}$. For example, the Jaumann rate of change of $\boldsymbol{\sigma}$ is

$$\overset{\circ}{\boldsymbol{\sigma}} = \dot{\boldsymbol{\sigma}} - \mathbf{W}\boldsymbol{\sigma} - \boldsymbol{\sigma}\mathbf{W}^{\mathrm{T}}. \tag{3.3.14}$$

3.4. GENERAL RESULTS ON OBJECTIVE STRESS RATES

In general, *the material rate of change of any stress tensor measured with respect to the Lagrangian triad is objective.* Consider, for example, the Kirchhoff stress,

$$\boldsymbol{\tau} = \sum_{a,\,b=1}^{3} \tau_{ab}\, \mathbf{n}^a \otimes \mathbf{n}^b, \tag{3.4.1}$$

and express it in the Lagrangian triad, as

$$\hat{\boldsymbol{\tau}} = \mathbf{R}^{\mathrm{T}} \boldsymbol{\tau} \mathbf{R} = \sum_{a,\,b=1}^{3} \tau_{ab}\, \mathbf{N}^a \otimes \mathbf{N}^b. \tag{3.4.2a,b}$$

Direct material time differentiation of (3.4.2) results in

$$\begin{aligned} \dot{\hat{\boldsymbol{\tau}}} &= \mathbf{R}^{\mathrm{T}} \dot{\boldsymbol{\tau}} \mathbf{R} - \mathbf{R}^{\mathrm{T}} \boldsymbol{\Omega}^{\mathrm{R}} \boldsymbol{\tau} \mathbf{R} + \mathbf{R}^{\mathrm{T}} \boldsymbol{\tau} \boldsymbol{\Omega}^{\mathrm{R}} \mathbf{R} \\ &= \dot{\hat{\boldsymbol{\tau}}} + \hat{\boldsymbol{\tau}} \hat{\boldsymbol{\Omega}}^{\mathrm{R}} - \hat{\boldsymbol{\Omega}}^{\mathrm{R}} \hat{\boldsymbol{\tau}}, \end{aligned} \tag{3.4.3a,b}$$

where (2.6.3) is also used. The stress rate $\dot{\hat{\boldsymbol{\tau}}}$ is an objective stress flux, measured in the Lagrangian triad. With respect to the Eulerian triad, this becomes

$$\overset{\Delta}{\boldsymbol{\tau}} \equiv \mathbf{R} \dot{\hat{\boldsymbol{\tau}}} \mathbf{R}^{\mathrm{T}} = \dot{\boldsymbol{\tau}} + \boldsymbol{\tau} \boldsymbol{\Omega}^{\mathrm{R}} - \boldsymbol{\Omega}^{\mathrm{R}} \boldsymbol{\tau} \tag{3.4.4a,b}$$

which is an objective rate of change for the Kirchhoff stress. Similar expressions can be obtained for the Cauchy stress (replace $\boldsymbol{\tau}$ in the above equations by $\boldsymbol{\sigma}$). It is interesting to note the relation between the stress flux $\overset{\Delta}{\boldsymbol{\tau}}$ and the corresponding Jaumann rate. A direct connection is obtained if (2.8.6) is substituted into (3.4.4) (Nemat-Nasser, 1983),

$$\overset{\Delta}{\boldsymbol{\tau}} = \overset{\circ}{\boldsymbol{\tau}} + \boldsymbol{\tau} \boldsymbol{\Delta} - \boldsymbol{\Delta} \boldsymbol{\tau}, \tag{3.4.5}$$

where $\boldsymbol{\Delta}$ is given by (2.8.6b). If the current state is chosen as the reference one, then, instantaneously, $\mathbf{V} = \mathbf{1}$, and hence, $\boldsymbol{\Delta} = \mathbf{0}$, and the stress flux (3.4.4) reduces to the Jaumann rate.

3.4.1. General Relation Connecting Objective Stress Rates

The material time derivative of a general stress measure $\mathbf{S}$ conjugate to any material strain measure $\mathbf{E}$ can be obtained by differentiating (3.1.14b) with respect to time, as follows:

$$\dot{S}^{P}_{ij} = \frac{\partial^2 E^{L}_{ij}}{\partial E_{mn}\, \partial E_{kl}}\, S^{P}_{kl}\, \dot{E}_{mn} + \frac{\partial E^{L}_{ij}}{\partial E_{kl}}\, \dot{S}^{P}_{kl}. \tag{3.4.6a}$$

For a coordinate-invariant expression, note that $\mathbf{h}(\mathbf{E}, \mathbf{S}^P)$ in (3.1.19b) is linear and homogeneous in $\mathbf{S}^P$. Take the material time derivative of (3.1.19b) to arrive at

$$\dot{\mathbf{S}} = \frac{\partial \mathbf{h}(\mathbf{E}, \mathbf{S}^P)}{\partial \mathbf{E}} : \dot{\mathbf{E}} + \mathbf{h}(\mathbf{E}, \dot{\mathbf{S}}^P). \tag{3.4.6b}$$

The second term in the right-hand side of this equation is obtained by simply substituting $\dot{\mathbf{S}}^P$ for $\mathbf{S}^P$ in (3.1.21)-(3.1.23). However, as can be seen from (3.1.21)-(3.1.23), the explicit expression for the first term will be rather complex.

The material time derivative of (3.1.29) yields the first-order approximate expression

$$\dot{\mathbf{S}} = \dot{\mathbf{S}}^P + \frac{1}{2}\{1 - f''(1)\}(\dot{\mathbf{E}}\mathbf{S}^P + \mathbf{S}^P\dot{\mathbf{E}}) + O(\mathbf{E}). \tag{3.4.6c}$$

A general relation between the rate of change of any two stress measures conjugate to the class of material strain measures discussed in Subsection 2.3 will now be established by considering two such strain measures, say, $\mathbf{E}$ and $\bar{\mathbf{E}}$. Let their corresponding conjugate stresses be $\mathbf{S}$ and $\bar{\mathbf{S}}$. Therefore, from (3.4.6c),

$$\begin{aligned} &\dot{\bar{\mathbf{S}}} - \frac{1}{2}\{1 - \bar{f}''(1)\}(\dot{\bar{\mathbf{E}}}\mathbf{S}^P + \mathbf{S}^P\dot{\bar{\mathbf{E}}}) + O(\bar{\mathbf{E}}) \\ &\qquad = \dot{\mathbf{S}} - \frac{1}{2}\{1 - f''(1)\}(\dot{\mathbf{E}}\mathbf{S}^P + \mathbf{S}^P\dot{\mathbf{E}}) + O(\mathbf{E}). \end{aligned} \tag{3.4.7a}$$

Since $\mathbf{E}$ and $\bar{\mathbf{E}}$ have the following general representation:

$$\begin{aligned} \mathbf{E} &= (\mathbf{U} - \mathbf{1}) + \frac{1}{2} f''(1)(\mathbf{U} - \mathbf{1})^2 + \dots, \\ \bar{\mathbf{E}} &= (\mathbf{U} - \mathbf{1}) + \frac{1}{2} \bar{f}''(1)(\mathbf{U} - \mathbf{1})^2 + \dots, \end{aligned} \tag{3.4.8a,b}$$

it follows that

$$\bar{\mathbf{E}} = \mathbf{E} + O(\mathbf{E}^2), \quad \dot{\bar{\mathbf{E}}} = \dot{\mathbf{E}} + O(\mathbf{E}). \tag{3.4.9a,b}$$

Also, from (3.1.29),

$$\mathbf{S}^P = \mathbf{S} + O(\mathbf{E}) = \bar{\mathbf{S}} + O(\bar{\mathbf{E}}) = \bar{\mathbf{S}} + O(\mathbf{E}). \tag{3.4.9c}$$

Substitution of (3.4.9) into (3.4.7a) yields

$$\begin{aligned} &\dot{\bar{\mathbf{S}}} + \frac{1}{2}\bar{f}''(1)(\bar{\mathbf{S}}\dot{\bar{\mathbf{E}}} + \dot{\bar{\mathbf{E}}}\bar{\mathbf{S}}) + O(\bar{\mathbf{E}}) \\ &\qquad = \dot{\mathbf{S}} + \frac{1}{2} f''(1)(\mathbf{S}\dot{\mathbf{E}} + \dot{\mathbf{E}}\mathbf{S}) + O(\mathbf{E}). \end{aligned} \tag{3.4.7b}$$

3.5. STRESS RATES WITH CURRENT STATE AS REFERENCE

If the current state is used as the reference one, then $\mathbf{F} = \mathbf{1}$, and $J = 1$. From Table 3.1.1 it then follows that all stress measures become identical with, say, the Cauchy stress $\boldsymbol{\sigma}$. Furthermore, the results of Sections 2.3 and 2.10 show that all considered strains are zero and their rates reduce to the deformation rate $\mathbf{D}$. Therefore, (3.4.6c) and (3.4.7b) then yield the following *exact* relations:

$$\dot{\mathbf{S}} = \dot{\mathbf{S}}^{\mathrm{P}} + \frac{1}{2}\{1 - \mathrm{f}''(1)\}\,(\mathbf{D}\boldsymbol{\sigma} + \boldsymbol{\sigma}\mathbf{D}),$$

$$\dot{\overline{\mathbf{S}}} + \frac{1}{2}\overline{\mathrm{f}}''(1)\,(\boldsymbol{\sigma}\mathbf{D} + \mathbf{D}\boldsymbol{\sigma}) = \dot{\mathbf{S}} + \frac{1}{2}\mathrm{f}''(1)\,(\boldsymbol{\sigma}\mathbf{D} + \mathbf{D}\boldsymbol{\sigma}). \qquad (3.5.1\text{a,b})$$

Consider the stress measure $\mathbf{S}^l$, which is conjugate to the logarithmic strain, $\mathbf{E}^l = ln\,\mathbf{U}$. Substitution of $\mathrm{f}(\lambda) = ln\,\lambda$ into (3.5.1a) yields

$$\dot{\mathbf{S}}^l = \dot{\mathbf{S}}^{\mathrm{P}} + \mathbf{D}\boldsymbol{\sigma} + \boldsymbol{\sigma}\mathbf{D}, \qquad (3.5.2\text{a})$$

when the current state is used as the reference one. Time differentiation of (3.1.30c) results in

$$\hat{\dot{\boldsymbol{\tau}}} = \mathbf{U}\dot{\mathbf{S}}^{\mathrm{P}}\mathbf{U} + \dot{\mathbf{U}}\mathbf{S}^{\mathrm{P}}\mathbf{U} + \mathbf{U}\mathbf{S}^{\mathrm{P}}\dot{\mathbf{U}} = \dot{\mathbf{S}}^{\mathrm{P}} + \mathbf{D}\boldsymbol{\sigma} + \boldsymbol{\sigma}\mathbf{D}. \qquad (3.5.2\text{b,c})$$

Therefore, when the current state is used as the reference one,

$$\hat{\dot{\boldsymbol{\tau}}} = \dot{\mathbf{S}}^l. \qquad (3.5.2\text{d})$$

Therefore, identify $\overline{\mathbf{S}}$ with $\mathbf{S}^l$, and substitute $\overline{\mathrm{f}}''(1) = -1$ into (3.5.1b) to arrive at the following exact relation:

$$\dot{\mathbf{S}} = \overset{\circ}{\boldsymbol{\tau}} - \frac{1}{2}\{\mathrm{f}''(1) + 1\}\,(\boldsymbol{\sigma}\mathbf{D} + \mathbf{D}\boldsymbol{\sigma}), \qquad (3.5.3)$$

where, since the current configuration is used as reference, $\mathbf{R} = \mathbf{1}$ which corresponds to $\overset{\triangle}{\boldsymbol{\tau}} = \overset{\circ}{\boldsymbol{\tau}} = \hat{\dot{\boldsymbol{\tau}}} = \dot{\mathbf{S}}^l$. The important relation (3.5.3), due to Hill (1968), shows that *a broad class of objective stress rates is equal to the Jaumann rate of the Kirchhoff stress plus objective terms which are linear in the deformation rate tensor and the Cauchy stress.*

3.5.1. Nominal Stress Rate

For the incremental formulation of large-deformation problems, the nominal stress and its rate play important roles and possess special properties. The rate of the nominal stress can be expressed in terms of the rate of any other stress measure, by direct material differentiation. For example, in terms of the Cauchy stress, $\mathbf{F}\,\mathbf{S}^{\mathrm{N}} = J\,\boldsymbol{\sigma}$, obtain, by time differentiation,

$$\dot{\mathbf{S}}^{\mathrm{N}} = \mathbf{F}^{-1}\{\dot{J}\boldsymbol{\sigma} + J(\dot{\boldsymbol{\sigma}} - \mathbf{L}\boldsymbol{\sigma})\}. \qquad (3.5.4)$$

Now, if the current configuration is used as reference, this becomes

$$\dot{\mathbf{S}}^{\mathrm{N}} = \dot{\boldsymbol{\sigma}} + \boldsymbol{\sigma}\,\mathrm{tr}\,\mathbf{D} - \mathbf{L}\boldsymbol{\sigma} = \dot{\boldsymbol{\tau}} - \mathbf{L}\boldsymbol{\tau}\;; \qquad (3.5.5\text{a,b})$$

the last expression follows from the fact that

$$\dot{\boldsymbol{\tau}} = \dot{\boldsymbol{\sigma}} + \boldsymbol{\sigma}\,\mathrm{tr}(\mathbf{D}), \tag{3.5.6}$$

when the current configuration is the reference one, so that $\boldsymbol{\sigma} = \boldsymbol{\tau}$. The nominal stress rate can now be related easily to other stress rates. For example, in terms of the Jaumann rate of the Kirchhoff stress, obtain

$$\dot{\mathbf{S}}^{\mathrm{N}} = \overset{\circ}{\boldsymbol{\tau}} - (\mathbf{D}\boldsymbol{\tau} + \boldsymbol{\tau}\mathbf{D}) + \boldsymbol{\tau}\mathbf{L}^{\mathrm{T}}. \tag{3.5.7}$$

Upon substitution into (3.5.3), find

$$\dot{\mathbf{S}}^{\mathrm{N}} = \dot{\mathbf{S}} + \frac{1}{2}\,\{f''(1) - 1\}\,(\mathbf{D}\mathbf{S} + \mathbf{S}\mathbf{D}) + \mathbf{S}\mathbf{L}^{\mathrm{T}}, \tag{3.5.8}$$

for any stress measure $\mathbf{S}$, conjugate to the strain measure $\mathbf{E}$; note that $\mathbf{S} = \boldsymbol{\tau} = \boldsymbol{\sigma} = \mathbf{S}^{\mathrm{N}}$, etc., since the current configuration is the reference one.

3.6. BALANCE RELATIONS AND BOUNDARY-VALUE PROBLEMS

Since basic balance laws are fully discussed in elementary books on continuum mechanics, only a brief summary is given here. Consider an arbitrary material element with initial volume V, bounded by surface ∂V, having a mass density (not necessarily uniform) ρ_0. The current (deformed) volume, its boundary, and the mass density are denoted, respectively, by v, ∂v, and ρ.

3.6.1. Conservation of Mass

The conservation of mass requires that

$$\int_{V} \rho_0\, dV = \int_{v} \rho\, dv \tag{3.6.1}$$

which, for suitably smooth fields, implies,

$$J \equiv \det \mathbf{F} = \rho_0 / \rho. \tag{3.6.2}$$

Time differentiation of this yields the usual spatial continuity equation,

$$\dot{\rho} + \rho\, \mathrm{tr}\,\mathbf{D} = 0. \tag{3.6.3}$$

3.6.2. Balance of Linear Momentum

The balance of linear momentum has already been given by (3.1.3) which leads to (3.1.4) when the current configuration is used as reference. From (3.1.35), (3.1.36), and (3.1.37),

$$\int_{\partial v} \mathbf{t}^{(\mathrm{n})}\, da = \int_{\partial V} \mathbf{T}^{(\mathrm{N})}\, dA, \tag{3.6.4a}$$

where

$$\mathbf{T}^{(N)} = \mathbf{N}\,\mathbf{S}^N \tag{3.6.4b}$$

is the applied traction measured per unit reference area of an element initially normal to the unit vector **N**. Substituting (3.6.4) and (2.2.24) into (3.1.3) results in

$$\int_{\partial V} \mathbf{T}^{(N)}\,dA = \int_V \rho_0\,(\mathbf{a}-\mathbf{b})\,dV\,. \tag{3.6.5}$$

From (3.6.4b), (3.6.5), the Gauss theorem, and the fact that the material element is arbitrary, it follows that

$$\boldsymbol{\nabla}^0 \cdot \mathbf{S}^N = \rho_0\,(\mathbf{a}-\mathbf{b}) \tag{3.6.6a}$$

or

$$S^N_{Ai,A} = \rho_0\,(a_i - b_i)\,. \tag{3.6.6b}$$

Note that $\boldsymbol{\nabla}^0 \equiv \mathbf{e}_A\,\partial/\partial X_A$ in (3.6.6a), and $\boldsymbol{\nabla} \equiv \mathbf{e}_i\,\partial/\partial x_i$ in (3.1.4a). Only for the nominal stress rate (and also, of course, for its transpose, *i.e.*, the first Piola-Kirchhoff stress) do the equations of motion take on the simple form given by (3.6.6) which is similar in form to the one in terms of the Cauchy stress.[3]

3.6.3. Balance of Angular Momentum

The conservation of angular momentum leads to the symmetry of the Cauchy stress, from which and Table 3.1.1, it follows that

$$\mathbf{F}\,\mathbf{S}^N = (\mathbf{S}^N)^T\,\mathbf{F}^T\,. \tag{3.6.7}$$

3.6.4. Conservation of Energy

Finally, observe that the local form of the conservation of energy may be expressed as

$$\rho\,\dot{e} = \mathrm{tr}\,(\boldsymbol{\sigma}\mathbf{D}) - \boldsymbol{\nabla}\cdot\mathbf{q} + \rho\,h\,, \tag{3.6.8}$$

where e is the internal energy per unit mass, **q** is the heat flux vector, and h is the energy added per unit mass, for instance, by radiation heating.

3.6.5. Boundary-value Problems

This section is completed by considering the significance of the nominal stress in formulating quasi-static boundary-value problems. Let the body forces,

[3] Indeed, the nominal stress and its rate, and the deformation gradient and its rate have many useful averaging properties which are discussed in Chapter 8.

b, and surface tractions, **t**, be prescribed such that

$$\boldsymbol{\nabla} \cdot \boldsymbol{\sigma} + \rho\, \mathbf{b} = \mathbf{0} \quad \text{in v}, \qquad \boldsymbol{\sigma} \cdot \mathbf{n} = \mathbf{t} \quad \text{on } \partial \mathrm{v}_1, \tag{3.6.9a,b}$$

where, on the remaining boundary, $\partial \mathrm{v} - \partial \mathrm{v}_1$, the displacement is prescribed. Suppose now, with the current configuration as reference, the body forces and the prescribed tractions are quasi-statically changed by increments associated with rates $\dot{\mathbf{b}}$ and $\dot{\mathbf{t}}$. The Cauchy stress changes at the rate $\dot{\boldsymbol{\sigma}}$, and the nominal stress changes at the rate $\dot{\mathbf{S}}^{\mathrm{N}}$. These two stress rates are related by (3.5.5a). Since now the nominal stress refers to the particles held fixed in their current positions, $\mathbf{X} = \mathbf{x}$,

$$\dot{\mathbf{S}}^{\mathrm{N}} = \frac{\partial}{\partial t}\, \mathbf{S}^{\mathrm{N}}(\mathbf{x}, t), \tag{3.6.10}$$

so that the rate equations of equilibrium take on the form

$$\boldsymbol{\nabla} \cdot \dot{\mathbf{S}}^{\mathrm{N}} + \rho\, \dot{\mathbf{b}} = \mathbf{0} \ \text{in v}, \qquad \mathbf{n} \cdot \dot{\mathbf{S}}^{\mathrm{N}} = \dot{\mathbf{t}} \quad \text{on } \partial \mathrm{v}_1, \tag{3.6.11a,b}$$

where $\dot{\mathbf{b}}$ is the traction rate measured per current area. The corresponding equations for the Cauchy stress are more complicated, and are obtained by substituting from (3.5.5) for $\dot{\mathbf{S}}^{\mathrm{N}}$.

3.7. PRINCIPLE OF VIRTUAL WORK

Consider the equations of motion (3.1.4) and, for simplicity, set

$$\rho\, (\mathbf{b} - \mathbf{a}) \equiv \mathbf{f} \tag{3.7.1}$$

in (3.1.4), to obtain

$$\boldsymbol{\nabla} \cdot \boldsymbol{\sigma} + \mathbf{f} = 0 \ \text{in v}, \qquad \boldsymbol{\sigma}\mathbf{n} = \mathbf{t} \ \text{on } \partial \mathrm{v}_1. \tag{3.7.2a,b}$$

Any stress field that satisfies (3.7.2), is called *statically admissible*. Refer to a velocity field, $\mathbf{v}^*$, which is suitably smooth and which does not violate any prescribed kinematical conditions, as a *kinematically admissible* velocity field. A kinematically admissible velocity field need not be the actual velocity field. Consider an arbitrary but kinematically admissible velocity field, $\mathbf{v}^*$, and taking its inner product with (3.7.2a), integrate the results over volume v, to obtain

$$\int_{\partial \mathrm{v}_1} \mathbf{t} \cdot \mathbf{v}^* \, da + \int_{\mathrm{v}} \mathbf{f} \cdot \mathbf{v}^* \, dv = \int_{\mathrm{v}} \mathrm{tr}\{\boldsymbol{\sigma}\, (\boldsymbol{\nabla} \otimes \mathbf{v}^*)\}\, dv, \tag{3.7.3}$$

where the Gauss theorem and (3.7.2b) are also used. This is an expression for the principle of virtual work which applies to any statically admissible stress field, $\boldsymbol{\sigma}$, and kinematically admissible velocity field, $\mathbf{v}^*$. These stress and velocity fields need not be related.

An equivalent expression in terms of the nominal stress can be obtained, by referring all quantities to the initial configuration. If the corresponding prescribed tractions are **T**, and the effective body forces are $\mathbf{f}^0$, then, for any kinematically admissible virtual velocity field, $\mathbf{V}^* = \mathbf{V}^*(\mathbf{X}, t)$, and any statically admissible nominal stress field, $\mathbf{S}^N$, it follows that

$$\nabla^0 \cdot \mathbf{S}^N + \mathbf{f}^0 = \mathbf{0} \quad \text{in } V, \qquad \mathbf{N} \cdot \mathbf{S}^N = \mathbf{T} \quad \text{on } \partial V_1, \tag{3.7.4a,b}$$

$$\int_{\partial V_1} \mathbf{T} \cdot \mathbf{V}^* \, dA + \int_V \mathbf{f}^0 \cdot \mathbf{V}^* \, dV = \int_V \mathrm{tr}\{\mathbf{S}^N (\nabla^0 \otimes \mathbf{V}^*)^T\} \, dV, \tag{3.7.5a}$$

or

$$S^N_{Ai,A} + f^0_i = 0 \quad \text{in } V, \qquad S^N_{Ai} N_A = T_i \quad \text{on } \partial V_1, \tag{3.7.4c,d}$$

$$\int_{\partial V_1} T_i V_i^* \, dA + \int_V f_i^0 V_i^* \, dV = \int_V S^N_{Ai} V^*_{i,A} \, dV, \tag{3.7.5b}$$

where

$$\mathbf{f}^0 \equiv \rho_0 (\mathbf{b} - \mathbf{a}). \tag{3.7.4e}$$

Note that ∇^0 in this equation is the del operator with respect to the initial particle position, $\mathbf{X}$, whereas ∇ in (3.7.2) and (3.7.3) represents the del operator with respect to the current particle position, $\mathbf{x}$.

The principle of virtual work in the form (3.7.5) also applies to the quasi-static rate problem defined by (3.6.11). This leads to

$$\int_{\partial v_1} \dot{\mathbf{t}} \cdot \mathbf{v}^* da + \int_v \rho \dot{\mathbf{b}} \cdot \mathbf{v}^* dv = \int_v \mathrm{tr}\{\dot{\mathbf{S}}^N (\nabla \otimes \mathbf{v}^*)^T\} \, dv. \tag{3.7.6}$$

Here, the current configuration, prior to the incremental loading, is used as the reference one. An expression in terms of the general stress measure $\mathbf{S}$ is obtained if (3.5.8) is substituted into (3.7.6) for $\dot{\mathbf{S}}^N$.

3.8. FINITE-ELEMENT FORMULATION

In finite-element solutions of initial, boundary-value problems,[4] volume V is divided into a finite number of subregions, usually hexahedral- or tetrahedral-shaped. A common point at which vertices of adjacent elements meet, is called a *node*. Within each element (subregion) the field quantities are represented by simple functions which include a number of free parameters (nodal values). These parameters are then calculated from the weak form (principle of virtual work) of the equations of motion and the boundary conditions. In view of (3.7.1), the virtual work equation (3.7.3) is expressed as

$$\int_V \rho (\mathbf{a} - \mathbf{b}) \cdot \mathbf{v}^* \, dv + \int_V \mathrm{tr}\{\boldsymbol{\sigma} (\nabla \otimes \mathbf{v}^*)\} dv - \int_{\partial v_1} \mathbf{t} \cdot \mathbf{v}^* \, da = 0, \tag{3.8.1}$$

[4] For modern accounts of finite-element modeling and computation of nonlinear solid mechanics and structural problems, see Belytschko *et al.* (2000), and Simo and Hughes (1998).

where $\mathbf{b}$ is the body force per unit volume, $\mathbf{a}$ is the acceleration, and $\mathbf{v}^*$ is any kinematically admissible virtual velocity field. This integral form of the equations of motion is used as a starting point to formulate the corresponding finite-element equations, as discussed in the sequel.

3.8.1. Weak Form of Equations of Motion in the Current Configuration

Let the current volume v be subdivided into NEL number of elements with NND number of nodes. The number of nodes of a typical element is denoted by n. Number the nodes and denote the node number of the αth node of an element e by N_α^e. Define a unit field, ϕ_α^e, which vanishes outside the element e and has unit value at node N_α^e. Then, the field quantities in a typical element, e, are represented by

$$\mathbf{u}(\mathbf{x};\, e) = \sum_{\alpha=1}^{n} \phi_\alpha^e(\mathbf{x})\, \mathbf{u}_\alpha^e, \quad \mathbf{v}(\mathbf{x};\, e) = \sum_{\alpha=1}^{n} \phi_\alpha^e(\mathbf{x})\, \mathbf{v}_\alpha^e,$$

$$\mathbf{a}(\mathbf{x};\, e) = \sum_{\alpha=1}^{n} \phi_\alpha^e(\mathbf{x})\, \mathbf{a}_\alpha^e, \quad \mathbf{v}^*(\mathbf{x};\, e) = \sum_{\alpha=1}^{n} \phi_\alpha^e(\mathbf{x})\, \mathbf{v}_\alpha^{*e}, \tag{3.8.2a-d}$$

where the subscript α and the superscript e denote the corresponding field quantity being evaluated at the αth node of the element e, and n is the total number of nodes of this element. Substituting (3.8.2) into (3.8.1), obtain

$$\sum_e \mathbf{v}_\alpha^{*e} \cdot \{\, \mathbf{a}_\beta^e \int_{v^e} \rho\, \phi_\beta^e \phi_\alpha^e\, dv - \int_{v^e} \rho\, \mathbf{b}\, \phi_\alpha^e\, dv + \int_{v^e} (\boldsymbol{\nabla}\, \phi_\alpha^e) \cdot \boldsymbol{\sigma}\, dv - \int_{\partial v^e \cap \partial v_1} \mathbf{t}\, \phi_\alpha^e\, da \,\} = 0, \tag{3.8.3a}$$

where $\partial v^e \cap \partial v_1$ denotes the intersection of the element boundary ∂v^e and the boundary of the body ∂v_1, where the tractions are prescribed, and the summation is over all elements, e = 1, 2, ..., NEL. This can be rewritten as

$$\sum_e \mathbf{v}_\alpha^{*e} \cdot (M_{\alpha\beta}^e\, \mathbf{a}_\beta - \mathbf{f}_{e\alpha}^b + \mathbf{f}_{e\alpha}^{in} - \mathbf{f}_{e\alpha}^{ex}) = 0, \tag{3.8.3b}$$

where $M_{\alpha\beta}^e$, $\mathbf{f}_{e\alpha}^b$, $\mathbf{f}_{e\alpha}^{in}$, and $\mathbf{f}_{e\alpha}^{ex}$ are the consistent mass matrix, body force vector, internal force vector, and external force vector of the αth node of element e, respectively. These are defined as follows:

$$M_{\alpha\beta}^e = \int_{v^e} \rho\, \phi_\alpha^e \phi_\beta^e\, dv, \quad \mathbf{f}_{e\alpha}^b = \int_{v^e} \rho\, \mathbf{b}\, \phi_\alpha^e\, dv,$$

$$\mathbf{f}_{e\alpha}^{in} = \int_{v^e} (\boldsymbol{\nabla}\, \phi_\alpha^e) \cdot \boldsymbol{\sigma}\, dv, \quad \mathbf{f}_{e\alpha}^{ex} = \int_{\partial v^e \cap \partial v_1} \mathbf{t}\, \phi_\alpha^e\, da. \tag{3.8.4a-d}$$

Note that, the interpolation functions are functions of material coordinates alone. Only the nodal quantities are time-dependent. From (3.8.2a), the deformation is approximated by

$$\mathbf{X} = \mathbf{x} - \mathbf{u}_\alpha^e\, \phi_\alpha^e(\mathbf{x}).$$

This is a one-to-one mapping. Denote the inverse transformation by

$$\mathbf{x} = \Psi(\mathbf{X}).$$

Therefore, the interpolation functions can be expressed in the initial configuration as follows:

$$\phi^e_\alpha(\mathbf{x}) = \phi^e_\alpha\{\Psi(\mathbf{X})\} \equiv \Phi^e_\alpha(\mathbf{X}). \tag{3.8.5}$$

Then (3.8.4) can be expressed in the initial configuration as

$$M^e_{\alpha\beta} = \int_{V^e} \rho_0 \, \Phi^e_\alpha \Phi^e_\beta \, dV, \quad \mathbf{f}^b_{e\alpha} = \int_{V^e} \rho_0 \, \mathbf{b} \, \Phi^e_\alpha \, dV,$$

$$\mathbf{f}^{in}_{e\alpha} = \int_{V^e} (\boldsymbol{\nabla}^0 \Phi^e_\alpha) \cdot \mathbf{S}^N \, dV, \quad \mathbf{f}^{ex}_{e\alpha} = \int_{\partial V^e \cap \partial V_1} \mathbf{T} \, \Phi^e_\alpha \, dA. \tag{3.8.6a-d}$$

The same expressions result when the following interpolations are used in (3.7.5):

$$\mathbf{u}(\mathbf{X}; e) = \Phi^e_\alpha(\mathbf{X}) \, \mathbf{u}^e_\alpha, \quad \mathbf{v}(\mathbf{X}; e) = \Phi^e_\alpha(\mathbf{X}) \, \mathbf{v}^e_\alpha,$$

$$\mathbf{V}^*(\mathbf{X}; e) = \Phi^e_\alpha(\mathbf{X}) \, \mathbf{V}^{*e}_\alpha, \quad \mathbf{a}(\mathbf{X}; e) = \Phi^e_\alpha(\mathbf{X}) \, \mathbf{a}^e_\alpha. \tag{3.8.7a-d}$$

The internal force vector $\mathbf{f}^{in}_{e\alpha}$ in (3.8.4c) or (3.8.6c) can be expressed in terms of the *Kirchhoff* stress as[5]

$$\mathbf{f}^{in}_{e\alpha} = \int_{v^e} \frac{1}{J} (\boldsymbol{\nabla}\phi^e_\alpha) \cdot \boldsymbol{\tau} \, dv = \int_{V^e} (\boldsymbol{\nabla}\phi^e_\alpha) \cdot \boldsymbol{\tau} \, dV. \tag{3.8.4e}$$

Note that, the del operator in the integrand of (3.8.4e) is $\boldsymbol{\nabla} = \mathbf{e}_i \, \partial/\partial x_i$, even though the domain of the integral is in the initial configuration.

3.8.2. Weak Form of the Rate Equations of Motion

Using the interpolations (3.8.2) in the principle of virtual work (3.7.6) for the rate problem, obtain

$$\sum_e \mathbf{v}^{*e}_\alpha \cdot (M^e_{\alpha\beta} \, \dot{\mathbf{a}}_\beta - \dot{\mathbf{f}}^b_{e\alpha} + \dot{\mathbf{f}}^{in}_{e\alpha} - \dot{\mathbf{f}}^{ex}_{e\alpha}) = 0, \tag{3.8.8}$$

$$M^e_{\alpha\beta} = \int_{v^e} \rho \, \phi^e_\alpha \phi^e_\beta \, dv, \quad \dot{\mathbf{f}}^b_{e\alpha} = \int_{v^e} \rho \, \dot{\mathbf{b}} \, \phi^e_\alpha \, dv,$$

$$\dot{\mathbf{f}}^{in}_{e\alpha} = \int_{v^e} (\boldsymbol{\nabla} \phi^e_\alpha) \cdot \dot{\mathbf{S}}^N \, dv, \quad \dot{\mathbf{f}}^{ex}_{e\alpha} = \int_{\partial v^e \cap \partial v_1} \dot{\mathbf{t}} \, \phi^e_\alpha \, da. \tag{3.8.9a-d}$$

The same expressions are obtained by taking the material time derivative of (3.8.3a,b) and (3.8.4).

---

[5] In general, formulation in terms of the Kirchhoff stress and its objective rates is preferred, since it leads to attractive results in constitutive modeling. This has an important consequence in finite-element formulations, as is discussed in connection with the symmetry of the element stiffness matrix; see Subsection 3.8.2 and comments which follow (3.8.13b).

For implicit time integration, in order to calculate the field quantities over a timestep, it is necessary to estimate the nodal forces first. These forces are usually obtained by linearization. They are expressed in terms of a *tangent stiffness matrix* which relates the nodal force rates to the nodal velocities.

Consider the rate of change of internal forces $\mathbf{f}^{\text{in}}_{\text{e}\alpha}$. The gradient of the interpolation function ϕ^{e}_{α} can be written as a vector $\boldsymbol{b}_{\text{e}\alpha}$,

$$\boldsymbol{b}_{\text{e}\alpha}(\mathbf{x}) \equiv \boldsymbol{\nabla}\phi^{\text{e}}_{\alpha}\,. \tag{3.8.10a}$$

Then the velocity gradient at any point in element e is related to the nodal velocities by

$$\mathbf{L}(\mathbf{x}) = \frac{\partial \mathbf{v}}{\partial \mathbf{x}} = \mathbf{v}^{\text{e}}_{\alpha} \otimes \boldsymbol{b}_{\text{e}\alpha}(\mathbf{x})\,. \tag{3.8.10b}$$

Now consider the constitutive equations of the material in the form which relates the Jaumann rate of the Kirchhoff stress to the deformation rate through a tangent modulus $\boldsymbol{\mathcal{L}}$,

$$\overset{\circ}{\boldsymbol{\tau}} = \boldsymbol{\mathcal{L}} : \mathbf{D}\,. \tag{3.8.11}$$

Substitute (3.8.10b) and (3.8.11) into (3.5.7), to obtain

$$\dot{\mathbf{S}}^{\text{N}} = \frac{1}{2}\,\{\boldsymbol{\mathcal{L}} : (\mathbf{v}^{\text{e}}_{\alpha} \otimes \boldsymbol{b}_{\text{e}\alpha} + \boldsymbol{b}_{\text{e}\alpha} \otimes \mathbf{v}^{\text{e}}_{\alpha}) - \boldsymbol{\tau}\,(\mathbf{v}^{\text{e}}_{\alpha} \otimes \boldsymbol{b}_{\text{e}\alpha} - \boldsymbol{b}_{\text{e}\alpha} \otimes \mathbf{v}^{\text{e}}_{\alpha})$$

$$- (\mathbf{v}^{\text{e}}_{\alpha} \otimes \boldsymbol{b}_{\text{e}\alpha} + \boldsymbol{b}_{\text{e}\alpha} \otimes \mathbf{v}^{\text{e}}_{\alpha})\,\boldsymbol{\tau}\}\,. \tag{3.8.12}$$

From (3.8.9c) and (3.8.12), the rate of the internal force vector, $\dot{\mathbf{f}}^{\text{in}}_{\text{e}\alpha}$, is related to the nodal velocities by

$$\dot{\mathbf{f}}^{\text{in}}_{\text{e}\alpha} = \mathbf{K}^{\text{e}}_{\alpha\beta}\,\mathbf{v}^{\text{e}}_{\beta}\,, \tag{3.8.13a}$$

where $\mathbf{K}^{\text{e}}_{\alpha\beta}$, $\beta = 1, 2, ..., n$, are the tangent stiffness matrices for the αth node of element e. The Cartesian components of these stiffness matrices are given in terms of the components of $\boldsymbol{\mathcal{L}}$, $\boldsymbol{\tau}$, and $\boldsymbol{b}_{\text{e}\beta}$, $\beta = 1, 2, ..., n$, by

$$\text{K}^{\text{e}}_{\alpha\beta ij} = \frac{1}{2}\int_{\text{V}} \{b_{\text{e}\alpha k}\,\mathcal{L}_{kilj}\,b_{\text{e}\beta l} + b_{\text{e}\alpha k}\,\mathcal{L}_{kijl}\,b_{\text{e}\beta l} - b_{\text{e}\alpha j}\,b_{\text{e}\beta k}\,\tau_{ki} - b_{\text{e}\alpha k}\,b_{\text{e}\beta k}\,\tau_{ji}$$

$$- b_{\text{e}\alpha k}\,\tau_{kj}\,b_{\text{e}\beta i} + b_{\text{e}\alpha k}\,\tau_{kl}\,b_{\text{e}\beta l}\,\delta_{ij}\}\text{dv}\,. \tag{3.8.13b}$$

The tangent stiffness, $\mathbf{K}^{\text{e}}_{\alpha\beta}$, *is always symmetric* in the following sense:

$$\text{K}_{\alpha\beta ij} = \text{K}_{\beta\alpha ji}\,, \tag{3.8.14a}$$

provided the constitutive modulus tensor $\boldsymbol{\mathcal{L}}$ has the following symmetry:

$$\mathcal{L}_{ijkl} = \mathcal{L}_{klij} = \mathcal{L}_{ijlk}\,. \tag{3.8.14b}$$

Indeed, $\mathbf{K}^{\text{e}}_{\alpha\beta}$ is symmetric, if a symmetric constitutive modulus tensor is used to relate *any objective rate (including the Jaumann rate) of the Kirchhoff stress* to the deformation rate. On the other hand, *if instead of the Jaumann rate of the Kirchhoff stress, the Jaumann rate of the Cauchy stress is used in (3.8.11) with a symmetric tangent modulus, then the resulting* $\mathbf{K}^e_{\alpha\beta}$ *will no longer be symmetric.*

From (3.5.6), when the current configuration is used as reference, the Jaumann rates of the Cauchy and Kirchhoff stresses are related by

$$\overset{\circ}{\boldsymbol{\tau}} = \overset{\circ}{\boldsymbol{\sigma}} + \boldsymbol{\sigma}\,\mathrm{tr}(\mathbf{D})\,. \tag{3.8.15}$$

Then the constitutive modulus tensor which relates the Jaumann rate of the Cauchy stress to the deformation rate is given by

$$\overset{\circ}{\boldsymbol{\sigma}} = \overline{\mathcal{L}} : \mathbf{D}\,, \quad \overline{\mathcal{L}} = \mathcal{L} - \boldsymbol{\sigma} \otimes \mathbf{1}\,,$$

$$\overline{\mathcal{L}}_{ijkl} = \mathcal{L}_{ijkl} - \sigma_{ij}\,\delta_{kl}\,. \tag{3.8.16a-c}$$

Therefore, if $\mathcal{L}$ has the symmetry defined by (3.8.14b), then $\overline{\mathcal{L}}$ will no longer be symmetric. Hence, in general, $\mathbf{K}^e_{\alpha\beta}$ will not be symmetric when $\mathcal{L}$ is symmetric and the Jaumann rate of the Cauchy stress is used (Belytschko, 1983). On the other hand, if the objective Cauchy stress rate,

$$\overset{*}{\boldsymbol{\sigma}} \equiv \dot{\boldsymbol{\sigma}} + \mathrm{tr}(\mathbf{D})\,\boldsymbol{\sigma} - \mathbf{L}\,\boldsymbol{\sigma} - \boldsymbol{\sigma}\,\mathbf{L}^{\mathrm{T}}, \tag{3.8.17a}$$

is used instead of $\overset{\circ}{\boldsymbol{\sigma}}$ in (3.8.16a), then $\mathbf{K}^e_{\alpha\beta}$ becomes symmetric. This stress rate has been introduced by Truesdell (1952), and Prager (1961, page 156) calls it the Truesdell rate. It is indeed the *convected rate* of the Kirchhoff stress when the current configuration is used as reference[6], *i.e.*,

$$\overset{*}{\boldsymbol{\sigma}} = \overset{\nabla}{\boldsymbol{\tau}} = \dot{\boldsymbol{\tau}} - \mathbf{L}\,\boldsymbol{\tau} - \boldsymbol{\tau}\,\mathbf{L}^{\mathrm{T}}. \tag{3.8.17b}$$

3.8.3. Discretized Equations of Motion

Denote the nodal displacements, velocities, and accelerations by $\mathbf{u}_I$, $\dot{\mathbf{u}}_I$, and $\ddot{\mathbf{u}}_I$, I = 1, 2, ..., NND, respectively. Then, for arbitrary nodal virtual velocities, (3.8.4) reduces to

$$M_{IJ}\,\ddot{\mathbf{u}}_J = \mathbf{F}^b_I + \mathbf{F}^{ex}_I - \mathbf{F}^{in}_I\,, \qquad I, J = 1, 2, \cdots, \mathrm{NND}\,, \tag{3.8.18}$$

$$M_{IJ} = \sum_e \sum_{\alpha,\beta=1}^{n} M^e_{\alpha\beta}\,\delta_{IN^e_\alpha}\,\delta_{JN^e_\beta}\,, \quad \mathbf{F}^b_I = \sum_e \sum_{\alpha=1}^{n} \mathbf{f}^b_{e\alpha}\,\delta_{IN^e_\alpha}\,,$$

$$\mathbf{F}^{ex}_I = \sum_e \sum_{\alpha=1}^{n} \mathbf{f}^{ex}_{e\alpha}\,\delta_{IN^e_\alpha}\,, \quad \mathbf{F}^{in}_I = \sum_e \sum_{\alpha=1}^{n} \mathbf{f}^{in}_{e\alpha}\,\delta_{IN^e_\alpha}\,, \tag{3.8.19a-d}$$

where N^e_α is the global node number of the αth node of a typical element e, and $\delta_{IN^e_\alpha}$ is the Kronecker delta.

Similarly, for quasi-static problems, the rate form of the discretized equations of equilibrium is obtained from (3.8.8), as

[6] Thus, complications of the type discussed by Belytschko (1983) disappear when an objective rate (*any* objective rate) of the Kirchhoff stress is used in constitutive relations (3.8.11) with a symmetric modulus tensor $\mathcal{L}$.

$$\mathbf{K}_{IJ}\,\dot{\mathbf{u}}_J + \dot{\mathbf{F}}_I^{ex} + \dot{\mathbf{F}}_I^{b} = \mathbf{0}, \tag{3.8.20}$$

$$\dot{\mathbf{F}}_I^{ex} = \sum_e \sum_{\alpha=1}^{n} \dot{\mathbf{f}}_{e\alpha}^{ex}\,\delta_{IN_\alpha^e}, \qquad \dot{\mathbf{F}}_I^{b} = \sum_e \sum_{\alpha=1}^{n} \dot{\mathbf{f}}_{e\alpha}^{b}\,\delta_{IN_\alpha^e},$$

$$\mathbf{K}_{IJ} = \sum_e \sum_{\alpha,\,\beta=1}^{n} \mathbf{K}_{\alpha\beta}^{e}\,\delta_{IN_\alpha^e}\,\delta_{JN_\beta^e}\,. \tag{3.8.21a-c}$$

3.9. BALANCE RELATIONS WITH SURFACES OF DISCONTINUITY

So far the treatment has been based on the assumption that all kinematical and dynamical quantities of motion are continuous, possessing continuous derivatives of any desired order, everywhere within and on the considered material volume. In some problems of practical and theoretical importance, however, it becomes necessary to allow for finite discontinuities to occur in certain quantities of motion across a propagating *singular surface of discontinuity*[7], called the *wave front*. For example, in fracture mechanics, one is concerned with displacement fields which are discontinuous across moving (or stationary) *fracture surfaces*. If, on the other hand, the velocity field or the displacement gradient suffers finite jumps across a moving surface S, while the displacement field remains continuous, this propagating surface of discontinuity is called the *first-order wave*: it is a *shock wave* if the normal derivative of the displacement field undergoes a jump as S is traversed, and it is a *vortex sheet* if the tangential derivative is discontinuous. If only the second derivatives of the displacement field possess discontinuities across S, the wave is called the *second-order* or the *acceleration wave*. Higher-order waves may also be defined but are not considered here.

Here, following Thomas (1957, 1961), various jump conditions that are involved in the study of the propagation of singular surfaces in continuous media are formulated.

3.9.1. Kinematics of Volume Integrals

Let $\psi(\mathbf{x}, t)$ be a field (geometrical or physical) quantity in the material volume v bounded by the surface ∂v in the current configuration C. Then the *material* time derivative of the volume integral of ψ is given by

$$\frac{d}{dt}\int_v \psi\,dv = \int_v \left\{ \frac{\partial \psi}{\partial t} + \frac{\partial \psi}{\partial \mathbf{x}} \cdot \dot{\mathbf{x}} + \psi\,\dot{J} \right\} dv$$

[7] The expression "singular surface" is used to convey the fact that a certain quantity of motion suffers a discontinuity as this surface is crossed. It does *not* imply that the surface itself is irregular. Indeed, it is here assumed that this surface has a continuously turning tangent plane.

$$= \int_{v} \left\{ \frac{\partial \psi}{\partial t} + \boldsymbol{\nabla} \cdot (\psi \, \mathbf{v}) \right\} dv$$

$$= \int_{v} \frac{\partial \psi}{\partial t} dv + \int_{\partial v} \psi \, \mathbf{v} \cdot d\mathbf{a} . \tag{3.9.1}$$

3.9.2. Transport Theorem

Let ϕ be a function of space $\mathbf{x}$ and time t, defined over a suitable region and a time interval. Denote by $I(v^*, t)$ the volume integral of this function over a spatial volume v^* bounded by a surface ∂v^* which is moving with a velocity $\mathbf{c}(t)$, *i.e.*, set

$$I(v^*, t) \equiv \int_{v^*} \phi(\mathbf{x}, t) \, dv . \tag{3.9.2a}$$

Now, to obtain the time derivative of the volume integral, consider a fictitious material volume v bounded by ∂v. If the fictitious material volume v, surface ∂v, and the surface velocity $\mathbf{v}$ coincide instantaneously with the spatial volume v^*, its surface ∂v^*, and the surface velocity $\mathbf{c}$, respectively, then it follows that

$$\frac{d}{dt} I(v^*, t) \equiv \frac{d}{dt} \int_{v} \phi \, dv . \tag{3.9.2b}$$

Therefore, from (3.9.1), obtain

$$\frac{d}{dt} \int_{v^*} \phi(\mathbf{x}, t) \, dv = \int_{v^*} \frac{\partial \phi}{\partial t} dv + \int_{\partial v^*} \phi \, \mathbf{c} \cdot d\mathbf{a} . \tag{3.9.3}$$

This *transport theorem* is named after O. Reynolds; see Truesdell and Toupin (1960).

3.9.3. Dynamical Conditions at Surfaces of Discontinuity

Consider now a material volume v, bounded by a regular surface $\partial v = \partial v_1 + \partial v_2$, and let v be divided into two parts, v_1^* and v_2^*, by a moving surface $S(t)$, where v_1^* is instantaneously bounded by $\partial v_1 + S_1$, and v_2^* by $\partial v_2 + S_2$. In this manner, refer to S as S_1 when it is viewed as a part of v_1^*, and[8] as S_2 when it is regarded as a part of v_2^*. Denote the moving velocity of $S(t)$ by $\mathbf{c}(t)$; see Figure 3.9.1. Note that, since $S(t)$ is not moving with the material velocity, v_1^* and v_2^* are not material volumes. Denote by $\mathbf{n}$ the unit normal to S_1 that points from v_1^* toward v_2^*; hence $-\,\mathbf{n}$ is the exterior unit normal to S_2. Write (3.9.3) for v_1^* and v_2^*, to obtain

$$\frac{d}{dt} \int_{v_1^*} \phi \, dv = \int_{v_1^*} \frac{\partial \phi}{\partial t} dv + \int_{\partial v_1} \phi \, \mathbf{v} \cdot d\mathbf{a} + \int_{S} \phi \, \mathbf{c} \cdot \mathbf{n} \, da$$

[8] Note that S_1 and S_2 are two faces of the same surface S.

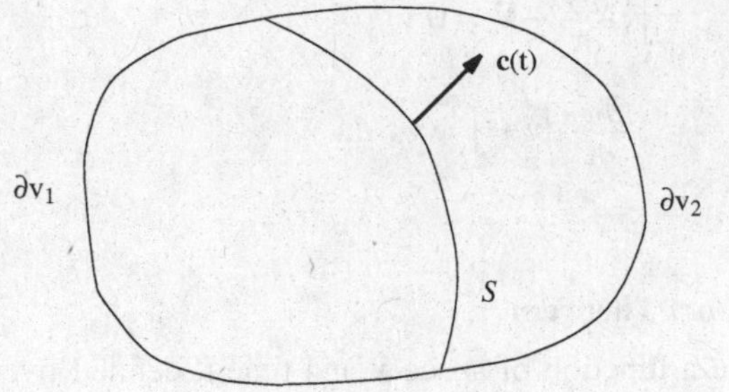

Figure 3.9.1

A surface of discontinuity S moving with a velocity $\mathbf{c}(t)$ in a material volume v bounded by $\partial v = \partial v_1 + \partial v_2$

$$= \int_{v_1^*} \left\{ \frac{\partial \phi}{\partial t} + \mathbf{v} \cdot \boldsymbol{\nabla} \phi \right\} dv + \int_S \phi (\mathbf{c} - \mathbf{v}) \cdot \mathbf{n} \, da, \tag{3.9.4a}$$

$$\frac{d}{dt} \int_{v_2^*} \phi \, dv = \int_{v_2^*} \frac{\partial \phi}{\partial t} dv + \int_{\partial v_2} \phi \mathbf{v} \cdot d\mathbf{a} - \int_S \phi \mathbf{c} \cdot \mathbf{n} \, da$$

$$= \int_{v_2^*} \left\{ \frac{\partial \phi}{\partial t} + \mathbf{v} \cdot \boldsymbol{\nabla} \phi \right\} dv - \int_S \phi (\mathbf{c} - \mathbf{v}) \cdot \mathbf{n} \, da. \tag{3.9.4b}$$

Now, assume that ϕ has a finite jump across S, it being continuous on either side of S. Denote by $\phi^{(1)}$ and $\phi^{(2)}$ the values (or the one-sided limits) of ϕ on S, as this surface is approached from the interior of v_1^* and v_2^*, respectively. From (3.9.4), obtain

$$\frac{d}{dt} \int_v \phi \, dv = \int_v \left\{ \frac{\partial \phi}{\partial t} + \mathbf{v} \cdot \boldsymbol{\nabla} \phi \right\} dv - \int_S [\phi \, U] \, da, \tag{3.9.5a}$$

where $[\phi \, U]$ is the jump in the quantity $\phi \, U$ defined by

$$[\phi \, U] = \phi^{(2)} U^{(2)} - \phi^{(1)} U^{(1)},$$

$$U^{(i)} \equiv (\mathbf{c} - \mathbf{v}^{(i)}) \cdot \mathbf{n}, \quad i = 1, 2. \tag{3.9.5b,c}$$

To obtain a general jump condition on S, let S be an arbitrary part of S that is contained in v, and consider the limiting value of (3.9.5a), as v approaches zero, passing at the limit into S. Therefore, from (3.9.5a), obtain

$$\lim_{v \to 0} \frac{d}{dt} \int_v \phi \, dv = 0 = -\int_S [\phi \, U] \, da$$

which must hold for an arbitrary S, yielding

$$[\phi \, U] = 0. \tag{3.9.6}$$

As an example, let ϕ stand for the mass density $\rho(\mathbf{x}, t)$ and, since mass is conserved, obtain

$$[\rho\,U] = 0 \tag{3.9.7a}$$

or

$$\rho^{(1)}\,(\mathbf{c} - \mathbf{v}^{(1)}) \cdot \mathbf{n} = \rho^{(2)}\,(\mathbf{c} - \mathbf{v}^{(2)}) \cdot \mathbf{n}\,. \tag{3.9.7b}$$

3.9.4. Balance Relations at Surfaces of Discontinuity

Now formulate a general jump condition at a surface of discontinuity for the following general balance equation:

$$\frac{d}{dt} \int_{v} F\,\rho\,dv = \int_{v} f\,\rho\,dv - \int_{\partial v} \mathbf{q} \cdot \mathbf{da} \tag{3.9.8a}$$

which expresses the instantaneous rate of change of a quantity of motion F, contained within the material volume v, in terms of the source f and the flux $\mathbf{q}$. Let ϕ in (3.9.5a) stand for $F\rho$. As the volume v is shrunk into S_0, the left-hand side (LHS) of (3.9.8a) yields

$$\text{LHS} = -\int_{S_0} \{\rho^{(2)}\,U^{(2)}\,F^{(2)} - \rho^{(1)}\,U^{(1)}\,F^{(1)}\}\,da\,. \tag{3.9.8b}$$

Next consider the right-hand side (RHS) of (3.9.8a). As v shrinks to zero, passing at the limit into S_0, the surface integral in (3.9.8a) reduces to,

$$\text{RHS} = \int_{\partial v} \mathbf{q} \cdot \mathbf{da} \;\rightarrow\; \int_{S_0} (\mathbf{q}^{(2)} - \mathbf{q}^{(1)}) \cdot \mathbf{n}\,da\,, \tag{3.9.8c}$$

while the volume integral in (3.9.8a) reduces to zero. From (3.9.8), and the fact that S_0 is arbitrary, conclude that

$$[\rho\,U\,F - \mathbf{q} \cdot \mathbf{n}] = 0 \tag{3.9.9a}$$

or

$$\rho^{(1)}\,\mathbf{n} \cdot (\mathbf{c} - \mathbf{v}^{(1)})\,(F^{(2)} - F^{(1)}) = (\mathbf{q}^{(2)} - \mathbf{q}^{(1)}) \cdot \mathbf{n}\,, \tag{3.9.9b}$$

where (3.9.5c) and (3.9.7b) are also used.

Equation (3.9.9a) is the general jump condition associated with the general balance law (3.9.8a). Note that, using Gauss' theorem, the global balance equation (3.9.8a) can be reduced to the following local one:

$$\rho\,\dot{F} = \rho\,f - \operatorname{div}\mathbf{q}\,. \tag{3.9.10}$$

Now, specialize (3.9.9a) to obtain various dynamical conditions on singular surfaces of discontinuity. As the first case, consider the conservation of linear momentum,

$$\frac{d}{dt} \int_{v} \mathbf{v}\,\rho\,dv = \int_{v} \rho\,\mathbf{b}\,dv + \int_{\partial v} \boldsymbol{\sigma} \cdot \mathbf{da}\,, \tag{3.9.11a}$$

and from (3.9.9a) obtain

$$[\rho\,U\,\mathbf{v} + \mathbf{t}^{(n)}] = \mathbf{0}\,, \tag{3.9.11b}$$

where $\mathbf{t}^{(n)} = \sigma_{ij}\, n_i\, \mathbf{e}_j$ is the traction vector on the element of S with the unit normal $\mathbf{n}$. Note that the jump in the traction vector is given by $[\mathbf{t}^{(n)}] = [\sigma_{ij}]\, n_i\, \mathbf{e}_j$.

As the second case, consider the conservation of angular momentum

$$\frac{d}{dt}\int_V \mathbf{x}\times\mathbf{v}\,\rho\, dv = \int_V \rho\,\mathbf{x}\times\mathbf{b}\, dv + \int_{\partial v} \mathbf{x}\times(\boldsymbol{\sigma}\cdot d\mathbf{a}), \tag{3.9.12a}$$

and obtain

$$\mathbf{x}\times[\rho\, U\,\mathbf{v}+\mathbf{t}^{(n)}] = \mathbf{0} \tag{3.9.12b}$$

which is also implied by (3.9.11b).

As the third case, consider the conservation of energy,

$$\frac{d}{dt}\int_V (e+\frac{1}{2}\mathbf{v}\cdot\mathbf{v})\,\rho\, dv = \int_V (\mathbf{b}\cdot\mathbf{v}+h)\,\rho\, dv + \int_{\partial v} (\mathbf{v}\cdot\boldsymbol{\sigma}-\mathbf{q})\cdot d\mathbf{a}, \tag{3.9.13a}$$

and obtain

$$[\rho\, U(e+\frac{1}{2}\mathbf{v}\cdot\mathbf{v})+\mathbf{t}^{(n)}\cdot\mathbf{v}-\mathbf{q}\cdot\mathbf{n}] = 0, \tag{3.9.13b}$$

where e is the internal energy density, and h is the density of energy added to the body by sources such as radiation.

3.10. GENERAL VARIATIONAL PRINCIPLES FOR HYPERELASTIC MATERIALS[9]

Variational methods gained prominence in the applied sciences with the work of Ritz (1908,1909); see also Rayleigh (1945). In linear elasticity the theorems of the minimum potential energy and the minimum complementary energy are well known; see, *e.g.*, Sokolnikoff (1956) and, for a more modern account, Nemat-Nasser and Hori (1993). In the application of these theorems, the displacement or the stress fields are sought on classes of admissible functions by the minimization of the corresponding functional.[10]

A variational theorem with displacement and stress as independent fields, is presented by Hellinger (1914; Section 7e), for finite elasticity problems; a more modern version of Hellinger's results with generalization and clarification, and in terms of a more modern notation, is found in Truesdell and Toupin (1960; Sections 231-238). Hellinger's variational statement does not account for the boundary conditions. Reissner (1953) presented a variational theorem for finite

---

[9] In Chapter 4 various continuum constitutive relations are summarized. For a brief account of hyperelasticity, see Section 4.1.1. The material in this section is a somewhat extended version of Chapter 5 in *Mechanics Today*, Vol. 1, 1972; see Nemat-Nasser (1974).

[10] An account of various mathematical implications of these theorems is found in Mikhlin (1964,1965).

elasticity which is based on the complementary energy, and explicitly includes boundary data, by allowing the independent variation of the surface tractions. This work is an extension of an earlier result (Reissner, 1950), where, for linear elasticity, a variational statement with independent displacement and stress fields which includes the boundary data, is given. Koppe (1956) reformulates Reissner's (1953) result, in terms of the field quantities originally employed by Hellinger, and arrives at a more general variational statement than that reported in Hellinger (1914). Reissner, in a later paper (1970; Section 4), gives a short historical account of the development in this area. Prange (1916) in an unpublished thesis discusses, for linear elasticity problems, a variational theorem involving independent displacement and stress fields, and explicitly includes the boundary conditions. Prange's work remained unknown for over fifty years. Hu (1955) published an English version of his earlier paper (1954) in which a variational theorem that includes independent displacement, stress, *and* strain fields, is presented. A similar result was independently obtained by Washizu (1955). Later on, more general statements for linear elasticity problems were given by Naghdi (1964) and Prager (1967). For linear elasticity, Prager includes not only independent variations of displacement, strain, and stress fields, but also the boundary and continuity conditions which are deduced as part of the variational theorem; see also Sandhu and Pister (1971), where various forms of variational methods (but excluding the discontinuities) are discussed. For finite elasticity, a variational statement with three independent fields is found in Washizu (1968).

For problems of small elastic deformations superimposed on large ones, variational methods have been proposed by Biot (1939,1965) and Washizu (1968, Chapter 5). Biot's work is confined to one independent field, *i.e.*, the displacement, whereas Washizu considers the influence of initial strain and initial stress.

A unified presentation of variational principles in mechanics and their relations with nonlinear programming, is given by Sewell (1969) who exploits the Legendre transformation, and emphasizes various dual principles; the genesis of Sewell's work can be found in a paper by Hill (1956) and one by Moreau (1967).

Nemat-Nasser (1972a, 1974) develops various general variational principles for nonlinear, incremental, and linear elasticity. These principles include the most general forms of discontinuities which may either be required as part of a solution (for example, when material properties change discontinuously, or when fracture occurs over an interior surface), or are introduced in order to enlarge the class of functions from which solutions are sought.[11]

---

[11] Perhaps the most significant and impressive use of these principles is found in Nemat-Nasser (1972b), Fu and Nemat-Nasser (1974), Nemat-Nasser *et al.* (1975), Minagawa and Nemat-Nasser (1975, 1976), Minagawa *et al.* (1981, 1984, 1992), Nemat-Nasser and Lee (1973), and Nemat-Nasser and Minagawa (1975, 1977), where methods giving extremely accurate eigenvalues of elastic waves in composites, and buckling and vibration of beams, are developed. In the numerical cal-

3.10.1. Preliminaries

Consider body B which, in its initial virgin state, has a volume V and a regular surface ∂V. Let it be deformed from this initial configuration C_0 to a deformed configuration C, by means of body forces **b** prescribed per unit mass and by the following boundary conditions: on surface ∂V some components of the surface tractions and the complementary components of the surface displacements are prescribed to be $\mathbf{T}^0$ and $\mathbf{u}^0$; the prescribed quantities are denoted by the superscript 0. Note that this includes cases in which surface tractions **T** are prescribed on a portion ∂V_1, and surface displacements **u** on the remaining part ∂V_2 of ∂V.

In this section, it is assumed that body B consists of an elastic material which admits a strain-energy-density function $\Phi = \Phi(\mathbf{E})$, and a complementary-energy-density function $\Psi = \Psi(\mathbf{S})$, where **E** is any of the general material strain measures introduced in Section 2.3, and **S** is its conjugate stress[12].

In a general variational statement, the displacement, the strain, and the stress may be considered as independent fields with arbitrary variation. When the strain **E** is obtained from the displacement field **u**, it is said to *correspond* to this displacement field. Then, the variation of the deformation-gradient is given, once the variation of the displacement field is defined. In such a case, the dependence of the strain, **E**, on the displacement, **u**, is denoted by $\mathbf{E}(\mathbf{u})$. On the other hand, when the strain is *not* calculated from the displacement field, it is denoted by **E**; it can be varied independently of the variation of the displacement field. Similarly, when **S** corresponds to the strain **E**, then $\mathbf{S} = \partial\Phi/\partial\mathbf{E}$, otherwise **S** is a field which can be varied independently. Other stress measures are viewed as functions of the independent stress field and the displacement field. For example, when the second Piola-Kirchhoff stress, $\mathbf{S}^\mathrm{P}$, is considered as the *independent stress field*, then the nominal stress, $\mathbf{S}^\mathrm{N}$, is viewed as a *function* of this independent stress *and* the displacement, **u**, *i.e.*, $\mathbf{S}^\mathrm{N} = \mathbf{S}^\mathrm{N}(\mathbf{S}^\mathrm{P}, \mathbf{u})$. From Table 3.1.1, page 99, $\mathbf{S}^\mathrm{N} = \mathbf{S}^\mathrm{P}\,\mathbf{F}^\mathrm{T}$, or, since $\mathbf{F} = \mathbf{1} + \partial\mathbf{u}/\partial\mathbf{X}$, it follows that

$$\mathbf{S}^\mathrm{N}(\mathbf{S}^\mathrm{P}, \mathbf{u}) = \mathbf{S}^\mathrm{P}\,(\mathbf{1} + \partial\mathbf{u}/\partial\mathbf{X})^\mathrm{T}. \tag{3.10.1a}$$

Similarly, the nominal stress corresponding to the independent strain *and* the displacement fields is denoted by $\mathbf{S}^\mathrm{N}(\partial\Phi/\partial\mathbf{E}, \mathbf{u})$. It is given by

$$\mathbf{S}^\mathrm{N}(\partial\Phi/\partial\mathbf{E}, \mathbf{u}) = (\partial\Phi/\partial\mathbf{E})\,(\mathbf{1} + \partial\mathbf{u}/\partial\mathbf{X})^\mathrm{T}. \tag{3.10.1b}$$

It follows from (3.10.1a) that, for a variation $\delta\mathbf{u}$ in the displacement field,

$$\mathbf{S}^\mathrm{N}(\mathbf{S}^\mathrm{P}, \mathbf{u}) : \frac{\partial\delta\mathbf{u}}{\partial\mathbf{X}} = \mathbf{S}^\mathrm{P}\,\{\mathbf{1} + \frac{\partial\mathbf{u}}{\partial\mathbf{X}}\}^\mathrm{T} : \frac{\partial\delta\mathbf{u}}{\partial\mathbf{X}} = \mathbf{S}^\mathrm{P} : \delta\mathbf{E}^\mathrm{L}(\mathbf{u}), \tag{3.10.2}$$

culations of large deformation problems, general variational principles play an important role. In special forms, they have been used by some authors in the past for linear problems (see Herrmann, 1965; Pian and Tong, 1969; and Tong, 1969, 1970) and for nonlinear problems (see Nemat-Nasser and Shatoff, 1971, 1973).

[12] Hence, **S** and **E** are related by $\mathbf{S} = \partial\Phi/\partial\mathbf{E}$ and $\mathbf{E} = \partial\Psi/\partial\mathbf{S}$.

where $\mathbf{E}^L$ is the Lagrangian strain tensor, conjugate to the second Piola-Kirchhoff stress; see Subsections 2.3.3 and 3.1.6. Similar results hold for other strain measures and their conjugate stresses.

3.10.2. General Variational Principles

Following Nemat-Nasser and Hori (1993), a new energy functional Π, with three independent fields is defined for the finite deformation, in line with the earlier work of Nemat-Nasser (1972a, 1974). To this end, the potential energy Π is regarded as a functional of the independent field variables $\mathbf{u}$, $\mathbf{E}$, and $\mathbf{S}$. Its stationary value is sought, subject to the following side conditions: in the volume V,

$$\nabla^0 \cdot \mathbf{S}^N(\mathbf{S},\ \mathbf{u}) + \rho_0\, \mathbf{b} = \mathbf{0}, \qquad \mathbf{E} = \mathbf{E}(\mathbf{u}); \tag{3.10.3a,b}$$

and on the boundary ∂V,

$$\begin{Bmatrix} \mathbf{u} = \mathbf{u}^0 \\ \mathbf{N} \cdot \mathbf{S}^N = \mathbf{T}^0 \end{Bmatrix} \text{ at points where } \begin{Bmatrix} \mathbf{u}^0 \\ \mathbf{T}^0 \end{Bmatrix} \text{ is prescribed}. \tag{3.10.4a-b}$$

To include these side conditions in the variational statement, introduce the following four Lagrange multipliers: a vector field $\boldsymbol{\lambda}^u$, and a second-order tensor field $\boldsymbol{\Lambda}^S$, both in V for the side conditions (3.10.3a,b); and two vector fields $\boldsymbol{\mu}^u$ and $\boldsymbol{\mu}^t$ on ∂V, for the boundary conditions (3.10.4a,b). Thus, define functional Π by

$$\begin{aligned}
\Pi &\equiv \Pi(\mathbf{u},\ \mathbf{E},\ \mathbf{S},\ \boldsymbol{\lambda}^u,\ \boldsymbol{\Lambda}^S,\ \boldsymbol{\mu}^u,\ \boldsymbol{\mu}^t;\ \mathbf{b},\ \mathbf{t}^0,\ \mathbf{u}^0) \\
&= \int_V \{\Phi(\mathbf{E}) + \Psi(\mathbf{S}) - \rho_0\, \mathbf{b} \cdot \mathbf{u}\}\, dV - \int_{\partial V'} \mathbf{T}^0 \cdot \mathbf{u}\, dA - \int_{\partial V''} \boldsymbol{\mu}^t \cdot \{\mathbf{u} - \mathbf{u}^0\}\, dA \\
&+ \int_V \Big\{ \rho_0\, \boldsymbol{\lambda}^u \cdot \mathbf{b} - \mathbf{E}(\boldsymbol{\lambda}^u) : \mathbf{S} + \boldsymbol{\Lambda}^S : \{\mathbf{E}(\mathbf{u}) - \mathbf{E}\} \Big\}\, dV \\
&+ \int_{\partial V''} \{\mathbf{N} \cdot \mathbf{S}^N(\mathbf{S},\ \boldsymbol{\lambda}^u)\} \cdot (\boldsymbol{\lambda}^u - \mathbf{u}^0)\, dA \\
&+ \int_{\partial V'} (\boldsymbol{\lambda}^u - \boldsymbol{\mu}^u) \cdot \{\mathbf{N} \cdot \mathbf{S}^N(\mathbf{S},\ \boldsymbol{\lambda}^u) - \mathbf{T}^0\}\, dA,
\end{aligned} \tag{3.10.5}$$

where the independent fields subject to variation, now are the field variables $\mathbf{u}$, $\mathbf{E}$, as well as $\mathbf{S}$, and the Lagrange multipliers $\boldsymbol{\lambda}^u$, $\boldsymbol{\Lambda}^S$, $\boldsymbol{\mu}^u$, and $\boldsymbol{\mu}^t$. In (3.10.5), $\partial V'$ is the part of the boundary ∂V, on which certain components of the tractions are prescribed, and $\partial V''$ is the portion on which the complementary components of the displacements are given.

The structure of functional Π in (3.10.5) should be noted. The first integral represents the strain energy *plus* the complementary energy stored in the elastic body, minus the work of the dead-load type body forces. The second integral is the work of the prescribed surface tractions, whereas the third integral represents the kinematical condition corresponding to the prescribed surface displacements. The first two terms in the fourth integral represent the difference in the work by the body force **b** and the stress **S** for the dual displacement field $\boldsymbol{\lambda}^u$. The third term in this integral represents the kinematical condition in the volume V. The Lagrange multiplier $\boldsymbol{\Lambda}^S$ then turns out to be the corresponding stress field conjugate to the strain field **E**. The fifth integral represents the kinematical condition for the dual displacement field, whereas the sixth integral corresponds to the traction boundary condition for the stress field **S**.

First consider the variation of $\mathbf{E}(\mathbf{u}) : \mathbf{S}$. It follows from (3.10.2) that[13]

$$\begin{aligned}\delta\{\mathbf{E}(\mathbf{u}) : \mathbf{S}\} &= \delta\mathbf{E}(\mathbf{u}) : \mathbf{S} + \mathbf{E}(\mathbf{u}) : \delta\mathbf{S} \\ &= S^N_{ij}(\mathbf{S},\, \mathbf{u})\, \partial\delta u_j / \partial X_i + \mathbf{E}(\mathbf{u}) : \delta\mathbf{S} \\ &= \boldsymbol{\nabla}^0 \cdot (\mathbf{S}^N \delta\mathbf{u}) - (\boldsymbol{\nabla}^0 \cdot \mathbf{S}^N) \cdot \delta\mathbf{u} + \mathbf{E}(\mathbf{u}) : \delta\mathbf{S}\,. \end{aligned} \tag{3.10.6}$$

Note that, $\mathbf{S}^N$ here is the nominal stress tensor relating to the considered stress **S** and its conjugate strain **E**(**u**), according to (3.10.1b). In view of (3.10.6) and Gauss' theorem, the variation of the functional Π results in

$$\delta\Pi = \int_V \delta\pi_1\, dV + \int_{\partial V'} \delta\pi_1\, dA + \int_{\partial V''} \delta\pi_2\, dA\,, \tag{3.10.7}$$

where the integrand $\delta\pi_1$ in the volume integral is given by

$$\begin{aligned}\delta\pi_1 = \; & \left\{\frac{\partial\Phi}{\partial\mathbf{E}} - \boldsymbol{\Lambda}^S\right\} : \delta\mathbf{E} && \text{(elasticity)} \\ + \; & \left\{\frac{\partial\Psi}{\partial\mathbf{S}} - \mathbf{E}(\boldsymbol{\lambda}^u)\right\} : \delta\mathbf{S} && \text{(compliance)} \\ + \; & \{\boldsymbol{\nabla}^0 \cdot \mathbf{S}^N(\mathbf{S},\, \boldsymbol{\lambda}^u) + \rho_0\mathbf{b}\} \cdot \delta\boldsymbol{\lambda}^u && \text{(equilibrium)} \\ + \; & \{\mathbf{E}(\mathbf{u}) - \mathbf{E}\} : \delta\boldsymbol{\Lambda}^S && \text{(strain–displacement)} \\ + \; & \{\boldsymbol{\nabla}^0 \cdot \mathbf{S}^N(\boldsymbol{\Lambda}^S,\, \mathbf{u}) + \rho_0\mathbf{b}\} \cdot (-\delta\mathbf{u}) && \text{(dual equilibrium)}\,; \end{aligned} \tag{3.10.8a}$$

the integrands $\delta\pi_2$ and $\delta\pi_3$ in the surface integrals are given by

$$\begin{aligned}\delta\pi_2 = \; & \{\mathbf{N} \cdot \mathbf{S}^N(\boldsymbol{\Lambda}^S,\, \mathbf{u}) - \mathbf{T}^0\} \cdot \delta\mathbf{u} && \text{(dual tractions B.C.)} \\ + \; & \{\mathbf{N} \cdot \mathbf{S}^N(\mathbf{S},\, \boldsymbol{\lambda}^u) - \mathbf{T}^0\} \cdot (-\delta\boldsymbol{\mu}^u) && \text{(tractions B.C.)} \\ + \; & (\boldsymbol{\lambda}^u - \boldsymbol{\mu}^u) \cdot \{\mathbf{N} \cdot \delta\mathbf{S}^N(\mathbf{S},\, \boldsymbol{\lambda}^u)\} && \text{(dual displacements B.C.)}\,; \end{aligned} \tag{3.10.8b}$$

$$\begin{aligned}\delta\pi_3 = \; & \{\mathbf{N} \cdot \mathbf{S}^N(\boldsymbol{\Lambda}^S,\, \mathbf{u}) - \boldsymbol{\mu}^t\} \cdot \delta\mathbf{u} && \text{(dual tractions B.C.)} \\ + \; & (\boldsymbol{\lambda}^u - \mathbf{u}^0) \cdot \{\mathbf{N} \cdot \delta\mathbf{S}^N(\mathbf{S},\, \boldsymbol{\lambda}^u)\} && \text{(dual displacements B.C.)} \\ + \; & (\mathbf{u} - \mathbf{u}^0) \cdot (-\delta\boldsymbol{\mu}^t) && \text{(displacements B.C.)}\,, \end{aligned} \tag{3.10.8c}$$

where B.C. stands for "boundary conditions". The consequence of the

---

[13] It follows from (3.1.12) and (3.1.34) that, by definition $\mathrm{tr}(\mathbf{S}\,\delta\mathbf{E}) = \mathrm{tr}(\mathbf{S}^N\,\delta\mathbf{F})$, whenever $\delta\mathbf{E}$ corresponds to $\delta\mathbf{F}$, where $\delta\mathbf{F} = \partial\delta\mathbf{u}/\partial\mathbf{X}$.

arbitrariness of the variation of each field is stated on the right of the corresponding term.

As is seen from (3.10.7) and (3.10.8a-c), $\delta\Pi = 0$ yields the proper kinematical and constitutive relations among the displacement, strain, and stress fields, and gives the appropriate field equations and boundary conditions. Furthermore, each Lagrange multiplier in Π is associated with a field variable, and indeed is the *corresponding dual field quantity*. This leads to the following correspondence between the field variables and their dual fields:

$$\mathbf{u} \Longleftrightarrow \boldsymbol{\lambda}^u, \quad \mathbf{S} \Longleftrightarrow \boldsymbol{\Lambda}^S, \quad \text{in } V,$$

$$\mathbf{u} \Longleftrightarrow \boldsymbol{\mu}^u, \quad \mathbf{N} \cdot \mathbf{S}^N(\mathbf{S}, \boldsymbol{\lambda}^u) \Longleftrightarrow \boldsymbol{\mu}^t, \quad \text{on } \partial V.$$

Therefore, the vanishing of the integrand of the right side of (3.10.8a) for arbitrary field variations $\delta\mathbf{u}$, $\delta\mathbf{E}$, and $\delta\mathbf{S}$, yields the field equations which govern the dual field variables $\boldsymbol{\lambda}^u$ and $\boldsymbol{\Lambda}^S$. Conversely, arbitrary field variations $\delta\boldsymbol{\lambda}^u$ and $\delta\boldsymbol{\Lambda}^S$ give the governing field equations for the fields $\mathbf{S}$ and $\mathbf{u}$. Similarly, the vanishing of the integrands of the right side of (3.10.8b,c) provides the corresponding boundary conditions.

3.10.3. General Variational Principle in Terms of Nominal Stress and Deformation Gradient

The simplest general variational statement for continuous fields in finite elasticity results if the *nominal stress tensor* $\mathbf{S}^N$, the deformation-gradient $\mathbf{F}$, and the displacement $\mathbf{u}$, are used as independent fields with arbitrary variation. In this case the functional Π becomes

$$\begin{aligned}
\Pi &\equiv \Pi(\mathbf{u}, \mathbf{F}, \mathbf{S}^N, \boldsymbol{\lambda}^u, \boldsymbol{\Lambda}^N, \boldsymbol{\mu}^u, \boldsymbol{\mu}^t; \mathbf{b}, \mathbf{t}^0, \mathbf{u}^0) \\
&= \int_V \{\Phi(\mathbf{F}) + \Psi(\mathbf{S}^N) - \rho_0\, \mathbf{b} \cdot \mathbf{u}\}\, dV - \int_{\partial V'} \mathbf{T}^0 \cdot \mathbf{u}\, dA - \int_{\partial V''} \boldsymbol{\mu}^t \cdot \{\mathbf{u} - \mathbf{u}^0\}\, dA \\
&+ \int_V \Big\{\boldsymbol{\lambda}^u \cdot \{\boldsymbol{\nabla}^0 \cdot \mathbf{S}^N + \rho_0\, \mathbf{b}\} + \boldsymbol{\Lambda}^N : \{\mathbf{F}(\mathbf{u}) - \mathbf{F}\}\Big\}\, dV \\
&- \int_{\partial V''} (\mathbf{N} \cdot \mathbf{S}^N) \cdot \mathbf{u}^0\, dA - \int_{\partial V'} \boldsymbol{\mu}^u \cdot \{\mathbf{N} \cdot \mathbf{S}^N - \mathbf{T}^0\}\, dA.
\end{aligned} \tag{3.10.9}$$

Then the variational results corresponding to (3.10.7) become,

$$\begin{aligned}
\delta\pi_1 = \;& \left\{\frac{\partial \Phi}{\partial \mathbf{F}^T} - \boldsymbol{\Lambda}^N\right\} : \delta\mathbf{F} && \text{(elasticity)} \\
+\;& \left\{\frac{\partial \Psi}{\partial (\mathbf{S}^N)^T} - \mathbf{F}(\boldsymbol{\lambda}^u)\right\} : \delta\mathbf{S}^N && \text{(compliance)} \\
+\;& \{\boldsymbol{\nabla}^0 \cdot \mathbf{S}^N + \rho_0\, \mathbf{b}\} \cdot \delta\boldsymbol{\lambda}^u && \text{(equilibrium)} \qquad (3.10.10\text{a}) \\
+\;& \{\mathbf{F}(\mathbf{u}) - \mathbf{F}\} : \delta\boldsymbol{\Lambda}^N && \text{(strain–displacement)} \\
+\;& \{\boldsymbol{\nabla}^0 \cdot \boldsymbol{\Lambda}^N + \rho_0\, \mathbf{b}\} \cdot (-\delta\mathbf{u}), && \text{(dual equilibrium)},
\end{aligned}$$

$$\begin{aligned}\delta\pi_2 = \quad & \{\mathbf{N}\cdot\mathbf{\Lambda}^N - \mathbf{T}^0\}\cdot\delta\mathbf{u} \quad \text{(dual tractions B.C.)}\\ + \; & \{\mathbf{N}\cdot\mathbf{S}^N - \mathbf{T}^0\}\cdot(-\delta\boldsymbol{\mu}^u) \quad \text{(tractions B.C.)}\\ + \; & (\boldsymbol{\lambda}^u - \boldsymbol{\mu}^u)\cdot\{\mathbf{N}\cdot\delta\mathbf{S}^N\} \quad \text{(dual displacements B.C.)},\end{aligned} \tag{3.10.10b}$$

$$\begin{aligned}\delta\pi_3 = \quad & \{\mathbf{N}\cdot\mathbf{\Lambda}^N - \boldsymbol{\mu}^t\}\cdot\delta\mathbf{u} \quad \text{(dual tractions B.C.)}\\ + \; & (\boldsymbol{\lambda}^u - \mathbf{u}^0)\cdot\{\mathbf{N}\cdot\delta\mathbf{S}^N\} \quad \text{(dual displacements B.C.)}\\ + \; & (\mathbf{u} - \mathbf{u}^0)\cdot(-\delta\boldsymbol{\mu}^t), \quad \text{(displacements B.C.)},\end{aligned} \tag{3.10.10c}$$

where

$$\mathbf{F}(\boldsymbol{\lambda}^u) = \mathbf{1} + \frac{\partial\boldsymbol{\lambda}^u}{\partial\mathbf{X}}, \quad \mathbf{F}(\mathbf{u}) = \mathbf{1} + \frac{\partial\mathbf{u}}{\partial\mathbf{X}}. \tag{3.10.10d}$$

3.10.4. Variational Principle with Two Independent Fields

With *two* continuous fields, two general variational statements can be developed, one in which the displacement and the strain, and the other in which the displacement and the stress, are given arbitrary variation. In each case, moreover, a different variational statement is obtained, depending on the particular strain and stress measures employed.

It can be seen from (3.10.8) that the independent strain field is related to the independent displacement field **u** and the Lagrange multipliers $\mathbf{\Lambda}^S$ and $\boldsymbol{\mu}^t$. On the other hand, the independent stress field is related to the independent displacement field **u** and the Lagrange multipliers $\boldsymbol{\lambda}^u$ and $\boldsymbol{\mu}^u$. Therefore, the functional Π can be split into two parts, a strain-energy-functional Π^E, and a complementary-energy-functional Π^S, as follows:

$$\begin{aligned}\Pi &= \Pi^E + \Pi^S\\ \Pi^E &= \Pi^E(\mathbf{u}, \mathbf{E}, \mathbf{\Lambda}^S, \boldsymbol{\mu}^t; \mathbf{b}, \mathbf{t}^0, \mathbf{u}^0)\\ \Pi^S &= \Pi^S(\mathbf{u}, \mathbf{S}, \boldsymbol{\lambda}^u, \boldsymbol{\mu}^u; \mathbf{b}, \mathbf{t}^0, \mathbf{u}^0).\end{aligned} \tag{3.10.11a-c}$$

Strain-energy Functional: First, consider the strain-energy-functional Π^E, when the displacement **u**, strain **E**, and the Lagrange multiplier $\mathbf{\Lambda}^S$ are regarded as the independent fields in volume V. From (3.10.5), the strain-energy-functional is given by

$$\begin{aligned}\Pi^E &= \Pi^E(\mathbf{u}, \mathbf{E}, \mathbf{\Lambda}^S, \boldsymbol{\mu}^t; \mathbf{b}, \mathbf{T}^0, \mathbf{u}^0)\\ &= \int_V \Big\{\Phi(\mathbf{E}) + \mathbf{\Lambda}^S : \{\mathbf{E}(\mathbf{u}) - \mathbf{E}\} - \rho_0\, \mathbf{b}\cdot\mathbf{u}\Big\}\, dV\\ &\quad - \int_{\partial V'} \mathbf{T}^0\cdot\mathbf{u}\, dA - \int_{\partial V''} \boldsymbol{\mu}^t\cdot\{\mathbf{u} - \mathbf{u}^0\}\, dA.\end{aligned} \tag{3.10.11d}$$

Then the variational statement produces

$$\delta\Pi^E = \int_V \delta\pi_1^E \, dV + \int_{\partial V'} \delta\pi_2 \, dA + \int_{\partial V'} \delta\pi_3 \, dA, \tag{3.10.12a}$$

$$\begin{aligned} \delta\pi_1^E = \; & \left\{ \frac{\partial\Phi}{\partial \mathbf{E}} - \boldsymbol{\Lambda}^S \right\} : \delta\mathbf{E} && \text{(elasticity)} \\ & + \{\mathbf{E}(\mathbf{u}) - \mathbf{E}\} : \delta\boldsymbol{\Lambda}^S && \text{(strain–displacement)} \\ & + \{\boldsymbol{\nabla}^0 \cdot \mathbf{S}^N(\boldsymbol{\Lambda}^S, \mathbf{u}) + \rho_0 \mathbf{b}\} \cdot (-\delta\mathbf{u}) && \text{(dual equilibrium)}, \end{aligned} \tag{3.10.12b}$$

$$\delta\pi_2^E = \{\mathbf{N} \cdot \mathbf{S}^N(\boldsymbol{\Lambda}^S, \mathbf{u}) - \mathbf{T}^0\} \cdot \delta\mathbf{u} \quad \text{(tractions B.C.)}, \tag{3.10.12c}$$

$$\begin{aligned} \delta\pi_3^E = \; & \{\mathbf{N} \cdot \mathbf{S}^N(\boldsymbol{\Lambda}^S, \mathbf{u}) - \boldsymbol{\mu}^t\} \cdot \delta\mathbf{u} && \text{(dual tractions B.C.)} \\ & + (\mathbf{u} - \mathbf{u}^0) \cdot (-\delta\boldsymbol{\mu}^t) && \text{(displacements B.C.)}. \end{aligned} \tag{3.10.12d}$$

When the deformation gradient $\mathbf{F}$ is considered as the independent strain measure $\mathbf{E}$, the strain-energy-functional Π^E and the variational statement reduce to those obtained by Nemat-Nasser (1972a, 1974), as follows:

$$\begin{aligned} \Pi^E &= \Pi^E(\mathbf{u}, \mathbf{F}, \boldsymbol{\Lambda}^N, \boldsymbol{\mu}^t; \mathbf{b}, \mathbf{T}^0, \mathbf{u}^0) \\ &= \int_V \left\{ \Phi(\mathbf{F}) + \boldsymbol{\Lambda}^N : \{\mathbf{F}(\mathbf{u}) - \mathbf{F}\} - \rho_0 \mathbf{b} \cdot \mathbf{u} \right\} dV \\ &\quad - \int_{\partial V'} \mathbf{T}^0 \cdot \mathbf{u} \, dA - \int_{\partial V''} \boldsymbol{\mu}^t \cdot \{\mathbf{u} - \mathbf{u}^0\} \, dA, \end{aligned} \tag{3.10.13a}$$

$$\begin{aligned} \delta\pi_1^E = \; & \left\{ \frac{\partial\Phi}{\partial \mathbf{F}^T} - \boldsymbol{\Lambda}^N \right\} : \delta\mathbf{F} && \text{(elasticity)} \\ & + \{\mathbf{F}(\mathbf{u}) - \mathbf{F}\} : \delta\boldsymbol{\Lambda}^N && \text{(strain–displacement)} \\ & + \{\boldsymbol{\nabla}^0 \cdot \boldsymbol{\Lambda}^N + \rho_0 \mathbf{b}\} \cdot (-\delta\mathbf{u}) && \text{(dual equilibrium)}, \end{aligned} \tag{3.10.13b}$$

$$\delta\pi_2^E = \{\mathbf{N} \cdot \boldsymbol{\Lambda}^N - \mathbf{T}^0\} \cdot \delta\mathbf{u} \quad \text{(dual tractions B.C.)}, \tag{3.10.13c}$$

$$\begin{aligned} \delta\pi_3 = \; & \{\mathbf{N} \cdot \boldsymbol{\Lambda}^N - \boldsymbol{\mu}^t\} \cdot \delta\mathbf{u} && \text{(dual tractions B.C.)} \\ & + (\mathbf{u} - \mathbf{u}^0) \cdot (-\delta\boldsymbol{\mu}^t) && \text{(displacements B.C.)}. \end{aligned} \tag{3.10.13d}$$

A variational statement of this kind which does not include the boundary conditions, was given by Hellinger (1914). Reissner (1953) used the complementary energy functional, together with the second Piola-Kirchhoff stress tensor $\mathbf{S}^P$ and the Lagrangian strain $\mathbf{E}^L$, to developed a variational statement which includes explicitly the boundary conditions. Koppe (1956) reformulated Reissner's result, giving a result close to (3.10.13).

Similarly when the Lagrangian strain measure $\mathbf{E}^L$ is considered as the independent strain field $\mathbf{E}$, the strain-energy-functional and the variational statement become

$$\begin{aligned} \Pi^E &= \Pi^E(\mathbf{u}, \mathbf{E}^L, \boldsymbol{\Lambda}^P, \boldsymbol{\mu}^t; \mathbf{b}, \mathbf{T}^0, \mathbf{u}^0) \\ &= \int_V \left\{ \Phi(\mathbf{E}^L) + \boldsymbol{\Lambda}^P : \{\mathbf{E}^L(\mathbf{u}) - \mathbf{E}^L\} - \rho_0 \mathbf{b} \cdot \mathbf{u} \right\} dV \\ &\quad - \int_{\partial V'} \mathbf{T}^0 \cdot \mathbf{u} \, dA - \int_{\partial V''} \boldsymbol{\mu}^t \cdot \{\mathbf{u} - \mathbf{u}^0\} \, dA, \end{aligned} \tag{3.10.14a}$$

$$\begin{aligned}\delta\pi_1^E = \; & \left\{\frac{\partial\Phi}{\partial\mathbf{E}^L} - \boldsymbol{\Lambda}^P\right\} : \delta\mathbf{E}^L && \text{(elasticity)} \\ & + \{\mathbf{E}^L(\mathbf{u}) - \mathbf{E}^L\} : \delta\boldsymbol{\Lambda}^P && \text{(strain–displacement)} \\ & + \{\boldsymbol{\nabla}^0 \cdot \mathbf{S}^N(\boldsymbol{\Lambda}^P,\ \mathbf{u}) + \rho_0\,\mathbf{b}\} \cdot (-\delta\mathbf{u}) && \text{(dual equilibrium)},\end{aligned} \tag{3.10.14b}$$

$$\delta\pi_2^E = \left\{\mathbf{N} \cdot \{\boldsymbol{\Lambda}^P\,\mathbf{F}^T(\mathbf{u})\} - \mathbf{T}^0\right\} \cdot \delta\mathbf{u} \;\; \text{(dual tractions B.C.)}, \tag{3.10.14c}$$

$$\begin{aligned}\delta\pi_3^E = \; & \left\{\mathbf{N} \cdot \{\boldsymbol{\Lambda}^P\,\mathbf{F}^T(\mathbf{u})\} - \boldsymbol{\mu}^t\right\} \cdot \delta\mathbf{u} && \text{(dual tractions B.C)} \\ & + (\mathbf{u} - \mathbf{u}^0) \cdot (-\delta\boldsymbol{\mu}^t) && \text{(displacements B.C.)}.\end{aligned} \tag{3.10.14d}$$

Complementary-energy Functional: Now consider the case where the stress **S**, the displacement **u**, and the Lagrange multiplier $\boldsymbol{\lambda}^u$ are the independent fields in V. Then, from (3.10.5), the complementary-energy-functional is given by

$$\begin{aligned}\Pi^S = \; & \Pi^S(\mathbf{u},\ \mathbf{S},\ \boldsymbol{\lambda}^u,\ \boldsymbol{\mu}^u;\ \mathbf{b},\ \mathbf{t}^0,\ \mathbf{u}^0) \\ & + \int_V \left\{\Psi(\mathbf{S}) + \rho_0\,\boldsymbol{\lambda}^u \cdot \mathbf{b} - \mathbf{E}(\boldsymbol{\lambda}^u) : \mathbf{S}\right\} dV - \int_{\partial V'} \mathbf{T}^0 \cdot \mathbf{u}\, dA \\ & + \int_{\partial V''} \{\mathbf{N} \cdot \mathbf{S}^N(\mathbf{S},\ \boldsymbol{\lambda}^u)\} \cdot (\boldsymbol{\lambda}^u - \mathbf{u}^0)\, dA \\ & + \int_{\partial V'} (\boldsymbol{\lambda}^u - \boldsymbol{\mu}^u) \cdot \{\mathbf{N} \cdot \mathbf{S}^N(\mathbf{S},\ \boldsymbol{\lambda}^u) - \mathbf{T}^0\}\, dA.\end{aligned} \tag{3.10.11e}$$

Then the variational statement leads to

$$\delta\Pi^S = \int_V \delta\pi_1^S\, dV + \int_{\partial V'} \delta\pi_2^S\, dA + \int_{\partial V''} \delta\pi_3^S\, dA, \tag{3.10.15a}$$

$$\begin{aligned}\delta\pi_1^S = \; & \left\{\frac{\partial\Psi}{\partial\mathbf{S}} - \mathbf{E}(\boldsymbol{\lambda}^u)\right\} : \delta\mathbf{S} && \text{(compliance)} \\ & + \{\boldsymbol{\nabla}^0 \cdot \mathbf{S}^N(\mathbf{S},\ \boldsymbol{\lambda}^u) + \rho_0\,\mathbf{b}\} \cdot \delta\boldsymbol{\lambda}^u && \text{(equilibrium)},\end{aligned} \tag{3.10.15b}$$

$$\begin{aligned}\delta\pi_2^S = \; & \{\mathbf{N} \cdot \mathbf{S}^N(\mathbf{S},\ \boldsymbol{\lambda}^u) - \mathbf{T}^0\} \cdot (-\delta\boldsymbol{\mu}^u) && \text{(tractions B.C.)} \\ & + (\boldsymbol{\lambda}^u - \boldsymbol{\mu}^u) \cdot \{\mathbf{N} \cdot \delta\mathbf{S}^N(\mathbf{S},\ \boldsymbol{\lambda}^u)\} && \text{(dual displacements B.C.)},\end{aligned} \tag{3.10.15c}$$

$$\delta\pi_3^S = (\boldsymbol{\lambda}^u - \mathbf{u}^0) \cdot \{\mathbf{N} \cdot \delta\mathbf{S}^N(\mathbf{S},\ \boldsymbol{\lambda}^u)\} \;\; \text{(dual displacements B.C.)}. \tag{3.10.15d}$$

When the nominal stress and the displacement are used as independent fields, the functional Π^S and the variational statement (3.10.15) become

$$\begin{aligned}\Pi^S = \; & \Pi^S(\mathbf{u},\ \mathbf{S}^N,\ \boldsymbol{\lambda}^u,\ \boldsymbol{\mu}^u;\ \mathbf{b},\ \mathbf{t}^0,\ \mathbf{u}^0) \\ = \; & \int_V \left\{\Psi(\mathbf{S}^N) + \boldsymbol{\lambda}^u \cdot \{\boldsymbol{\nabla}^0 \cdot \mathbf{S}^N + \rho_0\,\mathbf{b}\}\right\} dV - \int_{\partial V'} \mathbf{T}^0 \cdot \mathbf{u}\, dA\end{aligned}$$

$$- \int_{\partial V''} \{\mathbf{N} \cdot \mathbf{S}^{\mathrm{N}}\} \cdot \mathbf{u}^0 \, dA - \int_{\partial V'} \boldsymbol{\mu}^{\mathrm{u}} \cdot \{\mathbf{N} \cdot \mathbf{S}^{\mathrm{N}} - \mathbf{T}^0\} \, dA , \tag{3.10.16a}$$

$$\begin{aligned} \delta\pi_1^{\mathrm{S}} = & \left\{ \frac{\partial \Psi}{\partial \mathbf{S}^{\mathrm{N}}} - \mathbf{F}(\boldsymbol{\lambda}^{\mathrm{u}}) \right\} : \delta\mathbf{S}^{\mathrm{N}} \ \text{(compliance)} \\ & + \{\boldsymbol{\nabla}^0 \cdot \mathbf{S}^{\mathrm{N}} + \rho_0 \mathbf{b}\} \cdot \delta\boldsymbol{\lambda}^{\mathrm{u}} \ \text{(equilibrium)}, \end{aligned} \tag{3.10.16b}$$

$$\begin{aligned} \delta\pi_2^{\mathrm{S}} = & \{\mathbf{N} \cdot \mathbf{S}^{\mathrm{N}} - \mathbf{T}^0\} \cdot (-\delta\boldsymbol{\mu}^{\mathrm{u}}) \ \text{(tractions B.C.)} \\ & + (\boldsymbol{\lambda}^{\mathrm{u}} - \boldsymbol{\mu}^{\mathrm{u}}) \cdot \{\mathbf{N} \cdot \delta\mathbf{S}^{\mathrm{N}}\} \ \text{(dual displacements B.C.)}, \end{aligned} \tag{3.10.16c}$$

$$\delta\pi_3^{\mathrm{S}} = (\boldsymbol{\lambda}^{\mathrm{u}} - \mathbf{u}^0) \cdot \{\mathbf{N} \cdot \delta\mathbf{S}^{\mathrm{N}}\} \ \text{(dual displacements B.C.)}. \tag{3.10.16d}$$

Similarly, when the second Piola-Kirchhoff stress $\mathbf{S}^{\mathrm{P}}$, and the displacement $\mathbf{u}$ are used as independent fields, the functional Π and the variational statement become

$$\begin{aligned} \Pi^{\mathrm{S}} \equiv & \ \Pi^{\mathrm{S}}(\mathbf{u}, \mathbf{S}^{\mathrm{P}}, \boldsymbol{\lambda}^{\mathrm{u}}, \boldsymbol{\mu}^{\mathrm{u}}; \mathbf{b}, \mathbf{t}^0, \mathbf{u}^0) \\ = & \int_V \left\{ \Psi(\mathbf{S}^{\mathrm{P}}) + \rho_0 \boldsymbol{\lambda}^{\mathrm{u}} \cdot \mathbf{b} - \mathbf{E}^{\mathrm{L}}(\boldsymbol{\lambda}^{\mathrm{u}}) : \mathbf{S}^{\mathrm{P}} \right\} dV - \int_{\partial V'} \mathbf{T}^0 \cdot \mathbf{u} \, dA \\ & + \int_{\partial V''} \left\{ \mathbf{N} \cdot \{\mathbf{S}^{\mathrm{P}} \mathbf{F}^{\mathrm{T}}(\boldsymbol{\lambda}^{\mathrm{u}})\} \right\} \cdot (\boldsymbol{\lambda}^{\mathrm{u}} - \mathbf{u}^0) \, dA \\ & + \int_{\partial V'} (\boldsymbol{\lambda}^{\mathrm{u}} - \boldsymbol{\mu}^{\mathrm{u}}) \cdot \left\{ \mathbf{N} \cdot \{\mathbf{S}^{\mathrm{P}} \mathbf{F}^{\mathrm{T}}(\boldsymbol{\lambda}^{\mathrm{u}})\} - \mathbf{T}^0 \right\} dA , \end{aligned} \tag{3.10.17a}$$

$$\begin{aligned} \delta\pi_1^{\mathrm{S}} = & \left\{ \frac{\partial \Psi}{\partial \mathbf{S}^{\mathrm{P}}} - \mathbf{E}^{\mathrm{L}}(\boldsymbol{\lambda}^{\mathrm{u}}) \right\} : \delta\mathbf{S}^{\mathrm{P}} \quad \text{(compliance)} \\ & + \left\{ \boldsymbol{\nabla}^0 \cdot \{\mathbf{S}^{\mathrm{P}} \mathbf{F}^{\mathrm{T}}(\boldsymbol{\lambda}^{\mathrm{u}})\} + \rho_0 \mathbf{b} \right\} \cdot \delta\boldsymbol{\lambda}^{\mathrm{u}} \ \text{(equilibrium)}, \end{aligned} \tag{3.10.17b}$$

$$\begin{aligned} \delta\pi_2^{\mathrm{S}} = & \left\{ \mathbf{N} \cdot \{\mathbf{S}^{\mathrm{P}} \mathbf{F}^{\mathrm{T}}(\boldsymbol{\lambda}^{\mathrm{u}})\} - \mathbf{T}^0 \right\} \cdot (-\delta\boldsymbol{\mu}^{\mathrm{u}}) \ \text{(tractions B.C.)} \\ & + (\boldsymbol{\lambda}^{\mathrm{u}} - \boldsymbol{\mu}^{\mathrm{u}}) \cdot \left\{ \mathbf{N} \cdot \delta\{\mathbf{S}^{\mathrm{P}} \mathbf{F}^{\mathrm{T}}(\boldsymbol{\lambda}^{\mathrm{u}})\} \right\} \quad \text{(dual displacements B.C.)}, \end{aligned} \tag{3.10.17c}$$

$$\delta\pi_3^{\mathrm{S}} = (\boldsymbol{\lambda}^{\mathrm{u}} - \mathbf{u}^0) \cdot \left\{ \mathbf{N} \cdot \delta\{\mathbf{S}^{\mathrm{P}} \mathbf{F}^{\mathrm{T}}(\boldsymbol{\lambda}^{\mathrm{u}})\} \right\} \quad \text{(dual displacements B.C.)}, \tag{3.10.17d}$$

where

$$\delta\{\mathbf{S}^{\mathrm{P}} \mathbf{F}^{\mathrm{T}}(\boldsymbol{\lambda}^{\mathrm{u}})\} = \delta\mathbf{S}^{\mathrm{P}} \left[\mathbf{1} + \frac{\partial \boldsymbol{\lambda}^{\mathrm{u}}}{\partial \mathbf{X}} \right]^{\mathrm{T}} + \mathbf{S}^{\mathrm{P}} \left[\frac{\partial \delta\boldsymbol{\lambda}^{\mathrm{u}}}{\partial \mathbf{X}} \right]^{\mathrm{T}}. \tag{3.10.17e}$$

This reduces to Reissner's (1950) results when $\mathbf{u} = \boldsymbol{\lambda}^{\mathrm{u}} = \boldsymbol{\mu}^{\mathrm{u}}$ is substituted.

3.10.5. Variational Principles with Discontinuous Fields

Variational principles with discontinuous fields attain significance in connection with: (1) the statics and dynamics of composite materials in which material properties vary discontinuously; (2) fracture mechanics, where across some interfaces certain field quantities can admit finite jumps; and (3) finite-element formulation of problems in continuum mechanics, in which solutions are sought on classes of functions which, together with their derivatives, are continuous within an element, but vary discontinuously across two adjacent elements. A given variational statement can be augmented by a suitable jump condition in order to account for any one, or all of the above possibilities.

Let the initial configuration C_0 of the body B be divided into a finite number of regular regions which are separated by discontinuity surfaces. Divide these surfaces into three categories: portion S' on which jump $[\mathbf{T}^0]$ in certain components of the tractions is prescribed[14]; portion S'' on which jump $[\mathbf{u}^0]$ in certain components of the displacements is prescribed; and finally, portion $\bar{S}$ on which the test functions that are used for the purpose of obtaining solutions, admit discontinuities. Note that S, S', and $\bar{S}$ may have parts in common. Denote by S the collection of all these discontinuity surfaces, and let $\mathbf{N}^S$ be its unit normal which points outward from one domain of regularity, say domain 1, toward the adjacent subregion, say domain 2.

Define the jump [q] of a field quantity q at a point P on S by

$$[q] = q^{(1)} - q^{(2)}, \tag{3.10.18a}$$

where $q^{(1)}$ and $q^{(2)}$ are the limiting values of q at P, as this point is approached along $\mathbf{N}^S$ from the interior of domain 1 and 2, respectively. For equilibrium, the tractions across any interior surface must be continuous. When this requirement is relaxed, the jump in the tractions on a surface of discontinuity S is defined in terms of the jump in the nominal stress, as follows:

$$\begin{aligned}[\mathbf{T}] &\equiv \mathbf{N}^S \cdot [\mathbf{S}^N] = \mathbf{N}^S \cdot (\mathbf{S}^N)^{(1)} - \mathbf{N}^S \cdot (\mathbf{S}^N)^{(2)} \\ &= \mathbf{N}^1 \cdot (\mathbf{S}^N)^{(1)} + \mathbf{N}^2 \cdot (\mathbf{S}^N)^{(2)} = \mathbf{T}^{(1)} + \mathbf{T}^{(2)}.\end{aligned} \tag{3.10.18b}$$

Then the jump in the product $\mathbf{T} \cdot \mathbf{u}$, on the discontinuous surface can be expressed as follows:

$$[\mathbf{T} \cdot \mathbf{u}] = \mathbf{T}^{(1)} \cdot \mathbf{u}^{(1)} - \mathbf{T}^{(2)} \cdot \mathbf{u}^{(2)} = \bar{\mathbf{T}} \cdot [\mathbf{u}] + [\mathbf{T}] \cdot \bar{\mathbf{u}}, \tag{3.10.18c}$$

where $\bar{\mathbf{T}}$ and $\bar{\mathbf{u}}$ may be the weighted averages defined by (Nemat-Nasser, 1972a, 1974)

$$\bar{\mathbf{T}} = \alpha \mathbf{T}^{(1)} - (1-\alpha)\mathbf{T}^{(2)}, \quad \bar{\mathbf{u}} = (1-\alpha)\mathbf{u}^{(1)} + \alpha \mathbf{u}^{(2)}, \tag{3.10.19a,b}$$

where α is an arbitrary parameter. The weighted average traction $\bar{\mathbf{T}}$ can be

[14] Note that the prescribed jumps $[\mathbf{T}^0]$ and $[\mathbf{u}^0]$, at a common point on the discontinuity surface, must be *complementary*.

expressed as

$$\overline{\mathbf{T}} = \mathbf{N}^S \cdot \overline{\mathbf{S}^N}, \quad \overline{\mathbf{S}^N} = \alpha(\mathbf{S}^N)^{(1)} + (1-\alpha)(\mathbf{S}^N)^{(2)}. \tag{3.10.19c}$$

In view of (3.10.18), the functional Π in (3.10.9) can be augmented to include conditions which must hold at surfaces of discontinuity, as follows:

$$\Pi^* = \Pi - \int_{S'} \left\{ [\mathbf{T}^0] \cdot \overline{\mathbf{u}} + \boldsymbol{\mu}^{u*} \cdot \{\mathbf{N} \cdot [\mathbf{S}^N] - [\mathbf{T}^0]\} \right\} dA$$

$$- \int_{S''} \left\{ (\mathbf{N} \cdot \overline{\mathbf{S}^N}) \cdot [\mathbf{u}^0] + \boldsymbol{\mu}^{t*} \cdot \{[\mathbf{u}] - [\mathbf{u}^0]\} \right\} dA$$

$$- \int_{\overline{S}} \{\boldsymbol{\mu}^{t*} \cdot [\mathbf{u}] + \boldsymbol{\mu}^{u*} \cdot (\mathbf{N} \cdot [\mathbf{S}^N])\} \, dA, \tag{3.10.20}$$

where the two vector fields $\boldsymbol{\mu}^{t*}$ and $\boldsymbol{\mu}^{u*}$ on S are the Lagrange multipliers. Then the variation in Π^* is

$$\delta\Pi^* = \delta\Pi + \int_{S'} \delta\pi_1^* \, dA + \int_{S''} \delta\pi_2^* \, dA + \int_{\overline{S}} \delta\pi_3^* \, dA,$$

$$\begin{aligned}
\delta\pi_1^* = &\quad \{\mathbf{N} \cdot [\boldsymbol{\Lambda}^N] - [\mathbf{T}^0]\} \cdot \delta\overline{\mathbf{u}} && \text{(dual tractions J.C.)} \\
&- \{\mathbf{N} \cdot [\mathbf{S}^N] - [\mathbf{T}^0]\} \cdot \delta\boldsymbol{\mu}^{u*} && \text{(tractions J.C.)} \\
&+ (\overline{\boldsymbol{\lambda}}^u - \boldsymbol{\mu}^{u*}) \cdot \{\mathbf{N} \cdot \delta[\mathbf{S}^N]\} && \text{(dual displacements B.C.)}, \\
\delta\pi_2^* = &\quad \{\mathbf{N} \cdot \overline{\boldsymbol{\Lambda}}^N - \boldsymbol{\mu}^{t*}\} \cdot \delta[\mathbf{u}] && \text{(dual tractions B.C.)} \\
&+ \{[\boldsymbol{\lambda}^u] - [\mathbf{u}^0]\} \cdot (\mathbf{N} \cdot \delta\overline{\mathbf{S}^N}) && \text{(dual displacements J.C.)} \\
&- \{[\mathbf{u}] - [\mathbf{u}^0]\} \cdot \delta\boldsymbol{\mu}^{t*} && \text{(displacements J.C.)}, \\
\delta\pi_3^* = &\quad \mathbf{N} \cdot [\boldsymbol{\Lambda}^N] \cdot \delta\overline{\mathbf{u}} && \text{(dual traction continuity)} \\
&- \mathbf{N} \cdot [\mathbf{S}^N] \cdot \delta\boldsymbol{\mu}^{t*} && \text{(traction continuity)} \\
&+ (\overline{\boldsymbol{\lambda}}^u - \boldsymbol{\mu}^{u*}) \cdot \{\mathbf{N} \cdot \delta[\mathbf{S}^N]\} && \text{(dual displacement B.C.)} \\
&+ \{\mathbf{N} \cdot \overline{\boldsymbol{\Lambda}}^N - \boldsymbol{\mu}^{t*}\} \cdot \delta[\mathbf{u}] && \text{(dual traction B.C.)} \\
&+ [\boldsymbol{\lambda}^u] \cdot (\mathbf{N} \cdot \delta\overline{\mathbf{S}^N}) && \text{(dual displacement continuity)} \\
&- [\mathbf{u}] \cdot \delta\boldsymbol{\mu}^{t*} && \text{(displacement continuity)},
\end{aligned} \tag{3.10.21a-d}$$

where J.C. stands for "jump conditions."

3.10.6. Incremental Formulation

In actual calculations the incremental formulation of large deformation elasticity problems is more useful, since (1) for each increment of loading a linearized problem may be considered, and (2) an assessment can be made of the stability of a given configuration before proceeding to the next incremental loading (Hill, 1957). Calculations based on an incremental formulation, however, accumulate error, and the accuracy of the final results is highly dependent upon the size of each load increment. This shortcoming may be remedied to a certain extent if a variational formulation is used, which provides the incremental field equations, as well as the equations which define the finitely deformed state preceding the incremental load. Then for the numerical calculation a combination of the linearized incremental formulation and an iterative scheme may be used, where the numerical results obtained by means of the incremental method are employed as the initial values of the field quantities, in order to start the iteration[15].

Let the body be in an equilibrium configuration C, and consider the increments of body forces $\Delta\mathbf{b}$, of surface tractions $\Delta\mathbf{T}^0$, and of surface displacements $\Delta\mathbf{u}^0$, and denote the increment of the field quantity q by Δq.

The strain-energy-density and the comlementary-energy-density functions, after the incremental loading, become

$$\Phi(\mathbf{F}+\Delta\mathbf{F}) = \Phi(\mathbf{F}) + \frac{\partial\Phi}{\partial F_{iA}}\,\Delta F_{iA} + \frac{1}{2}\,\frac{\partial^2\Phi}{\partial F_{iA}\,\partial F_{jB}}\,\Delta F_{iA}\,\Delta F_{jB} + \ldots,$$

$$\Psi(\mathbf{S}^N+\Delta\mathbf{S}^N) = \Psi(\mathbf{S}^N) + \frac{\partial\Psi}{\partial S^N_{Ai}}\,\Delta S^N_{Ai} + \frac{1}{2}\,\frac{\partial^2\Psi}{\partial S^N_{Ai}\,\partial S^N_{Bj}}\,\Delta S^N_{Ai}\,\Delta S^N_{Bj} + \ldots, \tag{3.10.22a,b}$$

where it is assumed that Φ and Ψ are sufficiently smooth functions of their arguments, in order to permit the involved expansion.

For the incremental loading, new functionals can be obtained by replacing the field quantities by their new values in (3.10.9). For example, $\mathbf{u}$ by $\mathbf{u}+\Delta\mathbf{u}$. However, since only the variation in the incremental field quantities is considered, the terms which are not functions of the increments can be dropped. In this manner, consider the functional $\Delta\Pi$ defined by

$$\begin{aligned}
\Delta\Pi \equiv \int_V \Big\{ & \frac{\partial\Phi}{\partial F_{iA}}\,\Delta F_{iA} + \frac{1}{2}\,\frac{\partial^2\Phi}{\partial F_{iA}\,\partial F_{jB}}\,\Delta F_{iA}\,\Delta F_{jB} + \frac{\partial\Psi}{\partial S^N_{Ai}}\,\Delta S^N_{Ai} \\
& + \frac{1}{2}\,\frac{\partial^2\Psi}{\partial S^N_{Ai}\,\partial S^N_{Bj}}\,\Delta S^N_{Ai}\,\Delta S^N_{Bj} - \rho_0\,(\mathbf{b}+\Delta\mathbf{b})\cdot\Delta\mathbf{u}\Big\}\,dV \\
& - \int_{\partial V'} \{\mathbf{T}^0+\Delta\mathbf{T}^0\}\cdot\Delta\mathbf{u}\,dA - \int_{\partial V''} \Big\{\boldsymbol{\mu}^t\cdot\Delta\mathbf{u} + \Delta\boldsymbol{\mu}^t\cdot\{\mathbf{u}-\mathbf{u}^0+\Delta\mathbf{u}-\Delta\mathbf{u}^0\}\Big\}\,dA \\
& + \int_V \Big\{ \Delta\boldsymbol{\lambda}^u\cdot\{\boldsymbol{\nabla}^0\cdot(\mathbf{S}^N+\Delta\mathbf{S}^N) + \rho_0\,(\mathbf{b}+\Delta\mathbf{b})\} + \boldsymbol{\lambda}^u\cdot\boldsymbol{\nabla}^0\cdot\Delta\mathbf{S}^N \\
& + \Delta\boldsymbol{\Lambda}^N : \{\mathbf{F}(\mathbf{u}+\Delta\mathbf{u}) - \mathbf{F} - \Delta\mathbf{F}\} + \boldsymbol{\Lambda}^N : \{\mathbf{F}(\Delta\mathbf{u}) - \Delta\mathbf{F}\}\Big\}\,dV \\
& - \int_{\partial V''} (\mathbf{N}\cdot\Delta\mathbf{S}^N) : (\mathbf{u}^0+\Delta\mathbf{u}^0)\,dA \\
& - \int_{\partial V'} \Big\{\Delta\boldsymbol{\mu}^u\cdot\{\mathbf{N}\cdot(\mathbf{S}^N+\Delta\mathbf{S}^N) - \mathbf{T}^0 - \Delta\mathbf{T}^0\} + \boldsymbol{\mu}^u\cdot\mathbf{N}\cdot\Delta\mathbf{S}^N\Big\}\,dA\,.
\end{aligned} \tag{3.10.23}$$

---

[15] For illustration, see Nemat-Nasser and Shatoff (1971, 1973).

The corresponding variation becomes

$$\delta\Delta\Pi = \int_V \delta\Delta\pi_1\, dV + \int_{\partial V'} \delta\Delta\pi_2\, dA + \int_{\partial V''} \delta\Delta\pi_3\, dA, \tag{3.10.24}$$

where the integrands are given as follows:

$$\begin{aligned}
\delta\Delta\pi_1 = & \left\{\frac{\partial\Phi}{\partial\mathbf{F}^T} - \mathbf{\Lambda}^N\right\} : \delta\Delta\mathbf{F} + \left\{\frac{1}{2}\frac{\partial^2\Phi}{\partial\mathbf{F}^T\partial\mathbf{F}^T} : \Delta\mathbf{F} - \Delta\mathbf{\Lambda}^N\right\} : \delta\Delta\mathbf{F} \\
& + \left\{\frac{\partial\Psi}{\partial(\mathbf{S}^N)^T} - \mathbf{F}(\boldsymbol{\lambda}^u)\right\} : \delta\Delta\mathbf{S}^N + \left\{\frac{1}{2}\frac{\partial^2\Psi}{\partial(\mathbf{S}^N)^T\partial(\mathbf{S}^N)^T} : \Delta\mathbf{S}^N - \right. \\
& \left. \frac{\partial(\Delta\boldsymbol{\lambda}^u)}{\partial\mathbf{X}}\right\} : \delta\Delta\mathbf{S}^N \\
& + \{\boldsymbol{\nabla}^0 \cdot \mathbf{S}^N + \rho_0\,\mathbf{b}\} \cdot \delta\Delta\boldsymbol{\lambda}^u + \{\boldsymbol{\nabla}^0 \cdot \Delta\mathbf{S}^N + \rho_0\,\Delta\mathbf{b}\} \cdot \delta\Delta\boldsymbol{\lambda}^u \\
& + \{\mathbf{F}(\mathbf{u}) - \mathbf{F}\} : \delta\Delta\mathbf{\Lambda}^N + \{\partial(\Delta\mathbf{u})/\partial\mathbf{X} - \Delta\mathbf{F}\} : \delta\Delta\mathbf{\Lambda}^N \\
& + \{\boldsymbol{\nabla}^0 \cdot \mathbf{\Lambda}^N + \rho_0\,\mathbf{b}\} \cdot (-\delta\Delta\mathbf{u}),\ + \{\boldsymbol{\nabla}^0 \cdot \Delta\mathbf{\Lambda}^N + \rho_0\,\Delta\mathbf{b}\} \cdot (-\delta\Delta\mathbf{u}),
\end{aligned} \tag{3.10.25a}$$

$$\begin{aligned}
\delta\Delta\pi_2 = & \{\mathbf{N} \cdot \mathbf{\Lambda}^N - \mathbf{T}^0\} \cdot \delta\Delta\mathbf{u} + \{\mathbf{N} \cdot \Delta\mathbf{\Lambda}^N - \Delta\mathbf{T}^0\} \cdot \delta\Delta\mathbf{u} \\
& + \{\mathbf{N} \cdot \mathbf{S}^N - \mathbf{T}^0\} \cdot (-\delta\Delta\boldsymbol{\mu}^u) + \{\mathbf{N} \cdot \Delta\mathbf{S}^N - \Delta\mathbf{T}^0\} \cdot (-\delta\Delta\boldsymbol{\mu}^u) \\
& + (\boldsymbol{\lambda}^u - \boldsymbol{\mu}^u) \cdot \{\mathbf{N} \cdot \delta\Delta\mathbf{S}^N\} + (\Delta\boldsymbol{\lambda}^u - \Delta\boldsymbol{\mu}^u) \cdot \{\mathbf{N} \cdot \delta\Delta\mathbf{S}^N\},
\end{aligned} \tag{3.10.25b}$$

$$\begin{aligned}
\delta\Delta\pi_3 = & \{\mathbf{N} \cdot \mathbf{\Lambda}^N - \boldsymbol{\mu}^t\} \cdot \delta\Delta\mathbf{u} + \{\mathbf{N} \cdot \Delta\mathbf{\Lambda}^N - \Delta\boldsymbol{\mu}^t\} \cdot \delta\Delta\mathbf{u} \\
& + (\boldsymbol{\lambda}^u - \mathbf{u}^0) \cdot \{\mathbf{N} \cdot \delta\Delta\mathbf{S}^N\} + (\Delta\boldsymbol{\lambda}^u - \Delta\mathbf{u}^0) \cdot \{\mathbf{N} \cdot \delta\Delta\mathbf{S}^N\} \\
& + (\mathbf{u} - \mathbf{u}^0) \cdot (-\delta\Delta\boldsymbol{\mu}^t),\ + (\Delta\mathbf{u} - \Delta\mathbf{u}^0) \cdot (-\delta\Delta\boldsymbol{\mu}^t).
\end{aligned} \tag{3.10.25c}$$

It is observed that the quantity inside the *first braces in each line* in (3.10.25) yields the field equations that characterize the equilibrium condition of the finitely deformed configuration C, as can be seen by comparison to (3.10.10). The vanishing of the terms inside the second braces in each line, for arbitrary variations in the corresponding field quantities, on the other hand, yields all the field equations and boundary data for the small incremental deformations superimposed on configuration C.

It is clear that a similar incremental formulation can be given in terms of a general strain measure **E** and its conjugate stress **S**, similarly to that discussed in Subsection 3.10.2. Since the results can be read off in analogy with (3.10.25), they are not written down here.

In a manner similar to (3.10.20), the functional $\Delta\Pi$ in (3.10.23) can be augmented to include the conditions at surfaces of discontinuity, as follows:

$$\begin{aligned}
\Delta\Pi^* = \Delta\Pi - \int_S \Big\{ & [\mathbf{T}^0 + \Delta\mathbf{T}^0] \cdot \Delta\bar{\mathbf{u}} + \boldsymbol{\mu}^{u*} \cdot \mathbf{N} \cdot [\Delta\mathbf{S}^N] \\
& + \Delta\boldsymbol{\mu}^{u*} \cdot \{\mathbf{N} \cdot [\mathbf{S}^N + \Delta\mathbf{S}^N] - [\mathbf{T}^0 + \Delta\mathbf{T}^0]\}\Big\}\, dA
\end{aligned}$$

$$- \int_{S''} \Big\{ (\mathbf{N} \cdot \overline{\Delta \mathbf{S}^{N}}) \cdot [\mathbf{u}^0 + \Delta \mathbf{u}^0] + \boldsymbol{\mu}^{t*} \cdot [\Delta \mathbf{u}]$$

$$+ \Delta \boldsymbol{\mu}^{t*} \cdot \{[\mathbf{u} + \Delta \mathbf{u}] - [\mathbf{u}^0 + \Delta \mathbf{u}^0]\} \Big\} \, dA$$

$$- \int_{\bar{S}} \Big\{ \Delta \boldsymbol{\mu}^{t*} \cdot [\mathbf{u} + \Delta \mathbf{u}] + \boldsymbol{\mu}^{t*} \cdot [\Delta \mathbf{u}] + \boldsymbol{\mu}^{u*} \cdot (\mathbf{N} \cdot [\Delta \mathbf{S}^{N}])$$

$$+ \Delta \boldsymbol{\mu}^{u*} \cdot (\mathbf{N} \cdot [\mathbf{S}^{N} + \Delta \mathbf{S}^{N}]) \Big\} \, dA . \qquad (3.10.26)$$

The first variation of this gives

$$\delta \Delta \Pi^* = \delta \Delta \Pi + \int_{S'} \delta \Delta \pi_1^* \, dA + \int_{S''} \delta \Delta \pi_2^* \, dA + \int_{\bar{S}} \delta \Delta \pi_3^* \, dA , \qquad (3.10.27a)$$

where the integrands are

$$\begin{aligned} \delta \Delta \pi_1^* = \; & \{\mathbf{N} \cdot [\boldsymbol{\Lambda}^{N}] - [\mathbf{T}^0]\} \cdot \delta \Delta \bar{\mathbf{u}} + \{\mathbf{N} \cdot [\Delta \boldsymbol{\Lambda}^{N}] - [\Delta \mathbf{T}^0]\} \cdot \delta \Delta \bar{\mathbf{u}} \\ & - \{\mathbf{N} \cdot [\mathbf{S}^{N}] - [\mathbf{T}^0]\} \cdot \delta \Delta \boldsymbol{\mu}^{u*} - \{\mathbf{N} \cdot [\Delta \mathbf{S}^{N}] - [\Delta \mathbf{T}^0]\} \cdot \delta \Delta \boldsymbol{\mu}^{u*} \\ & + (\bar{\boldsymbol{\lambda}}^{u} - \boldsymbol{\mu}^{u*}) \cdot \{\mathbf{N} \cdot \delta [\Delta \mathbf{S}^{N}]\} + (\Delta \bar{\boldsymbol{\lambda}}^{u} - \Delta \boldsymbol{\mu}^{u*}) \cdot \{\mathbf{N} \cdot \delta [\Delta \mathbf{S}^{N}]\} , \end{aligned} \qquad (3.10.27b)$$

$$\begin{aligned} \delta \Delta \pi_2^* = \; & \{\mathbf{N} \cdot \bar{\boldsymbol{\Lambda}}^{N} - \boldsymbol{\mu}^{t*}\} \cdot \delta [\Delta \mathbf{u}] + \{\mathbf{N} \cdot \Delta \bar{\boldsymbol{\Lambda}}^{N} - \Delta \boldsymbol{\mu}^{t*}\} \cdot \delta [\Delta \mathbf{u}] \\ & + \{[\boldsymbol{\lambda}^{u}] - [\mathbf{u}^0]\} \cdot (\mathbf{N} \cdot \delta \overline{\Delta \mathbf{S}^{N}}) + \{[\Delta \boldsymbol{\lambda}^{u}] - [\Delta \mathbf{u}^0]\} \cdot (\mathbf{N} \cdot \delta \overline{\Delta \mathbf{S}^{N}}) \\ & - \{[\mathbf{u}] - [\mathbf{u}^0]\} \cdot \delta \Delta \boldsymbol{\mu}^{t*} - \{[\Delta \mathbf{u}] - [\Delta \mathbf{u}^0]\} \cdot \delta \Delta \boldsymbol{\mu}^{t*} , \end{aligned} \qquad (3.10.27c)$$

$$\begin{aligned} \delta \Delta \pi_3^* = \; & \mathbf{N} \cdot [\boldsymbol{\Lambda}^{N}] \cdot \delta \Delta \bar{\mathbf{u}} + \mathbf{N} \cdot [\Delta \boldsymbol{\Lambda}^{N}] \cdot \delta \Delta \bar{\mathbf{u}} \\ & - \mathbf{N} \cdot [\mathbf{S}^{N}] \cdot \delta \Delta \boldsymbol{\mu}^{t*} - \mathbf{N} \cdot [\Delta \mathbf{S}^{N}] \cdot \delta \Delta \boldsymbol{\mu}^{t*} \\ & + (\bar{\boldsymbol{\lambda}}^{u} - \boldsymbol{\mu}^{u*}) \cdot \{\mathbf{N} \cdot \delta [\Delta \mathbf{S}^{N}]\} + (\Delta \bar{\boldsymbol{\lambda}}^{u} - \Delta \boldsymbol{\mu}^{u*}) \cdot \{\mathbf{N} \cdot \delta [\Delta \mathbf{S}^{N}]\} \\ & + \{\mathbf{N} \cdot \bar{\boldsymbol{\Lambda}}^{N} - \boldsymbol{\mu}^{t*}\} \cdot \delta [\Delta \mathbf{u}] + \{\mathbf{N} \cdot \overline{\Delta \boldsymbol{\Lambda}}^{N} - \Delta \boldsymbol{\mu}^{t*}\} \cdot \delta [\Delta \mathbf{u}] \\ & + \{[\boldsymbol{\lambda}^{u}] \cdot (\mathbf{N} \cdot \delta \overline{\Delta \mathbf{S}^{N}}) + \{[\Delta \boldsymbol{\lambda}^{u}] \cdot (\mathbf{N} \cdot \delta \overline{\Delta \mathbf{S}^{N}}) \\ & - [\mathbf{u}] \cdot \delta \Delta \boldsymbol{\mu}^{t*} - [\Delta \mathbf{u}] \cdot \delta \Delta \boldsymbol{\mu}^{t*} . \end{aligned} \qquad (3.10.27d)$$

If it is assumed that the field equations corresponding to configuration C are identically satisfied, then expressions (3.10.23) and (3.10.26) can be simplified to yield the incremental field equations only. In this case, for example, the functional $\Delta\Pi$ in (3.10.23) reduces to

$$\Delta \Pi \equiv \int_{V} \Big\{ \frac{1}{2} \frac{\partial^2 \Phi}{\partial F_{iA} \, \partial F_{jB}} \Delta F_{iA} \, \Delta F_{jB} + \frac{1}{2} \frac{\partial^2 \Psi}{\partial S^{N}_{Ai} \, \partial S^{N}_{Bj}} \Delta S^{N}_{Ai} \, \Delta S^{N}_{Bj} - \rho_0 \, \Delta \mathbf{b} \cdot \Delta \mathrm{u} \Big\} \, dV$$

$$- \int_{\partial V'} \Delta \mathbf{T}^0 \cdot \Delta \mathbf{u} \, dA - \int_{\partial V''} \Delta \boldsymbol{\mu}^{t} \cdot \{\Delta \mathbf{u} - \Delta \mathbf{u}^0\} \, dA$$

$$+\int_V \left\{ \Delta\boldsymbol{\lambda}^u \cdot \{\boldsymbol{\nabla}^0 \cdot \Delta\mathbf{S}^N + \rho_0\,\Delta\mathbf{b}\} + \Delta\boldsymbol{\Lambda}^N : \{\partial(\Delta\mathbf{u})/\partial\mathbf{X} - \Delta\mathbf{F}\} \right\} dV$$

$$- \int_{\partial V^u} (\mathbf{N} \cdot \Delta\mathbf{S}^N) \cdot \Delta\mathbf{u}^0\, dA - \int_{\partial V^t} \Delta\boldsymbol{\mu}^u \cdot \{\mathbf{N} \cdot \Delta\mathbf{S}^N - \Delta\mathbf{T}^0\}\, dA\,. \qquad (3.10.28)$$

3.10.7. Linearized Formulation

The linearized formulation of the variational statements can be regarded as a special case of the incremental formulation, where (1) all quantities are referred to the given deformed state (Eulerian formulation), and (2) the field equations corresponding to the superimposed incremental deformation are linearized. It is therefore more useful to consider first the general case of small deformations superimposed on large ones, and then to develop the usual linearized equations as special cases.

From the field equations (3.10.25a), the incremental nominal stress and the deformation gradient are related by

$$\Delta S^N_{Ai} = \frac{1}{2}\,\frac{\partial^2\Phi}{\partial F_{iA}\,\partial F_{jB}}\,\Delta F_{jB}\,,$$

$$\Delta F_{iA} = \frac{1}{2}\,\frac{\partial^2\Psi}{\partial S^N_{Ai}\,\partial S^N_{Bj}}\,\Delta S^N_{Bj}\,. \qquad (3.10.29a,b)$$

Introduce the following field quantities and boundary data when the deformed configuration is used as reference:

$$d\mathbf{S}^N = \frac{1}{J}\,\mathbf{F}\,\Delta\mathbf{S}^N\,, \quad d\boldsymbol{\Lambda}^N = \frac{1}{J}\,\mathbf{F}\,\Delta\boldsymbol{\Lambda}^N\,, \quad d\mathbf{F} = \Delta\mathbf{F}\,\mathbf{F}^{-1}\,,$$

$$d\mathbf{t}^0 = \frac{dA}{da}\,\Delta\mathbf{T}^0\,, \quad d\boldsymbol{\mu}^t = \frac{dA}{da}\,\Delta\boldsymbol{\mu}^t\,, \quad d\mathbf{b} = \frac{\rho_0}{\rho}\,\Delta\mathbf{b}\,,$$

$$d\mathbf{u} = \Delta\mathbf{u}\,, \quad d\boldsymbol{\mu}^u = \Delta\boldsymbol{\mu}^u\,, \quad d\boldsymbol{\lambda}^u = \Delta\boldsymbol{\lambda}^u\,. \qquad (3.10.30a\text{-}i)$$

In terms of $d\mathbf{S}^N$ and $d\mathbf{F}$, (3.10.29a,b) can be expressed as

$$dS^N_{ij} = \mathcal{L}_{ijkl}\, dF_{lk}\,, \quad \mathcal{L}_{ijkl} = \frac{1}{J}\,F_{iA}\,\frac{\partial^2\Phi}{\partial F_{jA}\,\partial F_{lB}}\,F_{kB}\,,$$

$$dF_{ij} = \mathcal{M}_{ijkl}\, dS^N_{lk}\,, \quad \mathcal{M}_{ijkl} = J\,F^{-1}_{Aj}\,\frac{\partial^2\Psi}{\partial S^N_{Ai}\,\partial S^N_{Bk}}\,F^{-1}_{Bl}\,. \qquad (3.10.31a\text{-}d)$$

Then the functional $\Delta\Pi$ becomes

$$\Delta\Pi \equiv \int_v \left\{ \frac{1}{2}\,\mathcal{L}_{ijkl}\, dF_{ji}\, dF_{lk} + \frac{1}{2}\,\mathcal{M}_{ijkl}\, dS^N_{ji}\, dS^N_{lk} - \rho\, d\mathbf{b} \cdot d\mathbf{u} \right\} dv$$

$$- \int_{\partial v^t} d\mathbf{t}^0 \cdot d\mathbf{u}\, da - \int_{\partial v^u} d\boldsymbol{\mu}^t \cdot \{d\mathbf{u} - d\mathbf{u}^0\}\, da$$

$$+ \int_v \left\{ d\boldsymbol{\lambda}^u \cdot \{\boldsymbol{\nabla} \cdot d\mathbf{S}^N + \rho\, d\mathbf{b}\} + d\boldsymbol{\Lambda}^N : \{\partial(d\mathbf{u})/\partial\mathbf{x} - d\mathbf{F}\} \right\} dv$$

$$- \int_{\partial v''} (\mathbf{n} \cdot \mathrm{d}\mathbf{S}^{\mathrm{N}}) \cdot \mathrm{d}\mathbf{u}^{0} \,\mathrm{da} - \int_{\partial v'} \mathrm{d}\boldsymbol{\mu}^{\mathrm{u}} \cdot \{\mathbf{n} \cdot \mathrm{d}\mathbf{S}^{\mathrm{N}} - \mathrm{d}\mathbf{t}^{0}\} \,\mathrm{da}. \tag{3.10.32}$$

In the linearized formulation, the effective constitutive modulus tensors $\mathcal{L}$ and $\mathcal{M}$ are assumed to be constant during the incremental deformation. Then the corresponding variation becomes

$$\delta\Delta\Pi = \int_{v} \delta\Delta\pi_1 \,\mathrm{dv} + \int_{\partial v'} \delta\Delta\pi_2 \,\mathrm{da} + \int_{\partial v''} \delta\Delta\pi_3 \,\mathrm{da}, \tag{3.10.33a}$$

where the integrands are given as follows:

$$\begin{aligned} \delta\pi_1 = \quad & \{\mathcal{L} : \mathrm{d}\mathbf{F} - \mathrm{d}\boldsymbol{\Lambda}^{\mathrm{N}}\} : \delta\mathrm{d}\mathbf{F} && \text{(elasticity)} \\ + \quad & \{\mathcal{M} : \mathrm{d}\mathbf{S}^{\mathrm{N}} - \partial(\delta\mathrm{d}\boldsymbol{\lambda}^{\mathrm{u}})/\partial\mathbf{x}\} : \delta\mathrm{d}\mathbf{S}^{\mathrm{N}} && \text{(compliance)} \\ + \quad & \{\boldsymbol{\nabla} \cdot \mathrm{d}\mathbf{S}^{\mathrm{N}} + \rho\,\mathrm{d}\mathbf{b}\} \cdot \delta\mathrm{d}\boldsymbol{\lambda}^{\mathrm{u}} && \text{(equilibrium)} \\ + \quad & \{\partial(\mathrm{d}\mathbf{u})/\partial\mathbf{x} - \mathrm{d}\mathbf{F}\} : \delta\mathrm{d}\boldsymbol{\Lambda}^{\mathrm{N}} && \text{(strain–displacement)} \\ + \quad & \{\boldsymbol{\nabla} \cdot \mathrm{d}\boldsymbol{\Lambda}^{\mathrm{N}} + \rho\,\mathrm{d}\mathbf{b}\} \cdot (-\delta\mathrm{d}\mathbf{u}), && \text{(dual equilibrium)}, \end{aligned} \tag{3.10.33b}$$

$$\begin{aligned} \delta\pi_2 = \quad & \{\mathbf{n} \cdot \mathrm{d}\boldsymbol{\Lambda}^{\mathrm{N}} - \mathrm{d}\mathbf{t}^{0}\} \cdot \delta\mathrm{d}\mathbf{u} && \text{(dual tractions B.C.)} \\ + \quad & \{\mathbf{n} \cdot \mathrm{d}\mathbf{S}^{\mathrm{N}} - \mathrm{d}\mathbf{t}^{0}\} \cdot (-\delta\mathrm{d}\boldsymbol{\mu}^{\mathrm{u}}) && \text{(tractions B.C.)} \\ + \quad & (\mathrm{d}\boldsymbol{\lambda}^{\mathrm{u}} - \mathrm{d}\boldsymbol{\mu}^{\mathrm{u}}) \cdot \{\mathbf{n} \cdot \delta\mathrm{d}\mathbf{S}^{\mathrm{N}}\} && \text{(dual displacements B.C.)}, \end{aligned} \tag{3.10.33c}$$

$$\begin{aligned} \delta\pi_3 = \quad & \{\mathbf{n} \cdot \mathrm{d}\boldsymbol{\Lambda}^{\mathrm{N}} - \mathrm{d}\boldsymbol{\mu}^{\mathrm{t}}\} \cdot \delta\mathrm{d}\mathbf{u} && \text{(dual tractions B.C.)} \\ + \quad & (\mathrm{d}\boldsymbol{\lambda}^{\mathrm{u}} - \mathrm{d}\mathbf{u}^{0}) \cdot \{\mathbf{n} \cdot \delta\mathrm{d}\mathbf{S}^{\mathrm{N}}\} && \text{(dual displacements B.C.)} \\ + \quad & (\mathrm{d}\mathbf{u} - \mathrm{d}\mathbf{u}^{0}) \cdot (-\delta\mathrm{d}\boldsymbol{\mu}^{\mathrm{t}}), && \text{(displacements B.C.)}. \end{aligned} \tag{3.10.33d}$$

3.11. REFERENCES

Belytschko, T. (1983), An overview of semidiscretization and time integration procedures, in *Computational methods for transient analysis* (Belytschko, T. and Hughes, T. J. R., eds.), North-Holland, Amsterdam, 1-65.

Belytschko, T., Liu, W. K., and Moran, B. (2000), *Nonlinear Finite Elements for Continua and Structures,* Wiley & Sons, New York.

Biot, M. (1939), Non-linear theory of elasticity and the linearized case for a body under initial stress, *Phil. Mag.*, Vol. 27, Ser. 7, 468-489.

Biot, M. (1965), *Mechanics of incremental deformations*, John Wiley & Sons, New York.

Fu, F. C. L. and Nemat-Nasser, S. (1974), Harmonic waves in layered composites: bounds on frequencies, *J. Appl. Mech.*, Vol. 41, 288-290.

Hellinger, E. (1914), Die allgemeinen Ansätze der Mechanik der Kontinua, *Enz. math. Wis.*, Vol. 4, 602-694.

Herrmann, L. (1965), Elasticity equations for incompressible and nearly incompressible materials by a variational theorem, *AIAAJ*, Vol. 3, 1896-1901.

Hill, R. (1956), New horizons in the mechanics of solids, *J. Mech. Phys. Solids*, Vol. 5, 66-74.

Hill, R. (1957), On uniqueness and stability in the theory of finite elastic strain, *J. Mech. Phys. Solids*, Vol. 5, 229-241.

Hill, R. (1968), On constitutive inequalities for simple materials - I and II, *J. Mech. Phys. Solids*, Vol. 16, 229-242 and 315-322.

Hill, R. (1972), On constitutive macro-variables for heterogeneous solids at finite strain, *Proc. Roy. Soc. Lond. A*, Vol. 326, 131-147.

Hill, R. (1978), Aspects of invariance in solid mechanics, *Adv. Appl. Mech.*, Vol. 18, 1-75.

Hu, H.-C. (1955), On some variational principles in the theory of elasticity and the theory of plasticity, *Scientia Sinica*, Vol. 4, 33-54 (First published in Chinese in 1954).

Koppe, E. (1956), Die Ableitung der Minimalprinzipien der nichtlinearen Elastizitästheorie mittels kanonischer Transformation, *Nachrichten Ak. Wiss. Göttingen, Math.-Phys. Klasse*, Vol. 12, 259-266.

Mikhlin, S. G. (1964), *Variational methods in mathematical physics*, MacMillan, New York.

Mikhlin, S. G. (1965), *The problem of the minimum of a quadratic functional*, Holden Day, San Francisco.

Minagawa, S. and Nemat-Nasser, S. (1975), Harmonic waves in layered composites: comparison among several schemes, *J. Appl. Mech.*, Vol. 42, 699-704.

Minagawa, S. and Nemat-Nasser, S. (1976), Harmonic waves in three-dimensional elastic composites, *Int. J. Solids Struct.*, Vol. 12, 769-777.

Minagawa, S., Nemat-Nasser, S., and Yamada, M. (1981), Finite element analysis of harmonic waves in layered and fiber-reinforced composites, *Int. J. Num. Meth. Engrg*, Vol. 17, 1335-1353.

Minagawa, S., Nemat-Nasser, S., and Yamada, M. (1984), Dispersion of waves in two-dimensional layered, fiber-reinforced, and other elastic composites, *Computers & Structures*, Vol. 19, (Spec. Memorial Issue to Prof. K. Washizu), 119-128.

Minagawa, S., Yoshihara K., and Nemat-Nasser, S. (1992), Analysis of harmonic waves in a composite of materials with piezoelectric effect, *Int. J. Solids and Struct.*, Vol. 28, 1901-1906.

Moreau, J. J. (1967), One-sided constraint in hydrodynamics, *Nonlinear Programming*, North-Holland, Amsterdam, 259-279.

Naghdi, P. M. (1964), On a variational theory in elasticity and its application to shell theory, *J. Appl. Mech.*, Vol. 31, 647-653.

Nemat-Nasser, S. (1972a), On variational methods in finite and incremental elastic deformation problems with discontinuous fields, *Q. Appl. Math.*, Vol. 30, 143-156.

Nemat-Nasser, S. (1972b), General variational methods for waves in elastic composites, *J. Elasticity*, Vol. 2, 73-90.

Nemat-Nasser, S. (1974), General variational principles in nonlinear and linear elasticity with applications, *Mechanics Today*, Vol. 1, 1972, 214-261.

Nemat-Nasser, S. (1983), On finite plastic flow of crystalline solids and geomaterials, *J. Appl. Mech. (50th Anniversary Issue)*, Vol. 50, 1114-1126.

Nemat-Nasser, S. and Hori, M. (1993), *Micromechanics: overall properties of heterogeneous solids*, Elsevier, Amsterdam; second revised edition (1999), Elsevier, Amsterdam.

Nemat-Nasser, S. and Lee, K. N. (1973), Application of general variational methods with discontinuous fields to bending, buckling, and vibration of beams, *Comput. Meth. Appl. Mech. Eng.*, Vol. 2, 33-41.

Nemat-Nasser, S. and Minagawa, S. (1975), Harmonic waves in layered composites: comparison among several schemes, *J. Appl. Mech.*, Vol. 42, 699-704.

Nemat-Nasser, S. and Minagawa, S. (1977), On harmonic waves in layered composites, *J. Appl. Mech.*, Vol. 44, 689-695.

Nemat-Nasser, S. and Shatoff, H. (1971), A consistent numerical method for the solution of nonlinear elasticity problems at finite strains, *SIAM J. Appl. Math.*, Vol. 20, 462-481.

Nemat-Nasser, S. and Shatoff, H. (1973), Numerical analysis of pre-and post-critical response of elastic continua at finite strains, *Int. J. Comput. Struct.*, Vol. 3, 983-999.

Nemat-Nasser, S., Fu, F. C. L., and Minagawa, S. (1975), Harmonic waves in one-, two-, and three-dimensional composites: bounds for eigenfrequencies, *Int. J. Solids Struct.*, Vol. 11, 617-642.

Pian, T. H. and Tong, P. (1969), The basis of finite element methods for solid continua, *Int. J. Num. Meth. Engrg.*, Vol. 1, 3-28.

Prager, W. (1961), *Introduction to mechanics of continua*, Ginn, Boston.

Prager, W. (1967), Variational principles of linear elastostatics for discontinuous displacements, strains and stresses, in *Recent progress in applied mechanics, the Folke-Odqvist volume*, Almqvist and Wiksell, Stockholm, 463-474.

Prange, G. (1916), unpublished *Habilitationsschrift*, Hannover.

Rayleigh, J. W. (1945) *The theory of sound*, Dover, New York.

Reissner, E. (1950), On a variational theorem in elasticity, *J. Math. Phys.*, Vol. 29, 90-95.

Reissner, E. (1953), On a variational theorem for finite elastic deformations, *J. Math. Phys.*, Vol. 32, 129-135.

Reissner, E. (1970), Variational methods and boundary conditions in shell theory, in *Studies in optimization*, Proc. Symp. on Optimization, Vol. 1, Soc. Indus. Appl. Math., Philadelphia, 78-94.

Ritz, W. (1908), Über eine neue Methode zur Lösung gewisser Variationsprobleme der mathematischen Physik, *J. Reine Angew. Math.*, Vol. 135, 1-61.

Ritz, W. (1909), Theorie der Transversalschwingungen einer quadratischen Platte mit freien Rändern, *Annalen der Physik*, Vol. 38, 737-786.

Sandhu, R. S. and Pister, K. S. (1971), Variational principles for boundary value and initial boundary value problems in continuum mechanics, *Int. J. Solids Struct.*, Vol. 7, 639-654.

Sewell, M. J. (1969), On dual approximation principles and optimization in continuum mechanics, *Phil. Trans. Roy. Soc. Lond. A*, Vol. 265, 319-351.

Simo, J. C. and Hughes, T. J. R. (1998), *Computational inelasticity*, Springer Verlag, New York.

Sokolnikoff, I. S. (1956), *Mathematical theory of elasticity*, McGraw-Hill, New York.

Thomas, T. Y. (1957), Extended compatibility conditions for the study of surfaces of discontinuity in continuum mechanics, *J. Math. Mech.*, Vol. 6, 311-322.

Thomas, T. Y. (1961), *Plastic flow and fracture in solids*, Academic Press, New York.

Tong, P. (1969), An assumed stress hybrid finite element method for an incompressible and near-incompressible material, *Int. J. Solids Struct.*, Vol. 5, 455-461.

Tong, P. (1970), New displacement hybrid finite-element models for solid continua, *Int. J. Num. Meth. Eng.*, Vol. 2, 73-83.

Truesdell, C. (1952), The mechanical foundations of elasticity and fluid dynamics, *J. Rat. Mech. Anal.*, Vol. 1, 125-300, and (1953) Vol.2, 593-616.

Truesdell, C. and Toupin, R. (1960), The classical field theories, in *Encyclopedia of physics*, Vol. III/1, Springer-Verlag, Berlin.

Washizu, K. (1955), On the variational principles of elasticity and plasticity, Technical Report, 25-18, March.

Washizu, K. (1968), *Variational methods in elasticity and plasticity*, Pergamon, New York.

4

CONTINUUM THEORIES OF ELASTOPLASTICITY

In this chapter fundamental concepts underlying the continuum theories of elastic-plastic deformations at finite strains and rotations are presented. Certain commonly used plasticity theories are summarized. Particular attention is paid to the thermodynamic basis of inelastic deformation. Conditions for the existence of inelastic potentials are discussed. The results are presented in terms of a general material strain and its conjugate stress (see Chapters 2 and 3), and then specialized for particular applications, emphasizing quantities and theories which are reference- and strain measure-independent. Rate-independent and rate-dependent elastoplasticity relations are developed, starting from a finite deformation version of the von Mises or the J_2*-plasticity model with isotropic and kinematic hardening, and leading to theories which include dilatancy, pressure (mean-stress) sensitivity, frictional effects, and the noncoaxiality of the plastic strain rate and the stress deviator. Application of the results to soils and metals is considered and comparison with some experimental data is presented for illustration. A class of commonly used deformation plasticity theories is examined and their relation to nonlinear elasticity is discussed. For frictional granular materials, the load-induced fabric or texture is related to the kinematic hardening, and a continuum theory, capable of predicting dilatancy and densification in loading and unloading regimes, is developed. The question of plastic spin, and its relation to the decomposition of the deformation gradient into elastic and plastic constituents, is reviewed in some detail. It is shown that, for elastically isotropic materials, this decomposition yields explicit relations which uniquely define all spins in terms of the velocity gradient and the elastic and plastic deformation rates. The issue of possible overall elastic anisotropy in polycrystalline metals is examined in some detail, identifying certain inherent difficulties in its continuum modeling.*

More specifically, the thermodynamics of inelasticity, the concepts of elastic and inelastic potentials, and the basis of the plastic normality rule are examined in Section 4.1. The rate-independent theories of plasticity, with smooth yield surfaces (Section 4.2) and yield surfaces with corners (vertex models, Section 4.4) are presented in a general setting, in terms of an arbitrary strain measure and its conjugate stress measure. Several commonly used plasticity models with isotropic, kinematic, and combined isotropic-kinematic hardening, are developed, and explicit equations useful for computational implementation, are given in various subsections of Section 4.2. The questions of dilatancy, pressure-sensitivity, friction, and their constitutive modeling are examined in some detail in Section 4.3, providing specific illustrative examples and comparisons with experimental results. The vertex models are discussed in Section 4.4, and in Section 4.5, deformation theories of plasticity are examined, starting from nonlinear elasticity and systematically leading to the vertex model of elastoplasticity. The physical basis of the noncoaxiality of the plastic strain rate and the deviatoric stress is discussed in Section 4.6. It is shown in Section 4.7 that this property is an integral part of the response of all frictional materials. The rate- and temperature-dependent models are examined in some detail in Section 4.8. First, several empirical models are discussed. Then, based on the mechanisms of plastic flow by the motion of dislocations, physically-based models for the inelastic deformation of bcc, fcc, and hcp metals are developed and used to predict the flow stress (i.e., stress as a function of strain rate, temperature, and some internal structural parameters, e.g, the average dislocation density) of a class of metals, including refractory (bcc) metals such as commercially pure tantalum, molybdenum, niobium, and vanadium; various types of titanium (hcp); as well as stainless steel, DH-36 steel, HSLA65 steel; and oxygen-free high-conductivity (OFHC) copper (fcc), over a broad range of strains, strain rates, and temperatures, arriving at good correlation with experimental results. Experimentally obtained values of the constitutive parameters of these metals are tabulated and some illustrative comparative examples are presented. Section 4.9 is devoted to an examination of the consequences of the multiplicative decomposition of the total deformation gradient into a plastic and a nonplastic part. The kinematics of continuum elastoplastic deformation is examined in terms of the decomposition of the deformation gradient into purely inelastic and purely elastic constituents with no rigid-body rotation, and an accompanying rigid-body rotation. Objective spin tensors associated with elastic and inelastic strain rates are identified and formulae are given which explicitly express these spins as linear and homogeneous functions of the corresponding deformation rates. Included in this section are the examination of the requirements of the material frame indifference, the general elastic response in terms of an elastic potential, and a detailed description of how, for a given velocity gradient, all the rate quantities can be calculated explicitly, once the elastic potential and the constitutive relations for the plastic strain-rate tensor are given. The results for small elastic deformations and very small elastic strains are obtained as special cases. Also discussed in this section is the notion of backstress and its objective time rate of change, and an alternative additive decomposition of the deformation gradient. A number of controversial issues are critically addressed and clarified, in light of exact kinematical relations and commonly accepted notions of dislocation-induced plasticity of metals.

4.1. ELASTIC AND INELASTIC POTENTIALS

Continuum elastic-plastic constitutive models are considered. The elasticity stems from deformations which do not change the microstructure of the solid and hence do not introduce dislocations, twinning, slip, voids, cracks, breakage of molecular chains, and other irreversible microstructural changes, which are recognized as sources of inelastic deformation.

For a given material element, the microstructure changes in the course of deformation, as existing microdefects evolve and new microdefects emerge. To characterize the thermodynamic state of a material element, therefore, in addition to strain measures and a measure of temperature, parameters commonly referred to as *internal* (or *hidden*) *variables*, are introduced. In a certain sense, these internal variables quantify the microstructure of the material.[1]

More specifically, let $\mathbf{E}$ be a material strain measure,[2] θ be the temperature, and denote by ξ the set of parameters which measure the current microstructure of the material element. The quantity ξ may stand for a set of scalar internal variables, ξ_α, $\alpha = 1, 2, ..., n$, or they may be second-order tensors, or both,[3] which in a certain sense measure the existing defects. *Deformations which leave ξ unchanged are, by definition, elastic. Any inelastic deformation is accompanied by a change in ξ, and conversely, deformations which result in a change of ξ, must include inelastic contributions.* In this general setting, inelasticity is associated with a broad class of changes in the microdefects. Hence the results will apply to a broad class of materials. For example, they apply to ceramics where inelastic deformation may stem from the nucleation and growth of microcracks.[4]

4.1.1. Elastic Potentials

Let the thermodynamic state of the material element be defined by ($\mathbf{E}$, θ, ξ). When ξ is held fixed, the response of the material is elastic. Let $\phi = \phi(\mathbf{E}, \theta; \xi)$ be the Helmholtz free energy.[5] At constant, ξ, the response is thermo-elastic and, therefore,

---

[1] Biot (1954), Meixner (1954), Valanis (1966), Coleman and Gurtin (1967), Schapery (1968), Lubliner (1969), Onat (1970), Mandel (1971), Rice (1971, 1975), Nemat-Nasser (1975a,b, 1977, 1983), Halphen and Nguyen (1975), Kestin and Bataille (1980), and Germain *et al.* (1983).

[2] See Chapter 2.

[3] It is tacitly implied that some of the internal variables may be second-order (or possibly higher, even-order) tensors, although this is not explicitly displayed in the form of the equations. When any of the ξ's is a tensor, then the corresponding conjugate internal force, $\mathbf{\Lambda}$, is also the same rank tensor.

[4] See for example, Rice (1970), Talreja (1985), Krajcinovic (1989, 1996), and Nemat-Nasser and Hori (1990, 1993, 1999), for comments, references, and illustrations.

[5] The semicolon is used here to emphasize the special role of the internal variables ξ. More will be said about this later on.

$$\mathbf{S} = \partial\phi/\partial\mathbf{E} = \mathbf{S}(\mathbf{E}, \theta; \boldsymbol{\xi}),$$
$$\eta = -\partial\phi/\partial\theta = \eta(\mathbf{E}, \theta; \boldsymbol{\xi}), \tag{4.1.1a,b}$$

where $\mathbf{S}$ is the stress conjugate to the strain measure $\mathbf{E}$ in the sense that $\mathbf{S}:\dot{\mathbf{E}}$ is the rate of stress work and η is entropy, both measured per unit reference volume.[6]

The free energy need not be a differentiable function of the history $\boldsymbol{\xi}$. When $\boldsymbol{\xi}$ changes to $\boldsymbol{\xi}+d\boldsymbol{\xi}$, the free energy changes. At constant $\mathbf{E}$ and θ, this change is solely due to microstructural variations. It is purely inelastic. The notation,

$$d^{in}\phi \equiv \phi(\mathbf{E}, \theta; \boldsymbol{\xi}+d\boldsymbol{\xi}) - \phi(\mathbf{E}, \theta; \boldsymbol{\xi}), \tag{4.1.2}$$

is used by Hill and Rice (1973) for elastoplastic materials. These authors use $d^p\phi$ for the plastic change in ϕ. Here the superscript "in" is used instead of p to emphasize that inelasticity need not be confined to the classical rate-independent plasticity.[7]

When the free energy is regarded as a differentiable function of $\boldsymbol{\xi}$ which hence stands for a set of internal variables ξ_α, then *internal forces*,

$$\boldsymbol{\Lambda} = -\partial\phi/\partial\boldsymbol{\xi} = \boldsymbol{\Lambda}(\mathbf{E}, \theta; \boldsymbol{\xi}), \tag{4.1.1c}$$

are introduced such that

$$d^{in}\phi = \frac{\partial\phi}{\partial\xi_\alpha}\,d\xi_\alpha = -\Lambda_\alpha\,d\xi_\alpha \quad \alpha = 1, 2, \ldots, n, \quad (\alpha \text{ summed}). \tag{4.1.3a}$$

Associated with this is the inelastic rate of change of the free energy defined by

$$\dot{\phi}^{in} = -\Lambda_\alpha\,\dot{\xi}_\alpha = -\boldsymbol{\Lambda}\cdot\dot{\boldsymbol{\xi}}. \tag{4.1.3b}$$

The total rate of change of the free energy is given by

$$\dot{\phi} = \dot{\phi}^e + \dot{\phi}^{in} = \{\mathbf{S}:\dot{\mathbf{E}} - \eta\dot{\theta}\} - \boldsymbol{\Lambda}\cdot\dot{\boldsymbol{\xi}}. \tag{4.1.4}$$

Similarly, assuming suitable differentiability, the rate of change of the stress, entropy, and internal forces can be expressed, respectively, as

$$\dot{\mathbf{S}} = \dot{\mathbf{S}}^e + \dot{\mathbf{S}}^{in} = \left\{\frac{\partial\mathbf{S}}{\partial E_{AB}}\,\dot{E}_{AB} + \frac{\partial\mathbf{S}}{\partial\theta}\,\dot{\theta}\right\} + \frac{\partial\mathbf{S}}{\partial\xi_\alpha}\,\dot{\xi}_\alpha$$
$$= \left\{\boldsymbol{\mathcal{L}}^0:\dot{\mathbf{E}} - \frac{\partial\eta}{\partial\mathbf{E}}\,\dot{\theta}\right\} - \frac{\partial\Lambda_\alpha}{\partial\mathbf{E}}\,\dot{\xi}_\alpha, \tag{4.1.5}$$

$$\dot{\eta} = \dot{\eta}^e + \dot{\eta}^{in} = \left\{\frac{\partial\eta}{\partial\mathbf{E}}:\dot{\mathbf{E}} + \frac{\partial\eta}{\partial\theta}\,\dot{\theta}\right\} + \frac{\partial\eta}{\partial\xi_\alpha}\,\dot{\xi}_\alpha$$
$$= \left\{-\frac{\partial\mathbf{S}}{\partial\theta}:\dot{\mathbf{E}} + \frac{\partial\eta}{\partial\theta}\,\dot{\theta}\right\} + \frac{\partial\Lambda_\alpha}{\partial\theta}\,\dot{\xi}_\alpha, \tag{4.1.6}$$

---

[6] See Section 1.1, Chapter 1, for notation.

[7] See Rice (1970, 1971) and Nemat-Nasser (1975a,b, 1983).

$$\dot{\mathbf{\Lambda}} = \dot{\mathbf{\Lambda}}^e + \dot{\mathbf{\Lambda}}^{in} = \left\{ \frac{\partial \mathbf{\Lambda}}{\partial \mathbf{E}} : \dot{\mathbf{E}} + \frac{\partial \mathbf{\Lambda}}{\partial \theta} \dot{\theta} \right\} + \frac{\partial \mathbf{\Lambda}}{\partial \xi_\alpha} \dot{\xi}_\alpha$$

$$= \left\{ -\frac{\partial \mathbf{S}}{\partial \boldsymbol{\xi}} : \dot{\mathbf{E}} + \frac{\partial \eta}{\partial \boldsymbol{\xi}} \dot{\theta} \right\} + \frac{\partial \Lambda_\alpha}{\partial \boldsymbol{\xi}} \dot{\xi}_\alpha, \tag{4.1.7}$$

where all repeated indices are summed, with A, B = 1, 2, 3, and $\alpha = 1, 2, ..., n$. In (4.1.5),

$$\boldsymbol{\mathcal{L}}^0 \equiv \frac{\partial^2 \phi}{\partial \mathbf{E}\, \partial \mathbf{E}} \tag{4.1.8}$$

is the current elastic modulus tensor associated with strain measure **E**, measured relative to the chosen reference state. The elastic modulus tensor varies with the strain measure and the reference state. The governing relations are discussed by Hill (1978).

The quantity $\dot{\mathbf{\Lambda}}^{in}$ is the rate of change (relaxation) of the internal forces, with strain **E** and temperature θ held fixed. This rate of change is strictly due to the change of the internal variables associated with the microstructural changes, occurring with the current strain and temperature of the material element held constant. Since, in general,

$$\frac{\partial \Lambda_\alpha}{\partial \xi_\beta} = -\frac{\partial^2 \phi}{\partial \xi_\beta\, \partial \xi_\alpha} = \frac{\partial \Lambda_\beta}{\partial \xi_\alpha}, \tag{4.1.1d}$$

it follows that

$$\dot{\mathbf{\Lambda}}^{in} = \frac{\partial \mathbf{\Lambda}}{\partial \xi_\alpha} \dot{\xi}_\alpha = \frac{\partial \Lambda_\alpha}{\partial \boldsymbol{\xi}} \dot{\xi}_\alpha. \tag{4.1.3c}$$

By a *Legendre transformation*, the roles of **E** and **S** are reversed. To this end, the function $\psi = \psi(\mathbf{S}, \theta; \boldsymbol{\xi})$ is introduced such that

$$\phi + \psi = \mathbf{S} : \mathbf{E}. \tag{4.1.9}$$

It then follows that

$$\mathbf{E} = \frac{\partial \psi}{\partial \mathbf{S}}, \quad \eta = \frac{\partial \psi}{\partial \theta}, \quad \mathbf{\Lambda} = \frac{\partial \psi}{\partial \boldsymbol{\xi}}. \tag{4.1.10}$$

Assuming suitable differentiability, these now lead to

$$\dot{\mathbf{E}} = \dot{\mathbf{E}}^e + \dot{\mathbf{E}}^{in}$$

$$= \left\{ \frac{\partial \mathbf{E}}{\partial S_{AB}} \dot{S}_{AB} + \frac{\partial \mathbf{E}}{\partial \theta} \dot{\theta} \right\} + \frac{\partial \mathbf{E}}{\partial \xi_\alpha} \dot{\xi}_\alpha$$

$$= \left\{ \boldsymbol{\mathcal{M}}^0 : \dot{\mathbf{S}} + \frac{\partial \eta}{\partial \mathbf{S}} \dot{\theta} \right\} + \frac{\partial \Lambda_\alpha}{\partial \mathbf{S}} \dot{\xi}_\alpha, \tag{4.1.11}$$

$$\dot{\eta} = \dot{\eta}^e + \dot{\eta}^{in}$$

$$= \left\{ \frac{\partial \mathbf{E}}{\partial \theta} : \dot{\mathbf{S}} + \frac{\partial \eta}{\partial \theta} \dot{\theta} \right\} + \frac{\partial \Lambda_\alpha}{\partial \theta} \dot{\xi}_\alpha, \tag{4.1.12}$$

$$\dot{\mathbf{\Lambda}} = \dot{\mathbf{\Lambda}}^{e} + \dot{\mathbf{\Lambda}}^{in} = \left\{ \frac{\partial \mathbf{E}}{\partial \boldsymbol{\xi}} : \dot{\mathbf{S}} + \frac{\partial \eta}{\partial \boldsymbol{\xi}} \dot{\theta} \right\} + \frac{\partial \Lambda_\alpha}{\partial \boldsymbol{\xi}} \dot{\xi}_\alpha . \tag{4.1.13}$$

In (4.1.11), the fourth-order tensor

$$\mathcal{M}^0 \equiv \frac{\partial^2 \psi}{\partial \mathbf{S} \, \partial \mathbf{S}} \tag{4.1.14}$$

is the current elastic compliance tensor, the inverse of the elastic modulus tensor $\mathcal{L}^0$, $\mathcal{M}^0 = \mathcal{L}^{0-1}$, and similarly to the latter, it varies with the change of the strain measure and the reference state. In general, both $\mathcal{L}^0$ and $\mathcal{M}^0$ are functions of state and change with the change in the microstructure.[8]

Take the inner product of both sides of (4.1.7) and (4.1.13) with $\dot{\boldsymbol{\xi}}$ and obtain

$$\begin{aligned} \dot{\mathbf{\Lambda}} \cdot \dot{\boldsymbol{\xi}} &= (-\dot{\mathbf{S}}^{in} : \dot{\mathbf{E}} + \dot{\eta}^{in} \dot{\theta}) - \frac{\partial^2 \phi}{\partial \xi_\alpha \, \partial \xi_\beta} \dot{\xi}_\beta \, \dot{\xi}_\alpha \\ &= (\dot{\mathbf{E}}^{in} : \dot{\mathbf{S}} + \dot{\eta}^{in} \dot{\theta}) + \frac{\partial^2 \psi}{\partial \xi_\alpha \, \partial \xi_\beta} \dot{\xi}_\beta \, \dot{\xi}_\alpha . \end{aligned} \tag{4.1.15}$$

From this, it follows that

$$\dot{\mathbf{S}}^{in} : \dot{\mathbf{E}} + \dot{\mathbf{E}}^{in} : \dot{\mathbf{S}} = 0 . \tag{4.1.16a}$$

Also, comparing (4.1.7) and (4.1.13), observe that

$$\frac{\partial \mathbf{S}}{\partial \xi_\alpha} : \dot{\mathbf{E}} + \frac{\partial \mathbf{E}}{\partial \xi_\alpha} : \dot{\mathbf{S}} = 0 . \tag{4.1.16b}$$

Furthermore, in view of (4.1.1), (4.1.9), and (4.1.10), it follows that,

$$\dot{\phi}^{in} + \dot{\psi}^{in} = 0 . \tag{4.1.16c}$$

4.1.2. Inelastic Potential and Normality Rule

Constitutive relations are needed in order to define *fluxes* $\dot{\boldsymbol{\xi}}$ in terms of *forces* $\mathbf{\Lambda}$. These constitutive relations must reflect the corresponding micromechanisms which give rise to inelasticity and hence to the change in the internal variables. An example of such physically-based constitutive relations is in single-crystal plasticity, where the slip rate of a slip system is defined in terms of the corresponding resolved shear stress. The Schmid law for single crystals falls in this category. Another example is the linear relation between the fluxes and forces in classical nonequilibrium thermodynamics,

$$\dot{\xi}_\alpha = K_{\alpha\beta} \Lambda_\beta, \quad \mathbf{K} = \mathbf{K}^T , \tag{4.1.17a,b}$$

where the symmetry of the coefficient matrix $\mathbf{K}$ is the well-known Onsager

---

[8] For small deformations, comprehensive accounts are given by Nemat-Nasser and Hori (1993, 1999), and Torquato (2002).

reciprocal relation. When such a symmetry holds, then (4.1.17a) may be written as

$$\dot{\boldsymbol{\xi}} = \frac{\partial}{\partial \boldsymbol{\Lambda}}\left[\frac{1}{2}K_{\alpha\beta}\,\Lambda_\alpha\,\Lambda_\beta\right], \qquad \alpha, \beta = 1, 2, ..., n, \tag{4.1.17c}$$

where repeated indices are summed. In this case, the fluxes are derivable from the potential $\Omega = \frac{1}{2}K_{\alpha\beta}\,\Lambda_\alpha\,\Lambda_\beta$, *i.e.*, $\dot{\boldsymbol{\xi}} = \partial\Omega/\partial\boldsymbol{\Lambda}$.

A third example is the case where each flux is only a function of its corresponding force and no other forces, nor explicitly of any state variables. This corresponds to the classical hypothesis of the separation of individual processes, discussed by Thomson (1882) and restated and used by Li (1962), and Rice (1971). In this case, one has

$$\dot{\xi}_1 = h_1(\Lambda_1), \quad \dot{\xi}_2 = h_2(\Lambda_2), \; ..., \quad \dot{\xi}_n = h_n(\Lambda_n). \tag{4.1.18a}$$

Then

$$h_\alpha\, d\Lambda_\alpha = h_1(\Lambda_1)\, d\Lambda_1 + h_2(\Lambda_2)\, d\Lambda_2 + ... + h_n(\Lambda_n)\, d\Lambda_n$$

is a perfect differential $d\Omega$, and hence

$$\dot{\boldsymbol{\xi}} = \frac{\partial \Omega}{\partial \boldsymbol{\Lambda}}. \tag{4.1.18b}$$

In general, a broad class of materials may be considered, for which the fluxes are explicit functions of the forces but not explicitly of the state variables.[9] Then,

$$\dot{\boldsymbol{\xi}} = \mathbf{h}(\Lambda_1, \Lambda_2, ..., \Lambda_n) \equiv \mathbf{h}(\boldsymbol{\Lambda}). \tag{4.1.19a}$$

The question is, under what conditions does there exist an *integrating factor* $\lambda = \lambda(\boldsymbol{\Lambda})$ such that[10]

$$\dot{\boldsymbol{\xi}} = \mathbf{h}(\boldsymbol{\Lambda}) = \lambda\,\frac{\partial \Omega}{\partial \boldsymbol{\Lambda}}, \tag{4.1.19b}$$

where $\Omega = \Omega(\boldsymbol{\Lambda})$ is the *inelastic potential.* Before this general question is addressed, the consequences of the existence of such a potential for the fluxes is explored in the sequel.

4.1.3. Normality Rules

Suppose the fluxes are governed by (4.1.19b). Then, from (4.1.5) to (4.1.7) and the chain rule of differentiation, it follows that

$$\dot{\mathbf{S}}^{in} = -\,h_\alpha\,\frac{\partial \Lambda_\alpha}{\partial \mathbf{E}} = -\,\lambda\,\frac{\partial \overline{\Omega}}{\partial \mathbf{E}}, \tag{4.1.20a}$$

[9] They are, of course, implicit functions of the state variables since the forces are explicit functions of the state, (4.1.1c).

[10] With no loss in generality, $\lambda = \lambda(\boldsymbol{\Lambda})$ is assumed to be positive.

$$\dot{\eta}^{\text{in}} = h_\alpha \frac{\partial \Lambda_\alpha}{\partial \theta} = \lambda \frac{\partial \overline{\Omega}}{\partial \theta}, \tag{4.1.20b}$$

$$\dot{\boldsymbol{\Lambda}}^{\text{in}} = h_\alpha \frac{\partial \Lambda_\alpha}{\partial \xi} = \lambda \frac{\partial \overline{\Omega}}{\partial \xi}, \tag{4.1.20c}$$

where $\Omega = \Omega(\boldsymbol{\Lambda}(\mathbf{E}, \theta; \xi)) = \overline{\Omega}(\mathbf{E}, \theta; \xi)$ is the *inelastic potential.*

In the n+7-dimensional space of $(\mathbf{E}, \theta, \xi)$, $\overline{\Omega}$ = constant forms a hypersurface whose gradient defines the direction of the inelastic rates $(\dot{\mathbf{S}}^{\text{in}}, \dot{\eta}^{\text{in}}, \dot{\boldsymbol{\Lambda}}^{\text{in}})$. In particular, in the sub-**E**-space, the inelastic rate of change of the corresponding stress measure, $\dot{\mathbf{S}}^{\text{in}}$, is normal to the inelastic potential $\overline{\Omega}$.

A similar comment applies to the inelastic change of other quantities, as noted by Nemat-Nasser (1975a). In particular, from (4.1.11), the inelastic strain rate becomes,

$$\dot{\mathbf{E}}^{\text{in}} = \lambda \frac{\partial \hat{\Omega}}{\partial \mathbf{S}}, \tag{4.1.20d}$$

where[11] $\Omega = \hat{\Omega}(\mathbf{S}, \theta; \xi)$.

4.1.4. On the Existence of the Inelastic Potential

When there are only two scalar fluxes, then it is always possible to find an integrating factor $\lambda > 0$ such that $(h_1\, d\Lambda_1 + h_2\, d\Lambda_2)/\lambda = d\Omega$ is a perfect differential for suitably smooth functions $h_1(\Lambda_1, \Lambda_2)$ and $h_2(\Lambda_1, \Lambda_2)$. In general, however, and without additional assumptions, there seems no reason for the existence of such an integrating factor and the corresponding inelastic potential. However, if the functions **h** satisfy certain additional side conditions, then inelastic potentials would follow. For example, when the functions h_α in (4.1.19a) are suitably smooth and satisfy

$$\frac{\partial h_\alpha}{\partial \Lambda_\beta} = \frac{\partial h_\beta}{\partial \Lambda_\alpha}, \tag{4.1.21a}$$

or more generally,

$$h_\alpha (h_{\beta,\gamma} - h_{\gamma,\beta}) + h_\beta (h_{\gamma,\alpha} - h_{\alpha,\gamma}) + h_\gamma (h_{\alpha,\beta} - h_{\beta,\alpha}) = 0, \tag{4.1.21b}$$

where comma followed by an index denotes partial differentiation with respect to the corresponding component of $\boldsymbol{\Lambda}$, then it can be shown that the fluxes are derivable from a potential in accordance with (4.1.19b); Nemat-Nasser (1975a).

Observe that *assumption* (4.1.19a) *renders the fluxes* $\dot{\xi}$ *functions of state.* At a typical point in the state-space $(\mathbf{E}, \theta, \xi)$, the rates $\dot{\mathbf{E}}$ and $\dot{\theta}$ can be prescribed essentially arbitrarily. The same is *not* true for the fluxes $\dot{\xi}$. Hence, *the inelastic rates* $(\dot{\mathbf{S}}^{\text{in}}, \dot{\eta}^{\text{in}}, \dot{\boldsymbol{\Lambda}}^{\text{in}})$ *are all prescribed functions of state, once the*

[11] Based on (4.1.18a), Rice (1971) obtains (4.1.20a) and (4.1.20d). Nemat-Nasser (1975a, 1977) obtains (4.1.20a-d), starting with the less restrictive assumption (4.1.19a) and examining conditions under which (4.1.19b) holds.

functions $\mathbf{h}(\mathbf{\Lambda})$ *are given*, since $\mathbf{\Lambda} = \mathbf{\Lambda}(\mathbf{E}, \theta; \boldsymbol{\xi})$ is a function of the state variables. For example, $\dot{\mathbf{S}}^{in} = \{\partial^2\phi(\mathbf{E}, \theta; \boldsymbol{\xi})/\partial\xi_\alpha\,\partial\mathbf{E}\}\, \dot{h}_\alpha(\mathbf{\Lambda}(\mathbf{E}, \theta; \boldsymbol{\xi}))$ is a state function only. On the other hand, the elastic rates $(\dot{\mathbf{S}}^e, \dot{\eta}^e, \dot{\mathbf{\Lambda}}^e)$ are proportional to the rate of change of *observable* state variables $\mathbf{E}$ and θ, *i.e.*, proportional to $\dot{\mathbf{E}}$ and $\dot{\theta}$. The fact that all inelastic rates of the variables are state functions only, and the fact that all inelastic changes are dissipative, can be used to show that inelastic potentials exist when the number and nature of the internal variables are restricted; see Nemat-Nasser (1977).

In view of the above comments, it is clear that the equations defining the fluxes in terms of the internal forces, (4.1.19a), must be formulated on the basis of careful micromechanical observation and modeling. These equations contain the basic physics of the involved process, and cannot be prescribed in advance without due regard for the experimental observations. The second law of thermodynamics, in particular, only requires that the rate of dissipation be non-negative, *i.e.*,

$$\mathbf{\Lambda} \cdot \dot{\boldsymbol{\xi}} \geq 0, \tag{4.1.22}$$

and, hence, does not provide any other specific information. These points further emphasize the need for physically-based models of the inelastic deformation of materials, especially when realistic modeling of materials at various temperatures and deformation rates is considered; see Section 4.8.

There are other assumptions that may be used to ensure normality in plasticity theories, when the physics of the phenomenon warrants it. Two such postulates are that of Il'yushin (1961) and that of Drucker (1950, 1964). Various aspects of these specialized considerations have been examined in the literature; see Hill (1968), Hill and Rice (1973), and Nemat-Nasser (1983). As noted by Drucker (1964) and restated by Rice (1970), these postulates provide for classification of material behavior, but do not have universal applicability. Both postulates, however, lead to normality for *small deformations*, as originally intended by their authors; see also Hill (1968). For large plastic deformations, Drucker's postulate is not strain-measure invariant (Hill, 1968), and Il'yushin's postulate produces normality for *restricted* cycles of deformations only, as shown by Hill and Rice (1973) and re-emphasized by Nemat-Nasser (1983) and others.[12]

In the present context, the issue of plastic normality is viewed as stemming from specialized restrictions imposed on the internal variables and their evolution, leading to results (4.1.20a-d). These restrictions, however, do *not* limit the magnitude of either elastic or plastic strains and rotations. A similar comment does *not* apply to the above-mentioned postulates. Hill and Rice (1973), while dealing with finite elastic-plastic deformations, extend the normality implication of Il'yushin's postulate for "finite restricted cycles which contain

[12] Note that Carroll (1987) misstates Nemat-Nasser's discussion of this issue; see Nemat-Nasser (1983, Section 4.2, p.1123) and compare with Carroll (1987, p.20).

only infinitesimal additional plastic straining".

4.2. RATE-INDEPENDENT THEORIES

Classical rate-independent plasticity is based on the concept of the yield surface, either in stress space or in strain space. Traditionally, these theories have been developed for pressure-insensitive plastically incompressible solids, approximating in general the plastic response of metals. For application to porous metals and other materials such as granular masses, where pressure sensitivity and inelastic volumetric strain are significant and often dominating features, the classical theories have been modified to include pressure and volumetric effects. In this section, first, a general framework is laid out for a broad class of such theories, and then, various special cases which are commonly used in applications are considered. The basic results are first developed in terms of a general material strain measure **E**, with the conjugate stress measure **S**, reckoned with respect to an arbitrarily chosen reference state, and then specialized by choosing a convenient strain measure.[13]

The material is regarded to be elastoplastic and hence, when no microstructural changes occur, its response is purely elastic. The rate of change of the strain decomposes *exactly* into an elastic and a plastic contribution as

$$\dot{\mathbf{E}} = \dot{\mathbf{E}}^{e} + \dot{\mathbf{E}}^{p}, \tag{4.2.1}$$

independently of the particular material strain measure or the reference state, as long as the same reference state is used to measure the elastic and inelastic rates; Nemat-Nasser (1979, 1982). The stress rate relates to the elastic part of the strain rate by the elasticity relation[14]

$$\dot{\mathbf{S}} = \boldsymbol{\mathcal{L}}^{0} : \dot{\mathbf{E}}^{e}, \tag{4.2.2}$$

where $\dot{\mathbf{S}}$ is an objective stress rate, since it is referred to a reference material state. Let $\boldsymbol{\mathcal{M}}^{0}$ be the elastic compliance tensor, the inverse of $\boldsymbol{\mathcal{L}}^{0}$, and recall that both of these tensors depend on and vary with the strain measure and the reference state, and both are symmetric; *e.g.*, $\mathcal{L}^{0}_{ijkl} = \mathcal{L}^{0}_{klij} = \mathcal{L}^{0}_{jikl} = \mathcal{L}^{0}_{ijlk}$, when a fixed rectangular Cartesian coordinate system is used together with a proper symmetric material strain measure.

---

[13] Hill (1968, 1972, 1978), Havner (1982b), Nemat-Nasser (1983, 1992), and Mehrabadi and Nemat-Nasser (1987).

[14] In a consistent formulation, the elasticity tensor $\boldsymbol{\mathcal{L}}^{0}$ is defined as the second derivative of the elastic strain potential, taken with respect to the considered material strain measure, keeping all other measures fixed; see (4.1.8).

4.2.1. Yield Surface

An important ingredient in the classical theory of elastoplasticity is the concept of the yield surface. The yield surface, either in the stress space or the strain space, defines a region within which the material response is elastic and on which the response may be elastoplastic; Hill (1967). The shape and size of the yield surface vary with the history of the inelastic deformation of the material element. The yield surface may be smooth, having a continuously turning tangent plane or it may possess corners or vertices. It may consist of a set of smooth surfaces which intersect and hence form a corner or a vertex. The existence of such vertices is an integral part of the physics of crystal plasticity.[15] It plays a dominant role in modeling phenomena such as instability by localized deformation, necking, and shearbanding.[16] In the sequel, first, cases associated with a smooth yield surface are considered and illustrated, and then, the response at a vertex is discussed.

4.2.2. Constitutive Relations: Smooth Yield Surface

Rate-independent continuum plasticity essentially generalizes the results of the uniaxial and torsional deformation of metals, where over a range of stresses (or strains) the sample behaves essentially elastically. Once a critical stress (strain) is reached, the sample's response becomes elastoplastic in continued loading, and elastic upon unloading. The theory has been formulated traditionally, both in the stress and the strain space. In this regard there is duality. This dual development has been most clearly delineated by Hill (1967) based on his early work, Hill (1958, 1959); a concise description is given in a treatise by Hill (1978). In the sequel, the theory is presented in a general setting, both in the strain and the stress space, starting with the stress space.

4.2.3. Yield Surface in Stress Space

In the stress space the yield surface marks elastoplastic states. For stress points within the yield surface the material response is elastic. For points on the yield surface the response is elastoplastic. When the change in the stress state tends to move the stress point into the yield surface, the material element undergoes elastic *unloading*. When the stress point moves on the current yield surface, the material element undergoes *neutral loading*. When the change in the

[15] Kocks (1970), Bui (1970), Hutchinson (1970), Rice (1971), Zarka (1973, 1975), Havner and Shalaby (1977), Havner (1979), and Iwakuma and Nemat-Nasser (1984).

[16] Hill and Hutchinson (1975), Rudnicki and Rice (1975), Stören and Rice (1975), Rice (1977), Needleman and Rice (1978), Bassani *et al.* (1979), Nemat-Nasser *et al.* (1981), Iwakuma and Nemat-Nasser (1982), Needleman and Tvergaard (1982), Nemat-Nasser (1979, 1982, 1983), Pan and Rice (1983), Carroll and Carman (1985), Anand and Kalidindi (1994), and Kim *et al.* (1998). For an account of the physical bases and mathematical modeling of adiabatic shearbanding in metals, see Wright (2002).

stress state is such that the stress point tends to move out of the current yield surface, the material element is said to undergo *loading*. In this case the yield surface moves with the stress point such that the stress point always remains on the yield surface. The yield surface, therefore, may expand, or simply move, or do both, and this is called *workhardening*. When the yield surface expands self-similarly, the workhardening is called *isotropic hardening*. On the other hand, when the yield surface does not change shape and size but simply moves with the stress point, the material is said to be *kinematically hardening*. In general, a combined *isotropic-kinematic hardening* can be considered. It is pointed out that these are idealizations, since in general, the yield surface changes shape in a complex manner, as is the case for polycrystalline metals.[17]

Let $\boldsymbol{\mu}$ be normal to the yield surface in the *stress space*, and let $\mathbf{m}$ define the direction of the plastic strain rate; see Figure 4.2.1. In general, the second-order symmetric tensors $\boldsymbol{\mu}$ and $\mathbf{m}$ are not necessarily coincident. In view of (4.2.1), set

$$\begin{aligned}\dot{\mathbf{E}} &= \dot{\mathbf{E}}^e + \dot{\mathbf{E}}^p = \mathcal{M}^0 : \dot{\mathbf{S}} + \mathbf{m}\,(\boldsymbol{\mu} : \dot{\mathbf{S}}) \\ &= \overline{\mathcal{M}} : \dot{\mathbf{S}},\end{aligned} \tag{4.2.3}$$

where

$$\overline{\mathcal{M}} = \mathcal{M}^0 + \mathbf{m} \otimes \boldsymbol{\mu} \tag{4.2.4}$$

is the *instantaneous elastoplastic compliance tensor*.

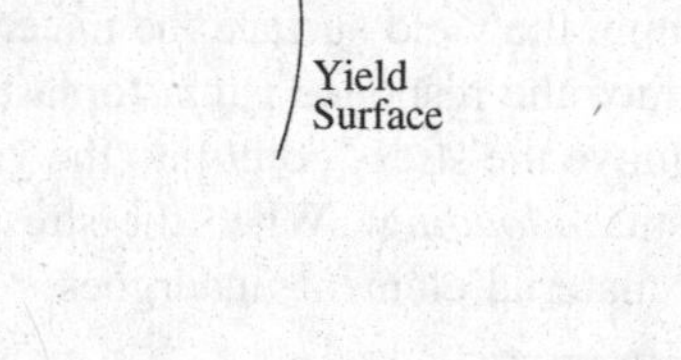

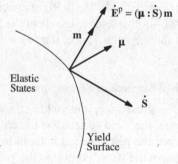

Figure 4.2.1

Yield surface in the *stress space*: plastic strain rate $\dot{\mathbf{E}}^p$ is given by $(\boldsymbol{\mu} : \dot{\mathbf{S}})\,\mathbf{m}$

[17] Mróz (1967), Hutchinson (1970), Argon (1975), Kocks (1975), Hecker (1976), Bassani (1977), Hill (1979), Canova *et al.* (1984), Iwakuma and Nemat-Nasser (1984), Nemat-Nasser and Obata (1986), and Parks and Ahzi (1990).

4.2.4. Yield Surface in Strain Space[18]

Let $\boldsymbol{\lambda}$ be normal to the yield surface at a point in the *strain space*, and denote by $\boldsymbol{l}$ the direction of the inelastic stress *decrement*; see Figure 4.2.2. In general, $\boldsymbol{\lambda}$ and $\boldsymbol{l}$ are distinct second-order symmetric tensors. The stress rate is now given by

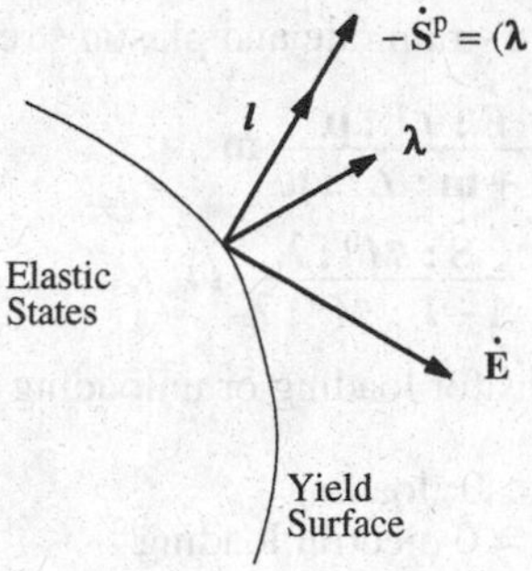

Figure 4.2.2

Yield surface in the *strain space*: plastic stress *decrement* is given by $(\boldsymbol{\lambda} : \dot{\mathbf{E}})\,\boldsymbol{l}$

$$\dot{\mathbf{S}} = \dot{\mathbf{S}}^e + \dot{\mathbf{S}}^p = \mathcal{L}^0 : \dot{\mathbf{E}} - \boldsymbol{l}\,(\boldsymbol{\lambda} : \dot{\mathbf{E}}) = \overline{\mathcal{L}} : \dot{\mathbf{E}}, \tag{4.2.5}$$

where[19]

$$\overline{\mathcal{L}} = \mathcal{L}^0 - \boldsymbol{l} \otimes \boldsymbol{\lambda} \tag{4.2.6}$$

is the *instantaneous elastoplastic modulus* tensor. It is required that $\overline{\mathcal{L}}$ and $\overline{\mathcal{M}}$ be each other's inverse. From this and the symmetry of the elastic modulus and compliance tensors embedded in their definitions (4.1.8) and (4.1.14), it follows that[20]

$$\boldsymbol{\lambda} = \mathcal{L}^0 : \boldsymbol{\mu}, \quad \boldsymbol{\mu} = \mathcal{M}^0 : \boldsymbol{\lambda},$$
$$\boldsymbol{l} = \frac{\mathcal{L}^0 : \mathbf{m}}{1 + \boldsymbol{\lambda} : \mathbf{m}}, \quad \mathbf{m} = \frac{\mathcal{M}^0 : \boldsymbol{l}}{1 - \boldsymbol{\mu} : \boldsymbol{l}},$$
$$(1 + \boldsymbol{\lambda} : \mathbf{m})\,(1 - \boldsymbol{\mu} : \boldsymbol{l}) = 1. \tag{4.2.7a-e}$$

Furthermore, the instantaneous elastoplastic and compliance tensors are given

---

[18] The significance of the strain space in continuum theories of plasticity has been emphasized by Naghdi and Trapp (1975), and Casey and Naghdi (1983). The formulation in this subsection follows Hill's (1967) pioneering contribution; see also Hill (1978). For a detailed account of contributions by Naghdi and coworkers, see Naghdi (1990) and relevant references cited there.

[19] The decomposition of strain rate, $\dot{\mathbf{E}}$, into the elastic, $\dot{\mathbf{E}}^e$, and plastic, $\dot{\mathbf{E}}^p$, contributions has operational significance. On the other hand, $\dot{\mathbf{S}}^e$ and $\dot{\mathbf{S}}^p$ merely stand for $\mathcal{L}^0 : \dot{\mathbf{E}}$ and $-\boldsymbol{l}\,(\boldsymbol{\lambda} : \dot{\mathbf{E}})$, respectively. The former is the stress rate produced by the strain rate $\dot{\mathbf{E}}$ in the absence of any plastic deformation, whereas $\dot{\mathbf{S}}^p$ is the stress decrement due to an accompanying plastic strain rate.

[20] From $(\mathcal{L}^0 - \boldsymbol{l} \otimes \boldsymbol{\lambda}) : (\mathcal{M}^0 + \mathbf{m} \otimes \boldsymbol{\mu}) = (\mathcal{M}^0 + \mathbf{m} \otimes \boldsymbol{\mu}) : (\mathcal{L}^0 - \boldsymbol{l} \otimes \boldsymbol{\lambda}) = \mathbf{1}^{(4s)}$ and the symmetry of $\mathcal{L}^0$ and $\mathcal{M}^0$, expressions (4.2.7a-e) are obtained.

by

$$\overline{\mathcal{M}} = \mathcal{M}^0 + \frac{(\mathcal{M}^0 : \boldsymbol{l}) \otimes (\mathcal{M}^0 : \boldsymbol{\lambda})}{1 - \boldsymbol{l} : \mathcal{M}^0 : \boldsymbol{\lambda}},$$

$$\overline{\mathcal{L}} = \mathcal{L}^0 - \frac{(\mathcal{L}^0 : \mathbf{m}) \otimes (\mathcal{L}^0 : \boldsymbol{\mu})}{1 + \mathbf{m} : \mathcal{L}^0 : \boldsymbol{\mu}}, \tag{4.2.8a,b}$$

and the plastic strain rate and plastic stress *decrement* are

$$\dot{\mathbf{E}}^{\mathrm{p}} = \frac{\dot{\mathbf{E}} : \mathcal{L}^0 : \boldsymbol{\mu}}{1 + \mathbf{m} : \mathcal{L}^0 : \boldsymbol{\mu}} \mathbf{m},$$

$$\dot{\mathbf{S}}^{\mathrm{p}} = -\frac{\dot{\mathbf{S}} : \mathcal{M}^0 : \boldsymbol{\lambda}}{1 - \boldsymbol{l} : \mathcal{M}^0 : \boldsymbol{\lambda}} \boldsymbol{l}. \tag{4.2.9a,b}$$

The conditions for loading or unloading then become

$$\boldsymbol{\lambda} : \dot{\mathbf{E}} \begin{cases} > 0 & \text{loading} \\ = 0 & \text{neutral loading} \\ < 0 & \text{unloading}. \end{cases} \tag{4.2.10a}$$

The corresponding conditions in the stress space are

$$\frac{\boldsymbol{\mu} : \dot{\mathbf{S}}}{1 - \boldsymbol{\mu} : \boldsymbol{l}} \begin{cases} > 0 & \text{loading} \\ = 0 & \text{neutral loading} \\ < 0 & \text{unloading}. \end{cases} \tag{4.2.10b}$$

In addition, the conditions for *strain-hardening, perfect plasticity*, or *strain-softening* can be expressed as

$$\boldsymbol{\mu} : \boldsymbol{l} \begin{cases} < 1 & \text{strain–hardening} \\ = 1 & \text{perfect plasticity} \\ > 1 & \text{strain–softening}. \end{cases} \tag{4.2.10c}$$

For neutral loading, $\dot{\mathbf{S}}$ is tangent to the yield surface, and, hence, $\boldsymbol{\mu} : \dot{\mathbf{S}} = 0$. For perfect plasticity, both the numerator and the denominator in the left expression in (4.2.10b) vanish. Since the yield surface remains unchanged for the perfectly plastic case, $1 - \boldsymbol{\mu} : \boldsymbol{l} \le 0$ which defines either elastic unloading ($1 - \boldsymbol{\mu} : \boldsymbol{l} < 0$) or neutral plastic loading ($1 - \boldsymbol{\mu} : \boldsymbol{l} = 0$). In this latter case, (4.2.7e) shows that the magnitude of the plastic strain rate is undefined and must be obtained through kinematical considerations.

The second-order tensors $\boldsymbol{\mu}$ and $\boldsymbol{l}$ depend on the choice of the strain measure and the reference state. Hence, in general, the notion of strain-hardening should be viewed relative to this choice; Hill (1978). As is seen, the formulation of elastoplasticity in the stress or strain space leads to equivalent results. This notwithstanding, there seems to be some confusion in the literature on this issue. Thus the reader is cautioned on this point.

4.2.5. Flow Potential and Associative Flow Rule

Let the yield surface in the stress space be defined by

$$f = f(\mathbf{S}, \ldots) = 0\,, \tag{4.2.11}$$

where dots stand for temperature and for internal variables that characterize material hardening. In addition, let there exist another surface

$$g = g(\mathbf{S}, \ldots) \tag{4.2.12}$$

whose gradient is in the direction of the plastic strain rate in the stress space,

$$\dot{\mathbf{E}}^p = \dot{\gamma}\, \frac{\partial g}{\partial \mathbf{S}}\,. \tag{4.2.13}$$

The function g is called the *flow potential*. Comparison with (4.2.3) now shows

$$\mathbf{m} = \frac{\partial g}{\partial \mathbf{S}}, \qquad \dot{\gamma} = \boldsymbol{\mu} : \dot{\mathbf{S}}\,. \tag{4.2.14}$$

For continued plastic deformation, furthermore, the stress point must remain on the yield surface and hence,

$$\dot{f} = \frac{\partial f}{\partial \mathbf{S}} : \dot{\mathbf{S}} + \ldots = 0\,, \tag{4.2.15}$$

from which it follows that

$$\dot{\gamma} = \frac{1}{H} \frac{\partial f}{\partial \mathbf{S}} : \dot{\mathbf{S}}\,,$$

$$\boldsymbol{\mu} = \frac{1}{H} \frac{\partial f}{\partial \mathbf{S}}\,, \tag{4.2.16a,b}$$

where H, the workhardening parameter, depends on the manner by which the yield surface changes in the course of plastic deformation. Hence,

$$\dot{\mathbf{E}} = \mathcal{M}^0 : \dot{\mathbf{S}} + \frac{1}{H} \frac{\partial g}{\partial \mathbf{S}} \left\{ \frac{\partial f}{\partial \mathbf{S}} : \dot{\mathbf{S}} \right\}. \tag{4.2.17a}$$

The inverse relation is easily obtained by calculating $\boldsymbol{\lambda}$ and $\boldsymbol{l}$ in terms of $\boldsymbol{\mu}$ and $\mathbf{m}$, using (4.2.7). This leads to

$$\dot{\mathbf{S}} = \mathcal{L}^0 : \left\{ \mathbf{1}^{(4s)} - \left[H + \frac{\partial f}{\partial \mathbf{S}} : \mathcal{L}^0 : \frac{\partial g}{\partial \mathbf{S}} \right]^{-1} \frac{\partial g}{\partial \mathbf{S}} \otimes \mathcal{L}^0 : \frac{\partial f}{\partial \mathbf{S}} \right\} : \dot{\mathbf{E}}\,, \tag{4.2.17b}$$

where $\mathbf{1}^{(4s)}$ is the fourth-order symmetric identity tensor with components

$$1^{(4s)}_{ijkl} = \frac{1}{2}(\delta_{ik}\delta_{jl} + \delta_{il}\delta_{jk})\,. \tag{4.2.17c}$$

Equations (4.2.17a,b) are general rate-constitutive relations for rate-independent finite elastoplastic deformation with *nonassociative* flow rule.[21]

---

[21] Rudnicki and Rice (1975), Stören and Rice (1975), Rice (1977), Christoffersen and Hutchinson (1979), Nemat-Nasser and Shokooh (1980), Anand (1980, 1983), Neale (1981), and Nemat-Nasser (1984).

They are valid for plastically compressible or incompressible cases. Both the strain measure and the reference state may be any convenient choice. In particular, the stress flux, $\dot{\mathbf{S}}$, can be chosen to suit any desired application. In addition, the dependence of the yield function and flow potential on the internal state variables can be general, including any required workhardening rules.[22]

When the functions f and g are identical, the tensors $\boldsymbol{\mu}$ and $\mathbf{m}$ become unidirectional in the stress space. In this case the plasticity is said to be governed by an *associative flow rule*, otherwise the flow rule is said to be *nonassociative*. Some examples are worked out in the following subsections.

4.2.6. The J_2-flow Theory with Isotropic Hardening

The most widely used continuum plasticity model is the so-called J_2-flow theory. The current configuration is used as the reference one, in which case all stresses coincide and all strain rates reduce to the deformation rate tensor, $\mathbf{D}$. It is convenient to use the Kirchhoff stress, $\boldsymbol{\tau}$, as the stress measure. The yield function, f, and the flow potential, g, in this theory are defined by

$$f \equiv g \equiv \tau - F(\gamma), \tag{4.2.18}$$

where[23]

$$\tau = (\tfrac{1}{2}\boldsymbol{\tau}' : \boldsymbol{\tau}')^{1/2} \equiv \sqrt{J_2},$$
$$\dot{\gamma} = (2\,\mathbf{D}^{p\prime} : \mathbf{D}^{p\prime})^{1/2}. \tag{4.2.19a,b}$$

In these expressions, prime denotes the deviatoric part, and $\mathbf{D}^p$ is the plastic part of the deformation rate,

$$\mathbf{D} = \mathbf{D}^e + \mathbf{D}^{p\prime}. \tag{4.2.19c}$$

In this theory, the plastic deformation is volume-preserving and pressure-independent. Hence, hydrostatic pressure or tension induces only elastic deformation. The deviatoric plastic deformation rate is given by

$$\mathbf{D}^{p\prime} = \dot{\gamma}\,\frac{\partial f}{\partial \boldsymbol{\tau}'} = \frac{1}{H}\left\{\frac{\boldsymbol{\mu}}{\sqrt{2}} \otimes \frac{\boldsymbol{\mu}}{\sqrt{2}}\right\} : \overset{\circ}{\boldsymbol{\tau}}{}', \tag{4.2.20a}$$

where $\overset{\circ}{\boldsymbol{\tau}}$ is the Jaumann rate of the Kirchhoff stress, and

$$\frac{\boldsymbol{\mu}}{\sqrt{2}} = \frac{\partial f}{\partial \boldsymbol{\tau}'} = \frac{\boldsymbol{\tau}'}{2\tau}, \qquad H = \frac{\partial F}{\partial \gamma}. \tag{4.2.20b,c}$$

In the deviatoric stress space, $f \equiv (\tfrac{1}{2}\boldsymbol{\tau}' : \boldsymbol{\tau}')^{1/2} - F = 0$ for F positive, is a sphere of radius $\sqrt{2}F$. The quantity τ is referred to as the *effective stress*, and the

---

22 *Ibid.*, and Dafalias and Popov (1976).

23 Many authors use $\sigma = ((3/2)\,\boldsymbol{\tau}' : \boldsymbol{\tau}')^{1/2}$ and $\dot{\varepsilon} = ((2/3)\,\mathbf{D}^{p\prime} : \mathbf{D}^{p\prime})^{1/2}$ as the *effective stress* and *effective plastic strain rate*, respectively. In this case, σ is the tensile stress in uniaxial tension. As is seen, $\sigma = \sqrt{3}\,\tau$ and $\dot{\gamma} = \sqrt{3}\,\dot{\varepsilon}$.

quantity γ is referred to as the *effective plastic strain*. In pure shearing, τ is the shear stress and γ is the corresponding (engineering[24]) shear strain. Then $d\tau/d\gamma = F'$ defines the slope of the shear stress versus the plastic shear strain curve. Depending on whether F' is positive, zero, or negative, the radius of the yield surface increases (workhardening), remains constant (perfect plasticity), or decreases (work-softening).

The final strain rate-stress rate relation now is

$$\mathbf{D} = \mathcal{M} : \overset{\circ}{\boldsymbol{\tau}} + \frac{\boldsymbol{\tau}'}{4\,\tau^2\,H}\,\boldsymbol{\tau}' : \overset{\circ}{\boldsymbol{\tau}}{}' = \left\{ \mathcal{M} + \frac{1}{H}\,\frac{\boldsymbol{\mu}}{\sqrt{2}} \otimes \frac{\boldsymbol{\mu}}{\sqrt{2}} \right\} : \overset{\circ}{\boldsymbol{\tau}}\,, \tag{4.2.21a}$$

with the following inverse:

$$\overset{\circ}{\boldsymbol{\tau}} = \mathcal{L} : \left\{ \mathbf{1}^{(4s)} - \frac{\boldsymbol{\tau}' \otimes \mathcal{L} : \boldsymbol{\tau}'}{4\,\tau^2 H + \boldsymbol{\tau}' : \mathcal{L} : \boldsymbol{\tau}'} \right\} : \mathbf{D}$$

$$= \mathcal{L} : \left\{ \mathbf{1}^{(4s)} - \frac{\boldsymbol{\mu} \otimes \mathcal{L} : \boldsymbol{\mu}}{2H + \boldsymbol{\mu} : \mathcal{L} : \boldsymbol{\mu}} \right\} : \mathbf{D}\,, \tag{4.2.21b}$$

where $\mathcal{L}$ and $\mathcal{M}$ are the elastic modulus and compliance tensors associated with the logarithmic strain measure, with the current state as the reference one. Since $\overset{\circ}{\boldsymbol{\tau}} = \mathcal{L} : \mathbf{D}^e = \mathcal{L} : (\mathbf{D} - \mathbf{D}^{p'})$, it follows from (4.2.21b) that (see Figure 4.2.3)

$$\mathbf{D}^{p'} = \frac{\mathbf{D} : \mathcal{L} : \boldsymbol{\mu}}{2H + \boldsymbol{\mu} : \mathcal{L} : \boldsymbol{\mu}}\,\boldsymbol{\mu} = \frac{\boldsymbol{\mu} \otimes \boldsymbol{\lambda}}{2H + \boldsymbol{\mu} : \boldsymbol{\lambda}} : \mathbf{D}\,. \tag{4.2.21c}$$

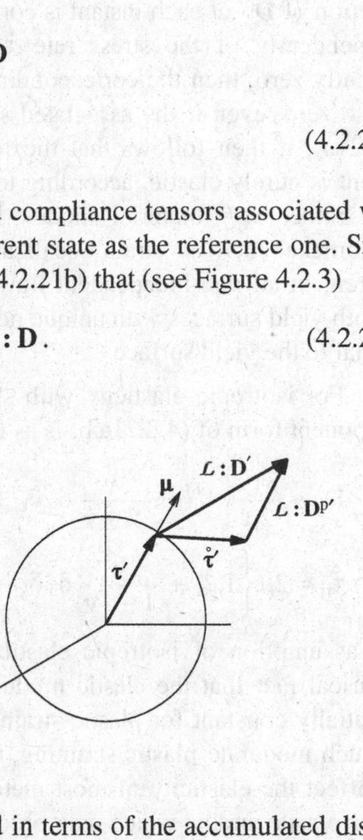

Figure 4.2.3
Isotropic yield surface
in the *deviatoric stress space*

Isotropic hardening is often expressed in terms of the accumulated distortional plastic work rather than the accumulated effective plastic strain γ, *i.e.*, function F is regarded to depend on the plastic distortional work instead of on γ. In this simple theory, the rate of distortional plastic work, measured per unit *reference* volume, is given by

$$\boldsymbol{\tau}' : \mathbf{D}^{p'} = \tau\dot{\gamma} \equiv \dot{\xi}\,. \tag{4.2.22}$$

[24] For pure shearing in the x_1, x_2-plane, $D^p_{12} = D^p_{21} \neq 0$, with all other components of the deformation-rate tensor, $\mathbf{D}^{p'}$, being zero. Then, $\dot{\gamma} = 2\,D^p_{12}$.

If the yield function is defined by

$$f \equiv \tau - \bar{F}(\xi), \tag{4.2.23a}$$

then it follows from (4.2.22), (4.2.20c), and (4.2.23a) that

$$H = \frac{\partial F}{\partial \gamma} = \tau \frac{\partial \bar{F}}{\partial \xi}. \tag{4.2.23b}$$

As noted before, this simple J_2-flow theory is based on a smooth yield surface (a sphere in the deviatoric stress space). The direction of the plastic strain rate $\mathbf{D}^{p\prime}$ is fixed by the normal to this surface at the corresponding stress point. This normal is, in this case, *coaxial* with the deviatoric stress $\boldsymbol{\tau}'$. Thus, the direction of $\mathbf{D}^{p\prime}$ at each instant is completely defined by the existing stress-state, independently of the stress rate $\dot{\boldsymbol{\tau}}'$. If any component of the stress τ_{ij}' is currently zero, then the corresponding component of the plastic strain rate $D^{p\prime}_{ij}$, is also zero, even if the associated stress-rate component $\dot{\tau}_{ij}'$ is nonzero. From (4.2.21a), it then follows that the response corresponding to that strain component is purely elastic, according to this flow theory of plasticity. This results in a rather *stiff* material model, incapable of predicting certain observed phenomena such as strain localization and shearbanding. This difficulty is inherent in all rate-independent flow theories of plasticity, which are based on smooth yield surfaces with unique normals, and in which the plastic strain rate is normal to the yield surface.

For isotropic elasticity with shear modulus μ and Poisson's ratio ν, the component form of (4.2.21a,b) is as follows:

$$D_{ij} = \frac{1}{2\mu}\left\{ 1^{(4s)}_{ijkl} - \frac{\nu}{1+\nu}\,\delta_{ij}\,\delta_{kl} + \frac{2\mu}{H}\,\frac{\mu_{ij}}{\sqrt{2}}\,\frac{\mu_{kl}}{\sqrt{2}} \right\} \overset{\circ}{\tau}_{kl},$$

$$\overset{\circ}{\tau}_{ij} = 2\mu\left\{ 1^{(4s)}_{ijkl} + \frac{\nu}{1-2\nu}\,\delta_{ij}\,\delta_{kl} - \frac{2\mu}{\mu+H}\,\frac{\mu_{ij}}{\sqrt{2}}\,\frac{\mu_{kl}}{\sqrt{2}} \right\} D_{kl}. \tag{4.2.24a,b}$$

The assumption of isotropic elasticity with constant moduli is based on the empirical fact that the elastic moduli of most initially isotropic metals remain essentially constant for plastic strains of up to 20 to 30%, or even greater; *i.e.*, for such moderate plastic straining, the presence of dislocations and twins does not affect the elasticity of most metals. The assumption ceases to be valid for larger plastic strains which may produce texture; see Chapter 8. It does not generally hold for geomaterials, even at small inelastic strains; see Section 4.7 and Chapter 7.

4.2.7. The J_2-flow Theory with Kinematic Hardening

For a purely kinematic hardening model, the radius of the yield surface in the deviatoric stress space remains constant, but the center of this surface moves, as plastic deformation takes place. Let $\boldsymbol{\beta}$ be a symmetric *deviatoric* second-order tensor defining the location of the center of the yield surface in the deviatoric stress space. The yield surface is then expressed as

$$f \equiv \tau - \tau_Y^0, \tag{4.2.25a}$$

where τ_Y^0 is a constant defining the yield stress in pure shear, and

$$\tau = \{\tfrac{1}{2}(\boldsymbol{\tau}' - \boldsymbol{\beta}) : (\boldsymbol{\tau}' - \boldsymbol{\beta})\}^{1/2}. \tag{4.2.25b}$$

The plastic part of the deformation rate is now given by

$$\mathbf{D}^{p'} = \dot{\gamma}\,\frac{\partial f}{\partial \boldsymbol{\tau}'} = \frac{1}{\tilde{H}}\left\{\frac{\boldsymbol{\mu}}{\sqrt{2}} \otimes \frac{\boldsymbol{\mu}}{\sqrt{2}}\right\} : \overset{\circ}{\boldsymbol{\tau}}{}', \tag{4.2.26a}$$

where

$$\frac{\boldsymbol{\mu}}{\sqrt{2}} = \frac{\partial f}{\partial \boldsymbol{\tau}'} = \frac{\boldsymbol{\tau}' - \boldsymbol{\beta}}{2\tau}, \tag{4.2.26b}$$

and $\tilde{H}$ is obtained from the condition that, in continued plastic deformation, the stress point remains on the yield surface, *i.e.*, from the *consistency condition* $f = 0$ which yields,[25]

$$\boldsymbol{\mu} : \overset{\circ}{\boldsymbol{\tau}}{}' = \boldsymbol{\mu} : \overset{\circ}{\boldsymbol{\beta}}. \tag{4.2.26c}$$

Hence, from (4.2.26a-c), the workhardening parameter $\tilde{H}$ is given by

$$\tilde{H} = \frac{\boldsymbol{\mu} : \overset{\circ}{\boldsymbol{\tau}}}{\sqrt{2}\,\dot{\gamma}} = \frac{\boldsymbol{\mu} : \overset{\circ}{\boldsymbol{\beta}}}{\sqrt{2}\,\dot{\gamma}}. \tag{4.2.26d}$$

The strain rate-stress rate relations then are

$$\mathbf{D} = \left\{\mathcal{M} + \frac{1}{\tilde{H}}\,\frac{\boldsymbol{\mu}}{\sqrt{2}} \otimes \frac{\boldsymbol{\mu}}{\sqrt{2}}\right\} : \overset{\circ}{\boldsymbol{\tau}},$$

$$\overset{\circ}{\boldsymbol{\tau}} = \mathcal{L} : \left\{\mathbf{1}^{(4s)} - \frac{\boldsymbol{\mu} \otimes \mathcal{L} : \boldsymbol{\mu}}{2\tilde{H} + \boldsymbol{\mu} : \mathcal{L} : \boldsymbol{\mu}}\right\} : \mathbf{D}, \tag{4.2.26e,f}$$

which also yield (see Figure 4.2.4).

$$\mathbf{D}^{p'} = \frac{\mathbf{D} : \mathcal{L} : \boldsymbol{\mu}}{2\tilde{H} + \boldsymbol{\mu} : \mathcal{L} : \boldsymbol{\mu}}\,\boldsymbol{\mu} = \frac{\boldsymbol{\mu} \otimes \boldsymbol{\lambda}}{2\tilde{H} + \boldsymbol{\mu} : \boldsymbol{\lambda}} : \mathbf{D}. \tag{4.2.21d}$$

Note that $\overset{\circ}{\boldsymbol{\beta}}$ is the Jaumann rate of change of $\boldsymbol{\beta}$, and hence, if γ, the accumulated effective plastic strain, is used as the time parameter, then $\dot{\gamma} = 1$ in the expression for $\tilde{H}$.

Additional assumptions are required to define the evolution of the parameter $\boldsymbol{\beta}$ and hence the workhardening quantity $\tilde{H}$. The parameter $\boldsymbol{\beta}$ is a macroscopic measure of the *backstress,* representing the effect of local residual stresses and strains on the overall plastic deformation of the material element. These residual stresses are developed by, *e.g.*, dislocation pile-ups against barriers, in the course of plastic deformation.[26]

---

[25] See, Ziegler (1959) and Tvergaard (1978).

[26] The reduction in the yield stress observed upon reverse uniaxial loading of some metals, is called the Bauschinger effect, after Bauschinger (1881). It has been studied both experimentally and theoretically by many investigators; see, *e.g.*, Masing (1922), Edelman and Drucker (1951),

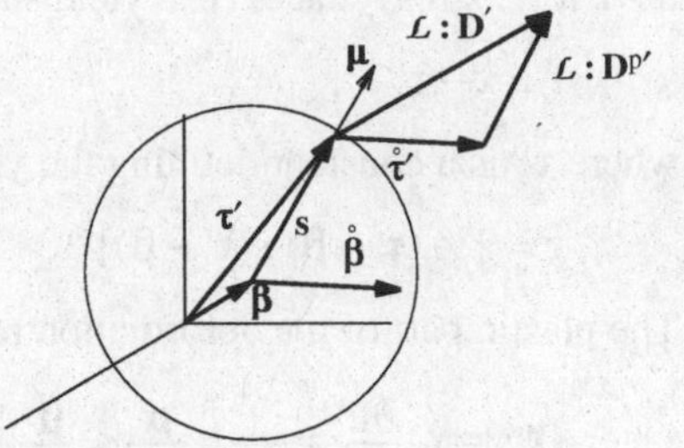

Figure 4.2.4

Yield surface in the *stress space* with kinematic hardening ($\boldsymbol{\mu}:\overset{\circ}{\boldsymbol{\tau}}' = \boldsymbol{\mu}:\overset{\circ}{\boldsymbol{\beta}}$)

When $\boldsymbol{\beta} = \mathbf{0}$, the material element is in a *plastically isotropic* state. Deviation from this state introduces plastic anisotropy through the location of the center of the yield surface defined by $\boldsymbol{\beta}$. Hence, the evolutionary equation for the kinematic hardening parameter $\boldsymbol{\beta}$ must reflect the microstructural changes in the material element in the course of its plastic deformation. In the literature, the flux of $\boldsymbol{\beta}$ has been assumed to be proportional to the deviatoric plastic strain rate, or equivalently, to be normal to the yield surface. Such an *isotropic* relation between the rate of change of $\boldsymbol{\beta}$ and the plastic strain rate may be reasonable if they are both measured relative to the state of plastic isotropy, where $\boldsymbol{\beta} = \mathbf{0}$ (Nemat-Nasser, 1983); this then is consistent with the anisotropy inherent in kinematic hardening.

To examine this, let $\tilde{\mathbf{F}}$ be the deformation gradient from the *reference state of* $\boldsymbol{\beta} = 0$, *to the current state.* With the aid of polar decomposition, $\tilde{\mathbf{F}} = \tilde{\mathbf{R}}\tilde{\mathbf{U}}$, introduce the objective flux of $\boldsymbol{\beta}$ as

$$\overset{\Delta}{\boldsymbol{\beta}} \equiv \tilde{\mathbf{R}}\,(\tilde{\mathbf{R}}^{T}\boldsymbol{\beta}\,\tilde{\mathbf{R}})^{\bullet}\,\tilde{\mathbf{R}}^{T} \equiv \overset{\circ}{\boldsymbol{\beta}} + \boldsymbol{\beta}\tilde{\boldsymbol{\varepsilon}} - \tilde{\boldsymbol{\varepsilon}}\boldsymbol{\beta}, \tag{4.2.27a}$$

where $\overset{\circ}{\boldsymbol{\beta}} = \dot{\boldsymbol{\beta}} - \tilde{\mathbf{W}}\boldsymbol{\beta} + \boldsymbol{\beta}\tilde{\mathbf{W}}$ is the Jaumann rate of $\boldsymbol{\beta}$, and $\tilde{\boldsymbol{\varepsilon}} = \dot{\tilde{\mathbf{R}}}\tilde{\mathbf{R}}^{T} - \tilde{\mathbf{W}}$ is linear and homogeneous in deformation rate tensor $\tilde{\mathbf{D}} = \frac{1}{2}(\tilde{\mathbf{L}} + \tilde{\mathbf{L}}^{T})$, where $\tilde{\mathbf{L}} = (\partial\tilde{\mathbf{F}}/\partial t)\,\tilde{\mathbf{F}}^{-1}$, and $\tilde{\mathbf{W}}$ is the associated spin; see Section 4.9. For the evolution of $\boldsymbol{\beta}$, assume, *e.g.*,

$$\overset{\Delta}{\boldsymbol{\beta}} = A\,\tilde{\mathbf{D}}, \qquad A = A(\gamma), \tag{4.2.27b,c}$$

from which it follows that

$$\overset{\circ}{\boldsymbol{\beta}} = A\,\tilde{\mathbf{D}} + \tilde{\boldsymbol{\varepsilon}}\boldsymbol{\beta} - \boldsymbol{\beta}\tilde{\boldsymbol{\varepsilon}}. \tag{4.2.27d}$$

While the model (4.2.27b) relates $\overset{\Delta}{\boldsymbol{\beta}}$ to $\tilde{\mathbf{D}}$ *isotropically,* (4.2.27d) is *anisotropic,* with its last two terms dominating at large shear strains, as follows from the structure of $\tilde{\boldsymbol{\varepsilon}}$; see Section 4.9.[27]

Woolley (1953), Orowan (1958), Wilson (1964), Abel and Ham (1966), Mróz (1969), Brown and Stobbs (1971), Daniels and Horn (1971), Abel and Muir (1972), Stoltz and Pelloux (1974, 1976), Kocks *et al.* (1975), Dafalias and Popov (1976), Moan and Embury (1979), Hidayetoglu *et al.* (1985), Bate and Wilson (1986), and Stout and Rollet (1990). Recently, the phenomenon has been shown to also exist at high strain rates; see Thakur (1994), and Thakur *et al.* (1996a,b).

[27] This has been examined by Nemat-Nasser (1983, 1990).

4.2.8. The J_2-flow Theory with Combined Isotropic-kinematic Hardening

For combined isotropic-kinematic hardening, the yield function and the flow potential in the J_2-flow theory become

$$f \equiv g \equiv \tau - F(\gamma), \quad \tau = \{(\boldsymbol{\tau}' - \boldsymbol{\beta}) : (\boldsymbol{\tau}' - \boldsymbol{\beta})/2\}^{1/2}, \tag{4.2.28a,b}$$

where the notation is the same as in the previous two subsections. The plastic part of the deformation rate then is

$$\mathbf{D}^{p\prime} = \dot{\gamma}\,\frac{\partial f}{\partial \boldsymbol{\tau}'} = \frac{1}{\overline{H}}\left\{\frac{\boldsymbol{\mu}}{\sqrt{2}} \otimes \frac{\boldsymbol{\mu}}{\sqrt{2}}\right\} : \overset{\circ}{\boldsymbol{\tau}}{}', \tag{4.2.29a}$$

where $\boldsymbol{\mu}$ is defined by (4.2.26b) and the hardening parameter $\overline{H}$ is given by

$$\overline{H} = H + \tilde{H} = \frac{\partial F}{\partial \gamma} + \frac{\boldsymbol{\mu} : \overset{\circ}{\boldsymbol{\beta}}}{\sqrt{2}\,\dot{\gamma}}. \tag{4.2.29b}$$

Moreover, the evolution of the backstress $\boldsymbol{\beta}$ may be assumed to be governed by (4.2.27b).

Substitution from (4.2.29a) into (4.2.19c) yields the total deformation rate and hence, with the aid of the elasticity relations, it follows that (see Figure 4.2.5),

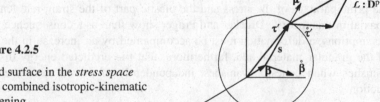

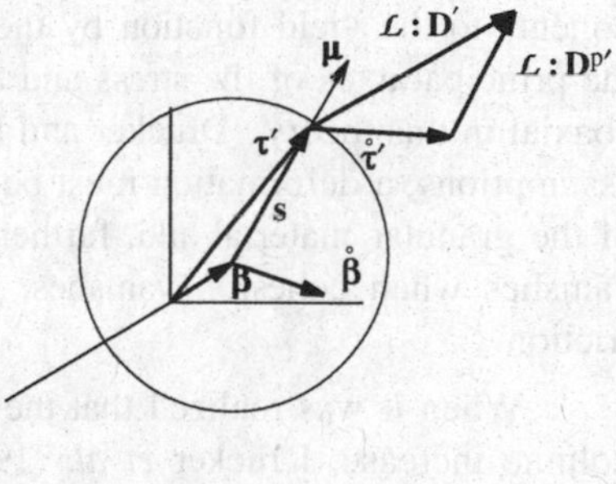

Figure 4.2.5

Yield surface in the *stress space* with combined isotropic-kinematic hardening

$$\mathbf{D} = \mathcal{M} : \overset{\circ}{\boldsymbol{\tau}} + \frac{(\boldsymbol{\tau}' - \boldsymbol{\beta})}{4\tau^2 \overline{H}}\,(\boldsymbol{\tau}' - \boldsymbol{\beta}) : \overset{\circ}{\boldsymbol{\tau}}{}' = \left\{\mathcal{M} + \frac{1}{\overline{H}}\,\frac{\boldsymbol{\mu}}{\sqrt{2}} \otimes \frac{\boldsymbol{\mu}}{\sqrt{2}}\right\} : \overset{\circ}{\boldsymbol{\tau}},$$

$$\overset{\circ}{\boldsymbol{\tau}} = \mathcal{L} : \left\{ \mathbf{1}^{(4s)} - \frac{\boldsymbol{\mu} \otimes \mathcal{L} : \boldsymbol{\mu}}{2\overline{H} + \boldsymbol{\mu} : \mathcal{L} : \boldsymbol{\mu}} \right\} : \mathbf{D},$$

$$\mathbf{D}^{p\prime} = \frac{\mathbf{D} : \mathcal{L} : \boldsymbol{\mu}}{2\overline{H} + \boldsymbol{\mu} : \mathcal{L} : \boldsymbol{\mu}}\,\boldsymbol{\mu} = \frac{\boldsymbol{\mu} \otimes \boldsymbol{\lambda}}{2\overline{H} + \boldsymbol{\mu} : \boldsymbol{\lambda}} : \mathbf{D}. \tag{4.2.30a-c}$$

4.3. DILATANCY AND PRESSURE SENSITIVITY

The importance of inelastic volumetric strain has long been recognized in the study of geomaterials, *e.g.*, Schofield and Wroth (1968). For materials of this kind, volumetric strains are usually closely related to the frictional effects and the pressure sensitivity of the material. These factors are central in the study of localized deformation in geomaterials, especially granular media. The volumetric strain is also of importance in porous metals. The pressure sensitivity and plastic volumetric strains may stem from a variety of possibly interacting micromechanisms, as discussed by Rice (1977) and Nemat-Nasser *et al.* (1981).

Classical plasticity has been modified for application to problems which involve inelastic volumetric strains, as well as pressure effects. This has been accomplished by including in the yield condition the dependence of the yield surface on the the first stress invariant; see, *e.g.*, Drucker and Prager (1952), and Berg (1970). The pressure effects and the volumetric strain, however, are not dominating factors in the response of most metals under moderate conditions. They are, on the other hand, of paramount importance in the proper description of the flow of frictional granules, ceramics, rocks, and concrete, under essentially all conditions.

The first yield-function-based, systematically developed, continuum theory of granular material behavior was presented by Drucker and Prager (1952) and applied by Shield (1953, 1954). The Drucker-Prager theory uses the Coulomb yield function as the plastic potential, and relates the strain-rate components to the yield function by the associated flow-rule of plasticity. Hence, the principal axes of the stress and the plastic part of the strain-rate tensors are coaxial in this theory. Drucker and Prager show that, as a consequence of these assumptions, a deformation must be accompanied by an increase in the volume of the granular material and, furthermore, that the predicted energy dissipation vanishes when cohesion vanishes, independently of the value of the internal friction.

When it was realized that the Drucker-Prager theory predicts too great a volume increase, Drucker *et al.* (1957), Roscoe *et al.* (1958), and Jenike and Shield (1959) proposed workhardening models with density-dependent cohesions which make more realistic volume-change predictions. In these theories, it is still assumed that the principal axes of the stress tensor coincide with those of the plastic strain-rate tensor. In general, the principal axes of the stress and the plastic strain rate tensors do not coincide (Mandl and Luque, 1970) in frictional deformation. This is called *noncoaxiality*. Its occurrence in frictional granules was supported by further experimental evidence given by Drescher (1976). It is an important feature of geomaterials, which stems from their frictional property; see Rudnicki and Rice (1975). This is discussed in Section 4.6.

A derivation of the constitutive relations based on the concepts of yield surface and flow potential, has been given by Nemat-Nasser and Shokooh (1980) for large-deformation problems, where the effects of inelastic dilatation (or densification) and friction (pressure sensitivity) are delineated and discussed; for small deformation, a similar model is examined by Wilde (1977). In this formulation, the J_2-plasticity model is used for both the yield function and the flow potential, but the radii of these surfaces in the deviatoric stress space are assumed to depend on pressure, and the volumetric and the effective plastic strains. The theoretical predictions have been compared with experimental results, showing reasonably good success.[28]

When a granular medium is monotonically sheared under confining pressure, a strong texture or anisotropy is induced, as the granules roll and slide over each other in response to the applied loads. For granular materials, the microstructure or texture is often called the *fabric*. Various measures are introduced to quantify the fabric in a granular mass, in terms of the distribution of the *contact unit normals*, and the *unit branches*. The unit branch is a unit vector in the direction which connects the centroids of two contacting granules. This and related issues are discussed in Section 4.7 and more thoroughly in Chapter 7, in connection with the micromechanics of the deformation of frictional granules.

From a continuum point of view, the induced anisotropy or fabric in a granular mass may be associated with the kinematic hardening which is strongly coupled with the dilatancy and densification, for this class of materials. This is unlike the kinematic hardening in metals which stems from dislocation pile-ups at various obstacles, essentially unrelated to volumetric strains which are negligibly small in metal plasticity. In the case of granular materials, the induced anisotropy may be so prominent that, upon unloading, reverse plastic shearing may take place against the still finite applied shear stress. For metals, on the other hand, this has never been observed experimentally. In the absence of volumetric strains, such reverse plastic deformation would violate the energy conservation law. For granular materials, however, a reverse plastic deformation against an applied shear stress is accompanied by volumetric contraction, where positive work is performed by the applied pressure, supplying the necessary energy. This and related topics are considered in Section 4.7 and in greater detail in Chapter 7.

4.3.1. Nonassociative Plasticity with Dilatancy and Friction

In this section, both isotropic and kinematic hardening are examined, focusing on monotonic loading; load reversal and cyclic shearing are examined in Section 4.7. The following yield surface and flow potential are considered:

$$f \equiv \tau - F(I, \Delta, \gamma), \qquad g \equiv \tau + G(I, \Delta, \gamma), \tag{4.3.1a,b}$$

where[29]

$$\Delta = \int_0^t \frac{\rho_0}{\rho} D^p_{kk}\, dt, \quad I = \tau_{kk}, \quad \gamma = \int_0^t \sqrt{2}\,\boldsymbol{\mu} : \mathbf{D}^{p\prime}\, dt, \tag{4.3.1c-e}$$

---

[28] See Dorris and Nemat-Nasser (1982) and Rowshandel and Nemat-Nasser (1987) for applications to soils and rocks, and Lade *et al.*(1987, 1988) who examines instability phenomena in granular materials.

[29] Here the yield surface and the flow potential are expressed in terms of the Kirchhoff stress $\boldsymbol{\tau}$. Nemat-Nasser and Shokooh (1980) use the Cauchy stress $\boldsymbol{\sigma}$ which relates to the Kirchhoff stress by $\boldsymbol{\sigma} = \boldsymbol{\tau}\rho/\rho_0$.

and where τ is defined by (4.2.25b), γ being, as before, the accumulated effective plastic strain. The quantity Δ is the total accumulated plastic volumetric strain, measured relative to a reference state with mass density ρ_0, the current mass density being ρ.

The plastic strain rate (not necessarily deviatoric) is given by

$$\mathbf{D}^p = \dot{\gamma}\,\frac{\partial g}{\partial \boldsymbol{\tau}}\,. \tag{4.3.2a}$$

From (4.3.1b) it follows that

$$\frac{\partial g}{\partial \boldsymbol{\tau}} = \frac{\boldsymbol{\mu}}{\sqrt{2}} + \frac{\partial G}{\partial I}\,\mathbf{1}\,, \tag{4.3.2b}$$

where $\boldsymbol{\mu}$ is defined by (4.2.26b). For continued plastic deformation, $\dot{f} = 0$ which yields

$$\dot{\gamma} = \frac{1}{\overline{H}}\left\{\frac{\boldsymbol{\mu}}{\sqrt{2}} - \frac{\partial F}{\partial I}\,\mathbf{1}\right\} : \overset{\circ}{\boldsymbol{\tau}}, \tag{4.3.2c}$$

where

$$\overline{H} = \frac{\partial F}{\partial \Delta}\,\frac{\dot{\Delta}}{\dot{\gamma}} + \frac{\partial F}{\partial \gamma} + \frac{\boldsymbol{\mu} : \overset{\circ}{\boldsymbol{\beta}}}{\sqrt{2}\,\dot{\gamma}}\,. \tag{4.3.2d}$$

It then follows that the plastic strain rate (including dilatancy or densification) is given by

$$\mathbf{D}^p = \frac{1}{\overline{H}}\left\{\frac{\boldsymbol{\mu}}{\sqrt{2}} + \frac{\partial G}{\partial I}\,\mathbf{1}\right\}\left\{\frac{\boldsymbol{\mu}}{\sqrt{2}} - \frac{\partial F}{\partial I}\,\mathbf{1}\right\} : \overset{\circ}{\boldsymbol{\tau}}\,. \tag{4.3.3a}$$

Note that the hardening parameter $\overline{H}$ may be expressed as

$$\overline{H} = 3\frac{\rho_0}{\rho}\,\frac{\partial G}{\partial I}\,\frac{\partial F}{\partial \Delta} + \frac{\partial F}{\partial \gamma} + \frac{\boldsymbol{\mu} : \overset{\circ}{\boldsymbol{\beta}}}{\sqrt{2}\,\dot{\gamma}}\,. \tag{4.3.3b}$$

When only isotropic hardening is involved, then $\boldsymbol{\beta} = \mathbf{0}$ and $\overline{H}$ is replaced by H which is then given by

$$H = 3\frac{\rho_0}{\rho}\,\frac{\partial G}{\partial I}\,\frac{\partial F}{\partial \Delta} + \frac{\partial F}{\partial \gamma}\,. \tag{4.3.3c}$$

In these equations, the quantity $-\partial F/\partial I$ is the pressure sensitivity parameter and $\partial G/\partial I$ relates to the volumetric straining produced by shearing, *i.e.*, it is the dilatancy parameter. If $I = -3p$, with p representing the pressure, then $-3\partial F/\partial I = \partial F/\partial p$, and $3\partial G/\partial I = -\partial G/\partial p$. With this change in the variable I, (4.3.3a,b) become,

$$\mathbf{D}^p = \frac{1}{\overline{H}}\left\{\frac{\boldsymbol{\mu}}{\sqrt{2}} \otimes \frac{\boldsymbol{\mu}}{\sqrt{2}} + \frac{1}{3}\left[\frac{\partial F}{\partial p}\,\frac{\boldsymbol{\mu}}{\sqrt{2}} \otimes \mathbf{1} - \frac{\partial G}{\partial p}\,\mathbf{1} \otimes \frac{\boldsymbol{\mu}}{\sqrt{2}}\right] - \frac{1}{9}\,\frac{\partial F}{\partial p}\,\frac{\partial G}{\partial p}\,\mathbf{1} \otimes \mathbf{1}\right\} : \overset{\circ}{\boldsymbol{\tau}}\,,$$

$$\overline{H} = -\frac{\rho_0}{\rho}\,\frac{\partial G}{\partial p}\,\frac{\partial F}{\partial \Delta} + \frac{\partial F}{\partial \gamma} + \frac{\boldsymbol{\mu} : \overset{\circ}{\boldsymbol{\beta}}}{\sqrt{2}\,\dot{\gamma}}\,. \tag{4.3.3d,e}$$

From (4.3.3d) now obtain

$$\mathbf{D}^{p\prime} = \frac{1}{H}\left\{\frac{\boldsymbol{\mu}}{\sqrt{2}} : \overset{\circ}{\boldsymbol{\tau}}' - \frac{\partial F}{\partial p}\dot{p}\right\}\frac{\boldsymbol{\mu}}{\sqrt{2}},$$

$$D^p_{kk} = \frac{1}{H}\frac{\partial G}{\partial p}\left\{\frac{\partial F}{\partial p}\dot{p} - \frac{\boldsymbol{\mu}}{\sqrt{2}} : \overset{\circ}{\boldsymbol{\tau}}'\right\}. \qquad (4.3.3f,g)$$

The pressure dependence of the yield surface leads to plastic distortion when the pressure is changed; *i.e., the term* $(\partial F/\partial p)\,\dot{p}$ *inside the braces in* (4.3.3f). *Similarly, the dependence of the flow potential on pressure results in an inelastic volumetric strain, being produced when the deviatoric stress is changed; i.e., the term* $\boldsymbol{\mu} : \overset{\circ}{\boldsymbol{\tau}}'$ *inside the braces in* (4.3.3g).

For geomaterials, $\partial F/\partial p$ *may be interpreted as a measure of the overall friction coefficient.* To see this, consider pure shearing in the x_1, x_2-plane, under hydrostatic pressure p; Figure 4.3.1.

The stress components now are,

$$\tau_{12} = \tau_{21} \equiv \tau,$$

$$\tau_{11} = \tau_{22} = \tau_{33} \equiv -p,$$

with all other components being zero. The yield function then becomes $\tau = F(p, \underline{\Delta}, \gamma)$. Thus,

$$\frac{\partial \tau}{\partial p} = \frac{\partial F}{\partial p}$$

which is the change of the resistance to shearing per unit applied compression. *In this sense,* $\partial F/\partial p$ *is a strictly positive quantity.*

The dilatancy parameter, $-\partial G/\partial p = D^p_{kk}/\dot{\gamma}$, measures the rate of change of the overall inelastic volumetric strain per unit rate of shear straining. *It may be positive, negative, or zero, depending on the history of the deformation and the current rate of loading.* This suggests that, within the framework of such continuum theories, the use of the associated flow rule may introduce a basic contradiction, because then $-\partial F/\partial p = \partial G/\partial p$. This then requires positive dilatancy throughout the entire deformation, contradicting experimental observations. The dilatancy and pressure sensitivity parameters can be related through the energy balance equation.[30]

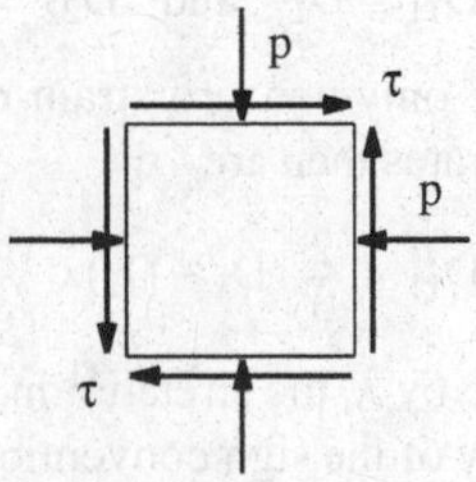

Figure 4.3.1

Plane strain, pure shearing under hydrostatic pressure

[30] See Nemat-Nasser (1980a,b), and Nemat-Nasser and Shokooh (1980).

In the following subsections, various aspects of the constitutive relations (4.3.3) are examined in relation to specific application to the mechanical response of soils and metals. In particular, the physical significance of various constitutive functions, *e.g.*, $\partial F/\partial p$ and $\partial G/\partial p$, is further discussed and clarified.

4.3.2. Application: Triaxial Test of Soils

Consider a possible application of the theory to the triaxial test commonly used to study the mechanical response of soils. Assume an initial homogeneous stress state defined by

$$\tau_{11} = \tau_1, \quad \tau_{22} = \tau_{33} = \tau_2, \quad \text{all other } \tau_{ij} = 0, \quad |\tau_1| > |\tau_2|. \tag{4.3.4a-d}$$

In line with the usual practice in soil mechanics, *regard compression and contraction positive and tension and extension negative.* The deviatoric components of the stress are,

$$\tau'_{11} = \frac{2}{3}(\tau_1 - \tau_2), \quad \tau'_{22} = \tau'_{33} = -\frac{1}{3}(\tau_1 - \tau_2). \tag{4.3.4e,f}$$

Hence,

$$I = \tau_1 + 2\tau_2, \quad \tau = \frac{1}{\sqrt{3}}(\tau_1 - \tau_2). \tag{4.3.4g,h}$$

Since, for granular materials consisting of very hard or almost rigid particles, the strains are essentially inelastic, consider only the plastic strain rates, D^p_{ij}. *For simplicity, drop out the superimposed* p in the sequel. Then, in view of (4.3.2a,b) and (4.3.4), it follows that

$$D_{11} \equiv D_1 \quad \text{and} \quad D_{22} = D_{33} \equiv D_2 \tag{4.3.5a,b}$$

are the only nonzero strain-rate components. The deviatoric components of the strain rates then are

$$D'_{11} = \frac{2}{3}(D_1 - D_2), \quad D'_{22} = D'_{33} = -\frac{1}{3}(D_1 - D_2). \tag{4.3.5c,d}$$

Denote by λ_i the stretch[31] in the x_i-direction, i = 1, 2, 3. Then, by definition and in view of the sign convention,

$$-D_i = (\ln \lambda_i)^{\cdot}, \quad D_1 - D_2 = \{\ln(\lambda_2/\lambda_1)\}^{\cdot} > 0. \tag{4.3.6a,b}$$

For monotonic loading, use the strain measure

$$\varepsilon = \frac{2}{3}\ln(\lambda_2/\lambda_1) \tag{4.3.7}$$

as the *time parameter*. Then, when $\lambda_1 = \lambda_2 = 1$ (no deformation), $\varepsilon = 0$. Since $\tau_1 > \tau_2$, then $\lambda_1 < 1$ and $\lambda_2 > 1$, *i.e.*, contraction in the x_1-direction, and

[31] λ_i is the length of an element in the current configuration, ds_i, divided by the corresponding initial length, dS_i.

expansion in the x_2- and x_3-directions. Hence, for this kind of loading, $\lambda_2/\lambda_1 > 1$, which implies that $\varepsilon > 0$, and, in view of (4.3.6), ε is monotone increasing for continued deformation. In what follows, only deformations of this kind are considered. The results must be modified if the sample extends in the x_1-direction and contracts in the x_2- and x_3-directions. In such a case, the roles of λ_1 and λ_2 on the one hand, and D_1 and D_2 on the other hand, are reversed. In terms of λ_1 and λ_2, the volumetric strain is

$$\Delta = 1 - \lambda_1 \lambda_2^2 . \tag{4.3.8}$$

In soil mechanics, the state of stress in a triaxial test is usually characterized by pressure, p, and differential stress, q, defined by

$$p = \frac{1}{3}(\tau_1 + 2\tau_2) = \frac{1}{3}I, \quad q = (\tau_1 - \tau_2) = \sqrt{3}\,\tau . \tag{4.3.9a,b}$$

The state of deformation is characterized by the rate of the volumetric strain, $\dot{v}/v$, measured per unit current volume, and the distortional strain, $\dot{\varepsilon}$,

$$\dot{v}/v = D_1 + 2D_2, \quad \dot{\varepsilon} = \frac{2}{3}(D_1 - D_2) \equiv D'_{11} . \tag{4.3.10a,b}$$

In the sequel (p, q) are used to represent the state of stress, and $(\dot{v}/v, \dot{\varepsilon})$ to represent the strain rate. Note that

$$\dot{\varepsilon} = \frac{1}{\sqrt{3}}\dot{\gamma} . \tag{4.3.10c}$$

The rate of distortional work per unit *reference* volume is

$$\dot{\xi} \equiv \frac{\rho_0}{\rho}\boldsymbol{\sigma}' : \mathbf{D} = \frac{2}{3}(\tau_1 - \tau_2)(D_1 - D_2) = q\dot{\varepsilon} , \tag{4.3.11}$$

where $\boldsymbol{\sigma}'$ is the deviatoric part of the Cauchy stress tensor. In terms of q, p, and ε, (4.3.2c) is expressed as

$$\frac{dq}{d\varepsilon} - \sqrt{3}\frac{\partial F}{\partial p}\frac{dp}{d\varepsilon} = 3H , \tag{4.3.12}$$

where (4.3.9) and (4.3.10c) are used. This equation relates q and p. If the work-hardening parameter H is established experimentally or by other means, then (4.3.12) can be used directly, as is illustrated in the sequel; see (4.3.23b) and the corresponding discussions.

The dilatancy factor,[32] $3\,\partial G/\partial I = \partial G/\partial p$, can be given a clear physical interpretation in the present special case. From (4.3.2a,b) and (4.3.10), it follows that

$$\sqrt{3}\,\frac{\partial G}{\partial p} = \frac{3}{2}\,\frac{D_1 + 2D_2}{D_1 - D_2} = \frac{1}{v}\,\frac{dv}{d\varepsilon} \tag{4.3.13a,b}$$

which indicates that $\sqrt{3}\,\partial G/\partial p$ is the rate of the volumetric strain, $\dot{v}/v$, per unit

[32] Note that pressure is positive.

rate of distortional strain, $\dot{\varepsilon}$. The dilatancy factor may be established experimentally, as suggested by (4.3.13); in the present case, the denominator in the right-hand side of (4.3.13a) is positive.

Let e denote the *void ratio*, *i.e.*, the ratio of the void volume, V_v, to the volume of the solid, V_s, contained within a sample of volume, $V_v + V_s$,

$$e = V_v / V_s . \tag{4.3.14a}$$

Then, by definition, the volumetric strain measure, Δ, is related to the void ratio by

$$\Delta = \frac{e_0 - e}{1 + e_0}, \tag{4.3.14b}$$

where e_0 is the initial value of the void ratio.[33]

From (4.3.1c) and (4.3.2a) obtain

$$\dot{\Delta} = \frac{\rho_0}{\rho} D_{kk} = \sqrt{3} \frac{\rho_0}{\rho} \frac{\partial G}{\partial p} \dot{\varepsilon} \tag{4.3.15a}$$

which can be integrated,

$$\Delta = \sqrt{3} \int_0^{\varepsilon} \frac{\rho_0}{\rho} \frac{\partial G}{\partial p} d\varepsilon', \tag{4.3.15b}$$

with the initial condition $\Delta = 0$ at $\varepsilon = 0$. Equation (4.3.15b) is valid in general. It can be further reduced for the triaxial state of strain considered here. To this end, from (4.3.8) and (4.3.13), obtain

$$\frac{d\Delta}{1 - \Delta} = \sqrt{3} \frac{\partial G}{\partial p} d\varepsilon \tag{4.3.16a}$$

which, upon integration, gives

$$\Delta = 1 - \exp\left\{ - \sqrt{3} \int_0^{\varepsilon} \frac{\partial G}{\partial p} d\varepsilon' \right\}. \tag{4.3.16b}$$

With the aid of (4.3.14) it now follows that

$$e = e_0 - (1 + e_0) \left[1 - \exp\left\{ - \sqrt{3} \int_0^{\varepsilon} \frac{\partial G}{\partial p} d\varepsilon' \right\} \right]. \tag{4.3.16c}$$

It is commonly assumed that a monotone and continuous shearing of cohesionless sand leads to a void ratio, e_c, called *critical*, which depends on the confining pressure and the grain size distribution and shape, but is independent of the initial value of the void ratio. This means that, in a continuous monotone deformation, the dilatancy factor $\partial G / \partial p$ must approach zero. Hence, set

$$K = \lim_{\varepsilon \to \infty} \exp\left\{ - \sqrt{3} \int_0^{\varepsilon} \frac{\partial G}{\partial p} d\varepsilon' \right\}, \tag{4.3.17a}$$

---

[33] In view of the sign convention (*i.e.*, contraction positive), Δ is defined by $(v_0 - v)/v_0$, where v is the current volume and v_0 is the initial volume of the sample.

and obtain

$$e_c = e_0 - (1 + e_0)(1 - K),$$

from which it follows that

$$K = \frac{1 + e_c}{1 + e_0}. \tag{4.3.17b}$$

Hence, arrive at

$$\sqrt{3}\int_0^\infty \frac{\partial G}{\partial p}\, d\varepsilon' = ln\left[\frac{1 + e_0}{1 + e_c}\right]. \tag{4.3.17c}$$

Thus, in continued monotone shearing, there is a net amount of densification or dilatancy depending on whether $e_0 > e_c$ (loose sand) or $e_0 < e_c$ (dense sand). Figure 4.3.2 sketches the observed variation of the void ratio e with the strain, ε.

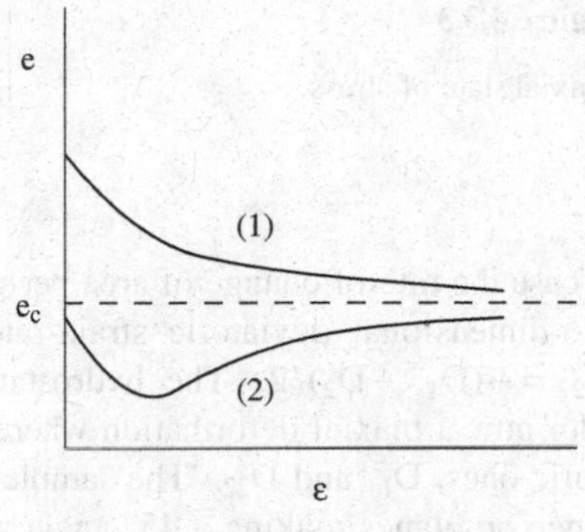

Figure 4.3.2
Variation of the void ratio e with strain ε in a triaxial test: Curve (1) is for loose, and Curve (2) is for dense sands

Curve (1) is for loose sand, $e_0 > e_c$, for which the void ratio e decreases monotonically with increasing ε, approaching e_c asymptotically. In this case, $\partial G/\partial p$ remains positive (continuous densification). Curve (2) of Figure 4.3.2, on the other hand, is for dense sand, $e_0 < e_c$. In this case, experiment shows that, similarly to the case of loose sand, there is an initial densification which is, however, followed by dilatancy. Hence, $\partial G/\partial p$ is initially positive, but for dense sand, it becomes zero at a certain ε, and then is negative as ε is increased, approaching zero asymptotically. For this case, there is always a net amount of dilatancy, as straining continues. It is shown in the sequel that these experimentally observed features are displayed by the present constitutive model.

4.3.3. A Model for Cohesionless Sands

A relation between the dilatancy factor, $\partial G/\partial p$, and the measure of the internal friction, $\partial F/\partial p$, is now sought for cohesionless sands. The total rate of plastic work, $\dot{w}_p$, measured per unit *reference* volume is

$$\dot{w}_p = p\,\dot{v}/v + q\,\dot{\varepsilon}\,. \tag{4.3.18a}$$

If the only source of dissipation is internal friction due to the intergranular sliding, the above rate of work must equal the rate of frictional loss.

The calculation of the rate of frictional loss requires consideration of the relative motion of the grains, and is a complicated problem; see Chapter 7. Here, consider a simple (approximate) approach instead, as follows.

As a motivation, first consider *biaxial* deformation of cylindrical rods (plane strain problem), where $D_{33} = 0$; see Figure 4.3.3.

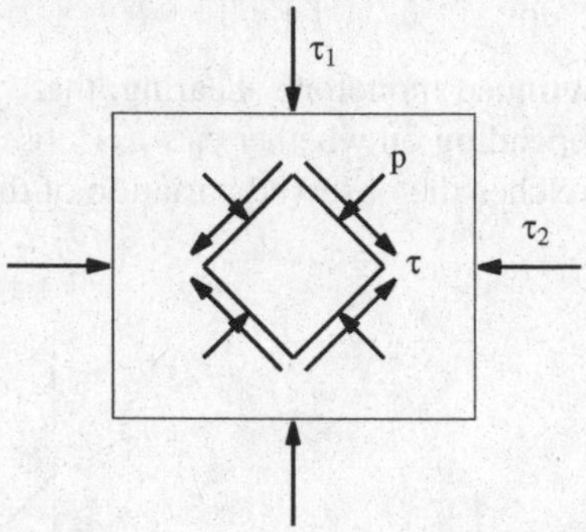

Figure 4.3.3

Biaxial state of stress

In this case the rate of change of area per unit current area is $\dot{a}/a = D_1 + D_2$, and the two-dimensional deviatoric strain-rate components are $D'_{11} = (D_1 - D_2)/2$ and $D'_{22} = -(D_1 - D_2)/2$. The hydrostatic pressure becomes $p = (\tau_1 + \tau_2)/2$. Consider now a biaxial deformation where the axial strain rates are equal to the deviatoric ones, D'_{11} and D'_{22}. The sample undergoes plane strain pure shearing and, since on planes making a 45° angle with the principal stress axes, the normal stress is equal to the hydrostatic pressure, the frictional stress on these planes would be $\tau_f = (\partial F/\partial p)p$, $\partial F/\partial p$ being the coefficient of friction. Hence, the rate of frictional work per unit reference volume becomes $\tau_f(|D'_{11}| + |D'_{22}|) = 2\,\tau_f D'_{11} = \tau_f(D_1 - D_2)$.

In the triaxial case with $D_{11} = D_1$ and $D_{22} = D_{33} = D_2$, add the contribution for sliding friction on the x_1, x_2- and x_3, x_1-planes, arriving at

$$\dot{w}_f \approx \frac{\partial F}{\partial p}\,p\,(|D'_{11} - D'_{22}| + |D'_{22} - D'_{33}| + |D'_{33} - D'_{11}|)$$

$$= 2\frac{\partial F}{\partial p}\,p\,(D_1 - D_2) = 3\,\frac{\partial F}{\partial p}\,p\,\dot{\varepsilon}\,. \tag{4.3.18b}$$

If now the rate of plastic work is equated with the rate of frictional loss, in view of (4.3.13), it follows that

$$\sqrt{3}\,\frac{\partial G}{\partial p} = 3\,\frac{\partial F}{\partial p} - \frac{q}{p}\,. \tag{4.3.19}$$

This relates the dilatancy factor to the coefficient of internal friction.[34] The

[34] Equation (4.3.19) is a generalization of Eq. (5.21a) of Schofield and Wroth (1968). It shows

relation (4.3.19) is an approximation. Although it models the triaxial test in monotone loading, it does not include the effects of unloading and, especially, of stress reversal. In Section 4.7, this expression is generalized to include the effect of induced anisotropy, through the introduction of texture- or fabric-related kinematic hardening; see (4.7.6a,b).

Since the friction coefficient, $\partial F/\partial p$, is positive, (4.3.19) shows that under a constant confining pressure, p = constant, the dilatancy factor, $\partial G/\partial p$, is initially, *i.e.,* at q = 0, positive. Hence, initially the sample tends to densify, as shown for both curves (*i.e.,* for both loose sand and dense sand) in Figure 4.3.2.

In the p, q-plane, the curve

$$3\,\frac{\partial F}{\partial p}\,p - q = 0 \tag{4.3.20}$$

is the locus of points on which $\partial G/\partial p = 0$. This is called the *critical curve* which passes through the origin; see Figure 4.3.4. If the average friction, $\partial F/\partial p$, is assumed to be constant, the critical *line* introduced by Schofield and Wroth (1968), is obtained.

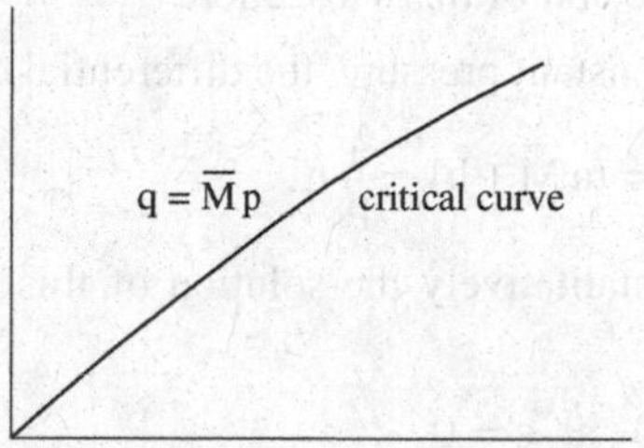

Figure 4.3.4
The critical curve in the q, p-plane for cohesionless sand

Assuming no kinematic hardening, consider now the hardening parameter (4.3.3c). In view of (4.3.16a) and (4.3.19), obtain

$$H = \frac{1}{\sqrt{3}}\left\{\left[3\,\frac{\partial F}{\partial p} - \frac{q}{p}\right](1-\Delta)\,\frac{\partial F}{\partial \Delta} + \frac{\partial F}{\partial \varepsilon}\right\}. \tag{4.3.21a}$$

The last term in the right-hand side represents hardening due to the distortional plastic work, *i.e.,* $\partial F/\partial\varepsilon = \sqrt{3}\,\partial F/\partial\gamma$; see (4.3.10c). The remaining terms on the right-hand side of (4.3.21a), on the other hand, represent hardening due to changes in the density. The quantity $(1-\Delta)\,\partial F/\partial\Delta$ is always positive, as it represents hardening due to volumetric contraction.[35]

---

that the concepts of internal friction and dilatancy are not compatible with the assumption of an associative flow rule.

[35] Note that F increases with increasing Δ, *i.e.,* with plastic *compaction*, and that $(1-\Delta) > 0$, in view of (4.3.8).

For simplicity set

$$3\,\frac{\partial F}{\partial p} \equiv \overline{M} > 0, \quad a \equiv \sqrt{3}\,(1-\Delta)\,\frac{\partial F}{\partial \Delta} > 0, \quad h \equiv \sqrt{3}\,\frac{\partial F}{\partial \varepsilon}, \tag{4.3.22a-c}$$

and reduce (4.3.21a) to

$$H = \frac{1}{3}\left\{ a\left[\overline{M} - \frac{q}{p}\right] + h \right\}. \tag{4.3.21b}$$

Here, a, $\overline{M}$, and h are, in general, functions of ε. Substitute from (4.3.21b) into (4.3.12), to arrive at

$$\frac{dq}{d\varepsilon} - \frac{1}{\sqrt{3}}\,\overline{M}\,\frac{dp}{d\varepsilon} = a\left[\overline{M} - \frac{q}{p}\right] + h. \tag{4.3.23a}$$

The general solution of the differential equation (4.3.23a) includes many commonly observed features of cohesionless sands in triaxial tests under monotone loading conditions. To show this, consider a triaxial test under constant confining pressure, p = constant. Qualitatively, the results are similar for a triaxial test with constant lateral confinement, *i.e.*, with τ_2 = constant; see comments at the end of this subsection.

At constant pressure, the differential equation (4.3.23a) becomes

$$\frac{dq}{d\varepsilon} = (a\,\overline{M} + h) - \frac{a}{p}\,q. \tag{4.3.23b}$$

Examine qualitatively the solution of this differential equation with the initial condition

$$q = 0 \quad \text{at} \quad \varepsilon = 0, \tag{4.3.23c}$$

for two possible variations of the workhardening function h = h(ε). These variations are sketched in Figure 4.3.5(a) by Curves (1) and (2). Assume that the positive quantity $a\overline{M}$ remains essentially constant.

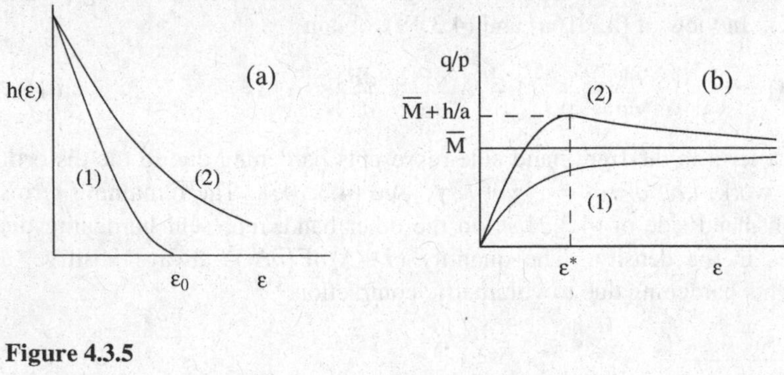

Figure 4.3.5

(a) Variation of the distortional hardening parameter h with strain ε; and (b) variation of the normalized differential stress q/p with strain ε: Curves (1) are for loose and Curves (2) are for dense samples

In the case of Curve (1), h drops to zero very quickly and is assumed to be essentially zero for $\varepsilon > \varepsilon_0$, with[36] $\varepsilon_0 << p/a$. Then from (4.3.23b) it is clear that, in the q, ε-plane, a solution curve similar to Curve (1) of Figure 4.3.5(b), is obtained, where q increases monotonically from zero, approaching $q/p = \overline{M}$ asymptotically. This is the response of loose sand, $e_0 > e_c$. From this and (4.3.16c), written as

$$e = e_0 - (1 + e_0)\left[1 - \exp\left\{-\int_0^{\varepsilon} (\overline{M} - q/p)d\varepsilon'\right\}\right], \tag{4.3.16d}$$

it then follows that the void ratio e decreases monotonically, approaching a constant value, e_c, as shown by Curve (1) in Figure 4.3.2.

The assumption that the material hardening due to distortional work, $h = \sqrt{3}\,(\partial F/\partial \varepsilon)$, should behave for loose sand in the manner sketched by Curve (1) of Figure 4.3.5a, may be justified. In this case, a substantial amount of densification occurs and, therefore, hardening due to the density changes is dominant. In fact, for a very loose sand, the effect of h may be ignored, leading to a model with only density hardening.

In Figure 4.3.5(a), Curve (2) shows a variation for h which, although monotonically decreasing with increasing ε, does not approach zero fast enough. Therefore, the function h will affect the qualitative behavior of the solution of (4.3.23b). For a and $\overline{M}$ positive constants, this is sketched by Curve (2) in Figure 4.3.5(b). Since h approaches zero for sufficiently large ε, the solution curve of (4.3.23b), in the q, ε-plane, again approaches asymptotically the line $q/p = \overline{M}$, as shown by Curve (2) in Figure 4.3.5(b). However, in this case, q/p first increases monotonically, becoming equal to $\overline{M}$ for a certain value of ε at which h(ε) is still finite. Hence, q/p continues to increase until the right-hand side of (4.3.23b) is zero at $\varepsilon = \varepsilon^*$. After this, because of the monotone decreasing nature of h, q/p decreases monotonically, approaching asymptotically *from above*, the value $q/p = \overline{M}$. As is well known, this is the observed behavior of dense sand. Moreover, with this variation of q/p with respect to ε, (4.3.16c) reveals that the void ratio in this case varies with ε in a manner shown by Curve (2) of Figure 4.3.2. Here, as q/p increases from zero, the void ratio decreases, attaining its minimum value when $q/p = \overline{M}$. After this, the void ratio begins to increase, attaining the initial value e_0 at the point where $\int_0^{\varepsilon} (\overline{M} - q/p)\, d\varepsilon' = 0$. After this point, e increases monotonically, attaining the critical value e_c asymptotically; see Curve (2) of Figure 4.3.2. For dense sand, therefore, both the density-hardening and the hardening due to plastic distortion play dominant roles. The model seems to adequately represent these observed behaviors of cohesionless sands.

In the illustration discussed above, the hydrostatic pressure is kept constant. In triaxial experiments, generally the lateral confining pressure, τ_2, is

---

36 Note the role of hydrostatic pressure.

constant. In such a case, from (4.3.9), it follows that

$$p = \tau_2 + \frac{q}{3}, \quad \frac{dp}{d\varepsilon} = \frac{1}{3}\frac{dq}{d\varepsilon}. \tag{4.3.24a,b}$$

The differential equation (4.3.23a) now becomes

$$(1 - \frac{1}{3\sqrt{3}}\,\overline{M})\,\frac{dq}{d\varepsilon} + \frac{a\,q}{\tau_2 + q/3} = a\,\overline{M} + h. \tag{4.3.24c}$$

This is a nonlinear, first-order, ordinary differential equation for the stress measure $q = \tau_1 - \tau_2$. Qualitatively, however, the solution curves are similar to those discussed for the constant-pressure case.

4.3.4. Model with Cohesion

For cohesionless granular materials, it is assumed that the resistance to intergranular sliding stems only from frictional effects. The model may be generalized somewhat by including the cohesive effects. As a simple approach, modify the frictional resistance by the addition of a cohesive force, $C/3$, which may depend on p, Δ, and ε, or it may be a constant, as the situation may dictate; the factor $1/3$ is introduced for convenience. The rate of frictional work, (4.3.18b), is modified to read,

$$\dot{w}_f = 3\left[\frac{\partial F}{\partial p}\,p + \frac{1}{3}\,C\right]\dot{\varepsilon}. \tag{4.3.18c}$$

Equations (4.3.19) and (4.3.20), respectively become

$$\sqrt{3}\,\frac{\partial G}{\partial p} = 3\,\frac{\partial F}{\partial p} - \frac{q - C}{p}, \tag{4.3.25}$$

and

$$3\,\frac{\partial F}{\partial p}\,p - (q - C) = 0. \tag{4.3.26}$$

The critical curve defined by (4.3.26) no longer passes through the origin in the p, q-plane, but intersects the q-axis at $q = C$; see Figure 4.3.6.

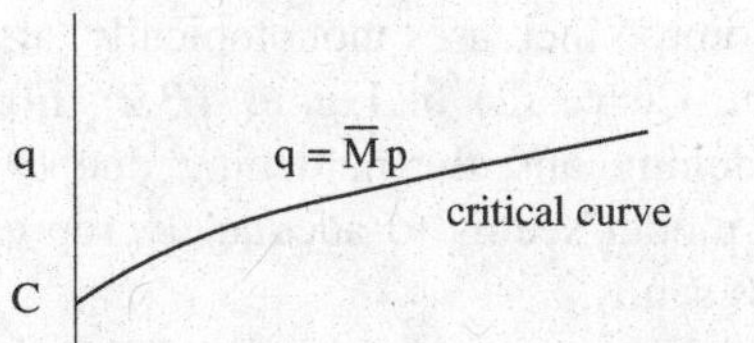

Figure 4.3.6

The critical curve in the q, p-plane for cohesive soils

This is compatible with experimental observations on cohesive soils.

In the present case the hardening parameter, H, becomes

$$H = \frac{1}{\sqrt{3}}\left\{\left[3\,\frac{\partial F}{\partial p} - \frac{q-C}{p}\right](1-\Delta)\,\frac{\partial F}{\partial \Delta} + \frac{\partial F}{\partial \varepsilon}\right\}. \tag{4.3.27}$$

The presence of cohesion therefore, has the effect of modifying the frictional coefficient $\overline{M} = 3\,\partial F/\partial p$. Depending on how the cohesion factor C, which is a non-negative quantity, varies as ε is increased, a modified response is obtained in the q, ε-plane, but qualitatively the results are similar to those already discussed for cohesionless sand.

4.3.5. Simple Shear

It is seen that the constitutive relations (4.3.3) seem to account relatively well for the observed mechanical response of granular materials in triaxial tests, under monotonic loading. A similar conclusion is obtained for the simple shear test. In fact, in this case, a more rigorous estimate for the rate of frictional loss can be obtained. Hence, if the shear stress is denoted by τ and the shear strain by γ, then, for the simple shear test, $D_{12} = D_{21} = (1/2)\,\dot{\gamma}$, $\dot{w} = p\,\dot{v}/v + \tau\,\dot{\gamma}$, and $\dot{w}_f = (\partial F/\partial p)\,p\,\dot{\gamma}$. Hence, instead of Eq. (4.3.19), obtain

$$\frac{\partial G}{\partial p} = \frac{\partial F}{\partial p} - \frac{\tau}{p}, \tag{4.3.28}$$

and (4.3.12) is replaced by

$$\frac{d\tau}{d\gamma} - \frac{\partial F}{\partial p}\frac{dp}{d\gamma} = H. \tag{4.3.29}$$

The workhardening parameter H is given by

$$\begin{aligned} H &= \left[\frac{\partial F}{\partial p} - \frac{\tau}{p}\right](1-\Delta)\,\frac{\partial F}{\partial \Delta} + \frac{\partial F}{\partial \gamma} \\ &= \frac{a}{\sqrt{3}}\left[\frac{1}{3}\,\overline{M} - \frac{\tau}{p}\right] + h, \end{aligned} \tag{4.3.30}$$

where a and $\overline{M}$ are as defined by (4.3.22a,b), and h is the tangent modulus in simple shear.

Except for slight modification of the coefficients, in simple shear, equations similar to those of the triaxial test are obtained. Therefore, the main features of the solution are the same as those of the triaxial test.

4.3.6. Comparison with Experimental Results

In light of the model features, some experimental results on triaxial tests of crushed Westerly granite and Ottawa sand, reported by Zoback and Byerlee (1976a,b), are examined in this subsection.

Figure 4.3.7 reproduces Figure 1 of Zoback and Byerlee (1976a, p.292), and shows the differential stress (q in the present notation) and the change in the void volume per unit initial void volume (here denoted by $\Delta\phi/\phi$), as functions of the axial strain $\Delta l/l$. These experimental results are replotted in Figures 4.3.8 and 4.3.9. Figure 4.3.8 represents the *normalized* differential stress, q/p, versus the strain, ε, while Figure 4.3.9 represents the void ratios as functions of the same strain measure.[37] As is seen from Figure 4.3.8, *all experimental results for crushed Westerly granite in this new representation, essentially fall on the same stress-strain curve.* Moreover, from Figure 4.3.9, the variation of the void ratios seems to follow a single pattern, their differences essentially being due to the initial value of the corresponding void ratio.

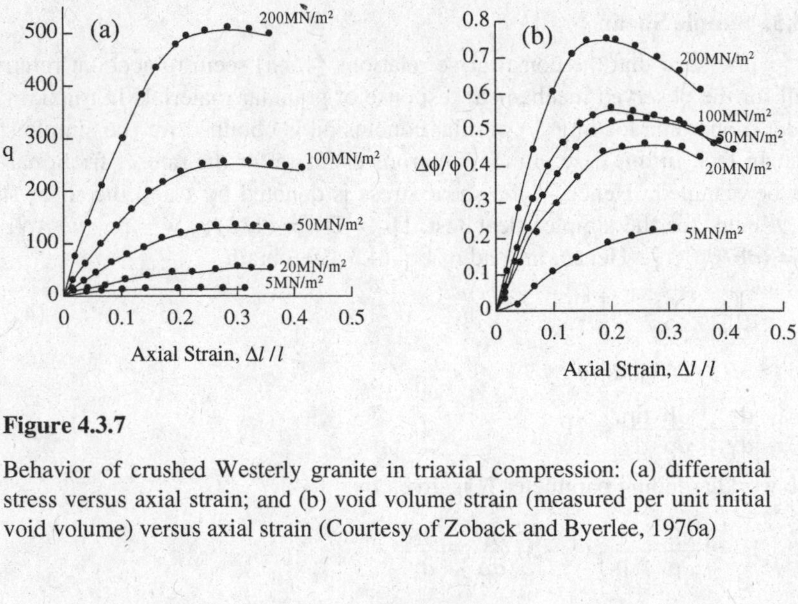

Figure 4.3.7

Behavior of crushed Westerly granite in triaxial compression: (a) differential stress versus axial strain; and (b) void volume strain (measured per unit initial void volume) versus axial strain (Courtesy of Zoback and Byerlee, 1976a)

To fit these experimental data with a minimum number of parameters, use the simplest possible form for the material functions, and assume in (4.3.23b) that a/p and $\overline{M}$ are both constants and that the hardening function h has the following simple exponential form:

$$h = h_0 e^{-\beta\varepsilon}, \tag{4.3.31}$$

β being a constant. Hence, (4.3.23b) can be integrated to yield

$$\frac{q}{p} = \overline{M}\left[1 - \exp\left\{-\frac{a}{p}\varepsilon\right\}\right] + \frac{h_0/p}{a/p - \beta}\left[\exp\{-\beta\varepsilon\} - \exp\left\{-\frac{a}{p}\varepsilon\right\}\right]. \tag{4.3.32a}$$

[37] Only those points for which all the needed data were available are replotted .

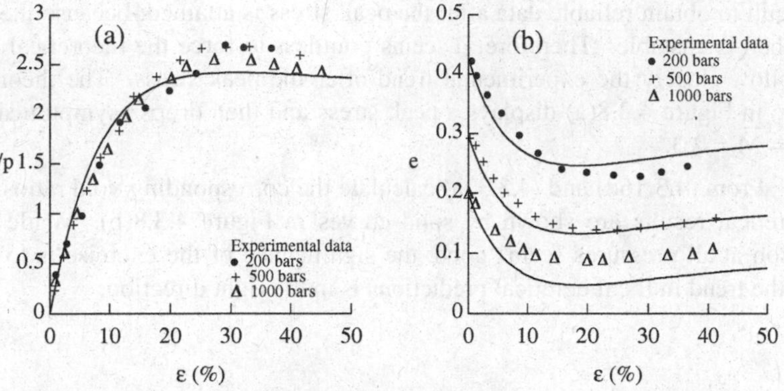

Figure 4.3.8

(a) Normalized differential stress, q/p, versus strain, $\varepsilon = (2/3)\,ln\,(\lambda_{(2)}/\lambda_{(1)})$, in triaxial tests of crushed Westerly granite; experimental data from Zoback and Byerlee (1976a); the theoretical (solid) curve displays (4.3.32a) with $\overline{M} = 2.3$, $B = (a/p)\,\overline{M} + h_0/p = 32$, $a/p = 9$, and $\beta = 8$; (b) the corresponding void ratio, e, versus strain ε; the theoretical (solid) curves display (4.3.16d) for the q/p, ε-relation of (a)

Now, the values of the parameters $\overline{M}$, a/p, h_0/p, and β are established for the experimental results reported in Figures 4.3.8(a,b). Since $\overline{M} = 3\,\partial F/\partial p$, and since $\partial F/\partial p$ is a measure of the coefficient of the overall friction in shear, the possible range of variation of $\overline{M}$ can be identified from the frictional strength of rocks, reported by Byerlee (1968, 1975), which has been reproduced as Figure 3 in Zoback and Byerlee (1976a). From this result, it is evident that, for the low-pressure range, $\partial F/\partial p$ varies somewhere between 0.6 and 1. On the other hand, the curve q/p, according to the theory, crosses the line $q/p = \overline{M}$ at the point where the void ratio has its minimum value. From the experimental results, the crossing occurs for $\overline{M}$ between 2 and 3.

Evaluate (4.3.23b) at ε = 0, to obtain the initial slope of the stress-strain curve, as

$$B = \left\{\frac{1}{p}\frac{dq}{d\varepsilon}\right\}_{\varepsilon=0} = \frac{a}{p}\overline{M} + \frac{h_0}{p}\,. \tag{4.3.33}$$

Comparison with the data reported in Figure 4.3.8 shows that $B \approx 32$. Moreover, a/p = 9 and β = 8 seem to be adequate estimates in the present context.[38]

With the above values of the coefficients, the curve in Figure 4.3.8(a) displays (4.3.32a), which seems to fit the experiments. It is experimentally very

[38] Nemat-Nasser and Shokooh (1980) use different values of the parameters than those used here.

difficult to obtain reliable data after the peak stress is attained, because the sample then is unstable. Therefore, it seems pointless to force the theoretical curve to follow closely the experimental trend after the peak stress. The theoretical curve in Figure 4.3.8(a) displays a peak stress and then drops asymptotically to $q/p = \overline{M} = 2.3$.

From (4.3.16d) and (4.3.32a) calculate the corresponding void ratios. The theoretical results are shown by solid curves in Figure 4.3.8(b). While comparison at all pressures seems good, the significance of the exercise is to show that the trend in the theoretical predictions is in the right direction.

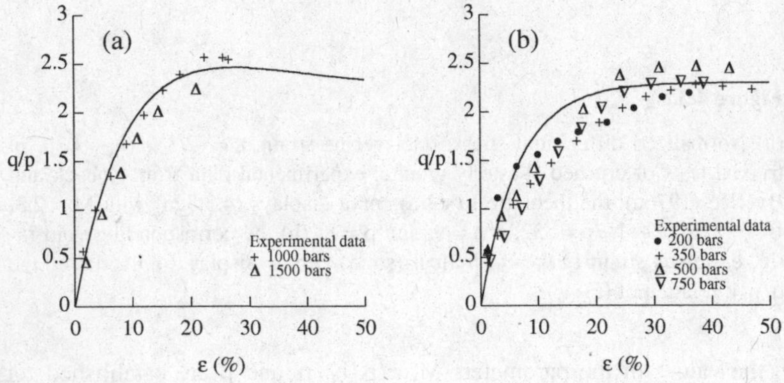

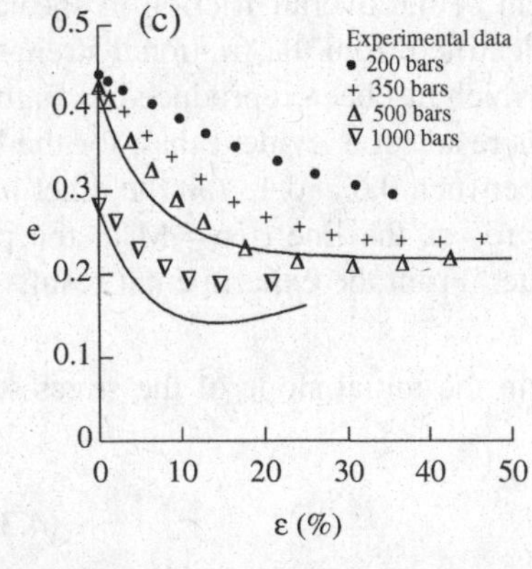

Figure 4.3.9

(a) Normalized differential stress, q/p, versus strain, $\varepsilon = (2/3)\, ln\, (\lambda_2/\lambda_1)$, in triaxial high-pressure tests of Ottawa sand; the theoretical (solid) curve displays (4.3.32a) with $\overline{M} = 2.2$, $B = (a/p)\overline{M} + (h_o/p) = 32$, $a/p = 8$, and $\beta = 7$; (b) normalized differential stress, q/p, versus strain, ε, in triaxial low-pressure tests of Ottawa sand; the theoretical (solid) curve displays (4.3.32b) with $\overline{M} = 2.3$ and $a/p = 15$; and (c) void ratio, e, versus strain, ε, corresponding to (a) and (b); the theoretical (solid) curves display (4.3.16d): the upper solid curve corresponds to the q/p, ε-relation of (a), and the other curve to that of (b); experimental data from Zoback and Byerlee (1976b)

The experimental results reported by Zoback and Byerlee (1976b) for Ottawa sand may be divided into two categories: those associated with $p = 1{,}000$ and 1,500 bars, and the others associated with lower pressures. Although it is difficult to precisely ascertain from the reported results, the higher pressure data seem to suggest the existence of a peak stress, whereas the lower pressure data do not. Figures 4.3.9(a,b) are the plots of q/p versus ε.

For the results in Figure 4.3.9(a), $\overline{M} = 2.2$, $B = 32$, $a/p = 8$, and $\beta = 7$ are used. From (4.3.32a), the theoretical results are plotted by a solid curve. This is a good fit to the experimental results, because not only are the experimental data for the stress-strain relation closely followed, but also the theory predicts nicely the variation in the corresponding void ratio for $p = 1{,}000$ bars, as shown in Figure 4.3.9(c).

The experimental results for lower pressures in Figure 4.3.9(b), are fitted by a curve associated with the theoretical results for loose sand with $h = 0$. In this case, from (4.3.32a),

$$q/p = \overline{M}\left\{1 - \exp\{-a\varepsilon/p\}\right\}, \qquad (4.3.32b)$$

which involves only two material parameters, and does not display a peak stress. To obtain the solid curve shown in Figure 4.3.9(b), $\overline{M} = 2.3$ and $a/p = 15$ are used. In Figure 4.3.9(c), the corresponding variation of the void ratio is shown by the solid curve: the upper curve. It follows the data for 350 and 500 bar pressure, but not those for 200 bar pressure.

The experiments used here were not conducted to test the present theory. *The suggested comparison therefore, should be viewed as an illustration of the general features that are included in the model, rather than as an experimental verification of the theory.*

4.3.7. Application to Porous Metals

For application to porous metals, the associative flow rule is commonly used. In this case the flow potential and the yield function are identical, so that $G \equiv -F$. Furthermore, special forms for the function F have been proposed. A relation has been suggested by Gurson (1977) based on a micromechanical analysis of an isolated spherical cavity in a concentric spherical ideally plastic matrix: a thick hollow sphere. The resulting yield function is

$$f \equiv (\sigma/\sigma_Y)^2 + 2\phi\cosh(\tfrac{1}{2} I/\sigma_Y) - (1+\phi^2),$$
$$\sigma = \sqrt{3}\,\tau = \left[(3/2)\,\boldsymbol{\tau}' : \boldsymbol{\tau}'\right]^{1/2}, \qquad I = \mathrm{tr}\,\boldsymbol{\tau}, \qquad (4.3.34a\text{-}c)$$

where σ_Y is the tensile yield stress of the matrix material, and ϕ is the void volume fraction. The relation between the rate of change of the void volume fraction and the rate of change of the volumetric strain is

$$\dot{\phi}/(1-\phi) = D^p_{kk}, \qquad (4.3.34d)$$

so that by comparing with (4.3.15a),

$$\rho\dot{\Delta}/\rho_0 = \dot{\phi}/(1-\phi)\,. \tag{4.3.34e}$$

Moreover, the equivalence of the overall rate of plastic work and that in the matrix material leads to

$$\boldsymbol{\tau} : \mathbf{D}^p = (1-\phi)\,\sigma_Y\,\dot{\varepsilon}_M\,, \tag{4.3.34f}$$

where $\dot{\varepsilon}_M$ is the effective strain rate of the matrix.

Equations (4.3.34) are based on a perfectly plastic matrix. When the matrix material workhardens, σ_Y is replaced by σ_F which is the flow stress of the matrix.

Comparison with numerical computations for periodically distributed cavities has shown that the model requires modification; Tvergaard (1982). Also, the model fails to predict the accelerated loss of the stress-bearing capacity observed experimentally during ductile failure. Based on these observations, additional free parameters have been introduced and adjusted, in order to bring the model into closer accord with computational and experimental results. The following version of the model has been extensively used by Tvergaard and Needleman (1984) and their coworkers for numerical computations:

$$f \equiv (\sigma/\sigma_Y)^2 + 2q_1\,\phi^*\cosh(\tfrac{1}{2}\,q_2\,I/\sigma_Y) - (1+\phi^{*2})\,,$$

$$\phi^* = \begin{cases} \phi & \text{for } \phi \le \phi_c \\ \phi_c + (\phi_u^* - \phi_c)\,(\phi-\phi_c)/(\phi_f-\phi_c) & \text{for } \phi > \phi_c\,. \end{cases} \tag{4.3.35a,b}$$

The fracture is assumed to occur for void volume fraction $\phi = \phi_f$ or $\phi^*(\phi_f) = \phi_u^* = 1/q_1$. The parameters ϕ_c and ϕ_f are adjusted on the basis of empirical data, whereas $q_1 > 1$ and q_2 are adjusted on the basis of comparison with numerical computations.

Mear and Hutchinson (1985) have suggested a modified version of Gurson's model to include the effect of kinematic hardening. The corresponding yield function is

$$f \equiv (\tilde{\sigma}/\sigma_F)^2 + 2\phi\cosh(\tfrac{1}{2}\,I/\sigma_F) - (1+\phi^2)\,,$$

$$\sigma_F = (1-b)\,\sigma_Y^0 + b\,\sigma_Y\,,$$

$$\tilde{\sigma} = \sqrt{3}\,\tau = \{(3/2)\,(\boldsymbol{\tau}'-\boldsymbol{\beta}) : (\boldsymbol{\tau}'-\boldsymbol{\beta})\,\}^{1/2}\,. \tag{4.3.36a-c}$$

Here, $\boldsymbol{\beta}$ defines the center of the yield surface, σ_F is the radius of the yield surface for the matrix material, σ_Y^0 is the initial and σ_Y the current flow stress for the matrix material, and the parameter b ranges from 0 to 1. The model is formulated in such a manner that, for b = 1, it reduces to the Gurson model with isotropic hardening, and for b = 0, purely kinematical hardening results.

When pressure sensitivity and volumetric strain are included in these models, an additional expression is obtained by equating the overall rate of plastic work to the rate of plastic work associated with the actual internal dissipation mechanisms. As discussed in the preceding subsection, for granular materials the internal mechanisms may be assumed to be dominated by the frictional effects; Nemat-Nasser (1980a,b) and Nemat-Nasser and Shokooh (1980). A similar energy balance equation has been used for porous metals. An approximate expression of this kind would be

$$\boldsymbol{\tau} : \mathbf{D}^{\mathrm{p}} = (1-\phi)\,\sigma_{\mathrm{F}}\,\dot{\varepsilon}_{\mathrm{M}}, \tag{4.3.37}$$

where $\dot{\varepsilon}_{\mathrm{M}}$ is the rate of effective plastic strain in the matrix material, which can be related to the current rate of change of the matrix flow stress, using experimental stress-strain relations for the matrix material. For example, if E and E_t are the elastic Young modulus and the tangent modulus associated with the true stress-natural strain curve in a uniaxial test of the matrix material, then set

$$\dot{\varepsilon}_{\mathrm{M}} = (1/E_t - 1/E)\,\dot{\sigma}_{\mathrm{Y}}, \tag{4.3.38a}$$

and hence, from (4.3.37) obtain

$$\dot{\sigma}_{\mathrm{Y}} = \frac{E_t E}{E - E_t}\,\frac{\boldsymbol{\tau} : \mathbf{D}^{\mathrm{p}}}{(1-\phi)\,\sigma_{\mathrm{F}}} \tag{4.3.38b}$$

which defines the evolution of the flow stress of the matrix material.

4.4. YIELD VERTEX MODELS

As pointed out before, vertices or corners are an integral part of crystal plasticity. They play a central role in the proper modeling of material instability by localization. Therefore, effective continuum models must include vertices or, at least account, in some sense, for the effect of the vertex structure in the elastoplastic response of such materials. In the sequel, a general framework is first presented for the elastoplastic response at a vertex of the yield surface. Then, some specific cases are briefly discussed. Since the vertices are part of single-crystal plasticity,[39] many ideas and concepts may be borrowed from the theory of single-crystal plasticity. A certain degree of arbitrariness however, is inherent in the continuum developments, since these are not as closely related to the physics and micromechanics of the phenomena, as are the concepts in slip-induced crystal plasticity. In this section, the concept of *yield surface with vertices* is presented in the general setting developed by Hill (1978). Representations in both the stress space and strain space are considered, using a conjugate pair of general material strain and stress measures; see Chapter 3.

---

[39] See Hill (1966), Hutchinson (1970), Hill and Rice (1972), and Havner (1979).

4.4.1. Strain-space Representation

A corner on the yield surface is formed by the intersection of several smooth surfaces, each characterizing a portion of the overall yield surface. In the strain space, consider a corner formed by the intersection of n surfaces, $\phi^\alpha = \phi^\alpha(\mathbf{E}, \boldsymbol{\gamma})$, $\alpha = 1, 2, ..., n$, where $\boldsymbol{\gamma}$ with components γ^α represents n scalar parameters characterizing the workhardening associated with each segment of the yield surface, and $\mathbf{E}$ is a general material strain measure. For continued plastic deformation the stress rate conjugate to $\mathbf{E}$ is expressed as

$$\dot{\mathbf{S}} = \mathcal{L}^0 : \dot{\mathbf{E}} - \sum_{\alpha=1}^{n} \dot{\gamma}^\alpha \boldsymbol{\lambda}^\alpha, \tag{4.4.1}$$

where $\boldsymbol{\lambda}^\alpha$ is normal to the α'th yield surface.

The quantity $\dot{\gamma}^\alpha$ is calculated from the consistency conditions which require $\dot{\phi}^\alpha = 0$ for continued plastic deformation with the state remaining at the corner. It thus follows that

$$-\boldsymbol{\lambda}^\alpha : \dot{\mathbf{E}} + \sum_{\beta=1}^{n} g^{\alpha\beta} \dot{\gamma}^\beta = 0, \quad \boldsymbol{\lambda}^\alpha = -\partial\phi^\alpha / \partial\mathbf{E},$$

$$g^{\alpha\beta} = \partial\phi^\alpha / \partial\gamma^\beta. \tag{4.4.2a-c}$$

If it is assumed that the $n \times n$ matrix with components $g^{\alpha\beta}$ admits an inverse with components $g_{-1}^{\alpha\beta}$, then (4.4.2a) can be solved to obtain

$$\dot{\gamma}^\alpha = \sum_{\beta=1}^{n} g_{-1}^{\alpha\beta} \boldsymbol{\lambda}^\beta : \dot{\mathbf{E}} \tag{4.4.2d}$$

and (4.4.1) becomes

$$\dot{\mathbf{S}} = (\mathcal{L}^0 - \sum_{\alpha,\beta=1}^{n} g_{-1}^{\alpha\beta} \boldsymbol{\lambda}^\alpha \otimes \boldsymbol{\lambda}^\beta) : \dot{\mathbf{E}}. \tag{4.4.3}$$

4.4.2. Stress-space Representation

In a similar manner, consider a typical corner of the yield surface in the stress space, being defined by the intersection of the n surfaces $\psi^\alpha = \psi^\alpha(\mathbf{S}, \boldsymbol{\gamma})$. Then the strain rate for continued plastic deformation, with the stress point remaining at the corner, becomes

$$\dot{\mathbf{E}} = \mathcal{M}^0 : \dot{\mathbf{S}} + \sum_{\alpha=1}^{n} \dot{\gamma}^\alpha \boldsymbol{\mu}^\alpha, \tag{4.4.4}$$

where $\boldsymbol{\mu}^\alpha$ is normal to the α'th surface at the corner. For continued plastic deformation,

$$\boldsymbol{\mu}^\alpha : \dot{\mathbf{S}} - \sum_{\beta=1}^{n} h^{\alpha\beta} \dot{\gamma}^\beta = 0, \quad \boldsymbol{\mu}^\alpha = -\partial\psi^\alpha / \partial\mathbf{S},$$

$$h^{\alpha\beta} = \partial\psi^\alpha / \partial\gamma^\beta. \tag{4.4.5a-c}$$

If the matrix of $h^{\alpha\beta}$ admits an inverse with components $h_{-1}^{\alpha\beta}$, from (4.4.5a) obtain

$$\dot{\gamma}^\alpha = \sum_{\beta=1}^{n} h_{-1}^{\alpha\beta}\,\boldsymbol{\mu}^\beta : \dot{\mathbf{S}}, \tag{4.4.5d}$$

and hence,

$$\dot{\mathbf{E}} = (\boldsymbol{\mathcal{M}}^0 + \sum_{\alpha,\beta=1}^{n} h_{-1}^{\alpha\beta}\,\boldsymbol{\mu}^\alpha \otimes \boldsymbol{\mu}^\beta) : \dot{\mathbf{S}}. \tag{4.4.6}$$

4.4.3. Consistency Conditions

For consistency, $\dot{\gamma}$ obtained from (4.4.2a) and (4.4.5a) must be the same. Adding these equations, and observing that

$$\boldsymbol{\lambda}^\alpha = \boldsymbol{\mathcal{L}}^0 : \boldsymbol{\mu}^\alpha, \quad \boldsymbol{\mu}^\alpha = \boldsymbol{\mathcal{M}}^0 : \boldsymbol{\lambda}^\alpha, \tag{4.4.7a,b}$$

it follows that

$$[(g^{\alpha\beta} - h^{\alpha\beta}) - (\boldsymbol{\mu}^\alpha : \boldsymbol{\mathcal{L}}^0 : \boldsymbol{\mu}^\beta)]\,\dot{\gamma}^\beta = 0. \tag{4.4.8a}$$

This is automatically satisfied if

$$g^{\alpha\beta} - h^{\alpha\beta} = \boldsymbol{\mu}^\alpha : \boldsymbol{\mathcal{L}}^0 : \boldsymbol{\mu}^\beta, \tag{4.4.8b}$$

in which case, while neither $g^{\alpha\beta}$ nor $h^{\alpha\beta}$ need to be symmetric, their sum must be.

In general, the matrix of $g^{\alpha\beta}$ may be singular and may not admit an inverse. The situation is similar to that of crystal plasticity. Here, however, there is little physically-based guidance for further development. Some simplifying assumptions can be made. For example, if each surface, ϕ^α or ψ^α, is regarded to depend only on *one* scalar parameter[40] γ^α, then the matrices of both $g^{\alpha\beta}$ and $h^{\alpha\beta}$ are diagonal and can easily be inverted, when they are nonsingular. Other assumptions may be made. These have been discussed in connection with crystal plasticity by Hill and Rice (1972) and Havner (1992), and in the continuum context by Sewell (1972, 1974), and summarized by Hill (1978).[41]

4.5. NONLINEAR ELASTICITY AND DEFORMATION THEORIES

Deformation theories of plasticity are essentially specialized versions of nonlinear elasticity; Budiansky *et al.* (1951), and Budiansky (1959). They have been used to analyze unstable deformations and bifurcation problems. Here some fundamental relations in nonlinear elasticity are first briefly reviewed and then a concise account of a class of deformation theories is given.

---

[40] *E.g.*, ϕ^1 is a function of γ^1 only, *etc.*

[41] For a development of other features of metal plasticity, especially plastic anisotropy, see Bassani (1977), Hill (1979), and Bassani *et al.* (1979).

4.5.1. Isotropic Elasticity

When there are no internal microstructural changes, the Helmholtz free energy for isothermal deformations depends only on the material strain measure. Equation (4.1.1a) then gives

$$\mathbf{S} = \partial\phi / \partial\mathbf{E}\,, \tag{4.5.1}$$

where $\mathbf{S}$ and $\mathbf{E}$ are conjugate stress and strain measures, and $\phi = \phi(\mathbf{E})$. The explicit form of (4.5.1) depends on the material symmetries, as well as on the form of the strain energy function, ϕ, which characterizes the material response. The simplest form is obtained for isotropic materials. In this case, a material line element taken from a material neighborhood in any arbitrary direction will have the same mechanical response. Hence the strain energy ϕ depends only on the basic invariants of the strain measure $\mathbf{E}$. Denoting these invariants by

$$\mathrm{I} \equiv \mathrm{tr}\,\mathbf{E}\,, \quad \mathrm{II} \equiv \frac{1}{2}\{(\mathrm{tr}\,\mathbf{E})^2 - \mathrm{tr}\,\mathbf{E}^2\}\,, \quad \mathrm{III} \equiv \det\mathbf{E}\,, \tag{4.5.2a-c}$$

from $\phi = \phi(\mathrm{I}, \mathrm{II}, \mathrm{III})$ and the chain rule of differentiation, it follows that

$$\begin{aligned}\mathbf{S} = {} & (\partial\phi/\partial\mathrm{I} + \mathrm{I}\,\partial\phi/\partial\mathrm{II} + \mathrm{II}\,\partial\phi/\partial\mathrm{III})\mathbf{1} \\ & - (\partial\phi/\partial\mathrm{II} + \mathrm{I}\,\partial\phi/\partial\mathrm{III})\,\mathbf{E} + (\partial\phi/\partial\mathrm{III})\,\mathbf{E}^2\,,\end{aligned} \tag{4.5.3a}$$

where the identities

$$\begin{aligned}&\partial\mathrm{I}/\partial\mathbf{E} = \mathbf{1}\,, \quad \partial\mathrm{II}/\partial\mathbf{E} = \mathrm{I}\,\mathbf{1} - \mathbf{E}\,, \\ &\partial\mathrm{III}/\partial\mathbf{E} = \mathrm{II}\,\mathbf{1} - \mathrm{I}\,\mathbf{E} + \mathbf{E}^2\,,\end{aligned} \tag{4.5.3b-d}$$

are used. Expression (4.5.3a) can be written as

$$\mathbf{S} = \phi_0\mathbf{1} + \phi_1\mathbf{E} + \phi_2\,\mathbf{E}^2\,, \tag{4.5.3e}$$

where the coefficients ϕ_i, i = 0, 1, 2, are functions of the basic invariants of $\mathbf{E}$; these coefficients can be expressed in terms of the derivatives of the strain energy ϕ, by comparison with (4.5.3a).

From (4.5.3e), stress-strain relations associated with other stress or strain measures are deduced. For example, for the Lagrangian strain $\mathbf{E} = \frac{1}{2}(\mathbf{C} - \mathbf{1})$, $\mathbf{S}$ is the second Piola-Kirchhoff stress. For the Kirchhoff stress, (4.5.3e) becomes

$$\boldsymbol{\tau} = \alpha_0\mathbf{1} + \alpha_1\mathbf{B} + \alpha_2\mathbf{B}^2\,, \tag{4.5.4a}$$

where $\mathbf{B} = \mathbf{F}\,\mathbf{F}^{\mathrm{T}}$, and α_i, i = 0, 1, 2, are functions of the basic invariants of $\mathbf{E}$; $\mathbf{F}$ being the deformation gradient. The Hamilton-Cayley relation may be used to reduce (4.5.4a) to

$$\boldsymbol{\tau} = \beta_0\mathbf{1} + \beta_1\mathbf{B} + \beta_{-1}\mathbf{B}^{-1}\,, \tag{4.5.4b}$$

where β_i, i = − 1, 0, 1, are functions of the invariants of $\mathbf{B}$. For incompressible materials, $\det\mathbf{B} = 1$, and the stress tensor is defined to within an additive isotropic (spherical) tensor which represents a uniform tension, say, $-p\,\mathbf{1}$. In this case, (4.5.4b) becomes

$$\boldsymbol{\tau} = -p\mathbf{1} + a_1\mathbf{B} + a_{-1}\mathbf{B}^{-1}, \tag{4.5.4c}$$

where a_1 and a_{-1}, depend on the first two basic invariants of **B**.

For the purposes of what follows, the above description of nonlinear elasticity suffices. Readers interested in other aspects of nonlinear elasticity are referred to a number of treatises written on the subject, *e.g.*, Treolar (1958), Green and Atkins (1960), and Ogden (1982, 1984); for an introductory account, see Atkins and Fox (1980).

4.5.2. The J_2-deformation Theory

Here, first a special version of the incompressible nonlinear isotropic elasticity model (4.5.4c) is presented, and then it is modified to obtain a class of deformation theories of plasticity for finite strains and rotations. Since, for the isotropic case, the principal directions of the Kirchhoff stress, $\boldsymbol{\tau}$, coincide with the Eulerian triad $\mathbf{n}^a$, a = 1, 2, 3,[42] it follows that

$$\boldsymbol{\tau} = \sum_{a=1}^{3} \tau_a \, \mathbf{n}^a \otimes \mathbf{n}^a, \tag{4.5.5}$$

where τ_a are the principal stresses. With the effective stress τ given by (4.2.19a), define an effective *total* strain,

$$\gamma_e = (2\varepsilon_a \varepsilon_a)^{1/2} \quad a = 1, 2, 3, \quad \text{(a summed)}, \tag{4.5.6a}$$

where

$$\varepsilon_a \equiv \ln \lambda_a, \tag{4.5.6b}$$

λ_a being the principal stretch.

Now consider a class of incompressible isotropic elastic *model* materials, for which the elastic strain energy density is a function of only the total effective strain γ_e. Denote this function by $\phi = \phi(\gamma_e)$, and introduce the complementary energy density function $\psi = \psi(\tau)$, such that

$$\tau = \partial\phi/\partial\gamma_e, \quad \gamma_e = \partial\psi/\partial\tau. \tag{4.5.7a,b}$$

The corresponding principal values of the deviatoric stress and the logarithmic strain are then related by

$$\tau_a' = \partial\phi/\partial\varepsilon_a = 2h_s\,\varepsilon_a, \tag{4.5.7c}$$

where

$$h_s = \tau/\gamma_e \tag{4.5.7d}$$

is the *secant modulus* which can be obtained from a simple shear test. Taking the time derivative of (4.5.7c),

---

[42] Chadwick (1976), Hill (1978), and Mehrabadi and Nemat-Nasser (1987). See also, Chapters 2 and 3.

$$\dot{\tau}'_a = 2h_s\,\dot{\varepsilon}_a - 2(h_s - h_t)\,\frac{\tau'_b\,\dot{\varepsilon}_b}{2\tau^2}\,\tau'_a\,, \tag{4.5.8a}$$

where[43]

$$h_t = d\tau/d\gamma_e \tag{4.5.8b}$$

is the *tangent modulus* in the corresponding simple shear test. In view of definition (4.5.8b), (4.5.8a) is actually an identity. Since $\dot{\varepsilon}_a = \dot{\lambda}_a/\lambda_a = D_{aa}$ (no sum), and since $\boldsymbol{\tau}'$ and hence $\boldsymbol{\mu}$ are diagonal in the Eulerian triad, it follows that

$$\frac{\tau'_b\,\dot{\varepsilon}_b}{\sqrt{2}\,\tau} = \boldsymbol{\mu} : \mathbf{D}\,. \tag{4.5.8c}$$

4.5.3. Generalization to Three-dimensional Vertex Model

Consider now (4.5.5), and taking the material time derivative,[44] obtain

$$\dot{\boldsymbol{\tau}} = \sum_{a,b=1}^{3} \{\dot{\tau}_a\,\delta_{ab} + (\tau_a - \tau_b)\,(W_{ba} + D_{ba}\coth(\varepsilon_b - \varepsilon_a))\}\;\mathbf{n}^a \otimes \mathbf{n}^b\,, \tag{4.5.9a}$$

where W_{ab} and D_{ab} are the components of the spin and deformation-rate tensors in the Eulerian triad. Hence,

$$\overset{\circ}{\boldsymbol{\tau}} \equiv \dot{\boldsymbol{\tau}} - \mathbf{W}\,\boldsymbol{\tau} - \boldsymbol{\tau}\,\mathbf{W}^T$$
$$= \sum_{a,b=1}^{3} \{\dot{\tau}_a\,\delta_{ab} + (\tau_a - \tau_b)\,D_{ab}\,\coth(\varepsilon_b - \varepsilon_a)\}\;\mathbf{n}^a \otimes \mathbf{n}^b\,. \tag{4.5.9b}$$

Now substitute from (4.5.8a) into (4.5.9b), and obtain

$$\overset{\circ}{\boldsymbol{\tau}}{}' = \overset{\circ}{\boldsymbol{\tau}} + \dot{p}\mathbf{1} = 2h_s\Big\{\mathbf{1}^{(4s)} - \frac{h_s - h_t}{h_s}\,\boldsymbol{\mu} \otimes \boldsymbol{\mu}\Big\} : \mathbf{D} + \mathbf{Q} : \mathbf{D}\,, \tag{4.5.10a}$$

where the hydrostatic pressure rate, $\dot{p}$, is positive,

$$\boldsymbol{\mu} = \frac{\boldsymbol{\tau}'}{\sqrt{2}\,\tau} \tag{4.5.10b}$$

and

$$\mathbf{Q} : \mathbf{D} = \sum_{a,b=1}^{3} \{(\tau_a - \tau_b)\coth(\varepsilon_b - \varepsilon_a) - 2h_s\,(1 - \delta_{ab})\}D_{ab}\,\mathbf{n}^a \otimes \mathbf{n}^b\,. \tag{4.5.10c}$$

The finite-strain J_2-deformation theory proposed by Stören and Rice (1975) and Rudnicki and Rice (1975), is obtained if the term $\mathbf{Q} : \mathbf{D}$ in (4.5.10a) is ignored and the resulting equation is used to define the plastic part of the

---

43 To obtain the last expression in (4.5.8a), observe that $\dot{\gamma}_e = \tau'_b\,\dot{\varepsilon}_b/\tau$, and that $2\,\varepsilon_a\,h_s = 2\,\varepsilon_a\,(d\tau/d\gamma_e - h_s)\,\dot{\gamma}_e/\gamma_e = 2\,(h_t - h_s)\,\tau'_a\,\tau'_b\,\dot{\varepsilon}_b/(2\tau^2)$, where b = 1 , 2 , 3 is summed.

44 Hill (1978), Nemat-Nasser (1983), Mehrabadi and Nemat-Nasser (1987), and Chapter 2.

deformation rate, $\mathbf{D}^p$, in terms of the Jaumann rate of the Kirchhoff stress,

$$\overset{\circ}{\boldsymbol{\tau}}' = 2h_s\left\{ \mathbf{1}^{(4s)} - \frac{h_s - h_t}{h_s}\boldsymbol{\mu}\otimes\boldsymbol{\mu}\right\} : \mathbf{D}^{p\prime} \tag{4.5.11a}$$

which can be inverted to give

$$\mathbf{D}^{p\prime} = \left\{\frac{1}{2h_s}\mathbf{1}^{(4s)} + \left[\frac{1}{h_t} - \frac{1}{h_s}\right]\frac{\boldsymbol{\mu}}{\sqrt{2}}\otimes\frac{\boldsymbol{\mu}}{\sqrt{2}}\right\} : \overset{\circ}{\boldsymbol{\tau}}'. \tag{4.5.11b}$$

Now, with the elastic part of the deformation rate given by

$$\mathbf{D}^e = \boldsymbol{\mathcal{M}} : \overset{\circ}{\boldsymbol{\tau}}, \tag{4.5.11c}$$

the total deformation rate tensor is obtained by adding this to (4.5.11b),

$$\mathbf{D} = \boldsymbol{\mathcal{M}} : \overset{\circ}{\boldsymbol{\tau}} + \left\{\frac{1}{2h_s}\mathbf{1}^{(4s)} + \left[\frac{1}{h_t} - \frac{1}{h_s}\right]\frac{\boldsymbol{\mu}}{\sqrt{2}}\otimes\frac{\boldsymbol{\mu}}{\sqrt{2}}\right\} : \overset{\circ}{\boldsymbol{\tau}}'. \tag{4.5.11d}$$

The plastic part of the deformation rate, given by (4.5.11b), may be viewed as consisting of two parts, one *coaxial* with the deviatoric Kirchhoff stress $\boldsymbol{\tau}'$, and the other *normal* to this tensor. This can be seen by rewriting (4.5.11b) as

$$\begin{aligned}\mathbf{D}^{p\prime} &= \frac{1}{2h_t}(\boldsymbol{\mu} : \overset{\circ}{\boldsymbol{\tau}}')\boldsymbol{\mu} + \frac{1}{2h_s}\{\overset{\circ}{\boldsymbol{\tau}}' - (\boldsymbol{\mu} : \overset{\circ}{\boldsymbol{\tau}}')\boldsymbol{\mu}\} \\ &= \frac{1}{2h_t}(\boldsymbol{\mu} : \overset{\circ}{\boldsymbol{\tau}}')\boldsymbol{\mu} + \frac{\tau}{\sqrt{2}\,h_s}\overset{\circ}{\boldsymbol{\mu}}.\end{aligned} \tag{4.5.12a}$$

Furthermore, by definition, $\sqrt{2}\dot{\tau} = \boldsymbol{\mu} : \overset{\circ}{\boldsymbol{\tau}}'$ and hence, (4.5.12a) takes on the following simple form:

$$\mathbf{D}^{p\prime} = \frac{\dot{\tau}}{\sqrt{2}\,h_t}\boldsymbol{\mu} + \frac{\tau}{\sqrt{2}\,h_s}\overset{\circ}{\boldsymbol{\mu}} \tag{4.5.12b}$$

which can be expressed simply as,

$$\mathbf{D}^{p\prime} = \frac{1}{\sqrt{2}}(\gamma_e\boldsymbol{\mu})^{\circ}, \tag{4.5.12c}$$

where definitions (4.5.7d), (4.5.8b), and the Jaumann rate of (4.5.10b) are used. While (4.5.12c) is the logical outcome of the basic assumptions of the theory, *in application*, (4.5.11d) *is used with* h_t and h_s as unrelated free constitutive parameters.

Since from (4.5.10b) $\boldsymbol{\mu} : \boldsymbol{\mu} = 1$, then $\boldsymbol{\mu} : \overset{\circ}{\boldsymbol{\mu}} = 0$, and while the first term in the right-hand side of (4.5.12a) is coaxial with $\boldsymbol{\mu}$ and hence with $\boldsymbol{\tau}'$, the last term is normal to this direction. In particular, the rate of plastic work is given by

$$\boldsymbol{\tau}' : \mathbf{D}^{p\prime} = \tau\dot{\tau}/h_t = \tau\dot{\gamma}_e. \tag{4.5.13}$$

4.5.4. Relation to Granular-material Models and Summary of Equations

As noted by Nemat-Nasser (1983), a term proportional to $\overset{\circ}{\boldsymbol{\mu}}$ emerges in a natural way in the double-sliding theory of granular materials proposed by Mandel (1947), Spencer (1964), and Mehrabadi and Cowin (1978). A similar term also arises in the double-slip model of single crystals; Asaro (1979, 1983), and Nemat-Nasser *et al.* (1981). Therefore, while the derivation of (4.5.12) has been motivated on the basis of isotropic nonlinear elasticity, the results are apparently in some accord with physically-based models of the finite deformation of crystals, granular materials, and jointed rocks.[45] In these physically-based models, the moduli h_s and h_t do not necessarily have the same meaning as those arising in the deformation theory presented above. For example, the coefficient of the noncoaxiality term in the case of granular materials is directly defined in terms of the coefficient of friction, and for the single crystal, it relates to the crystal structure, as well as to the crystal's physical properties. Representation (4.5.12c) is a direct consequence of the underlying nonlinear elasticity structure which is used to obtain (4.5.12b). This structure is *not* shared by the slip theory of single crystals nor the micromechanical theory of the deformation of granular materials; see Nemat-Nasser (1983, 2000a), and Balendran and Nemat-Nasser (1993a). This issue is further discussed in the following two sections and in Chapter 7.

This section is closed by recording the final constitutive relations of the J_2-deformation theory,

$$\mathbf{D} = \mathcal{M} : \overset{\circ}{\boldsymbol{\tau}} + \frac{1}{\sqrt{2}} (\gamma_e \boldsymbol{\mu})^o = \mathcal{M} : \overset{\circ}{\boldsymbol{\tau}} + \frac{\dot{\tau}}{\sqrt{2}\, h_t} \boldsymbol{\mu} + \frac{\tau}{\sqrt{2}\, h_s} \overset{\circ}{\boldsymbol{\mu}}. \tag{4.5.14a}$$

When isotropic hypoelasticity is assumed,

$$\mathcal{M} = \frac{1}{2\mu}\left[\mathbf{1}^{(4s)} - \frac{1}{3}\mathbf{1} \otimes \mathbf{1}\right] + \frac{1}{3\kappa}\left[\frac{1}{3}\mathbf{1} \otimes \mathbf{1}\right] \tag{4.5.15a}$$

with an inverse given by

$$\mathcal{L} = 2\mu\left[\mathbf{1}^{(4s)} - \frac{1}{3}\mathbf{1} \otimes \mathbf{1}\right] + 3\kappa\left[\frac{1}{3}\mathbf{1} \otimes \mathbf{1}\right], \tag{4.5.15b}$$

where μ is the shear modulus and κ is the bulk modulus. Then (4.5.14a) becomes

$$\mathbf{D} = \left\{ \mathcal{M}^* + \left[\frac{1}{h_t} - \frac{1}{h_s}\right] \left[\frac{\boldsymbol{\mu}}{\sqrt{2}} \otimes \frac{\boldsymbol{\mu}}{\sqrt{2}}\right] \right\} : \overset{\circ}{\boldsymbol{\tau}},$$

$$\mathcal{M}^* = \frac{1}{2\mu^*}\left[\mathbf{1}^{(4s)} - \frac{1}{3}\mathbf{1} \otimes \mathbf{1}\right] + \frac{1}{3\kappa}\left[\frac{1}{3}\mathbf{1} \otimes \mathbf{1}\right],$$

[45] See also Rudnicki and Rice (1975), Stören and Rice (1975), Christoffersen *et al.* (1981), Anand (1983), Lance and Nemat-Nasser (1986), Balendran (1993), and Balendran and Nemat-Nasser (1993a,b).

$$\frac{1}{\mu^*} = \frac{1}{\mu} + \frac{1}{h_s}\,. \tag{4.5.14b-d}$$

The inverse of (4.5.14c) is

$$\mathcal{L}^* = 2\mu^*\left[\mathbf{1}^{(4s)} - \frac{1}{3}\mathbf{1}\otimes\mathbf{1}\right] + 3\kappa\left[\frac{1}{3}\mathbf{1}\otimes\mathbf{1}\right]. \tag{4.5.14e}$$

In view of (4.5.11d) and (4.2.8), the inverse of (4.5.14c) becomes

$$\overset{\circ}{\boldsymbol{\tau}} = \mathcal{L}^* : \left\{\mathbf{1}^{(4s)} - \frac{\boldsymbol{\mu}\otimes\mathcal{L}^* : \boldsymbol{\mu}}{2h_t h_s/(h_s - h_t) + \boldsymbol{\mu} : \mathcal{L}^* : \boldsymbol{\mu}}\right\} : \mathbf{D}\,. \tag{4.5.14f}$$

For isotropic hypoelasticity, $\boldsymbol{\mu}\otimes\mathcal{L}^* : \boldsymbol{\mu} = 2\mu^*\,\boldsymbol{\mu}\otimes\boldsymbol{\mu}$ and $\boldsymbol{\mu} : \mathcal{L}^* : \boldsymbol{\mu} = 2\mu^*$. Thus, the deviatoric part in (4.5.14f) reduces to

$$\overset{\circ}{\boldsymbol{\tau}}' = 2\mu^*\left\{\mathbf{1}^{(4s)} - \frac{2\mu\,(h_s - h_t)}{h_s\,(\mu + h_t)}\,\frac{\boldsymbol{\mu}}{\sqrt{2}}\otimes\frac{\boldsymbol{\mu}}{\sqrt{2}}\right\} : \mathbf{D}'\,. \tag{4.5.14g}$$

4.6. THE J_2-FLOW THEORY WITH NONCOAXIALITY

From the physically-based models (see Chapter 7), and guided by the J_2-*deformation* theory presented in the preceding section, the J_2-*flow* theory of plasticity may be modified to account for the effect of the vertex structure of the yield surface, while maintaining the simplicity associated with the smooth yield surface.[46] In this section, first the notion of *noncoaxiality* is discussed. Then the resulting model is generalized to include dilatancy.

4.6.1. Nondilatant Model with Noncoaxiality

Let the plastic contribution to the deformation rate tensor, $\mathbf{D}^p$, consist of two constituents (both deviatoric): one, denoted by $\mathbf{D}_1^p$, is defined by the classical plastic potential and the yield function in the manner discussed for various cases in subsections of this chapter, and the other, denoted by $\mathbf{D}_2^p$, is *normal* to the deviatoric part of the Kirchhoff stress; *i.e.*, $\mathbf{D}_2^p : \boldsymbol{\tau}' = 0$. Since $\boldsymbol{\mu} : \boldsymbol{\mu} = 1$, it follows that $\boldsymbol{\mu} : \overset{\circ}{\boldsymbol{\mu}} = 0$. Hence, the part of the deformation-rate tensor which is normal to the deviatoric part of the Kirchhoff stress, is often represented as

$$\mathbf{D}_2^p = \alpha\,\tau\,\frac{\overset{\circ}{\boldsymbol{\mu}}}{\sqrt{2}}\,, \tag{4.6.1}$$

where α is viewed as a material parameter which, in general, depends on the

[46] Mandel (1947), Spencer (1964, 1982), Rudnicki and Rice (1975), Stören and Rice (1975), Mehrabadi and Cowin (1978), Nemat-Nasser *et al.* (1981), Anand (1983), Nemat-Nasser (1983), Lance and Nemat-Nasser (1986), and, more recently, Balendran and Nemat-Nasser (1993a,b) and Nemat-Nasser (2000a).

state of stress, the temperature, and on the history of plastic deformation. Since $\boldsymbol{\tau}' : \mathbf{D}_2^p = 0$, there is no contribution to the rate of the plastic work by the term (4.6.1).

As an example, the vertex model for the J_2-flow theory with isotropic hardening, Subsection 4.2.3, becomes, by identifying $\mathbf{D}_1^p$ with (4.2.20a),

$$\mathbf{D}^p = \dot{\gamma}\,\frac{\boldsymbol{\mu}}{\sqrt{2}} + \alpha\tau\,\frac{\overset{\circ}{\boldsymbol{\mu}}}{\sqrt{2}}, \qquad \dot{\gamma} = \frac{1}{H}\,\frac{\boldsymbol{\mu}}{\sqrt{2}} : \overset{\circ}{\boldsymbol{\tau}}', \tag{4.6.2a,b}$$

H being defined by (4.2.20c). Here, α is not necessarily related to γ which itself is a quantity depending on the deformation path; see definition (4.2.19b). For single crystals and for granular materials which support the external loads through contact friction, the noncoaxiality parameter α can be given a physical basis. This is mentioned in the following section for granular materials, and further discussed in Chapters 6 and 7 for single crystals and granular materials, respectively. The discussion in this chapter remains macroscopic, whereas that in Chapters 6 and 7 is micromechanical. Note from (4.5.10b) that (see also equations (4.8.25a-e)), in general,

$$\tau\,\frac{\overset{\circ}{\boldsymbol{\mu}}}{\sqrt{2}} = \left\{ \frac{1}{2}\left[\mathbf{1}^{(4s)} - \frac{1}{3}\,\mathbf{1}\otimes\mathbf{1}\right] - \frac{\boldsymbol{\mu}}{\sqrt{2}}\otimes\frac{\boldsymbol{\mu}}{\sqrt{2}} \right\} : \overset{\circ}{\boldsymbol{\tau}}. \tag{4.6.3a}$$

Hence, the total (deviatoric) plastic deformation rate is

$$\mathbf{D}^p = \left\{ \frac{\alpha}{2}\left[\mathbf{1}^{(4s)} - \frac{1}{3}\,\mathbf{1}\otimes\mathbf{1}\right] + \left[\frac{1}{H} - \alpha\right]\frac{\boldsymbol{\mu}}{\sqrt{2}}\otimes\frac{\boldsymbol{\mu}}{\sqrt{2}} \right\} : \overset{\circ}{\boldsymbol{\tau}}. \tag{4.6.3b}$$

4.7. MODELS FOR FRICTIONAL GRANULAR MATERIALS

As is mentioned in Section 4.3, the response of frictional granules is strongly dominated by anisotropy or fabric which is invariably induced upon shearing under confinement. Concomitant features of this characteristic of frictional granules are their dilatancy in monotone shearing and their densification upon unloading. A further remarkable property of materials of this kind is that *they can actually undergo reverse inelastic shearing* **against** *an applied shear stress, when unloading follows a monotone shearing under confining pressure. The energy required to effect plastic deformation against the applied shear stress is supplied by the work done by the confining pressure going through the accompanying shear-induced volumetric contraction.*

Figure 4.7.1(a) is a schematic of the two-dimensional deformation of a frictional granular mass in simple shearing. The confining pressure is supplied by the normal stress, σ, which is kept constant, while inelastic deformation is induced by the application of the monotonically increasing shear stress, τ. This results in the shear stress τ, shear strain γ-curve, OAB, shown in Figure 4.7.1(b). The unloading from stress state B, is shown by curve BCD which includes

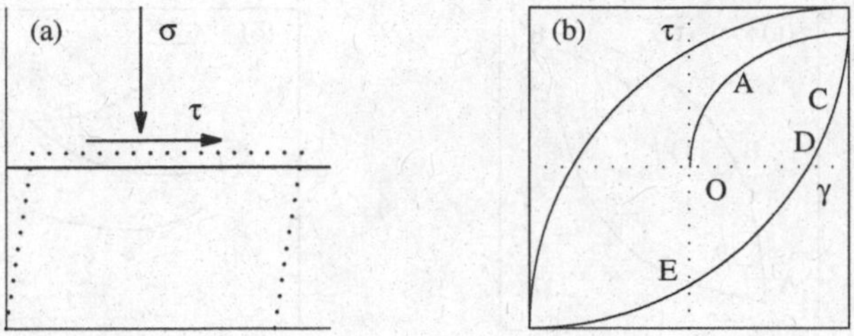

Figure 4.7.1

(a) Simple shearing of frictional granules under constant normal compression σ; (b) when monotone shearing, OAB, is followed by unloading, BCDE, plastic deformation occurs *against* the applied shear stress τ along BCD; and (c) initial densification, oa, is followed by dilatancy, ab, and substantial densification during unloading, along bcde

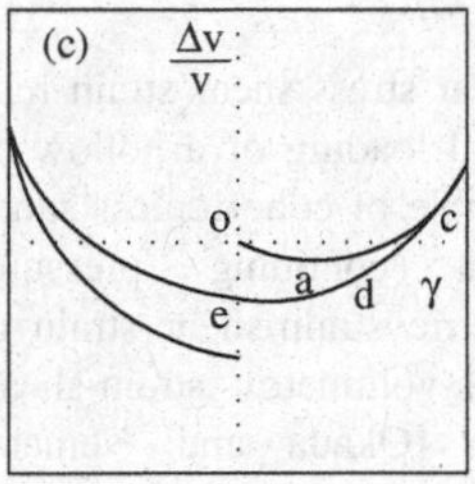

inelastic shearing *against* the shear stress τ, all along the unloading curve BCD. This process is accompanied by volumetric contraction which is schematically shown in Figure 4.7.1(c). In this figure, the curve oab displays an initial densification (from o to a), followed by dilatancy (from a to b), in monotone shearing along OAB of Figure 4.7.1(b). The curve bcd represents continued densification which accompanies the corresponding unloading curve BCD of Figure 4.7.1(b).

Figure 4.7.2 shows the results of an actual experiment performed on a large sample of cohesionless sand under a constant confining pressure. The sample is a hollow circular cylinder with 20cm inside and 25cm outside diameter and 25cm height.[47] The shear stress τ is applied through torsional loading of the sample.

Figure 4.7.2(a) shows the shear stress τ as a function of the shear strain γ for a strain-controlled cyclic deformation of 2% strain amplitude. A strain path, from O to A to A′ to B to B′, is shown. In the experiment, this path is continued from B′ and followed by further shearing (not shown) which is finally terminated at zero shear strain on the τ-axis. The volumetric strain, $\frac{\Delta v}{v}$, corresponding to the straining portion which terminates at point B′ in Figure 4.7.2(a), is shown as a function of shear strain, γ, in Figure 4.7.2(b), and as a

[47] See Chapter 9, Sections 9.5 and 9.6, as well as Okada and Nemat-Nasser (1994) and Nemat-Nasser and Okada (2001) for detailed description of the experimental setup, procedure, and results.

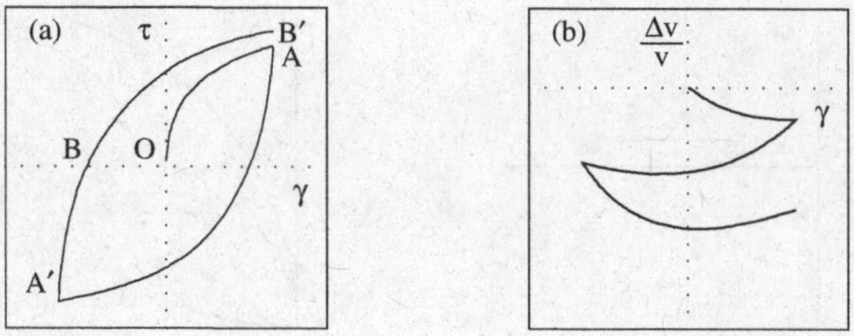

Figure 4.7.2

(a) Shear stress-shear strain relation in torsional loading of a hollow cylindrical sample of cohesionless sand, under constant confining pressure; (b) volumetric strain-shear strain relation; and (c) volumetric strain-shear stress relation (Okada and Nemat-Nasser, 1994)

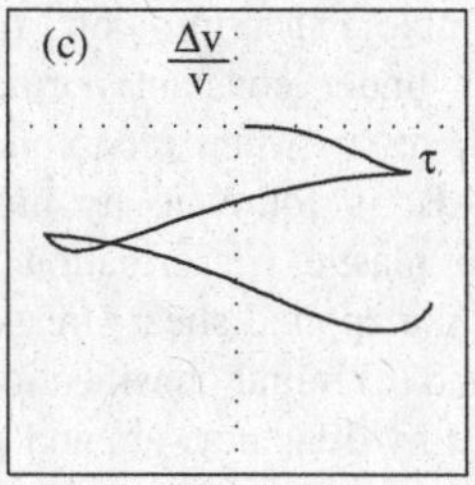

function of shear stress, τ, in Figure 4.7.2(c).[48]

Figure 4.7.3 shows the results of the further shearing of the same sample, now at 3% shear strain amplitude. Because of the densification in the initial two cycles of 2% strain amplitude, and since now the strain amplitude is larger, the sample shows greater dilatancy in each quarter of the cycle, and the net amount of densification at the end of each half-cycle (at zero shear strain) is progressively smaller.

The above results and numerous other experiments[49] show that, under relatively small confining pressures, the following features are among the essential characteristics which must be captured in the modeling of the inelastic deformation of granular masses which carry the applied loads through frictional contact (sheared under relatively small confining pressures):

- an initial densification (decrease in void volume), the magnitude of which decreases as the void ratio (the ratio of void volume to the volume of the solid) approaches a minimum value;

---

[48] As is seen, the dependence of the volumetric strain on the shear stress is more complex than that on the corresponding shear strain. This fact has been documented experimentally by Nemat-Nasser and Tobita (1982) and Nemat-Nasser and Takahashi (1984) who show that the dilatancy vanishes in this kind of experiment, close to a state at which the shear strain is almost zero, where the induced anisotropy (or fabric) is zero.

[49] See, *e.g.*, Nemat-Nasser and Tobita (1982) and Nemat-Nasser and Takahashi (1984) and references cited therein.

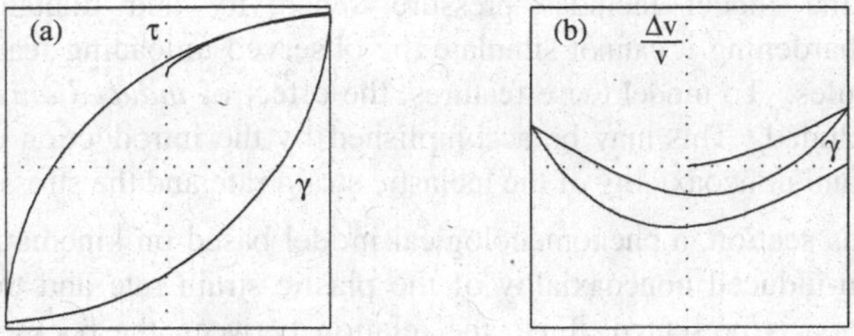

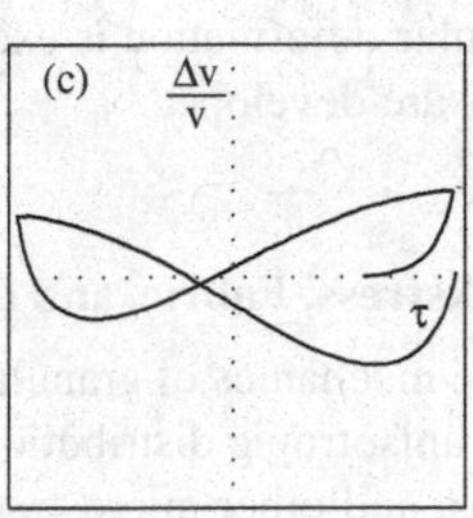

Figure 4.7.3

Shear stress-shear strain relation in torsional loading of a hollow cylindrical sample of cohesionless sand, under constant confining pressure: (a) the third cycle of shearing (at 3% strain amplitude) after two cycles of 2% strain amplitude (the first cycle is shown in Figure 4.7.2(a)); (b) volumetric strain-shear strain relation; and (c) volumetric strain-shear stress relation (Okada and Nemat-Nasser, 1994)

- if the sample is dense (*i.e.*, its void ratio is close to the corresponding minimum value), then the initial (small) densification is followed by dilatancy (increase in void volume) which continues until a critical[50] void ratio is attained asymptotically (see curve (2) in Figure 4.3.2, p.171);
- if at a certain stage during the course of dilatancy, stated above, the shearing is reversed and the shear strain is gradually decreased to its initial zero value (completing half of a strain cycle), then there is always a net amount of densification, this amount decreasing as the initial void ratio approaches its minimum value; and
- if the sample is loose, *i.e.*, the initial void ratio is larger than the critical value, then the sample may densify continuously until the critical void ratio is reached asymptotically (see curve (1) in Figure 4.3.2).

The plasticity model of Section 4.3 includes in equations (4.3.2a-d) both isotropic and kinematic hardening. Its application to triaxial tests of soils, discussed in Subsections 4.3.1 to 4.3.5, however, is based on isotropic hardening.

---

[50] This critical value, as well as the minimum void ratio, in general, depend on the value of the confining pressure, on the size and shape distribution of the grains, and on other relevant parameters; see Subsection 4.7.5.

Although the model includes pressure sensitivity and dilatancy, without kinematic hardening it cannot simulate the observed unloading features of frictional granules. To model these features, the effect of *induced anisotropy* must also be included. This may be accomplished by the introduction of kinematic hardening and noncoaxiality of the inelastic strain rate and the stress tensors.

In this section, a phenomenological model based on kinematic hardening and friction-induced noncoaxiality of the plastic strain rate and the deviatoric stress tensors, is presented. First, the relation between the backstress and the induced anisotropy or fabric is briefly examined. Then, the corresponding constitutive relations are developed and discussed. In Chapter 7, the micromechanics of granular deformation is examined in detail, and physically-based constitutive models are developed.

4.7.1. Backstress, Fabric, and Dilatancy

In the mechanics of granular deformation, the term *fabric* is used to define the overall anisotropic distribution of the granules, their contact forces, the associated voids, and other microstructural parameters which are responsible for the anisotropic behavior of the granular mass. Two kinds of fabric are usually considered: (1) *inherent fabric* which is due to the initial packing of the particles; and (2) *induced fabric* which is developed during the course of deformation.[51] To explain the induced fabric and its relation to the backstress (kinematic hardening) of phenomenological plasticity, consider again the two-dimensional simple shearing of Figure 4.7.1(a). In response to the applied shear stress τ, the macroscopic shearing occurs in the horizontal direction. The *actual* motion of the individual grains (or families of grains), however, consists of sliding and rolling against one another, which results in both shear deformation and volumetric change; the dotted lines in Figure 4.7.1(a). If the sample is not very densely packed, initial shearing invariably results in the rearrangement of particles, which, under a confining pressure, produces an initial densification.

Consider the schematic representation of Figure 4.7.4. As a typical particle, say, A, comes in contact with its neighbor, say, B, its further movement relative to its neighbor will require it to slide and roll relative to B over their contact area. The direction of the motion of particle A relative to particle B, lies on the tangent plane at their common contact point. This direction makes an angle ν with the macroscopic shearing direction which, in the present case, is horizontal and from left to right.

The angle ν is called the *dilatancy angle*. When ν is positive, A tends to rise during the course of deformation (dilatancy), and when ν is negative, the

[51] In general, the anisotropy which is induced through shearing is distinguished from the inherent anisotropy that may exist due to, say, sample preparation; see, *e.g.*, Arthur and Menzies (1972), Oda (1972), Konishi *et al.* (1982, 1983), Ishihara and Towhata (1983), and references cited there, as well as Chapter 7.

particle moves down (densification). When, on the average, there are more active particles with positive dilatancy angles, then the overall shearing is accompanied by volumetric expansion, *against* the applied normal stress σ of Figure 4.7.1(a). Therefore, in this case, the local shear stress due to the applied loads, must overcome both the interparticle frictional resistance and the local compression.

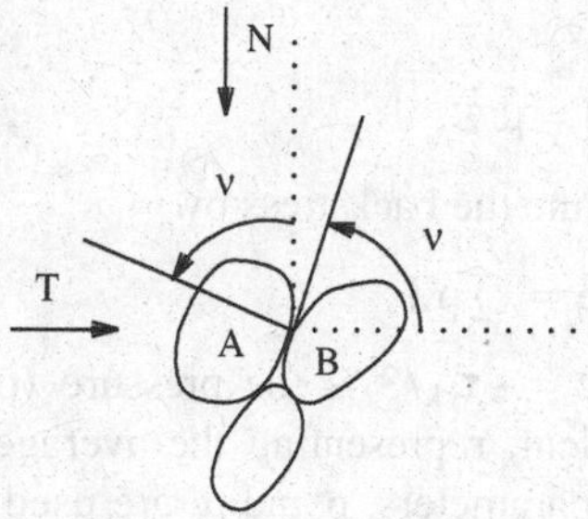

Figure 4.7.4

Locally, shearing in the direction T requires sliding and rolling of granule A relative to granule B in a direction which lies on the tangent plane at their common contact point; ν is the dilatancy angle

Thus, at the local level, the effective force T that moves particle A relative to particle B of Figure 4.7.4, must also overcome the associated normal force, N, which helps to resist the relative motion of the two contacting particles. When the dilatancy angle ν is suitably large, and if the applied shear stress in Figure 4.7.1(a) is *reduced*, leading to a reduction in the magnitude of T, particle A may tend to move down relative to B, under the action of the normal force, N. When large numbers of particles undergo reverse relative motions of this kind, then the granular mass experiences reverse plastic deformation against an applied shear stress. This asymmetric influence of the pressure on the behavior of frictional granules is the root of the resulting *backstress,* and the associated volumetric straining.

The backstress, therefore, should depend directly on the distribution of the dilatancy angles ν of the active granules. This distribution of the dilatancy angles may be measured in terms of the *distribution of the density of the orientation of the contact unit normals.*[52] This distribution is a measure of the *fabric* of the granular mass.

Denote by **n** a typical contact unit normal, and let E(**n**) be its *distribution-density function.* Hence, E(**n**) dΩ represents the fractional number (per unit volume) of contact unit normals whose directions fall within the *solid angle* dΩ. Various aspects of this representation of the fabric of a granular mass are discussed in Chapter 7. Here, it suffices to note that a second-order

---

[52] Note that ν also measures the angle of the contact normal with the vertical axis; see Figure 4.7.4.

approximation of E(**n**) is given by[53]

$$\mathrm{E}(\mathbf{n}) = \frac{1}{2\pi(\mathrm{r}-1)}(1+\hat{\mathcal{E}}):(\mathbf{n}\otimes\mathbf{n}), \tag{4.7.1a}$$

where r = 2 in two dimensions, r = 3 in three dimensions, and $\hat{\mathcal{E}}$ is a second-order symmetric and deviatoric tensor. In the present continuum model, this tensor, $\hat{\mathcal{E}}$, is used as a measure of the fabric of the material.

The backstress must vanish when the fabric is isotropic. To ensure this, set

$$\mathbf{J} = \hat{\mu}\,\hat{\mathcal{E}}, \tag{4.7.1b}$$

and define the backstress by

$$\boldsymbol{\beta} = \sqrt{2}\,\mathrm{p}\,\mathbf{J}. \tag{4.7.1c}$$

Here, $\mathrm{p} = -\tau_{kk}/3$ is the pressure (positive) and $\hat{\mu}$ is a microscopic frictional coefficient, representing the average particle-to-particle coefficient of friction. These parameters, p and $\hat{\mu}$, are used as normalizing factors. In this representation, the magnitude of the backstress for a given microstructure, is assumed to increase with increasing frictional resistance of the contacting granules, which, in turn, increases with increasing pressure and with the friction coefficient. This is reasonable, since, in the absence of internal friction, the granular mass would deform as an inviscid fluid. With nonzero internal friction, furthermore, the frictional resistance increases with increasing pressure. In this phenomenological model, however, the fabric tensor $\hat{\mathcal{E}}$ and the frictional coefficient $\hat{\mu}$ are not used directly. Instead, the fabric tensor **J** is used as an internal variable to represent the effects of the microstructure. The evolution of the microstructure is thus measured in terms of the evolution of the fabric tensor, **J**.

4.7.2. Yield Criterion

First, the isotropic response of the material is expressed in terms of a frictional yield condition similar to that discussed in Section 4.3. In particular, the yield function (4.3.1a) is written as

$$\mathrm{f}_1 \equiv \tau - \tau_y, \quad \tau_y \equiv \mathrm{F}(\mathrm{p}, \Delta, \gamma); \tag{4.7.2a,b}$$

where p is the presure, Δ is the inelastic volumetric strain defined by (4.3.1c), and γ is the total effective plastic strain defined by (4.3.1e).

Second, the effect of the fabric anisotropy is introduced through the definition of the *effective stress difference*, τ, as follows:

$$\tau = (\mathbf{s}:\mathbf{s}/2)^{1/2}, \quad \mathbf{s} = \boldsymbol{\tau}' - \boldsymbol{\beta}, \tag{4.7.2c,d}$$

[53] See expression (7.3.24), Section 7.3 of Chapter 7, and Mehrabadi *et al.* (1982), and Kanatani (1984).

where $\mathbf{s}$ is the *stress difference,* and $\boldsymbol{\beta}$ is the backstress, both being deviatoric tensors; see (4.2.25b) and (4.2.26a,b).

Finally, and for completeness, a yield condition is provided for isotropic compaction, as follows:

$$f_2 = p - p_c\,, \tag{4.7.2e}$$

where p_c is the critical value of the pressure at which *shear-independent* isotropic inelastic compaction takes place, *i.e.,* when the pressure is increased beyond this critical value, the granular mass is compacted isotropically.

4.7.3. Plastic Strain Rate with Dilatancy and Noncoaxiality

It is assumed that the inelastic deformation of granular materials stems from two mechanisms:

- microscopic shearing along wavy surfaces consisting of collections of granules; and
- collapse of voids when the pressure exceeds a critical value.

It is assumed that the first mechanism occurs when the yield condition $f_1 = 0$ is satisfied by the stress states at points on the sliding surfaces. The deformation corresponding to this mechanism is primarily shearing, accompanied by volumetric straining which may be dilatancy or densification, depending in part on the orientation[54] of the fabric tensor, $\mathbf{J}$, relative to the direction of the stress difference, $\mathbf{s}$, which is represented by the normal to the yield surface, $\boldsymbol{\mu}$, *i.e.,* depending in part on the sign and magnitude of $\boldsymbol{\mu} : \mathbf{J}$. The second mechanism, the collapse of voids, is assumed to occur when the yield condition $f_2 = 0$ is satisfied.

Hence, in this model, the inelastic part of the deformation rate tensor has the general form

$$\mathbf{D}^{p\prime} = \frac{1}{\sqrt{2}}\,\{\dot{\gamma}\,\boldsymbol{\mu} + u\,\alpha\,\mathring{\boldsymbol{\mu}}\}\,, \quad D^p_{kk} = \dot{\gamma}\,B + \dot{\theta}\,, \tag{4.7.3a,b}$$

where B is the dilatancy parameter which relates the *shear-induced* volumetric strain rate to the shear strain rate;[55] $\alpha \geq 0$ is the noncoaxiality coefficient; $u = \pm 1$ denotes the direction of the noncoaxial plastic deformation rate relative to the yield surface, chosen such that the corresponding rate of plastic work is nonnegative (see (4.7.7)); $\boldsymbol{\mu}$ is the normal to the yield surface, $f_1 = 0$, in the deviatoric stress space, *i.e.,*

---

[54] The orientation of a second-order deviatoric tensor (*e.g.,* $\mathbf{s}$) is given by the orientation of the associated unit tensor (*e.g.,* $\boldsymbol{\mu}$). In the present context, two such tensors are called coaxial when their associated unit tensors are the same.

[55] Note that B stands for $\partial G/\partial p$ of Section 4.3. Here, however, volumetric *expansion* is positive and $p = -\tau_{kk}/3$, whereas the opposite sign convention is used in Section 4.3.

$$\frac{\boldsymbol{\mu}}{\sqrt{2}} = \frac{\partial f_1}{\partial \mathbf{s}} = \frac{\mathbf{s}}{2\,\tau}\,; \tag{4.7.3c}$$

and $\dot{\theta}$ is the volumetric strain rate due to isotropic compaction.

The conditions for the plastic deformation are

$$\dot{\gamma},\ \alpha \begin{cases} = 0 \text{ for } f_1 < 0 \text{ or } \dot{f}_1 < 0 \\ > 0 \text{ for } f_1 = 0 \text{ and } \dot{f}_1 = 0, \end{cases} \quad \dot{\theta} \begin{cases} = 0 \text{ for } f_2 < 0 \text{ or } \dot{f}_2 < 0 \\ < 0 \text{ for } f_2 = 0 \text{ and } \dot{f}_2 = 0\,. \end{cases} \tag{4.7.4a,b}$$

4.7.4. Energy Balance Equation

Now, consider the rate of plastic work as follows:

$$\dot{w}_p \equiv \boldsymbol{\tau} : \mathbf{D}^p = \frac{1}{\sqrt{2}}\,(\boldsymbol{\beta} + \sqrt{2}\,\tau\,\boldsymbol{\mu}) : (\dot{\gamma}\,\boldsymbol{\mu} + u\,\alpha\,\mathring{\boldsymbol{\mu}}) - p\,(\dot{\gamma}\,B + \dot{\theta})$$

$$= \dot{w}_{pd} + \dot{w}_\alpha + \dot{w}_\theta\,, \tag{4.7.5a}$$

where, in view of (4.7.1c),

$$\dot{w}_{pd} = \dot{\gamma}\,(\tau/p + \mathbf{J} : \boldsymbol{\mu} - B)\,p \tag{4.7.5b}$$

which is the rate of work due to *plastic distortion* $\dot{\gamma}\,\boldsymbol{\mu}/\sqrt{2}$ which is coaxial with the stress difference, **s**. In (4.7.5a),

$$\dot{w}_\alpha = u\,\alpha\,p\,\mathbf{J} : \mathring{\boldsymbol{\mu}} \quad \text{and} \quad \dot{w}_\theta = -p\,\dot{\theta}\,. \tag{4.7.5c,d}$$

The first, $\dot{w}_\alpha$, is the rate of plastic work due to the stress component tangent to the yield surface, and the second, $\dot{w}_\theta$, is due to the pressure-induced isotropic compaction.

It is now necessary to relate the distortional rate of the plastic work to the rate of frictional loss due to intergranular sliding and rolling. One possible model is given in Subsection 4.3.2, for triaxial loading of cohesionless granules. Based on this model, assume that the stress τ_f, conjugate to the effective distortional plastic strain rate, $\dot{\gamma}$, can be approximated by $\tau_f = M_f\,p$, where M_f is an *effective* frictional resistance to distortion, *i.e.*, an *effective overall internal friction coefficient*, and p is the pressure. The parameter M_f may be considered to be representing the intergranular friction, but since $\dot{\gamma}$ is an *effective* shear strain rate, M_f must similarly be an *effective*, rather than the actual, friction coefficient. M_f is, at least a function of the void ratio and possibly some fabric measures.

In Subsection 4.3.2, following Nemat-Nasser and Shokooh (1980), the effective friction coefficient is taken to be proportional to $\partial F/\partial p$ which represents the rate of change of the radius of the yield surface with respect to the pressure; see (4.2.18) and (4.7.2a,b). The model in Subsection 4.3.2, however, does not include a backstress, and hence $\tau = F(p, \Delta, \gamma)$ is the total effective stress during plastic deformation. In the present section, on the other hand,

$\tau = \tau_y = F(p, \Delta, \gamma)$ in (4.7.2a,b) is the effective stress *difference* defined by (4.7.2c,d), representing only the size of the yield surface. Therefore, M_f here is not interpreted as the gradient of τ_y with respect to the pressure. Instead, M_f here is regarded as representing an average overall frictional resistance of the material, responsible for energy dissipation through the intergranular sliding and rolling. To identify this and separate it from the anisotropic material resistance due to the induced fabric, M_f will be called the *overall frictional coefficient.*

The rate of frictional loss, $\dot{w}_f$, due to internal sliding and rolling of the granules is thus approximated by

$$\dot{w}_f = \tau_f \dot{\gamma} = M_f p \dot{\gamma}. \tag{4.7.5e}$$

In general, and as in Section 4.3, M_f, the overall frictional coefficient, is not constant. It varies as the void ratio varies, and, in general, is a function of the plastic strain, the plastic strain rate, and other parameters. Its functional form must be established either by direct experiments on given materials, or through micromechanical modeling. For a very loose soil sample, M_f is initially small, but increases as the sample density increases in monotone deformations.

4.7.5. Dilatancy, Friction, and Fabric

An expression for the dilatancy parameter B is obtained by setting the rate of frictional loss, $\dot{w}_f$, equal to the rate of distortional plastic work, $\dot{w}_{pd}$. In view of (4.7.5b) and (4.7.5e), this leads to

$$-B = M_f - (\tau/p + \mu_f), \quad \mu_f = \mathbf{J} : \boldsymbol{\mu}, \tag{4.7.6a,b}$$

where μ_f is viewed as the instantaneous *anisotropic frictional coefficient* due to the microstructure and the fabric of the granular mass, measured by the orientation of the fabric tensor, $\mathbf{J}$, relative to the deviatoric part of the stress difference tensor, $\boldsymbol{\mu} = \mathbf{s}/(\sqrt{2}\tau)$. Note that, since $\dot{\gamma}$, M_f, and p are all nonnegative, $\dot{w}_f$ is always nonnegative; therefore, $\dot{w}_{pd} \geq 0$. Similarly, since $\dot{\theta}$, the volumetric strain rate due to *hydrostatic pressure*, is nonpositive, $\dot{w}_\theta$ is also nonnegative.[56] To ensure that $\dot{w}_\alpha$ is also nonnegative for positive α, the sign of the noncoaxial plastic deformation is taken as

$$u = \mathrm{sign}(\mathbf{J} : \mathring{\boldsymbol{\mu}}). \tag{4.7.7}$$

Consider now some of the implications of this model, by examining (4.7.6). For an isotropic virgin sample, $\mathbf{J} = \mathbf{0}$. This means that initially $\mu_f = 0$ for such a sample. Since $M_f > 0$, it follows from (4.7.6a) that the dilatancy parameter B is initially negative (densification)[57] for a virgin sample, until τ/p becomes suitably large. Hence, (4.7.6) captures the fact that when an

---

[56] Note that $\dot{\theta}$ is the rate of volume contraction due to void collapse. The shear-induced volumetric strain is represented by the dilatancy parameter B which may be positive, negative, or zero.

[57] Here, volumetric expansion is taken to be positive, whereas in Section 4.3, compaction is po-

isotropically consolidated virgin sample is subjected to shearing under confining pressure, it initially densifies.

During the course of continued monotone shearing, an anisotropic microstructure is developed, as the fabric of the material evolves. It will thus be required that $\boldsymbol{\mu}$ and $\mathbf{J}$ tend to become coaxial in the course of monotone loading; see Subsection 4.7.6. As a result, μ_f increases from zero monotonically, and may eventually attain a saturation value, μ_s. As τ/p and μ_f increase, the dilatancy parameter B decreases. However, as the sample densifies, the overall frictional resistance of the material also increases, leading to an increasing M_f. For sufficiently loose samples, $\tau/p+\mu_f$ may never exceed the frictional resistance, measured by (an increasing) M_f. In such cases, the dilatancy parameter B remains negative (continued densification), approaching zero asymptotically.

For dense samples, on the other hand, the effective stress τ, required to produce plastic deformation, is initially large and continues to increase, as the material initially densifies, and as the granules are further engaged by their neighbors, during their relative sliding and rolling. This also leads to an increasing μ_f. Hence, the dilatancy B becomes positive, and the initial densification is followed by dilatancy. As discussed in Subsection 4.3.2, this dilatancy gradually diminishes, approaching zero upon continued monotone shearing.[58]

When the direction of shearing is reversed, μ_f changes its sign and attains a negative value; see Figure 4.7.5. Hence, in reversing the direction of shearing, the dilatancy parameter B becomes negative, predicting that the sample would densify. This is in agreement with essentially all experimental observations; see Figure 4.7.2.

As the reversed shearing continues, the fabric anisotropy evolves and $\mathbf{J}$ changes with the applied stress. For the model to predict positive dilatancy, the evolution of $\mathbf{J}$ must be such that $\mathbf{J}$ tends to become coaxial with $\boldsymbol{\mu}$, rendering $\mathbf{J}:\boldsymbol{\mu}$ positive again. The evolution equations for the fabric tensor $\mathbf{J}$, as well as for the overall effective friction coefficient, M_f, and the radius of the yield surface, defined by $F(p, \Delta, \gamma)$, must thus reflect these basic physical requirements.

In many situations, the elastic range for frictional granules is rather small under small or moderate confining pressures. This means that yielding may begin soon upon loading. Resistance to the deformation, however, may increase in the course of plastic deformation because of the microstructural changes. In cases of this kind, the magnitude of $\tau_y = F(p, \Delta, \gamma)$ in (4.7.2a) may be rather small. The resistance to plastic deformation of the granular mass then stems essentially from the induced fabric anisotropy. Such a material may display reverse plastic deformation immediately upon unloading. This is examined in Subsection 4.7.10.

---

sitive. Hence, the negative sign in the left-hand side of (4.7.6a).

[58] Homogeneous deformation of frictional granular masses which are sheared under pressure, usually becomes unstable, leading to strain localization, after 5 to 6% shear strain; see Nemat-Nasser and Okada (2001).

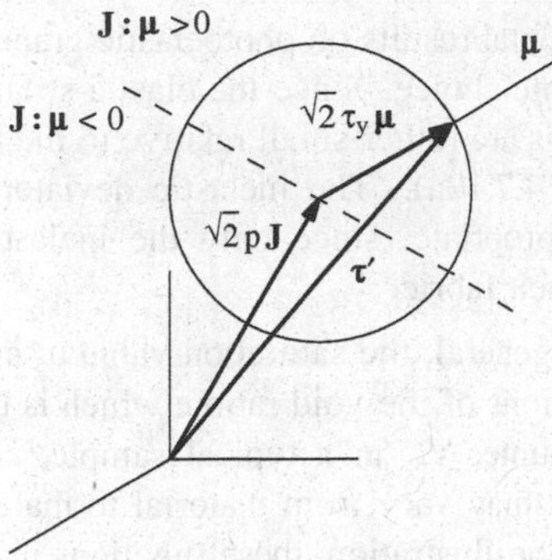

Figure 4.7.5

Yield surface for frictional granular materials with kinematic hardening, in the deviatoric stress space

4.7.6. Evolution Equations

Now, consider the evolution equations which describe the changes in the material resistance to inelastic deformation, measured by the overall and anisotropic frictional parameters, M_f and $\mu_f = \mathbf{J} : \boldsymbol{\mu}$, respectively. It has been observed from the microscopic experiments on photoelastic granules that the principal axes of the fabric tensor tend to move toward those of the stress tensor, as the granular mass is being sheared.[59] In order to include this feature in the evolution equation for the fabric tensor **J**, first split its objective rate of change into two parts: one which is coaxial with the stress-difference tensor, $\mathbf{s} = \boldsymbol{\tau}' - \boldsymbol{\beta}$, *i.e.*, it is coaxial with $\boldsymbol{\mu}$; and the other, which is normal to $\boldsymbol{\mu}$. Since, from (4.7.6b), the time derivative of μ_f is $\dot{\mu}_f = \mathring{\mathbf{J}} : \boldsymbol{\mu} + \mathbf{J} : \mathring{\boldsymbol{\mu}}$, the evolution of **J** may be expressed by

$$\mathring{\mathbf{J}} = (\dot{\mu}_f - \mathbf{J} : \mathring{\boldsymbol{\mu}})\,\boldsymbol{\mu} + \rho\,\frac{\mathring{\boldsymbol{\mu}}}{\sqrt{2}}\,, \tag{4.7.8}$$

where ρ is a material parameter. Consider a simple model in which $\mathring{\mathbf{J}}$ is coaxial with $\boldsymbol{\mu}$,

$$\mathring{\mathbf{J}} = \Lambda\,\dot{\gamma}\,\boldsymbol{\mu}, \qquad \Lambda = \Lambda(\mu_f/\mu_s, \;\cdots) \geq 0\,. \tag{4.7.9a,b}$$

In (4.7.9b), μ_s is the *saturation* value of μ_f, or the value of μ_f just before a load reversal. In general, Λ is expected to depend on the history of the deformation, and this is denoted in (4.7.9b) by the dots. In cyclic shearing, Nemat-Nasser and Zhang (2002) find this function to depend on the value of the strain attained just before each load reversal.

An alternative to the model (4.7.9a) is to take $\mathring{\mathbf{J}}$ proportional to the deviatoric part of the deformation rate tensor, $\mathbf{D}^{p\prime}$,

$$\mathring{\mathbf{J}} = \sqrt{2}\,\hat{\Lambda}\,\mathbf{D}^{p\prime}, \qquad \hat{\Lambda} = \hat{\Lambda}(\mu_f/\mu_s, \;\cdots) \geq 0\,. \tag{4.7.10a,b}$$

[59] See Oda *et al.* (1982, 1985), Mehrabadi *et al.* (1988, 1993), and Subhash *et al.* (1991). See also Chapter 7.

Experimental results on photoelastic granules suggest that either alternative may be a viable choice. Since the elastic strains in a granular mass under moderate conditions are rather small relative to the inelastic ones, either $\mathbf{D}'$ or $\mathbf{D}^{p\prime}$ may be used in (4.7.10a). The inelastic deviatoric deformation rate, $\mathbf{D}^{p\prime}$, however, is more appropriate, since only the inelastic deformation of frictional granules affects their fabric.

In general, the saturation value μ_s and the overall frictional coefficient M_f are functions of the void ratio e which is the ratio of the void volume, V_v, to the solid volume, V_s, in a typical sample, *i.e.*, $e = V_v/V_s$. Expressions for these functions may vary from material to material and can be obtained from experiments. For illustration, these functions may be expressed as power laws of their arguments, as follows:

$$\mu_s = \mu_M - b\,(e - e_m)^{N_1}, \quad M_f = [M_m + c\,(e_M - e)^{N_2}]\, f(\mu_f)\,, \tag{4.7.11a,b}$$

where μ_M, b, e_m, N_1, M_m, N_2, c, and e_M, are material constants, and f is some function of μ_f. These quantities must be determined empirically from experiments; see Nemat-Nasser and Zhang (2002) for an illustration and comparison with experimental results. Here, μ_M and e_M are viewed as the maximum values of the corresponding quantities, and e_m as the minimum value of the void ratio. The last two quantities, e_M and e_m, are commonly used to characterize sands.

It is finally necessary to define the evolution of the yield surfaces, $f_1 = 0$ and $f_2 = 0$, and to provide an explicit expression for the plastic part of the deformation tensor, $\mathbf{D}^{p\prime}$; see (4.7.2a,e) and (4.7.3a,b). This can be accomplished following the procedure of Subsection 4.3.2, though the calculations are now a bit more involved due to the coupling that exists between the deviatoric and dilatational deformation rates.

First, note that

$$\overset{\circ}{\boldsymbol{\mu}} = (\mathbf{1}^{(4s)} - \boldsymbol{\mu}\otimes\boldsymbol{\mu}) : \frac{\overset{\circ}{\mathbf{s}}}{\sqrt{2}\,\tau}\,, \quad \boldsymbol{\mu} : \overset{\circ}{\boldsymbol{\mu}} = 0\,, \quad (\mathbf{1}^{(4s)} - \boldsymbol{\mu}\otimes\boldsymbol{\mu}) : \overset{\circ}{\boldsymbol{\mu}} = \overset{\circ}{\boldsymbol{\mu}}\,,$$
$$(\mathbf{1}^{(4s)} - \boldsymbol{\mu}\otimes\boldsymbol{\mu}) : \boldsymbol{\mu} = \mathbf{0}\,, \quad (\mathbf{1}^{(4s)} - \boldsymbol{\mu}\otimes\boldsymbol{\mu}) : (\mathbf{1}^{(4s)} - \boldsymbol{\mu}\otimes\boldsymbol{\mu}) = (\mathbf{1}^{(4s)} - \boldsymbol{\mu}\otimes\boldsymbol{\mu})\,, \tag{4.7.12a-e}$$

and setting

$$M = \frac{\partial F}{\partial p}\,, \quad B = -\frac{\partial G}{\partial p}\,, \quad \frac{1}{h_1} = \frac{u\,\alpha}{\tau}\,, \tag{4.7.13a-c}$$

from (4.3.3d,e), (4.7.2a,b), and (4.7.3a,b), obtain

$$\mathbf{D}^{p\prime} = \frac{1}{H}\left\{\frac{\boldsymbol{\mu}}{\sqrt{2}} \otimes \frac{\boldsymbol{\mu}}{\sqrt{2}} + \frac{M}{3}\,\frac{\boldsymbol{\mu}}{\sqrt{2}} \otimes \mathbf{1}\right\} : \overset{\circ}{\boldsymbol{\tau}}$$
$$+ \frac{1}{h_1}\left\{\frac{1}{2}\,\mathbf{1}^{(4s)} - \frac{\boldsymbol{\mu}}{\sqrt{2}} \otimes \frac{\boldsymbol{\mu}}{\sqrt{2}}\right\} : (\overset{\circ}{\boldsymbol{\tau}}' - \overset{\circ}{\boldsymbol{\beta}})\,,$$
$$D^p_{kk} = \frac{B}{H}\left\{\frac{\boldsymbol{\mu}}{\sqrt{2}} + \frac{M}{3}\,\mathbf{1}\right\} : \overset{\circ}{\boldsymbol{\tau}} + \dot{\theta}\,. \tag{4.7.14a,b}$$

The workhardening parameter $\overline{H}$ in these equations is defined by (4.3.3e), which, in the present notation, becomes

$$\overline{H} = \frac{\rho_0}{\rho} B \frac{\partial F}{\partial \Delta} + \frac{\partial F}{\partial \gamma} + \frac{\boldsymbol{\mu} : \mathring{\boldsymbol{\beta}}}{\sqrt{2}\,\dot{\gamma}} . \tag{4.7.14c}$$

In (4.7.14a), the term involving the Jaumann rate of the backstress, $\mathring{\boldsymbol{\beta}}$, may be rendered explicit using either (4.7.9a), or (4.7.10a), or another suitable model for the evolution of the fabric tensor, $\mathbf{J}$.

From (4.7.1c), first note that

$$\mathring{\boldsymbol{\beta}} = \sqrt{2}\,\dot{p}\,\mathbf{J} + \sqrt{2}\,p\,\mathring{\mathbf{J}} . \tag{4.7.15}$$

Then, for model (4.7.9a), $\mathring{\mathbf{J}} = \Lambda \dot{\gamma}\, \boldsymbol{\mu}$, obtain

$$\frac{1}{h_1}\left\{ \frac{1}{2} \mathbf{1}^{(4s)} - \frac{\boldsymbol{\mu}}{\sqrt{2}} \otimes \frac{\boldsymbol{\mu}}{\sqrt{2}} \right\} : (\mathring{\boldsymbol{\tau}}' - \mathring{\boldsymbol{\beta}}) = \frac{1}{3 h_1}\left\{ \frac{\mathbf{J}}{\sqrt{2}} - \mu_f \frac{\boldsymbol{\mu}}{\sqrt{2}} \right\} \otimes \mathbf{1} : \mathring{\boldsymbol{\tau}}$$

$$+ \frac{1}{h_1}\left\{ \frac{1}{2} \mathbf{1}^{(4s)} - \frac{\boldsymbol{\mu}}{\sqrt{2}} \otimes \frac{\boldsymbol{\mu}}{\sqrt{2}} \right\} : \mathring{\boldsymbol{\tau}}' . \tag{4.7.16a}$$

The last term in the workhardening parameter $\overline{H}$, in (4.7.14c), becomes

$$\frac{\boldsymbol{\mu} : \mathring{\boldsymbol{\beta}}}{\sqrt{2}\,\dot{\gamma}} = \mu_f \frac{\dot{p}}{\dot{\gamma}} + \Lambda\, p . \tag{4.7.16b}$$

Similarly, for model (4.7.10a), $\mathring{\mathbf{J}} = \sqrt{2}\,\hat{\Lambda}\,\mathbf{D}^{p\prime}$, and simple calculation shows that

$$\frac{1}{h_1}\left\{ \frac{1}{2} \mathbf{1}^{(4s)} - \frac{\boldsymbol{\mu}}{\sqrt{2}} \otimes \frac{\boldsymbol{\mu}}{\sqrt{2}} \right\} : (\mathring{\boldsymbol{\tau}}' - \mathring{\boldsymbol{\beta}}) = \frac{1}{3\,(h_1 + \hat{\Lambda}\,p)}\left\{ \frac{\mathbf{J}}{\sqrt{2}} - \mu_f \frac{\boldsymbol{\mu}}{\sqrt{2}} \right\} \otimes \mathbf{1} : \mathring{\boldsymbol{\tau}}$$

$$+ \frac{1}{h_1 + \hat{\Lambda}\,p}\left\{ \frac{1}{2} \mathbf{1}^{(4s)} - \frac{\boldsymbol{\mu}}{\sqrt{2}} \otimes \frac{\boldsymbol{\mu}}{\sqrt{2}} \right\} : \mathring{\boldsymbol{\tau}}' , \tag{4.7.17a}$$

and that

$$\frac{\boldsymbol{\mu} : \mathring{\boldsymbol{\beta}}}{\sqrt{2}\,\dot{\gamma}} = \mu_f \frac{\dot{p}}{\dot{\gamma}} + \hat{\Lambda}\, p . \tag{4.7.17b}$$

Hence, for both models, only the pressure rate, $\dot{p}$, affects the term associated with $\mathring{\boldsymbol{\beta}}$ in (4.7.14a) as well as in (4.7.14c). Since (4.7.16a) is obtained from (4.7.17a) by simply setting $\hat{\Lambda}$ in the latter expression equal to zero, in what follows, only model (4.7.10a) is further examined.

For future use, it is convenient to recast expressions (4.7.14a-c) into the general form (4.2.3) so that the inverse relations can be obtained from expressions (4.2.8) by simple inspection. The presence of the terms associated with $\mathring{\boldsymbol{\mu}}$ makes this a bit involved. However, as pointed out by Rudnicki and Rice (1975), redefinition of the parameters may be effectively used to simplify this procedure. In the present context, first define

$$\overline{h} = h_1 + \hat{\Lambda}\,p , \quad \frac{1}{\hat{H}} = \frac{1}{\overline{H}} - \frac{1}{\overline{h}} ,$$

$$\hat{B} = \frac{B\hat{H}}{\overline{H}}, \quad \hat{M} = \frac{M\hat{H}}{\overline{H}} - \frac{\mu_f \hat{H}}{\overline{h}}. \tag{4.7.18a-d}$$

Then observe that (4.7.14a,b) can be combined to yield

$$\mathbf{D}^p = \frac{1}{\hat{H}}\left\{\frac{\boldsymbol{\mu}}{\sqrt{2}} + \frac{\hat{B}}{3}\mathbf{1}\right\}\left\{\frac{\boldsymbol{\mu}}{\sqrt{2}} + \frac{\hat{M}}{3}\mathbf{1}\right\} : \overset{\circ}{\boldsymbol{\tau}}$$
$$+ \frac{1}{\overline{h}}\left\{\frac{1}{2}(\mathbf{1}^{(4s)} - \frac{1}{3}\mathbf{1}\otimes\mathbf{1}) + \frac{1}{3}\frac{\mathbf{J}}{\sqrt{2}}\otimes\mathbf{1} + \frac{1}{9\hat{A}}\mathbf{1}\otimes\mathbf{1}\right\} : \overset{\circ}{\boldsymbol{\tau}} + \frac{1}{3}\dot{\theta}\,\mathbf{1}, \tag{4.7.19}$$

where $1/\hat{A}$ is defined by

$$\frac{1}{\hat{A}} = (M - \mu_f)\,B\,\{(\overline{H}/\overline{h}) - 1\}^{-1}. \tag{4.7.18e}$$

In this form, expression (4.7.19) can be combined with the elasticity relations of the following subsection to obtain a complete set of rate constitutive relations.

Finally, for void collapse, the critical pressure, p_c, is, in general, a function of the grain size and distribution, the particle to particle friction coefficient, pressure, and other microstructural parameters. A simple model is obtained by assuming that p_c is a function of the void ratio alone, $p_c = p_c(e)$. Then,

$$\dot{p}_c = (B\,\dot{\gamma} + \dot{\theta})\,h, \quad h = (1 + e)\frac{\partial p_c}{\partial e}. \tag{4.7.20a,b}$$

4.7.7. Elasticity Relations

The deformation of most granular materials includes both elastic and inelastic parts. The elastic deformation generally stems from the elasticity of the constituent granules as well as that of the fluids which may fill the intergranular voids. When the granules are cemented to one another, then the elasticity of the bonding material also contributes to the elastic deformation of the composite. In most cases, however, the deformation due to sliding and rolling of the granules may be so large relative to that due to the elasticity of the particles, that this elasticity may be neglected in the description of the overall deformation of the granular mass. For the calculation of the overall stresses, however, the elasticity of the material may have to be included in order to obtain reasonably accurate estimates of the stress field.

In the continuum models, the elastic properties are expressed in terms of the elastic parameters which, in general, do not remain constant in the course of deformation of granular masses. While it is possible to estimate these parameters in terms of the properties of the granules and the microstructure, here they will be assumed to be given.[60]

---

[60] For a comprehensive account of the micromechanical formulation of heterogeneous elastic solids, see Nemat-Nasser and Hori (1993, 1999).

Let $\mathcal{L}$ be the instantaneous elasticity tensor, and $\mathcal{M}$ be its inverse, the compliance tensor. In general, these tensors are not isotropic. If $\mathcal{L}_{ijkl}$, i, j, k, l = 1, 2, 3, are the components of $\mathcal{L}$ with respect to a fixed rectangular Cartesian coordinate system, then, in general, the following symmetry holds for these components: $\mathcal{L}_{ijkl} = \mathcal{L}_{klij} = \mathcal{L}_{jikl} = \mathcal{L}_{ijlk}$, leading to a maximum of twenty-one elasticity parameters. As shown by Nemat-Nasser and Hori (1993, p.562), a fourth-order symmetric tensor, say, $\mathcal{L} = \mathcal{L}^T$, can be expressed in terms of its six principal values, $\mathcal{L}^I$, I = 1, 2, ..., 6, and the corresponding principal second-order symmetric tensors, $\mathbf{E}^I$, as follows:

$$\mathcal{L} = \sum_{I=1}^{6} \mathcal{L}^I\, \mathbf{E}^I \otimes \mathbf{E}^I, \tag{4.7.21a}$$

where the principal second-order symmetric tensors are orthonormal in the sense that

$$\mathbf{E}^I : \mathbf{E}^J = \delta_{IJ}, \tag{4.7.21b}$$

where δ_{IJ} is the Kronecker delta. Hence, for an arbitrary second-order symmetric tensor $\mathbf{D}$,

$$\mathcal{L} : \mathbf{D} = \sum_{I=1}^{6} \mathcal{L}^I\, (\mathbf{D} : \mathbf{E}^I)\, \mathbf{E}^I. \tag{4.7.21c}$$

The principal values and principal second-order symmetric tensors of $\mathcal{L}$ are *related* to the eigenvalues and eigenvectors of the corresponding six by six matrix $[\mathcal{L}_{ab}]$. The spectral decomposition of a fourth-order symmetric tensor is examined in some detail by Nemat-Nasser and Hori (1993). Here, the spectral representation (4.7.21) is used as a tool to clarify the decomposition of the elasticity and compliance tensors, $\mathcal{L}$ and $\mathcal{M}$, into their deviatoric and dilatational components, as discussed in the sequel.

Consider $\mathcal{M}$ with components $\mathcal{M}_{ijkl}$; a similar representation applies to $\mathcal{L}$. Define a second-order symmetric tensor $\mathbf{M}$ such that

$$M_{ij} = \mathcal{M}_{ijkk} = \mathcal{M}_{kkij}. \tag{4.7.22a}$$

In terms of the spectral representation of $\mathcal{M}$,

$$\mathcal{M} = \sum_{I=1}^{6} \mathcal{M}^I\, \mathbf{E}^I \otimes \mathbf{E}^I \quad \text{or} \quad \mathcal{M}_{ijkl} = \sum_{I=1}^{6} \mathcal{M}^I\, E^I_{ij} E^I_{kl}, \tag{4.7.22b,c}$$

and tensor $\mathbf{M}$ is given by

$$\mathbf{M} = \sum_{I=1}^{6} \mathcal{M}^I\, \mathrm{tr}(\mathbf{E}^I)\, \mathbf{E}^I, \quad \mathrm{tr}(\mathbf{E}^I) = E_{ii}. \tag{4.7.22d,e}$$

Decompose $\mathbf{M}$ into its deviatoric and dilatational (spherical) parts, as follows:

$$\mathbf{M} = \mathbf{M}' + \frac{1}{3\kappa}\mathbf{1}, \quad \frac{1}{\kappa} = \mathrm{tr}(\mathbf{M}) = \mathcal{M}_{iijj}. \tag{4.7.22f,g}$$

Based on the above results, now set

$$\mathcal{M} = \mathcal{M}' + \frac{1}{3}(\mathbf{M} \otimes \mathbf{1} + \mathbf{1} \otimes \mathbf{M}) - \frac{1}{9\kappa}\mathbf{1} \otimes \mathbf{1}$$

$$= \mathcal{M}' + \frac{1}{3}(\mathbf{M}' \otimes \mathbf{1} + \mathbf{1} \otimes \mathbf{M}') + \frac{1}{9\kappa}\mathbf{1} \otimes \mathbf{1}, \qquad (4.7.23a,b)$$

where $\mathcal{M}'$ is deviatoric, defined such that $\mathcal{M}'_{iijk} = \mathcal{M}'_{jkii} = 0$, *i.e.*,

$$\mathcal{M}' = \sum_{I=1}^{6} \mathcal{M}^I \mathbf{E}^{I'} \otimes \mathbf{E}^{I'}, \quad \mathbf{E}^{I'} = \mathbf{E}^I - \frac{1}{3}\mathrm{tr}(\mathbf{E}^I)\,\mathbf{1}. \qquad (4.7.23c,d)$$

Observe that, since (4.7.21a) and (4.7.22b) are each other's inverse, their eigenvalues are related by $\mathcal{M}^I \mathcal{L}^I = 1$, $I = 1, 2, ..., 6$, (I not summed).

Let $\mathbf{D}^e$ be the elastic part of the total deformation rate tensor, $\mathbf{D}$, *i.e.*, $\mathbf{D} = \mathbf{D}^e + \mathbf{D}^p$. Consider the elasticity relation

$$\mathbf{D}^e = \mathcal{M} : \mathring{\boldsymbol{\tau}}$$

$$= \{\mathcal{M}' + \frac{1}{3}(\mathbf{M}' \otimes \mathbf{1} + \mathbf{1} \otimes \mathbf{M}') + \frac{1}{9\kappa}\mathbf{1} \otimes \mathbf{1}\} : \mathring{\boldsymbol{\tau}}. \qquad (4.7.24a,b)$$

These equations can be split into dilatational and deviatoric parts, as follows:

$$\mathbf{D}^{e\prime} = \mathcal{M}' : \mathring{\boldsymbol{\tau}}' + \frac{1}{3}\mathbf{M}' \mathring{\tau}_{kk} = \{\mathcal{M}' + \frac{1}{3}\mathbf{M}' \otimes \mathbf{1}\} : \mathring{\boldsymbol{\tau}},$$

$$D^e_{kk} = \mathbf{M} : \mathring{\boldsymbol{\tau}} = \frac{1}{3\kappa}\mathring{\tau}_{kk} + \mathbf{M}' : \mathring{\boldsymbol{\tau}}. \qquad (4.7.24c\text{-}f)$$

For isotropic elasticity, the compliance tensor, $\mathcal{M}$, takes on the following form:

$$\mathcal{M} = \frac{1}{2\mu}(\mathbf{1}^{(4s)} - \frac{1}{3}\mathbf{1} \otimes \mathbf{1}) + \frac{1}{3\kappa}(\frac{1}{3}\mathbf{1} \otimes \mathbf{1}). \qquad (4.7.23e)$$

Hence, in this case,

$$\mathbf{M}' = \mathbf{0}, \quad \mathcal{M}' = \frac{1}{2\mu}(\mathbf{1}^{(4s)} - \frac{1}{3}\mathbf{1} \otimes \mathbf{1}). \qquad (4.7.23f,g)$$

The inverse for the isotropic case is simply given by

$$\mathcal{L} = 2\mu(\mathbf{1}^{(4s)} - \frac{1}{3}\mathbf{1} \otimes \mathbf{1}) + 3\kappa(\frac{1}{3}\mathbf{1} \otimes \mathbf{1}). \qquad (4.7.21d)$$

4.7.8. General Constitutive Relations

Combining (4.7.19) with (4.7.24a), first define an *effective compliance tensor* by

$$\mathcal{M}^* = \mathcal{M} + \frac{1}{h}\left\{\frac{1}{2}(\mathbf{1}^{(4s)} - \frac{1}{3}\mathbf{1} \otimes \mathbf{1}) + \frac{1}{9\hat{A}}\mathbf{1} \otimes \mathbf{1}\right\}, \qquad (4.7.25a)$$

where $\hat{A}$ is defined by (4.7.18e). For isotropic elasticity, this becomes

$$\mathcal{M}^* = \frac{1}{2\mu^*}(\mathbf{1}^{(4s)} - \frac{1}{3}\mathbf{1} \otimes \mathbf{1}) + \frac{1}{9\kappa^*}\mathbf{1} \otimes \mathbf{1}, \qquad (4.7.25b)$$

where

$$\frac{1}{\mu^*} = \frac{1}{\mu} + \frac{1}{\bar{h}}, \quad \frac{1}{\kappa^*} = \frac{1}{\kappa} + \frac{1}{\bar{h}\hat{A}}. \tag{4.7.25c,d}$$

The inverse then is

$$\mathcal{L}^* = 2\mu^* (\mathbf{1}^{(4s)} - \frac{1}{3}\mathbf{1}\otimes\mathbf{1}) + 3\kappa^* (\frac{1}{3}\mathbf{1}\otimes\mathbf{1}). \tag{4.7.25e}$$

Now, referring to (4.7.19), set

$$\bar{\mathbf{D}} = \mathbf{D} - \frac{1}{3}\dot{\theta}\,\mathbf{1} - \frac{1}{3\bar{h}}\frac{\mathbf{J}}{\sqrt{2}}\,\overset{\circ}{\tau}_{kk}, \tag{4.7.26a}$$

where $\bar{h}$ is defined by (4.7.18a). The rate constitutive relation then is

$$\bar{\mathbf{D}} = \mathcal{M}^* : \overset{\circ}{\boldsymbol{\tau}} + \frac{1}{\hat{H}}\left\{\frac{\boldsymbol{\mu}}{\sqrt{2}} + \frac{\hat{B}}{3}\mathbf{1}\right\}\otimes\left\{\frac{\boldsymbol{\mu}}{\sqrt{2}} + \frac{\hat{M}}{3}\mathbf{1}\right\} : \overset{\circ}{\boldsymbol{\tau}} \tag{4.7.26b}$$

which can be inverted using the method outlined in Section 4.2. To complete the process, it is necessary to eliminate $\overset{\circ}{\tau}_{kk}$ in the expression for $\bar{\mathbf{D}}$, as is discussed in the sequel.

To obtain the inverse of (4.7.26b) for general anisotropic elasticity, denote by $\mathcal{L}^*$ the inverse of the general effective anisotropic compliance tensor $\mathcal{M}^*$. Then, in view of (4.2.4) and (4.2.7), the inverse of (4.7.26b) becomes

$$\overset{\circ}{\boldsymbol{\tau}} = (\mathcal{L}^* - \boldsymbol{l}\otimes\boldsymbol{\lambda}) : \left\{\mathbf{D} - \frac{1}{3}\dot{\theta}\,\mathbf{1} - \frac{1}{3\bar{h}}\frac{\mathbf{J}}{\sqrt{2}}\overset{\circ}{\tau}_{kk}\right\}, \tag{4.7.27a}$$

where the second-order tensors, $\boldsymbol{l}$ and $\boldsymbol{\lambda}$, are given in terms of

$$\mathbf{m} = \frac{1}{\hat{H}}\left\{\frac{\boldsymbol{\mu}}{\sqrt{2}} + \frac{\hat{B}}{3}\mathbf{1}\right\} \quad \text{and} \quad \hat{\boldsymbol{\mu}} = \frac{\boldsymbol{\mu}}{\sqrt{2}} + \frac{\hat{M}}{3}\mathbf{1}, \tag{4.7.27b,c}$$

by the following expressions:

$$\boldsymbol{\lambda} = \mathcal{L}^* : \hat{\boldsymbol{\mu}}, \quad \boldsymbol{l} = \frac{\mathcal{L}^* : \mathbf{m}}{1 + \boldsymbol{\lambda} : \mathbf{m}}. \tag{4.7.27d,e}$$

Consider now eliminating $\overset{\circ}{\tau}_{kk}$ from the left-hand side of (4.7.26b), *i.e.*, in expression (4.7.26a) for $\bar{\mathbf{D}}$. To simplify notation, set

$$\bar{\mathcal{L}} = \mathcal{L}^* - \boldsymbol{l}\otimes\boldsymbol{\lambda}, \quad \boldsymbol{\eta} = \mathbf{M}' + \frac{B}{\hat{H}}\frac{\boldsymbol{\mu}}{\sqrt{2}}, \quad A_0 = \frac{1}{3\kappa} + \frac{BM}{3H},$$

$$A_1 = \frac{\boldsymbol{\eta} : \bar{\mathcal{L}} : \mathbf{J}}{3\sqrt{2}\,\bar{h}}. \tag{4.7.28a-d}$$

Then, in view of (4.7.14b) and (4.7.24f), obtain

$$D_{kk} = D^e_{kk} + D^p_{kk} = \boldsymbol{\eta} : \overset{\circ}{\boldsymbol{\tau}}' + A_0\,\overset{\circ}{\tau}_{kk} + \dot{\theta}. \tag{4.7.29}$$

To find $\overset{\circ}{\tau}_{kk}$, multiply (4.7.27a) from the left by $\boldsymbol{\eta}:$, substitute the result into (4.7.29), and solve for $\overset{\circ}{\tau}_{kk}$,

$$\overset{\circ}{\tau}_{kk} = \{D_{kk} - \dot{\theta} - \boldsymbol{\eta} : \bar{\mathcal{L}} : (\mathbf{D} - \frac{1}{3}\dot{\theta}\,\mathbf{1})\}\,(A_0 - A_1)^{-1}. \tag{4.7.30}$$

4.7.9. The Case of Isotropic Elasticity

Explicit expressions for $\boldsymbol{l}$, $\boldsymbol{\lambda}$, and the stress rate, $\overset{\circ}{\boldsymbol{\tau}}$, result when the elastic response of the material can be assumed to be isotropic. Then, the elasticity tensor $\boldsymbol{\mathcal{L}}$ is given by (4.7.21d) with μ and κ as the shear and bulk moduli, and $\boldsymbol{\mathcal{L}}^*$, the inverse of $\boldsymbol{\mathcal{M}}^*$, is given explicitly by (4.7.25e), involving simple modification of the elasticity moduli, as defined by (4.7.25c,d). With $\boldsymbol{\mathcal{L}}^*$ known, (4.7.27d,e) yield

$$\boldsymbol{\lambda} = 2\mu^* \left\{ \frac{\boldsymbol{\mu}}{\sqrt{2}} + \frac{M^*}{3} \mathbf{1} \right\}, \quad \boldsymbol{l} = \frac{1}{H^*} \left\{ \frac{\boldsymbol{\mu}}{\sqrt{2}} + \frac{B^*}{3} \mathbf{1} \right\}, \tag{4.7.31a,b}$$

where

$$M^* = \frac{3\hat{M}\kappa^*}{2\mu^*}, \quad B^* = \frac{3\hat{B}\kappa^*}{2\mu^*}, \quad H^* = \frac{3\hat{H} + 3\mu^* + 2\mu^* M^* \hat{B}}{6\mu^*}. \tag{4.7.31c-e}$$

From these, it now follows that

$$\begin{aligned} \overline{\boldsymbol{\mathcal{L}}} = {} & 2\mu^* \{ \mathbf{1}^{(4s)} - \frac{1}{3} \mathbf{1} \otimes \mathbf{1} \} + 3\kappa^* \{ \frac{1}{3} \mathbf{1} \otimes \mathbf{1} \} \\ & - \frac{2\mu^*}{H^*} \left\{ \frac{\boldsymbol{\mu}}{\sqrt{2}} + \frac{B^*}{3} \mathbf{1} \right\} \otimes \left\{ \frac{\boldsymbol{\mu}}{\sqrt{2}} + \frac{M^*}{3} \mathbf{1} \right\}. \end{aligned} \tag{4.7.31f}$$

Now, to obtain an expression for $\overset{\circ}{\tau}_{kk}$, first note that $\mathbf{M}' = \mathbf{0}$ in this case. From (4.7.28a-d) then obtain

$$\boldsymbol{\eta} : \overline{\boldsymbol{\mathcal{L}}} = \frac{2\mu^* B}{\overline{H} N} \left\{ \frac{\boldsymbol{\mu}}{\sqrt{2}} + \frac{\tilde{M}}{3} \mathbf{1} \right\}, \quad A_1 = \frac{\mu^* B \mu_f}{3\bar{h}\overline{H} N},$$

$$\tilde{M} = -\frac{M^*}{2H^*} N, \quad N = \frac{3\hat{H} + 3\mu^* + 2\mu^* M^* \hat{B}}{3\hat{H} + 2\mu^* \hat{M} \hat{B}}. \tag{4.7.32a-d}$$

With this, expression (4.7.30) now gives $\overset{\circ}{\tau}_{kk}$ explicitly, where A_0 is given by (4.7.28c).

4.7.10. Granular Materials with Vanishing Elastic Range in Shearing

Under moderate conditions, certain granular materials, *e.g.*, cohesionless sands, often exhibit negligibly small elasticity, even in unloading. This behavior may be modeled by assigning a small value to the radius of the yield surface, *e.g.*, to F/p. Such a direct approach invariably leads to numerical problems in the computational implementation of the theory. In this subsection, the model of the previous subsection is modified and simplified for application to cases where the elastic response of the frictional granular mass can be neglected.

To this end, first introduce a unit deviatoric tensor, $\boldsymbol{\nu}$, coaxial with the deviatoric deformation rate tensor, $\mathbf{D}'$,

$$\boldsymbol{\nu} = \frac{\sqrt{2}\mathbf{D}'}{d}, \quad d = (2\mathbf{D}' : \mathbf{D}')^{1/2}, \quad \mathbf{D}' = \frac{d}{\sqrt{2}} \boldsymbol{\nu}. \tag{4.7.33a-c}$$

Now, noting that, for *any* nonzero symmetric, deviatoric, second-order tensor $\mathbf{J}$, the second-order deviatoric tensor $(\mathbf{1}^{(4s)} - \boldsymbol{\nu}\otimes\boldsymbol{\nu}) : \mathbf{J}$ is normal to $\boldsymbol{\nu}$, decompose the inelastic deformation rate tensor, $\mathbf{D}^p$, as follows:

$$\mathbf{D}^{p\prime} = \dot{\gamma}\,\frac{\boldsymbol{\nu}}{\sqrt{2}} + \dot{\lambda}\,(\mathbf{1}^{(4s)} - \boldsymbol{\nu}\otimes\boldsymbol{\nu}) : \frac{\mathbf{J}}{\sqrt{2}}, \qquad D^p_{kk} = \dot{\gamma}\,B, \tag{4.7.34a,b}$$

where the volumetric straining associated with $\dot{\theta}$ in (4.7.3b) is neglected; this can easily be included in the model which follows, but is not considered any further in this subsection. In (4.7.34a,b), $\dot{\gamma}$ and $\dot{\lambda}$ are yet to be calculated.

As before, let $\mathbf{J}$ denote the fabric tensor, relating to the backstress, $\boldsymbol{\beta}$, by (4.7.1c). Since, the stress difference, $\mathbf{s} = \boldsymbol{\tau}' - \boldsymbol{\beta}$, is negligibly small, it follows from (4.7.2c) and (4.7.1c) that

$$\boldsymbol{\tau}' = \sqrt{2}\,p\,\mathbf{J}. \tag{4.7.35a}$$

Hence, taking the Jaumann rate, arrive at

$$\overset{\circ}{\boldsymbol{\tau}}{}' = \sqrt{2}\,\dot{p}\,\mathbf{J} + \sqrt{2}\,p\,\overset{\circ}{\mathbf{J}}. \tag{4.7.35b}$$

For the evolution of the fabric tensor $\mathbf{J}$, consider model (4.7.9a), and since the elastic range is small, set

$$\overset{\circ}{\mathbf{J}} = \Lambda\,\dot{\gamma}\,\boldsymbol{\nu}. \tag{4.7.35c}$$

Also, redefine the anisotropic friction coefficient, $\mu_f = \mathbf{J} : \boldsymbol{\mu}$, by

$$\nu_f = \mathbf{J} : \boldsymbol{\nu}. \tag{4.7.35d}$$

To calculate $\dot{\gamma}$, the effective plastic strain rate coaxial with the deformation rate, and $\dot{\lambda}$, the effective plastic strain rate normal to the deformation rate, assume isotropic elasticity for the vanishingly small elastic range, and write

$$\overset{\circ}{\tau}_{kk} = 3\,\kappa\,(D_{kk} - \dot{\gamma}\,B),$$

$$\overset{\circ}{\boldsymbol{\tau}}{}' = 2\,\mu\left\{\mathbf{D}' - \dot{\gamma}\,\frac{\boldsymbol{\nu}}{\sqrt{2}} - \dot{\lambda}\,(\mathbf{1} - \boldsymbol{\nu}\otimes\boldsymbol{\nu}) : \frac{\mathbf{J}}{\sqrt{2}}\right\}. \tag{4.7.36a,b}$$

Now, combining (4.7.35b,c) with (4.7.36b), consider

$$\sqrt{2}\,\dot{p}\,\mathbf{J} + \sqrt{2}\,p\,\Lambda\,\dot{\gamma}\,\boldsymbol{\nu} = 2\,\mu\left\{\mathbf{D}' - \dot{\gamma}\,\frac{\boldsymbol{\nu}}{\sqrt{2}} - \dot{\lambda}\,(\mathbf{1} - \boldsymbol{\nu}\otimes\boldsymbol{\nu}) : \frac{\mathbf{J}}{\sqrt{2}}\right\}. \tag{4.7.37a}$$

From the double contraction of this equation, first with $\boldsymbol{\nu}$ and then with $(\mathbf{1}^{(4s)} - \boldsymbol{\nu}\otimes\boldsymbol{\nu}) : \mathbf{J}$, arrive at

$$\dot{p}\,\nu_f + \Lambda\,p\,\dot{\gamma} = \mu\,d - \mu\,\dot{\gamma}, \qquad \dot{p} = -\mu\,\dot{\lambda}. \tag{4.7.37b,c}$$

Since $\dot{p} = -\kappa\,(D_{kk} - \dot{\gamma}\,B)$, the above two equations now yield

$$\dot{\gamma} = \frac{\mu\,d + \kappa\,\nu_f\,D_{kk}}{\mu + \Lambda\,p + \kappa\,\nu_f\,B},$$

$$\dot{\lambda} = \frac{\kappa}{\mu}\,(D_{kk} - B\,\dot{\gamma}). \tag{4.7.38a,b}$$

Note that, since there is no elastic range in shearing, $\dot{\gamma}$ is always nonzero for a nonzero deformation rate.

To obtain an expression for the dilatancy parameter, B, examine the rate of inelastic distortional work. From (4.7.35a), the stress is given by

$$\boldsymbol{\tau} = \sqrt{2}\,\mathrm{p}\,\mathbf{J} - \mathrm{p}\,\mathbf{1}\,. \tag{4.7.39}$$

The rate of distortional work corresponding to $\dot{w}_{pd}$ in (4.7.5b) now becomes

$$\begin{aligned}\dot{w}_{pd} &= \{\sqrt{2}\,\mathrm{p}\,\mathbf{J} - \mathrm{p}\,\mathbf{1}\} : \{\dot{\gamma}\,(\boldsymbol{\nu}/\sqrt{2} + \mathrm{B}/3\,\mathbf{1})\}\\ &= \dot{\gamma}\,\mathrm{p}\,(\nu_f - \mathrm{B})\,.\end{aligned} \tag{4.7.40a,b}$$

Equating this with the rate of frictional dissipation, approximated by (4.7.5e), the dilatancy B now becomes

$$-\mathrm{B} = \mathrm{M}_f - \nu_f\,. \tag{4.7.41}$$

As in Subsection 4.7.6, it is also necessary to define the functional form of the parameter Λ in (4.7.35c), as well as that of the saturation value of ν_f. These functions must be selected empirically, based on the experimental results.

4.8. RATE-DEPENDENT THEORIES

Inelastic deformation of crystalline solids and geomaterials is, in general, rate-dependent. Rate-independent plasticity theories, therefore, represent idealizations which, in general, have limited applicability. The rate-dependency becomes especially dominant for high-strain-rate deformation problems.

To deal with rate effects within a phenomenological framework, two classes of constitutive models may be identified:

(1) fully rate-dependent plasticity; and

(2) viscoplasticity.

In the first approach, the deformation rate consists of an elastic and an inelastic constituent throughout the entire deformation history, *i.e., there is no yield surface.* In the second approach, on the other hand, a yield surface is considered, within which the response is elastic and on which the response may be elastic-viscoplastic, where the plastic deformation is rate-dependent. In this latter case, loading and unloading, analogous to the rate-independent theories, are included.

In the rate-dependent phenomenological approach, the effective plastic strain rate, $\dot{\gamma}$, is expressed in terms of the effective stress, temperature, and some internal variables which represent the effect of the thermomechanical loading history on the current microstructure and, hence, on the response of the material. In many models for metals, the total accumulated plastic strain is used as the only variable representing the history effects. Though more realistic than the classical rate-independent theories, this representation still has only a limited predictive capability. Indeed, it is often necessary to evaluate the parameters in constitutive models of this kind from experiments which are conducted at strains, strain rates, and temperatures close to the range of the values to which the model will eventually be applied.

There are other models which are based on rational analysis of the underlying mechanisms of the inelastic response of the material. For applications, however, simplifying assumptions are generally inevitable, and this kind of model also eventually relies on empirical evaluation of its constitutive parameters.

In this section, some of the commonly used rate-dependent models for metals, are examined, focusing on the temperature range where inelasticity is essentially due to the motion of dislocations. First, several empirical models are discussed. Then, based on the mechanisms of plastic deformation by the motion of dislocations, equations are developed that relate the effective stress, τ, to the effective strain rate, $\dot{\gamma}$, and the temperature, T, as well as to the dislocation densities (both total and mobile) and parameters that characterize their distribution and hence the microstructure. In this representation, the effective plastic strain, γ, enters only as a *load parameter*, since plastic strain or its gradient is *not* a proper microstructural (thermodynamic) state variable. Interestingly enough, the densities and distribution of dislocations introduce natural length scales in this physics-based plasticity model, requiring constitutive relations to describe their evolution in the course of a given thermomechanical loading history. The basic pillar of this model is a vast body of experimental data on various metals over broad ranges of strain rates, from quasi-static to 10^4/s and greater, and temperatures, from 77 to 1,300K and greater. Indeed, the dislocation-based formulation given in the present section and in Section 6.5 of Chapter 6, was originally suggested by the experimental data obtained for direct measurement of the strain hardening, the strain rate effect, and the temperature softening in tantalum and tantalum-tungsten alloys (Nemat-Nasser and Isaacs, 1997a,b) and subsequently applied to model experimental data on oxygen-free high-conductivity copper (OFHC) (Nemat-Nasser and Li, 1998) and many other metals, with remarkable success. In Chapter 6, this dislocation-based model is modified and applied to develop constitutive relations for single-crystal plasticity, which can be employed to simulate deformation of either polycrystalline metals (using certain averaging techniques, see Chapter 8), or several interacting crystals. As discussed in Section 6.5, when there are only a few crystals in a deforming aggregate, geometric and textural incompatibilities will most likely manifest themselves through a *size effect*, and may affect the overall material's resistance to deformation (flow stress[61]) in a major way. This size effect should be distinguished from the length scales in plasticity. While these dislocation-based length scales naturally enter plasticity constitutive relations, the size effect is a problem-dependent phenomenon that must be examined in each case using a relevant material-specific, possibly dislocation-based, crystal plasticity model. For a few interacting crystals, the slip-induced crystal plasticity (Chapter 6) should adequately account for any such size effects.

---

[61] The term *flow stress* is used to refer to the effective stress as a function of the effective strain rate and the temperature in general loading, with the effective strain as the load parameter. The ex-

It may be helpful for readers who may not have been directly involved in the experimental characterization of materials, to note the following comments. Most often, small variations in composition and alloying can have profound effects on the resulting metal's thermomechanical response. This response seldom is strain-rate and temperature independent. Early development of classical plasticity has focused, to a large extent, on the mathematical foundation of rate-independent models of materials, possibly suitable for application to certain structural steels of the 1930-1950's, particularly those of the WWII technology. Modern metals (even steels) can seldom be modeled adequately by a simple rate-independent yield function and some assumed workhardening rules. For example, various versions of tantalum, obtained through minor changes in their composition and processing, have surprisingly distinct mechanical responses critical to their intended applications.[62] Over the past 15 years, modern tools have been developed to directly characterize materials, particularly metals, for their rate, temperature, and strain dependencies, thus providing an opportunity for a more physics-based constitutive modeling than that provided by the classical rate- and temperature-independent formalism.[63]

4.8.1. Flow Stress: Empirical Models

Uniaxial experiments are often used to obtain the stress-strain response of a material at various strain rates and temperatures. The results are then generalized and used to formulate the three-dimensional constitutive relations for that material, as is illustrated in Subsections 4.8.10, 4.8.11, and 4.8.12. Most empirical models express the one-dimensional stress, say, τ, in terms of the corresponding strain rate, $\dot{\gamma}$, strain, γ, and temperature, T, as follows:

$$\tau/\tau_0 = F(\dot{\gamma}, \gamma, T), \tag{4.8.1a}$$

and seek to identify the function F based on the experimental results; here, τ_0 is some normalizing stress, *e.g.*, the initial yield stress of a virgin sample at some reference temperature and strain rate. Then, for application to three-dimensional problems, the stress, strain, and strain rate, τ, γ, and $\dot{\gamma}$, are interpreted as the *effective stress, effective strain, and effective strain rate*, respectively.

The form of the function F is often such that it can be solved explicitly for $\dot{\gamma}$ in terms of the stress, strain, and temperature,

perimental data that are used are for uniaxial loading which renders the effective stress and strain (strain rate) the same as the measured axial stress and strain (strain rate).

62 See, *e.g.*, Chen *et al.* (1996) and other articles in the same volume on tantalum, as well as Nemat-Nasser *et al.* (1998a).

63 See, *e.g.*, Nemat-Nasser *et al.* (1991), Rashid (1990), Rashid *et al.*, (1992), Nemat-Nasser and Isaacs (1997a,b), Kapoor and Nemat-Nasser (1998), Kim *et al.* (1998), and Chapter on **High Strain Testing** in *ASM Volume 8, Mechanical Testing and Evaluation Handbook* (2000).

$$\dot{\gamma}/\dot{\gamma}_0 = g(\tau, \gamma, T), \tag{4.8.1b}$$

where $\dot{\gamma}_0$ is some normalizing reference strain rate. Indeed, most existing empirical flow-stress models have the following form:

$$\tau/\tau_0 = f_1(\dot{\gamma})\, f_2(\gamma)\, f_3(T), \tag{4.8.1c}$$

with the inverse relation for $\dot{\gamma}$ in the form of

$$\dot{\gamma}/\dot{\gamma}_0 = g_1(\tau)\, g_2(\gamma)\, g_3(T), \tag{4.8.1d}$$

where f_i and g_i, $i = 1, 2, 3$, are material functions to be determined by experiments. In most cases, this is accomplished empirically by seeking to use functional forms which may make the model effective over the range of strains, strain rates, and temperatures of interest. Some of these functional forms are also based on certain theoretical consideration of the kinematics and kinetics of the micromechanisms that are responsible for the corresponding inelastic material behavior.[64]

As an illustration, consider the power law,

$$\dot{\gamma}/\dot{\gamma}_0 = (\tau/\tau_r)^m, \tag{4.8.2a}$$

where m is a fixed parameter, usually very large for metals at relatively low temperatures, *e.g.*, it is of the order of 100 at room temperature and at strain rates less than about 10^4/s. In (4.8.2a), $\dot{\gamma}_0$ and τ_r are the reference shear strain rate and shear stress, respectively. In this representation, the reference stress, τ_r, includes the essential physics of the deformation. In general, τ_r is assumed to depend on the accumulated plastic strain, on the temperature, and on other parameters which define the current state of the microstructure[65]. A simple representation, often used, is

$$\tau_r = \tau_0\,(1 + \gamma/\gamma_0)^N \exp\{-\lambda_0(T - T_0)\}, \tag{4.8.2b}$$

where τ_0 and γ_0 are normalizing stress and strain measures, N and λ_0 are material parameters, and T is temperature, with T_0 being its reference value. Here, N is the strain-hardening parameter, whereas λ_0 is a parameter characterizing thermal softening effects. Hence, the various functions in (4.8.1d) are identified as

$$g_1(\tau) \equiv (\tau/\tau_0)^m, \quad g_2(\gamma) \equiv (1 + \gamma/\gamma_0)^{-Nm},$$
$$g_3(T) \equiv \exp\{m\,\lambda_0(T - T_0)\}. \tag{4.8.2c}$$

Since λ_0 is not dimensionless, it may be preferable to use $g_3 \equiv \exp\{m\,\lambda_1(T - T_0)/(T_M - T_0)\}$, where T_M is the material's melting temperature, and λ_1 is a dimensionless parameter. The flow stress, τ, in this model is given by

64 See Kocks *et al.* (1975), Follansbee and Kocks (1988), Zerilli and Armstrong (1987), and Subsections 4.8.2 to 4.8.9.

65 See, *e.g.*, Nemat-Nasser *et al.* (1989) and references cited therein. As pointed out before, the accumulated plastic strain, however, is not a microstructural state variable.

$$\tau = \tau_0 (\dot{\gamma}/\dot{\gamma}_0)^{1/m} (1+\gamma/\gamma_0)^N \exp\{-\lambda_1 (T-T_0)/(T_M-T_0)\}. \qquad (4.8.2d)$$

From (4.8.2a), with m of the order of 100, it is clear that the plastic strain rate is small if the stress, τ, is less than the reference stress, τ_r. Once τ reaches the value of τ_r, and especially when it tends to exceed this value, the plastic strain rate dominates, and therefore, the overall response for large values of m resembles that of the elastic-plastic rate-independent model.

An alternative to representation (4.8.2a) is the exponential form,

$$\dot{\gamma}/\dot{\gamma}_0 = \exp\{a_0 \frac{\tau-\tau_r}{\tau_r}\}, \qquad (4.8.3a)$$

proposed by Johnson and Cook (1983, 1985), where τ_r is defined as

$$\tau_r = \tau_0 \{1 + A(\gamma/\gamma_0)^n\}\Big\{1 - \{(T-T_0)/(T_M-T_0)\}^{m_0}\Big\}. \qquad (4.8.3b)$$

The Johnson-Cook model can be written as

$$\tau = \tau_0 \{1 + A(\gamma/\gamma_0)^n\}\{1 + B\, ln(\dot{\gamma}/\dot{\gamma}_0)\}\Big\{1 - \{(T-T_0)/(T_M-T_0)\}^{m_0}\Big\}, \qquad (4.8.3c)$$

where τ_0, A, $B = 1/a_0$, n, and m_0 are regarded as material constants to be fixed by comparing with experimental results.[66] This model has five free parameters, whereas the power-law model in (4.8.2) has four free parameters.

Another often used model is that proposed by Zerilli and Armstrong (1987, 1990, 1992). This model distinguishes between fcc and bcc crystal structures. For fcc crystals, these authors suggest

$$\sigma = \Delta\sigma'_G + c_2\, \varepsilon^{1/2} \exp\{-c_3 T + c_4 T\, ln\, \dot{\varepsilon}\} + k\, d_G^{-1/2}, \qquad (4.8.4a)$$

and for bcc, they propose

$$\sigma = \Delta\sigma'_G + c_1 \exp\{-c_3 T + c_4 T\, ln\, \dot{\varepsilon}\} + c_5 \varepsilon^n + k\, d_G^{-1/2}, \qquad (4.8.4b)$$

where c_i's, i = 1,2, ..., 5, n, and k are viewed as material parameters; $\Delta\sigma'_G$ represents the additional stress due to the influence of solute atoms and original dislocations on the yield stress; and σ and ε stand for τ and γ, respectively. In these equations, d_G represents the average grain size in the considered polycrystal. Since the parameters in this model are established empirically through direct comparison with experimental results, to relate these expressions to the empirical models considered in this section, rewrite (4.8.4a) in the following equivalent form:[67]

---

[66] The Johnson-Cook model is usually expressed as $\sigma = \sigma_0 \{A + B(\varepsilon/\varepsilon_0)^n\}\ \{1 + C\, ln(\dot{\varepsilon}/\dot{\varepsilon}_0)\} \times \{1 - \{(T-T_0)/(T_M-T_0)\}^{m_0}\}$. However, since σ_0, A, B, and C are to be obtained from experimental data, this form is exactly equivalent to the slightly simpler form given by (4.8.3c), where, in addition, σ is replaced by τ, and ε by γ, in order to conform with the notation used in this book.

$$\tau = \tau_0 + C_1 (\gamma/\gamma_0)^{1/2} (\dot{\gamma}/\dot{\gamma}_0)^{C_2 T} \exp\{-C_3 T\}, \tag{4.8.4c}$$

where τ stands for σ, γ stands for ε, $\tau_0 = \Delta\sigma_G + k\, d_G^{-1/2}$, and the constants C_i, $i = 1,3$, are adjusted so as to account for the normalizing reference values of the strain and strain rate, γ_0 and $\dot{\gamma}_0$.[68] Similarly, (4.8.4b) can be expressed in the following equivalent form:

$$\tau = \tau_0 + C_4 (\dot{\gamma}/\dot{\gamma}_0)^{C_2 T} \exp\{-C_3 T\} + C_5 (\gamma/\gamma_0)^n. \tag{4.8.4d}$$

In comparing this model with (4.8.2d), note that: (1) T here is the absolute temperature; (2) the exponent of the strain-rate term is proportional to the absolute temperature; (3) for the fcc case, the exponent of the strain term is fixed at 1/2; and (4) for the bcc case, the strain effect is included additively. These expressions are motivated by Zerilli and Armstrong on the basis of the kinetics of dislocation dynamics and the thermally activated dislocation motion. The final expression includes the history effects through the accumulated plastic strain. This is similar to the other models mentioned in this and subsequent subsections.

4.8.2. Physically-based Models

For metals which deform plastically essentially through dislocation motion and accumulation, models have been developed based on the notion of thermally activated dislocation kinetics, for moderate strain rates, and the notion of a dislocation-drag mechanism for deformations at greater strain rates.[69]

The motion of dislocations through the crystals of a polycrystalline alloy is a complex phenomenon with various features which *cannot* be described by simple mathematical models similar to the empirical equations discussed in the preceding subsection. While aspects of the kinetics and kinematics of dislocation motion have been understood and modeled,[70] their incorporation into tractable and computationally useful constitutive relations is not yet fully resolved. In the following development, therefore, the concept of the motion of dislocations and the barriers that they must overcome in their motion, is used as an underlying *motivation* to obtain general expressions which include a number of free constitutive parameters. These parameters must then be evaluated by direct comparison with experimental data, as will be illustrated for several metals. For

---

67 Note that $\exp\{a + b\, ln\, x\} = \exp\{a\}\exp\{ln\, x^b\} = x^b \exp\{a\}$.

68 With ε replaced by γ, the constants in (4.8.4c) and (4.8.4a) are related as follows: $C_1 = c_2 \gamma_0^{1/2}$, $C_2 = c_4$, $C_3 = c_3 - c_4\, ln\, \dot{\varepsilon}_0$, $C_4 = c_1$, and $C_5 = c_5 \gamma_0^n$.

69 See, Orowan (1940, 1958), Dorn *et al.* (1949), Conrad (1964), Gilman (1969), Kocks *et al.* (1975), Argon and East (1979), Clifton (1983), Follansbee (1986), Klepaczko and Chiem (1986), Regazzoni *et al.* (1987), Follansbee and Kocks (1988), Klepaczko (1989), Follansbee and Gray (1991), Nemat-Nasser and Isaacs (1997a,b), Kothari and Anand (1998), Nemat-Nasser and Li (1998), Kapoor and Nemat-Nasser (2000), Nemat-Nasser and Guo (2000a,b), Cheng and Nemat-Nasser (2000), Cheng *et al.* (2001) and Nemat-Nasser *et al.* (1999a,b, 2001a,b).

70 See, *e.g.*, Hirth and Lothe (1992) and Hull and Bacon (1992).

certain cases where several mechanisms may be involved in providing resistance to the plastic deformation, this approach may present a useful guidance for identifying the influence of each mechanism on the resulting material response and, hence, a better understanding of the existing experimental data. This would also be helpful in planning new experiments.

Plastic deformation in the range of temperatures and strain rates where diffusion and creep are not dominant, occurs basically by the motion of dislocations. The flow stress, therefore, is essentially defined by the resistance to dislocation motion. The motion of dislocations is opposed by both short-range and long-range obstacles. In addition, the dislocations must overcome any drag forces that may act on them as they move from one set of short-range barriers to the next.

The short-range barriers may include the lattice resistance itself, which is commonly called Peierls' resistance or Peierls' stress, the resistance due to point defects such as vacancies and self-interstitials, other dislocations which intersect the slip plane, alloying elements, and solute atoms (interstitial and substitutional). The dislocations can overcome their short-range barriers partly by their thermal activation, and partly by the action of the net shear stress due to the externally applied forces.

The long-range barriers may include the elastic-stress field due to grain boundaries, farfield forests of dislocations, and other structural defects. The resistance of the long-range barriers is often referred to as the *athermal* component of the flow stress. Being due to the elastic field of the dislocations and defects, its dependence on temperature is through the temperature-dependence of the elastic moduli, especially the shear modulus, and the temperature-history dependence of the microstructure, *e.g.*, dislocation density.

In the simplest setting, the flow stress, τ, is expressed as[71]

$$\tau = \tau^* + \tau_a\,, \tag{4.8.5a}$$

where τ^* and τ_a are the thermal and athermal parts of the resistance to the dislocation motion, respectively. The stress τ^* is a decreasing function of temperature, T, and an increasing function of strain rate, $\dot{\gamma}$. The athermal part, τ_a, increases with increasing accumulated dislocations whose elastic field hinders the motion of mobile dislocations. While this elastic field does not *directly* depend on the temperature, it is *affected* by the temperature in at least two ways: (1) through the elastic moduli and their dependence on the temperature; and (2) through the effect of the *temperature history* on the density of the dislocation forests. At suitably high temperatures, metals anneal, leading to a reduction in

[71] In general, τ depends on the density and distribution of dislocations (often empirically represented by the plastic strain γ), as well as on the strain rate, $\dot{\gamma}$, and temperature, T, *i.e.*, $\tau = \tau(\rho, \rho_m, \dot{\gamma}, T)$, where ρ is the total dislocation density, and ρ_m is the density of the mobile dislocations. To simplify the expressions, this dependency is not always explicitly shown in all expressions. It is, however, implied, unless stated otherwise.

the dislocation density and hence in the corresponding elastic stress field. It may be necessary to account for this effect of the temperature history when the model is applied in high-temperature regimes. The athermal part of the flow stress is, however, independent of the strain rate, although the strain-rate history does affect the temperature history and the total dislocation density. In the following subsections, these and related topics are examined and illustrated.

4.8.3. Thermal Activation and Flow Stress

Consider the temperature- and strain rate-dependent part, τ^*, of the flow stress, τ. Let the change in the Helmholtz free energy, measured per barrier, due to an isothermal crossing of a set of barriers by a dislocation line, be denoted by ΔF. As is sketched in Figure 4.8.1(a), the corresponding mechanical work, ΔW, at *constant stress*, τ^*, is given by

$$\Delta W = \tau^* V^*, \qquad V^* = b\lambda l, \tag{4.8.5b,c}$$

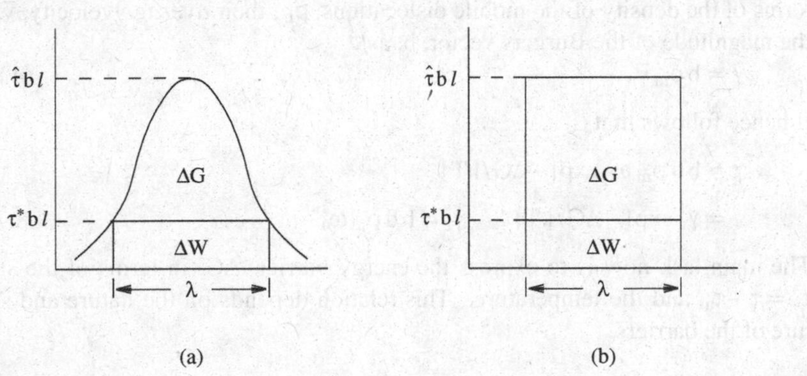

Figure 4.8.1

Profile of the resistance force to dislocation motion: (a) a general profile; and (b) simplified constant profile

where b is the magnitude of the Burgers vector, l is some length parameter that characterizes the structure of the barriers, λ is the width of the short-range barrier at the stress level τ^*, and $V^* = b\lambda l$ is the corresponding *activation volume*. The length parameter l may be the average spacing of the barriers along the dislocation line (mostly for fcc and hcp metals), or it may represent the critical length of the double kink necessary for screw dislocations to overcome the lattice resistance (Peierls' resistance, mostly for bcc metals). The energy difference between the mechanical work, $\tau^* V^*$, and the change in the free energy, ΔF,

$$\Delta G = \Delta F - \tau^* V^*, \tag{4.8.5d}$$

must be overcome by the dislocation through its thermal activation. The frequency of the occurrence of such an event is given by[72]

$$\omega = \omega_0 \exp\{-\Delta G/kT\}, \tag{4.8.6a}$$

where k is the Boltzmann constant, T is the absolute temperature, and ω_0 is the attempt frequency which, among other factors, depends on the structure and composition of the crystal, as well as on the core structure of the dislocation. Once the barriers are crossed, the dislocation line moves to the next set of barriers under the action of the net force $(\tau - \tau_a)\,b$ (per unit dislocation length), balancing the *drag resistance*,[73] τ_D. For dislocations of slow (running) velocities, the effect of τ_D may be insignificant; see, however, Subsection 4.8.8. The waiting time for the crossing of a barrier is usually (assumed to be) much longer than the time required to travel between barriers. Hence, the average dislocation velocity may be estimated by

$$\bar{v} = d\,\omega_0 \exp\{-\Delta G/kT\}, \tag{4.8.6b}$$

where d is the average distance the dislocations move between the barriers.

It is shown in Subsection 6.1.10 that the slip rate, $\dot{\gamma}$, may be written in terms of the density of the mobile dislocations, ρ_m, their *average* velocity, $\bar{v}$, and the magnitude of the Burgers vector, b, as[74]

$$\dot{\gamma} = b\,\rho_m\,\bar{v}. \tag{4.8.7a}$$

It hence follows that

$$\begin{aligned}\dot{\gamma} &= b\,d\,\rho_m\,\omega_0 \exp\{-\Delta G/kT\}\\ &= \dot{\gamma}_r \exp\{-\Delta G/kT\}, \quad \dot{\gamma}_r = b\,d\,\rho_m\,\omega_0.\end{aligned} \tag{4.8.7b-d}$$

The main task now is to express the energy barrier, ΔG, in terms of the stress, $\tau^* = \tau - \tau_a$, and the temperature. This relation depends on the nature and structure of the barriers.

---

72 To obtain this expression, assume there are ω successful events (measured per second) in ω_0 possible attempts. Let the energy associated with a successful attempt be ΔG. Then $\mathcal{F} = \omega\,\Delta G - T\,\mathcal{S}$ is the corresponding free energy, where $\mathcal{S} = k\,ln\,g(\omega, \omega_0)$ is the *entropy*, with $g(\omega, \omega_0)$ being the probability of ω successes out of ω_0 attempts. For statistically independent events, $g(\omega, \omega_0) = \omega_0!/\{(\omega_0 - \omega)!\,\omega!\}$. When $\omega_0 >> \omega >> 1$, minimization of $\mathcal{F}$ with respect to ω leads to (4.8.6a); see, *e.g.*, Kittel (1969).

73 This is an extreme simplification of a complex process involving dislocation bowing, kinking, jogging, and cross slip, to mention only a few important well-studied phenomena; see Hirth and Lothe (1992) and Hull and Bacon (1992). This simple interpretation is, however, adequate for the present purposes. Note that the *average* dislocation velocity and the *average* running velocity of a dislocation between barriers are not the same. A better estimate of the average dislocation velocity is given in Subsection 4.8.8.

74 Transition from crystal slip to the deformation of a polycrystal introduces a parameter, M, known as the Taylor factor, so that, *e.g.*, $\dot{\gamma} = b\,\rho_m\,\bar{v}/M$, with $M \approx 3$; see Taylor (1934, 1938), Bishop and Hill (1951a,b), Cottrell (1963, page 116), and Section 6.5. In (4.8.7a) and in the expression for the flow stress, this factor is absorbed in the involved parameters.

4.8.4. A Simple Model

The simplest model is to assume barriers with constant energy profile at a *threshold stress,* $\hat{\tau}$, over the barrier length λ; see Figure 4.8.1(b). Here, $\hat{\tau}$ is the mechanical stress required to cross the barrier without any thermal activation, *i.e.,* at 0K temperature. If the stress required for the dislocation to jump this barrier at temperature T is denoted by $\tau^*(\dot{\gamma}, T)$, then, it follows that[75]

$$\Delta G = G_0\left\{1 - \frac{\tau^*(\dot{\gamma}, T)}{\hat{\tau}}\right\}, \qquad G_0 = \hat{\tau}\, V^*, \qquad \hat{\tau} = \tau^*(\dot{\gamma}, 0), \tag{4.8.8a-c}$$

where the activation volume, V^*, is defined by (4.8.5c). From (4.8.8a) and (4.8.7c) obtain

$$\tau^*(\dot{\gamma}, T) = \hat{\tau}\left\{1 + \frac{kT}{G_0}\, ln\frac{\dot{\gamma}}{\dot{\gamma}_r}\right\}. \tag{4.8.9a}$$

As sketched in Figure 4.8.2, for temperatures less than a critical value, T_c, the resistance to the thermally activated motion of the dislocations decreases linearly (for this simple model) with the increasing temperature, T.

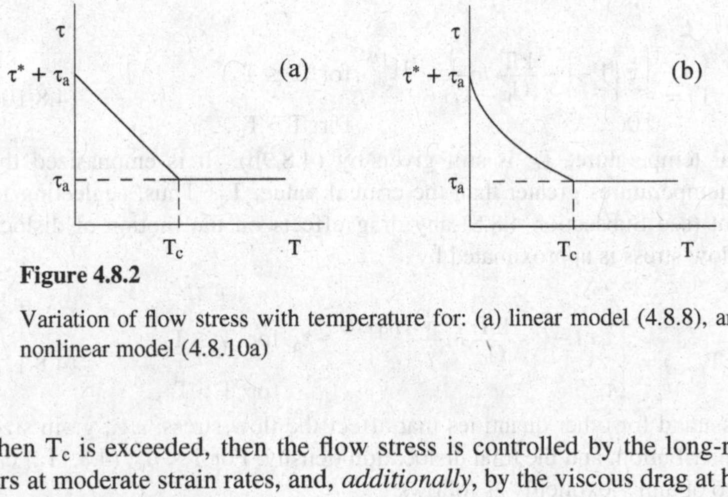

Figure 4.8.2

Variation of flow stress with temperature for: (a) linear model (4.8.8), and (b) nonlinear model (4.8.10a)

When T_c is exceeded, then the flow stress is controlled by the long-range barriers at moderate strain rates, and, *additionally*, by the viscous drag at high strain rates. The critical temperature, T_c, depends on the strain rate. For the present simple illustrative model, this critical temperature, T_c, is given by

$$T_c = -\frac{G_0}{k}\left\{ln\frac{\dot{\gamma}}{\dot{\gamma}_r}\right\}^{-1}. \tag{4.8.9b}$$

Note that, expression (4.8.9a) holds only for temperatures less than the critical value, T_c. For $T > T_c$, the thermally activated part of the stress, τ^*, is zero.

---

[75] Note that τ^* also depends on the dislocation density; see Subsection 4.8.6.

In general, the barrier profile depends on the structure of the crystal and on the nature of the barrier itself. For example, the Peierls barrier is not expected to be the same as a dislocation barrier, or as a solute atom barrier. A study of this and other related issues is found in Kocks *et al.* (1975); see also Ono (1968).

4.8.5. More General Models

The constant barrier profile discussed above illustrates the concept of the thermally activated part of the flow stress. In general, however, it is often necessary to consider more general relations between the thermally activated flow stress, $\tau^*(\dot{\gamma}, T)$, and the activation energy, ΔG. The following form has been considered in the literature; see Kocks *et al.* (1975):

$$\Delta G = G_0 \left\{1 - \left[\frac{\tau^*(\dot{\gamma}, T)}{\hat{\tau}}\right]^p\right\}^q, \tag{4.8.10a}$$

where, as suggested by Ono (1968), a number of barrier profiles of interest can be represented by assuming $0 < p \leq 1$ and $1 \leq q \leq 2$. This model gives (Figure 4.8.2(b))

$$\tau^*(\dot{\gamma}, T) = \begin{cases} \hat{\tau}\left\{1 - \left[-\frac{kT}{G_0} \ln\frac{\dot{\gamma}}{\dot{\gamma}_r}\right]^{1/q}\right\}^{1/p} & \text{for } T \leq T_c \\ 0 & \text{for } T > T_c\,. \end{cases} \tag{4.8.10b}$$

The critical temperature, T_c, is still given by (4.8.9b). It is emphasized that $\tau^* \equiv 0$ for temperatures greater than the critical value, T_c. Thus, neglecting for the moment (see Subsection 4.8.8) any drag effects on the motion of dislocations, the flow stress is approximated by

$$\tau(\dot{\gamma}, T, \ldots) = \begin{cases} \hat{\tau}\left\{1 - \left[-\frac{kT}{G_0} \ln\frac{\dot{\gamma}}{\dot{\gamma}_r}\right]^{1/q}\right\}^{1/p} + \tau_a & \text{for } T \leq T_c \\ \tau_a & \text{for } T > T_c\,, \end{cases} \tag{4.8.11a}$$

where dots stand for other quantities that affect the flow stress, *e.g.*, grain sizes and their distribution, and the total dislocation density. For $T < T_c$, (4.8.11a) can be solved to obtain $\dot{\gamma}$ explicitly as follows:

$$\dot{\gamma} = \dot{\gamma}_r \exp\left\{-\frac{G_0}{kT}\left[1 - \left(\frac{\tau - \tau_a}{\hat{\tau}}\right)^p\right]^q\right\}. \tag{4.8.11b}$$

4.8.6. Dislocations as Short-range Barriers

From (4.8.5b,c) and (4.8.8b), $\hat{\tau}$ is given by

$$\hat{\tau} = \frac{G_0}{V^*} = \frac{G_0}{b\lambda l}\,. \tag{4.8.12a,b}$$

Here b is a constant and λ is dependent on the profile of the barrier. The distance, l, on the other hand, depends on the dislocation density, ρ, when the dislocations which intersect the slip planes are actually the barriers to the motion of mobile dislocations which are lying on the slip planes. For the lattice or Peierls barriers, λ is of the order of the lattice dimensions, and l is a critical double-kink length which depends on the lattice structure and the structure of the dislocation core. Many bcc crystals are assumed to possess this feature, although in actuality the thermally activated component of the resistance to the dislocation motion will also include the effects of other than the Peierls barriers; *e.g.*, other dislocations, substitutional atoms, vacancies, interstitials, and alloying elements. For fcc metals, on the other hand, the energy required to cross the Peierls barrier is often rather small (less than 0.2eV), so that only at very low temperatures does the Peierls mechanism provide significant resistance to the dislocation motion. In this case, the dislocation forests which intersect the slip planes may be the essential barriers to the motion of the dislocations lying on the slip planes.

Consider a case where the dislocation forests intersecting the slip planes are the essential barriers to the slip. Then the average spacing of the barriers along the dislocation line, l, and the average distance the dislocation travels before encountering the next barrier, d, are basically of the same order of magnitude, relating to the current density of the dislocations, ρ, by

$$l \approx \mathrm{d} \approx \rho^{-1/2}. \tag{4.8.12c}$$

In general, the current density of dislocations, ρ, does not directly relate to the accumulated plastic deformation, γ, which is the result of the motion of *all mobile dislocations throughout the entire history of the plastic deformation.* Indeed, integrating (4.8.7a) over the time-history of the plastic deformation, it follows that

$$\gamma = \mathrm{b} \int_0^t \rho_m(t')\, \bar{v}(t')\, dt' = \mathrm{b}\, \bar{\rho}\, \bar{x}, \tag{4.8.12d}$$

where $\bar{\rho}$ is an *intermediate value* of the mobile dislocation density at an instant between 0 and the current time t, and $\bar{x}$ is the travel distance corresponding to the average dislocation velocity $\bar{v}(t')$, over the time interval $0 \le t' \le t$.[76] During the initial stages of the plastic deformation, ρ and $\bar{\rho}$ may be closely related. However, as the plastic deformation proceeds, new dislocations are generated and some existing ones are annihilated. Many dislocations may pile up against grain boundaries and other barriers, or they may react with each other, forming locked dislocations which remain immobile. Hence, in actuality, both the deformation and temperature *histories* must be considered in estimating the average current dislocation density, ρ, and hence the average dislocation spacing, l.

In certain cases, a simple empirical model may be considered, where the average dislocation spacing is assumed to be a decreasing function of the

[76] This follows from the mean-value theorem.

accumulated plastic strain and an increasing function of temperature, *i.e.*,

$$l = d = \frac{l_0}{f(\gamma, T)}, \quad f(\gamma, T) > 0, \quad f(0, T_0) = 1,$$
$$\frac{\partial f(\gamma, T)}{\partial \gamma} \geq 0, \quad \frac{\partial f(\gamma, T)}{\partial T} \leq 0, \qquad (4.8.12e\text{-}j)$$

where f is some nondecreasing dimensionless function of the plastic strain, γ, and a nonincreasing function of temperature, T; T_0 is the initial temperature; and l_0 is the *initial average dislocation spacing.* As illustrations, consider

$$l = \frac{l_0}{1 + a(T)\,\gamma^{n_0}}, \quad \text{or} \quad l = \frac{l_0}{[1 + a(T)\,\gamma]^{n_0}},$$
$$a(T),\ n_0 \geq 0, \quad \frac{\partial a(T)}{\partial T} \leq 0. \qquad (4.8.12k\text{-}n)$$

The first example (with constant a) leads to a strain-hardening model similar to the Johnson and Cook model (1983), and the second is similar to the power-law model used by Nemat-Nasser *et al.* (1994). In these expressions, a(T) and n_0 are viewed as adjustable constitutive parameters, to be established empirically. Since workhardening is associated with the increasing dislocation density and, hence, with the decreasing average dislocation spacing, the function a(T) is assumed to be non-negative. Furthermore, since the density of dislocations is expected to decrease with increasing temperature, a(T) is expected to be a decreasing function of the temperature. For application to OFHC Cu, for example, it turns out that a(T) may be chosen to have the following simple form (Nemat-Nasser and Li, 1998):

$$a(T) = a_0 \{1 - (T/T_m)^2\}, \qquad (4.8.12o)$$

where T_m is the melting temperature ($\approx$ 1,350K, for copper), and a_0 depends on the initial average dislocation spacing. For the annealed samples, a_0 is expected to be an order of magnitude greater than that for the as-received samples. In general, a_0 is an adjustable parameter to be fixed empirically. In the OFHC Cu case, $a_0 = 20$ for the annealed samples and $a_0 = 1.8$ for the as-received samples, fit a large body of high-strain-rate experimental data, over a broad range of temperatures, strain rates, and strains; see Subsection 4.8.10. For this application, l is approximated by the first alternative in (4.8.12k-n), with a(T) given by (4.8.12o). Expression (4.8.12c) suggests that n_0 may be taken equal to 1/2. While, in general, n_0 is treated as a free constitutive parameter, for application to OFHC copper, $n_0 = 1/2$ seems to fit the experimental data well, as is shown in Subsection 4.8.10. For a more general setting, however, choose

$$l = l_0/(1 + a(T)\,\gamma^{n_0}), \quad a(T) = a_0\{1 - (T/T_m)^2\}, \qquad (4.8.12p,q)$$

and fix the value of n_0 empirically; see Table 4.8.1, page 235, for experimentally obtained values.

It is important to bear in mind that (4.8.12p,q) do not have general validity, and that the average spacing of the dislocations must be calculated based on the thermomechanical history of the deformation, including the strain-rate history that affects the density of the mobile dislocations, especially at high strain rates. Hence, for a general case (*e.g.*, for fcc, bcc, and hcp metals), consider

$$\hat{\tau} = \frac{l_0}{l}\hat{\tau}_0, \quad \hat{\tau}_0 = \frac{G_0}{b\lambda l_0}, \tag{4.8.13a,b}$$

where $\hat{\tau}_0$ is now material-dependent and can be estimated directly from experimental results; see Tables 4.8.1, page 235, and 4.8.2, page 238.

From (4.8.7d), it is seen that $\dot{\gamma}_r$, for dislocation barriers, is also dependent on the dislocation density, ρ, being proportional to the average dislocation spacing, d. Here, set $d = l_s$ and consider the following model:

$$\dot{\gamma}_r = \frac{l_s}{l_0}\dot{\gamma}_0, \quad \dot{\gamma}_0 = b\,\rho_m\,\omega_0\,l_0. \tag{4.8.14a,b}$$

Note that the average density of mobile dislocations also introduces a length scale in plasticity, *i.e.*, $l_m = \rho_m^{-1/2}$, which may be included in the model directly, *e.g.*, by redefining l_s and $\dot{\gamma}_0$ as follows:

$$l_s = \frac{\rho_m d}{\rho_m^0}, \quad \dot{\gamma}_0 = b\,\rho_m^0\,\omega_0\,l_0, \tag{4.8.14c,d}$$

where ρ_m^0 is a reference value of the mobile dislocation density. Hence, for dislocation barriers, or in general,

$$ln\frac{\dot{\gamma}}{\dot{\gamma}_r} = ln\frac{\dot{\gamma}}{\dot{\gamma}_0} + ln\frac{l_0}{l_s}. \tag{4.8.14e}$$

Combining (4.8.13) and (4.8.14), and in view of (4.8.10a), the thermal part of the flow stress, τ^*, when dislocations are the barriers to the motion of other dislocations, becomes

$$\tau^*(\dot{\gamma}, T) = \begin{cases} \tau^0\left\{1 - \left[-\frac{kT}{G_0}\left[ln\frac{\dot{\gamma}}{\dot{\gamma}_0} + ln\frac{l_0}{l_s}\right]\right]^{1/q}\right\}^{1/p}\frac{l_0}{l} & \text{for } T \le T_c \\ 0 & \text{for } T > T_c. \end{cases} \tag{4.8.15a}$$

Note that, in this case, the critical temperature, T_c, at which the dislocation barriers cease to be significant, is dependent on the dislocation density, $\rho = l^{-2}$. This temperature is given by

$$T_c = -\frac{G_0}{k}\left\{ln\frac{\dot{\gamma}}{\dot{\gamma}_0} + ln\frac{l_0}{l}\right\}^{-1}. \tag{4.8.15b}$$

When both the Peierls barriers and the existing dislocations with density ρ, are resisting slip due to the motion of the mobile dislocations, then τ^* in (4.8.10b) represents both mechanisms. Since the spacing of the Peierls barriers is of the order of the lattice dimensions, while the average dislocation spacing is orders of magnitude greater, it may be assumed that the two mechanisms are essentially uncoupled, writing

$$\tau^*(\dot\gamma, T, ...) = \tau_p^*(\dot\gamma, T) + \tau_d^*(\dot\gamma, T, ...) = \hat\tau_p^0\left\{1 - \left[-\frac{kT}{G_0^p}\, ln\frac{\dot\gamma}{\dot\gamma_0^p}\right]^{1/q}\right\}^{1/p}$$

$$+ \hat\tau_d^0\left\{1 - \left[-\frac{kT}{G_0^d}\left[ln\frac{\dot\gamma}{\dot\gamma_0^d} + ln\frac{l_0}{l_s}\right]\right]^{1/q}\right\}^{1/p}\frac{l_0}{l}, \tag{4.8.16}$$

where the corresponding quantities are denoted by subscript or superscript p for Peierls, and d for dislocation, respectively. Note that $\tau_p^* \equiv 0$ for $T \geq T_c^p$ and $\tau_d^* \equiv 0$ for $T \geq T_c^d$. These restrictions must be imposed when (4.8.16) is used to fit experimental data. Also, although the lattice resistance and the dislocations as barriers involve different mechanisms, usually the same $p = 2/3$ and $q = 2$ may be used here, as commented on before; see, however, Tables 4.8.1, page 235, and 4.8.2, page 238.

4.8.7. Effect of Long-range Barriers

The athermal part, τ_a, of the flow stress, τ, represents the resistance to the motion of a dislocation provided by the potential fields of all other dislocations, grain boundaries, and defects. A dislocation must overcome this resistance through the applied resolved shear stress. This was recognized by Taylor (1934) who noted that workhardening can result from the interaction between nonintersecting dislocations. Consider, for illustration, two parallel edge dislocations on two different parallel slip planes, a distance l apart. From Subsections 6.1.5 and 6.1.9, the interaction force has the form

$$F_{in} \approx \mu\, b^{(1)} b^{(2)}/l\,, \tag{4.8.17}$$

where μ is the shear modulus, and $b^{(1)}$ and $b^{(2)}$ are the magnitudes of the Burgers vectors. This suggests that τ_a should be proportional to the square root of the dislocation density, as well as being proportional to the shear modulus. This is the kind of model commonly considered in rate-independent theories of the plastic deformation of metals.

The athermal stress, τ_a, however, represents the long-range influence of all the existing dislocations, as well as the effects of the grain boundaries and other inhomogeneities and defects that help to produce an elastic stress field. Thus, in addition to its dependence on the dislocation density ρ, as inferred from expression (4.8.17) through its dependence on l, τ_a is also a function of the average grain size, d_G, and other parameters which represent the overall microstructural inhomogeneities that affect the elastic stress field. Based on this observation, it may thus be reasonable to write

$$\tau_a = \tau_a^0\,\hat g(\rho, d_G, ...)\,, \quad \tau_a^0 = A_0\mu(T)/\mu_0\,, \tag{4.8.18a,b}$$

where τ_a^0 and A^0 have the dimensions of stress, g is a dimensionless function of the indicated arguments, and $\mu(T)$ is the shear modulus at temperature T, with μ_0 being its reference value. If the range of temperatures of interest is not very wide, then $\mu(T)/\mu_0$ may be replaced by an average (constant) value. The dependence on the dislocation density, ρ, is often empirically represented by the dependence on the plastic strain, γ. While this may be a useful approximation, it is fundamentally questionable, since plastic strain in *not* a microstructural state variable, while the dislocation density is.[77] On the other hand, the effective plastic strain is always monotonically increasing and may be used as a load parameter. Thus, (4.8.18a) may be expressed as

$$\tau_a = \tau_a^0\,\hat{g}(\rho(\gamma),\, d_G,\, \ldots) = \tau_a^0\, g(\gamma,\, d_G,\, \ldots)\,. \qquad (4.8.18c)$$

As an example of such representation, consider

$$\tau_a = (a_1 + \gamma^{n_1} + k_0\, d_G^{-1/2})\,\tau_a^0\,, \qquad (4.8.18d)$$

where a_1, n_1, and k_0 are constants. The first two terms in the parentheses represent the influence of the farfield dislocations, and the second term is due to the so-called Hall-Petch effect. Note that (4.8.18d) is similar to some of the terms in the Zerilli-Armstrong (1987) model; see (4.8.4d).

4.8.8. Drag-controlled Plastic Flow

A simple model is often used to explain the flow stress of metals at high strain rates, where the viscous drag is the dominant resisting force to the motion of dislocation.[78] Here, again, the model is used to motivate the derivation of a general constitutive expression which is then viewed in a phenomenological context.

As a first step, it is assumed that the average time required for dislocations to move from one set of barriers to the next set, consists of two parts, a waiting period, t_w, and a running period, t_r. Hence, the *average* dislocation velocity is estimated by

$$\bar{v} = \frac{d+\lambda}{t_w + t_r}\,, \qquad (4.8.19a)$$

where d is the average travel distance between barriers and λ is the average barrier width, usually assumed to be much smaller than d, *i.e.*, $\lambda \ll d$; this assumption is not used in what follows. As a second step, it is assumed that the force resisting the dislocation motion, $\tau_D^* b$, measured per unit dislocation length, is linearly related to the *average running*[79] *velocity* of the dislocation, $\bar{v}_r$,

$$\tau_D^* b = D\,\bar{v}_r\,, \quad \bar{v}_r = \frac{d}{t_r}\,, \qquad (4.8.19b,c)$$

---

[77] At a given plastic strain, a metal sample can be subjected to a temperature cycle such that the plastic strain (the sample dimensions) remains unchanged, while the dislocation density is changed substantially. This profoundly affects the subsequent material response; see Figure 4.8.11.

[78] See, *e.g.*, Kocks *et al.* (1975), Clifton (1983), Follansbee *et al.* (1984), Regazzoni *et al.* (1987), and references cited therein.

[79] In crystals with strong Peierls resistance, this drag may be associated with the lattice friction. Then the drag coefficient D in (4.8.19b) may be a function of the average running velocity of the

where D is the *drag* coefficient, and, in general, $\bar{v} \neq \bar{v}_r$. Between the barriers, the frictional stress, $\tau_D^* = D\,\bar{v}_r/b$, must be overcome by the net driving stress, $\tau_D^* = \tau - \tau_a$. In a more general setting, this driving net force may exceed the drag force, in which case the dislocation accelerates to the next barrier, and may even "overshoot" that barrier. A systematic analysis of this phenomenon requires consideration of the "mass" of a dislocation and other subtle concepts which are not considered in the present work; see Kocks *et al.* (1975).

Now, if the waiting period, t_w, is estimated by (4.8.6a),

$$t_w = \omega_0^{-1} \exp\{\Delta G/kT\}, \tag{4.8.19d}$$

and the running time, t_r, is obtained from (4.8.19c), then, in view of (4.8.7a) and (4.8.19a,b), it follows that

$$\frac{\dot{\gamma}}{\dot{\gamma}_0} = \frac{l_s'}{l_0}\left\{\exp\{\Delta G/kT\} + \frac{\tau_D^0}{\tau - \tau_a}\,\frac{d}{l_0}\right\}^{-1},$$

$$l_s' = \frac{\rho_m\,(d + \lambda)}{\rho_m^0}, \qquad \tau_D^0 = \omega_0\, D\, l_0/b, \tag{4.8.20a-c}$$

where $\dot{\gamma}_0$ is defined by (4.8.14d), *i.e.*, $\dot{\gamma}_0 = b\,\rho_m^0\,\omega_0\, l_0$, ΔG is defined in (4.8.10a), and τ_D^0 is a reference drag stress. When $\tau > \hat{\tau} = \tau^*(\dot{\gamma}, T = 0, ...)$, then the resistance to the deformation is solely due to the athermal, τ_a, and frictional, τ_D^*, parts of the process, so that $t_w \approx 0$. In this case, only the viscous drag and the stress field due to farfield inhomogeneities resist the motion of dislocations. Then, each dislocation vibrational "attempt" is successful. Thus, $t_w = \omega_0^{-1} << t_r$ and (4.8.20a) may be replaced by

$$\frac{\dot{\gamma}}{\dot{\gamma}_0} = B\,\frac{\tau - \tau_a}{\tau_D^0}, \qquad B = \frac{\rho_m}{\rho_m^0}, \tag{4.8.20d,e}$$

where τ_D^0 is defined by (4.8.20c).

At very high deformation rates, common in shocks, the density of the mobile dislocations increases with increasing deformation rate, since the maximum velocity that dislocations can attain is essentially limited by the elastic shear-wave velocity. Thus, the parameter B is expected to increase with increasing deformation rate. On the other hand, for moderate strain rates, B may be assumed to be a constant, depending on the material. Experimental results for strain rates less than 10^4/s suggest that the drag resistance varies exponentially with the strain rate, attaining a constant value for strain rates greater than a few thousand per second.[80] Based on this observation, the following alternative representation of the flow stress may be considered:

---

dislocations, $\bar{v}_r$. When the drag is due to the interaction of the moving dislocation with phonons and, at low temperatures, with electrons, D may be assumed to be a constant; see Kocks *et al.* (1975).

80 See Kapoor and Nemat-Nasser (2000), Nemat-Nasser *et al.* (2001a), and the following two subsections.

$$\tau = \tau_a + \tau^* + \tau_d, \qquad \tau_d = m_0[1 + \exp(-\alpha\dot{\gamma})], \tag{4.8.21a,b}$$

where m_0 is a constant, and, in the range of the considered strain rates, $\alpha = O(10^{-4})$s. In this case, expansion of the exponential term in Taylor series gives,

$$\tau_d \approx m_0 \alpha \dot{\gamma}, \tag{4.8.21c}$$

so that, in view of (4.8.20d,e), it follows that[81]

$$\alpha \approx \frac{D}{m_0 b^2 \rho_m}. \tag{4.8.21d}$$

4.8.9. Application: Flow Stress of Commercially Pure Tantalum

Consider as illustration, applying the physically-based model (4.8.11) with τ_a defined by (4.8.18b) to tantalum (Ta) which is a bcc metal. To simplify the analysis, neglect the influence of the grain size, and use, for the athermal part of the flow stress,

$$\tau_a = c_0 + c_1 \gamma^{n_1}, \tag{4.8.18e}$$

where c_0 and c_1 are to be fixed empirically.

For bcc metals, the lattice itself provides the short-range barrier (the Peierls barrier) to the motion of dislocations. Generally, a double kink is formed with the assistance of thermal vibrations, and the kinks then move sideways, leading to the advancement of the dislocation line. In this process, the kinks may be pinned down by other defects or alloying elements, thereby increasing the resistance to the plastic deformation. For tantalum-tungsten (Ta-W) alloys, the (substitutional) tungsten atoms may pin down the kinked dislocations in their lateral motion. Hence, the flow stress for Ta-W is generally higher than that for commercially pure tantalum (Nemat-Nasser and Kapoor, 2001). In what follows, the thermally activated part of the flow stress is assumed to correspond solely to the Peierls resistance to the dislocation motion. The results, therefore, are applicable to, for example, commercially pure tantalum, molybdenum, niobium, and vanadium; see Table 4.8.1, page 235, for the values of their constitutive parameters.

Modified Hopkinson experimental techniques have been used to measure the flow stress of ductile materials over a broad range of strains, strain rates, and temperatures, in uniaxial stress.[82] Figure 4.8.3 gives the adiabatic flow stress of commercially pure tantalum tested in compression at a 5,000/s strain rate and at

---

[81] If m_0 is identified with the material's yield stress at a suitably high temperature, say, about 1,000K, and the Taylor factor M is also introduced, then equation (4.8.21d) corresponds to equation $(3.6)_2$ of Nemat-Nasser *et al.* (2001a).

[82] Nemat-Nasser *et al.* (1991), Nemat-Nasser and Isaacs (1997a,b), Nemat-Nasser (2000b), and *ASM Volume 8* (2000).

the indicated initial temperatures. The isothermal curves at temperatures of 25, 125, 225, 325°C (*i.e.*, 298, 398, 498, 598K) are shown in Figure 4.8.4 by the dashed curves.

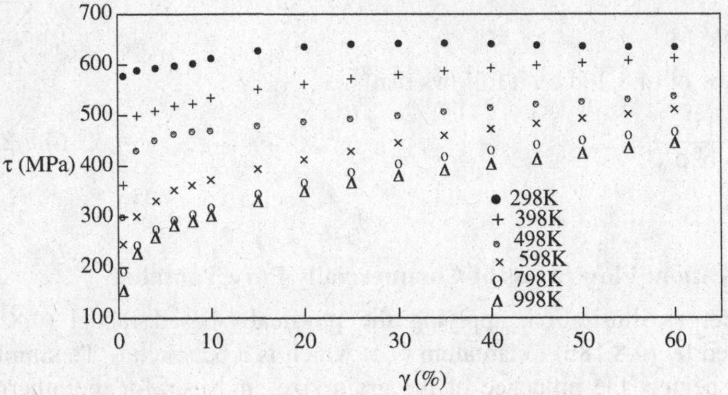

Figure 4.8.3

Adiabatic flow stress of commercially pure Ta for indicated initial temperatures, at a 5,000/s strain-rate

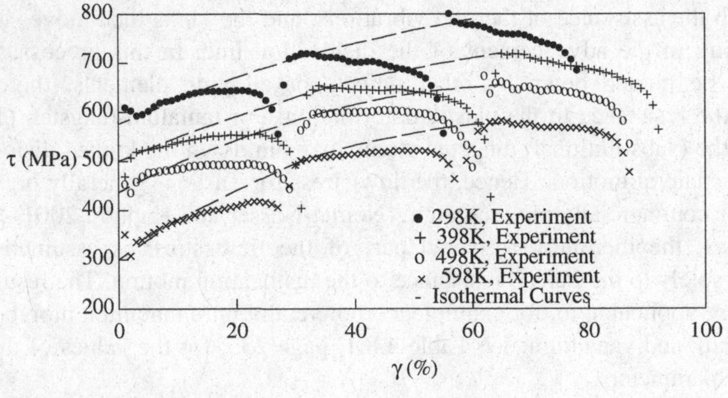

Figure 4.8.4

Isothermal (dashed curves) flow stress of commercially pure Ta for indicated temperatures, at a 5,000/s strain-rate

The results presented in Figure 4.8.4 are obtained by an incremental loading using a modified version of the Hopkinson bar technique, in which a sample can be subjected to a single stress pulse of predetermined profile and then recovered without it having been subjected to any additional loading; see Chapter 9. This is called the *recovery Hopkinson technique*.

To construct an isothermal flow stress at a high strain rate and a given initial temperature, the sample is heated to the required temperature in the furnace attached to the recovery Hopkinson bar, and then loaded incrementally. After the application of each load increment, the sample is unloaded without being subjected to any additional stress pulses. The sample is then allowed to return to the furnace temperature before the application of the next strain increment.

Since the unloading, the cooling of the sample to its initial temperature, and the reloading, may affect the microstructure and hence the thermomechanical properties of the material, it is necessary to check this in each case. To this end, a sample which has been loaded, unloaded, and cooled to its initial temperature, may then be reheated to the temperature that it had just prior to its unloading. If there are no substantial changes in the microstructure that affect the flow-stress properties, then the flow stress, upon reloading, should follow the previous stress-strain curve. This fact has been verified to be the case for all data presented in this book. This verification also allows checking whether or not essentially all the involved plastic work has been converted to heat, at high strain rates and for suitably large strains.

Data Analysis: The measured flow stresses shown in Figures 4.8.3 and 4.8.4, are replotted in Figure 4.8.5, as stress versus absolute temperature, for indicated plastic strains. This figure also includes estimates obtained using the adiabatic flow stress curves in Figure 4.8.3, for high temperatures, *e.g.*, 998 and 798K. At such high temperatures, the flow stress is rather low. Hence, the change in the temperature due to the plastic work is small relative to the total temperature and can be estimated directly from the corresponding stress-strain curves without introducing significant errors.[83] These adjusted points are also included in Figure 4.8.5.

Effect of Long-range Barriers: Examination of the trend in Figure 4.8.5 suggests that the flow stress becomes essentially independent of the temperature, close to 1,000K. These high-temperature limiting values of the flow stress are plotted in Figure 4.8.6, as stress versus the corresponding plastic strain. The points in Figure 4.8.6 nicely fit a simple power law,

$$\tau_a = 473\,\gamma^{1/5}\text{MPa}\,. \tag{4.8.22a}$$

This is taken as an empirical[84] representation of the athermal part, τ_a, of the flow stress for this material, writing the total flow stress as the sum of the athermal and a thermally-activated part, τ^*, as in (4.8.5a).

---

[83] The temperature change is computed using $\Delta T = \hat{\rho}\, C_v \int_0^{\gamma} \tau(\gamma')\, d\gamma'$, where $\hat{\rho}\, C_v = 0.433\text{K/MPa}$ for Ta, $\hat{\rho}$ being the mass density. This assumes that all the plastic work is converted to heat and is used to increase the temperature of the sample. As discussed in Chapter 9, this is reasonable for the large plastic strains involved in the present case (Kapoor and Nemat-Nasser, 1998). It is also assumed that the heat capacity C_v is constant within the considered range of temperatures. It is, however, more appropriate to include the effect of temperature on C_v and on the elastic moduli.

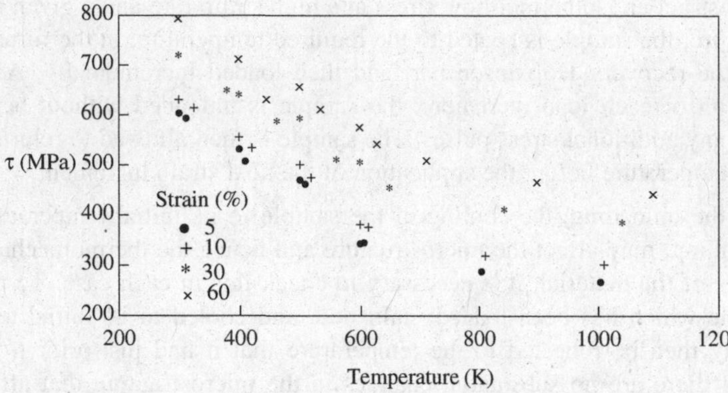

Figure 4.8.5

Flow stress of commercially pure Ta for indicated plastic strains, at a 5,000/s strain-rate

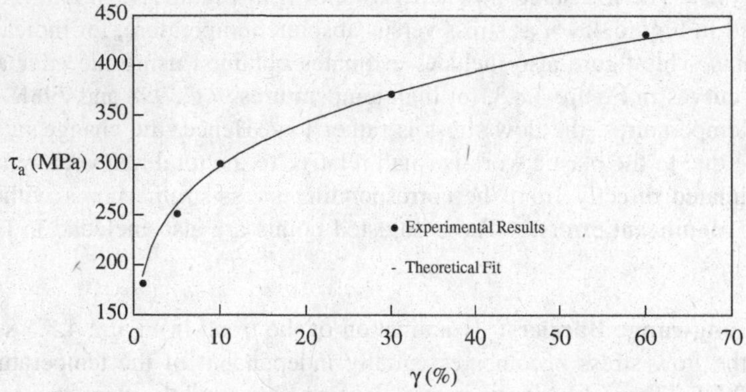

Figure 4.8.6

Limiting high-temperature flow stress (athermal stress) for commercially pure Ta at a 5,000/s strain-rate

Based on (4.8.5a) and (4.8.22a), the results of Figure 4.8.5 are now replotted in Figure 4.8.7, as τ^* versus the temperature, T. As is seen, all the experimental points collapse into basically a single curve which may be represented by the following simple expression:

$$\tau^* = 1{,}100\,\{1 - (0.001T)^{1/2}\}^{3/2}\,. \tag{4.8.22b}$$

---

84 Note that τ_a is a function of dislocation density and *not* the plastic strain. The empirical relation (4.8.22a), as well as other empirical equations, must be used with caution.

This equation is shown by a solid curve in Figure 4.8.7. The choice of the exponents in this equation is based on the nonlinear model of Subsection 4.8.5. The coefficient 0.001, of the temperature, T, is selected by noting that the influence of thermal activation disappears once the temperature reaches a critical value of about $T_c = 1{,}000$K. The exponents 1/2 and 3/2 are obtained empirically here, but they are in accord with Ono's results (Ono, 1968).

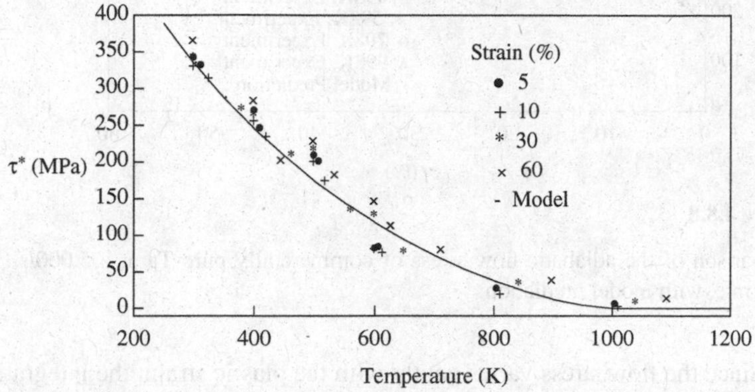

Figure 4.8.7

Thermally-activated stress for commercially pure Ta at a 5,000/s strain rate

Comparison with Data used for Modeling: Figure 4.8.8 compares the flow stress given by

$$\tau = 473\,\gamma^{1/5} + 1{,}100\,\{1 - (0.001\mathrm{T})^{1/2}\}^{3/2} \tag{4.8.22c}$$

with the experimental data. The temperature, T, is calculated by assuming that all the plastic work is converted into heat. As pointed out before, our experiments show this to be essentially the case for large plastic strains. Initially, for small strains, *e.g.*, 5 to 10%, and especially when the sample has been annealed, a portion of the plastic work is stored in the material in the form of the elastic energy of the dislocations. This is, however, quickly saturated, and, as is shown on the basis of experimental results in Chapter 9 and theoretical estimates,[85] it does not seem to have any significance in the present case. Hence, the temperature is calculated incrementally from

$$\hat{\rho}\, C_v \frac{d\mathrm{T}}{d\gamma} = \tau(\gamma, \dot{\gamma}, \mathrm{T})\,, \tag{4.8.22d}$$

with $\mathrm{T} = \mathrm{T}_0$ at $\gamma = 0$, where $\tau = \tau(\gamma, \dot{\gamma}, \mathrm{T})$ is defined by (4.8.22c), T_0 is the initial temperature, $\hat{\rho}$ is the mass density, and C_v is the heat capacity at constant

[85] Stroh (1954), Moore and Kuhlmann-Wilsdorf (1970), Zehnder (1991), and Kapoor and Nemat-Nasser (1998).

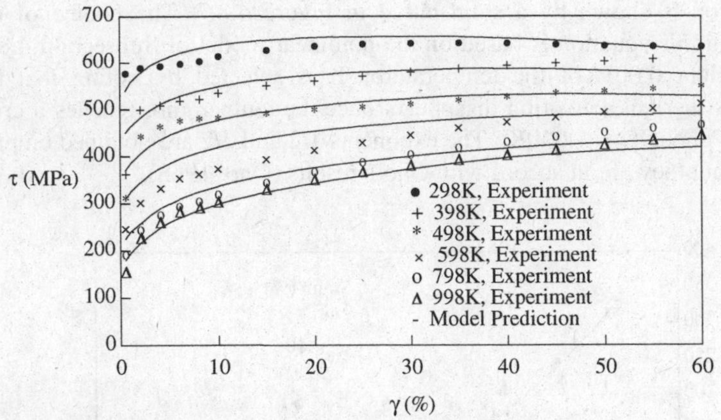

Figure 4.8.8

Comparison of the adiabatic flow stress of commercially pure Ta at a 5,000/s strain rate, with model prediction

volume. Since the flow stress varies gently with the plastic strain, the integration of (4.8.22d) can be performed using the mean-value theorem or even the simple Euler method, without introducing any noticeable errors.

Discussion of the Model: The general constitutive equation for commercially pure tantalum, obtained semi-empirically, is now expressed as

$$\tau(\gamma, \dot{\gamma}, T) = \hat{\tau}\left\{1-\left[-\frac{kT}{G_0}\, ln\frac{\dot{\gamma}}{\dot{\gamma}_r}\right]^{1/2}\right\}^{3/2} + c_0 + c_1\gamma^{n_1}, \qquad (4.8.22e)$$

where the dependence on the dislocation density is represented by the dependence on γ. Here,[86] the orientation (or Taylor) factor is absorbed into the definition of stress, τ, and plastic strain, γ. In (4.8.22e), G_0 is the energy of the Peierls barrier per atom, and will be taken as 1eV/atom for the present application to tantalum. The reference strain rate, $\dot{\gamma}_r$, on the other hand, must be estimated by direct comparison with experiment. An order-of-magnitude estimate is obtained by noting that $d = b \approx 3.31\text{Å}$, $\omega_0 = 10^{12}$/s, and $\rho_m = 10^{15}\text{m}^{-2}$, leading to $\dot{\gamma}_r = O(10^8)$/s. With $G_0 = 1$eV/atom, it follows that $k/G_0 = 8.62\times10^{-5}$/K, and $\dot{\gamma}_r = 5.46\times10^8$/s. With these values of the parameters, and setting $c_0 = 0$, Nemat-Nasser and Isaacs (1997a,b) show that the model fits essentially all existing high-strain-rate data, by simply viewing $\hat{\tau}$, c_1, and n_1 as free parameters. The same model also applies to other bcc metals at high strain

[86] The stress τ and the strain γ are the *uniaxial* stress and strain values in Figure 4.8.8. Hence, when the same equations are used in three-dimensional relations, or when they are used to model slip in single crystals (as in Chapter 6), the distinction becomes important. For example, if $\tau_{eff} = (\frac{1}{2}\boldsymbol{\tau}' : \boldsymbol{\tau}')^{\frac{1}{2}}$ and $\dot{\gamma}_{eff} = (2\mathbf{D}' : \mathbf{D}')^{1/2}$ are the effective stress and strain rate, then $\tau = \sqrt{3}\,\tau_{eff}$ and $\dot{\gamma} = \sqrt{1/3}\,\dot{\gamma}_{eff}$; in these expressions, a prime denotes the deviatoric.

rates. The constitutive parameters of several refractory metals that have been experimentally characterized, are listed in Table 4.8.1.

Finally, note that, in (4.8.22e), the thermal component of the flow stress is *non-negative*, and should be set equal to zero when the temperature exceeds a corresponding critical value which is strain-rate dependent, and which is given by

$$T_c = \frac{k}{G_0}\left\{ln\frac{\dot{\gamma}_r}{\dot{\gamma}}\right\}^{-1}. \tag{4.8.22f}$$

For $\dot{\gamma} = 10^{-3}$, 10^{-1}, 1,000, 5,000, and 40,000/s, this gives, $T_c =$ 430, 520, 880, 1,000, and 1,220K, respectively.

Table 4.8.1

Values of constitutive parameters in (4.8.22e) for indicated commercially pure metals[a]

	p	q	k/G_0 10^{-5}K	$\dot{\gamma}_r$ 10^7	$\hat{\tau}$ MPa	c_1 MPa	n_1
Ta	2/3	2	8.62	54.6	1,100	473	1/5
Va	2/3	2	12.72	0.357	1,050	305	1/5
Nb	2/3	2	12.4	0.35	1,680	440	1/4
Mo	2/3	2	8.62	1.45	2,450	720	1/4

[a] It is assumed that $c_0 \approx 0$ in all cases.

Application to Molybdenum: As pointed out before, the model predicts experimental results obtained for commercially pure molybdenum, as reported by Nemat-Nasser *et al.* (1999b). The corresponding parameters are given in Table 4.8.1. Figure 4.8.9 shows some of the results obtained by these authors for a 3,100/s strain rate and indicated initial temperatures. For the model calculation, it is assumed that the entire plastic work is used to heat the sample. The sample temperature is estimated using (4.8.22d), with $(\hat{\rho}\, C_v)^{-1} \approx 0.39$K/MPa. Figure 4.8.10 compares the experimental results for an 8,000/s strain rate and indicated initial temperatures, with the corresponding model predictions.

4.8.10. Application to OFHC Copper

Consider applying the model of (4.8.5a) with the thermally activated part of the flow stress given by (4.8.15a) and the athermal part defined by (4.8.18d), to predict the response of OFHC copper deformed in compression at various strain rates and temperatures. To simplify the modeling, neglect the grain-size effect, use the approximation (4.8.12p,q) with $n_0 = 1/2$, $q = 2$, and $p = 2/3$, and consider the following semi-empirical form of the basic equations:

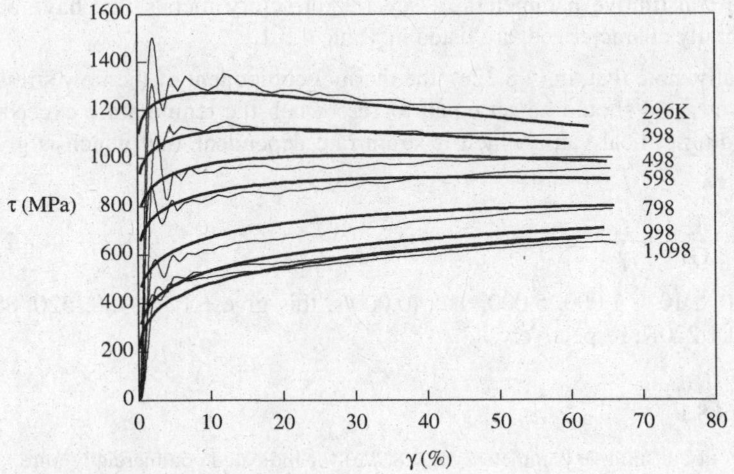

Figure 4.8.9

Comparison of the adiabatic flow stress of commercially pure Mo at a 3,100/s strain rate, with model prediction

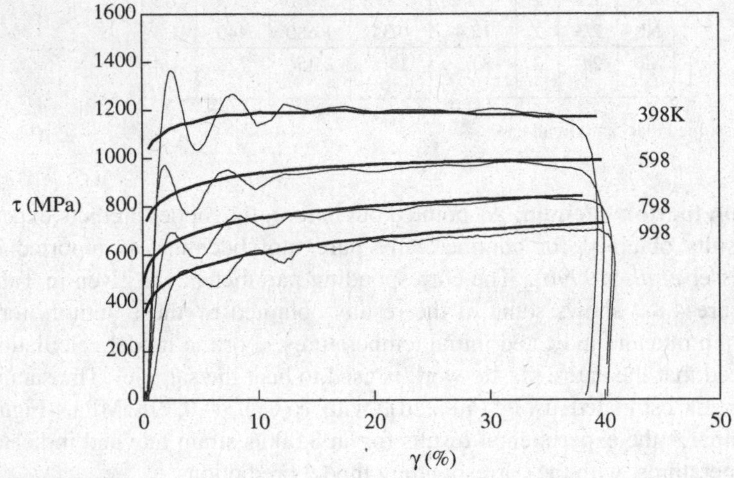

Figure 4.8.10

Comparison of the adiabatic flow stress of commercially pure Mo at an 8,000/s strain rate, with model prediction

$$\tau(\dot{\gamma}, \gamma, T) = \tau^0 \left\{ 1 - \left[-\frac{k\,T}{G_0} \left[ln \frac{\dot{\gamma}}{\dot{\gamma}_0} + ln\,(1 + a(T)\,\gamma^{n_0}) \right] \right]^{1/q} \right\}^{1/p} (1 + a(T)\,\gamma^{n_0})$$

$$+ c_0 + c_1 \gamma^{n_1}, \quad \text{for } T \leq T_c,$$

$$\tau^0 = \frac{G_0}{b\lambda l_0}, \quad \dot{\gamma}_0 = b\,\rho_m\,\omega_0\,l_0, \quad a(T) = a_0\{1-(T/T_m)^2\}. \qquad (4.8.23a\text{-}d)$$

With $\tau^0 = \hat{\tau}$, $a_0 = 0$, and $l_0 = b$, this constitutive relation reduces to the expression (4.8.22e) used for commercially pure tantalum, as outlined in Subsection 4.8.10, leading to good correlation with a broad range of experimental data. As is shown in this subsection, (4.8.23) fits a large body of high- and low-strain-rate experimental results for both annealed and as-received OFHC Cu samples.

First, estimate some of the constitutive parameters on the basis of the physics of the plastic deformation, for dislocation barriers. Consider k/G_0 and with G_0 slightly less than 2eV, set $k/G_0 = 5\times10^{-5}$/K. This energy barrier, G_0, is about twice the value used for commercially pure tantalum and molybdenum, for which the lattice stress field provides the barrier. For a dislocation as a barrier, the core of the dislocation involves a few surrounding atoms, and hence is expected to involve greater total energy. The actual value of k/G_0 is established empirically, once other parameters are fixed. To estimate $\dot{\gamma}_0$, note that $b \approx 2\times10^{-8}$cm, $\omega_0 = O(10^{11}/s)$, and if it is assumed that $\rho_m = O(10^{15}m^{-2})$ and l_0 is about 500 lattice spacings, then it follows that $\dot{\gamma}_0 \approx 2\times10^{10}$/s. This value of $\dot{\gamma}_0$ is used to model the OFHC copper, adjusting k/G_0 empirically in order to fit the data, leading to $k/G_0 = 4.9\times10^{-5}$/K. In this estimate, the value of ω_0 is that which has been suggested and justified by Kocks *et al.* (1975, page 124).

Table 4.8.2 summarizes the values of the constitutive parameters, for both the annealed and as-received OFHC Cu samples, as well as for several other metals. Note that greater values of $\dot{\gamma}_0$ and G_0 correspond to smaller strain-rate and temperature sensitivity, respectively. As is seen, many of the parameters have essentially the same values. The basic differences are in the parameters for the athermal part and in the values for τ^0. Comparing the results for OFHC Cu with those of Ta, note that, in the case of the commercially pure tantalum, the lattice is assumed to be the source of resistance to the motion of dislocations. In this case, l is of the order of the lattice spacing, *i.e.*, $l \approx b$. It is for this reason that $\dot{\gamma}_0$ for copper is about two orders of magnitude greater than that for tantalum (Table 4.8.1, page 235). This is a reflection of the fact that copper shows considerably greater strain-rate sensitivity than does tantalum. On the other hand, G_0 for copper is about twice that of tantalum, and, consequently, copper is less temperature-sensitive than tantalum. Similar comments apply to other metals listed in Table 4.8.2.

Experimental Data: Experiments are performed on both annealed and as-received OFHC Cu samples, over the temperature range from 77 to 1,100K, at stain rates from 0.001 to 8,000/s, achieving strains of 100% and greater. The quasi-static tests are performed using an Instron machine. In what follows, the test results are presented together with the corresponding theoretical predictions.

Effect of Long-range Barriers: As before, the high-temperature results are used to fix the constitutive parameters, c_0, c_1, and n_1, which define the athermal part of the flow stress in (4.8.23a). Since heating a sample to high-temperatures

Table 4.8.2

Values of various constitutive parameters in (4.8.23a-d) for indicated metals[b]

Parameter	p	q	k/G_0 $10^{-5}K^{-1}$	$\dot{\gamma}_0$ $10^{10}/s$	τ^0 MPa	a_0	n_0	c_1 MPa	n_1
An[c], Cu	2/3	2	4.9	2	46	20	1/2	220	0.3
Ar[d], Cu	2/3	2	4.9	2	400	1.8	1/2	220	0.3
AL-6XN	2/3	2	6.6	5.46	630	5	1/2	900	0.35
DH36	2/3	2	6.6	2	1,500	0	--	750	0.25
HSLA-65	2/3	2	10.6	0.04	1,450	0	--	760	0.15
Ti–1[e]	1	2	6.2	1.3	1,560	2.4	1	685	0.05
Ti–2[f]	1	2	6.2	1.3	1,900	2.4	1	710	0.03
Ti–3[g]	1	2	6.2	1.3	1,620	2.4	1	680	0.04

[b] It is assumed that $c_0 \approx 0$ in all cases. For AL-6XN, m_0 = 140MPa and $\alpha = 3 \times 10^{-4}$s. These parameters are not available for other listed metals.
[c] Annealed OFHC; [d] As-received OFHC; [e] Commercial; [f] Rapidly solidified Ti-6Al-4V powder is milled and then HIPed; and [g] Rapidly solidified Ti-6Al-4V powder is HIPed.

automatically anneals it, the high-temperature results should be independent of the initial condition of the sample, *i.e.*, whether it is annealed or workhardened. The experimental results support this observation. Indeed, repeated incremental loading of a sample at 600K and greater initial temperatures, reproduces the same stress-strain relation, as is illustrated in Figure 4.8.11, for a 696K initial temperature and a 4,000/s strain rate.[87] The curves shown in this figure are obtained by straining the sample by about 28% true strain and then allowing the sample to cool to room temperature, before it is reheated to the same initial temperature of 696K and then restrained by an additional 30% strain increment at the same strain rate of 4,000/s. The three curves are essentially the same and almost coincide with one another if they are translated by their initial plastic strain to a common origin. Thus, the sample in each of these incremental loadings should be viewed as a virgin and annealed one. Except for a translation along the strain axis, the constitutive relation (4.8.23a) should (and does) fit all three curves whether the parameters for the annealed or those for the as-received cases are used.

To obtain the tentative estimates of the constitutive parameters, follow the procedure outlined for tantalum, and plot the experimentally obtained flow stress as a function of the absolute temperature, for fixed strains and a constant strain rate. Figure 4.8.12 shows these results for the as-received material, at a 4,000/s strain rate. As is seen, the curves tend to limiting values which are more or less independent of the temperature. These limiting values fit a simple power law (see Figure 4.8.13),

[87] This also shows that plastic strain is not a microstructural state variable, as was pointed out before.

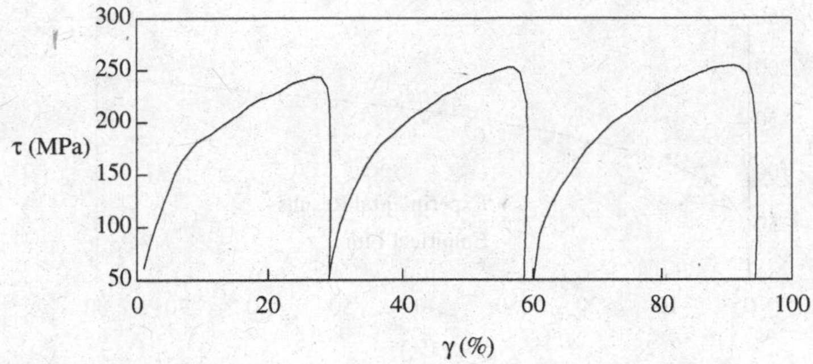

Figure 4.8.11

Stress-strain relations for an OFHC copper sample incrementally deformed at a 4,000/s strain rate and 696K initial temperature; note that the three curves are essentially the same

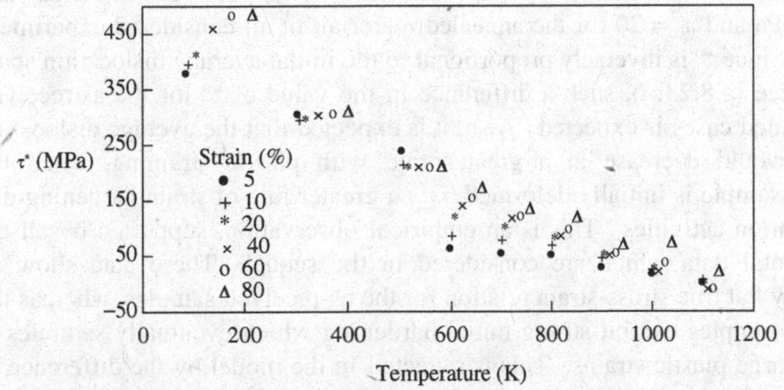

Figure 4.8.12

Experimentally obtained flow stress for as-received OFHC copper sample at a 4,000/s strain rate and indicated constant strain, as functions of the absolute temperature; note the tendency of the curves towards limiting, athermal values

$$\tau_a = 208\,\gamma^{0.27}, \qquad (4.8.23e)$$

suggesting that,[88] in (2.21), $c_0 \approx 0$, $c_1 \approx 210$MPa, and $n_1 \approx 0.3$. It turns out that $c_1 = 220$MPa fits the data slightly better. These are the values listed in Table 4.8.2.

---

[88] It turns out that the model is not too sensitive to the exact values of the parameters which best fit the experimental data. For this reason, we round all the parameters to within 10%, using instead of $n_1 = 0.27$, the rounded value of $n_1 = 0.3$. In general, the accuracy of the experimental evaluation of the flow stress of metals, does not warrant more accurate estimates of the model parameters.

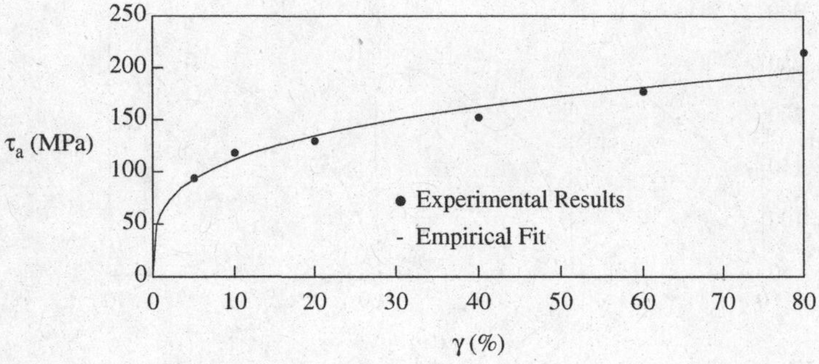

Figure 4.8.13

Athermal stress as a function of strain for OFHC copper

As pointed out before, $k/G_0 = 4.9 \times 10^{-5}$/K and $\dot{\gamma}_0 = 2 \times 10^{10}$/s are suitable values for OFHC Cu, both for annealed and as-received samples. The remaining parameters in (4.8.23) are τ^0 and a_0. Examination of the experimental results reveals that, $\tau^0 = 400$MPa and $a_0 = 1.8$ for the as-received material, and $\tau^0 = 46$MPa and $a_0 = 20$ for the annealed material, fit all considered experimental data. Since τ^0 is inversely proportional to the initial average dislocation spacing, l_0, (see (4.8.23b)), such a difference in the value of τ^0 for the as-received and annealed cases is expected. Also, it is expected that the average dislocation spacing would decrease at a greater rate with plastic straining, when the annealed sample is initially deformed, *i.e.*, a greater rate of strain hardening due to dislocation activities. This is an empirical observation, supported by all the experimental data which are considered in the sequel. These data show an essentially flat true stress-strain relation for the as-received samples, whereas the annealed samples exhibit strong initial hardening which eventually saturates at suitably large plastic strains. This is reflected in the model by the difference in the values of a_0 for the as-received and annealed cases.

Figures 4.8.14 and 4.8.15 compare the model predictions with the experimental curves. The first figure is for the as-received samples, whereas the second one is for the annealed samples. Only the samples tested at low temperatures were pre-annealed. As pointed out before, for initial temperatures of 696K and greater, the sample anneals while it is being brought to the required temperature in the furnace. As is seen, the model fits these data, whether the parameters for the annealed or those for the as-received case are used.

In addition, experiments have been performed at an 8,000/s strain rate and initial temperatures of 77 and 296K, for both the as-received and annealed samples. These are shown in Figures 4.8.16 and 4.8.17 which also include the quasi-static results obtained at room temperature. The theoretical results are shown by the solid curves in these figures. Both the high-strain-rate predictions and the quasi-static ones are in good accord with the experimental results.

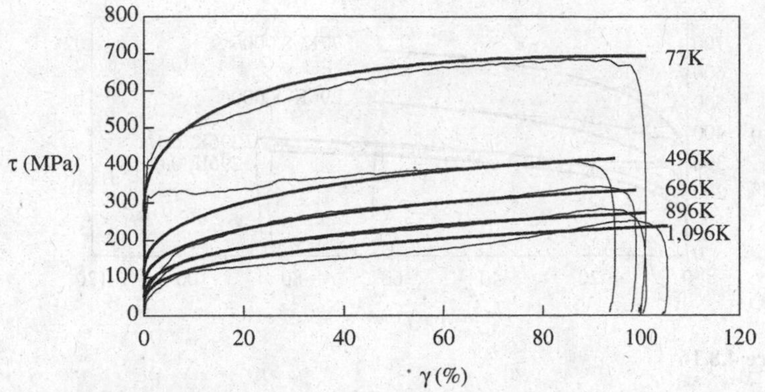

Figure 4.8.14

Comparison of model predictions (thick curves) with experimental results (thin curves) for as-received OFHC copper at a 4,000/s strain rate and indicated initial temperatures

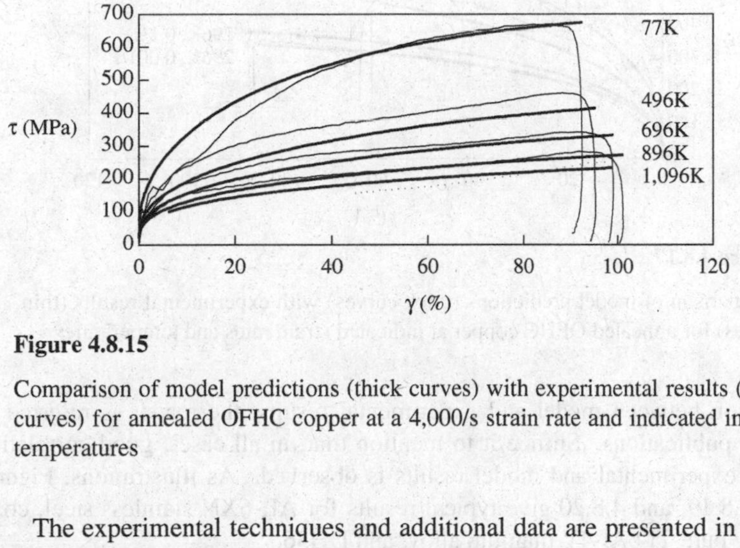

Figure 4.8.15

Comparison of model predictions (thick curves) with experimental results (thin curves) for annealed OFHC copper at a 4,000/s strain rate and indicated initial temperatures

The experimental techniques and additional data are presented in Chapter 9, where the quality of the data is also discussed.

Other Metals: Table 4.8.2 lists the values of the constitutive parameters obtained through experimental characterization of three different titanium alloys (Nemat-Nasser *et al.*, 2001b), Al-6XN stainless steel (Nemat-Nasser *et al.*, 2001a), and DH36 steel (Nemat-Nasser and Guo, 2003); included in this table are also the values of constitutive parameters for HSLA-65 steel(Nemat-Nasser and Guo, 2004). For DH36, these authors also give the experimentally evaluated constitutive parameters of the corresponding Johnson-Cook model (4.8.3c), and discuss the relative merits of the two models. For data analysis and

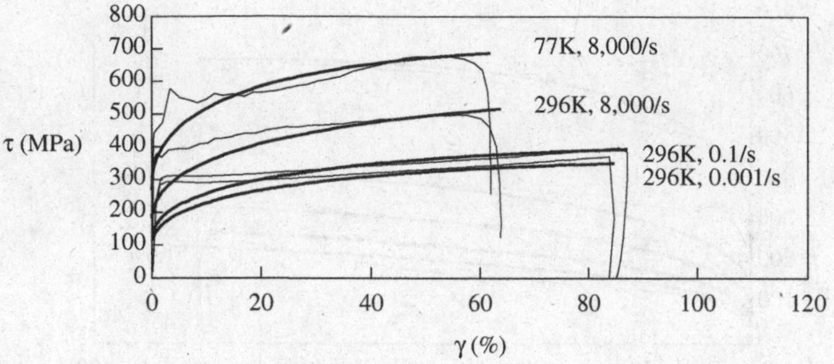

Figure 4.8.16

Comparison of model predictions (thick curves) with experimental results (thin curves) for as-received OFHC copper at indicated strain rates and temperatures

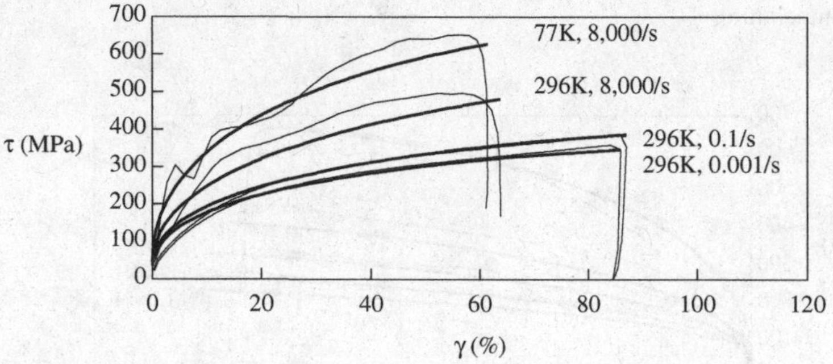

Figure 4.8.17

Comparison of model predictions (thick curves) with experimental results (thin curves) for annealed OFHC copper at indicated strain rates and temperatures

comparison between model and experimental results, the reader is referred to the cited publications. Suffice it to mention that, in all cases, good correlation between experimental and model results is observed. As illustrations, Figures 4.8.18, 4.8.19, and 4.8.20 give typical results for AL-6XN stainless steel, commercially pure Ti-6Al-4V titanium alloy, and DH36.

4.8.11. Rate-dependent J_2-plasticity

The results of the one-dimensional experiments are used to obtain the flow stress of the material, as discussed in the preceding subsections. With the strain rate $\dot{\gamma}$ expressed as a function of stress, τ, strain, γ, and temperature, T, the results are then generalized for application to three-dimensional problems. This is examined in the remaining subsections of the present section.

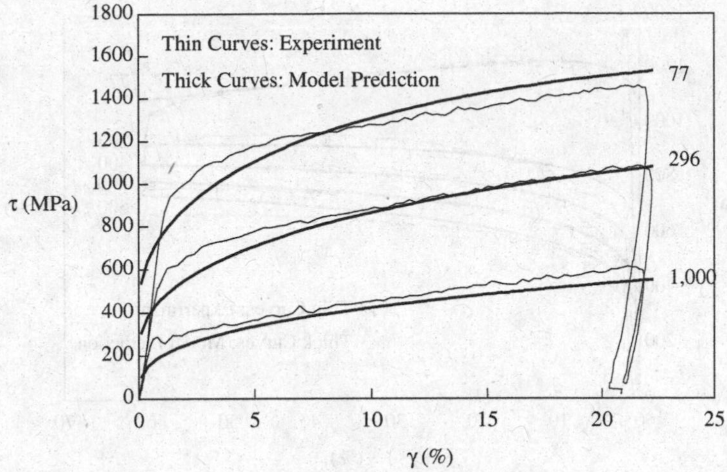

Figure 4.8.18

Comparison between experimental and model results for AL-6XN stainless steel at indicated initial temperatures and at a 1,000/s strain rate

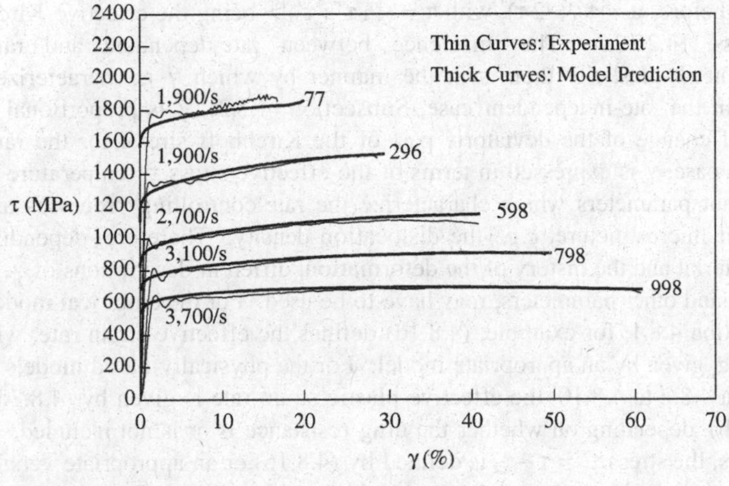

Figure 4.8.19

Comparison between experimental and model results for commercial Ti-6AL-4V at indicated initial temperatures and strain rates

The simplest model is obtained when the plastic part of the deformation rate is expressed as

$$\mathbf{D}^p = \frac{1}{\sqrt{2}} \dot{\gamma}\, \boldsymbol{\mu}, \tag{4.8.24a}$$

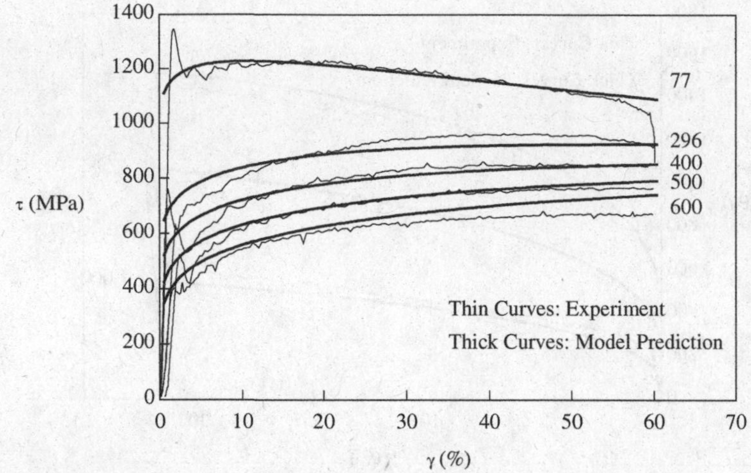

Figure 4.8.20

Comparison between experimental and model results for DH36 steel at indicated initial temperatures and at a 3,000/s strain rate

where, as before, $\boldsymbol{\mu} = \boldsymbol{\tau}'/(\sqrt{2}\,\tau)$, with $\tau = (\frac{1}{2}\boldsymbol{\tau}' : \boldsymbol{\tau}')^{1/2}$ being the effective Kirchhoff stress, (4.2.19a). The difference between rate-dependent and rate-independent J_2-plasticity then is in the manner by which $\dot{\gamma}$ is characterized. Whereas in the rate-independent case, Subsection (4.2.2), $\dot{\gamma}$ is proportional to the rate of change of the deviatoric part of the Kirchhoff stress, for the rate-dependent case, $\dot{\gamma}$ is expressed in terms of the effective stress τ, temperature T, and relevant parameters which characterize the rate-controlling processes and the current microstructure, *e.g.,* the dislocation density. Therefore, depending on the material and the history of the deformation, different descriptions of $\dot{\gamma}$, in terms of τ and other parameters, may have to be used. For the empirical models of Subsection 4.8.1, for example, (4.8.1d) defines the effective strain rate, with functions g_i given by an appropriate model. For the physically-based models of Subsection 4.8.2 to 4.8.10, the effective plastic strain rate is given by (4.8.20a) or (4.8.11b), depending on whether the drag resistance is or is not included. In these cases, the stress $\tau^* = \tau - \tau_a$ is defined by (4.8.16) or an appropriate version of this expression, depending on the material. As pointed out before, the experimental results reported in this section are for uniaxial compression. Hence, τ and γ in all these experiments must be divided by $\sqrt{3}$ and $1/\sqrt{3}$, respectively, in order to obtain the effective stress and strain used in this book for the J_2-plasticity models.

Once an appropriate expression for $\dot{\gamma}$ is obtained, then the total (deviatoric)[89] deformation rate is the sum of the elastic and plastic parts,

---

[89] As pointed out before, plastic deformation of most metals is isochoric. The elastic volume changes are also generally very small. Hence, expressions (4.8.24) to (4.8.26) are limited to devia-

$$\mathbf{D} = \mathcal{M} : \overset{\circ}{\boldsymbol{\tau}} + \dot{\gamma}\,\frac{\boldsymbol{\mu}}{\sqrt{2}}\,. \tag{4.8.24b}$$

In Chapter 5, numerical algorithms are given for effective integration of this kind of model which does not involve a yield surface. These and other numerical-integration techniques generally require an expression for the stress rate in terms of the deformation-rate tensor. From (4.8.24b), this is given by

$$\overset{\circ}{\boldsymbol{\tau}} = \mathcal{L} : \left[\mathbf{D} - \dot{\gamma}\,\frac{\boldsymbol{\mu}}{\sqrt{2}}\right], \tag{4.8.24c}$$

where $\mathcal{L}$ is the inverse of $\mathcal{M}$.

In the model (4.8.24b), the inelastic strain rate is coaxial with the deviatoric stress. In certain applications, it may be necessary to include a component for the inelastic strain rate which is noncoaxial with $\boldsymbol{\mu}$, *i.e.*,

$$\mathbf{D} = \mathcal{M} : \overset{\circ}{\boldsymbol{\tau}} + \dot{\gamma}\,\frac{\boldsymbol{\mu}}{\sqrt{2}} + \alpha\,\tau\,\frac{\overset{\circ}{\boldsymbol{\mu}}}{\sqrt{2}}\,. \tag{4.8.25a}$$

In this case, and for computational purposes, it is more effective to express $\overset{\circ}{\boldsymbol{\mu}}$ directly in terms of $\mathbf{D}$. Since $\boldsymbol{\mu} : \overset{\circ}{\boldsymbol{\mu}} = \mathbf{0}$, assume

$$\overset{\circ}{\boldsymbol{\mu}} = a\,(\mathbf{1}^{(4s)} - \boldsymbol{\mu} \otimes \boldsymbol{\mu}) : \mathbf{D} \tag{4.8.25b}$$

and calculate a from (4.8.25a,b) to be

$$\frac{\tau\,a}{\sqrt{2}} = \frac{\mathbf{D} : (\mathcal{L} - \boldsymbol{\lambda} \otimes \boldsymbol{\mu}) : \mathbf{D} - (\dot{\gamma}/\sqrt{2})\,\{\boldsymbol{\lambda} - (\boldsymbol{\lambda} : \boldsymbol{\mu})\,\boldsymbol{\mu}\} : \mathbf{D}}{\alpha\,\mathbf{D} : \mathcal{L}^{**} : \mathbf{D} + 2\,\mathbf{D} : (\mathbf{1}^{(4s)} - \boldsymbol{\mu} \otimes \boldsymbol{\mu}) : \mathbf{D}}\,, \tag{4.8.25c}$$

where, as before, all second-order tensors are *deviatoric*, and

$$\mathcal{L}^{**} = \mathcal{L} - (\boldsymbol{\lambda} \otimes \boldsymbol{\mu} + \boldsymbol{\mu} \otimes \boldsymbol{\lambda}) + (\boldsymbol{\lambda} : \boldsymbol{\mu})\,\boldsymbol{\mu} \otimes \boldsymbol{\mu}\,, \quad \boldsymbol{\lambda} = \mathcal{L} : \boldsymbol{\mu}\,. \tag{4.8.25d,e}$$

Hence, (4.8.25a) can now be expressed as

$$\overset{\circ}{\boldsymbol{\tau}} = \left[(1 - \bar{\alpha})\mathcal{L} + 2\,\bar{\alpha}\,\frac{\boldsymbol{\lambda}}{\sqrt{2}} \otimes \frac{\boldsymbol{\mu}}{\sqrt{2}}\right] : \mathbf{D} - \dot{\gamma}\,\frac{\boldsymbol{\lambda}}{\sqrt{2}}\,, \quad \bar{\alpha} = \alpha\,\tau\,a/\sqrt{2}\,. \tag{4.8.26a,b}$$

Note that, for the isotropic elasticity, (4.8.25c) simplifies considerably, and hence $\bar{\alpha}$ reduces to

$$\bar{\alpha} = \frac{\mu\,\alpha}{1 + \mu\,\alpha}\,. \tag{4.8.25f}$$

4.8.12. Viscoplastic J_2-flow Theory

In the preceding subsection, the plastic part of the deformation rate is always nonzero, although it may be negligibly small in certain regimes.[90] A viscoplasticity model is obtained if a rate-dependent yield stress is introduced, so that, whenever the effective stress τ falls below the current value of the yield

---

toric parts of all tensorial quantities, *without designating them with the usual prime*. These expressions can readily be modified to include the volume strain and strain rate, as in Section 4.7.

stress, the plastic strain rate vanishes. There are a number of phenomenological models of this kind discussed in the literature; see, for example, Perzyna (1966, 1980, 1984).

As an illustration, consider the plastic deformation rate given by (4.8.24a), with $\dot{\gamma}$ defined by any of the models discussed in this section. The viscoplastic version of this simple model would be

$$\mathbf{D}^{\mathrm{p}} = \begin{cases} \dot{\gamma}\, \boldsymbol{\mu}/\sqrt{2} & \text{when } \tau > \tau_{\mathrm{Y}} \\ 0 & \text{when } \tau < \tau_{\mathrm{Y}}, \end{cases} \tag{4.8.27a}$$

where τ_{Y} may be defined by a yield function, or it may be given as a function of the effective plastic strain, γ, effective plastic strain rate, $\dot{\gamma}$, temperature, T, and other parameters.

4.8.13. Viscoplastic J_2-flow Theory with Noncoaxiality

The viscoplastic model of (4.8.27a) may be modified to include the effect of the yield surface vertex through a noncoaxiality term, similar to (4.8.25a). The plastic strain rate then is given by

$$\mathbf{D}^{\mathrm{p}} = \begin{cases} \dfrac{1}{\sqrt{2}}\{\dot{\gamma}\, \boldsymbol{\mu} + \alpha\, \tau\, \mathring{\boldsymbol{\mu}}\} & \text{when } \tau > \tau_{\mathrm{Y}} \\ 0 & \text{when } \tau < \tau_{\mathrm{Y}}, \end{cases} \tag{4.8.27b}$$

where τ_{Y}, $\dot{\gamma}$, and α are defined in an appropriate way for the considered application.

4.9. GENERAL ANISOTROPIC ELASTOPLASTICITY

In this section, the consequences of the multiplicative decomposition of the deformation gradient $\mathbf{F}$ into a plastic constituent, $\mathbf{F}^{\mathrm{p}}$, and an elastic part, $\mathbf{F}^*$, are examined.[91] This decomposition is

$$\mathbf{F} = \mathbf{F}^* \mathbf{F}^{\mathrm{p}}. \tag{4.9.1}$$

Each constituent may include a rigid-body rotation. The plastic deformation is

---

[90] For example, for the power-law model (4.8.2b), with large values of m, $\dot{\gamma}$ is almost zero when $\tau < \tau_{\mathrm{r}}$.

[91] This decomposition was introduced independently by Lee (1969) and Willis (1969).

assumed to admit a general *flow potential*, $g = g(\boldsymbol{\tau}; \boldsymbol{\beta})$, such that the plastic part of the deformation-rate tensor, $\mathbf{D}^p$, is given by[92]

$$\mathbf{D}^p = \dot{\gamma}\,\frac{\partial g}{\partial \boldsymbol{\tau}}\,, \tag{4.9.2}$$

where $\boldsymbol{\beta}$ stands for a set of variables which measure the changes in the plastic response of the material due to the microstructural evolution, induced in the course of inelastic deformation. For example, they may represent workhardening parameters (isotropic, kinematic, or both), or other parameters which characterize the density and distribution of dislocations. They may represent the changes in the plastic response caused by distortional inelastic strain accumulation, *e.g.*, dislocation pileups. It will be assumed that the evolution of these variables is defined by a set of differential equations, *e.g.*,

$$\overset{\circ}{\boldsymbol{\beta}} = \boldsymbol{\psi}(\boldsymbol{\beta}; \ldots)\,, \tag{4.9.3}$$

where dots, ..., stand for possible dependence on the stress and temperature. The number and nature of these evolution equations are such that they can be (at least, in principle) integrated to yield expressions for the internal variables in terms of a given loading history. This, of course, must be done in conjunction with the other constitutive relations which are considered in this section. In what follows, *assume there are n independent internal variables that may be scalar- or tensor-valued,* $\boldsymbol{\beta}_\alpha$, $\alpha = 1, 2, \ldots, n$, and there are commensurately n corresponding evolution equations in (4.9.3), which characterize their changes.

The plastic deformation rate, given by (4.9.2), may include volumetric strains.[93] However, in order to simplify the discussion and focus on the implications of the decomposition (4.9.1) in metal plasticity, *inelastic volumetric strains are excluded in what follows.* The plastic potential g may depend on all deviatoric stress components, τ'_{ij}, $i, j = 1, 2, 3$. If there are any specializations (*e.g.*, plastic isotropy), these can be included once the general theory is laid out.

In (4.9.2), $\dot{\gamma}$ may relate to a *yield surface* or it may be defined in terms of the material's *flow stress*, in a manner similar to the special cases discussed in the preceding sections of this chapter; see Section 4.8 and also Subsection 4.9.7. At this point, $\dot{\gamma}$ is a scalar-valued function of all state variables, including $\boldsymbol{\beta}$'s. Hence, (4.9.2) describes *a general anisotropic (but isochoric) plastic deformation.*

While the plastic potential may be affected by the microstructural changes, it is commonly understood that large dimensional changes can and do occur by plastic deformation without affecting substantially the elasticity of the

---

[92] The flow potential, g, also depends on temperature.

[93] In application to metal plasticity, and in line with experimental observations, the plastic deformation and its rate are generally regarded to be volume preserving (isochoric). This type of plastic deformation is caused by dislocation motion and the concomitant crystallographic slips that produce large inelastic distortions with generally negligibly small (or, essentially no) volumetric strains.

material.[94] This is in line with crystal plasticity, where plastic flow by slip is known to leave the lattice unaffected. Therefore, in general, it may safely be assumed that, for most metals, plastic deformations defined by (4.9.2) and (4.9.3), do not affect the elastic structure or the elastic response of the material. Therefore, it may be assumed that the internal variables, $\boldsymbol{\beta}$'s, in the expression for the flow potential, affect only the plastic response and not the elasticity of the material. While this is not necessary in what follows, it is in line with most experimental results on metal plasticity.

The elastic deformation which may accompany the plastic deformation, is assumed to admit an *elastic potential*, $\phi = \tilde{\phi}(\tilde{\mathbf{E}}^e; ...)$, where dots, ..., stand for temperature and internal parameters which characterize the possible changes in the elasticity of the material.[95] The quantity $\tilde{\mathbf{E}}^e$ is an elastic material strain (*e.g.*, the Lagrangian strain), measured with respect to a suitable fixed reference (unstressed) configuration of the material, and ϕ is also measured per unit volume of this configuration. It is assumed, with no loss in generality, that $\tilde{\phi}(\mathbf{0}; ...) = 0$. The stress, $\tilde{\mathbf{S}}$, conjugate to the elastic material-strain measure, $\tilde{\mathbf{E}}^e$, is given by

$$\tilde{\mathbf{S}} = \frac{\partial \tilde{\phi}}{\partial \tilde{\mathbf{E}}^e} . \tag{4.9.4}$$

Note that function ϕ and the stress measure $\tilde{\mathbf{S}}$ depend on the reference state, with respect to which the elastic material strain, $\tilde{\mathbf{E}}^e$, is measured; see Subsections 4.9.5 and 4.9.6 for details. For the general anisotropic elastic response, the elastic potential, ϕ, depends on all components of the strain tensor $\tilde{\mathbf{E}}^e$. In this case, the principal directions of the stress tensor, $\tilde{\mathbf{S}}$, and strain tensor, $\tilde{\mathbf{E}}^e$, are noncoincident. When there are any elastic symmetries, then simplification becomes possible. In the isotropic case, for example, the dependence of the elastic potential, ϕ, on the elastic strain $\tilde{\mathbf{E}}^e$ is only through the basic invariants of this strain

---

[94] This is evident from the experimental results shown in Figure 4.8.11, page 239, where sequential large plastic deformation (about 30% each) of a copper sample at 696K temperature and a 4,000/s strain rate, produces essentially the same stress-strain relation. It thus appears that metal-plasticity models that assume the free energy to depend on plastic strain measures, are not supported by experimental evidence; see, *e.g.*, Green and Naghdi (1965, 1971), and for further justification of such a model, Naghdi (1990). See also comments in the footnote on page 479, regarding Gurtin (2000).

[95] The elastic moduli of most metals are temperature dependent. On the other hand, for many metals, the presence of dislocations may not change the instantaneous elastic response by any significant amount. Thus, it may safely be assumed that, for most metals, ϕ is unaffected by dislocation-induced purely plastic large deformations. In geomaterials, however, induced fabric due to inelastic deformation may change the elasticity of the material substantially, as shown in Section 7.5 of Chapter 7; see expressions (7.5.34a,b), page 582, where the fabric tensor associated with the backstress directly enters the instantaneous elasticity tensor. The elastic strains in the case of frictional granular materials, as well as most other geomaterials (and also most metals), are generally infinitesimally small, as is discussed in Section 4.7; see also Chapter 7. Moreover, most materials of this kind cannot sustain large inelastic deformations without strain localization and shearbanding; see Section 9.6, Chapter 9. The present discussion therefore, addresses metal plasticity only.

tensor. Then, the principal directions of the conjugate stress and strain tensors coincide.

Based on the above minimal assumptions, a general anisotropic elastoplasticity theory is considered in this section. No restriction is imposed on the magnitude of the elastic or plastic deformations, although metals, in reality, can support only infinitesimally small elastic strains, even at extremely large dislocation-induced plastic strains. Large strains and rotations are included. The aim is to precisely identify the number and the nature of constitutive relations that are necessary for a self-consistent and complete anisotropic elastoplasticity theory, *based on decomposition (4.9.1)*. For example, based on Mandel's work (Mandel, 1973), it has been suggested in the literature (Dafalias, 1983 and Loret, 1983) that separate constitutive relations are required to define the *plastic spin*, $\mathbf{W}^p$, similarly to those necessary to define the *plastic strain-rate tensor*, $\mathbf{D}^p$, and that identification of the *elastic symmetry axes* (if any), requires some "plastic rotation", and hence constitutive relations for the corresponding plastic spin (Wright, 2002, page 60).

In this section, first *kinematical* implications of decomposition (4.9.1) are examined, independently of the associated *dynamical* issues. It is shown that, (4.9.1) implies the existence of three independent and unique kinematical variables. These are: the total purely plastic deformation measure, $\mathbf{U}^p$; a unique rigid-body rotation tensor, $\mathbf{Q}$; and a purely elastic deformation measure, $\mathbf{V}^e$ (or equivalently, $\mathbf{U}^e = \mathbf{Q}\,\mathbf{V}^e\,\mathbf{Q}^T$), each being fully defined by a corresponding rate quantity. Indeed, starting from a given known state of the material, the time-rate-of-change of all kinematical variables (*i.e.*, $\dot{\mathbf{V}}^e$, $\dot{\mathbf{U}}^e$, $\dot{\mathbf{U}}^p$, and $\dot{\mathbf{Q}}$) can be computed through explicit equations, when the velocity gradient, $\mathbf{L} = \dot{\mathbf{F}}\,\mathbf{F}^{-1}$, and the plastic deformation rate, $\mathbf{D}^p$, are given. This is a purely kinematical conclusion that directly stems from decomposition (4.9.1), independently of any dynamical assumptions.

The stress rate is calculated using the elastic potential, ϕ, referred to a suitable fixed configuration, with respect to which the elastic structure of the material is assumed to be known, The fixed frame in this configuration is similar to the lattice vectors in a crystal; see Figures 6.2.2 and 6.2.3, pages 405 and 407, respectively. In general, the plastic deformation gradient which transforms the material from its initial undeformed configuration C_0, to this fixed reference configuration, would involve both pure deformation and a rigid-body rotation. When the material is elastically isotropic, no such rotation need be included. On the other hand, for anisotropic elasticity, a rotation must necessarily accompany pure plastic deformation. In crystal plasticity, the plastic spin corresponding to this rotation is completely defined, once the slip rates which give the associated plastic deformation-rate tensor are known. Hence, there is no need for additional constitutive relations to define the plastic spin, in crystal plasticity; see, (6.2.10a-d), page 411. The dynamic issues relating to elastic anisotropy are examined in Subsections 4.9.5 and 4.9.6, where a possible approach to calculate the plastic spin in a continuum setting is also described.

If it is assumed that the material elasticity is affected by plastic deformations, then this must be accounted for through the dependence of the elastic potential ϕ on some internal variables, in an appropriate manner. In what follows, this possibility is not examined since there are few experimental results to support and guide a continuum approach, and that deformation-induced texturing in polycrystals, for example, is essentially due to rigid-body rotation of the constituent crystals that should and can be modeled directly by crystal plasticity; see Chapter 6.

4.9.1. Decomposition of Deformation Gradient

For a single crystal with slip-induced plasticity, the kinematical quantity $\mathbf{F}^*$ corresponds to the elastic distortion and rigid-body rotation of the lattice, while $\mathbf{F}^p$ represents the plastic deformation associated with slips on crystallographic planes; this slip-induced simple shearing also includes rigid-body rotation of the material relative to the lattice, but does not distort or rotate the lattice. In the continuum approach, $\mathbf{F}^p$, and hence $\mathbf{F}^*$, may not have such a clear geometric interpretation. As is seen from (4.9.1), by the introduction of an arbitrary rotation $\mathbf{R}$, the deformation gradients $\mathbf{F}^*$ and $\mathbf{F}^p$ can be changed to $(\mathbf{F}^*\,\mathbf{R})$ and $(\mathbf{R}^T\,\mathbf{F}^p)$, without affecting the final deformation,

$$\mathbf{F} = (\mathbf{F}^*\,\mathbf{R})\,(\mathbf{R}^T\,\mathbf{F}^p), \quad \mathbf{R}^T = \mathbf{R}^{-1}, \quad \det \mathbf{R} = 1. \tag{4.9.5a-c}$$

Thus, $\mathbf{F}^*$ and $\mathbf{F}^p$ are understood to be defined to within an arbitrary rigid-body rotation.

Since $\mathbf{F}^*$ and $\mathbf{F}^p$ are deformation gradients, they may be decomposed by polar decomposition as follows:

$$\mathbf{F}^* = \mathbf{V}^e\,\mathbf{R}^*, \quad \mathbf{F}^p = \mathbf{R}^{**}\,\mathbf{U}^p, \tag{4.9.6a,b}$$

where $\mathbf{R}^*$ and $\mathbf{R}^{**}$ are proper orthogonal tensors, representing rigid-body rotations; $\mathbf{V}^e$ and $\mathbf{U}^p$ are the left *elastically-induced* and the right *plastically-induced* stretch tensors, respectively. In general, $\mathbf{R}^* \neq \mathbf{R}^{**}$. From (4.9.6), it follows that

$$\mathbf{F} = (\mathbf{V}^e\,\mathbf{R}^*)\,(\mathbf{R}^{**}\,\mathbf{U}^p) = \mathbf{V}^e\,\mathbf{Q}\,\mathbf{U}^p, \quad \mathbf{Q} = \mathbf{R}^*\,\mathbf{R}^{**}, \tag{4.9.7a-c}$$

and it is convenient to define

$$\mathbf{U}^e \equiv \mathbf{Q}^T\,\mathbf{V}^e\,\mathbf{Q} \tag{4.9.7d}$$

as the right elastically-induced stretch tensor. While $\mathbf{R}^*$ and $\mathbf{R}^{**}$ in (4.9.6a,b) are defined to within an arbitrary rotation $\mathbf{R}$, *the rotation tensor* $\mathbf{Q}$ *is unique.*[96]

---

[96] In the case of single crystals, the rigid-body rotation $\mathbf{R}^{**}$ is defined through the process of keeping the lattice orientations in an *intermediate configuration,* parallel to their respective orientations in the initial undeformed configuration of the crystal, while plastic deformation occurs by simple shearing on specific slip planes, through the motion of dislocations in specific slip directions.

The decomposition (4.9.7b) is schematically shown in Figure 4.9.1, where *two intermediate configurations* of the material neighborhood are denoted by C_p and C_R, respectively.

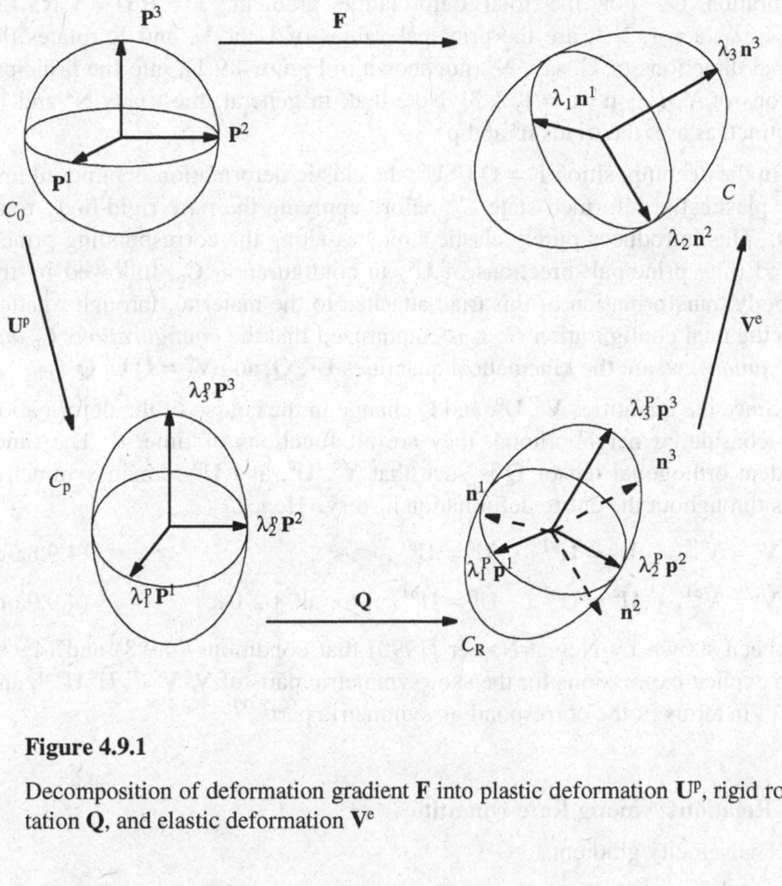

Figure 4.9.1

Decomposition of deformation gradient **F** into plastic deformation $\mathbf{U}^p$, rigid rotation **Q**, and elastic deformation $\mathbf{V}^e$

The configuration C_R is realized by *pure elastic unloading* from the current configuration, without any rigid-body rotation. This is an *elastically relaxed* state. The subscript R in C_R stands for elastically "relaxed." In the language of crystal plasticity, this unloading occurs *conceptually* with all active or potentially active slip systems and all mobile or potentially mobile dislocations considered *locked*; see also Section 6.2, and Figure 6.2.2. In the context of internal variables, discussed in Section 4.1, the unloading is realized with all internal variables held fixed at their current values. The intermediate configuration C_p is realized by *pure rotation* $\mathbf{Q}^T$ from configuration C_R, or by *pure plastic deformation* with no rigid-body rotation, from the initial undeformed configuration C_0.

In Figure 4.9.1, $\mathbf{P}^a$, a = 1, 2, 3, are the principal directions (eigenvectors) of the purely plastic right-stretch tensor, $\mathbf{U}^p$, and λ_a^p, a = 1, 2, 3, are the corresponding principal stretches. The proper orthogonal tensor, $\mathbf{Q}$, rotates the unit triad $\mathbf{P}^a$ into the unit triad $\mathbf{p}^a$. Then, the purely elastic distortion defined by $\mathbf{V}^e$, deforms the rotated ellipsoid shown in configuration C_R, into its final configuration C. For the total deformation gradient, $\mathbf{F} = \mathbf{R}\,\mathbf{U} = \mathbf{V}\,\mathbf{R}$, the stretches λ_a, a = 1, 2, 3, are the principal values of $\mathbf{U}$ or $\mathbf{V}$, and $\mathbf{R}$ rotates the principal directions of $\mathbf{U}$, say, $\mathbf{N}^a$ (not shown in Figure 4.9.1), into the principal directions of $\mathbf{V}$, *i.e.*, $\mathbf{n}^a$, a = 1, 2, 3. Note that, in general, the triads $\mathbf{N}^a$ and $\mathbf{P}^a$ are distinct, as are the triads $\mathbf{n}^a$ and $\mathbf{p}^a$.

In the decomposition $\mathbf{F} = \mathbf{Q}\,\mathbf{U}^e\,\mathbf{U}^p$, the elastic deformation is imposed first on the plastically deformed state C_p, before applying the pure rigid-body rotation $\mathbf{Q}$. This introduces purely elastic stretches along the corresponding principal triad (*i.e.,* principal directions of $\mathbf{U}^e$) in configuration C_p, followed by the rigid-body transformation of this triad attached to the material, through rotation $\mathbf{Q}$ into the final configuration C. It is emphasized that the *configurations* C_p *and* C_R *are unique,* as are the kinematical quantities $\mathbf{U}^p$, $\mathbf{Q}$, and $\mathbf{V}^e = \mathbf{Q}\,\mathbf{U}^e\,\mathbf{Q}^T$.

Since the quantities $\mathbf{V}^e$, $\mathbf{U}^p$, and $\mathbf{Q}$ change in the course of the deformation of the considered neighborhood, they are all functions of time, t. The time-dependent orthogonal tensor $\mathbf{Q}$ is such that $\mathbf{V}^e$, $\mathbf{U}^e$, and $\mathbf{U}^p$ remain symmetric tensors throughout the entire deformation history. Hence,

$$\mathbf{V}^e = \mathbf{V}^{eT}, \quad \mathbf{U}^e = \mathbf{U}^{eT}, \quad \mathbf{U}^p = \mathbf{U}^{pT}, \tag{4.9.8a-c}$$

$$\dot{\mathbf{V}}^e = \dot{\mathbf{V}}^{eT}, \quad \dot{\mathbf{U}}^e = \dot{\mathbf{U}}^{eT}, \quad \dot{\mathbf{U}}^p = \dot{\mathbf{U}}^{pT}, \quad \text{for all } t \geq 0. \tag{4.9.9a-c}$$

It has been shown by Nemat-Nasser (1990) that conditions (4.9.8) and (4.9.9) lead to explicit expressions for the skewsymmetric parts of $\dot{\mathbf{V}}^e\,\mathbf{V}^{e-1}$, $\dot{\mathbf{U}}^e\,\mathbf{U}^{e-1}$, and $\dot{\mathbf{U}}^p\,\mathbf{U}^{p-1}$, in terms of the corresponding symmetric parts.[97]

4.9.2. Relations Among Rate Quantities

The velocity gradient

$$\mathbf{L} = \dot{\mathbf{F}}\,\mathbf{F}^{-1}, \tag{4.9.10}$$

may be expressed as

$$\begin{aligned}\mathbf{L} &= \mathbf{Q}\,\dot{\mathbf{U}}^e\,\mathbf{U}^{e-1}\mathbf{Q}^T + \dot{\mathbf{Q}}\,\mathbf{Q}^T + \mathbf{V}^e\,\mathbf{Q}\,\dot{\mathbf{U}}^p\,\mathbf{U}^{p-1}\,\mathbf{Q}^T\,\mathbf{V}^{e-1} \\ &= \mathbf{Q}\,\hat{\mathbf{L}}^e\,\mathbf{Q}^T + \boldsymbol{\Omega} + \mathbf{V}^e\,\mathbf{Q}\,\hat{\mathbf{L}}^p\,\mathbf{Q}^T\,\mathbf{V}^{e-1} \\ &= \mathbf{L}^e + \boldsymbol{\Omega} + \mathbf{V}^e\,\bar{\mathbf{L}}^p\,\mathbf{V}^{e-1},\end{aligned} \tag{4.9.11a-c}$$

where (4.9.7b,d) and the following *notation* are used:

---

[97] Note that $\dot{\mathbf{V}}^e$ is not objective, since it is affected by rigid-body spin. Indeed, if $\overset{\circ}{\mathbf{V}}{}^e$ is the Jaumann rate of $\mathbf{V}^e$, corotational with a frame of spin $\boldsymbol{\Omega}$, then $\dot{\mathbf{V}}^e = \overset{\circ}{\mathbf{V}}{}^e + \boldsymbol{\Omega}\,\mathbf{V}^e - \mathbf{V}^e\,\boldsymbol{\Omega}$, and $\dot{\mathbf{U}}^e = \mathbf{Q}^T\,\overset{\circ}{\mathbf{V}}{}^e\,\mathbf{Q}$.

$$\hat{\mathbf{L}}^e = \dot{\mathbf{U}}^e \mathbf{U}^{e-1}, \quad \hat{\mathbf{L}}^p = \dot{\mathbf{U}}^p \mathbf{U}^{p-1}, \quad \mathbf{L}^e = \mathbf{Q}\hat{\mathbf{L}}^e\mathbf{Q}^T,$$
$$\overline{\mathbf{L}}^p = \mathbf{Q}\hat{\mathbf{L}}^p\mathbf{Q}^T, \quad \boldsymbol{\Omega} = \dot{\mathbf{Q}}\mathbf{Q}^T. \tag{4.9.12a-e}$$

Since $\hat{\mathbf{L}}^p$ in (4.9.11b) is a purely plastic velocity gradient, consider the following identification:

$$\mathbf{L} = \mathbf{L}^{e\Omega} + \mathbf{L}^p, \quad \mathbf{L}^{e\Omega} = \mathbf{L}^e + \boldsymbol{\Omega}, \quad \mathbf{L}^p = \mathbf{V}^e\overline{\mathbf{L}}^p\mathbf{V}^{e-1}, \tag{4.9.13a-c}$$

where, for convenience only, the total rigid-body spin, $\boldsymbol{\Omega}$, is combined with the purely elastic (and *objective*) velocity gradient, $\mathbf{L}^e$, to define $\mathbf{L}^{e\Omega}$, so that,

$$\mathbf{L}^{e\Omega} = \mathbf{D}^e + \mathbf{W}^{e\Omega}, \quad \mathbf{W}^{e\Omega} = \mathbf{W}^e + \boldsymbol{\Omega}. \tag{4.9.13d,e}$$

The quantity $\boldsymbol{\Omega}$ is the rigid-body spin associated with rotation $\mathbf{Q}$. The *objective* quantities $\mathbf{L}^e$ and $\overline{\mathbf{L}}^p$, represent the elastically and plastically induced velocity gradients, with respect to the current C and relaxed C_R configurations, respectively.[98] The quantity $\hat{\mathbf{L}}^p$ is the plastically induced objective velocity gradient, measured with respect to the plastically deformed but unrotated configuration C_p. Table 4.9.1 summarizes the above kinematical relations. The velocity gradients may be divided into symmetric and skewsymmetric parts (denoted by letters $\mathbf{D}$ and $\mathbf{W}$, respectively),[99]

$$\hat{\mathbf{L}}^e = \hat{\mathbf{D}}^e + \hat{\mathbf{W}}^e, \quad \hat{\mathbf{L}}^p = \hat{\mathbf{D}}^p + \hat{\mathbf{W}}^p,$$
$$\mathbf{L}^e = \mathbf{D}^e + \mathbf{W}^e, \quad \overline{\mathbf{L}}^p = \overline{\mathbf{D}}^p + \overline{\mathbf{W}}^p. \tag{4.9.14a-d}$$

A key observation now is that, the symmetry of $\mathbf{U}^e$ and hence its rate $\dot{\mathbf{U}}^e$, implies that the objective spin tensor $\hat{\mathbf{W}}^e$ *is a linear and homogeneous function of the elastic deformation-rate tensor,* $\hat{\mathbf{D}}^e$, *with coefficients in this relation being defined in terms of the total elastic stretch tensor,* $\mathbf{U}^e$. Similar comments apply to other objective spin tensors, *i.e.*, to $\hat{\mathbf{W}}^p$, $\mathbf{W}^e$, and $\overline{\mathbf{W}}^p$, as is shown in the following subsection. These are *purely kinematical results*, embedded in decomposition (4.9.1). They hold if and when (4.9.1) holds, independently of the associated material properties.[100] Thus, none of the objective spins that stem from (4.9.1) is an independent quantity to have an independent constitutive description.

The elastic and plastic spins, $\mathbf{W}^e$ and $\overline{\mathbf{W}}^p$, are real material spins, produced by the elastic and plastic deformation, respectively. *They are unique* kinematical quantities, fully defined by the corresponding deformation-rate tensor. The remaining spin $\boldsymbol{\Omega}$ is a rigid-body spin. *It is also unique* and independent of any arbitrariness that the rotation tensors $\mathbf{R}^*$ and $\mathbf{R}^{**}$ in (4.9.6a,b) may

---

[98] Note that $\mathbf{V}^{e-1}\mathbf{L}^e\mathbf{V}^e = \mathbf{L}^{e^T}$ is the objective elastic velocity gradient, measured with respect to C_R.

[99] Note that, $\mathbf{D}^p \neq \mathbf{V}^e\overline{\mathbf{D}}^p\mathbf{V}^{e-1}$ and $\mathbf{W}^p \neq \mathbf{V}^e\overline{\mathbf{W}}^p\mathbf{V}^{e-1}$, although $\mathbf{L}^p = \mathbf{D}^p + \mathbf{W}^p = \mathbf{V}^e(\overline{\mathbf{D}}^p + \overline{\mathbf{W}}^p)\mathbf{V}^{e-1}$. On the other hand, $\overline{\mathbf{D}}^p = \mathbf{Q}\hat{\mathbf{D}}^p\mathbf{Q}^T$ and $\overline{\mathbf{W}}^p = \mathbf{Q}\hat{\mathbf{W}}^p\mathbf{Q}^T$.

[100] *E.g.*, whether or not the elastic response is isotropic, or whether or not the plastic response is rate- and temperature-dependent.

Table 4.9.1

Summary of kinematical relations

Decomposition $\mathbf{F} = \mathbf{F}^* \mathbf{F}^p = \mathbf{V}^e \mathbf{Q} \mathbf{U}^p = \mathbf{Q} \mathbf{U}^e \mathbf{U}^p$		
Reference Configuration	Velocity Gradients	Definition & Relations
Current, C	Total	$\mathbf{L} = \mathbf{D} + \mathbf{W} = \dot{\mathbf{F}} \mathbf{F}^{-1}$ $= \mathbf{L}^e + \boldsymbol{\Omega} + \mathbf{V}^e \bar{\mathbf{L}}^p \mathbf{V}^{e-1} = \mathbf{L}^{e\Omega} + \mathbf{L}^p$
	Elastic (Objective)	$\mathbf{L}^e = \mathbf{D}^e + \mathbf{W}^e$
	Elastic & Rigid Spin	$\mathbf{L}^{e\Omega} = \mathbf{L}^e + \boldsymbol{\Omega} = \mathbf{D}^e + \mathbf{W}^{e\Omega}$
	Plastic (Objective)	$\mathbf{L}^p = \mathbf{V}^e \bar{\mathbf{L}}^p \mathbf{V}^{e-1} = \mathbf{D}^p + \mathbf{W}^p$
	Rigid Spin, Total	$\boldsymbol{\Omega} = \dot{\mathbf{Q}} \mathbf{Q}^T$
	Spin, Elastic & Rigid	$\mathbf{W}^{e\Omega} = \mathbf{W}^e + \boldsymbol{\Omega}$
	Spin, Total	$\mathbf{W} = \mathbf{W}^e + \boldsymbol{\Omega} + \mathbf{W}^p$
Elastically Relaxed, C_R	Total	$\bar{\mathbf{L}} \equiv \mathbf{V}^{e-1} \mathbf{L} \mathbf{V}^e = \mathbf{L}^{eT} + \bar{\boldsymbol{\Omega}} + \bar{\mathbf{L}}^p$
	Elastic (Objective)	$\mathbf{L}^{eT} \equiv \mathbf{V}^{e-1} \mathbf{L}^e \mathbf{V}^e = \mathbf{D}^e - \mathbf{W}^e$
	Spin	$\bar{\boldsymbol{\Omega}} = \mathbf{V}^{e-1} \boldsymbol{\Omega} \mathbf{V}^e$
	Plastic (Objective)	$\bar{\mathbf{L}}^p = \mathbf{Q} \dot{\mathbf{U}}^p \mathbf{U}^{p-1} \mathbf{Q}^T$
Plastically Deformed but Unrotated, C_p	Total	$\hat{\mathbf{L}} \equiv \mathbf{Q}^T \bar{\mathbf{L}} \mathbf{Q}$
	Elastic (Objective)	$\hat{\mathbf{L}}^e \equiv \dot{\mathbf{U}}^e \mathbf{U}^{e-1} = \hat{\mathbf{D}}^e + \hat{\mathbf{W}}^e$
	Plastic (Objective)	$\hat{\mathbf{L}}^p \equiv \dot{\mathbf{U}}^p \mathbf{U}^{p-1} = \hat{\mathbf{D}}^p + \hat{\mathbf{W}}^p$

introduce into the definition of $\mathbf{F}^*$ and $\mathbf{F}^p$. These are exact mathematical results. They are necessary for clear identification of objective and independent kinematical quantities that require constitutive description. Most importantly, the partitioning[101] of the rate quantities shows that *only* $\mathbf{D}^e$ and $\mathbf{D}^p$ *are independent deformation rates*, and that *only* $\boldsymbol{\Omega}$ *is a unique rigid-body spin*.

The above observations suggest direct relations for incrementally calculating $\mathbf{U}^e = \mathbf{Q}^T \mathbf{V}^e \mathbf{Q}$, $\mathbf{Q}$, and $\mathbf{U}^p$, as functions of time, in terms of the velocity gradient $\mathbf{L}$ and $\mathbf{D}^p$ or in terms of $\mathbf{L}$ and $\bar{\mathbf{D}}^p$ (or equivalently, in terms of $\mathbf{L}$ and $\hat{\mathbf{D}}^p$), as is shown in the sequel.[102] All other (kinematical) rate quantities are then *calculated* through *explicit* expressions. This observation also shows the decomposition (4.9.1) to be more restrictive than it may first appear.[103]

---

101 See Lubarda and Shih (1994, Concluding Remarks, page 528) for an alternative comments.

102 An elasticity constitutive relation, *i.e.*, the instantaneous elastic modulus tensor, is required for the calculation of the stress rate; see Subsection 4.9.5.

103 Indeed, modified or alternative versions of (4.9.1) have been proposed in order to better account for physical phenomena that may not readily be represented through the (4.9.1) formalism; see, for example, Kratochvil (1971), Davison (1995), Butler and McDowell (1998), Lion (2000), Kalidindi (2001), and Clayton and McDowell (2003). See also Chapters 6, 7, and 8.

4.9.3. Explicit Expressions for Objective Spin Tensors $\hat{\mathbf{W}}^e$, $\hat{\mathbf{W}}^p$, $\mathbf{W}^e$ and $\overline{\mathbf{W}}^p$

Explicit expressions are obtained in this subsection, for the spin tensors, $\hat{\mathbf{W}}^e$, $\hat{\mathbf{W}}^p$, $\mathbf{W}^e$, and $\overline{\mathbf{W}}^p$, in terms of the corresponding deformation-rate tensors, showing also that these spin tensors are in fact objective quantities.

Consider first (4.9.9b). By definition,

$$\hat{\mathbf{L}}^e \equiv \dot{\mathbf{U}}^e \mathbf{U}^{e-1} = \hat{\mathbf{D}}^e + \hat{\mathbf{W}}^e\,, \tag{4.9.15a}$$

and, from the symmetry of $\dot{\mathbf{U}}^e$,

$$(\hat{\mathbf{D}}^e + \hat{\mathbf{W}}^e)\,\mathbf{U}^e = \mathbf{U}^e\,(\hat{\mathbf{D}}^e - \hat{\mathbf{W}}^e)\,. \tag{4.9.15b}$$

Hence, spin tensor $\hat{\mathbf{W}}^e$ is the solution of

$$\mathbf{U}^e\,\hat{\mathbf{W}}^e + \hat{\mathbf{W}}^e\,\mathbf{U}^e = \mathbf{U}^e\,\hat{\mathbf{D}}^e - \hat{\mathbf{D}}^e\,\mathbf{U}^e\,. \tag{4.9.15c}$$

Since $\mathbf{U}^e$ and $\hat{\mathbf{D}}^e$ are both symmetric and $\hat{\mathbf{W}}^e$ is skewsymmetric, (4.9.15c) can be solved explicitly.[104] The solution is

$$(\mathrm{I}_{\mathbf{U}^e}\,\mathrm{II}_{\mathbf{U}^e} - \mathrm{III}_{\mathbf{U}^e})\,\hat{\mathbf{W}}^e = \mathrm{I}_{\mathbf{U}^e}{}^2\,(\mathbf{U}^e\,\hat{\mathbf{D}}^e - \hat{\mathbf{D}}^e\,\mathbf{U}^e) - \mathrm{I}_{\mathbf{U}^e}\,(\mathbf{U}^{e2}\,\hat{\mathbf{D}}^e - \hat{\mathbf{D}}^e\,\mathbf{U}^{e2})$$
$$+\,\mathbf{U}^e\,(\mathbf{U}^e\,\hat{\mathbf{D}}^e - \hat{\mathbf{D}}^e\,\mathbf{U}^e)\,\mathbf{U}^e\,, \tag{4.9.16a}$$

where $\mathrm{I}_{\mathbf{U}^e}$, $\mathrm{II}_{\mathbf{U}^e}$, and $\mathrm{III}_{\mathbf{U}^e}$ are the basic invariants of $\mathbf{U}^e$. Hence, $\hat{\mathbf{W}}^e$ is objective, but it is not an independent kinematical variable.

Similarly, consider

$$\hat{\mathbf{L}}^p \equiv \dot{\mathbf{U}}^p \mathbf{U}^{p-1} = \hat{\mathbf{D}}^p + \hat{\mathbf{W}}^p\,, \tag{4.9.17a}$$

and from the symmetry of $\mathbf{U}^p$ and $\dot{\mathbf{U}}^p$, obtain

$$\mathbf{U}^p\,\hat{\mathbf{W}}^p + \hat{\mathbf{W}}^p\,\mathbf{U}^p = \mathbf{U}^p\,\hat{\mathbf{D}}^p - \hat{\mathbf{D}}^p\,\mathbf{U}^p\,. \tag{4.9.17b}$$

The solution of (4.9.17b) is

$$(\mathrm{I}_{\mathbf{U}^p}\,\mathrm{II}_{\mathbf{U}^p} - \mathrm{III}_{\mathbf{U}^p})\,\hat{\mathbf{W}}^p = \mathrm{I}_{\mathbf{U}^p}{}^2\,(\mathbf{U}^p\,\hat{\mathbf{D}}^p - \hat{\mathbf{D}}^p\,\mathbf{U}^p) - \mathrm{I}_{\mathbf{U}^p}\,(\mathbf{U}^{p2}\,\hat{\mathbf{D}}^p - \hat{\mathbf{D}}^p\,\mathbf{U}^{p2})$$
$$+\,\mathbf{U}^p\,(\mathbf{U}^p\,\hat{\mathbf{D}}^p - \hat{\mathbf{D}}^p\,\mathbf{U}^p)\,\mathbf{U}^p\,, \tag{4.9.18a}$$

where $\mathrm{I}_{\mathbf{U}^p}$, $\mathrm{II}_{\mathbf{U}^p}$, and $\mathrm{III}_{\mathbf{U}^p}$ are the basic invariants of $\mathbf{U}^p$.

To obtain expressions for $\mathbf{W}^e$ and $\overline{\mathbf{W}}^p$ in terms of $\mathbf{D}^e$ and $\overline{\mathbf{D}}^p$, respectively, use $\mathbf{V}^e = \mathbf{Q}\,\mathbf{U}^e\,\mathbf{Q}^T$, let $\overline{\mathbf{U}}^p$ define the left plastic stretch tensor,

$$\overline{\mathbf{U}}^p \equiv \mathbf{Q}\,\mathbf{U}^p\,\mathbf{Q}^T\,, \tag{4.9.19}$$

and note that the basic invariants of $\mathbf{U}^e$ and $\mathbf{V}^e$ are the same, as are those of $\mathbf{U}^p$ and $\overline{\mathbf{U}}^p$. Since symmetry is preserved by rigid-body transformation, it follows from (4.9.16a) and (4.9.18a) that

---

[104] Mehrabadi and Nemat-Nasser (1987), and Chapter 1, Section 1.4. See also Dienes (1979), Nemat-Nasser (1990), and Obata *et al.* (1990).

$$(\mathrm{I}_{\mathbf{U}^e}\,\mathrm{II}_{\mathbf{U}^e} - \mathrm{III}_{\mathbf{U}^e})\,\mathbf{W}^e = \mathrm{I}_{\mathbf{U}^e}^{\,2}\,(\mathbf{V}^e\,\mathbf{D}^e - \mathbf{D}^e\,\mathbf{V}^e) - \mathrm{I}_{\mathbf{U}^e}\,(\mathbf{V}^{e2}\,\mathbf{D}^e - \mathbf{D}^e\,\mathbf{V}^{e2}) + \mathbf{V}^e\,(\mathbf{V}^e\,\mathbf{D}^e - \mathbf{D}^e\,\mathbf{V}^e)\,\mathbf{V}^e\,, \tag{4.9.20a}$$

$$(\mathrm{I}_{\mathbf{U}^p}\,\mathrm{II}_{\mathbf{U}^p} - \mathrm{III}_{\mathbf{U}^p})\,\overline{\mathbf{W}}^p = \mathrm{I}_{\mathbf{U}^p}^{\,2}\,(\overline{\mathbf{U}}^p\,\overline{\mathbf{D}}^p - \overline{\mathbf{D}}^p\,\overline{\mathbf{U}}^p) - \mathrm{I}_{\mathbf{U}^p}\,(\overline{\mathbf{U}}^{p2}\,\overline{\mathbf{D}}^p - \overline{\mathbf{D}}^p\,\overline{\mathbf{U}}^{p2}) + \overline{\mathbf{U}}^p\,(\overline{\mathbf{U}}^p\,\overline{\mathbf{D}}^p - \overline{\mathbf{D}}^p\,\overline{\mathbf{U}}^p)\,\overline{\mathbf{U}}^p\,. \tag{4.9.21a}$$

It is seen from (4.9.16a), (4.9.18a), (4.9.20a), and (4.9.21a) that, $\hat{\mathbf{W}}^e$, $\hat{\mathbf{W}}^p$, $\mathbf{W}^e$, and $\overline{\mathbf{W}}^p$ are linear and homogeneous functions of their corresponding deformation-rate tensors, $\hat{\mathbf{D}}^e$, $\hat{\mathbf{D}}^p$, $\mathbf{D}^e$ and $\overline{\mathbf{D}}^p$, respectively. The coefficients in these linear relations are explicit functions of the associated stretch tensor. These spin tensors are thus all objective, but they are not independent kinematical variables. To express these relations in a unified manner, introduce a fourth-order *tensor operator* $\boldsymbol{\mathcal{K}}(...)$, as follows: for two second-order tensors, $\mathbf{A} = \mathbf{A}^T$ (a stretch tensor) and $\mathbf{B} = \mathbf{B}^T$ (any tensor), the operation $\boldsymbol{\mathcal{K}}(\mathbf{A}) : \mathbf{B}$ yields a second-order skewsymmetric tensor $\mathbf{C} = -\,\mathbf{C}^T$, in the following manner:

$$\begin{aligned}\mathbf{C} &= \boldsymbol{\mathcal{K}}(\mathbf{A}) : \mathbf{B} \\ &\equiv (\mathrm{I}_\mathbf{A}\,\mathrm{II}_\mathbf{A} - \mathrm{III}_\mathbf{A})^{-1}\,\{\mathrm{I}_\mathbf{A}^{\,2}\,(\mathbf{A}\,\mathbf{B} - \mathbf{B}\,\mathbf{A}) - \mathrm{I}_\mathbf{A}\,(\mathbf{A}^2\,\mathbf{B} - \mathbf{B}\,\mathbf{A}^2) \\ &\quad + \mathbf{A}\,(\mathbf{A}\,\mathbf{B} - \mathbf{B}\,\mathbf{A})\,\mathbf{A}\,\}\,.\end{aligned} \tag{4.9.22a}$$

In component form, the operator $\boldsymbol{\mathcal{K}}(\mathbf{A})$ becomes

$$\begin{aligned}\mathcal{K}_{ijkl}(\mathbf{A}) \equiv{}& (\mathrm{I}_\mathbf{A}\,\mathrm{II}_\mathbf{A} - \mathrm{III}_\mathbf{A})^{-1}\,\{\mathrm{I}_\mathbf{A}^{\,2}\,(A_{ik}\delta_{jl} - \delta_{ik}\,A_{lj}) \\ &- \mathrm{I}_\mathbf{A}\,(A_{im}\,A_{mk}\,\delta_{jl} - \delta_{ik}\,A_{lm}\,A_{mj}) \\ &+ A_{im}\,(A_{mk}\,\delta_{nl} - \delta_{mk}\,A_{ln})\,A_{nj}\}\,, \quad i, j, k, l, m, n = 1, 2, 3\,.\end{aligned} \tag{4.9.22b}$$

With this notation, (4.9.16a), (4.9.18a), (4.9.20a), and (4.9.21a), respectively become,

$$\hat{\mathbf{W}}^e = \boldsymbol{\mathcal{K}}(\mathbf{U}^e) : \hat{\mathbf{D}}^e\,, \tag{4.9.16b}$$

$$\hat{\mathbf{W}}^p = \boldsymbol{\mathcal{K}}(\mathbf{U}^p) : \hat{\mathbf{D}}^p\,, \tag{4.9.18b}$$

$$\mathbf{W}^e = \boldsymbol{\mathcal{K}}(\mathbf{V}^e) : \mathbf{D}^e\,, \tag{4.9.20b}$$

$$\overline{\mathbf{W}}^p = \boldsymbol{\mathcal{K}}(\overline{\mathbf{U}}^p) : \overline{\mathbf{D}}^p\,. \tag{4.9.21b}$$

Hence, the corresponding (objective) velocity gradients are

$$\hat{\mathbf{L}}^e = \{\boldsymbol{\mathcal{K}}(\mathbf{U}^e) + \mathbf{1}^{(4s)}\} : \hat{\mathbf{D}}^e\,, \tag{4.9.23a}$$

$$\hat{\mathbf{L}}^p = \{\boldsymbol{\mathcal{K}}(\mathbf{U}^p) + \mathbf{1}^{(4s)}\} : \hat{\mathbf{D}}^p\,, \tag{4.9.24a}$$

$$\mathbf{L}^e = \{\boldsymbol{\mathcal{K}}(\mathbf{V}^e) + \mathbf{1}^{(4s)}\} : \mathbf{D}^e\,, \tag{4.9.25a}$$

$$\overline{\mathbf{L}}^p = \{\boldsymbol{\mathcal{K}}(\overline{\mathbf{U}}^p) + \mathbf{1}^{(4s)}\} : \overline{\mathbf{D}}^p\,. \tag{4.9.26a}$$

Introduce the operator

$$\mathcal{H}(...) \equiv \mathcal{K}(...) + \mathbf{1}^{(4s)}, \tag{4.9.22c}$$

and rewrite the above expressions in the following more compact form:

$$\hat{\mathbf{L}}^e = \mathcal{H}(\mathbf{U}^e) : \hat{\mathbf{D}}^e, \tag{4.9.23b}$$

$$\hat{\mathbf{L}}^p = \mathcal{H}(\mathbf{U}^p) : \hat{\mathbf{D}}^p, \tag{4.9.24b}$$

$$\mathbf{L}^e = \mathcal{H}(\mathbf{V}^e) : \mathbf{D}^e, \tag{4.9.25b}$$

$$\overline{\mathbf{L}}^p = \mathcal{H}(\overline{\mathbf{U}}^p) : \overline{\mathbf{D}}^p. \tag{4.9.26b}$$

In component form, (4.9.22c) becomes

$$\mathcal{H}_{ijkl}(...) = \mathcal{K}_{ijkl}(...) + \tfrac{1}{2}(\delta_{ik}\,\delta_{jl} + \delta_{il}\,\delta_{jk}). \tag{4.9.22d}$$

It is emphasized that (4.9.16) to (4.9.26) are exact mathematical results, valid independently of the specific nature of the deformation gradients $\mathbf{F}^*$ and $\mathbf{F}^p$, as long as (4.9.1) and polar decompositions (4.9.6a,b) are assumed to hold. Whatever the origin and physical meaning of these gradients, each of them can then be decomposed into a rotation and a pure stretch by polar decomposition. The two rotations, $\mathbf{R}^*$ of $\mathbf{F}^*$ and $\mathbf{R}^{**}$ of $\mathbf{F}^p$, can always be combined to yield a *unique rotation* $\mathbf{Q}$. Thus, $\mathbf{F} = \mathbf{V}^e\,\mathbf{Q}\,\mathbf{U}^p = \mathbf{Q}\,\mathbf{U}^e\,\mathbf{U}^p$ can always be established uniquely, with the two stretch tensors, $\mathbf{U}^e$ and $\mathbf{U}^p$ (or, equivalently, $\mathbf{V}^e$ and $\mathbf{U}^p$), symmetric and positive-definite. Hence, *no matter what constitutive model is used, as long as* $\mathbf{F} = \mathbf{F}^*\,\mathbf{F}^p$ *is the starting point, the skewsymmetric parts of* $\dot{\mathbf{U}}^e\,\mathbf{U}^{e-1}$ and $\dot{\mathbf{U}}^p\,\mathbf{U}^{p-1}$ *are explicitly given by their corresponding symmetric parts*, through (4.9.16) and (4.9.18), respectively.[105]

Suppose that, at an instant in the course of a deformation history, $\mathbf{U}^e$ (or $\mathbf{V}^e$), $\mathbf{U}^p$, and $\mathbf{Q}$ are known. Consider a procedure to calculate the rate quantities $\dot{\mathbf{U}}^e$ (or $\dot{\mathbf{V}}^e$), $\dot{\mathbf{U}}^p$, and $\dot{\mathbf{Q}}$, given the velocity gradient $\mathbf{L}$ and one of the plastic deformation rate tensors. To this end, first assume that $\overline{\mathbf{D}}^p$ is given. Then, (4.9.21b) yields $\overline{\mathbf{W}}^p$, (4.9.26b) yields $\overline{\mathbf{L}}^p$, and from (4.9.13c), $\mathbf{L}^p = \mathbf{D}^p + \mathbf{W}^p$ is obtained. Now use (4.9.12d) to calculate $\hat{\mathbf{L}}^p$, and hence obtain $\dot{\mathbf{U}}^p = \hat{\mathbf{L}}^p\,\mathbf{U}^p$. The symmetric part of $\mathbf{L}$ equals $\mathbf{D}^e$, (4.9.20b) yields $\mathbf{W}^e$, and hence $\mathbf{L}^e$ and $\boldsymbol{\Omega} = \mathbf{L} - \mathbf{L}^e - \mathbf{L}^p$ are calculated, leading to $\dot{\mathbf{Q}} = \boldsymbol{\Omega}\,\mathbf{Q}$. Finally, set $\hat{\mathbf{D}}^e = \mathbf{Q}^T\,\mathbf{D}^e\mathbf{Q}$ and use (4.9.23b) to find $\hat{\mathbf{L}}^e$, and $\dot{\mathbf{U}}^e = \hat{\mathbf{L}}^e\,\mathbf{U}^e$.

When $\mathbf{L}$ and $\mathbf{D}^p$ are given, (4.9.13c) and (4.9.26b) yield,

$$\mathbf{D}^p = \{\mathbf{V}^e\,\mathcal{H}(\overline{\mathbf{U}}^p) : \overline{\mathbf{D}}^p\,\mathbf{V}^{e-1}\}_{\mathrm{sym}}, \tag{4.9.27a}$$

where the subscript sym denotes the symmetric part. Since $\mathbf{V}^e$ is never zero or unbounded, and $\mathcal{H}$ is of the order of one, (4.9.27a) can be solved to obtain $\overline{\mathbf{D}}^p$ in terms of $\mathbf{D}^p$, and the above procedure applied. In component form, (4.9.27a) becomes

---

105 For historical reasons, Nemat-Nasser (1990) has chosen to interpret $\mathbf{F}^*$ as the elastic, and $\mathbf{F}^p$ as the plastic part of the total deformation gradient $\mathbf{F}$. In application to crystal plasticity, $\mathbf{F}^*$ defines the lattice elastic distortion accompanied by the lattice rigid-body rotation; see Section 6.2.

$$D^p_{ij} = \tfrac{1}{2}(V^e_{ik}V^{e-1}_{jl} + V^e_{jk}V^{e-1}_{il})\,\mathcal{H}_{klmn}\,\bar{D}^p_{mn}. \tag{4.9.27b}$$

4.9.4. Material Frame Indifference

Constitutive relations must remain invariant under any rigid-body rotation of the reference coordinate system. This is called *objectivity* or the *material frame indifference*.[106] It is often automatically guaranteed when material strain measures, their rates, and objective stress rates are used. These conditions are satisfied for conjugate stress and strain tensors which are referenced with respect to the fixed initial configuration C_0, as discussed in Chapters 2 and 3. By their very definition, material strain measures and their rates remain unchanged when the material neighborhood undergoes rigid-body motions. The same comment applies to all objective stress rates.

In the context of the present application, the question now is whether objectivity must be enforced with respect to both the current configuration and the intermediate configurations, or whether the former automatically guarantees the latter. It is shown in the sequel that objectivity with respect to the current state, C, also guarantees objectivity with respect to the intermediate state, C_R.

Since configuration C_R relates to the current configuration through pure elastic deformations, $\mathbf{V}^e$, any rigid-body motion superimposed on the current configuration C will also equally apply to the relaxed configuration C_R. Indeed, suppose the current configuration is subjected to an arbitrary rotation, $\mathbf{R}$, $\mathbf{R}^{-1} = \mathbf{R}^T$, $\det\mathbf{R} = +1$. Then, with respect to the *new* configuration, say, C_{new}, the deformation gradient is $\mathbf{F}_{new} = \mathbf{R}\mathbf{F}$. Hence, $\mathbf{F}_{new} = (\mathbf{R}\mathbf{V}^e\mathbf{R}^T)(\mathbf{R}\mathbf{Q})\mathbf{U}^p = \mathbf{R}\mathbf{Q}\mathbf{U}^e\mathbf{U}^p$, and only the rotation tensor changes from $\mathbf{Q}$ to the new rotation $\mathbf{Q}_{new} = \mathbf{R}\mathbf{Q}$. The quantity $\mathbf{R}\mathbf{V}^e\mathbf{R}^T$ is the same elastic stretch tensor with its principal directions $\mathbf{n}^a$, a = 1, 2, 3, now rotated by rotation tensor, $\mathbf{R}$, in both the new relaxed and the new current configurations, while $\mathbf{U}^e$ remains unchanged. These comments also show that, if a rigid rotation is imposed on the relaxed configuration C_R, then the same rotation will necessarily be imposed on the current configuration.

The superimposed time-dependent rotation $\mathbf{R}$ also changes the velocity gradient $\mathbf{L}$ to $\mathbf{L}_{new} = \dot{\mathbf{R}}\mathbf{R}^T + \mathbf{R}\mathbf{L}\mathbf{R}^T$. This then only changes the spin $\boldsymbol{\Omega}$ by the added spin, *i.e.*, $\dot{\mathbf{Q}}_{new}\mathbf{Q}^T_{new} = \dot{\mathbf{R}}\mathbf{R}^T + \mathbf{R}\boldsymbol{\Omega}\mathbf{R}^T$. Therefore, the spin $\boldsymbol{\Omega} = \dot{\mathbf{Q}}\mathbf{Q}^T$ is *uniquely* defined in terms of the total spin $\mathbf{W}$. To change $\boldsymbol{\Omega}$, an additional rotation must be imposed on the *current* configuration, *relative* to the undeformed state. Hence, *objectivity imposed with respect to rigid rotations superimposed on the current deformed state also ensures objectivity with respect to the intermediate configurations, C_p and C_R. Indeed, since $\mathbf{W}^e$ and $\overline{\mathbf{W}}^p$ are objective*, the only way that the rigid spin $\boldsymbol{\Omega}$ can be changed is by imposing rigid spin on the

[106] The issue of objectivity or material frame indifference received considerable attention by many researchers in the 1950's and early 1960's; see, *e.g.*, Truesdell and Noll (1965).

current state. It is thus sufficient to apply the requirement of material frame indifference to the constitutive relations in the current configuration C. No other objectivity should be necessary in the present context.[107]

4.9.5. Elastic Response

In the preceding sections of this chapter, the elastic and plastic deformation-rate tensors, $\mathbf{D}^e$ and $\mathbf{D}^p$, are prescribed through constitutive relations in terms of the Kirchhoff stress and its rate, $\boldsymbol{\tau}$ and $\dot{\boldsymbol{\tau}}$. In general, constitutive relations are used to define the elastic and inelastic strain-rate measures in terms of the conjugate stress and objective stress rate.[108]

As a starting point, assume first that plastic deformations do not affect the elastic structure and hence the elastic potential. Therefore, the elastic response is the same whether the elastic strain is imposed on the material in configuration C_0 or in a new configuration, say, $C_{\mathbf{F}^p}$, obtained from C_0 by imposing the plastic deformation defined by $\mathbf{F}^p$. As discussed in the next subsection, in general, plastic deformations do not affect the elastic structure and elastic symmetries of polycrystalline metals, and that it is only the rigid-body lattice rotations of the individual crystals that produce textures and hence elastic, as well as plastic, anisotropy. These lattice rotations are not the same for all crystals within a continuum volume element, and, thus, the overall elasticity of the volume element evolves with elastoplastic deformations. These and related issues are examined in Subsection 4.9.6. *In what follows, it is assumed that* $\mathbf{A}_i$, $i = 1, 2, 3$, are known overall elastic symmetry axes of the material volume element, and that these axes remain the same as the plastic deformation, $\mathbf{F}^p$, is imposed on this material element. Figure 4.9.2 schematically shows these overall continuum elastic symmetry axes, in the initial C_0, in the plastically deformed configuration $C_{\mathbf{F}^p}$, in the elastoplastically deformed configuration C_{pe}, and finally in the current configuration C, respectively.

Consider now the elastic relation (4.9.4), and examine its consequences. For illustration, let the elastic potential $\phi = \tilde{\phi}(\tilde{\mathbf{E}}^e)$ be a function of the Lagrangian elastic strain,[109]

---

[107] Since this is an exact mathematical consequence of the basic kinematical decomposition $\mathbf{F} = \mathbf{F}^* \mathbf{F}^p$, contrary conclusions that are also based on the same starting kinematical decomposition, must necessarily be incorrect; for comments, see Casey and Naghdi (1980), Xiano *et al.* (2000), and, for additional references, see Lubarda (2002).

[108] Hill (1978), Havner (1982a), and Nemat-Nasser (1983).

[109] If the continuum model is to account for the evolution of the material element's elasticity, which may be produced by variable rigid-body rotations of its individual crystal constituents, then the elastic potential must similarly evolve, and, hence, be also dependent on certain internal variables that represent such evolutions. There are no experimental results to guide such a formulation. Therefore, generally, either direct crystal plasticity together with some averaging scheme are used (Chapters 6 and 8), or a direct rate-problem is formulated (Sections 4.2 and 4.8) without reference to possible decompositions similar to (4.9.1).

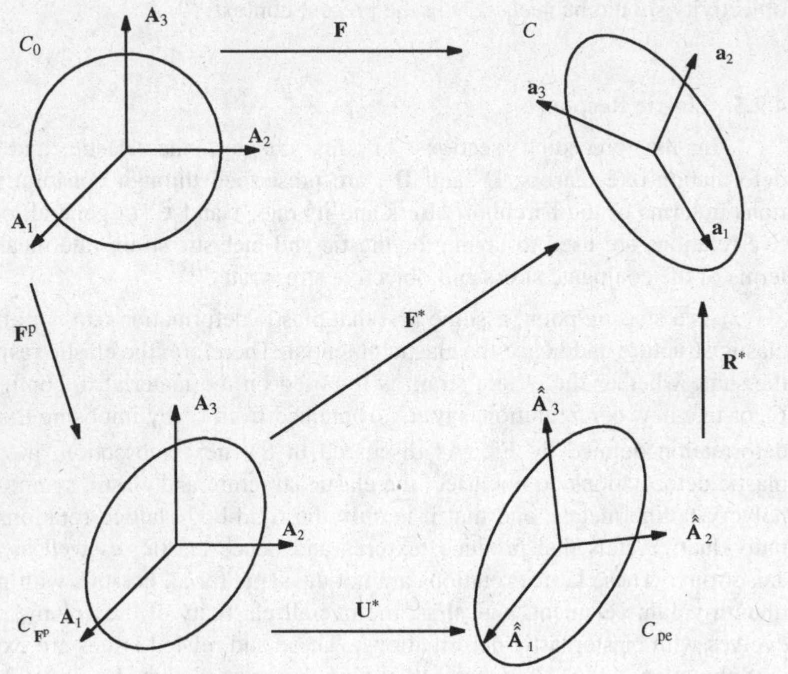

Figure 4.9.2

Decomposition of deformation gradient **F** into plastic part, $\mathbf{F}^p$, purely elastic distortion part, $\mathbf{U}^*$, and rigid-body rotation part, $\mathbf{R}^*$, such that $\mathbf{F} = \mathbf{R}^* \mathbf{U}^* \mathbf{R}^{**} \mathbf{U}^p$; $\mathbf{A}_i$, i = 1, 2, 3, represent the axes of elastic symmetry, which are unaffected by plastic deformation, $\mathbf{F}^p$, elastically distorted by $\mathbf{U}^*$ into $\hat{\hat{\mathbf{A}}}_i$, and then rotated together with the material into their final state, $\mathbf{a}_i$

$$\tilde{\mathbf{E}}^e = \tfrac{1}{2}(\mathbf{F}^{*T}\mathbf{F}^* - \mathbf{1}), \quad \mathbf{F}^* = \mathbf{R}^*\mathbf{U}^* = \mathbf{V}^e\mathbf{R}^*, \quad \mathbf{F}^{*T}\mathbf{F}^* = \mathbf{U}^{*2}, \quad (4.9.28a\text{-}d)$$

measured relative to the plastically deformed,[110] but elastically relaxed, stress-free configuration $C_{\mathbf{F}^p}$. The conjugate stress then is the Piola-Kirchhoff stress, defined by (see Table 3.1.1, page 99),

$$\tilde{\mathbf{S}}^p = \mathbf{F}^{*-1}\,\boldsymbol{\tau}\,\mathbf{F}^{*-T}, \qquad (4.9.29a)$$

again, measured relative to configuration $C_{\mathbf{F}^p}$. In terms of the potential ϕ, this becomes

$$\tilde{\mathbf{S}}^p = \frac{\partial\tilde{\phi}}{\partial\tilde{\mathbf{E}}^e}. \qquad (4.9.29b)$$

---

[110] Note that $\mathbf{U}^* = \mathbf{R}^{**}\mathbf{U}^e\mathbf{R}^{**T}$, where $\mathbf{U}^e = \mathbf{Q}^T\mathbf{V}^e\mathbf{Q}$, with $\mathbf{Q}$ given by (4.9.7c).

Take the time derivative to obtain

$$\dot{\tilde{\mathbf{S}}}^{\mathrm{p}} = \frac{\partial^2 \tilde{\phi}}{\partial \tilde{\mathbf{E}}^{\mathrm{e}} \partial \tilde{\mathbf{E}}^{\mathrm{e}}} : \dot{\tilde{\mathbf{E}}}^{\mathrm{e}} \equiv \tilde{\mathcal{L}}(\mathbf{U}^*) : \dot{\tilde{\mathbf{E}}}^{\mathrm{e}}, \tag{4.9.29c,d}$$

where the elasticity tensor $\tilde{\mathcal{L}}(\mathbf{U}^*)$, viewed as a function of $\mathbf{U}^*$, is defined by

$$\tilde{\mathcal{L}} \equiv \frac{\partial^2 \tilde{\phi}}{\partial \tilde{\mathbf{E}}^{\mathrm{e}} \partial \tilde{\mathbf{E}}^{\mathrm{e}}}, \quad \tilde{\mathcal{L}}_{ijkl} = \frac{\partial^2 \tilde{\phi}}{\partial \tilde{E}^{\mathrm{e}}_{ij} \partial \tilde{E}^{\mathrm{e}}_{kl}}, \quad i, j, k, l = 1, 2, 3. \tag{4.9.29e,f}$$

Since the elastic strain rate, $\dot{\tilde{\mathbf{E}}}^{\mathrm{e}}$, and the elastic deformation rate $\mathbf{D}^{\mathrm{e}} = ½(\mathbf{L}^* + \mathbf{L}^{*\mathrm{T}})$, are related by

$$\dot{\tilde{\mathbf{E}}}^{\mathrm{e}} = \mathbf{F}^{*\mathrm{T}} \mathbf{D}^{\mathrm{e}} \mathbf{F}^*, \tag{4.9.28e}$$

in view of (4.9.29a), obtain

$$\begin{aligned} \dot{\boldsymbol{\tau}} &= \mathbf{L}^* \boldsymbol{\tau} + \boldsymbol{\tau} \mathbf{L}^{*\mathrm{T}} + \mathbf{F}^* \dot{\tilde{\mathbf{S}}}^{\mathrm{p}} \mathbf{F}^{*\mathrm{T}} \\ &\equiv \mathbf{L}^* \boldsymbol{\tau} + \boldsymbol{\tau} \mathbf{L}^{*\mathrm{T}} + \mathcal{L}(\mathbf{U}^*) : \mathbf{D}^{\mathrm{e}}, \end{aligned} \tag{4.9.30a,b}$$

where $\mathbf{L}^*$ is now defined by[111]

$$\dot{\mathbf{F}}^* \mathbf{F}^{*-1} \equiv \mathbf{L}^* = \mathbf{L}^{\mathrm{e}} + \boldsymbol{\Omega}^*, \quad \boldsymbol{\Omega}^* = \dot{\mathbf{R}}^* \mathbf{R}^{*\mathrm{T}}, \tag{4.9.30c,d}$$

and $\mathcal{L}$ is the instantaneous elasticity tensor, referred to the current configuration C, having the following components:

$$\mathcal{L}_{ijkl}(\mathbf{U}^*) \equiv F^*_{ia} F^*_{jb} F^*_{kc} F^*_{ld} \tilde{\mathcal{L}}_{abcd}(\mathbf{U}^*), \quad i, j, k, l, a, b, c, d = 1, 2, 3. \tag{4.9.30e}$$

Introduce the stress rate, convected with the deformation rate, $\mathbf{L}^*$, as follows:

$$\overset{\nabla}{\boldsymbol{\tau}}{}^* \equiv \dot{\boldsymbol{\tau}} - \mathbf{L}^* \boldsymbol{\tau} - \boldsymbol{\tau} \mathbf{L}^{*\mathrm{T}}, \quad \overset{\nabla}{\boldsymbol{\tau}}{}^* = \mathcal{L} : \mathbf{D}^{\mathrm{e}}. \tag{4.9.31a,b}$$

Denote the inverse of the elasticity tensor $\mathcal{L}$ by $\mathcal{M}$, and rewrite (4.9.31b) as

$$\mathbf{D}^{\mathrm{e}} = \mathcal{M} : \overset{\nabla}{\boldsymbol{\tau}}{}^*. \tag{4.9.31c}$$

Both $\mathcal{M}$ and $\mathcal{L}$ are functions of the elastic stretch tensor $\mathbf{U}^*$, *i.e.*,

$$\mathcal{L} = \mathcal{L}(\mathbf{U}^*), \quad \mathcal{M} = \mathcal{M}(\mathbf{U}^*). \tag{4.9.32a,b}$$

To separate the purely spin-induced stress rate from that caused by elastic straining, consider the Jaumann rate of Kirchhoff stress, $\overset{\circ}{\boldsymbol{\tau}}{}^*$, corotational with a frame of spin $\boldsymbol{\Omega}^*$, as follows:

$$\overset{\circ}{\boldsymbol{\tau}}{}^* = \dot{\boldsymbol{\tau}} - \boldsymbol{\Omega}^* \boldsymbol{\tau} + \boldsymbol{\tau} \boldsymbol{\Omega}^*, \tag{4.9.33a}$$

and from (4.9.31a,b) and (4.9.25b), obtain

$$\overset{\circ}{\boldsymbol{\tau}}{}^* = \{\mathcal{H}(\mathbf{U}^*) : \mathbf{D}^{\mathrm{e}}\} \boldsymbol{\tau} + \boldsymbol{\tau} \{\mathcal{H}(\mathbf{U}^*) : \mathbf{D}^{\mathrm{e}}\}^{\mathrm{T}} + \mathcal{L}(\mathbf{U}^*) : \mathbf{D}^{\mathrm{e}}. \tag{4.9.33b}$$

As noted, the coefficients in this relation are functions of the elastic stretch

---

[111] Note that $\boldsymbol{\Omega}^* = \boldsymbol{\Omega} = \dot{\mathbf{Q}} \mathbf{Q}^{\mathrm{T}}$ only when $\mathbf{R}^{**} = \mathbf{1}$; see (4.9.12e) and (4.9.13b). Hence, in general, it is necessary to calculate $\mathbf{R}^{**}$, as discussed in Subsection 4.9.6.

tensor, $\mathbf{U}^*$. This is a fully anisotropic relation. No symmetry is assumed. Furthermore, no restriction is placed on the magnitude of the elastic strains, although, in reality, these strains are insignificantly small. Now, based on the right-hand side of (4.9.33b), define the modulus tensor $\boldsymbol{\mathcal{C}}$ and its inverse (assuming the inverse exists) $\boldsymbol{\mathcal{D}}$ such that

$$\overset{\circ}{\boldsymbol{\tau}}{}^* = \boldsymbol{\mathcal{C}} : \mathbf{D}^e, \quad \mathbf{D}^e = \boldsymbol{\mathcal{D}} : \overset{\circ}{\boldsymbol{\tau}}{}^*. \tag{4.9.34a,b}$$

In component form, the modulus tensor $\boldsymbol{\mathcal{C}}$ is given by

$$\mathcal{C}_{ijkl} = \mathcal{H}_{imkl}\,\tau_{mj} + \mathcal{H}_{jmkl}\,\tau_{im} + \mathcal{L}_{ijkl}. \tag{4.9.34c}$$

In the case of *elastic isotropy*, use the decomposition of Figure 4.9.1 and expression (4.9.7b), with the elastic stretch tensor defined by (4.9.7d). Then (4.9.28e) becomes

$$\dot{\bar{\mathbf{E}}}^e = (\mathbf{V}^e\,\mathbf{Q})^T\,\mathbf{D}^e\,(\mathbf{V}^e\,\mathbf{Q}), \tag{4.9.28f}$$

and (4.9.31a) changes to

$$\overset{\nabla}{\boldsymbol{\tau}}{}^{e\Omega} \equiv \dot{\boldsymbol{\tau}} - \mathbf{L}^{e\Omega}\boldsymbol{\tau} - \boldsymbol{\tau}\,\mathbf{L}^{e\Omega T}, \quad \overset{\nabla}{\boldsymbol{\tau}}{}^{e\Omega} = \boldsymbol{\mathcal{L}} : \mathbf{D}^e, \tag{4.9.31d,e}$$

where $\mathbf{L}^{e\Omega}$ now includes the total rigid spin, as defined by (4.9.13d); see also Table 4.9.1. Similarly, (4.9.33a,b) may now be replaced by,

$$\overset{\circ}{\boldsymbol{\tau}}{}^{\Omega} \equiv \dot{\boldsymbol{\tau}} - \boldsymbol{\Omega}\boldsymbol{\tau} + \boldsymbol{\tau}\boldsymbol{\Omega},$$

$$\overset{\circ}{\boldsymbol{\tau}}{}^{\Omega} = \{\boldsymbol{\mathcal{H}}(\mathbf{U}^e) : \mathbf{D}^e\}\,\boldsymbol{\tau} + \boldsymbol{\tau}\,\{\boldsymbol{\mathcal{H}}(\mathbf{U}^e) : \mathbf{D}^e\}^T + \boldsymbol{\mathcal{L}}(\mathbf{U}^e) : \mathbf{D}^e, \tag{4.9.33c,d}$$

where $\overset{\circ}{\boldsymbol{\tau}}{}^{\Omega}$ is the Jaumann rate of Kirchhoff stress, corotational with a frame of the total rigid-body spin, $\boldsymbol{\Omega}$. Equations (4.9.34a-c) remain the same, except for $\overset{\circ}{\boldsymbol{\tau}}{}^*$ that now is replaced by $\overset{\circ}{\boldsymbol{\tau}}{}^{\Omega}$.

4.9.6. Comments on Elastic Anisotropy

In general, elastic anisotropy in polycrystalline metals stems from biased orientational distribution of the constituent crystals which themselves must also be elastically anisotropic. Random orientational distributions lead to overall isotropic elasticity and plasticity of the polycrystalline metals. In addition, since large plastic deformations invariably would involve large rigid-body rotations of the crystals, even initially randomly distributed crystal orientations may evolve into biased configurations, and, hence, an anisotropic overall elastoplastic response. It is important to bear in mind that *it is the rigid-body rotation of the lattice of the crystals that may lead to their biased distributions and hence elastic anisotropy, and not their plastic deformations by simple shearing due to slip on crystallographic slip planes.* Such plastic deformations do not affect the lattice orientation and, hence, cannot produce elastic anisotropy in polycrystalline metals.

It would be rather complicated to account for the above-mentioned physical facts within a continuum setting. Therefore, continuum characterizations of large elastoplastic deformations of crystalline metals, usually tend to become formal and nonspecific, addressing mostly the general structure of the continuum relations.[112] Most models of actual metal elastoplasticity generally ignore the elastic deformations, as being insignificantly small, and focus only on plastic deformations and texturing due to lattice rotations.[113] In general, these models are based on the physical observation that plastic deformations relative to crystal lattice occur by simple shearing due to slip on crystallographic planes that produces no lattice distortions nor any lattice rotations. Such simple shearing, however, does produce material spin relative to the lattice. In crystal plasticity, this spin is directly calculated as a kinematical result, and constitutive relations are reserved for dislocation-induced crystallographic slip rates (Sections 6.4 and 6.5; see, also Section 4.8). In view of these observations, it is reasonable to require also that, in a continuum approach:

(1) any *overall* anisotropic elasticity be due to certain biased distributions of the lattice orientations of the constituent crystals;

(2) the plastic deformation associated with the deformation gradient $\mathbf{F}^p$ should not affect the overall elastic structure of the continuum, and, hence, its overall elastic symmetry axes (if any);

(3) such plastic deformations should generally include material rotation, say, $\mathbf{R}^{**}$, relative to the overall elastic symmetry axes of the continuum; and

(4) in addition to a rigid-body rotation of the elastic axes by, say, $\mathbf{R}^*$, the elasticity should be allowed to change due to possible texturing and texture evolution.

Therefore, it follows that, in general, $\mathbf{F}^p = \mathbf{R}^{**}\,\mathbf{U}^p$, where $\mathbf{R}^{**}$ need now be computed, in order to be able to calculate the rigid-body rotation, say, $\mathbf{R}^*$, of the overall elastic-symmetry frame.[114] Also, as a consequence of (2), it must be assumed that the plastic deformation defined by $\mathbf{F}^p$ does not affect the elastic potential, ϕ, so that the elastic response is the same whether the elastic strain is imposed on the material in configuration C_0 or in a new configuration $C_{\mathbf{F}^p}$, obtained by imposing the plastic deformation $\mathbf{F}^p$ on the initial undeformed material, as is assumed in Subsection 4.9.5.

---

[112] See, *e.g.*, Scheidler and Wright (2001, 2003) and references cited therein.

[113] See Taylor (1934, 1938), Bishop and Hill (1951a,b), Canova *et al.* (1984), Havner (1992), and, more recent work by Garmestani *et al.* (2001) and references cited therein.

[114] It is tempting to resolve this complication in a continuum approach based on decomposition (4.9.1), by simply resorting to some possible *constitutive* characterization of the plastic spin. As is evident from expression (4.9.36d), this characterization cannot be arbitrary. Hence, in the absence of any theoretical or experimental guidance, it would be difficult to quantify such a constitutive relation, consistent with the constitutive relation for the corresponding deformation rate tensor. Furthermore, in principle, rigid-body spin, whether it is induced as a consequence of elastic or plastic deformations of crystalline metals, should not be viewed as a *constitutive response*. In a physically well-posed problem, all rigid-body spins must simply be calculated as accompanying kinematical quantities.

Consider now a possible continuum modeling that is motivated by the physics of polycrystal plasticity, where each continuum material element is assumed to consist of a large number of single crystals with an initially known orientational distribution. Knowing also the crystal structure, *e.g.*, body-centered cubic (bcc) or face-centered cubic (fcc), the initial distribution of the slip systems is therefore also known.

A slip system consists of a slip-plane unit normal, say, **n**, which defines the plane of slip, and a slip-direction unit vector, say, **s**, which gives the direction of slip on the slip plane. In the plastically deformed configuration, C_{F^p}, let the orientation of a typical slip-plane unit normal, **n**, be defined by a pair of angles (θ, ψ), measured relative to a fixed frame, as shown in Figure 7.2.4, page 521, where $0 \geq \theta > 2\pi$ and $0 \geq \psi > \pi$. For a fixed **n**, say, $\mathbf{n} = \mathbf{n}_0$ with $\theta = \theta_0$ and $\psi = \psi_0$, choose a reference direction on the corresponding slip plane, and measure the slip direction **s** by an angle α relative to this direction.

Now, let the distribution of the slip-plane unit normals be defined by a distribution function, $\hat{E}(\mathbf{n})$, and, for a slip plane of unit normal, say, $\mathbf{n}_0$, assume that the distribution of the slip directions is defined by a distribution function $\tilde{E}(\mathbf{n}_0; \alpha) = E(\theta_0, \psi_0; \alpha)$, where $0 \geq \alpha > 2\pi$. It follows that

$$\frac{1}{2\pi}\int_0^{2\pi} \tilde{E}(\mathbf{n}_0; \alpha)\, d\alpha = \int_0^{2\pi} E(\theta_0, \psi_0; \alpha)\, d\alpha = 1\,,$$

$$\frac{1}{8\pi^2}\int_0^{2\pi} E(\theta, \psi; \alpha)\, d\alpha \int_0^{\pi} \sin\psi\, d\psi \int_0^{2\pi} d\theta = 1\,. \tag{4.9.35a,b}$$

Symmetry requires that

$$\tilde{E}(\mathbf{n}; \alpha) = \tilde{E}(\mathbf{n}; \alpha + \pi)\,, \qquad \tilde{E}(\mathbf{n}; \alpha) = \tilde{E}(-\mathbf{n}; \alpha)\,, \tag{4.9.35c,d}$$

or, equivalently,

$$E(\theta, \psi; \alpha) = E(\theta, \psi; \alpha + \pi)\,,$$

$$E(\theta, \psi; \alpha) = E(\theta + \pi, \pi - \psi; \alpha)\,. \tag{4.9.35e,f}$$

Let $\dot{\gamma}(\theta, \psi; \alpha)$ denote the slip rate[115] on the (θ, ψ)-slip plane, in the α-direction. In general, this slip rate has the sign of the corresponding shear stress, $\tau(\theta, \psi; \alpha) \equiv \mathbf{s} \cdot (\tilde{\mathbf{S}}^P \mathbf{n})$. Therefore, in general,

$$\dot{\gamma}(\theta, \psi; \alpha) = -\dot{\gamma}(\theta, \psi; \alpha + \pi)$$

$$= -\dot{\gamma}(\theta + \pi, \pi - \psi; \alpha)\,. \tag{4.9.36a,b}$$

The slip rate is generally defined in terms of the driving shear stress, temperature, and some parameters that characterize the material's resistance to plastic deformation. This is discussed in detail in Section 4.8 of this chapter and in

115 Note that the same notation, namely $\dot{\gamma}$, is used to represent the effective plastic strain rate in (4.9.2) and below in (4.9.38a,b). The context should make the distinction clear. Similar comments apply to other notations with different physical meanings in different contexts.

Chapter 6. Thus, in general, $\dot{\gamma} = \tilde{g}(\tau, T; ...)$, where $\tilde{g}$ is a known function, T is the absolute temperature, and dots, ..., stand for some relevant microstructural hardening parameters that represent, for example, dislocation accumulation and pileups on the slip plane.

Now, the velocity gradient due to plastic deformation,

$$\tilde{\mathbf{L}}^p = \dot{\mathbf{F}}^p\,\mathbf{F}^{p-1} = \tilde{\mathbf{D}}^p + \tilde{\mathbf{W}}^p, \tag{4.9.36c}$$

is given by

$$\tilde{\mathbf{L}}^p = \frac{1}{2\pi^2}\int_0^{\pi} E(\theta, \psi; \alpha)\,\dot{\gamma}(\theta, \psi; \alpha)\,\mathbf{s}(\theta, \psi; \alpha)\otimes\mathbf{n}(\theta, \psi)\,d\alpha$$

$$\times\int_0^{\pi/2}\sin\psi\,d\psi\int_0^{2\pi} d\theta\,, \tag{4.9.36d}$$

subject to the restriction that the rate of plastic work is given by[116]

$$\mathrm{tr}(\tilde{\mathbf{S}}^p\,\tilde{\mathbf{L}}^p) = \frac{1}{2\pi^2}\int_0^{\pi} E(\theta, \psi; \alpha)\,\dot{\gamma}(\theta, \psi; \alpha)\,\tau(\theta, \psi; \alpha)\,d\alpha\int_0^{\pi/2}\sin\psi\,d\psi\int_0^{2\pi} d\theta\,,$$

$$\tau(\theta, \psi; \alpha) \equiv \mathbf{s}(\theta, \psi; \alpha)\cdot\{\tilde{\mathbf{S}}^p\,\mathbf{n}(\theta, \psi)\}\,. \tag{4.9.37a,b}$$

If there are structural changes that occur in the course of deformation that change the elastic symmetry axes, $\mathbf{A}_i$, these changes must stem from different rigid-body rotations of different crystal orientations. The rigid-body rotation $\mathbf{R}^*$ cannot account for this kind of change in the material's texture or anisotropy. Hence, this change must be represented by the evolution of the distribution function, $E(\theta, \psi; \alpha)$, in an appropriate manner. In such a case, the elastic potential, ϕ, must also be modified commensurately, through appropriate dependence on some internal state variables, say, ξ's, which are by definition unaffected by the plastic deformation. However, there are no physics-based continuum models that include these complex behaviors of crystalline metals. Hence, texture analysis and elastoplastic anisotropy are usually treated by direct application of crystal plasticity and some appropriate averaging scheme, as discussed in Chapters 6 and 8.

Note that, the plastic spin is a result in the present case, independently of the structure of the distribution function E. In fact, whether there is only one[117] active slip system or infinitely many, no constitutive relation here is required for the corresponding rigid-body spin. This is similar to the single crystal case, where even though there may be many slip systems involved in producing the overall (rate-dependent) plastic deformation of the crystal, in the final analysis,

---

[116] Note that, $\mathrm{tr}(\boldsymbol{\tau}\,\mathbf{L}^p) = \mathrm{tr}(\mathbf{U}^{*2}\,\tilde{\mathbf{S}}^p\,\tilde{\mathbf{L}}^p)$, where $\boldsymbol{\tau}$ is the Kirchhoff stress, and $\mathbf{L}^p = \mathbf{F}^*\,\tilde{\mathbf{L}}^p\,\mathbf{F}^{*-1}$.

[117] The case of a single active slip system corresponds to a distribution function, $\tilde{E}(\mathbf{n}; \alpha) = 8\pi^2\,\delta(\mathbf{n} - \mathbf{n}_0)\,\delta(\alpha - \alpha_0)$. Choosing the fixed reference coordinates such that $\mathbf{n}_0 = \mathbf{e}_2$ and $\mathbf{s}_0 = \mathbf{e}_1$, it follows that $\tilde{D}^p_{12} = \tilde{D}^p_{21} = \frac{1}{2}\dot{\gamma}$ and $\tilde{W}^p_{12} = -\tilde{W}^p_{21} = \frac{1}{2}\dot{\gamma}$, where $\dot{\gamma}$ is the corresponding slip rate; see Figure 6.2.1, page 403. This also shows that $\tilde{\mathbf{W}}^p$ is necessarily related to $\tilde{\mathbf{D}}^p$ and should not be prescribed arbitrarily.

no more that five independent slip rates are calculated, from which all other slip rates are obtained.[118] In a fully developed plastic deformation of an fcc crystal, for example, there may be six or eight slip systems that are active at an instant, and in the case of bcc crystals, there may be many more simultaneously active slip systems. The slip rates of all these active slip systems can be computed explicitly from the values of the five (and not eight) independent ones, as is exemplified in Chapter 6; see also Okinaka (1995), and Nemat-Nasser *et al.* (1998b,c). Hence, the corresponding spin is calculated as a kinematical result accompanying the plastic deformation rate tensor.

4.9.7. Calculation of Stress Rate

In view of the complexity in treating elastic anisotropy and its evolution in the course of elastoplastic deformations of polycrystalline metals within a continuum framework, based on decomposition (4.9.1), elastic isotropy is generally assumed. Hence, *assume elastic isotropy in what follows* and consider decompositions in (4.9.7a-c), as well as Figure 4.9.1, page 251. Use the constitutive relation (4.9.2) to rewrite (4.9.34a) as

$$\overset{\nabla}{\boldsymbol{\tau}}{}^{*} = \mathcal{L} : (\mathbf{D} - \dot{\gamma}\frac{\partial g}{\partial \boldsymbol{\tau}}), \quad \dot{\gamma} = (2\,\mathbf{D}^{p\prime} : \mathbf{D}^{p\prime})^{1/2}. \qquad (4.9.38a,b)$$

For rate-dependent plasticity models, the effective plastic strain rate, $\dot{\gamma}$, is a given function of the stress and other parameters, as discussed in Section 4.8. When thermal effects are included, then the energy-balance equation, (3.6.8), must be used in order to account for the change in the temperature of the material due to plastic dissipation and heat conduction; additional equations are required to account for heat conduction and the related energy exchanges.

For rate-independent models of elastoplasticity, $\dot{\gamma}$ is defined in terms of a yield surface, as discussed in Sections 4.2 to 4.7. As an illustration, consider a general *anisotropic* yield surface defined by

$$f \equiv f(\boldsymbol{\tau};\, \gamma,\, \boldsymbol{\beta}), \qquad (4.9.39a)$$

where γ is the accumulated effective plastic strain calculated according to (4.9.38b) and used as a load parameter, and $\boldsymbol{\beta}$ stands for other relevant parameters which affect the yield function. These parameters may include the back-stress. For continued plastic deformation, $f = 0$ and $\dot{f} = 0$, from which it follows that

[118] It is easy to see why there are at most only five independent slip systems in any crystal, even though there may be many more simultaneously active slip systems. Slip occurs on slip *planes*, and there are only three independent planes in the three-dimensional space. On each plane, there are only two independent directions. Hence, there are only six independent slip systems. Since simple shearing produces isochoric deformations, the resulting deformation rate tensor is traceless. This leads to one more restriction, and, thus, in general, there are only five independent slip systems in a single crystal.

$$\dot{\gamma} = \frac{1}{H}\frac{\partial f}{\partial \boldsymbol{\tau}} : \dot{\boldsymbol{\tau}}, \tag{4.9.39b}$$

where the workhardening parameter H is defined by

$$H = -\left\{\frac{\partial f}{\partial \gamma} + \frac{\partial f}{\partial \boldsymbol{\beta}} \cdot \frac{d\boldsymbol{\beta}}{d\gamma}\right\}. \tag{4.9.39c}$$

Given $\mathbf{D}$ and $\mathbf{W}$, (4.9.38a) must be integrated together with the evolution equations (4.9.3), to obtain the stress increment and the increments in the internal variables $\boldsymbol{\beta}$, over a suitable time interval, as discussed in Chapter 5. For this, it is necessary to calculate $\mathbf{L}^{e\Omega} = \mathbf{L}^e + \boldsymbol{\Omega}$, in order to obtain an expression for the objective stress rate $\overset{\nabla}{\boldsymbol{\tau}}{}^{e\Omega}$, (4.9.31d). In principle, this is always possible, although the presence of the elastic deformation, *i.e.*, the term $\mathbf{V}^e$, in the kinematical equations complicates the details. The elastic strains are, however, usually very small. If this fact is used, then the expressions simplify considerably, as is discussed in the sequel, after the general case is outlined.

Without regard for the smallness of the elastic strains, based on (4.9.7a-d) note that $\mathbf{W}^p$ is given as a homogeneous and linear function of $\mathbf{D}^p$. Indeed, from $\overline{\mathbf{D}}^p + \overline{\mathbf{W}}^p = \mathbf{V}^{e-1}(\mathbf{D}^p + \mathbf{W}^p)\mathbf{V}^e$ and the relation between $\overline{\mathbf{W}}^p$ and $\overline{\mathbf{D}}^p$, *i.e.*, (4.9.21a), obtain

$$\{\mathbf{V}^{e-1}(\mathbf{D}^p + \mathbf{W}^p)\mathbf{V}^e\}_{\text{skew}} = \boldsymbol{\mathcal{K}}(\overline{\mathbf{U}}^p) : \{\mathbf{V}^{e-1}(\mathbf{D}^p + \mathbf{W}^p)\mathbf{V}^e\}_{\text{sym}}, \tag{4.9.40}$$

where the subscripts skew and sym denote the skewsymmetric and the symmetric parts of the corresponding expression. Since $\mathbf{D}^p$ is defined by the constitutive relation (4.9.2), expression (4.9.40) gives $\mathbf{W}^p$ as a linear function of $\dot{\gamma}\,\partial g/\partial\boldsymbol{\tau}$. Once $\mathbf{W}^p$ is defined in this manner, $\mathbf{W}^{e\Omega} = \mathbf{W} - \mathbf{W}^p$ and hence, $\mathbf{L}^{e\Omega} = \mathbf{W}^{e\Omega} + (\mathbf{D} - \mathbf{D}^p)$, are expressed as linear functions of $\dot{\gamma}\,\partial g/\partial\boldsymbol{\tau}$. With these expressions, the left-hand side of (4.9.38a) is defined in terms of $\dot{\boldsymbol{\tau}}$, $\boldsymbol{\tau}$, and $\dot{\gamma}\,\partial g/\partial\boldsymbol{\tau}$. The resulting differential equation can hence be integrated numerically to obtain the stress increment.

4.9.8. Small Elastic Deformations

To make the above results transparent, examine the consequences of the fact that the elastic strains are usually very small, say, less than 1% for most metals. Set

$$\mathbf{V}^e = \mathbf{1} + \boldsymbol{\varepsilon}, \quad \mathbf{V}^{e-1} = \mathbf{1} - \boldsymbol{\varepsilon} + O(\boldsymbol{\varepsilon}^2), \tag{4.9.41a,b}$$

where $O(\boldsymbol{\varepsilon}^2)$ denotes terms of the order of $\boldsymbol{\varepsilon}^2$ and greater. Note that, if $|\boldsymbol{\varepsilon}|$ is of the order of 0.01, then $|\boldsymbol{\varepsilon}^2|$ would be of the order of 0.0001. The plastic strains and rigid-body rotations of material elements, on the other hand, are usually rather large. From $\overline{\mathbf{L}} = \mathbf{V}^{e-1}\mathbf{L}^p\mathbf{V}^e$ and (4.9.41a,b), it follows that

$$\overline{\mathbf{W}}^p = \mathbf{W}^p - \boldsymbol{\varepsilon}\,\mathbf{D}^p + \mathbf{D}^p\,\boldsymbol{\varepsilon} + O(\boldsymbol{\varepsilon}^2),$$

$$\overline{\mathbf{D}}^p = \mathbf{D}^p - \boldsymbol{\varepsilon}\,\mathbf{W}^p + \mathbf{W}^p\,\boldsymbol{\varepsilon} + O(\boldsymbol{\varepsilon}^2). \tag{4.9.42a,b}$$

Since $\overline{\mathbf{W}}^p$ is given in terms of $\overline{\mathbf{D}}^p$ by (4.9.21b), these equations yield $\mathbf{W}^p$ in terms of $\mathbf{D}^p$, and, hence, $\overline{\mathbf{D}}^p$ in terms of $\mathbf{D}^p = \dot{\gamma}\, \partial g/\partial \boldsymbol{\tau}$. Indeed, substitution from these equations into (4.9.21b), or direct evaluation of (4.9.42a), leads to

$$\mathbf{W}^p + \boldsymbol{\mathcal{K}}(\overline{\mathbf{U}}^p) : (\boldsymbol{\varepsilon}\, \mathbf{W}^p - \mathbf{W}^p\, \boldsymbol{\varepsilon}) = \boldsymbol{\mathcal{K}}(\overline{\mathbf{U}}^p) : \mathbf{D}^p + \boldsymbol{\varepsilon}\, \mathbf{D}^p - \mathbf{D}^p\, \boldsymbol{\varepsilon} + O(\boldsymbol{\varepsilon}^2)\,. \quad (4.9.42c)$$

In component form, this becomes[119]

$$\{\delta_{jm}\, \delta_{in} + (\mathcal{K}_{ijml} + \mathcal{K}_{ijlm})\, \varepsilon_{ln}\}\, W^p_{nm} = (\mathcal{K}_{ijmn} + \delta_{mj}\, \varepsilon_{in} - \delta_{mi}\, \varepsilon_{jn})\, D^p_{nm} + O(\varepsilon_{il}\, \varepsilon_{lj})\,, \quad (4.9.42d)$$

where repeated indices are summed, and all indices range over 1, 2, 3. Hence, W^p_{ij} is given by

$$W^p_{ij} = \{\mathcal{K}_{ijrs} + \delta_{ir}\, \varepsilon_{js} - \delta_{jr}\, \varepsilon_{is} - (\mathcal{K}_{ijml} + \mathcal{K}_{ijlm})\, \varepsilon_{ln}\, \mathcal{K}_{nmrs}\}\, D^p_{sr} + O(\varepsilon_{il}\, \varepsilon_{lj})\,. \quad (4.9.42e)$$

When terms of the order of $\boldsymbol{\varepsilon}$ are also regarded as negligible, as is usually the case for most metals, then rewrite (4.9.42a,b) as

$$\overline{\mathbf{D}}^p = \mathbf{D}^p\ (= \dot{\gamma}\, \partial g/\partial \boldsymbol{\tau}) + O(\boldsymbol{\varepsilon})\,, \qquad \overline{\mathbf{W}}^p = \mathbf{W}^p + O(\boldsymbol{\varepsilon})\,. \quad (4.9.43a,b)$$

Again, consider isotropic elasticity with very small elastic strains. With terms of the order of $\boldsymbol{\varepsilon}$ neglected, use decomposition $\mathbf{F} = \mathbf{V}^e\, \mathbf{Q}\, \mathbf{U}^p$ and from (4.9.31a,b) and (4.9.33a,b) obtain

$$\overset{\nabla}{\boldsymbol{\tau}}{}^{e\Omega} \approx \overset{\circ}{\boldsymbol{\tau}}{}^{\Omega} = \dot{\boldsymbol{\tau}} - \boldsymbol{\Omega}\, \boldsymbol{\tau} + \boldsymbol{\tau}\, \boldsymbol{\Omega} = \boldsymbol{\mathcal{L}} : \mathbf{D}^e\,. \quad (4.9.44a\text{-}c)$$

This then leads to the following elastoplasticity equation:

$$\overset{\circ}{\boldsymbol{\tau}} \approx \boldsymbol{\mathcal{L}} : \mathbf{D}^e + \boldsymbol{\tau}\, \mathbf{W}^p - \mathbf{W}^p\, \boldsymbol{\tau}\,, \quad (4.9.44d)$$

where $\overset{\circ}{\boldsymbol{\tau}}$ is the Jaumann rate of the Kirchhoff stress,

$$\overset{\circ}{\boldsymbol{\tau}} = \dot{\boldsymbol{\tau}} - \mathbf{W}\, \boldsymbol{\tau} + \boldsymbol{\tau}\, \mathbf{W}\,. \quad (4.9.44e)$$

The plastically-induced spin, $\mathbf{W}^p$, is of the order of the plastic strain rate, $\mathbf{D}^p$, and the final constitutive equations now are

$$\overset{\circ}{\boldsymbol{\tau}} \approx \boldsymbol{\mathcal{L}} : (\mathbf{D} - \dot{\gamma}\, \frac{\partial g}{\partial \boldsymbol{\tau}}) + \boldsymbol{\tau}\, \mathbf{W}^p - \mathbf{W}^p\, \boldsymbol{\tau}\,,$$

$$\mathbf{W}^p \approx \boldsymbol{\mathcal{K}}(\overline{\mathbf{U}}^p) : (\dot{\gamma}\, \frac{\partial g}{\partial \boldsymbol{\tau}})\,, \quad (4.9.44f,g)$$

where $\boldsymbol{\mathcal{K}}(\overline{\mathbf{U}}^p)$ is defined by (4.9.22a,b). These are the usual elastoplasticity equations which, however, include the effect of the plastic spin (Nemat-Nasser; 1979, 1982).

---

[119] Note that $\mathcal{K}_{ijkl}$ is skewsymmetric with respect to its first two indices, ij, and symmetric with respect to the second two indices, kl; see (4.9.21b).

4.9.9. Evolution of Backstress

As discussed before, the rate equation for the evolution of the backstress, $\boldsymbol{\beta}$, must be defined on the basis of the physics of the process. The plastic deformation is defined by the stretch tensor, $\mathbf{U}^{\mathrm{p}}$. It therefore seems reasonable to measure the rate of change of the backstress, corotationally with the principal triad of $\mathbf{U}^{\mathrm{p}}$, if the backstress develops as a result of plastic deformation.

Let the backstress, $\boldsymbol{\beta}$, be measured relative to the current configuration C; *i.e.,* this backstress is the counterpart of the Kirchhoff stress $\boldsymbol{\tau}$, so that the stress difference is given by $\mathbf{s} = \boldsymbol{\tau} - \boldsymbol{\beta}$.

Now, in line with the definition of the Kirchhoff stress rate, (4.9.31a), introduce

$$\overset{\nabla}{\boldsymbol{\beta}}{}^{e\Omega} \equiv \dot{\boldsymbol{\beta}} - \mathbf{L}^{e\Omega}\boldsymbol{\beta} - \boldsymbol{\beta}\,\mathbf{L}^{e\Omega T}. \tag{4.9.45a}$$

The corresponding rate of change of the stress difference becomes

$$\overset{\nabla}{\mathbf{s}}{}^{e\Omega} = \dot{\mathbf{s}} - \mathbf{L}^{e\Omega}\mathbf{s} - \mathbf{s}\,\mathbf{L}^{e\Omega T}. \tag{4.9.45b}$$

It is often assumed that

$$\overset{\nabla}{\boldsymbol{\beta}}{}^{e\Omega} = A\,\mathbf{D}^{\mathrm{p}\prime}, \tag{4.9.46a}$$

where A is a material parameter which may depend on the accumulated plastic strain, γ, and other variables. Then, for small elastic strains, terms of the order of $\boldsymbol{\varepsilon}$ and higher may be neglected, leading to

$$\overset{\circ}{\boldsymbol{\beta}} \approx A\,\mathbf{D}^{\mathrm{p}\prime} + \boldsymbol{\beta}\,\mathbf{W}^{\mathrm{p}} - \mathbf{W}^{\mathrm{p}}\boldsymbol{\beta}, \tag{4.9.46b}$$

where $\mathbf{W}^{\mathrm{p}}$ is given by (4.9.44g) and $\mathbf{D}^{\mathrm{p}} = \dot{\gamma}\,\dfrac{\partial g}{\partial \boldsymbol{\tau}}$, requiring no additional constitutive relations.

4.9.10. Alternative Decomposition of Deformation Gradient

The decomposition of the deformation gradient in the manner of (4.9.1) has only a conceptual value, since, in actuality, the elastically relaxed configurations C_{p} and C_{R} can seldom be realized. In view of this, alternative kinematical decompositions, with similar conceptual value, may be considered. One such decomposition is (Clifton, 1972)

$$\mathbf{F} = \tilde{\mathbf{F}}^{\mathrm{p}}\,\tilde{\mathbf{F}}^{*}. \tag{4.9.47}$$

Here, the elastic deformation gradient, $\tilde{\mathbf{F}}^{*}$, which may include rigid-body rotation, is implemented first, followed by the remaining deformation which is the plastic deformation gradient, $\tilde{\mathbf{F}}^{\mathrm{p}}$, including the remaining rigid-body rotation.

The deformation gradients $\tilde{\mathbf{F}}^{*}$ in (4.9.47) and $\mathbf{F}^{\mathrm{p}}$ in (4.9.1) are both measured relative to the initial undeformed configuration C_0. Moreover, for homogeneous deformations, they can be interpreted as the gradients of linear fields, *i.e.,*

$$\tilde{\mathbf{F}}^* = \left[\frac{\partial \tilde{\mathbf{x}}^*}{\partial \mathbf{X}}\right]^T, \qquad \mathbf{F}^p = \left[\frac{\partial \mathbf{x}^p}{\partial \mathbf{X}}\right]^T, \tag{4.9.48a,b}$$

where $\tilde{\mathbf{x}}^* = \tilde{\mathbf{F}}^* \cdot \mathbf{X}$ and $\mathbf{x}^p = \mathbf{F}^p \cdot \mathbf{X}$. Hence, given $\mathbf{F}^p$, it is always possible to identify $\tilde{\mathbf{F}}^*$ such that

$$\mathbf{x}(\mathbf{X}) = \mathbf{x}^p(\mathbf{X}) + \tilde{\mathbf{x}}^*(\mathbf{X}) - \mathbf{X}. \tag{4.9.49a}$$

Indeed, for this homogeneous deformation,

$$\tilde{\mathbf{F}}^* = \mathbf{F} - \mathbf{F}^p + \mathbf{1}, \tag{4.9.49b}$$

and

$$\mathbf{x} = \mathbf{F}^p \cdot \mathbf{X} + \tilde{\mathbf{F}}^* \cdot \mathbf{X} - \mathbf{X}. \tag{4.9.49c}$$

In this manner, an additive decomposition of the deformation gradient is obtained,

$$\mathbf{F} = \mathbf{F}^p + \tilde{\mathbf{F}}^* - \mathbf{1}. \tag{4.9.50}$$

Since, in continuum mechanics, all deformations are considered to be *locally* homogeneous, (4.9.50) holds locally even when the global deformation is nonhomogeneous. Various implications of the decomposition (4.9.50) are discussed by Nemat-Nasser (1979, 1982). More recently, Lubarda (1999) has examined various features of decomposition (4.9.47) in relation to (4.9.1).

4.10. REFERENCES

Abel, A. and Ham, R. K. (1966), The cyclic strain behaviour of crystals aluminum-4 wt.% copper - I, the Bauschinger effect, *Acta Metall.*, Vol. 14, 1489-1494.

Abel, A. and Muir, H. (1972), The Bauschinger effect and stacking fault energy, *Phil. Mag.*, Vol. 27, 585-594.

Anand, L. (1980), Constitutive equations for rate-independent, isotropic, elastic-plastic solids exhibiting pressure-sensitive yielding and plastic dilatancy, *J. Appl. Mech.*, Vol. 47, 439-441.

Anand, L. (1983), Plane deformations of ideal granular materials, *J. Mech. Phys. Solids*, Vol. 31, 105-122.

Anand, L. and Kalidindi, S. R. (1994), The process of shearband formation in plane strain compression of fcc metals: effects of crystallographic texture, *Mech. Mat.*, Vol. 17, 223-243.

Argon, A. S. (1975), *Constitutive equations in plasticity*, MIT Press, Cambridge, Massachusetts.

Argon, A. and East, G. (1979), Forest dislocation intersections in stage I deformation of copper single crystals, (Haasen, P., Gerold, V., and Kostorz, G. eds.), *Proceedings of the Fifth International Conference on the Strength of Metals and Alloys*, Pergamon Press, Oxford, 9-16.

Arthur, J. R. F. and Menzies, B. K. (1972), Inherent anisotropy in a sand, *Géotechnique*, Vol. 22, 115-128.

Asaro, R. J. (1979), Geometrical effects in the inhomogeneous deformation of ductile single crystals, *Acta Metall.*, Vol. 27, 445-453.

Asaro, R. J. (1983), Micromechanics of crystals and polycrystals, *Advanced Solid Mechanics*, Vol. 23, 1-115.

ASM Volume 8, Mechanical Testing and Evaluation Handbook (2000), Chapter on High Strain Testing, page 427.

Atkins, R. J. and Fox, N. (1980), *An introduction to the theory of elasticity*, Longman, London.

Balendran, B. (1993), Physically based constitutive modeling of granular materials, *Ph.D. Thesis*, University of California, San Diego.

Balendran, B. and Nemat-Nasser, S. (1993a), Double sliding model cyclic deformation of granular materials, including dilatancy effects, *J. Mech. Phys. Solids*, Vol. 41, 573-612.

Balendran, B. and Nemat-Nasser, S. (1993b), Viscoplastic flow of planar granular materials, *Mech. Mat.*, Vol 16, 1-12.

Bassani, J. L. (1977), Yield characterization of metals with transversely isotropic plastic properties, *Int. J. Mech. Sci.*, Vol. 19, 651-660.

Bassani, J. L., Hutchinson, J. W., and Neale, K. W. (1979), On the prediction of necking in anisotropic sheets, in *Metal forming plasticity* (Lippmann, H. ed.), Springer-Verlag, Berlin, 1-13.

Bate, P. S. and Wilson, D. V. (1986), Analysis of the Bauschinger effect, *Acta Metall.*, Vol. 34, 1097-1105.

Bauschinger, J. (1881), Tension prisma Stabe, *Civilingenieur*, Vol. 27, 289-301.

Berg, C. A. (1970), Plastic dilation and void interaction, *Inelastic behavior of solids* (Kanninen, M. F. *et. al.* eds.), McGraw-Hill, New York, 171-210.

Biot, M. A. (1954), Theory of stress-strain relations in anisotropic viscoelasticity and relaxation phenomena, *J. Appl. Phys.*, Vol. 25, 1385-1391.

Bishop, J. F. W. and Hill, R. (1951a), A theory of plastic distortion of a polycrystalline aggregate under combined stresses, *Phil. Mag.*, Vol. 42, 414-427.

Bishop, J. F. W. and Hill, R. (1951b), A theoretical derivation of the plastic properties of a polycrystalline face-centered metal, *Phil. Mag.,* Vol. 42, 1290-1307.

Brown, L. M. and Stobbs, W. M. (1971), The workhardening of copper-silica - I, a model based on internal stresses, with no plastic relaxation, *Phil. Mag.*, Vol. 23, 1185-1199.

Budiansky, B. (1959), A reassessment of deformation theories of plasticity, *J. Appl. Mech.*, Vol. 26, 259-264.

Budiansky, B., Dow, N. F., Peters, R. W., and Shepherd, R. (1951), Experimental studies of polyaxial stress-strain laws of plasticity, in *Proc. 1st US Nat. Cong. Appl. Mech.*, ASME, New York, 503-512.

Bui, H. D. (1970), Évolution de la frontière du domaine élastique des métaux avec l'écrouissage plastique et comportement élastoplastique d'un agrégat de cristaux cubiques, *Mem. Artillerie Franc.: Sci. Tech. Armement*, Vol. 1, 141-165.

Butler, G. C. and McDowell, D. L. (1998), Polycrystal constraint and grain subdivision, *Int. J. Plasticity*, Vol. 14, 703-717.

Byerlee, J. D. (1968), Brittle-ductile transition in rocks, *J. Geophys. Res.*, Vol. 73, 4741-4750.

Byerlee, J. D. (1975), The fracture strength and frictional strength of Weber sandstone, *Int. J. Rock Mech. Min. Sci. & Geomech. Abstr.*, Vol. 12, 1-4.

Canova, G. R., Kocks, U. F., and Jonas, J. J. (1984), Theory of torsion texture development, *Acta Metall.*, Vol. 32, 211-226.

Carroll, M. M. (1987), A rate independent constitutive theory for finite inelastic deformation, *J. Appl. Mech.*, Vol. 54, 15-21.

Carroll, M. M. and Carman, R. A. (1985), Influence of yield surface curvature on flow localization in dilatant plasticity, *Mech. Mat.*, Vol. 4, 409-415.

Casey, J. and Naghdi, P. M. (1980), A remark on the use of the decomposition $\mathbf{F} = \mathbf{F}^{e}\mathbf{F}^{p}$ in plasticity, *J. Appl. Mech.*, Vol. 47, 672-675.

Casey, J. and Naghdi, P. M. (1983), On the nonequivalence of the stress space and strain space formulation of plasticity theory, *J. Appl. Mech.*, Vol. 50, 350-354.

Chadwick, P. (1976), *Continuum mechanics*, Allen and Unwin, London.

Chen, S. R., Gray, G. T., and Bingert, S. R. (1996), Mechanical properties and constitutive relations for tantalum and tantalum alloys under high-rate deformation, in *Tantalum, 1996 TMS annual meeting*.

Cheng, J. Y. and Nemat-Nasser, S. (2000), A model for experimentally-observed high-strain-rate dynamic strain aging in titanium, *Acta Mater.*, Vol. 48, 3131-3144.

Cheng, J. Y., Nemat-Nasser, S., and Guo, W. G. (2001), A unified constitutive model for strain-rate and temperature dependent behavior of molybdenum, *Mech. Mat.*, Vol. 33, 603-616.

Christoffersen, J. and Hutchinson, J. W. (1979), A class of phenomenological corner theories of plasticity, *J. Mech. Phys. Solids*, Vol. 27, 465-487.

Christoffersen, J., Mehrabadi, M. M., and Nemat-Nasser, S. (1981), A micromechanical description of granular material behavior, *J. Appl. Mech.*, Vol. 48, 339-344.

Clayton, J. D. and McDowell, D. L. (2003), A multiscale multiplicative decomposition for elastoplasticity of polycrystals, *Int. J. Plasticity*, Vol. 19, 1401-1444.

Clifton, R. J. (1972), On the equivalence of $\mathbf{F}^{e}\mathbf{F}^{p}$ and $\overline{\mathbf{F}}^{p}\overline{\mathbf{F}}^{e}$, *J. Appl. Mech.*, Vol. 39, pp 287-289.

Clifton, R. J. (1983), Dynamic plasticity, *J. Appl. Mech.*, Vol. 50, 941-952.

Coleman, B. and Gurtin, M. (1967), Thermodynamics with internal state variables, *J. Chem. Phys.*, Vol. 47, 597-613.

Conrad, H. (1964), Thermally activated deformation of metals, *J. Metals*, 16, 582-588.

Cottrell, A. H. (1963), *Dislocations and plastic flow in crystals,* Oxford

University Press; first edition (1953).
Dafalias, Y. F. (1983), Corotational rates for kinematic hardening at large plastic deformations, *J. Appl. Mech.*, Vol. 50, 561-565.
Dafalias, Y. F. and Popov, E. P. (1976), Plastic internal variables formalism of cyclic plasticity, *J. Appl. Mech.*, Vol. 43, 645-651.
Daniels, R. C. and Horn, G. T. (1971), The Bauschinger effect and cyclic hardening in copper, *Metall. Trans.*, Vol. 2, 1161-1172.
Davison, L. (1995), Kinematics of finite elastoplastic deformation, *Mech. Mat.*, Vol.21, 73-88.
Dienes, J. K. (1979), On the analysis of rotation and stress rate in deforming bodies, *Acta Mech.*, Vol. 32, 217-232.
Dorn, J., Goldberg, A., and Tietz, T. (1949), The effect of thermal-mechanical history on the strain hardening of metals, *Trans. AIME*, Vol. 180, 205-224.
Dorris, J. F. and Nemat-Nasser, S. (1982), A plasticity model for flow of granular materials under triaxial stress states, *Int. J. Solids Struct.*, Vol. 18, 497-531.
Drescher, A. (1976), An experimental investigation of flow rules for granular materials using optically sensitive glass particles, *Géotechnique*, Vol. 26, 591-601.
Drucker, D. (1950), Some implications of work hardening and ideal plasticity, *Q. Appl. Math.*, Vol. 7, 411-418.
Drucker, D. (1964), Stress-strain-time relations and irreversible thermodynamics, in *Proc. Int. Symp. Second-Order Effects in Elasticity, Plasticity and Fluid dynamics* (Reiner, M. and Abir, D. eds.), Pergamon, Oxford, 331-351.
Drucker, D. and Prager, W. (1952), Soil mechanics and plastic analysis for limit design, *Q. Appl. Math.*, Vol. 10, 157-165.
Drucker, D., Gibson, R. E., and Henkel, D. J. (1957), Soil mechanics and work-hardening theories of plasticity, *Trans. ASCE*, Vol. 122, 338-346.
Edelman, F. and Drucker, D. (1951), Some extensions of elementary plasticity theory, *J. Franklin Inst.*, Vol. 251, 581-605.
Follansbee, P. S. (1986), High-strain-rate deformation of fcc metals and alloys, in *Metallurgical applications of shock-wave and high-strain-rate phenomena,* (Murr, L. E., Staudhammer, K. P., and Meyers, M. A. eds.), Marcel Dekker, New York, 451-480.
Follansbee, P. S. and Gray, G. T. (1991), The response of single crystal and polycrystal nickel to quasistatic and shock deformation, *Int. J. Plasticity*, Vol. 7, 651-660.
Follansbee, P. S. and Kocks, U. F. (1988), A constitutive description of the deformation of copper based on the use of the mechanical threshold stress as an internal state variable, *Acta Metall.*, Vol. 36, 81-93.
Follansbee, P. S., Regazzoni, G., and Kocks, U. F. (1984), The transition to drag-controlled deformation in copper at high strain rates, in *Mechanical properties at high rates of strain,* (Harding, J. ed), Institute of Physics, London, 71-80.
Garmestani, H., Lin, S., Adams, B. L., and Ahzi, S. (2001), Statistical

continuum theory for large plastic deformation of polycrystalline materials, *J. Mech. Phys. Solids*, Vol. 49, 589-607.

Germain, P., Nguyen, Q. S., and Suquet, P. (1983), Continuum thermodynamics, *J. Appl. Mech.*, Vol. 50, 1010-1020.

Gilman, J. J. (1969), A unified view of flow mechanisms in materials, *Strength and plasticity* (Argon, A. S. ed.), MIT Press, Cambridge, Massachusetts, 3-14.

Green, A. E. and Atkins, J. E. (1960), *Large elastic deformations and non-linear continuum mechanics*, Clarendon, Oxford.

Green, A. E. and Naghdi, P. M. (1965), A general theory of an elastic-plastic continuum, *Arch. Rat. Mech. Anal.*, Vol. 18, 251-281.

Green, A. E. and Naghdi, P. M. (1971), Some remarks on elastic-plastic deformation at finite strain, *Int. J. Engng. Sci.*, Vol. 9, 1219-1229.

Gurson, A. L. (1977), Continuum theory of ductile rupture by void nucleation and growth - I - yield criteria and flow rules for porous ductile media, *J. Eng. Mat. Tech.*, Vol. 99, 2-15.

Gurtin, M. E., (2000), On the plasticity of single crystals: free energy, microforces, plastic strain gradients, *J. Mech. Phys. Solids*, Vol. 48, 989-1036.

Halphen, B. and Nguyen, Q. S. (1975), Sur les matériaux généralisés, *J. de Mécanique*, Vol. 14, 39-63.

Havner, K. S. (1979), The kinematics of double slip with application to cubic crystals in the compression test, *J. Mech. Phys. Solids*, Vol. 27, 415-429.

Havner, K. S. (1982a), Minimum plastic work selects the highest symmetry deformation in axially-loaded fcc crystals, *Mech. Mat.*, Vol. 1, 97-111.

Havner, K. S. (1982b), The theory of finite plastic deformation of crystalline solids, in *Mech. of Solids* (Hopkins, H. G. and Sewell, M. J. eds.), Pergamon, New York, 265-302.

Havner, K. S. (1992), *Finite plastic deformation of crystalline solids*, Cambridge University Press, Cambridge.

Havner, K. S. and Shalaby, A. H. (1977), A simple mathematical theory of finite distortional latent hardening in single crystals, *Proc. R. Soc. Lond. A*, Vol. 358, 47-70.

Hecker, S. S. (1976), Experimental studies of yield phenomena in biaxially loaded metals, in *Constitutive equations in viscoplasticity: computational and engineering aspects* (Stricklin, J. A. and Saczalski, K. J. eds.), ASME AMD-20, 1-33.

Hidayetoglu, U. K., Pica, P. N., and Haworth, W. L. (1985), Aging dependence of the Bauschinger effect in aluminum alloy 2024, *Mat. Sci. Eng.*, Vol. 73, 65-76.

Hill, R. (1958), A general theory of uniqueness and stability in elastic-plastic solids, *J. Mech. Phys. Solids*, Vol. 6, 236-249.

Hill, R. (1959), Some basic principles in the mechanics of solids without a natural time, *J. Mech. Phys. Solids*, Vol. 7, 209-225.

Hill, R. (1966), Generalized constitutive relations for incremental deformation of metals crystals by multislip, *J. Mech. Phys. Solids*, Vol. 14, 95-102.

Hill, R. (1967), On the classical constitutive relations for elastic-plastic solids, in *Recent progress in applied mechanics*, Folke Odqvist Volume (Broberg,, B., Hult, J., and Niordson, F. eds.), Almqvist and Wiksell, Stockholm, 241-249.

Hill, R. (1968), On constitutive inequalities for simple materials - I and II, *J. Mech. Phys. Solids*, Vol. 16, 229-242 and 315-322.

Hill, R. (1972), On constitutive macro-variables for heterogeneous solids at finite strain, *Proc. Roy. Soc. Lond. A*, Vol. 326, 131-147.

Hill, R. (1978), Aspects of invariance in solid mechanics, *Advances in applied mechanics* (Yih, C.-S. ed.), Academic Press, New York, Vol. 18, 1-75.

Hill, R. (1979), Theoretical plasticity of textured aggregates, *Math. Proc. Cambridge Phil. Soc.*, Vol. 85, 179-191.

Hill, R. and Hutchinson, J. W. (1975), Bifurcation phenomena in the plane tension test, *J. Mech. Phys. Solids*, Vol. 23, 179-191.

Hill, R. and Rice, J. R. (1972), Constitutive analysis of elastic-plastic crystals at arbitrary strain, *J. Mech. Phys. Solids*, Vol. 20, 401-413.

Hill, R. and Rice, J. R. (1973), Elastic potentials and the structure of inelastic constitutive laws, *SIAM J. Appl. Math.*, Vol. 25, 448-461.

Hirth, J. P. and Lothe, J. (1992), *Theory of dislocation*, Krieger, Melbourne, Florida.

Hull, D. and Bacon, D. J. (1992), *Introduction to dislocations*, Pergamon, Oxford.

Hutchinson, J. W. (1970), Elastic-plastic behavior of polycrystalline metals and composites, *Proc. Roy. Soc. Lond. A*, Vol. 319, 247-272.

Il'yushin, A. A. (1961), On the postulate of plasticity, *J. Appl. Math. Mech.*, Vol. 25, 746-752.

Ishihara, K. and Towhata, I. (1983), Sand response to cyclic rotation of principal stress directions as induced by wave loads, *Soils and Foundations* Vol. 23, 11-26.

Iwakuma, T. and Nemat-Nasser, S. (1982), An analytical estimate of shear band initiation in a necked bar, *Int. J. Solids Struct.*, Vol. 18, 69-83.

Iwakuma, T. and Nemat-Nasser, S. (1984), Finite elastic-plastic deformation of polycrystalline metals, *Proc. Roy. Soc. Lond. A*, Vol. 23, 87-119.

Jenike, A. W. and Shield, R. T. (1959), On the plastic flow of Coulomb solids beyond original failure, *J. Appl. Mech.*, Vol. 26, 599-602.

Johnson, G. R. and Cook, W. H. (1983), A constitutive model and data for metals subjected to large strains, high strain rates, and high temperatures, in *Proc. 7th Int. Symp. Ballistics*, The Hague, The Netherlands, 1-7.

Johnson, G. R. and Cook, W. H. (1985), Fracture characteristics of three metals subjected to various strains, strain rates, temperatures and pressures, *Eng. Frac. Mech.*, Vol. 21, 31-48.

Kalidindi, S. (2001), Modeling anisotropic strain hardening and deformation textures in low stacking fault energy fcc metals, *Int. J. Plasticity*, Vol. 17, 837-860.

Kanatani, K. (1984), Distribution of directional data and fabric tensors, *Int. J. Eng. Sci.*, Vol. 22, 149-164.

Kapoor, R. and Nemat-Nasser, S. (1998), Determination of temperature rise during high strain rate deformation, *Mech. Mat.*, Vol. 27, 1-12.

Kapoor, R. and Nemat-Nasser, S. (2000), Comparison between high strain-rate and low strain-rate deformation of tantalum, *Metall. and Mat. Trans.*, Vol. 31A, 815-823.

Kestin, J. and Bataille, J. (1980), Thermodynamics of solids, in *Continuum models of discrete systems* (Kroner, E. and Anthony, K. H. eds.), University of Waterloo Press, Ontario, 99-147.

Kim, D. S., Nemat-Nasser, S., Isaacs, J. B., and Lischer, D. (1998), Adiabatic shearband in WHA in high-strain-rate compression, *Mech. Mat.*, Vol. 28, 227-236.

Kittel, C. (1969), *Thermal Physics*, John Wiley and Sons, Inc.

Klepaczko, J. (1989), Discussion of microstructural effects and their modeling at high rates of strain, in *Proc. Int. Conf. Mech. Prop. Mat. High Rates of Strain*, Oxford, 183-298.

Klepaczko, J. and Chiem, C. Y. (1986), On rate sensitivity of fcc metals, instantaneous rate sensitivity and rate sensitivity of strain hardening, *J. Mech. Phys. Solids*, Vol. 34, 29-34.

Kocks, U. F. (1970), The relation between polycrystal deformation and single-crystal deformation, *Metall. Trans.*, Vol. 1, 1121-1142.

Kocks, U. F. (1975), Constitutive relations for slip, in *Constitutive equations in plasticity* (Argon, A. S. ed.), MIT Press, Cambridge, Massachusetts, 81-115.

Kocks, U. F., Argon, A. S., and Ashby, M. F. (1975), Thermodynamics and kinetics of slip, in *Progress in Materials Science*, Pergamon, New York, Vol. 19, 68-170.

Konishi, J., Oda, M., and Nemat-Nasser, S. (1982), Inherent anisotropy and shear strength of assemblies of oval cross-sectional rods, *Deformation and Failure of Granular Materials*, Proceedings of the IUTAM Symposium, Delft, The Netherlands, August 31 - September 3, 1982, P.A. Vermeer and H.J. Luger (eds.), A.A. Balkema/Rotterdam, 403-412.

Konishi, J., Oda, M., and Nemat-Nasser, S. (1983), Induced anisotropy in oval cross-sectional rods in biaxial compression, in *Proc. US-Japan Sem. Mech. Granular Mat.: New Models and Constitutive Relations* (Jenkins, J. T. and Satake, M. eds.), 31-39.

Kothari, M. and Anand, L. (1998), Elasto-viscoplastic constitutive equations for polycrystalline metals: application to tantalum, *J. Mech. Phys. Solids*, 46, 51-83.

Krajcinovic, D. (1989), Damage mechanics, *Mech. Mat.*, Vol. 8, 117-197.

Krajcinovic, D. (1996), *Damage mechanics*, North-Holland, New York.

Kratochvil, J. (1971), Finite-strain theory of crystalline elastic-inelastic materials, *J. Appl. Phys.*, Vol. 42, 1104-1108.

Lade, P. V., Nelson, R. B., and Ito, Y. M. (1987), Nonassociated flow and stability of granular materials, *J. Eng. Mech.*, ASCE, Vol. 113, 1302-1328.

Lade, P. V., Nelson, R. B., and Ito, Y. M., (1988), Instability of granular materials with nonassociated flow, *J. Eng. Mech.*, ASCE, Vol. 114, 2173-2191.

Lance, G. and Nemat-Nasser, S. (1986), Slip-induced plastic flow of geomaterials and crystals, *Mech. Mat.*, Vol. 5, 1-11.

Lee, E. H. (1969), Elastic-plastic deformations at finite strains, *J. Appl. Mech.*, Vol. 36, 1-6.

Li, J. C. M. (1962), Thermodynamics for nonequilibrium systems; The principle of macroscopic separability and the thermokinetic potential, *J. Appl. Phys.*, Vol. 33, 616-624.

Lion, A. (2000), Constitutive modelling in finite thermoviscoplasticity: a physical approach based on nonlinear rheological models, *Int. J. Plasticity*, Vol. 16, 469-494.

Loret, B. (1983), On the effect of plastic rotation in the finite deformation of anisotropic elastoplastic materials, *Mech. Mat.*, Vol. 2, 287-304.

Lubarda, V. A. (1999), Duality in constitutive formulation of finite-strain elastoplasticity based on $\mathbf{F} = \mathbf{F}_e \mathbf{F}_p$ and $\mathbf{F} = \mathbf{F}^p \mathbf{F}^e$ decompositions, *J. Plasticity*, Vol. 15, 1277-1290.

Lubarda, V. A. (2002), *Elastoplasticity theory*, CRC Press, New York.

Lubarda, V. A. and Shih, C. F. (1994), Plastic spin and related issues in phenomenological plasticity, *J. Appl. Mech.*, Vol. 61, 524-529.

Lubliner, J. (1969), On fading memory in materials of evolutionary type, *Acta Mech.*, Vol. 8, 75-81.

Mandel, J. (1947), Sur les lignes de glissement et le calcul des déplacements dans la déformation plastique, *Comptes rendus de l'Académie des Sciences*, Vol. 225, 1272-1273.

Mandel, J. (1971), Courses and Lectures, *Plasticité classique et viscoplasticité*, CIDM, Udine, Springer, New York.

Mandel, J. (1973), Equations constitutives et directeurs dans les milieux plastiques et viscoplastiques, *Int. J. Solids Struct.*, Vol. 9, 725-740.

Mandl, G. and Luque, F. R. (1970), Fully developed plastic shear flow of granular materials, *Géotechnique*, Vol. 20, 277-307.

Masing, G. (1922), Zur Theorie der Warmespannungen, *Z. Tech. Phys.*, Vol. 3, 167-174.

Mear, M. E. and Hutchinson, J. W. (1985), Influence of yield surface curvature on flow localization in dilatant plasticity, *Mech. Mat.*, Vol. 4, 395-407.

Mehrabadi, M. M. and Cowin, S. C. (1978), Initial planar deformation of dilatant granular materials, *J. Mech. Phys. Solids*, Vol. 26, 269-284.

Mehrabadi, M. M. and Nemat-Nasser, S. (1987), Some basic kinematical relations for finite deformations of continua, *Mech. Mat.*, Vol. 6, 127-138.

Mehrabadi, M. M, Loret, B., and Nemat-Nasser, S. (1993), Incremental constitutive relations for granular materials based on micromechanics, *Proc. Roy. Soc. Lond. A.*, Vol. 441, 433-463.

Mehrabadi, M. M., Nemat-Nasser, S., and Oda, M. (1982), On statistical description of stress and fabric in granular materials, *Int. J. Num. Anal. Methods in Geomechanics*, Vol. 6, 95-108.

Mehrabadi, M. M., Nemat-Nasser, S., Shodja, H. M., and Subhash, G. (1988), Some basic theoretical and experimental results on micromechanics of granular flow, *Micromechanics of Granular Materials*: Proc. 2nd US-Japan Seminar on Mechanics of Granular Materials, October 26-30, 1987, in Sendai-Zao, Japan (Jenkins, J.T. and Satake, M. eds.), Elsevier Science Publishers, Amsterdam, 253-262.

Meixner, J. (1954), Thermodynamische Theorie der elastischen Relaxation, *Z. Naturf.*, Vol. 9(a), 654-663.

Moan, G. D. and Embury, J. D. (1979), A study of the Bauschinger effect in Al-Cu alloys, *Acta Mat.*, Vol. 27, 903-914.

Moore, J. T. and Kuhlmann-Wilsdorf, D. (1970), The rate of energy storage in workhardened metals, In *2nd International Conference on the strength of metals and alloys*, Vol. 2, 484-488.

Mróz, Z. (1967), On the description of anisotropic work-hardening, *J. Mech. Phys. Solids*, Vol. 15, 163-175.

Mróz, Z. (1969), An attempt to describe the behavior of metals under cyclic loads using a more general workhardening model, *Acta Mech.*, Vol. 7, 199-212.

Mróz, Z. (1980), Deformation and flow of granular materials, in *Theoretical and applied mechanics* (Rimrott, F. P. J. and Tabarrok, B. eds.), North-Holland, Amsterdam, 119-132.

Naghdi, P. M. (1990), A critical review of the state of finite plasticity, *J. Appl. Math. Phys. (ZAMP)*, Vol. 41, 315-394.

Naghdi, P. M. and Trapp, J. A. (1975), On the nature of normality of plastic strain rate and convexity of yield surfaces in plasticity, *J. Appl. Mech.*, Vol. 42, 61-66.

Neale, K. W. (1981), Phenomenological constitutive laws in finite plasticity, *SM Arch.*, Vol. 6, 79-128.

Needleman, A. and Rice, J. R. (1978), Limit to ductility set by plastic localization, *Mechanics of sheet metal forming*, (Koistinen, D. P. and Wang, N.-M., eds.), Plenum Press, 237-267.

Needleman, A. and Tvergaard, V. (1982), Aspects of plastic postbuckling behavior, *Mechanics of Solids, the Rodney Hill 60th anniversary volume* (Hopkins, H. G. and Sewell, M. J., eds.), Pergamon, Oxford, 453-498.

Nemat-Nasser, S. (1975a), On nonequilibrium thermodynamics of viscoelasticity: inelastic potentials and normality conditions, in *Mechanics of viscoelastic media and bodies* (Hult, J. ed.), Springer-Verlag, New York, 375-391.

Nemat-Nasser, S. (1975b), On nonequilibrium thermodynamics of continua, in *Mechanics Today* (Nemat-Nasser, S. ed.), Pergamon, New York, Vol. 2, Chpt. 2, 94-158.

Nemat-Nasser, S. (1977), On nonequilibrium thermodynamics of continua. Addendum, in *Mechanics Today* (Nemat-Nasser, S. ed.), Pergamon, New York, Vol. 4, 391-405.

Nemat-Nasser, S. (1979), Decomposition of strain measures and their rates in finite deformation elastoplasticity, *Int. J. Solids Struct.*, Vol. 15, 155-166.

Nemat-Nasser, S. (1980a), On behavior of granular materials in simple shear,

Soils and Foundations, Vol. 20, 59-73.

Nemat-Nasser, S. (1980b), On constitutive behavior of fault materials, *Solid Earth Geophysics and Geotechnology*, Proceedings of the ASME Symposium, AMD-Vol. 42, ASME, New York, 31-37.

Nemat-Nasser, S. (1982), On finite deformation elasto-plasticity, *Int. J. Solids Struc.*, Vol. 18, 857-872.

Nemat-Nasser, S. (1983), On finite plastic flow of crystalline solids and geomaterials, *J. Appl. Mech.* (50th Anniversary Issue), Vol. 50, 1114-1126.

Nemat-Nasser, S. (1984), Theoretical foundations of plasticity, *Theoretical foundation for large-scale computations of nonlinear material behavior*, Nemat-Nasser, S., Asaro, R. J., and Hegemier, G. A. eds.), Martinus Nijhoff Publishers, Dordrecht, 7-24.

Nemat-Nasser, S. (1990), Certain basic issues in finite-deformation continuum plasticity, *Meccanica*, Vol. 25, 223-229.

Nemat-Nasser, S. (1992), Phenomenological theories of elastoplasticity and strain localization at high strain rates, *J. Appl. Mech. Rev.*, Vol. 45, Part 2, 519-545.

Nemat-Nasser, S., (2000a), A micromechanically-based constitutive model for frictional deformation of granular materials, *J. Mech. Phys. Solids*, Vol. 38, 1541-1563.

Nemat-Nasser, S., (2000b), Recovery Hopkinson bar techniques, *ASM Volume 8, Mechanical Testing and Evaluation Handbook*, 477-487.

Nemat-Nasser, S. and Guo, W. G. (2000a), High-strain-rate response of commercially pure vanadium, *Mech. Mat.*, Vol. 32, 243-260.

Nemat-Nasser, S. and Guo, W. G. (2000b), Flow stress of commercially pure niobium over a broad range of temperatures and strain rates, *Mat. Sci. & Eng. A*, Vol. 284, 202-210.

Nemat-Nasser, S. and Guo, W. G. (2003), Thermomechanical response of DH-36 structural steel over a wide range of strain rates and temperature, *Mech. Mat*, Vol. 35, 1023-1047.

Nemat-Nasser, S. and Guo, W. G. (2004), Thermomechanical response of HSLA-65 steel plates: experiments and modeling, to be published.

Nemat-Nasser, S. and Hori, M. (1990), Elastic solids with microdefects, *Micromechanics and inhomogeneity; the Toshio Mura 50th anniversary volume* (Weng, G. R., Taya, M., and Abe, H. eds.), 297-320.

Nemat-Nasser, S. and Hori, M. (1993), *Micromechanics: overall properties of heterogeneous solids*, Elsevier, Amsterdam; second revised edition (1999), Elsevier, Amsterdam.

Nemat-Nasser, S. and Isaacs, J. B. (1997a), High strain-rate, high-temperature response of tantalum and tantalum alloys, *Tantalum* (Chen, E., Crowson, A., Lavernia, E., Ebihara, W., and Kumar, P. eds.), The Minerals, Metals & Materials Society, Warrendale, Pennsylvania, 135-138.

Nemat-Nasser, S. and Isaacs J. B. (1997b), Direct measurement of isothermal flow stress of metals at elevated temperatures and high strain rates with application to Ta and Ta-W alloys, *Acta Mat.*, Vol. 45, 907-919.

Nemat-Nasser, S. and Kapoor, R. (2001), Deformation behavior of tantalum and a tantalum tungsten alloy, *Int. J. Plasticity*, Vol. 17, 1351-1366.

Nemat-Nasser, S. and Li, Y. F. (1998), Flow stress of fcc polycrystals with application to OFHC Cu, *Acta Mat.*, Vol. 46, 565-577.

Nemat-Nasser, S. and Obata, M. (1986), Rate dependent, finite elasto-plastic deformation of polycrystals, *Proc. Roy. Soc. Lond. A*, Vol. 407, 343-375.

Nemat-Nasser, S. and Okada, N. (2001), Radiographic and microscopic observation of shear bands in granular materials, *Géotechnique*, Vol. 51, 753-765.

Nemat-Nasser, S. and Shokooh, A. (1980), On finite plastic flow of compressible materials with internal friction, *Int. J. Solids Struc.*, Vol. 16, 495-514.

Nemat-Nasser, S. and Takahashi, K. (1984), Liquefaction and fabric of sand, *J. Geotech. Eng.*, ASCE, Vol. 110, 1291-1306.

Nemat-Nasser, S. and Tobita, Y. (1982), Influence of fabric on liquefaction and densification potential of cohesionless sand, *Mech. Mat.*, Vol. 1, 43-62.

Nemat-Nasser, S., and Zhang, J. (2002), Constitutive relations for cohesionless frictional granular materials, *J. Plasticity*, Special Issue for Tim Wright, Vol. 18, 531-547.

Nemat-Nasser, S., Chung, D.-T., and Taylor, L. M. (1989), Phenomenological modeling of rate-dependent plasticity for high strain rate problems, *Mech. Mat.*, Vol. 7, 319-344.

Nemat-Nasser, S., Guo, W. G., and Cheng, J. Y. (1999a), Mechanical properties and deformation mechanisms of a commercially pure titanium, *Acta Mat.*, Vol. 47, 3705-3720.

Nemat-Nasser, S., Guo, W. G., and Kihl, D. (2001a), Thermomechanical response of a AL6-XN stainless steel over a wide range of strain rates and temperatures, *J. Mech. Phys. Solids*, Vol. 49, 1823-1846.

Nemat-Nasser, S., Guo, W. G., and Liu, M. (1999b), Experimentally-based micromechanical modeling of dynamic response of molybdenum, *Scripta Mat.,* Vol. 40, 859-872.

Nemat-Nasser, S., Isaacs, J. B., and Liu, M. (1998a), Microstructure of high-strain, high-strain-rate deformed tantalum, *Acta Mat.*, Vol. 46, 1307-1325.

Nemat-Nasser, S., Isaacs, J. B., and Starrett, J. E. (1991), Hopkinson techniques for dynamic recovery experiments, *Proc. Royal Soc. London, A*, Vol. 435, 371-391.

Nemat-Nasser, S., Li, Y. F., and Isaacs, J. B. (1994), Experimental/ computational evaluation of flow stress at high strain rates with application to adiabatic shearbanding, *Mech. Mat.*, Vol. 17, 111-134.

Nemat-Nasser, S., Mehrabadi, M. M., and Iwakuma, T. (1981), On certain macroscopic and microscopic aspects of plastic flow of ductile materials, in *Three-dimensional constitutive relations and ductile fracture*, North-Holland, Amsterdam, 157-172.

Nemat-Nasser, S., Ni, L., and Okinaka, T. (1998b), A constitutive model for FCC crystals with application to polycrystalline OFHC copper, *Mech. Mat.*, Vol. 30, 325-341.

Nemat-Nasser, S., Okinaka, T., and Ni. L. (1998c), A physically-based constitutive model for BCC crystals with application to polycrystalline tantalum, *J. Mech. Phys. Solids* , Vol. 46, 1009-1038.

Nemat-Nasser, S., Guo, W. G., Nesterenko, V., Indrakanti, S. S., and Gu, Y. (2001b), Dynamic response of conventional and hot isostatically pressed Ti-6Al-4V Alloys: experiments and modeling, *Mech. Mat.*, Vol. 33, 425-439.

Obata, M., Goto, Y., and Matsura, S. (1990), Micromechanical consideration on the theory of elasto-plasticity at finite deformations, *Int. J. Eng. Sci.* Vol. 28, 241-252.

Oda, M. (1972), Initial fabrics and their relations to mechanical properties of granular materials, *Soils and Foundations*, Vol. 12, 17-36.

Oda, M., Nemat-Nasser, S., and Konishi, J. (1985), Stress induced anisotropy in granular masses, *Soils and Foundations*, Vol. 25, 85-97.

Oda, M., Nemat-Nasser, S., and Mehrabadi, M. M. (1982), A statistical study of fabric in a random assembly of spherical granules, *Int. J. Num. Anal. Methods Geomech.* Vol. 6, 77-94.

Ogden, R. W. (1982), Elastic deformation of rubberlike solids, *Mechanics of Solids, the Rodney Hill 60th anniversary volume* (Hopkins, H. G. and Sewell, M. J., eds.), Pergamon, Oxford, 499-537.

Ogden, R. W. (1984), *Non-linear elastic deformations*, Halsted Press, New York.

Okada, N. and Nemat-Nasser, S. (1994), Energy dissipation in inelastic flow of saturated cohesionless granular media, *Géotechnique*, Vol. 44, 1-19.

Okinaka, T. (1995), *A new computational approach to crystal plasticity: fcc crystals*, Ph.D. thesis, University of California, San Diego.

Onat, E. T. (1970), Representation of inelastic mechanical behavior by means of state variables, in *Proc. IUTAM Symp. Thermoinelasticity* (Boley, B. A. ed.), Springer-Verlag, Berlin, 213-224.

Ono, K. (1968), Temperature dependence of dispersed barrier hardening, *J. Appl. Phys.*, Vol. 39, 1803-1806.

Orowan, E. (1940), Problems of plastic gliding, *Phil. Trans. R. Soc. London A*, Vol.52, 8-22.

Orowan, E. (1958), Internal stresses and fatigue in metals, in *Proc. Symp. on Internal Stresses and Fatigue in Metals* (Rassweiler, G. M. and Grube, W. L. eds.), Detroit and Warren, MI, 1958 and 1959, Elsevier, New York, 59-80.

Pan, J. and Rice, J. R. (1983), Rate sensitivity of plastic flow and implications for yield-surface vertices, *Int. J. Solids Struct.*, Vol. 19, 973-987.

Parks, D. M. and Ahzi, S. (1990), Polycrystalline plastic deformation and texture evolution for crystals lacking five independent slip systems, *J. Mech. Phys. Solids*, Vol. 38, 701-724.

Perzyna, P. (1966), Fundamental problems in viscoplasticity, *Advances in Applied Mechanics*, Vol. 9, Academic Press, 243-377.

Perzyna, P. (1980), Modified theory of viscoplasticity. Application to advanced flow and instability phenomena, *Arch. Mech.*, Vol. 32, 403-420.

Perzyna, P. (1984), Constitutive modeling of dissipative solids for postcritical behavior and fracture, *J. Eng. Mat. Tech.*, 410-419.

Rashid, M. M. (1990), High-strain-rate flow in metallic single crystals, *Ph.D. Thesis*, University of California, San Diego.

Rashid, M. M., Gray, G. T., and Nemat-Nasser, S. (1992), Heterogeneous deformations in copper single crystals at high and low strain rates, *Phil. Mag. A*, Vol. 65, 707-735.

Regazzoni G., Kocks, U. F., and Follansbee, P. S. (1987), Dislocation kinetics at high strain rates, *Acta Metall.*, Vol. 35, 2865-2875.

Rice, J. R. (1970), On the structure of stress-strain relations for time-dependent plastic deformation in metals, *J. Appl. Mech.*, Vol. 37, 728-737.

Rice, J. R. (1971), Inelastic constitutive relations for solids: an internal-variable theory and its application to metal plasticity, *J. Mech. Phys. Solids*, Vol. 19, 433-455.

Rice, J. R. (1975), Continuum mechanics and thermodynamics of plasticity in relation to microscale deformation mechanisms, *Constitutive equations in plasticity* (Argon, A. S. ed.), MIT Press, Cambridge, Massachusetts, 23-79.

Rice, J. R. (1977), The localization of plastic deformation, *Theoretical and applied mechanics*, (Koiter, W. T. ed.), North-Holland, 207-220.

Roscoe, K. H., Schofield, A. N., and Wroth, C. P. (1958), On the yielding of soils, *Géotechnique*, Vol. 8, 22-53.

Rowshandel, B. and Nemat-Nasser, S. (1987), Finite strain rock plasticity: Stress triaxiality, pressure and temperature effects, *Soil Dyn. Earthquake Eng.*, Vol. 6, 203-219.

Rudnicki, J. W. and Rice, J. R. (1975), Conditions for the localization of deformation in pressure-sensitive dilatant materials, *J. Mech. Phys. Solids*, Vol. 23, 371-394.

Schapery, R. A. (1968), On a thermodynamic constitutive theory and its application to various nonlinear materials, in *Proc. IUTAM Symp. Irreversible Aspects of Continuum Mech.* (Parkus, H. and Sedov, L. I. eds.), Springer-Verlag, Berlin, 259-285.

Scheidler, M. and Wright, T. W. (2001), A continuum framework for finite viscoplasticity, *Int. J. Plasticity*, Vol. 17, 1033-1085.

Scheidler, M. and Wright, T. W. (2003), Classes of flow rules for finite viscoplasticity, *Int. J. Plasticity*, Vol. 19, 1119-1165.

Schofield, A. N. and Wroth, P. (1968), *Critical state soil mechanics*, McGraw-Hill, London.

Sewell, M. J. (1972), A survey of plastic buckling, *Stability* (Leipholz, H. ed.), University of Waterloo Press, Ontario, 85-197.

Sewell, M. J. (1974), On applications of saddle-shaped and convex generating functionals, *Physical structure in systems theory,* (Van Dixhoorn, J. J. and Evans, F. J. eds.), Academic Press, London, 219-245.

Shield, R. T. (1953), Mixed boundary value problems in soil mechanics, *Q. Appl. Math.*, Vol. 11, 81.

Shield, R. T. (1954), Stress and velocity fields in soil mechanics, *J. Math. Phys.*,

Vol. 33, 144-156.

Spencer, A. J. M. (1964), A theory of kinematics of ideal soil under plane strain conditions, *J. Mech. Phys. Solids*, Vol. 12, 337-351.

Spencer, A. J. M. (1982), Deformation of ideal granular materials, *Mechanics of Solids, the Rodney Hill 60th anniversary volume* (Hopkins, H. G. and Sewell, M. J., eds.), Pergamon, Oxford, 607-652.

Stoltz, R. E. and Pelloux, R. M. (1974), Cyclic deformation and Bauschinger effect in Al-Cu-Mg alloys, *Scripta Metall.*, Vol. 8, 269-276.

Stoltz, R. E. and Pelloux, R. M. (1976), The Bauschinger effect in precipitation strengthened aluminum alloys, *Metall. Trans. A.*, Vol. 7A, 1295-1306.

Stören, S. and Rice, J. R. (1975), Localized necking in thin sheets, *J. Mech. Phys. Solids*, Vol. 23, 421-441.

Stout, M. G. and Rollet, A. D. (1990), *Metall. Trans. A*, Vol. 21A, 3201-3213.

Stroh, A. N. (1954), A theoretical calculation of the stored energy in a work-hardened material, *Proc. Royal Soc.*, 218A, 391-400.

Subhash, G., Nemat-Nasser, S., Mehrabadi, M. M., and Shodja, H. M. (1991), Experimental investigation of fabric-stress relations in granular materials, *Mech. Mat.*, Vol. 11, 87-106.

Talreja, R. (1985), A continuum mechanics characterization of damage in composite materials, *Proc. R. Soc. Lond. A*, Vol. 399, 195-216.

Taylor, G. I. (1934), The mechanism of plastic deformation of crystals - I, theoretical, *Proc. R. Soc. Lond. A*, Vol. 145, 362-387.

Taylor, G. I. (1938), Plastic strain in metals, *J. Inst. Metals*, Vol 62, 307-325.

Thakur, A. (1994), Deformation behavior and Bauschinger effect in super alloy, *Ph.D. Thesis*, University of California, San Diego.

Thakur, A., Nemat-Nasser, S., and Vecchio, K. S. (1996a), Dynamic Bauschinger effect, *Acta Metall.*, Vol. 44, 2797-2807

Thakur, A., Vecchio, K. S., and Nemat-Nasser, S. (1996b), Bauschinger effect in Haynes 230 alloys: influence of strain rate and temperature, *Metall. and Mat. Trans.*, Vol. 27A, 1739-1748.

Thomson, J. (1882), On the jointed prismatic structure in balsatic rocks, *Glasgow Geol. Soc. Trans.*, Vol. 6, 95-110.

Torquato, S. (2002), *Random heterogeneous materials*, Springer.

Treolar, L. R. G. (1958), *Physics of rubber elasticity*, Oxford University Press.

Truesdell, C. and Noll, W. (1965), The non-linear field theories of mechanics, in *Handbuch der Physik* (Flügge, S. ed.), Springer-Verlag, Berlin, Vol. III/3, 1-602.

Tvergaard, V. (1978), Effect of kinematic hardening on localized necking in biaxially stretched sheets, *Int. J. Mech. Sci.*, Vol. 20, 651-658.

Tvergaard, V. (1982), On localization in ductile materials containing spherical voids, *Int. J. Fract.*, Vol. 18, 237-252.

Tvergaard, V. and Needleman, A. (1984), Analysis of the cup-cone fracture in a round tensile bar, *Acta Metall.*, Vol. 32, 157-169.

Valanis, K. C. (1966), Thermodynamics of large viscoelastic deformations, *J. Math. Phys.*, Vol. 45, 197-212.

Wilde, P. (1977), Two invariant-dependent models of granular media, *Arch. Mech.*, Vol. 29, 799-809.

Willis, J. R. (1969), Some constitutive equations applicable to problems of large dynamic plastic deformation, *J. Mech. Phys. Solids*, Vol. 17, 359-369.

Wilson, D. V. (1964), Reversible work hardening in alloys of cubic metals, *Acta Metall.*, Vol. 13, 807-814.

Woolley, R. L. (1953), The Bauschinger effect in some face-centered and body-centered cubic metals, *Phil. Mag.*, Vol. 44, 597-618.

Wright, T. W. (2002), *The physics and mathematics of adiabatic shear bands*, Cambridge University Press.

Xiano, H., Bruhns, O. T., and Meyers, A. (2000), A consistent finite elastoplasticity theory combining additive and multiplicative decomposition of stretching and the deformation gradient, *Int. J. Plasticity*, Vol. 16, 143-177.

Zarka, J. (1973), Étude du comportement des monocristaux métalliques: application à la traction du monocristal CFC, *J. Mécanique*, Vol. 12, 275-318.

Zarka, J. (1975), Modelling of changes of dislocation structures in monotonically deformed single phase crystals, in *Constitutive equations in plasticity* (Argon, A. S. ed.), MIT Press, Cambridge, Massachusetts, 359-386.

Zehnder, A. T. (1991), A model for the heating due to plastic work, *Mech. Res. Comm.*, Vol. 18, 23-28.

Zerilli, F. J. and Armstrong, R. W. (1987), Dislocation-mechanics-based constitutive relations for material dynamics calculations, *J. Appl. Phys.*, Vol. 61, 1816-1825.

Zerilli, F. J. and Armstrong, R. W. (1990), Description of tantalum deformation behavior by dislocation mechanics based constitutive relations, *J. Appl. Phys.*, Vol. 68, 1580-1591.

Zerilli, F. J. and Armstrong, R. W. (1992), The effect of dislocation drag on the stress-strain behavior of fcc metals, *Acta Metall.*, Vol. 40, 1803-1808.

Ziegler, H. (1959), A modification of Prager's hardening rule, *Q. Appl. Math.*, Vol. 17, 55-65.

Zoback, M. D. and Byerlee, J. D. (1976a), A note on the deformational behavior and permeability of crushed granite, *Int. J. Rock Mech. Min. Sci. & Geomech. Abstr.*, Vol. 13, 291-294.

Zoback, M. D. and Byerlee, J. D. (1976b), Effect of high-pressure deformation on permeability of Ottawa sand, *Bull. Am. Assoc. Petro. Geolog.*, Vol. 60, 1531-1542.

5

INTEGRATION OF CONTINUUM CONSTITUTIVE EQUATIONS

This chapter focuses on numerical integration of continuum constitutive equations. First, the relation between the incremental deformation gradient and the velocity gradient is examined. Then explicit algorithms are developed and detailed for various constitutive models which are given in Chapter 4. The chapter is organized as follows.

After some brief historical comments in Section 5.1, the kinematical assumptions used in the numerical computation of the incremental deformations are reviewed in Section 5.2. The constant incremental velocity gradient and the incrementally unidirectional stretch tensor are examined in relation to other kinematical quantities, and various numerical approximations are discussed. Exact, explicit, coordinate-independent relations between the velocity gradient and the incremental deformation gradient, as well as the incremental stretch and rotation tensors, are developed. The rate-constitutive equations for the J_2-*flow theory, with rate-independent and rate-dependent flow conditions, are presented in Section 5.3, including the effect of thermal softening on the flow stress. Section 5.4 examines the basis of the plastic-predictor/elastic-corrector method, and provides error estimates. Section 5.5 is devoted to a detailed account of the application of the singular perturbation method to solve the stiff constitutive equations which govern the variation of the stress magnitude. This section is closed by presenting a modified outer solution which is explicit, simple, accurate, and can be used directly in large-scale computations. Section 5.6 provides a detailed account of computational algorithms for proportional loading and unloading. In the algorithm, the deformation-rate tensor is assumed to remain codirectional with the stress-difference tensor which is defined as the deviatoric part of the Kirchhoff stress less the (deviatoric) backstress. Various integration methods are discussed, illustrated, and compared, for both the rate-dependent and rate-independent flow stress. The computational steps are outlined in tables, through specific numerical examples. Section 5.7 outlines the computational algorithms when the incremental deformation corresponds to the unidirectional stretch assumption. Both elastic, perfectly-plastic and workhardening models are considered. For the perfectly plastic case, an exact integration procedure to obtain the stress tensor, is given. For the rate-dependent or rate-independent models with workhardening, the radial-return method and its generalization which directly includes the spin effect, are developed, illustrated, and compared. Important results are summarized in terms of computational steps which would allow direct implementation into numerical algorithms. Numerical results are given, which may be used to check computational implementation of the steps.*

Section 5.8 provides details similar to those covered in Section 5.7, but based on the kinematical assumption that the velocity gradient is constant over the time increment. Examples are given to demonstrate the effectiveness of several novel integration techniques. This includes cases where the spin or the stretching dominates the incremental deformation, or where the spin and stretching are of the same order of magnitude. In this section, again, important results are summarized in terms of computational steps, and the results are illustrated. Unlike the commonly used radial-return method, the generalized radial-return methods presented in Sections 5.7 and 5.8, for both the rate-dependent and rate-independent cases, directly and accurately include the effects of the spin and the noncoaxiality of the strain-rate tensor and the stress-deviatoric tensor. In the computational methods presented here, any strain and strain-rate hardening and thermal softening flow-stress model may be used, following the indicated computational steps.

A number of basic identities and mathematical results for exact integration of some of the constitutive relations, are summarized in Appendices 5.A and 5.B.

5.1. INTRODUCTION

Explicit finite-element codes are generally used for large-scale computations of dynamic elastoplastic problems. At each timestep, these codes integrate the equations of motion to obtain the nodal displacements at the end, and the nodal velocities in the middle of the timestep. These are then used along with the interpolation functions (shape functions) to obtain the incremental deformation gradient and the mid-point velocity gradient at the Gauss points in each element. The task of the constitutive algorithm then is to use the velocity gradient (or the incremental deformation gradient) at a typical Gauss point and the corresponding current values of the stress, deformation measure, temperature, and other state variables, to calculate all the associated incremental quantities over a timestep which is prescribed by the code. An important requirement is that this task be performed accurately and efficiently. Techniques which necessitate subdivision of the time increment, or the solution of nonlinear equations, or iteration with conditional statements, tend to compromise efficiency.

For elastoviscoplasticity, the issue of an explicit algorithm has been discussed by a number of authors; a brief discussion of this with references, is given by Rashid and Nemat-Nasser (1990). The simplest approach is a one-step Eulerian explicit integration, requiring unreasonably small timesteps for stiff elastoviscoplasticity; Zienkiewicz and Cormeau (1974). Improved techniques, based on the forward gradient method, have been proposed.[1] These methods employ truncated Taylor series, and are thus limited in accuracy by the time increment associated with the series expansion.

For rate-independent elastoplasticity, most large-scale finite-element codes use the radial-return method of Wilkins (1964) to determine the stress orientation. Explicit Euler integrations are used to determine the yield surface; see Hughes (1984), and Dodds (1986). Krieg and Krieg (1977) consider the perfect-plasticity model and obtain an exact expression for the stress orientation. Recently Nemat-Nasser and coworkers have developed a technique which combines the exact expression for the stress orientation by Krieg and Krieg and a *plastic-predictor/elastic-corrector* method to calculate the yield surface.[2] In this chapter, these methods are reviewed and where necessary generalized to include nonclassical effects of noncoaxiality of the plastic strain rate and the stress deviator, and large spins.

The plastic-predictor/elastic-corrector method exploits the physical fact that most of the deformation in incremental elastoplasticity or elastoviscoplasticity of many metals, their alloys, and even geomaterials, is due to inelastic

[1] See for example Kanchi *et al.* (1978), Hughes and Taylor (1978), Marques and Owen (1983), Brockman (1984), Peirce *et al.* (1984), Chen and Krempl (1986), Yoshimura *et al.* (1987), and Szabo (1990).

[2] See Nemat-Nasser (1991), Nemat-Nasser and Chung (1992), Nemat-Nasser and Li (1992, 1994), Nemat-Nasser and Ni (1994), and Balendran and Nemat-Nasser (1995).

deformation. The incremental elastic contribution is generally very small. The method, therefore, assigns *all the incremental deformation to be plastic and then computes the elastic contribution as a corrector.* To this end, the rate-constitutive equations are directly integrated over the time increment, and the various resulting integrals are estimated. This avoids Taylor series expansions of nonlinear expressions, providing considerable efficiency and accuracy.

It turns out that the plastic-predictor/elastic-corrector method is related to the singular-perturbation solution of the stiff rate-constitutive relations, when a single timestep is used, with the elastic strain being the *small parameter*; Nemat-Nasser and Ni (1994). The *outer* solution then corresponds to the *plastic-predictor* estimate of the field quantities, and the *inner* solution can be replaced by an estimate which exploits the physics of the elastoplasticity.

This chapter provides a detailed account of the incremental computation of the elastoplastic constitutive relations, emphasizing both rate-independent and rate-dependent models which are based on the J_2-flow theory of plasticity, also including the noncoaxiality, dilatancy, and pressure sensitivity. For the calculation of the stress magnitude, both the plastic-predictor/elastic-corrector method, and methods based on the forward-gradient technique, are examined and illustrated. A brief account of the error estimate for the plastic-predictor/elastic-corrector method is also included; see Nemat-Nasser and Ni (1994) for details. For the calculation of the orientation of the stress tensor, the radial-return method and its modified and generalized versions, the method of Krieg and Krieg and its modified and generalized versions, as well as other direct and essentially exact methods, are presented.

5.2. INCREMENTAL KINEMATICS

In finite-element calculations, equations of motion, (3.8.18), are solved numerically to obtain the *nodal* displacements at the end of each timestep, and the nodal velocities at the middle of the timestep. Then the displacement *field* at the end of each timestep, and the velocity *field* at the middle of the timestep, are obtained using the interpolation formulae, *e.g.,* (3.8.7a,b).

It is usually assumed that the configuration (particle positions) at time $t+\Delta t$, and the velocity field (particle velocities) at time $t+\Delta t/2$, can be expressed in terms of the configuration at time t,

$$\mathbf{x}(t+\Delta t;\, \mathbf{x}(t)) = \mathbf{\Phi}(\mathbf{x}(t)),$$

$$\mathbf{v}(t+\Delta t/2;\, \mathbf{x}(t)) = \mathbf{\Psi}(\mathbf{x}(t)), \tag{5.2.1a,b}$$

where the vector fields $\mathbf{\Phi}$ and $\mathbf{\Psi}$ are related to the nodal displacements and velocities through the interpolation functions[3] by

[3] Φ^e_α, $\alpha = 1, 2, ..., n$, are the interpolation functions of the element e, corresponding to the parti-

$$\boldsymbol{\Phi}(\mathbf{x}(t)) = \mathbf{x}(t) + \sum_{\alpha=1}^{n} \Phi_\alpha^e(\mathbf{x}(t))\, \mathbf{u}_\alpha^e(t+\Delta t)\,,$$

$$\boldsymbol{\Psi}(\mathbf{x}(t)) = \sum_{\alpha=1}^{n} \Phi_\alpha^e(\mathbf{x}(t))\, \mathbf{v}_\alpha^e(t+\Delta t/2)\,. \qquad (5.2.1c,d)$$

Then the incremental deformation gradient becomes

$$\mathbf{F}_t(\Delta t) \equiv \left[\frac{\partial \mathbf{x}(t+\Delta t)}{\partial \mathbf{x}(t)}\right]^T = [\boldsymbol{\nabla}_t \otimes \boldsymbol{\Phi}(\mathbf{x}(t))]^T\,, \qquad (5.2.2a)$$

where the del operator is $\boldsymbol{\nabla}_t \equiv \mathbf{e}_i\, \partial/\partial x_i(t)$, and *the subscript t emphasizes that the configuration at time t is the reference one.* Similarly, the midpoint velocity gradient at any material point is given by

$$\mathbf{L}(t+\Delta t/2) = \{\boldsymbol{\nabla}_{t+\Delta t/2} \otimes \boldsymbol{\Psi}(\mathbf{x}(t))\}^T\,, \qquad (5.2.2b)$$

where $\boldsymbol{\nabla}_{t+\Delta t/2} \equiv \mathbf{e}_i\, \partial/\partial x_i(t+\Delta t/2)$. The midpoint configuration is approximated by

$$\mathbf{x}(t+\Delta t/2) = \frac{1}{2}\{\mathbf{x}(t) + \mathbf{x}(t+\Delta t)\}$$

$$= \mathbf{x}(t) + \frac{1}{2} \sum_{\alpha=1}^{n} \Phi_\alpha^e(\mathbf{x}(t))\, \mathbf{u}_\alpha^e(t+\Delta t)\,. \qquad (5.2.2c,d)$$

Most computational codes use a constant velocity gradient, $\mathbf{L}(t+\xi) \equiv \mathbf{L}(t+\Delta t/2)$, $0 \le \xi \le \Delta t$, given by (5.2.2b), over the timestep, to calculate the corresponding stress increment. The relations between the incremental deformation gradient (5.2.2a) and the velocity gradient (5.2.2b) are examined in the sequel under different kinematical restrictions. Exact coordinate-independent results are given in Subsection (5.2.3).

5.2.1. Constant Velocity Gradient

In general, a constant velocity gradient $\mathbf{L}(t+\Delta t/2)$ given by (5.2.2b) over the timestep Δt, results in a deformation gradient which is not the same as that given by (5.2.2a), and therefore

$$\mathbf{L}(t+\Delta t/2) \neq \frac{1}{\Delta t}\, ln\, (\mathbf{F}_t(\Delta t))\,; \qquad (5.2.3)$$

see (5.2.6b) below. Therefore, the assumption that the velocity gradient is constant and equal to that given by (5.2.2b) is inconsistent with the incremental deformation gradient (5.2.2a). This inconsistency is examined in the sequel.

The *central-difference* method is used in most explicit finite-element codes, where it is assumed that

---

cle position $\mathbf{x}(t)$, n is the number of nodes in the element, and $\mathbf{u}_\alpha^e$ and $\mathbf{v}_\alpha^e$, $\alpha = 1, 2, ..., n$, are the nodal displacements and velocities, respectively.

$$\mathbf{x}(t+\Delta t) = \mathbf{x}(t) + \mathbf{v}(t+\Delta t/2)\,\Delta t\,. \tag{5.2.4}$$

In this case, and for small incremental deformations, the midpoint velocity gradient and the incremental deformation gradient are related by[4]

$$\mathbf{L}(t+\Delta t/2) = \frac{2}{\Delta t}\,[\mathbf{F}_t(\Delta t) - \mathbf{1}]\,[\mathbf{F}_t(\Delta t) + \mathbf{1}]^{-1}$$
$$= \frac{1}{\Delta t}\,\{\mathbf{A} - \frac{1}{2}\,\mathbf{A}^2 + \frac{1}{4}\,\mathbf{A}^3 - \frac{1}{8}\,\mathbf{A}^4 + \ldots\}\,, \tag{5.2.5a}$$

where

$$\mathbf{A} = \mathbf{F}_t(\Delta t) - \mathbf{1}\,. \tag{5.2.5b}$$

Since

$$ln\,(\mathbf{F}_t(\Delta t)) = \mathbf{A} - \frac{1}{2}\,\mathbf{A}^2 + \frac{1}{3}\,\mathbf{A}^3 - \frac{1}{4}\,\mathbf{A}^4 + \ldots\,, \tag{5.2.6a}$$

it follows that

$$\mathbf{L}(t+\Delta t/2) = \frac{1}{\Delta t}\,ln\,(\mathbf{F}_t(\Delta t)) + O(\mathbf{A}^3)\,. \tag{5.2.6b}$$

Therefore, for small deformations where **A** corresponds to infinitesimally small strains and rotations, the constant velocity gradient given by (5.2.2b) is consistent with the incremental deformation gradient (5.2.2a). It should be noted that, in the localized deformation zones (adiabatic shearbands), the deformation in each timestep may be large even if the timestep is very small. In such a case, the above expansions are not valid. The constant velocity gradient may then be defined by

$$\mathbf{L} = \frac{1}{\Delta t}\,ln\,(\mathbf{F}_t(\Delta t))\,. \tag{5.2.7}$$

Closed-form, coordinate-independent expressions for the logarithm of a non-symmetric tensor are obtained from Subsections 1.5.2 to 1.5.5, by simply substituting $f(\lambda) = ln\,(\lambda)$; see Subsection 5.2.3.

5.2.2. Unidirectional Stretch

Alternative to the assumption that the velocity gradient is constant over a timestep, is the assumption that the stretch is unidirectional over the same time increment. The incremental rotation is then implemented at the end of this timestep. In order to examine this kinematical assumption, first consider the polar decomposition of the incremental deformation gradient into an incremental stretch tensor $\mathbf{U}_t$ and a rotation tensor $\mathbf{R}_t$, as follows:

$$\mathbf{F}_t(\xi) = \mathbf{R}_t(\xi)\,\mathbf{U}_t(\xi)\,, \quad 0 \le \xi \le \Delta t\,. \tag{5.2.8}$$

---

[4] Note that $\mathbf{L}(t+\Delta t/2) = \dot{\mathbf{F}}_t(t+\Delta t/2)\,\mathbf{F}_t^{-1}(t+\Delta t/2)$, with $\dot{\mathbf{F}}_t(t+\Delta t/2) = \frac{1}{\Delta t/2}\,[\mathbf{F}_t(t+\Delta t/2) - 1] = \frac{2}{\Delta t}\,\{½\,[\mathbf{F}_t(t+\Delta t) + 1] - 1\} = \frac{1}{\Delta t}[\mathbf{F}_t(t+\Delta t) - 1]$, and, similarly, $\mathbf{F}_t^{-1}(t+\Delta t/2) = 2\,[\mathbf{F}_t(t+\Delta t) + 1]^{-1}$.

The material time derivative of (5.2.8) yields

$$\mathbf{L}(\xi) \equiv \dot{\mathbf{F}}_t(\xi)\,\mathbf{F}_t^{-1}(\xi) = \mathbf{\Omega}_t(\xi) + \mathbf{R}_t(\xi)\,\dot{\mathbf{U}}_t(\xi)\,\mathbf{U}_t^{-1}(\xi)\,\mathbf{R}_t^{\mathrm{T}}(\xi)\,, \tag{5.2.9a}$$

where $\mathbf{\Omega}_t(\xi)$ is the incremental rigid spin given by

$$\mathbf{\Omega}_t(\xi) = \dot{\mathbf{R}}_t(\xi)\,\mathbf{R}_t^{\mathrm{T}}(\xi)\,. \tag{5.2.9b}$$

Now, assume that $\mathbf{U}_t(\xi)$ is unidirectional for $0 \le \xi \le \Delta t$, arriving at

$$\mathbf{U}_t(\xi) = \sum_{a=1}^{3} \lambda_a(\xi)\,\mathbf{N}_t^a \otimes \mathbf{N}_t^a\,,$$

$$\mathbf{U}_t(\Delta t) = \sum_{a=1}^{3} \lambda_a(\Delta t)\,\mathbf{N}_t^a \otimes \mathbf{N}_t^a\,,$$

$$\dot{\mathbf{U}}_t(\xi) = \sum_{a=1}^{3} \dot{\lambda}_a(\xi)\,\mathbf{N}_t^a \otimes \mathbf{N}_t^a\,, \tag{5.2.10a-c}$$

where $\lambda_a(\xi)$ and $\mathbf{N}_t^a$, a = 1, 2, 3, are the principal values and directions of $\mathbf{U}_t(\xi)$. Since in this particular case, $\dot{\mathbf{U}}_t\,\mathbf{U}_t^{-1}$ is a symmetric tensor, the *incremental rigid spin* and the material spin are equal,

$$\mathbf{\Omega}_t(\xi) = \mathbf{W}(\xi)\,, \quad \dot{\mathbf{U}}_t(\xi)\,\mathbf{U}_t^{-1}(\xi) = \hat{\mathbf{D}}(\xi) = \mathbf{R}_t^{\mathrm{T}}(\xi)\,\mathbf{D}(\xi)\,\mathbf{R}_t(\xi)\,. \tag{5.2.11a,b}$$

Even under these assumptions, neither $\hat{\mathbf{D}}$ nor $\mathbf{D}$ is constant.[5] However, an additional assumption that $\dot{\lambda}_a/\lambda_a$ (a not summed) is constant, renders $\hat{\mathbf{D}}$ a constant during the timestep, it then being given by[6]

$$\hat{\mathbf{D}} = \frac{1}{\Delta t}\,ln\,(\mathbf{U}_t(\Delta t))\,. \tag{5.2.12}$$

The polar decomposition of $\mathbf{F}_t(\Delta t)$ and the calculation of $\hat{\mathbf{D}}$, need not be based on the *eigenvectors* $\mathbf{N}_t^a$, a = 1, 2, 3. Following the procedure discussed in Section 2.2, incremental deformation gradient $\mathbf{F}_t(\Delta t)$ can be *exactly* decomposed to obtain $\mathbf{U}_t(\Delta t)$ and $\mathbf{R}_t(\Delta t)$, without any *eigenvector* calculation. A closed-form expression for $\hat{\mathbf{D}}$ is then obtained, using the method described in Subsection 2.3.1, by setting $f(\lambda) = ln\,(\lambda)$; see Subsection 5.2.3.

Hughes and Winget (1980) have proposed,

$$\mathbf{R}_t^{\mathrm{HW}}(\Delta t) = (\mathbf{1} - \tfrac{1}{2}\,\boldsymbol{\omega})^{-1}\,(\mathbf{1} + \tfrac{1}{2}\,\boldsymbol{\omega}) = (\mathbf{1} + \tfrac{1}{2}\,\boldsymbol{\omega})\,(\mathbf{1} - \tfrac{1}{2}\,\boldsymbol{\omega})^{-1}\,, \tag{5.2.13}$$

where

$$\boldsymbol{\omega} \equiv \tfrac{1}{2}\,(\mathbf{G} - \mathbf{G}^{\mathrm{T}})\,,$$

$$\mathbf{G} \equiv \frac{\partial \mathbf{x}(t+\Delta t)}{\partial \mathbf{x}(t+\Delta t/2)} - \frac{\partial \mathbf{x}(t)}{\partial \mathbf{x}(t+\Delta t/2)}\,,$$

---

[5] Note that all quantities with a superimposed caret, ^, are measured with respect to the unrotated frame, *e.g.*, $\hat{\mathbf{D}} = \mathbf{R}_t^{\mathrm{T}}\mathbf{D}\mathbf{R}_t$, with similar definitions of $\hat{\boldsymbol{\tau}}$, $\hat{\mathbf{s}}$, and $\hat{\boldsymbol{\mu}}$, and other second-order tensors. This notation is used throughout this chapter.

[6] See ABAQUS Manual (1989; Subsection 2.7.2). A numerical approach to calculate $ln\,\mathbf{U}$ is given by Bažant (1998).

$$\mathbf{x}(t+\Delta t/2) = \tfrac{1}{2}\{\mathbf{x}(t)+\mathbf{x}(t+\Delta t)\}. \qquad (5.2.14\text{a-c})$$

Superscript HW is used to denote the approximate expression by Hughes and Winget (1980). Note that $\mathbf{R}_t^{HW}(\Delta t)$ is a proper orthogonal tensor.[7] Hughes (1984) proposes the following approximate expression for $\hat{\mathbf{D}}$:

$$\hat{\mathbf{D}}^H = \frac{1}{\Delta t}(\mathbf{R}_t^H(\Delta t/2))^T \Delta\boldsymbol{\varepsilon}\, \mathbf{R}_t^H(\Delta t/2), \qquad (5.2.15\text{a})$$

where

$$\Delta\boldsymbol{\varepsilon} \equiv \tfrac{1}{2}(\mathbf{G}+\mathbf{G}^T),$$

$$\mathbf{R}_t^H(\Delta t/2) \equiv \sqrt{\mathbf{R}_t^{HW}(\Delta t)}. \qquad (5.2.15\text{b,c})$$

Substituting (5.2.2a) into (5.2.14b,c), tensor $\mathbf{G}$ is expressed in terms of the incremental deformation gradient as

$$\mathbf{G} = 2[\mathbf{F}_t - \mathbf{1}][\mathbf{F}_t + \mathbf{1}]^{-1}. \qquad (5.2.16)$$

Note that, if the incremental deformation is pure rotation, *i.e.*, if $\mathbf{F}_t = \mathbf{R}_t$ and $\mathbf{U}_t = \mathbf{1}$, then the deformation rate is identically zero, $\hat{\mathbf{D}}^H = \mathbf{0}$. Moreover, if it is pure deformation, *i.e.*, if $\mathbf{F}_t = \mathbf{F}_t^T = \mathbf{U}_t$, then the rotation is zero and $\mathbf{R}_t^H = \mathbf{1}$. On the other hand, if the deformation includes both rotation and distortion, then $\hat{\mathbf{D}}^H$ may deviate from its actual value by an error of the first order in the incremental strain.

Rashid (1993) observes the deficiency of the above approximations and calls the result *weakly objective*. He then proposes an algorithm which does not involve any calculation of *eigenvectors* or *eigenvalues*, but yields a constant deformation rate and a rotation tensor such that the error in the deformation rate is of the third order in the strain increment. Rashid's results are as follows:

$$(\mathbf{R}_t^R)_{ij} = \delta_{ij}\cos\theta + \frac{(1-\cos\theta)}{4Q}\alpha_i\alpha_j - \frac{\sin\theta}{2\sqrt{Q}} e_{ijk}\alpha_k, \qquad (5.2.17\text{a})$$

where

$$\alpha_i = e_{ijk}(\mathbf{F}_t)_{jk}, \quad Q = 1/4\,\alpha_i\alpha_i,$$

$$\cos^2\theta = P + \frac{3P^2\{1-(P+Q)\}}{(P+Q)^2} - \frac{2P^3\{1-(P+Q)\}}{(P+Q)^3},$$

$$P = 1/4[\mathrm{tr}(\mathbf{F}_t)-1]^2. \qquad (5.2.17\text{b-d})$$

For the constant $\hat{\mathbf{D}}$, (5.2.12) is approximated as

$$\hat{\mathbf{D}}^R = \frac{1}{2\Delta t}\{(\mathbf{C}_t - 1) - \tfrac{1}{2}(\mathbf{C}_t - 1)^2\},$$

$$\mathbf{C}_t = \mathbf{F}_t^T\mathbf{F}_t. \qquad (5.2.18\text{a,b})$$

Superscript R is used to denote the approximations proposed by Rashid (1993).

---

[7] *I.e.*, its transpose is its inverse and its determinant is +1.

These expressions yield

$$\hat{\mathbf{D}}^R = \mathbf{0}, \quad \mathbf{R}_t^R = \mathbf{R}_t \quad \text{for } \mathbf{F}_t = \mathbf{R}_t,$$

$$\mathbf{R}_t^R = \mathbf{1}, \quad \text{for } \mathbf{F}_t = \mathbf{F}_t^T. \tag{5.2.19a-c}$$

Exact expressions for $\mathbf{R}_t$ and $\hat{\mathbf{D}}$ are given in the following subsection.

5.2.3. Exact Coordinate-independent Relations Between Velocity and Incremental Deformation Gradients

In this subsection, the exact relations between the velocity and incremental deformation gradients are given for the two kinematical conditions considered above, namely, for constant velocity gradient and for unidirectional stretch. The final expressions are given by (5.2.24) and (5.2.28c) of this subsection, for these two cases. These expressions include three coefficients, a_0, a_1, and a_2, which relate the required constant velocity gradient, $\mathbf{L}$, to the constant deformation gradient, $\mathbf{F}_t$; or, in the case of the unidirectional stretch, the required constant deformation rate, $\hat{\mathbf{D}}$, to the constant deformation measure $\mathbf{C}_t = \mathbf{F}_t^T \mathbf{F}_t$. In what follows, these coefficients are given explicitly.

Constant Velocity Gradient: First consider the kinematical condition of a constant velocity gradient. In this case the velocity gradient is related to the incremental deformation gradient by (5.2.7). In order to obtain the logarithm of the nonsymmetric tensor $\mathbf{F}_t$, first introduce the following expressions:

$$I_F = \mathrm{tr}(\mathbf{F}_t), \quad a = \frac{1}{6}\,\mathrm{tr}(\mathbf{A}^2),$$

$$b = \frac{1}{6}\,\mathrm{tr}(\mathbf{A}^3), \quad \mathbf{A} = \mathbf{F}_t - \frac{1}{3}\,I_F\,\mathbf{1}. \tag{5.2.20a-d}$$

Then evaluate the coefficients a_0, a_1, and a_2, depending on the sign of $b^2 - a^3$, as follows (see Subsection 1.5.2):

(i) $b^2 < a^3$

$$\phi = \cos^{-1}(b/a^{3/2}),$$

$$\lambda_{n+1} = \frac{I_F}{3} + 2\sqrt{a}\cos\left(\frac{2n\pi + \phi}{3}\right), \quad n = 0,\ 1,\ 2,$$

$$\Delta = -(\lambda_1 - \lambda_2)(\lambda_2 - \lambda_3)(\lambda_3 - \lambda_1),$$

$$\begin{bmatrix} a_0 \\ -a_1 \\ a_2 \end{bmatrix} = \frac{1}{\Delta}\begin{bmatrix} \lambda_2\lambda_3 & \lambda_3\lambda_1 & \lambda_1\lambda_2 \\ \lambda_2+\lambda_3 & \lambda_3+\lambda_1 & \lambda_1+\lambda_2 \\ 1 & 1 & 1 \end{bmatrix}\begin{bmatrix} \ln(\lambda_1)(\lambda_2-\lambda_3) \\ \ln(\lambda_2)(\lambda_3-\lambda_1) \\ \ln(\lambda_3)(\lambda_1-\lambda_2) \end{bmatrix}. \tag{5.2.21a-d}$$

(ii) $b^2 = a^3 \neq 0$

$$\lambda_1 = 1/3\, I_F + 2\sqrt{a}, \quad \lambda_2 = \lambda_3 = 1/3\, I_F - \sqrt{a},$$

$$a_0 = \frac{\lambda_1\, ln\,(\lambda_2) - \lambda_2\, ln\,(\lambda_1)}{\lambda_1 - \lambda_2} + a_2\, \lambda_1\, \lambda_2,$$

$$a_1 = \frac{ln\,(\lambda_1) - ln\,(\lambda_2)}{\lambda_1 - \lambda_2} - a_2\,(\lambda_1 + \lambda_2),$$

$$a_2 = \frac{ln\,(\lambda_1) - ln\,(\lambda_2) - (\lambda_1 - \lambda_2)/\lambda_2}{(\lambda_1 - \lambda_2)^2}. \tag{5.2.22a-e}$$

(iii) $a = b = 0$

$$\lambda = \frac{I_F}{3}, \quad a_0 = ln\,(\lambda) - \frac{3}{2},$$

$$a_1 = \frac{2}{\lambda}, \quad a_2 = -\frac{1}{2\,\lambda^2}. \tag{5.2.23a-d}$$

Note that the case $b^2 > a^3$ is excluded, since it corresponds to complex-valued stretches. Now, the constant velocity gradient is obtained from

$$\mathbf{L} = \frac{1}{\Delta t}(a_0\, \mathbf{1} + a_1\, \mathbf{F}_t + a_2\, \mathbf{F}_t^2). \tag{5.2.24}$$

Unidirectional Stretch: For this kinematical condition, the incremental rotation and stretch tensors are obtained from the polar decomposition of $\mathbf{F}_t(\Delta t)$, as follows (see Subsection 2.2.2):

$$\mathbf{C}_t = \mathbf{F}_t^T\, \mathbf{F}_t, \quad k = \frac{1}{3}\, \mathrm{tr}(\mathbf{C}_t),$$

$$a = \frac{1}{6}\, \{\mathrm{tr}(\mathbf{C}_t^2) - 3\, k^2\} \quad b = \frac{1}{6}\, \{\mathrm{tr}(\mathbf{C}_t^3) - 18\, a\, k - 3\, k^3\},$$

$$\phi = \begin{cases} \cos^{-1}(b/a^{3/2}) & \text{for} \quad a \neq 0 \\ 0 & \text{for} \quad a = 0 \end{cases},$$

$$\lambda_{n+1}^2 = k + 2\sqrt{a}\cos\left(\frac{2\, n\, \pi + \phi}{3}\right), \quad n = 0,\ 1,\ 2,$$

$$I_U = \lambda_1 + \lambda_2 + \lambda_3,$$

$$II_U = \lambda_1\, \lambda_2 + \lambda_2\, \lambda_3 + \lambda_3\, \lambda_1, \quad III_U = \lambda_1\, \lambda_2\, \lambda_3,$$

$$\mathbf{U}_t = \frac{1}{III_U - I_U\, II_U}\, \{\mathbf{C}_t^2 - (I_U^2 - II_U)\, \mathbf{C}_t - I_U\, III_U\, \mathbf{1}\},$$

$$\mathbf{U}_t^{-1}(\Delta t) = \frac{1}{III_U}\, (\mathbf{C}_t - I_U\, \mathbf{U}_t + II_U\, \mathbf{1}),$$

$$= \frac{1}{III_U\,(III_U - I_U\, II_U)}\, \{-I_U\, \mathbf{C}_t^2 + (I_U^3 - 2\, I_U\, II_U + III_U)\, \mathbf{C}_t$$

$$+ (I_U^2\, III_U + II_U\, III_U - I_U\, II_U^2)\, \mathbf{1}\},$$

$$\mathbf{R}_t = \mathbf{F}_t\,\mathbf{U}_t^{-1}. \tag{5.2.25a-l}$$

Note that, the evaluation of the tensors $\mathbf{C}_t$ and $\mathbf{C}_t^2$ and the coefficients k, a, and b, requires 88 floating point operations. With an additional 14 floating point operations, and 1, 3, and 4, evaluations of *arc cosine*, *cosine*, and *square root*, respectively, the principal stretches can be obtained. Then the invariants of the stretch tensor $\mathbf{U}_t$ require 9 more floating point operations. Evaluation of $\mathbf{U}_t^{-1}$ and $\mathbf{R}_t$ requires 21 and 45 floating point operations, respectively. In total, 186 floating point operations, and 1, 3, and 4 evaluations of *arc cosine*, *cosine*, and *square root*, respectively, are required to determine the rotation tensor *exactly*. Rashid's (1993) *approximate* estimate requires 105 floating point operations and 2 evaluations of *square root*.

Using (5.2.12), the constant corotational deformation rate is evaluated *exactly* as follows:

(i) a = 0

$$\hat{\mathbf{D}} = \frac{ln\,(\lambda_1)}{\Delta t}\,\mathbf{1}. \tag{5.2.26}$$

(ii) $a^3 = b^2 \neq 0$

$$\hat{\mathbf{D}} = \frac{ln\,(\lambda_1^2) - ln\,(\lambda_2^2)}{(\lambda_1^2 - \lambda_2^2)\,\Delta t}\,\mathbf{C}_t + \frac{\lambda_1^2\,ln\,(\lambda_2) - \lambda_2^2\,ln\,(\lambda_1)}{(\lambda_1^2 - \lambda_2^2)\,\Delta t}\,\mathbf{1}. \tag{5.2.27}$$

(iii) $a^3 \neq b^2$

$$\Delta = -\,(\lambda_1^2 - \lambda_2^2)\,(\lambda_2^2 - \lambda_3^2)\,(\lambda_3^2 - \lambda_1^2),$$

$$\begin{bmatrix} a_0 \\ -a_1 \\ a_2 \end{bmatrix} = \frac{1}{\Delta}\begin{bmatrix} \lambda_2^2\lambda_3^2 & \lambda_3^2\lambda_1^2 & \lambda_1^2\lambda_2^2 \\ \lambda_2^2+\lambda_3^2 & \lambda_3^2+\lambda_1^2 & \lambda_1^2+\lambda_2^2 \\ 1 & 1 & 1 \end{bmatrix}\begin{bmatrix} ln\,(\lambda_1)\,(\lambda_2^2-\lambda_3^2) \\ ln\,(\lambda_2)\,(\lambda_3^2-\lambda_1^2) \\ ln\,(\lambda_3)\,(\lambda_1^2-\lambda_2^2) \end{bmatrix},$$

$$\hat{\mathbf{D}} = \frac{1}{\Delta t}\,(a_0\,\mathbf{1} + a_1\,\mathbf{C}_t + a_2\,\mathbf{C}_t^2). \tag{5.2.28a-c}$$

This requires a maximum of 32 floating point operations and 3 evaluations of *logarithm*. The approximate expression in Rashid's procedure requires an additional 52 floating point operations.

Summarizing, the *exact* evaluation of $\mathbf{R}_t$ and $\hat{\mathbf{D}}$ for a given $\mathbf{F}_t$, requires 218 floating point operations and 1, 3, 4, and 3, evaluations of *arc cosine*, *cosine*, *square root*, and *logarithm*, respectively. The *approximate* evaluation of $\mathbf{R}_t$ and $\hat{\mathbf{D}}$ for a given $\mathbf{F}_t$, requires 157 floating point operations and 2 evaluations of *square root*. As shown by Rashid (1993), the approximate expressions $\hat{\mathbf{D}}^R$ and $\mathbf{R}_t^R$ are accurate only to the second and first order of $\mathbf{U}_t - \mathbf{1}$, respectively.

5.3. J_2-FLOW THEORIES

In general, the continuum constitutive equations can be expressed as

$$\overset{\circ}{\boldsymbol{\tau}} = \boldsymbol{\mathcal{L}} : (\mathbf{D} - \mathbf{D}^p), \tag{5.3.1a}$$

where $\overset{\circ}{\boldsymbol{\tau}}$ is the objective Jaumann rate of the Kirchhoff stress, $\mathbf{D}$ is the deformation rate tensor, $\mathbf{D}^p$ is the inelastic part of the deformation rate, and $\boldsymbol{\mathcal{L}}$ is the fourth-order instantaneous elasticity tensor. The material time derivative of the Kirchhoff stress is given by

$$\dot{\boldsymbol{\tau}} = \boldsymbol{\mathcal{L}} : (\mathbf{D} - \mathbf{D}^p) + \mathbf{W}\boldsymbol{\tau} - \boldsymbol{\tau}\mathbf{W}, \tag{5.3.1b}$$

where $\mathbf{W}$ is the spin tensor. To integrate this constitutive equation from time t to $t + \Delta t$, the time variation of the velocity gradient during this timestep needs to be specified. In most finite-element codes, it is assumed that the velocity gradient is constant over each timestep. Because of the presence of $\boldsymbol{\tau}$ on the right-hand side, the integration of (5.3.1b) is complicated even when $\boldsymbol{\mathcal{L}}$ is constant.

Alternatively, the constitutive rate equation (5.3.1) can be expressed in terms of the Kirchhoff stress defined with respect to the deformed configuration at time t, as follows:

$$\dot{\hat{\boldsymbol{\tau}}} = \hat{\boldsymbol{\mathcal{L}}} : (\hat{\mathbf{D}} - \hat{\mathbf{D}}^p),$$

$$\hat{\boldsymbol{\tau}} = \mathbf{R}_t^T \boldsymbol{\tau} \mathbf{R}_t,$$

$$\hat{\mathbf{D}} = \mathbf{R}_t^T \mathbf{D} \mathbf{R}_t,$$

$$\hat{\mathbf{D}}^p = \mathbf{R}_t^T \mathbf{D}^p \mathbf{R}_t, \tag{5.3.2a-d}$$

$$\hat{\mathcal{L}}_{ijkl} = (R_t)_{pi} (R_t)_{qj} (R_t)_{rk} (R_t)_{sl} \mathcal{L}_{pqrs}, \tag{5.3.3}$$

$$\mathbf{R}_t(\xi) = \exp\{ \int_t^{t+\xi} \mathbf{W}(\zeta)\, d\zeta \}. \tag{5.3.4}$$

When the elastic response of the material is *isotropic*, then $\hat{\boldsymbol{\mathcal{L}}} = \boldsymbol{\mathcal{L}}$. For constant $\hat{\mathbf{D}}$, the integration of (5.3.2a) is then similar to that for small deformations; see Hughes (1984).

In this chapter, numerical algorithms are presented for integration of the continuum constitutive equations discussed in Chapter 4. For simplicity, only isotropic elasticity is considered. The elasticity tensor in coordinate-independent and index notation takes on the following form:

$$\hat{\boldsymbol{\mathcal{L}}} = 2\mu [\mathbf{1}^{(4s)} - 1/3\, \mathbf{1} \otimes \mathbf{1}] + 3\kappa [1/3\, \mathbf{1} \otimes \mathbf{1}],$$

$$\hat{\mathcal{L}}_{ijkl} = \mathcal{L}_{ijkl} = \mu (\delta_{ik}\delta_{jl} + \delta_{il}\delta_{jk}) + (\kappa - 2\mu/3)\delta_{ij}\delta_{kl}, \tag{5.3.5a,b}$$

where μ and κ are the shear and bulk moduli of the material.[8]

---

[8] This approximation of *elasticity* is justified on the ground that the elastic moduli of most metals remain constant for moderate plastic strains, *i.e.*, they are essentially unaffected by moderate prestraining. Assumption (5.3.5) is a special case of hypoelasticity.

In the J_2-flow theory, the plastic deformation is volume-preserving and pressure-independent. Hence, only the deviatoric part of the constitutive equations is considered in the remaining part of this section. Then, the constitutive equation for the *deviatoric* part of the stress rate is

$$\overset{\circ}{\boldsymbol{\tau}} = 2\mu\,(\mathbf{D} - \mathbf{D}^p)\,. \tag{5.3.6}$$

Here and in the sequel, the usual superscript, ′, that is used to denote the deviatoric part, is omitted for the J_2*-flow theory.*

5.3.1. Rate-independent Model

For the rate-independent plasticity model, the plastic part of the deformation rate considered here is in the form

$$\mathbf{D}^p = \begin{cases} \dot{\gamma}\,\boldsymbol{\mu}/\sqrt{2} + \alpha\,\tau\,\overset{\circ}{\boldsymbol{\mu}}/\sqrt{2} & \text{if } \tau = \tau_y \text{ and } \boldsymbol{\mu}:\mathbf{D} > 0 \\ 0 & \text{if } \tau < \tau_y \text{ and } \boldsymbol{\mu}:\mathbf{D} > 0 \\ 0 & \text{if } \tau \le \tau_y \text{ and } \boldsymbol{\mu}:\mathbf{D} < 0\,, \end{cases} \tag{5.3.7a}$$

where $\dot{\gamma}$ is the effective plastic strain rate, α is the noncoaxiality coefficient, and τ and $\boldsymbol{\mu}$ are defined by

$$\tau = (\tfrac{1}{2}\,\mathbf{s}:\mathbf{s})^{1/2}\,, \quad \boldsymbol{\mu} = \frac{\mathbf{s}}{\sqrt{2}\,\tau}\,, \quad \mathbf{s} \equiv \boldsymbol{\tau} - \boldsymbol{\beta}\,. \tag{5.3.8a-c}$$

In (5.3.8c), $\boldsymbol{\beta}$ is the (deviatoric) backstress whose evolution is assumed to be given by

$$\overset{\circ}{\boldsymbol{\beta}} = \sqrt{2}\,\Lambda\,\dot{\gamma}\,\boldsymbol{\mu}\,, \tag{5.3.9}$$

where Λ is a given function of the effective plastic strain, γ. For illustration, the yield stress τ_y is assumed to be a function of γ,

$$\tau_y = F(\gamma)\,. \tag{5.3.10a}$$

Both Λ and τ_y, in general, are also functions of the temperature. This is considered in Subsection 5.3.3.

5.3.2. Rate-dependent Model

For the rate-dependent deformation model, the plastic part (deviatoric) of the deformation rate considered here is given by

$$\mathbf{D}^p = \dot{\gamma}\,\frac{\boldsymbol{\mu}}{\sqrt{2}} + \alpha\,\tau\,\frac{\overset{\circ}{\boldsymbol{\mu}}}{\sqrt{2}}\,. \tag{5.3.7b}$$

For illustration, a general flow rule (flow stress) is considered in the form[9]

$$\tau = F(\dot{\gamma}, \gamma)\,, \tag{5.3.10b}$$

[9] Several commonly used rate- and temperature-dependent flow-stress models are examined in Chapter 4, Section 4.8, where physically-based models for metal plasticity are also developed, and the corresponding experimentally obtained constitutive parameters are given for a number of metals

or, equivalently, in the form

$$\dot{\gamma} = g(\tau, \gamma)\,. \tag{5.3.10c}$$

When there is plastic deformation, constitutive equation (5.3.6) is expressed in terms of the Jaumann rate of the stress difference **s**, as follows:

$$\overset{\circ}{\mathbf{s}} = 2\mu\mathbf{D} - 2(\mu+\Lambda)\dot{\gamma}\,\frac{\boldsymbol{\mu}}{\sqrt{2}} - 2\mu\tau\alpha\,\frac{\overset{\circ}{\boldsymbol{\mu}}}{\sqrt{2}}\,. \tag{5.3.11a}$$

This is further simplified into

$$\overset{\circ}{\mathbf{s}} = 2\mu\{\mathbf{D} - \bar{\alpha}(\mathbf{1}^{(4s)} - \boldsymbol{\mu}\otimes\boldsymbol{\mu}):\mathbf{D}\} - 2(\mu+\Lambda)\dot{\gamma}\,\frac{\boldsymbol{\mu}}{\sqrt{2}}\,, \tag{5.3.11b}$$

where, based on expression (4.8.25f), for isotropic hypoelasticity,

$$\bar{\alpha} = \frac{\mu\alpha}{1+\mu\alpha}\,. \tag{5.3.12}$$

In a coordinate system rotating with the material element, (5.3.11b) becomes

$$\dot{\hat{\mathbf{s}}} = 2\mu\{\hat{\mathbf{D}} - \bar{\alpha}(\mathbf{1}^{(4s)} - \hat{\boldsymbol{\mu}}\otimes\hat{\boldsymbol{\mu}}):\hat{\mathbf{D}}\} - 2(\mu+\Lambda)\dot{\gamma}\,\frac{\hat{\boldsymbol{\mu}}}{\sqrt{2}}\,,$$

$$\hat{\boldsymbol{\mu}} = \frac{\hat{\mathbf{s}}}{\sqrt{2}\tau}\,, \qquad \tau = (\tfrac{1}{2}\hat{\mathbf{s}}:\hat{\mathbf{s}})^{1/2}\,. \tag{5.3.13a-c}$$

The evolution equation for the backstress then takes on the form

$$\dot{\hat{\boldsymbol{\beta}}} = \sqrt{2}\Lambda\dot{\gamma}\,\hat{\boldsymbol{\mu}}\,. \tag{5.3.13d}$$

Take the scalar product of (5.3.13a) with $\hat{\boldsymbol{\mu}}$ to arrive at

$$\dot{\tau} = \sqrt{2}\mu(\hat{\mathbf{D}}:\hat{\boldsymbol{\mu}}) - (\mu+\Lambda)\dot{\gamma}\,, \tag{5.3.14a}$$

and from (5.3.13a,b) obtain

$$\dot{\hat{\boldsymbol{\mu}}} = \frac{\sqrt{2}\mu(1-\bar{\alpha})}{\tau}(\mathbf{1} - \hat{\boldsymbol{\mu}}\otimes\hat{\boldsymbol{\mu}}):\hat{\mathbf{D}}\,. \tag{5.3.14b}$$

5.3.3. Thermal Softening with Isotropic Hardening

For materials which undergo thermal softening, the yield (flow) stress depends on the temperature[10]. Since the temperature change is related to the plastic work, the yield stress is an implicit function of the plastic strain (and strain rate). In continuum theories with thermal softening, the yield stress is usually expressed in the form

$$\tau_y = F(\gamma, T) = F_0(\gamma)\,/\,F_T(T)\,, \qquad \text{for rate–independent flow}\,,$$

in Tables 4.8.1 and 4.8.2, pages 235 and 238. As in Chapter 4, γ is viewed as a "load parameter".

[10] Throughout this chapter, the temperature is denoted by T which is the absolute temperature minus a reference temperature. The reference temperature may be adjusted to suit a given model; see Section 4.8. For physically-based models, T is the absolute temperature.

$$\tau = F(\dot{\gamma}, \gamma, T) = F_0(\dot{\gamma}, \gamma) / F_T(T), \quad \text{for rate–dependent flow}, \quad (5.3.15a,b)$$

where F_T is some given (monotonically increasing) function of temperature T. When there is plastic deformation, the rate of change of the temperature is related to the rate of change of the plastic work per unit mass, as follows:

$$\dot{T} = \frac{\eta}{\hat{\rho}\, C_v}\, \tau \dot{\gamma}\,, \tag{5.3.16a}$$

where $\hat{\rho}$ is the density of the material; C_v is the constant-volume heat capacity of the material; and η, a dimensional parameter, defines the fraction of plastic work which is converted into heat[11]. Equation (5.3.16a) can be rewritten as

$$\tau = \frac{1}{C_0}\frac{dT}{d\gamma}\,, \quad C_0 = \frac{\eta}{\hat{\rho}\, C_v}\,. \tag{5.3.16b}$$

In general, C_0 is a function of temperature, $C_0 = C_0(T)$. Substitute (5.3.16b) into (5.3.15) and integrate to arrive at

$$\int_{T_t}^{T_t + \Delta T} \frac{F_T(T)}{C_0(T)}\, dT = \int_{\gamma_t}^{\gamma_t + \Delta\gamma} F_0\, d\gamma\,. \tag{5.3.17a}$$

Usually, C_0 changes slowly with temperature. Hence, consider the following approximation:

$$\Delta T = \frac{C_0(T_t) + C_0(T_t + \Delta T_A)}{F_T(T_t) + F_T(T_t + \Delta T_A)} \int_{\gamma_t}^{\gamma_t + \Delta\gamma} F_0\, d\gamma\,,$$

$$\Delta T_A = \frac{C_0(T_t)}{F_T(T_t)} \int_{\gamma_t}^{\gamma_t + \Delta\gamma} F_0\, d\gamma\,, \tag{5.3.17b,c}$$

where T_t is the temperature at time t.[12] Note that, in many applications C_0 may be assumed to remain constant over each time increment.

In continuum theories, one-parameter models are often used to express $F_T(T)$. For example, in the Johnson-Cook model,[13]

$$F_T = \frac{(T_M)^{m_0}}{(T_M)^{m_0} - T^{m_0}}\,, \tag{5.3.18a}$$

where the power m_0 is the thermal softening exponent, and T_M is the melting temperature. Here, F_T is normalized such that $F_T = 1$ when the material is at the

[11] Expressions (5.3.16a,b) are valid when the backstress is zero. In the presence of kinematic hardening, these expressions do not apply, unless the plastic work associated with backstress does not contribute to heating, *e.g.*, in the case of fabric anisotropy in plastic deformation of frictional granules; see Section 4.7 and Chapter 7.

[12] In this chapter, the subscript t is used to denote the value of the corresponding quantity at time t, and the subscript A is used to denote the first (or starting) estimate (or approximation) of the associated variable (or quantity).

[13] See Johnson and Cook (1983); also see Section 4.8 where various rate-dependent models, including those based on the physics of dislocation motion, are discussed.

reference temperature, *i.e.*, when $T = 0$. Note that, in the Johnson-Cook model, the yield stress goes to 0 as the material approaches the melting temperature.

When the function F_T is such that

$$F_T(T) = \frac{dG_T}{dT}, \tag{5.3.19a}$$

where G_T is a function of T with an inverse relation

$$T = \chi(G_T), \tag{5.3.19b}$$

then substitution of (5.3.19) into (5.3.17a) results in

$$T_t + \Delta T = \chi(G_T(T_t + \Delta T)),$$

$$G_T(T_t + \Delta T) \approx G_T(T_t) + \tfrac{1}{2}\{C_0(T_t) + C_0(T_t + \Delta T_A)\} \int_{\gamma_t}^{\gamma_t + \Delta\gamma} F_0 \, d\gamma, \tag{5.3.20a,b}$$

where T_t is the temperature at time t, the start of the timestep. For example, if F_T is an exponential function, as in the power-law model,

$$F_T = e^{\lambda_0 T}, \tag{5.3.18b}$$

where λ_0 is the thermal softening parameter, assumed constant, then

$$G_T(T) = \frac{1}{\lambda_0} e^{\lambda_0 T}, \quad \chi(G_T) = \frac{1}{\lambda_0} \, ln\,(\lambda_0 G_T). \tag{5.3.21a,b}$$

5.4. PLASTIC-PREDICTOR/ELASTIC-CORRECTOR METHOD WITH ERROR ESTIMATE

To simplify the analysis, kinematic hardening is excluded in what follows; see, however, Subsection 5.6.3. With $\Lambda = 0$, from (5.3.14a) and (5.3.10b), the governing constitutive relations are

$$\dot{\tau}(\xi) + \mu\,\dot{\gamma}(\xi) = \mu\, d(\xi),$$

$$\tau(\xi) = F(\dot{\gamma}(\xi), \gamma(\xi)), \quad \xi \geq 0. \tag{5.4.1a,b}$$

The stress, $\tau(\xi)$, the plastic strain, $\gamma(\xi)$, and the plastic strain rate, $\dot{\gamma}(\xi)$, are unknown, with their initial values at $\xi = 0$ given. The total rate of deformation, $d(\xi) = \sqrt{2}\,\hat{\mathbf{D}} : \hat{\boldsymbol{\mu}}$, and the function $F(\dot{\gamma}, \gamma)$ are assumed to be prescribed. As before, the function F depends on $\dot{\gamma}(\xi)$ and $\gamma(\xi)$ for rate-dependent elastovisco-plasticity, and only on $\gamma(\xi)$ for rate-independent elastoplasticity.

The system (5.4.1a,b) is very stiff for problems of practical interest, in the sense that $\dot{\gamma}(\xi)$ is essentially zero until $\tau(\xi)$ attains a certain threshold value, after which $\dot{\gamma}(\xi)$ becomes increasingly large. For example, consider the power-law model,

$$\tau = F(\dot{\gamma}, \gamma) = \tau_c (1 + \gamma/\gamma_0)^N (\dot{\gamma}/\dot{\gamma}_0)^{1/m}, \tag{5.4.2}$$

where τ_c is the initial yield stress, $\dot{\gamma}_0$ and γ_0 are the reference strain rate and strain, and N and m are material parameters. For most metals, m is large, between 20 and 200. Therefore, $\dot{\gamma}$ is rather small for τ smaller than the reference stress, $\tau_r \equiv \tau_c (1 + \gamma/\gamma_0)^N$. Once τ exceeds this reference stress, then $\dot{\gamma}$ becomes very large.

5.4.1. Stiff Systems

It is convenient to non-dimensionalize (5.4.1) and (5.4.2), using the initial yield stress τ_c. Without loss of generality, assume that τ is measured in τ_c as the unit of stress, and set $\varepsilon = \tau_c/\mu$. Rewrite (5.4.1) in the *non-dimensional* form as

$$\varepsilon \dot{\tau}(\xi) + \dot{\gamma}(\xi) = d(\xi), \qquad \tau(\xi) = F(\dot{\gamma}(\xi), \gamma(\xi)), \tag{5.4.3a,b}$$

where the same notation, *i.e.*, τ and F, is used for the non-dimensional stress. Note that (5.4.2) now is given by

$$F = (1 + \gamma/\gamma_0)^N (\dot{\gamma}/\dot{\gamma}_0)^{1/m}. \tag{5.4.4}$$

The inverse of (5.4.3b) is defined by

$$\dot{\gamma}(\xi) = g(\tau, \gamma), \tag{5.4.3c}$$

where $\dot{\gamma}$ is measured in $\dot{\gamma}_0$ as the unit of the strain rate.

Since $\varepsilon = \tau_c/\mu$ is in general a small parameter in the range 10^{-4} to 10^{-2}, (5.4.3a,b) defines a typical boundary-layer problem, with initial conditions

$$\gamma(0) = \gamma_t, \qquad \tau(0) = \tau_t. \tag{5.4.3d,e}$$

Recently, Nemat-Nasser and Ni (1994) have investigated the solution of this problem and have related it to the plastic-predictor/elastic-corrector method of Nemat-Nasser (1991) and Nemat-Nasser and Chung (1992). These authors show that the plastic-predictor/elastic-corrector method is a simplified singular perturbation method, where the outer solution and the boundary-layer correction are constructed by exploiting the physics of elastoviscoplasticity in the form of an inelastic predictor followed by an estimate of the remaining part of the solution as a whole. They also provide computable error bounds for the solutions obtained by this algorithm, and, in this manner, assess the effectiveness and accuracy of the method. What follows in the remaining part of this section is a summary of these authors' results.

5.4.2. The Constitutive Algorithm

The algorithm is based on an inelastic predictor, followed by a corrector for the associated elastic increment, which also gives the value of the stress increment. Integrating both sides of (5.4.3a), use the initial conditions, (5.4.3d,e), to obtain

$$\varepsilon\{\tau(\xi)-\tau_t\}+\gamma(\xi)-\gamma_t=\int_0^{\xi} d(\zeta)\, d\zeta\,. \tag{5.4.5}$$

Define the *inelastic predictor* by

$$\gamma_A(\xi)=\gamma_t+\int_0^{\xi} d(\zeta)\, d\zeta\,, \quad \dot{\gamma}_A(\xi)=d(\xi)\,, \quad \gamma_A(0)=\gamma_t\,. \tag{5.4.6a-c}$$

Combine (5.4.5) and (5.4.6) with (5.4.3b) to obtain the main equation for the correction term, $\gamma(\xi)-\gamma_A(\xi)$, *i.e.*, the *error* in the inelastic predictor $\gamma_A(\xi)$,

$$\gamma(\xi)-\gamma_A(\xi)=\varepsilon\{\tau_t-F(\dot{\gamma}(\xi),\,\gamma(\xi))\}\,, \tag{5.4.7}$$

in which $\tau_t=\tau(0)$ and $\gamma_A(\xi)$ are known. Note that, for positive workhardening, the right-hand side of (5.4.7) is always negative and small.

To construct an explicit solution for the nonlinear differential equation (5.4.7), consider the linearization of the function $F(\dot{\gamma},\,\gamma)$. The novelty here is that, since (5.4.7) is considered an equation for the correction term, $\gamma(\xi)-\gamma_A(\xi)$, *the linearization is performed with respect to the deviation from quantities* $(\dot{\gamma}_A(\xi),\gamma_A(\xi))$. This exploits the important physical fact that, for this kind of nonlinear elastoplastic deformation problem, essentially all deformation is due to plasticity with a very small elastic contribution. In view of (5.4.3a), (5.4.6) implies that, tentatively, the elastic contribution is taken to be zero.

The linearization with respect to $\dot{\gamma}_A(\xi)$ and $\gamma_A(\xi)$ yields,

$$\begin{aligned} F(\dot{\gamma}(\xi),\,\gamma(\xi)) &\approx F(\dot{\gamma}_A(\xi),\,\gamma_A(\xi))+\frac{\partial F(\dot{\gamma}_A,\,\gamma_A)}{\partial\dot{\gamma}}\{\dot{\gamma}(\xi)-\dot{\gamma}_A(\xi)\} \\ &\quad+\frac{\partial F(\dot{\gamma}_A,\,\gamma_A)}{\partial\gamma}\{\gamma(\xi)-\gamma_A(\xi)\}\,. \end{aligned} \tag{5.4.8}$$

Based on the *exact* relation

$$\int_0^{\xi}\{\dot{\gamma}(\zeta)-\dot{\gamma}_A(\zeta)\}\, d\zeta=\gamma(\xi)-\gamma_A(\xi)\,, \tag{5.4.9a}$$

the following approximation is made:

$$\dot{\gamma}(\xi)-\dot{\gamma}_A(\xi)\approx\frac{1}{\xi}\{\gamma(\xi)-\gamma_A(\xi)\}\,. \tag{5.4.9b}$$

Substitute (5.4.8) and (5.4.9b) into (5.4.7), to obtain

$$\begin{aligned} &\gamma_A(\xi)-\gamma(\xi)\approx\varepsilon\,\eta(\xi)\{\gamma(\xi)-\gamma_A(\xi)\}+\varepsilon\{\tau_A(\xi)-\tau_t\}\,, \\ &\eta(\xi)=\frac{1}{\xi}\frac{\partial F(\dot{\gamma}_A(\xi),\,\gamma_A(\xi))}{\partial\dot{\gamma}}+\frac{\partial F(\dot{\gamma}_A(\xi),\,\gamma_A(\xi))}{\partial\gamma}\,, \\ &\tau_A(\xi)=F(\dot{\gamma}_A(\xi),\,\gamma_A(\xi))\,. \end{aligned} \tag{5.4.10a-c}$$

Then, from (5.4.9), (5.4.10), and (5.4.8), the approximate *final* value of the plastic strain is obtained,[14]

---

[14] Here and in the remaining part of this section, the subscript f is used to denote the *final* estimate of the corresponding quantity by the constitutive algorithm.

$$\gamma_f(\xi) = \gamma_A(\xi) - \frac{\varepsilon\,\{\tau_A(\xi) - \tau_t\}}{1 + \varepsilon\,\eta(\xi)}\,. \tag{5.4.11a}$$

Combining (5.4.11a) with (5.4.9a) and (5.4.7), the plastic strain rate and the stress are given, respectively, by

$$\dot\gamma_f(\xi) = \dot\gamma_A(\xi) - \frac{\varepsilon\,\{\tau_A(\xi) - \tau_t\}}{\{1 + \varepsilon\,\eta(\xi)\}\xi}\,, \tag{5.4.11b}$$

$$\tau_f(\xi) = \frac{\tau_A(\xi) + \varepsilon\,\eta(\xi)\tau_t}{1 + \varepsilon\,\eta(\xi)}\,, \tag{5.4.11c}$$

for any desired ξ, provided that

$$\dot\gamma_f(\xi) \geq 0\,, \tag{5.4.12a}$$

or, equivalently, that

$$\xi \geq \frac{\varepsilon\,(\tau_A(\xi) - \tau_t)}{(1 + \varepsilon\,\eta(\xi))\,d(\xi)}\,. \tag{5.4.12b}$$

Note that (5.4.12b) places a *lower bound* on the value of the time increment that should be used. This is in contrast to the algorithms which are based on the forward-gradient methods, where the timestep must not exceed an *upper bound*. For the rate-independent elastoplasticity, inequality (5.4.12b) always holds, since F is independent of $\dot\gamma$.

If (5.4.12) is not satisfied, then the plastic contribution to the total deformation is very small. It then may be assumed that the stress rate and the plastic strain rate are constant during the time interval t to $t + \xi$. Hence,

$$\dot\tau = \varepsilon^{-1}\,\{d_0 - g(\tau_t, \gamma_t)\}\,, \qquad \dot\gamma = g(\tau_t, \gamma_t)\,. \tag{5.4.13a,b}$$

The stress and the plastic strain then are,

$$\tau(\xi) = \tau_t + \varepsilon^{-1}\,\{d_0 - g(\tau_t, \gamma_t)\}\,\xi\,,$$

$$\gamma(\xi) = \gamma_t + g(\tau_t, \gamma_t)\,\xi\,, \quad d_0 = d(0)\,. \tag{5.4.13c-e}$$

In summary, the approximate solutions for the non-dimensional plastic strain, plastic strain rate, and stress, by the constitutive algorithm of Nemat-Nasser (1991) and Nemat-Nasser and Chung (1992), are given as follows.

(A) Rate-independent elastoplasticity

$$\gamma_f(\xi) = \gamma_A(\xi) - \frac{\varepsilon\,\{\tau_A(\xi) - \tau_t\}}{1 + \varepsilon\,\eta(\xi)}\,, \quad \dot\gamma_f(\xi) = \dot\gamma_A(\xi) - \frac{\varepsilon\,\{\tau_A(\xi) - \tau_t\}}{\xi\,\{1 + \varepsilon\,\eta(\xi)\}}\,,$$

$$\tau_f(\xi) = \tau_A(\xi) - \varepsilon\,\eta(\xi)\,\frac{\{\tau_A(\xi) - \tau_t\}}{1 + \varepsilon\,\eta(\xi)}\,, \tag{5.4.14a-c}$$

where

$$\tau_A(\xi) = F(\gamma_A(\xi)), \qquad \eta(\xi) = \frac{\partial F(\gamma_A(\xi))}{\partial\gamma}\,,$$

$$\gamma_A(\xi) = \gamma_t + \int_0^{\xi} d(\zeta)\,d\zeta\,. \tag{5.4.14d,f}$$

(B) Rate-dependent elastoviscoplasticity

$$\gamma_f(\xi) = \gamma_A(\xi) - \frac{\varepsilon\,\{\tau_A(\xi) - \tau_t\}}{1 + \varepsilon\,\eta(\xi)}, \quad \dot{\gamma}_f(\xi) = \dot{\gamma}_A(\xi) - \frac{\varepsilon\,\{\tau_A(\xi) - \tau_t\}}{\xi\,\{1 + \varepsilon\,\eta(\xi)\}},$$

$$\tau_f(\xi) = \tau_A(\xi) - \frac{\varepsilon\,\eta(\xi)\,\{\tau_A(\xi) - \tau_t\}}{1 + \varepsilon\,\eta(\xi)}, \tag{5.4.14g-i}$$

provided that $\dot{\gamma}_f(\xi) > 0$, where

$$\tau_A(\xi) = F(d(\xi),\, \gamma_A(\xi)),$$

$$\eta(\xi) = \frac{1}{\xi}\,\frac{\partial F(\dot{\gamma}_A(\xi),\, \gamma_A(\xi))}{\partial \dot{\gamma}} + \frac{\partial F(\dot{\gamma}_A(\xi),\, \gamma_A(\xi))}{\partial \gamma},$$

$$\gamma_A(\xi) = \gamma_t + \int_0^{\xi} d(\zeta)\, d\zeta. \tag{5.4.14j-l}$$

When $\dot{\gamma}_f(\xi) \le 0$, then

$$\dot{\gamma}_A(\xi) = g(\tau_t,\, \gamma_t), \quad \gamma(\xi) = \gamma_t + g(\tau_t,\, \gamma_t)\,\xi,$$

$$\tau(\xi) = \tau_t + \varepsilon^{-1}\,\{d_0 - g(\tau_t,\, \gamma_t)\}\,\xi, \quad d_0 = d(0), \tag{5.4.14m-p}$$

where g is the inverse function of F.

The numerical examples show that this method is quite accurate. Consider the case when the function F is described by the power-law (5.4.4). Let the non-dimensional parameters be given by $(\varepsilon, \gamma_0, \dot{\gamma}_0) = 10^{-2}$, $d(\xi) = 0.008$ per microsecond (8,000 per second), and $N = 0.08$, and assume the initial values are $\gamma_t = \tau_t = 0$. The corresponding numerical example is shown in Figure 5.4.1, for $m = 200$. It is emphasized that, the curve designated as "PP/EC Solution" is obtained *nonincrementally*, where, for each value of $\xi = t > 0$, the value of $\tau_f(t)$ is computed directly from either (5.4.14i) or (5.4.14o), depending on whether $\dot{\gamma}_f(t) > 0$ or $\dot{\gamma}_f(t) \le 0$, each time starting from the initial conditions $\gamma_t = \tau_t = 0$. On the other hand, the curve designated as "Exact Solution" is obtained incrementally, using the following rate equation which results when $\dot{\gamma}(t)$ and $\gamma(t)$ are eliminated between (5.4.3a) and (5.4.4):

$$\dot{\tau}(t) = \frac{d}{\varepsilon} - \frac{\dot{\gamma}_0}{\varepsilon}\,\{\tau(t)\}^m \left[1 + \frac{t\,d - \varepsilon\,\tau(t)}{\gamma_0}\right]^{Nm}. \tag{5.4.15}$$

5.4.3. Error Estimate

To assess the accuracy of the plastic-predictor/elastic-corrector algorithm, Nemat-Nasser and Ni (1994) consider the computable (*a posteriori*) error bounds for the one-step algorithm described in Subsection 5.4.2. This is briefly reviewed in this subsection.

Errors are caused only by inaccuracies in (5.4.8) and (5.4.9b). From physical considerations, Nemat-Nasser and Ni (1994) assume that:

(a) The prescribed total deformation rate, $d(\xi)$, the plastic strain, $\gamma(\xi)$, the plastic strain rate, $\dot{\gamma}(\xi)$, and the stress rate, $\dot{\tau}(\xi)$, are all non-negative;

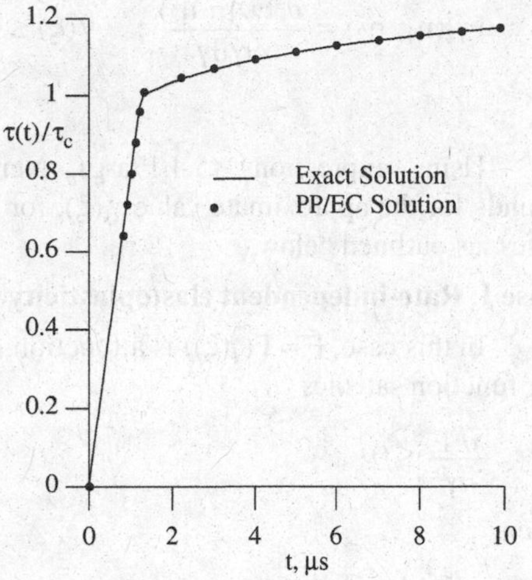

Figure 5.4.1

Comparison between the one-step (from t = 0 to any t) results of the plastic-predictor/elastic-corrector (denoted by PP/EC Solution) algorithm and the exact results, for rate-dependent plasticity; N = 0.08, m = 200, d = 8,000/s, $(\gamma_t, \tau_t) = 0$, $(\varepsilon, \gamma_0, \dot{\gamma}_0) = 10^{-2}$

(b) $F(\dot{\gamma}, \gamma)$ is a non-decreasing class C^2 function with respect to both its arguments;

(c) $d(\xi)$ is of class C^1, and $\gamma(\xi)$ and $\tau(\xi)$ are of class C^2 functions;

(d) The stress $\tau(\xi)$ is a concave function, *i.e.*, $\ddot{\tau}(\xi)$ is non-positive.

Based on these assumptions and in view of (5.4.11a-c), Nemat-Nasser and Ni then show that,

$$\dot{\gamma}_A(\xi) \geq \dot{\gamma}(\xi), \quad \gamma_A(\xi) \geq \gamma(\xi);$$
$$\dot{\gamma}_A(\xi) \geq \dot{\gamma}_f(\xi), \quad \gamma_A(\xi) \geq \gamma_f(\xi). \tag{5.4.16a-d}$$

Furthermore, they prove that the following relation always holds:

$$\gamma(\xi) - \gamma_f(\xi) = -\frac{\varepsilon}{1+\varepsilon F_2(\xi)} \{F_1(\xi)(\dot{\gamma}(\xi) - \dot{\gamma}_f(\xi)) + R(\dot{\gamma}(\xi), \gamma(\xi), \eta_1, \eta_2)\}, \tag{5.4.17a}$$

where

$$R(\dot{\gamma}(\xi), \gamma(\xi), \eta_1, \eta_2) = \frac{1}{2} F_{11}(\eta_1, \eta_2)\{\dot{\gamma}_A(\xi) - \dot{\gamma}(\xi)\}^2 + F_{12}(\eta_1, \eta_2)\{\dot{\gamma}_A(\xi) - \dot{\gamma}(\xi)\}\{\gamma_A(\xi) - \gamma(\xi)\} + \frac{1}{2} F_{22}(\eta_1, \eta_2)\{\gamma_A(\xi) - \gamma(\xi)\}^2,$$

$$F_1(\xi) = \frac{\partial F(\dot{\gamma}_A(\xi), \gamma_A(\xi))}{\partial \dot{\gamma}}, \quad F_2(\xi) = \frac{\partial F(\dot{\gamma}_A(\xi), \gamma_A(\xi))}{\partial \gamma},$$

$$F_{11}(\eta_1, \eta_2) = \frac{\partial^2 F(\eta_1, \eta_2)}{\partial \dot{\gamma}^2}, \quad F_{22}(\eta_1, \eta_2) = \frac{\partial^2 F(\eta_1, \eta_2)}{\partial \gamma^2},$$

$$F_{12}(\eta_1, \eta_2) = \frac{\partial^2 F(\eta_1, \eta_2)}{\partial\dot{\gamma}\,\partial\gamma}, \quad \dot{\gamma}(\xi) \le \eta_1 \le \dot{\gamma}_A(\xi), \quad \gamma(\xi) \le \eta_2 \le \gamma_A(\xi). \tag{5.4.17b-g}$$

Using expressions (5.4.17a-g), Nemat-Nasser and Ni establish error bounds for the approximate value $\gamma_f(\xi)$, for rate-independent and rate-dependent cases, as outlined below.

Case I: Rate-independent elastoplasticity

In this case, $F = F(\gamma(\xi))$ is a function of $\gamma(\xi)$. In general, this workhardening function satisfies

$$\frac{\partial^2 F}{\partial \gamma^2} \le 0. \tag{5.4.18a}$$

Then,

$$\gamma_f(\xi) \le \gamma(\xi) \le \gamma_A(\xi), \tag{5.4.18b}$$

and, an error bound for $\gamma_f(\xi)$ is

$$|\gamma(\xi) - \gamma_f(\xi)| \le \frac{1}{2} \frac{\varepsilon^3 \{\tau_A(\xi) - \tau_t\}^2}{\{1 + \varepsilon F_1(\xi)\}^3} \max\{F_{22}(\eta_2)\}. \tag{5.4.18c}$$

As an example, when $F(\gamma)$ is given by

$$F(\gamma) = (1 + \gamma/\gamma_0)^N, \tag{5.4.18d}$$

the error bound is

$$|\gamma(\xi) - \gamma_f(\xi)| \le \frac{N(1-N)}{2} \, \frac{\varepsilon^3 \tau_A(\xi)\,(\tau_A(\xi) - \tau_t)^2}{\{1 + \varepsilon F_2(\xi)\}^3 \{\gamma_f(\xi) + \gamma_0\}^2}. \tag{5.4.18e}$$

With $N = 0.08$, and when the prescribed constants are such that

$$\varepsilon\,\tau_A(\xi) \approx 10^{-2}, \tag{5.4.18f}$$

then the error bound is

$$|\gamma(\xi) - \gamma_f(\xi)| \le 10^{-7}. \tag{5.4.18g}$$

Case II: Rate-dependent elastoviscoplasticity

In this case, Nemat-Nasser and Ni (1994) first note that, when the matrix

$$M = \begin{bmatrix} \dfrac{\partial^2 F}{\partial \dot{\gamma}^2} & \dfrac{\partial^2 F}{\partial \dot{\gamma}\,\partial \gamma} \\[2ex] \dfrac{\partial^2 F}{\partial \dot{\gamma}\,\partial \gamma} & \dfrac{\partial^2 F}{\partial \gamma^2} \end{bmatrix} \tag{5.4.19}$$

is negative-definite, then $\dot{\gamma}(\xi)$ has the bounds

$$\dot{\gamma}_f(\xi) \le \dot{\gamma}(\xi) \le \dot{\gamma}_A(\xi). \tag{5.4.20}$$

As an example, consider the power law (5.4.4). A direct calculation confirms that matrix M is negative-definite, provided $N \le 1/2$ and $m \ge 2$, which are always satisfied for the class of considered problems. Hence, (5.4.20) holds when F is given by the power-law relation (5.4.4).

For the error estimate, consider two mutually exclusive possibilities: either $\gamma(\xi) \le \gamma_f(\xi)$ or $\gamma(\xi) > \gamma_f(\xi)$. For the first case, obtain

$$\gamma_f(\xi) - \gamma(\xi) \le \frac{\varepsilon^2 F_1(\xi)\{\tau_A(\xi) - \tau_t\}}{\xi\{1 + \varepsilon\eta(\xi)\}}. \tag{5.4.21a}$$

When F satisfies the power-law relation (5.4.4), this gives

$$\gamma_f(\xi) - \gamma(\xi) \le \frac{\varepsilon^2 \tau_A(\xi)\{\tau_A(\xi) - \tau_t\}}{m\,\xi\, d(\xi)\{1 + \varepsilon\eta(\xi)\}}. \tag{5.4.21b}$$

When, on the other hand, $\gamma(\xi) > \gamma_f(\xi)$, Nemat-Nasser and Ni obtain

$$\gamma(\xi) - \gamma_f(\xi) \le \frac{\varepsilon}{\{1 + \varepsilon\eta(\xi)\}} \operatorname{Sup} |R(\dot\gamma_f(\xi), \gamma_f(\xi), \eta_1, \eta_2)|, \tag{5.4.22a}$$

where the supremum is taken over $\dot\gamma_f(\xi) \le \eta_1 \le \dot\gamma_A(\xi)$, and $\gamma_f(\xi) \le \eta_2 \le \gamma_A(\xi)$. It can be shown that this case dominates when ξ is near zero, *i.e.*, $\gamma(\xi) > \gamma_f(\xi)$ when ξ approaches zero. When F satisfies the power-law relation (5.4.4), then simple calculation gives

$$\begin{aligned}&\operatorname{Sup} |R(\dot\gamma_f(\xi), \gamma_f(\xi), \eta_1, \eta_2)| \\ &\quad \le \frac{\varepsilon^2 \{\tau_A(\xi) - \tau_t\}^2 \tau_A(\xi)}{2\xi^2\{1 + \varepsilon\eta(\xi)\}^2} \left[\frac{1}{m\{\dot\gamma_f(\xi)\}^2} + \frac{\xi^2 N}{\{\gamma_0 + \gamma_f(\xi)\}^2}\right].\end{aligned} \tag{5.4.22b}$$

Hence, in this case,

$$\gamma(\xi) - \gamma_f(\xi) \le \frac{\varepsilon^3 \{\tau_A(\xi) - \tau_t\}^2 \tau_A(\xi)}{2\{1 + \varepsilon\eta(\xi)\}^3} \left[\frac{1}{m\xi^2\{\dot\gamma_f(\xi)\}^2} + \frac{N}{\{\gamma_0 + \gamma_f(\xi)\}^2}\right]. \tag{5.4.22c}$$

Therefore, under assumptions (a)-(d), and when matrix M in negative, the following results are obtained for the rate-dependent case:

$$\begin{aligned}&|\gamma(\xi) - \gamma_f(\xi)| \le \\ &\max\left\{\frac{\varepsilon^2 F_1(\xi)\{\tau_A(\xi) - \tau_t\}}{\xi\{1 + \varepsilon\eta(\xi)\}}, \ \frac{\varepsilon}{\{1 + \varepsilon\eta(\xi)\}} \operatorname{Sup} |R(\dot\gamma_f(\xi), \gamma_f(\xi), \eta_1, \eta_2)|\right\},\end{aligned} \tag{5.4.23a}$$

where the supremum is taken over $\dot\gamma_f(\xi) \le \eta_1 \le \dot\gamma_A(\xi)$, and $\gamma_f(\xi) \le \eta_2 \le \gamma_A(\xi)$.

In particular, when F satisfies the power-law relation (5.4.4), the *a posteriori* error bound is

$$\begin{aligned}|\gamma(\xi) - \gamma_f(\xi)| \le \max&\left\{\frac{\varepsilon^2 \tau_A(\xi)(\tau_A(\xi) - \tau_t)}{m\,\xi\, d(\xi)\{1 + \varepsilon\eta(\xi)\}},\right. \\ &\left.\frac{\varepsilon^3 \{\tau_A(\xi) - \tau_t\}^2 \tau_A(\xi)}{2\{1 + \varepsilon\eta(\xi)\}^3} \left\{\frac{m}{\xi^2\{\dot\gamma_f(\xi)\}^2} + \frac{N}{\{\gamma_0 + \gamma_f(\xi)\}^2}\right\}\right\}.\end{aligned} \tag{5.4.23b}$$

As an example, suppose that

$$\varepsilon\,\tau_A(\xi) \approx 10^{-2}\,,$$

$$0.005 \leq \gamma_f(\xi) + \gamma_0 \leq 1\,, \qquad (5.4.23c,d)$$

then after simple calculation, the following rough numerical error bounds can be established. When $\gamma_f(\xi)+\gamma_0$ is near 0.005, the maximum in (5.4.23b) is generally given by the second term inside the braces. Hence

$$|\,\gamma(\xi) - \gamma_f(\xi)\,| \leq 1.6\times 10^{-3} \qquad (5.4.23e)$$

for $\gamma_f(\xi)+\gamma_0$ near 0.005. When $\gamma_f(\xi)+\gamma_0$ is near 1, the maximum in (5.4.23b) is generally given by the first term inside the braces. Hence

$$|\,\gamma(\xi) - \gamma_f(\xi)\,| \leq 10^{-6}\,. \qquad (5.4.23f)$$

for $\gamma_f(\xi)+\gamma_0$ near 1. Note that *the error decreases as the strain increment increases.* This is in contrast with the results of algorithms which are based on the forward-gradient method.

The error bound for the approximate value of the stress, $\tau_f(\xi)$, is obtained using (5.4.11a,c),

$$\tau(\xi) - \tau_f(\xi) = \varepsilon^{-1}\,\{\gamma_f(\xi) - \gamma(\xi)\}\,. \qquad (5.4.24)$$

Thus, the error bound of the approximate value of the non-dimensional stress, $\varepsilon\,\tau(\xi)$, is the same as the error bound of the approximate value of the plastic strain, $\gamma(\xi)$.

5.5. SINGULAR PERTURBATION METHOD FOR CONSTITUTIVE ALGORITHMS

The stiff system of equations (5.4.3a,b) is a boundary-layer type. The small parameter ε, in general, has a value between 10^{-4} to 10^{-2}. The smaller this parameter, the stiffer is the system (5.4.3a,b). There are several techniques available for treating boundary-layer problems of this kind, *e.g.,* the method of matched asymptotic expansions, the method of composite scales, and the method of multiple scales.[15] Nemat-Nasser and Ni (1994) employ a two-scale singular perturbation technique to solve this stiff system of equations. What follows in this section is an account of the singular perturbation method and a modified outer solution given by these authors for system (5.4.3a,b).

---

[15] See, for example, O'Malley (1971a,b), Hoppensteadt (1971), Van Dyke (1975), Nayfeh (1980), Miranker (1981), and Smith (1985).

5.5.1. Singular Perturbation Solution

Consider system (5.4.3a,b) with initial conditions (5.4.3d,e). Examine the *asymptotic expansions* of the unknown functions $\gamma(\xi)$ and $\tau(\xi)$, in the following form:

$$\gamma(\xi) \approx \gamma^*(\xi) + \Gamma(s), \qquad \tau(\xi) \approx \tau^*(\xi) + T(s), \tag{5.5.1a,b}$$

where $s = \xi/\varepsilon$ is the *stretched variable.* In (5.5.1), $\gamma^*(\xi)$ and $\tau^*(\xi)$ are the *outer solutions,* and $\Gamma(s)$ and $T(s)$ are the *boundary-layer corrections.* These functions are expressed as,

$$\gamma^*(\xi) = \sum_{k=0}^{\infty} \gamma_k^*(\xi)\,\frac{\varepsilon^k}{k!}, \qquad \tau^*(\xi) = \sum_{k=0}^{\infty} \tau_k^*(\xi)\,\frac{\varepsilon^k}{k!},$$

$$\Gamma(s) = \sum_{k=0}^{\infty} \Gamma_k(s)\,\frac{\varepsilon^k}{k!}, \qquad T(s) = \sum_{k=0}^{\infty} T_k(s)\,\frac{\varepsilon^k}{k!}. \tag{5.5.2a-d}$$

The initial conditions (5.4.3d,e) require that

$$\gamma_0^*(0) + \Gamma_0(0) = \gamma_t, \qquad \tau_0^*(0) + T_0(0) = \tau_t,$$

$$\gamma_k^*(0) + \Gamma_k(0) = 0, \qquad \tau_k^*(0) + T_k(0) = 0, \qquad k \geq 1. \tag{5.5.3a-d}$$

In addition, the following asymptotic conditions are imposed:

$$\Gamma_k(s) \to 0, \qquad T_k(s) \to 0;$$

$$\frac{d}{ds}\Gamma_k(s) \to 0, \qquad \frac{d}{ds}T_k(s) \to 0, \tag{5.5.3e-h}$$

when $s \to \infty$, for $k \geq 0$.

Substituting (5.5.1a,b) into (5.4.3a), letting s go to infinity, and then using (5.5.3e-h), obtain

$$\dot{\gamma}_0^*(\xi) = d(\xi), \qquad k\,\dot{\tau}_{k-1}^*(\xi) + \dot{\gamma}_k^*(\xi) = 0, \qquad \text{for } k \geq 1. \tag{5.5.4a,b}$$

Integrate (5.5.4) to obtain the required relations between γ_k^* and τ_k^*, for $k \geq 0$,

$$\gamma_0^*(\xi) = \gamma_0^*(0) + \int_0^{\xi} d(\zeta)\,d\zeta,$$

$$\gamma_k^*(\xi) = k\,\{\tau_{k-1}^*(0) - \tau_{k-1}^*(\xi)\} + \gamma_k^*(0), \qquad \text{for } k \geq 1. \tag{5.5.5a,b}$$

To evaluate $\tau_k^*(\xi)$, substitute (5.5.1) and (5.5.2) into the flow rule (5.4.3b), fix ξ, and take the limit as $\varepsilon \to 0$. Using a Taylor expansion of the function $F(\dot{\gamma}(\xi), \gamma(\xi))$ for small ε, and in view of (5.5.3e-h), arrive at

$$\tau_0^*(\xi) = F_0(\xi), \qquad \tau_1^*(\xi) = F_1(\xi)\,\dot{\gamma}_1^*(\xi) + F_2(\xi)\,\gamma_1^*(\xi),$$

$$\tau_2^*(\xi) = F_{11}(\xi)\,\{\dot{\gamma}_1^*(\xi)\}^2 + 2\,F_{12}(\xi)\,\dot{\gamma}_1^*(\xi)\,\gamma_1^*(\xi)$$

$$+ F_{22}(\xi)\,\{\gamma_1^*(\xi)\}^2 + F_1(\xi)\,\dot{\gamma}_2^*(\xi) + F_2(\xi)\,\gamma_2^*(\xi), \tag{5.5.6a-c}$$

where

$$F_0(\xi) = F(\dot{\gamma}_0^*(\xi), \gamma_0^*(\xi)), \qquad F_1(\xi) = \frac{\partial F(\dot{\gamma}_0^*(\xi), \gamma_0^*(\xi))}{\partial \dot{\gamma}},$$

$$F_2(\xi) = \frac{\partial F(\dot{\gamma}_0^*(\xi), \gamma_0^*(\xi))}{\partial \gamma}, \quad F_{11}(\xi) = \frac{\partial^2 F(\dot{\gamma}_0^*(\xi), \gamma_0^*(\xi))}{\partial \dot{\gamma}^2},$$

$$F_{12}(\xi) = \frac{\partial^2 F(\dot{\gamma}_0^*(\xi), \gamma_0^*(\xi))}{\partial \dot{\gamma}\, \partial \gamma}, \quad F_{22}(\xi) = \frac{\partial^2 F(\dot{\gamma}_0^*(\xi), \gamma_0^*(\xi))}{\partial \gamma^2}. \tag{5.5.6d-i}$$

Combining (5.5.5) with (5.5.6), it now follows that

$$\gamma_0^*(\xi) = \gamma_0^*(0) + \int_0^\xi d(\zeta) d\zeta, \quad \gamma_1^*(\xi) = F_0(0) - F_0(\xi) + \gamma_1^*(0),$$

$$\gamma_2^*(\xi) = \gamma_2^*(0) + 2\Big[F_1(\xi) F_0'(\xi)$$

$$- F_2(\xi)\{F_0(0) - F_0(\xi) + \gamma_1^*(0)\} + \tau_1^*(0)\Big], \tag{5.5.7a-c}$$

where, here and in the sequel, prime denotes differentiation with respect to the corresponding argument. This process may be continued to obtain terms up to any desired order.

Constants $\gamma_0^*(0)$, $\gamma_1^*(0)$, $\gamma^*{}_2(0)$, and $\tau_1^*(0)$ are to be determined by (5.5.3). From (5.4.3a) and (5.5.1), note that

$$\varepsilon\, T'(s) = -\Gamma'(s). \tag{5.5.8}$$

Using (5.5.2c,d), from (5.5.8) obtain

$$\Gamma_0(s) = \Gamma_0(0), \quad (k+1)\{T_k(s) - T_k(0)\} = \Gamma_{k+1}(0) - \Gamma_{k+1}(s). \tag{5.5.9a,b}$$

From (5.5.3e-h) and (5.5.9), note that

$$\Gamma_0(s) = 0, \quad \Gamma_{k+1}(0) = -(k+1)\, T_k(0),$$

$$\Gamma_{k+1}(s) = -(k+1)\, T_k(s), \quad \text{for } k \geq 0. \tag{5.5.10a-c}$$

Then, combining (5.5.10) with (5.5.3), it follows that

$$\gamma_0^*(0) = \gamma_t, \quad \gamma_{k+1}^*(0) = -(k+1)\, \tau_k^*(0), \quad \text{for } k \geq 0. \tag{5.5.11a,b}$$

In particular,

$$\gamma_1^*(0) = \tau_t - \tau_0^*(0) = \tau_t - F_0(0), \quad \gamma_2^*(0) = -2\tau_1^*(0). \tag{5.5.11c,d}$$

Therefore, the outer solution is,

$$\gamma^*(\xi) = \gamma_t + \int_0^\xi d(\zeta)\, d\zeta + \varepsilon\{\tau_t - F_0(\xi)\}$$

$$+ \varepsilon^2\{F_1(\xi) F_0'(\xi) + F_2(\xi)(F_0(\xi) - \tau_t)\} + O(\varepsilon^3),$$

$$\tau^*(\xi) = F_0(\xi) - \varepsilon\{F_1(\xi) F_0'(\xi) + F_2(\xi)(F_0(\xi) - \tau_t)\} + O(\varepsilon^2). \tag{5.5.12a,b}$$

The boundary-layer correction terms are obtained by substituting (5.5.1) and (5.5.2) into (5.4.3b). Set $\xi = \varepsilon s$, fix s, and expand the resulting equation in terms of ε. Equating coefficients of like powers of ε, obtain equations for the boundary-layer correction terms, $T_k(s)$ and $\Gamma_k(s)$, $k \geq 0$. The leading term $T_0(s)$, in the expansion of T(s), is defined by the following nonlinear equation and the initial conditions:

$$T_0(s) = F(d_0 - T_0'(s),\ \gamma_t) - F_0(0)\,, \qquad T_0(0) = \tau_t - F_0(0)\,, \tag{5.5.13a,b}$$

where $d_0 = d(0)$ and $F_0(0) = F(d_0, \gamma_t)$.

Since $\Gamma_1(s) = -T_0(s)$ by (5.5.10c), similar equations hold for $\Gamma_1(s)$,

$$\Gamma_1(s) = -F(d_0 + \Gamma_1'(s)\,,\ \gamma_t) + F(0)\,, \qquad \Gamma_1(0) = F_0(0) - \tau_t\,. \tag{5.5.14a,b}$$

For more details and results, see Nemat-Nasser and Ni (1994).

In the following two subsections, the basic elements of the constitutive algorithm, based on the singular perturbation technique, are summarized for rate-independent and rate-dependent elastoplasticity, respectively.

5.5.2. Rate-independent Elastoplasticity

For rate-independent plasticity the algorithm is explicit. The approximate solutions of the non-dimensional plastic strain $\gamma(\xi)$ and stress $\tau(\xi)$ are given by a direct evaluation of the elementary functions, as follows:

$$\gamma(\xi) = \gamma_t + \int_0^\xi d(\zeta)\,d\zeta + \varepsilon\,\{\tau_t - F_0(\xi)\} + \varepsilon^2 F_2(\xi)\,\{\tau_t - F_0(\xi)\} + O(\varepsilon^3)\,,$$

$$\tau(\xi) = F_0(\xi) - \varepsilon\,F_2(\xi)\,\{\tau_t - F_0(\xi)\} + O(\varepsilon^2)\,, \tag{5.5.15a,b}$$

where

$$F_0(\xi) = F(\gamma_0^*(\xi))\,, \qquad F_2(\xi) = \frac{\partial F}{\partial \gamma}\{\gamma_0^*(\xi)\}\,,$$

$$\gamma_0^*(\xi) = \gamma_0(\xi) + \int_0^\xi d(\zeta)\,d\zeta\,. \tag{5.5.15c-e}$$

For the rate-independent case, there are no boundary-layer contributions. The conventional perturbation technique gives the desired results. The perturbation solution (5.5.15a,b) and the solution (5.4.14a,c) which is based on the algorithm suggested by Nemat-Nasser (1991), are exactly the same up to the ε-terms for the stress, $\tau(\xi)$, and the ε^2-terms for the plastic strain, $\gamma(\xi)$.

When the function F is given by (5.4.18d) for $\gamma_0 = 10^{-2}$, $d(\xi) = 0.008$ per microsecond (or 8,000 per second), the initial values $\gamma(0) = \gamma_t = 0$, $\tau(0)/\tau_c = \tau_t/\tau_c = 1$, and the parameters $N = 0.08$, $\varepsilon = 10^{-2}$, the numerical result is shown in Figure 5.5.1. Note that, here again, both the perturbation and the plastic-predictor/elastic-corrector solutions (designated as "PP/EC Solution") are obtained by a *one-step algorithm, i.e.,* the approximate value of the physical quantity at time t is obtained using a *one-step time increment* always starting from the *initial condition,* $(\gamma(0),\ \tau(0)/\tau_c) = (0,\ 1)$ at time $t = 0$.

In Figure 5.5.1, the curve referred to as the "exact solution", on the other hand, is obtained by applying incrementally the simple Euler scheme to the following equation, obtained by combining (5.4.3a) and (5.4.18d):

Figure 5.5.1

Comparison between the one-step (from t = 0 to any t) results of the perturbation and the plastic-predictor/elastic-corrector (denoted by PP/EC Solution) solutions with the exact results, for rate-independent plasticity; N = 0.08, d = 8,000/s, $(\gamma(0), \tau(0)/\tau_c) = (0, 1)$, and $(\varepsilon, \gamma_0) = 10^{-2}$

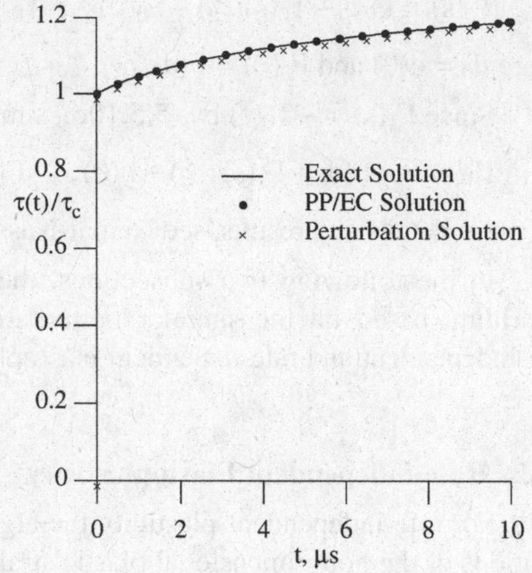

$$\dot{\tau} = N\, d \left[N\varepsilon + \gamma_0\, \tau^{(1-N)/N} \right]^{-1}, \tag{5.5.16}$$

and using very small timesteps.

5.5.3. Rate-dependent Elastoviscoplasticity

The approximate solutions of the non-dimensional plastic strain and stress are given by

$$\begin{aligned} \gamma(\xi) &= \gamma_t + \int_0^{\xi} d(\zeta)\, d\zeta + \varepsilon\, \{\tau_t - F_0(\xi) - T_0(\xi/\varepsilon)\} \\ &\quad + \varepsilon^2 \{F_1(\xi)\, F_0'(\xi) + F_2(\xi)(\tau_t - F_0(\xi)) - T_1(\xi/\varepsilon)\} + O(\varepsilon^3), \\ \tau(\xi) &= F_0(\xi) + T_0(\xi/\varepsilon) - \varepsilon\, \{F_1(\xi)\, F_0'(\xi) \\ &\quad + F_2(\xi)\,(\tau_t - F_0(\xi)) - T_1(\xi/\varepsilon)\} + O(\varepsilon^2), \end{aligned} \tag{5.5.17a,b}$$

$$\begin{aligned} F_0(\xi) &= F(\dot{\gamma}_0^*(\xi), \gamma_0^*(\xi)), \quad F_1(\xi) = \frac{\partial F(\dot{\gamma}_0^*(\xi), \gamma_0^*(\xi))}{\partial \dot{\gamma}}, \\ F_2(\xi) &= \frac{\partial F(\dot{\gamma}_0^*(\xi), \gamma_0^*(\xi))}{\partial \gamma}, \\ \gamma_0^*(\xi) &= \gamma(0) + \int_0^{\xi} d(\zeta) d\zeta, \quad \dot{\gamma}_0^*(\xi) = d(\xi). \end{aligned} \tag{5.5.17c-g}$$

In (5.5.17a,b), the boundary-layer correction terms $T_0(\xi/\varepsilon)$ and $T_1(\xi/\varepsilon)$ are given by (5.5.13) and (5.5.14), respectively; see Nemat-Nasser and Ni (1994) for details.

Comparing the plastic-predictor/elastic-corrector method of Section 5.4, with the singular perturbation method of this section, note from (5.5.5a) and (5.4.6a) that, the inelastic predictor $\gamma_A(\xi)$ coincides exactly with the leading term $\gamma_0^*(\xi)$ of the outer solution $\gamma^*(\xi)$. A similar correspondence exists for the estimates of the stress $\tau(\xi)$. Instead of finding the remaining terms of the outer solution and the boundary-layer correction, the plastic-predictor/elastic-corrector method obtains an approximation for the entire remaining part of the solution, using the linearization (5.4.8) and simplification (5.4.9b). In essence, the plastic-predictor/elastic-corrector method of Nemat-Nasser and Chung is a simplified perturbation method which exploits the physics of the elastoplasticity of metals.

The approximate solutions (5.5.17a,b) are very accurate. The error estimate of the two-scale singular perturbation method is discussed by Smith (1985). For the function F given by the power-law (5.4.4) with $(\gamma_0, \dot{\gamma}_0, \varepsilon) = 10^{-2}$, $d(\xi) = 0.008$, $N = 0.08$, $m = 20$, $\gamma_t = \gamma(0) = 0$, and $\tau_t = \tau(0) = 0$, numerical results are shown in Figure 5.5.2. Similar results are obtained for m = 200, as shown by Nemat-Nasser and Ni (1994).

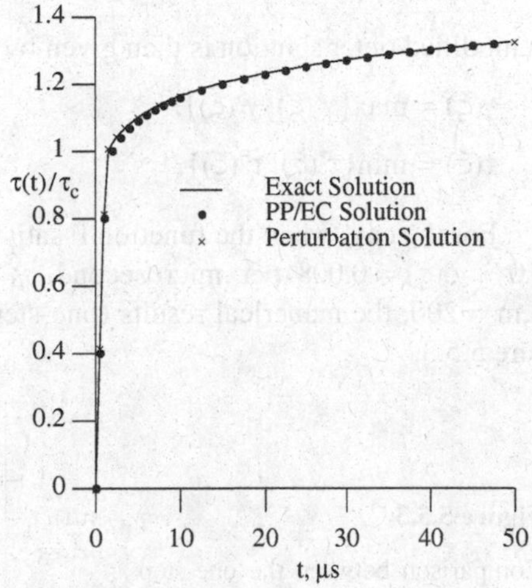

Figure 5.5.2

Comparison between the one-step (from t = 0 to any t) results of the perturbation and the plastic-predictor/elastic-corrector (denoted by PP/EC Solution) solutions with the exact results, for rate-dependent plasticity; N = 0.08, m = 20, d = 8,000/s, $(\gamma_t, \tau_t) = 0$, and $(\varepsilon, \gamma_0, \dot{\gamma}_0) = 10^{-2}$

From Figures 5.5.1 and 5.5.2, it is seen that the singular perturbation algorithm is very accurate for both the rate-independent and rate-dependent cases. For the former case, the algorithm is explicit with no boundary layer, so that the singular perturbation reduces to the conventional perturbation scheme. For the latter case, the singular perturbation requires the solution of nonlinear equations, which unfortunately is no less complicated than the original problem. In view of this, direct implementation of the singular perturbation technique into computer codes as an explicit constitutive algorithm for rate-dependent elastoviscoplasticity, seems impractical. The following modified outer solution is, therefore proposed by Nemat-Nasser and Ni (1994), which provides a very accurate, uniformly valid, one-step alternative.[16]

[16] Another alternative is given in Section 5.6; see (5.6.24) to (5.6.27).

5.5.4. Modified Outer Solution Method

Consider combining the singular perturbation method with the basic ideas underlying the plastic predictor/elastic corrector method, for rate-dependent elastoviscoplasticity. To this end, seek to obtain an explicit outer solution which is accurate outside the boundary layer. For this, construct an approximate solution for $\tau(\xi)$ by using a linear relation for the outer solution near the boundary layer. The resulting modified outer solution is then given by

$$\gamma^*(\xi) = \gamma_t + \int_0^\xi d(\zeta)\, d\zeta - \varepsilon F_0(\xi) + \varepsilon^2 \{F_1(\xi) F_0'(\xi) + F_2(\xi)(F_0(\xi) - \tau_t)\}\,,$$
$$\tau^*(\xi) = F_0(\xi) - \varepsilon \{F_1(\xi) F_0'(\xi) + F_2(\xi)(F_0(\xi) - \tau_t)\}\,, \qquad (5.5.18\text{a-b})$$

where the notation of (5.5.17c-g) is used. Within the boundary-layer region, set

$$\gamma^b(\xi) = \gamma_t + \dot{\gamma}_t \xi,$$
$$\tau^b(\xi) = \tau_t + \frac{1}{\varepsilon}\{d(0) - \dot{\gamma}_t\}\,\xi\,. \qquad (5.5.19\text{a,b})$$

The thickness, ξ^b, of the boundary layer is estimated by the following criterion:

$$\xi^b = \max\,\{\xi : \tau^b(\xi) - \tau^*(\xi) \le 0\}. \qquad (5.5.20)$$

The modified outer solution is then given by

$$\gamma(\xi) = \max\{\gamma^*(\xi),\, \gamma^b(\xi)\},$$
$$\tau(\xi) = \min\{\tau^*(\xi),\, \tau^b(\xi)\}. \qquad (5.5.21\text{a,b})$$

For the case when the function F satisfies the power law (5.5.4), $(\varepsilon, \gamma_0, \dot{\gamma}_0) = 10^{-2}$, $d(\xi) = 0.008$ per microsecond, $\gamma_t = \gamma(0) = 0$, $\tau_t = \tau(0) = 0$, $N = 0.08$, and $m = 200$, the numerical results (one-step, from t = 0 to any t) are shown in Figure 5.5.3.

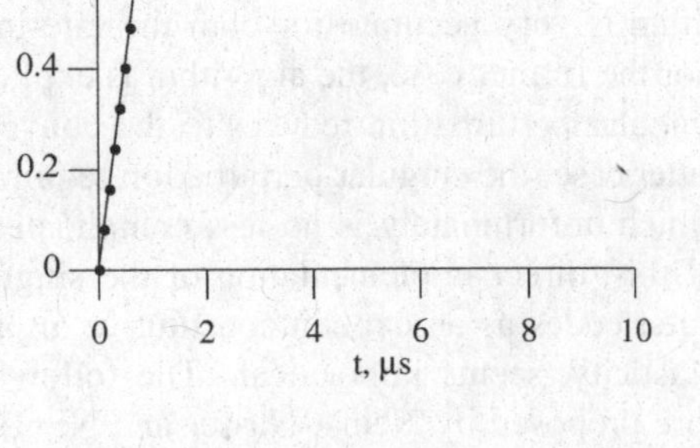

Figure 5.5.3

Comparison between the one-step (from t = 0 to any t) results of the modified outer solution algorithm and the exact results, for rate-dependent plasticity; $N = 0.08$, $m = 200$, $d = 8{,}000$/s, $(\gamma_t, \tau_t) = 0$, $(\varepsilon, \gamma_0, \dot{\gamma}_0) = 10^{-2}$

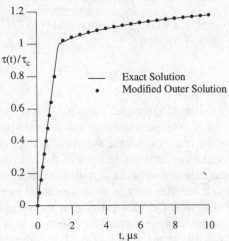

5.6. PROPORTIONAL LOADING

In Sections 5.4 and 5.5, the set of scalar constitutive equations (5.4.1a,b) is solved, assuming the functions $d(\xi)$ and $F(\dot{\gamma}, \gamma)$ are prescribed. For an arbitrarily defined $d(\xi)$, and using reasonable restrictions on the function $F(\dot{\gamma}, \gamma)$, computable error bounds are obtained which verify the effectiveness of the plastic-predictor/elastic-corrector method. In application to large-scale computations, a constant velocity gradient is generally used over each timestep, as discussed in Section 5.2. For d constant over a timestep, the solution of the basic constitutive relations is examined in the remaining part of this chapter, including both isotropic and kinematic hardening.

In this section, several numerical integration methods are detailed for a case where, in the coordinate system which rotates with the material element, the deformation rate tensor, $\hat{\mathbf{D}} = \mathbf{R}^T{}_t\mathbf{D}\mathbf{R}_t$, is coaxial with the stress-difference tensor, $\hat{\mathbf{s}}(t) = \mathbf{R}_t{}^T\mathbf{s}(t)\mathbf{R}_t$; see (5.3.8c). In Section 5.7, the results presented in this section are then generalized to cases of nonproportional loading, in which stretches remain unidirectional over a timestep. Finally, in Section 5.8, integration methods are presented for any constant velocity gradient, including the effect of large spins.

When $\hat{\mathbf{D}}$ is *proportional* to $\hat{\boldsymbol{\mu}}(t) = \hat{\mathbf{s}}(t)/(\sqrt{2}\tau_t)$ during the timestep from t to $t+\Delta t$, *i.e.*, when (see (5.3.2) and (5.3.13) for definition of terms)

$$\hat{\mathbf{D}} = d\,\frac{\hat{\boldsymbol{\mu}}(t)}{\sqrt{2}}, \quad d = (2\,\hat{\mathbf{D}} : \hat{\mathbf{D}})^{1/2}, \tag{5.6.1a,b}$$

then the loading is called proportional. During such a timestep, (5.3.14a) reduces to

$$\dot{\tau} = \mu\, d - (\mu + \Lambda)\,\dot{\gamma}, \tag{5.6.1c}$$

exactly. Since d is constant, the time integration of (5.6.1c) from t to $t+\Delta t$ results in the following *exact* relations:

$$\tau(t+\Delta t) = \tau_t + \mu\,(d\,\Delta t - \Delta\gamma) - \Delta\beta, \quad \Delta\beta = \int_{\gamma_t}^{\gamma_t+\Delta\gamma} \Lambda\, d\gamma. \tag{5.6.2a,b}$$

For the considered proportional loading, the final values of the stress difference and the backstress are then given by

$$\hat{\mathbf{s}}(t+\Delta t) = \sqrt{2}\,\tau(t+\Delta t)\,\hat{\boldsymbol{\mu}}(t), \quad \hat{\boldsymbol{\beta}}(t+\Delta t) = \hat{\boldsymbol{\beta}}(t) + \sqrt{2}\,\Delta\beta\,\hat{\boldsymbol{\mu}}(t). \tag{5.6.3a,b}$$

Therefore, in the case of proportional loading, only the scalar incremental equations (5.6.2) need to be solved in conjunction with the flow rule (5.3.10), in order to obtain the final values of the tensors **s** and $\hat{\boldsymbol{\beta}}$.

5.6.1. Rate-independent Model

First consider the solution of (5.6.2) for rate-independent plasticity with the flow rule (5.3.10a). For the final stress state to be on the yield surface, it is necessary that

$$\tau(t+\Delta t) = \tau_y(t+\Delta t) = F(\gamma_t + \Delta\gamma)\,.$$

Substitution of this into (5.6.2a) results in

$$\tau_y(t+\Delta t) = F(\gamma_t+\Delta\gamma) = \tau^{tr} - (\mu\,\Delta\gamma + \Delta\beta)\,,$$
$$\tau^{tr} = \tau(t) + \mu\, d\,\Delta t\,. \qquad (5.6.2c,d)$$

Several methods for solving the nonlinear equation (5.6.2c) are now discussed.

Forward-gradient Method: The simplest and first-order accurate approach is the forward-gradient method. For this, the yield stress and the increment in the backstress at the end of the timestep are approximated by

$$\tau_y(t+\Delta t) = \tau_y(t) + H(\gamma_t)\,\Delta\gamma\,, \quad \Delta\beta = \Lambda(\gamma_t)\,\Delta\gamma\,,$$
$$H(\gamma_t) \equiv \frac{dF(\gamma_t)}{d\gamma}\,. \qquad (5.6.4a\text{-}c)$$

Substitution of this into (5.6.2c) results in

$$\Delta\gamma = \max\left\{0\,, \ \frac{\tau^{tr} - \tau_y(t)}{\mu + H + \Lambda}\right\}. \qquad (5.6.5)$$

Hence, $\tau(t+\Delta t)$ and $\Delta\beta$ are given by

$$\tau(t+\Delta t) = \tau^{tr} - \mu\,\Delta\gamma - \Delta\beta\,, \quad \Delta\beta = \Lambda(\gamma_t)\,\Delta\gamma\,. \qquad (5.6.6a,b)$$

For nonlinear hardening, the final stress state obtained by this method, will not, in general, be on the yield surface, *i.e.*,

$$\tau(t+\Delta t) \neq F(\gamma_t + \Delta\gamma)\,. \qquad (5.6.6c)$$

To illustrate this, consider the following example. For simplicity, assume isotropic hardening, *i.e.*, $\Lambda = 0$, with yield function

$$F(\gamma) = \tau_c\,(1 + \gamma / \gamma_0)^N\,, \qquad (5.6.7)$$

and set $N = 0.1$, $\gamma_0 = 0.1$, and $\tau_c = \mu / 100$. The *forward-gradient* method involves evaluation of τ^{tr}, $\Delta\gamma$, and $\tau(t+\Delta t)$, using (5.6.2c,d), (5.6.5), and (5.6.6), respectively. These are given in Table 5.6.1 for four different values of $d\,\Delta t$, *each starting with initial conditions* $\tau_t = 0$ *and* $\gamma_t = 0$. In order to check whether the stress state is on the yield surface[17], values of $F(\gamma_t + \Delta\gamma)$ are also given in

[17] In Tables 5.6.1 to 5.6.9, *for each indicated strain increment* $d\,\Delta t$, *a one-step solution, starting from the zero initial conditions, is used.* As is well known, the forward-gradient method should be used with very small strain increments. The results of large strain increments in Table 5.6.1 illustrate this.

Table 5.6.1. The solution becomes progressively more inaccurate as the magnitude of the timestep increases.

Iterative Method: An iterative method may be used to improve the results of the forward-gradient method; see Dodds (1986). After the ith iteration, the plastic strain increment and the effective stress are obtained from (5.6.6a), as follows:

$$\Delta\gamma^{i+1} = \frac{1}{\mu}\{\tau^{tr} - \tau_y^i(t+\Delta t) - \Delta\beta^i\},$$

$$\tau^{i+1}(t+\Delta t) = \tau_y^i(t+\Delta t) = F(\gamma_t + \Delta\gamma^i), \quad \Delta\beta^i = \int_{\gamma_t}^{\gamma_t+\Delta\gamma^i} \Lambda\, d\gamma. \tag{5.6.8a-d}$$

Table 5.6.1

Computational steps for a single time increment: Forward-gradient Method

$d\,\Delta t$	0.02	0.2	0.4	1.0
τ^{tr}/τ_c	2.0	20.0	40.0	100.0
$\Delta\gamma$	0.009901	0.188119	0.386139	0.980198
$\tau(t+\Delta t)/\tau_c$	1.0099	1.1881	1.3861	1.9802
$F(\gamma_t+\Delta\gamma)/\tau_c$	1.0095	1.1116	1.1713	1.2687

The $\Delta\gamma$ obtained by the *forward-gradient* method is used as the first trial in (5.6.8c,d) to obtain $\tau_y^1(t+\Delta t)$ and $\Delta\beta^1$. These are then substituted into (5.6.8a,b), to obtain $\Delta\gamma$ and $\tau^2(t+\Delta t)$. The integration is repeated until $\Delta\gamma$ converges to a final value to within a desired accuracy. The final values of the effective and yield stresses of the previous example are given in Table 5.6.2 for three iterations. This table shows that in the second iteration the stresses are almost exact, *i.e.*, the error is less than 0.1% even for a 50% strain increment.

Plastic-predictor/Elastic-corrector Method: This technique, proposed by Nemat-Nasser (1991), exploits the fact that is detailed in Section 5.4, for a general function $d(\xi)$, using isotropic hardening. Here, a constant function $d(\xi) = d$ = constant is considered, and kinematic hardening is included. As pointed out before, the method exploits the fact that, in finite elastoplastic deformations, elastic strains are negligibly small in comparison to the total strains. Thus, the nonlinear hardening or softening equations are integrated explicitly, first assuming that the plastic strain increment is equal to the total strain increment, and then correcting for the error which is introduced by this assumption. For most materials with moderate hardening (or softening), this yields an almost exact result in one step.

Denote by $\Delta\gamma_A$ and $\Delta\gamma_{er}$, respectively, the first estimate of the plastic strain increment and the error which is introduced by neglecting the elastic strain,

$$\Delta\gamma = \Delta\gamma_A - \Delta\gamma_{er}. \tag{5.6.9a}$$

Table 5.6.2

Computational steps for a single time increment: Iterative Method

$d\,\Delta t$	0.02	0.2	0.4	1.0
τ^{tr}/τ_c	2.0	20.0	40.0	100.0
$\Delta\gamma^1$	0.009901	0.188119	0.386139	0.980198
$\tau^1(t+\Delta t)/\tau_c$	1.0099	1.1881	1.3861	1.9802
$\tau_y^1(t+\Delta t)/\tau_c$	1.0095	1.1116	1.1713	1.2687
$\Delta\gamma^2$	0.009905	0.188884	0.388287	0.987313
$\tau^2(t+\Delta t)/\tau_c$	1.0095	1.1116	1.1713	1.2686
$\tau_y^2(t+\Delta t)/\tau_c$	1.0095	1.1119	1.1718	1.2695
$\Delta\gamma^3$	0.009905	0.188881	0.388282	0.987305
$\tau^3(t+\Delta t)/\tau_c$	1.0095	1.1119	1.1718	1.2695
$\tau_y^3(t+\Delta t)/\tau_c$	1.0095	1.1119	1.1718	1.2695

For the rate-independent plastic deformation, the first step is to estimate the plastic strain increment, using (5.6.2d) for τ^{tr},

$$\Delta\gamma_A = \max\left\{0,\quad \frac{\tau^{tr} - \tau_y(t)}{\mu} \le d\,\Delta t\right\}. \tag{5.6.9b}$$

Then the yield stress and the increment in the backstress are estimated as follows:

$$\gamma_A \equiv \gamma_t + \Delta\gamma_A\,, \qquad \tau_A = F(\gamma_A)\,, \qquad \Delta\beta_A = \int_{\gamma_t}^{\gamma_A} \Lambda\, d\gamma\,. \tag{5.6.10a-c}$$

With these tentative values, the final yield stress and the increment in the backstress are expressed as

$$\tau_y(t+\Delta t) = \tau_A - H_A\,\Delta\gamma_{er}\,, \qquad \Delta\beta = \Delta\beta_A - \Lambda_A\,\Delta\gamma_{er}\,, \tag{5.6.11a,b}$$

where, from (5.6.4),

$$H_A \equiv H(\gamma_A)\,, \qquad \Lambda_A \equiv \Lambda(\gamma_A)\,. \tag{5.6.11c,d}$$

For the stress state to be on the yield surface, set $\tau(t+\Delta t) = \tau_y(t+\Delta t)$ in (5.6.2a) and substitute in the result from (5.6.9) and (5.6.11), to obtain

$$\Delta\gamma_{er} = \frac{\tau_A - \tau_y(t) + \Delta\beta_A}{\mu + H_A + \Lambda_A}\,. \tag{5.6.12a}$$

This "error" expression is the same as $\gamma_A(\Delta t) - \gamma_f(\Delta t)$, given by (5.4.14a) at $\xi = \Delta t$, except that the kinematic hardening is now included. Now, the final value of the effective stress becomes

$$\tau(t+\Delta t) = \tau^{tr} - \mu\,\Delta\gamma - \Delta\beta\,, \tag{5.6.12b}$$

where τ^{tr}, $\Delta\gamma$, and $\Delta\beta$ are defined in (5.6.2d), (5.6.9a), and (5.6.11b), respectively. When $\Delta\gamma_A$ given by (5.6.9b) is zero, from (5.6.10), $\tau_A = \tau_y(t)$, $\Delta\beta_A = 0$, and, hence, $\Delta\gamma_{er} = 0$. In that case, $\tau(t+\Delta t) = \tau^{tr}$. Table 5.6.3 shows the results obtained by this method for the same example as in Tables 5.6.1 and 5.6.2.

Table 5.6.3

Computational steps for a single time increment: Plastic-predictor/Elastic-corrector Method

$d\,\Delta t$	0.02	0.2	0.4	1.0
τ^{tr}/τ_c	2.0	20.0	40.0	100.0
$\Delta\gamma_A$	0.01	0.19	0.39	0.99
τ_A	1.0096	1.1123	1.1722	1.2698
H_A	0.9178	0.3836	0.2392	0.1165
$\Delta\gamma_{er}$	0.000095	0.001119	0.001718	0.002695
$\Delta\gamma$	0.009905	0.188881	0.388282	0.987305
$\tau(t+\Delta t)/\tau_c$	1.0095	1.1119	1.1718	1.2695
$F(\gamma_t+\Delta\gamma)/\tau_c$	1.0095	1.1119	1.1718	1.2695

5.6.2. Rate-dependent Model

Now, consider the solution of (5.6.2a,b) for the rate-dependent case with the flow rule (5.3.10b).

Tangent-modulus Method: A forward-gradient method known as the *tangent-modulus method* had been proposed by Peirce *et al.* (1984) for rate-dependent plastic deformation. For the case of proportional loading, this is considered here. In this method, a linear interpolation is employed to relate the plastic strain increment to the plastic strain rates at the beginning and at the end of the timestep, as follows:

$$\Delta\gamma = \Delta t\,\{(1-\theta)\,\dot{\gamma}(t) + \theta\,\dot{\gamma}(t+\Delta t)\}\,, \tag{5.6.13a}$$

where the interpolation parameter θ may vary from 0 to 1. The flow rule (5.3.10c) is approximated by

$$\dot{\gamma}(t+\Delta t) \approx \dot{\gamma}_t + \frac{\partial g(\tau_t, \gamma_t)}{\partial \tau}\,\Delta\tau + \frac{\partial g(\tau_t, \gamma_t)}{\partial \gamma}\,\Delta\gamma\,,$$

$$\Delta\tau = \tau(t+\Delta t) - \tau_t\,. \tag{5.6.13b,c}$$

Substitute (5.6.13b) into (5.6.13a) to arrive at

$$\Delta\gamma = \Delta t\,\{\dot{\gamma}(t) + \theta\,\frac{\partial g(\tau_t, \gamma_t)}{\partial \tau}\,\Delta\tau + \theta\,\frac{\partial g(\tau_t, \gamma_t)}{\partial \gamma}\,\Delta\gamma\}\,. \tag{5.6.13d}$$

From (5.6.13c,d) and (5.6.2a), now obtain

$$\Delta\gamma = \frac{\Delta t}{1+\zeta}\{\dot{\gamma}_t + \mu\, d\,\zeta / h\}\,, \tag{5.6.14a}$$

where

$$h = \mu + \Lambda(\gamma_t) - \frac{\partial g(\tau_t, \gamma_t)}{\partial \gamma}\left[\frac{\partial g(\tau_t, \gamma_t)}{\partial \tau}\right]^{-1},$$

$$\zeta = \theta\, h \frac{\partial g(\tau_t, \gamma_t)}{\partial \tau} \Delta t. \qquad (5.6.14b,c)$$

The final values of the effective stress difference, plastic strain rate, and the increment in the backstress are then calculated from

$$\tau(t+\Delta t) = \tau^{tr} - \mu\,\Delta\gamma - \Delta\beta, \quad \dot{\gamma}(t+\Delta t) = \frac{1}{\theta}\{\frac{\Delta\gamma}{\Delta t} - (1-\theta)\,\dot{\gamma}(t)\},$$

$$\Delta\beta = \Lambda(\gamma_t)\,\Delta\gamma. \qquad (5.6.15a\text{-}c)$$

In general, these final values will not satisfy the flow rule (5.3.10b). To illustrate this, consider a rate-dependent power-law model

$$\tau = F(\dot{\gamma}, \gamma) = \tau_c\,(1+\gamma/\gamma_0)^N\,(\dot{\gamma}/\dot{\gamma}_0)^m,$$

or, equivalently, in the form

$$\dot{\gamma} = g(\tau, \gamma) = \dot{\gamma}_0\,(\tau/\tau_c)^{1/m}\,(1+\gamma/\gamma_0)^{-N/m},$$

and use the following parameters: $\tau_c = \mu/100$, $N = 0.1$, $m = 0.01$, and $\gamma_0 = 0.1$. Consider an initial-value problem[18] with $\tau_t/\tau_c = 0.1$, $\gamma_t = 0$, and $d = 200\,\dot{\gamma}_0$. The forward-gradient method involves evaluations of $\dot{\gamma}(t)$, τ^{tr}, $\partial g/\partial\gamma$, $\partial g/\partial\tau$, h, ζ, $\Delta\gamma$, and $\tau(t+\Delta t)$. These values are given in Table 5.6.4 for four different strain increments, $d\,\Delta t$, *each starting from the initial conditions* $\tau_t = 0.1$, $\gamma_t = 0$. In order to check whether the final values of τ, γ and $\dot{\gamma}$ satisfy the flow rule, the quantities $\dot{\gamma}(t+\Delta t)$ and $F(\dot{\gamma}(t+\Delta t), \Delta\gamma)$ are also given in Table 5.6.4[19].

Iterative Method: The iterative method introduced for the rate-independent case can be used to satisfy the flow rule. After the ith iteration, the plastic strain increment is as follows:

$$\Delta\gamma^{i+1} = \frac{1}{\mu}\max\{0,\ (\tau^{tr} - \tau_y^i - \Delta\beta^i)\}, \qquad (5.6.16a)$$

where

$$\tau_y^i = F(\dot{\gamma}^i(t+\Delta t), \gamma_t + \Delta\gamma^i), \quad \Delta\beta^i = \int_{\gamma_t}^{\gamma_t+\Delta\gamma^i} \Lambda\, d\gamma,$$

$$\dot{\gamma}^i(t+\Delta t) = \frac{1}{\theta}\left\{\frac{\Delta\gamma^i}{\Delta t} - (1-\theta)\,\dot{\gamma}_t\right\}. \qquad (5.6.16b\text{-}d)$$

---

[18] If the initial value of the effective shear stress is zero, then $\dot{\gamma}_t = \partial\dot{\gamma}/\partial\gamma = \partial\dot{\gamma}/\partial\tau = 0$. Therefore, for $\tau_t = 0$, this method gives $\tau(t+\Delta t) = \tau^{tr} = \mu\, d\,\Delta t$, regardless of the value of $d\,\Delta t$. Hence, this method is not applicable when large strain increments are applied at $\tau_t = 0$.

[19] Here, again, such large strain increments are used to illustrate the procedure, since only very small strain increments must be used with this method.

Table 5.6.4

Computational steps for a time increment: Tangent-modulus Method, $\theta = 0.5$

$d\,\Delta t$	0.01	0.02	0.2	0.4	1.0
$\dot{\gamma}(t) / \dot{\gamma}_0$	0.0000	0.0000	0.0000	0.0000	0.0000
$\frac{\tau_c}{\dot{\gamma}_0}\frac{\partial g}{\partial \tau}$	977.2372	977.2372	977.2372	977.2372	977.2372
$\frac{1}{\dot{\gamma}_0}\frac{\partial g}{\partial \gamma}$	- 9.7724	- 9.7724	- 9.7724	- 9.7724	- 9.7724
τ^{tr} / τ_c	1.1000	2.1000	20.1000	40.1000	100.1000
h / τ_c	100.01	100.01	100.01	100.01	100.01
ζ	0.002500	0.004999	0.049995	0.099990	0.249975
$\Delta\gamma$	0.000025	0.000100	0.010000	0.040000	0.250000
$\tau(t+\Delta t) / \tau_c$	1.0975	2.0900	19.1000	36.1000	75.1000
$\dot{\gamma}(t+\Delta t) / \dot{\gamma}_0$	0.0000	2.0000	20.0000	40.0000	100.0000
$F(\dot{\gamma}(t+\Delta t), \Delta\gamma) / \tau_c$	1.0000	1.0071	1.0403	1.0731	1.1869

The $\Delta\gamma$ obtained from the *tangent-modulus* method is used as the first trial, $\Delta\gamma^1$, and the iteration is repeated until $\Delta\gamma$ converges to a final value to within a desired accuracy. Though the iterative method converges to a final solution in a couple of iterations, the accuracy of the solution is limited by the validity of the approximation (5.6.13a). To illustrate this, the results from the iterative method using $\theta = 0.5$ and 1 are given in Tables 5.6.5 and 5.6.6, respectively.

As seen from Tables 5.6.5 and 5.6.6, for large strain increments and for both values of θ, the final values of $\Delta\gamma$ converge to the same values. However, because of the approximation (5.6.13), the converged values of $\dot{\gamma}(t+dt)$ for different values of θ, differ considerably. This results in different values for $\tau(t+\Delta t)$. It should be noted that, in this example, the strain-rate sensitivity parameter m is set to 0.01 which is very small. For highly rate-sensitive materials, the variation in the value of the final strain rate would give significant variation in the final values of $\tau(t+\Delta t)$. The correct value of the $\tau(t+\Delta t)$ can be obtained by subdividing the time increment Δt into very small time increments. For very small strain increments per subdivision, the *forward-gradient method* gives the same results for any value of θ. The values of τ, γ, and $\dot{\gamma}$ obtained using a strain increment of 10^{-5} per subdivision are given in Table 5.6.7. It is seen from Tables 5.6.5, 5.6.6, and 5.6.7, that when the iterative method is used, the best value for θ is one.

Plastic-predictor/Elastic-corrector Method: For large strain increments, the *tangent-modulus* method is inaccurate while the iterative method is inefficient. The *plastic-predictor/elastic-corrector* method introduced by Nemat-Nasser and Chung (1992) is an accurate, one-step method. In the original work, and for constant $d(\xi)$, it is proposed that the first estimate of the plastic strain rate be a constant equal to d, as discussed in Section 5.4. This generally works well, but can be improved with little additional computational effort, as proposed by Fotiu

Table 5.6.5

Computational steps for a single time increment: Iterative Method, $\theta = 0.5$

$d\,\Delta t$	0.01	0.02	0.2	0.4	1.0
$\dot{\gamma}(t)/\dot{\gamma}_0$	0.0000	0.0000	0.0000	0.0000	0.0000
$\frac{\tau_c}{\dot{\gamma}_0}\frac{\partial\dot{\gamma}}{\partial\tau}$	977.2372	977.2372	977.2372	977.2372	977.2372
$\frac{1}{\dot{\gamma}_0}\frac{\partial\dot{\gamma}}{\partial\gamma}$	- 9.7724	- 9.7724	- 9.7724	- 9.7724	- 9.7724
τ^{tr}/τ_c	1.1000	2.1000	20.1000	40.1000	100.1000
h/τ_c	100.01	100.01	100.01	100.01	100.01
ζ	0.002500	0.004999	0.049995	0.099990	0.249975
$\Delta\gamma^1$	0.000025	0.000100	0.010000	0.040000	0.250000
$\tau^1(t+\Delta t)/\tau_c$	1.0975	2.0900	19.1000	36.1000	75.1000
$\dot{\gamma}^1(t+\Delta t)/\dot{\gamma}_0$	0.0000	2.0000	20.0000	40.0000	100.0000
$\tau_y^1(t+\Delta t)/\tau_c$	1.0000	1.0071	1.0403	1.0731	1.1869
$\Delta\gamma^2$	0.001000	0.010929	0.190597	0.390269	0.989131
$\tau^2(t+\Delta t)/\tau_c$	1.0000	1.0071	1.0403	1.0731	1.1869
$\dot{\gamma}^2(t+\Delta t)/\dot{\gamma}_0$	39.9900	218.5887	381.1945	390.2692	355.6625
$\tau_y^2(t+\Delta t)/\tau_c$	1.0386	1.0664	1.1807	1.2444	1.3465
$\Delta\gamma^3$	0.000614	0.010336	0.189193	0.388556	0.987535
$\tau^3(t+\Delta t)/\tau_c$	1.0386	1.0664	1.1807	1.2444	1.3465
$\dot{\gamma}^3(t+\Delta t)/\dot{\gamma}_0$	24.5569	206.7294	378.3860	388.5561	370.4569
$\tau_y^3(t+\Delta t)/\tau_c$	1.0332	1.0652	1.1800	1.2439	1.3469
$\Delta\gamma^4$	0.000668	0.010348	0.189200	0.388561	0.987531
$\tau^4(t+\Delta t)/\tau_c$	1.0332	1.0652	1.1800	1.2439	1.3469
$\dot{\gamma}^4(t+\Delta t)/\dot{\gamma}_0$	26.7360	206.9625	378.3991	388.5610	368.2764
$\tau_y^4(t+\Delta t)/\tau_c$	1.0341	1.0652	1.1800	1.2439	1.3468

and Nemat-Nasser (1995); see also Balendran and Nemat-Nasser (1995). What follows, therefore, is this improved version of the original method, including the kinematic hardening effect.

Denote the first estimate (yet to be calculated)[20] of the (variable) plastic strain rate by $\dot{\gamma}_A$ and its deviation from the exact value, *i.e.*, the error, by $\dot{\gamma}_{er}$. Hence set,

$$\dot{\gamma}(t+\xi) = \dot{\gamma}_A(\xi) - \dot{\gamma}_{er}, \quad \gamma(t+\xi) - \gamma_t \equiv \Delta\gamma(\xi) = \Delta\gamma_A(\xi) - \Delta\gamma_{er},$$

$$\Delta\gamma_A(\xi) = \int_0^{\xi} \dot{\gamma}_A(\zeta)\, d\zeta, \quad 0 \le \xi \le \Delta t. \tag{5.6.17a-c}$$

Unlike the similar relations in Section 5.4, none of these functions are yet known. They are all calculated in what follows; see (5.6.21) to (5.6.23). Since,

[20] Note that, in Section 5.4, $\dot{\gamma}_A(\xi) = d(\xi)$ is assumed; see (5.4.6b). Here, $\dot{\gamma}_A(\xi)$ is calculated; see (5.6.22b).

Table 5.6.6

Computational steps for a time increment: Iterative Method, $\theta = 1$

$d\,\Delta t$	0.01	0.02	0.2	0.4	1.0
$\dot{\gamma}(t)/\dot{\gamma}_0$	0.0000	0.0000	0.0000	0.0000	0.0000
$\frac{\tau_c}{\dot{\gamma}_0}\frac{\partial\dot{\gamma}}{\partial\tau}$	977.2372	977.2372	977.2372	977.2372	977.2372
$\frac{1}{\dot{\gamma}_0}\frac{\partial\dot{\gamma}}{\partial\gamma}$	- 9.7724	- 9.7724	- 9.7724	- 9.7724	- 9.7724
τ^{tr}/τ_c	1.1000	2.1000	20.1000	40.1000	100.1000
h/τ_c	100.01	100.01	100.01	100.01	100.01
ζ	0.004999	0.009999	0.099990	0.199980	0.499950
$\Delta\gamma^1$	0.000050	0.000200	0.020000	0.080000	0.500000
$\tau^1(t+\Delta t)/\tau_c$	1.0950	2.0800	18.1000	32.1000	50.1000
$\dot{\gamma}^1(t+\Delta t)/\dot{\gamma}_0$	0.0000	2.0000	20.0000	40.0000	100.0000
$\tau_y^1(t+\Delta t)/\tau_c$	1.0000	1.0072	1.0494	1.1004	1.2526
$\Delta\gamma^2$	0.001000	0.010928	0.190506	0.389996	0.988474
$\tau^2(t+\Delta t)/\tau_c$	1.0000	1.0072	1.0494	1.1004	1.2526
$\dot{\gamma}^2(t+\Delta t)/\dot{\gamma}_0$	19.9900	109.2843	190.5063	194.9980	197.6948
$\tau_y^2(t+\Delta t)/\tau_c$	1.0314	1.0590	1.1725	1.2357	1.3386
$\Delta\gamma^3$	0.000686	0.010410	0.189275	0.388643	0.987614
$\tau^3(t+\Delta t)/\tau_c$	1.0314	1.0590	1.1725	1.2357	1.3386
$\dot{\gamma}^3(t+\Delta t)/\dot{\gamma}_0$	13.7139	104.1015	189.2750	194.3214	197.5229
$\tau_y^3(t+\Delta t)/\tau_c$	1.0272	1.0580	1.1719	1.2353	1.3385
$\Delta\gamma^4$	0.000728	0.010420	0.189281	0.388647	0.987615
$\tau^4(t+\Delta t)/\tau_c$	1.0272	1.0580	1.1719	1.2353	1.3385
$\dot{\gamma}^4(t+\Delta t)/\dot{\gamma}_0$	14.5537	104.2025	189.2807	194.3233	197.5231
$\tau_y^4(t+\Delta t)/\tau_c$	1.0279	1.0580	1.1719	1.2353	1.3385

Table 5.6.7

Exact values of τ, γ and $\dot{\gamma}$, obtained by subdividing the time increments and using the Tangent-modulus Method

$d\,\Delta t$	0.01	0.02	0.2	0.4	1.0
τ	1.0547	1.0647	1.1725	1.2357	1.3386
γ	0.000453	0.010352	0.189272	0.388609	0.987577
$\dot{\gamma}$	195.96	198.09	199.19	199.49	199.75

in general, $\dot{\gamma}_{er}$ is orders of magnitude smaller than $\dot{\gamma}_A$, it is assumed to be a constant. Denote by $\tau^e(\xi)$ the estimate of $\tau(\xi)$ corresponding to the approximation $\dot{\gamma}(\xi) = \dot{\gamma}_A(\xi)$. Thus, from (5.6.1c), obtain

$$\dot{\tau}^e(\xi) = \dot{\tau}^e = \mu\,(d - \dot{\gamma}_A(\xi)), \quad 0 \le \xi \le \Delta t. \tag{5.6.18}$$

Then, the final value of τ is given by

$$\tau(t+\Delta t) = \tau^e(\Delta t) - \Delta\beta_A + (\mu + \Lambda_A)\,\Delta\gamma_{er}(\Delta t)\,, \quad \Delta\beta_A \equiv \int_{\gamma_t}^{\gamma_A(t+\Delta t)} \Lambda(\gamma)\,d\gamma\,,$$

$$\gamma_A(t+\Delta t) \equiv \gamma_t + \Delta\gamma_A(\Delta t)\,, \quad \Lambda_A \equiv \Lambda(\gamma_A(t+\Delta t))\,. \tag{5.6.19a-d}$$

From the flow rule, on the other hand, the final value of τ is approximated by[21]

$$\tau(t+\Delta t) = F(\dot\gamma(t+\Delta t),\ \gamma_t + \Delta\gamma(\Delta t)) = \tau_A - \eta_A\,\Delta\gamma_{er}(\Delta t)\,,$$

$$\eta_A \equiv \frac{1}{\Delta t}\,\frac{\partial F(\dot\gamma_A(\Delta t),\ \gamma_A(t+\Delta t))}{\partial\dot\gamma} + \frac{\partial F(\dot\gamma_A(\Delta t),\ \gamma_A(t+\Delta t))}{\partial\gamma}\,,$$

$$\tau_A \equiv F(\dot\gamma_A(\Delta t),\ \gamma_A(t+\Delta t))\,. \tag{5.6.20a-c}$$

Substitute (5.6.20a) into (5.6.19a), to obtain

$$\Delta\gamma_{er} = \frac{\tau_A - \tau^e(\Delta t) + \Delta\beta_A}{\mu + \Lambda_A + \eta_A}\,. \tag{5.6.20d}$$

To complete the solution, it is now necessary to solve (5.6.18), to obtain $\dot\gamma_A(\xi)$, $\Delta\gamma_A(\xi)$, and $\tau^e(\xi)$, as functions of ξ for $0 \le \xi \le \Delta t$.

To solve (5.6.18), assume that $\dot\gamma_A$ varies linearly with τ^e from $\dot\gamma(t)$ to d, as τ^e varies from[22] $\tau(t)$ to τ_d,

$$\tau^e(\xi) = \tau_d + \rho\,(\dot\gamma_A(\xi) - d)\,,$$

$$\rho = (\tau_d - \tau_t)/(d - \dot\gamma_t)\,,$$

$$\tau_d \equiv F(d,\ \gamma_t)\,. \tag{5.6.21a-c}$$

Substitute (5.6.21a) into (5.6.18) and integrate, to obtain

$$\tau^e(\xi) = \tau(t)\,e^{-k\xi} + \tau_d\,(1 - e^{-k\xi})$$

$$= \tau_t + \mu\,\frac{d - \dot\gamma_t}{k}\,(1 - e^{-k\xi})\,, \tag{5.6.22a}$$

where $k = \mu/\rho$. Substitute (5.6.22a) into (5.6.21a) to arrive at

$$\dot\gamma_A(\xi) = d - (d - \dot\gamma_t)\,e^{-k\xi}\,,$$

$$\Delta\gamma_A(\xi) = \xi d - \frac{d - \dot\gamma_t}{k}\,(1 - e^{-k\xi})\,. \tag{5.6.22b,c}$$

Note that, as ξ goes to zero, $\dot\gamma_A(\xi)$ and $\Delta\gamma_A(\xi)$ go to $\dot\gamma_t = \dot\gamma_t$ and $\dot\gamma_t\,\xi$, respectively. However, in general, $\Delta\gamma_{er}$ does not go to zero as Δt goes to zero. Therefore, in general, the plastic strain increment resulting from (5.6.17b) may not be

---

[21] In the expression for η_A, the differentiation with respect to $\dot\gamma$ and γ is first performed and then the results are evaluated at the indicated values of these arguments. Similar notation is used throughout this chapter.

[22] As pointed out before, the subscript A is used to denote the first (or starting) estimate (or approximation) of the associated variable (or quantity).

positive for very small time increments. However, in these cases the plastic strain increment is negligible. To account for such cases, (5.6.20d) is modified to read

$$\Delta\gamma_{er} = \max\left\{\Delta\gamma_A,\ \frac{\tau_A - \tau^e(\Delta t) + \Delta\beta_A}{\mu + \Lambda_A + \eta_A}\right\}. \tag{5.6.22d}$$

If ρ in (5.6.21a) is set to zero, then (5.6.22a-c) reduces to

$$\tau^e(\xi) = \tau_t, \quad \dot{\gamma}_A(\xi) = d, \quad \Delta\gamma_A(\xi) = \xi d. \tag{5.6.23a-c}$$

These also result if the estimate $\dot{\gamma}_A$ is assumed to be a constant over the time-step; see Nemat-Nasser and Chung (1992) and Section 5.4.

In Table 5.6.8 the results obtained by this method for the same problem considered before, are summarized. In this table, $\Delta\gamma_A(\Delta t)$, $\dot{\gamma}_A(\Delta t)$, $\tau^e(\Delta t)$, are obtained from (5.6.22), while τ_A, η_A, $\Delta\gamma_{er}$ are from (5.6.20). The final values of $\dot{\gamma}$ and $\Delta\gamma$ are then obtained from (5.6.17). The final value of τ is obtained from (5.6.19a). In order to check whether the final values of τ, $\dot{\gamma}$, and γ satisfy the flow rule (5.3.10b), the values of $F(\dot{\gamma}(t+\Delta t), \Delta\gamma)$ (last row) are also given in Table 5.6.8. As is seen, for very small strains there is a small error, whereas essentially exact results are obtained at large strain increments.

Because of the remarkable accuracy of the above-detailed plastic-predictor/elastic-corrector technique, the necessary steps are summarized below for the convenience of the reader.

Table 5.6.8

Computational steps for a single time increment: Plastic-predictor/Elastic-corrector Method

$d\Delta t$	0.01	0.02	0.2	0.4	1.0
$\dot{\gamma}(t)/\dot{\gamma}_0$	0.0000	0.0000	0.0000	0.0000	0.0000
τ_d/τ_c	1.0471	1.0471	1.0471	1.0471	1.0471
$k/\dot{\gamma}_0$	21116.46	21116.46	21116.46	21116.46	21116.46
$\Delta\gamma_A$	0.003824	0.011675	0.190529	0.390529	0.990529
$\dot{\gamma}_A/\dot{\gamma}_0$	130.4188	175.7923	200.0000	200.0000	200.0000
τ^e/τ_c	0.7176	0.9325	1.0471	1.0471	1.0471
τ_A/τ_c	1.053860	1.064745	1.173084	1.236166	1.338980
η_A/τ_c	2.631164	1.559114	0.462430	0.282911	0.136172
$\Delta\gamma_{er}$	0.003276	0.001302	0.001254	0.001885	0.002915
$\Delta\gamma$	0.000548	0.010373	0.189275	0.388644	0.987614
$\dot{\gamma}(t+\Delta t)/\dot{\gamma}_0$	64.8941	162.7698	198.7462	199.0575	199.4171
$\tau(t+\Delta t)/\tau_c$	1.0452	1.0627	1.1725	1.2356	1.3386
$F(\dot{\gamma}(t+\Delta t),\Delta\gamma)/\tau_c$	1.0432	1.0627	1.1725	1.2356	1.3386

Computational Steps in Plastic-predictor/Elastic-corrector Method:

Step 1

From the initial values $\tau_t \equiv \tau_t$, γ_t, and the flow rule (5.3.10b), calculate

$$\tau_d = F(d,\ \gamma_t), \quad \dot{\gamma}_t \equiv \dot{\gamma}_t = g(\tau_t,\ \gamma_t),$$

$$k = \mu \frac{(d - \dot{\gamma}_t)}{(\tau_d - \tau_t)}. \tag{5.6.24a-c}$$

Step 2

Estimate the plastic strain, plastic strain rate, and the effective stress at time $t + \Delta t$ from

$$\dot{\gamma}_A = d - (d - \dot{\gamma}_t)\, e^{-k\Delta t},$$

$$\gamma_A = \gamma_t + d\,\Delta t - \frac{d - \dot{\gamma}_t}{k}(1 - e^{-k\Delta t}).$$

$$\tau^e = \tau_t + \mu \frac{d - \dot{\gamma}_t}{k}(1 - e^{-k\Delta t}). \tag{5.6.25a-c}$$

Step 3

Estimate the flow stress, τ_A, the tangent modulus, η_A, and the kinematic-hardening parameters (when relevant), using the above approximate values of the plastic strain and its rate,

$$\tau_A = F(\dot{\gamma}_A,\ \gamma_A),$$

$$\eta_A = \frac{1}{\Delta t}\frac{\partial F(\dot{\gamma}_A,\ \gamma_A)}{\partial \dot{\gamma}} + \frac{\partial F(\dot{\gamma}_A,\ \gamma_A)}{\partial \gamma},$$

$$\Delta\beta_A = \int_{\gamma_t}^{\gamma_A} \Lambda(\gamma)\, d\gamma, \quad \Lambda_A = \Lambda(\gamma_A). \tag{5.6.26a-d}$$

Step 4

Calculate the error in the estimated plastic strain, using the above approximation,

$$\Delta\gamma_{er} = \max\left\{\gamma_A - \gamma_t,\ \frac{\tau_A - \tau^e + \Delta\beta_A}{\mu + \Lambda_A + \eta_A}\right\}. \tag{5.6.27a}$$

Step 5

Obtain the final values of the plastic strain and effective stress

$$\gamma(t + \Delta t) = \gamma_A - \Delta\gamma_{er}, \quad \Delta\beta = \Delta\beta_A - \Lambda\,\Delta\gamma_{er},$$

$$\tau(t + \Delta t) = \tau^e - \Delta\beta - \mu\,\Delta\gamma_{er}. \tag{5.6.27b-d}$$

These steps are illustrated in Table 5.6.8.

5.6.3. Asymptotic Analysis

Section 5.5 presents the singular perturbation analysis used by Nemat-Nasser and Ni (1993) to investigate the integration of (5.6.1) for small τ/μ. Subsection 5.5.4 gives their proposed modified outer-solution method which is simple and accurate. In this modified method, the outer solution (for large time-steps) is approximated by

$$\gamma^o(t+\Delta t) = \gamma_t + \Delta\gamma_A - \frac{1}{\mu}(\tau_A + \Delta\beta_A - \tau_t)$$
$$+ \frac{1}{\mu^2}\left\{\eta_A\,[(h_A+\Lambda_A)\,\dot{\gamma}_A + \eta_A\,\ddot{\gamma}_A] + (h_A+\Lambda_A)\,(\tau_A+\Delta\beta_A-\tau_t)\right\},$$
$$\tau^o(t+\Delta t) = \tau_A - \frac{1}{\mu}\left\{\eta_A\,[(h_A+\Lambda_A)\,\dot{\gamma}_A + \eta_A\,\ddot{\gamma}_A] + (h_A+\Lambda_A)\,(\tau_A+\Delta\beta_A-\tau_t)\right\},$$
$$\Delta\beta^o = \Delta\beta_A - \frac{\Lambda_A}{\mu}[\tau_A + \Delta\beta_A - \tau_t], \tag{5.6.28a-c}$$

where

$$\tau_A = F(\dot{\gamma}_A,\, \gamma_A), \quad \eta_A = \eta(\dot{\gamma}_A,\, \gamma_A), \quad h_A = h(\dot{\gamma}_A,\, \gamma_A), \quad \Lambda_A = \Lambda(\gamma_A),$$
$$\gamma_A = \gamma_t + \Delta\gamma_A, \quad \Delta\beta_A = \int_{\gamma_t}^{\gamma_A} \Lambda(\gamma)\,d\gamma,$$
$$\dot{\gamma}_A = d(\Delta t), \quad \ddot{\gamma}_A = \dot{d}(\Delta t), \quad \Delta\gamma_A = \int_0^{\Delta t} d(\zeta)\,d\zeta,$$
$$\eta(\dot{\gamma},\, \gamma) = \frac{\partial F}{\partial \dot{\gamma}}(\dot{\gamma},\, \gamma), \quad h(\dot{\gamma},\, \gamma) = \frac{\partial F}{\partial \gamma}(\dot{\gamma},\, \gamma), \tag{5.6.29a-k}$$

where $\dot{d}(\Delta t)$ is the derivative of function $d(\xi)$ with respect to ξ, evaluated at $\xi = \Delta t$. The inner solution (small timestep) is approximated by

$$\gamma^i(t+\Delta t) = \gamma_t + \dot{\gamma}_t\,\Delta t, \quad \tau^i(t+\Delta t) = \tau_t + \mu\,d\,\Delta t - (\mu+\Lambda_t)\,\dot{\gamma}_t\,\Delta t,$$
$$\Delta\beta(t+\Delta t) = \Lambda_t\,\dot{\gamma}_t\,\Delta t. \tag{5.6.30a-c}$$

For arbitrary timesteps, these two solutions are combined to obtain

$$\gamma(t+\Delta t) = \max\{\gamma^i(t+\Delta t),\ \gamma^o(t+\Delta t)\},$$
$$\tau(t+\Delta t) = \min\{\tau^i(t+\Delta t),\ \tau^o(t+\Delta t)\},$$
$$\Delta\beta(t+\Delta t) = \min\{\Delta\beta^i(t+\Delta t),\ \Delta\beta^o(t+\Delta t)\}. \tag{5.6.31a-c}$$

Table 5.6.9 illustrates this for the same problem that is examined in Tables 5.6.1 to 5.6.8. Comparing with the "exact" results of Table 5.6.7, it is seen that the asymptotic method provides uniformly accurate results for all strain increments.

Table 5.6.9

Computational steps for a single time increment: Asymptotic Method

$d\,\Delta t$	0.01	0.02	0.20	0.40	1.00
$\dot\gamma(t)/\dot\gamma_0$	0.0000	0.0000	0.0000	0.0000	0.0000
$\Delta\gamma_A$	0.0100	0.0200	0.2000	0.4000	1.0000
$\dot\gamma_A/\dot\gamma_0$	200.0000	200.0000	200.0000	200.0000	200.0000
τ_A/τ_c	1.0645	1.0738	1.1769	1.2385	1.3401
$\eta_A\dot\gamma_0/\tau_c$	0.000053	0.000054	0.000059	0.000062	0.000067
h_A/τ_c	0.9677	0.8948	0.3923	0.2477	0.1218
$\Delta\gamma^o(t+\Delta t)$	-.000551	0.009350	0.188274	0.387643	0.986614
$\tau^o(t+\Delta t)/\tau_c$	1.0551	1.0650	1.1726	1.2357	1.3386
$\Delta\gamma^i(t+\Delta t)$	0.000000	0.000000	0.000000	0.000000	0.000000
$\tau^i(t+\Delta t)/\tau_c$	1.1000	2.1000	20.1000	40.1000	100.1000
$\Delta\gamma(t+\Delta t)$	0.000000	0.009350	0.188274	0.387643	0.986614
$\tau(t+\Delta t)$	1.0551	1.0650	1.1726	1.2357	1.3386

5.6.4. Reverse Loading

So far, in this section, *proportional loading* is considered only for continued loading. In this subsection, consider the case of proportional reverse loading, *i.e.*, when

$$\hat{\mathbf{D}} = -d\,\frac{\hat{\boldsymbol{\mu}}(t)}{\sqrt{2}}, \quad d = (2\,\hat{\mathbf{D}} : \hat{\mathbf{D}})^{1/2}, \quad \hat{\boldsymbol{\mu}}(t) = \frac{\hat{\mathbf{s}}(t)}{\sqrt{2}\,\tau_t}. \tag{5.6.32a-c}$$

In this case, the effective stress difference τ will first decrease from τ_t to zero, and then will increase to the flow stress. Therefore, it is convenient to use a new variable s defined by

$$s(t+\xi) = -\,\hat{\mathbf{s}}(t+\xi) : \hat{\boldsymbol{\mu}}(t), \tag{5.6.33}$$

instead of the effective stress difference τ. Then (5.3.13a) reduces to

$$\dot{s} = \mu\,d - (\mu+\Lambda)\,\dot\varepsilon^p, \quad \dot\varepsilon^p = \mathrm{sgn}(s)\,\dot\gamma. \tag{5.6.34a,b}$$

This case is similar to (5.6.2a). The initial value of s is $-\tau_t$. Since $\dot\gamma$ is always positive, $\dot\varepsilon^p$ is initially negative and then becomes positive as s becomes positive.

Rate-independent Model: From the flow rule (5.3.7a), $\dot\varepsilon^p$ is zero until s reaches the value $\tau_y(t)$, after which (5.6.34b) becomes

$$\dot\varepsilon^p = \dot\gamma. \tag{5.6.34c}$$

Integration of (5.6.34a), along with the flow rule (5.3.10a), can be performed using any of the methods discussed in Subsection 5.4.1. For example, using the *plastic-predictor/elastic-corrector* method, obtain

$$s^{tr} = -\tau_t + 2\,\mu\,d\,\Delta t,$$

$$\Delta\gamma_A = \max\left\{0,\ \frac{s^{tr}-\tau_y(t)}{\mu}\right\},$$

$$\gamma_A = \gamma_t + \Delta\gamma_A\,, \quad s_A = F(\gamma_A)\,, \quad \Delta\beta_A = \int_{\gamma_t}^{\gamma_A} \Lambda\, d\gamma,$$

$$H_A = H(\gamma_A)\,, \quad \Lambda_A = \Lambda(\gamma_A)\,,$$

$$\Delta\gamma_{er} = \frac{s_A - \tau_y(t) + \Delta\beta_A}{\mu + H_A + \Lambda_A}\,,$$

$$\Delta\gamma = \Delta\gamma_A - \Delta\gamma_{er}\,,$$

$$\Delta\beta = \Delta\beta_A - \Lambda_A\,\Delta\gamma_{er}$$

$$s(t+\Delta t) = s^{tr} - \mu\,\Delta\gamma - \Delta\beta\,. \tag{5.6.35a-k}$$

These expressions also provide the necessary steps in the numerical implementation of the method.

Rate-dependent Model: Now, consider the integration of (5.6.34a) for rate-dependent plastic deformation. In this case, from (5.6.33), (5.6.34b), and (5.3.10c), the initial values of s and $\dot{\varepsilon}^p$ are

$$s(t) = -\tau_t\,, \quad \dot{\varepsilon}^p(t) = -\,g(\tau_t,\, \gamma_t)\,. \tag{5.6.36a,b}$$

With these initial values, (5.6.34a) can be integrated using the methods discussed in Subsection 5.4.2. The computational steps in using the *plastic-predictor/elastic-corrector* method are as follows:

Step 1.

From the initial values given by (5.6.36a,b) and the flow rule, calculate

$$\tau_d = F(d,\, \gamma_t)\,, \quad \dot{\varepsilon}^p(t) = -\,g(\tau_t,\, \gamma_t)\,,$$

$$k = \mu\,\frac{(d - \dot{\varepsilon}^p(t))}{(\tau_d + \tau_t)}\,. \tag{5.6.37a-c}$$

Step 2.

Estimate the plastic strain, plastic strain rate, and the effective stress at time $t + \Delta t$, from

$$\dot{\varepsilon}^p_A = d - (d - \dot{\varepsilon}^p(t))\, e^{-k\Delta t}\,,$$

$$\Delta\varepsilon^p_A = d\,\Delta t - \frac{d - \dot{\varepsilon}^p(t)}{k}\,(1 - e^{-k\Delta t})\,,$$

$$s^e = -\tau_t + \mu\,\frac{d - \dot{\varepsilon}^p(t)}{k}\,(1 - e^{-k\Delta t})\,. \tag{5.6.38a-c}$$

Step 3.

Estimate the time at which $\dot{\varepsilon}^p_A$ becomes zero, and the plastic strain increment during unloading,[23]

[23] For rate-dependent models, inelastic strains always accompany a deformation.

$$\Delta t_0 = \frac{1}{k}\, ln\,(1 - \dot{\varepsilon}^p(t)/d)\,, \qquad \Delta\varepsilon_0^p = \Delta t_0 d + \dot{\varepsilon}^p(t)/k\,. \tag{5.6.39a,b}$$

Step 4.

Estimate the increments in the plastic strain and the backstress during the time-step (unloading followed by reversed loading)

$$\gamma_A = \gamma_t + \int_0^{\Delta t} |\dot{\varepsilon}_A^p(\zeta)|\, d\zeta = \begin{cases} \gamma_t - \Delta\varepsilon_A^p & \text{for } \dot{\varepsilon}_A^p \le 0 \\ \gamma_t + \Delta\varepsilon_A^p - 2\,\Delta\varepsilon_0^p & \text{for } \dot{\varepsilon}_A^p > 0\,, \end{cases}$$

$$\Delta\beta_A = \begin{cases} \displaystyle\int_{\gamma_t}^{\gamma_A} -\Lambda(\gamma)\, d\gamma & \text{for } \dot{\varepsilon}_A^p \le 0 \\ \displaystyle\int_{\gamma_t}^{\gamma_t - \Delta\varepsilon_0^p} -\Lambda(\gamma)\, d\gamma + \int_{\gamma_t - \Delta\varepsilon_0^p}^{\gamma_A} \Lambda(\gamma)\, d\gamma & \text{for } \dot{\varepsilon}_A^p > 0\,. \end{cases} \tag{5.6.40a,b}$$

Step 5.

Estimate the flow stress and the tangent modulus η_A, using the approximate values of the plastic strain and its rate,

$$\dot{\gamma}_A(\Delta t) = |\dot{\varepsilon}_A^p(\Delta t)|\,, \qquad \tau_A = F(\dot{\gamma}_A,\ \gamma_A)\,,$$

$$s_A = \text{sgn}(\dot{\varepsilon}_A^p)\,\tau_A\,, \qquad \eta_A = \frac{1}{\Delta t}\,\frac{\partial F(\dot{\gamma}_A,\ \gamma_A)}{\partial \dot{\gamma}} + \frac{\partial F(\dot{\gamma}_A,\ \gamma_A)}{\partial \gamma}\,. \tag{5.6.41a-d}$$

Step 6.

Calculate the error in the estimated plastic strain,

$$\Delta\varepsilon_{er}^p = \frac{s_A - s^e + \Delta\beta_A}{\mu + \Lambda_A + \eta_A}\,. \tag{5.6.42a}$$

Step 7.

Obtain the final values of the plastic strain and effective stress,

$$\gamma(t+\Delta t) = \gamma_A - \text{sgn}(\dot{\varepsilon}_A^p)\,\Delta\varepsilon_{er}^p\,, \qquad s(t+\Delta t) = s_A - \eta_A\,\Delta\varepsilon_{er}^p\,. \tag{5.6.42b,c}$$

5.7. INTEGRATION FOR UNIDIRECTIONAL STRETCH

The results of the preceding section are generalized in this section to cases of nonproportional loading, in which the stretches remain unidirectional; see Subsection 5.2.2. Then, the deformation rate tensor, $\hat{\mathbf{D}} = \mathbf{R}_t^T\,\mathbf{D}\,\mathbf{R}_t$, measured corotationally with the material element, is constant over the time increment.

For small deformations, there are several numerical algorithms for integrating the inelastic constitutive equations. These can be adapted for integrating (5.3.13) in finite deformations, provided that $\hat{\mathbf{D}}$ is constant, as is detailed in this section.

5.7.1. Elastic, Perfectly-plastic Model

Consider the case of an *elastic, perfectly-plastic material model without noncoaxiality and backstress.* Then, $\Lambda = \alpha = 0$, and $F(\gamma) = \tau_y$ is a constant for all γ. For the J_2-flow theory, constitutive equation (5.3.13a) becomes

$$\overset{\bullet}{\hat{\boldsymbol{\tau}}} = \overset{\bullet}{\hat{\mathbf{s}}} = 2\,\mu\left[\,\hat{\mathbf{D}} - \dot{\gamma}\,\frac{\hat{\boldsymbol{\mu}}}{\sqrt{2}}\right], \quad \hat{\boldsymbol{\mu}} = \frac{\hat{\boldsymbol{\tau}}}{\sqrt{2}\tau}. \tag{5.7.1a,b}$$

Now, seek to integrate (5.7.1) from time t to $t+\Delta t$ along with the flow rule (5.3.7a) for a constant tensor $\hat{\mathbf{D}}$, with $\alpha = \Lambda = 0$ and τ_y = constant. Inside the yield surface, the value of τ is less than τ_y. Denote by Δt^e the time required for the stress state to reach the yield surface. This is given by the solution of[24]

$$\hat{\boldsymbol{\tau}}(t+\Delta t^e) : \hat{\boldsymbol{\tau}}(t+\Delta t^e) = 2\,\tau_y^2, \quad \hat{\boldsymbol{\tau}}(t+\Delta t^e) = \hat{\boldsymbol{\tau}}(t) + 2\,\mu\,\hat{\mathbf{D}}\,\Delta t^e. \tag{5.7.2a,b}$$

If the timestep Δt is less than Δt^e, then the entire deformation increment is elastic. If Δt is greater than Δt^e, the deformation is elastic from time t to $t+\Delta t^e$ and then plastic from $t+\Delta t^e$ to $t+\Delta t$. Therefore, the stress state moves from $\hat{\boldsymbol{\tau}}(t)$ to $\hat{\boldsymbol{\tau}}(t+\Delta t^e)$ on a straight line in the deviatoric stress space and then along the yield surface from $\hat{\boldsymbol{\tau}}(t+\Delta t^e)$ to $\hat{\boldsymbol{\tau}}(t+\Delta t)$ which is the unknown to be calculated. Therefore, the plastic part of the deformation rate varies from

$$\hat{\mathbf{D}}^p(t+\Delta t^e) = \dot{\gamma}(t+\Delta t^e)\,\frac{\hat{\boldsymbol{\mu}}(t+\Delta t^e)}{\sqrt{2}}, \quad \hat{\boldsymbol{\mu}}(t+\Delta t^e) = \frac{\hat{\boldsymbol{\tau}}(t+\Delta t^e)}{\sqrt{2}\,\tau_y}, \tag{5.7.3a,b}$$

to

$$\hat{\mathbf{D}}^p(t+\Delta t) = \dot{\gamma}(t+\Delta t)\,\frac{\hat{\boldsymbol{\mu}}(t+\Delta t)}{\sqrt{2}}, \quad \hat{\boldsymbol{\mu}}(t+\Delta t) = \frac{\hat{\boldsymbol{\tau}}(t+\Delta t)}{\sqrt{2}\,\tau_y}. \tag{5.7.3c,d}$$

Radial-return Method: In the *radial-return method*, the required time integral is approximated as follows (Wilkins, 1964):

$$\int_{t+\Delta t^e}^{t+\Delta t} \hat{\mathbf{D}}^p(\xi)\,d\xi \approx \Delta\gamma\,\frac{\hat{\boldsymbol{\mu}}(t+\Delta t)}{\sqrt{2}}, \tag{5.7.4}$$

where $\Delta\gamma$ is the effective plastic strain increment. In view of (5.7.4), integration of (5.7.1) results in

$$\hat{\boldsymbol{\tau}}(t+\Delta t) \approx \hat{\boldsymbol{\tau}}(t) + 2\,\mu\,\hat{\mathbf{D}}\,\Delta t - 2\,\mu\,\Delta\gamma\,\frac{\hat{\boldsymbol{\mu}}(t+\Delta t)}{\sqrt{2}}. \tag{5.7.5a}$$

For the final stress orientation this method considers,

---

[24] As explained in the sequel, in the *radial-return method*, it is not necessary to consider (5.7.2a,b), thereby simplifying the numerical procedure.

$$\hat{\boldsymbol{\mu}}(t+\Delta t) = \hat{\boldsymbol{\mu}}^{tr} \equiv \frac{\hat{\boldsymbol{\tau}}^{tr}}{\sqrt{2}\,\tau^{tr}}, \qquad \tau^{tr} = (\tfrac{1}{2}\hat{\boldsymbol{\tau}}^{tr} : \hat{\boldsymbol{\tau}}^{tr})^{1/2},$$
$$\hat{\boldsymbol{\tau}}^{tr} \equiv \hat{\boldsymbol{\tau}}(t) + 2\mu\,\hat{\mathbf{D}}\,\Delta t, \tag{5.7.6a-c}$$

where superscript tr stands for a *trial* value of the corresponding variable. If τ^{tr} is less than τ_y, then Δt^e is greater than Δt and the deformation is elastic. Then, the final stress state is given by

$$\hat{\boldsymbol{\tau}}(t+\Delta t) = \hat{\boldsymbol{\tau}}^{tr} - \mu\,\Delta\gamma\,\frac{\hat{\boldsymbol{\tau}}^{tr}}{\tau^{tr}}, \qquad \Delta\gamma = \max\left\{\frac{\tau^{tr}-\tau_y}{\mu},\ 0\right\}. \tag{5.7.5b,c}$$

It can be seen from (5.7.5) that, in this method, *it is not required to calculate the exact time required to reach the yield surface. The method gives exact results if the deformation is elastic. In large-scale numerical applications, considerable efficiency is thus obtained. The strain increment must, however, be small enough to ensure the necessary accuracy.*

Exact Integration: Now consider the exact integration of (5.7.1) from time[25] t^e to $t^e+\Delta t^p$, assuming perfect plasticity, where $\Delta t^p = \Delta t - \Delta t^e$, and $t^e = t + \Delta t^e$. Since $\tau = \tau_y$ is constant during the timestep Δt^p, it follows that

$$\hat{\boldsymbol{\tau}} : \dot{\hat{\boldsymbol{\tau}}} = 0. \tag{5.7.7a}$$

Substitute (5.7.1) into (5.7.7a) to obtain

$$\dot{\gamma} = \frac{\hat{\boldsymbol{\tau}} : \hat{\mathbf{D}}}{\tau_y}, \qquad \dot{\hat{\boldsymbol{\tau}}} + \frac{\mu}{\tau_y^2}(\hat{\boldsymbol{\tau}} : \hat{\mathbf{D}})\,\hat{\boldsymbol{\tau}} = 2\mu\,\hat{\mathbf{D}}. \tag{5.7.7b,c}$$

Integration of (5.7.7c) from t^e to $t^e+\xi$, results in

$$\hat{\boldsymbol{\tau}}(t^e+\xi)\,z(\xi) = \hat{\boldsymbol{\tau}}(t^e) + 2\mu\,\hat{\mathbf{D}}\int_0^{\xi} z(\zeta)\,d\zeta, \tag{5.7.8a}$$

where

$$z(\xi) \equiv \exp\left\{\frac{\mu}{\tau_y^2}\int_0^{\xi}\hat{\boldsymbol{\tau}}(t^e+\zeta) : \hat{\mathbf{D}}\,d\zeta\right\}. \tag{5.7.8b}$$

Take the scalar product of (5.7.8a) with $\hat{\mathbf{D}}$ and then differentiate with respect to ξ, to obtain

$$\ddot{z} = A^2 z, \qquad A = \mu\,d/\tau_y, \qquad d = (2\,\hat{\mathbf{D}} : \hat{\mathbf{D}})^{1/2}. \tag{5.7.9a-c}$$

The initial values of z and $\dot{z}$ are

$$z(0) = 1, \qquad \dot{z}(0) = \frac{\mu}{\tau_y^2}\,\hat{\boldsymbol{\tau}}(t^e) : \hat{\mathbf{D}}. \tag{5.7.9d,e}$$

Then, the solution of (5.7.9a) is

$$z(\xi) = \cosh(A\,\xi) + x_0\sinh(A\,\xi), \qquad x_0 = \frac{\hat{\boldsymbol{\tau}}(t^e) : \hat{\mathbf{D}}}{d\,\tau_y}. \tag{5.7.10a,b}$$

---

[25] See Subsection 5.7.4 for the evaluation of Δt^e.

Substitute (5.7.10a) into (5.7.8a) to obtain

$$\hat{\boldsymbol{\tau}}(t^e+\xi)\,z(\xi) = \hat{\boldsymbol{\tau}}(t^e) + \frac{2\,\tau_y}{d}\,\{\sinh(A\,\xi) + x_0\cosh(A\,\xi) - x_0\}\,\hat{\mathbf{D}}\,. \qquad (5.7.10c)$$

These results are obtained by Krieg and Krieg (1977).

5.7.2. Rate-independent Model

Consider the integration of constitutive equations (5.3.13a) and (5.3.13d) for the rate-independent flow rule (5.3.7a) with the yield stress given by (5.3.10a). *Both the noncoaxiality and the backstress are included.* Hence, *neither Λ nor α is assumed to be zero.*

Generalized Radial-return Method: Consider the generalization of the radial-return method to the J_2-flow theory that includes hardening or softening, the backstress and the noncoaxiality of the plastic strain rate and the stress difference. The isotropic and kinematic hardening moduli, $H = dF/d\gamma$ and Λ, are *not* necessarily constants.

For this case, in addition to the approximation (5.7.4), the evolution equation (5.3.13d) for the backstress and the noncoaxial term must be integrated. These are approximated as follows:

$$\overset{\circ}{\hat{\boldsymbol{\beta}}} = \sqrt{2}\,\Lambda\,\dot{\gamma}\,\hat{\boldsymbol{\mu}}(t+\Delta t)\,,$$

$$\int_0^{\Delta t} \bar{\alpha}\,\{\mathbf{1} - \hat{\boldsymbol{\mu}}(t+\xi)\otimes\hat{\boldsymbol{\mu}}(t+\xi)\} : \hat{\mathbf{D}}\,d\xi = \bar{\alpha}\,\{\mathbf{1} - \hat{\boldsymbol{\mu}}^{\alpha}\otimes\hat{\boldsymbol{\mu}}^{\alpha}\} : \hat{\mathbf{D}}\,\Delta t^p\,, \qquad (5.7.11a,b)$$

where

$$\hat{\boldsymbol{\mu}}^{\alpha} \equiv \frac{\hat{\mathbf{s}}^{\alpha}}{\sqrt{2}\,\tau^{\alpha}}\,, \qquad \hat{\mathbf{s}}^{\alpha} \equiv \hat{\mathbf{s}}(t) + 2\,\mu\,\hat{\mathbf{D}}\,\Delta t\,, \qquad \tau^{\alpha} \equiv (\tfrac{1}{2}\,\hat{\mathbf{s}}^{\alpha} : \hat{\mathbf{s}}^{\alpha})^{1/2}\,. \qquad (5.7.11c\text{-}e)$$

The part of the timestep, Δt^p, during which plastic deformation takes place, may be approximated by

$$\Delta t^p = \max\left\{\frac{\tau^{\alpha} - \tau_y(t)}{\tau^{\alpha} - \tau(t)}\,\Delta t\,,\ \ 0\right\}. \qquad (5.7.12)$$

Then, integration of (5.3.13a) and (5.3.13d) results in

$$\hat{\mathbf{s}}(t+\Delta t) = \hat{\mathbf{s}}^{tr} - 2\,(\mu\,\Delta\gamma + \Delta\beta)\,\frac{\hat{\boldsymbol{\mu}}(t+\Delta t)}{\sqrt{2}}\,,$$

$$\hat{\mathbf{s}}^{tr} \equiv \hat{\mathbf{s}}^{\alpha} + 2\,\mu\,\bar{\alpha}\,(\mathbf{1}^{(4s)} - \hat{\boldsymbol{\mu}}^{\alpha}\otimes\hat{\boldsymbol{\mu}}^{\alpha}) : \hat{\mathbf{D}}\,\Delta t^p\,,$$

$$\hat{\boldsymbol{\beta}}(t+\Delta t) = \hat{\boldsymbol{\beta}}(t) + \sqrt{2}\,\Delta\beta\,\hat{\boldsymbol{\mu}}(t+\Delta t)\,,$$

$$\Delta\beta = \int_{\gamma(t)}^{\gamma(t+\Delta t)} \Lambda\,d\gamma\,. \qquad (5.7.13a\text{-}d)$$

For the final stress state to be on the yield surface, substitute

$$\hat{\mathbf{s}}(t+\Delta t) = \sqrt{2}\,\tau_y(t+\Delta t)\,\hat{\boldsymbol{\mu}}(t+\Delta t)$$

into (5.7.13a), to arrive at

$$\tau_y(t+\Delta t) = F(\gamma(t)+\Delta\gamma) = \tau^{tr} - (\mu\,\Delta\gamma + \Delta\beta)\,,$$

$$\hat{\boldsymbol{\mu}}(t+\Delta t) = \hat{\boldsymbol{\mu}}^{tr}\,, \qquad (5.7.14a,b)$$

where the trial quantities $\boldsymbol{\mu}^{tr}$ and τ^{tr} are given by

$$\hat{\boldsymbol{\mu}}^{tr} = \frac{\hat{\mathbf{s}}^{tr}}{\sqrt{2}\,\tau^{tr}}\,, \qquad \tau^{tr} = (\tfrac{1}{2}\,\hat{\mathbf{s}}^{tr} : \hat{\mathbf{s}}^{tr})^{1/2}\,. \qquad (5.7.14c,d)$$

For isotropic hardening, and in the absence of noncoaxiality, (5.7.14a) reduces to the equation obtained by Dodds (1986).

Note that, in (5.7.12), $\Delta t^p = 0$ if $\tau^\alpha \le \tau_y(t)$. Since the term due to noncoaxiality in (5.7.13b) is normal to $\hat{\mathbf{s}}^\alpha$, it follows that $\tau^{tr} \ge \tau^\alpha$. Therefore, if $\tau^\alpha \ge \tau_y(t)$, then $\tau^{tr} \ge \tau_y(t)$.

The nonlinear equation (5.7.14a) is exactly the same as (5.6.2c). It is pointed out in Subsection 5.4.1 that the *plastic-predictor/elastic-corrector* is the preferred method for solving (5.6.2c). Therefore, only the steps involved in the *plastic-predictor/elastic-corrector* method for solving (5.7.14a) are given here.

Computational Steps: For the convenience of the reader, the necessary computational steps in the *generalized radial-return method* are summarized .

Step 1

Obtain the initial values of the stress difference and the yield stress,

$$\hat{\mathbf{s}}(t) = \hat{\boldsymbol{\tau}}(t) - \hat{\boldsymbol{\beta}}(t)\,, \quad \tau(t) = \{\tfrac{1}{2}\,\hat{\mathbf{s}}(t) : \hat{\mathbf{s}}(t)\}^{1/2}\,, \quad \tau_y(t) = F(\gamma(t))\,. \quad (5.7.15a\text{-}c)$$

Step 2

Obtain the trial stress difference, and its magnitude,

$$\hat{\mathbf{s}}^\alpha = \hat{\mathbf{s}}(t) + 2\,\mu\,\hat{\mathbf{D}}\,\Delta t\,, \qquad \tau^\alpha = (\tfrac{1}{2}\,\hat{\mathbf{s}}^\alpha : \hat{\mathbf{s}}^\alpha)^{1/2}\,. \qquad (5.7.16a,b)$$

Step 3

Estimate the fraction of the timestep during which the deformation is plastic and correct the trial stress difference for noncoaxiality,

$$\Delta t^p = \max\left\{\frac{\tau^\alpha - \tau_y(t)}{\tau^\alpha - \tau(t)}\,\Delta t\,,\ 0\right\},$$

$$\hat{\mathbf{s}}^{tr} = \hat{\mathbf{s}}^\alpha + \mu\,\overline{\alpha}\,\{2\,\hat{\mathbf{D}} - (\hat{\mathbf{s}}^\alpha : \hat{\mathbf{D}})\,\hat{\mathbf{s}}^\alpha/(\tau^\alpha)^2\}\,\Delta t^p\,,$$

$$\tau^{tr} = (\tfrac{1}{2}\,\hat{\mathbf{s}}^{tr} : \hat{\mathbf{s}}^{tr})^{1/2}\,. \qquad (5.7.16c\text{-}e)$$

Step 4

Estimate the values of the plastic strain, yield stress, increment in the backstress, and the isotropic and kinematic hardening moduli at the *end* of the timestep,

$$\Delta\gamma_A = \max\left\{0\,,\ \frac{\tau^{tr} - \tau_y(t)}{\mu}\right\}, \qquad \gamma_A = \gamma(t) + \Delta\gamma_A\,,$$

$$\tau_A = F(\gamma_A)\,, \quad \Delta\beta_A = \int_{\gamma(t)}^{\gamma_A} \Lambda(\gamma)\,d\gamma\,,$$

$$H_A = H(\gamma_A)\,, \quad \Lambda_A = \Lambda(\gamma_A)\,. \tag{5.7.17a-f}$$

Step 5

Estimate the error in the plastic strain increment and correct,

$$\Delta\gamma_{er} = \frac{\tau_A - \tau_y(t) + \Delta\beta_A}{\mu + H_A + \Lambda_A}\,, \quad \Delta\gamma = \Delta\gamma_A - \Delta\gamma_{er}\,. \tag{5.7.18a,b}$$

Step 7

Obtain the final values of the equivalent plastic strain, total stress, and the back-stress,

$$\gamma(t+\Delta t) = \gamma(t) + \Delta\gamma\,,$$

$$\hat{\boldsymbol{\tau}}(t+\Delta t) = \left\{1 - \frac{2\mu\Delta\gamma}{\tau^{tr}}\right\}\hat{\mathbf{s}}^{tr} + \hat{\boldsymbol{\beta}}(t)\,,$$

$$\hat{\boldsymbol{\beta}}(t+\Delta t) = \hat{\boldsymbol{\beta}}(t) + (\Delta\beta_A - \Lambda_A\Delta\gamma_{er})\frac{\hat{\mathbf{s}}^{tr}}{\tau^{tr}}\,. \tag{5.7.19a-c}$$

Note that, if $\tau^\alpha \le \tau_y(t)$, then from (5.7.12), $\Delta t^p = 0$ and $\hat{\mathbf{s}}^{tr} = \hat{\mathbf{s}}^\alpha$. Hence, from (5.7.17a), $\Delta\gamma_A = 0$. This yields $\Delta\beta_A = 0$ and $\tau_A = \tau_y(t)$, and from (5.7.18a), $\Delta\gamma_{er} = 0$. In this case, the final values are given by

$$\hat{\boldsymbol{\tau}}(t+\Delta t) = \hat{\mathbf{s}}^\alpha + \hat{\boldsymbol{\beta}}(t)\,, \quad \hat{\boldsymbol{\beta}}(t+\Delta t) = \hat{\boldsymbol{\beta}}(t)\,. \tag{5.7.19d,e}$$

Thus, the method produces exact results for elastic deformation. The time required to reach the yield surface is not explicitly calculated. This fact renders the *generalized radial-return method* very attractive for implementation in large-scale computer codes. *The generalization outlined above ensures that the objectivity is satisfied in the presence of isotropic and kinematic hardening, including the noncoaxiality effect. Furthermore, the use of the plastic-predictor/elastic-corrector method, ensures that the stress magnitude is calculated almost exactly in one step.*

Example: As an illustration consider an isotropic hardening with yield function

$$F(\gamma) = \tau_c\,(1 + \gamma/\gamma_0)^N\,, \tag{5.7.20a}$$

where $\tau_c = \mu/100$, $\gamma_0 = 0.005$, and $N = 0.1$. As an example consider the following time history for the corotational deformation rate $\hat{\mathbf{D}}$:

$$\hat{\mathbf{D}} = \begin{bmatrix} 1 & 0 & 0 \\ 0 & -1 & 0 \\ 0 & 0 & 0 \end{bmatrix} \times 10^{-2}\,, \quad 0 \le t < 50\,,$$

$$\hat{\mathbf{D}} = \begin{bmatrix} 1 & 1 & 0 \\ 1 & -1 & 0 \\ 0 & 0 & 0 \end{bmatrix} \times 10^{-2}\,, \quad 50 \le t < 100\,,$$

$$\hat{\mathbf{D}} = \begin{bmatrix} 0 & 0 & 0 \\ 0 & 0 & 1 \\ 0 & 1 & 0 \end{bmatrix} \times 10^{-2}, \qquad 100 \le t < 150. \tag{5.7.20b-d}$$

For this example with the initial conditions $\tau(0) = 0$ and $\gamma(0) = 0$, the components of the deviatoric stress $\boldsymbol{\tau}'$ are given by the *generalized radial-return method* using various numbers of timesteps, as shown in Figure 5.7.1.

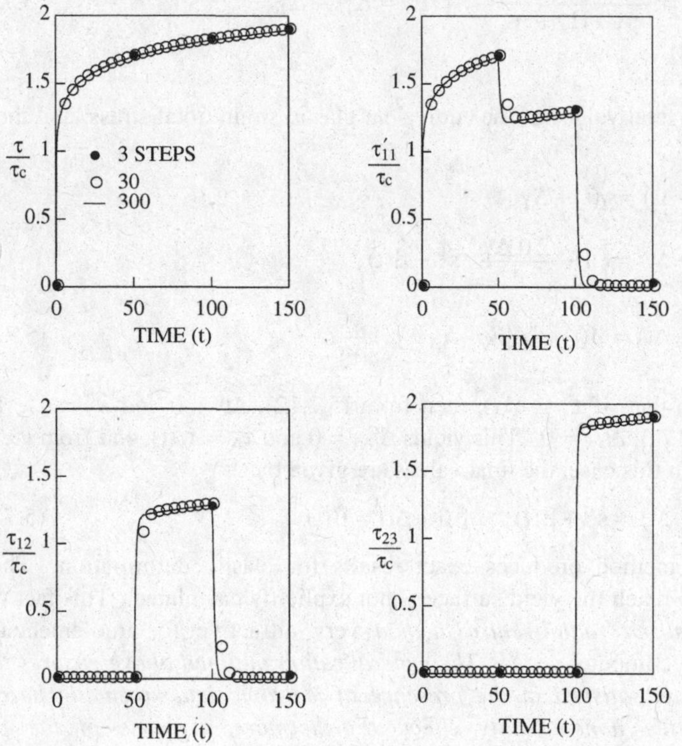

Figure 5.7.1

Time history of deviatoric stress components, obtained using the *generalized radial-return method*, for prescribed deformation history (5.7.20b-d) and isotropic hardening (5.7.20a)

As is seen in this figure, the effective stress τ is very accurately estimated, even when a 50% strain increment is used in one timestep. However, the results for the stress components are not similarly accurate when there is a rotation in the stress direction. This is due to the approximation for the final stress orientation. To improve the calculation of the stress orientation, consider the following alternative to the radial-return method.

Perfect-plasticity Path: In the radial-return method, the final orientation of the stress difference is obtained approximately. Krieg and Krieg (1977) show that this final orientation is almost exact if the strain increment is either infinitesimally small or infinitely large. The error in the final orientation is significant for moderate strain increments. Even though real materials show hardening and thermal softening during plastic deformation, the hardening moduli are generally orders of magnitude smaller than the elastic moduli. In this case, the final orientation can be obtained more accurately than that given by the *radial-return* method, if a perfect-plasticity path is assumed and the exact results of Krieg and Krieg (1977) are used.

Integration of (5.3.13a) and (5.3.13d) from time $t^e = t + \Delta t^e$ to $t^e + \Delta t^p$, is considered now, assuming a perfect-plasticity path. Note that, the time Δt^e required to reach the yield surface is obtained from (5.7.2). Hence, here it is assumed that

$$\hat{\boldsymbol{\tau}}(t^e) : \hat{\boldsymbol{\tau}}(t^e) = 2\,\tau_y^2(t)\,. \tag{5.7.21}$$

Take the scalar product of (5.3.13a) with $\hat{\boldsymbol{\mu}}/\sqrt{2}$, to obtain

$$\dot{\tau} = \mu\, x(\xi)\, d - (\mu + \Lambda)\,\dot{\gamma}\,, \tag{5.7.22}$$

where

$$x(\xi) \equiv \hat{\boldsymbol{\mu}}(t^e + \xi) : \hat{\mathbf{v}}\,, \qquad d \equiv (2\,\hat{\mathbf{D}} : \hat{\mathbf{D}})^{1/2}\,, \qquad \hat{\mathbf{v}} \equiv \frac{\sqrt{2}\,\hat{\mathbf{D}}}{d}\,. \tag{5.7.23a-c}$$

Differentiate (5.3.13b) with respect to time and substitute into (5.3.13a), to obtain

$$\dot{\hat{\boldsymbol{\mu}}} = A\,(\mathbf{1} - \hat{\boldsymbol{\mu}} \otimes \hat{\boldsymbol{\mu}}) : \hat{\mathbf{v}}\,, \tag{5.7.24a}$$

where

$$A \equiv \mu\, d\,(1 - \overline{\alpha})/\tau\,. \tag{5.7.24b}$$

Equations (5.7.22) and (5.7.24a) are coupled differential equations for the rate of change of the magnitude and orientation of the stress difference $\hat{\mathbf{s}}$. To integrate these equations, first assume τ is constant and integrate (5.7.24a) analytically. Then use this result and the *plastic-predictor/elastic-corrector* method to integrate (5.7.22).

For a constant A, it follows from (5.7.10) that

$$\hat{\boldsymbol{\mu}}(t^e + \xi) = \frac{1}{z(\xi)}\,\hat{\boldsymbol{\mu}}(t^e) + \frac{1}{z(\xi)}\,\{\sinh(A\xi) + k\cosh(A\xi) - k\}\,\hat{\mathbf{v}}\,,$$

$$z(\xi) = \exp\left\{ A \int_0^{\xi} \hat{\boldsymbol{\mu}}(t^e + \zeta) : \hat{\mathbf{v}}\, d\zeta \right\} = \cosh(A\xi) + k\sinh(A\xi)\,, \tag{5.7.25a,b}$$

where

$$k = \hat{\boldsymbol{\mu}}(t^e) : \hat{\mathbf{v}}\,. \tag{5.7.25c}$$

Then, integration of (5.7.22) results in

$$\tau_y(t+\Delta t) = \tau_y(t) + \frac{\mu\, d}{A}\, ln\,(z(\Delta t^p)) - (\mu\,\Delta\gamma + \Delta\beta)\,,$$
$$\tau_y(t+\Delta t) = F(\gamma(t) + \Delta\gamma)\,,$$
$$\tau_y(t) = F(\gamma(t))\,. \qquad (5.7.26a\text{-}c)$$

These nonlinear equations are also similar to (5.6.2). Hence, they can be solved using the *plastic-predictor/elastic-corrector* method. First estimate the plastic strain increment by

$$\Delta\gamma_A = \frac{d}{A}\, ln\,(z(\Delta t^p))\,. \qquad (5.7.27)$$

Now, calculate the approximate values γ_A, $\Delta\beta_A$, H_A, Λ_A, and τ_A from (5.7.17) and the error in the estimated plastic strain increment, $\Delta\gamma_{er}$, from (5.7.18). Then the final value of the stress difference is given by

$$\hat{\mathbf{s}}\,(t^e + \Delta t^p) = \sqrt{2}\,\tau(t^e + \Delta t^p)\,\hat{\boldsymbol{\mu}}(t^e + \Delta t^p)\,,$$
$$\tau(t^e + \Delta t^p) = \tau_y(t) - \Delta\beta_A + (\mu + \Lambda_A)\,\Delta\gamma_{er}\,. \qquad (5.7.28a,b)$$

The backstress may be approximated by

$$\hat{\boldsymbol{\beta}}(t+\Delta t) = \hat{\boldsymbol{\beta}}(t) + \sqrt{2}\,(\Delta\beta_A - \Lambda_A\,\Delta\gamma_{er})\,\hat{\boldsymbol{\mu}}(t+\Delta t)\,. \qquad (5.7.29)$$

Computational Steps: In the following, the necessary computational steps are summarized for the perfect-plasticity path method.

Step 1

Obtain the initial values of the stress difference and yield stress,

$$\hat{\mathbf{s}}\,(t) = \hat{\boldsymbol{\tau}}(t) - \hat{\boldsymbol{\beta}}(t)\,, \quad \tau(t) = \{\tfrac{1}{2}\,\hat{\mathbf{s}}\,(t) : \hat{\mathbf{s}}\,(t)\}^{1/2}\,,$$
$$\tau_y(t) = F(\gamma(t))\,. \qquad (5.7.30a\text{-}c)$$

Step 2

Calculate the elastic-response time increment and the stress difference at the end of the elastic deformation (Subsection 5.7.4),

$$\Delta t^e = \min\left\{\Delta t\,,\ \frac{\sqrt{(\hat{\mathbf{s}}_t : \hat{\mathbf{D}})^2 + 2\,(\hat{\mathbf{D}} : \hat{\mathbf{D}})\,(\tau_y^2 - \tau_t^2)} - \hat{\mathbf{s}}_t : \hat{\mathbf{D}}}{2\,\mu\,(\hat{\mathbf{D}} : \hat{\mathbf{D}})}\right\}\,,$$
$$\hat{\mathbf{s}}\,(t^e) = \hat{\mathbf{s}}\,(t) + 2\,\mu\,\hat{\mathbf{D}}\,\Delta t^e\,. \qquad (5.7.31a,b)$$

Step 3

Calculate the final orientation of the stress difference by assuming perfect-plasticity,

$$d = (2\,\hat{\mathbf{D}} : \hat{\mathbf{D}})^{1/2}\,, \quad A = \mu\, d\,(1 - \bar{\alpha})/\tau_y(t)\,, \quad k = \frac{\hat{\mathbf{s}}\,(t^e) : \hat{\mathbf{D}}}{\tau_y(t)\, d}\,,$$
$$\Delta t^p = \Delta t - \Delta t^e\,, \quad z = \cosh(A\,\Delta t^p) + k\sinh(A\,\Delta t^p)\,,$$
$$z_1 = \sinh(A\,\Delta t^p) + k\cosh(A\,\Delta t^p) - k\,,$$

$$\hat{\boldsymbol{\mu}}(t+\Delta t) = \frac{1}{\sqrt{2}\, z\, \tau_y}\, \hat{\mathbf{s}}(t^e) + \frac{\sqrt{2}\, z_1}{z\, d}\, \hat{\mathbf{D}}\,. \quad (5.7.32a\text{-}g)$$

Step 4

Estimate the plastic strain, yield stress, increment in the back stress, and the isotropic and kinematic hardening moduli at the end of the timestep,

$$\Delta\gamma_A = \frac{d}{A}\, ln\,(z)\,, \quad \gamma_A = \gamma(t) + \Delta\gamma_A\,,$$

$$\tau_A = F(\gamma_A)\,, \quad \Delta\beta_A = \int_{\gamma(t)}^{\gamma_A} \Lambda(\gamma)\, d\gamma\,,$$

$$H_A = H(\gamma_A)\,, \quad \Lambda_A = \Lambda(\gamma_A)\,. \quad (5.7.33a\text{-}f)$$

Step 5

Estimate the error in the estimated plastic strain increment,

$$\Delta\gamma_{er} = \frac{\tau_A - \tau_y(t) + \Delta\beta_A}{\mu + H_A + \Lambda_A}\,. \quad (5.7.34)$$

Step 6

Obtain the final values of the equivalent plastic strain, the stress difference, the backstress, and the total stress,

$$\gamma(t+\Delta t) = \gamma_A - \Delta\gamma_{er}\,,$$

$$\tau(t+\Delta t) = \tau_y(t) - \Delta\beta_A + (\mu + \Lambda_A)\, \Delta\gamma_{er}\,,$$

$$\hat{\mathbf{s}}(t+\Delta t) = \sqrt{2}\, \tau(t+\Delta t)\, \hat{\boldsymbol{\mu}}(t+\Delta t)\,,$$

$$\hat{\boldsymbol{\beta}}(t+\Delta t) = \hat{\boldsymbol{\beta}}(t) + \sqrt{2}\, (\Delta\beta_A - \Lambda_A\, \Delta\gamma_{er})\, \hat{\boldsymbol{\mu}}(t+\Delta t)\,,$$

$$\hat{\boldsymbol{\tau}}(t+\Delta t) = \hat{\mathbf{s}}(t+\Delta t) + \hat{\boldsymbol{\beta}}(t+dt)\,. \quad (5.7.35a\text{-}f)$$

The time history of the components of the deviatoric stress $\hat{\boldsymbol{\tau}}'$ for the example (5.7.20) using the *perfect-plasticity-path method* is shown in Figure 5.7.2 for various timesteps. Unlike the *generalized radial-return method*, the *perfect-plasticity-path method* yields very accurate results for both the magnitude (effective stress) and the orientation (components) of $\hat{\boldsymbol{\tau}}'$.

5.7.3. Rate-dependent Model

Consider the integration of constitutive equations (5.3.13a) and (5.3.13d) for the rate-dependent flow rule (5.3.10b). Let the inverse relation for the flow rule be given by[26]

$$\dot{\gamma} = g(\tau,\, \gamma)\,. \quad (5.7.36)$$

[26] See Section 4.8 for a discussion of various models and Chapter 9 for their experimental evaluation.

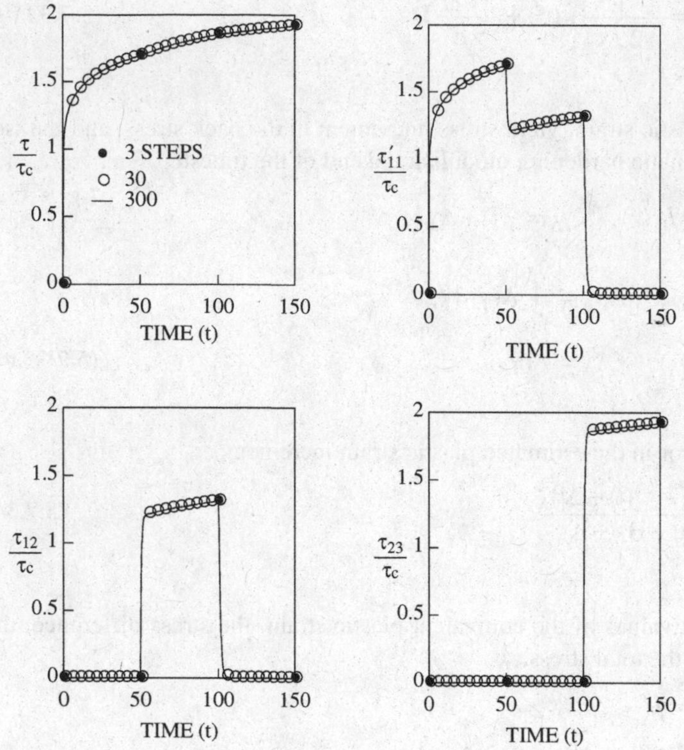

Figure 5.7.2

Time history of deviatoric stress components using the *perfect-plasticity-path method*, for prescribed deformation history (5.7.20b-d) and isotropic hardening (5.7.20a)

For example, if the flow rule is

$$\tau = F(\dot{\gamma},\ \gamma) = \tau_0 \left[\frac{\dot{\gamma}}{\dot{\gamma}_0}\right]^m \left[1 + \frac{\gamma}{\gamma_0}\right]^N, \tag{5.7.37a}$$

then the inverse relation is

$$\dot{\gamma} = g(\tau,\ \gamma) = \dot{\gamma}_0 \left[\frac{\tau}{\tau_0}\right]^{\frac{1}{m}} \left[1 + \frac{\gamma}{\gamma_0}\right]^{-\frac{N}{m}}. \tag{5.7.37b}$$

Here, m and N are material parameters (assumed to be constant) and τ_0, γ_0, and $\dot{\gamma}_0$ are the reference values of the stress, strain, and strain rate, respectively.

Generalized Radial-return Method: Consider the generalization of the radial-return method for application to the rate-dependent case which includes combined isotropic-kinematic hardening and noncoaxiality. Following the procedure outlined in Section 5.7.2 for the rate-independent flow, obtain

$$\hat{\mathbf{s}}(t+\Delta t) = \hat{\mathbf{s}}^{tr} - 2(\mu\,\Delta\gamma + \Delta\beta)\,\frac{\hat{\boldsymbol{\mu}}(t+\Delta t)}{\sqrt{2}},$$

$$\hat{\boldsymbol{\beta}}(t+\Delta t) = \hat{\boldsymbol{\beta}}(t) + \sqrt{2}\,\Delta\beta\,\hat{\boldsymbol{\mu}}(t+\Delta t), \qquad (5.7.38a,b)$$

where

$$\hat{\mathbf{s}}^{tr} = \hat{\mathbf{s}}^{\alpha} + 2\mu\,\bar{\alpha}\,(\mathbf{1} - \hat{\boldsymbol{\mu}}^{\alpha}\otimes\boldsymbol{\mu}^{\alpha}) : \hat{\mathbf{D}}\,\Delta t, \quad \Delta\beta = \int_{\gamma(t)}^{\gamma(t+\Delta t)} \Lambda\, d\gamma,$$

$$\hat{\mathbf{s}}^{\alpha} = \hat{\mathbf{s}}(t) + 2\mu\,\hat{\mathbf{D}}\,\Delta t, \quad \hat{\boldsymbol{\mu}}^{\alpha} = \frac{\hat{\mathbf{s}}^{\alpha}}{\sqrt{2}\,\tau^{\alpha}}, \quad \tau^{\alpha} = (\tfrac{1}{2}\hat{\mathbf{s}}^{\alpha} : \hat{\mathbf{s}}^{\alpha})^{1/2}. \qquad (5.7.39a\text{-}e)$$

Substitute

$$\hat{\mathbf{s}}(t+\Delta t) = \sqrt{2}\,\tau(t+\Delta t)\,\hat{\boldsymbol{\mu}}(t+\Delta t) \qquad (5.7.40)$$

into (5.7.38a), and take the double scalar product with $\hat{\boldsymbol{\mu}}(t+\Delta t)$, to obtain

$$\tau(t+\Delta t) = F\{\dot{\gamma}(t+\Delta t),\ \gamma(t)+\Delta\gamma\} = \tau^{tr} - (\mu\,\Delta\gamma + \Delta\beta),$$

$$\hat{\boldsymbol{\mu}}(t+\Delta t) = \hat{\boldsymbol{\mu}}^{tr} = \frac{\hat{\mathbf{s}}^{tr}}{\sqrt{2}\,\tau^{tr}}, \quad \tau^{tr} = (\tfrac{1}{2}\hat{\mathbf{s}}^{tr} : \hat{\mathbf{s}}^{tr})^{1/2}. \qquad (5.7.41a\text{-}c)$$

The scalar incremental equation (5.7.41a) can be treated as a rate equation for proportional loading in the direction of $\hat{\boldsymbol{\mu}}^{tr}$; see Balendran and Nemat-Nasser (1995). The effective initial stress s_t and the deformation rate $\dot{\varepsilon}$ are given by

$$s_t = \hat{\mathbf{s}}(t) : \hat{\boldsymbol{\mu}}^{tr}/\sqrt{2}, \quad \dot{\varepsilon} = (\tau^{tr} - s_t)/(\mu\,\Delta t). \qquad (5.7.42a,b)$$

It should be noted that, as Δt goes to zero, $\hat{\boldsymbol{\mu}}^{tr}$ goes to $\hat{\boldsymbol{\mu}}(t)$ and s_t goes to $\tau(t)$. Therefore, this approximation will converge to the correct solution as the time-step is reduced.

For continuous loading, $\dot{\varepsilon}$ and $s_t > 0$. On the other hand, $\dot{\varepsilon} < 0$ corresponds to unloading, while $s_t < 0$ and $\dot{\varepsilon} > 0$ correspond to unloading followed by reversed loading. In these cases, the scalar equation (5.7.41a) can be solved using the method discussed in Subsection 5.6.4. However, for most cases, the plastic strain is negligible in unloading. Therefore, to retain the simplicity of the radial-return method, set

$$\tau(t+\Delta t) = \tau^{tr}, \quad \Delta\gamma = \Delta\beta = 0, \quad \text{for } \dot{\varepsilon} < 0. \qquad (5.7.43a,b)$$

In the cases of continuous loading, and unloading followed by reversed loading, solve the scalar equation (5.7.41a), using one of the methods outlined in Subsection 5.4.2 for proportional loading.

Computational Steps: Since the *plastic-predictor/elastic-corrector* method is accurate and efficient, the necessary computational steps for the generalized radial-return method with plastic-predictor/elastic-corrector are given below.

Step 1

Obtain the initial values of the stress difference,

$$\hat{\mathbf{s}}(t) = \hat{\boldsymbol{\tau}}(t) - \hat{\boldsymbol{\beta}}(t), \quad \tau(t) = \{\tfrac{1}{2}\hat{\mathbf{s}}(t) : \hat{\mathbf{s}}(t)\}^{1/2}. \qquad (5.7.44a,b)$$

Step 2

In the absence of noncoaxiality, obtain the trial stress difference, and its magnitude and then correct for noncoaxiality,

$$\hat{\mathbf{s}}^{\alpha} = \hat{\mathbf{s}}(t) + 2\mu\,\hat{\mathbf{D}}\,\Delta t, \qquad \tau^{\alpha} = (\tfrac{1}{2}\,\hat{\mathbf{s}}^{\alpha} : \hat{\mathbf{s}}^{\alpha})^{1/2}.$$

$$\hat{\mathbf{s}}^{\mathrm{tr}} = \hat{\mathbf{s}}^{\alpha} + \mu\,\bar{\alpha}\,\{2\,\hat{\mathbf{D}} - (\hat{\mathbf{s}}^{\alpha} : \hat{\mathbf{D}})\,\hat{\mathbf{s}}^{\alpha}/(\tau^{\alpha})^2\}\,\Delta t,$$

$$\tau^{\mathrm{tr}} = (\tfrac{1}{2}\,\hat{\mathbf{s}}^{\mathrm{tr}} : \hat{\mathbf{s}}^{\mathrm{tr}})^{1/2}. \tag{5.7.45a-d}$$

Step 3

Calculate the effective initial stress and deformation rate,

$$s_t = \frac{\hat{\mathbf{s}}(t) : \hat{\boldsymbol{\mu}}^{\mathrm{tr}}}{\sqrt{2}}, \qquad \dot{\varepsilon} = \frac{\tau^{\mathrm{tr}} - s_t}{\mu\,\Delta t}, \qquad \hat{\boldsymbol{\mu}}^{\mathrm{tr}} = \frac{\hat{\mathbf{s}}^{\mathrm{tr}}}{\sqrt{2}\,\tau^{\mathrm{tr}}}. \tag{5.7.46a-c}$$

Step 4

If $\dot{\varepsilon} < 0$, then set $\Delta\gamma = \Delta\beta = 0$ and go to Step 8.

Step 5

Calculate the initial effective estimated plastic strain rate, and the final flow stress from the flow rule,

$$\dot{\varepsilon}^{p}_{t} = \mathrm{sign}(s_t)\,g(\,|\,s_t\,|,\ \gamma(t)),$$

$$\tau_d = F(\dot{\varepsilon},\ \gamma(t)). \tag{5.7.47a,b}$$

Step 6

Estimate the plastic strain increment, plastic strain rate, and the effective stress at the end of the timestep,

$$k = \mu\,\frac{\dot{\varepsilon} - \dot{\varepsilon}^{p}_{t}}{\tau_d - s_t},$$

$$\Delta\gamma_A = \dot{\varepsilon}\,\Delta t - \frac{\tau_d - s_t}{G}\,(1 - e^{-k\Delta t}),$$

$$\dot{\gamma}_A = \dot{\varepsilon} - (\dot{\varepsilon} - \dot{\varepsilon}^{p}_{t})\,e^{-k\Delta t},$$

$$\tau^{e} = s_t + (\tau_d - s_t)\,(1 - e^{-k\Delta t}). \tag{5.7.48a-d}$$

Step 7

Estimate the flow stress, the increment in the backstress, and the hardening moduli,

$$\gamma_A = \gamma(t) + \Delta\gamma_A, \qquad \tau_A = F(\dot{\gamma}_A,\ \gamma_A), \qquad \Delta\beta_A = \int_{\gamma(t)}^{\gamma_A} \Lambda(\gamma)\,d\gamma,$$

$$\eta_A = \frac{1}{\Delta t}\,\frac{\partial F}{\partial \dot{\gamma}} + \frac{\partial F}{\partial \gamma}, \qquad \Lambda_A = \Lambda(\gamma_A). \tag{5.7.49a-e}$$

Step 8

Estimate the error in the plastic strain increment and correct the increments in the plastic strain and the backstress,

$$\Delta\gamma_{er} = \frac{\tau_A - \tau^e + \Delta\beta_A}{\mu + \Lambda_A + \eta_A},$$

$$\Delta\gamma = \max\{0, \quad \Delta\gamma_A - \Delta\gamma_{er}\},$$

$$\Delta\beta = \begin{cases} 0 & \text{for } \Delta\gamma = 0 \\ \Delta\beta_A - \Lambda_A \Delta\gamma_{er} & \text{for } \Delta\gamma \neq 0 \end{cases} \tag{5.7.50a-c}$$

Step 9

Obtain the final values of the equivalent plastic strain, total stress, and the backstress,

$$\gamma(t+\Delta t) = \gamma(t) + \Delta\gamma,$$

$$\hat{\boldsymbol{\tau}}(t+\Delta t) = \left[1 - \frac{2\mu\Delta\gamma}{\tau^{tr}}\right]\hat{\mathbf{s}}^{tr} + \hat{\boldsymbol{\beta}}(t),$$

$$\hat{\boldsymbol{\beta}}(t+\Delta t) = \hat{\boldsymbol{\beta}}(t) + \Delta\beta\,\frac{\hat{\mathbf{s}}^{tr}}{\tau^{tr}}. \tag{5.7.51a-c}$$

Example: As an illustration, consider the power law for the isotropic flow-stress

$$\tau = \tau_c(1+\gamma/\gamma_0)^N(\dot{\gamma}/\dot{\gamma}_0)^m, \tag{5.7.52a}$$

with $\tau_c = \mu/100$, $\gamma_0 = 0.005$, $N = 0.1$, and $\dot{\gamma}_0 = 0.0001$. As an example consider the following time history for the corotational deformation rate $\hat{\mathbf{D}}$:

$$\hat{\mathbf{D}} = \begin{bmatrix} 1 & 0 & 0 \\ 0 & -1 & 0 \\ 0 & 0 & 0 \end{bmatrix} \times 10^{-2}, \quad 0 \le t < 50,$$

$$\hat{\mathbf{D}} = \frac{1}{\sqrt{2}}\begin{bmatrix} 1 & 1 & 0 \\ 1 & -1 & 0 \\ 0 & 0 & 0 \end{bmatrix} \times 10^{-4}, \quad 50 \le t < 5050,$$

$$\hat{\mathbf{D}} = \frac{1}{\sqrt{2}}\begin{bmatrix} -1 & 0 & 0 \\ 0 & 1 & 1 \\ 0 & 1 & 0 \end{bmatrix} \times 10^{-2}, \quad 5050 \le t < 6000. \tag{5.7.52b-d}$$

The components of the deviatoric stress $\hat{\boldsymbol{\tau}}'$ obtained using the *generalized radial-return method* with various numbers of timesteps are shown in Figures 5.7.3 and 5.7.4 for $m = 0.01$ and $m = 0.05$, respectively. It is seen in Figure 5.7.3 that when the rate sensitivity is low, the *generalized radial-return method* gives very accurate results for the effective stress. However, the stress components are not very accurate when there is any rotation in the stress orientation. To improve the calculation of the stress orientation, consider the following alternative to the radial-return method.

Perfect-plasticity Path: Alternative to the *radial-return* method is the perfect-

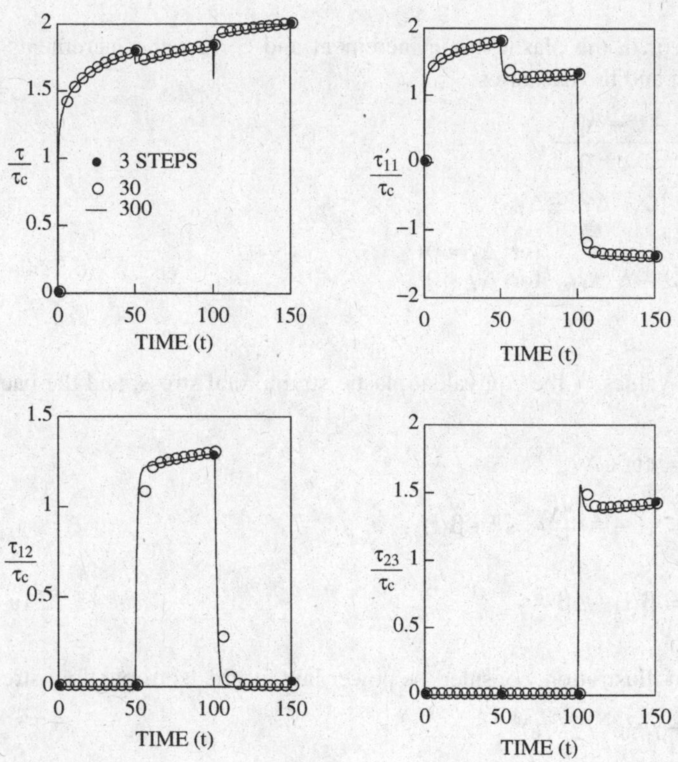

Figure 5.7.3

Time history of deviatoric stress components for rate-dependent plasticity model (5.7.52a) with m = 0.01, and deformation history (5.7.52b-d), using the *generalized radial-return method*

plasticity assumption. However, unlike for the rate-independent flow, there may be no yield surface for the rate-dependent flow[27]. On the other hand, most rate-dependent materials undergo very little plastic deformation when the effective shear stress is less than a reference stress. Hence, in addition to the flow rule (5.7.36), it may be assumed that

$$\dot{\gamma} << d \quad \text{for } \tau < \tau_d \equiv F(d, \gamma)\,. \tag{5.7.53a}$$

This also implies that

$$\dot{\gamma} >> d\,, \quad \text{for } \tau > \tau_d\,. \tag{5.7.53b}$$

Condition (5.7.53) shows that the elastic deformation dominates when $\tau < \tau_d$

[27] Some authors have introduced rate-dependent plastic deformation models with a yield condition, as discussed in Subsection 4.3.2 of Chapter 4.

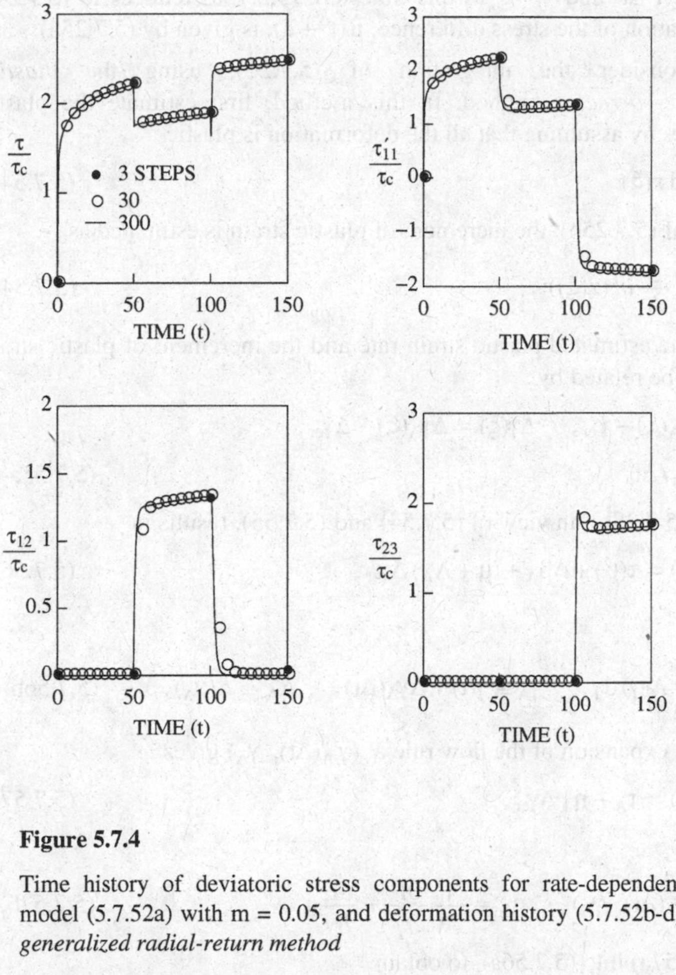

Figure 5.7.4

Time history of deviatoric stress components for rate-dependent plasticity model (5.7.52a) with m = 0.05, and deformation history (5.7.52b-d), using the *generalized radial-return method*

and hence, the stress state moves mostly parallel to the direction of $\hat{\mathbf{D}}$ in the deviatoric stress space. On the other hand, when $\tau > \tau_d$, the plastic deformation dominates, and hence the stress state moves mostly in the direction of $\hat{\boldsymbol{\mu}}$ in the deviatoric stress space. Only if $\tau \approx \tau_d$, then the stress state moves in the direction normal to $\hat{\boldsymbol{\mu}}$. Therefore, the perfect-plasticity-path assumption is reasonable only after the stress magnitude attains the value τ_d.

Denote by Δt^e the time that is taken for the effective stress to reach the value τ_d. This depends on whether the initial effective stress $\tau(t)$ is greater or less than τ_d. Note that, even if $\tau(t)$ is greater than τ_d, because of the dominant plastic deformation, the effective stress drops to τ_d, regardless of whether it is loading ($\hat{\mathbf{D}} : \hat{\boldsymbol{\mu}} > 0$) or unloading ($\hat{\mathbf{D}} : \hat{\boldsymbol{\mu}} < 0$).

Consider the integration of (5.3.13a) and (5.3.13d) from time $t^e = t + \Delta t^e$ to $t^e + \Delta t^p$, where $\Delta t^p = \Delta t - \Delta t^e$. Assume that $\tau(t^e) \approx \tau_d$. See Subsection 5.7.4 for the evaluation of Δt^e and $\boldsymbol{\tau}(t^e)$. In this timestep, (5.3.13a) reduces to (5.7.24). Then the orientation of the stress difference, $\hat{\boldsymbol{\mu}}(t^e+\xi)$, is given by (5.7.25a).

Now, consider the integration of (5.7.22), using the *plastic-predictor/elastic-corrector* method. In this method, first estimate the plastic deformation rate by assuming that all the deformation is plastic,

$$\dot{\gamma}_A(\xi) = d\,x(\xi)\,. \tag{5.7.54a}$$

Then, in view of (5.7.25b), the increment of plastic strain is estimated as

$$\Delta\gamma_A(\xi) = \frac{d}{A}\,ln\,(z(\xi))\,. \tag{5.7.54b}$$

The errors in the estimated plastic strain rate and the increment of plastic strain are assumed to be related by

$$\dot{\gamma}(\xi) = \dot{\gamma}_A(\xi) - \dot{\gamma}_{er}\,, \quad \Delta\gamma(\xi) = \Delta\gamma_A(\xi) - \Delta\gamma_{er}\,,$$

$$\dot{\gamma}_{er} = \Delta\gamma_{er}/\Delta t\,. \tag{5.7.55a-c}$$

Integration of (5.7.28b), in view of (5.7.54) and (5.7.55), results in

$$\tau(t^e+\Delta t^p) = \tau(t^e) + \Delta\beta_A + (\mu + \Lambda_A)\,\Delta\gamma_{er}\,, \tag{5.7.56a}$$

where

$$\Delta\beta_A \equiv \int_{\gamma(t)}^{\gamma_A} \Lambda(\gamma)\,d\gamma\,, \quad \gamma_A \equiv \gamma(t) + \Delta\gamma_A(\Delta t)\,, \quad \Lambda_A \equiv \Lambda(\gamma_A)\,. \tag{5.7.56b-d}$$

A Taylor series expansion of the flow rule at $(\dot{\gamma}_A(\Delta t),\ \gamma_A)$ gives

$$\tau(t^e+\Delta t^p) = \tau_A - \eta_A\,\Delta\gamma_{er}\,, \tag{5.7.57a}$$

where

$$\tau_A \equiv F(\dot{\gamma}_A(\Delta t),\ \gamma_A)\,, \quad \eta_A \equiv \frac{1}{\Delta t}\,\frac{\partial F}{\partial \dot{\gamma}} + \frac{\partial F}{\partial \gamma}\,. \tag{5.7.57b,c}$$

Substitute (5.7.57a) into (5.7.56a), to obtain

$$\Delta\gamma_{er} = \frac{\tau_A - \tau(t^e) + \Delta\beta_A}{\mu + \Lambda_A + \eta_A}\,. \tag{5.7.58}$$

Finally, the values of the stress difference and the backstress are given by

$$\hat{\mathbf{s}}(t^e+\Delta t^p) = \tau(t^e+\Delta t^p)\,\hat{\boldsymbol{\mu}}(t^e+\Delta t^p)\,,$$

$$\hat{\boldsymbol{\beta}}(t^e+\Delta t^p) = \hat{\boldsymbol{\beta}}(t^e) + \sqrt{2}\,(\Delta\beta_A - \Lambda_A\,\Delta\gamma_{er})\,\hat{\boldsymbol{\mu}}(t^e+\Delta t^p)\,. \tag{5.7.59a,b}$$

Computational Steps: The necessary computational steps are summarized below.

Step 1

Obtain the initial values of the stress difference and the dynamic critical stress,

$$\hat{\mathbf{s}}(t) = \hat{\boldsymbol{\tau}}(t) - \hat{\boldsymbol{\beta}}(t)\,, \quad \tau_t \equiv \tau(t) = \{\tfrac{1}{2}\,\hat{\mathbf{s}}(t) : \hat{\mathbf{s}}(t)\}^{\frac{1}{2}}\,,$$

$$d = \{2\hat{\mathbf{D}} : \hat{\mathbf{D}}\}^{1/2}, \quad \tau_d = F(d, \gamma_t), \quad \dot{\gamma}_t = g(\tau_t, \gamma_t), \tag{5.7.60a-e}$$

where $\gamma_t \equiv \gamma(t)$.

Step 2

If $\tau_t \le \tau_d$, set $\Delta t_1^e = 0$, $\hat{\mathbf{s}}(t_1) = \hat{\mathbf{s}}(t)$, and $\tau(t_1) = \tau(t) \equiv \tau_t$, and go to Step 4.

Step 3 (Subsection 5.7.4)

Find the time increment required for the effective stress difference, τ, to reach the τ_d-value, and evaluate the stress difference at the end of this timestep,

$$k = (\mu + \Lambda)\frac{\dot{\gamma}_t - d}{\tau_t - \tau_d},$$

$$\dot{\gamma}_A = \max\{\dot{\gamma}_t(1 - e^{-k\Delta t})/(k\Delta t), \quad (\dot{\gamma}_t - d)/ln(\dot{\gamma}_t/d)\},$$

$$\mathbf{K}_1 = 2\mu(1 - \bar{\alpha})\hat{\mathbf{D}} + \{\mu\bar{\alpha}(\hat{\mathbf{s}}_t : \hat{\mathbf{D}})/\tau_t^2 - (\mu + \Lambda_t)\dot{\gamma}_A/\tau_t\}\hat{\mathbf{s}}(t),$$

$$\Delta t_1^e = \min\left\{\Delta t, \frac{-\sqrt{(\hat{\mathbf{s}}_t : \mathbf{K}_1)^2 - 2(\mathbf{K}_1 : \mathbf{K}_1)(\tau_t^2 - \tau_d^2)} - \hat{\mathbf{s}}_t : \mathbf{K}_1}{(\mathbf{K}_1 : \mathbf{K}_1)}\right\},$$

$$\hat{\mathbf{s}}(t_1) = \hat{\mathbf{s}}(t) + \mathbf{K}_1\Delta t_1^e, \quad \tau(t_1) = \{\tfrac{1}{2}\hat{\mathbf{s}}(t_1) : \hat{\mathbf{s}}(t_1)\}^{1/2}. \tag{5.7.61a-f}$$

Step 4

If $\tau(t_1) \ge \tau_d$ and $\hat{\mathbf{s}}(t_1) : \hat{\mathbf{D}} \ge 0$, set $\Delta t_2^e = 0$, $\hat{\mathbf{s}}(t^e) = \hat{\mathbf{s}}(t_1)$, and $\tau(t^e) = \tau(t_1)$, and go to Step 6.

Step 5 (Subsection 5.7.4)

Find the time required for the effective stress difference to reach the τ_d-value and evaluate the stress difference at the end of this timestep,

$$\mathbf{K}_2 = 2\mu(1 - \bar{\alpha})\hat{\mathbf{D}} + \{\mu\bar{\alpha}(\hat{\mathbf{s}}_{t_1} : \hat{\mathbf{D}})/\tau_{t_1}^2\}\hat{\mathbf{s}}(t_1),$$

$$\Delta t_2^e = \min\left\{\Delta t - \Delta t_1^e, \frac{\sqrt{(\hat{\mathbf{s}}_{t_1} : \mathbf{K}_2)^2 + 2(\mathbf{K}_2 : \mathbf{K}_2)(\tau_d^2 - \tau_{t_1}^2)} - \hat{\mathbf{s}}_{t_1} : \mathbf{K}_2}{(\mathbf{K}_2 : \mathbf{K}_2)}\right\},$$

$$\hat{\mathbf{s}}_A(t^e) = \hat{\mathbf{s}}(t_1) + \mathbf{K}_2\Delta t_2^e, \quad \tau_A = \{\tfrac{1}{2}\hat{\mathbf{s}}_A(t^e) : \hat{\mathbf{s}}_A(t^e)\}^{1/2},$$

$$\dot{\gamma}_A = g(\tau_A, \gamma_t), \quad k = (\mu + \Lambda)\left[\frac{\partial F}{\partial \dot{\gamma}}\right]^{-1},$$

$$\tau(t^e) = \tau_A - \dot{\gamma}_A(\partial F/\partial\dot{\gamma})(1 - e^{-k\Delta t_2^e}), \quad \hat{\mathbf{s}}(t^e) = \frac{\tau(t^e)}{\tau_A}\hat{\mathbf{s}}_A(t^e). \tag{5.7.62a-h}$$

Step 6

Calculate the final orientation of the stress difference by assuming perfect plasticity,

$$A = \mu d(1 - \bar{\alpha})/\tau(t^e), \quad k = \frac{\hat{\mathbf{s}}(t^e) : \hat{\mathbf{D}}}{\tau(t^e)\, d},$$

$$\Delta t^p = \Delta t - \Delta t_1^e - \Delta t_2^e, \quad z = \cosh(A\Delta t^p) + k\sinh(A\Delta t^p),$$

$$z_1 = \sinh(A\,\Delta t^p) + k\cosh(A\,\Delta t^p) - k\,,$$

$$\hat{\boldsymbol{\mu}}(t+\Delta t) = \frac{1}{\sqrt{2}\,z\,\tau_y}\,\hat{\mathbf{s}}(t^e) + \frac{\sqrt{2}\,z_1}{z\,d}\,\hat{\mathbf{D}}\,. \tag{5.7.63a-f}$$

Step 7

Estimate the plastic strain rate, plastic strain, flow stress, increment in the back stress, and the hardening moduli at the end of the timestep,

$$\dot{\gamma}_A = \sqrt{2}\,\hat{\boldsymbol{\mu}}(t+\Delta t) : \hat{\mathbf{D}}\,, \quad \Delta\gamma_A = \frac{1}{A}\,d\,ln(z)\,, \quad \gamma_A = \gamma(t) + \Delta\gamma_A\,,$$

$$\tau_A = F(\dot{\gamma}_A,\ \gamma_A)\,, \quad \Delta\beta_A = \int_{\gamma(t)}^{\gamma_A} \Lambda(\gamma)\,d\gamma\,,$$

$$\Lambda_A = \Lambda(\gamma_A)\,, \quad \eta_A = \frac{1}{\Delta t}\frac{\partial F}{\partial \dot{\gamma}}(\dot{\gamma}_A,\ \gamma_A) + \frac{\partial F}{\partial \gamma}(\dot{\gamma}_A,\ \gamma_A)\,. \tag{5.7.64a-g}$$

Step 8

Estimate the error in the estimated plastic strain increment, and obtain the final values of the equivalent plastic strain, the stress difference, the backstress, and the total stress,

$$\Delta\gamma_{er} = \frac{\tau_A - \tau(t^e) + \Delta\beta_A}{\mu + \eta_A + \Lambda_A}\,,$$

$$\gamma(t+\Delta t) = \gamma_A - \Delta\gamma_{er}\,,$$

$$\tau(t+\Delta t) = \tau(t^e) - \Delta\beta_A + (\mu + \Lambda_A)\,\Delta\gamma_{er}\,,$$

$$\hat{\mathbf{s}}(t+\Delta t) = \sqrt{2}\,\tau(t+\Delta t)\,\hat{\boldsymbol{\mu}}(t+\Delta t)\,,$$

$$\hat{\boldsymbol{\beta}}(t+\Delta t) = \hat{\boldsymbol{\beta}}(t) + \sqrt{2}\,(\Delta\beta_A - \Lambda_A\,\Delta\gamma_{er})\,\hat{\boldsymbol{\mu}}(t+\Delta t)\,,$$

$$\hat{\boldsymbol{\tau}}(t+\Delta t) = \hat{\mathbf{s}}(t+\Delta t) + \hat{\boldsymbol{\beta}}(t+\Delta t)\,. \tag{5.7.65a-f}$$

The stress components obtained using the *perfect-plasticity-path method* for the same example in Figures 5.7.3 and 5.7.4, with various numbers of timesteps, are shown in Figures 5.7.5 and 5.7.6, for $m = 0.01$ and $m = 0.05$ respectively. Essentially exact results are obtained for any number of timesteps, from 3 to 300.

5.7.4. Elasticity-dominated Deformation

In the previous two subsections, the *perfect-plasticity-path* method is used when the deformation is dominated by plastic deformation. In rate-independent plasticity, the deformation is regarded elastic in unloading and when the effective stress difference is less than the current value of the yield stress. In the rate-dependent case with strain-rate softening or hardening, the stress path deviates from the perfect-plasticity path even in loading. In that case, both elastic and plastic deformations are involved essentially in all time increments.

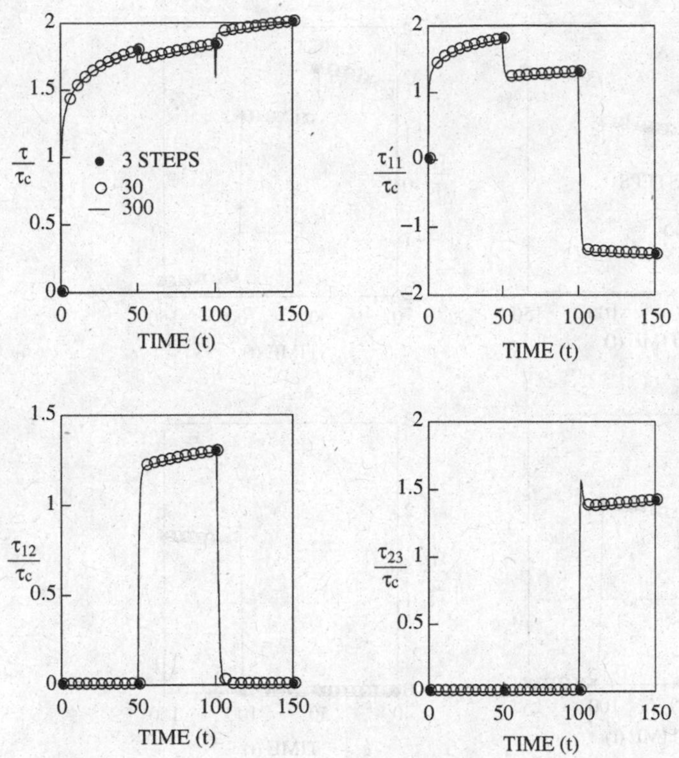

Figure 5.7.5

Time history of deviatoric stress components for rate-dependent plasticity model (5.7.52a) with m = 0.01, and deformation history (5.7.52b-d), using the *perfect-plasticity-path method*

Rate-independent Model: For rate-independent materials, the deformation is purely elastic if the effective stress is less than the current yield stress. Consider a time increment Δt, and denote by Δt^e, the portion of this time which it takes to reach the yield surface. In this range, the rate of change of the stress difference (5.3.13a) is

$$\mathbf{K} \equiv \dot{\hat{\mathbf{s}}} = 2\,\mu\,\hat{\mathbf{D}}\,. \tag{5.7.66a}$$

Therefore, the duration of the elastic deformation, Δt^e, is obtained from

$$(\hat{\mathbf{s}}_t + \mathbf{K}\,\Delta t^e) : (\hat{\mathbf{s}}_t + \mathbf{K}\,\Delta t^e) = 2\,\tau_y^2 \tag{5.7.66b}$$

which yields

$$\Delta t^e = \min\left\{ \Delta t\,,\ \frac{\sqrt{(\hat{\mathbf{s}}_t : \mathbf{K})^2 + 2\,\mathbf{K} : \mathbf{K}\,(\tau_y^2 - \tau_t^2)} - \hat{\mathbf{s}}_t : \mathbf{K}}{\mathbf{K} : \mathbf{K}} \right\}, \tag{5.7.66c}$$

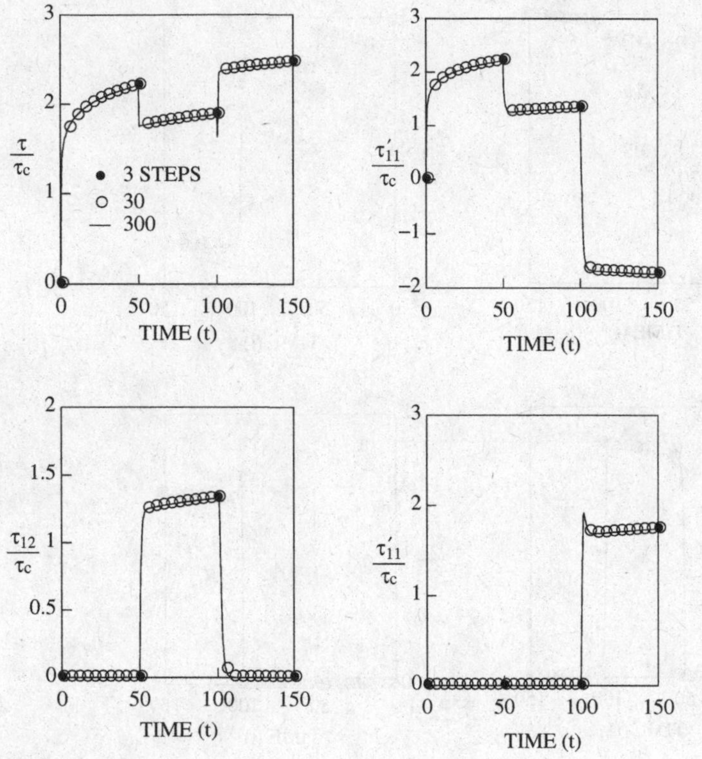

Figure 5.7.6

Time history of deviatoric stress components for rate-dependent plasticity model (5.7.52a) with m = 0.05, and deformation history (5.7.52b-d), using the *perfect-plasticity-path method*

where $\hat{\mathbf{s}}_t = \hat{\mathbf{s}}(t)$. The stress difference at the end of the period of the elastic deformation is obtained as follows:

$$\hat{\mathbf{s}}(t+\Delta t^e) = \hat{\mathbf{s}}(t) + \mathbf{K}\,\Delta t^e\,. \tag{5.7.66d}$$

The timestep during which plastic deformation is dominant is then given by

$$\Delta t^p = \Delta t - \Delta t^e\,. \tag{5.7.66e}$$

Rate-dependent Model: In the case of rate-dependent materials, there is no yield surface. However, if a material element is subjected to a constant deformation rate for a relatively long time, in the absence of strain hardening, the effective stress difference would reach a value τ_d, given by

$$\tau_d \equiv F(d,\ \gamma(t))\,. \tag{5.7.67}$$

(i) Strain-rate softening: If the initial (*i.e.*, at the start of the time increment) effective stress difference, $\tau_t \equiv \tau(t)$, is higher than τ_d, then the initial plastic strain rate $\dot{\gamma}(t) \equiv \dot{\gamma}_t$ will be much higher than the effective deformation rate d. In this case, the effective stress difference τ will decrease regardless of the sign of $\hat{\mathbf{D}} : \hat{\boldsymbol{\mu}}$, until the effective stress difference reaches the value τ_d.

In this case, equation (5.3.13a) may be first approximated by

$$\dot{\hat{\mathbf{s}}} = -\sqrt{2}\,(\mu + \Lambda_t)\,\dot{\gamma}_t\,\hat{\boldsymbol{\mu}}\,, \quad \Lambda_t = \Lambda(\gamma(t))\,. \tag{5.7.68a}$$

This yields a scalar equation

$$\dot{\tau} = -(\mu + \Lambda)\,\dot{\gamma}\,. \tag{5.7.68b}$$

Thus $\dot{\gamma}$ varies sharply from $\dot{\gamma}_t$ defined by (5.7.37b), to d, as τ varies from τ_t to τ_d. Assume the variation of τ in $\dot{\gamma}$ is linear, and set

$$\tau(t+\xi) = \tau_d + \rho\,(\dot{\gamma} - d)\,,$$
$$\rho = (\tau_t - \tau_d)/(\dot{\gamma}_t - d)\,. \tag{5.7.69a,b}$$

Substitute (5.7.69a) into (5.7.68b) and integrate, to obtain

$$\tau(t+\xi) = \tau_t - \rho\,\dot{\gamma}_t\,(1 - e^{-k\xi})\,,$$
$$k = (\mu + \Lambda_t)/\rho\,. \tag{5.7.69c,d}$$

Therefore, the time that it takes for the effective stress difference to reach the τ_d-value, is given by

$$\Delta t_A^e = \frac{\rho}{\mu + \Lambda_t}\, ln\,(\dot{\gamma}_t/d)\,, \tag{5.7.70}$$

where $\Lambda_t = \Lambda(\gamma(t))$. From (5.7.68b), the average plastic strain rate over this time interval may be approximated by

$$\dot{\gamma}_A = \max\left\{\frac{\tau_t - \tau(t+\Delta t)}{(\mu + \Lambda_t)\,\Delta t}\,,\ \frac{\tau_t - \tau_d}{(\mu + \Lambda_t)\,\Delta t_A^e}\right\}. \tag{5.7.71a}$$

Substitute (5.7.69) and (5.7.70) into (5.7.71a), to obtain

$$\dot{\gamma}_A = \max\,\{\dot{\gamma}_t\,(1 - e^{-k\Delta t})/(k\,\Delta t)\,,\ (\dot{\gamma}_t - d)/ln\,(\dot{\gamma}_t/d)\}\,. \tag{5.7.71b}$$

Using this as an approximate expression for $\dot{\gamma}$, the time it takes for the effective stress difference to reach the τ_d-value, is more accurately obtained by solving the quadratic equation

$$\hat{\mathbf{s}}(t_1) : \hat{\mathbf{s}}(t_1) = 2\,\tau_d^2\,, \quad \hat{\mathbf{s}}(t_1) = \hat{\mathbf{s}}(t) + \mathbf{K}_1\,\Delta t_1^e\,, \quad t_1 = t + \Delta t_1^e\,,$$
$$\mathbf{K}_1 = 2\,\mu\,\hat{\mathbf{D}} - 2\,\mu\,\bar{\alpha}\,(\mathbf{1} - \hat{\boldsymbol{\mu}}_t \otimes \hat{\boldsymbol{\mu}}_t) : \hat{\mathbf{D}} - \sqrt{2}\,(\mu + \Lambda_t)\,\dot{\gamma}_A\,\hat{\boldsymbol{\mu}}_t\,. \tag{5.7.72a-d}$$

This yields

$$\Delta t_1^e = \min\left\{\Delta t\,,\ \frac{-\sqrt{(\hat{\mathbf{s}}_t : \mathbf{K}_1)^2 - 2\,\mathbf{K}_1 : \mathbf{K}_1\,(\tau_t^2 - \tau_d^2)} - \hat{\mathbf{s}}_t : \mathbf{K}_1}{\mathbf{K}_1 : \mathbf{K}_1}\right\}. \tag{5.7.72e}$$

Note that, since $\hat{\mathbf{s}}_t : \mathbf{K}_1 < 0$, and $\tau_t > \tau_d$ for this case, Δt_1^e given by (5.7.72e) is always nonnegative.

(ii) Strain-rate hardening and unloading: If the effective stress difference $\tau(t+\Delta t_1^e)$ is less than τ_d, then the plastic strain rate will be orders of magnitude smaller than d until τ reaches the τ_d-value. Also for $\tau(t+\Delta t_1^e) = \tau_d$ and $\hat{\mathbf{s}}(t+\Delta t_1^e) : \hat{\mathbf{D}} < 0$, the plastic deformation will be negligible until τ reaches the τ_d-value again. Therefore, the stress difference at the end of such a timestep must first be estimated by solving the quadratic equation

$$\hat{\mathbf{s}}_A(t^e) : \hat{\mathbf{s}}_A(t^e) = 2\,\tau_d^2\,,$$
$$\hat{\mathbf{s}}_A(t^e) = \hat{\mathbf{s}}(t_1) + \mathbf{K}_2\,\Delta t_2^e\,, \quad t^e = t_1 + \Delta t_2^e\,,$$
$$\mathbf{K}_2 = 2\,\mu\,\hat{\mathbf{D}} - 2\,\mu\,\bar{\alpha}\,(\mathbf{1} - \hat{\boldsymbol{\mu}} \otimes \hat{\boldsymbol{\mu}}) : \hat{\mathbf{D}}\,. \tag{5.7.73a-d}$$

This yields

$$\Delta t_2^e = \min\left\{\Delta t - \Delta t_1^e\,,\ \frac{\sqrt{(\hat{\mathbf{s}}_{t_1} : \mathbf{K}_2)^2 + 2\,\mathbf{K}_2 : \mathbf{K}_2\,(\tau_d^2 - \tau_{t_1}^2)} - \hat{\mathbf{s}}_{t_1} : \mathbf{K}_2}{\mathbf{K}_2 : \mathbf{K}_2}\right\}. \tag{5.7.73e}$$

Note that, since $\tau_{t_1} \leq \tau_d$, for this case, Δt_2^e given by (5.7.73e) is alway nonnegative. Though the plastic strain rate is negligible during most of this timestep, this may cease to be the case when $\tau_A = \tau_d$. Therefore, when very small timesteps are used, the approximation based on neglecting the plastic strain rate may not lead to the correct solution. Thus, the first estimate of the stress difference, $\hat{\mathbf{s}}_A(t^e)$, is corrected in order to account for the plastic strain increment, as follows:

$$\hat{\mathbf{s}}(t^e) = \hat{\mathbf{s}}_A(t^e) - \sqrt{2}\,(\mu + \Lambda_t)\,\Delta\gamma_A(\Delta t_2^e)\,\hat{\boldsymbol{\mu}}_A(t^e)\,, \tag{5.7.74a}$$

where $\Lambda_t = \Lambda(\gamma(t))$. The plastic strain rate is assumed to vary in this timestep, as

$$\dot{\gamma}_A(t_1 + \xi) = \dot{\gamma}_A(t^e)\,e^{k(\xi - \Delta t_2^e)}\,,$$
$$\dot{\gamma}_A(t^e) = g(\tau_A(t^e),\ \gamma(t))\,, \tag{5.7.75a,b}$$

where k is assumed to be

$$k = (\mu + \Lambda_t)\left[\frac{\partial F}{\partial \dot{\gamma}}\right]^{-1}. \tag{5.7.75c}$$

Then, the plastic strain increment is given by

$$\Delta\gamma_A(t_1 + \xi) = \dot{\gamma}_A(t^e)\,(1 - e^{-k\xi})/k\,. \tag{5.7.74b}$$

5.8. INTEGRATION FOR CONSTANT VELOCITY GRADIENT

The integration of constitutive equation (5.3.11b), and the evolution equation for the backstress, (5.3.7a), for a constant velocity gradient, $\mathbf{L} = \mathbf{D} + \mathbf{W}$, is considered in this section. From (5.3.11b), (5.3.7a), and (5.3.9), the material time derivatives of the stress difference and that of the backstress are

$$\dot{\mathbf{s}} = 2\mu\{\mathbf{D} - \bar{\alpha}(\mathbf{1} - \boldsymbol{\mu}\otimes\boldsymbol{\mu}) : \mathbf{D}\} - \sqrt{2}(\mu + \Lambda)\dot{\gamma}\,\boldsymbol{\mu} + \mathbf{W}\,\mathbf{s} - \mathbf{s}\,\mathbf{W}, \tag{5.8.1a}$$

$$\dot{\boldsymbol{\beta}} = \sqrt{2}\Lambda\dot{\gamma}\,\boldsymbol{\mu} + \mathbf{W}\boldsymbol{\beta} - \boldsymbol{\beta}\mathbf{W}. \tag{5.8.1b}$$

Consider integrating these over the time interval t to $t + \Delta t$, when **D** and **W** are the given constant deformation rate and spin tensors. In this section, algorithms are presented for performing this integration with various degrees of accuracy. Integration steps are outlined for the convenience of numerical implementation. Various models, discussed before, are examined, starting with the perfect-plasticity model.

5.8.1. Elastic, Perfectly-plastic Model

Consider the case of the elastic, perfectly-plastic model without noncoaxiality, *i.e.*, $\Lambda = \bar{\alpha} = 0$ and $F(\gamma) = \tau_y$, a constant. In this case, (5.8.1a) becomes

$$\dot{\boldsymbol{\tau}} = \dot{\mathbf{s}} = 2\mu\{\mathbf{D} - \dot{\gamma}\,\boldsymbol{\mu}/\sqrt{2}\} + \mathbf{W}\boldsymbol{\tau} - \boldsymbol{\tau}\mathbf{W}. \tag{5.8.2}$$

Generalized Radial-return Method: Integration of (5.8.2) from time t to $t + \Delta t$ with the approximation

$$\int_0^{\Delta t}\{\mathbf{W}\boldsymbol{\tau}(t+\xi) - \boldsymbol{\tau}(t+\xi)\mathbf{W}\}d\xi$$

$$\approx \mathbf{W}\{\theta\,\boldsymbol{\tau}_A(t+\Delta t) + (1-\theta)\,\boldsymbol{\tau}(t)\}\,\Delta t - \{\theta\,\boldsymbol{\tau}_A(t+\Delta t) + (1-\theta)\,\boldsymbol{\tau}(t)\}\mathbf{W}\,\Delta t,$$

results in

$$\boldsymbol{\tau}_A(t+\Delta t) - \theta\{\mathbf{W}\,\boldsymbol{\tau}_A(t+\Delta t) - \boldsymbol{\tau}_A(t+\Delta t)\,\mathbf{W}\}\,\Delta t$$

$$= \boldsymbol{\tau}(t) + 2\mu\mathbf{D}\,\Delta t - \sqrt{2}\mu\int_0^{\Delta t}\dot{\gamma}\,\boldsymbol{\mu}\,d\xi + (1-\theta)\{\mathbf{W}\boldsymbol{\tau}(t) - \boldsymbol{\tau}(t)\,\mathbf{W}\}\,\Delta t, \tag{5.8.3}$$

where the subscript A in $\boldsymbol{\tau}_A(t+\Delta t)$ emphasizes that this is an initial approximation, which will be improved, and the parameter θ may be assigned a value between 0 and 1. Equation (5.8.3) can be rewritten as

$$\boldsymbol{\tau}^\theta(t+\Delta t) = \boldsymbol{\tau}^{tr} - \sqrt{2}\mu\int_0^{\Delta t}\dot{\gamma}\,\boldsymbol{\mu}\,d\xi, \tag{5.8.4a}$$

$$\boldsymbol{\tau}^\theta(t+\Delta t) \equiv \boldsymbol{\tau}_A(t+\Delta t) - \theta\{\mathbf{W}\,\boldsymbol{\tau}_A(t+\Delta t) - \boldsymbol{\tau}_A(t+\Delta t)\,\mathbf{W}\}\,\Delta t,$$

$$\boldsymbol{\tau}^{tr} = \boldsymbol{\tau}^\theta(t) + 2\mu\mathbf{D}\,\Delta t,$$

$$\boldsymbol{\tau}^\theta(t) \equiv \boldsymbol{\tau}(t) + (1-\theta)\{\mathbf{W}\boldsymbol{\tau}(t) - \boldsymbol{\tau}(t)\,\mathbf{W}\}\,\Delta t. \tag{5.8.4b-d}$$

The radial-return method in (5.7.4) can be generalized and applied to this case as follows:

$$\int_0^{\Delta t} \dot{\gamma}\,\boldsymbol{\mu}\,d\xi \approx \Delta\gamma\,\boldsymbol{\mu}^\theta(t+\Delta t)\,, \qquad \boldsymbol{\mu}^\theta(t+\Delta t) = \frac{\boldsymbol{\tau}^\theta(t+\Delta t)}{\sqrt{2}\,\tau^\theta(t+\Delta t)}\,,$$

$$\tau^\theta(t+\Delta t) = \{\tfrac{1}{2}\boldsymbol{\tau}^\theta(t+\Delta t) : \boldsymbol{\tau}^\theta(t+\Delta t)\}^{1/2}\,. \qquad (5.8.5a\text{-}c)$$

Substitution of (5.8.5a) into (5.8.4a) results in

$$\boldsymbol{\tau}^\theta(t+\Delta t) = \boldsymbol{\tau}^{tr} - \sqrt{2}\,\mu\,\Delta\gamma\,\boldsymbol{\mu}^\theta(t+\Delta t)\,. \qquad (5.8.6a)$$

This shows that $\boldsymbol{\tau}^\theta(t+\Delta t)$ and $\boldsymbol{\tau}^{tr}$ are coaxial, *i.e.*,

$$\boldsymbol{\mu}^\theta(t+\Delta t) = \boldsymbol{\mu}^{tr} = \frac{\boldsymbol{\tau}^{tr}}{\sqrt{2}\,\tau^{tr}}\,, \qquad \tau^{tr} = (\tfrac{1}{2}\boldsymbol{\tau}^{tr} : \boldsymbol{\tau}^{tr})^{1/2}\,. \qquad (5.8.6b,c)$$

Then (5.8.6a) reduces to

$$\boldsymbol{\tau}^\theta(t+\Delta t) = \boldsymbol{\tau}^{tr} - \mu\,\Delta\gamma\,\boldsymbol{\tau}^{tr}/\tau^{tr}\,. \qquad (5.8.6d)$$

In order to obtain $\Delta\gamma$, it is assumed that $\tau^\theta(t+\Delta t) \leq \tau_y$. This yields

$$\Delta\gamma = \max\{(\tau^{tr} - \tau_y)/\mu,\ 0\}\,. \qquad (5.8.6e)$$

Once $\boldsymbol{\tau}^\theta(t+\Delta t)$ is known, $\boldsymbol{\tau}_A(t+\Delta t)$ is obtained by solving (5.8.4b). An estimate is,

$$\boldsymbol{\tau}_A(t+\Delta t) \approx \boldsymbol{\tau}^\theta(t+\Delta t) + \theta\,\{\mathbf{W}\,\boldsymbol{\tau}^\theta(t+\Delta t) - \boldsymbol{\tau}^\theta(t+\Delta t)\,\mathbf{W}\}\,\Delta t\,. \qquad (5.8.7a)$$

This is equivalent to treating the constant velocity gradient $\mathbf{L}$ over a period of time Δt by a pure rotation $(1-\theta)\,\mathbf{W}\,\Delta t$, followed by a constant deformation rate $\mathbf{D}$ over a period of time Δt, and followed by a pure rotation $\theta\,\mathbf{W}\,\Delta t$. Note that, when the material element is subjected to a rigid body rotation $(1-\theta)\,\mathbf{W}\,\Delta t$, at time t, the stress changes from $\boldsymbol{\tau}(t)$ to $\boldsymbol{\tau}^\theta(t)$, as follows:

$$\boldsymbol{\tau}^\theta(t) = \exp\{(1-\theta)\,\mathbf{W}\,\Delta t\}\,\boldsymbol{\tau}(t)\exp\{-(1-\theta)\,\mathbf{W}\,\Delta t\}\,. \qquad (5.8.7b)$$

Here, this is approximated by (5.8.4d) to the first order in $\mathbf{W}\,\Delta t$. For the constant deformation rate, the stress changes from $\boldsymbol{\tau}^\theta(t)$ to $\boldsymbol{\tau}^\theta(t+\Delta t)$. For the rotation $\theta\,\mathbf{W}\,\Delta t$ at time $t+\Delta t$, the stress changes from $\boldsymbol{\tau}^\theta(t+\Delta t)$ to $\boldsymbol{\tau}(t+\Delta t)$, as follows:

$$\boldsymbol{\tau}(t+\Delta t) = \exp\{\theta\,\mathbf{W}\,\Delta t\}\,\boldsymbol{\tau}^\theta(t+\Delta t)\exp\{-\theta\,\mathbf{W}\,\Delta t\}\,. \qquad (5.8.7c)$$

This is approximated to the first order in $\mathbf{W}\,\Delta t$ by (5.8.7a). The subscript A stands for approximate expression. For $\theta = 1/2$, this method corresponds to the algorithm used in DYNA; see Benson (1992).

The *exact* solution of (5.8.4b) is given by (see Appendix 5.B)

$$\boldsymbol{\tau}_A(t+\Delta t) = \boldsymbol{\tau}_1^\theta + \frac{1}{1+p^2}\,(\boldsymbol{\tau}_3^\theta - p\,\boldsymbol{\tau}_5^\theta) + \frac{1}{1+4\,p^2}\,(\boldsymbol{\tau}_2^\theta - p\,\boldsymbol{\tau}_4^\theta)\,, \qquad (5.8.7d)$$

$$\boldsymbol{\tau}_1^\theta = \{\boldsymbol{\tau}^\theta(t+\Delta t) : \overline{\mathbf{W}}^2\}\,(3/2\,\overline{\mathbf{W}}^2 + \mathbf{1})\,,$$

$$\boldsymbol{\tau}_2^\theta = \overline{\mathbf{W}}\,\{\boldsymbol{\tau}^\theta(t+\Delta t) - \boldsymbol{\tau}_1^\theta\}\,\overline{\mathbf{W}}\,, \qquad \boldsymbol{\tau}_3^\theta = \boldsymbol{\tau}^\theta(t+\Delta t) - \boldsymbol{\tau}_1^\theta - \boldsymbol{\tau}_2^\theta\,,$$

$$\boldsymbol{\tau}_4^\theta = \boldsymbol{\tau}_2^\theta \overline{\mathbf{W}} - \overline{\mathbf{W}} \boldsymbol{\tau}_2^\theta, \quad \boldsymbol{\tau}_5^\theta = \boldsymbol{\tau}_3^\theta \overline{\mathbf{W}} - \overline{\mathbf{W}} \boldsymbol{\tau}_3^\theta,$$
$$\overline{\mathbf{W}} = \mathbf{W}/\omega, \quad \omega = \sqrt{W_{12}^2 + W_{13}^2 + W_{23}^2}, \quad p = \theta\,\omega\,\Delta t. \tag{5.8.8a-h}$$

When $p = 0$, ($\theta = 0$ and/or $\omega = 0$), this results in

$$\boldsymbol{\tau}_A(t+\Delta t) = \boldsymbol{\tau}_1^\theta + \boldsymbol{\tau}_2^\theta + \boldsymbol{\tau}_3^\theta = \boldsymbol{\tau}^\theta(t+\Delta t). \tag{5.8.9a}$$

For small rotations, the estimate (5.8.7a) is accurate. Since the approximation (5.8.3) is also valid only for small rotations, the exact solution (5.8.7d) of (5.8.4b) is not necessary. In order to retain the simplicity of the *radial-return method*, only the approximate expression (5.8.7a) will be used in the remainder of this section.

In general, $\boldsymbol{\tau}_A(t+\Delta t)$ will not be on the yield surface for $p \neq 0$. To ensure that $\boldsymbol{\tau}(t+\Delta t)$ is on the yield surface, normalize $\boldsymbol{\tau}_A(t+\Delta t)$, as follows:

$$\boldsymbol{\tau}(t+\Delta t) = \frac{\tau^\theta}{\tau_A}\,\boldsymbol{\tau}_A(t+\Delta t), \quad \tau_A = \{\tfrac{1}{2}\boldsymbol{\tau}_A(t+\Delta t) : \boldsymbol{\tau}_A(t+\Delta t)\}^{1/2}. \tag{5.8.9b,c}$$

Note that, in the absence of the spin, *i.e.*, when $\mathbf{W} = \mathbf{0}$, this method is identical to the *radial-return method* discussed in Subsection 5.7.1.

It is seen in Section 5.5 that, in continued plastic flow, the stress orientation rotates to become coaxial with the corotational deformation rate. However, when the velocity gradient is constant, the corotational deformation rate is not a constant. Hence, when a constant velocity gradient is prescribed, the orientation of the stress varies as the corotational deformation rate varies. To illustrate this, first consider isotropic perfect plasticity with $\tau_y = \mu/100$. Consider the following problem where the material is initially subjected to biaxial deformation up to 2% shear strain,

$$\mathbf{L} = \begin{bmatrix} 1 & 0 & 0 \\ 0 & -1 & 0 \\ 0 & 0 & 0 \end{bmatrix} \times 10^{-2}, \quad 0 \le t < 2, \tag{5.8.10a}$$

and then is subjected to a simple shearing

$$\mathbf{L} = \begin{bmatrix} 0 & d+\omega & 0 \\ d-\omega & 0 & 0 \\ 0 & 0 & 0 \end{bmatrix}, \quad \omega = 1, \quad 2 \le t < 7. \tag{5.8.10b}$$

The results for a wide range of d/ω using the *generalized radial-return method* with very small timesteps are shown in Figure 5.8.1. The results show that the effect of spin on the stress orientation is negligible for $d/\omega \ge 1$. On the other hand, the effect of the deformation rate is negligible for $d/\omega < 0.001$. Both effects are significant in the range $0.001 < d/\omega < 1$. However, for $d/\omega \ge 0.01$, the stress orientation reaches its final value exponentially with negligible oscillation. For $d/\omega < 0.01$, the effect of spin dominates and therefore the stress orientation does not reach a final value, but oscillates. As the stress orientation oscillates, $\boldsymbol{\mu} : \mathbf{D}$ becomes negative for a certain period of time after a 90° rotation. The deformation is elastic during this time.

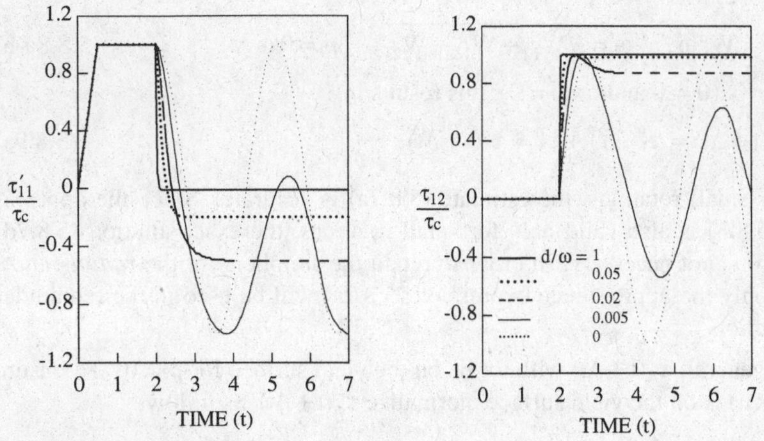

Figure 5.8.1

Time history of deviatoric stress components for indicated wide range of d/ω

Approximate Integration: Consider now a method which combines the approximation (5.8.3) for the spin effect and the exact results of Krieg and Krieg (1977) for the stress orientation. Hence, this method would yield exact results for the cases where $\mathbf{W} = \mathbf{0}$. Consider the integration of (5.8.2) from time t^e to $t^e + \Delta t^p$, where $\Delta t^p = \Delta t - \Delta t^e$, and Δt^e is the time it takes for the stress state to reach the yield surface. Hence $\tau(t^e) = \tau_y$. In this case, (5.8.3) is expressed as

$$\boldsymbol{\tau}^\theta(t^e+\xi) = \boldsymbol{\tau}^\theta(t^e) + 2\mu\,\mathbf{D}\,\xi - \sqrt{2}\,\mu\int_0^\xi \dot{\gamma}\,\boldsymbol{\mu}(t^e+\zeta)\,d\zeta\,,$$

$$\boldsymbol{\tau}^\theta(t^e) = \boldsymbol{\tau}(t^e) + (1-\theta)\,\{\mathbf{W}\,\boldsymbol{\tau}(t^e) - \boldsymbol{\tau}(t^e)\,\mathbf{W}\}\,\xi\,,$$

$$\boldsymbol{\tau}^\theta(t^e+\xi) = \boldsymbol{\tau}_A(t^e+\xi) - \theta\,\{\mathbf{W}\,\boldsymbol{\tau}_A(t^e+\xi) - \boldsymbol{\tau}_A(t^e+\xi)\,\mathbf{W}\}\,\xi\,, \qquad (5.8.11\text{a-c})$$

where $\boldsymbol{\tau}_A(t^e+\xi)$ is the (approximate) estimate of $\boldsymbol{\tau}$ at time $t^e+\xi$. Furthermore, the stress orientation $\boldsymbol{\mu}$ in the integrand of (5.8.11a) is approximated by

$$\boldsymbol{\mu} \approx \boldsymbol{\mu}^\theta(t^e+\xi) \equiv \frac{\boldsymbol{\tau}^\theta(t^e+\xi)}{\sqrt{2}\,\tau^\theta(t^e+\xi)}\,, \quad \tau^\theta \equiv \{\tfrac{1}{2}\,\boldsymbol{\tau}^\theta(t^e+\xi) : \boldsymbol{\tau}^\theta(t^e+\xi)\}^{1/2}\,. \qquad (5.8.12)$$

Then for perfect plasticity, (5.8.11a) is solved, to obtain (see (5.7.10c))

$$\boldsymbol{\tau}^\theta(t^e+\xi)\,z(\xi) = \boldsymbol{\tau}^\theta(t^e) + \frac{2\,\tau_y}{d}\,\{\sinh(A\,\xi) + x_0\cosh(A\,\xi) - x_0\}\,\mathbf{D}\,, \qquad (5.8.13\text{a})$$

$$z(\xi) = \cosh(A\,\xi) + x_0\sinh(A\,\xi)\,, \quad x_0 = \frac{\boldsymbol{\tau}^\theta(t^e) : \mathbf{D}}{d\,\tau_y}\,,$$

$$A = \mu\,d/\tau_y\,, \quad d = (2\,\mathbf{D} : \mathbf{D})^{1/2}\,. \qquad (5.8.13\text{b-e})$$

Once $\boldsymbol{\tau}^\theta(t^e+\xi)$ is known, $\boldsymbol{\tau}_A(t^e+\xi)$ is obtained by solving (5.8.11c); see (5.8.7) and (5.8.8). Then the final stress is estimated by

$$\boldsymbol{\tau}(t^e+\xi) = \sqrt{2}\,\tau_y\,\boldsymbol{\mu}_A(t^e+\xi)\,, \qquad \boldsymbol{\mu}_A(t^e+\xi) = \frac{\boldsymbol{\tau}_A(t^e+\xi)}{\sqrt{2}\,\tau_A(t^e+\xi)}\,,$$

$$\tau_A(t^e+\xi) = \{\tfrac{1}{2}\,\boldsymbol{\tau}_A(t^e+\xi) : \boldsymbol{\tau}_A(t^e+\xi)\}^{1/2}. \tag{5.8.14a-c}$$

A similar method has been presented by Nemat-Nasser and Li (1992) for both rate-dependent and rate-independent flows; see also Li (1993). They also use the approximation (5.8.3) and their formulation reduces to that of Krieg and Krieg (1977) for $\mathbf{W} = \mathbf{0}$. The above method is a modified version of Nemat-Nasser and Li (1992) but is consistent with the *generalized radial-return* method discussed above in this subsection. Note that, the formulation given by Krieg and Krieg (1977) for small deformations, *i.e.*, for $\mathbf{W} = \mathbf{0}$, is exact for perfect-plasticity models. Therefore, the results obtained by Nemat-Nasser and Li (1992) are very accurate for finite deformation with small rotations. It has been shown by Balendran and Nemat-Nasser (1995), that the effect of spin on the final results is significant, only if the spin $\mathbf{W}$ is orders of magnitude larger than the deformation rate $\mathbf{D}$. In such situations the approximation (5.8.3) is not accurate.

For large rotation, rewrite (5.8.2) as

$$\dot{\hat{\boldsymbol{\tau}}} = 2\,\mu\,(\hat{\mathbf{D}} - \dot{\gamma}\,\hat{\boldsymbol{\mu}}/\sqrt{2})\,, \tag{5.8.15a}$$

where, with $\mathbf{X}$ standing for any one of the second-order tensors in (5.8.15a),

$$\hat{\mathbf{X}}(\xi) = \mathbf{R}^T(\xi)\,\mathbf{X}\,\mathbf{R}(\xi)\,, \qquad \mathbf{R}(\xi) = \exp\{\mathbf{W}\,\xi\}\,. \tag{5.8.15b,c}$$

Nemat-Nasser and Li (1994) approximately integrate (5.8.15a), as follows:

$$\hat{\boldsymbol{\tau}}(t^e+\Delta t^p) = \hat{\boldsymbol{\tau}}(t^e) + 2\,\mu\left\{\overline{\mathbf{D}}\,\Delta t^p - \Delta\gamma \sum_{i=1}^{n} w_i\,\hat{\boldsymbol{\mu}}_A(t^e+\xi_i)\right\}, \tag{5.8.16a}$$

where[28]

$$\hat{\boldsymbol{\mu}}_A(t^e+\xi_i) = \mathbf{R}^T(\xi_i)\,\boldsymbol{\mu}_A(t^e+\xi_i)\,\mathbf{R}(\xi_i)\,, \quad 0 \le \xi_i \le \Delta t^p\,,$$

$$\overline{\mathbf{D}} = \frac{1}{\Delta t^p}\int_0^{\Delta t^p} \mathbf{R}^T(\xi)\,\mathbf{D}\,\mathbf{R}(\xi)\,d\xi$$

$$= \frac{1}{\Delta t^p}\{\mathbf{R}^T(\Delta t^p)\,\mathbf{X}\,\mathbf{R}(\Delta t^p) - \mathbf{X}\} + \frac{\mathrm{tr}(\mathbf{W}^2\,\mathbf{D})}{6\,\omega^4}\{3\,\mathbf{W}^2 + 2\,\omega^2\,\mathbf{1}\}\,,$$

$$\mathbf{X} = \frac{3}{4\,\omega^4}\,\mathbf{W}\,(\mathbf{D}\,\mathbf{W} - \mathbf{W}\,\mathbf{D})\,\mathbf{W} - \frac{1}{\omega^2}\,(\mathbf{D}\,\mathbf{W} - \mathbf{W}\,\mathbf{D})\,, \tag{5.8.16b-d}$$

w_i is the weighting function for the Gauss quadrature corresponding to ξ_i, and $\boldsymbol{\mu}_A(t^e+\xi_i)$ is an approximation of the stress orientation, obtained from (5.8.14b). The plastic strain increment $\Delta\gamma$ is obtained from the scalar equation

$$\tau(t^e+\Delta t^p) = \tau(t^e) + \mu\,(d^*\,\Delta t^p - \Delta\gamma)\,, \tag{5.8.17a}$$

---

28 The expression for $\overline{\mathbf{D}}$ which is simpler than the one given by Nemat-Nasser and Li (1994) follows from Appendix 5.A

where

$$d^* = \frac{1}{\Delta t^p} \int_0^{\Delta t^p} \sqrt{2}\,\mathbf{D} : \boldsymbol{\mu}(t^e + \xi)\, d\xi \approx \sum_{i=1}^{n} w_i \sqrt{2}\,\mathbf{D} : \boldsymbol{\mu}_A(t^e + \xi_i)\,. \tag{5.8.17b}$$

For this perfect-plasticity model, $\Delta\gamma$ is then given by

$$\Delta\gamma = d^* \, \Delta t^p\,. \tag{5.8.17c}$$

Nemat-Nasser and Li (1994) observe that one Gauss point is sufficient for accurate results, *i.e.*, $n = 1$, $\xi_1 = \Delta t^p/2$, and $w_1 = 1$. Once, $\hat{\boldsymbol{\tau}}(t^e + \Delta t^p)$ is known, $\boldsymbol{\tau}(t^e + \Delta t^p)$ is obtained from

$$\boldsymbol{\tau}(t^e + \Delta t^p) = \mathbf{R}(\Delta t^p)\, \hat{\boldsymbol{\tau}}(t^e + \Delta t^p)\, \mathbf{R}^T(\Delta t^p)\,. \tag{5.8.18}$$

The approximate expression for the final value of the stress tensor (5.8.14a) is accurate only for small rotations. For large rotations this is improved in (5.8.16)-(5.8.18). However, since the expression for $\boldsymbol{\mu}_A$ may be inaccurate in (5.8.16), the accuracy of (5.8.18) is limited. In order to develop an accurate and effective method, it is important to obtain the exact solution and then make reasonable approximations to simplify the numerical procedure. In the remainder of this subsection, first the exact integration for the perfect-plasticity model is developed, and based on this, an approximate but effective method is developed.

Exact Integration: In line with Balendran and Nemat-Nasser (1995), consider the exact integration of (5.8.2), from time t^e to $t^e + \Delta t^p$, with $\tau(t^e) = \tau_y$. In this timestep,

$$\boldsymbol{\tau}(t^e + \xi) = \sqrt{2}\, \tau_y\, \boldsymbol{\mu}(t^e + \xi)\,. \tag{5.8.19}$$

In a coordinate system which rotates with spin $\mathbf{W}$, (5.8.2) reduces to

$$\dot{\hat{\boldsymbol{\tau}}} = 2\,\mu\,(\hat{\mathbf{D}} - \dot{\gamma}\, \hat{\boldsymbol{\mu}}/\sqrt{2})\,, \tag{5.8.20a}$$

where

$$\hat{\boldsymbol{\tau}} = \mathbf{R}^T(\xi)\, \boldsymbol{\tau}\, \mathbf{R}(\xi), \quad \hat{\mathbf{D}} = \mathbf{R}^T(\xi)\, \mathbf{D}\, \mathbf{R}(\xi), \quad \hat{\boldsymbol{\mu}} = \mathbf{R}^T(\xi)\, \boldsymbol{\mu}\, \mathbf{R}(\xi)\,,$$

$$\mathbf{R}(\xi) = \exp\{\mathbf{W}\,\xi\}\,. \tag{5.8.21a-d}$$

In view of (5.8.20a), the time derivative of $2\,\tau^2 = \boldsymbol{\tau} : \boldsymbol{\tau} = \hat{\boldsymbol{\tau}} : \hat{\boldsymbol{\tau}}$ results in

$$\dot{\tau} = \mu\,(\sqrt{2}\,\mathbf{D} : \hat{\boldsymbol{\mu}} - \dot{\gamma})\,. \tag{5.8.20b}$$

Then, in view of (5.8.20a,b), the time derivative of $\hat{\boldsymbol{\mu}} = \dfrac{\hat{\boldsymbol{\tau}}}{\sqrt{2}\,\tau}$ yields

$$\dot{\hat{\boldsymbol{\mu}}} = A\,\{\hat{\mathbf{v}} - (\hat{\mathbf{v}} : \hat{\boldsymbol{\mu}})\,\hat{\boldsymbol{\mu}}\}\,, \tag{5.8.20c}$$

$$\hat{\mathbf{v}} = \frac{\sqrt{2}\,\hat{\mathbf{D}}}{d}, \quad d = (2\,\hat{\mathbf{D}} : \hat{\mathbf{D}})^{1/2} = (2\,\mathbf{D} : \mathbf{D})^{1/2}, \quad A = \mu\, d/\tau_y\,. \tag{5.8.21e-g}$$

Differential equation (5.8.20c) can be rearranged as

$$\frac{d(z\,\hat{\boldsymbol{\mu}})}{d\xi} = A\, z\, \hat{\mathbf{v}} \tag{5.8.22}$$

which can be integrated to obtain

$$z(\xi)\,\hat{\boldsymbol{\mu}}(t^e+\xi) = \hat{\boldsymbol{\mu}}(t^e) + A\int_0^{\xi} z(\zeta)\,\hat{\mathbf{v}}(\zeta)\,d\zeta\,, \tag{5.8.23}$$

$$z(\xi) = \exp\left\{A\int_0^{\xi} \hat{\mathbf{v}}(\zeta) : \hat{\boldsymbol{\mu}}(t^e+\zeta)\,d\zeta\right\}. \tag{5.8.24}$$

Taking the scalar product of (5.8.23) with $\hat{\mathbf{v}}(\xi)$, obtain

$$\frac{1}{A}\,\dot{z} = x(\xi) + A\int_0^{\xi} z(\zeta)\,y(\xi-\zeta)\,d\zeta\,,$$

$$y(\xi-\zeta) = \hat{\mathbf{v}}(\xi) : \hat{\mathbf{v}}(\zeta) = \mathbf{v} : \hat{\mathbf{v}}(\xi-\zeta)\,, \quad x(\xi) = \boldsymbol{\mu}(t^e) : \hat{\mathbf{v}}(\xi)\,. \tag{5.8.25a-c}$$

Applying the Laplace transform to (5.8.25a) and since $z(0) = 1$, arrive at

$$\tilde{z}(s) = \frac{A\,\tilde{x}+1}{s-A^2\,\tilde{y}}\,, \tag{5.8.26}$$

where the super tilde denotes the Laplace transform of the corresponding function.

The inverse transform of (5.8.26) gives $z(\xi)$. Substitution of that into (5.8.23) results in an expression for $\hat{\boldsymbol{\mu}}(\xi)$. Substituting

$$\mathbf{R}(\xi) = \exp\{\mathbf{W}\,\xi\} = \mathbf{1} + \overline{\mathbf{W}}^2(1-\cos\omega\,\xi) + \overline{\mathbf{W}}\sin\omega\,\xi\,,$$

$$\overline{\mathbf{W}} = \mathbf{W}/\omega\,, \tag{5.8.27a,b}$$

into (5.8.21b), arrive at

$$\hat{\mathbf{v}}(\xi) = \mathbf{v}_1 + \mathbf{v}_2\cos 2\omega\,\xi + \mathbf{v}_3\cos\omega\,\xi + \frac{1}{2}\,\mathbf{v}_4\sin 2\omega\,\xi + \mathbf{v}_5\sin\omega\,\xi\,,$$

$$\mathbf{v}_1 = (\mathbf{v} : \overline{\mathbf{W}}^2)\,(3/2\,\overline{\mathbf{W}}^2 + \mathbf{1})\,,$$

$$\mathbf{v}_2 = \overline{\mathbf{W}}\,(\mathbf{v}-\mathbf{v}_1)\,\overline{\mathbf{W}}\,, \quad \mathbf{v}_3 = \mathbf{v} - \mathbf{v}_1 - \mathbf{v}_2\,,$$

$$\mathbf{v}_4 = \mathbf{v}_2\,\overline{\mathbf{W}} - \overline{\mathbf{W}}\,\mathbf{v}_2\,, \quad \mathbf{v}_5 = \mathbf{v}_3\,\overline{\mathbf{W}} - \overline{\mathbf{W}}\,\mathbf{v}_3\,. \tag{5.8.28a-f}$$

Substituting (5.8.28) into (5.8.25b,c) and applying the Laplace transform, obtain

$$\tilde{x} = \frac{x_{10}}{s} + \frac{s\,x_{20}}{s^2+4\,\omega^2} + \frac{s\,x_{30}}{s^2+\omega^2} + \frac{2\,\omega\,x_{40}}{s^2+4\,\omega^2} + \frac{\omega\,x_{50}}{s^2+\omega^2}\,,$$

$$\tilde{y} = \frac{\alpha_1}{s} + \frac{s\,\alpha_2}{s^2+4\,\omega^2} + \frac{s\,\alpha_3}{s^2+\omega^2}\,, \tag{5.8.29a,b}$$

$$x_{i0} = \boldsymbol{\mu}(t^e) : \mathbf{v}_i\,, \quad \alpha_i = \mathbf{v}_i : \mathbf{v}_i\,, \quad i = 1, 2, \ldots, 5\,. \tag{5.8.30a,b}$$

Substituting (5.8.29) into (5.8.26), arrive at

$$\tilde{z} = \frac{f_2(s^2) + s\,f_3(s^2)}{f_1(s^2)}\,, \tag{5.8.31a}$$

where the functions f_1, f_2, and f_3 are defined by

$$f_1(s^2) \equiv (s^2-A^2)\,(s^2+4\,\omega^2)\,(s^2+\omega^2)$$

$$+ A^2 \omega^2 \alpha_3 (s^2 + 4\omega^2) + 4A^2 \omega^2 \alpha_2 (s^2 + \omega^2) = 0,$$

$$f_2(s^2) \equiv A x_{10} (s^2 + \omega^2)(s^2 + 4\omega^2)$$
$$+ A x_{30} s^2 (s^2 + 4\omega^2) + A x_{20} s^2 (s^2 + \omega^2),$$

$$f_3(s^2) \equiv A \omega x_{50} (s^2 + 4\omega^2)$$
$$+ A \omega x_{40} (s^2 + \omega^2) + (s^2 + \omega^2)(s^2 + 4\omega^2). \qquad (5.8.31b\text{-}d)$$

Since α_1, α_2, $\alpha_3 \geq 0$ and $\alpha_1 + \alpha_2 + \alpha_3 = 1$, (Appendix 5.A), $f_1(s^2)$ can be factored,

$$f_1(s^2) = (s^2 - \hat{\mu}^2)(s^2 + \lambda^2)(s^2 + \theta^2), \qquad (5.8.32a)$$

such that

$$4\omega^2 > \lambda^2 > \omega^2, \quad \omega^2 > \theta^2 > 0, \quad 0 \leq \hat{\mu}^2 \leq A^2. \qquad (5.8.32b\text{-}d)$$

Then the inverse transform of (5.8.26) yields

$$z(\xi) = a_1 \sinh \hat{\mu}\xi / \hat{\mu} + a_2 \sin \lambda\xi / \lambda + a_3 \sin \theta\xi / \theta$$
$$+ a_4 \cosh \hat{\mu}\xi + a_5 \cos \lambda\xi + a_6 \cos \theta\xi,$$

$$a_1 = f_2(\hat{\mu}^2) / \{(\hat{\mu}^2 + \lambda^2)(\hat{\mu}^2 + \theta^2)\},$$

$$a_2 = \begin{cases} f_2(-\lambda^2)(\theta^2 + \hat{\mu}^2)/\Delta & \text{for } \lambda \neq \theta \\ A\omega^2 (x_{20} - 3x_{10})/(\hat{\mu}^2 + \omega^2) & \text{for } \lambda = \theta, \end{cases}$$

$$a_3 = \begin{cases} -f_2(-\theta^2)(\hat{\mu}^2 + \lambda^2)/\Delta & \text{for } \lambda \neq \theta \\ 0 & \text{for } \lambda = \theta, \end{cases}$$

$$a_4 = f_3(\hat{\mu}^2) / \{(\hat{\mu}^2 + \lambda^2)(\hat{\mu}^2 + \theta^2)\},$$

$$a_5 = \begin{cases} f_3(-\lambda^2)(\theta^2 + \hat{\mu}^2)/\Delta & \text{for } \lambda \neq \theta \\ \omega(3\omega - 2A x_{40})/(\hat{\mu}^2 + \omega^2) & \text{for } \lambda = \theta, \end{cases}$$

$$a_6 = \begin{cases} -f_3(-\theta^2)(\hat{\mu}^2 + \lambda^2)/\Delta & \text{for } \lambda \neq \theta \\ 0 & \text{for } \lambda = \theta, \end{cases}$$

$$\Delta = (\hat{\mu}^2 + \lambda^2)(\hat{\mu}^2 + \theta^2)(\lambda^2 - \theta^2). \qquad (5.8.33a\text{-}h)$$

It follows from Appendix 5.A that

$$\mathbf{v}_1 \overline{\mathbf{W}} - \overline{\mathbf{W}} \mathbf{v}_1 = \mathbf{0},$$
$$\mathbf{v}_2 \overline{\mathbf{W}} - \overline{\mathbf{W}} \mathbf{v}_2 = \mathbf{v}_4, \quad \mathbf{v}_3 \overline{\mathbf{W}} - \overline{\mathbf{W}} \mathbf{v}_3 = \mathbf{v}_5,$$
$$\mathbf{v}_4 \overline{\mathbf{W}} - \overline{\mathbf{W}} \mathbf{v}_4 = -4\mathbf{v}_2, \quad \mathbf{v}_5 \overline{\mathbf{W}} - \overline{\mathbf{W}} \mathbf{v}_5 = -\mathbf{v}_3. \qquad (5.8.34a\text{-}e)$$

The time derivative of

$$\hat{\mathbf{v}}_i(\xi) \equiv \mathbf{R}^T(\xi)\, \mathbf{v}_i\, \mathbf{R}(\xi) \qquad (5.8.35a)$$

results in

$$\dot{\hat{\mathbf{v}}}_i = \mathbf{R}^T (\mathbf{v}_i \mathbf{W} - \mathbf{W} \mathbf{v}_i) \mathbf{R}. \qquad (5.8.35b)$$

Hence, it follows from (5.8.34) and (5.8.35), that

$$\dot{\hat{\mathbf{v}}}_1 = \mathbf{0}, \quad \dot{\hat{\mathbf{v}}}_2 = \omega\,\hat{\mathbf{v}}_4, \quad \dot{\hat{\mathbf{v}}}_4 = -4\,\omega\,\hat{\mathbf{v}}_2,$$
$$\dot{\hat{\mathbf{v}}}_3 = \omega\,\hat{\mathbf{v}}_5, \quad \dot{\hat{\mathbf{v}}}_5 = -\omega\,\hat{\mathbf{v}}_3. \tag{5.8.36a-e}$$

Using these relations, the following integral can be easily evaluated:

$$\int_0^{\xi} e^{\alpha\zeta}\,\hat{\mathbf{v}}(\zeta)\,d\zeta = \int_0^{\xi} e^{\alpha\zeta}\,\{\hat{\mathbf{v}}_1(\zeta) + \hat{\mathbf{v}}_2(\zeta) + \hat{\mathbf{v}}_3(\zeta)\}\,d\zeta$$
$$= e^{\alpha\xi}\left\{\frac{1}{\alpha}\,\hat{\mathbf{v}}_1(\xi) + \frac{\alpha}{\alpha^2+4\,\omega^2}\,\hat{\mathbf{v}}_2(\xi) + \frac{\alpha}{\alpha^2+\omega^2}\,\hat{\mathbf{v}}_3(\xi) - \frac{\omega}{\alpha^2+4\,\omega^2}\,\hat{\mathbf{v}}_4(\xi) - \frac{\omega}{\alpha^2+\omega^2}\,\hat{\mathbf{v}}_5(\xi)\right\}$$
$$- \left\{\frac{1}{\alpha}\,\mathbf{v}_1 + \frac{\alpha}{\alpha^2+4\,\omega^2}\,\mathbf{v}_2 + \frac{\alpha}{\alpha^2+\omega^2}\,\mathbf{v}_3 - \frac{\omega}{\alpha^2+4\,\omega^2}\,\mathbf{v}_4 - \frac{\omega}{\alpha^2+\omega^2}\,\mathbf{v}_5\right\}. \tag{5.8.36f}$$

In a similar manner, the resulting integral when (5.8.33) is substituted into (5.8.23), is evaluated, to arrive at

$$\hat{\boldsymbol{\mu}}(t^e+\xi) = \frac{1}{z(\xi)}\,\{\boldsymbol{\mu}(t^e) + A\,z_i(\xi)\,\hat{\mathbf{v}}_i - A\,z_i(0)\,\mathbf{v}_i\}, \tag{5.8.37}$$

where $i = 1, 2, ..., 5$ (i summed), and z_i's are given by

$$\begin{Bmatrix} z_1(\xi) \\ z_2(\xi) \\ z_3(\xi) \end{Bmatrix} = \begin{bmatrix} \frac{1}{\hat{\mu}^2} & \frac{1}{\lambda^2} & \frac{1}{\theta^2} \\ \frac{1}{\hat{\mu}^2+4\,\omega^2} & \frac{1}{\lambda^2-4\,\omega^2} & \frac{1}{\theta^2-4\,\omega^2} \\ \frac{1}{\hat{\mu}^2+\omega^2} & \frac{1}{\lambda^2-\omega^2} & \frac{1}{\theta^2-\omega^2} \end{bmatrix} \begin{Bmatrix} a_1 \cosh\hat{\mu}\,\xi + a_4\,\hat{\mu}\sinh\hat{\mu}\,\xi \\ -a_2\cos\lambda\,\xi + a_5\,\lambda\sin\lambda\,\xi \\ -a_3\cos\theta\,\xi + a_6\,\theta\sin\theta\,\xi \end{Bmatrix},$$

$$\begin{Bmatrix} z(\xi) \\ z_4(\xi) \\ z_5(\xi) \end{Bmatrix} = \begin{bmatrix} 1 & 1 & 1 \\ \frac{-\omega}{\hat{\mu}^2+4\,\omega^2} & \frac{\omega}{\lambda^2-4\,\omega^2} & \frac{\omega}{\theta^2-4\,\omega^2} \\ \frac{-\omega}{\hat{\mu}^2+\omega^2} & \frac{\omega}{\lambda^2-\omega^2} & \frac{\omega}{\theta^2-\omega^2} \end{bmatrix} \begin{Bmatrix} a_1 \sinh\hat{\mu}\,\xi/\hat{\mu} + a_4\cosh\hat{\mu}\,\xi \\ a_2\sin\lambda\,\xi/\lambda + a_5\cos\lambda\,\xi \\ a_3\sin\theta\,\xi/\theta + a_6\cos\theta\,\xi \end{Bmatrix}.$$

$$\tag{5.8.38a-d}$$

Approximate but Effective Integration Procedure: The lengthy procedure presented above for an exact solution is inevitable due to the presence of $\hat{\mathbf{v}}$, a variable, in (5.8.20c). When $\dot{\hat{\mathbf{v}}}$ is constant, the exact integration method of Subsection 5.7.1, which is rather simple, can be used. It is evident from (5.8.23) that the actual solution $\hat{\boldsymbol{\mu}}(t^e+\xi)$ lies on a plane formed by $\hat{\boldsymbol{\mu}}(t^e)$ and the time average of $z\,\hat{\mathbf{v}}$. Therefore, an approximate but accurate solution is obtained, if a suitable approximate expression for z is used in (5.8.23). The exact solution for z in (5.8.33a) consists of an exponential and two sinusoidal terms. For a large time increment, ξ, the effect of sinusoidal terms may be neglected in comparison with the exponential terms. This is also true for small time increments when $\hat{\mu}$ is very large compared to ω. Then $\hat{\mu}$ and z can be approximated by

$$\hat{\mu} = A\left[1 - \frac{4\,\alpha_2\,\omega^2}{A^2 + 4\,\omega^2} - \frac{\alpha_3\,\omega^2}{A^2 + \omega^2}\right]^{1/2},$$

$$z(\xi) = \cosh\hat{\mu}\,\xi + k\sinh\hat{\mu}\,\xi, \quad k = A\,x_0/\hat{\mu}, \quad x_0 = \boldsymbol{\nu} : \boldsymbol{\mu}(t^e). \tag{5.8.39a-d}$$

Substitute (5.8.39) into (5.8.23) and use (5.8.35), to obtain

$$\boldsymbol{\mu}(t^e + \Delta t^p) = \frac{1}{z(\Delta t^p)}\Big\{\tilde{\boldsymbol{\mu}}_{t^e} + A\,(\sinh\hat{\mu}\,\Delta t^p + k\cosh\hat{\mu}\,\Delta t^p)\,\mathbf{X}_1 - A\,k\,\tilde{\mathbf{X}}_1$$

$$- B\,(\cosh\hat{\mu}\,\Delta t^p + k\sinh\hat{\mu}\,\Delta t^p)\,\mathbf{X}_2 + B\,\tilde{\mathbf{X}}_2\Big\}, \tag{5.8.40}$$

$$\tilde{\boldsymbol{\mu}}_{t^e} = \mathbf{R}(\Delta t^p)\,\boldsymbol{\mu}(t^e)\,\mathbf{R}^T(\Delta t^p),$$

$$\tilde{\mathbf{X}}_i = \mathbf{R}(\Delta t^p)\,\mathbf{X}_i\,\mathbf{R}^T(\Delta t^p), \quad i = 1, 2,$$

$$\mathbf{X}_1 = \boldsymbol{\nu}_1/\hat{\mu} + \hat{\mu}\,\boldsymbol{\nu}_2/(\hat{\mu}^2 + 4\,\omega^2) + \hat{\mu}\,\boldsymbol{\nu}_3/(\hat{\mu}^2 + \omega^2),$$

$$\mathbf{X}_2 = \mathbf{X}_1\,\overline{\mathbf{W}} - \overline{\mathbf{W}}\,\mathbf{X}_1, \quad B = A\,\omega/\hat{\mu}. \tag{5.8.41a-e}$$

Note that, if $\omega = 0$, (5.8.40) reduces to the exact solution (5.7.10c). When $d = 0$, (5.8.40) reduces to $\boldsymbol{\mu}(t^e + \Delta t^p) = \tilde{\boldsymbol{\mu}}_{t^e}$ which is also exact. Therefore, even though (5.8.40) is approximate, it yields the exact solutions for both extreme cases, *i.e.*, pure distortion and rigid body rotation.

5.8.2. Rate-independent Model

Given a constant velocity gradient, $\mathbf{L} = \mathbf{D} + \mathbf{W}$, over a timestep t to $t + \Delta t$, consider the integration of the constitutive equations (5.8.1) for the rate-independent flow rule (5.3.7a), with the yield stress given by (5.3.10a).

Computational Steps for Generalized Radial-return Method: The simplest and most efficient way is to approximate the final orientation of the stress difference by the *generalized radial-return method* of Subsection 5.8.1, and then obtain the stress magnitude using the *plastic-predictor/elastic-corrector method* of Subsection 5.6.1. The necessary computational steps are summarized below.

Step 1

Obtain the initial values of the stress difference and the yield stress,

$$\mathbf{s}(t) = \boldsymbol{\tau}(t) - \boldsymbol{\beta}(t), \quad \tau(t) = \{\tfrac{1}{2}\,\mathbf{s}(t) : \mathbf{s}(t)\}^{1/2}, \quad \tau_y(t) = F(\gamma(t)). \tag{5.8.42a-c}$$

Step 2

Calculate the stress difference and backstress for the initial rotation $(1 - \theta)\,\mathbf{W}\,\Delta t$,

$$\mathbf{s}^\theta(t) = \mathbf{s}(t) + (1 - \theta)\,\{\mathbf{W}\,\mathbf{s}(t) - \mathbf{s}(t)\,\mathbf{W}\}\,\Delta t, \quad \tau^\theta = (\mathbf{s}^\theta : \mathbf{s}^\theta/2)^{1/2},$$

$$\boldsymbol{\beta}^\theta(t) = \boldsymbol{\beta}(t) + (1-\theta)\,\{\mathbf{W}\,\boldsymbol{\beta}(t) - \boldsymbol{\beta}(t)\,\mathbf{W}\}\,\Delta t\,. \tag{5.8.43a-c}$$

Step 3

Obtain the trial stress difference and its magnitude

$$\mathbf{s}^\alpha = \mathbf{s}^\theta(t) + 2\,\mu\,\mathbf{D}\,\Delta t\,, \quad \tau^\alpha = (\tfrac{1}{2}\,\mathbf{s}^\alpha : \mathbf{s}^\alpha)^{1/2}\,. \tag{5.8.44a,b}$$

Step 4

Estimate the fraction of the timestep during which the deformation is plastic and correct the trial stress difference for noncoaxiality,

$$\Delta t^p = \max\left\{\frac{\tau^\alpha - \tau_y(t)}{\tau^\alpha - \tau^\theta}\,\Delta t,\ 0\right\},$$

$$\mathbf{s}^{tr} = \mathbf{s}^\alpha + \mu\,\bar{\alpha}\,\{2\,\mathbf{D} - (\mathbf{s}^\alpha : \mathbf{D})\,\mathbf{s}^\alpha/(\tau^\alpha)^2\}\,\Delta t^p\,,$$

$$\tau^{tr} = (\tfrac{1}{2}\,\mathbf{s}^{tr} : \mathbf{s}^{tr})^{1/2}\,. \tag{5.8.44c-e}$$

Step 5

Estimate the plastic strain, yield stress, increment in the backstress, and the isotropic and kinematic hardening moduli at the end of the timestep,

$$\Delta\gamma_A = \max\left\{0,\ \frac{\tau^{tr} - \tau_y(t)}{\mu}\right\}, \quad \gamma_A = \gamma(t) + \Delta\gamma_A\,, \quad \tau_A = F(\gamma_A)\,,$$

$$\Delta\beta_A = \int_{\gamma(t)}^{\gamma_A} \Lambda(\gamma)\,d\gamma\,, \quad H_A = H(\gamma_A)\,, \quad \Lambda_A = \Lambda(\gamma_A)\,. \tag{5.8.45a-f}$$

Step 6

Estimate the error in the estimated plastic strain increment and correct it,

$$\Delta\gamma_{er} = \frac{\tau_A - \tau_y(t) + \Delta\beta_A}{\mu + H_A + \Lambda_A}\,, \quad \Delta\gamma = \Delta\gamma_A - \Delta\gamma_{er}\,. \tag{5.8.46a,b}$$

Step 7

Obtain the final values of the equivalent plastic strain, stress difference, and the backstress,

$$\gamma(t+\Delta t) = \gamma(t) + \Delta\gamma\,, \quad \mathbf{s}^\theta(t+\Delta t) = \{1 - (\mu\,\Delta\gamma + \Delta\beta_A - \Lambda_A\,\Delta\gamma_{er})/\tau^{tr}\}\,\mathbf{s}^{tr}\,,$$

$$\boldsymbol{\beta}^\theta(t+\Delta t) = \boldsymbol{\beta}(t) + \{(\Delta\beta_A - \Lambda_A\,\Delta\gamma_{er})/\tau^{tr}\}\,\mathbf{s}^{tr}\,. \tag{5.8.47a-c}$$

Step 8

Calculate the stress difference, backstress, and the total stress for the rotation $\theta\,\mathbf{W}\,\Delta t$,

$$\mathbf{s}_A(t+\Delta t) = \mathbf{s}^\theta(t+\Delta t) + \theta\,\{\mathbf{W}\,\mathbf{s}^\theta(t+\Delta t) - \mathbf{s}^\theta(t+\Delta t)\,\mathbf{W}\}\,,$$

$$\boldsymbol{\beta}(t+\Delta t) = \boldsymbol{\beta}^\theta(t+\Delta t) + \theta\,\{\mathbf{W}\,\boldsymbol{\beta}^\theta(t+\Delta t) - \boldsymbol{\beta}^\theta(t+\Delta t)\,\mathbf{W}\}\,,$$

$$\tau_A(t+\Delta t) = \{\mathbf{s}_A(t+\Delta t) : \mathbf{s}_A(t+\Delta t)/2\}^{1/2}\,.$$

$$\mathbf{s}(t+\Delta t) = \frac{\tau^{\theta}(t+\Delta t)}{\tau_A(t+\Delta t)}\,\mathbf{s}_A(t+\Delta t)\,,$$

$$\boldsymbol{\tau}(t+\Delta t) = \mathbf{s}(t+\Delta t) + \boldsymbol{\beta}(t+\Delta t)\,. \qquad (5.8.48a\text{-}e)$$

As an illustration, consider the isotropic hardening with the yield function given by (5.6.7). As an example for a piecewise constant velocity gradient, consider the following time histories:

$$\mathbf{L} = \begin{bmatrix} 1 & 0 & 0 \\ 0 & -1 & 0 \\ 0 & 0 & 0 \end{bmatrix} \times 10^{-2}, \quad 0 \le t < 50,$$

$$\mathbf{L} = \begin{bmatrix} 1 & 101 & 0 \\ -99 & -1 & 0 \\ 0 & 0 & 0 \end{bmatrix} \times 10^{-4}, \quad 50 \le t < 100,$$

$$\mathbf{L} = \begin{bmatrix} 0 & 0 & 0 \\ 0 & 0 & 2 \\ 0 & 0 & 0 \end{bmatrix} \times 10^{-2}, \quad 100 \le t < 150. \qquad (5.8.49a\text{-}c)$$

These velocity gradients are chosen such that in the first stage a pure distortion is followed by a simultaneous large rotation and small distortion which is again followed by a simple shearing. The components of the deviatoric stress $\boldsymbol{\tau}'$ obtained using the *generalized radial-return method* with various numbers of timesteps are shown in Figure 5.8.2. As is seen, the generalized radial-return method gives very accurate results for the stress magnitude (effective stress) even when large timesteps are used. However, when the rotation is orders of magnitude larger than the distortion, small timesteps are needed to achieve accurate results.

Though the *generalized radial-return method* is simple and efficient, it is not accurate for large strain increments and rotations. In these cases, an accurate method is to obtain the final orientation of the stress difference using the *perfect-plasticity path*, and then the magnitude of the stress using the *plastic-predictor/elastic-corrector method*. An efficient method to evaluate the final orientation of the stress difference is to use the *approximate but effective integration procedure* of Subsection 5.8.1. The necessary computational steps for this method are summarized below.

Computational Steps for the Perfect-plasticity Path Method:

Step 1

Obtain the initial values of the stress difference and yield stress,

$$\mathbf{s}(t) = \boldsymbol{\tau}(t) - \boldsymbol{\beta}(t)\,, \quad \tau(t) = \{\tfrac{1}{2}\,\mathbf{s}(t) : \mathbf{s}(t)\}^{1/2}\,, \quad \tau_y(t) = F(\gamma(t))\,. \qquad (5.8.50a\text{-}c)$$

Step 2

Calculate the elastic-deformation time increment and the stress difference at the end of the elastic deformation (Subsection 5.7.4),

$$\mathbf{K} = 2\mu\,\mathbf{D} + \mathbf{W}\,\mathbf{s}(t) - \mathbf{s}(t)\,\mathbf{W}\,,$$

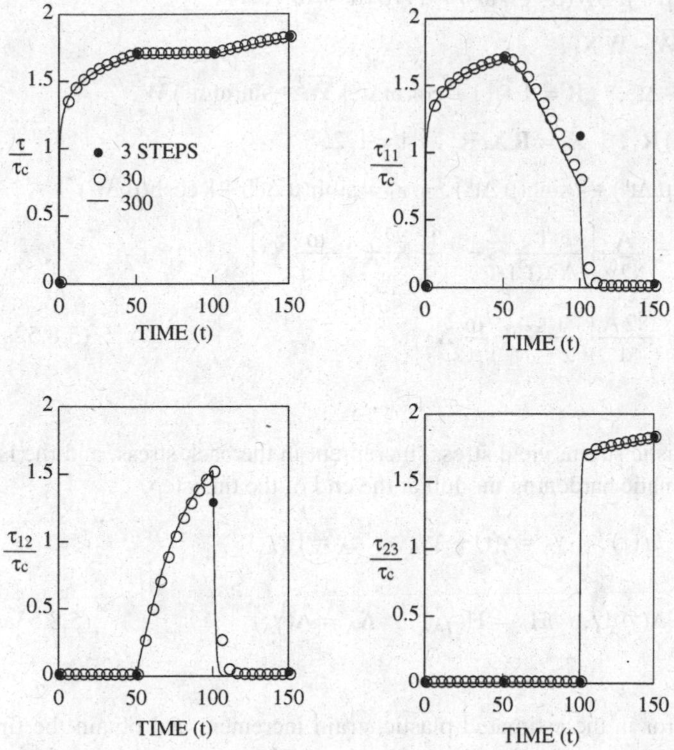

Figure 5.8.2

Time history of deviatoric stress components using the *generalized radial-return method*

$$\Delta t^e = \min\left\{ \Delta t,\ \frac{\sqrt{(\mathbf{s}_t : \mathbf{K})^2 + 2\,(\mathbf{K} : \mathbf{K})\,(\tau_y^2 - \tau_t^2)} - \mathbf{s}_t : \mathbf{K}}{(\mathbf{K} : \mathbf{K})} \right\},$$

$$\mathbf{s}(t^e) = \mathbf{s}(t) + \mathbf{K}\,\Delta t^e . \qquad (5.8.51a\text{-}c)$$

Step 3

Calculate the final orientation of the stress difference by assuming perfect plasticity,

$$d = (2\,\mathbf{D} : \mathbf{D})^{1/2}, \quad \omega = (-\,\mathbf{W} : \mathbf{W}/2)^{1/2} = \{W_{12}^2 + W_{23}^2 + W_{31}^2\}^{1/2},$$

$$\overline{\mathbf{W}} = \mathbf{W}/\omega, \quad \mathbf{D}_1 = (\mathbf{D} : \overline{\mathbf{W}}^2)\,(3/2\,\overline{\mathbf{W}}^2 + \mathbf{1}),$$

$$\mathbf{D}_2 = \overline{\mathbf{W}}\,(\mathbf{D} - \mathbf{D}_1)\,\overline{\mathbf{W}}, \quad \mathbf{D}_3 = \mathbf{D} - \mathbf{D}_1 - \mathbf{D}_2,$$

$$\alpha_2 = 2\,\mathbf{D}_2 : \mathbf{D}_2/d^2, \quad \alpha_3 = 2\,\mathbf{D}_3 : \mathbf{D}_3/d^2, \quad A = \mu\, d\,(1 - \overline{\alpha})/\tau(t^e),$$

$$\hat{\mu} = A\left[1 - \frac{4\,\alpha_2\,\omega^2}{A^2 + 4\,\omega^2} - \frac{\alpha_3\,\omega^2}{A^2 + \omega^2}\right]^{1/2}, \quad k = \frac{A\,\mathbf{D} : \mathbf{s}(t^e)}{d\,\tau(t^e)\,\hat{\mu}},$$

$$\mathbf{X}_1 = \mathbf{D}_1/\hat{\mu} + \hat{\mu}\,\mathbf{D}_2/(\hat{\mu}^2 + 4\,\omega^2) + \hat{\mu}\,\mathbf{D}_3/(\hat{\mu}^2 + \omega^2)\,,$$
$$\mathbf{X}_2 = \mathbf{X}_1\,\overline{\mathbf{W}} - \overline{\mathbf{W}}\,\mathbf{X}_1\,,$$
$$\Delta t^p = \Delta t - \Delta t^e, \qquad \mathbf{R} = \mathbf{1} + \{1 - \cos\omega\Delta t^p\}\,\overline{\mathbf{W}}^2 + \sin(\omega\Delta t^p)\,\overline{\mathbf{W}}\,,$$
$$\tilde{\mathbf{s}} = \mathbf{R}\,\mathbf{s}(t^e)\,\mathbf{R}^T, \qquad \tilde{\mathbf{X}}_i = \mathbf{R}\,\mathbf{X}_i\,\mathbf{R}^T, \quad i = 1,2,$$
$$z = \cosh(\hat{\mu}\,\Delta t^p) + k\sinh(\hat{\mu}\,\Delta t^p)\,, \qquad z_1 = \sinh(\hat{\mu}\,\Delta t^p) + k\cosh(\hat{\mu}\,\Delta t^p)\,,$$
$$\boldsymbol{\mu}(t+\Delta t) = \frac{A}{\sqrt{2}\,z}\left[\frac{1}{A\,\tau(t^e)}\tilde{\mathbf{s}} - 2\,\frac{k}{d}\,\tilde{\mathbf{X}}_1 + 2\,\frac{\omega}{\hat{\mu}\,d}\,\tilde{\mathbf{X}}_2\right]$$
$$+ \frac{\sqrt{2}\,A}{d}\left[\frac{z_1}{z}\,\mathbf{X}_1 - \frac{\omega}{\mu}\,\mathbf{X}_2\right]. \qquad (5.8.52a\text{-}t)$$

Step 4

Estimate the plastic strain, yield stress, increment in the back stress, and the isotropic and kinematic hardening moduli at the end of the timestep,

$$\Delta\gamma_A = \frac{d}{A}\,ln\,(z)\,, \qquad \gamma_A = \gamma(t) + \Delta\gamma_A\,, \qquad \tau_A = F(\gamma_A)\,,$$
$$\Delta\beta_A = \int_{\gamma(t)}^{\gamma_A} \Lambda(\gamma)\,d\gamma, \qquad H_A = H(\gamma_A)\,, \qquad \Lambda_A = \Lambda(\gamma_A)\,. \qquad (5.8.53a\text{-}f)$$

Step 5

Estimate the error in the estimated plastic strain increment and obtain the final values of the equivalent plastic strain, the stress difference, the backstress, and the total stress,

$$\Delta\gamma_{er} = \frac{\tau_A - \tau_y(t) + \Delta\beta_A}{\mu + H_A + \Lambda_A}\,,$$
$$\gamma(t+\Delta t) = \gamma_A - \Delta\gamma_{er}\,,$$
$$\tau(t+\Delta t) = \tau_y(t) - \Delta\beta_A + (\mu + \Lambda_A)\,\Delta\gamma_{er}\,,$$
$$\mathbf{s}(t+\Delta t) = \sqrt{2}\,\tau(t+\Delta t)\,\boldsymbol{\mu}(t+\Delta t)\,,$$
$$\boldsymbol{\beta}(t+\Delta t) = \boldsymbol{\beta}(t) + \sqrt{2}\,(\Delta\beta_A - \Lambda_A\,\Delta\gamma_{er})\,\boldsymbol{\mu}(t+\Delta t)\,,$$
$$\boldsymbol{\tau}(t+\Delta t) = \mathbf{s}(t+\Delta t) + \boldsymbol{\beta}(t+\Delta t)\,. \qquad (5.8.54a\text{-}f)$$

The components of the deviatoric stress $\boldsymbol{\tau}'$, obtained using the *perfect-plasticity-path method* with various numbers of timesteps, are shown in Figure 5.8.3. The results are very accurate for both stress magnitude and stress components, even when there is a large rotation.

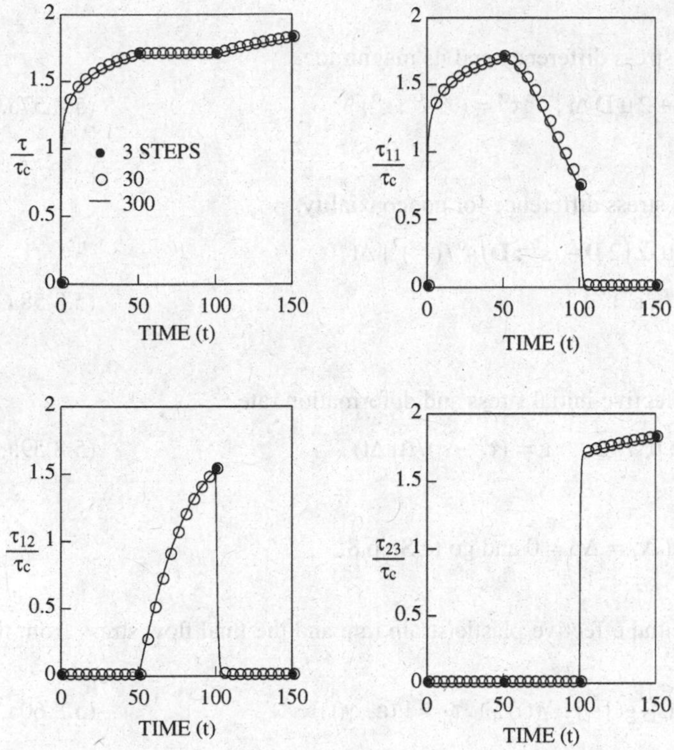

Figure 5.8.3

Time history of deviatoric stress components using the *perfect-plasticity-path method*

5.8.3. Rate-dependent Model

Now, consider the integration of the constitutive equations (5.8.1) for the rate-dependent flow rule (5.3.10b).

Computational Steps for Generalized Radial-return Method:

Step 1

Obtain the initial values of the stress difference,

$$\mathbf{s}(t) = \boldsymbol{\tau}(t) - \boldsymbol{\beta}(t), \quad \tau(t) = \{\tfrac{1}{2}\,\mathbf{s}(t) : \mathbf{s}(t)\}^{1/2}. \qquad (5.8.55a,b)$$

Step 2

Calculate the stress difference and backstress for the initial rotation $(1-\theta)\,\mathbf{W}\,\Delta t$,

$$\mathbf{s}^{\theta}(t) = \mathbf{s}(t) + (1-\theta)\,\{\mathbf{W}\,\mathbf{s}(t) - \mathbf{s}(t)\,\mathbf{W}\}\,\Delta t,$$

$$\boldsymbol{\beta}^{\theta}(t) = \boldsymbol{\beta}(t) + (1-\theta)\,\{\mathbf{W}\,\boldsymbol{\beta}(t) - \boldsymbol{\beta}(t)\,\mathbf{W}\}\,\Delta t\,. \qquad (5.8.56a,b)$$

Step 3

Obtain the trial stress difference and its magnitude,

$$\mathbf{s}^{\alpha} = \mathbf{s}^{\theta}(t) + 2\,\mu\,\mathbf{D}\,\Delta t\,, \quad \tau^{\alpha} = (\tfrac{1}{2}\,\mathbf{s}^{\alpha} : \mathbf{s}^{\alpha})^{1/2}\,. \qquad (5.8.57a,b)$$

Step 4

Correct the trial stress difference for noncoaxiality,

$$\mathbf{s}^{\mathrm{tr}} = \mathbf{s}^{\alpha} + \mu\,\bar{\alpha}\,\{2\,\mathbf{D} - (\mathbf{s}^{\alpha} : \mathbf{D})\,\mathbf{s}^{\alpha}/(\tau^{\alpha})^2\}\,\Delta t\,,$$

$$\tau^{\mathrm{tr}} = (\tfrac{1}{2}\,\mathbf{s}^{\mathrm{tr}} : \mathbf{s}^{\mathrm{tr}})^{1/2}\,. \qquad (5.8.58a,b)$$

Step 5

Calculate the effective initial stress and deformation rate,[29]

$$s_t = \mathbf{s}^{\theta}(t) : \boldsymbol{\mu}^{\mathrm{tr}}/\sqrt{2}\,, \quad \dot{\varepsilon} = (\tau^{\mathrm{tr}} - s_t)/(\mu\,\Delta t)\,. \qquad (5.8.59a,b)$$

Step 6

If $\dot{\varepsilon} < 0$, then set $\Delta\gamma = \Delta\beta = 0$ and go to Step 8.

Step 7

Calculate the initial effective plastic strain rate and the final flow stress from the flow rule,

$$\dot{\varepsilon}_t^p = \mathrm{sign}(s_t)\,g(\,|\,s_t\,|\,,\ \gamma(t))\,, \quad \tau_d = F(\dot{\varepsilon},\ \gamma(t))\,. \qquad (5.8.60a,b)$$

Step 8

Estimate the plastic strain increment, plastic strain rate, and the effective stress at the end of the timestep,

$$k = \mu\,\frac{\dot{\varepsilon} - \dot{\varepsilon}_t^p}{\tau_d - s_t}\,, \quad \Delta\gamma_A = \dot{\varepsilon}\,\Delta t - \frac{\tau_d - s_t}{\mu}\,(1 - e^{-k\Delta t})\,,$$

$$\dot{\gamma}_A = \dot{\varepsilon} - (\dot{\varepsilon} - \dot{\varepsilon}_t^p)\,e^{-k\Delta t}\,, \quad \tau^{e} = s_t + (\tau_d - s_t)\,(1 - e^{-k\Delta t})\,. \qquad (5.8.61a\text{-}d)$$

Step 9

Estimate the flow stress, the increment in the backstress, and the hardening moduli,

$$\gamma_A = \gamma(t) + \Delta\gamma_A\,, \quad \tau_A = F(\dot{\gamma}_A,\ \gamma_A)\,, \quad \Delta\beta_A = \int_{\gamma(t)}^{\gamma_A} \Lambda(\gamma)\,d\gamma\,,$$

$$\eta_A = \frac{1}{\Delta t}\,\frac{\partial F}{\partial \dot{\gamma}} + \frac{\partial F}{\partial \gamma}\,, \quad \Lambda_A = \Lambda(\gamma_A)\,. \qquad (5.8.62a\text{-}e)$$

---

[29] As before, the subscript t denotes the value of the corresponding quantity at time t, *e.g.*, $s_t \equiv s(t)$.

Step 10

Estimate the error in the estimated plastic strain increment and correct the increments in the plastic strain and the backstress,

$$\Delta\gamma_{er} = \frac{\tau_A - \tau^e + \Delta\beta_A}{\mu + \Lambda_A + \eta_A}, \qquad \Delta\gamma = \max\{0,\ \Delta\gamma_A - \Delta\gamma_{er}\},$$

$$\Delta\beta = \begin{cases} 0 & \text{for } \Delta\gamma = 0 \\ \Delta\beta_A - \Lambda_A\,\Delta\gamma_{er} & \text{for } \Delta\gamma \neq 0 \end{cases}. \tag{5.8.63a-c}$$

Step 11

Obtain the final values of the equivalent plastic strain, total stress, and the backstress,

$$\gamma(t+\Delta t) = \gamma(t) + \Delta\gamma, \qquad \boldsymbol{\tau}(t+\Delta t) = \left[1 - \frac{2\mu\,\Delta\gamma}{\tau^{tr}}\right]\mathbf{s}^{tr} + \boldsymbol{\beta}(t),$$

$$\boldsymbol{\beta}(t+\Delta t) = \boldsymbol{\beta}(t) + \Delta\beta\,\frac{\mathbf{s}^{tr}}{\tau^{tr}}. \tag{5.8.64a-c}$$

As an illustration, consider the isotropic power-law hardening model (5.7.38), and consider the following piecewise constant velocity gradient:

$$\mathbf{L} = \begin{bmatrix} 1 & 0 & 0 \\ 0 & -1 & 0 \\ 0 & 0 & 0 \end{bmatrix} \times 10^{-2}, \qquad 0 \le t < 50,$$

$$\mathbf{L} = \frac{1}{\sqrt{2}}\begin{bmatrix} 1 & -99 & 0 \\ 101 & -1 & 0 \\ 0 & 0 & 0 \end{bmatrix} \times 10^{-4}, \qquad 50 \le t < 100,$$

$$\mathbf{L} = \frac{1}{\sqrt{2}}\begin{bmatrix} -1 & 0 & 0 \\ 0 & -1 & 2 \\ 0 & 0 & 0 \end{bmatrix} \times 10^{-2}, \qquad 100 \le t < 150. \tag{5.8.65a-c}$$

In this example, the strain-rate sensitivity parameter $m = 0.05$ is used. The components of the deviatoric stress $\boldsymbol{\tau}'$, obtained by using the *generalized radial-return method* with various numbers of timesteps, are shown in Figure 5.8.4. As is seen, the *generalized radial-return method* gives accurate results even with large timesteps when the rotation is of the order of the distortion. However, when the rotation is orders of magnitude larger than the distortion, small timesteps are required for accurate results. To improve the results for large rotations, as an alternative to the *generalized radial-return method*, consider the *perfect-plasticity path method.* The computational steps for this method are given in the following:

Computational Steps for the Perfect-plasticity Path Method:

Step 1

Obtain the initial values of the stress difference and the dynamic critical stress,

$$\mathbf{s}(t) = \boldsymbol{\tau}(t) - \boldsymbol{\beta}(t), \qquad \tau(t) = \{\tfrac{1}{2}\,\mathbf{s}(t) : \mathbf{s}(t)\}^{1/2},$$

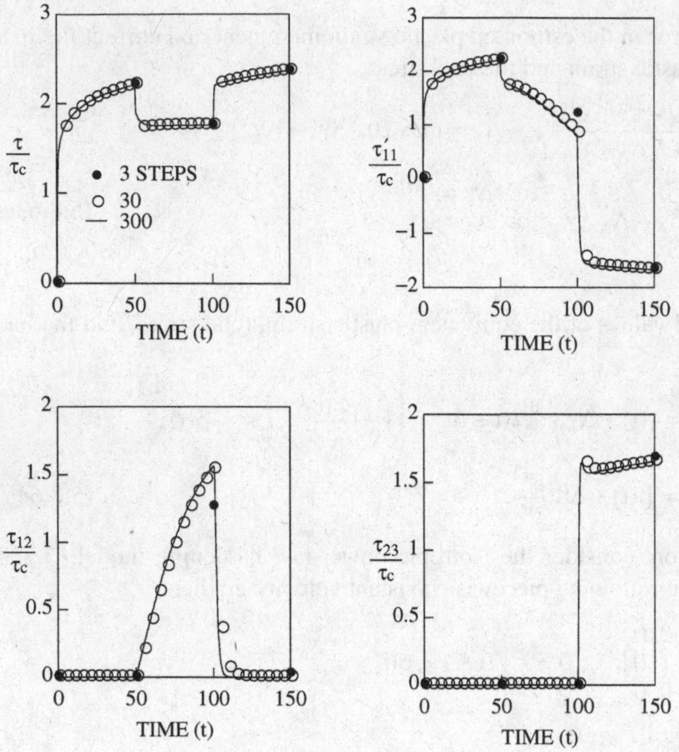

Figure 5.8.4

Time history of deviatoric stress components using the *generalized radial-return method*

$$d = \{2\,\mathbf{D} : \mathbf{D}\}^{1/2}, \quad \tau_d = F(d,\ \gamma(t)), \quad \dot{\gamma}_t = g(\tau_t,\ \gamma_t). \qquad (5.8.66a\text{-}e)$$

Step 2

If $\tau_t \le \tau_d$, set $\Delta t_1^e = 0$, $\mathbf{s}(t_1) = \mathbf{s}(t)$, and $\tau(t_1) = \tau(t)$, and go to Step 4.

Step 3

Find the time required for the effective stress difference τ, to reach the τ_d-value and evaluate the stress difference at the end of this timestep (Subsection 5.7.4),

$$k = (\mu + \Lambda)\,\frac{\dot{\gamma}_t - d}{\tau_t - \tau_d},$$

$$\dot{\gamma}_A = \max\{\dot{\gamma}_t\,(1 - e^{-k\Delta t})/(k\,\Delta t),\ (\dot{\gamma}_t - d)/\mathit{ln}\,(\dot{\gamma}_t/d)\},$$

$$\mathbf{K}_1 = 2\,\mu\,(1 - \overline{\alpha})\,\mathbf{D} + \{\mu\,\overline{\alpha}\,(\mathbf{s}_t : \mathbf{D})/\tau_t^2 - (\mu + \Lambda_t)\,\dot{\gamma}_A/\tau_t\}\,\mathbf{s}(t)$$
$$+\,\mathbf{W}\,\mathbf{s}(t) - \mathbf{s}(t)\,\mathbf{W},$$

$$\Delta t_1^e = \min\left\{\Delta t,\ \frac{-\sqrt{(\mathbf{s}_t : \mathbf{K}_1)^2 - 2\,(\mathbf{K}_1 : \mathbf{K}_1)\,(\tau_t^2 - \tau_d^2)} - \mathbf{s}_t : \mathbf{K}_1}{(\mathbf{K}_1 : \mathbf{K}_1)}\right\},$$

$$\mathbf{s}(t_1) = \mathbf{s}(t) + \mathbf{K}_1\,\Delta t_1^e, \quad \tau(t_1) = \{\tfrac{1}{2}\,\mathbf{s}(t_1) : \mathbf{s}(t_1)\}^{1/2}. \tag{5.8.67a-f}$$

Step 4

If $\tau(t_1) \geq \tau_d$ and $\mathbf{s}(t_1) : \mathbf{D} \geq 0$, set $\Delta t_2^e = 0$, $\mathbf{s}(t^e) = \mathbf{s}(t_1)$, and $\tau(t^e) = \tau(t_1)$, and go to Step 6.

Step 5

Find the time required for the effective stress difference to reach the τ_d-value and evaluate the stress difference at the end of this timestep (Subsection 5.7.4),

$$\mathbf{K}_2 = 2\,\mu\,(1-\bar{\alpha})\,\mathbf{D} + \{\mu\,\bar{\alpha}\,(\mathbf{s}_{t_1} : \mathbf{D})/\tau_{t_1}^2\}\,\mathbf{s}(t_1) + \mathbf{W}\,\mathbf{s}(t_1) - \mathbf{s}(t_1)\,\mathbf{W},$$

$$\Delta t_2^e = \min\left\{\Delta t - \Delta t_1^e,\ \frac{\sqrt{(\mathbf{s}_{t_1} : \mathbf{K}_2)^2 + 2\,(\mathbf{K}_2 : \mathbf{K}_2)\,(\tau_d^2 - \tau_{t_1}^2)} - \mathbf{s}_{t_1} : \mathbf{K}_2}{(\mathbf{K}_2 : \mathbf{K}_2)}\right\},$$

$$\mathbf{s}_A(t^e) = \mathbf{s}(t_1) + \mathbf{K}_2\,\Delta t_2^e,$$

$$\tau_A = \{\tfrac{1}{2}\,\mathbf{s}_A(t^e) : \mathbf{s}_A(t^e)\}^{1/2},$$

$$\dot{\gamma}_A = g(\tau_A,\ \gamma(t)), \quad k = (\mu + \Lambda)\left[\frac{\partial F}{\partial \dot{\gamma}}\right]^{-1},$$

$$\tau(t^e) = \tau_A - \dot{\gamma}_A\,(\partial F/\partial \dot{\gamma})\,(1 - e^{-k\Delta t_2^e}),$$

$$\mathbf{s}(t^e) = \frac{\tau(t^e)}{\tau_A}\,\mathbf{s}_A(t^e). \tag{5.8.68a-h}$$

Step 6

Calculate the final orientation of the stress difference by assuming perfect-plasticity,

$$d = (2\,\mathbf{D} : \mathbf{D})^{1/2}, \quad \omega = (-\mathbf{W} : \mathbf{W}/2)^{1/2} = \{W_{12}^2 + W_{23}^2 + W_{31}^2\}^{1/2},$$

$$\overline{\mathbf{W}} = \mathbf{W}/\omega, \quad \mathbf{D}_1 = (\mathbf{D} : \overline{\mathbf{W}}^2)\,(3/2\,\overline{\mathbf{W}}^2 + \mathbf{1}),$$

$$\mathbf{D}_2 = \overline{\mathbf{W}}\,(\mathbf{D} - \mathbf{D}_1)\,\overline{\mathbf{W}}, \quad \mathbf{D}_3 = \mathbf{D} - \mathbf{D}_1 - \mathbf{D}_2,$$

$$\alpha_2 = 2\,\mathbf{D}_2 : \mathbf{D}_2/d^2, \quad \alpha_3 = 2\,\mathbf{D}_3 : \mathbf{D}_3/d^2, \quad A = \mu\,d\,(1-\bar{\alpha})/\tau(t^e),$$

$$\hat{\mu} = A\left[1 - \frac{4\,\alpha_2\,\omega^2}{A^2 + 4\,\omega^2} - \frac{\alpha_3\,\omega^2}{A^2 + \omega^2}\right]^{1/2}, \quad k = \frac{A\,\mathbf{D} : \mathbf{s}(t^e)}{d\,\tau(t^e)\,\hat{\mu}},$$

$$\mathbf{X}_1 = \mathbf{D}_1/\hat{\mu} + \hat{\mu}\,\mathbf{D}_2/(\hat{\mu}^2 + 4\,\omega^2) + \hat{\mu}\,\mathbf{D}_3/(\hat{\mu}^2 + \omega^2),$$

$$\mathbf{X}_2 = \mathbf{X}_1\,\overline{\mathbf{W}} - \overline{\mathbf{W}}\,\mathbf{X}_1,$$

$$\Delta t^p = \Delta t - \Delta t_1^e - \Delta t_2^e, \quad \mathbf{R} = \mathbf{1} + \{1 - \cos\omega\Delta t^p\}\,\overline{\mathbf{W}}^2 + \sin(\omega\Delta t^p)\,\overline{\mathbf{W}},$$

$$\tilde{\mathbf{s}} = \mathbf{R}\,\mathbf{s}(t^e)\,\mathbf{R}^T, \quad \tilde{\mathbf{X}}_i = \mathbf{R}\,\mathbf{X}_i\,\mathbf{R}^T, \quad i = 1, 2,$$

$$z = \cosh(\hat{\mu}\,\Delta t^p) + k\sinh(\hat{\mu}\,\Delta t^p)\,, \qquad z_1 = \sinh(\hat{\mu}\,\Delta t^p) + k\cosh(\hat{\mu}\,\Delta t^p)\,,$$

$$\boldsymbol{\mu}(t+\Delta t) = \frac{A}{\sqrt{2}\,z}\left[\frac{1}{A\,\tau(t^e)}\tilde{\mathbf{s}} - \frac{2k}{d}\tilde{\mathbf{X}}_1 + \frac{2\omega}{\hat{\mu}\,d}\tilde{\mathbf{X}}_2\right]$$

$$+\frac{\sqrt{2}\,A}{d}\left[\frac{z_1}{z}\mathbf{X}_1 - \frac{\omega}{\hat{\mu}}\mathbf{X}_2\right]. \qquad (5.8.69a\text{-}t)$$

Step 7

Estimate the plastic strain rate, plastic strain, flow stress, increment in the backstress, and the hardening moduli at the end of the timestep,

$$\dot{\gamma}_A = \sqrt{2}\,\boldsymbol{\mu}(t+\Delta t) : \mathbf{D}\,, \qquad \Delta\gamma_A = \frac{d}{A}\,ln\,(z)\,,$$

$$\gamma_A = \gamma(t) + \Delta\gamma_A\,, \qquad \tau_A = F(\dot{\gamma}_A,\ \gamma_A)\,,$$

$$\Delta\beta_A = \int_{\gamma(t)}^{\gamma_A} \Lambda(\gamma)\,d\gamma\,, \qquad \Lambda_A = \Lambda(\gamma_A)\,,$$

$$\eta_A = \frac{1}{\Delta t}\frac{\partial F}{\partial\dot{\gamma}}(\dot{\gamma}_A,\ \gamma_A) + \frac{\partial F}{\partial\gamma}(\dot{\gamma}_A,\ \gamma_A)\,. \qquad (5.8.70a\text{-}g)$$

Step 8

Estimate the error in the estimated plastic strain increment and obtain the final values of the equivalent plastic strain, the stress difference, the backstress, and the total stress,

$$\Delta\gamma_{er} = \frac{\tau_A - \tau(t^e) + \Delta\beta_A}{\mu + \eta_A + \Lambda_A}\,,$$

$$\gamma(t+\Delta t) = \gamma_A - \Delta\gamma_{er}\,,$$

$$\tau(t+\Delta t) = \tau(t^e) - \Delta\beta_A + (\mu + \Lambda_A)\,\Delta\gamma_{er}\,,$$

$$\mathbf{s}(t+\Delta t) = \sqrt{2}\,\tau(t+\Delta t)\,\boldsymbol{\mu}(t+\Delta t)\,,$$

$$\boldsymbol{\beta}(t+\Delta t) = \boldsymbol{\beta}(t) + \sqrt{2}\,(\Delta\beta_A - \Lambda_A\,\Delta\gamma_{er})\,\boldsymbol{\mu}(t+\Delta t)\,,$$

$$\boldsymbol{\tau}(t+\Delta t) = \mathbf{s}(t+\Delta t) + \boldsymbol{\beta}(t+\Delta t)\,. \qquad (5.8.71a\text{-}f)$$

The components of the deviatoric stress $\boldsymbol{\tau}'$, obtained using the *perfect-plasticity-path* method with various numbers of timesteps for the example (5.8.65), are shown in Figure 5.8.5. This figure shows that this method gives very accurate results for both stress magnitude and orientation even when the rotations are large, and when there is a sudden change in the effective deformation rate.

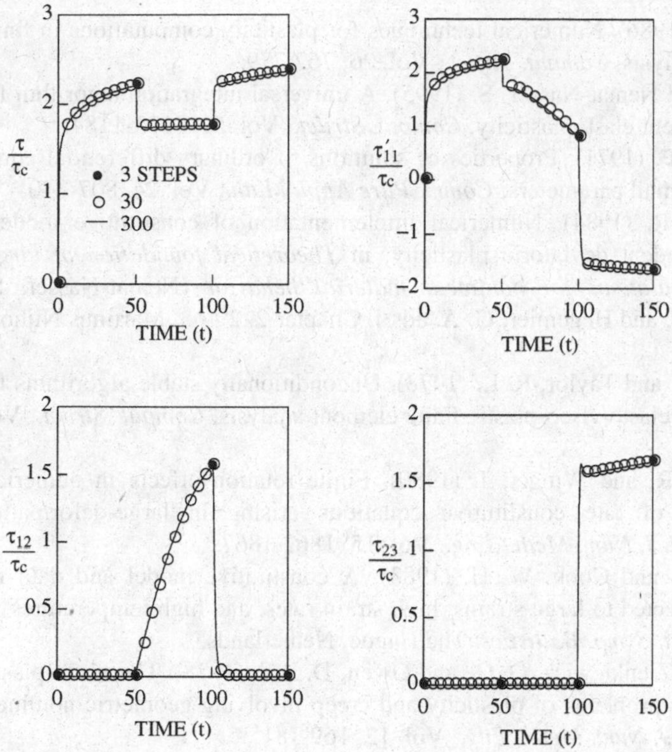

Figure 5.8.5

Time history of deviatoric stress components using the *perfect-plasticity path method*

5.9. REFERENCES

Balendran, B. and Nemat-Nasser, S. (1995), Integration of inelastic constitutive equations for constant velocity gradient with large rotation, *Appl. Math. Comput.*, Vol. 67, 161-195.

Bažant, Z. P. (1998), Easy-to-compute tensors with symmetric inverse approximating Hencky finite strain and its rate, *J. Eng. Mat. Tech.*, Vol. 120, 131-136.

Benson, D. J. (1992), Computational methods in Lagrangian and Eulerian hydrocodes, *Comput. Meth. Appl. Mech. Eng.*, Vol. 99, 235-394.

Brockman, R. A. (1984), Explicit forms for the tangent modulus tensor in viscoplastic stress analysis, *Int. J. Numer. Meth. Eng.*, Vol. 20, 315-319.

Chen, H. and Krempl, E. (1986), An adaptive time-stepping scheme for the viscoplasticity theory based on over stress, *Comput. Struct.*, Vol. 22, 573-

578.

Dodds, R. H. (1986), Numerical techniques for plasticity computations in finite element analysis, *Comput. Struct.*, Vol. 26, 767-779.

Fotiu, P. A. and Nemat-Nasser, S. (1995), A universal integration algorithm for rate-dependent elastoplasticity, *Comput. Struct.*, Vol. 59, 1173-1184.

Hoppensteadt, F. (1971), Properties of solutions of ordinary differential equations with small parameters, *Comm. Pure Appl. Math.*, Vol. 24, 807-840.

Hughes, T. J. R. (1984), Numerical implementation of constitutive models, rate-independent deviatoric plasticity, in *Theoretical foundation of large-scale computations for nonlinear material behavior* (Nemat-Nasser, S., Asaro, R. J., and Hegemier, G. A. eds.), Chapter 2, 29-63, Martinus Nijhoff, Dordrecht.

Hughes, T. J. R and Taylor, R. L. (1978), Unconditionally stable algorithms for quasi-static elasto/visco-plastic finite element analysis, *Comput. Struct.*, Vol. 8, 169-173.

Hughes, T. J. R. and Winget, J. (1980), Finite rotation effects in numerical integration of rate constitutive equations arising in large-deformation analysis, *Int. J. Num. Meth. Eng.*, Vol. 15, 1862-1867.

Johnson, G. R. and Cook, W. H. (1983), A constitutive model and data for metals subjected to large strains, high strain rates, and high temperatures, in *Proc. 7th Int. Symp. Ballistics*, The Hague, Netherlands.

Kanchi, M. B., Zienkiewicz, O. C., and Owen, D. R. J, (1978), The visco-plastic approach to problems of plasticity and creep involving geometric nonlinear effects, *Int. J. Num. Meth. Eng.*, Vol. 12, 169-181.

Krieg, R. D. and Krieg, D. B. (1977), Accuracies of numerical solution methods for the elastic, perfectly-plastic model, *ASME J. Press. Vessel Tech.*, Vol. 99, 510-515.

Li, Y.-F. (1993), Constitutive algorithm, constitutive modeling and simulation of high-strain, high-strain-rate finite deformation of heavy metals, Ph.D. thesis, University of California, San Diego.

Marques, J. M. M. C. and Owen, D. R. J. (1983), Strain hardening representation for implicit quasistatic elasto-viscoplastic algorithms, *Comput. Struct.*, Vol. 17, 301-304.

Miranker, W. L. (1981), *Numerical methods for stiff equations and singular perturbation problems*, D. Reidel, Dordrecht.

Nayfeh, A. H. (1980), *Introduction to perturbation techniques*, John Willey & Sons, New York.

Nemat-Nasser, S. (1991), Rate-independent finite-deformation elastoplasticity: a new explicit algorithm, *Mech. Mat.*, Vol. 11, 235-249.

Nemat-Nasser, S. and Chung, D.-T. (1992), An explicit constitutive algorithm for large-strain-rate elastic-viscoplasticity, *Comput. Meth. Appl. Mech. Eng.*, Vol. 95, 205-219.

Nemat-Nasser, S. and Li, Y. F. (1992), A new explicit algorithm for finite-deformation elastoplasticity and elastoviscoplasticity: performance evaluation, *Comput. Struct.*, Vol. 44, 937-963.

Nemat-Nasser, S. and Li, Y. F. (1994), An algorithm for large-scale computational finite-deformation plasticity, *Mech. Mat.*, Vol. 18, 231-264.

Nemat-Nasser, S. and Ni, L. (1993), Mechanics of interface fracture of anisotropic bimaterials, in *Ultrasonic characterization and mechanics of interfaces* (Rokhlin, S. I., Datta, S. K., and Rajapakse, Y. G. S. eds.), AMD Vol. 177, 65-77, The American Society of Mechanical Engineers, New York.

Nemat-Nasser, S. and Ni, L. (1994), Effective constitutive algorithms in elastoplasticity and elastoviscoplasticity, *European J. Appl. Math.*, Vol. 5, Part 3, 313-336.

O'Malley, R. E. Jr. (1971a), Boundary layer methods for nonlinear initial value problems, SIAM *Review*, Vol. 13, 425-434.

O'Malley, R. E. Jr. (1971b), On initial value problems for nonlinear system of differential equations with two small parameters, *Arch. Rat. Mech. Anal.*, Vol. 40, 209-222.

Peirce, D., Shih, C. F., and Needleman, A. (1984), A tangent modulus method for rate dependent solids, *Comput. Struct.*, Vol 18, 875-887.

Rashid, M. M. (1993), Incremental kinematics for finite element applications, *Int. J. Num. Meth. Eng.*, Vol. 36, 3937-3956.

Rashid, M. M. and Nemat-Nasser, S. (1990), Modelling very large plastic flows at very large strain rates for large scale computation, *Comput. Struct.*, Vol. 37, 119-132.

Smith, D. R. (1985), *Singular-perturbation theory*, Cambridge University Press, Cambridge.

Szabo, L. (1990), Tangent modulus tensors for elastic-viscoplastic solids, *Comput. Struct.*, Vol. 34, 401-419.

Van Dyke, M. (1975), *Perturbation methods in fluid mechanics*, The Parabolic Press, Stanford, California.

Wilkins, M. L. (1964), Calculation of elastic-plastic flow, *Meth. Comput. Phys.*, Vol. 3B, 211-263.

Yoshimura, S., Chen, K. L., and Atluri, S. N. (1987), A study of two alternate tangent modulus formulations and attendant implicit algorithms for creep as well as high-strain-rate plasticity, *Int. J. Plasticity*, Vol 3, 391-413.

Zienkiewicz, O. C. and Cormeau, I. C. (1974), Visco-plasticity-plasticity and creep in elastic solids-a unified numerical solution approach, *Int. J. Num. Meth. Eng.*, Vol. 8, 821-845.

APPENDIX 5.A. IDENTITIES FOR SECOND-ORDER DEVIATORIC AND SKEWSYMMETRIC TENSORS

Some useful relations involving a second-order symmetric deviatoric tensor **D**, and a second-order skewsymmetric tensor **W** are developed in this Appendix.

Since **W** is skewsymmetric, *i.e.*, $W_{ij} = -W_{ji}$, it follows that

$$\mathrm{tr}(\mathbf{W}) = \mathrm{tr}(\mathbf{W}^3) = 0, \quad \mathrm{tr}(\mathbf{W}^2) = -2\,(W_{12}^2 + W_{23}^2 + W_{31}^2) = -2\,\omega^2,$$
$$\omega \equiv \sqrt{W_{12}^2 + W_{23}^2 + W_{31}^2}\,. \tag{5.A.1a-c}$$

Hence, from (1.2.7b-d), the basic invariants of **W** are

$$I_W = 0, \quad II_W = \omega^2, \quad III_W = 0\,. \tag{5.A.2a-c}$$

The Hamilton-Cayley theorem (1.2.11) now yields

$$\mathbf{W}^3 + \omega^2\,\mathbf{W} = \mathbf{0}\,. \tag{5.A.3a}$$

Identities (1.4.3)-(1.4.5) become

$$\mathbf{W}^2\,\mathbf{D} + \mathbf{D}\,\mathbf{W}^2 + \mathbf{W}\,\mathbf{D}\,\mathbf{W} + \omega^2\,\mathbf{D} = \mathrm{tr}(\mathbf{W}^2\,\mathbf{D})\,\mathbf{1}\,,$$
$$\mathbf{W}^2\,\mathbf{D}\,\mathbf{W} + \mathbf{W}\,\mathbf{D}\,\mathbf{W}^2 = \mathrm{tr}(\mathbf{W}^2\,\mathbf{D})\,\mathbf{W}\,,$$
$$\mathbf{W}^2\,\mathbf{D}\,\mathbf{W}^2 - \omega^2\,\mathbf{W}\,\mathbf{D}\,\mathbf{W} = \mathrm{tr}(\mathbf{W}^2\,\mathbf{D})\,\mathbf{W}^2\,. \tag{5.A.3b-d}$$

Split the symmetric deviatoric tensor **D** into a part $\mathbf{D}_1$, coaxial with $\mathbf{W}^2$, and a remaining part **H**, as follows:

$$\mathbf{D} = \mathbf{D}_1 + \mathbf{H}, \quad \mathrm{tr}(\mathbf{H}) = \mathrm{tr}(\mathbf{W}\,\mathbf{H}) = \mathrm{tr}(\mathbf{W}^2\,\mathbf{H}) = 0\,,$$
$$\mathbf{D}_1 = (\mathbf{D} : \overline{\mathbf{W}}^2)(3\,\overline{\mathbf{W}}^2 + 2\,\mathbf{1})\,/\,2, \quad \overline{\mathbf{W}} = \mathbf{W}\,/\,\omega\,. \tag{5.A.4a-d}$$

Use **H** in place of **D** in (5.A.3d), and in view of (5.A.4b), obtain

$$\mathbf{W}^2\,\mathbf{H}\,\mathbf{W}^2 = \omega^2\,\mathbf{W}\,\mathbf{H}\,\mathbf{W}\,. \tag{5.A.5}$$

Now, split the tensor **H** into two parts,

$$\mathbf{H} = \mathbf{D}_2 + \mathbf{D}_3, \quad \mathbf{D}_2 = \overline{\mathbf{W}}\,\mathbf{H}\,\overline{\mathbf{W}}\,. \tag{5.A.6a,b}$$

Hence, it follows from (5.A.5) that the symmetric deviatoric tensors $\mathbf{D}_2$ and $\mathbf{D}_3$ are orthogonal, *i.e.*,

$$\mathrm{tr}(\mathbf{D}_2\,\mathbf{D}_3) = \mathrm{tr}(\mathrm{H}\,\overline{\mathbf{W}}\,\mathbf{H}\,\overline{\mathbf{W}}) - \mathrm{tr}(\overline{\mathbf{W}}\,\mathbf{H}\,\overline{\mathbf{W}}^2\,\mathbf{H}\,\overline{\mathbf{W}})$$
$$= \mathrm{tr}(\mathrm{H}\,\overline{\mathbf{W}}\,\mathbf{H}\,\overline{\mathbf{W}}) - \mathrm{tr}(\mathbf{H}\,\overline{\mathbf{W}}^2\,\mathbf{H}\,\overline{\mathbf{W}}^2) = 0\,. \tag{5.A.7a}$$

Therefore,

$$\mathbf{D} : \mathbf{D} = \alpha_1 + \alpha_2 + \alpha_3, \quad \alpha_i = (\mathbf{D}_i : \mathbf{D}_i), \quad (\text{i not summed})\,. \tag{5.A.7b,c}$$

Since $\mathbf{D}_2$ and $\mathbf{D}_3$ are orthogonal to $\mathbf{D}_1$, they can be expressed in the form

$$\mathbf{D}_2 = \mathbf{X}_2\,\mathbf{W} - \mathbf{W}\,\mathbf{X}_2, \quad \mathbf{D}_3 = \mathbf{X}_3\,\mathbf{W} - \mathbf{W}\,\mathbf{X}_3\,. \tag{5.A.8a,b}$$

In view of (1.4.21c), $\mathbf{X}_2$ and $\mathbf{X}_3$ are given by

$$\mathbf{X}_2 = \frac{3}{4\,\omega^4}\,\mathbf{W}\,(\mathbf{D}_2\,\mathbf{W} - \mathbf{W}\,\mathbf{D}_2)\,\mathbf{W} - \frac{1}{\omega^2}\,(\mathbf{D}_2\,\mathbf{W} - \mathbf{W}\,\mathbf{D}_2)\,,$$
$$\mathbf{X}_3 = \frac{3}{4\,\omega^4}\,\mathbf{W}\,(\mathbf{D}_3\,\mathbf{W} - \mathbf{W}\,\mathbf{D}_3)\,\mathbf{W} - \frac{1}{\omega^2}\,(\mathbf{D}_3\,\mathbf{W} - \mathbf{W}\,\mathbf{D}_3)\,. \tag{5.A.9a,b}$$

Substitute (5.A.5) and (5.A.6) into (5.A.9), to obtain

$$\mathbf{X}_2 = -\frac{1}{4\omega}\mathbf{D}_4, \quad \mathbf{X}_3 = -\frac{1}{\omega}\mathbf{D}_5, \tag{5.A.10a,b}$$

where

$$\mathbf{D}_4 = \mathbf{D}_2\overline{\mathbf{W}} - \overline{\mathbf{W}}\mathbf{D}_2, \quad \mathbf{D}_5 = \mathbf{D}_3\overline{\mathbf{W}} - \overline{\mathbf{W}}\mathbf{D}_3. \tag{5.A.11a,b}$$

Hence, it follows from (5.A.8) and (5.A.10) that

$$\mathbf{D}_4\overline{\mathbf{W}} - \overline{\mathbf{W}}\mathbf{D}_4 = -4\,\mathbf{D}_2, \quad \mathbf{D}_5\overline{\mathbf{W}} - \overline{\mathbf{W}}\mathbf{D}_5 = -\mathbf{D}_3. \tag{5.A.12a,b}$$

Summarizing these, for a given skewsymmetric tensor $\mathbf{W}$, any second-order symmetric deviatoric tensor $\mathbf{D}$ can be split into three orthogonal parts

$$\begin{aligned}
&\mathbf{D} = \mathbf{D}_1 + \mathbf{D}_2 + \mathbf{D}_3, \\
&\mathbf{D}_1 = (\mathbf{D} : \overline{\mathbf{W}}^2)(3\,\overline{\mathbf{W}}^2 + 2\,\mathbf{1})/2, \\
&\mathbf{D}_2 = \overline{\mathbf{W}}(\mathbf{D} - \mathbf{D}_1)\overline{\mathbf{W}}, \quad \mathbf{D}_3 = \mathbf{D} - \mathbf{D}_1 - \mathbf{D}_2,
\end{aligned} \tag{5.A.13a-d}$$

such that

$$\begin{aligned}
&\mathbf{D}_2\overline{\mathbf{W}} - \overline{\mathbf{W}}\mathbf{D}_2 = \mathbf{D}_4, \quad \mathbf{D}_4\overline{\mathbf{W}} - \overline{\mathbf{W}}\mathbf{D}_4 = -4\,\mathbf{D}_2, \\
&\mathbf{D}_3\overline{\mathbf{W}} - \overline{\mathbf{W}}\mathbf{D}_3 = \mathbf{D}_5, \quad \mathbf{D}_5\overline{\mathbf{W}} - \overline{\mathbf{W}}\mathbf{D}_5 = -\mathbf{D}_3.
\end{aligned} \tag{5.A.14a-d}$$

APPENDIX 5.B. SOLUTION OF $\mathbf{A} + \theta(\mathbf{A}\mathbf{W} - \mathbf{W}\mathbf{A}) = \mathbf{B}$

Consider the equation

$$\mathbf{A} + \theta(\mathbf{A}\mathbf{W} - \mathbf{W}\mathbf{A}) = \mathbf{B}, \tag{5.B.1}$$

where the second-order tensors $\mathbf{A}$ and $\mathbf{B}$ are symmetric deviatoric, while $\mathbf{W}$ is skewsymmetric. Now, seek to solve for $\mathbf{A}$ when the tensors $\mathbf{W}$ and $\mathbf{B}$, and the constant θ are given. As discussed in Appendix 5.A, split the symmetric deviatoric tensors $\mathbf{A}$ and $\mathbf{B}$ into three orthogonal tensors as follows:

$$\mathbf{A} = \mathbf{A}_1 + \mathbf{A}_2 + \mathbf{A}_3, \quad \mathbf{B} = \mathbf{B}_1 + \mathbf{B}_2 + \mathbf{B}_3, \tag{5.B.2a,b}$$

$$\begin{aligned}
&\mathbf{A}_1 = \mathrm{tr}(\mathbf{A} : \overline{\mathbf{W}}^2)(3\,\overline{\mathbf{W}}^2 + 2\,\mathbf{1})/2, \quad \mathbf{B}_1 = \mathrm{tr}(\mathbf{B} : \overline{\mathbf{W}}^2)(3\,\overline{\mathbf{W}}^2 + 2\,\mathbf{1})/2, \\
&\mathbf{A}_2 = \overline{\mathbf{W}}(\mathbf{A} - \mathbf{A}_1)\overline{\mathbf{W}}, \quad \mathbf{B}_2 = \overline{\mathbf{W}}(\mathbf{B} - \mathbf{B}_1)\overline{\mathbf{W}}, \\
&\mathbf{A}_3 = \mathbf{A} - \mathbf{A}_1 - \mathbf{A}_2, \quad \mathbf{B}_3 = \mathbf{B} - \mathbf{B}_1 - \mathbf{B}_2, \\
&\overline{\mathbf{W}} = \mathbf{W}/\omega, \quad \omega = \sqrt{W_{12}^2 + W_{23}^2 + W_{31}^2}.
\end{aligned} \tag{5.B.3a-h}$$

Define the symmetric deviatoric tensors $\mathbf{A}_4$, $\mathbf{A}_5$, $\mathbf{B}_4$, and $\mathbf{B}_5$, by

$$\mathbf{A}_4 = \mathbf{A}_2\overline{\mathbf{W}} - \overline{\mathbf{W}}\mathbf{A}_2, \quad \mathbf{B}_4 = \mathbf{B}_2\overline{\mathbf{W}} - \overline{\mathbf{W}}\mathbf{B}_2,$$

$$\mathbf{A}_5 = \mathbf{A}_3\,\overline{\mathbf{W}} - \overline{\mathbf{W}}\,\mathbf{A}_3\,, \qquad \mathbf{B}_5 = \mathbf{B}_3\,\overline{\mathbf{W}} - \overline{\mathbf{W}}\,\mathbf{B}_3\,. \tag{5.B.4a-d}$$

Hence, (5.B.1) is rewritten as

$$\mathbf{A}_1 + \mathbf{A}_2 + \mathbf{A}_3 + \mathrm{p}\,(\mathbf{A}_4 + \mathbf{A}_5) = \mathbf{B}_1 + \mathbf{B}_2 + \mathbf{B}_3\,, \qquad \mathrm{p} = \theta\,\omega\,. \tag{5.B.5a,b}$$

Now, introduce the function $\boldsymbol{\Phi}(\overline{\mathbf{W}}, \mathbf{X})$,

$$\boldsymbol{\Phi}(\overline{\mathbf{W}}, \mathbf{X}) = (\mathbf{X}\,\overline{\mathbf{W}} - \overline{\mathbf{W}}\,\mathbf{X})\,\overline{\mathbf{W}} - \overline{\mathbf{W}}\,(\mathbf{X}\,\overline{\mathbf{W}} - \overline{\mathbf{W}}\,\mathbf{X})\,. \tag{5.B.6}$$

From the results of Appendix 5.A it follows that

$$\begin{aligned}
&\boldsymbol{\Phi}(\overline{\mathbf{W}}, \mathbf{A}_1) = \boldsymbol{\Phi}(\overline{\mathbf{W}}, \mathbf{B}_1) = \mathbf{0}\,,\\
&\boldsymbol{\Phi}(\overline{\mathbf{W}}, \mathbf{A}_2) = -4\,\mathbf{A}_2\,, \quad \boldsymbol{\Phi}(\overline{\mathbf{W}}, \mathbf{A}_4) = -4\,\mathbf{A}_4\,,\\
&\boldsymbol{\Phi}(\overline{\mathbf{W}}, \mathbf{A}_3) = -\mathbf{A}_3\,, \quad \boldsymbol{\Phi}(\overline{\mathbf{W}}, \mathbf{A}_5) = -\mathbf{A}_5\,,\\
&\boldsymbol{\Phi}(\overline{\mathbf{W}}, \mathbf{B}_2) = -4\,\mathbf{B}_2\,, \quad \boldsymbol{\Phi}(\overline{\mathbf{W}}, \mathbf{B}_4) = -4\,\mathbf{B}_4\,,\\
&\boldsymbol{\Phi}(\overline{\mathbf{W}}, \mathbf{B}_3) = -\mathbf{B}_3\,, \quad \boldsymbol{\Phi}(\overline{\mathbf{W}}, \mathbf{B}_5) = -\mathbf{B}_5\,.
\end{aligned} \tag{5.B.7a-i}$$

Hence, from (5.B.6), obtain

$$\mathbf{A}_1 = \mathbf{B}_1\,, \qquad \mathbf{A}_2 + \mathrm{p}\,\mathbf{A}_4 = \mathbf{B}_2\,, \qquad \mathbf{A}_3 + \mathrm{p}\,\mathbf{A}_5 = \mathbf{B}_3\,. \tag{5.B.8a-c}$$

It follows from (5.A.12) that

$$\mathbf{A}_4\,\overline{\mathbf{W}} - \overline{\mathbf{W}}\,\mathbf{A}_4 = -4\,\mathbf{A}_2\,, \qquad \mathbf{A}_5\,\overline{\mathbf{W}} - \overline{\mathbf{W}}\,\mathbf{A}_5 = -\mathbf{A}_3\,. \tag{5.B.9a,b}$$

Substitute (5.B.9) into (5.B.8b,c), to obtain

$$\mathbf{A}_4 - 4\,\mathrm{p}\,\mathbf{A}_2 = \mathbf{B}_4\,, \qquad \mathbf{A}_5 - \mathrm{p}\,\mathbf{A}_3 = \mathbf{B}_5\,. \tag{5.B.8d,e}$$

Hence, (5.B.8) can be solved, to obtain $\mathbf{A}_i$, i = 1, 2, ..., 5 in terms of $\mathbf{B}_i$, i = 1, 2, ..., 5, as follows:

$$\begin{aligned}
&\mathbf{A}_1 = \mathbf{B}_1\,, \quad \mathbf{A}_2 = \frac{1}{1+4\,\mathrm{p}^2}\,(\mathbf{B}_2 - \mathrm{p}\,\mathbf{B}_4)\,, \quad \mathbf{A}_3 = \frac{1}{1+\mathrm{p}^2}\,(\mathbf{B}_3 - \mathrm{p}\,\mathbf{B}_5)\,,\\
&\mathbf{A}_4 = \frac{1}{1+4\,\mathrm{p}^2}\,(4\,\mathrm{p}\,\mathbf{B}_2 + \mathbf{B}_4)\,, \quad \mathbf{A}_3 = \frac{1}{1+\mathrm{p}^2}\,(\mathrm{p}\,\mathbf{B}_3 + \mathbf{B}_5)\,,
\end{aligned} \tag{5.B.9c-g}$$

and the solution to (5.B.1) is

$$\mathbf{A} = \mathbf{B}_1 + \frac{1}{1+4\,\mathrm{p}^2}\,(\mathbf{B}_2 - \mathrm{p}\,\mathbf{B}_4) + \frac{1}{1+\mathrm{p}^2}\,(\mathbf{B}_3 - \mathrm{p}\,\mathbf{B}_5)\,. \tag{5.B.10}$$

Note that, (5.B.10) is the coordinate-invariant solution of the tensor equation (5.B.1). Alternatively, (5.B.1) can be solved in component form. In view of the symmetry and skewsymmetry of the tensors **D** and **W**, (5.B.1) reduces to 6 linear equations as follows:

$$\mathrm{K}_{ij}\,\mathrm{u}_j = \mathrm{v}_i\,, \quad i, j = 1, 2, \cdots, 6, \quad (j \text{ summed})\,, \tag{5.B.11a}$$

where

$$\mathrm{u}_1 = \mathrm{A}_{11}\,, \quad \mathrm{u}_2 = \mathrm{A}_{22}\,, \quad \mathrm{u}_3 = \mathrm{A}_{33}\,,$$

$$u_4 = A_{23}, \quad u_5 = A_{31}, \quad u_6 = A_{12},$$

$$v_1 = B_{11}, \quad v_2 = B_{22}, \quad v_3 = B_{33},$$

$$v_4 = B_{23}, \quad v_5 = B_{31}, \quad v_6 = B_{12}. \tag{5.B.11b-m}$$

Nemat-Nasser (1991) expresses the 6×6 matrix $\mathbf{K}$ in terms of two 3×3 matrices as follows:

$$\mathbf{K} = \begin{bmatrix} \mathbf{1} & \mathbf{w} \\ -\tfrac{1}{2}\mathbf{w}^T & -\tfrac{1}{2}\mathbf{n} \end{bmatrix}, \tag{5.B.12a}$$

where $\mathbf{1}$ is the 3×3 identity matrix, and the 3×3 nonsymmetric matrices $\mathbf{w}$ and $\mathbf{n}$ are defined by

$$\mathbf{w} = \begin{bmatrix} 0 & \omega_2 & -\omega_3 \\ -\omega_1 & 0 & \omega_3 \\ \omega_1 & -\omega_2 & 0 \end{bmatrix}, \quad \mathbf{n} = \begin{bmatrix} -2 & -\omega_3 & \omega_2 \\ \omega_3 & -2 & -\omega_1 \\ -\omega_2 & \omega_1 & -2 \end{bmatrix},$$

$$\omega_1 = 2\theta W_{23}, \quad \omega_2 = 2\theta W_{31}, \quad \omega_3 = 2\theta W_{12}. \tag{5.B.12b-f}$$

In terms of matrices $\mathbf{w}$ and $\mathbf{n}$, Nemat-Nasser (1991) expresses the inverse of the matrix $\mathbf{K}$ as follows:

$$\mathbf{K}^{-1} = \begin{bmatrix} \mathbf{a} & \mathbf{b} \\ \mathbf{c} & \mathbf{d} \end{bmatrix}, \tag{5.B.13a}$$

where

$$\mathbf{a} = \mathbf{1} - \tfrac{1}{2}\,\mathbf{w}\,\mathbf{d}\,\mathbf{w}^T, \quad \mathbf{b} = -\mathbf{w}\,\mathbf{d},$$

$$\mathbf{c} = \tfrac{1}{2}\,\mathbf{d}\,\mathbf{w}^T, \quad \mathbf{d} = 2\,(\mathbf{w}^T\mathbf{w} - \mathbf{n})^{-1}. \tag{5.B.13b-e}$$

This way, the problem of obtaining the inverse of a 6×6 matrix is reduced to that of a 3×3 matrix. Following Nemat-Nasser (1991), the closed-form expressions for the components of the matrix $\mathbf{d}$ are

$$d_{11} = 2\,\{3\,\omega_2^2\,\omega_3^2 + 4\,(\omega_2^2 + \omega_3^2 + 1) + \omega_1^2\}\,/\,\Delta,$$

$$d_{22} = 2\,\{3\,\omega_3^2\,\omega_1^2 + 4\,(\omega_3^2 + \omega_1^2 + 1) + \omega_2^2\}\,/\,\Delta,$$

$$d_{33} = 2\,\{3\,\omega_1^2\,\omega_2^2 + 4\,(\omega_1^2 + \omega_2^2 + 1) + \omega_3^2\}\,/\,\Delta,$$

$$d_{12},\ d_{21} = -2\,\{\pm 2\,\omega_3^3 - 3\,\omega_1\,\omega_2\,\omega_3^2 \mp \omega_3\,(\omega_1^2 + \omega_2^2 - 2) - 3\,\omega_1\,\omega_2\}\,/\,\Delta,$$

$$d_{23},\ d_{32} = -2\,\{\pm 2\,\omega_1^3 - 3\,\omega_1^2\,\omega_2\,\omega_3 \mp \omega_1\,(\omega_2^2 + \omega_3^2 - 2) - 3\,\omega_2\,\omega_3\}\,/\,\Delta,$$

$$d_{31},\ d_{13} = -2\,\{\pm 2\,\omega_2^3 - 3\,\omega_1\,\omega_2^2\,\omega_3 \mp \omega_2\,(\omega_3^2 + \omega_1^2 - 2) - 3\,\omega_3\,\omega_1\}\,/\,\Delta,$$

$$\Delta = 2\,(\omega_1^4 + \omega_2^4 + \omega_3^4) + 4\,(\omega_1^2\,\omega_2^2 + \omega_2^2\,\omega_3^2 + \omega_3^2\,\omega_1^2) + 10\,(\omega_1^2 + \omega_2^2 + \omega_3^2) + 8. \tag{5.B.14a-g}$$

Once the inverse of $\mathbf{K}$ is known, the unknown $\mathbf{u}$ is obtained from

$$\mathbf{u} = \mathbf{K}^{-1}\,\mathbf{v}. \tag{5.B.15}$$

6

FINITE ELASTOPLASTIC DEFORMATION OF SINGLE CRYSTALS

Certain fundamentals of elastoplastic finite deformation of crystalline solids are discussed from a microscopic point of view. Physically-based constitutive relations for single crystals are formulated on the basis of slip-induced plastic deformation, and the accompanying elastic lattice distortion. The slip is produced through the motion of dislocations. Due account is taken of the crystal structure and the barriers which the dislocations must overcome in their motion through the lattice structure. A brief description of the physical structure of the crystals is given in Section 6.1. Starting with a review of the crystal systems and outlining the fourteen Bravais lattices, the notion of the flow of material through the lattice by crystallograghic slip, is introduced, leading to the concept of dislocation. Then certain elementary topics in the theory of dislocations are considered, including the characterization of dislocations, their elastic interactions which underlie certain aspects of strain hardening in metals, and other relevant issues. This section is concluded by developing an explicit expression for the slip rate in terms of the density of the associated mobile dislocations, their average velocity, and the magnitude of the Burgers vector. The kinematics of the finite deformation of single crystals is addressed in Section 6.2. Various decompositions of the deformation gradient and its rate are discussed and the results are compared. In particular, alternative representations of the deformation and spin tensors corresponding to various possible reference states, are developed and their relations and equivalence are discussed. The elasticity of crystals is examined in Section 6.3, where explicit general rate-constitutive relations are presented, again using possible alternative reference states. Various objective stress rates are considered, the associated constitutive relations are produced, and their equivalence is established. The notion of self- and latent-hardening is examined in Section 6.4, and various aspects of the commonly used assumption that latent-hardening exceeds self-hardening in magnitude, are discussed. The slip models which directly account for both the temperature- and strain-rate effects, are presented in Section 6.5. The short- and long-range barriers that the dislocations must overcome in their motion, are identified. The hardening issue, associated with the long-range athermal resistance to the motion of the dislocations, is reexamined. Illustrative physically-based models of the flow stress associated with slip, are developed in this section. Explicit results for bcc and fcc crystals are produced, taking into account the temperature, strain rate, and the long-range hardening effects. As an illustration, explicit results for a commercially pure tantalum (bcc) and OFHC copper (fcc) are given and compared with the experimental data. Many of the results presented in this chapter, particularly in this section, are of recent contribution, and some are new. As in Chapter 4, Section 4.8, dislocation-based crystal plasticity naturally involves several length scales relating to the dislocation densities and various microstructural characteristics of the material. In addition, when there are only a few interacting crystals within a small sample, geometric and textural incompatibilities may profoundly affect the sample deformation. These and related issues are briefly examined at the end of this chapter. Finally, for the convenience of the reader, a brief account of Miller indices is given in Appendix 6.A.

6.1. PHYSICS OF CRYSTAL PLASTICITY

Finite plasticity addresses mechanical processes that involve large plastic deformations. Examples are wire drawing and sheet metal forming. The continuum theories of elastoplastic deformation of crystalline solids do not directly examine the *micromechanisms* which give rise to their permanent plastic deformations and the accompanying recoverable elastic strains. These issues are considered in the theories of *crystal plasticity*, where the corresponding kinematics and dynamics are explored in terms of the atomic structure of crystals. At the microscale and at suitable temperatures, crystals deform dissipatively, generally by rate-dependent slip over crystallographic planes, accompanied by reversible elastic lattice distortion. In this section, the structure of crystals is first briefly examined. Then, the deformation of crystals is discussed in terms of the lattice distortion which is the basis of crystal elasticity, and the dislocation motion which leads to slip-induced crystal plasticity.

6.1.1. Crystal Structure and Elasticity

A *lattice* is a three-dimensional configuration of points, which can be repeated periodically in all three directions, maintaining the same surroundings for all lattice points. The basic structure is defined by a *unit cell*. A crystal is formed when the lattice points are occupied by atoms or a collection of atoms, while satisfying the required periodicity. The atoms within the crystal structure are held together through *interatomic bonds* which may have various origins, depending on the composition of the constituent elements. The *primary* interatomic bonds are very strong and may consist of *ionic, covalent*, and *metallic* bonding. The resulting substance has a relatively high melting point (1,000 to 5,000K) and high elastic moduli. Ceramics with ionic and covalent bonding, and metals with metallic and covalent bonding, are in this category. The *secondary* bonding with a melting point of 100 to 500K, may consist of *van der Waals* bonding (induced dipoles; bonding of polar molecules) and *hydrogen* bonding. Many atoms in a crystal are bound together by a mixture of bonding forces.[1]

The *elastic* deformation of crystalline solids is associated with the distortion of the lattice, while the atoms and molecules that occupy the lattice points remain in their original cell, only the cell being distorted. The slight relative displacement of the atoms within the cell in the presence of the bonding forces, creates the macroscopic stresses. Thus, the stresses in a quasi-statically deformed crystal, stem solely from lattice distortion and are elastic in nature. When the applied forces which *elastically* distort a crystal are removed, the atoms and molecules of the crystal, within each unit cell, return to their

[1] For elementary accounts, see, *e.g.*, Kittel (1986) and Reed-Hill and Abbaschian (1992). For a more comprehensive discussion, see Burns (1990).

equilibrium positions, hence restoring the crystal to its initial configuration. A lattice distortion of this kind involves very small displacements, and does not break the interatomic bonds. Thus, the elastic deformation of crystalline solids, by its very nature, is generally infinitesimally small. It is essential to distinguish between the *lattice* which consists of points in space arranged periodically and a *crystal* which has atoms and molecules occupying the lattice points.[2]

The crystals are classified into *seven systems*, according to the arrangement of the three sets of planes that produce the corresponding unit cell. The unit cell is defined by three independent vectors (usually denoted by **a**, **b**, and **c** or by $\mathbf{a}_i$, i = 1, 2, 3) and the associated three angles (usually denoted by α, β, and γ, or by α_i, i = 1, 2, 3). *The seven crystal systems are cubic, tetragonal, orthorhombic, monoclinic, triclinic, hexagonal, and rhombohedral;* see Table 6.1.1.

Table 6.1.1

Seven crystal systems and fourteen Bravais lattices

System	**Cell Geometry**	**Bravais Lattice**
Cubic	Equal axes; right angles $a = b = c$, $\alpha = \beta = \gamma = 90°$	Simple Body-centered Face-centered
Tetragonal	Two equal axes; right angles $a = b \neq c$, $\alpha = \beta = \gamma = 90°$	Simple Body-centered
Orthorhombic	Unequal axes; right angles $a \neq b \neq c$, $\alpha = \beta = \gamma = 90°$	Simple Body-centered Base-centered Face-centered
Monoclinic	Unequal axes; one non-orthogonal pair $a \neq b \neq c$, $\alpha = \gamma = 90° \neq \beta$	Simple Base-centered
Triclinic	Unequal axes; unequal and non-orthogonal angles $a \neq b \neq c$, $\alpha \neq \beta \neq \gamma \neq 90°$	Simple
Hexagonal	Two equal axes at 120°; third axis normal to them $a = b \neq c$, $\alpha = \beta = 90°$, $\gamma = 120°$	Simple
Rhombohedral (or 'Trigonal')	Equal non-orthogonal axes $a = b = c$, $\alpha = \beta = \gamma \neq 90°$,	Simple

The corresponding seven lattice structures are obtained by placing points at the corners of the unit cells. In a lattice, each point is required to have the same surroundings. This can be fulfilled in other ways than just putting points at

[2] See, *e.g.*, Chapter 2 of Cullity (1978).

lattice corners. It is shown that *there are only fourteen different ways that lattice points can be arranged, with all points having the same surroundings. These are called Bravais lattices,* summarized in Table 6.1.1 and sketched in Figure 6.1.1. Some of the most common and extensively studied crystal structures are *face-centered cubic (fcc), body-centered cubic (bcc), and hexagonal close-packed (hcp);* see Figure 6.1.2.

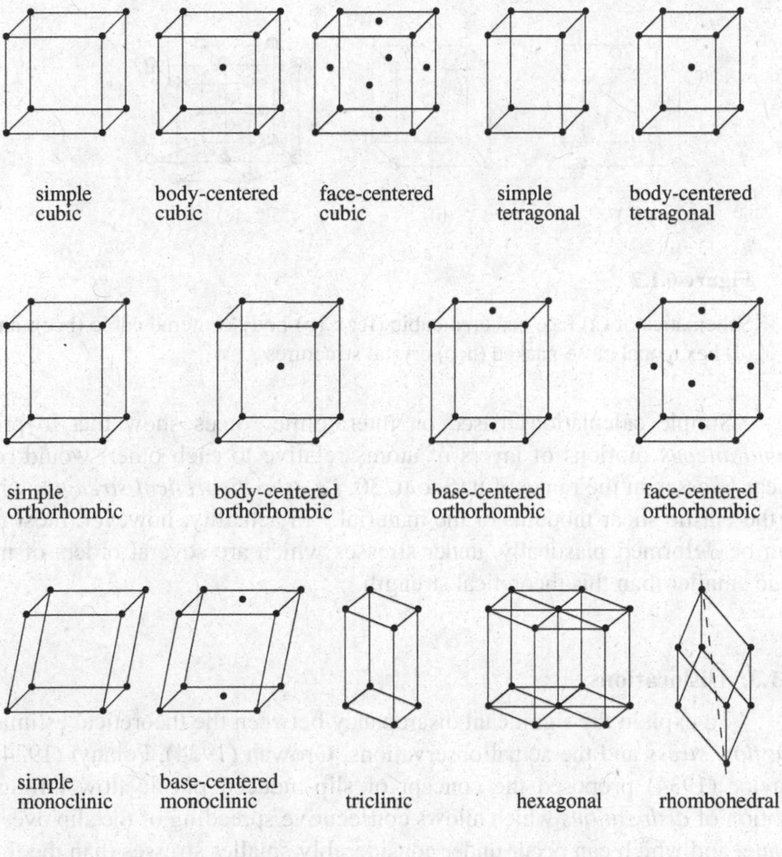

Figure 6.1.1

Fourteen Bravais lattices; see also Table 6.1.1

6.1.2. Plasticity and Slip

Under applied loads, crystals may undergo permanent deformation, accompanied by recoverable elastic lattice distortion. It is well established that the permanent deformation is the result of the relative movement of atoms, while the lattice structure is preserved. This is feasible because of the periodic arrangement of the atoms. If such movements were to occur by a collection of atoms sliding in unison relative to other atoms across a *slip plane*, then a permanent deformation could result without disturbing the basic lattice structure. Plastic deformation of metals by slip was pointed out by Mügge (1883), and was later established by Ewing and Rosenhain (1899, 1900) who, using optical microscopy, identify the terraced surfaces of plastically deformed polycrystalline lead samples as *slip bands*.

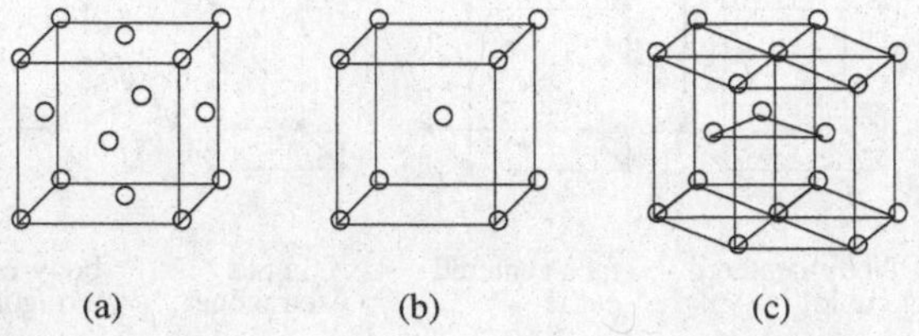

Figure 6.1.2

Schematics of (a) face-centered cubic (fcc); (b) body-centered cubic (bcc); and (c) hexagonal close-packed (hcp) crystal structures

Simple calculations based on interatomic forces show that to produce *simultaneous* motions of layers of atoms relative to each other, would require shear stresses in the range of $\mu/5$ to $\mu/30$, *i.e., the theoretical strength*, where μ is the elastic shear modulus of the material.[3] In actuality, however, most metals can be deformed plastically, under stresses which are several orders of magnitude smaller than this theoretical strength.

6.1.3. Dislocations

To explain the significant discrepancy between the theoretical estimates of the *flow stress* and the actual observations, Orowan (1934), Polanyi (1934), and Taylor (1934) proposed the concept of slip-induced plastic flow through the motion of *dislocations* which allows consecutive spreading of the slip over a slip plane, and which can occur under considerably smaller stresses than the theoretical strength.[4] Since these early works, the presence of dislocations and their prominent role in producing plastic flow, have been established by numerous observations. Figure 6.1.3 shows dislocation lines in tantalum which has been deformed at a high strain rate.[5]

---

[3] Frenkel (1926); see also Cottrell (1953).

[4] For comprehensive discussions of various aspects of dislocations and dislocation theories, see, *e.g.*, Cottrell (1953), Read (1953), Nabarro (1967), Hirth and Lothe (1992), Hull and Bacon (1992), and Weertman and Weertman (1992). Nabarro (1979-1989) has edited eight volumes on *Dislocations in Solids*, which also provide references to many other contributions.

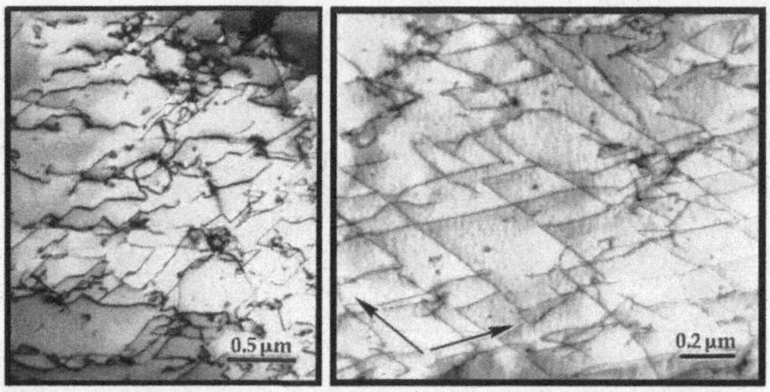

(a) (b)

Figure 6.1.3

TEM of Ta with dislocations: (a) after moderate high strain-rate deformation; (b) after extensive high strain-rate, high-temperature deformation

6.1.4. Burgers' Vector

When atoms in a crystal move from one equilibrium configuration to a neighboring one by slip on crystallographic planes, the crystal structure remains intact, except at the boundary of the slipped region, which is a *dislocation line.* This slip is characterized by the *Burgers vector*, **b**, which lies on the slip plane, defines the direction of the slip, and has one lattice-spacing length. Figure 6.1.4(a) shows a dislocation loop, l, in a crystal, on a plane with unit normal **n**. It is assumed that the region within the loop has slipped by one lattice spacing in the direction of the Burgers vector, *i.e.*, in the **b**-direction, such that atoms above the plane are moved to the left of the figure, in the **b**-direction, in relation to the atoms below the plane. If the corresponding displacement vector is denoted by **u**, then the Burgers vector is defined such that

$$\mathbf{u}^+ - \mathbf{u}^- = \mathbf{b}. \tag{6.1.1a}$$

Here, superscript + refers to the quantity evaluated just above the slip plane, and superscript - to that just below the plane. Thus, **b** *is the displacement of the upper plane measured relative to the lower plane.*[6] Figure 6.1.4(b) shows the AA′ cross section, indicating the relative displacement of the atoms. At B there is an extra layer of atoms above the slip plane, whereas at B′ the extra layer is below this plane. As is suggested by the sketch of Figure 6.1.4(b), outside and inside the dislocation loop, the crystal structure remains intact.

---

[5] Tantalum is a bcc crystal, and a highly ductile material. The dislocations are generally screw-type, which may jog out of their planes, creating segments of mixed screw and edge dislocations. These segments have different velocities and are left as dislocation loops; see Nemat-Nasser *et al.* (1998a) for a detailed account.

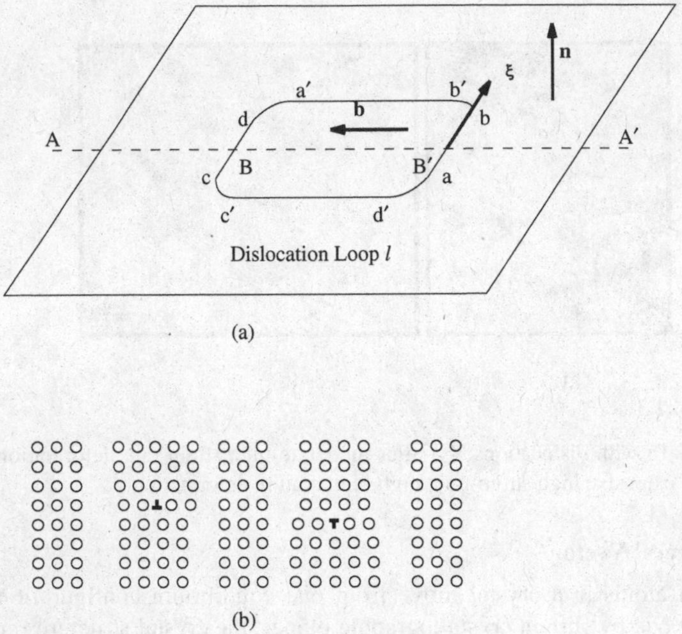

Figure 6.1.4

(a) A dislocation loop l, on a slip plane with unit normal **n**: **b** is the Burgers vector, showing the direction of slip; ab and cd are edge, a′b′ and c′d′ are screw; and the remaining part of the loop is mixed dislocation; and (b) cross section AA′ showing the effect of one lattice slip in the direction of **b**

When the Burgers vector is normal to the dislocation line, the dislocation is called an *edge dislocation,* and when it is parallel to the dislocation line, the dislocation is called a *screw dislocation.* In Figure 6.1.4(a), segments ab and cd are edge, and a′b′ and c′d′ are screw dislocations. In general, the Burgers vector will have a normal and a tangential component, which vary along the dislocation line, *e.g.*, segment bb′ in Figure 6.1.4(a). In such a case, the Burgers vector may be decomposed into a screw, $\mathbf{b}_s$, and an edge, $\mathbf{b}_e$, component, as follows:

$$\mathbf{b} = \mathbf{b}_s + \mathbf{b}_e = (\mathbf{b} \cdot \boldsymbol{\xi})\,\boldsymbol{\xi} + (\boldsymbol{\xi} \times \mathbf{b}) \times \boldsymbol{\xi}, \tag{6.1.1b}$$

where $\boldsymbol{\xi}$ is a unit vector tangent to the dislocation loop l at the considered point. The direction of $\boldsymbol{\xi}$ is chosen such that when an observer moves along the loop in the direction of $\boldsymbol{\xi}$, then the region enclosed by the loop is to the observer's left;

6 An alternative is to define the Burgers vector as the displacement of the lower plane relative to the upper plane; see Hirth and Lothe (1992). The sign of some of the corresponding quantities will then change accordingly; *e.g.,* the expression for the force acting on a dislocation, (6.1.2a), then will change to $\mathbf{F} = (\mathbf{b} \cdot \boldsymbol{\sigma}^0) \times \boldsymbol{\xi}$.

see Figure 6.1.4(a). Note that the dislocation must be a closed loop when totally inside a crystal, or else must terminate at the crystal boundaries. Figure 6.1.5 includes sketches of edge and screw dislocations.

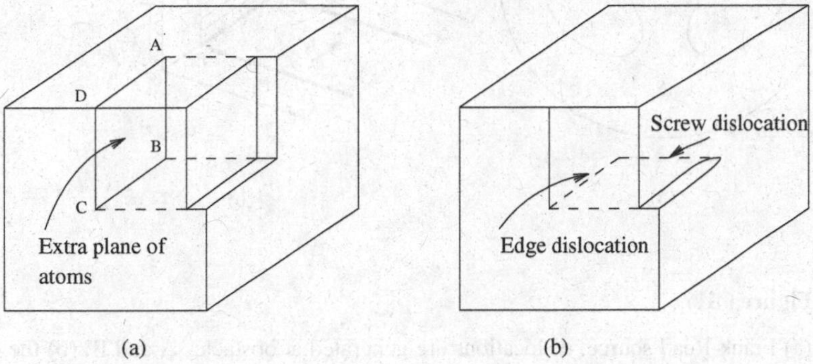

Figure 6.1.5

(a) An edge dislocation with extra plane of atoms on ABCD; (b) a dislocation composed of edge and screw segments

The plane formed by the Burgers vector, **b**, and the unit vector tangent to the dislocation line, $\boldsymbol{\xi}$, is called the *glide plane*. This defines a unique plane for an edge dislocation, but any plane containing the unit vector $\boldsymbol{\xi}$, or equivalently the Burgers vector **b**, is a glide plane for a screw dislocation. For a mixed dislocation, the edge component of the Burgers vector, $\mathbf{b}_e$, and $\boldsymbol{\xi}$ define the corresponding glide plane.

In a crystal under stress, dislocations can be generated from pre-existing defects, and they readily multiply in response to continued loading. Figure 6.1.6(a) shows loops generated from two neighboring obstacles, A and B. These obstacles may be part of a dislocation line that is not on the considered slip plane, or they may be two other dislocations, or other obstacles. As the segment between A and B is "bowed out" under the action of the applied stress field, it may go through a sequence of motions shown in the figure, generating dislocation loops one after another. In this manner, dislocations can readily multiply and produce large plastic deformations. This mechanism is known as the Frank-Read source; Frank and Read (1950). Dislocations may also *cross slip* from one glide plane to another, as is illustrated in Figure 6.1.6(b). Here, the screw segment AB on glide plane P_1 has *jugged* out into plane P_2 and then has moved into the neighboring glide plane P_3 which is parallel to the original glide plane P_1.

Dislocations can interact with each other (dislocation reactions), and can combine or break down into *partial dislocations*, whenever it is energetically

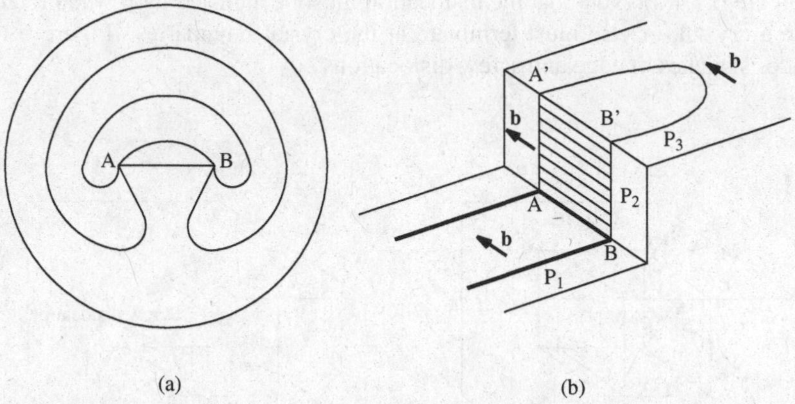

Figure 6.1.6

(a) Frank-Read source: dislocations are generated at obstacles A and B; (b) the screw segment AB has cross slipped into plane P_2 and then into the neighboring glide plane P_3

favorable. In this manner, plastic deformation can proceed at stresses that are orders of magnitude smaller than the elastic shear modulus of the material. The dislocations may also combine to form a new dislocation on a plane which may not be a slip plane, leading to *locked dislocations*. They may also *pile up* against point defects such as precipitates and impurities or grain boundaries, or they may become entangled. In this manner, the material's resistance to plastic flow may increase, and this is called *workhardening*.

6.1.5. Action of a Stress Field on a Dislocation

Consider a dislocation with Burgers' vector **b** in a stress field $\boldsymbol{\sigma}^o$. The force **F** acting on this dislocation at a typical point on the dislocation line with unit tangent vector $\boldsymbol{\xi}$, is given by

$$\mathbf{F} = \boldsymbol{\xi} \times (\mathbf{b} \cdot \boldsymbol{\sigma}^o), \quad F_i = e_{ijl}\, \xi_j\, b_k\, \sigma^o_{kl}, \tag{6.1.2a,b}$$

where e_{ijl} is the permutation symbol, and i, j, l, k = 1, 2, 3, with repeated indices being summed.[7] This force is measured per unit length of the dislocation line. Expression (6.1.2) can be established by considering the interaction energy between the stress field and the dislocation. The negative of the gradient of this energy functional, taken in a given direction, **r**, at a typical point on the dislocation line, then yields **F**.

[7] See Section 1.1 for notation.

Since the elastic lattice distortion is generally infinitesimally small, and since the presence of dislocations does not noticeably affect the elastic response of most crystals, the theory of linear elasticity may be used to establish the force of a stress field on a dislocation line.

To this end, denote the displacement field due to the presence of the dislocation by **u**, and note that the interaction elastic energy is given by

$$\phi = \int_V \boldsymbol{\sigma}^o : (\boldsymbol{\nabla} \otimes \mathbf{u})\, dV = \int_V \sigma^o_{ij}\, u_{j,i}\, dV\,, \tag{6.1.2c}$$

where V is the volume of the crystal, and where the symmetry of the stress tensor is also used. The integrand in (6.1.2c) is the work done per unit volume by the stress field $\boldsymbol{\sigma}^o$ going through the displacement (deformation) induced by the dislocation. It is assumed that $\boldsymbol{\sigma}^o$ is a pre-existing (*i.e., it exists before the dislocation is created), self-equilibrating stress field, so that*

$$\boldsymbol{\nabla} \cdot \boldsymbol{\sigma}^o = \mathbf{0}\,, \quad \text{or} \quad \sigma^o_{ij,i} = 0 \quad \text{in } V\,,$$

i, j, = 1, 2, 3. *Since the displacement field,* $\mathbf{u}(\mathbf{x})$, *produced by the dislocation loop,* l, *vanishes far away from the loop, the integral is finite.*

For an unbounded medium containing a dislocation loop, l, which encloses a surface, S, the integral in the above expression can be written as

$$\phi = \int_{S^+ + S^-} \mathbf{n} \cdot \boldsymbol{\sigma}^o \cdot \mathbf{u}\, dS = \int_{S^+ + S^-} \sigma^o_{ij}\, n_i\, u_j\, dS\,, \tag{6.1.2d}$$

where S^+ and S^- are the two faces of the area enclosed by the dislocation loop (see Figure 6.1.7); and where the divergence theorem is used. Let the unit normal to S^+ be $\mathbf{n}^+ \equiv \mathbf{n}$, and choose its sense as shown in the figure.[8] Then that of S^- is $\mathbf{n}^- = -\mathbf{n}$. Identify S^+ with S, and obtain

$$\begin{aligned} \phi &= -\int_S (\mathbf{n} \cdot \boldsymbol{\sigma}^o) \cdot (\mathbf{u}^+ - \mathbf{u}^-)\, dS \\ &= -\int_S (\mathbf{n} \cdot \boldsymbol{\sigma}^o) \cdot \mathbf{b}\, dS \\ &= -\int_S (\mathbf{b} \cdot \boldsymbol{\sigma}^o) \cdot \mathbf{n}\, dS\,, \end{aligned} \tag{6.1.2e}$$

where the symmetry of $\boldsymbol{\sigma}^o$ and (6.1.1a) are used.

Consider now the change in this energy when at a typical point on the dislocation line, surface S is increased by an increment $\mathbf{n}\, dS = d\mathbf{r} \times \boldsymbol{\xi}\, d\eta$; see Figure 6.1.7(b). Then the increment in the interaction energy, $d\phi$, becomes

$$\begin{aligned} d\phi &= -(\mathbf{b} \cdot \boldsymbol{\sigma}^o) \cdot (d\mathbf{r} \times \boldsymbol{\xi})\, d\eta \\ &= -\boldsymbol{\xi} \times (\mathbf{b} \cdot \boldsymbol{\sigma}^o) \cdot d\mathbf{r}\, d\eta\,. \end{aligned} \tag{6.1.2f}$$

---

[8] Since within the dislocation loop, l, the crystal structure remains intact, surface S can be chosen arbitrarily within the intact crystal. For a planar dislocation loop, use the area on the glide plane enclosed by the loop. Note that the *exterior unit normal* of S^+ is $\mathbf{n}^-$, and that of S^- is $\mathbf{n}^+$; hence, the minus sign in the right-hand side of (6.1.2e).

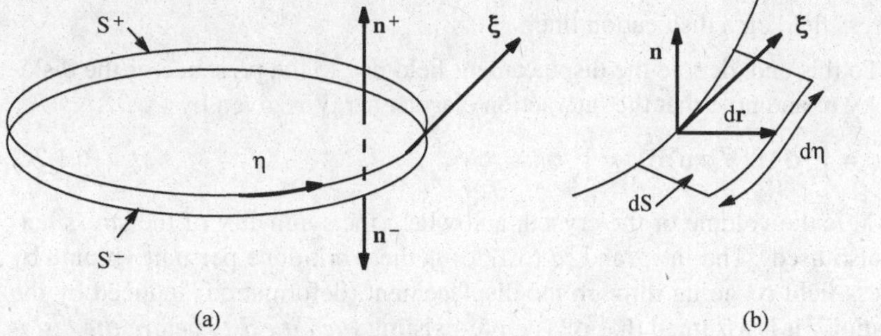

Figure 6.1.7

(a) S^+ and S^-, with (interior) unit normals $\mathbf{n}^+$ and $\mathbf{n}^-$, are the two faces of the surface S which is bounded by the dislocation loop: η measures length along the loop, and $\boldsymbol{\xi}$ is the unit tangent vector; (b) the surface S is increased by dS which can be represented by $d\mathbf{S} = \mathbf{n}\, dS = d\mathbf{r} \times \boldsymbol{\xi}\, d\eta$

Here, η measures length along the dislocation loop. Hence, obtain

$$-\frac{d\phi}{d\mathbf{r}}\frac{1}{d\eta} \equiv \mathbf{F} = \boldsymbol{\xi} \times (\mathbf{b} \cdot \boldsymbol{\sigma}^o). \tag{6.1.2g}$$

This expression is called the Peach-Koehler equation, after Peach and Koehler (1950).

As an illustration, choose the coordinate unit triad, $\mathbf{e}_i$, i = 1, 2, 3, such that

$$\boldsymbol{\xi} = \mathbf{e}_3, \quad \mathbf{b} = b_1\,\mathbf{e}_1 + b_2\,\mathbf{e}_2 + b_3\,\mathbf{e}_3, \tag{6.1.3a,b}$$

and obtain

$$\begin{aligned}\mathbf{F} &= \mathbf{e}_3 \times (\sigma^o_{ij}\,\mathbf{e}_i \otimes \mathbf{e}_j) \cdot (b_k\,\mathbf{e}_k)\\ &= \mathbf{e}_3 \times \mathbf{e}_i\,(\sigma^o_{i1}\,b_1 + \sigma^o_{i2}\,b_2 + \sigma^o_{i3}\,b_3) = -(\sigma^o_{2k}\,b_k)\,\mathbf{e}_1 + (\sigma^o_{1k}\,b_k)\,\mathbf{e}_2\\ &= -(\sigma^o_{21}\,b_1 + \sigma^o_{22}\,b_2 + \sigma^o_{23}\,b_3)\,\mathbf{e}_1 + (\sigma^o_{11}\,b_1 + \sigma^o_{12}\,b_2 + \sigma^o_{13}\,b_3)\,\mathbf{e}_2.\end{aligned} \tag{6.1.3c}$$

If $\mathbf{b} = b\,\mathbf{e}_1$, an edge dislocation with its glide plane normal to the $\mathbf{e}_2$-direction, then (6.1.3c) becomes

$$\begin{aligned}\mathbf{F} &= F_1\,\mathbf{e}_1 + F_2\,\mathbf{e}_2\\ &= -(\sigma^o_{21}\,b)\,\mathbf{e}_1 + (\sigma^o_{11}\,b)\,\mathbf{e}_2.\end{aligned} \tag{6.1.3d}$$

The component $F_1 = -\sigma^o_{21}\,b$ tends to produce glide, while the component $F_2 = \sigma^o_{11}\,b$ leads to the dislocation *climb*; see Figure 6.1.8. Note that b may be positive or negative, depending on the sense of **b**.

If, on the other hand, $\mathbf{b} = b\,\mathbf{e}_3$, a screw dislocation, then (6.1.3c) leads to

$$\mathbf{F} = -\sigma^o_{23}\, b\, \mathbf{e}_1 + \sigma^o_{13}\, b\, \mathbf{e}_2 . \tag{6.1.3e}$$

For a stress field with only a nonzero anti-plane component, $\sigma^o_{13} \neq 0$, the force is $\mathbf{F} = \sigma^o_{13}\, b\, \mathbf{e}_2$. While the slip for this screw dislocation occurs in the $\mathbf{e}_3$-direction, the applied force tends to move the dislocation line in the $\mathbf{e}_2$-direction.

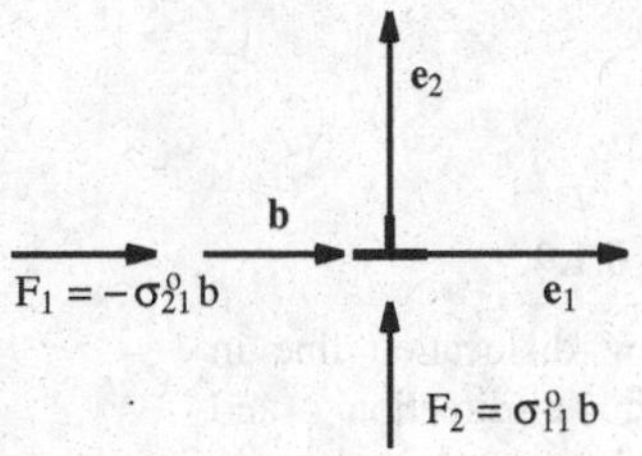

Figure 6.1.8

A dislocation line in the $\boldsymbol{\xi} = \mathbf{e}_3$-direction and Burgers' vector $\mathbf{b} = b\,\mathbf{e}_1$, is subjected to a gliding force F_1 and a climbing force F_2, under the action of the stress field $\boldsymbol{\sigma}^o$

The stress field in the above results may correspond to an external load, or it may be due to the action of other dislocations. In this manner the interaction of a dislocation with various sources can be studied. Note that, the force acting on the dislocation line, measured per unit length of the line, is always proportional to the corresponding component of the Burgers vector.

6.1.6. Elastic Field of a Dislocation

The stress field of a dislocation line can be calculated by viewing the dislocation as a displacement discontinuity in an elastic continuum.[9] The use of linear elasticity is justified since the displacement field associated with the corresponding lattice distortion is infinitesimally small, except at the dislocation core.

Screw Dislocation: As an illustration, consider the stress field of a screw dislocation with the Burgers vector $\mathbf{b} = -b\,\mathbf{e}_3$, and dislocation line $\boldsymbol{\xi} = \mathbf{e}_3$, in an isotropic continuum, where $b = |\mathbf{b}| > 0$; see Figure 6.1.9.

The equilibrium equations,

$$\sigma_{ij,j} = 0 , \quad i, j = 1, 2, 3 , \quad (\text{j summed}) , \tag{6.1.4}$$

reduce to

[9] The dislocation as a displacement discontinuity was introduced into linear elasticity theory by Timpe (1905), Volterra (1907), Somigliana (1914), and Love (1927), who examined certain elastic displacement fields which are not single-valued.

$$\frac{\partial \sigma_{13}}{\partial x_1} + \frac{\partial \sigma_{23}}{\partial x_2} = 0\,. \tag{6.1.5}$$

For a screw dislocation, $u_3 = u_3(x_1, x_2)$ is the only nonzero component of the displacement field. Then, the Hooke law yields

$$\mu \frac{\partial u_3}{\partial x_1} = \sigma_{13} \quad \text{and} \quad \mu \frac{\partial u_3}{\partial x_2} = \sigma_{23}\,. \tag{6.1.6a,b}$$

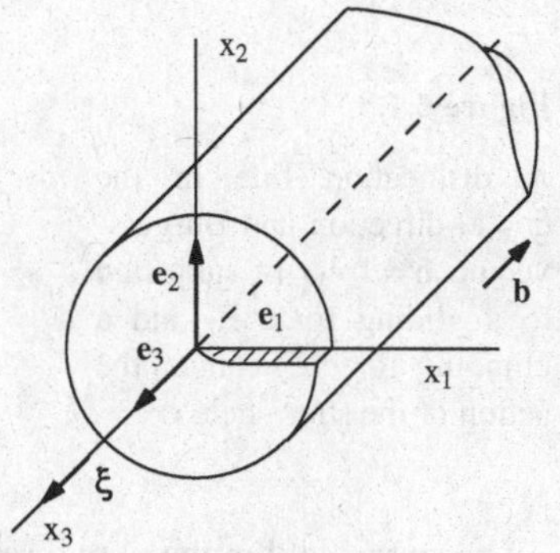

Figure 6.1.9

A screw dislocation line in the $\xi = \mathbf{e}_3$-direction and Burgers' vector $\mathbf{b} = -b\,\mathbf{e}_3$, in an isotropic linearly elastic solid

Hence, it follows that u_3 satisfies

$$\nabla^2 u_3 = 0\,. \tag{6.1.7}$$

Using the cylindrical coordinates, (r, θ, x_3), note[10] that $u_3 = \hat{u}_3(\theta) = c_0\,\theta + c_1$, where c_0 and c_1 are the integration constants. From (6.1.1a), it follows that $u_3(0) - u_3(2\pi) = -b$. Hence,

$$\hat{u}_3 = \frac{b\,\theta}{2\pi} \quad \text{or} \quad u_3 = \frac{b}{2\pi}\tan^{-1}(x_2/x_1)\,. \tag{6.1.8a,b}$$

The resulting stress field in cylindrical coordinates then is

$$\sigma_{\theta 3} = \sigma_{3\theta} = \frac{\mu\,b}{2\pi\,r}\,, \tag{6.1.9}$$

with all other stress components being zero. From (6.1.9), it is seen that the stress field is $1/r$ singular at the origin, or within the core of the dislocation. Linear elasticity, however, is not applicable within the core of the dislocation, since a discrete atomistic model is then necessary for a proper analysis of the deformation of the core region.

Assuming a core of radius r_0, the elastic energy of this screw dislocation, contained within the annulus $r_0 < r < R$, of unit length in the x_3-direction, then is given by

---

[10] Note that $\nabla^2(...) \equiv (1/r)\,\partial(r\,\partial(...)/\partial r)/\partial r + (1/r^2)\,\partial^2(...)/\partial\theta^2 + \partial^2(...)/\partial x_3^2$. Hence, $\nabla^2 u_3 = 0$ yields $d^2\hat{u}_3/d\theta^2 = 0$, or $\hat{u}_3 = c_0\,\theta + c_1$.

$$W = \frac{\mu b^2}{4\pi} \ln(R/r_0). \qquad (6.1.10)$$

Edge Dislocation: Expressions (6.1.8) to (6.1.10) are estimates for a screw dislocation, based on linear elasticity and an isotropic continuum. Similar results are obtained for an edge dislocation. In this case also, the elastic fields all depend only on x_1 and x_2. The equilibrium equations, (6.1.4), now take on the form

$$\sigma_{i1,1} + \sigma_{i2,2} = 0, \quad i = 1, 2, 3. \qquad (6.1.11a)$$

The i = 3 corresponds to the screw dislocation which has already been considered. For i = 1, 2, and isotropic elasticity, these equations are satisfied by introducing the Airy stress function, $U(x_1, x_2)$, such that

$$\sigma_{11} = \frac{\partial^2 U}{\partial x_2^2}, \quad \sigma_{22} = \frac{\partial^2 U}{\partial x_1^2}, \quad \sigma_{12} = -\frac{\partial^2 U}{\partial x_1 \partial x_2}. \qquad (6.1.11b\text{-}d)$$

Then, with

$$\varepsilon_{ij} = \frac{1}{2}\left[\frac{\partial u_i}{\partial x_j} + \frac{\partial u_j}{\partial x_i}\right], \quad i, j = 1, 2, 3, \qquad (6.1.12)$$

denoting the linearized strain components, compatibility of the strain field would necessarily require,[11]

$$\frac{\partial^2 \varepsilon_{11}}{\partial x_2^2} + \frac{\partial^2 \varepsilon_{22}}{\partial x_1^2} = 2\frac{\partial^2 \varepsilon_{12}}{\partial x_1 \partial x_2}. \qquad (6.1.13)$$

For isotropic elasticity, the strain-stress relation is

$$\varepsilon_{ij} = \frac{1}{2\mu}\left[\sigma_{ij} - \delta_{ij}\frac{\lambda}{3\lambda + 2\mu}\sigma_{kk}\right]; \qquad (6.1.14a)$$

μ and λ being the Lamé constants. The inverse relation is,

$$\sigma_{ij} = 2\mu\,\varepsilon_{ij} + \lambda\,\delta_{ij}\,\varepsilon_{kk}. \qquad (6.1.14b)$$

The compatibility condition is satisfied if the Airy stress function satisfies[12]

$$\nabla^4 U = 0. \qquad (6.1.15)$$

The general solution of this equation is obtained in terms of two complex potentials, known as Muskhelishvili's stress potentials. Here, a special solution, valid for $r_0 < r < R$, with suitable boundary data prescribed on $r = r_0$, R, is considered, as follows:

---

[11] See, for example, Timoshenko and Goodier (1970) and Sokolnikoff (1956). It can be shown that (6.1.13) is also sufficient for the compatibility of the strain tensor, in the sense that a suitably smooth strain field which satisfies this condition is necessarily the symmetric part of the gradient of a single-valued displacement field.

[12] See Muskhelishvili (1956); a concise account, including results for anisotropic elasticity, is given by Nemat-Nasser and Hori (1993 and 1999, Section 21).

$$U = D\, r \sin\theta\, ln\, r, \quad D = \frac{\mu b}{2\pi(1-\nu)}, \tag{6.1.16a,b}$$

ν being the Poisson ration. The corresponding stresses then are[13]

$$\sigma_{rr} = \sigma_{\theta\theta} = \frac{D}{r}\sin\theta, \quad \sigma_{r\theta} = \sigma_{\theta r} = -\frac{D}{r}\cos\theta. \tag{6.1.16c-f}$$

These stress components vanish at infinity (*i.e.*, far away from the dislocation), and become unbounded close to the core of the dislocation, where the solution in no longer valid. Thus, the solution is reasonable in the range $r_0 << r << R$, where r_0 is of the order of the Burgers vector and R is of the order of, say, one micron.

The Cartesian stress components corresponding to (6.1.16) are

$$\sigma_{11} = D\frac{x_2(3x_1^2+x_2^2)}{(x_1^2+x_2^2)^2}, \quad \sigma_{22} = -D\frac{x_2(x_1^2-x_2^2)}{(x_1^2+x_2^2)^2},$$

$$\sigma_{12} = -D\frac{x_1(x_1^2-x_2^2)}{(x_1^2+x_2^2)^2}, \quad \sigma_{33} = 2\nu D\frac{x_2}{x_1^2+x_2^2}. \tag{6.1.17a-d}$$

The associated strain field is given by

$$\varepsilon_{11} = D'\left\{\frac{x_2(3x_1^2+x_2^2)}{(x_1^2+x_2^2)^2} - 2\nu\frac{x_2}{x_1^2+x_2^2}\right\},$$

$$\varepsilon_{22} = -D'\left\{\frac{x_2(x_1^2-x_2^2)}{(x_1^2+x_2^2)^2} + 2\nu\frac{x_2}{x_1^2+x_2^2}\right\},$$

$$\varepsilon_{12} = -D'\left\{\frac{x_1(x_1^2-x_2^2)}{(x_1^2+x_2^2)^2}\right\}, \quad D' = \frac{b}{4\pi(1-\nu)}. \tag{6.1.18a-d}$$

Integrating these expressions, the displacement field is obtained as follows:

$$u_1 = D'\left\{-2(1-\nu)\tan^{-1}(x_2/x_1) - \frac{x_1 x_2}{x_1^2+x_2^2}\right\} + A x_2 + B,$$

$$u_2 = D'\left\{\frac{1-2\nu}{2}\, ln\,(x_1^2+x_2^2) + \frac{x_1^2}{x_1^2+x_2^2}\right\} - A x_1 + C, \tag{6.1.19a,b}$$

where the integration constants, A, B , C, correspond to rigid-body displacement, and can be set equal to zero. Let the positive direction of the dislocation line be defined by $\boldsymbol{\xi} = \mathbf{e}_3$. As θ is changed from zero to 2π, counterclockwise, the displacement component u_1 changes by $|\mathbf{b}| = b$, where $\mathbf{b}$ is the Burgers vector, $\mathbf{b} = b\,\mathbf{e}_1$, lying on, and pointing toward the positive x_1-axis.

The elastic energy per unit length of an edge dislocation, contained within the annulus $r_0 < r < R$, is given by

---

[13] In polar coordinates, expressions (6.1.11b-d) become, $\sigma_{rr} = \partial^2 U/(r^2\partial\theta^2) + \partial U/(r\partial r)$, $\sigma_{\theta\theta} = \partial^2 U/\partial r^2$, and $\sigma_{r\theta} = -\partial/\partial r\,(\partial U/(r\partial\theta))$.

$$W = \frac{\mu b^2}{4\pi(1-\nu)} \ln(R/r_0). \tag{6.1.20}$$

Comparing with (6.1.10), and since, in general, $\frac{1}{2} \geq \nu > 0$, observe that this energy of the edge dislocation is greater than that of a screw dislocation.

A Dislocation in an Anisotropic Solid: Modified expressions can be obtained for general anisotropic media; see Stroh (1958) and Barnett and Lothe (1973). Consider a case where the displacement field, $\mathbf{u}$, and hence the strain, $\boldsymbol{\varepsilon}$, and stress, $\boldsymbol{\sigma}$, fields are functions of only two space variables, say, the rectangular Cartesian coordinates, x_1 and x_2. In such a case, the equilibrium equations, in the absence of body forces, reduce to (6.1.11a).[14] These equations are identically satisfied if the stress components are expressed as suitable gradients of a vector potential, $\boldsymbol{\phi}$, as follows:

$$\sigma_{i1} = \frac{\partial \phi_i}{\partial x_2}, \quad \sigma_{i2} = -\frac{\partial \phi_i}{\partial x_1}, \tag{6.1.21a,b}$$

where ϕ_i, $i = 1,2,3$, are the components of $\boldsymbol{\phi}$.

Now, from the constitutive relations and the equations of equilibrium,

$$\sigma_{ij} = C_{ijkl}\,\varepsilon_{kl} = C_{ijkl}\,u_{k,l},$$
$$\sigma_{i\beta,\beta} = C_{i\beta k\alpha}\,u_{k,\alpha\beta} = 0, \tag{6.1.22a,b}$$

where the Latin subscripts $i, j, k, l = 1, 2, 3$, and the Greek subscripts $\alpha, \beta = 1, 2$, it follows that

$$\frac{\partial \phi_i}{\partial x_2} = C_{i1k1}\frac{\partial u_k}{\partial x_1} + C_{i1k2}\frac{\partial u_k}{\partial x_2},$$
$$-\frac{\partial \phi_i}{\partial x_1} = C_{i2k1}\frac{\partial u_k}{\partial x_1} + C_{i2k2}\frac{\partial u_k}{\partial x_2}. \tag{6.1.23a,b}$$

Therefore, setting

$$\mathbf{Q} \equiv [C_{i1k1}], \quad \mathbf{R} \equiv [C_{i1k2}], \quad \mathbf{T} \equiv [C_{i2k2}], \tag{6.1.24a-c}$$

define

$$\mathcal{N} = \begin{bmatrix} \mathbf{N}_{11} & \mathbf{N}_{12} \\ \mathbf{N}_{21} & \mathbf{N}_{11}^T \end{bmatrix}, \tag{6.1.25a}$$

where the following notation is used:[15]

---

[14] It is assumed that all quantities, *e.g.*, the coordinate variables, the displacement vector, the stress and elasticity tensors, and the stress-potential vector, are rendered physically dimensionless, using suitable length, stress, and force scales.

[15] Note that $\mathbf{N}_{12}$ and $\mathbf{N}_{21}$ are symmetric, $\mathbf{N}_{12} = \mathbf{N}_{12}^T$ and $\mathbf{N}_{21} = \mathbf{N}_{21}^T$. The matrix $\mathcal{N}$ was introduced by Ingebrigtsen and Tonning (1969). Its eigenvalue problem for anisotropic elasticity is examined by Malen (1971), and Barnett and Lothe (1973); see also Ni and Nemat-Nasser (1996) and Ting (1996) who gives a comprehensive account and review. The six-dimensional elastic field equations in terms of $\mathcal{N}$ are studied by Chadwick and Smith (1977).

$$\mathbf{N}_{11} = -\mathbf{T}^{-1}\mathbf{R}^T, \quad \mathbf{N}_{21} = \mathbf{Q} - \mathbf{R}\,\mathbf{T}^{-1}\mathbf{R}^T,$$

$$\mathbf{N}_{12} = -\mathbf{T}^{-1}. \tag{6.1.25b-d}$$

Then, the basic field equations (6.1.22) to (6.1.25) reduce to

$$\frac{\partial \boldsymbol{\eta}}{\partial x_2} = \boldsymbol{\mathcal{N}} \frac{\partial \boldsymbol{\eta}}{\partial x_1}, \tag{6.1.26a}$$

where

$$\boldsymbol{\eta} = [\mathbf{u}, \boldsymbol{\phi}]^T \tag{6.1.26b}$$

is a six-dimensional vector field with components $(u_1, u_2, u_3, \phi_1, \phi_2, \phi_3)$.

The six by six matrix $\boldsymbol{\mathcal{N}}$ is called the *fundamental elasticity matrix*. It can be shown that the fundamental elasticity matrix $\boldsymbol{\mathcal{N}}$ has no real-valued eigenvalues; see Chadwick and Smith (1977). Hence, since it is real-valued, it has three pairs of complex conjugate eigenvalues. Denote by p_k, $k = 1, 2, 3$, its first three eigenvalues with positive imaginary parts. Then, the other three eigenvalues are $p_{k+3} = \bar{p}_k$, $k = 1, 2, 3$.

Since $\boldsymbol{\mathcal{N}}$ is not symmetric, it has two sets of eigenvectors, the right- and the left-eigenvectors. Let the right-eigenvector $\boldsymbol{\zeta}_k$ correspond to the eigenvalue p_k, $k = 1, 2, ..., 6$. These eigenvalues (assumed to be distinct) and eigenvectors are defined by the following system of homogeneous linear equations:

$$\boldsymbol{\mathcal{N}}\boldsymbol{\zeta}_k = p_k\,\boldsymbol{\zeta}_k, \quad \text{(k not summed)}. \tag{6.1.27a}$$

Now, denote the left-eigenvectors by $\boldsymbol{\xi}_k$, and observe that these are defined by

$$\boldsymbol{\mathcal{N}}^T\boldsymbol{\xi}_k = p_k\,\boldsymbol{\xi}_k, \quad \text{(k not summed)}. \tag{6.1.27b}$$

The first three components of the right-eigenvector $\boldsymbol{\zeta}_k$ correspond to the displacement components, (u_1, u_2, u_3), while the second three are associated with the components of the stress-potential vector, (ϕ_1, ϕ_2, ϕ_3). Set

$$\boldsymbol{\zeta}_k = \begin{bmatrix} \mathbf{a}_k \\ \boldsymbol{l}_k \end{bmatrix}. \tag{6.1.27c}$$

Then, it can be shown that[16]

$$\boldsymbol{\xi}_k = \begin{bmatrix} \boldsymbol{l}_k \\ \mathbf{a}_k \end{bmatrix}. \tag{6.1.27d}$$

Hence, the first three components of the left-eigenvector, $\boldsymbol{\xi}_k$, correspond to the stress-potential vector, (ϕ_1, ϕ_2, ϕ_3), while the second three are associated with the components of the displacement field, (u_1, u_2, u_3), *i.e.*, $\boldsymbol{\zeta}_k$ and $\boldsymbol{\xi}_k$ are each other's dual.

---

[16] For a thorough exposition, see Ni and Nemat-Nasser (1996) and Ting (1996), where many illustrative examples, comments, and references are presented; see also Nemat-Nasser and Hori (1993, 1999), Section 21.6. The cases where there are repeated eigenvalues are examined in the first two references.

Define the six by six matrix $\mathcal{Z}$ by

$$\mathcal{Z} \equiv [\zeta_1, \zeta_2, \zeta_3, \zeta_4, \zeta_5, \zeta_6] = \begin{bmatrix} \mathbf{A} & \overline{\mathbf{A}} \\ \mathbf{L} & \overline{\mathbf{L}} \end{bmatrix}, \tag{6.1.27e}$$

where **A** and **L** are given by

$$\mathbf{A} = [\mathbf{a}_1, \mathbf{a}_2, \mathbf{a}_3], \quad \mathbf{L} = [\boldsymbol{l}_1, \boldsymbol{l}_2, \boldsymbol{l}_3]. \tag{6.1.27f,g}$$

Assume now that a straight dislocation with Burgers vector **b** is situated at the origin of the coordinate system, having an infinitely long dislocation line in the x_3-direction. This dislocation represents a discontinuity in the displacement field, described by the Burgers vector **b**. For a pure screw dislocation, the only nonzero component of **b** is b_3, while for a pure edge dislocation, $b_3 = 0$, and the Burgers vector **b** is orthogonal to the x_3-axis. The stress field is continuous everywhere (except at the dislocation core, where it is not defined). Thus the following two auxiliary conditions must be satisfied:

$$\mathbf{u}(x_1, 0^+) - \mathbf{u}(x_1, 0^-) = H(x_1)\,\mathbf{b},$$
$$\sigma_{i2}(x_1, 0^+) = \sigma_{i2}(x_1, 0^-), \quad i = 1, 2, 3, \tag{6.1.28a,b}$$

where the superscripts + and - denote the values of the corresponding quantity evaluated at the upper and lower faces of the plane $x_2 = 0$, respectively; the upper face at $x_2 = 0$ is defined by $x_2 > 0$, $x_2 \to 0$, and the lower face by $x_2 < 0$, $x_2 \to 0$; and H is the Heaviside step function.

With the aid of a Fourier transform, the basic six-dimensional equation (6.1.26) with the auxiliary conditions (6.1.28) can be solved explicitly, leading to (Stroh, 1958; Barnett and Lothe, 1973)

$$\mathbf{u}(x_1, x_2) = -\frac{1}{\pi}\, Im\{\mathbf{A}[\sum_{k=1}^{3} \log(x_1 + p_k x_2)\,\mathbf{J}_k]\,\mathbf{L}^T\}\,\mathbf{b},$$
$$\boldsymbol{\phi}(x_1, x_2) = -\frac{1}{\pi}\, Im\{\mathbf{L}[\sum_{k=1}^{3} \log(x_1 + p_k x_2)\,\mathbf{J}_k]\,\mathbf{L}^T\}\,\mathbf{b}, \tag{6.1.29a,b}$$

where the principal value of the logarithmic function $\log z$ is defined by $\log z = \log|z| + i\arg(z)$, with $0 \le \arg(z) < 2\pi$, and $\mathbf{J}_k = [\delta_{ik}\,\delta_{jk}]$ (k not summed) is a three by three matrix with zero elements except for the k-th diagonal element which is one. The corresponding stress field can be obtained from the stress potential, $\boldsymbol{\phi}$, by simple differentiation, according to (6.1.21a,b).

The periodicity of the atomic structure of a crystal provides a periodic potential field and hence, the stress field for a dislocation. The estimates of the self-stress and self-energy outlined above, account neither for this periodicity nor for the discrete structure of the crystal. Furthermore, the elasticity solution breaks down close to the center of the dislocation and hence, does not provide information about the nature and size of the dislocation core. Efforts have been made to include these effects in the estimates of the properties of dislocations, beginning with the pioneering work of Peierls (1940) and subsequently, Nabarro (1947). The corresponding stress and elastic energy are referred to as the

Peierls-Nabarro stress and energy. The issue is reexamined by Movchan *et al.* (1998), providing more rigorous detailed treatment.

6.1.7. Slip Systems

A *slip system* in a crystal is defined by a pair of unit vectors, **s** and **n**, which, respectively, define the slip direction on the slip plane and the normal to the slip plane. In general, the primary slip planes are crystallographic planes with the highest density of atoms, and the slip directions on the slip planes are the directions along which the greatest atomic packing occurs.

If a crystal model with atoms as spherical balls in contact and with their centers at the lattice points, is considered, then the octahedral planes, {111}, are the slip planes in an fcc crystal, with <110> as the slip directions.[17] For a bcc crystal, on the other hand, the primary slip planes of such a model crystal would be the {110} planes, with <111> as the slip directions. In many bcc crystals, however, slip lines are often wavy and not clearly defined. This may be attributed to simultaneous slip over several slip planes, or it may be attributed to slip on the {112} and {123}, in addition to the {110} planes. For an hcp crystal, the basal plane (0001) is usually the primary slip plane, although slip has been observed on $\{10\bar{1}1\}$ and $\{10\bar{1}2\}$, at higher temperatures; see, *e.g.*, Hirth and Lothe (1992) for details.

Many slip systems may participate in a continued plastic deformation of a crystalline solid, depending on the loading history, the temperature, and the rate of deformation. A basic analysis of the plasticity of crystalline solids, therefore, must include the kinetics and kinematics of the dislocation generation and evolution, which result in crystallographic slips and the accompanying plastic strains and rotations. In crystal plasticity, however, the plastic strains and rotations are often cast in terms of crystallographic slips, and empirical continuum models are used to relate the slip rates to the driving stress fields. The kinetics and kinematics of the dislocations are often used to *motivate* and qualitatively understand the mechanisms of slip. In Subsection 6.1.10, a relation is established, giving a slip rate in terms of certain parameters which in an average sense characterize the corresponding mobile dislocations; see expression (6.1.34a), page 402.

6.1.8. Slip Systems in fcc and bcc Crystals

For an fcc crystal, there are four {111} type slip planes, with three <110> slip directions on each plane, leading to a total of *twelve slip systems*. A transparent designation of these systems would be to define the four slip planes by their unit normals, $\mathbf{n}^{\mathrm{a}}$, $\mathrm{a} = 1, 2, 3, 4$, with the slip directions designated by $\mathbf{s}^{\mathrm{a}\alpha}$,

[17] See Appendix 6.A for a brief description of the Miller indices used here and elsewhere in this book to identify directions and planes in a crystal lattice.

$\alpha = 1, 2, 3$. Thus, slip systems of this kind are collectively denoted by ($\mathbf{n}^a$, $\mathbf{s}^{a\alpha}$). Usually, however, a single superscript (or sometimes, subscript) is used with the range 1, 2, ..., 12, to represent the slip systems which are then denoted by ($\mathbf{n}^\alpha$, $\mathbf{s}^\alpha$; $\alpha = 1, 2, ..., 12$). This latter notation is used in most of this chapter, although the former notation may be considered more transparent; see Nemat-Nasser and Okinaka (1996). In Subsection 6.5.6, the slip systems in a bcc crystal are examined in some detail. The fcc crystal is discussed in Subsections 6.5.7 to 6.5.9. An interesting example of anisotropic crystal deformation under isotropic loads, is examined in Subsections 6.5.13 and 6.5.14 for fcc crystals.

For a bcc crystal, the most densely packed directions are defined by the <111>-family. The planes {110}, {112}, and {123} are generally considered as the primary slip planes. This provides a total of 48 potential slip systems for a bcc crystal. Table 6.5.2 of Subsection 6.5.6 lists these systems in terms of the Miller indices.

6.1.9. Dislocation-induced Distortion

Consider Figure 6.1.7, and assume that the dislocation loop, l, is contained within an otherwise intact crystal of volume V. Denote the displacement field resulting from the slip, $\mathbf{b} = b\,\mathbf{s}$, of this dislocation, by $\mathbf{u}(\mathbf{x})$. The average of the gradient of the displacement field is given by

$$\begin{aligned}\overline{\nabla\otimes\mathbf{u}} &= \frac{1}{V}\int_V \nabla\otimes\mathbf{u}\,dV \\ &= \frac{1}{V}\int_{\partial V}\mathbf{n}\otimes\mathbf{u}\,dS + \frac{1}{V}\int_{S^+ + S^-}\mathbf{n}\otimes\mathbf{u}\,dS\,, \end{aligned} \qquad (6.1.30a,b)$$

where the Gauss theorem is used. The last integral can be written as

$$\begin{aligned}\int_{S^+ + S^-}\mathbf{n}\otimes\mathbf{u}\,dS &= -\int_S \mathbf{n}\otimes[\mathbf{u}]\,dS \\ &= -b\,\mathbf{n}\otimes\mathbf{s}\,S\,. \end{aligned} \qquad (6.1.31a,b)$$

The quantity, $b\,\mathbf{n}\otimes\mathbf{s}\,S/V$, therefore, is the contribution of the slip-induced distortion to the overall average distortion of the crystal.

6.1.10. Dislocation Motion and Plastic Distortion Rate

When the dislocation line AB in Figure 6.1.10 moves by an increment, $d\mathbf{r} = \mathbf{v}\,dt$, it contributes to the inelastic strain rate. Consider the dislocation loop of Figure 6.1.7(a), and let it move in its glide plane of unit normal $\mathbf{n}$. Denote the velocity of the loop at a typical point η by $v(\eta)$. This velocity is measured in the glide plane, normal to the dislocation loop. The rate of change of the dislocated area, S, is

$$\dot{S} = \int_l v(\eta)\,d\eta\,. \qquad (6.1.32a)$$

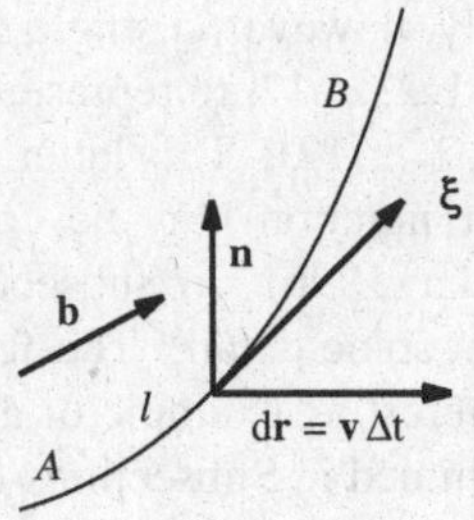

Figure 6.1.10

Dislocation line AB moves in the direction $\mathbf{dr} = \mathbf{v}\,dt$, contributing to the overall plastic distortion

Hence, from (6.1.30), the contribution of the motion of this loop to the average distortion rate becomes[18]

$$\frac{1}{V}\,b\,\mathbf{s}\otimes\mathbf{n}\,\dot{S} = \frac{1}{V}\,b\,\mathbf{s}\otimes\mathbf{n}\int_{l} v(\eta)\,d\eta\,. \tag{6.1.32b}$$

Consider now all slip planes with unit normal $\mathbf{n}$, and identify a slip direction of these planes by the unit vector $\mathbf{s}$. Seek to define the slip rate $\dot{\gamma}$ of this system in terms of the motion of the corresponding *mobile* dislocations. Let there be N mobile dislocation loops of this kind, l^a, a = 1, 2, ..., N. Denote the velocity of the dislocation loop l^a at its typical point η, by $v^a(\eta)$. This velocity is normal to the loop and lies on the corresponding glide plane. The *plastic distortion rate*, $\boldsymbol{l}^p$, associated with the motion of these dislocation loops then is

$$\boldsymbol{l}^p = \frac{1}{V}\sum_{a}^{N}\{b\int_{l^a} v^a(\eta)\,d\eta\}\,\mathbf{s}\otimes\mathbf{n}\,. \tag{6.1.33a}$$

Using the mean-value theorem, the integral can be written as

$$\int_{l^a} v^a(\eta)\,d\eta = l^a\,\bar{v}^a\,, \tag{6.1.33b}$$

where $\bar{v}^a$ is the average velocity of the dislocation loop l^a in the slip direction, $\mathbf{s}$. Hence, (6.1.33a) becomes

$$\boldsymbol{l}^p = \frac{1}{V}\sum_{a}^{N} b\,l^a\,\bar{v}^a\,\mathbf{s}\otimes\mathbf{n}\,. \tag{6.1.33c}$$

Define the slip rate of this slip system by

$$\dot{\gamma} = \frac{1}{V}\sum_{a}^{N} b\,l^a\,\bar{v}^a \approx b\,\rho_m\,\bar{v}\,, \tag{6.1.34a}$$

where ρ_m is the density of the mobile dislocations,[19] and $\bar{v}$ is an average velocity associated with the corresponding dislocations. Hence, (6.1.33c) becomes

$$\boldsymbol{l}^p = (b\,\rho_m\,\bar{v})\,\mathbf{s}\otimes\mathbf{n} = \dot{\gamma}\,\mathbf{s}\otimes\mathbf{n}\,. \tag{6.1.34b}$$

---

[18] Note that the *velocity gradient* is defined by $(\nabla\otimes\mathbf{v})^T$.

[19] This description of slip rate was proposed by Orowan (1940). Regarding (6.1.34a), since, in general, the average of the products is *not* equal to the product of the averages, (6.1.34a) is viewed as defining the average dislocation velocity, $\bar{v}$. This type of characterization of the slip rates and hence the plastic strain rate is common in the literature. The velocity $\bar{v}$ is thus some *effective* aver-

The plastic deformation rate, $\mathbf{D}^p$, and the associated spin tensor, $\mathbf{W}^p$, are now given by the symmetric and skewsymmetric parts of the above plastic distortion rate, summed over all active slip systems, as discussed in Section 6.2 below.

6.2. KINEMATICS OF FINITE DEFORMATION OF SINGLE CRYSTALS

As discussed in the preceding section, most metals sustain plastic deformation by the motion of dislocations. Dislocations tend to move on closed-packed crystallographic planes and in closed-packed directions. The activation energies associated with these crystallographic *slip systems* are lower than for any others. In general, therefore, there are a finite number of slip systems which are *active*. The plastic flow takes place by *simple shearing* over the slip planes in the slip directions. This is illustrated in Figure 6.2.1 which shows simple shearing in the $\mathbf{s}_0$-direction on the slip plane with unit normal $\mathbf{n}_0$. The slip rate is $\dot{\gamma}$ and a layer at the distance x_2 from the x_1-axis moves at velocity $\mathbf{v} = \dot{\gamma}\, x_2\, \mathbf{e}_1$; in Subsection 6.1.10, expression (6.1.34a), page 402, the slip rate $\dot{\gamma}$ is related to the density and the average velocity of the mobile dislocations.

Figure 6.2.1

Plastic flow occurs by simple shearing on crystallographic slip planes, along densely packed slip directions, as illustrated in this figure; here, $(\partial\mathbf{v}/\partial\mathbf{x})^T = \dot{\gamma}\, \mathbf{e}_1 \otimes \mathbf{e}_2 = \dot{\gamma}\, \mathbf{s}_0 \otimes \mathbf{n}_0$

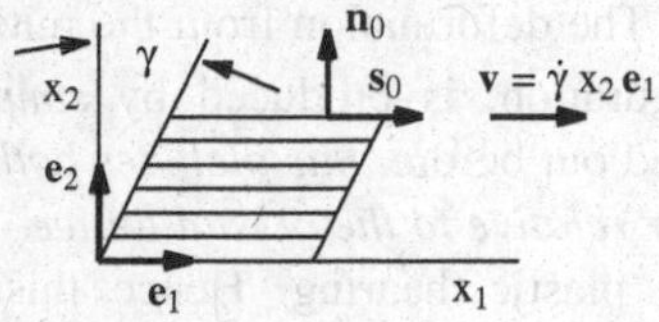

For each crystal structure, there is a well-defined set of slip systems similar to that sketched in Figure 6.2.1, which may be activated under suitable loading conditions. For example, room-temperature plastic deformation in face-centered cubic metals (*e.g.*, Al, Cu, Ni) is basically by dislocation motion on {111} crystallographic planes and in <110> directions.

For the fcc crystals, the {111}<110> are the *primary slip systems*. In certain cases, slip on these primary systems may be accompanied by slip on *secondary slip systems*. For example, at high temperatures (*e.g.*, above 450°C), aluminum slips also on {100} and other planes; see Schmid and Boas (1935, 1950). The slip direction always is the most closely packed <110> family. In body-centered crystals, {110}, {112}, and {123} are all often viewed as the primary slip planes, with <111> being the slip directions. The primary slip planes in the hexagonal closed-packed crystals, are the family of the basal planes

age measure of the velocity of the mobile dislocation lines.

{0001}, as well as the pyramidal, $\{10\bar{1}1\}$, and prismatic, $\{10\bar{1}2\}$, planes.

The inelastic deformation by crystallographic slip, leaves the lattice structure unaffected. The lattice distortion is due to elastic deformation only. The atoms and molecules are viewed to flow through the lattice in this picture of crystal plasticity. This flow consists of simple shearing on a finite number of active slip systems, resulting in both *plastic deformation and rigid-body rotation of the material relative to the lattice*.

6.2.1. Decomposition of Deformation Gradient

Consider a finitely deformed crystal with initial configuration C_0 and current deformed configuration C. Introduce two intermediate configurations, $C_{\mathbf{F}^p}$ and C_R. This is similar to those discussed in Subsection 4.9.5, page 259. The configuration C_R is obtained from the current deformed configuration C, by *relaxing the lattice elastic deformation without any rotation*; the subscript R stands for "relaxed". This is similar to that in Subsection 4.9, where the same notation is also used. Here, the mapping from C_R to the current configuration C, is due to a purely elastic lattice distortion. *Denote this purely elastic lattice distortion by* $\mathbf{V}^e$. In general, the elastic strains and rotations are infinitesimally small. Hence, the current configuration C and the elastically relaxed configuration C_R are, in most cases, essentially indistinguishable.

The deformation from the initial undeformed configuration C_0, to the $C_{\mathbf{F}^p}$ configuration, is produced by *simple shearing*, or crystallographic slip. As pointed out before, *this includes both rigid rotation and pure deformation of the matter relative to the crystal lattice.* The lattice remains unchanged in the course of this plastic shearing. Hence, this is *not* similar to that in Figure 4.9.1, page 251, where the intermediate configuration C_p is obtained by *pure plastic deformation*, $\mathbf{U}^p$, without any rigid-body rotation. It is similar to Figure 4.9.2, page 260, with the lattice vectors now defining the crystal's axes of elastic symmetry. *Let* $\mathbf{F}^p$ *denote the deformation gradient corresponding to the plastic flow of matter through the lattice by slip-induced simple shearing.*[20]

Finally, let the mapping from configuration $C_{\mathbf{F}^p}$ to C_R be a *purely rigid-body rotation*, denoted by $\mathbf{R}^*$. This then rotates the lattice from its initial orientation which is common to both C_0 and $C_{\mathbf{F}^p}$, together with the material points which are now fixed to the lattice, into its final but elastically relaxed configuration C_R; see Figure 6.2.2.

The total deformation gradient, $\mathbf{F}$, therefore, is decomposed as[21]

$$\mathbf{F} = \mathbf{F}^* \mathbf{F}^p . \tag{6.2.1a}$$

[20] Subscript $\mathbf{F}^p$ in $C_{\mathbf{F}^p}$ emphasizes the fact that both rotation and pure deformation are involved, since $\mathbf{F}^p = \mathbf{R}^{**} \mathbf{U}^p$; see (6.2.2a).

[21] In the context of crystal plasticity, this decomposition was introduced by Rice (1971) who examined the physical meaning of the slip-induced plastic deformation which is quantified by $\mathbf{F}^p$, and the rigid rotation and the elastic distortion of the lattice which are embedded in $\mathbf{F}^*$; see also Hill and

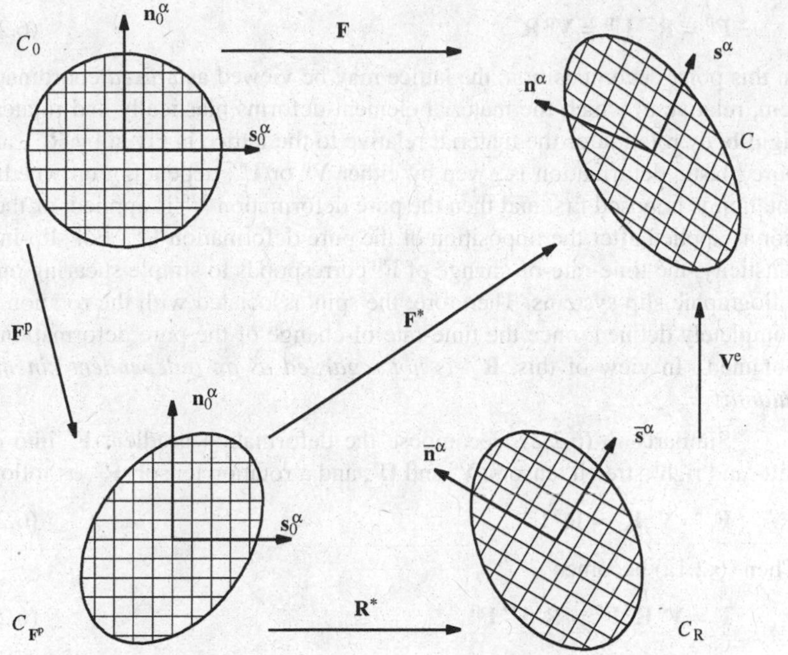

Figure 6.2.2

Decomposition of deformation gradient **F** into plastic deformation $\mathbf{F}^p$, lattice rotation $\mathbf{R}^*$, and lattice distortion (elastic deformation) $\mathbf{V}^e$, $\mathbf{F}^* = \mathbf{V}^e\,\mathbf{R}^* = \mathbf{R}^*\,\mathbf{U}^*$ and $\mathbf{F} = \mathbf{F}^*\,\mathbf{F}^p$

In general, neither $\mathbf{F}^*$ nor $\mathbf{F}^p$ may be compatible, *i.e.*, they may *not* be gradients of some smooth displacement fields. Their product, **F**, is compatible. Indeed, if **x**(**X**, t) denotes the position vector in the current configuration of a material point which is at point **X** in the initial configuration,[22] then

$$\mathbf{F} = (\partial\mathbf{x}/\partial\mathbf{X})^T\,, \tag{6.2.1b}$$

whereas a similar representation of $\mathbf{F}^*$ and $\mathbf{F}^p$ in terms of partial derivatives can only have a *symbolic* meaning, *i.e.,* without implying an actual partial differentiation.

Rice (1972), Hill and Havner (1982), and Havner (1992).

[22] The material point is identified by its initial position.

The plastic part of the deformation gradient, $\mathbf{F}^p$, may be decomposed into a rigid-body rotation, $\mathbf{R}^{**}$, and a pure deformation, $\mathbf{U}^p$ or $\mathbf{V}^p$, as follows:

$$\mathbf{F}^p = \mathbf{R}^{**}\,\mathbf{U}^p = \mathbf{V}^p\,\mathbf{R}^{**}. \tag{6.2.2a,b}$$

In this polar decomposition, the lattice may be viewed as a *fixed* coordinate system, relative to which the material element deforms plastically and rotates. The rigid-body rotation of the material relative to the lattice is given by $\mathbf{R}^{**}$, and the pure plastic deformation is given by either $\mathbf{V}^p$ or $\mathbf{U}^p$, depending on whether the rotation is imposed first and then the pure deformation $\mathbf{V}^p$ is applied, or the rotation is applied after the imposition of the pure deformation $\mathbf{U}^p$. For slip-induced plasticity, the time-rate-of-change of $\mathbf{F}^p$ corresponds to simple shearing on crystallographic slip systems. Therefore, the spin associated with the rotation $\mathbf{R}^{**}$ is completely defined, once the time-rate-of-change of the pure deformation $\mathbf{U}^p$ is obtained. In view of this, $\mathbf{R}^{**}$ *is not regarded as an independent kinematical quantity.*[23]

Similarly to (6.2.2), decompose the deformation gradient $\mathbf{F}^*$ into elastic left- and right-stretch tensors, $\mathbf{V}^e$ and $\mathbf{U}^*$, and a rotation tensor, $\mathbf{R}^*$, as follows:

$$\mathbf{F}^* = \mathbf{V}^e\,\mathbf{R}^* = \mathbf{R}^*\,\mathbf{U}^*. \tag{6.2.3a,b}$$

Then (6.2.1a) becomes

$$\mathbf{F} = \mathbf{V}^e\,\mathbf{R}^*\,\mathbf{F}^p = \mathbf{R}^*\,\mathbf{U}^*\,\mathbf{F}^p. \tag{6.2.1c,d}$$

Decomposition (6.2.1c) is shown in Figure 6.2.2, and decomposition (6.2.1d) is shown in Figure 6.2.3. In Figure 6.2.3, configuration C_{pe} is obtained from $C_{\mathbf{F}^p}$ by *pure elastic distortion*, $\mathbf{U}^*$, *of the lattice*. Since elastic deformation by lattice distortion is generally quite small, configurations $C_{\mathbf{F}^p}$ and C_{pe} are essentially indistinguishable. However, rotation $\mathbf{R}^*$ is generally finite.

From (6.2.1c) and (6.2.2a), it follows that[24]

$$\begin{aligned} \mathbf{F} &= \mathbf{V}^e\,\mathbf{R}^*\,\mathbf{R}^{**}\,\mathbf{U}^p \\ &= \mathbf{V}^e\,\mathbf{Q}\,\mathbf{U}^p = \mathbf{Q}\,\mathbf{U}^e\,\mathbf{U}^p, \\ \mathbf{Q} &\equiv \mathbf{R}^*\,\mathbf{R}^{**}, \quad \mathbf{V}^e\,\mathbf{Q} = \mathbf{Q}\,\mathbf{U}^e. \end{aligned} \tag{6.2.1e-i}$$

Here, $\mathbf{Q}$ is the total rigid-body rotation of the *material* relative to its *initial undeformed* configuration. Since the lattice orientation in the current configuration, relative to that in the initial state, is defined by the rotation $\mathbf{R}^*$, it is necessary to calculate this rotation in order to be able to establish the current orientation of the lattice.

---

[23] This can be seen from Figure 6.2.1, where the single kinematical quantity $\dot{\gamma}$ defines both the pure shearing and the accompanying spin, *i.e.*, $\dot{\gamma}\,(\mathbf{s}\otimes\mathbf{n}+\mathbf{n}\otimes\mathbf{s})/2$ and $\dot{\gamma}\,(\mathbf{s}\otimes\mathbf{n}-\mathbf{n}\otimes\mathbf{s})/2$, respectively.

[24] Note that the purely elastic stretch measures, $\mathbf{U}^e = \mathbf{Q}^T\,\mathbf{V}^e\,\mathbf{Q}$ and $\mathbf{U}^* = \mathbf{R}^*\,\mathbf{V}^e\,\mathbf{R}^{*T}$, are related by, $\mathbf{U}^* = \mathbf{R}^{**}\,\mathbf{U}^e\,\mathbf{R}^{**T}$.

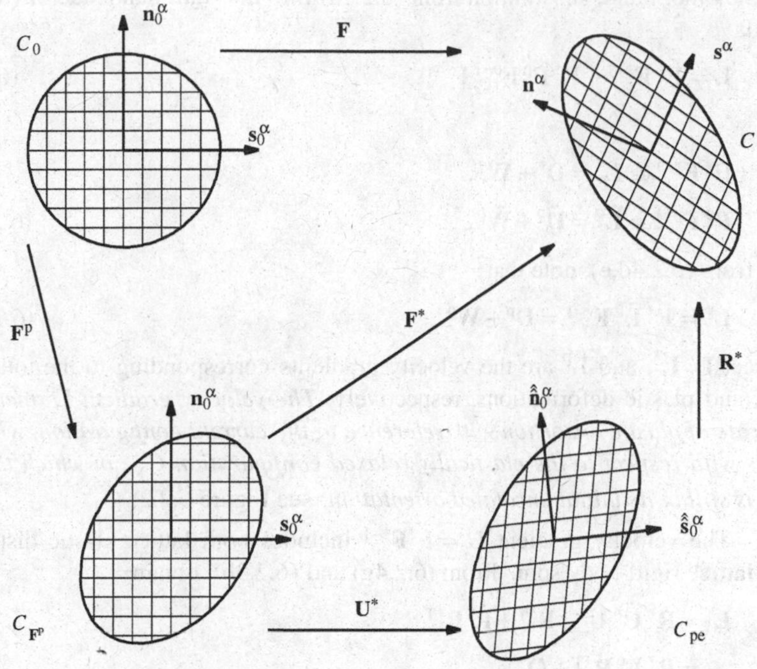

Figure 6.2.3

Decomposition of deformation gradient $\mathbf{F}$ into plastic deformation $\mathbf{F}^p$, lattice distortion (elastic deformation) $\mathbf{U}^*$, and lattice rotation $\mathbf{R}^*$, $\mathbf{F} = \mathbf{R}^* \mathbf{U}^* \mathbf{F}^p$

6.2.2. Decomposition of Velocity Gradient

The material time derivative of (6.2.1b) yields

$$\dot{\mathbf{F}} = \left[\frac{\partial \dot{\mathbf{x}}}{\partial \mathbf{X}} \right]^T. \tag{6.2.4a}$$

This is the gradient of the particle velocity with respect to the initial particle position, $\mathbf{X}$. To obtain the Eulerian velocity gradient, post multiply (6.2.4a) by $\mathbf{F}^{-1}$, arriving at

$$\mathbf{L} \equiv \dot{\mathbf{F}}\, \mathbf{F}^{-1} = \left[\frac{\partial \dot{\mathbf{x}}}{\partial \mathbf{x}} \right]^T. \tag{6.2.4b,c}$$

This velocity gradient may be decomposed into a part due to lattice rotation and lattice elastic distortion, and a part due to slip-induced plastic deformation, as follows:

$$\mathbf{L} = \mathbf{L}^* + \mathbf{L}^{\mathrm{p}}. \tag{6.2.4d}$$

On the other hand, substitution from (6.2.1a) into the right-hand side of (6.2.4b) yields

$$\mathbf{L} = \dot{\mathbf{F}}^* \mathbf{F}^{*-1} + \mathbf{F}^* \dot{\mathbf{F}}^{\mathrm{p}} \mathbf{F}^{\mathrm{p}-1} \mathbf{F}^{*-1}. \tag{6.2.4e}$$

Set

$$\dot{\mathbf{F}}^* \mathbf{F}^{*-1} \equiv \mathbf{L}^* = \mathbf{D}^* + \mathbf{W}^*,$$
$$\dot{\mathbf{F}}^{\mathrm{p}} \mathbf{F}^{\mathrm{p}-1} \equiv \tilde{\mathbf{L}}^{\mathrm{p}} = \tilde{\mathbf{D}}^{\mathrm{p}} + \tilde{\mathbf{W}}^{\mathrm{p}}, \tag{6.2.4f-i}$$

and from (6.2.4d,e), note that[25]

$$\mathbf{L}^{\mathrm{p}} = \mathbf{F}^* \tilde{\mathbf{L}}^{\mathrm{p}} \mathbf{F}^{*-1} = \mathbf{D}^{\mathrm{p}} + \mathbf{W}^{\mathrm{p}}. \tag{6.2.4j,k}$$

Hence, $\mathbf{L}$, $\mathbf{L}^*$, and $\tilde{\mathbf{L}}^{\mathrm{p}}$ are the velocity gradients corresponding to the total, lattice, and plastic deformations, respectively. *The velocity gradient* $\mathbf{L}^{\mathrm{p}}$ *measures the rate of plastic distortion with reference to the current configuration, whereas* $\tilde{\mathbf{L}}^{\mathrm{p}}$ *is with respect to the elastically relaxed configuration* $C_{\mathbf{F}^{\mathrm{p}}}$*, in which the lattice is still in its initial unrotated orientation;* see Figure 6.2.2.

The velocity gradient $\mathbf{L}^* = \dot{\mathbf{F}}^*\mathbf{F}^{*-1}$ includes both lattice elastic distortion and lattice rigid-body spin. From (6.2.4g) and (6.2.3b), obtain[26]

$$\begin{aligned}\mathbf{L}^* &= \mathbf{R}^* \dot{\mathbf{U}}^* \mathbf{U}^{*-1} \mathbf{R}^{*\mathrm{T}} + \dot{\mathbf{R}}^* \mathbf{R}^{*\mathrm{T}} \\ &= \mathbf{R}^* \hat{\hat{\mathbf{L}}}{}^{\mathrm{e}} \mathbf{R}^{*\mathrm{T}} + \boldsymbol{\Omega}^* \\ &= \mathbf{L}^{\mathrm{e}} + \boldsymbol{\Omega}^*,\end{aligned} \tag{6.2.5a-c}$$

where

$$\mathbf{L}^{\mathrm{e}} = \mathbf{R}^* \hat{\hat{\mathbf{L}}}{}^{\mathrm{e}} \mathbf{R}^{*\mathrm{T}}, \qquad \hat{\hat{\mathbf{L}}}{}^{\mathrm{e}} = \dot{\mathbf{U}}^* \mathbf{U}^{*-1},$$
$$\boldsymbol{\Omega}^* = \dot{\mathbf{R}}^* \mathbf{R}^{*\mathrm{T}}; \tag{6.2.5d,f}$$

here, $\hat{\hat{\mathbf{L}}}{}^{\mathrm{e}}$ is the elastic (objective) lattice distortion rate (which also includes spin), measured with respect to the unrotated configuration in the intermediate state, C_{pe}. The purely elastic (also objective) distortion rate, $\mathbf{L}^{\mathrm{e}}$, includes both elastic strain rate, $\mathbf{D}^{\mathrm{e}}$, and (objective) elastic spin, $\mathbf{W}^{\mathrm{e}}$,

$$\mathbf{L}^{\mathrm{e}} = \mathbf{D}^{\mathrm{e}} + \mathbf{W}^{\mathrm{e}}. \tag{6.2.5g}$$

As shown in Subsection 4.9.3, $\mathbf{W}^{\mathrm{e}}$ is proportional to $\mathbf{D}^{\mathrm{e}}$; see (4.9.20b), page 256. The contribution of the elastic spin $\mathbf{W}^{\mathrm{e}}$ to the lattice rotation, however, is insignificantly small, as the elastic strains are infinitesimally small. Note also that $\mathbf{D}^{\mathrm{e}}$ is the lattice deformation-rate tensor, *i.e.*, $\mathbf{D}^* \equiv \mathbf{D}^{\mathrm{e}}$, as is restated below in expression (6.2.8b).

---

[25] Quantities referred to configuration $C_{\mathbf{F}^{\mathrm{p}}}$ are denoted by superimposed tilde, ˜.

[26] Note that $\mathbf{L}^* = \overset{\circ}{\mathbf{V}}{}^{\mathrm{e}} \mathbf{V}^{\mathrm{e}-1} + \boldsymbol{\Omega}^*$, where $\overset{\circ}{\mathbf{V}}{}^{\mathrm{e}} = \dot{\mathbf{V}}^{\mathrm{e}} + \mathbf{V}^{\mathrm{e}} \boldsymbol{\Omega}^* - \boldsymbol{\Omega}^* \mathbf{V}^{\mathrm{e}}$.

Use the elastically relaxed configuration C_R as the reference one (see Figure 6.2.2), and obtain[27]

$$\begin{aligned}\bar{\mathbf{L}}^* &= \mathbf{V}^{e-1}\,\mathbf{L}^*\,\mathbf{V}^e \\ &= \mathbf{L}^{eT} + \mathbf{V}^{e-1}\,\boldsymbol{\Omega}^*\,\mathbf{V}^e \\ &= \mathbf{D}^e - \mathbf{W}^e + \bar{\boldsymbol{\Omega}}^*, \qquad \bar{\boldsymbol{\Omega}}^* = \mathbf{V}^{e-1}\,\boldsymbol{\Omega}^*\,\mathbf{V}^e.\end{aligned} \tag{6.2.5h-k}$$

The velocity gradient $\tilde{\mathbf{L}}^p = \dot{\mathbf{F}}^p\,\mathbf{F}^{p-1}$ includes both pure plastic distortion and rigid-body spin relative to the lattice. From (6.2.2a) and (6.2.4i), obtain

$$\begin{aligned}\tilde{\mathbf{L}}^p &= \tilde{\mathbf{D}}^p + \tilde{\mathbf{W}}^p \\ &= \dot{\mathbf{R}}^{**}\,\mathbf{R}^{**T} + \mathbf{R}^{**}\,\dot{\mathbf{U}}^p\,\mathbf{U}^{p-1}\,\mathbf{R}^{**T} \\ &= \boldsymbol{\Omega}^{**} + \mathbf{R}^{**}\,\hat{\mathbf{L}}^p\,\mathbf{R}^{**T} \\ &= \mathbf{R}^{**}\,\hat{\mathbf{D}}^p\,\mathbf{R}^{**T} + (\boldsymbol{\Omega}^{**} + \mathbf{R}^{**}\,\hat{\mathbf{W}}^p\,\mathbf{R}^{**T}),\end{aligned} \tag{6.2.6a-d}$$

where

$$\boldsymbol{\Omega}^{**} = \dot{\mathbf{R}}^{**}\,\mathbf{R}^{**T}, \quad \hat{\mathbf{L}}^p = \dot{\mathbf{U}}^p\,\mathbf{U}^{p-1} = \hat{\mathbf{D}}^p + \hat{\mathbf{W}}^p, \tag{6.2.6e-g}$$

and, hence, it follows that

$$\tilde{\mathbf{D}}^p = \mathbf{R}^{**}\,\hat{\mathbf{D}}^p\,\mathbf{R}^{**T}, \qquad \tilde{\mathbf{W}}^p = \boldsymbol{\Omega}^{**} + \mathbf{R}^{**}\,\hat{\mathbf{W}}^p\,\mathbf{R}^{**T}, \tag{6.2.6h,i}$$

where $\hat{\mathbf{W}}^p$ is objective and is proportional to $\hat{\mathbf{D}}^p$; see (4.9.18b), page 256.

With respect to the elastically relaxed configuration C_R, the plastic deformation rate and spin become,

$$\begin{aligned}\bar{\mathbf{L}}^p &= \bar{\mathbf{D}}^p + \bar{\mathbf{W}}^p \\ &= \mathbf{R}^*\,\tilde{\mathbf{D}}^p\,\mathbf{R}^{*T} + \mathbf{R}^*\,\tilde{\mathbf{W}}^p\,\mathbf{R}^{*T} \\ &= \mathbf{Q}\,\hat{\mathbf{D}}^p\,\mathbf{Q}^T + (\bar{\boldsymbol{\Omega}}^{**} + \mathbf{Q}\,\hat{\mathbf{W}}^p\,\mathbf{Q}^T),\end{aligned} \tag{6.2.6j-l}$$

where $\mathbf{Q} = \mathbf{R}^*\,\mathbf{R}^{**}$. Hence,

$$\begin{aligned}\bar{\mathbf{D}}^p &= \mathbf{R}^*\,\tilde{\mathbf{D}}^p\,\mathbf{R}^{*T} = \mathbf{Q}\,\hat{\mathbf{D}}^p\mathbf{Q}^T, \\ \bar{\mathbf{W}}^p &= \bar{\boldsymbol{\Omega}}^{**} + \mathbf{Q}\,\hat{\mathbf{W}}^p\,\mathbf{Q}^T, \qquad \bar{\boldsymbol{\Omega}}^{**} = \mathbf{R}^*\,\boldsymbol{\Omega}^{**}\,\mathbf{R}^{*T}.\end{aligned} \tag{6.2.6m-p}$$

The total velocity gradient in the elastically relaxed configuration C_R, therefore, takes on the form,

$$\begin{aligned}\bar{\mathbf{L}} &\equiv \mathbf{V}^{e-1}\,\mathbf{L}\,\mathbf{V}^e \\ &= \mathbf{L}^{eT} + \bar{\boldsymbol{\Omega}} + \mathbf{Q}\,\hat{\mathbf{L}}^p\,\mathbf{Q}^T, \qquad \bar{\boldsymbol{\Omega}} = \bar{\boldsymbol{\Omega}}^* + \bar{\boldsymbol{\Omega}}^{**}.\end{aligned} \tag{6.2.7a-c}$$

The kinematical results presented in this subsection are summarized in Table 6.2.1. These are all *exact* results.

---

[27] Quantities referred to configuration C_R are denoted by superimposed bar, -.

Table 6.2.1

Summary of kinematical relations

Decomposition $\mathbf{F} = \mathbf{F}^* \mathbf{F}^p = \mathbf{V}^e \mathbf{R}^* \mathbf{F}^p = \mathbf{V}^e \mathbf{R}^* \mathbf{R}^{**} \mathbf{U}^p = \mathbf{V}^e \mathbf{Q} \mathbf{U}^p$		
Reference Configuration	Velocity Gradients	Definition & Relations
Current, C	Total	$\mathbf{L} = \dot{\mathbf{F}} \mathbf{F}^{-1} = \mathbf{L}^* + \mathbf{L}^p$
	Lattice	$\mathbf{L}^* = \dot{\mathbf{F}}^* \mathbf{F}^{*-1} = \mathbf{D}^* + \mathbf{W}^*$
		$= \mathbf{R}^* \hat{\hat{\mathbf{L}}}^e \mathbf{R}^{*T} + \boldsymbol{\Omega}^*$
		$= \mathbf{L}^e + \boldsymbol{\Omega}^*, \quad \hat{\hat{\mathbf{L}}}^e = \dot{\mathbf{U}}^* \mathbf{U}^{*-1}$
	Plastic	$\mathbf{L}^p = \mathbf{F}^* \dot{\mathbf{F}}^p \mathbf{F}^{p-1} \mathbf{F}^{*-1} = \mathbf{D}^p + \mathbf{W}^p$
Elastically Relaxed,	Total	$\bar{\mathbf{L}} = \mathbf{V}^{e-1} \mathbf{L} \mathbf{V}^e = \bar{\mathbf{D}} + \bar{\mathbf{W}}$
Lattice Rotated		$= \bar{\mathbf{L}}^* + \bar{\mathbf{L}}^p$
but Undeformed, C_R	Lattice, Total	$\bar{\mathbf{L}}^* = \mathbf{L}^{eT} + \bar{\boldsymbol{\Omega}}^* = \mathbf{D}^e + (-\mathbf{W}^e + \bar{\boldsymbol{\Omega}}^*)$
	Lattice, Elastic (Objective)	$\mathbf{L}^{eT} = \mathbf{R}^* \mathbf{U}^{*-1} \dot{\mathbf{U}}^* \mathbf{R}^{*T} = \mathbf{D}^e - \mathbf{W}^e$
	Lattice, Rigid Spin	$\bar{\boldsymbol{\Omega}}^* = \mathbf{V}^{e-1} \boldsymbol{\Omega}^* \mathbf{V}^e, \quad \boldsymbol{\Omega}^* = \dot{\mathbf{R}}^* \mathbf{R}^{*T}$
	Lattice, Total Spin	$\bar{\mathbf{W}}^* = -\mathbf{W}^e + \bar{\boldsymbol{\Omega}}^*$
	Lattice, Deformation Rate	$\mathbf{D}^e = \mathbf{D}^*$
	Material Spin, Total	$\bar{\mathbf{W}} = -\mathbf{W}^e + \bar{\boldsymbol{\Omega}}^* + \bar{\mathbf{W}}^p$
	Plastic, Total	$\bar{\mathbf{L}}^p = \mathbf{R}^* \tilde{\mathbf{L}}^p \mathbf{R}^{*T}, \quad \tilde{\mathbf{L}}^p = \dot{\mathbf{F}}^p \mathbf{F}^{p-1}$
	Plastic, Deformation Rate	$\bar{\mathbf{D}}^p = \mathbf{R}^* \tilde{\mathbf{D}}^p \mathbf{R}^{*T} = \mathbf{Q} \hat{\mathbf{D}}^p \mathbf{Q}^T$
	Plastic, Spin	$\bar{\mathbf{W}}^p = \mathbf{R}^* \tilde{\mathbf{W}}^p \mathbf{R}^{*T}$
		$= \bar{\boldsymbol{\Omega}}^{**} + \mathbf{Q} \hat{\mathbf{W}}^p \mathbf{Q}^T$
		$\hat{\mathbf{W}}^p = \frac{1}{2}(\dot{\mathbf{U}}^p \mathbf{U}^{p-1} - \mathbf{U}^{p-1} \dot{\mathbf{U}}^p)$
	Total, Rigid Spin	$\bar{\boldsymbol{\Omega}} = \bar{\boldsymbol{\Omega}}^* + \bar{\boldsymbol{\Omega}}^{**}$
Plastically Deformed,	Total	$\tilde{\mathbf{L}} = \mathbf{F}^{*-1} \dot{\mathbf{F}}^* + \dot{\mathbf{F}}^p \mathbf{F}^{p-1} = \tilde{\mathbf{L}}^* + \tilde{\mathbf{L}}^p$
Lattice Unrotated, $C_{\mathbf{F}^p}$	Lattice, Total	$\tilde{\mathbf{L}}^* = \mathbf{F}^{*-1} \dot{\mathbf{F}}^* = \tilde{\mathbf{D}}^* + \tilde{\mathbf{W}}^*$
	Plastic, Total	$\tilde{\mathbf{L}}^p = \dot{\mathbf{F}}^p \mathbf{F}^{p-1} = \tilde{\mathbf{D}}^p + \tilde{\mathbf{W}}^p$
		$= \boldsymbol{\Omega}^{**} + \mathbf{R}^{**} \hat{\mathbf{L}}^p \mathbf{R}^{**T}$
	Slip-induced, Rigid Spin	$\boldsymbol{\Omega}^{**} = \dot{\mathbf{R}}^{**} \mathbf{R}^{**T}$
	Plastic (Objective)	$\hat{\mathbf{L}}^p = \dot{\mathbf{U}}^p \mathbf{U}^{p-1}$

The velocity gradient $\mathbf{L}^* = \dot{\mathbf{F}}^* \mathbf{F}^{*-1}$ may be decomposed into a symmetric and a skewsymmetric part,

$$\mathbf{L}^* = \mathbf{D}^* + \mathbf{W}^*, \qquad \mathbf{D}^* = \mathbf{D}^e, \qquad \mathbf{W}^* = \mathbf{W}^e + \boldsymbol{\Omega}^* . \tag{6.2.8a-c}$$

For small elastic strains, set $\mathbf{V}^e = \mathbf{1} + \boldsymbol{\varepsilon}$, and from (6.2.4j,k) and (6.2.6f,g), obtain

$$\begin{aligned} \mathbf{L}^p &= \mathbf{V}^e (\bar{\mathbf{D}}^p + \bar{\mathbf{W}}^p) \mathbf{V}^{e-1} \\ &= (\mathbf{1} + \boldsymbol{\varepsilon})(\bar{\mathbf{D}}^p + \bar{\mathbf{W}}^p)(\mathbf{1} - \boldsymbol{\varepsilon}) + \mathrm{O}(\boldsymbol{\varepsilon}^2) \\ &= (\bar{\mathbf{D}}^p + \boldsymbol{\varepsilon} \bar{\mathbf{W}}^p - \bar{\mathbf{W}}^p \boldsymbol{\varepsilon}) + (\bar{\mathbf{W}}^p + \boldsymbol{\varepsilon} \bar{\mathbf{D}}^p - \bar{\mathbf{D}}^p \boldsymbol{\varepsilon}) + \mathrm{O}(\boldsymbol{\varepsilon}^2) . \end{aligned} \tag{6.2.9a-c}$$

Therefore, it follows that

$$\mathbf{D}^p = \overline{\mathbf{D}}^p + \boldsymbol{\varepsilon}\,\overline{\mathbf{W}}^p - \overline{\mathbf{W}}^p\,\boldsymbol{\varepsilon} + O(\boldsymbol{\varepsilon}^2)\,.$$

$$\mathbf{W}^p = \overline{\mathbf{W}}^p + \boldsymbol{\varepsilon}\,\overline{\mathbf{D}}^p - \overline{\mathbf{D}}^p\,\boldsymbol{\varepsilon} + O(\boldsymbol{\varepsilon}^2)\,. \tag{6.2.9d,e}$$

Even though the elastic strain, $\boldsymbol{\varepsilon}$, may be small, the rigid-body spin, $\overline{\boldsymbol{\Omega}}^{**}$, and the plastic deformation rate, $\overline{\mathbf{D}}^p$, may be large, so that the corresponding terms in the right-hand side of (6.2.9b,c) may actually be significant.

6.2.3. Plastic Distortion

Denote the slip direction and slip normal of a typical slip system in the *undeformed configuration* by $\mathbf{s}_0^\alpha$ and $\mathbf{n}_0^\alpha$, respectively. As stated before, *the plastic deformation leaves the lattice structure unaffected. Hence, the slip direction and slip normal in the intermediate configuration, C_{F^p}, are the same as those in the initial undeformed configuration, C_0*; see Figure 6.2.2. In view of this fact, the plastic part of the velocity gradient due to crystallographic slip may be expressed by (see Figure 6.2.1)

$$\tilde{\mathbf{L}}^p = \sum_{\alpha=1}^{n} \dot{\gamma}^\alpha\, \mathbf{s}_0^\alpha \otimes \mathbf{n}_0^\alpha\,, \quad \tilde{L}_{ij} = \sum_{\alpha=1}^{n} \dot{\gamma}^\alpha\, s_{0i}{}^\alpha\, n_{0j}{}^\alpha\,, \tag{6.2.10a,b}$$

where $\dot{\gamma}^\alpha$ is the slip rate of the α'th slip system measured relative to the undeformed lattice, n is the total number of active slip systems in the single crystal, and i, j = 1, 2, 3. If a slip system is inactive, then the corresponding slip rate would be zero. Hence, n can be set equal to the number of all *potentially* active slip systems. For example, in the case of fcc single crystals, n = 12.

For a typical slip system, define the following symmetric and skewsymmetric second-order tensors (α not summed):

$$\mathbf{p}_0^\alpha \equiv \frac{1}{2}\,(\mathbf{s}_0^\alpha \otimes \mathbf{n}_0^\alpha + \mathbf{n}_0^\alpha \otimes \mathbf{s}_0^\alpha)\,,$$

$$\mathbf{r}_0^\alpha \equiv \frac{1}{2}\,(\mathbf{s}_0^\alpha \otimes \mathbf{n}_0^\alpha - \mathbf{n}_0^\alpha \otimes \mathbf{s}_0^\alpha)\,. \tag{6.2.11a,b}$$

The symmetric and skewsymmetric parts of the velocity gradient (6.2.10a), then become

$$\tilde{\mathbf{D}}^p = \sum_{\alpha=1}^{n} \dot{\gamma}^\alpha\, \mathbf{p}_0^\alpha\,,$$

$$\tilde{\mathbf{W}}^p = \sum_{\alpha=1}^{n} \dot{\gamma}^\alpha\, \mathbf{r}_0^\alpha\,. \tag{6.2.10c,d}$$

To obtain an expression for the plastic part of the velocity gradient, measured relative to the current configuration, $\mathbf{L}^p$, substitute (6.2.10a) into (6.2.4j), arriving at

$$\mathbf{L}^p = \sum_{\alpha=1}^{n} \dot{\gamma}^\alpha\, \mathbf{s}^\alpha \otimes \mathbf{n}^\alpha\,, \quad L_{ij} = \sum_{\alpha=1}^{n} \dot{\gamma}^\alpha\, s_i{}^\alpha\, n_j{}^\alpha\,, \tag{6.2.12a,b}$$

where

$$\mathbf{s}^\alpha = \mathbf{F}^* \mathbf{s}_0^\alpha, \quad \mathbf{n}^\alpha = \mathbf{F}^{*-T} \mathbf{n}_0^\alpha. \tag{6.2.13a,b}$$

As is shown in Figure 6.2.2, $\mathbf{F}^* = \mathbf{V}^e \mathbf{R}^*$ first rotates by $\mathbf{R}^*$ the lattice from configuration C_{F^p}, and then it deforms the lattice elastically by $\mathbf{V}^e$ into the final state C. The elastic deformation changes the lattice only slightly, but the rigid-body rotation by $\mathbf{R}^*$ may be finite. *In general,* $\mathbf{s}^\alpha$ and $\mathbf{n}^\alpha$ are *not* unit vectors, although they remain normal to each other. However, the deviation of their lengths from unity is of the order of the elastic strains only, and is generally neglected. Here, these vectors are not regarded as unit vectors.

As noted before, $\dot{\gamma}^\alpha$ is the slip rate of the α'th slip system, measured with respect to the elastically relaxed (undeformed) lattice. Since the elastic deformation, defined by $\mathbf{V}^e$ (or $\mathbf{U}^*$), stretches the lattice, the actual slip rate differs from $\dot{\gamma}^\alpha$ by the factor $\lambda^*_{(s)}/\lambda^*_{(n)}$, where $\lambda^*_{(s)}$ is the elastic lattice stretch in the slip direction, and $\lambda^*_{(n)}$ is that in the direction normal to the slip plane. Normalize the deformed vectors, $\mathbf{s}^\alpha$ and $\mathbf{n}^\alpha$, defined by (6.2.13a,b), to have unit lengths, and rewrite (6.2.12a) as

$$\mathbf{L}^p = \sum_{\alpha=1}^{n} (\dot{\gamma}^\alpha \lambda^*_{(s)}/\lambda^*_{(n)}) (\mathbf{s}^\alpha \otimes \mathbf{n}^\alpha \lambda^*_{(n)}/\lambda^*_{(s)}), \tag{6.2.12c}$$

where $(\dot{\gamma}^\alpha \lambda^*_{(s)}/\lambda^*_{(n)})$ now is the slip rate measured in the deformed lattice, using orthogonal unit vectors. Observe that the orthogonal vectors, $\mathbf{s}^\alpha$ and $\mathbf{n}^\alpha$, form a pair of reciprocal base vectors; see Section 1.3, page 18.

The symmetric and skewsymmetric parts of $\mathbf{L}^p$ are expressed by

$$\mathbf{D}^p = \sum_{\alpha=1}^{n} \dot{\gamma}^\alpha \mathbf{p}^\alpha,$$

$$\mathbf{W}^p = \sum_{\alpha=1}^{n} \dot{\gamma}^\alpha \mathbf{r}^\alpha, \tag{6.2.12d,e}$$

where

$$\mathbf{p}^\alpha = \frac{1}{2} (\mathbf{s}^\alpha \otimes \mathbf{n}^\alpha + \mathbf{n}^\alpha \otimes \mathbf{s}^\alpha),$$

$$\mathbf{r}^\alpha = \frac{1}{2} (\mathbf{s}^\alpha \otimes \mathbf{n}^\alpha - \mathbf{n}^\alpha \otimes \mathbf{s}^\alpha). \tag{6.2.14a,b}$$

These expressions measure the corresponding quantities with respect to the current configuration, C.

It is often more convenient to use the *rotated but elastically relaxed configuration* C_R as the reference one. As in Subsection 4.9, denote quantities referred to configuration C_R by superimposed bar. The slip directions and the unit vectors normal to the slip planes then become

$$\bar{\mathbf{s}}^\alpha = \mathbf{R}^* \mathbf{s}_0^\alpha, \quad \bar{\mathbf{n}}^\alpha = \mathbf{R}^* \mathbf{n}_0^\alpha, \tag{6.2.15a,b}$$

where $\bar{\mathbf{s}}^\alpha$ and $\bar{\mathbf{n}}^\alpha$ are *orthogonal unit vectors*. The plastically-induced velocity gradient, $\bar{\mathbf{L}}^p$, measured relative to C_R, now is

$$\bar{\mathbf{L}}^p = \sum_{\alpha=1}^{n} \dot{\gamma}^\alpha \bar{\mathbf{s}}^\alpha \otimes \bar{\mathbf{n}}^\alpha. \tag{6.2.16a}$$

The corresponding deformation rate and spin tensors then become

$$\overline{\mathbf{D}}^{\mathrm{p}} = \sum_{\alpha=1}^{n} \dot{\gamma}^{\alpha} \overline{\mathbf{p}}^{\alpha},$$

$$\overline{\mathbf{W}}^{\mathrm{p}} = \sum_{\alpha=1}^{n} \dot{\gamma}^{\alpha} \overline{\mathbf{r}}^{\alpha}, \qquad (6.2.16b,c)$$

where

$$\overline{\mathbf{p}}^{\alpha} = \frac{1}{2}(\overline{\mathbf{s}}^{\alpha} \otimes \overline{\mathbf{n}}^{\alpha} + \overline{\mathbf{n}}^{\alpha} \otimes \overline{\mathbf{s}}^{\alpha}),$$

$$\overline{\mathbf{r}}^{\alpha} = \frac{1}{2}(\overline{\mathbf{s}}^{\alpha} \otimes \overline{\mathbf{n}}^{\alpha} - \overline{\mathbf{n}}^{\alpha} \otimes \overline{\mathbf{s}}^{\alpha}), \quad \alpha \text{ not summed.} \qquad (6.2.17a,b)$$

In most cases, the associated "barred" and "unbarred" quantities are essentially indistinguishable, as they differ from each other by terms of the order of the elastic strains. Here, however, the barred and the unbarred quantities are viewed as distinct measures.

6.2.4. Elastic Lattice Distortion Rate and Spin

The elastic lattice distortion rate and spin can be described in terms of the time-rate-of-change of the vectors $\mathbf{s}^{\alpha}$ and $\mathbf{n}^{\alpha}$, which define the current direction of the α'th slip on the slip plane of normal $\mathbf{n}^{\alpha}$. These orthogonal vectors define both the rotation and the elastic distortion of the corresponding lattice system. The lattice spin, on the other hand, can be expressed in terms of the time-rate-of-change of $\overline{\mathbf{s}}^{\alpha}$ and $\overline{\mathbf{n}}^{\alpha}$. As pointed out before, the orthogonal $\mathbf{s}^{\alpha}$ and $\mathbf{n}^{\alpha}$ are not unit vectors, whereas $\overline{\mathbf{s}}^{\alpha}$ and $\overline{\mathbf{n}}^{\alpha}$ are, although the deviation from unity for the first set of vectors, is of the order of the elastic strains only.

Consider first the current configuration as reference and from (6.2.13a) obtain,

$$\dot{\mathbf{s}}^{\alpha} = \dot{\mathbf{F}}^{*} \mathbf{s}_0^{\alpha} = (\dot{\mathbf{F}}^{*} \mathbf{F}^{*-1}) \mathbf{F}^{*} \mathbf{s}_0^{\alpha} = \mathbf{L}^{*} \mathbf{s}^{\alpha}. \qquad (6.2.18a\text{-}c)$$

Similarly, from (6.2.13b) it follows that[28]

$$\dot{\mathbf{n}}^{\alpha} = -\mathbf{L}^{*\mathrm{T}} \mathbf{n}^{\alpha}. \qquad (6.2.18d)$$

Using these results, observe that

$$(\mathbf{s}^{\alpha} \otimes \mathbf{n}^{\alpha})^{\boldsymbol{\cdot}} = \mathbf{L}^{*} (\mathbf{s}^{\alpha} \otimes \mathbf{n}^{\alpha}) - (\mathbf{s}^{\alpha} \otimes \mathbf{n}^{\alpha}) \mathbf{L}^{*}. \qquad (6.2.19a)$$

Hence, the material time derivative of (6.2.14) becomes

$$\dot{\mathbf{p}}^{\alpha} = \mathbf{W}^{*} \mathbf{p}^{\alpha} - \mathbf{p}^{\alpha} \mathbf{W}^{*} + (\mathbf{D}^{*} \mathbf{r}^{\alpha} - \mathbf{r}^{\alpha} \mathbf{D}^{*}),$$

$$\dot{\mathbf{r}}^{\alpha} = \mathbf{W}^{*} \mathbf{r}^{\alpha} - \mathbf{r}^{\alpha} \mathbf{W}^{*} + (\mathbf{D}^{*} \mathbf{p}^{\alpha} - \mathbf{p}^{\alpha} \mathbf{D}^{*}), \qquad (6.2.19b,c)$$

---

[28] Since $\mathbf{F}\mathbf{F}^{-1} = \mathbf{1}$, it follows that $\mathbf{F}(\mathbf{F}^{-1})^{\boldsymbol{\cdot}} = -\dot{\mathbf{F}}\mathbf{F}^{-1} = -\mathbf{L}$. Hence, $(\mathbf{F}^{-\mathrm{T}})^{\boldsymbol{\cdot}}\mathbf{F}^{\mathrm{T}} = -\mathbf{F}^{-\mathrm{T}}\dot{\mathbf{F}}^{\mathrm{T}} = -\mathbf{L}^{\mathrm{T}}$. Similar expressions hold for the "starred" quantities.

where (6.2.8a) is also used; note that $\mathbf{D}^*$ is the symmetric part, while $\mathbf{W}^*$ is the skewsymmetric part of the velocity gradient $\mathbf{L}^*$. The last two terms (within parentheses) in the right-hand side of (6.2.19b,c) are due to the elastic distortion of the lattice. These terms are hence of the order of the elastic strains; see (6.2.8).

Consider now the lattice spin with reference to the elastically relaxed configuration C_R. First note from (6.2.15a,b) that

$$\dot{\bar{\mathbf{s}}}^\alpha = \boldsymbol{\Omega}^* \bar{\mathbf{s}}^\alpha, \quad \dot{\bar{\mathbf{n}}}^\alpha = \boldsymbol{\Omega}^* \bar{\mathbf{n}}^\alpha, \tag{6.2.20a,b}$$

where $\boldsymbol{\Omega}^*$, defined by (6.2.5f), is the lattice rigid-body spin. Now, in view of (6.2.15a,b), obtain

$$(\bar{\mathbf{s}}^\alpha \otimes \bar{\mathbf{n}}^\alpha)^\bullet = \boldsymbol{\Omega}^* (\bar{\mathbf{s}}^\alpha \otimes \bar{\mathbf{n}}^\alpha) - (\bar{\mathbf{s}}^\alpha \otimes \bar{\mathbf{n}}^\alpha)\, \boldsymbol{\Omega}^*. \tag{6.2.20c}$$

Then, the time rates-of-change of $\bar{\mathbf{p}}^\alpha$ and $\bar{\mathbf{r}}^\alpha$ become

$$\dot{\bar{\mathbf{p}}}^\alpha = \boldsymbol{\Omega}^* \bar{\mathbf{p}}^\alpha - \bar{\mathbf{p}}^\alpha \boldsymbol{\Omega}^*,$$
$$\dot{\bar{\mathbf{r}}}^\alpha = \boldsymbol{\Omega}^* \bar{\mathbf{r}}^\alpha - \bar{\mathbf{r}}^\alpha \boldsymbol{\Omega}^*. \tag{6.2.20d,e}$$

Comparing (6.2.20d,e) with (6.2.19b,c), note that the difference between them is of the order of the elastic strain rates only.

Consider now the decompositions (6.2.1c,d) and the corresponding configurations C_0, $C_{\mathbf{F}^\mathrm{p}}$, C_pe, and C, as shown in Figure 6.2.3, page 407. Based on (6.2.3b), $\mathbf{L}^* = \dot{\mathbf{F}}^* \mathbf{F}^{*-1}$ yields (6.2.5a) which is written as,

$$\mathbf{L}^* = \boldsymbol{\Omega}^* + \mathbf{R}^* \dot{\mathbf{U}}^* \mathbf{U}^{*-1} \mathbf{R}^{*\mathrm{T}}, \tag{6.2.21}$$

where $\boldsymbol{\Omega}^*$ is defined by (6.2.5f).

Consider the configuration C_pe, in which *the lattice is elastically distorted but is unrotated.* Denote quantities referred to this configuration, C_pe, by superimposed double carets, $\hat{\hat{}}$. The total velocity gradient becomes

$$\hat{\hat{\mathbf{L}}} = \mathbf{R}^{*\mathrm{T}} \mathbf{L} \mathbf{R}^* = \hat{\hat{\mathbf{L}}}{}^* + \hat{\hat{\mathbf{L}}}{}^\mathrm{p},$$
$$\hat{\hat{\mathbf{L}}}{}^* = \mathbf{R}^{*\mathrm{T}} \mathbf{L}^* \mathbf{R}^*,$$
$$\hat{\hat{\mathbf{L}}}{}^\mathrm{p} \equiv \mathbf{R}^{*\mathrm{T}} \mathbf{L}^\mathrm{p} \mathbf{R}^* = \mathbf{U}^* \tilde{\mathbf{L}}^\mathrm{p} \mathbf{U}^{*-1}. \tag{6.2.22a-e}$$

Also, in view of (6.2.10a), note that

$$\hat{\hat{\mathbf{s}}}{}_0^\alpha = \mathbf{U}^* \mathbf{s}_0^\alpha, \quad \hat{\hat{\mathbf{n}}}{}_0^\alpha = \mathbf{U}^{*-1} \mathbf{n}_0^\alpha. \tag{6.2.23a,b}$$

These are orthogonal but not unit vectors. From (6.2.22) and (6.2.23), now obtain

$$\hat{\hat{\mathbf{L}}}{}^\mathrm{p} = \sum_{\alpha=1}^{n} \dot{\gamma}^\alpha \hat{\hat{\mathbf{s}}}{}_0^\alpha \otimes \hat{\hat{\mathbf{n}}}{}_0^\alpha. \tag{6.2.24a}$$

The symmetric and skewsymmetric parts of $\hat{\hat{\mathbf{L}}}{}^\mathrm{p}$ are

$$\hat{\hat{\mathbf{D}}}{}^\mathrm{p} = \frac{1}{2}(\hat{\hat{\mathbf{L}}}{}^\mathrm{p} + \hat{\hat{\mathbf{L}}}{}^{\mathrm{pT}}) = \sum_{\alpha=1}^{n} \dot{\gamma}^\alpha \hat{\hat{\mathbf{p}}}{}_0^\alpha,$$

$$\hat{\hat{\mathbf{W}}}^{\mathrm{p}} = \frac{1}{2}(\hat{\hat{\mathbf{L}}}^{\mathrm{p}} - \hat{\hat{\mathbf{L}}}^{\mathrm{pT}}) = \sum_{\alpha=1}^{n} \dot{\gamma}^{\alpha} \hat{\hat{\mathbf{r}}}_0^{\alpha}, \tag{6.2.24b,c}$$

$$\hat{\hat{\mathbf{p}}}_0^{\alpha} = \frac{1}{2}(\hat{\hat{\mathbf{s}}}_0^{\alpha} \otimes \hat{\hat{\mathbf{n}}}_0^{\alpha} + \hat{\hat{\mathbf{n}}}_0^{\alpha} \otimes \hat{\hat{\mathbf{s}}}_0^{\alpha}),$$

$$\hat{\hat{\mathbf{r}}}_0^{\alpha} = \frac{1}{2}(\hat{\hat{\mathbf{s}}}_0^{\alpha} \otimes \hat{\hat{\mathbf{n}}}_0^{\alpha} - \hat{\hat{\mathbf{n}}}_0^{\alpha} \otimes \hat{\hat{\mathbf{s}}}_0^{\alpha}) \quad (\alpha \text{ not summed}). \tag{6.2.23c,d}$$

In view of (6.2.22) and (6.2.23), note that

$$\dot{\hat{\hat{\mathbf{s}}}}_0^{\alpha} = (\hat{\mathbf{L}}^* - \hat{\mathbf{\Omega}}^*)\, \hat{\hat{\mathbf{s}}}_0^{\alpha},$$

$$\dot{\hat{\hat{\mathbf{n}}}}_0^{\alpha} = -(\hat{\mathbf{L}}^* - \hat{\mathbf{\Omega}}^*)^{\mathrm{T}}\, \hat{\hat{\mathbf{n}}}_0^{\alpha},$$

$$\hat{\mathbf{\Omega}}^* = \mathbf{R}^{*\mathrm{T}} \mathbf{\Omega}^* \mathbf{R}^*. \tag{6.2.25a-c}$$

From this and by taking the material time derivative of (6.2.23c,d), obtain

$$\dot{\hat{\hat{\mathbf{p}}}}_0^{\alpha} = (\hat{\mathbf{W}}^* - \hat{\mathbf{\Omega}}^*)\, \hat{\hat{\mathbf{p}}}_0^{\alpha} - \hat{\hat{\mathbf{p}}}_0^{\alpha} (\hat{\mathbf{W}}^* - \hat{\mathbf{\Omega}}^*) + \hat{\mathbf{D}}^* \hat{\hat{\mathbf{r}}}_0^{\alpha} - \hat{\hat{\mathbf{r}}}_0^{\alpha} \hat{\mathbf{D}}^*,$$

$$\dot{\hat{\hat{\mathbf{r}}}}_0^{\alpha} = (\hat{\mathbf{W}}^* - \hat{\mathbf{\Omega}}^*)\, \hat{\hat{\mathbf{r}}}_0^{\alpha} - \hat{\hat{\mathbf{r}}}_0^{\alpha} (\hat{\mathbf{W}}^* + \hat{\mathbf{\Omega}}^*) + \hat{\mathbf{D}}^* \hat{\hat{\mathbf{p}}}_0^{\alpha} - \hat{\hat{\mathbf{p}}}_0^{\alpha} \hat{\mathbf{D}}^*. \tag{6.2.26a,b}$$

6.2.5. Small Lattice Distortion

The plastic deformation and lattice rotation are generally large. On the other hand, the lattice distortion (elastic strain) is very small for essentially all metals and many other materials. Hence, the configurations C and C_{R} in Figure 6.2.2, may be viewed as essentially indistinguishable. The current elastically distorted lattice vectors, $\mathbf{s}^{\alpha}$ and $\mathbf{n}^{\alpha}$, may hence be identified by their rotated but undeformed image, $\bar{\mathbf{s}}^{\alpha}$ and $\bar{\mathbf{n}}^{\alpha}$, in configuration C_{R}, *i.e.*,

$$\mathbf{s}^{\alpha} = \mathbf{F}^* \mathbf{s}_0^{\alpha} \approx \mathbf{R}^* \mathbf{s}_0^{\alpha} = \bar{\mathbf{s}}^{\alpha}, \quad \mathbf{n}^{\alpha} = \mathbf{F}^{*-\mathrm{T}} \mathbf{n}_0^{\alpha} \approx \mathbf{R}^* \mathbf{n}_0^{\alpha} = \bar{\mathbf{n}}^{\alpha}. \tag{6.2.27a,b}$$

The corresponding rate quantities then become

$$\dot{\mathbf{s}}^{\alpha} = \mathbf{L}^* \mathbf{s}^{\alpha} \approx \mathbf{W}^* \mathbf{s}^{\alpha},$$

$$\dot{\mathbf{n}}^{\alpha} = -\mathbf{L}^{*\mathrm{T}} \mathbf{n}^{\alpha} \approx \mathbf{W}^* \mathbf{n}^{\alpha},$$

$$\dot{\mathbf{p}}^{\alpha} \approx \mathbf{W}^* \mathbf{p}^{\alpha} - \mathbf{p}^{\alpha} \mathbf{W}^*,$$

$$\dot{\mathbf{r}}^{\alpha} \approx \mathbf{W}^* \mathbf{r}^{\alpha} - \mathbf{r}^{\alpha} \mathbf{W}^*, \tag{6.2.28a-d}$$

where $\mathbf{\Omega}^* \approx \mathbf{W}^*$.

6.3. CONSTITUTIVE EQUATIONS FOR SINGLE CRYSTALS

In this section, constitutive relations are developed for the stress rates in terms of the elastic and inelastic deformation measures and their rates. First, exact results are presented, without any assumption regarding the magnitude of the elastic distortion of the lattice. Then, these results are simplified for cases in which the elastic strains are negligibly small. No restriction is placed on the anisotropic nature of the elasticity of the crystal lattice.

6.3.1. Crystal Elasticity

The *elastic* distortion of the lattice produces the stress field in the crystal. This elastic distortion may be measured by the right-stretch tensor, $\mathbf{U}^*$, relative to the undistorted and unrotated lattice in configuration $C_{\mathbf{F}^p}$; see Figure 6.2.3, page 407. The elastically relaxed configuration $C_{\mathbf{F}^p}$ is also a natural state, with respect to which the elastic stress potential, $\tilde{\phi}(\tilde{\mathbf{E}}^e)$, may be measured.[29] The elastic potential, $\tilde{\phi}$, depends on the structure of the crystal, and may change if this structure changes (*e.g.*, phase transformation), although the elasticity of most metals remains essentially unaffected by the accumulated dislocation-induced plastic deformations, up to rather large strains. Hence, in general, the elastic potential, $\tilde{\phi}$, may depend parametrically on internal variables that account for possible structural changes.

Consider crystals which deform plastically through the motion of dislocations. Experiments show that their elasticity is basically unaffected by the presence of the dislocations, up to rather large dislocation densities. To account for the crystal's elastic anisotropy, it is most convenient to define the elasticity of the crystal directly in the undeformed and unrotated lattice coordinate system in configuration $C_{\mathbf{F}^p}$, and then transfer the results to other suitable reference coordinates.

While the strain measure can be arbitrary, a possible choice is the Lagrangian (elastic) strain,

$$\tilde{\mathbf{E}}^e = \tfrac{1}{2}(\mathbf{F}^{*T}\mathbf{F}^* - \mathbf{1}), \quad \mathbf{F}^* = \mathbf{R}^*\mathbf{U}^* = \mathbf{V}^e\mathbf{R}^*, \quad \mathbf{F}^{*T}\mathbf{F}^* = \mathbf{U}^{*2}, \tag{6.3.1a-c}$$

with an associated stress (see (4.9.29a)),

$$\tilde{\mathbf{S}}^p = \mathbf{F}^{*-1}\,\boldsymbol{\tau}\,\mathbf{F}^{*-T} = \frac{\partial\tilde{\phi}}{\partial\tilde{\mathbf{E}}^e}\,. \tag{6.3.2a,b}$$

The corresponding elasticity tensor, $\tilde{\mathcal{L}}$, measured relative to $C_{\mathbf{F}^p}$, then becomes

$$\tilde{\mathcal{L}} = \frac{\partial^2\tilde{\phi}}{\partial\tilde{\mathbf{E}}^e\,\partial\tilde{\mathbf{E}}^e}\,,$$

$$\tilde{\mathcal{L}}_{ijkl} = \frac{\partial^2\tilde{\phi}}{\partial\tilde{E}^e_{ij}\,\partial\tilde{E}^e_{kl}}\,, \quad i, j, k, l = 1, 2, 3\,, \tag{6.3.3a,b}$$

and the stress rate relates to the strain rate by[30]

[29] As before, quantities which are measured with respect to the elastically relaxed and unrotated lattice in configuration $C_{\mathbf{F}^p}$, are denoted by superimposed tilde, ˜.

[30] As pointed out before, the elasticity tensor, $\tilde{\mathcal{L}}_{ijkl}$, may conveniently be first defined in the (unrotated and elastically relaxed) lattice coordinates where crystal anisotropy can be naturally

$$\dot{\tilde{\mathbf{S}}}^{\mathrm{p}} = \tilde{\mathcal{L}} : \dot{\tilde{\mathbf{E}}}^{\mathrm{e}}, \tag{6.3.4}$$

where, from (6.3.1) and (6.3.2), it follows that

$$\dot{\tilde{\mathbf{E}}}^{\mathrm{e}} = \mathbf{F}^{*\mathrm{T}} \mathbf{D}^{*} \mathbf{F}^{*}, \tag{6.3.5}$$

$$\dot{\tilde{\mathbf{S}}}^{\mathrm{p}} = \mathbf{F}^{*-1} (\dot{\boldsymbol{\tau}} - \mathbf{L}^{*} \boldsymbol{\tau} - \boldsymbol{\tau} \mathbf{L}^{*\mathrm{T}}) \mathbf{F}^{*-\mathrm{T}}. \tag{6.3.6a}$$

From (6.3.6a), now obtain

$$\dot{\boldsymbol{\tau}} = \mathbf{L}^{*} \boldsymbol{\tau} + \boldsymbol{\tau} \mathbf{L}^{*\mathrm{T}} + \mathbf{F}^{*} \dot{\tilde{\mathbf{S}}}^{\mathrm{p}} \mathbf{F}^{*\mathrm{T}}, \tag{6.3.6b}$$

and, in view of (6.3.4), arrive at

$$\dot{\boldsymbol{\tau}} = \mathbf{L}^{*} \boldsymbol{\tau} + \boldsymbol{\tau} \mathbf{L}^{*\mathrm{T}} + \mathcal{L}(\mathbf{U}^{*}) : \mathbf{D}^{*}, \tag{6.3.7a}$$

where

$$\mathcal{L}_{\mathrm{ijkl}} = \mathrm{F}^{*}_{\mathrm{ia}} \mathrm{F}^{*}_{\mathrm{jb}} \mathrm{F}^{*}_{\mathrm{kc}} \mathrm{F}^{*}_{\mathrm{ld}} \tilde{\mathcal{L}}_{\mathrm{abcd}}, \tag{6.3.7b}$$

is the instantaneous elasticity tensor, in general, a function of elastic strain, $\mathbf{U}^{*}$.

6.3.2. Crystal Elastoplasticity Relative to Current Configuration

The goal is to express an objective time-rate-of-change of the Kirchhoff stress, $\boldsymbol{\tau}$, in terms of the total and plastic deformation rates, $\mathbf{D}$ and $\mathbf{D}^{\mathrm{p}}$. The choice of the objective stress rate depends on the choice of the strain measure. As a starting point, consider the objective rate of change of the Kirchhoff stress,

$$\overset{\nabla}{\boldsymbol{\tau}}{}^{*} \equiv \dot{\boldsymbol{\tau}} - \mathbf{L}^{*} \boldsymbol{\tau} - \boldsymbol{\tau} \mathbf{L}^{*\mathrm{T}}, \tag{6.3.8a}$$

with[31] $\mathbf{L}^{*} = \dot{\mathbf{F}}^{*} \mathbf{F}^{*-1}$. It relates *exactly* to the elastic deformation rate tensor, $\mathbf{D}^{*}$, by

$$\overset{\nabla}{\boldsymbol{\tau}}{}^{*} = \mathcal{L} : \mathbf{D}^{*} = \mathcal{L} : (\mathbf{D} - \mathbf{D}^{\mathrm{p}}), \tag{6.3.8b,c}$$

where $\mathcal{L}$ is defined in terms of $\tilde{\mathcal{L}}$ by (6.3.7b), $\mathbf{D}^{\mathrm{p}}$ is given by (6.2.12d),[32] and $\mathbf{D}^{*}$ is defined by (6.2.8a,b).

It is important to note the practical usefulness of the constitutive relation (6.3.8c) for rate-dependent plasticity. Given the current orientation of the lattice, ($\mathbf{s}^{\alpha}$, $\mathbf{n}^{\alpha}$, $\alpha = 1, 2, ..., n$), and the constitutive relations for the slip rates, $\dot{\gamma}^{\alpha}$, the plastic parts of the deformation rate and spin tensors, $\mathbf{D}^{\mathrm{p}}$ and $\mathbf{W}^{\mathrm{p}}$, are defined by (6.2.12d,e). Then, if the velocity gradient, $\mathbf{L} = \dot{\mathbf{F}} \mathbf{F}^{-1}$, is prescribed, $\mathbf{L}^{*} = \mathbf{L} - \mathbf{L}^{\mathrm{p}}$ can be used in the left-hand side of (6.3.8b) to express $\overset{\nabla}{\boldsymbol{\tau}}{}^{*}$ in terms of the slip rates, $\dot{\gamma}^{\alpha}$, the Kirchhoff stress, $\boldsymbol{\tau}$, and its rate, $\dot{\boldsymbol{\tau}}$. This leads to a set of

identified, and then the results transferred to other suitable coordinate systems.

[31] The stress rate $\overset{\nabla}{\boldsymbol{\tau}}{}^{*}$ here is defined by $\overset{\nabla}{\boldsymbol{\tau}}{}^{*} = \mathbf{F}^{*} (\mathbf{F}^{*-1} \boldsymbol{\tau} \mathbf{F}^{*-\mathrm{T}})^{\boldsymbol{\cdot}} \mathbf{F}^{*\mathrm{T}}$.

[32] The calculation of (6.3.8a-d) follows the steps detailed in Subsection 4.9.5; see expressions (4.9.31a,b), page 261.

differential equations for the stress rate $\dot{\boldsymbol{\tau}}$, to be integrated over a given time increment.

As an example, assume that $\dot{\gamma}^\alpha$'s depend only on the stress,[33] $\boldsymbol{\tau}$, and some internal variables, $\boldsymbol{\beta}$, for which constitutive relations are known, *e.g.*, (4.9.3), page 247. Then, based on the above-explained procedure, the following differential equations are obtained from (6.3.8a-c), for the stress rate tensor, $\dot{\boldsymbol{\tau}}$:

$$\dot{\boldsymbol{\tau}} = \boldsymbol{\mathcal{L}} : (\mathbf{D} - \sum_{\alpha=1}^{n} \dot{\gamma}^\alpha \mathbf{p}^\alpha) + (\mathbf{L} - \sum_{\alpha=1}^{n} \dot{\gamma}^\alpha \mathbf{s}^\alpha \otimes \mathbf{n}^\alpha) \boldsymbol{\tau}$$
$$+ \boldsymbol{\tau} (\mathbf{L} - \sum_{\alpha=1}^{n} \dot{\gamma}^\alpha \mathbf{s}^\alpha \otimes \mathbf{n}^\alpha)^T . \tag{6.3.8d}$$

If, at an instant, $\mathbf{L}$ and hence $\mathbf{D} = \frac{1}{2}(\mathbf{L} + \mathbf{L}^T)$ are given, ($\mathbf{s}^\alpha$, $\mathbf{n}^\alpha$, $\alpha = 1, 2,..., n$) and $\boldsymbol{\tau}$ are known, and $\dot{\gamma}^\alpha$'s are prescribed as functions of $\boldsymbol{\tau}$ and internal variables, $\boldsymbol{\beta}$, then equations (6.3.8d) and (4.9.3) can be integrated (at least numerically) over a time increment, to obtain the new values of the stress tensor and the internal variables.

While expression (6.3.8d) is *exact*, the actual computations are usually simplified using the fact that the elastic distortions and their rates are generally small, and that, for most metals, the elastic shear modulus is orders of magnitude greater than the yield stress. This means that $\boldsymbol{\mathcal{L}}$ in (6.3.8d) may be replaced by $\bar{\boldsymbol{\mathcal{L}}}$, since $\boldsymbol{\mathcal{L}} = \bar{\boldsymbol{\mathcal{L}}} + O(\boldsymbol{\varepsilon})$, where $\mathbf{V}^e = \mathbf{1} + \boldsymbol{\varepsilon}$, and

$$\bar{\mathcal{L}}_{ijkl} = R^*_{ia} R^*_{jb} R^*_{kc} R^*_{ld} \tilde{\mathcal{L}}_{abcd} . \tag{6.3.7c}$$

In addition, $\mathbf{L}^*$ in (6.3.8a) may be replaced by $\mathbf{W}^*$, or, indeed, to the first order in $\boldsymbol{\varepsilon}$, by $\boldsymbol{\Omega}^* = \dot{\mathbf{R}}^* \mathbf{R}^{*T}$. In principle, however, the exact expression (6.3.8d) may be equally effective. At any rate, it provides a framework from which various approximate expressions may be deduced systematically.

Consider now the stress rate convected with the *total deformation gradient*, $\mathbf{F}$,

$$\overset{\nabla}{\boldsymbol{\tau}} = \mathbf{F} (\mathbf{F}^{-1} \boldsymbol{\tau} \mathbf{F}^{-T})^{\boldsymbol{\cdot}} \mathbf{F}^T = \dot{\boldsymbol{\tau}} - \mathbf{L} \boldsymbol{\tau} - \boldsymbol{\tau} \mathbf{L}^T$$
$$= \overset{\nabla}{\boldsymbol{\tau}}{}^* - \mathbf{L}^p \boldsymbol{\tau} - \boldsymbol{\tau} \mathbf{L}^{pT} . \tag{6.3.9a-c}$$

Constitutive relation (6.3.8c) then takes on the form,

$$\overset{\nabla}{\boldsymbol{\tau}} = \boldsymbol{\mathcal{L}} : \mathbf{D} - \sum_{\alpha=1}^{n} \dot{\gamma}^\alpha \boldsymbol{\lambda}^\alpha , \tag{6.3.10a}$$

where $\boldsymbol{\lambda}^\alpha$ is defined by

$$\boldsymbol{\lambda}^\alpha \equiv \boldsymbol{\mathcal{L}} : \mathbf{p}^\alpha + (\mathbf{s}^\alpha \otimes \mathbf{n}^\alpha \boldsymbol{\tau} + \boldsymbol{\tau} \mathbf{n}^\alpha \otimes \mathbf{s}^\alpha) . \tag{6.3.10b}$$

Compare with (4.4.1) of Chapter 4, page 184, and note that (6.3.10a) is formulated with reference to the current configuration, whereas the reference state in

---

[33] See Sections 6.4. and 6.5, where various constitutive relations for slip rates are discussed.

Subsection 4.4.1 is arbitrary.

Expression (6.3.10a) can be inverted to yield the deformation rate in terms of the stress rate, as follows:

$$\mathbf{D} = \boldsymbol{\mathcal{M}} : \overset{\nabla}{\boldsymbol{\tau}} + \sum_{\alpha=1}^{n} \dot{\gamma}^{\alpha} \boldsymbol{\mu}^{\alpha},$$

$$\boldsymbol{\mu}^{\alpha} \equiv \boldsymbol{\mathcal{M}} : \boldsymbol{\lambda}^{\alpha} = \mathbf{p}^{\alpha} + \boldsymbol{\mathcal{M}} : (\mathbf{s}^{\alpha} \otimes \mathbf{n}^{\alpha} \boldsymbol{\tau} + \boldsymbol{\tau} \mathbf{n}^{\alpha} \otimes \mathbf{s}^{\alpha}), \tag{6.3.10c,d}$$

where $\boldsymbol{\mathcal{M}} = \boldsymbol{\mathcal{L}}^{-1}$ is the current elastic compliance tensor.

In equation (6.3.10c), the last term in the right-hand side, $\sum_{\alpha=1}^{n} \dot{\gamma}^{\alpha} \boldsymbol{\mu}^{\alpha}$, is due to the plastic slip contribution to the deformation rate tensor, $\mathbf{D}$. The quantity $\sum_{\alpha=1}^{n} \dot{\gamma}^{\alpha} \boldsymbol{\mu}^{\alpha}$, therefore, is the effective plastic deformation rate tensor associated with the Lagrangian strain measure, when the current state is used as the reference one. This quantity is strain measure-dependent; see, *e.g.*, (6.3.15d), page 421. In (6.3.10a), $\boldsymbol{\mathcal{L}} : \mathbf{D}$ would be the stress rate if the crystal response to the imposed deformation rate, $\mathbf{D}$, were purely elastic. Hence, $-\sum_{\alpha=1}^{n} \dot{\gamma}^{\alpha} \boldsymbol{\lambda}^{\alpha}$ is *the decrement in the stress rate* $\overset{\nabla}{\boldsymbol{\tau}}$, *due to the plastic deformation rate which accompanies the elastic deformation rate.* For the sake of referencing, this decrement in the stress rate is *denoted* by

$$\overset{\nabla}{\boldsymbol{\tau}}{}^{\text{in}} \equiv - \sum_{\alpha=1}^{n} \dot{\gamma}^{\alpha} \boldsymbol{\lambda}^{\alpha}. \tag{6.3.10e}$$

Expression (6.3.10a) may now be rewritten as

$$\overset{\nabla}{\boldsymbol{\tau}} = \boldsymbol{\mathcal{L}} : \mathbf{D} + \overset{\nabla}{\boldsymbol{\tau}}{}^{\text{in}}. \tag{6.3.10f}$$

It is emphasized that $\overset{\nabla}{\boldsymbol{\tau}}{}^{\text{in}}$ is just a notation since there is no such physical quantity as "the plastic stress rate". On the other hand, $\mathbf{D}^{\text{p}}$ is indeed the rate of deformation of the crystal due to plastic slip on crystallographic slip planes, a physical quantity that actually can be observed and related to the motion of dislocations. Note that $\overset{\nabla}{\boldsymbol{\tau}}{}^{\text{in}}$ *is independent of the chosen strain measure.*

The quantity[34]

$$\tau^{\alpha} = \boldsymbol{\tau} : \mathbf{p}^{\alpha} \tag{6.3.11a}$$

is a measure of the *resolved shear stress* acting on the α'th slip system. This quantity, τ^{α}, is invariant under a change of the strain measure. It is called the *Taylor-Schmid stress*; Havner (1992). Its rate of change is given by

$$\dot{\tau}^{\alpha} = \overset{\nabla}{\boldsymbol{\tau}}{}^{*} : \mathbf{p}^{\alpha} + (\mathbf{s}^{\alpha} \otimes \mathbf{n}^{\alpha} \boldsymbol{\tau} + \boldsymbol{\tau} \mathbf{n}^{\alpha} \otimes \mathbf{s}^{\alpha}) : \mathbf{D}^{*}, \tag{6.3.11b}$$

---

[34] Since $\mathbf{s}^{\alpha}$ and $\mathbf{n}^{\alpha}$ are *not* unit vectors, $\mathbf{p}^{\alpha}$ must be normalized, in order for $\boldsymbol{\tau} : \mathbf{p}^{\alpha}$ to be the resolved shear stress of the α'th slip system. This correction, however, is of the order of the elastic strains, and, indeed, τ^{α} is essentially the resolved shear stress of the α'th slip system.

as can be verified using (6.2.18a-d), where $\mathbf{D}^* = \mathbf{D} - \sum_{\beta=1}^{n} \dot{\gamma}^\beta \mathbf{p}^\beta$. From (6.3.11b) and (6.3.8b), obtain

$$\dot{\tau}^\alpha = \{\boldsymbol{\mathcal{L}} : \mathbf{p}^\alpha + (\mathbf{s}^\alpha \otimes \mathbf{n}^\alpha \boldsymbol{\tau} + \boldsymbol{\tau}\, \mathbf{n}^\alpha \otimes \mathbf{s}^\alpha)\} : \mathbf{D}^* = \boldsymbol{\lambda}^\alpha : \mathbf{D}^*, \tag{6.3.11c,d}$$

where $\boldsymbol{\lambda}^\alpha$ is given by (6.3.10b). Furthermore, from (6.3.11c) and (6.3.8b), obtain

$$\dot{\tau}^\alpha = \boldsymbol{\lambda}^\alpha : \mathbf{D}^* = \boldsymbol{\lambda}^\alpha : \boldsymbol{\mathcal{M}} : \overset{\nabla}{\boldsymbol{\tau}}{}^* = \boldsymbol{\mu}^\alpha : \overset{\nabla}{\boldsymbol{\tau}}{}^*, \tag{6.3.11e-g}$$

where $\boldsymbol{\mu}^\alpha$ is defined in (6.3.10d). Expressions (6.3.10) and (6.3.11) are in accord with the general results of Hill and Rice (1972), leading to the normality rule which is discussed in the next section.

Consider now the Jaumann rate of the Kirchhoff stress, $\overset{\circ}{\boldsymbol{\tau}}$, *corotational with the material element,*

$$\overset{\circ}{\boldsymbol{\tau}} = \dot{\boldsymbol{\tau}} - \mathbf{W}\boldsymbol{\tau} + \boldsymbol{\tau}\mathbf{W}, \tag{6.3.12a}$$

and define the Jaumann rate of the Kirchhoff stress, $\overset{\circ}{\boldsymbol{\tau}}{}^*$, *corotational with the lattice,* by[35]

$$\overset{\circ}{\boldsymbol{\tau}}{}^* = \dot{\boldsymbol{\tau}} - \mathbf{W}^*\boldsymbol{\tau} + \boldsymbol{\tau}\mathbf{W}^*. \tag{6.3.12b}$$

Hence, since $\mathbf{W} = \mathbf{W}^* + \mathbf{W}^p$,

$$\overset{\circ}{\boldsymbol{\tau}} = \overset{\circ}{\boldsymbol{\tau}}{}^* - \mathbf{W}^p\boldsymbol{\tau} + \boldsymbol{\tau}\mathbf{W}^p. \tag{6.3.12c}$$

Now, in view of (6.3.8a), note that

$$\begin{aligned} \overset{\nabla}{\boldsymbol{\tau}}{}^* &= \overset{\circ}{\boldsymbol{\tau}}{}^* - (\mathbf{D}^*\boldsymbol{\tau} + \boldsymbol{\tau}\mathbf{D}^*) \\ &= \overset{\circ}{\boldsymbol{\tau}} + (\mathbf{W}^p\boldsymbol{\tau} - \boldsymbol{\tau}\mathbf{W}^p) - (\mathbf{D}^*\boldsymbol{\tau} + \boldsymbol{\tau}\mathbf{D}^*). \end{aligned} \tag{6.3.13a,b}$$

Therefore, constitutive relation (6.3.8c) becomes

$$\overset{\circ}{\boldsymbol{\tau}} = \boldsymbol{\mathcal{L}} : (\mathbf{D} - \mathbf{D}^p) - (\mathbf{W}^p\boldsymbol{\tau} - \boldsymbol{\tau}\mathbf{W}^p) + (\mathbf{D}^*\boldsymbol{\tau} + \boldsymbol{\tau}\mathbf{D}^*). \tag{6.3.14a}$$

This equation may be written as

$$\overset{\circ}{\boldsymbol{\tau}} = \boldsymbol{\mathcal{L}} : (\mathbf{D} - \mathbf{D}^p) + (\mathbf{D}\boldsymbol{\tau} + \boldsymbol{\tau}\mathbf{D}) - (\mathbf{L}^p\boldsymbol{\tau} + \boldsymbol{\tau}\mathbf{L}^{pT}), \tag{6.3.14b}$$

or, equivalently, as

$$\overset{\circ}{\boldsymbol{\tau}} = \boldsymbol{\mathcal{L}}' : \mathbf{D} - \boldsymbol{\mathcal{L}}' : \mathbf{D}^p - (\mathbf{W}^p\boldsymbol{\tau} - \boldsymbol{\tau}\mathbf{W}^p), \tag{6.3.15a}$$

where

$$\mathcal{L}'_{ijkl} = \mathcal{L}_{ijkl} + \frac{1}{2}(\delta_{il}\tau_{jk} + \delta_{jl}\tau_{ik} + \delta_{ik}\tau_{jl} + \delta_{jk}\tau_{il}). \tag{6.3.15b}$$

Here, $\boldsymbol{\mathcal{L}}'$ is the instantaneous elasticity tensor associated with the logarithmic

[35] Note that $\mathbf{W}^*$ includes spin due to the elastic distortion of the lattice, see (6.2.8a-f); however, this elastic contribution is generally very small, and, hence, $\mathbf{W}^* \approx \boldsymbol{\Omega}^*$ which is pure rigid lattice spin.

strain, as is shown in the next subsection. Constitutive relation (6.3.15a) may be written as[36]

$$\overset{\circ}{\boldsymbol{\tau}} = \mathcal{L}' : \mathbf{D} - \sum_{\alpha=1}^{n} \dot{\gamma}^{\alpha} \boldsymbol{\lambda}^{\alpha}, \tag{6.3.15c}$$

where $\boldsymbol{\lambda}^{\alpha}$ is defined by (6.3.10b). As shown by Hill and Rice (1972), the difference, $\overset{\circ}{\boldsymbol{\tau}} - \mathcal{L}' : \mathbf{D} = -\sum_{\alpha=1}^{n} \dot{\gamma}^{\alpha} \boldsymbol{\lambda}^{\alpha}$, is independent of the chosen strain measure. This is further discussed in the following subsection.

The inverse of (6.3.15c) is

$$\mathbf{D} = \mathcal{M}' : \overset{\circ}{\boldsymbol{\tau}} + \sum_{\alpha=1}^{n} \dot{\gamma}^{\alpha} \boldsymbol{\mu}'^{\alpha},$$

$$\boldsymbol{\mu}'^{\alpha} = \mathcal{M}' : \boldsymbol{\lambda}^{\alpha} = \mathbf{p}^{\alpha} + \mathcal{M}' : (\mathbf{r}^{\alpha} \boldsymbol{\tau} - \boldsymbol{\tau} \mathbf{r}^{\alpha}), \tag{6.3.15d-f}$$

where $\mathcal{M}' = \mathcal{L}'^{-1}$. As is evident from (6.3.15e,f), unlike $\boldsymbol{\lambda}^{\alpha}$, the symmetric second-order tensor $\boldsymbol{\mu}'^{\alpha}$ is strain measure-dependent.

Similarly to expression (6.3.10e), here it will also prove convenient to introduce the *notation*

$$\overset{\circ}{\boldsymbol{\tau}}{}^{\text{in}} \equiv - \sum_{\alpha=1}^{n} \dot{\gamma}^{\alpha} \boldsymbol{\lambda}^{\alpha} \tag{6.3.15g}$$

for *the stress rate decrement* which results from the plastic strain rate, $\mathbf{D}^{\text{p}}$, accompanying the elastic deformation rate, $\mathbf{D}^{*}$. Note that *this decrement is the same no matter which objective stress rate is employed*, as is further discussed in the following subsection.

Finally, in view of (6.3.12c) and (6.3.14b), obtain

$$\overset{\circ}{\boldsymbol{\tau}}{}^{*} = \mathcal{L}' : \mathbf{D}^{*} = \mathcal{L}' : (\mathbf{D} - \mathbf{D}^{\text{p}})$$

$$= \mathcal{L}' : (\mathbf{D} - \sum_{\alpha=1}^{n} \dot{\gamma}\, \mathbf{p}^{\alpha}). \tag{6.3.16a-c}$$

The inverse relation is

$$\mathbf{D} = \mathcal{M}' : \overset{\circ}{\boldsymbol{\tau}}{}^{*} + \sum_{\alpha=1}^{n} \dot{\gamma}^{\alpha} \mathbf{p}^{\alpha}, \tag{6.3.16d}$$

and the time rate of change of the Taylor-Schmid stress, $\dot{\tau}^{\alpha}$, becomes

$$\dot{\tau}^{\alpha} = \boldsymbol{\lambda}^{\alpha} : \mathbf{D}^{*} = \boldsymbol{\mu}'^{\alpha} : \overset{\circ}{\boldsymbol{\tau}}{}^{*}. \tag{6.3.17a,b}$$

Note that $(\mathbf{D}^{*} \boldsymbol{\tau} + \boldsymbol{\tau} \mathbf{D}^{*})$ in (6.3.13a) is usually neglected, since it is generally insignificantly small, being of the order of the elastic strain rate times the yield stress. For application to cases where such approximations are not allowed, constitutive relation (6.3.15a) provides an attractive exact expression.

---

[36] Note that $\boldsymbol{\lambda}^{\alpha} = \mathcal{L} : \mathbf{p}^{\alpha} + (\mathbf{s}^{\alpha} \otimes \mathbf{n}^{\alpha} \boldsymbol{\tau} + \boldsymbol{\tau} \mathbf{n}^{\alpha} \otimes \mathbf{s}^{\alpha}) = \mathcal{L}' : \mathbf{p}^{\alpha} + (\mathbf{r}^{\alpha} \boldsymbol{\tau} - \boldsymbol{\tau} \mathbf{r}^{\alpha})$, where $\mathcal{L}' : \mathbf{p}^{\alpha} = \mathcal{L} : \mathbf{p}^{\alpha} + (\mathbf{p}^{\alpha} \boldsymbol{\tau} + \boldsymbol{\tau} \mathbf{p}^{\alpha})$.

6.3.3. Crystal Elastoplasticity with General Strain and Stress Measures

The strain measure with its conjugate stress measure can be general, *i.e.*, any of the quantities discussed in Chapter 2, Section 2.3, and Chapter 3, Section 3.1. The relations among the corresponding objective stress-rate measures are discussed in Section 3.5. In particular the general relation (3.5.3), due to Hill (1968), provides a natural means of connecting the elasticity relations expressed in terms of any general strain measure with that expressed in terms of the logarithmic strain measure, $ln\,\mathbf{U}$. When the current configuration is the reference one, the Cauchy stress, $\boldsymbol{\sigma}$, equals the Kirchhoff stress, $\boldsymbol{\tau}$, and (3.5.3) becomes

$$\dot{\mathbf{S}} = \overset{\circ}{\boldsymbol{\tau}} - \frac{1}{2}(\mathrm{f}''(1)+1)(\mathbf{D}\boldsymbol{\tau}+\boldsymbol{\tau}\mathbf{D}), \tag{6.3.18a}$$

where $\overset{\circ}{\boldsymbol{\tau}}$ is the Jaumann rate of the Kirchhoff stress given by (6.3.12a), and $\dot{\mathbf{S}}$ is a stress-rate measure conjugate to the chosen material strain measure, $\mathbf{E}$. If the logarithmic strain measure is chosen, then $\mathrm{f}(x) = ln\,x$ and hence $\mathrm{f}''(1) = -1$, showing that the Jaumann rate of the Kirchhoff stress, $\overset{\circ}{\boldsymbol{\tau}}$, indeed corresponds to the logarithmic strain measure. Choosing the Lagrangian strain measure, $\mathbf{E}^{L}$, the right-hand side of (6.3.18a) gives

$$\overset{\circ}{\boldsymbol{\tau}} - (\mathbf{D}\boldsymbol{\tau}+\boldsymbol{\tau}\mathbf{D}) = \overset{\nabla}{\boldsymbol{\tau}}, \tag{6.3.18b}$$

where the convected time derivative of the Kirchhoff stress, $\overset{\nabla}{\boldsymbol{\tau}}$, is defined by (6.3.9a,b). Thus, rewriting (6.3.18a) as

$$\dot{\mathbf{S}} = \overset{\nabla}{\boldsymbol{\tau}} - \frac{1}{2}(\mathrm{f}''(1)-1)(\mathbf{D}\boldsymbol{\tau}+\boldsymbol{\tau}\mathbf{D}), \tag{6.3.19a}$$

and substituting from (6.3.10a) for $\overset{\nabla}{\boldsymbol{\tau}}$, obtain the following general relation for the stress tensor $\mathbf{S}$, conjugate to an arbitrary material strain measure, $\mathbf{E}$, of the class discussed in Chapter 2:

$$\dot{\mathbf{S}} = \boldsymbol{\mathcal{L}}:\mathbf{D} - \frac{1}{2}(\mathrm{f}''(1)-1)(\mathbf{D}\boldsymbol{\tau}+\boldsymbol{\tau}\mathbf{D}) - \sum_{\alpha=1}^{n}\dot{\gamma}^{\alpha}\boldsymbol{\lambda}^{\alpha}, \tag{6.3.19b}$$

where $\boldsymbol{\mathcal{L}}$ and $\boldsymbol{\lambda}^{\alpha}$ are defined by (6.3.8d) and (6.3.10b), respectively, $\boldsymbol{\mathcal{L}}$ corresponding to the Lagrangian strain measure.

Note that expressions (6.3.18a) and (6.3.19a) also apply when only the elastic strains are involved. In such a case, (6.3.18a) becomes

$$\dot{\mathbf{S}}^{*} = \overset{\circ}{\boldsymbol{\tau}}{}^{*} - \frac{1}{2}(\mathrm{f}''(1)+1)(\mathbf{D}^{*}\boldsymbol{\tau}+\boldsymbol{\tau}\mathbf{D}^{*}), \tag{6.3.18c}$$

with a similar expression for the equation corresponding to (6.3.19a). With the Lagrangian strain, $\mathrm{f}(x) = (x^2-1)/2$, and (6.3.18c) reduces to (6.3.13a).

6.3.4. Crystal Elastoplasticity Relative to Elastically Relaxed Lattice

Equivalent *exact* formulations are obtained with respect to the elastically relaxed lattice in either the unrotated configuration $C_{\mathbf{F}^p}$ or the rotated configuration $C_{\mathbf{R}}$, in both of which the material is plastically deformed (stretched and rotated by $\mathbf{U}^p$ and $\mathbf{R}^{**}$, respectively); see Figure 6.2.3, page 407. As before, denote quantities measured with respect to $C_{\mathbf{F}^p}$ by superimposed tilde, ˜, and those relative to C_R by superimposed bar, -. Then, the Lagrangian elastic strain relative to the configuration C_R becomes

$$\overline{\mathbf{E}}^e = \frac{1}{2}(\mathbf{F}^* \mathbf{F}^{*T} - \mathbf{1}) = \mathbf{R}^* \tilde{\mathbf{E}}^e \mathbf{R}^{*T} = \frac{1}{2}(\mathbf{V}^{e2} - \mathbf{1}). \tag{6.3.20a-c}$$

The associated stress tensor is

$$\begin{aligned}\overline{\mathbf{S}}^p &= \mathbf{R}^* \tilde{\mathbf{S}}^p \mathbf{R}^{*T} \\ &= \mathbf{R}^* \frac{\partial \tilde{\phi}}{\partial \tilde{\mathbf{E}}^e} \mathbf{R}^{*T} = \frac{\partial \overline{\phi}}{\partial \overline{\mathbf{E}}^e},\end{aligned} \tag{6.3.21a-c}$$

where $\overline{\phi} = \overline{\phi}(\overline{\mathbf{E}}^e) = \tilde{\phi}(\tilde{\mathbf{E}}^e)$; note that the mass densities in configurations C_R and $C_{\mathbf{F}^p}$ are the same. The relation between the stress rate and the elastic strain rate takes on the form

$$\overset{\circ}{\overline{\mathbf{S}}}{}^p = \overline{\mathcal{L}} : \overset{\circ}{\overline{\mathbf{E}}}{}^e, \tag{6.3.22a}$$

where the objective stress and strain rates, $\overset{\circ}{\overline{\mathbf{S}}}{}^p$ and $\overset{\circ}{\overline{\mathbf{E}}}{}^e$ are respectively given by[37]

$$\overset{\circ}{\overline{\mathbf{S}}}{}^p = \dot{\overline{\mathbf{S}}}{}^p - \boldsymbol{\Omega}^* \overline{\mathbf{S}}^p + \overline{\mathbf{S}}^p \boldsymbol{\Omega}^*, \quad \overset{\circ}{\overline{\mathbf{E}}}{}^e = \dot{\overline{\mathbf{E}}}{}^e - \boldsymbol{\Omega}^* \overline{\mathbf{E}}^e + \overline{\mathbf{E}}^e \boldsymbol{\Omega}^*, \tag{6.3.22b,c}$$

and the elasticity tensor $\overline{\mathcal{L}}_{ijkl}$ is defined in terms of $\tilde{\mathcal{L}}_{abcd}$ by expression (6.3.7c), which is equivalent to

$$\begin{aligned}\overline{\mathcal{L}}_{ijkl} &= \frac{\partial^2 \overline{\phi}}{\partial \overline{E}^e_{ij} \partial \overline{E}^e_{kl}} \\ &= R^*_{ia} R^*_{jb} R^*_{kc} R^*_{ld} \frac{\partial^2 \tilde{\phi}}{\partial \tilde{E}^e_{ab} \partial \tilde{E}^e_{cd}} \\ &= R^*_{ia} R^*_{jb} R^*_{kc} R^*_{ld} \tilde{\mathcal{L}}_{abcd}.\end{aligned} \tag{6.3.22d-f}$$

The stress rate $\dot{\tilde{\mathbf{S}}}{}^p$ and the strain rate $\dot{\tilde{\mathbf{E}}}{}^e$ are both objective. These quantities relate to the (also objective) measures $\overset{\circ}{\overline{\mathbf{S}}}{}^p$ and $\overset{\circ}{\overline{\mathbf{E}}}{}^e$ by $\dot{\tilde{\mathbf{S}}}{}^p = \mathbf{R}^{*T} \overset{\circ}{\overline{\mathbf{S}}}{}^p \mathbf{R}^*$ and $\dot{\tilde{\mathbf{E}}}{}^e = \mathbf{R}^{*T} \overset{\circ}{\overline{\mathbf{E}}}{}^e \mathbf{R}^*$, respectively.

Consider configuration $C_{\mathbf{F}^p}$ and observe that

$$\dot{\tilde{\mathbf{S}}}{}^p = \tilde{\mathcal{L}} : (\tilde{\mathbf{D}} - \sum_{\alpha=1}^{n} \dot{\gamma}^\alpha \tilde{\mathbf{p}}_0^\alpha), \tag{6.3.23a}$$

where

$$\tilde{\mathbf{D}} = \mathbf{F}^{*T} \mathbf{D} \mathbf{F}^*, \quad \tilde{\mathbf{p}}_0^\alpha = \frac{1}{2}(\mathbf{C}^* \mathbf{s}_0^\alpha \otimes \mathbf{n}_0^\alpha + \mathbf{n}_0^\alpha \otimes \mathbf{s}_0^\alpha \mathbf{C}^*), \quad \mathbf{C}^* = \mathbf{F}^{*T} \mathbf{F}^*. \tag{6.3.23b-d}$$

Expression (6.3.23a) may also be cast in the framework of Hill and Rice (1972),

[37] A deformation-rate measure is *objective* if it is free of any rigid-body spin.

as follows:

$$\dot{\tilde{\mathbf{S}}}^{\mathrm{P}} = \tilde{\mathcal{L}} : \tilde{\mathbf{D}} - \sum_{\alpha=1}^{n} \dot{\gamma}^{\alpha} \tilde{\boldsymbol{\lambda}}^{\alpha}, \quad \tilde{\boldsymbol{\lambda}}^{\alpha} = \tilde{\mathcal{L}} : \tilde{\mathbf{p}}_0^{\alpha},$$

$$\tilde{\mathbf{D}} = \tilde{\mathcal{M}} : \dot{\tilde{\mathbf{S}}}^{\mathrm{P}} + \sum_{\alpha=1}^{n} \dot{\gamma}^{\alpha} \tilde{\boldsymbol{\mu}}^{\alpha}, \quad \tilde{\boldsymbol{\mu}}^{\alpha} = \tilde{\mathcal{M}} : \tilde{\boldsymbol{\lambda}}^{\alpha}. \tag{6.3.24a-d}$$

Finally, consider the measure of the resolved shear stress of the $(\mathbf{s}_0, \mathbf{n}_0)$'th slip system, defined by[38]

$$\tilde{\tau}^{\alpha} = \tilde{\mathbf{S}}^{\mathrm{P}} : \tilde{\mathbf{p}}_0^{\alpha}, \tag{6.3.25a}$$

and, noting that

$$\dot{\tilde{\tau}}^{\alpha} = \dot{\tilde{\mathbf{S}}}^{\mathrm{P}} : \tilde{\mathbf{p}}_0^{\alpha} + \tilde{\mathbf{S}}^{\mathrm{P}} : \dot{\tilde{\mathbf{p}}}_0^{\alpha}, \tag{6.3.25b}$$

obtain,

$$\dot{\tilde{\tau}}^{\alpha} = \{ \tilde{\mathcal{L}} : \tilde{\mathbf{p}}_0^{\alpha} + \mathbf{s}_0^{\alpha} \otimes \mathbf{n}_0^{\alpha} \tilde{\mathbf{S}}^{\mathrm{P}} + \tilde{\mathbf{S}}^{\mathrm{P}} \mathbf{n}_0^{\alpha} \otimes \mathbf{s}_0^{\alpha} \} : \dot{\tilde{\mathbf{E}}}^{\mathrm{e}}. \tag{6.3.25c}$$

6.3.5. Crystal Elastoplasticity Relative to Undeformed Configuration

To obtain the constitutive relation with the undeformed state as the reference configuration, examine the material time derivative of the second Piola-Kirchhoff stress, $\mathbf{S}^{\mathrm{P}}$, expressed in terms of the Kirchhoff stress, $\boldsymbol{\tau}$, as follows:

$$\mathbf{S}^{\mathrm{P}} = \mathbf{F}^{-1} \boldsymbol{\tau} \mathbf{F}^{-\mathrm{T}}. \tag{6.3.26a}$$

This material derivative is given by

$$\dot{\mathbf{S}}^{\mathrm{P}} = \mathbf{F}^{-1} \overset{\nabla}{\boldsymbol{\tau}} \mathbf{F}^{-\mathrm{T}}, \tag{6.3.26b}$$

as can be verified by direct calculation. Substitution from (6.3.10a) for $\overset{\nabla}{\boldsymbol{\tau}}$ then yields

$$\dot{\mathbf{S}}^{\mathrm{P}} = \mathbf{F}^{-1} \{ \mathcal{L} : \mathbf{D} - \sum_{\alpha=1}^{n} \dot{\gamma}^{\alpha} \boldsymbol{\lambda}^{\alpha} \} \mathbf{F}^{-\mathrm{T}}$$

$$= \mathcal{L}^0 : \dot{\mathbf{E}}^{\mathrm{L}} - \sum_{\alpha=1}^{n} \dot{\gamma}^{\alpha} \boldsymbol{\lambda}_0^{\alpha}, \tag{6.3.27a,b}$$

where

$$\dot{\mathbf{E}}^{\mathrm{L}} = \mathbf{F}^{\mathrm{T}} \mathbf{D} \mathbf{F}, \quad \boldsymbol{\lambda}_0^{\alpha} = \mathbf{F}^{-1} \boldsymbol{\lambda}^{\alpha} \mathbf{F}^{-\mathrm{T}},$$

$$\mathcal{L}^0_{ABCD} = F^{-1}_{Ai} F^{-1}_{Bj} F^{-1}_{Ck} F^{-1}_{Dl} \mathcal{L}_{ijkl}, \quad i, j, k, l, A, B, C, D = 1, 2, 3. \tag{6.3.27c-e}$$

In expression (6.3.27c), $\dot{\mathbf{E}}^{\mathrm{L}}$ is the time derivative of the Lagrangian strain, $\mathbf{E}^{\mathrm{L}} = \frac{1}{2}(\mathbf{F}^{\mathrm{T}}\mathbf{F} - \mathbf{1})$. With $\boldsymbol{\lambda}^{\alpha}$ defined by (6.3.10b), $\boldsymbol{\lambda}_0^{\alpha}$ becomes

[38] The measure $\tilde{\tau}^{\alpha}$ is in fact equal to the generalized Taylor-Schmid stress, $\tau^{\alpha} = \boldsymbol{\tau} : \mathbf{p}^{\alpha}$, defined by (6.3.11a). This is easily verified as follows: $\boldsymbol{\tau} : \mathbf{p}^{\alpha} = (\mathbf{F}^{*-1} \boldsymbol{\tau} \mathbf{F}^{*-\mathrm{T}}) : (\mathbf{F}^{*\mathrm{T}} \mathbf{p}^{\alpha} \mathbf{F}^{*}) = \frac{1}{2} \tilde{\mathbf{S}}^{\mathrm{P}} : (\mathbf{C}^{*} \mathbf{s}_0^{\alpha} \otimes \mathbf{n}_0^{\alpha} + \mathbf{n}_0^{\alpha} \otimes \mathbf{s}_0^{\alpha} \mathbf{C}^{*})$.

$$\boldsymbol{\lambda}_0^\alpha = \mathcal{L}^0 : \mathbf{q}_0^\alpha + (\boldsymbol{m}_0^\alpha \, \mathbf{S}^\mathrm{P} + \mathbf{S}^\mathrm{P} \, \boldsymbol{m}_0^{\alpha\mathrm{T}}),$$
$$\mathbf{q}_0^\alpha = \mathbf{F}^\mathrm{T} \, \mathbf{p}^\alpha \, \mathbf{F} = \mathbf{F}^{\mathrm{pT}} \, \tilde{\mathbf{p}}_0^\alpha \, \mathbf{F}^\mathrm{p},$$
$$\boldsymbol{m}_0^\alpha = \mathbf{F}^{-1} \, \mathbf{s}^\alpha \otimes \mathbf{n}^\alpha \, \mathbf{F} = \mathbf{F}^{\mathrm{p}-1} \, \mathbf{s}_0^\alpha \otimes \mathbf{n}_0^\alpha \, \mathbf{F}^\mathrm{p}. \qquad (6.3.27\text{f-i})$$

6.3.6. Rate of Stress Work and Measure-invariance of Plastic Dissipation

The kinetics of crystal elastoplasticity can be examined in terms of any material strain measure and the corresponding work-conjugate stress, using any desired reference configuration, as is illustrated in the preceding subsections.[39] The generalized Taylor-Schmid stress, τ^α, in particular, is the conjugate measure for the corresponding plastic slip rate, $\dot{\gamma}^\alpha$, and remains invariant with the change of the reference state or the strain measure, or both.

To illustrate this fact, consider the decomposition of the total velocity gradient, $\mathbf{L}$, defined by (6.2.4), and rewrite it for convenience in the following alternative forms:

$$\mathbf{L} = \dot{\mathbf{F}}^* \mathbf{F}^{*-1} + \mathbf{F}^* \, \dot{\mathbf{F}}^\mathrm{p} \, \mathbf{F}^{\mathrm{p}-1} \, \mathbf{F}^{*-1} = \mathbf{L}^* + \mathbf{L}^\mathrm{p}$$
$$= \mathbf{L}^* + \mathbf{F}^* \sum_{\alpha=1}^{n} \dot{\gamma}^\alpha \, \mathbf{s}_0^\alpha \otimes \mathbf{n}_0^\alpha \, \mathbf{F}^{*-1}$$
$$= \mathbf{L}^* + \sum_{\alpha=1}^{n} \dot{\gamma}^\alpha \, \mathbf{s}^\alpha \otimes \mathbf{n}^\alpha. \qquad (6.3.28\text{a-d})$$

Then observe that the rate of plastic dissipation (plastic stress work),

$$(\mathbf{L} - \mathbf{L}^*) : \boldsymbol{\tau} = \mathbf{L}^\mathrm{p} : \boldsymbol{\tau} = \mathbf{D}^\mathrm{p} : \boldsymbol{\tau}$$
$$= \sum_{\alpha=1}^{n} \dot{\gamma}^\alpha \, \tau^\alpha, \qquad (6.3.29\text{a-c})$$

is invariant under the change of the stress measure and the reference state, where the symmetry of the Kirchhoff stress, $\boldsymbol{\tau}$, is used, and the generalized Taylor-Schmid stress, τ^α, is given by (6.3.11a).

By direct calculation, it is verified that

$$\tau^\alpha = \mathbf{p}^\alpha : \boldsymbol{\tau} = \tilde{\mathbf{p}}_0^\alpha : \tilde{\mathbf{S}}^\mathrm{P} = \mathbf{q}_0^\alpha : \mathbf{S}^\mathrm{P}, \qquad (6.3.30\text{a-c})$$

where $\tilde{\mathbf{p}}_0^\alpha$ and $\mathbf{q}_0^\alpha$ are defined in (6.3.23c) and (6.3.27g), respectively, and $\tilde{\mathbf{S}}^\mathrm{P}$ and $\mathbf{S}^\mathrm{P}$ are the Piola-Kirchhoff stress tensors (both symmetric) referred to configurations $C_{\mathbf{F}^\mathrm{p}}$ and C_0, respectively. Furthermore, let $\tilde{\mathbf{S}}^{\mathrm{N}*} = \mathbf{F}^{*-1} \boldsymbol{\tau}$ be the

[39] See Hill and Rice (1972) and Hill and Havner (1982); also, Rice (1971), Teodosiu and Sidoroff (1976), and Havner (1973, 1974, 1992). Comparison of results presented in Subsections 6.3.1 to 6.3.5 suggests that the decomposition of the deformation gradient according to Figure 6.2.2 provides the simplest representation of both plasticity and elasticity of the crystal with reference to the current state, where elastic anisotropy may most naturally be accounted for directly in the rotated but undistorted lattice coordinate system.

nominal stress, conjugate to the elastic velocity gradient $\dot{\mathbf{F}}^*$, referred to the plastically deformed configuration $C_{\mathbf{F}^p}$. Then

$$\tau^\alpha = \mathbf{p}^\alpha \mathbf{F}^* : \mathbf{F}^{*-1}\boldsymbol{\tau} = \mathbf{s}^\alpha \otimes \mathbf{n}_0^\alpha : \tilde{\mathbf{S}}^{N*}$$
$$= \mathbf{s}^\alpha . (\mathbf{n}_0^\alpha \tilde{\mathbf{S}}^{N*}) . \tag{6.3.30d,e}$$

Here, $\mathbf{n}_0^\alpha \tilde{\mathbf{S}}^{N*}$ is the traction vector acting on the glide plane with unit normal $\mathbf{n}_0^\alpha$ in the plastically deformed configuration $C_{\mathbf{F}^p}$.

6.3.7. Crystal Elastoplasticity with Small Elastic Strains

As pointed out before, the pure (elastic) deformation of the lattice is generally negligibly small and can be neglected in many applications, without noticeably affecting the results. The general results presented in the preceding sections apply with no restrictions on the elastic strains. However, the term $\mathbf{D}^*\boldsymbol{\tau} + \boldsymbol{\tau}\mathbf{D}^*$ in (6.3.14a) is usually negligible in comparison with the remaining terms, since $\mathcal{L}$ is of the order of the shear modulus which, for most metals, generally is orders of magnitude greater than their yield stress, and the plastic spin, $\mathbf{W}^p$, is generally much greater than the the elastic deformation rate, $\mathbf{D}^*$. Hence, it is usually assumed that (6.3.14a) can be written as

$$\overset{\circ}{\boldsymbol{\tau}} = \mathcal{L} : (\mathbf{D} - \mathbf{D}^p) - (\mathbf{W}^p \boldsymbol{\tau} - \boldsymbol{\tau}\mathbf{W}^p) + O(\boldsymbol{\varepsilon})$$
$$= \mathcal{L} : \mathbf{D} - \sum_{\alpha=1}^{n} \dot{\gamma}^\alpha \boldsymbol{\lambda}^\alpha + O(\boldsymbol{\varepsilon}) , \tag{6.3.31a,b}$$

where $\boldsymbol{\lambda}^\alpha$ is now given by

$$\boldsymbol{\lambda}^\alpha \approx \mathcal{L} : \mathbf{p}^\alpha + \mathbf{r}^\alpha \boldsymbol{\tau} - \boldsymbol{\tau}\mathbf{r}^\alpha , \tag{6.3.31c}$$

since the elasticity tensors $\mathcal{L}$ and $\mathcal{L}'$ are no longer distinguishable.

With $\mathcal{M}$ denoting the elastic compliance, $\mathcal{M} = \mathcal{L}^{-1}$, constitutive relation (6.3.31b) can be inverted to yield

$$\mathbf{D} \approx \mathcal{M} : \overset{\circ}{\boldsymbol{\tau}} + \sum_{\alpha=1}^{n} \dot{\gamma}^\alpha \boldsymbol{\mu}^\alpha , \tag{6.3.31d}$$

where

$$\boldsymbol{\mu}^\alpha = \mathbf{p}^\alpha + \mathcal{M} : (\mathbf{r}^\alpha \boldsymbol{\tau} - \boldsymbol{\tau}\mathbf{r}^\alpha) . \tag{6.3.31e}$$

Relation (6.3.31e) is in the same form as the continuum constitutive relation (4.4.7).

Finally, recall that, for negligible elastic strains and strain rates, the lattice parameters in the current configuration are related to their initial state by the following expressions:

$$\mathbf{s}^\alpha \approx \mathbf{R}^* \mathbf{s}_0^\alpha , \quad \mathbf{n}^\alpha \approx \mathbf{R}^* \mathbf{n}_0^\alpha ,$$
$$\mathbf{p}^\alpha \approx \mathbf{R}^* \mathbf{p}_0^\alpha , \quad \mathbf{r}^\alpha \approx \mathbf{R}^* \mathbf{r}_0^\alpha , \tag{6.3.32a-d}$$

where, in this approximation, $\mathbf{s}^\alpha$ and $\mathbf{n}^\alpha$ are orthogonal unit vectors.

Note that, with the new definition of $\boldsymbol{\lambda}^\alpha$, given by (6.3.31c), the important results (6.3.11d) and (6.3.11g), *i.e.*, $\dot{\tau}^\alpha = \boldsymbol{\lambda}^\alpha : \mathbf{D}^*$ and $\dot{\tau}^\alpha = \boldsymbol{\mu}^\alpha : \overset{\nabla}{\boldsymbol{\tau}}{}^*$, remain valid, taking on the equivalent form

$$\dot{\tau}^\alpha = \boldsymbol{\lambda}^\alpha : \mathbf{D}^* \approx \boldsymbol{\mu}^\alpha : \overset{\circ}{\boldsymbol{\tau}}{}^*. \tag{6.3.33a,b}$$

6.4. RESISTANCE TO SLIP AND WORKHARDENING: RATE-INDEPENDENT MODELS

Crystallographic slip occurs by the motion of dislocations on slip planes. In this motion, the dislocations must overcome various barriers. The collective effect of these barriers is the resistance to the motion of dislocations and hence to the corresponding slip. Models have been proposed to quantify the rate of slip of the active slip systems in terms of the stress, dislocation density and distribution, temperature, and various parameters which are introduced to represent the microstructure. Physically-based models of this kind are developed in Section 6.5. Rate-independent models are examined in the present section, starting with the *Schmid law* which also leads to the normality of the plastic strain rate to the yield surface of the crystal, as well as the normality of the plastically-induced stress-rate decrement to the yield surface in the strain space.

6.4.1. Schmid Law

Early experiments by Taylor and Elam (1925) and Schmid (1926) suggest that, for a slip system to become active, the shear stress acting on the slip plane in the slip direction must attain a critical value, and that this critical value tends to increase with continued plastic deformation. The driving shear stress is called *the resolved shear stress* and the resulting resistance to shearing in the considered slip direction is called *the critical shear strength.*

If $\boldsymbol{\sigma}$ is the Cauchy stress, then the resolved shear stress of the α'th slip system becomes

$$\sigma^\alpha = \boldsymbol{\sigma} : \mathbf{p}^\alpha \frac{\lambda^*_{(n)}}{\lambda^*_{(s)}}, \tag{6.4.1a}$$

where $\mathbf{p}^\alpha$ is defined by (6.2.14a), and $\lambda^*_{(s)}$ and $\lambda^*_{(n)}$ are the elastic stretches in the slip direction and in the direction normal to the slip plane, respectively; see comments associated with (6.2.12c). As pointed out before, the elastic lattice deformation in general is negligibly small, and hence $\lambda^*_{(n)}/\lambda^*_{(s)}$ can be set equal to one, without affecting the results. Indeed, essentially all experimental measurements of the resistance to slip have neglected the effects of the lattice elastic deformation. Since plastic deformation by crystallographic slip is volume preserving, $\det|\mathbf{F}^{\mathrm{p}}| = 1$ and it follows that $J = \det|\mathbf{F}| = \det|\mathbf{F}^*| \approx 1$. The Cauchy and Kirchhoff stress tensors, $\boldsymbol{\sigma}$ and $\boldsymbol{\tau}$, are then related by

$$\boldsymbol{\tau} = J\boldsymbol{\sigma} = \boldsymbol{\sigma} + O(\boldsymbol{\varepsilon}), \tag{6.4.2a,b}$$

where, as before, $\boldsymbol{\varepsilon} = \mathbf{V}^{\mathrm{e}} - \mathbf{1}$, and $\mathbf{F}^* = \mathbf{V}^{\mathrm{e}}\mathbf{R}^*$; see (6.2.1c). Thus, the actual resolved shear stress, σ^α, defined by (6.4.1a), differs from the generalized Taylor-Schmid stress, $\tau^\alpha = \boldsymbol{\tau} : \mathbf{p}^\alpha$, by terms of the order of the elastic strain. Indeed, since

$$\tau^\alpha = \boldsymbol{\tau} : \mathbf{p}^\alpha = J\boldsymbol{\sigma} : \mathbf{p}^\alpha, \tag{6.4.1b,c}$$

it follows that

$$\tau^\alpha = \sigma^\alpha + O(|\boldsymbol{\varepsilon}|). \tag{6.4.1d}$$

The use of the Taylor-Schmid stress, τ^α, on the other hand, leads to a precise general theory of elastoplastic crystals, with a number of attractive features, as presented by Rice (1971), Hill and Rice (1972), Hill and Havner (1982), and Havner (1992). Therefore, in what follows, the crystallographic flow models are all expressed in terms of the Kirchhoff stress, $\boldsymbol{\tau}$, and its various appropriate components.

The Schmid law states that the α'th slip system is potentially active if the resolved shear stress, σ^α, attains a critical value, σ_c^α. Hill and Rice (1972) restate this in terms of the Kirchhoff stress. Hence, according to this extended version of the Schmid law, the slip system defined by $(\mathbf{s}^\alpha, \mathbf{n}^\alpha)$ is potentially *active*, when the generalized Taylor-Schmid stress, τ^α, attains a critical value, τ_c^α, which is a function of the current density and the distribution of dislocations and other microstructural parameters.[40] All slip systems for which $\tau^\alpha = \tau_c^\alpha$, are called *critical*. For a slip system to remain critical, the Taylor-Schmid stress, τ^α, must continue to retain its critical value. This critical stress is assumed to change according to the following *flow rule*:

$$\dot{\tau}^\alpha = \dot{\tau}_c^\alpha = \sum_{\beta=1}^{n} h^{*\alpha\beta}\dot{\gamma}^\beta, \quad \dot{\gamma}^\alpha > 0. \tag{6.4.3a-c}$$

Here, (6.4.3a) states that the driving stress rate, $\dot{\tau}^\alpha$, should keep up with the rate of increase in the resistance to slip, and (6.4.3b) gives the rate by which the resistance shear stress, τ_c^α, changes with continued slipping.[41] A critical slip system with $\tau^\alpha = \tau_c^\alpha$, becomes *inactive* when

$$\dot{\tau}^\alpha \le \dot{\tau}_c^\alpha = \sum_{\beta=1}^{n} h^{*\alpha\beta}\dot{\gamma}^\beta, \quad \dot{\gamma}^\alpha = 0. \tag{6.4.4a-c}$$

---

[40] In general, the flow stress associated with each slip system depends on the temperature, strain rate, the density and distribution of dislocations, grain boundaries, and other microstructural factors; see Section 6.5.

[41] Actually, it is the rate of change of the dislocation density affecting the hardening of the α'th system that should be used in (6.4.3b), and not the slip rate.

A slip system is called *noncritical* or *inactive*, if

$$\tau^\alpha < \tau_c^\alpha, \quad \dot{\gamma}^\alpha = 0. \tag{6.4.5a,b}$$

In expressions (6.4.3b) and (6.4.4b), $h^{*\alpha\beta}$, $\alpha, \beta = 1, 2, ..., n$, are the elements of the *workhardening* matrix, $[h^{*\alpha\beta}]$, often *assumed* to be a constant matrix but, in reality, it changes with the change in the microstructure of the crystal. The diagonal elements, $h^{*\alpha\alpha}$ (α not summed), define the *self-hardening* rates, and the off-diagonal terms, $h^{*\alpha\beta}$, $\alpha \neq \beta$, are the *cross-hardening* or the *latent-hardening* terms. To simplify the counting procedure, usually all slip rates, $\dot{\gamma}^\alpha$, $\alpha = 1, 2, ..., n$, are regarded either *positive* (when active) or *zero* (when inactive).[42] Forward and reverse slips of the α'th system are then viewed as two separate cases, *i.e.*, $(\mathbf{s}^\alpha, \mathbf{n}^\alpha)$ and $(-\mathbf{s}^\alpha, \mathbf{n}^\alpha)$ are regarded as two separate, but mutually exclusive, slip systems. For an fcc crystal, for example, this counting procedure yields twenty-four possible slip systems, no more than twelve of which can possibly be potentially active at each instance. Since there are only four slip planes in an fcc crystal, no more than eight slip systems are active at each instant; see Section 6.5.

At this point, no restriction is placed on the hardening matrix $[h^{*\alpha\beta}]$. Its elements are functions of, for example, dislocation density and dislocation distribution, both of which change with the plastic deformation of the crystal. In general, the matrix $[h^{*\alpha\beta}]$ is not constant. Indeed, as discussed in Subsections 6.4.5 and 6.4.6, existing experimental data do not support the assumption of a constant hardening matrix. Furthermore, the latent-hardening terms, $h^{*\alpha\beta}$, $\alpha \neq \beta$, need not be symmetric, *i.e.*, in general, there is no reason to expect that $h^{*\alpha\beta} = h^{*\beta\alpha}$, $\alpha \neq \beta$, in all cases.[43] This is particularly so when secondary slip systems interact with the primary ones; *e.g.*, possible slip systems {112}<111>, or {123}<111> interact with the primary slip systems {110}<111> in bcc crystals, or slip systems on prismatic and pyramidal planes interact with the primary basal slip systems in hcp crystals.

6.4.2. Yield Surface and Plastic Normality in Stress Space

In the (deviatoric) stress space, the critical conditions,

$$\tau^\alpha \equiv \boldsymbol{\tau} : \mathbf{p}^\alpha = \tau_c^\alpha, \quad \alpha = 1, 2, 3, ..., n, \tag{6.4.6a,b}$$

define a set of surfaces (planes) which constitute the yield surface of the crystal. As pointed out before, this extended version of the Schmid law leads to the normality of the plastic strain rate to the yield surface. For the slip rate $\dot{\gamma}^\alpha$ to be positive, it is first necessary that the corresponding stress state satisfy (6.4.6b), *i.e.*, the stress state must lie on the corresponding plane of the yield locus.

---

[42] It will be explicitly stated wherever this convention is used.

[43] These facts have been emphasized by Zarka (1968, 1975) and, more recently by Wu *et al.* (1991) and Bassani and Wu (1991); see also the discussion in Subsection 6.4.6.

Furthermore, since, in general, $\dot{\tau}_c^\alpha = \sum_{\beta=1}^{n} h^{*\alpha\beta}\dot{\gamma}^\beta \geq 0$ when one or several slip systems are active, any stress change which results in $\dot{\tau}^\alpha \leq 0$ renders $\dot{\gamma}^\alpha = 0$ and, hence, produces a purely elastic response in the corresponding slip system. Hence, in view of (6.3.11g), with the stress point on the yield surface, the response associated with the α'th slip system is elastic, if

$$\dot{\tau}^\alpha = \boldsymbol{\mu}^\alpha : \overset{\nabla}{\boldsymbol{\tau}}{}^* \leq 0. \tag{6.4.7a,b}$$

From (6.3.10c),

$$\mathbf{D} - \mathbf{D}^* = \mathbf{D} - \boldsymbol{\mathcal{M}} : \overset{\nabla}{\boldsymbol{\tau}} = \sum_{\alpha=1}^{n} \dot{\gamma}^\alpha \boldsymbol{\mu}^\alpha. \tag{6.4.8a,b}$$

According to (6.4.7b), $\boldsymbol{\mu}^\alpha$ is the outward normal to the yield plane of the α'th slip system, in the stress space. Hence, in view of (6.4.8), *the contribution of the α'th slip system to the plastic part of the deformation rate,* $\dot{\gamma}^\alpha \boldsymbol{\mu}^\alpha$, *is normal to the corresponding yield plane.* This normality rule is a consequence of the extended Schmid law.

In general, several slip systems may be active simultaneously and, hence, contributing to the plastic part of the deformation rate tensor. The stress point then lies on the yield planes of these active slip systems. These yield planes form a vertex of the yield surface, and the stress point falls on this yield vertex. For each contributing active slip system of this kind, the *flow rule*

$$\boldsymbol{\mu}^\alpha : \overset{\nabla}{\boldsymbol{\tau}}{}^* = \sum_{\beta=1}^{n} h^{*\alpha\beta}\dot{\gamma}^\beta, \quad \dot{\gamma}^\alpha > 0, \tag{6.4.9a,b}$$

must hold. If a slip system is not active, then it follows that

$$\boldsymbol{\mu}^\alpha : \overset{\nabla}{\boldsymbol{\tau}}{}^* \leq \sum_{\beta=1}^{n} h^{*\alpha\beta}\dot{\gamma}^\beta, \quad \dot{\gamma}^\alpha = 0, \tag{6.4.10a,b}$$

for that slip system.

Elastic unloading from a potentially active state (*i.e.*, a state with the stress point on one or several yield planes), results in

$$\left[\sum_{\alpha=1}^{n} \dot{\gamma}^\alpha \boldsymbol{\mu}^\alpha\right] : \overset{\nabla}{\boldsymbol{\tau}}{}^* < 0, \tag{6.4.11}$$

whereas, continued plastic flow requires

$$\left[\sum_{\alpha=1}^{n} \dot{\gamma}^\alpha \boldsymbol{\mu}^\alpha\right] : \overset{\nabla}{\boldsymbol{\tau}}{}^* > 0. \tag{6.4.12}$$

Thus, the plastic strain rate, $\sum_{\alpha=1}^{n} \dot{\gamma}^\alpha \boldsymbol{\mu}^\alpha$, is along the outward normal of the yield plane at a regular yield point, and falls within a pyramid formed by the outward normals of the yield planes at a yield vertex.

Expressions (6.4.9) and (6.4.10) can be written in the following alternative forms, based on the relation (6.3.9c) which connects the objective stress rates $\overset{\nabla}{\boldsymbol{\tau}}$ and $\overset{\nabla}{\boldsymbol{\tau}}{}^*$; the first, $\overset{\nabla}{\boldsymbol{\tau}}$, is convected with the *total* deformation, and the second, $\overset{\nabla}{\boldsymbol{\tau}}{}^*$, with the lattice deformation:

$$\boldsymbol{\mu}^\alpha : \overset{\nabla}{\boldsymbol{\tau}} = \sum_{\beta=1}^{n} h^{\alpha\beta}\,\dot{\gamma}^\beta, \quad \dot{\gamma}^\alpha > 0, \tag{6.4.13a,b}$$

$$\boldsymbol{\mu}^\alpha : \overset{\nabla}{\boldsymbol{\tau}} \le \sum_{\beta=1}^{n} h^{\alpha\beta}\,\dot{\gamma}^\beta, \quad \dot{\gamma}^\alpha = 0, \tag{6.4.14a,b}$$

$$\begin{aligned} h^{\alpha\beta} &= h^{*\alpha\beta} - \boldsymbol{\mu}^\alpha : (\mathbf{s}^\beta \otimes \mathbf{n}^\beta\,\boldsymbol{\tau} + \boldsymbol{\tau}\,\mathbf{n}^\beta \otimes \mathbf{s}^\beta) \\ &= h^{*\alpha\beta} - (\mathbf{s}^\alpha \otimes \mathbf{n}^\alpha\,\boldsymbol{\tau} + \boldsymbol{\tau}\,\mathbf{n}^\alpha \otimes \mathbf{s}^\alpha) : \boldsymbol{\mathcal{M}} : (\mathbf{s}^\beta \otimes \mathbf{n}^\beta\,\boldsymbol{\tau} + \boldsymbol{\tau}\,\mathbf{n}^\beta \otimes \mathbf{s}^\beta) \\ &\quad - \mathbf{p}^\alpha : (\mathbf{s}^\beta \otimes \mathbf{n}^\beta\,\boldsymbol{\tau} + \boldsymbol{\tau}\,\mathbf{n}^\beta \otimes \mathbf{s}^\beta). \end{aligned} \tag{6.4.15a,b}$$

Furthermore, the Taylor-Schmid stress rate, $\dot{\tau}^\alpha$, and the flow rule can be expressed as

$$\dot{\tau}^\alpha = \tau_c^\alpha = \sum_{\beta=1}^{n} \{h^{\alpha\beta} + \boldsymbol{\mu}^\alpha : (\mathbf{s}^\beta \otimes \mathbf{n}^\beta\,\boldsymbol{\tau} + \boldsymbol{\tau}\,\mathbf{n}^\beta \otimes \mathbf{s}^\beta)\}\,\dot{\gamma}^\beta. \tag{6.4.16}$$

Similar results are obtained in terms of the Jaumann rates of the Kirchhoff stress, $\overset{\circ}{\boldsymbol{\tau}}{}^*$ and $\overset{\circ}{\boldsymbol{\tau}}$; see (6.3.12) and (6.3.17). In particular, from (6.3.17b), it follows that

$$\boldsymbol{\mu}'^\alpha : \overset{\circ}{\boldsymbol{\tau}}{}^* = \sum_{\beta=1}^{n} h^{*\alpha\beta}\,\dot{\gamma}^\beta, \quad \dot{\gamma}^\alpha > 0, \tag{6.4.17a,b}$$

$$\boldsymbol{\mu}'^\alpha : \overset{\circ}{\boldsymbol{\tau}}{}^* \le \sum_{\beta=1}^{n} h^{*\alpha\beta}\,\dot{\gamma}^\beta, \quad \dot{\gamma}^\alpha = 0. \tag{6.4.18a,b}$$

In addition, in view of (6.3.12e), obtain

$$\boldsymbol{\mu}'^\alpha : \overset{\circ}{\boldsymbol{\tau}} = \sum_{\beta=1}^{n} h'^{\alpha\beta}\,\dot{\gamma}^\beta, \quad \dot{\gamma}^\alpha > 0, \tag{6.4.19a,b}$$

$$\boldsymbol{\mu}'^\alpha : \overset{\circ}{\boldsymbol{\tau}} \le \sum_{\beta=1}^{n} h'^{\alpha\beta}\,\dot{\gamma}^\beta, \quad \dot{\gamma}^\alpha = 0, \tag{6.4.20a,b}$$

$$\begin{aligned} h'^{\alpha\beta} &= h^{*\alpha\beta} - \boldsymbol{\mu}'^\alpha : (\mathbf{r}^\beta\,\boldsymbol{\tau} - \boldsymbol{\tau}\,\mathbf{r}^\beta) \\ &= h^{*\alpha\beta} - (\mathbf{r}^\alpha\,\boldsymbol{\tau} - \boldsymbol{\tau}\,\mathbf{r}^\alpha) : \boldsymbol{\mathcal{M}} : (\mathbf{r}^\beta\,\boldsymbol{\tau} - \boldsymbol{\tau}\,\mathbf{r}^\beta) - \mathbf{p}^\alpha : (\mathbf{r}^\beta\,\boldsymbol{\tau} - \boldsymbol{\tau}\,\mathbf{r}^\beta). \end{aligned} \tag{6.4.21a,b}$$

The Taylor-Schmid stress rate, $\dot{\tau}^\alpha$, and the flow rule now become

$$\dot{\tau}^\alpha = \tau_c^\alpha = \sum_{\beta=1}^{n} \{h'^{\alpha\beta} + \boldsymbol{\mu}'^\alpha : (\mathbf{r}^\beta\,\boldsymbol{\tau} + \boldsymbol{\tau}\,\mathbf{r}^\beta)\}\,\dot{\gamma}^\beta. \tag{6.4.22}$$

As discussed in Subsection 6.3.2, the convected stress rate $\overset{\nabla}{\boldsymbol{\tau}}$ corresponds to the Lagrangian strain, $\mathbf{E}^L = ½\,(\mathbf{F}^T\mathbf{F} - \mathbf{1})$, whereas the Jaumann stress rate $\overset{\circ}{\boldsymbol{\tau}}$, corotational with the total material spin, is associated with the logarithmic strain, $\mathbf{E}^l = \ln \mathbf{U}$. Similarly, the second-order symmetric tensors, $\boldsymbol{\mu}^\alpha$ and $\boldsymbol{\mu}'^\alpha$, as well as the hardening matrices, $[h^{\alpha\beta}]$ and $[h'^{\alpha\beta}]$, correspond to the same respective strain measures. These quantities, therefore, are *not* invariant under a change in the strain measure, even if the same (here the current) configuration is used as the reference one. In contrast, certain invariance properties are preserved in the strain-based formulation, as is discussed in the next subsection.

6.4.3. Yield Surface and Plastic Normality in Strain Space

It has been noted in Subsection 6.3.2 that the time-rate-of-change of the Taylor-Schmid stress, $\dot{\tau}^\alpha$, is invariant under a change in the strain measure. With respect to the current state as the reference, $\dot{\tau}^\alpha = \boldsymbol{\lambda}^\alpha : \mathbf{D}^*$, no matter which material strain measure is used, where $\boldsymbol{\lambda}^\alpha$ is defined by (6.3.10b). The flow rules are

$$\dot{\tau}^\alpha = \boldsymbol{\lambda}^\alpha : \mathbf{D}^* = \sum_{\beta=1}^{n} h^{*\alpha\beta} \dot{\gamma}^\beta, \quad \dot{\gamma}^\alpha > 0, \tag{6.4.23a-c}$$

$$\dot{\tau}^\alpha = \boldsymbol{\lambda}^\alpha : \mathbf{D}^* \leq \sum_{\beta=1}^{n} h^{*\alpha\beta} \dot{\gamma}^\beta, \quad \dot{\gamma}^\alpha = 0, \tag{6.4.24a-c}$$

where $[h^{*\alpha\beta}]$ is the hardening matrix, defined in (6.4.3); see (6.3.11e) and (6.4.9).

The slip rate $\dot{\gamma}^\alpha$ is zero for elastic loading or unloading of the α'th slip system. Also, with the α'th system at the critical state, $\dot{\tau}^\alpha = \boldsymbol{\lambda}^\alpha : \mathbf{D}^* < 0$ leads to elastic unloading of this system, *i.e.*, leads to $\dot{\gamma}^\alpha = 0$. It thus follows from (6.3.10a) that

$$\left[\boldsymbol{\mathcal{L}} : \mathbf{D} - \overset{\nabla}{\boldsymbol{\tau}}\right] : \mathbf{D}^* = \left[\sum_{\alpha=1}^{n} \dot{\gamma}^\alpha \boldsymbol{\lambda}^\alpha\right] : \mathbf{D}^* < 0 \tag{6.4.25a,b}$$

corresponds to elastic unloading from a state on the yield surface with n critical slip systems. This ensures that, at a regular point on the yield surface in the strain space, the plastically-induced stress-rate decrement, $-\dot{\gamma}^\alpha \boldsymbol{\lambda}^\alpha$, is directed along the outward normal to the yield plane. At a vertex where several yield planes intersect, this decrement, $\overset{\nabla}{\boldsymbol{\tau}}{}^{\text{in}} \equiv -\sum_{\alpha=1}^{n} \dot{\gamma}^\alpha \boldsymbol{\lambda}^\alpha$, falls within the pyramid formed by the outward normals of the intersecting yield planes. Each corresponding slip system then makes the contribution, $-\dot{\gamma}^\alpha \boldsymbol{\lambda}^\alpha$, $\alpha = 1, 2, \ldots, n$, to this stress-rate decrement. As pointed out before, $\boldsymbol{\lambda}^\alpha : \mathbf{D}^*$ is invariant under a change in the strain measure. For example, from (6.3.10e) or (6.3.15g) it is seen that the decrement in the stress-rate is given by $-\sum_{\alpha=1}^{n} \dot{\gamma}^\alpha \boldsymbol{\lambda}^\alpha$, whether $\overset{\nabla}{\boldsymbol{\tau}}$ or $\overset{\circ}{\boldsymbol{\tau}}$ is used as the objective stress rate. As noted before, this observation holds for other objective stress rates associated with other material strain measures. The normality condition (6.4.25), therefore, is unaffected by a change in the strain measure, where the current state is used as the reference one.

The flow rules (6.4.23) and (6.4.24) may be expressed in the following equivalent form:

$$\boldsymbol{\lambda}^\alpha : \mathbf{D} = \sum_{\beta=1}^{n} g^{\alpha\beta} \dot{\gamma}^\beta, \quad \dot{\gamma}^\alpha > 0, \tag{6.4.26a,b}$$

$$\boldsymbol{\lambda}^\alpha : \mathbf{D} \le \sum_{\beta=1}^{n} g^{\alpha\beta}\dot{\gamma}^\beta, \quad \dot{\gamma}^\alpha = 0, \tag{6.4.27a,b}$$

$$g^{\alpha\beta} = h^{*\alpha\beta} + \boldsymbol{\lambda}^\alpha : \mathbf{p}^\beta. \tag{6.4.28a}$$

In view of the definition of $\boldsymbol{\lambda}^\alpha$, given by (6.3.10b), this expression becomes

$$g^{\alpha\beta} = h^{*\alpha\beta} + \mathbf{p}^\alpha : \mathcal{L} : \mathbf{p}^\beta + (\mathbf{s}^\alpha \otimes \mathbf{n}^\alpha \boldsymbol{\tau} + \boldsymbol{\tau}\mathbf{n}^\alpha \otimes \mathbf{s}^\alpha) : \mathbf{p}^\beta. \tag{6.4.28b}$$

Note that the elasticity tensor $\mathcal{L}$ corresponds to the Lagrangian strain measure, referred to the current configuration. With the logarithmic strain measure, (6.4.28b) becomes

$$g^{\alpha\beta} = h^{*\alpha\beta} + \mathbf{p}^\alpha : \mathcal{L}' : \mathbf{p}^\beta + (\mathbf{r}^\alpha \boldsymbol{\tau} - \boldsymbol{\tau}\mathbf{r}^\alpha) : \mathbf{p}^\beta, \tag{6.4.28c}$$

where $\mathcal{L}'$ and $\mathcal{L}$ are related by (6.3.15b); see also footnote 36, page 421. Finally, the Taylor-Schmid stress rate, $\dot{\tau}^\alpha$, and the flow rule become,

$$\dot{\tau}^\alpha = \tau_c^\alpha = \sum_{\beta=1}^{n} \{g^{\alpha\beta} - \boldsymbol{\lambda}^\alpha : \mathbf{p}^\beta\} : \dot{\gamma}^\beta. \tag{6.4.29}$$

Note that, while neither $[g^{\alpha\beta}]$ nor $[h^{\alpha\beta}]$ need be symmetric, their difference, $[g^{\alpha\beta} - h^{\alpha\beta}]$, always is, whether or not the hardening matrix $[h^{*\alpha\beta}]$ is symmetric. This follows directly from (6.4.15b) and (6.4.28b).

6.4.4. Slip Rates

In general, the number of active slip systems depends on the stress state, the crystal structure, the hardening mechanisms, and the hardening history of the slip systems. For crystals with a high degree of symmetry, *e.g.*, fcc and bcc crystals, many slip systems can be active at the same time. Indeed, for fcc crystals, out of twelve possible cases, eight slip systems can be active simultaneously; see Subsections 6.5.8 and 6.5.9. For bcc crystals there are, in general, forty-eight slip systems which can be potentially activated, and many of them actually are, *i.e.*, as many as thirty-two slip systems; see Nemat-Nasser *et al.* (1998a). However, not all the corresponding equations in, say, (6.4.28c) are linearly independent. Hence, in general, it may not be possible to invert the system of equations (6.4.26a), in order to express the slip rates, $\dot{\gamma}^\alpha$'s, in terms of the deformation rate tensor, $\mathbf{D}$, as has been noted by Hill (1966), Hill and Rice (1972), and others; see Havner (1992) for references.

To examine this in greater detail, use (6.4.28c) to rewrite (6.4.26a) as

$$\sum_{\beta=1}^{n} \{h^{*\alpha\beta} + \mathbf{p}^\alpha : \mathcal{L}' : \mathbf{p}^\beta + (\mathbf{r}^\alpha \boldsymbol{\tau} - \boldsymbol{\tau}\mathbf{r}^\alpha) : \mathbf{p}^\beta\}\, \dot{\gamma}^\beta = \boldsymbol{\lambda}^\alpha : \mathbf{D},$$

$$\alpha = 1, 2, \ldots, n. \tag{6.4.30a}$$

Now, for illustration, assume isotropic elasticity, and note that $\mathbf{p}^\alpha : \mathcal{L}' : \mathbf{p}^\beta = 2\,\mu\,\mathbf{p}^\alpha : \mathbf{p}^\beta$, where μ is the shear modulus. The shear modulus is orders of magnitude greater than the yield stress. Furthermore, the shear modulus also dominates the hardening parameters. Therefore, the matrix $[\mathbf{p}^\alpha : \mathcal{L}' : \mathbf{p}^\beta]$ dominates the matrix of the coefficients of $\dot{\gamma}^{\beta}$'s in the left-hand side of (6.4.30a). However, only five of the twelve (for fcc) or of the forty-eight (for bcc) deviatoric tensors, $\mathbf{p}^{\alpha}$, ($\alpha = 1, 2, ..., 12$, for fcc, or $\alpha = 1, 2, ..., 48$, for bcc), are linearly independent; see Subsections 6.5.6 and 6.5.7. Hence, if $n > 5$, then the matrix $[\mathbf{p}^{\alpha} : \mathcal{L}' : \mathbf{p}^{\beta}]$ becomes singular, and equations (6.4.30a) may no longer be solved for the slip rates. On the other hand, for $n \leq 5$, and when linearly independent $\mathbf{p}^{\alpha}$'s are chosen, then slip rates can be computed from (6.4.30a).

In a more recent work, Wu *et al.* (1991) and Bassani and Wu (1991) have reexamined the rate-independent slip model of crystal plasticity and the associated supporting experiments. Supplementing the experimental results by some of their own, these authors develop a new interpretation for the rate-independent models, which appears to have important implications, as discussed in Subsection 6.4.6.

Assume, at this point, that five (or fewer) suitable linearly independent slip systems are chosen from (6.4.26a), and denote the inverse of the matrix $[g^{\alpha\beta}]$ by $[g_{-1}^{\alpha\beta}]$. From (6.4.30a) then obtain,

$$\dot{\gamma}^{\alpha} = \sum_{\beta=1}^{n} g_{-1}^{\alpha\beta} \boldsymbol{\lambda}^{\beta} : \mathbf{D}, \quad \alpha = 1, 2, ..., n \leq 5. \tag{6.4.30b}$$

Constitutive relation (6.3.15c) now becomes

$$\overset{\circ}{\boldsymbol{\tau}} = (\mathcal{L}' - \sum_{\alpha, \beta=1}^{n} g_{-1}^{\alpha\beta} \boldsymbol{\lambda}^{\alpha} \otimes \boldsymbol{\lambda}^{\beta}) : \mathbf{D}, \tag{6.4.31}$$

with a similar expression for other objective stress rates. Note that, since matrix $[g^{\alpha\beta}]$ and tensor $\boldsymbol{\lambda}^{\alpha}$ are independent of the chosen strain measure, the term $\sum_{\alpha, \beta=1}^{n} g_{-1}^{\alpha\beta} \boldsymbol{\lambda}^{\alpha} \otimes \boldsymbol{\lambda}^{\beta}$ is strain measure-independent. Hence, *only the elasticity tensor, $\mathcal{L}'$, in (6.4.31), changes with a change of the strain measure.*

6.4.5. Critical Shear Stress and Hardening Matrix

At each instant during the course of the deformation of a crystal, the value of the critical shear strength, τ_c^{α}, associated with the slip system α, will depend on the existing microstructural barriers hindering the motion of dislocations which produce the slip rate $\dot{\gamma}^{\alpha}$. As the microstructure changes in the course of deformation of the crystal, so does the critical shear strength, τ_c^{α}. The element $h^{*\alpha\beta}$ of the hardening matrix $[h^{*\alpha\beta}]$ represents the rate of change of the critical shear strength τ_c^{α}, with respect to the plastic slip of the β'th slip system, *i.e.*,

$$h^{*\alpha\beta} = \frac{\partial \tau_c^{\alpha}}{\partial \gamma^{\beta}}, \quad \alpha, \beta = 1, 2, ..., n. \tag{6.4.32}$$

The diagonal elements of the hardening matrix are called the *self-hardening* elements. For $\alpha \neq \beta$, and when the α'th system is active while the β'th system is inactive, $h^{*\alpha\beta}$ represents the hardening of the β'th slip system produced by the dislocation activity of the α'th slip system. For the purposes of referencing, the

α'th (active) system is called[44] the *primary* slip system and the β'th (inactive) one is called the *latent* system. Once the loading of the crystal is changed such that the α'th system becomes inactive and the β'th one is activated, the latent shear strength of the β'th slip system, τ_c^β, can be measured. However, direct measurement of the latent hardening parameter, $h^{*\alpha\beta}$, $\alpha \neq \beta$, is very difficult.

Since the early work of Taylor and Elam (1925, 1926) and Schmid (1926), considerable experimental effort has been devoted to measuring the critical shear strength, τ_c^α, $\alpha = 1, 2, ..., n$. The literature in this area has been reviewed by Kocks (1970), Basinski and Basinski (1979), and, more recently, by Wu *et al.* (1991), Bassani and Wu (1991), and Havner (1992). Essentially all (quasi-static, isothermal) experimental results suggest that slip on a given slip system results in an *increase* in the critical shear strength of all (active or inactive) slip systems. In addition, many researchers have concluded that the latent hardening exceeds that of self-hardening.[45] Furthermore, the hardening matrix, $[h^{*\alpha\beta}]$, in general, is not constant, and varies with the microstructural changes.

As has been explained and illustrated in Section 4.8, plastic deformation of most metals is caused by the motion of dislocations, which leads to crystallographic slip. This is a highly rate- and temperature-dependent process. On the other hand, the above-mentioned classical results are generally discussed within the rate-independent plasticity theory, presumably under quasi-static and isothermal loading conditions. Once the rate and temperature effects are included, and particularly when the effects of the long-range and short-range barriers to the motion of dislocations are examined separately, a considerably simpler and clearer picture of crystal plasticity emerges. This and related issues are examined in some detail in Section 6.5; see also Nemat-Nasser (1997) and Nemat-Nasser *et al.* (1998b,c). The remaining part of this section is devoted to quasi-static rate-independent slip models.

---

44 The term *primary* here refers to the currently active system which becomes *latent*, once inactive. The same term has been used to denote the slip planes of the highest potential for becoming active, as compared with the *secondary* slip planes; *e.g.*, in hcp crystals, the basal plane is the primary and the pyramidal planes are the secondary slip planes.

45 See Kocks (1964), Bell (1965), Basinski and Jackson (1965a,b), Ramaswami *et al.* (1965), Kocks and Brown (1966), Nakada and Keh (1966), Bell and Green (1967), Jackson and Basinski (1967), Nowacki and Zarka (1971), Francoisi *et al.* (1980), and Wu *et al.* (1991). Note that experimental results of Nowacki and Zarka (1971) show larger critical shear strength for the *active* slip system than for the previously *inactive* ones, and those of Wu *et al.* (1991) confirm this latter conclusion. Calculations by Stroh (1953) and Zarka (1968) also show that the latent hardening due to the elastic field of the dislocations can never exceed the corresponding self-hardening; see Subsection 6.5.1. High strain-rate experiments by Kapoor and Nemat-Nasser (1999) on single-crystal tantalum, over a broad range of temperatures, show greater hardening in the athermal part of the flow stress associated with the latent slip system, but not in the thermal part; see Section 6.5.

6.4.6. Latent Hardening

Many experiments in this area consist of first quasi-statically loading an annealed crystal in such a manner as to activate essentially one slip system (referred to as the *primary* system), and then preparing samples from this deformed parent crystal for further tests in which other *latent* slip systems are individually (quasi-statically) activated in separate tests.[46] Most experiments of this kind show for a number of metals, that the critical shear strengths of the latent systems, obtained by backward extrapolation, are greater than those of the primary system. For fcc (copper, aluminum, gold, and silver) crystals, the critical shear strengths of the latent systems which are coplanar with the primary system, are reported to be essentially the same as those of the primary system, whereas for bcc (iron) crystals, the latent coplanar systems are reported to have hardened more than the primary system.

Figure 6.4.1 is from Jackson and Basinski (1967) for copper single crystals which have been first prestrained in single glide along the primary slip system $[\bar{1}01](111)$, *e.g.*, the parent crystal G4. Then, samples have been obtained from these parent crystals in such orientations that subsequent loadings activate only one intersecting latent system.[47] These latent systems are $[01\bar{1}](\bar{1}11)$ and $[0\bar{1}\bar{1}](1\bar{1}1)$ which are designated by Jackson and Basinski as A2 and D1, respectively; see also Havner (1992) for a detailed discussion. In addition, Figure 6.4.1 shows the stress-strain relation for a coplanar latent system, $[0\bar{1}1](111)$, designated as B2.

As is seen, the *level* of the latent critical shear strengths of the two intersecting systems is considerably greater than that of the primary system, with no measurable change being observed for the coplanar slip system.[48] Furthermore, the hardening rate, *i.e.*, the slope of the stress-strain curve, of the latent systems, after some initial plastic flow, is very small, sometimes actually zero, and then begins to increase after considerable straining.[49] This result is interpreted as showing that the latent hardening rate has been greater than that of the primary (active) slip system, leading to greater critical shear stress for the latent system, once it is activated.

Wu *et al.* (1991) have reexamined this question experimentally, using single-crystal copper. These authors report that the *initial flow stress* of a latent system, once it is activated, is actually *smaller* than the final flow stress of the

---

[46] As is illustrated in Subsections 4.8.9 and 4.8.10, the flow stress of metals (*e.g.*, tantalum and copper) can change several-fold when the test temperature and strain rate are appropriately changed. Thus, the classical (quasi-static and isothermal) rate-independent results can have only a limited range of validity; see Section 6.5 and, also, Kapoor and Nemat-Nasser (1998, 1999).

[47] A latent system is called *intersecting* when its plane intersects the glide plane of the primary slip system.

[48] Note that experiments by Nowacki and Zarka (1971) on single-crystal copper do not support this conclusion.

[49] Jackson and Basinski (1967) report that this response is accompanied by an inhomogeneous slip pattern.

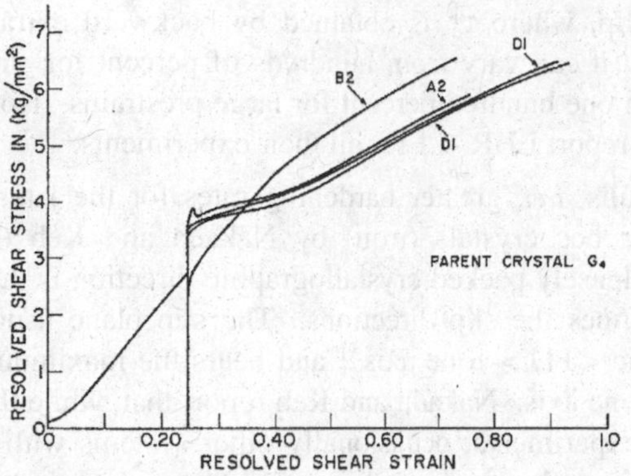

Figure 6.4.1

Latent hardening in copper crystals, prestrained in single glide on [$\bar{1}$01](111), and then on latent systems [01$\bar{1}$]($\bar{1}$11), [01$\bar{1}$](1$\bar{1}$1), and [01$\bar{1}$](111), (denoted by A2, D1, and B2, respectively); the first two are intersecting and the last one is a coplanar latent system (from Jackson and Basinski, 1967)

associated primary system, and that it is the dislocation activities which follow the initial yielding of the activated-latent system,[50] that quickly harden this system to a flow stress higher than the final flow stress of the then inactivated-primary system. This observation appears to be in accord with the analysis presented in Subsection 6.5. As is shown, the athermal resistance to the motion of dislocations, produced by the elastic field of the dislocations of the primary slip, is never greater for the latent systems than that of the primary system; Stroh (1953) and Zarka (1968). This resistance may define the *initial* flow stress of the latent system, provided the effects of other defects and impurities are excluded. Moreover, the dislocation forests which have been generated during the plastic flow of the primary slip, and which are piercing the slip plane of the latent system, serve as barriers to the motion of the dislocations of the activated-latent system, may create pileups, and may cause rapid hardening of the activated-latent slip system.

The self-hardening parameter, $h^{*\beta\beta}$ (β not summed), of the latent system β, is *not* constant. This fact has been emphasized by, *e.g.*, Kocks (1964) and Kocks and Brown (1966). Furthermore, the *latent hardening ratio*, defined by

$$\mathrm{LHR} \equiv \frac{\tau_c^\beta}{\tau_c^\alpha}, \quad \alpha\,(\text{primary}) \neq \beta\,(\text{latent}), \tag{6.4.33}$$

[50] To simplify the discussion, refer to an active but previously latent system, as an "activated-latent system" and to the associated primary system which has become inactive, as an "inactivated-primary system."

depends on the extent of the primary plastic slip, and decreases with increasing primary plastic slip, where τ_c^β is obtained by backward extrapolation. It has been reported that it can vary from hundreds of percent for small prestrains to slightly more than one hundred percent for large prestrains. However, Nowacki and Zarka (1971) report LHR < 1 for all their experiments.

Similar results, *i.e.,* greater hardening rates for the latent systems, have been reported for bcc crystals (iron) by Nakada and Keh (1966). For bcc metals, the most densely packed crystallographic direction is the $<111>$-family which usually defines the slip directions. The slip plane usually is the plane which contains the $<111>$-zone axis[51] and bears the maximum resolved shear stress along the zone axis. Nakada and Keh report that, while this was observed in most of their experiments, occasionally other systems with smaller Schmid factors, and hence smaller resolved shear stress, become active and contribute to the deformation; these cases are, however, excluded from consideration of the latent hardening.

Figure 6.4.2 shows some of the results reported by Nakada and Keh (1966). The curve marked P is the primary (parent) case. The others, one to four in Figure 6.4.2, are for the latent directions which were activated (each in a separate test) after prestraining in the primary glide direction. Curve 4 in this figure corresponds to a case where double slip was activated, even though the sample was oriented to activate only a single secondary slip system.

Figure 6.4.3 is Nakada and Keh's room-temperature results for the latent system $[\bar{1}11](101)$, *coplanar* with the primary system $[11\bar{1}](101)$. As is seen, unlike for the fcc metals, here a latent hardening ratio greater than one is observed, but the stress-strain curve of the latent system quickly joins that of the primary system. An LHR of 1.2 for this case, is among the smallest values reported by these authors. Figure 6.4.3 shows results for a case in which the resolved shear stress on the latent system (Q) was zero during the primary slip, and was also zero on the primary system (P) during the slip of the latent system; curve Q_0 is the virgin response of the system Q (separate test), with the origin shifted along the strain axis. The magnitude of the latent hardening was the smallest that these authors report to have observed.

The results for bcc crystals, similarly to those for fcc ones, show greater levels of flow stress for the latent system, once it has undergone some plastic flow. The corresponding slope of the stress-strain curve, however, may be rather small, *e.g.,* curve 3 of Figure 6.4.2. Nakada and Keh (1966) seek to discuss this in light of various potential mechanisms. In their experiments, the flow stress was observed to be temperature-independent for temperatures above 300K, where the athermal resistance to the dislocation motion must be operating. For low-temperature tests, down to 77K, these authors observe temperature dependence for both the primary and latent systems, with the latter showing greater temperature sensitivity. Similarly, both the primary and latent systems

[51] When several planes intersect on a common line, the line is called their *zone axis*.

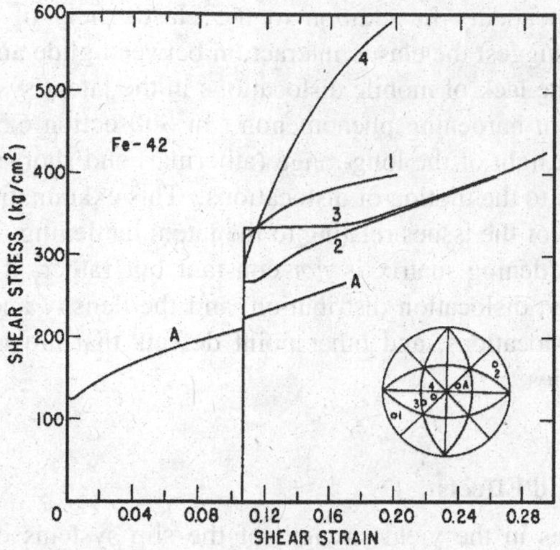

Figure 6.4.2

Latent hardening in single-crystal iron: curve A is the primary and curves 1 to 4 are the latent systems; double slip occurred for curve 4 which shows a very high hardening rate (from Nakada and Keh, 1966)

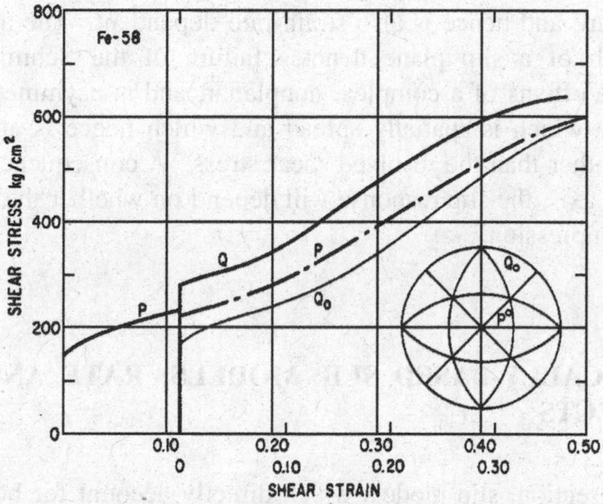

Figure 6.4.3

Latent hardening in single-crystal iron: curve P is the primary and curves Q_0 (virgin) and Q are the latent systems; the resolved shear stress on the latent (primary) system was zero during the primary (secondary) deformation (from Nakada and Keh, 1966) are strain-rate-dependent. In addition to the elastic field of all dislocations, Nakada and Keh suggest the elastic interaction between glide and forest dislocations, as well as the lack of mobile dislocations in the latent system, as possible causes of the latent hardening phenomenon. In Subsection 6.5.2, these results are reexamined in light of the long-range (athermal) and short-range (thermally activated) barriers to the motion of dislocations. This examination should somewhat clarify some of the issues relating to the latent hardening. In paricular, it is noted that the hardening matrix is *not* constant but rather, it varies with the dislocation density, dislocation distribution, and the density and distribution of debris, locked dislocations, and other point defects that are generated by the dislocation activities.

6.4.7. Non-Schmid Effects

Asymmetries in the yield strength of the slip systems on the *same* slip plane have been observed for many bcc metals, *e.g.*, iron and silicon-iron alloys, tungsten, niobium, tantalum, and molybdenum,[52] revealing non-Schmid responses in this class of crystals. This has been discussed by Christian (1983) who notes that the asymmetry in the shear resistance to slip in the {112}-planes of bcc metals is due to the intrinsic anisotropic lattice resistance to dislocation motion and *not* because of nonlattice obstacles such as solute atoms, precipitates, and forest dislocations. This strength asymmetry increases with decreasing temperature and hence is also strain-rate dependent. The anisotropy of the shear strength of a slip plane denotes failure of the Schmid law, since it represents the effects of a complex, nonplanar, and nonsymmetric, dislocation-core structure which is spatially spread and which hence is affected by stress components other than the resolved shear stress. A consequence is that, for the same loading axis, the slip response will depend on whether the sample is under tension or compression.

6.5. PHYSICALLY-BASED SLIP MODELS: RATE AND TEMPERATURE EFFECTS

In this section, slip models which directly account for both the temperature and strain-rate effects, are presented. Unlike the rate-independent models which have received considerable attention within the continuum framework, the rate- and temperature-dependent models of crystal plasticity have not been examined extensively in the literature, although a general rate-dependent theory has been proposed by Rice (1971). A power-law rate-dependent model is used

[52] See Taoka *et al.* (1964), Sestak and Zarubova (1965), Argon and Maloof (1966), Bowen *et al.* (1967), Foxall *et al.* (1967), Sherwood *et al.* (1967), Stein (1967), and Hull *et al.* (1967).

by Hutchinson (1976), and furthermore, by Pan and Rice (1983), Peirce *et al.* (1983, 1984), and Nemat-Nasser and Obata (1986), and, more recently, by Rashid (1990), Zikry (1990), Zikry and Nemat-Nasser (1990), Rashid and Nemat-Nasser (1992), Anand and Kalidindi (1994), Nemat-Nasser and Okinaka (1996), and Nemat-Nasser *et al.* (1998c), where references to other related work can be found. This type of model is similar to the empirical models discussed in Section 4.8 of Chapter 4. In particular, the slip rate of each slip system is assumed to be defined by an empirical relation similar to (4.8.1), with (4.8.2) representing an illustration, *i.e.*, a power law, as follows:

$$\dot{\gamma}^\alpha = \dot{\gamma}_0^\alpha \,\mathrm{sgn}(\tau^\alpha)\left[\frac{|\tau^\alpha|}{\tau_c^\alpha}\right]^m, \quad \dot{\gamma}^\alpha \geq 0, \quad (\alpha \text{ not summed}), \tag{6.5.1a}$$

where the evolution of τ_c^α is defined by

$$\dot{\tau}_c^\alpha = \sum_{\beta=1}^{n} h^{*\alpha\beta}\dot{\gamma}^\beta, \quad \dot{\gamma}^\beta \geq 0, \tag{6.5.1b}$$

and m is assumed to be a material constant, usually very large for most metals, $m \gg 1$. In most of these considerations, either a constant hardening matrix, $[h^{*\alpha\beta}]$, is used, in which case it follows that

$$\tau_c^\alpha = \sum_{\beta=1}^{n} h^{*\alpha\beta}\gamma^\beta + \tau_0^\alpha, \tag{6.5.1c}$$

or it is assumed that[53]

$$\tau_c^\alpha = \tau_c^\alpha(\gamma), \tag{6.5.1d}$$

from which the matrix $[h^{*\alpha\beta}]$ is obtained using (6.4.32), where γ is the accumulated total plastic slip of all active slip systems,

$$\gamma = \sum_{\alpha=1}^{n} \int_0^t |\dot{\gamma}^\alpha(t')|\, dt'. \tag{6.5.1e}$$

A latent hardening may also be implemented in the structure of the hardening matrix; see Peirce *et al.* (1983).

Experimental results show that most metals are both rate- and temperature-sensitive, and that generally the temperature dependence is of great significance. All physically-based models of the flow stress of metals include thermal activation and temperature sensitivity; see, Conrad (1964), Kocks *et al.* (1975), Argon (1995), and references cited there. Therefore, a somewhat more effective (yet inadequate) approach than the one mentioned above, is to identify the critical shear stress, τ_c^α, with one of the empirical relations given in Section 4.8 for the reference effective stress, τ_r, with γ now defined by (6.5.1e). For example, from (4.8.2b), obtain

---

[53] See, however, Rashid *et al.* (1992), where the hardening moduli are assumed to be functions of both the plastic strain and its rate, and Nemat-Nasser *et al.* (1998c) who use a hardening model that increases with the effective plastic strain and eventually becomes constant.

$$\tau_c^\alpha = \tau_0^\alpha (1 + \gamma/\gamma_0)^N \exp\{-\lambda_0(T - T_0)\}. \tag{6.5.2}$$

Similarly, the Johnson-Cook model (4.8.3) or the Zerilli-Armstrong model (4.8.4) may be modified and used accordingly.

Such empirical approaches, while useful and more powerful than the classical models, are still of limited applicability. It is more effective to develop models based on the physics of the dislocation motion associated with each slip system, and to integrate the basic crystal structure and the parameters which influence the dislocation motion, into the constitutive relations, somewhat similar to the models discussed in Subsections 4.8.2 to 4.8.10. This is the subject of the present section. For illustration, and since direct experimental results are not available, only two cases are examined in detail. The first is for bcc crystals in which it is assumed that the lattice resistance (Peierls stress) to the motion of dislocations is the only dominant thermally activated feature, and the second is for fcc crystals in which the forests of dislocations (Argon and East, 1979) are assumed to serve as the essential barriers to the motion of the mobile dislocations. The general approach, however, can be used for other metals, once temperature and rate dependency are established experimentally. In what follows, first the notion of long-range barriers to the motion of dislocations, is examined, particularly in relation to the latent hardening. Then the thermally-activated short-range barriers, as well as the drag effects, are considered.

6.5.1. Long-range Barriers and Latent Hardening

Stroh (1952) has calculated the root-mean-square shear stress on a slip plane due to an array of dislocations of an arbitrary distribution lying on a primary slip plane. The results show that this shear stress on a latent slip plane and in a latent direction, cannot be greater than that acting on the primary slip system. Stroh's calculations apply to both edge and screw dislocations. His conclusion does not depend on the distribution of the dislocations. Thus, according to this calculation, *the stress field of the dislocations of a primary slip system, hardens that system more than it hardens any other system.* These dislocations pierce through an intersecting latent slip plane, and provide sources for dislocation generation on the latent plane, once the latent system is activated, leading to rapid hardening of the activated-latent system. These comments are in accord with the results presented by Zarka (1968) and Wu *et al.* (1991). The Stroh results are summarized in this subsection, and the effects of the short-range barriers are examined subsequently in this section.

Consider a crystal, and choose the x_1-axis along the slip direction on the primary x_1, x_2-slip plane normal to the x_3-direction. Let the x_1', x_2'-plane be a latent slip plane normal to the x_3'-direction. Consider first the screw dislocations. The Burgers vector then is in the x_1-direction. The stresses at the origin due to a dislocation situated at a point with coordinates (0, x_2, x_3), are

$$\tau_{12} = -\frac{\mu b \sin\theta}{2\pi r}, \quad \tau_{13} = \frac{\mu b \cos\theta}{2\pi r}, \tag{6.5.3a,b}$$

where $r^2 = x_2^2 + x_3^2$ and $\theta = \tan^{-1}(x_3/x_2)$. The shear stresses acting on the x_1', x_2'-plane then are[54]

$$\tau_{12}' = (c_{11}\,c_{22} + c_{12}\,c_{21})\,\tau_{12} + (c_{11}\,c_{32} + c_{12}\,c_{31})\,\tau_{13}\,, \tag{6.5.4a}$$

with a similar expression for τ_{13}', where $c_{ij} = \mathbf{e}_i \cdot \mathbf{e}_j'$, i, j = 1, 2, 3, are the direction cosines of the primed coordinate system of unit base vectors $\mathbf{e}_i'$, measured relative to the unprimed system with unit base vectors $\mathbf{e}_i$. It thus follows that

$$\tau_{12}' = \frac{\mu\,b}{2\pi\,r}\,(-m_1 \sin\theta + m_2 \cos\theta)\,, \tag{6.5.4b}$$

where

$$m_1 = c_{11}\,c_{22} + c_{12}\,c_{21}\,, \qquad m_2 = c_{11}\,c_{32} + c_{12}\,c_{31}\,. \tag{6.5.4c,d}$$

For an array of such dislocations with the i'th dislocation at (r_i, θ_i), the resulting shear stress is

$$\tau_{12}' = \frac{\mu\,b}{2\pi}\sum_i \frac{-m_1 \sin\theta_i + m_2 \cos\theta_i}{r_i^2}\,. \tag{6.5.5}$$

The squared shear stress becomes

$$(\tau_{12}')^2 = \left\{\frac{\mu\,b}{2\pi}\right\}^2 \Big\{ \sum_i \pm \frac{1}{r_i^2}\,(m_1^2 \sin^2\theta_i - m_1\,m_2 \sin 2\theta_i + m_2^2 \cos^2\theta_i)$$
$$+ \sum_{i>j} \frac{1}{r_i\,r_j}\,[(m_1^2 + m_2^2)\cos(\theta_i - \theta_j)$$
$$-(m_1^2 + m_2^2)\cos(\theta_i + \theta_j) - 2m_1\,m_2 \sin(\theta_i + \theta_j)]\Big\}\,. \tag{6.5.6a}$$

Consider now all dislocation arrays associated with the primary slip system, and note that all values of θ_i and $(\theta_i + \theta_j)/2$ are equally probable. Hence, upon averaging (6.5.6a) over all such dislocations, $\sin^2\theta_i = (1 - \cos 2\theta_i)/2$ and $\cos^2\theta_i = (1 + \cos 2\theta_i)/2$ can be replaced by 1/2, while terms involving $\sin 2\theta_i$, $\sin(\theta_i + \theta_j)$, and $\cos(\theta_i + \theta_j)$ vanish. Denote the mean of $(\tau_{12}')^2$ by $<(\tau_{12}')^2>$ and observe that,

$$<(\tau_{12}')^2> \;=\; \tau_0^2\,(m_1^2 + m_2^2)\,, \quad \text{and}$$
$$\tau_0^2 = \left\{\frac{\mu\,b}{2\pi}\right\}^2 <\sum_i \frac{1}{r_i^2} + 2\sum_{i>j}\frac{\cos(\theta_i - \theta_j)}{r_i\,r_j}>\,, \tag{6.5.6b,c}$$

are the mean-square shear stresses on the latent and primary slip systems.

As is seen from (6.5.6b,c), the ratio τ_{12}'/τ_0 depends only on the relative orientation of the latent and the primary slip systems. For fcc crystals, for example, let $(111)[01\bar{1}]$ be the primary slip system, and note that $\tau_{12}'/\tau_0 \leq 1$ for all slip systems. On the slip plane $(\bar{1}11)$, for example, this ratio is 1, 0.5, and

54 This follows from transformation, $\tau_{ij}' = c_{ki}\,c_{lj}\,\tau_{kl}$, i, j, k, l = 1, 2, 3.

0.5, for the slip directions [011], [101], and [110], respectively. Table 6.5.1 gives the values of this ratio for both edge and screw dislocations.

Table 6.5.1

Values of τ'_{12}/τ_0 in percent due to the stress field of dislocations of the primary slip system $(111)[01\bar{1}]$ (from Stroh, 1952)

Slip Plane	Slip Direction	Screw Directions	Edge Dislocations, $\nu = 1/4$
(111)	$[01\bar{1}]$	100	100
	$[\bar{1}01]$	50	50
	$[1\bar{1}0]$	50	50
$(\bar{1}11)$	$[01\bar{1}]$	100	33.3
	$[101]$	50	25.2
	$[110]$	50	25.2
$(1\bar{1}1)$	$[01\bar{1}]$	81.7	69.3
	$[\bar{1}01]$	50	66.7
	$[110]$	50	90.2
$(11\bar{1})$	$[01\bar{1}]$	81.7	69.3
	$[101]$	50	90.2
	$[1\bar{1}0]$	50	67.7

For edge dislocations lying on a primary slip plane, similar results are given by Stroh (1952). Unlike for the screw dislocations, the stress ratio now depends on the Poisson ratio, ν. The calculation is straightforward and follows the same procedure as outlined above for the screw case. The mean-square shear stress acting on a latent slip system, *e.g.*, τ_{12}, now is given by

$$< (\tau'_{12})^2 > = (n_1^2 + n_3^2)\,\tau_0^2 + 4(2n_2^2 + n_1 n_2)\,\tau_1^2\,,$$

$$n_1 = c_{21}c_{22} - c_{11}c_{12}\,, \quad n_2 = c_{11}c_{12} + \nu\, c_{31}c_{32}\,, \quad n_3 = c_{11}c_{22} + c_{12}c_{21}\,,$$

$$\tau_0^2 = \frac{D^2}{4} < \sum_i \frac{1}{r_i^2} + 2\sum_{i>j} \frac{\cos(\theta_i - \theta_j)\cos 2(\theta_i - \theta_j)}{r_i r_j} >\,,$$

$$\tau_1^2 = \frac{D^2}{4} < \sum_i \frac{1}{r_i^2} + 2\sum_{i>j} \frac{\cos(\theta_i - \theta_j)}{r_i r_j} >\,, \quad D = \frac{b\mu}{2\pi(1-\nu)}\,. \qquad (6.5.7a\text{-}g)$$

For randomly distributed dislocations, the second sums in (6.5.7e,f) are zero. Stroh suggests that $\tau_1 = \tau_0$ may still be a good approximation when the pile-up of dislocations occurs. In this case, the terms in the considered sums may be divided into two groups according to whether or not the i'th and j'th dislocations belong to the same pile-up. When from different pile-ups, they are regarded as random, with zero net contributions. When from the same pile-up, then $\theta_i - \theta_j$ would generally be small so that $\cos(\theta_i - \theta_j) \approx 1$ is a good approximation. Hence, $\tau_1 \approx \tau_0$ is an acceptable approximation, especially since the coefficient of τ_1^2 is found to be generally somewhat smaller than that of τ_0^2.

With such approximations, Stroh (1952) obtains

$$(\tau'_{12}/\tau_0)^2 = (1 - c_{31}^2)(1 - c_{32}^2) - 4(1 - 2\nu)\, c_{31}\, c_{32}\,. \tag{6.5.7h}$$

In Table 6.5.1, the above estimate of τ'_{12}/τ_0 is given for $\nu = 1/4$ and indicated latent slip systems. As is seen, the stress field of the dislocations producing a primary slip, cannot harden a latent system more that it hardens the primary system.

Zarka (1968) arrives at a similar conclusion, using a more detailed calculation which also involves the distribution of the dislocations. Zarka groups randomly distributed segments of dislocations of parallel Burgers vectors and parallel slip planes, into a mechanism, say, mechanism i, and seeks to calculate the resulting force that these dislocations apply to any dislocation segment, measured per its unit length. Denoting this force on a j'th segment, by F^j, Zarka obtains,

$$F^j = \mu\, b^j \sum_i K_i^j\, (N^i\, l^i)^{1/2}\,, \tag{6.5.8a}$$

where N^i and l^i are the number per unit volume and the mean length of the i'th segment (hence, $N^i\, l^i = \rho^i$ is the dislocation density of the i'th segment), and K_i^j is the interaction coefficient which Zarka estimates to have the following values (no sum on repeated indices, $i \neq j$):

$$K_i^i = 1/8 \quad \text{(self–hardening)}\,,$$

$$K_i^j = 1/16 \quad \text{(parallel slip planes)}\,,$$

$$K_i^j = 1/20 \quad \text{(same Burgers vector)}\,,$$

$$K_i^j = 1/12 \quad \text{(all other mechanisms)}\,. \tag{6.5.8b-e}$$

Based on such calculations, it may be inferred that *the latent hardening due to the long-range effects of dislocations cannot exceed the corresponding self-hardening.* If the initial yielding is associated with the resistive stress field of all dislocations and defects, *i.e.,* with the athermal, rate-independent, long-range effects, then the starting value of the critical shear stress of a just-activated latent system cannot be expected to exceed the current value of the critical shear stress of the primary system, as suggested by Zarka (1968) and Wu *et al.* (1991), as well as earlier by Stroh (1952).

6.5.2. Self Hardening of a Previously Latent System

Experiments show that the flow stress of a previously latent slip system, estimated by backward extrapolation of the corresponding stress-strain relation, is generally greater than the final value of the flow stress of the primary system; see Section 6.4 and references cited therein. Wu *et al.* (1991) and Bassani and Wu (1991) suggest that this is caused by the extensive dislocation activities which follow, once a latent system is activated. This suggestion appears to be in accord with the observation that the dislocation forests of the primary slip system, piercing through the plane of the latent system, serve as sources of dislocation generation, which rapidly pile up, leading to rapid saturation and extensive self-hardening of such an activated-latent system. Indeed, Campbell *et al.* (1961) show by calculations that the activation time to nucleate a dislocation under a stress of $5 \times 10^{-4}\mu$ (μ is the shear modulus), is less than one nanosecond for aluminum. The piercing dislocations may act as, *e.g.*, Frank-Read sources; see Figure 6.1.6, page 390. The Wu *et al.* conclusion appears to apply independently of the nature of the short-range barriers, *i.e.*, to both bcc and fcc crystals.

Figure 6.5.1 illustrates this conclusion. In Figure 6.5.1a, the initial yielding of the latent system is assumed to occur at point Y_l. This is then followed by a rapid self-hardening portion which finally levels off due to dislocation saturation. Figure 6.5.1b shows an experimental result from Wu *et al.* (1991), which seems to be in line with this conclusion. The experiment has been performed on single-crystal copper.

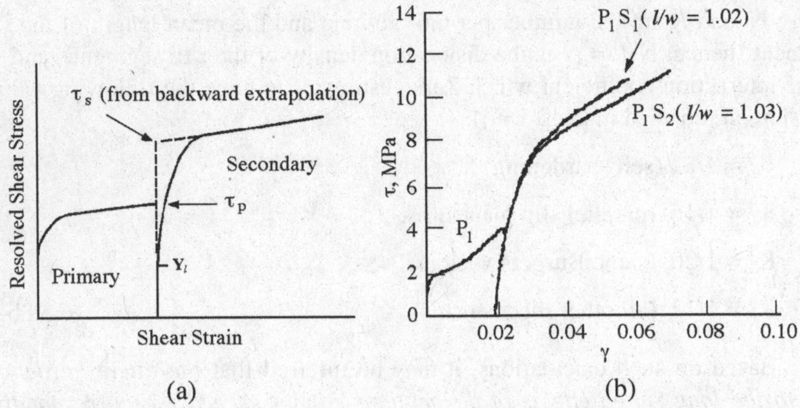

Figure 6.5.1

Primary and latent yield stresses for the rate-independent slip model, (a) initial yielding of the latent system occurs at Y_l, followed by rapid self-hardening; and (b) experimental data on copper (from Wu *et al.*, 1991)

As has been shown in Section 4.8, metals are generally both rate- and temperature-sensitive. Thus the thermally-activated part of the flow stress must be formulated in terms of the short-range resistance to the motion of dislocations.[55] A thorough modeling requires estimates of the dislocation generation

[55] The classical empirical rate-independent or rate-dependent models combine the short- and long-range parts of the resistance to the motion of dislocations into a single hardening matrix. This perhaps is the root of some of the confusion on latent- versus self-hardening. This comment notwithstanding, recent high strain-rate experimental data of Kapoor and Nemat-Nasser (1999) show, over a broad range of temperatures, 10% greater long-range (temperature-independent) hardening for an intersecting latent system in tantalum single crystals. The experiments, performed using a

and annihilation as functions of time. This is a complex phenomenon which requires a treatise in its own right, outside the scope of the present work. Instead, here a semi-empirical approach is adopted. Following the approach of Section 4.8, general expressions for the slip rates are first developed based on the physics of the phenomenon, and then the corresponding constitutive parameters are adjusted using the experimental results. For this approach to be successful, it is necessary to consider specific material classes. For illustration, commercially pure tantalum (bcc crystal) and copper (fcc crystal) are examined in Subsections 6.5.6 to 6.5.11, since a large body of experimental data is now available for these materials, over a broad range of temperatures and strain rates; see Nemat-Nasser and Isaacs (1997) and Nemat-Nasser and Li (1998).

6.5.3. Resistance to Crystalline Slip: Long-range Barriers

Divide the resistance to the motion of dislocations associated with the plastic slip of, say, the α'th slip system, into two parts, the short-range, $\tau^{*\alpha}$, which is rate- and temperature-dependent, and the long-range, τ_a^α, which is due to the elastic stress field of all other dislocations and defects. Let there be a total of n potential slip systems, and set

$$\tau^\alpha = \tau^{*\alpha} + \tau_a^\alpha, \quad (\alpha = 1, 2, ..., n). \tag{6.5.9}$$

As a first approximation, let the athermal stress, τ_a^α, of the α'th slip system depend solely on the stress field of all the existing dislocations, and represent this by

$$\tau_a^\alpha \approx \sum_{\beta=1}^{n} K_\beta^\alpha \frac{1}{l^\beta} = \sum_{\beta=1}^{n} K_\beta^\alpha \left[\rho^\beta\right]^{1/2}, \tag{6.5.10a}$$

where l^β is the average spacing and ρ^β is the average density of dislocations corresponding to the β'th slip system, and K_β^α's are parameters which can be estimated based on the crystal structure and the distribution of dislocations; *e.g.*, the estimates given by Zarka (1968) and summarized in (6.5.8) may be used. In general, ρ^β's are difficult to estimate. Since γ given by (6.5.1e) is a monotonically increasing parameter, it may be used as a load parameter. Hence, consider the following representation:

$$\tau_a^\alpha \approx c_0^\alpha + \sum_{\beta=1}^{n} K_\beta^\alpha \left\{\rho^\beta(\gamma)\right\}^{1/2} = \tau_a^0\, g^\alpha(\gamma, ...), \tag{6.5.10b}$$

where τ_a^0 is a normalizing parameter with physical dimensions of stress,[56] and dots stand for all other variables that also affect the athermal stress. A

recovery Hopkinson technique (see Chapter 9), do not, however, provide information relating to the initial yield stresses, since Kapoor and Nemat-Nasser compare the flow stress of the primary and the intersecting latent systems at 5% strain, at which Hopkinson data are reliable.

[56] This stress should include the effect of temperature on the elastic moduli, especially on the shear modulus, similarly to (4.8.18b), page 226.

simplified model of this kind is obtained by using a two-term approximation of the functions g^α and replacing (6.5.10b) by

$$\tau_a^\alpha \approx c_0^\alpha + c_1^\alpha \gamma^{n_1^\alpha}, \tag{6.5.10c}$$

where c_0^α, c_1^α, and n_1^α are empirical constants which may be taken to be independent of a particular slip system, by setting

$$\tau_a^\alpha \approx c_0 + c_1 \gamma^{n_1}, \tag{6.5.10d}$$

and estimating the values of c_0, c_1, and n_1 empirically, as is exemplified later on in this section.[57] Note again that, while the dislocation densities, ρ^β's, qualify as microstructural state variables, the plastic strain γ, does not, and is used only as a load parameter. Furthermore, expression (6.5.10d) is an empirical relation with limited applicability; for a systematic calculation of plastic deformation based on dislocation kinetics, see Zikry and Kao (1996a,b).

6.5.4. Resistance to Crystalline Slip: Short-range Barriers

Consider a typical slip system, α, and examine the associated slip rate, $\dot{\gamma}^\alpha$, produced by the dislocations which lie on the slip plane and move in the slip direction. The average shear resistance to the motion of these dislocations, $\tau^{*\alpha}$, provided by the short-range barriers, is examined below for two cases: (1) Peierls' barriers which are inherent in the lattice periodic structure; and (2) dislocation forests which intersect the slip plane. The general features of these models have been discussed in Section 4.8. Here, the basic idea is applied to define the slip rate associated with each slip system, starting with the Peierls barriers.

Slip Rate in bcc Crystals: Based on (6.1.34a), page 402, define the slip rate of the α'th slip system by[58]

$$\dot{\gamma}^\alpha = b^\alpha \rho_m^\alpha \bar{v}^\alpha, \tag{6.5.11a}$$

where b^α, ρ_m^α, and $\bar{v}^\alpha$, respectively, are the Burgers vector, the density of the mobile dislocations, and the average dislocation velocity, corresponding to the α'th slip system. As in Section 4.8, equation (4.8.6b), set

$$\bar{v}^\alpha = d^\alpha \omega_0^\alpha \exp\{-\Delta G^\alpha/kT\}, \tag{6.5.11b}$$

where d^α is the average distance the dislocations move on the slip plane, between the barriers; k is the Boltzmann constant; T is the absolute temperature; ω_0^α is the attempt frequency which depends on the structure and composition of the crystal, on the core structure of the dislocations, as well as on the particular

---

[57] In Subsection 6.5.7, this model and the Taylor averaging (see Chapter 8) are used to simulate the flow stress of commercially pure tantalum and OFHC copper over a broad range of strain rates, temperatures, and strains.

[58] Here and in the sequel, repeated indices are *not* summed unless otherwise explicitly stated.

slip system; and ΔG^α is the energy barrier to be overcome by the mobile dislocations through their thermal activation. Some or all of these parameters may take on different values depending on the particular slip system. For example, there is no reason to assume that any of these parameters should have the same value for a basal and a pyramidal slip system in an hcp crystal. In application, however, many of these parameters may be given the same constant values for all slip systems.

The slip rate of the α'th slip system is obtained using the procedure outlined for the continuum model in Subsection 4.8.8, as follows:

$$\dot{\gamma}^\alpha = b^\alpha d^\alpha \rho_m^\alpha \omega_0^\alpha \exp\{-\Delta G^\alpha / kT\}$$
$$= \dot{\gamma}_r^\alpha \exp\{-\Delta G^\alpha / kT\}, \quad \dot{\gamma}_r^\alpha = b^\alpha d^\alpha \rho_m^\alpha \omega_0^\alpha. \tag{6.5.11c-e}$$

As in (4.8.10a), set

$$\Delta G^\alpha = G_0^\alpha \left\{1 - \left[\frac{\tau^{*\alpha}(\dot{\gamma}^\alpha, T)}{\hat{\tau}^\alpha}\right]^p\right\}^q, \tag{6.5.12}$$

where $0 < p \le 1$ and $1 \le q \le 2$, and $\hat{\tau}^\alpha$ is the shear stress required for the dislocation to cross the barrier without the assistance of the thermal activation.

Finally, in view of (6.5.9) to (6.5.12), obtain

$$\dot{\gamma}^\alpha = \dot{\gamma}_r^\alpha \exp\left\{-\frac{G_0^\alpha}{kT}\left[1 - \left[\frac{\tau^\alpha - \tau_a^\alpha}{\hat{\tau}^\alpha}\right]^p\right]^q\right\}, \tag{6.5.13a}$$

for $T \le T_c^\alpha$, where T_c^α is the temperature above which the dislocations can overcome their barriers through their thermal activation without the help of the applied shear stress. This critical temperature is given by

$$T_c^\alpha = -\frac{G_0^\alpha}{k}\left\{ln\frac{\dot{\gamma}^\alpha}{\dot{\gamma}_r^\alpha}\right\}^{-1}. \tag{6.5.13b}$$

If the viscous drag resisting the motion of the dislocations is neglected, then, for $T \ge T_c^\alpha$, the flow stress is rate- and temperature-independent, being then given by any one of (6.5.10a-c). On the other hand, when the drag forces are also included (Subsection 6.5.5), then the resistance to slip is always rate-dependent.

The flow stress of the α'th slip system is obtained by solving (6.5.13a) for τ^α. This gives

$$\tau^\alpha = \tau_a^\alpha + \hat{\tau}^\alpha\left\{1 - \left[-\frac{kT}{G_0^\alpha}\left[ln\frac{\dot{\gamma}^\alpha}{\dot{\gamma}_r^\alpha}\right]\right]^{1/q}\right\}^{1/p}, \quad \dot{\gamma}^\alpha > 0, \tag{6.5.13c}$$

where τ_a^α is given by either (6.5.10b) or (6.5.10c), depending on the choice of the approximate expression for the athermal part of the flow stress.

When the short-range energy barrier is due to the lattice stress field, the Peierls barrier, then $\dot{\gamma}_r^\alpha$ is defined by (6.5.11e) and can be estimated from the lattice properties and the density of the mobile dislocations, as in Subsection 4.8.9; see comments relating to expression (4.8.22e), page 234, and Table 4.8.1, page 235.

Slip Rate in fcc Crystals: Consider now the case in which the barriers are due to the dislocation forests which intersect the slip plane (Argon and East, 1979). By necessity, these dislocations must all lie on intersecting slip planes. Denote by l^α the average spacing of such dislocations that pierce the α'th slip plane, and, from (4.8.12a,b), note that

$$\hat{\tau}^\alpha = \frac{G_0^\alpha}{b^\alpha \lambda^\alpha l^\alpha}, \tag{6.5.14a}$$

where λ^α is the average width of the energy barrier. The spacing l^α is given by

$$l^\alpha = (\rho^\alpha)^{-1/2}, \quad \rho^\alpha = \sum_{\beta=1}^{n} \rho_c^\beta, \quad (\beta \neq \alpha), \tag{6.5.15a,b}$$

where the summation is over all slip systems whose dislocations intersect the α'th slip plane, and ρ_c^β is the current density of the dislocations associated with the β'th slip system. As a rough approximation of the above expression, consider

$$l^\alpha = \frac{l_0}{1 + \left[\sum_{\beta=1}^{n} a^\beta(T)\, \gamma^\beta \right]^n}, \quad (\beta \neq \alpha), \tag{6.5.15c}$$

or, even a rougher approximation,

$$l^\alpha = \frac{l_0}{1 + a(T) \left[\sum_{\beta=1}^{n} \gamma^\beta \right]^n}, \quad (\beta \neq \alpha), \tag{6.5.15d}$$

where the parameters $a^\beta(T)$ and $a(T)$ are functions of the temperature, T, the length scale l_0 is the corresponding reference average dislocation spacing, and n is a constitutive parameter that may be set equal to 1/2. Generally, workhardening is associated with the increasing dislocation density and, hence, with the decreasing average dislocation spacing. Hence, with a suitable choice of the reference l_0, the free constitutive parameters in (6.5.15c,d) are non-negative. Furthermore, since the density of dislocations is expected to decrease with increasing temperature, these parameters are expected to be *decreasing* functions of the temperature. In application, these parameters are obtained empirically. For the simple case (6.5.15d), this is illustrated in Subsection 6.5.11.

In a more general setting, and in view of (4.8.13a,b), consider expressing (6.5.14a) in a form that also transparently displays the involved length scales, as follows:

$$\hat{\tau}^\alpha = \frac{l_0}{l^\alpha} \hat{\tau}_0^\alpha, \quad \hat{\tau}_0^\alpha = \frac{G_0^\alpha}{b^\alpha \lambda^\alpha l_0}. \tag{6.5.14b,c}$$

Similarly, the reference strain rate, (6.5.11e), is expressed as

$$\dot{\gamma}_r^\alpha = \frac{l_s^\alpha}{l_0} \dot{\gamma}_0^\alpha, \quad l_s^\alpha = \frac{\rho_m^\alpha d^\alpha}{\rho_m^0},$$

$$\dot{\gamma}_0^\alpha = b^\alpha \rho_m^0 \omega_0^\alpha l_0. \tag{6.5.16a-c}$$

The flow stress for the α'th slip system now becomes

$$\tau^{\alpha} = \tau_a^{\alpha} + \hat{\tau}_0^{\alpha}\left\{1 - \left[-\frac{kT}{G_0^{\alpha}}\left[ln\frac{\dot{\gamma}^{\alpha}}{\dot{\gamma}_0^{\alpha}} + ln\frac{l_0}{l_s^{\alpha}}\right]\right]^{1/q}\right\}^{1/p}\frac{l_0}{l^{\alpha}}. \tag{6.5.17}$$

Thus, in addition to the average spacing of the total dislocations that affects the athermal part of the flow stress, as in (6.5.10a), expression (6.5.17) includes two length scales for each slip system, l^{α} and l_s^{α}. The first one relates to the structure of the short-range barriers, and the second, to their spacing and the density of the mobile dislocations. These expressions clearly display the influence of the initial density of the dislocations through the presence of the l_0-term. It is thus expected that, *e.g.*, τ^{α} should decrease by more than an order of magnitude when a plastically deformed sample is annealed, since annealing can reduce the dislocation density by a factor 1,000 and greater.

For actual applications, it may be convenient to assume that $l_s^{\alpha} \approx l^{\alpha}$, and approximate this by, say, (6.5.15d). Then, as in the case of the continuum model discussed in Subsection 4.8.6, page 222, the following empirical relation may be used for a(T):

$$a(T) = a_0\{1 - (T/T_m)^2\}, \tag{6.5.15e}$$

where a_0 is fixed empirically, and T_m is the melting temperature. In Subsection 6.5.16, this model is used to simulate experimental data on the flow stress of OFHC copper. There, again, the continuum estimates of $k/G_0^{\alpha} = k/G_0 = 4.9\times10^{-5}/K$ and $\dot{\gamma}_0^{\alpha} = \dot{\gamma}_0 = 2\times10^{10}/s$ are used, leaving c_0, c_1, a_0, and $\hat{\tau}_0^{\alpha} = \hat{\tau}_0$ as free constitutive parameters, where c_0 and c_1 define the athermal part of the flow stress; see (6.5.10d). With (6.5.10d) and (6.5.17), the final approximate expression which is used in Subsection 6.5.16, becomes

$$\tau^{\alpha} = c_0 + c_1\gamma^{n_1} + \hat{\tau}_0^{\alpha}\left\{1 - \left[-\frac{kT}{G_0}\left[ln\frac{\dot{\gamma}^{\alpha}}{\dot{\gamma}_0} + ln\left\{1 + a(T)\left[\sum_{\beta=1}^{n}\gamma^{\beta}\right]^{n}\right\}\right]\right]^{1/q}\right\}^{1/p}$$

$$\times\left\{1 + a(T)\left[\sum_{\beta=1}^{n}\gamma^{\beta}\right]^{n}\right\}, \tag{6.5.18}$$

where a(T) is given by (6.5.15e). Note that, q = 2 and p = 2/3 are generally suitable values that can be used in (6.5.18).

6.5.5. Effect of Viscous Drag

At high strain rates, the effect of the viscous drag on the motion of dislocations may become important. This can be included by modifying the average dislocation velocity, $\overline{v}^{\alpha}$, as follows:

$$\overline{v}^{\alpha} = \frac{d^{\alpha} + \lambda^{\alpha}}{t_w^{\alpha} + t_r^{\alpha}}. \tag{6.5.19a}$$

Following now the procedure of Section 4.8, obtain

$$\dot{\gamma}^{\alpha} = \dot{\gamma}_0^{\alpha}\frac{l_s^{\alpha}}{l_0}\left\{\exp\{\Delta G^{\alpha}/kT\} + \frac{\tau_D^{0\alpha}}{\tau^{\alpha} - \tau_a^{\alpha}}\frac{d}{l_0}\right\}^{-1}, \tag{6.5.19b}$$

where $\dot{\gamma}_0^\alpha$ is given by (6.5.16c), ΔG^α is defined by (6.5.12), and l_s^α and the reference drag stress, $\tau_D^{0\alpha}$, are defined by

$$l_s^\alpha = \frac{\rho_m^\alpha(d^\alpha + \lambda^\alpha)}{\rho_m^0}, \quad \tau_D^{0\alpha} = \omega_0^\alpha D\, l_0/b, \tag{6.5.19c,d}$$

with D being the drag coefficient. Here again, simplifications similar to (4.8.21a-d) may be considered, but expression (6.5.19b) is most suited for direct inclusion into the expression for $\mathbf{D}^p$, *e.g.,* to be used in (6.3.8d), page 418.

6.5.6. bcc Crystals

For bcc crystals, it is often assumed that {110}, {112}, and {123} are the families of slip planes, with <111> providing the corresponding family of slip directions. This results in a total of forty-eight potential slip systems. Table 6.5.2 summarizes these systems, numbering them from 1 to 48.

Interdependency of Slip Systems: The forty-eight slip systems in a bcc crystal are not all independent. The slip systems can be divided into four families, each family having a common slip direction. These directions are

$$[111], \quad [\bar{1}11], \quad [1\bar{1}1], \quad [11\bar{1}].$$

Hence, twelve slip systems belong to each family. The slip-plane normals, $\mathbf{n}^\alpha$, of the slip systems of each family all lie on a plane which is perpendicular to the corresponding slip direction. It thus follows that only two of twelve slip systems in each family are linearly independent, since a proper linear combination of two arbitrarily chosen slip-plane normals can represent slip-plane normals of the other ten slip systems within that family. Hence, ten conditions for $\mathbf{p}^\alpha$'s are obtained in each of the families.[59] These forty conditions are listed in Table 6.5.3, according to the numbering system of Table 6.5.2.

Since the forty-eight slip-system tensors, $\mathbf{p}^\alpha$'s, are symmetric and traceless, in addition to the forty relations listed in Table 6.5.3, there are three more independent linear relations among them. It is easy to directly obtain these additional relations. The results are

$$\mathbf{p}^1 + \mathbf{p}^7 - \mathbf{p}^9 = \mathbf{0},$$

$$\mathbf{p}^2 - \mathbf{p}^5 + \mathbf{p}^{11} = \mathbf{0},$$

$$\mathbf{p}^3 + \mathbf{p}^6 + \mathbf{p}^{10} = \mathbf{0}. \tag{6.5.20a-c}$$

---

[59] The same conditions hold for both the symmetric part, $\mathbf{p}^\alpha$ and the skewsymmetric part, $\mathbf{r}^\alpha$ of the slip systems $\mathbf{s}^\alpha \otimes \mathbf{n}^\alpha$. Indeed, on each slip plane in a three-dimensional space, there are only two independent slip directions, and, overall, there are only three independent non-coplanar slip-plane unit normals. Thus, there are only six independent $\mathbf{s}^\alpha \otimes \mathbf{n}^\alpha$ for *any crystal structure*. Furthermore, since $\mathrm{tr}(\mathbf{s}^\alpha \otimes \mathbf{n}^\alpha) = 0$, at most, there are only five independent slip systems, in general.

Table 6.5.2

Forty-eight slip systems for bcc crystals

1	$(110)[\bar{1}11]$	13	$(112)[11\bar{1}]$	25	$(123)[11\bar{1}]$	37	$(231)[1\bar{1}1]$
2	$(110)[1\bar{1}1]$	14	$(\bar{1}12)[1\bar{1}1]$	26	$(\bar{1}23)[1\bar{1}1]$	38	$(\bar{2}31)[11\bar{1}]$
3	$(1\bar{1}0)[111]$	15	$(1\bar{1}2)[1\bar{1}1]$	27	$(1\bar{2}3)[\bar{1}11]$	39	$(2\bar{3}1)[111]$
4	$(1\bar{1}0)[11\bar{1}]$	16	$(11\bar{2})[111]$	28	$(12\bar{3})[111]$	40	$(23\bar{1})[\bar{1}11]$
5	$(101)[11\bar{1}]$	17	$(121)[1\bar{1}1]$	29	$(132)[1\bar{1}1]$	41	$(312)[\bar{1}11]$
6	$(101)[\bar{1}11]$	18	$(\bar{1}21)[11\bar{1}]$	30	$(\bar{1}32)[11\bar{1}]$	42	$(\bar{3}12)[111]$
7	$(10\bar{1})[111]$	19	$(1\bar{2}1)[111]$	31	$(1\bar{3}2)[111]$	43	$(3\bar{1}2)[11\bar{1}]$
8	$(10\bar{1})[1\bar{1}1]$	20	$(12\bar{1})[\bar{1}11]$	32	$(13\bar{2})[\bar{1}11]$	44	$(31\bar{2})[1\bar{1}1]$
9	$(011)[11\bar{1}]$	21	$(211)[\bar{1}11]$	33	$(213)[11\bar{1}]$	45	$(321)[\bar{1}11]$
10	$(011)[1\bar{1}1]$	22	$(\bar{2}11)[111]$	34	$(\bar{2}13)[1\bar{1}1]$	46	$(\bar{3}21)[111]$
11	$(01\bar{1})[111]$	23	$(2\bar{1}1)[11\bar{1}]$	35	$(2\bar{1}3)[\bar{1}11]$	47	$(3\bar{2}1)[11\bar{1}]$
12	$(01\bar{1})[\bar{1}11]$	24	$(21\bar{1})[1\bar{1}1]$	36	$(21\bar{3})[111]$	48	$(32\bar{1})[1\bar{1}1]$

Table 6.5.3

Forty linear relations among the forty-eight slip-system tensors, $\mathbf{p}^{\alpha}$'s, for bcc crystals (from Nemat-Nasser *et al.*, 1998a)

1	$\mathbf{p}^3-\mathbf{p}^7+\mathbf{p}^{11}=0$	2	$\mathbf{p}^3-2\,\mathbf{p}^7+\sqrt{3}\,\mathbf{p}^{16}=0$	3	$-2\,\mathbf{p}^3+\mathbf{p}^7+\sqrt{3}\,\mathbf{p}^{19}=0$
4	$\mathbf{p}^3+\mathbf{p}^7+\sqrt{3}\,\mathbf{p}^{22}=0$	5	$2\,\mathbf{p}^3-3\,\mathbf{p}^7+\sqrt{7}\,\mathbf{p}^{28}=0$	6	$-3\,\mathbf{p}^3+2\,\mathbf{p}^7+\sqrt{7}\,\mathbf{p}^{31}=0$
7	$\mathbf{p}^3-3\,\mathbf{p}^7+\sqrt{7}\,\mathbf{p}^{36}=0$	8	$-3\,\mathbf{p}^3+\mathbf{p}^7+\sqrt{7}\,\mathbf{p}^{39}=0$	9	$\mathbf{p}^3+2\,\mathbf{p}^7+\sqrt{7}\,\mathbf{p}^{42}=0$
10	$2\,\mathbf{p}^3+\mathbf{p}^7+\sqrt{7}\,\mathbf{p}^{46}=0$	11	$-\mathbf{p}^1+\mathbf{p}^6+\mathbf{p}^{12}=0$	12	$\mathbf{p}^1-2\,\mathbf{p}^6+\sqrt{3}\,\mathbf{p}^{15}=0$
13	$-2\,\mathbf{p}^1+\mathbf{p}^6+\sqrt{3}\,\mathbf{p}^{20}=0$	14	$-\mathbf{p}^1-\mathbf{p}^6+\sqrt{3}\,\mathbf{p}^{21}=0$	15	$2\,\mathbf{p}^1-3\,\mathbf{p}^6+\sqrt{7}\,\mathbf{p}^{27}=0$
16	$-3\,\mathbf{p}^1+2\,\mathbf{p}^6+\sqrt{7}\,\mathbf{p}^{32}=0$	17	$\mathbf{p}^1-3\,\mathbf{p}^6+\sqrt{7}\,\mathbf{p}^{35}=0$	18	$-3\,\mathbf{p}^1+\mathbf{p}^6+\sqrt{7}\,\mathbf{p}^{40}=0$
19	$-\mathbf{p}^1-2\,\mathbf{p}^6+\sqrt{7}\,\mathbf{p}^{41}=0$	20	$-2\,\mathbf{p}^1-\mathbf{p}^6+\sqrt{7}\,\mathbf{p}^{45}=0$	21	$-\mathbf{p}^2+\mathbf{p}^8+\mathbf{p}^{10}=0$
22	$-\mathbf{p}^2+2\,\mathbf{p}^8+\sqrt{3}\,\mathbf{p}^{14}=0$	23	$-2\,\mathbf{p}^2-\mathbf{p}^8+\sqrt{3}\,\mathbf{p}^{17}=0$	24	$-\mathbf{p}^2-\mathbf{p}^8+\sqrt{3}\,\mathbf{p}^{24}=0$
25	$-2\,\mathbf{p}^2+3\,\mathbf{p}^8+\sqrt{7}\,\mathbf{p}^{26}=0$	26	$-3\,\mathbf{p}^2+2\,\mathbf{p}^8+\sqrt{7}\,\mathbf{p}^{29}=0$	27	$-\mathbf{p}^2+3\,\mathbf{p}^8+\sqrt{7}\,\mathbf{p}^{34}=0$
28	$-3\,\mathbf{p}^2+\mathbf{p}^8+\sqrt{7}\,\mathbf{p}^{37}=0$	29	$-\mathbf{p}^2-2\,\mathbf{p}^8+\sqrt{7}\,\mathbf{p}^{44}=0$	30	$-2\,\mathbf{p}^2-\mathbf{p}^8+\sqrt{7}\,\mathbf{p}^{48}=0$
31	$\mathbf{p}^4-\mathbf{p}^5+\mathbf{p}^9=0$	32	$\mathbf{p}^4-2\,\mathbf{p}^5+\sqrt{3}\,\mathbf{p}^{13}=0$	33	$2\,\mathbf{p}^4-\mathbf{p}^5+\sqrt{3}\,\mathbf{p}^{18}=0$
34	$-\mathbf{p}^4-\mathbf{p}^5+\sqrt{3}\,\mathbf{p}^{23}=0$	35	$2\,\mathbf{p}^4-3\,\mathbf{p}^5+\sqrt{7}\,\mathbf{p}^{25}=0$	36	$-3\,\mathbf{p}^4-2\,\mathbf{p}^5+\sqrt{7}\,\mathbf{p}^{30}=0$
37	$\mathbf{p}^4-3\,\mathbf{p}^5+\sqrt{7}\,\mathbf{p}^{33}=0$	38	$3\,\mathbf{p}^4-\mathbf{p}^5+\sqrt{7}\,\mathbf{p}^{38}=0$	39	$-\mathbf{p}^4-2\,\mathbf{p}^5+\sqrt{7}\,\mathbf{p}^{43}=0$
40	$-2\,\mathbf{p}^4-\mathbf{p}^5+\sqrt{7}\,\mathbf{p}^{47}=0$				

6.5.7. fcc Crystals

The fcc crystals have been most extensively studied, since the pioneering work of Taylor (1934, 1938).[60] Early efforts have focused on rate-independent theories. More recent rate-dependent models have been mostly empirical, being based on some variance of the power-law description of the critical shear stresses; see, *e.g.*, Hutchinson (1976), Asaro (1983), Pan and Rice (1983), Peirce *et al.* (1983), Rashid *et al.* (1992), and Nemat-Nasser and Okinaka

(1996). In the remaining part of this chapter, some recent advances in physically-based modeling of fcc crystals are summarized, starting with an account of the general properties and response of fcc crystals.

As pointed out before, the {111}-family are the crystallographic slip planes for fcc single crystals. In Figure 6.5.2, the crystallographic plane (111) is identified by the dashed lines. There are four such slip planes in an fcc crystal. In each slip plane there are three slip directions of the < 110 >-type.

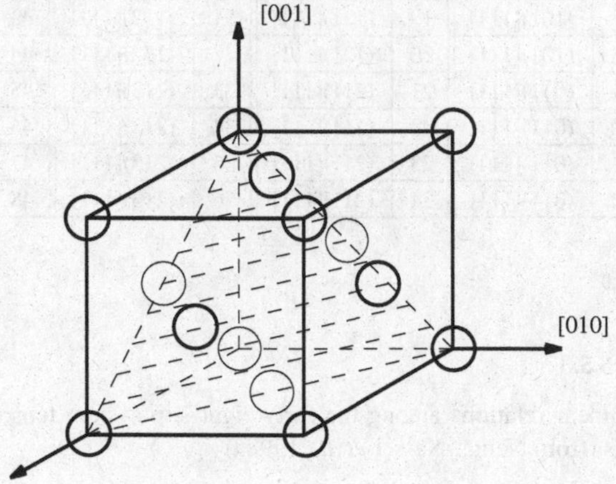

Figure 6.5.2

An fcc unit cell and the (111)-crystallographic plane

Consider a typical unit cell, and choose the x_1-, x_2-, and x_3-coordinate axes, with the corresponding unit base vectors, $\mathbf{e}_i$, i = 1, 2, 3, along the crystallographic directions, [100], [010], and [001], respectively. The slip-plane normals and the corresponding components in the $\mathbf{e}_i$-triad then are,

$$\mathbf{n}_0^1 = \mathbf{n}_0^2 = \mathbf{n}_0^3 = \frac{1}{\sqrt{3}}(1, 1, 1),$$

$$\mathbf{n}_0^4 = \mathbf{n}_0^5 = \mathbf{n}_0^6 = \frac{1}{\sqrt{3}}(-1, 1, 1),$$

$$\mathbf{n}_0^7 = \mathbf{n}_0^8 = \mathbf{n}_0^9 = \frac{1}{\sqrt{3}}(-1, -1, 1),$$

$$\mathbf{n}_0^{10} = \mathbf{n}_0^{11} = \mathbf{n}_0^{12} = \frac{1}{\sqrt{3}}(1, -1, 1). \qquad (6.5.21\text{a-d})$$

[60] For a lucid and thorough discussion of Taylor's contributions, and an account of rate-independent crystal theories, see Havner (1992).

The associated slip directions are similarly expressed as

$$\mathbf{s}_0^1 = \frac{1}{\sqrt{2}}(-1,1,0), \quad \mathbf{s}_0^2 = \frac{1}{\sqrt{2}}(0,-1,1), \quad \mathbf{s}_0^3 = \frac{1}{\sqrt{2}}(1,0,-1),$$
$$\mathbf{s}_0^4 = \frac{1}{\sqrt{2}}(-1,-1,0), \quad \mathbf{s}_0^5 = \frac{1}{\sqrt{2}}(1,0,1), \quad \mathbf{s}_0^6 = \frac{1}{\sqrt{2}}(0,1,-1),$$
$$\mathbf{s}_0^7 = \frac{1}{\sqrt{2}}(1,-1,0), \quad \mathbf{s}_0^8 = \frac{1}{\sqrt{2}}(0,1,1), \quad \mathbf{s}_0^9 = \frac{1}{\sqrt{2}}(-1,0,-1),$$
$$\mathbf{s}_0^{10} = \frac{1}{\sqrt{2}}(1,1,0), \quad \mathbf{s}_0^{11} = \frac{1}{\sqrt{2}}(-1,0,1), \quad \mathbf{s}_0^{12} = \frac{1}{\sqrt{2}}(0,-1,-1).$$
(6.5.22a-l)

The resulting twelve slip systems are shown in Figure 6.5.3.

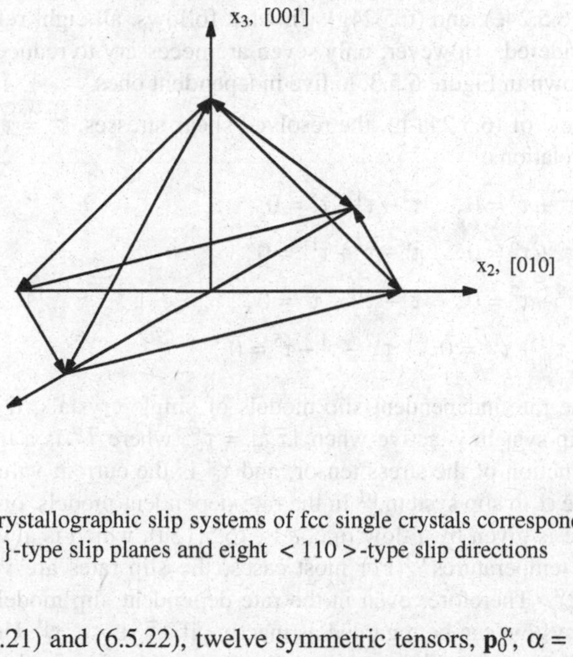

Figure 6.5.3

The twelve crystallographic slip systems of fcc single crystals corresponding to the four {111}-type slip planes and eight < 110 >-type slip directions

From (6.5.21) and (6.5.22), twelve symmetric tensors, $\mathbf{p}_0^\alpha$, $\alpha = 1, 2, ..., 12$, are now formed, according to (6.2.11a). As in the case of bcc crystals, only five of these tensors are linearly independent. A simple way to establish this fact is to note that, any second-order symmetric tensor in a three-dimensional Euclidean space, can be expressed as a linear combination of six symmetric second-order base tensors, $\mathbf{b}_a$, a = 1, 2, ..., 6, defined as follows (Nemat-Nasser and Hori, 1993, page 558):

$$\mathbf{b}_1 = \mathbf{e}_{11}^s, \quad \mathbf{b}_2 = \mathbf{e}_{22}^s, \quad \mathbf{b}_3 = \mathbf{e}_{33}^s,$$
$$\mathbf{b}_4 = \mathbf{e}_{23}^s, \quad \mathbf{b}_5 = \mathbf{e}_{31}^s, \quad \mathbf{b}_6 = \mathbf{e}_{12}^s,$$
(6.5.23a-f)

where the $\mathbf{e}_{ij}^s$'s are given by

$$\mathbf{e}_{ij}^{s} = \tfrac{1}{2}\,(\mathbf{e}_i \otimes \mathbf{e}_j + \mathbf{e}_j \otimes \mathbf{e}_i)\,. \tag{6.5.23g}$$

Furthermore, since $\mathbf{p}_0^{\alpha}$'s are traceless, they all belong to the five-dimensional deviatoric subspace of the linear space defined by the base tensors $\mathbf{b}_a$. Hence, there are only five linearly-independent $\mathbf{p}_0^{\alpha}$'s. From (6.5.21) and (6.5.22), the $\mathbf{p}_0^{\alpha}$'s satisfy the following linear relations:

$$\begin{aligned}
&\mathbf{p}_0^1 + \mathbf{p}_0^2 + \mathbf{p}_0^3 = \mathbf{0}, \quad && \mathbf{p}_0^3 - \mathbf{p}_0^4 + \mathbf{p}_0^8 = \mathbf{0},\\
&\mathbf{p}_0^4 + \mathbf{p}_0^5 + \mathbf{p}_0^6 = \mathbf{0}, \quad && \mathbf{p}_0^6 - \mathbf{p}_0^7 + \mathbf{p}_0^{11} = \mathbf{0},\\
&\mathbf{p}_0^7 + \mathbf{p}_0^8 + \mathbf{p}_0^9 = \mathbf{0}, \quad && \mathbf{p}_0^9 - \mathbf{p}_0^{10} + \mathbf{p}_0^2 = \mathbf{0},\\
&\mathbf{p}_0^{10} + \mathbf{p}_0^{11} + \mathbf{p}_0^{12} = \mathbf{0}, \quad && \mathbf{p}_0^{12} - \mathbf{p}_0^1 + \mathbf{p}_0^5 = \mathbf{0}.
\end{aligned} \tag{6.5.24a-h}$$

These are eight linear relations. Therefore, one relation is redundant. For example, $\mathbf{p}_0^6 - \mathbf{p}_0^7 + \mathbf{p}_0^{11} = \mathbf{0}$ can be reduced to (6.5.24a) by linear combination with (6.5.24c), (6.5.24e), and (6.5.24g). In what follows, all eight relations (6.5.24a-h) are considered. However, only seven are necessary to reduce the twelve slip systems shown in Figure 6.5.3, to five independent ones.

In view of (6.5.24a-h), the resolved shear stresses, $\tau^{\alpha} = \boldsymbol{\tau} : \mathbf{p}^{\alpha}$, satisfy the following relations:

$$\begin{aligned}
&\tau^1 + \tau^2 + \tau^3 = 0, \quad && \tau^3 - \tau^4 + \tau^8 = 0,\\
&\tau^4 + \tau^5 + \tau^6 = 0, \quad && \tau^6 - \tau^7 + \tau^{11} = 0,\\
&\tau^7 + \tau^8 - \tau^9 = 0, \quad && \tau^9 - \tau^{10} + \tau^2 = 0,\\
&\tau^{10} + \tau^{11} + \tau^{12} = 0, \quad && \tau^{12} - \tau^1 + \tau^5 = 0.
\end{aligned} \tag{6.5.25a-h}$$

In the rate-independent slip models of single crystals, it is assumed that the α'th slip system is active when $|\bar{\tau}^{\alpha}| = \tau_c^{\alpha}$, where $\bar{\tau}^{\alpha}$ is a linear and homogeneous function of the stress tensor, and τ_c^{α} is the current value of the critical stress of the α'th slip system.[61] In the rate-dependent models, on the other hand, the slip rate is given by a flow rule, *e.g.*, (6.5.13a), which is always nonzero for subcritical temperatures.[62] For most cases, the slip rates are very small when $|\bar{\tau}^{\alpha}| \ll \tau_c^{\alpha}$. Therefore, even in the rate-dependent slip model of single crystals, a slip system can be regarded as inactive if $|\bar{\tau}^{\alpha}| \ll \tau_c^{\alpha}$. Hence, only some (no more than eight) of the twelve slip systems in fcc single crystals can be active at any time. This and related issues are examined in the sequel.

---

61 When *non-Schmid* effects are included, then $\bar{\tau}^{\alpha}$ may differ from the corresponding *resolved shear* stress.

62 When the critical temperature is exceeded, the model reduces to the rate-independent case. On the other hand, if the more complete model (6.5.19b) which includes the dislocation-drag effects, is used, then $\dot{\gamma}^{\alpha}$ is always nonzero.

6.5.8. Schmid Rule

Consider the *Schmid rule*, and set $\overline{\tau}^{\alpha} = \tau^{\alpha}$, the resolved shear stress of the α'th slip system. Now seek to identify all possible combinations of active slip systems in fcc single crystals. If the slip systems 1 and 2 are active, then[63]

$$|\tau^1| \approx \tau_c^1, \quad |\tau^2| \approx \tau_c^2. \tag{6.5.26a,b}$$

It follows from (6.5.25a) then, that

$$|\tau^3| \approx \tau_c^1 + \tau_c^2 \quad \text{or} \quad |\tau^3| \approx |\tau_c^1 - \tau_c^2|. \tag{6.5.26c,d}$$

Now, consider the case where the critical shear stresses τ_c^{α} in all slip systems are of the same order of magnitude. Then,

$$|\tau_c^1 - \tau_c^2| << \tau_c^3 << \tau_c^1 + \tau_c^2. \tag{6.5.26e}$$

Hence (6.5.26c,d) become

$$|\tau^3| >> \tau_c^3, \quad \text{or} \quad |\tau^3| << \tau_c^3. \tag{6.5.26f,g}$$

However, only (6.5.26g) is admissible. Therefore, if slip systems 1 and 2 are active, then slip system 3 must be inactive, and if $\tau^1 \approx \pm\tau_c^1$, then $\tau^2 \approx \mp\tau_c^2$. From a similar argument, it follows from (6.5.25) that in each of the following sets of three slip systems, only two slip systems can be active at the same time:

$$\{1,2,3\}, \quad \{4,5,6\}, \quad \{7,8,9\}, \quad \{10,11,12\}. \tag{6.5.27a-d}$$

Therefore, *at most eight slip systems can be active at any time in fcc single crystals.* It also follows from (6.5.25) that only two slip systems can be active in each of the following sets of three slip systems:

$$\{3,4,8\}, \quad \{6,7,11\}, \quad \{9,10,2\}, \quad \{12,1,5\}. \tag{6.5.28a-d}$$

6.5.9. Fully Developed Plastic Flow

In general, the elastic shear moduli are orders of magnitude greater than any typical critical shear stress, τ_c^{α}. Therefore, when a single crystal undergoes large plastic deformations, the resolved shear stresses either reach their critical values or remain close to zero. This condition is readily achieved even when the overall strains are of the order of τ_c^{α}/μ, where μ is a typical shear modulus. This is a very small strain. Therefore, large plastic deformations are obtained through crystallographic slip on active slip systems. In the remainder of this section, deformation regimes of this kind are referred to as *fully developed plastic flow*. For the fully developed plastic flow, the resolved shear stresses are either almost zero or equal to the critical shear stresses. Therefore, each set in (6.5.27) and (6.5.28) will have either zero or two active slip systems. If all sets have two active slip systems, then there are a total of eight active slip systems. If

63 The slip systems are designated by their numbers, according to (6.5.24) and (6.5.25).

all slip systems are inactive in any of the sets in (6.5.27) or (6.5.28), then there can be a maximum of six active slip systems.

Eight Active Slip Systems: First consider the case where in each set in (6.5.27) and (6.5.28), one slip system is inactive and the other two are active. For example, let in the first set in (6.5.27), slip system 1 be inactive, while 2 and 3 are active, *i.e.*,

$$\tau^1 \approx 0, \quad \tau^2 \approx \mp \tau_c^2, \quad \tau^3 \approx \pm \tau_c^3 .$$

Then from (6.5.25b), either $\tau^4 \approx 0$ and $\tau^8 \approx \mp \tau_c^8$, or $\tau^4 \approx \pm \tau_c^4$ and $\tau^8 \approx 0$. Then, based on (6.5.25c-h), several combinations of possible active slip systems can be identified. This is displayed in Figure 6.5.4 in the form of a flow chart. The chart shows that, there are only *four combinations of eight slip systems* which can be active when slip system 1 is inactive and 2 and 3 are active. These are

$$\{\pm 3, \mp 2, \pm 6, \mp 5, \pm 9, \mp 8, \pm 12, \mp 11\},$$
$$\{\pm 3, \mp 2, \mp 6, \pm 5, \pm 9, \mp 8, \mp 12, \pm 11\},$$
$$\{\pm 3, \mp 2, \pm 6, \mp 5, \pm 7, \mp 8, \mp 10, \pm 12\},$$
$$\{\pm 3, \mp 2, \pm 4, \mp 5, \mp 7, \pm 9, \pm 12, \mp 11\}. \tag{6.5.29a-d}$$

In a similar manner, it can be shown that there are *four combinations of eight slip systems* which can be active when slip system 2 or 3 is inactive.

$$\{\pm 1, \mp 3, \mp 6, \pm 5, \mp 9, \pm 8, \mp 10, \pm 11\},$$
$$\{\pm 1, \mp 3, \mp 4, \pm 6, \pm 7, \mp 9, \mp 10, \pm 12\},$$
$$\{\pm 1, \mp 3, \mp 4, \pm 5, \pm 7, \mp 9, \mp 10, \pm 11\},$$
$$\{\pm 1, \mp 3, \mp 4, \pm 5, \mp 7, \pm 9, \pm 10, \mp 11\}. \tag{6.5.29e-h}$$

When slip system 3 is inactive, the four combinations are

$$\{\pm 1, \mp 2, \mp 4, \pm 6, \pm 9, \mp 8, \pm 12, \mp 11\},$$
$$\{\pm 1, \mp 2, \pm 4, \mp 6, \mp 7, \pm 8, \mp 10, \pm 12\},$$
$$\{\pm 1, \mp 2, \mp 4, \pm 6, \pm 7, -8, \mp 10, \pm 12\},$$
$$\{\pm 1, \mp 2, \mp 4, \pm 5, \pm 7, \mp 8, \mp 10, \pm 11\}. \tag{6.5.29i-l}$$

Six Active Slip Systems: If all slip systems in any set in (6.5.27) are inactive, then there can be a maximum of six active slip systems. First consider the case that all slip systems in the set (6.5.27a) are inactive, *i.e.*, $\tau^1 \approx \tau^2 \approx \tau^3 \approx 0$. Then from (6.5.25a,b), either $\tau^4 \approx \pm \tau_c^4$ and $\tau^8 \approx \pm \tau_c^8$, or $\tau^4 \approx \tau^8 \approx 0$. This procedure can be continued, as shown in Figure 6.5.5, to obtain all possible combinations of six active slip systems.

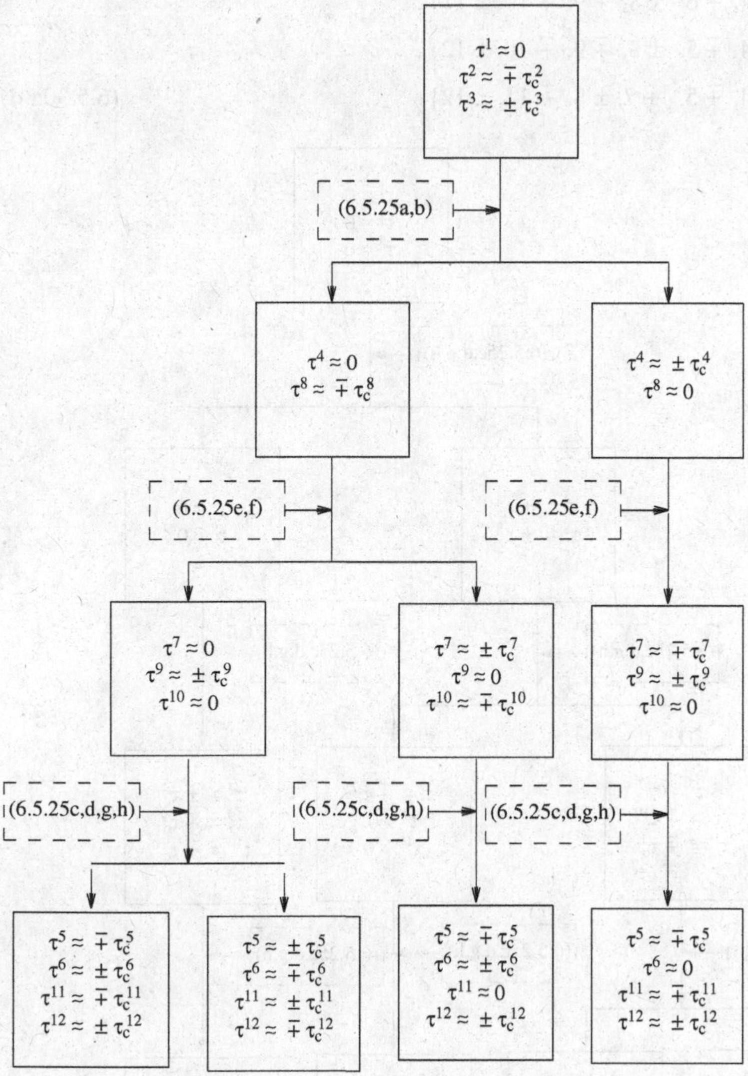

Figure 6.5.4

Flow chart for identifying various combinations of possible active slip systems when slip systems 2 and 3 are active

It is seen in Figure 6.5.5 that only *four combinations of six slip systems* can possibly be active when slip systems 1 , 2, and 3 are inactive. These are

$$\{\pm 5,\ \mp 6,\ \mp 7, \pm 9, \pm 10, \mp 12\},$$

$$\{\pm 4,\ \mp 6,\ \pm 8,\ \mp 9,\ \mp 10, \pm 11\},$$
$$\{\pm 4,\ \mp 5,\ \pm 8,\ \mp 9,\ \mp 10, \pm 12\},$$
$$\{\pm 4,\ \mp 5,\ \mp 7, \pm 8,\ \mp 11, \pm 12\}. \qquad (6.5.30a\text{-}d)$$

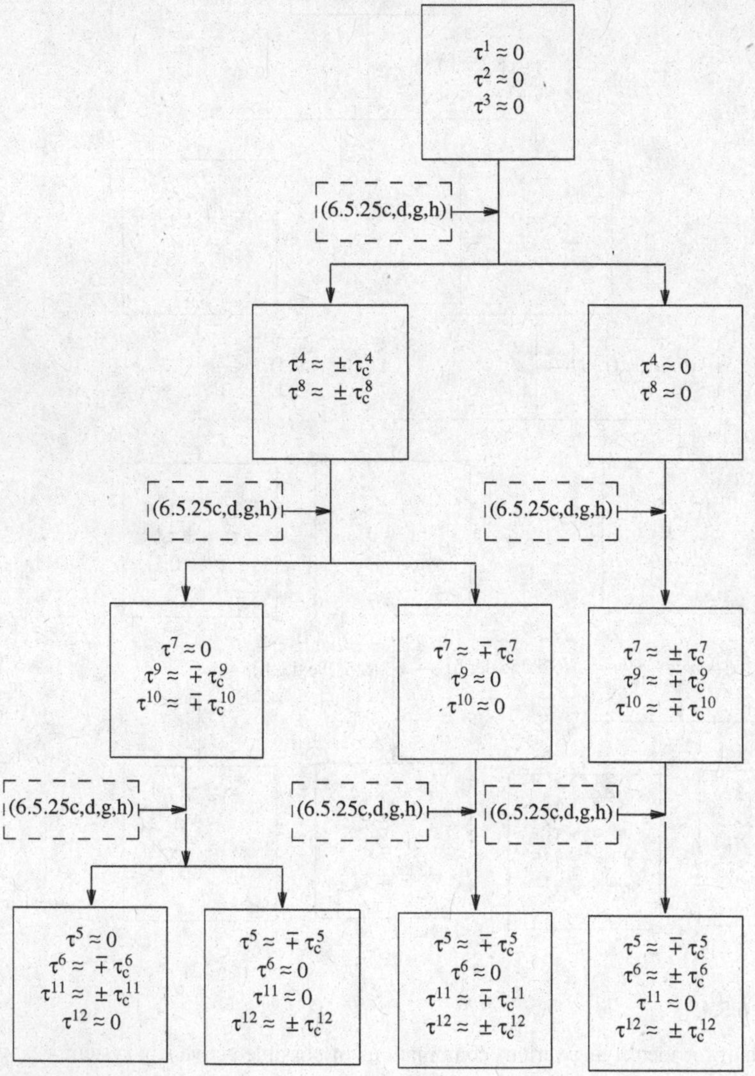

Figure 6.5.5

Flow chart for identifying various combinations of possible active slip systems when slip systems 1, 2, and 3 are inactive

Similarly, when slip systems 4, 5, and 6 are inactive, the possible combinations of 6 active slip systems are

$$\{\pm 2, \mp 3, \mp 7, \pm 8, \pm 10, \mp 11\},$$
$$\{\pm 1, \mp 3, \pm 8, \mp 9, \mp 10, \pm 12\},$$
$$\{\pm 1, \mp 3, \mp 7, \pm 8, \mp 11, \pm 12\},$$
$$\{\pm 1, \mp 2, \mp 7, \pm 9, \mp 11, \pm 12\}. \qquad (6.5.31a\text{-}d)$$

When slip systems 7, 8, and 9 are inactive, the possible combinations of 6 active slip systems are

$$\{\pm 1, \mp 3, \mp 4, \pm 6, \mp 11, \pm 12\},$$
$$\{\pm 2, \mp 3, \mp 4, \pm 5, \pm 10, \mp 12\},$$
$$\{\pm 2, \mp 3, \mp 4, \pm 6, \pm 10, \mp 11\},$$
$$\{\pm 1, \mp 2, \pm 5, \mp 6, \mp 10, \pm 11\}. \qquad (6.5.32a\text{-}d)$$

When slip systems 10, 11, and 12 are inactive, the possible combinations of 6 active slip systems are

$$\{\pm 2, \mp 3, \mp 4, \pm 6, \pm 7, \mp 9\},$$
$$\{\pm 1, \mp 3, \pm 5, \mp 6, \mp 7, \pm 8\},$$
$$\{\pm 1, \mp 2, \pm 5, \mp 6, \mp 7, \pm 9\},$$
$$\{\pm 1, \mp 2, \mp 4, \pm 5, \pm 8, \mp 9\}. \qquad (6.5.33a\text{-}d)$$

Note that the set of active slip systems is unique, in the sense that, six simultaneously active slip systems can belong to only one of the above sets.

6.5.10. Numerical Simulation of Single Crystals

In this subsection, computational algorithms necessary for accurate and efficient calculation of plastic deformation of bcc and fcc crystals are outlined and a number of illustrative examples are presented. For the constitutive model governing the slip rates, the power-law model (6.5.1a,b), as well as the dislocation-based models (6.5.13a) and (6.5.18) are used. First, general results applicable to all three cases are presented, and then specific examples are considered. To simplify the calculations, isotropic elasticity is used in all cases.

Kinematical Relations: For most metals, the elastic strains are generally very small, of the order of 10^{-3}. The particular applications considered in the sequel are to commercially pure tantalum and OFHC copper at large plastic strains. Hence, in the kinematical relations, only first-order terms in the elastic strains need to be retained. In view of this, and noting (6.2.4) and (6.2.8), together with (6.2.12d,e), it follows that the deformation rate and spin tensors, **D** and **W**, for the crystal can be written as

$$\mathbf{D} = \mathbf{D}^* + \sum_{\alpha=1}^{N} \dot{\gamma}^\alpha \mathbf{p}^\alpha, \quad \mathbf{W} = \mathbf{W}^* + \sum_{\alpha=1}^{N} \dot{\gamma}^\alpha \mathbf{r}^\alpha, \tag{6.5.34a,b}$$

where N is the number of slip systems (48 for bcc and 12 for fcc), and $\mathbf{p}^\alpha$ and $\mathbf{r}^\alpha$ are the symmetric and skewsymmetric parts of the rotated slip-system tensors, $\mathbf{s}^\alpha \otimes \mathbf{n}^\alpha$, which are now related to the initial slip-plane unit normals, $\mathbf{n}_0^\alpha$, and slip direction unit vectors, $\mathbf{s}_0^\alpha$, through the lattice rotation tensor $\mathbf{R}^*$, as follows:

$$\mathbf{s}^\alpha \otimes \mathbf{n}^\alpha = \mathbf{R}^* \mathbf{s}_0^\alpha \otimes \mathbf{n}_0^\alpha \mathbf{R}^{*T}. \tag{6.5.34c}$$

Hence, $\mathbf{s}^\alpha$ and $\mathbf{n}^\alpha$ are orthogonal unit vectors. Furthermore, the "nonplastic" (*i.e.*, the elastic plus the rigid-body spin) quantities $\mathbf{D}^*$ and $\mathbf{W}^*$, are approximated by

$$\mathbf{D}^* = \dot{\boldsymbol{\varepsilon}} + \boldsymbol{\varepsilon}\boldsymbol{\Omega}^* - \boldsymbol{\Omega}^*\boldsymbol{\varepsilon}$$

$$\mathbf{W}^* = \boldsymbol{\Omega}^* = \dot{\mathbf{R}}^* \mathbf{R}^{*T}, \tag{6.5.34d-f}$$

where $\mathbf{V}^e = \mathbf{1} + \boldsymbol{\varepsilon}$ defines the elastic strain $\boldsymbol{\varepsilon}$; see footnote 26, page 408. Finally, the slip rates are given by (6.5.1a,b), or (6.5.13a), or (6.5.18), depending on which model is being considered. For the application to commercially pure tantalum or OFHC copper, the athermal part, τ_a^α, is approximated by (6.5.10d), with γ being defined by (6.5.1e).

Calculation of Temperature: Depending on the rate of loading and the boundary conditions, a part of the heat which is generated by the plastic work, may remain in the material, resulting in an increase in its temperature. The actual calculation of the temperature, therefore, requires solving the corresponding boundary-value problem. For a single-crystal sample which is being deformed macroscopically uniformly, it may be assumed that a fraction of the plastic work is used to increase the sample temperature. This fraction, in general, will depend on the strain rate, strain, temperature, and the boundary conditions of the sample. The temperature at time t may be estimated using

$$T = T_0 + \sum_{\alpha=1}^{N} \int_0^t \eta\, \tau^\alpha(t')\, \dot{\gamma}^\alpha(t')\, dt', \tag{6.5.35}$$

where η is a temperature-dependent, dimensional parameter, and τ^α is the shear stress of the α'th slip system, *e.g.*, given by $\tau^{*\alpha} + \tau_a^\alpha$ or τ_a^α, depending on whether $T < T_c^\alpha$ or $T \geq T_c^\alpha$.

6.5.11. Computational Algorithm

The incremental equations of the rate-dependent elastoplastic deformations of crystals, in general, are very stiff.[64] Because of this stiffness, most

---

[64] Roughly speaking, when the solution curve of a differential equation suffers an abrupt variation in a small neighborhood of a point, the differential equation is called "stiff"; see Aiken (1985) for a systematic account.

conventional implicit or explicit time-integration schemes would require very small timesteps in order to ensure both accuracy and solution stability. Hence, excessive computational time is necessary to obtain reasonable solutions by these methods. In a recent work, Nemat-Nasser and Okinaka (1996) have proposed a computational method for fcc crystals, which is both efficient and accurate. It is based on the *plastic-predictor/elastic-corrector* method combined with the conventional forward gradient technique, both of which have been detailed in Chapter 5. The resulting algorithm is remarkably efficient and accurate at temperatures below the critical temperature, T_c. This computational strategy has been applied to solve finite-deformation problems of bcc or fcc single crystals (Nemat-Nasser *et al.*, 1998a,b). The solution is obtained incrementally, assuming that the (constant) value of the velocity gradient, **L**, is given at the start of each timestep. The calculation then starts with the tentative assumption that the total deformation increment is solely due to plastic slip, and afterwards corrects the results by including the accompanying elastic contribution when and if necessary. Since, in general, the elastic contribution is much smaller than the plastic one, the method efficiently yields accurate results. On the other hand, during the initial deformation, or when there is a sudden change in the loading condition, the elastic contribution may become equally important, as the change in the shear stress of inactive systems may not be small during the timestep. For this case, the explicit Euler method is employed; however, any other forward-gradient method may be used.

The deformation history under a given constant loading condition is thus divided into three regimes: (1) Transition Regime, in which fewer than five slip systems are active; (2) Rapidly-changing Regime, in which new slip systems become active; and (3) Steady-state (or fully developed plastic flow) Regime, in which more than five slip systems are active (Okinaka, 1995). In what follows, the computational methods used to deal with these regimes are outlined.

As a starting point, consider the inner product of the total deformation rate, **D**, with the symmetric part of the rotated slip-system tensor, $\mathbf{p}^\alpha$, defined in (6.2.14a), to obtain

$$\mathbf{D} : \mathbf{p}^\alpha = \frac{1}{2\mu}\,\dot{\tau}^\alpha + \sum_{\beta=1}^{N_s} \dot{\gamma}^\beta H^{\alpha\beta}, \qquad H^{\alpha\beta} = \mathbf{p}^\alpha : \mathbf{p}^\beta, \tag{6.5.36a,b}$$

$$\dot{\tau}^\alpha = 2\mu\,\mathbf{D}^* : \boldsymbol{\mathcal{L}} : \mathbf{p}^\alpha, \tag{6.5.37}$$

where $2\mu\,\boldsymbol{\mathcal{L}}$ is the elasticity tensor, and μ is a shear modulus used here to render the tensor $\boldsymbol{\mathcal{L}}$ dimensionless; in application to tantalum and OFHC copper, an elastically isotropic crystal is assumed, in which case, $\boldsymbol{\mathcal{L}} = \mathbf{1}^{(4s)}$, even though these metals have cubic symmetry with two shear moduli that are not quite the same. Modifications required to include the effects of elastic anisotropy are straightforward.

Given the current state, say, at time t_0, and constant deformation rate and spin tensors, **D** and **W**, over a time interval t_0 to $t_0 + \Delta t$, (6.5.37) and (6.5.36a), and (6.2.5d) are to be integrated to obtain, respectively, the increments $\Delta\tau^\alpha$'s and $\Delta\gamma^{\alpha}$'s, and the corresponding increment of lattice rotation, $\Delta\mathbf{R}^*$, all at time $t_0 + \Delta t$. To this end, two different computational strategies are used, depending on the deformation state of the material. In a continuous plastic slip of, say, the α'th slip system, the corresponding elastic distortion rate would be infinitesimally small. Then, $\dot{\tau}^{\alpha} \approx 0$. This is called the *stable flow condition*. It is assumed to hold *provisionally* for all *active* slip systems, over the considered time increment. The α'th slip system is *provisionally* called *active* if its resolved shear stress satisfies $|\tau^{\alpha}| \geq \tau_c^{\alpha}$ when the power-law model (6.5.1a) is used, or $|\tau^{\alpha}| \geq \tau_a^{\alpha}$ when the dislocation-based models (6.5.13c) or (6.5.18) are considered. Otherwise, it is regarded *inactive*. When the number of active slip systems, N_a, is greater than five, then the deformation is called *fully developed plastic flow* and the plastic-predictor technique is applied. Otherwise, a forward-gradient method, *e.g.*, the explicit Euler method, is used.[65]

Plastic Predictor and Transition Regime: Consider the case when fewer than five slip systems are active. Equation (6.5.36a) is solved to obtain approximate slip rates of the active slip systems with the tentative assumption that the deformation is accompanied by no elastic distortion in the active slip systems, and that slip rates of inactive slip systems are zero.[66] The resulting expressions are then corrected to include the elastic lattice deformation. Tentatively, it is also assumed that the increment of the rigid-body rotation in the concerned timestep is sufficiently small, so that

$$\mathbf{p}_A^{\alpha}(t_0 + \xi) \approx \mathbf{p}^{\alpha}(t_0), \quad 0 \leq \xi \leq \Delta t, \tag{6.5.38}$$

where the subscript A stands for the approximated value, since the lattice elastic strains and the variation in the spin during the timestep are not yet included. A more accurate lattice rotation is then implemented once the incremental slip rates and shear stresses are computed. In light of the above comments, equation (6.5.36a) is approximated by

$$\mathbf{D} : \mathbf{p}^{\alpha}(t_0) = \sum_{\beta=1}^{N_a} \dot{\gamma}_A^{\beta}(t_0 + \Delta t)\, H^{\alpha\beta}(t_0), \tag{6.5.36c}$$

where $\dot{\gamma}_A^{\beta}(t_0 + \Delta t)$ is the estimate of the β'th slip rate at $t_0 + \Delta t$, and the summation is performed only on active slip systems; this is automatic, since $\dot{\gamma}_A^{\beta}$'s are set equal to zero for inactive slip systems. Since there are fewer than five slip systems and these are linearly independent, (6.5.36c) can be solved to obtain

$$\dot{\gamma}_A^{\alpha}(t_0 + \Delta t) = \sum_{\beta=1}^{N_a} K^{\alpha\beta}(t_0)\, \{\mathbf{D} : \mathbf{p}^{\beta}(t_0)\}, \tag{6.5.36d}$$

---

[65] It is, however, possible to use the plastic-predictor technique even when there are fewer than five active slip systems, as has been commented on by Nemat-Nasser and Okinaka (1996).

[66] For rate-dependent models, a slip rate is non-zero when the corresponding resolved shear stress is non-zero. However, the slip rate of an "inactive" system is infinitesimally small compared to that of an "active" one.

where $[K^{\alpha\beta}(t_0)]$ is the inverse of $[H^{\alpha\beta}(t_0)]$. The corresponding increment in the slip is estimated by

$$\Delta\gamma_A^\alpha(t_0+\Delta t) = \tfrac{1}{2}\,\mathrm{sgn}(\dot{\gamma}^\alpha(t_0))\,\{\dot{\gamma}^\alpha(t_0) + \dot{\gamma}_A^\alpha(t_0+\Delta t)\}\Delta t, \qquad (6.5.36e)$$

and the temperature change by

$$\Delta T_A(t_0+\Delta t) = \eta \sum_{\alpha=1}^{N_a} \tau^\alpha(t_0)\,\Delta\gamma_A^\alpha(t_0+\Delta t). \qquad (6.5.36f)$$

Elastic Corrector: To estimate the involved error in the above calculation, use (6.5.38), integrate (6.5.36a) from t_0 to $t_0+\Delta t$, and obtain

$$\mathbf{D} : \mathbf{p}^\alpha(t_0)\,\Delta t = \frac{1}{2\mu}\,\Delta\tau^\alpha(t_0+\Delta t) + \sum_{\beta=1}^{N_a} \Delta\gamma^\beta H^{\alpha\beta}(t_0). \qquad (6.5.39a)$$

Also integrate (6.5.36c) over the same time interval, and combine the result with (6.5.39a), to obtain the following representation for the estimate of the resolved shear stress increments:

$$\frac{1}{2\mu}\,\Delta\tau_A^\alpha(t_0+\Delta t) = \frac{1}{2\mu}\,\tau_{er}^\alpha(t_0+\Delta t) + \sum_{\beta=1}^{N_a} \gamma_{er}^\beta(t_0+\Delta t)\,H^{\alpha\beta}(t_0). \qquad (6.5.39b)$$

In this equation, the following notation is used:

$$\Delta\tau_A^\alpha(t_0+\Delta t) = \tau_A^\alpha(t_0+\Delta t) - \tau^\alpha(t_0),$$

$$\tau_{er}^\alpha(t_0+\Delta t) = \tau_A^\alpha(t_0+\Delta t) - \tau^\alpha(t_0+\Delta t),$$

$$\Delta\gamma_A^\alpha(t_0+\Delta t) = \gamma_A^\alpha(t_0+\Delta t) - \gamma^\alpha(t_0),$$

$$\dot{\gamma}_{er}^\alpha(t_0+\Delta t) = \dot{\gamma}_A^\alpha(t_0+\Delta t) - \dot{\gamma}^\alpha(t_0+\Delta t). \qquad (6.5.39c\text{-}f)$$

The approximate values of the resolved shear stresses, $\tau_A^\alpha(t_0+\Delta t)$, are obtained using the constitutive relations that relate these stresses to the slip rates, temperature, and other parameters. In order to work within a general setting, consider a class of models that can be represented by

$$\tau^\alpha = F^\alpha(\dot{\gamma}^\alpha, T, \gamma^\beta;\ \beta = 1, 2, \ldots). \qquad (6.5.40a)$$

The power-law and both the fcc and bcc models that have been developed in this section, are included in this class. Hence, evaluate $\tau_A^\alpha(t_0+\Delta t)$ by direct substitution,

$$\tau_A^\alpha(t_0+\Delta t) = F^\alpha(\dot{\gamma}_A^\alpha(t_0+\Delta t),\ T_A(t_0+\Delta t),\ \gamma_A^\beta(t_0+\Delta t);\ \beta = 1, 2, \ldots), \qquad (6.5.40b)$$

where $\dot{\gamma}_A^\alpha(t_0+\Delta t)$ is given by (6.5.36d), and from (6.5.36e,f), $T_A(t_0+\Delta t) = T(t_0) + \Delta T_A(t_0+\Delta t)$ and $\gamma_A^\alpha(t_0+\Delta t) = \gamma^\alpha(t_0) + \Delta\gamma_A^\alpha(t_0+\Delta t)$ are calculated.

Now, establish a relation between the error terms τ_{er}^α's and γ_{er}^α's, using (6.5.40a),

$$\tau_{er}^\alpha(t_0+\Delta t) = \sum_{\beta=1}^{N_a} \frac{\partial F^\alpha}{\partial\gamma^\beta}(t_0+\Delta t)\,\gamma_{er}^\beta(t_0+\Delta t)$$

$$+\frac{\partial F^\alpha}{\partial T}(t_0+\Delta t)\,T_{er}(t_0+\Delta t)+\frac{\partial F^\alpha}{\partial \dot{\gamma}^\alpha}(t_0+\Delta t)\,\dot{\gamma}^\alpha_{er}(t_0+\Delta t), \quad (6.5.41a)$$

where, as indicated, all partial derivatives in the above equation are evaluated for the approximate values of the arguments at $t_0+\Delta t$. For γ^α_{er} and T_{er}, use the following interpolations:

$$\gamma^\alpha_{er}(t_0+\Delta t) = \tfrac{1}{2}\,\dot{\gamma}^\alpha_{er}(t_0+\Delta t)\,\Delta t,$$

$$T_{er}(t_0+\Delta t) = \tfrac{1}{2}\,\eta \sum_{\alpha=1}^{N_a} \tau^\alpha(t_0)\,\dot{\gamma}^\alpha_{er}(t_0+\Delta t)\,\Delta t. \quad (6.5.41b,c)$$

Finally, substitute (6.5.40b), (6.5.39c), and (6.5.41a-c) into (6.5.39b), to obtain a set of (nonsingular) linear equations for $\dot{\gamma}^\alpha_{er}(t_0+\Delta t)$'s. From the solution of these equations, all other error quantities are then calculated and the corresponding values are corrected.

Lattice Rotation: The lattice spin tensor over the time interval t_0 to $t_0+\Delta t$ is given by

$$\mathbf{\Omega}^*(t) = \mathbf{W} - \sum_{\alpha=1}^{N_a} \dot{\gamma}^\alpha(t)\mathbf{r}^\alpha(t), \quad (6.5.42a)$$

where $\mathbf{W}$ is constant over the timestep. Tentatively assume,

$$\mathbf{r}^\alpha(t_0+\xi) \approx \mathbf{r}^\alpha(t_0), \qquad 0 \le \xi \le \Delta t, \quad (6.5.42b)$$

and integrating (6.5.42a) over the timestep, obtain

$$\int_{t_0}^{t_0+\Delta t} \mathbf{\Omega}^*(\xi)\,d\xi \approx \mathbf{W}\,\Delta t - \sum_{\alpha=1}^{N_a} \Delta\gamma^\alpha(t_0+\Delta t)\,\mathbf{r}^\alpha(t_0), \quad (6.5.42c)$$

where $\Delta\gamma^\alpha(t_0+\Delta t) = \gamma(t_0+\Delta t) - \gamma(t_0)$. Expression (6.5.42a) shows that $\mathbf{\Omega}^*$ is not constant over the time interval. However, since

$$\Delta\gamma^\alpha(t_0+\Delta t) \approx \Delta\gamma^\alpha_A(t_0+\Delta t) - \gamma^\alpha_{er}(t_0+\Delta t), \quad (6.5.42d)$$

the right-hand side of (6.5.42c) can be computed rather accurately, and the incremental lattice rotation tensor may be approximated by

$$\Delta\mathbf{R}^*(\Delta t) = \mathbf{1} + \mathbf{W}\,\Delta t - \sum_{\alpha=1}^{N_a} \{\Delta\gamma^\alpha_A(t_0+\Delta t) - \gamma^\alpha_{er}(t_0+\Delta t)\}\,\mathbf{r}^\alpha(t_0). \quad (6.5.42e)$$

Then, the rotation tensor at $t_0+\Delta t$ becomes

$$\mathbf{R}^*(t_0+\Delta t) = \Delta\mathbf{R}^*(\Delta t)\,\mathbf{R}^*(t_0). \quad (6.5.42f)$$

This is a good estimate if the time increment is small.[67]

---

[67] For large timesteps, the solution can be improved using integration procedures similar to those given in Section 5.8.

Resolved Shear Stresses of Inactive Slip Systems: The changes in the shear stresses acting on inactive slip systems can be computed based on the corresponding elastic deformation. From (6.5.39a), it follows that

$$\Delta\tau^{\alpha}(t_0+\Delta t) \approx 2\,\mu\,\{\mathbf{D} : \mathbf{p}^{\alpha}(t_0)\,\Delta t - \sum_{\beta=1}^{N_a} \Delta\gamma^{\beta}\,H^{\alpha\beta}(t_0)\}\,. \tag{6.5.43a}$$

Hence, denoting by subscript "in", the stresses corresponding to inactive systems, calculate

$$\tau^{\alpha}_{in}(t_0+\Delta t) = \tau^{\alpha}_{in}(t_0) + \Delta\tau^{\alpha}_{in}(t_0+\Delta t)\,. \tag{6.5.43b}$$

An inactive slip system may become active during the timestep Δt. The timestep must then be modified accordingly. Note that more than one inactive slip system may become active during a timestep. When the total number of active slip systems is fewer than five, a new appropriate timestep is chosen and the computation is continued. On the other hand, when the number of active slip systems exceeds five, then a new computational technique is used, as discussed below.

Rapidly Changing Regime: In this regime, the elastic contributions are important and must be included directly. This occurs at the initial loading where the response is essentially elastic until a critical state is reached. It also occurs when there is an abrupt change in the loading regime. Deformation of this kind is classified as the "rapidly changing regime". The forward gradient method explained in Chapter 5 can be used in this regime, and will not be discussed in detail again. Since no more than five slip systems are involved, the solution is straightforward.

Fully Developed Flow: When there are more than five active slip systems, equations (6.5.36c) cannot be directly solved since the coefficient matrix, $[H^{\alpha\beta}]$, of the slip rates, $\dot{\gamma}^{\beta}_{A}$'s, is then singular. In such a case, this system of linear equations is supplemented by equations which represent the interdependency of the resolved shear stresses. In view of (6.4.6a), the results summarized in (6.5.20a-c) are expressed as

$$\tau^1 + \tau^7 - \tau^9 = 0\,, \quad \tau^2 - \tau^5 + \tau^{11} = 0\,,$$
$$\tau^3 + \tau^6 - \tau^{10} = 0\,. \tag{6.5.44a-c}$$

Similarly, the forty conditions that are given in Table 6.5.3, page 453, are also rewritten in terms of τ^{α}'s, leading to a total of forty-three conditions for the resolved shear stresses, in the bcc-case. For the fcc-case, seven conditions out of the eight relations (6.5.25a-h) are used. *Since the calculation is incremental, the number and combination of active slip systems are known at the start of each increment.* Hence, the forty-three (bcc) or seven (fcc) conditions can be screened and reduced to include only the active slip systems. The reduced $N_a - 5$ conditions are then collectively written in matrix form, as

$$\sum_{\alpha=1}^{N_a} M_{i\alpha}\tau^{\alpha} = 0\,, \quad (1 \le i \le N_a - 5). \tag{6.5.45}$$

These equations must now be solved together with (6.5.36c), to obtain the first estimate of the slip rates. The remaining calculations are as discussed before.

In what follows, examples are worked out for specific models, starting with the power-law model.

6.5.12. Power-law Model for fcc Crystals

Consider the model (6.5.1a,c) and assume that the hardening matrix is constant; the case of a variable hardening matrix is examined subsequently. The coefficients in (6.5.41a) can be calculated by direct differentiation, leading to

$$\tau^{\alpha}_{er}(t_0+\Delta t) = \sum_{\beta=1}^{N_s} \{ \delta_{\alpha\beta} \frac{\tau^{\beta}_{A}(t_0+\Delta t)}{m\,\dot{\gamma}^{\beta}(t_0+\Delta t)} + \mathrm{sgn}(\gamma^{\beta}_{A}(t_0+\Delta t))\, h_{\alpha\beta}$$

$$\times \frac{\tau^{\alpha}_{A}(t_0+\Delta t)}{\tau^{\alpha}_{c}(t_0+\Delta t)} \Delta t \,\}\, \dot{\gamma}^{\beta}_{er}(t_0+\Delta t)\,. \tag{6.5.46}$$

Consider now specific examples where a constant velocity gradient, **L**, is prescribed and it is required to calculate the resulting time variation of all resolved shear stresses, and the slips of active slip systems, as well as the overall effective stress and strain. To fix the example parameters, set m = 101, $\tau^{\alpha}_{c} = \frac{1}{2}\,\tau_0$, $\dot{\gamma}^{\alpha}_{0} = 1.0$/s, and $\mu = 100\,\tau_0$, where τ_0 is a normalizing unit of stress that we set equal to one. In the calculation, the timestep is adjusted to allow for a 5% increment in the equivalent plastic strain in both the transition and steady-state deformation regimes, whereas a 0.02% increment of the equivalent strain is used for the forward-gradient method in the Rapidly-changing Regime. To identify the slip systems, they are numbered according to their slip plane and slip direction. For example, System (43) denotes the slip plane with the unit normal $\mathbf{n}^4$ and slip direction $\mathbf{s}^3$; see Figure 6.5.2 and equations (6.5.21a-d) and (6.5.22a-l). To check the results, the explicit Euler method with 8,000 timesteps is used to obtain accurate reference results that are shown by solid lines in the figure. With 5,500 steps, an oscillating unstable solution is obtained by the Euler method, but not for 8,000 or more steps, which then yield indistinguishable results.

Example, fcc Crystal: Consider the following constant velocity gradient:

$$\mathbf{L} = \begin{bmatrix} 0 & 2{,}000 & 0 \\ 2{,}000 & 0 & 4{,}000 \\ 0 & 4{,}000 & 0 \end{bmatrix}, \tag{6.5.47}$$

measured per second. With a workhardening matrix of common diagonal elements, $h^{*\alpha\alpha} = 0.4\,\tau_0$, and off-diagonal elements $h^{*\alpha\beta} = 0.48\,\tau_0$, the corresponding equivalent stress and slip rates are calculated and plotted in Figures 6.5.6a,b, where geometric symbols represent the results obtained by the plastic-predictor method. The transition and steady-state regimes take place for

$0.59\% \leq \gamma_{eq} \leq 1.55\%$ and $1.93\% \leq \gamma_{eq}$, respectively, and the rapidly-changing regime occurs for $0\% \leq \gamma_{eq} \leq 0.59\%$ and $1.55\% \leq \gamma_{eq} \leq 1.93\%$. Note from Figure 6.5.6b that eight slip systems become active in this example, and that the algorithm efficiently and accurately predicts the results. Other worked-out examples (with and without workhardening) are given by Nemat-Nasser and Okinaka (1996) and Okinaka (1995).

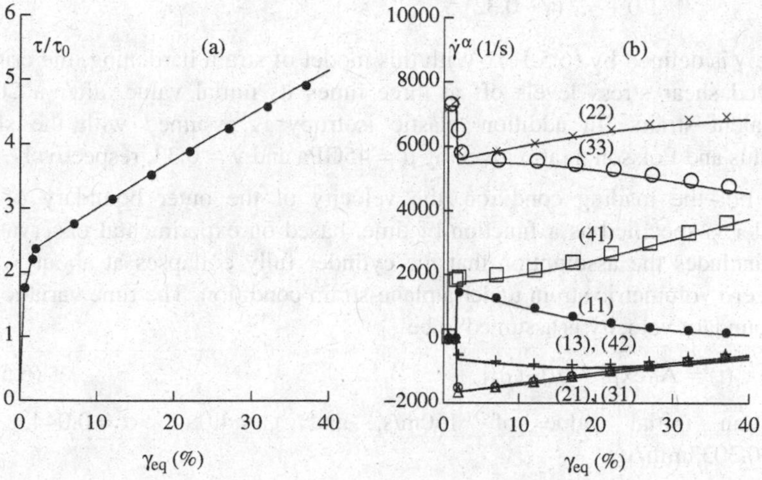

Figure 6.5.6

Equivalent stress (a), and slip rates (b) versus equivalent strain; results represented by geometric symbols are obtained using the new algorithm, and the solid lines are obtained using the Euler method with 8,000 timesteps (from Nemat-Nasser and Okinaka, 1996)

6.5.13. Dynamic Collapse of Single-crystal Hollow Cylinder

The computational algorithm has been implemented into the large-scale computer program DYNA2D, and has been used to simulate the dynamic collapse of thick-walled cylindrical single-crystal copper specimens subjected to explosive loading on their lateral surfaces.[68]

For the constitutive relation defining the slip rate in terms of the resolved shear stress and the resistance to shearing, the power-law model (6.5.1a-d) is used, with m = 101. The initial value of the critical resolved shear stress is set at

[68] See Nemat-Nasser *et al.* (1998c) where the experimental procedure is described and extensive numerical-simulation results are presented, showing that the initial configuration of the lattice profoundly affects the final deformed state and the flow localization phenomena.

0.25% of the shear modulus, or 112.5MPa. The hardening parameters, $h^{*\alpha\beta}$, used in the calculations, are assumed to vary with the plastic strain in the following manner:

$$h^{*\alpha\beta} = h(\gamma)\,[\delta_{\alpha\beta} + r(1-\delta_{\alpha\beta})]\,,$$

$$h(\gamma) = \begin{cases} 0.003125\,\mu & \text{for } \gamma \le 0.32 \\ \dfrac{0.003125\,\mu}{1.0+3.7\,(\gamma-0.32)} & \text{for } \gamma > 0.32 \end{cases}, \tag{6.5.48a,b}$$

where γ is defined by (6.5.1e). With this model of strain hardening, the critical resolved shear stress levels off at three times its initial value, after a 200% equivalent strain. In addition, elastic isotropy is assumed with the shear modulus and Poisson's ratio given by μ = 45GPa and ν = 0.33, respectively.

For the loading condition, the velocity of the outer boundary of the cylinder is specified as a function of time, based on experimental observation. This includes the assumption that the cylinder fully collapses at about 8.5μs, with zero volumetric strain under a plane-strain condition. The time variation of the boundary velocity is assumed to be

$$v_r(t) = A_0 \exp[-\alpha(t-t_0)^2]\,, \tag{6.5.49}$$

with an initial value of 150m/s, and $t_0 = 4.0\mu s$, $\alpha = 0.044$, and $A_0 = 0.3033 mm/\mu s$.

The specimen consists of a cylindrical tube of 11mm inside and 17mm outside diameter, cut from a single-crystal copper rod and encased in a polycrystal copper jacket with 30mm outside diameter. The initial lattice orientation of the single crystal is established, using X-ray diffraction. The axis of the cylinder is in the [134]-direction. The sample is loaded by detonating an explosive which surrounds the polycrystal jacket. Both full and partial collapse of the cylinder have been achieved. Figure 6.5.7a is the optical micrograph of a partially collapsed specimen. The outer boundary of the crystal has attained a non-circular shape. The initially circular inner boundary has deformed into a rectangular shape, and cracks are initiated at four corners of the rectangle. Shearbands have developed around the four cracks, and also, from the outer boundary to the upper and bottom segments of the inner boundary of the cylinder. Since the mechanism of crack initiation and propagation is not revealed by these final experimental facts, numerical simulations are used to examine the deformation process and to develop a detailed understanding of the phenomenon. Figure 6.5.7b is a typical example. As is seen, the final configuration of the partially collapsed sample is accurately predicted computationally. Good correlation between the simulation and the experimental results is obtained in both the structure of the flow localization around the inner boundary, and in the shape of the outer boundary. These results reveal that the final geometry is caused by the anisotropic inelastic response of the crystal, and not by any asymmetry in the explosive loading that may have existed.

Another interesting feature of this experiment is the formation of tension cracks at four locations in the interior surface of the collapsed cylinder. Dynamic void collapse and void growth in single crystals under uniaxial loads have been studied analytically by Nemat-Nasser and Hori (1987), revealing that tension cracks may be produced (even in a ductile material) during unloading, in a direction normal to the applied compression, even if the sample has not been subjected to any externally applied tensile stresses. This prediction has been verified experimentally by Nemat-Nasser and Chang (1990) for single-crystal copper as well as other metals, *e.g.*, mild steel and iron, using a split Hopkinson bar dynamic loading technique. A similar phenomenon occurs in the cylinder-collapse experiment. This is also verified by computational modeling which shows that very high tensile stresses can develop at certain critical points on and near the inside surface of the collapsed cylinder. This and many other issues relating to anisotropic deformation of thick-walled cylinders of single-crystal copper, under initially uniform dynamic loading, are discussed in Nemat-Nasser *et al.* (1998d), where the effect of the initial crystal orientation on the final collapsed geometry is studied by numerical simulation.

6.5.14. Dislocation-based Model for bcc and fcc Crystals

In this case, use constitutive relation (6.5.18) for the fcc-case, and then simply set $a_0 = 0$ to obtain the necessary equations for the bcc-case. To simplify calculations, rewrite (6.5.18) as

$$\tau^\alpha = \tau_a + \Psi^\alpha, \qquad \Psi^\alpha = \hat{\tau}_0 f [1 - X^{1/q}]^{1/p},$$

$$f = \begin{cases} 1 + a_0 [1 - (T/T_m)^2] \gamma^n & \text{for fcc} \\ 1 & \text{for bcc} \end{cases},$$

$$X = -\frac{kT}{G_0} \, ln \frac{\dot{\gamma}_\alpha}{\dot{\gamma}_0 f}. \tag{6.5.50a-d}$$

Hence, it is assumed that the athermal stress is the same for all slip systems, and that its dependence on slip is only through the total accumulated slip, γ, defined by (6.5.1e). Now, using the chain rule of differentiation, coefficients in (6.5.41a) are easily evaluated.

To this end, note that

$$\frac{\partial \tau_a}{\partial \gamma} = \frac{n_1}{\gamma}(\tau_a - c_0), \qquad \frac{\partial f}{\partial \gamma} = \frac{n}{\gamma}(f - 1),$$

$$\frac{\partial f}{\partial T} = -2a_0 (T/T_m^2) \gamma^n, \qquad \frac{\partial X}{\partial \gamma} = \frac{kT}{G_0 f} \frac{\partial f}{\partial \gamma},$$

$$\frac{\partial X}{\partial T} = \frac{X}{T} + \frac{kT}{G_0 f} \frac{\partial f}{\partial T}, \qquad \frac{\partial X}{\partial \dot{\gamma}^\alpha} = -\frac{kT}{G_0 \dot{\gamma}^\alpha}.$$

$$\frac{\partial \Psi^\alpha}{\partial X} = -\frac{\hat{\tau}_0 f}{pq} [1 - X^{1/q}]^{\frac{1-p}{p}} X^{\frac{1-q}{q}}. \tag{6.5.51a-g}$$

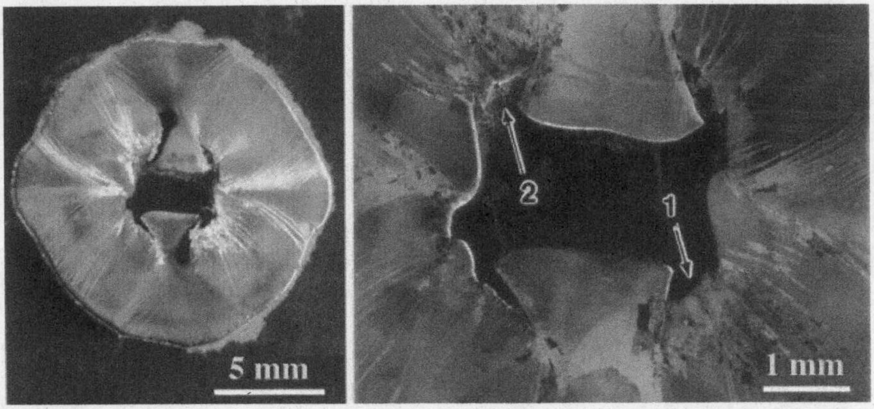

(a) (b)

Figure 6.5.7a

Optical micrograph of the partially collapsed single-crystal specimen of initially circular cylinder with 11mm inside and 17mm outside diameter; the axis of the cylinder is in the [134]-direction (from Nemat-Nasser *et al.*, 1998d)

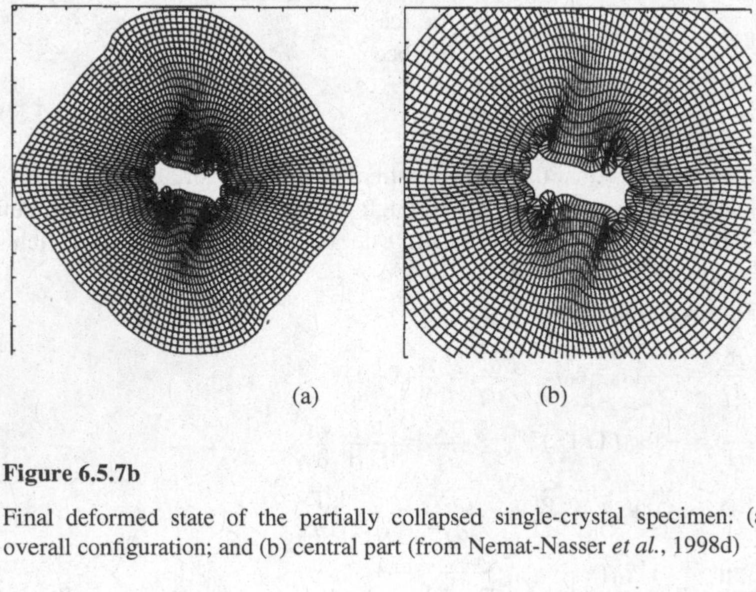

(a) (b)

Figure 6.5.7b

Final deformed state of the partially collapsed single-crystal specimen: (a) overall configuration; and (b) central part (from Nemat-Nasser *et al.*, 1998d)

The remaining calculations are the same as before.

Example, bcc Crystal: As an illustration, consider a constant velocity gradient, given by

$$\mathbf{L} = \begin{bmatrix} 1000 & 0 & 0 \\ 0 & 1000 & 0 \\ 0 & 0 & -2000 \end{bmatrix} (\mathrm{s}^{-1}) . \tag{6.5.52}$$

The corresponding equivalent stress is plotted in Figure 6.5.8, where solid dots represent the results obtained by the plastic-predictor method and the solid line is given by the explicit Euler method with 5,000 steps. In the considered algorithm, the explicit Euler method with small time increments is applied until $\gamma_{eq} = 0.19\%$, when the stable condition is satisfied, and then, the plastic-predictor method is applied until $\gamma_{eq} = 50.0\%$. The values of the equivalent stress obtained by the considered algorithm and by the explicit Euler method show good agreement.

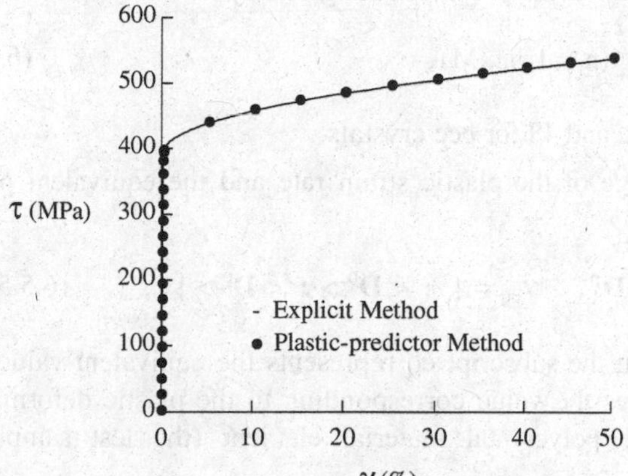

Figure 6.5.8

Comparison of the plastic-predictor results and those of the explicit Euler method with 5,000 timesteps (from Nemat-Nasser *et al.*, 1998b)

6.5.15. bcc and fcc Polycrystals

The constitutive model and computational algorithm of the preceding subsections are applied to predict the high strain-rate response of polycrystalline tantalum and OFHC copper at various initial temperatures. Computational results are compared with the experimental data reported by Nemat-Nasser and Issacs (1997) and Nemat-Nasser and Li (1998).

Polycrystal Calculations: The Taylor averaging model is employed, where it is assumed that each grain of the polycrystal undergoes the same deformation and deformation rate. Hence, in all grains, $\mathbf{L} = <\mathbf{L}>$, where $\mathbf{L}$ and $<\mathbf{L}>$ denote the local and overall velocity gradients, respectively. The global temperature of the material element is also assumed to have a value common to all grains, which must be calculated at each time increment.

The volume average of the Cauchy stress tensor, $\boldsymbol{\sigma}$, which is essentially the same as that of the Kirchhoff stress,[69] $\boldsymbol{\tau}$, is given by

$$\boldsymbol{\Sigma} = \frac{1}{V}\int_V \boldsymbol{\sigma}\, dV = \frac{1}{M}\sum_{n=1}^{M} \boldsymbol{\sigma}_n, \tag{6.5.53a,b}$$

where M is the number of grains, and all grains are assumed to have the same volume. An initially uniform distribution of the crystal orientations is assumed for illustration. The equivalent stress associated with uniaxial stress is then calculated as

$$\Sigma_{eq} = \{3/2\, \boldsymbol{\Sigma} : \boldsymbol{\Sigma}\}^{1/2}, \tag{6.5.54}$$

where the subscript eq represents the equivalent value. The plastic part of the strain rate for the n'th grain of the polycrystal is given by

$$\mathbf{D}_n^p = \sum_{\alpha=1}^{N} \dot{\gamma}_n^{\alpha} \mathbf{p}_n^{\alpha}, \quad (n = 1, ..., M), \tag{6.5.55}$$

where N equals 12 for fcc and 48 for bcc crystals.

The volume average of the plastic strain rate and the equivalent plastic strain rate are defined by

$$<\mathbf{D}^p> = \frac{1}{M}\sum_{n=1}^{M} \mathbf{D}_n^p, \quad \dot{\gamma}_{eq} = \{2/3 <\mathbf{D}^p> : <\mathbf{D}^p>\}^{1/2}, \tag{6.5.56a,b}$$

respectively, where again the subscript eq represents the equivalent value, and the superscript p indicates the value corresponding to the plastic deformation. The temperature of the polycrystal material element (the test sample) is estimated from

$$T_A = T_0 + \eta \int_{t_0}^{t_0+\Delta t} \Sigma_{eq}(\xi)\, \dot{\gamma}_{eq}(\xi)\, d\xi, \tag{6.5.57}$$

where η includes the fraction of the plastic work converted into heat, as well as the conversion factor; for a discussion of the calculation details, see Nemat-Nasser *et al.* (1998a). For large deformations at very high strain rates, essentially all the plastic work is converted into heat; see Kapoor and Nemat-Nasser (1998). Then $\eta \approx 0.433$ for tantalum, and $\eta \approx 0.29$ for copper. These are the values used below.

---

[69] See equation (6.4.2) and the related comments, page 428.

Simulation of Response of Tantalum: Figure 6.5.9 shows the experimentally obtained flow stress at a 5,000/s strain rate and initial temperatures of 300, 400, 500, 600, 800, and 1,000K, together with the computational results, where $\hat{\tau} = 440$MPa, $c_0 = 75$MPa, $c_1 = 65$MPa, and $n_1 = 1/2$ are used; all other parameters are the same as in the continuum case, listed in Table 4.8.1, page 235. As is seen, the model calculations nicely simulate the experimental results.

To obtain the overall strain rates used in the compression tests, model calculations are based on the following imposed constant overall velocity gradient, where $\dot{\gamma}_{eq}$ is set equal to the imposed uniaxial strain rate:

$$< \mathbf{L} > = \mathbf{L} = \begin{bmatrix} \dot{\gamma}_{eq} & 0 & 0 \\ 0 & -\frac{1}{2}\dot{\gamma}_{eq} & 0 \\ 0 & 0 & -\frac{1}{2}\dot{\gamma}_{eq} \end{bmatrix} (s^{-1}) . \qquad (6.5.58a)$$

This gives, when the elastic strains are neglected,

$$\dot{\gamma}_{eq} = \{\tfrac{2}{3}\,\mathbf{D} : \mathbf{D}\}^{\frac{1}{2}} . \qquad (6.5.58b)$$

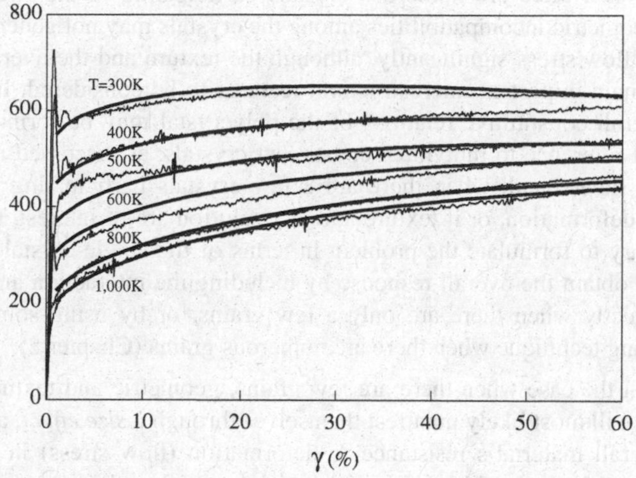

Figure 6.5.9

Comparison between experimental and model results for tantalum at 5,000/s strain rate and initial temperatures of 300, 400, 500, 600, 800, and 1,000K; light curves are experimental data of Nemat-Nasser and Isaacs (1997), and dark curves are model results with Taylor averaging (from Nemat-Nasser *et al.*, 1998b)

6.5.16. Numerical Simulation of Polycrystal Copper

The same computational modeling is applied to simulate fcc crystals and polycrystals, by Nemat-Nasser *et al.* (1998c), and the results are compared with experimental results reported by Nemat-Nasser and Li (1998) for OFHC copper. The basic approach is the same as that for the bcc crystals and polycrystals. For the polycrystal calculation, again the Taylor-averaging method is used. All constitutive parameters are the same as those used for the continuum model of Section 4.8, Table 4.8.2, page 238, except for the coefficients c_1 and $\hat{\tau}_0$ which are now reduced to $c_1 = 50, 48$ and $\hat{\tau}_0 = 9, 95$, for the annealed and as-received copper, respectively; these values are obtained empirically and c_0 is set equal to zero. Extensive comparison with experimental results over strain rates from 10^{-3} to 8×10^3/s and temperatures from 77 to 1,100K, with strains up to 100%, is given in Nemat-Nasser *et al.* (1998c). As illustrations, Figure 6.5.10 presents the results for several indicated strain rates and the indicated temperatures, for both, annealed (a), and as-received (b), samples. Similar correlations are obtained for high temperatures.

6.5.17. Length Scales and Size Effect in Crystal Plasticity

When there are numerous crystals participating in the overall deformation, geometric incompatibilities among the crystals may not generally affect the overall flow stress significantly, although the texture and the average grain size still remain important microstructural features to be considered. In many cases, the overall constitutive relations of the polycrystal may be formulated directly without reference to individual constituent crystals, as discussed in Section 4.8. On the other hand, when there are a few crystals participating in the overall plastic deformation, or if texture and its evolution are of interest, then it may be necessary to formulate the problem in terms of the single-crystal plasticity and seek to obtain the overall response by including the interaction among the crystals directly when there are only a few grains, or by using some appropriate averaging technique when there are numerous grains (Chapter 8).

In the case when there are few grains, geometric and textural incompatibilities will most likely manifest themselves through a *size effect*, and may affect the overall material's resistance to deformation (flow stress) in a major way. This size effect should be distinguished from the length scales in plasticity. As has been shown in this section, dislocation-based metal plasticity naturally involves several length scales associated with the densities of dislocations (mobile and total), as well as the microstructure of the material, whether or not there is also a size effect. While these length scales naturally enter plasticity constitutive relations, the size effect is a problem-dependent phenomenon that must be examined in each case, using a relevant material-specific model. For a few interacting crystals, the slip-induced crystal plasticity may well adequately account for any such size effects.

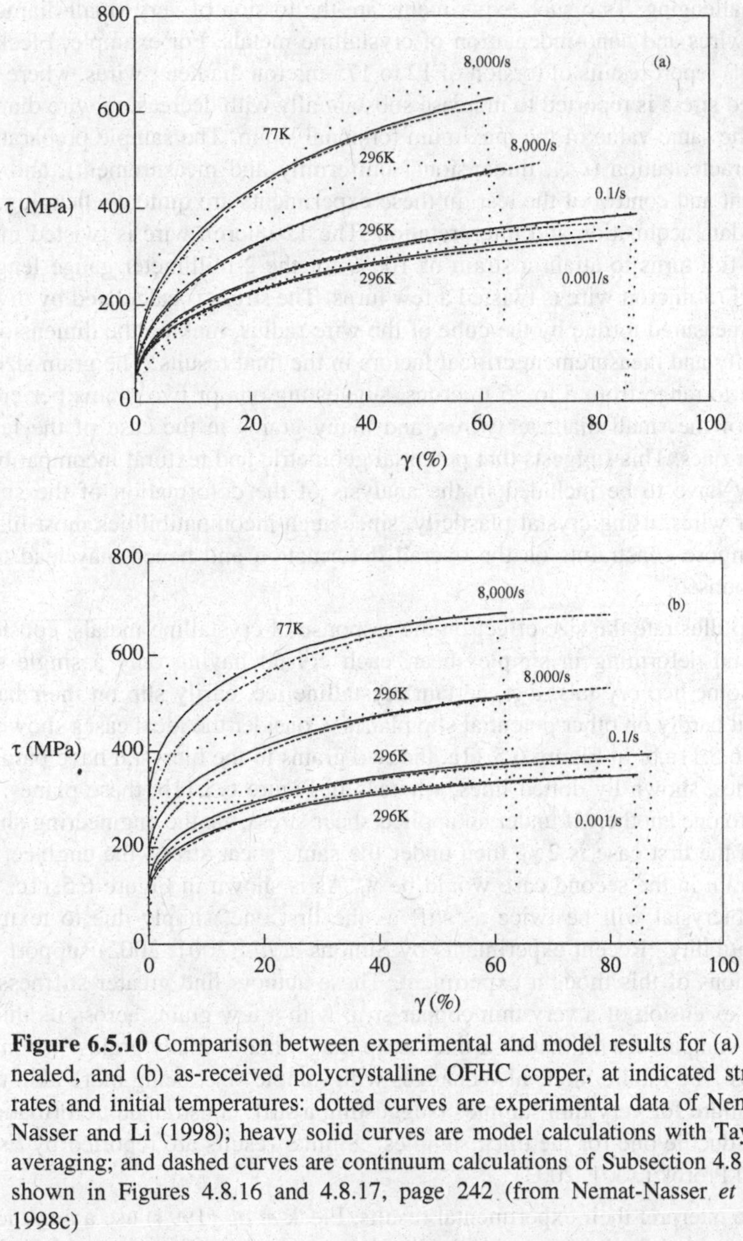

Figure 6.5.10 Comparison between experimental and model results for (a) annealed, and (b) as-received polycrystalline OFHC copper, at indicated strain rates and initial temperatures: dotted curves are experimental data of Nemat-Nasser and Li (1998); heavy solid curves are model calculations with Taylor averaging; and dashed curves are continuum calculations of Subsection 4.8.10, shown in Figures 4.8.16 and 4.8.17, page 242 (from Nemat-Nasser *et al.*, 1998c)

The size effects are most often realized in small-scale experiments that employ micron-dimension samples consisting of only a few crystals. Experiments of this class are extremely challenging and must be carried out under highly controlled conditions. Moreover, the interpretation of the results is also quite challenging. Two such experiments are the torsion of very small diameter copper wires and nano-indentation of crystalline metals. For example, Fleck *et al.* (1994) report results of torsion of 12 to 175 micron diameter wires, where the measured stress is reported to increase substantially with decreasing wire diameter for the same value of the maximum torsional strain. The sample preparation and characterization (*e.g.*, dimensional uniformity and measurement), and the alignment and control of the test, in these experiments are quite challenging, as are the data acquisition and interpretation. The 12-micron wire is twisted more than 50 full turns to attain a strain of 100% for the 2-millimeter gauge length, and the 175-micron wire is twisted a few turns. The stress is calculated by dividing the measured torque by the cube of the wire radius, making the dimensional uniformity and measurement critical factors in the final results. The grain size is reported to range from 5 to 25 microns, suggesting one or two grains per cross section of the small diameter wires, and many grains in the case of the large diameter ones. This suggests that potential geometric and textural incompatibilities may have to be included in the analysis of the deformation of the small diameter wires, using crystal plasticity, since such incompatibilities most likely could impose constraints on the overall deformation and hence may lead to a stiff response.

To illustrate the size effect on the response of crystalline metals, consider a bicrystal deforming in simple shear, each crystal having only a single slip plane. Some hcp crystals, *e.g.*, certain crystalline ice, easily slip on their basal plane but hardly on other potential slip planes. Consider the ideal cases shown in Figures 6.5.11a,b. In Figure 6.5.11a, the two grains in the bicrystal have parallel slip planes, shown by dotted lines, whereas in Figure 6.5.11b, these planes are normal to one another. If under an applied shear stress, τ_0, the engineering shear strain in the first case is $2\,\gamma_0$, then under the same shear stress, the engineering shear strain in the second case would be γ_0. As is shown in Figure 6.5.11c, the second bicrystal will be twice as stiff as the first one, simply due to textural incompatibility. Recent experiments by Simons *et al.* (2001, 2002) support the implications of this thought experiment. These authors find greater stiffness in uniaxial extension of a very thin copper strip with a few grains across its thickness, as compared with that of a thick strip that contains many grains within its thickness. The failure mode also changes with sample size, being more akin to a brittle failure for very thin samples (suggesting a stiff, constrained deformation) and to a ductile one for the thick samples. Similar results are reported by Espinosa and Prorok (2001, 2003).

To interpret their experimental results, Fleck *et al.* (1994) use a nonlinear, small deformation elasticity model (the Hencky-type deformation model) with the stress depending on the strain, as well as on the strain gradient in a certain way. The introduction of the strain gradient produces an additional free parameter that is then used to fit the test results. More recently, the theory has been modified and placed within the framework of incremental continuum plasticity, using the plastic strain rate and its gradient, involving three free length parameters (Fleck and Hutchinson, 2001). This kind of model lends itself to challenging

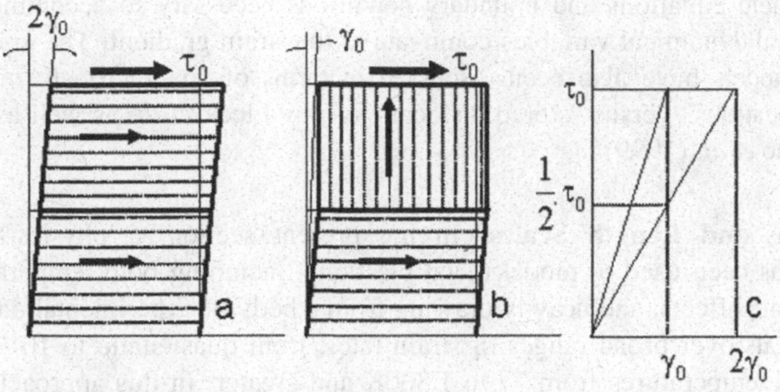

Figure 6.5.11 A bicrystal with parallel (a), and normal (b) slip planes; and the resulting response (c)

numerical model simulations, since, in addition to the usual stress and strain fields, it is now necessary to compute numerically accurate spatial derivatives of these fields. The presence of the strain (or strain-rate) gradients in the constitutive relations provides new constitutive parameters for continuum-type modeling of the experimental results. Thus, the theory has been attractive to many researchers interested in computational modeling, resulting in many new papers focused on computational simulations. It has also been applied to fit the nano-indentation test results, based on the usual engineering interpretation of the indentation data in terms of an effective stress as a function of the indentation depth. A formulation of the continuum gradient theory based on slip-induced crystal plasticity and continuum thermodynamics, is given by Gurtin (2000, 2002).[70] An alternative approach has been suggested by Acharya and Bassani (2000) and Bassani (2001), in which the strain gradient is embedded into the material's hardening parameters. This model circumvents the question of

---

70 In Gurtin (2000), it is assumed that the free energy of the crystal is a function of both an elastic deformation measure, denoted by $\mathbf{F}^e$, and a plastic deformation measure, denoted by $\mathbf{F}^p$, with the total deformation being characterized by $\mathbf{F} = \mathbf{F}^e\mathbf{F}^p$, as in finite-deformation crystal plasticity developed in Section 6.2, equation (6.2.1a), page 404. Since dislocations can traverse a crystal from one free face to another, producing large plastic deformations but no change in the internal structure, it is difficult to relate this assumption to experimental observations, unless the symbol $\mathbf{F}^p$ represents the plastic deformation of the crystal associated with the dislocations that are currently present in the crystal. In this case, it is then a microstructural state parameter and can occur in the expression for the free energy. As is discussed later on in this subsection, with this latter interpretation, proposed in Cermelli and Gurtin (2001) and Gurtin (2002), then the elastic and plastic deformations are of the same order of magnitude, hence infinitesimally small for most crystals. This then is a variant of the classical concepts proposed by Kondo (1952, 1955), Nye (1953), Kröner (1960), and others, as discussed later on in this section. Within the framework of linear elasticity, numerical simulations of discrete dislocations by Van der Giessen and Needleman (1995), and Cleveringa *et al.* (1997, 1998) also suggest that the plastic and elastic strains in a dislocated crystal are of the same order of magnitude and infinitesimally small.

additional field equations and boundary conditions necessary to accommodate the additional dynamical variables conjugate to the strain gradient. The gradient plasticity models have also been discussed in terms of "geometrically necessary[71] dislocations" versus "stored dislocations", by Fleck *et al.* as well as others, *e.g.*, Gao *et al.* (1999).

Dislocations and Length Scales: In the present section, a physics-based approach has been used to model metal plasticity, including both temperature- and strain-rate effects, and heavily drawing from a body of experimental data on various metals over broad ranges of strain rates, from quasi-static to 10^4/s and greater, and temperatures from 77 to 1,300K and greater. In this approach, the role of the strain gradient is embedded in the nature of the dislocations, their density and distribution, and the manner by which they produce slip in crystal plasticity and affect the overall flow stress. As is seen from equations (6.5.14) to (6.5.19), the length scales in this approach are directly related to the dislocation densities and microstructure of the material, and hence change with temperature and the strain-rate histories. This kind of model can be used to calculate the force-deformation relations at micron to continuum dimensions.[72]

As is well known, the core of a dislocation involves only a few lattices, within which immense lattice distortion and strain gradient are present. Since this strain gradient dies inversely with the square of the distance from the dislocation line,[73] each dislocation line carries with it a concentrated strain gradient field at the nano-scale, and hence, large strain gradients exist within each cluster of dislocations, as well as across various such clusters, on each slip plane of each crystal.

In general, it seems difficult to establish relations between the macroscopic and microscopic strain gradients. Indeed, intense microscopic strain gradient fields always accompany any plastic deformation, whether it is macroscopically uniform or nonuniform. Figure 6.5.12a shows a highly deformed shearband (910% shear strain) in tantalum, obtained at a strain rate in excess of 10^5/s (Nemat-Nasser *et al.*, 1998a). The temperature is estimated to have reached 1,900K inside the (about 100μm width) shearband in less than 100μs and then cooled to less than 1,000K in less than 100ms, quenching and preserving the microstructure. As is seen, within the shearband, the *macroscopic strain field is essentially uniform.* Figure 6.5.12b is a micrograph of the center of the shearband, showing a highly heterogeneous distribution of dislocations and hence

---

[71] Since the elastic distortion that the lattice can support is infinitesimally small, dislocations as geometric imperfections are generated in crystalline materials in response to internal incompatibilities and externally imposed deformations, in order to minimize the internal elastic energy, and, in this sense, they all seem to be geometrically necessary.

[72] To model deformation at sub nano-scales, it would be necessary to account for the actual lattice distortion and possible quantum effects.

[73] Since the strain components given by, say, (6.1.18a-d), die off inversely with distance from the dislocation line, their spatial gradients die off inversely with the square of the distance.

immense strain gradient variations, as well as dynamic recrystallization.

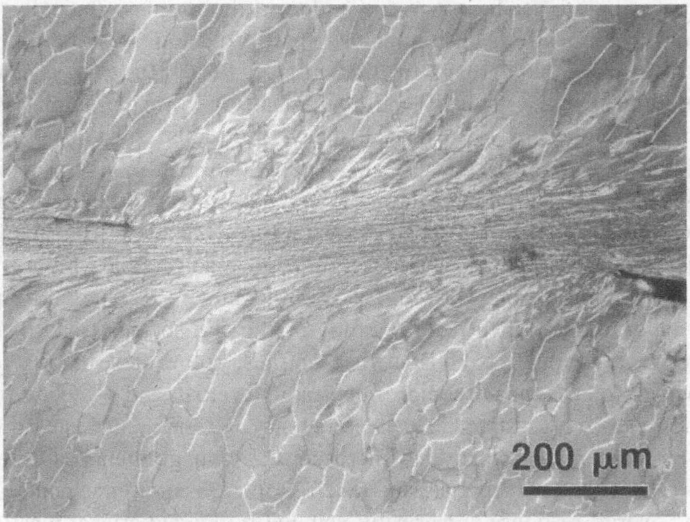

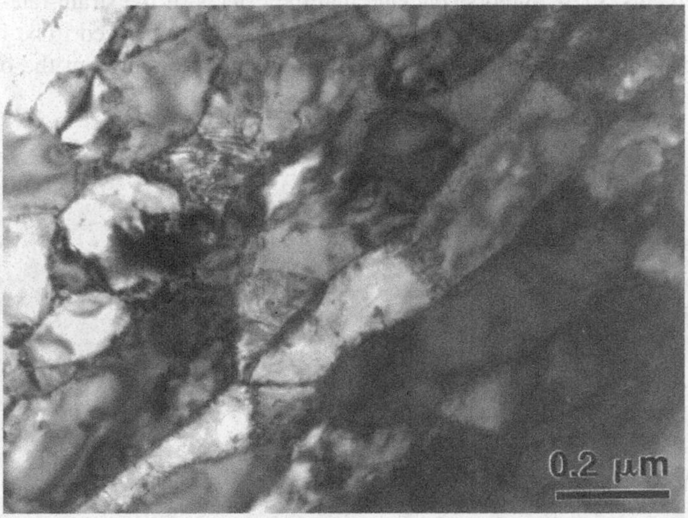

Figure 6.5.12

Highly deformed shearband (910% shear strain) in tantalum (a), obtained at a strain rate in excess of 10^5/s; and (b) resulting microstructure at the center of the shearband where macroscopic strain is essentially uniform, showing intense heterogeneous distribution of strain gradients at submicron scale (from Nemat-Nasser *et al.*, 1998a)

Plastic deformation of crystals is always highly heterogeneous on the micro-scale. This has been shown most vividly by Rashid *et al.* (1992). Thin samples of single-crystal copper are indented both quasi-statically (Figure 6.5.13) and dynamically in such a manner that in-plane deformation is induced by activating only two through-the-thickness {111}<110> slip systems. The microstructure of quasi-statically and dynamically deformed samples is highly heterogeneous, consisting of extensive dislocation cells and microbands, suggesting intense strain gradients basically independent of the macroscopic strain gradients and even at locations where the macroscopic deformation is uniform. Figures 6.5.14a,b,c show two such microstructures far away from the indenter (a, quasi-static; b, dynamic), where there are no macroscopic strain gradients, and one (c, dynamic) somewhat close to the indenter, where there are macroscopic strain gradients. Their detailed study of the slip structure and dislocation morphology at various locations in the samples, some close to the indented regions where intense macroscopic strain gradients exist, and some away from it where the macroscopic strain field is essentially uniform, suggests no obvious relation between macroscopic and microscopic strain gradients. On the other hand, correlating the microscopic and macroscopic observations, and seeking to model the overall response numerically using slip-induced crystal plasticity and finite elements, these authors conclude that the effects of the strain-rate-history-dependence of the evolution of the dislocation substructure need to be included in the constitutive model of slip, in order to adequately represent the observed results.

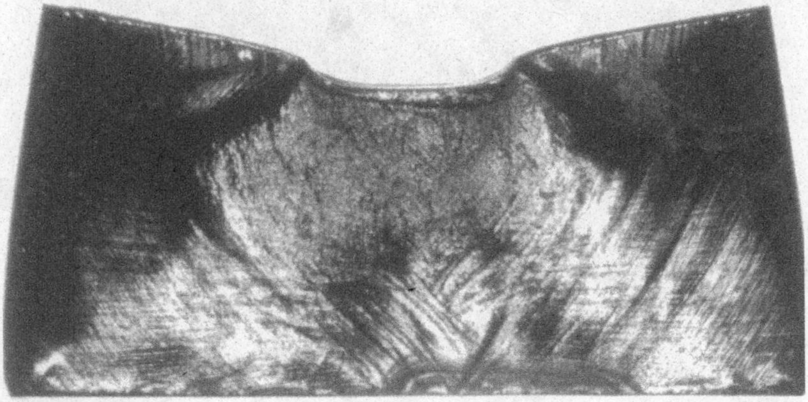

Figure 6.5.13

The deformed shape of the quasi-statically indented specimen of single-crystal copper; the dynamically indented specimen has a similar deformed shape (from Rashid *et al.*, 1992)

Similar, highly heterogeneous deformation patterns are observed at sub-millimeter scales in polycrystals. Figure 6.5.15a shows the initial state of a

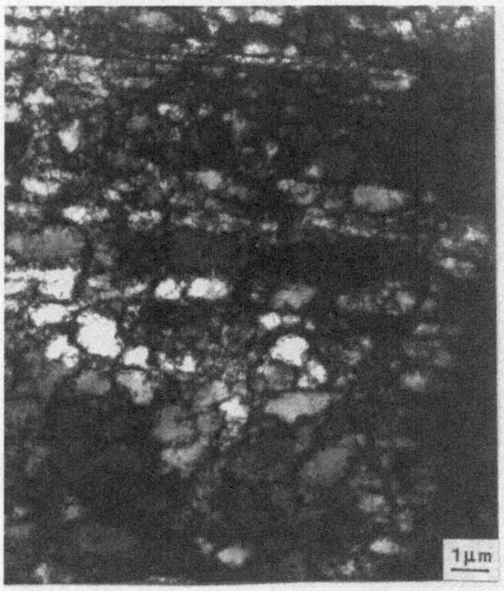

(a)

(b)

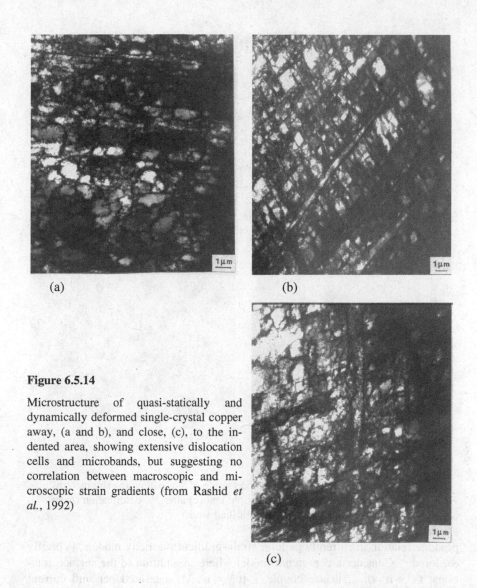

(c)

Figure 6.5.14

Microstructure of quasi-statically and dynamically deformed single-crystal copper away, (a and b), and close, (c), to the indented area, showing extensive dislocation cells and microbands, but suggesting no correlation between macroscopic and microscopic strain gradients (from Rashid *et al.*, 1992)

Waspalloy sample which was then incrementally deformed in uniaxial extension, and the deformation history of the same set of crystals was then documented. Figures 6.5.15b,c are slip patterns at 3% and 5% axial strains, respectively. Highly heterogeneous deformation with accompanying deformation gradients is observed in this macroscopically homogeneously deformed specimen.

A Simple Couple-stress Theory: For the sake of historical perspective, a brief account of a simple couple-stress theory in continuum mechanics is presented here first. Then the concept of continuously distributed dislocations and its

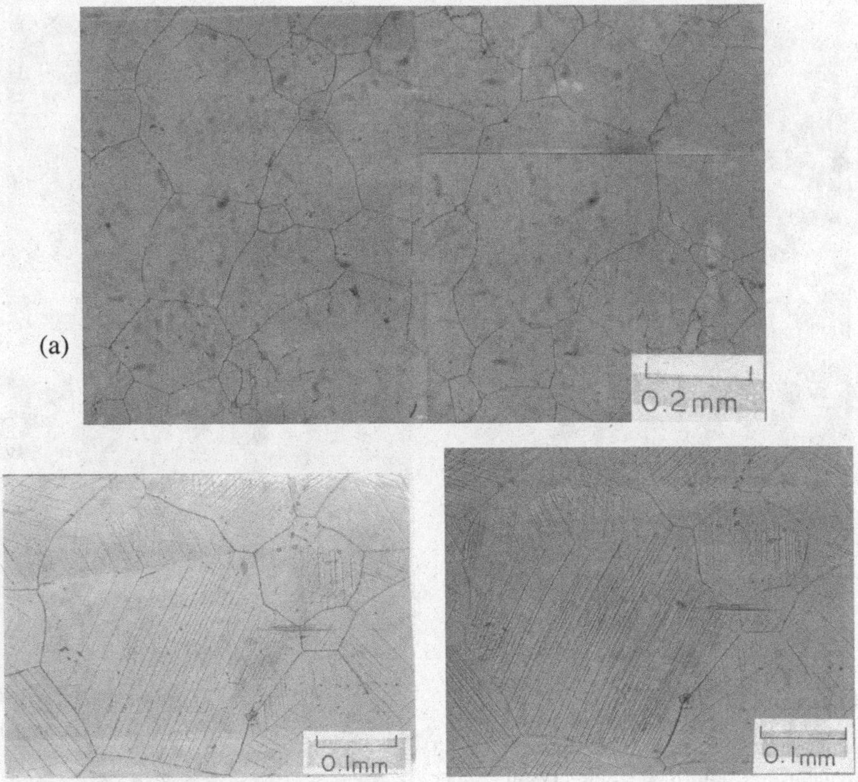

Figure 6.5.15

Microstructure of (a) initial undeformed sample of Waspalloy; (b) slip patterns after 3% uniaxial quasi-static deformation; and (c) after 5% deformation (from Nemat-Nasser and Sakane, 1980, unpublished work)

possible relation to currently popular strain-gradient plasticity models, is briefly explored.[74] Consider a continuum model where in addition to the surface tractions, $\mathbf{t}^{(n)} = \mathbf{n} \cdot \boldsymbol{\sigma}$, a surface couple,[75] $\boldsymbol{\mu}^{(n)} = \mathbf{n} \cdot \mathbf{M}$, measured per unit current area, is transmitted across a surface element da with unit normal $\mathbf{n}$, where $\mathbf{M} = M_{ij}\, \mathbf{e}_i \otimes \mathbf{e}_j$ is the *couple-stress tensor*, a counterpart of the Cauchy stress tensor $\boldsymbol{\sigma} = \sigma_{ij}\, \mathbf{e}_i \otimes \mathbf{e}_j$.

[74] The couple-stress theory was explored rather extensively in the 50's and 60's, as a special case of multi-polar continua. Interested readers may find elegant presentations by Toupin (1962) for finite elastic deformations and by Mindlin (1964) for the linearly elastic case, and an introductory account by Koiter (1964). The material presented here is from the author's class notes, Nemat-Nasser (1968-1970).

[75] Using the conservation of angular momentum, (6.5.59a), it is easy to show that $\boldsymbol{\mu}^{(n)}$ at any point within a continuum, can be expressed as a linear and homogeneous function of $\mathbf{n}$. The coefficients in this linear relation are defined in terms of the couple-stress tensor.

The balance of angular momentum for an elementary volume v of boundary ∂v requires that

$$\frac{d}{dt}\int_v \mathbf{x}\times\mathbf{v}\,\rho\,dv = \int_v \rho\,\{\mathbf{x}\times(\mathbf{a}-\mathbf{b})+\mathbf{c}\}\,dv + \int_{\partial v}\{\mathbf{x}\times\mathbf{t}^{(n)}+\boldsymbol{\mu}^{(n)}\}\,da\,, \tag{6.5.59a}$$

where ρ is the mass density, **a** is the acceleration, (2.4.3), **b** is the body force, and **c** is the body couple.[76] Since (6.5.59a) must hold for any *arbitrary* volume v, then using the Gauss theorem obtain,

$$\nabla\cdot\mathbf{M}+\rho\,\mathbf{c}+\boldsymbol{\zeta}=0\,,$$
$$M_{ji,j}+\rho\,c_i+\zeta_i=0\,, \tag{6.5.59b,c}$$

where $\boldsymbol{\zeta}$ is the dual vector of the skewsymmetric part of the Cauchy stress, defined by

$$\boldsymbol{\zeta}=\zeta_k\,\mathbf{e}_k=e_{ijk}\,\sigma_{ij}\,\mathbf{e}_k\,, \tag{6.5.59d}$$

as the Cauchy stress may now be unsymmetric in the presence of couple stresses.

Consider now the conservation of energy for the same volume element. As in Section 3.6 of Chapter 3, let e, **q**, and h respectively denote the internal energy density per unit mass, the heat flux vector per unit current area, and heat sources per unit mass. Then the rate of change of the sum of the internal and kinetic energies is balanced by the rate of applied work and heat exchange,

$$\frac{d}{dt}\int_v \rho\,[e+\tfrac{1}{2}\,\mathbf{v}\cdot\mathbf{v}]\,dv=\mathcal{F}+\mathcal{Q}\,, \tag{6.5.60a}$$

where $\mathcal{F}$ and $\mathcal{Q}$ are the rate of work and heat exchange, respectively, given by

$$\mathcal{F}=\int_{\partial v}[\mathbf{t}^{(n)}\cdot\mathbf{v}+\boldsymbol{\mu}^{(n)}\cdot\boldsymbol{\omega}]\,da+\int_v \rho\,[\mathbf{b}\cdot\mathbf{v}+\mathbf{c}\cdot\boldsymbol{\omega}]\,dv\,;$$
$$\mathcal{Q}=\int_v[-\nabla\cdot\mathbf{q}+\rho\,h]\,dv\,. \tag{6.5.60b,c}$$

In (6.5.60b), $\boldsymbol{\omega}$ is the spin vector, relating to the spin tensor, **W**, by (2.5.10), *i.e.*,

$$\boldsymbol{\omega}=\omega_i\,\mathbf{e}_i=-\tfrac{1}{2}\,e_{ijk}\,W_{jk}\,\mathbf{e}_i\,. \tag{6.5.60d}$$

From (6.5.60a) and the fact the this rate form of the conservation of energy is invariant under rigid-body motion of the observer, the balance laws in continuum mechanics are extracted. Indeed, the left-hand side of (6.5.60a) may be written as

$$\frac{d}{dt}\int_v \rho\,[e+\tfrac{1}{2}\,\mathbf{v}\cdot\mathbf{v}]\,dv=\int_v\{\frac{d}{dt}(\rho dv)\,[e+\tfrac{1}{2}\,\mathbf{v}\cdot\mathbf{v}]$$
$$+\rho\,dv\,\frac{d}{dt}[e+\tfrac{1}{2}\,\mathbf{v}\cdot\mathbf{v}]\}\,. \tag{6.5.61a}$$

---

[76] In a more general setting, a local relative deformation may also be considered. For example, a local angular momentum in excess of the moment of the linear momentum may be added.

Considering now arbitrary but constant translational velocity and spin of the observer, expressions for the balance of linear and angular momenta, energy, as well as mass are obtained. To see this, evaluate and arrange (6.5.60a) as,

$$\int_{\mathrm{v}} \{ [\boldsymbol{\nabla} \cdot \mathbf{M} + \rho\,\mathbf{c} + \boldsymbol{\zeta}] \cdot \boldsymbol{\omega} + [\boldsymbol{\nabla} \cdot \boldsymbol{\sigma} + \rho\,\mathbf{b} - \rho\,\dot{\mathbf{v}}] \cdot \mathbf{v}$$

$$+ [-\rho\,\dot{e} + \mathrm{tr}((\boldsymbol{\nabla} \otimes \boldsymbol{\omega})^{\mathrm{T}} \mathbf{M}) + \mathrm{tr}(\mathbf{D}\,\boldsymbol{\sigma}) + -\boldsymbol{\nabla} \cdot \mathbf{q} + \rho\,h]\}\, d\mathrm{v}$$

$$= \int_{\mathrm{v}} \frac{d}{dt}(\rho d\mathrm{v})\, [e + \tfrac{1}{2}\mathbf{v} \cdot \mathbf{v}], \tag{6.5.61b}$$

where (6.5.61a) is also used. As is seen, since $\boldsymbol{\omega}$ and $\mathbf{v}$ can be changed arbitrarily by the (constant) motion of the observer, and that the volume v is also arbitrary, (6.5.59b,c) and (3.1.4a,b) now follow, as does the following balance of energy:

$$\rho\,\dot{e} = \mathrm{tr}((\boldsymbol{\nabla} \otimes \boldsymbol{\omega})^{\mathrm{T}} \mathbf{M}) + \mathrm{tr}(\mathbf{D}\,\boldsymbol{\sigma}) - \boldsymbol{\nabla} \cdot \mathbf{q} + \rho\,h,$$

$$\rho\,\dot{e} = \omega_{j,i} M_{ij} + D_{ij}\,\sigma_{ij} - q_{i,i} + \rho\,h. \tag{6.5.61c,d}$$

Inclusion of local "micro-deformation and micro-spin", in a manner similar to those proposed, *e.g.*, by Mindlin (1964), introduces additional stress and inertia terms with certain length scales, into the relevant balance relations.[77] For example, as in Toupin (1962), it may be assumed that there is a micro-spin field associated with an excess angular momentum above the moment of linear momentum. The corresponding term is then added to the right-hand side of (6.5.59b,c), accordingly.

Microstructure and Continuously Distributed Dislocations: The interaction force between two dislocations in an elastic solid, is inversely proportional to their spacing, as may be inferred from (6.1.16a-f). From Stroh's calculations, reported in Subsection 6.5.1, it also follows that the elastic field of dislocations, hindering the motion of mobile dislocations in crystalline solids, may be assumed to be proportional to the inverse of the average dislocation spacing; see *e.g.*, equations (6.5.8a) and (6.5.10a). In addition, the forest of dislocations that intersects a slip plane (Argon and East, 1979), provides the short-range barrier to the motion of dislocations that lie on that plane and, hence, affects the flow stress, as has been discussed before. Therefore, the dislocation density and its spatial variation play a role in defining the flow stress of most crystals and crystalline materials.

To represent within a continuum framework, the dislocation density and its spatial variation, the notion of continuously distributed dislocations was introduced and extensively studied in the 50's and 60's, by Kondo (1952), Nye (1953), Kröner (1955, 1956, 1960), Bilby *et al.* (1955), and Bilby (1960); see also Lardner (1974) and Mura (1987). The presence of dislocations as

[77] See, *e.g.*, equations (4.1) and (4.2), page 57, of Mindlin (1964).

geometric line defects in crystals, produces inelastic strains accompanied by an accommodating elastic lattice distortion, leading to a compatible total deformation. Indeed, since interatomic forces cannot support large elastic distortions of the lattice, dislocations are generated to relieve the elastic deformation and to minimize the associated elastic energy of the crystal. In essentially all known crystalline materials, the lattice elastic strains are infinitesimally small, being of the order of 10^{-3}. The plastic strains associated with the dislocations that are present in a crystal, are also very small. Therefore, large plastic deformations of crystals occur by the motion of dislocations that are constantly generated and annihilated, as atoms rearrange and matter *flows* through the lattice, and not by a continuous storage of dislocations. For example, an annealed cylindrical sample of copper of centimeter dimensions, can be deformed, say, at a 600K temperature, to a thin sheet of submillimeter thickness, while its final dislocation density is about the same as that of its initial undeformed state, *i.e.*, about $10^7\,\mathrm{cm}^{-2}$. A three-orders of magnitude increase in this dislocation density, means an average dislocation spacing of about 0.1μm, which is rather small, but does not correspond to significant plastic strains.

It is therefore essential to clearly distinguish between the *total plastic deformation* of a crystalline solid, which can be hundreds or even thousands of percent, and *the plastic strains due to the presence of dislocations within a crystalline solid*, which generally is very small, of the order of the associated elastic strains. While the analysis of the total finite deformation of crystals must be performed within a fully nonlinear setting, as discussed in Section 6.2, small deformation, linearized kinematics is generally sufficient to deal with the elastoplastic deformation due to the existing dislocations within crystals or polycrystals.[78]

In view of the above comments, it is generally unnecessary to distiguish between the configurations of a crystal without and with dislocations, *i.e.*, the particle positions in an *undislocated* and in the corresponding *dislocated crystal* are indistinguishable. The displacement gradient field, $\boldsymbol{\nabla}\otimes\mathbf{u} = u_{i,j}(\mathbf{x})\,\mathbf{e}_j\otimes\mathbf{e}_i$, of a dislocated crystal may thus be additively split into and an elastic, $\boldsymbol{\beta}^{e}$, and a plastic, $\boldsymbol{\beta}^{p}$ (due to the presence of dislocations), part,

$$(\boldsymbol{\nabla}\otimes\mathbf{u})^{T} = \boldsymbol{\beta}^{e} + \boldsymbol{\beta}^{p}\,,$$

$$u_{i,j} = \beta^{e}_{ij} + \beta^{p}_{ij}\,. \tag{6.5.62a}$$

The plastic part, $\boldsymbol{\beta}^{p}$, is associated with the dislocations that are present on various slip planes of the crystal or the crystals in a polycrystal. The dislocation density tensor is now defined by Kröner's equation,[79]

---

78 The shear strain that produced the shearband in Figure 6.5.12a is about 910%, while the residual plastic (and elastic) strains due to the dislocations that remain in the material (Figure 6.5.12b), are fractions of one percent.

79 See Mura (1987, page 53). Note that Mura defines the positive direction of the normal to the S^{+} towards the S^{-}, whereas in the present work the opposite direction is used; see Figure 6.1.7, page 392. Hence, the minus sign in Mura's formula.

$$\alpha_{ij} = e_{ilk}\beta^{p}_{jl,k}\,. \tag{6.5.62b}$$

The requirement of compatibility of the total displacement gradient $\nabla\otimes\mathbf{u}$, then yields,

$$\alpha_{ij} = -e_{ilk}\beta^{e}_{jl,k}\,. \tag{6.5.62c}$$

Since, at each instant in the course of the finite deformation of a crystal, there are mobile and stationary dislocations present within the crystal, the crystal is dislocated. In line with equations (6.5.62a,b), but following Subsection 6.1.10, equations (6.1.33a-c) and (6.1.34a,b), consider the plastic distortion *rate* of a specific slip system, $\mathbf{s}\otimes\mathbf{n}$. Assume that the mobile dislocations of this slip system are all parallel, having a common direction, given by $\boldsymbol{\xi}$, and a common velocity, given by $\mathbf{v}$, which lies on the slip plane and is normal to the dislocation lines. The plastic distortion rate of this slip system can be expressed in terms of a third-order tensor, $\boldsymbol{\mathcal{V}}$, given by

$$\begin{aligned}\boldsymbol{\mathcal{V}} &= b\,\rho_m\,\mathbf{s}\otimes\mathbf{v}\otimes\boldsymbol{\xi}\\ &= b\,\rho_m\,s_i\,v_j\,\xi_k\,\mathbf{e}_i\otimes\mathbf{e}_j\otimes\mathbf{e}_k\,,\end{aligned} \tag{6.5.63a,b}$$

which will be called the *dislocation flux density tensor* of slip system $\mathbf{s}\otimes\mathbf{n}$. In this equation, b is the magnitude of the Burgers vector of direction $\mathbf{s}$, and ρ_m is the density of the associated mobile dislocations. Now, it follows that the corresponding plastic distortion rate, $\boldsymbol{l}^p$, is given in terms of this dislocation flux density tensor, by

$$\begin{aligned}\boldsymbol{l}^p &= l^p_{ij}\,\mathbf{e}_i\otimes\mathbf{e}_j = e_{jlk}\mathcal{V}_{ilk}\,\mathbf{e}_i\otimes\mathbf{e}_j\\ &= (b\,\rho_m)\,\mathbf{s}\otimes(\mathbf{v}\times\boldsymbol{\xi}) = (b\,\rho_m\,v)\,\mathbf{s}\otimes\mathbf{n}\,.\end{aligned} \tag{6.5.63c}$$

Consider now a mobile dislocation loop of the same slip system, with the dislocation velocity vector, $\mathbf{v}(\eta)$, lying on the slip plane for all values of η which measures length along the loop. Then, $\mathbf{v}\times\boldsymbol{\xi} = v(\eta)\,\mathbf{n}$ at all points of the loop. Thus, (6.5.62c) may be generalized by replacing v with an average effective value, say, $\bar{v}$, as commented on in Subsection 6.1.10; see footnote following equation (6.1.34a), page 402.

It may be of interest to note that

$$\begin{aligned}\mathrm{tr}(\boldsymbol{l}^p\boldsymbol{\sigma}) = l^p_{ji}\,\sigma_{ij} &= \rho_m\,[\boldsymbol{\xi}\times(\mathbf{b}\cdot\boldsymbol{\sigma})]\cdot\mathbf{v}\\ &= \rho_m\,\mathbf{F}\cdot\mathbf{v}\,,\end{aligned} \tag{6.5.63d}$$

where $\mathbf{F} = \boldsymbol{\xi}\times(\mathbf{b}\cdot\boldsymbol{\sigma})$ is the force on the dislocations of Burgers' vector $\mathbf{b}$ and direction $\boldsymbol{\xi}$, due to a stress field $\boldsymbol{\sigma}$; see Peach-Koehler equation (6.1.2g), page 392. Thus, as should be expected, $\mathrm{tr}(\boldsymbol{l}^p\,\boldsymbol{\sigma})$ is the associated rate of work per unit current volume.

While (6.5.62b) corresponds to linearized kinematics, (6.5.63c) can be associated with a fully nonlinear formulation, since it is in terms of the rate variables. In line with this model, define the rate of change of the dislocation density tensor, $\dot{\boldsymbol{\alpha}}$, for the general nonlinear case, by

$$\dot{\boldsymbol{\alpha}} = \dot{\alpha}_{ij}\,\mathbf{e}_i \otimes \mathbf{e}_j = e_{ikl}\, l^p_{jl,k}\, \mathbf{e}_i \otimes \mathbf{e}_j$$
$$= (\nabla\dot{\gamma} \times \mathbf{n}) \otimes \mathbf{s}\,. \qquad (6.5.64a,b)$$

With the slip system $\mathbf{s}\otimes\mathbf{n}$ fixed, the axial vector, $\dot{\mathbf{a}}$, of the rate of change of the dislocation density tensor, $\dot{\boldsymbol{\alpha}}$, *is the* **s***-direction gradient of the plastic distortion rate*,

$$\dot{\mathbf{a}} = e_{ijk}\dot{\alpha}_{ij}\,\mathbf{e}_k = (\mathbf{s} \cdot \nabla\dot{\gamma})\,\mathbf{n}\,,$$
$$\dot{\gamma} = b\,\rho_m\,\overline{v}\,. \qquad (6.5.64c,d)$$

In view of the above observation, define the gradient of the plastic distortion rate of the slip system $\mathbf{s}\otimes\mathbf{n}$, by

$$\dot{\boldsymbol{\pi}} = (\nabla\dot{\gamma}) \otimes \mathbf{n} \otimes \mathbf{s}\,, \qquad \dot{\pi}_{ijk} = \dot{\gamma}_{,i}\, n_j\, s_k, \qquad (6.5.65a,b)$$

and observe that

$$\dot{\alpha}_{kl} = e_{ijk}\,\dot{\gamma}_{,i}\, n_j\, s_l\,. \qquad (6.5.64e)$$

Thus, define a general *distortion-rate gradient vector*, $\dot{\mathbf{g}}$, as follows:

$$\dot{\mathbf{g}} = (\nabla\dot{\gamma}) \times \mathbf{n} \times \mathbf{s}$$
$$= (\mathbf{s} \cdot \nabla\dot{\gamma})\,\mathbf{n} - (\mathbf{n} \cdot \nabla\dot{\gamma})\,\mathbf{s}\,. \qquad (6.5.65c)$$

The **n**- and **s**-components of $\dot{\mathbf{g}}$ are the rates of change of dislocation activities in the **s**- and **n**-direction, respectively. These are the components of the axial vectors of the associated rate-of-change of the dislocation density tensors. As such, they do indeed characterize the spatial variation of the time-rate-of-change of the microstructure in the direction of slip and in the direction normal to the slip plane. To each slip system α, therefore, a distortion-rate gradient vector, $\dot{\mathbf{g}}^{\alpha}$, may be assigned.

It is worth emphasizing that the dislocation density tensor is in fact a microstructural parameter, and in this way the spatial gradient of the plastic distortion rate may enter the plasticity constitutive relations of crystals as well as those of polycrystals. From the application point of view, however, the introduction of the gradient of any distortion tensor, adds immense computational difficulties in finite-element or finite-difference simulations of real problems. As discussed in Chapter 5, even the calculation of deformation gradient, **F**, requires special attention in large-scale codes, in order to ensure accuracy and stability of the solution. Therefore, in the absence of clear and compelling reasons, it seems reasonable that this additional complication should be avoided. In fact, the dislocation-based models of plasticity presented in this and in Section 4.8 appear adequate and effective for any such numerical applications with essentially minimal requirements. They can be directly integrated into existing large-scale finite-element codes. In addition, they are backed by an immense body of experimental data.

Phenomenological Strain-gradient Plasticity: Most of the developments in this area have been for small deformations and hence outside the scope of the present treatise. However, in view of recent activities by many authors on this subject, a brief account appears warranted. There are various versions of the gradient plasticity, some by the same authors. The idea has its origin in high-order continuum mechanics theories that have been mentioned in this subsection. Specific application to plasticity problems dates to the work by Aifantis and coworkers; see Bammann and Aifantis (1982), Aifantis (1984), and a recent review by Aifantis (2003). Through introduction of the gradient of the (small deformation, linear) plastic strain or its rate, length scales emerge as free constitutive parameters that may be adjusted to fit various experimental results, particularly those associated with micron-size deformation events. As pointed out before, sample-size effects and length scales are distinct, though not necessarily mutually exclusive.

A conceptually simple gradient-plasticity model builds on the classical J_2-plasticity, and only modifies the definition of the effective plastic strain (in the case of deformation theory) or the effective plastic strain rate (in the case of incremental theory), as well as the associated yield condition and hardening rule. All models of this kind are rate- and temperature-independent, not applicable to experiments at various strain rates and temperatures. Here, to maintain notational consistency, consider the deformation rate tensor, **D**, and split it into the elastic and plastic parts, $\mathbf{D} = \mathbf{D}^e + \mathbf{D}^{p\prime}$, as has been discussed before. Assume that the elastic part relates to the stress rate through Hooke's law, and the plastic part is given by, say,

$$\mathbf{D}^{p\prime} = \dot{\varepsilon}_p \frac{\boldsymbol{\mu}}{\sqrt{2/3}}, \quad \boldsymbol{\mu} = \frac{\boldsymbol{\sigma}}{\sqrt{2/3}\,\sigma},$$
$$\sigma = (3/2\,\boldsymbol{\sigma}' : \boldsymbol{\sigma}')^{1/2}, \quad \dot{\varepsilon}_p = (2/3\,\mathbf{D}' : \mathbf{D}')^{1/2}. \tag{6.5.66a-d}$$

Here, $\boldsymbol{\sigma}$ is the Cauchy stress tensor, which remains symmetric, and other quantities are as defined before. The essential difference from the classical J_2-plasticity appears in the definition of the yield condition, now given by

$$Q = \sigma + \tau_{i,i} = Q_Y,$$
$$\dot{Q}_Y = H(E_p)\,[(1+C)\,\dot{\varepsilon}_p + 1/2\,B_i\,\dot{\varepsilon}_{p,i}],$$
$$\dot{\tau}_i = H(E_p)\,(A_{ij}\,\dot{\varepsilon}_{p,j} + 1/2\,B_i\,\dot{\varepsilon}_p),$$
$$\dot{E}_p^2 = (1+C)\,\dot{\varepsilon}_p^2 + A_{ij}\,\dot{\varepsilon}_{p,i}\,\dot{\varepsilon}_{p,j} + B_i\,\dot{\varepsilon}_{p,i}\,\dot{\varepsilon}_p, \tag{6.5.67a-d}$$

where H is the workhardening parameter, and the quantity τ_i is a force measure, work conjugate of the gradient of the plastic strain-rate measure, $\dot{\varepsilon}_{p,i}$. E_p is a generalized effective plastic strain for this model, obtained by integrating $\dot{E}_p(t)$ over time parameter, t; the model is rate-independent and hence t is any monotonically increasing load parameter. The coefficient A_{ij} in (6.5.67d) is a quadratic function of $\boldsymbol{\mu}$, and B_i and C are quadratic functions of its spatial gradient, $\boldsymbol{\nabla} \otimes \boldsymbol{\mu}$. These are given in Appendix A of Fleck and Hutchinson (2001) and will not be reproduced here. It is however noted that, since the plastic deformation-rate tensor is symmetric with zero trace, $\mathbf{D}^p = (\mathbf{D}^p)^T$ and $\mathrm{tr}(\mathbf{D}^p) = 0$, its gradient, a third-order tensor, can be split into two non-zero orthogonal constituents, as discussed by Toupin (1962, equation (4.43)). Then $\dot{E}_p^2$ is expressed in terms of $\dot{\varepsilon}_p^2$ and only three non-zero basic invariants of these orthogonal components.

The equilibrium, in the absence of body and inertia forces, becomes

$$\boldsymbol{\nabla} \cdot \boldsymbol{\sigma} = \mathbf{0}, \quad \boldsymbol{\sigma} = \boldsymbol{\sigma}^T, \quad \text{in } v,$$
$$\mathbf{n} \cdot \boldsymbol{\sigma} = \mathbf{t}^0, \quad \mathbf{n} \cdot \boldsymbol{\tau} = \tau^0, \quad \text{on } \partial v, \qquad (6.5.68a\text{-}d)$$

where $\mathbf{n}$ is the exterior unit normal to the surface ∂v that bounds the volume v, and $\mathbf{t}^0$ and τ^0 are the "surface data" to be prescribed on ∂v. For the displacement boundary data, in addition to the ususal surface velocity, the model requires prescription of the plastic strain-rate measure, $\dot{\varepsilon}_p$. It remains to identify the physics that should guide the assignment of the additional boundary conditions, *e.g.*, (6.5.68d), since the results of the solution of the corresponding boundary-value problems will be affected by this assignment.

6.6. REFERENCES

Acharya, A. and Bassani, J. L. (2000), Incompatibility and crystal plasticity, *J. Mech. Phys. Solids* Vol. 48, 1565-1595.

Aifantis, E. C. (1984), On the microstructural origin of certain inelastic models, *Trans. ASME, J. Eng. Mater. Technol.* Vol. 106, 326-330.

Aifantis, E. C., (2003), Update on a class of gradient theories, *Mech. Mat.*, Vol. 35, 259-280.

Aiken, R. C., (1985), *Stiff computation*, Oxford University Press.

Anand, L. and Kalidindi, S. R. (1994), The process of shearband formation in plane strain compression of fcc metals: effects of crystallographic texture, *Mech. Mat.*, Vol. 17, No. 2-3, 223-243.

Argon, A., (1995), Mechanical properties of single phase crystalline media: deformation at low temperatures, (Cahn, R., Haasen, P. eds.), *Physical Metallurgy*, Elsevier, Amsterdam.

Argon, A. and East, G. (1979), Forest dislocation intersections in stage I deformation of copper single crystals, (Haasen, P., Gerold, V., and Kostorz, G. eds.), *Proceedings of the Fifth International Conference on the Strength of Metals and Alloys*, Pergamon Press, Oxford, 9-16.

Argon, A. S. and Maloof, S. R. (1966), Plastic deformation of tungsten single crystal at low temperatures, *Acta Metall.*, Vol. 14, 1449-1962.

Asaro R. J. (1983), Crystal plasticity, *J. Appl. Mech.*, Vol. 50, 921-934.

Bammann, D. J. and Aifantis, E. C. (1982), On a proposal for a continuum with microstructure, *Acta Mech.* 45, 91-121.

Barnett, D. M. and Lothe, J. (1973), Synthesis of the sextic and the integral formalism for dislocations, Green's functions, and surface waves in anisotropic elastic solids, *Physica Norvegica*, Vol. 7, 13-19.

Basinski, S. J. and Basinski, Z. S. (1979), Plastic deformation and work hardening, *Dislocations in solids* (Nabarro, F. R. N. ed.) North-Holland, Amsterdam.

Basinski, Z. S. and Jackson, P. J. (1965a), The effect of extraneous deformation on strain hardening in Cu single crystals, *Appl. Phys. Lett.*, Vol. 6, 148-150.

Basinski, Z. S. and Jackson, P. J. (1965b), The instability of the work-hardened state — I; Slip in extraneously deformed crystals, *Phys. Stat. Solidi*, Vol. 9, 805-823.

Bassani, J. L. (2001), Incompatibility and a simple gradient theory of plasticity, *J. Mech. Phys. Solids,* Vol. 49, 1983-1996.

Bassani, J. L. and Wu, T. Y. (1991), Latent hardening in single crystals - II, analytical characterization and predictions, *Proc. R. Soc. Lond. A*, Vol. 435, 21-41.

Bell, J. F. (1965), A generalized large deformation behaviour for face-centred cubic solids: nickel, aluminium, gold, silver and lead, *Phil. Mag.*, Ser. 8, Vol. 11, 1135-1156.

Bell, J. F. and Green, R. E. Jr. (1967), An experimental study of the double slip deformation hypothesis for face-centered cubic single crystals, *Phil. Mag. Ser. 8*, Vol. 15, 469-476.

Bilby, B. A. (1960), Continuous distributions of dislocations, *Progress in Solid Mechanics* Vol. 1, 329-398.

Bilby, B. A., Bullough, R., and Smith, E. (1955), Continuous distributions of dislocations: a new application of the methods of non-Riemannian geometry, *Proc. R. Soc. Lond. A*, Vol. 231, 263-273.

Bowen, D. K., Christian, J. W., and Taylor, G. (1967), Deformation properties of niobium single crystals, *Can. J. Phys.*, Vol. 45, 903-938.

Burns, G. (1990), *Solid state physics*, Academic Press, New York.

Campbell, J. D., Simmons, J. A., and Dorn, J. E. (1961), On the dynamic behavior of a Frank-Read source, *J. Appl. Mech.*, Vol. 28, 447-453.

Cermelli, P. and Gurtin, M. E. (2001), On characterization of geometrically necessary dislocations in finite plasticity, *J. Mech. Phys. Solids*, Vol. 49, 1539-1568.

Chadwick, P. and Smith, G. D. (1977), Foundations of the theory of surface waves in anisotropic elastic materials, in *Advances in applied mechanics* (Yih, C.-S. ed.) Vol. 17, 303-376, Academic Press, New York.

Christian, J. W. (1983), Some surprising features of the plastic deformation of body-centered cubic metals and alloys, *Metall. Trans.*, Vol. 14A, 1237-1256.

Cleveringa, H. H. M., Van der Giessen, E., Needleman, A. (1997), Comparison of discrete dislocation and continuum plasticity predictions for a composite material, *Acta Mater.*, Vol. 45, 3163-3179.

Cleveringa, H. H. M., Van der Giessen, E., Needleman, A. (1998), Discrete dislocation simulations and size dependent hardening in single slip, *J. Phys.*, IV, 83-92.

Conrad, H. (1964), Thermally activated deformation of metals, *J. Metals*, Vol.

16, 582-588.

Cottrell, A. H. (1953), *Dislocations and plastic flow in crytals*, Oxford University Press, London.

Cullity, B. D. (1978), *Elements of x-ray diffraction*, Addison-Wesley, New York.

Espinosa, H. D. and Porok, B. C. (2001), Effects of film thickness on the yielding behavior of polycrystalline gold films, *Mater. Res. Soc. Proc.* 688, 365-372.

Espinosa, H. D. and Porok, B. C. (2003), Plasticity size effects in free-standing submicron polycrystalline FCC films subjected to pure tension, *J. Mech. Phys. Solids*, in press.

Ewing, J. A. and Rosenhain, W. (1899), Experiments in micro-metallurgy: effects of strain, preliminary notice, *Proc. R. Soc. Lond. A*, Vol. 65, 85-90.

Ewing, J. A. and Rosenhain, W. (1900), The crystalline structure of metals, *Phil. Trans. R. Soc. Lond.*, Vol. 193, 353-375.

Fleck, N. A., Muller, G. M., Ashby, M. F., and Hutchinson, J. W., (1994), Strain gradient plasticity: theory and experiment, *Acta Metall.* Vol. 42, 475-487.

Fleck, N. A. and Hutchinson, J. W., (2001), A reformulation of a class of strain gradient plasticity theories, *J. Mech. Phys. Solids*, Vol. 49, 2245-2271.

Foxall, R. A., Duesbery, M. S. and Hirsch, P. B. (1967), The deformation of niobium single crystals, *Can. J. Phys.*, Vol. 45, 607-629.

Francoisi, P., Berveiller, M. and Zaoui, A. (1980), Latent hardening in copper and aluminium single crystals, *Acta Metall.*, Vol. 28, 273-283.

Frank, F. C. and Read, W. T., Jr. (1950), Multiplication processes for slow moving dislocations, *Phys. Rev.*, Vol. 79, 722-724.

Frenkel, J. (1926), Zur theorie der Elastizitätsgrenze und der Festigheit kristallinischer Körper, *Z. Phys.*, Vol. 37, 572-609.

Gao, H., Huang, Y., Nix, W. D., and Hutchinson, J. W. (1999), Mechanism-based strain gradient plasticity- I. Theory, *J. Mech. Phys. Solids*, Vol. 47, 1239-1263.

Gurtin, M. E., (2000), On the plasticity of single crystals: free energy, microforces, plastic strain gradients, *J. Mech. Phys. Solids*, Vol. 48, 989-1036.

Gurtin, M. E. (2002), A gradient theory of single-crystal viscoplasticity that accounts for geometrically necessary dislocations, *J. Mech. Phys. Solids*, Vol. 50, 5 -32.

Havner K. S. (1973), On the mechanics of crystalline solids, *J. Mech. Phys. Solids*, Vol. 21, 383-394.

Havner, K. S. (1974), Aspects of theoretical plasticity at finite deformation and large pressure, *Z. angew. Math. Phys.*, Vol. 25, 765-781.

Havner, K. S. (1992), *Finite plastic deformation of crystalline solids*, Cambridge University Press, Cambridge.

Hill, R. (1966), Generalized constitutive relations for incremental deformation of metal crystals by multislip, *J. Mech. Phys. Solids*, Vol. 14, 95-102.

Hill, R. (1968), On constitutive inequalities for simple materials - I, II, *J. Mech. Phys. Solids*, Vol. 16, 229-242, 315-322.

Hill, R. and Havner, K. S. (1982), Perspectives in the mechanics of elastoplastic crystals, *J. Mech. Phys. Solids*, Vol. 30, 5-22.

Hill, R. and Rice, J. R. (1972), Constitutive analysis of elastic-plastic crystals at arbitrary strain, *J. Mech. Phys. Solids*, Vol. 20, 401-413.

Hirth, J. P. and Lothe, J. (1992), *Theory of dislocation*, Krieger, Melbourne, Florida.

Hull, D. and Bacon, D. J. (1992), *Introduction to dislocation*, Pergamon, Oxford.

Hull, D., Byron, J. F., and Noble, F. W. (1967), Orientation dependence of yield in body-centered cubic metals, *Can. J. Phys.*, Vol. 45, 1091-1099.

Hutchinson J. W. (1976), Bounds and self-consistent estimates for creep of polycrystalline materials, *Proc. R. Soc. Lond. A*, Vol. 348, 101-127.

Ingebrigtsen, K. A. and Tonning, A. (1969), Elastic surface waves in crystals, *Phys. Rev.* Vol. 184, 942-951.

Jackson, P. J. and Basinski, Z. S. (1967), Latent hardening and the flow stress in copper single crystals, *Can. J. Phys.*, Vol. 45, 707-735.

Kapoor, R. and Nemat-Nasser, S. (1998), Determination of temperature rise during high strain rate deformation, *Mech. Mat.*, Vol. 27, 1-12.

Kapoor, R. and Nemat-Nasser, S. (1999), High-rate deformation of single crystal tantalum: temperature dependence and latent hardening, *Scripta Mat.*, Vol. 40, 159-164.

Kittel, C. (1986), *Introduction to solid state physics*, John Wiley & Sons, New York.

Kocks, U. F. (1964), Latent hardening and secondary slip in aluminum and silver, *Trans. Metall. Soc. AIME*, Vol. 230, 1160-1167.

Kocks, U. F. (1970), The relation between polycrystal deformation and single-crystal deformation, *Metall. Trans.*, Vol. 1, 1121-1143.

Kocks, U. F. and Brown, T. J. (1966), Latent hardening in aluminum, *Acta Metall.*, Vol. 14, 87-98.

Kocks, U. F., Argon, A. S., and Ashby, M. F. (1975), Thermodynamics and kinetics of slip, *Prog. Mat. Sci.*, Vol. 19, 1-288.

Koiter, W. T. (1964), Couple stresses in the theory of elasticity, I and II, *Proc. K. Ned. Akad. Wet. (B)*, Vol. 56, 17-44.

Kondo, K. (1952), On the geometrical and physical foundations of the theory of yielding, *Proc. Jpn. Nat. Cong. Appl. Mech.* Vol. 2, 41-47.

Kondo, K. (1955), Non-Riemannian geometry of imperfect crystals from a macroscopic viewpoint. In: K. Kondo (Ed.) *RAAG Memoirs of the Unifying Study of Basic Problems in Engineering and Physical Science by Means of Geometry* Vol. 1, Gakuyusty Bunken Fukin-Kay, Tokyo.

Kröner, E. (1955), Der fundamentale Zusammenhang zwischen Versetzungsdichte und Spannungsfunktionen, *Z. Phys.*, Vol. 142, 463-475.

Kröner, E. (1956), Die Versetzung als elementare Eigenspannungsquelle, *Z. Naturforschung*, 2a, 969-985.

Kröner, E. (1960), Allgemeine Kontinuumstheorie der Versetzungen und Eigenspannungen, *Arch. Rational Mech. Anal.* Vol. 4, 273-334.

Lardner, R. W. (1974), *Mathematical theory of dislocations and fracture*, University of Toronto Press.

Love, A. E. H. (1927), *A Treatise on the mathematical theory of elasticity*, Cambridge University Press, Cambridge.

Malen, K. (1971), A unified six-dimensional treatment of elastic Green's functions and dislocations, *Phys. Stat. Solidi*, Vol. B44, 661-672.

Mindlin, R. D. (1964) Micro-structure in linear elasticity, *Arch. Rat. Mech. Anal.* Vol. 16, 51-78.

Movchan, A. B., Bullough, R., and Willis, J. R. (1998), Stability of a dislocation: Discrete model, *Eur. J. Appl. Math.*, Vol. 9, PT4, 373-396.

Mügge, O. (1883), Über neue Structurflachen an den Krystallen der gediegenen Metalle, *Neues Jahrb. Min.*, Vol. 13, 56-63.

Mura, T. (1987), *Micromechanics of defects in solids*, Martinus Nijhoff Publishers, Boston; First edition, 1982.

Muskhelishvili, N. I. (1956), *Some basic problems of the mathematical theory of elasticity,* Translated from the 3rd. Russian edition by J. R. M. Radok, Noordhoff, Groningen.

Nabarro, F. R. N. (1947), Dislocations in a simple cubic lattice, *Proc. Phys. Soc.*, Vol. 59, 256-272.

Nabarro, F. R. N. (1967), *Theory of crystal dislocation*, Clarendon, Oxford.

Nabarro, F. R. N. (1979-1989), *Dislocations in solids*, North-Holland, Amsterdam.

Nakada, Y. and Keh, A. S. (1966), Latent hardening in iron single crystals, *Acta Metall.*, Vol. 14, 961-973.

Nemat-Nasser, S. (1968-1970), *Notes on elasticity*, University of California, San Diego.

Nemat-Nasser, S. (1997), Plasticity: Inelastic flow of heterogeneous solids at finite strains and rotations, *Proceeding of the ICTAM*, Kyoto, Japan, August 25-31, 1996, Elsevier Science, 25-31.

Nemat-Nasser, S. and Chang S. N. (1990), Compression-induced high strain rate void collapse, tensile cracking, and recrystallization in ductile single and polycrystals, *Mech. Mat.*, Vol. 10, 1-17.

Nemat-Nasser, S. and Hori, M. (1987), Void collapse and void growth in crystalline solids, *J. Appl. Phys.*, Vol. 62, 2746-2757.

Nemat-Nasser, S. and Hori, M. (1993), *Micromechanics: overall properties of heterogeneous solids,* Elsevier, Amsterdam; second revised edition (1999), Elsevier, Amsterdam.

Nemat-Nasser, S. and Isaacs, J. B. (1997), Direct measurement of isothermal flow stress of metals at elevated temperatures and high strain rates with application to Ta and Ta-W alloys, *Acta Mater.*, Vol. 45, No. 3, 907-919.

Nemat-Nasser, S. and Li, Y. (1998), Flow stress of fcc polycrystals with application to OFHC Cu, *Acta Mater.*, Vol. 46, No. 2, 565-577.

Nemat-Nasser, S. and Obata, M. (1986), Rate dependent, finite elasto-plastic deformation of polycrystals, *Proc. R. Soc. Lond. A*, Vol. 407, 343-375.

Nemat-Nasser, S. and Okinaka, T. (1996), A new computational approach to crystal plasticity: fcc single crystals, *Mech. Mat.*, Vol. 24, 43-58.

Nemat-Nasser, S. and Sakane, M. (1980), Unpublished work, Northwestern University.

Nemat-Nasser, S., Isaacs, J. B., and Liu, M. (1998a), Microstructure of high-strain, high-strain-rate deformed tantalum, *Acta Mater.*, Vol. 46, No. 4, 1305-1325.

Nemat-Nasser, S., Okinaka, T., and Ni, L. (1998b), A physically-based constitutive model for bcc crystals with application to polycrystalline tantalum, *J. Mech. Phys. Solids*, Vol. 46, No. 6, 1009-1038.

Nemat-Nasser, S., Okinaka, T., and Ni, L. (1998c), A constitutive model for fcc crystals with application to polycrystalline OFHC copper, *Mech. Mat.*, Vol. 30, 325-341.

Nemat-Nasser, S., Okinaka, T., Nesterenko, V., and Liu, M. (1998d), Dynamic void collapse in crystals: computational modeling and experiments, *Phil. Mag. A*, Vol. 78, No. 5, 1151-1174.

Ni, L. and Nemat-Nasser, S. (1996), A general duality principle in elasticity, *Mech. Mat.*, Vol. 24, 87-123.

Nowacki, W. K. and Zarka, J. (1971), Étude de la frontiere élastique des monocristaux d'aluminum, *Int. J. Solids Struct.*, Vol. 7, 1277-1287.

Nye, J.F. (1953), Some geometrical relations in dislocated solids, *Acta Metall.* Vol. 1, 153-162.

Okinaka, T. (1995), *A new computational approach to crystal plasticity: fcc crystals*, Ph.D. thesis, University of California, San Diego.

Orowan, E. (1934), Zur Kristallplastizität-I, II, III, *Zeit. Phys.*, Vol. 89, 605-613, 614-633, 634-659.

Orowan, E. (1940), Problems of plastic gliding, *Philos. Trans. R. Soc. London A*, Vol.52, 8-22.

Pan, J. and Rice, J. R. (1983), Rate sensitivity of plastic flow and implications for yield-surface vertices, *Int. J. Solids Struct.*, Vol. 19, 973-987.

Peach, M. and Koehler, J. S. (1950), The forces exerted on dislocations and the stress fields produced by them, *Phys. Rev.*, Vol. 80, 436-439.

Peierls, R. (1940), The size of a dislocation, *Proc. Phys. Soc.*, Vol. 52, 34-37.

Peirce, D., Asaro, R. J., and Needleman, A. (1983), Material rate dependence and localized deformation in crystalline solids, *Acta Metall.*, Vol. 31, 1951-1976.

Peirce, D., Shih, C. F., and Needleman, A. (1984), A tangent modulus method for rate dependent solids, *Comput. Struct.*, Vol. 18, 875-887.

Polanyi, M. (1934), Über eine Art Gitterstörung, die einem Kristall plastisch machen konnte, *Zeit. Phys.*, Vol. 89, 660-664.

Ramaswami, B., Kocks, U. F., and Chalmers, B. (1965), Latent hardening in silver and an Ag-Au alloy, *Trans. Metall. Soc. AIME,* Vol. 233, 927-931.

Rashid, M. M. (1990), High-strain-rate flow in metallic single crystals, *Ph.D. Thesis*, University of California, San Diego.

Rashid, M. M. and Nemat-Nasser, S. (1992), A constitutive algorithm for rate-dependent crystal plasticity, *Comput. Meth. Appl. Mech. Eng.*, Vol. 94, 201-

228.
Rashid, M. M., Gray, G. T., and Nemat-Nasser, S., (1992), Heterogeneous deformations in copper single crystals at high and low strain rates, *Phil. Mag.*, Vol. 65, No. 3, 707-735.
Read, W. T. (1953), *Dislocations in crystals*, McGraw-Hill, New York.
Reed-Hill, R. E. and Abbaschian, R. (1992), *Physical metallurgy principles*, PWS-Kent, Boston.
Rice, J. R. (1971), Inelastic constitutive relations for solids: an internal-variable theory and its application to metal plasticity, *J. Mech. Phys. Solids*, Vol. 19, 433-455.
Schmid, E. (1926), Über die Schubverfestigung von Einkristallen bei plasticher Deformation, *Z. Phys.*, Vol. 40, 54-74.
Schmid, E. and Boas, W. (1935), *Kristallplastizität*, Springer-Verlag, Berlin.
Schmid, E. and Boas, W. (1950), *Plasticity of crystals*, Hughes, London (originally published in 1935, *ibid.*).
Sestak, B. and Zarubova, N. (1965), *Phys. Stat. Solidi*, Vol. 10, 239-250.
Sherwood, P. J., Guiu, F., Kim, H. C., and Pratt, P. L. (1967), Plastic anisotropy of tantalum, niobium, and molybdenum, *Can. J. Phys.*, Vol. 45, 1075-1089.
Simons, G., Weippert, Ch., Dual, J., and Villain, J. (2001), Investigating size effects on mechanical properties: preliminary work and results for thin copper foils, *Proceedings of Materialsweek 2001*, 1-4 Oct., International Congress Centre, Munich.
Simons, G., Weippert, Ch., Dual, J., and Villain, J. (2002), Experimental investigations of size effects in thin copper foils, *Proceedings of IUTAM Symposium on Multiscale Modeling and Characterization of Elastic-Inelastic Behavior of Engineering Materials*, Marrakech, 20-25 Oct.
Sokolnikoff, I. S. (1956), *Mathematical theory of elasticity*, McGraw-Hill, New York.
Somigliana, C. (1914), Sulla teoria delle distorsioni elastiche, *Rend. C. Accad. Lincei*, Vol. 23, 463-472.
Stein, D. F. (1967), The effect of orientation and impurities on the mechanical properties of molybdenum single crystals, *Can. J. Phys.*, Vol. 45, 1063-1074.
Stroh, A. N. (1952), The mean shear stress in an array of dislocations and latent hardening, *Proc. Phys. Soc. B*, Vol. 66, 2-6.
Stroh, A. N. (1953), A theoretical calculation of the stored energy in a work-hardened material, *Proc. Phys. Soc. B*, Vol. 67, 391-400.
Stroh, A. N. (1958), Dislocations and cracks in anisotropic elasticity, *Phil. Mag.*, Vol. 3, 625-646.
Taoka, T., Takeuchi, S., and Furubayashi, E. (1964), Slip systems and their critical shear stress in 3-percent silicon iron, *J. Phys. Soc. Japan* , Vol. 19, 701-711.
Taylor, G. I. (1934), The mechanism of plastic deformation of crystals - I , II, *Proc. R. Soc. Lond. A*, Vol. 145, 362-387, 388-404.
Taylor, G. I. (1938), Plastic strain in metals, *J. Inst. Metals*, Vol. 62, 307-324.

Taylor, G. I. and Elam, C. F. (1925), The plastic extension and fracture of aluminium crystals, *Proc. R. Soc. Lond. A*, Vol. 108, 28-51.

Taylor, G. I. and Elam, C. F. (1926), The distortion of iron crystals, *Proc. R. Soc. Lond. A*, Vol. 112, 337-361.

Teodosiu, C. and Sidoroff, F. (1976), A theory of finite elastoviscoplasticity of single crystals, *Int. J. Eng. Sci.*, 14, 165-176.

Timoshenko, S. P. and Goodier, J. N. (1970), *Theory of elasticity*, McGraw-Hill, New York.

Timpe, A. (1905), *Zeit. Math. Phys.*, Vol. 52, 348.

Ting, T. C. T. (1996), *Anisotropic elasticity*, Oxford University Press, Oxford.

Toupin, R. A. (1962), Elastic materials with couple stresses, *Arch. Rat. Mech. Anal.*, Vol. 11, 385-414.

Van der Giessen, E., Needleman, A. (1995), Discrete dislocation plasticity: a simple planar model, *Modell. Simul. Mat. Sci. Engin.*, Vol. 3, 689-735.

Volterra, V. (1907), Sur l'équilibre des corps élastiques multiplement connexes, *Ann. Sci. l'École Normale Super. Paris*, Vol. 24, 401-517; see also Volterra, V. and Volterra E., Sur les dislocations des corps élastiques, in Mémorial des Sciences Mathématiques, fasc. 147, Gauthier-Villars, Paris (1960).

Weertman, J. and Weertman, J. R. (1992), *Elementary dislocation theory*, Oxford University Press, London.

Wu, T. Y., Bassani, J. L. and Laird, C. (1991), Latent hardening in single crystals I. theory and experiments, *Proc. R. Soc. Lond. A*, Vol. 435, 1-19.

Zarka, J. (1968), Sur la viscoplasticité des métaux, *PhD Thesis*, Paris.

Zarka, J. (1975), Modelling of changes of dislocation structures in monotonically deformed single phase crystals, in *Constitutive equations in plasticity* (Argon, A. S. ed.), MIT Press, Cambridge, Massachusetts.

Zikry, M. A. (1990), High strain-rate deformation and failure of crystalline materials, *Ph.D. Thesis*, University of California, San Diego.

Zikry, M. A. and Kao, M. (1996a), Dislocation based multiple-slip crystalline constitutive formulation for finite-strain plasticity, *Scripta Mat.*, Vol. 34, 1115-1121.

Zikry, M. A. and Kao, M. (1996b), Inelastic microstructural failure mechanics in crystalline materials with high angle grain boundaries, *J. Mech. Phys. Solids*, Vol. 44, 1765-1798.

Zikry, M. A. and Nemat-Nasser, M. (1990), High strain rate localization and failure of crystalline materials, *Mech. Mat.*, Vol. 10, No. 3, 215-237.

APPENDIX 6.A: MILLER INDICES

The unit cell of a periodic lattice is defined by three noncoplanar lattice vectors, $\mathbf{a}_i$, i = 1, 2, 3; see Figure 6.A.1. The reciprocal lattice is defined by the corresponding three reciprocal vectors, say, $\mathbf{b}^i$; see Subsection 1.3.1, page 19. The lattice vectors and their reciprocals are related by

$$\mathbf{a}_i \cdot \mathbf{b}^j = \delta_i^j . \tag{6.A.1a}$$

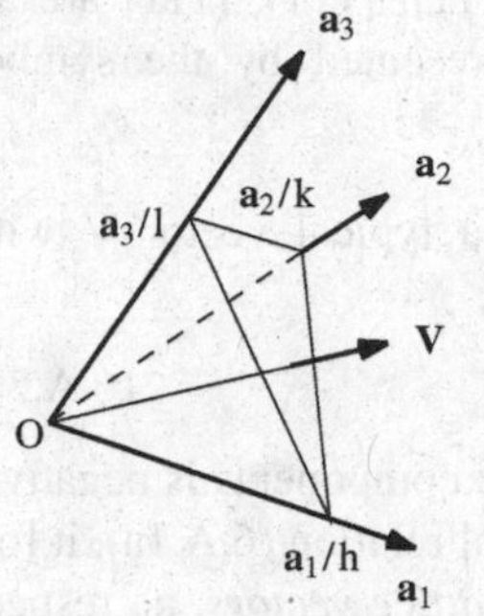

Figure 6.A.1

A unit cell defined by lattice vectors $\mathbf{a}_i$, i = 1, 2, 3, and the plane (hkl) normal to $\mathbf{V} = h\,\mathbf{b}^1 + k\,\mathbf{b}^2 + l\,\mathbf{b}^3$

Hence, it follows that

$$\mathbf{b}^1 = \frac{\mathbf{a}_2 \times \mathbf{a}_3}{V},$$

$$\mathbf{b}^2 = \frac{\mathbf{a}_3 \times \mathbf{a}_1}{V},$$

$$\mathbf{b}^3 = \frac{\mathbf{a}_1 \times \mathbf{a}_2}{V}, \tag{6.A.1b-d}$$

where $V = \mathbf{a}_1 \cdot \mathbf{a}_2 \times \mathbf{a}_3$ is the volume of the unit cell formed by the lattice vectors. With the aid of permutation symbol, e_{ijk}, (6.A.1b-d) are more concisely written as,

$$\mathbf{b}^i = \frac{1}{2\,V}\, e_{ijk}\, \mathbf{a}_j \times \mathbf{a}_k ,$$

$$\mathbf{a}_i \times \mathbf{a}_j = V\, e_{ijk}\, \mathbf{b}_k . \tag{6.A.1e,f}$$

Miller Indices for Directions: A direction can be defined by a vector, say **v**, through the origin of the unit cell. Let v_i define the components of this vector in the lattice coordinates,

$$\mathbf{v} = v_1\, \mathbf{a}_1 + v_2\, \mathbf{a}_2 + v_3\, \mathbf{a}_3 . \tag{6.A.2a}$$

In general, these components need not be integers or even rational numbers. In application to crystals, however, it is generally assumed that the directions of interest can be defined with rational components. Hence, multiplying the vector **v** by the least common denominator of its (rational) components, say m, a new vector m **v** is obtained which has integer components,

$$m\, v_1 = u , \quad m\, v_2 = v , \quad m\, v_3 = w , \tag{6.A.2b-d}$$

and defines the same direction. It is customary to represent this direction by [uvw]. If any of these integers is negative, then it is denoted by a superimposed bar. For example, if $m\,v_3 = -\,|m\,v_3|$, then set $w = |m\,v_3|$ and define this direction by $[uv\bar{w}]$.

Often a *class* of directions with certain characteristics may be of interest. For example, in bcc crystals, the directions [111], $[\bar{1}11]$, $[1\bar{1}1]$, $[11\bar{1}]$ are all potential slip directions. These are collectively represented by the symbol $<111>$, in the Miller notation.

Miller Indices for Planes: Let the components of a typical vector **V** with respect to the *reciprocal lattice* be given by (h, k,l), *i.e.,*

$$\mathbf{V} = h\,\mathbf{b}^1 + k\,\mathbf{b}^2 + l\,\mathbf{b}^3\,. \tag{6.A.3a}$$

Assume that h, k, and l are integers; if any one of these components is negative, then it must carry a superposed bar. From the reciprocal relation (6.A.1a), it follows that **V** is normal to a plane which intersects the *lattice vectors*, $\mathbf{a}_i$, respectively at $\mathbf{a}_1/h$, $\mathbf{a}_2/k$, and $\mathbf{a}_3/l$. This plane is denoted by (hkl), in Miller notation. Hence, in general, a crystal plane is identified by three integers, h, k, and l, which are the reciprocals of the intercepts of the plane with the lattice vectors. According to this notation, the octahedral plane shown in Figure 6.5.2, page 454, is designated as the (111)-plane.

From Figure 6.A.1, the distance, d, from O to the (hkl)-plane is obtained by taking the inner product of, say, $\mathbf{a}_1/h$ with the unit normal to this plane. Since **V** is normal to the (hkl)-plane, it follows from (6.A.1b) and (6.A.3a) that,

$$d = \frac{1}{|\mathbf{V}|}, \tag{6.A.3b}$$

where $|\mathbf{V}|$ is the length of **V**. Furthermore, the angle, ϕ, between this and another plane, say, a plane with Miller indices (uvw), is readily calculated from

$$\cos\phi = \frac{\mathbf{V}\cdot\mathbf{W}}{|\mathbf{V}|\;|\mathbf{W}|},$$
$$\mathbf{W} = u\,\mathbf{b}^1 + v\,\mathbf{b}^2 + w\,\mathbf{b}^3\,. \tag{6.A.3c,d}$$

As in the case of directions, a class of planes often may need to be identified, *e.g.*, the class of all octahedral planes in an fcc crystal. This is then denoted by {111} which represents all four octahedral planes, namely, (111), $(\bar{1}11)$, $(1\bar{1}1)$, and $(11\bar{1})$. Hence, {hkl} defines the set of all planes whose reciprocal intercepts are a combination of any three distinct components taken from $(\pm h,\ \pm k,\ \pm l)$.

hcp Crystals: The basal plane of an hcp crystal is represented by two lattice vectors, $\mathbf{a}_1$ and $\mathbf{a}_2$, making a 60° angle with one another. The direction normal to the basal plane is the so-called c-axis. The three vectors, $\mathbf{a}_1$, $\mathbf{a}_2$, and **c** define the lattice of an hcp crystal.

It is, however, customary to include a third vector, $\mathbf{a}_3 = -(\mathbf{a}_1 + \mathbf{a}_2)$, to the basal plane, and to represent planes in an hcp crystal by four indices. This additional index is therefore, the negative of the sum of the first two indices. For example, the plane of Miller indices (hkl) with respect to the ($\mathbf{a}_1$, $\mathbf{a}_2$, $\mathbf{c}$)-coordinate vectors, is represented by (hkil), where $i = -(h + k)$.

Zone Axis: When several planes intersect on a common line, this line is called their *zone axis*. Let the direction of this line be defined by [uvw]. Since a vector normal to a plane in this set of planes, is also normal to the zone axis, it follows that the Miller indices, (hkl), of any of these planes must satisfy

$$uh + vk + wl = 0. \tag{6.A.4}$$

7

FINITE PLASTIC DEFORMATION OF GRANULAR MATERIALS

The fundamentals of finite elastoplastic deformation of densely packed granular materials which carry the applied loads through contact friction are discussed, integrating the material behavior at three distinct length scales, i.e., the micro-scale, the meso-scale, and the macro-scale. Considering the contact forces, contact normals, and branches (which are vectors connecting the centroids of two adjacent contacting granules) at the micro-scale, expressions for the overall stress tensor and the tractions transmitted across interior macroscopic planes, are developed. Macroscopic parameters which characterize the microstructure and its evolution in the course of deformation, are identified in terms of fabric tensors and the distribution density functions of unit branch vectors and contact normals, leading to explicit relations between the stress and fabric tensors. At the meso-scale, the deformation is assumed to consist of dilatant simple shearing over several interacting sliding planes. Resistance to such simple shearing is provided by friction due to rolling and sliding of granules at the micro-scale, and restructuring through redistribution of the contact normals, i.e., the change in the fabric. An explicit expression is obtained for the critical value of the resistive shear stress, by integrating the corresponding contact forces over a typical sliding plane. Finally, physically-based continuum constitutive relations are obtained, based on the double sliding-plane model which includes dilatancy (densification), pressure sensitivity, and — most importantly — the anisotropy, measured by a fabric tensor which enters into the resulting continuum constitutive relations through a backstress.

This chapter is organized as follows. Some basic physical features of the response of frictional granular materials, when cyclically sheared under confining pressures, are examined in Section 7.1, illustrating the influence of the induced anisotropy or fabric on the accompanying dilatancy or densification. An expression for the overall continuum stress tensor is developed in Section 7.2, in terms of the contact forces and microstructural parameters which characterize the fabric of the granular mass. Various fabric measures are examined in Section 7.3, including second- and fourth-order tensors, which define the distribution of unit contact normals, unit branch vectors, their properly weighted representations, and other relevant vectorial quantities which define the interaction between contacting granules. A number of fabric measures are then presented for densely-packed rigid granules. Experimental results on rod-like photoelastic granules, subjected to cyclic shearing, as well as under mono-

tonically applied biaxial compression, are examined in Section 7.4, and various fabric measures are correlated with the corresponding stress and strain tensors.

These introductory sections are followed in Section 7.5 by a detailed development of physically-based constitutive relations, bringing into focus the interrelation between models of granular flow and plastic deformation of single crystals, wherever appropriate. Guided by various contributions in these areas, a unified constitutive formulation at the meso-scale is presented and some of its implications are discussed, including the close relation between the double-slip theory of single crystals and that of granular materials. While the elastoplasticity of crystalline solids and that of frictional granules have many common features in their kinematics, dynamics, and kinetics, there are also a number of fundamental differences which are also examined in this section which includes new results on the double-sliding model of frictional granules. Crystals in crystalline solids consist of a well-defined, periodic arrangement of atoms, held together by interatomic bonding forces. Granular masses, on the other hand, support the applied loads through intergranular contact forces which may diminish to zero in the absence of a confining pressure. In an assembly of cohesionless particulates, the intergranular frictional forces are developed to resist sliding and rolling, and hence, the granular medium bears the imposed stresses through frictional resistance. These and related issues are studied in this section.

7.1. INELASTIC DEFORMATION OF GRANULAR MASSES

Some features peculiar to the deformation of granular masses which carry the applied loads through contact friction, are considered in Section 4.7, from a macroscopic (continuum) point of view. In the present section, these and related features are reexamined within a micromechanical framework which serves to highlight the basic role that the microstructure and its evolution play in defining the overall response and failure modes of this class of materials.

Photoelastic granules have been used to demonstrate the interaction among granules, as a granular mass is deformed in a biaxial loading, or in simple shearing under a confining pressure. The results of model experiments of this kind are discussed in Section 7.3. Here, consider Figures 7.1.1a,b which show the initial and a deformed configuration of a randomly packed collection of rod-like photoelastic granules of oval cross section. (b)

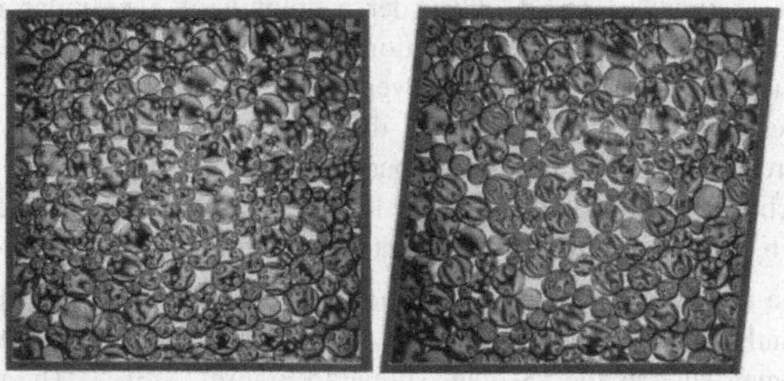

Figure 7.1.1

Simple shearing of rod-like photoelastic granules of oval cross section, under constant confining pressure and their own weight; (a) initial configuration; (b) configuration after a continuous simple shearing

In Figure 7.1.1(a), the granules are under a uniform confining pressure (as well as their own weight). Upon simple shearing, the granular mass deforms into the configuration shown in Figure 7.1.1(b). As is discussed in Section 7.3, the experimental setup allows the granular mass to expand or contract as it is sheared under an applied constant confining pressure. The fringe patterns seen in these figures, illustrate how the distribution of the orientation and magnitude of the contact forces change with the applied loads. Figure 7.1.2 shows the *rose diagram* of the distribution of the unit contact normals corresponding to the configurations shown in Figures 7.1.1a,b. As is seen, shearing produces a biased distribution, as the granules tend to slide and roll on one another. This *induced anisotropy or fabric*, is at the basis of the micromechanics of the deformation of granular media. In the sequel, through a simple micromechanical model, the profound influence of the induced fabric on dilatancy or densification of frictional granules, is quantified.

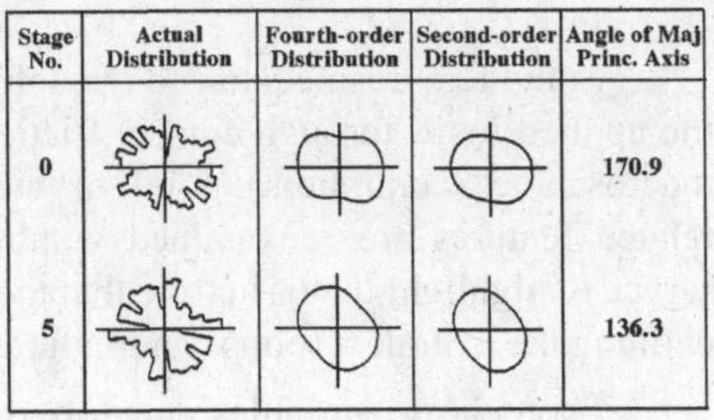
Stage No.
Actual Distribution
Fourth-order Distribution
Second-order Distribution
Angle of Maj Princ. Axis
0
170.9
5
136.3

Figure 7.1.2

Rose diagram of the distribution of the unit contact normals corresponding to the configurations shown in Figures 7.1.1a,b

7.1.1. Dilatancy, Densification, and Fabric in Simple Shearing

Examine plane-strain simple shearing of a layer of frictional granules under a constant normal stress, σ, as sketched in Figure 7.1.3(a). The assumption of plane strain is appropriate to the deformation of the rod-like photoelastic granules of Figures 7.1.1a,b.[1] Consider a column of a granular material of height h, measured along the x_2-direction, and of area a, being sheared in the x_1-direction; Figure 7.1.3(b). For a layer which extends to infinity in all horizontal directions, shearing in the x_1-direction does not change the cross-sectional area of a typical column. Therefore, volumetric changes occur only due to the change in the column's height h. The volumetric strain rate, $\dot{v}/v$, then is $\dot{v}/v = \dot{h}/h$, where the superimposed dot denotes time-rate-of-change.

While the overall shearing is in the x_1-direction, the overall deformation of the granular mass actually occurs by sliding and rolling of granules over inclined planes, such as the SS-plane shown in Figure 7.1.3(c). The motion of a typical granule relative to its neighboring granules, contributes to the overall volumetric expansion, if its *dilatancy angle,* ν, is positive. The dilatancy angle ν defines the direction of the motion of this granule, along the SS-plane, in relation to its neighboring granules.[2]

Since the local movements of granules must be accommodated by all the neighboring granules, it is appropriate to *divide active contacts into* Q *classes, according to their dilatancy angles.* Thus, *all contacts whose dilatancy angles are in the range* $\nu^\alpha \pm \Delta\nu$, *belong to class* α, *being identified by the dilatancy angle* ν^α, *where* $\Delta\nu$ *is some suitable pre-assigned small angle.* At each instant, the macroscopic volume v, contains a large number, N, of active contacts per its unit volume, each belonging to one of the Q classes. Let N^α be the number of such contacts per unit volume, in class α, and set

---

[1] Simple shearing of granular materials can be attained in a large hollow circular cylindrical sample of relatively thin wall thickness. The shearing is produced by the torsion of the sample about its cylindrical axis; see Chapter 9 and also Okada and Nemat Nasser (1994).

[2] For shearing in the *negative* x_1-direction, active contacts with *negative* dilatancy angles produce volumetric expansion.

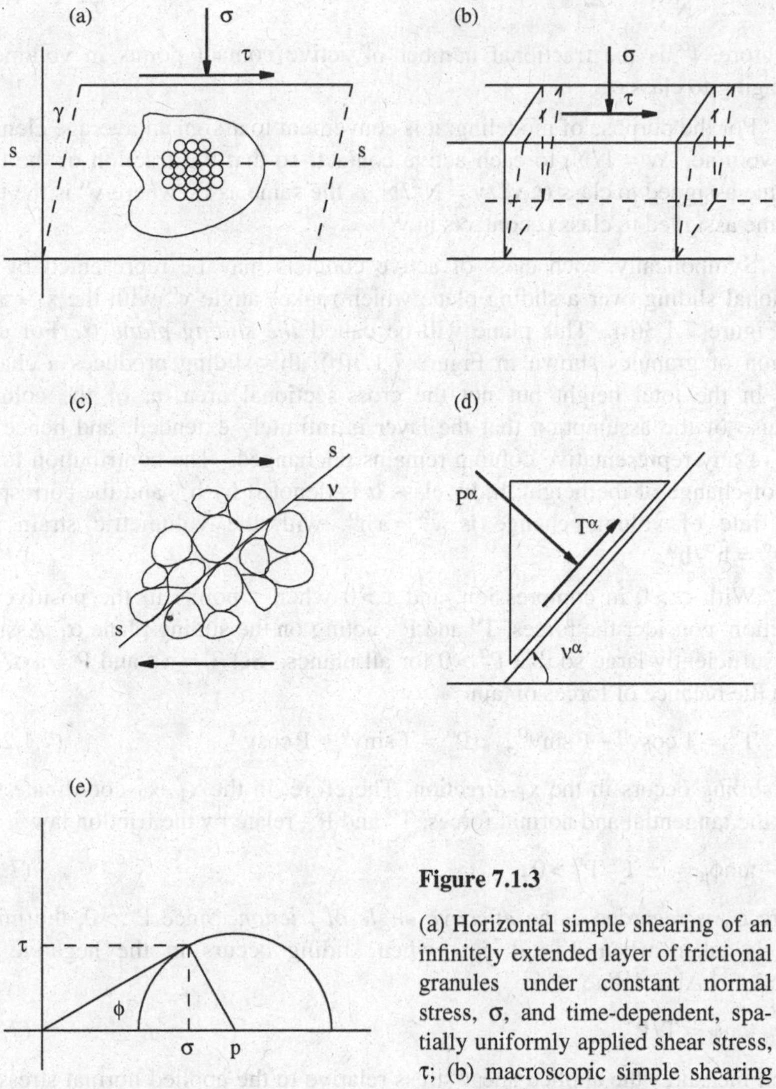

Figure 7.1.3

(a) Horizontal simple shearing of an infinitely extended layer of frictional granules under constant normal stress, σ, and time-dependent, spatially uniformly applied shear stress, τ; (b) macroscopic simple shearing of a typical column of cross-sectional area a and height h, being sheared in the x_1-direction, resulting in volumetric strain rate $\dot{v}/v = \dot{h}/h$; (c) microscopic flow of granules occurring over inclined planes, such as SS, with ν being the corresponding *dilatancy angle*; (d) a class of active contacts being represented by frictional sliding plane α, with dilatancy angle ν^α, and subjected over its cross-sectional area to resultant shear force T^α and normal force P^α; and (e) the Mohr-Coulomb failure criterion, relating the pressure, p, to the shear stress, τ, by $p = \tau/(\sin\phi \cos\phi)$

$$f^\alpha = \frac{N^\alpha}{N}, \qquad \sum_{\alpha=1}^{Q} f^\alpha = 1. \tag{7.1.1a,b}$$

Therefore, f^α is the fractional number of active contact points in volume v, belonging to class α.

For the purpose of modeling, it is convenient to assign an average elementary volume, $\Delta v = 1/N$, to each active contact, so that the fraction of the total volume assigned to class α, $v^\alpha/v = N^\alpha/N$, is the same as f^α, where v^α is the total volume assigned to class α contacts in v.

Symbolically, each class of active contacts may be represented by the frictional sliding over a sliding plane which makes angle ν^α with the x_1 – axis; see Figure 7.1.3(d). This plane will be called *the sliding plane* α. For each column of granules shown in Figure 7.1.3(b), this sliding produces a change only in the total height but not the cross-sectional area, a, of the column, because of the assumption that the layer is infinitely extended, and hence the area of any representative column remains unchanged. The contribution to the rate-of-change of the height, $\dot{h}$, by class α is denoted by $\dot{h}^\alpha$, and the corresponding rate of volume change is $\dot{v}^\alpha = a\,\dot{h}^\alpha$, with the volumetric strain rate $\dot{v}^\alpha/v^\alpha = \dot{h}^\alpha/h^\alpha$.

With $\sigma > 0$ in compression, and $\tau > 0$ when τ points in the positive x_1-direction, consider the forces, T^α and P^α, acting on the sliding plane α. Assume σ is sufficiently large so that $P^\alpha > 0$ for all planes. Set $T = a\tau$ and $P = a\sigma$, and from the balance of forces obtain

$$T^\alpha = T\cos\nu^\alpha - P\sin\nu^\alpha, \qquad P^\alpha = T\sin\nu^\alpha + P\cos\nu^\alpha. \tag{7.1.2a,b}$$

The sliding occurs in the x_1^*-direction. Therefore, in the x_1^*, x_2^*-coordinate system, the tangential and normal forces, T^α and P^α, relate by the friction law,

$$\tan\phi_\mu = \pm T^\alpha/P^\alpha > 0, \tag{7.1.3}$$

where ϕ_μ is regarded as an *effective angle of friction*. Since $P^\alpha > 0$, the minus sign is taken when $T^\alpha < 0$, *i.e.*, when sliding occurs in the negative x_1^*-direction.[3] Also define

$$\tan\phi = T/P \tag{7.1.4}$$

which measures the applied shear stress relative to the applied normal stress. It characterizes the applied loads.

From (7.1.2a,b), (7.1.3), and (7.1.4), now obtain (Nemat-Nasser, 1980)

$$\tan\phi_\mu = \pm\tan(\phi - \nu^\alpha) \quad \text{or} \quad \phi_\mu = \pm(\phi - \nu^\alpha) > 0. \tag{7.1.5a,b}$$

This is the critical condition, defining the loading state at which class α contacts are activated. It shows that, *for shearing in the positive (negative)* x_1*-direction, contacts with negative (positive) dilatancy angles attain the critical state for*

[3] Even for $\tau > 0$, this may occur for suitably large σ and $\nu^\alpha > 0$.

sliding first; this is also inherent in a model developed by Rowe (1962). It also shows that, under large enough vertical pressure, *reverse sliding can occur against an applied shear stress.*

Consider the rate of work done as the sliding plane α, slips at a rate denoted by $\dot{l}^\alpha$, in the x_1^*-direction. From the geometry, $\dot{l}^\alpha = \dot{h}^\alpha/\sin\nu^\alpha$. In view of (7.1.2a,b) and (7.1.5b), it follows that the rate of work per unit volume associated with the α'th class of active contacts, is

$$\dot{w}^\alpha = \frac{T^\alpha \dot{l}^\alpha}{v^\alpha} = \frac{\tau \sin\phi_\mu}{\sin(\phi_\mu \pm \nu^\alpha)\sin\nu^\alpha}\,\frac{\dot{v}^\alpha}{v^\alpha}, \tag{7.1.6a,b}$$

where the minus sign corresponds to the critical condition $\phi_\mu = -(\phi - \nu^\alpha) > 0$.

To relate the local pressure associated with the α'th sliding plane, to the applied shear and normal stresses, use the Mohr-Coulomb failure criterion, shown in Figure 7.1.3 (e). As is seen, the pressure, p, relates to the shear stress, τ, by $p = \tau/(\sin\phi\cos\phi)$. Therefore, if $\dot{\gamma}^\alpha$ is the associated rate of shearing, the rate of work per unit volume becomes[4]

$$\dot{w}^\alpha = \tau\dot{\gamma}^\alpha - p\,\dot{v}^\alpha/v^\alpha = \tau\dot{\gamma}^\alpha - \frac{\tau}{\sin\phi\cos\phi}\,\frac{\dot{v}^\alpha}{v^\alpha}. \tag{7.1.7a,b}$$

From (7.1.6b) and (7.1.7b), it follows that

$$\frac{\dot{v}^\alpha}{v^\alpha} = \frac{\cos(\phi_\mu \pm \nu^\alpha)\sin\nu^\alpha}{\cos\phi_\mu}\,\dot{\gamma}^\alpha \tag{7.1.8a}$$

The total rate of volumetric strain rate is

$$\frac{\dot{v}}{v} = \sum_{\alpha=1}^{Q} \frac{\dot{v}^\alpha}{v} = \sum_{\alpha=1}^{Q} f^\alpha\,\frac{\dot{v}^\alpha}{v^\alpha}, \tag{7.1.8b,c}$$

leading to the following dilatancy equation:

$$\frac{1}{v}\,\frac{\dot{v}}{\dot{\gamma}} = \sum_{\alpha=1}^{Q} \frac{\dot{\gamma}^\alpha}{\dot{\gamma}}\,f^\alpha\,\frac{\cos(\phi_\mu \pm \nu^\alpha)\sin\nu^\alpha}{\cos\phi_\mu}. \tag{7.1.9}$$

For the simple shearing of Figure 7.1.3(a), and based on the sliding model of Figure 7.1.3(d), expression (7.1.9) is a reasonable micromechanical representation. The basic assumptions leading to this result are: (1) the Mohr-Coulomb failure criterion applies to each class of sliding planes; and (2) the effective friction angle, ϕ_μ, is constant. Both assumptions are reasonable.

Consider now a third assumption which has been extensively used in the calculation of the mechanical behavior of polycrystalline solids and composites, and which dates back to the work of Taylor (1934, 1938); see Chapter 8. This assumption is that the shear strain rate, $\dot{\gamma}^\alpha$, of each class, α, of the sliding plane, is equal to the overall average strain rate, $\dot{\gamma}$. Hence, with $\dot{\gamma}^\alpha \approx \dot{\gamma}$, (7.1.9)

---

[4] Note that pressure, p, is positive, and volumetric strain rate, $\dot{v}^\alpha/v^\alpha$, is positive when it represents expansion.

becomes (Nemat-Nasser, 1980)

$$\frac{1}{v}\frac{\dot{v}}{\dot{\gamma}} = \sum_{\alpha=1}^{Q} f^{\alpha}\frac{\cos(\phi_\mu \pm \nu^{\alpha})\sin\nu^{\alpha}}{\cos\phi_\mu}. \tag{7.1.10}$$

In a sample of a granular mass, there are a large number of granules. It is thus appropriate to consider a continuous distribution of the dilatancy angles, characterized by a distribution density function, $f(\nu)$, with $\nu^- \leq \nu \leq \nu^+$, and replace the right-hand side of (7.1.10) by the following integral representation:

$$\frac{1}{v}\frac{\dot{v}}{\dot{\gamma}} = \int_{\nu^-}^{\nu^+} f(\nu)\frac{\cos(\phi_\mu \pm \nu)\sin\nu}{\cos\phi_\mu}\,d\nu, \tag{7.1.11a}$$

where $f(\nu)\,d\nu$ is the volume fraction of active granules whose dilatancy angles are between ν and $\nu + d\nu$, so that

$$\int_{\nu^-}^{\nu^+} f(\nu)\,d\nu = 1. \tag{7.1.11b}$$

In (7.1.11a), the positive sign is used when $\dot{\gamma} > 0$, and the negative sign when $\dot{\gamma} < 0$. Note that, $f(\nu)$ relates only to the active contact points. For the plane strain deformation of cylindrical granules, ν actually equals the orientation of the corresponding contact normal, measured from the vertical x_2-axis.

Konishi (1978) has measured the distribution of the orientation of *all* contact normals in simple shearing of photoelastic cylindrical granules. Some of his results which have also been reported in Oda *et al.* (1980), are reproduced in Figure 7.1.4; note that $\tau < 0$ and $\dot{\gamma} < 0$ in these figures. His measurements show that the distribution of contact normals is initially isotropic, but develops strong bias as the sample is sheared in the negative x_1-direction. Experimental results by the present author and coworkers support Konishi's results; see Figures 7.1.1 and 7.1.2, and Section 7.4.

Consider shearing in the positive x_1-direction, with $\tau > 0$ and $\dot{\gamma} > 0$. As has been pointed out before, sliding planes with negative dilatancy angles are initially activated, and, therefore, the distribution density function, $f(\nu)$, is initially biased toward negative dilatancy angles. Physically, this is because the normal force P hinders the sliding of granules with positive dilatancy angles, whereas it assists the granules with negative dilatancy angles. This explains the generally observed initial densification when granular masses are sheared under confining pressures. Note, however that, even if a distribution density function $f(\nu)$ is symmetric with respect to $\nu = 0$, and, hence, for which $|\nu^-| = \nu^+$, still the integral in (7.1.11a) is negative for $\dot{\gamma} > 0$ (with the plus sign) and positive for $\dot{\gamma} < 0$ (with the minus sign). Thus, (7.1.11a) incorporates, in a natural manner, the asymmetric influence of the normal stress on "upgoing" and "downgoing" granules, *i.e.,* on granules with *positive* and *negative dilatancy angles.*

As the shearing proceeds in the positive x_1-direction, the distribution function $f(\nu)$ becomes more biased toward the positive dilatancy angles.[5]

[5] This is also borne out by Konishi's results which show that larger fractions of contact normals

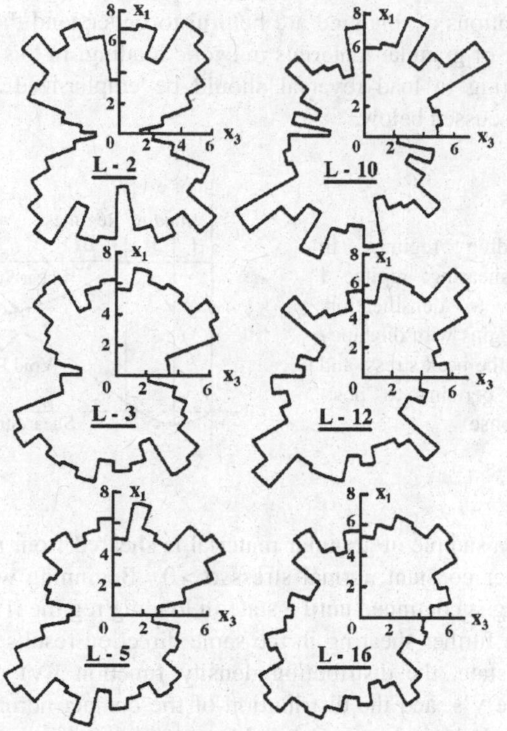

Figure 7.1.4

Changes of fabric during two-dimensional simple shearing (from right to left, *i.e.*, $\tau < 0$ and $\dot{\gamma} < 0$); symbols L-2, L-3...L-16 show the positions where photographs are taken (from Konishi, 1978)

Eventually, when a suitable bias toward positive dilatancy angles is attained, the corresponding integral in (7.1.11a) vanishes, as the granular mass attains a minimum void ratio (or maximum density). After this state, continual shearing in the same direction results in volume expansion, until the peak stress is reached, which, according to the theoretical considerations of Nemat-Nasser and Shokooh (1980), must correspond to a maximum rate of dilatancy. Upon further shearing, the rate of dilatancy begins to decrease in the post-failure regime, presumably becoming zero asymptotically at a critical state.

For monotonic shearing, the three loading regimes, initial densification (I), leading to dilatancy up to the peak stress (II), and then continuing in the

---

take on orientations in the first and third quadrants in Figure 7.1.4. Note that x_3 in Figure 7.1.4 is the same as the x_2-axis in Figure 7.1.3, and that, while the distribution function f(v) relates directly to the distribution of contact normals, it does not equal it.

post-failure behavior to a critical state (III), are shown schematically in Figure 7.1.5. Classifications of this kind are helpful to understand the effect of history on the behavior of granular materials in cyclic loading. In this regard, an important point relating to load reversal should be emphasized in the context of (7.1.11a), as discussed below.

Figure 7.1.5

Three loading regimes in monotone shearing: regime I corresponds to densification, regime II begins with dilatancy and ends at the peak stress, and regime III pertains to post-failure response

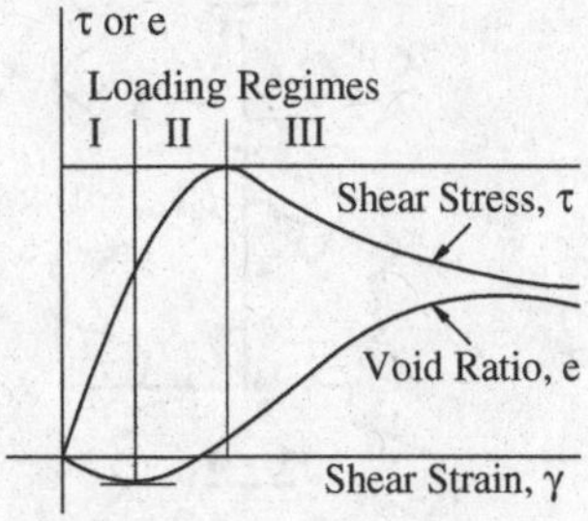

Suppose a sample of granular material is sheared from *right to left* ($\dot{\gamma} < 0$ and $\tau \leq 0$) under constant normal stress $\sigma > 0$. Beginning with $\tau = 0$, assume that the shearing is continued until a state in loading regime II of Figure 7.1.5, is attained, where further shearing in the same direction results in volume expansion. At this state, the distribution density function $f(\nu)$ is strongly biased toward negative ν's, and the distribution of the contact normals may resemble that denoted by L-16 in Figure 7.1.4. Many subcritical potential sliding planes now exist at this stage of loading; see Figure 7.1.6(a). Suppose now that the magnitude of the shear stress is gradually reduced, keeping the normal stress constant; Figure 7.1.6(b). It is possible that some sliding planes with large negative dilatancy angles may begin to slide down as the applied shear stress is reduced. Thus, if $|\nu^{\alpha}|$ is large enough and T is reduced, while P is kept constant, classes of contacts with suitable dilatancy angles, may undergo reverse sliding, under the action of P, *against an applied shear stress*. In this case, $\tau < 0$, while $\dot{\gamma} > 0$ and $\dot{v}/v < 0$. Because of this, it is expected that the distribution of contact normals, as well as the dilatancy angles, change somewhat as the shear stress is reduced to zero. However, it is reasonable to assume, as Konishi's experiments suggest, that even with these changes, when $\tau = T/a$ in Figure 7.1.6(a) is reduced to zero, a strong bias toward *negative dilatancy angles for shearing from right to left* still remains. Now, upon load reversal, namely as τ is gradually increased from zero (*from left to right*), considerable densification is expected and is generally observed. As shown in Figure 7.1.6(c), then, both the normal stress and the shear stress contribute to the sliding of active sliding planes with negative dilatancy angles. As is discussed in Section 7.3, this means that, for example, *a prestraining of saturated sands into the loading regime II under drained conditions, can lead to immediate liquefaction (loss of load-bearing strength), if load reversal is implemented under undrained conditions.*

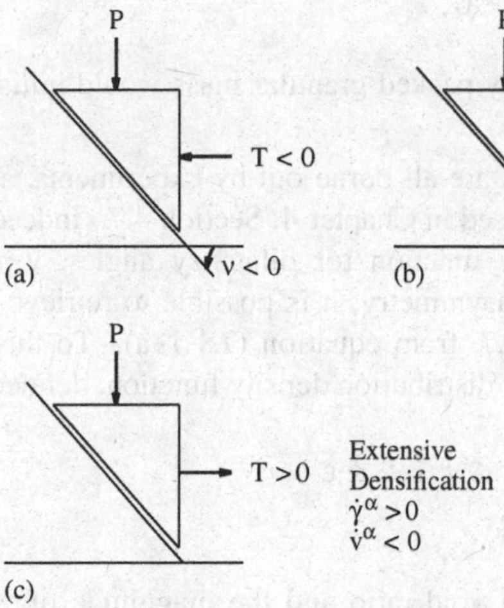

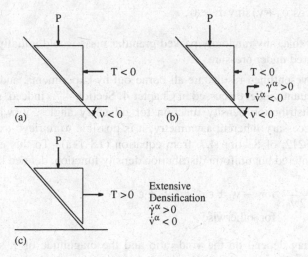

Figure 7.1.6

(a) Simple shearing from right to left, $\tau < 0$, into the loading regime II of Figure 7.1.5, produces a large number of subcritical, $\phi_\mu > \phi - \nu^\alpha > 0$, potential sliding planes (one is shown) with large negative dilatancy angles (the distribution of their unit normals is similar to L-16 of Figure 7.1.4); (b) reverse sliding under the action of normal force, P, when shear force, T, is reduced; and (c) extensive downward sliding (densification) upon load reversal

7.1.2. Relation to Continuum Models

To examine the relevance of expression (7.1.11a), to the continuum model of Section 4.7, consider first a special case where the distribution density function $f(\nu)$ is uniform over the range $\pm\,\nu_0$, and zero otherwise,

$$f(\nu) = \begin{cases} \dfrac{1}{2\,\nu_0} & \text{for } -\nu_0 \le \nu \le \nu_0 \\ 0 & \text{for } |\nu| > \nu_0\,, \end{cases} \tag{7.1.12a}$$

where ν_0 is a positive constant angle less than $\pi/2$. Assume that shearing occurs in the positive x_1-direction, so that $\tau > 0$ and $\dot{\gamma} > 0$. Then substitution into the corresponding (plus sign) expression in (7.1.11a), yields

$$\frac{1}{v}\,\frac{\dot{v}}{\dot{\gamma}} = -\frac{1}{2}\,\tan\phi_\mu\left\{1 - \frac{\sin(2\,\nu_0)}{2\nu_0}\right\}. \tag{7.1.12b}$$

This interesting result shows that: (1) if the internal friction is zero, then the granular mass flows without any volume change, as an incompressible fluid; (2) if there is no induced anisotropy, then, in the presence of internal friction, the granular mass can only densify;[6] and (3) since for any symmetric distribution density function, $f(\nu)$,

[6] Since this is an impossibility, induced anisotropy is an integral part of the deformation of frictional granules.

$$\int_{\nu^-}^{\nu^+} f(\nu)\cos(\phi_\mu+\nu)\sin\nu\,d\nu < 0\,, \qquad (7.1.12c)$$

it is concluded that, any randomly packed granular mass would initially densify when it is sheared under pressure.

The above-outlined results are all borne out by experiments, and do support the continuum model discussed in Chapter 4, Section 4.7. Indeed, selecting the simplest distribution density function for dilatancy angles, ν, which still displays the necessary inherent asymmetry, it is possible to retrieve expression (4.7.41), page 212, of Section 4.7, from equation (7.1.11a). To this end, consider an off-centered but uniform distribution density function, defined by

$$f(\nu) = \begin{cases} \dfrac{1}{2\,\nu_0} & \text{for } -\nu_0 \pm \varepsilon \le \nu \le \nu_0 \pm \varepsilon \\ 0 & \text{for otherwise}, \end{cases} \qquad (7.1.12d)$$

where $\varepsilon \ge 0$ may depend on the void ratio and the magnitude of γ, such that $\varepsilon = 0$ for $\gamma = 0$.[7] The plus sign is for $\dot{\gamma} > 0$ and the minus is for $\dot{\gamma} < 0$. Now, from (7.1.11a), obtain

$$\frac{1}{v}\frac{\dot{v}}{\dot{\gamma}} = \mp\frac{1}{2}\Big\{\tan\phi_\mu\Big\{1-\frac{\sin 2\nu_0}{2\nu_0}\cos 2\varepsilon\Big\} - \frac{\sin 2\nu_0}{2\nu_0}\sin 2\varepsilon\Big\}. \qquad (7.1.12e)$$

To compare with (4.7.41a), assume that $0 < \varepsilon << 1$, so that $\sin 2\varepsilon \approx 2\varepsilon$ and $\cos 2\varepsilon \approx 1$, and obtain, for $\dot{\gamma} > 0$,

$$-B = M_f - \nu_f = \frac{1}{2}\tan\phi_\mu\Big\{1-\frac{\sin 2\nu_0}{2\nu_0}\Big\} - \varepsilon\,\frac{\sin 2\nu_0}{2\nu_0}\,. \qquad (7.1.13a,b)$$

Therefore, M_f is proportional to $\tan\phi_\mu$, the effective internal friction, and ν_f, defined in (4.7.35d), page 211, is proportional to ε, with both proportionality factors being positive. Similar comments apply to (7.1.12e).

7.2. STRESS MEASURES IN GRANULAR MASSES

Granular masses transmit applied loads through contacts at discrete contact regions which are commonly called *contact points*. In many applications, the deformation of individual granules is small relative to the overall deformation which is produced in the granular mass by sliding and rolling of granules relative to one another. Therefore, for all practical purposes, the granules may be viewed as essentially *rigid*. For the calculation of the resulting stress field, it is thus necessary to relate the forces which act at contact points, to the overall stresses, in a systematic and rigorous manner. This then leads to relations

[7] Note that, here, in simple shearing, γ may be positive or negative.

between the parameters which characterize the microstructure of the granular assembly, and the associated contact forces. In this section, these relations are developed and discussed. In particular, expressions are obtained for various stress measures, and their physical implications are examined.

7.2.1. Overall Cauchy Stress in Granular Media

Stress is a continuum concept. For an assembly of granules, it must thus be defined in an average sense. Consider a collection of granules within an overall volume v, bounded by an imagined suitably smooth surface ∂v. The volume v, therefore, consists of solid particles of volume v_s, and the associated void volume v_v, where $v = v_s + v_v$. Let this assembly be subjected to self-equilibrating tractions, **t**(**x**), distributed on its boundary surface, ∂v. This results in contact forces developing at the contacting granules, to support the applied tractions. Assume v contains a large number of granules, so that a variable stress field, $\boldsymbol{\sigma}(\mathbf{x})$, in equilibrium with the applied tractions, **t**, can be defined in v. In the absence of body forces, equilibrium requires

$$\nabla \cdot \boldsymbol{\sigma} = \mathbf{0} \quad \text{or} \quad \sigma_{ji,j} = 0 \quad \text{in v},$$

$$\mathbf{n} \cdot \boldsymbol{\sigma} = \mathbf{t} \quad \text{or} \quad \sigma_{ji}\, n_j = t_i \quad \text{on } \partial\text{v}, \quad i, j = 1, 2, 3, \tag{7.2.1a-d}$$

where **n** with rectangular Cartesian components n_i, is the exterior unit normal to the *surface* ∂v.

In general, surface ∂v touches a finite (albeit very large) number of granules, at which external tractions are actually applied to the granules. This is schematically shown in Figure 7.2.1. Consider an elementary area, Δs, of the boundary surface ∂v. At a typical (continuum) point **x** on Δs, the traction vector, **t**(**x**), is defined as follows:

$$\mathbf{t}(\mathbf{x}) = \frac{1}{\Delta s} \sum_{\alpha=1}^{n} \mathbf{f}^{\alpha}, \tag{7.2.2}$$

where $\mathbf{f}^{\alpha}$ is the resultant force externally applied at the α'th point on Δs which is in contact with a granule, and **x** is any interior point of Δs. This definition is meaningful if and only if the traction **t**(**x**) is essentially independent of the choice of the elementary surface Δs which contains **x** as an interior point. At a minimum, it must thus be required that: (1) relative to the macroscopic length scale used to measure the tractions, the granules are so small and numerous, that any such (macroscopic) elementary area touches a very large number of granules at which external loads are actually applied; and that (2) the applied tractions are suitably regular. Note that, the nature and properties of the granules do not affect the tractions **t** defined in the above manner. Indeed, this is true also for the *average stress field* defined in v, as is discussed below.

Theorem I: *Whatever the nature and composition of the granules and the properties of their contact interfaces, the overall Cauchy stress, $\bar{\boldsymbol{\sigma}}$, defined by the unweighted volume average*

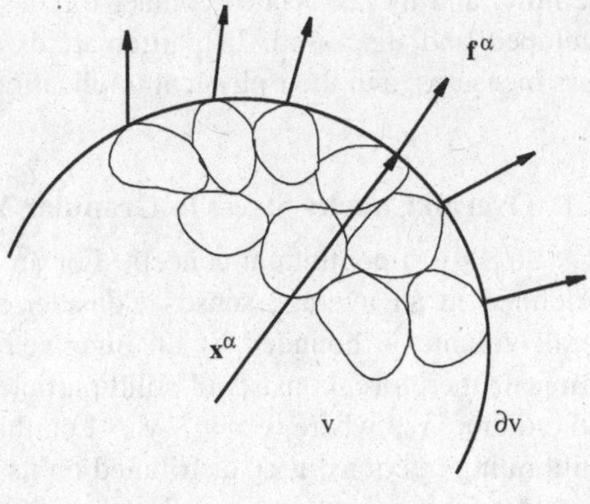

Figure 7.2.1

Volume v containing granules of total volume v_s, and voids of total volume v_v, where $v = v_s + v_v$; on boundary, ∂v, of v, resultant forces, $\mathbf{f}^\alpha$, are applied on granules which touch ∂v

$$\bar{\boldsymbol{\sigma}} = \frac{1}{v} \int_v \boldsymbol{\sigma}(\mathbf{x})\, dv\,, \tag{7.2.3a}$$

is given by

$$\bar{\boldsymbol{\sigma}} = \frac{1}{v} \sum_{\alpha=1}^{N_s} \mathbf{x}^\alpha \otimes \mathbf{f}^\alpha\,. \tag{7.2.3b}$$

Here, N_S is the total number of points on the boundary surface ∂v where the external loads are actually applied to the granules, and $\mathbf{x}^\alpha$'s are the corresponding position vectors; see Figure 7.2.1.

The proof of Theorem I follows from the equilibrium equations and the Gauss theorem. Indeed, from the following direct calculation:

$$\int_v \sigma_{ij}\, dv = \int_v (x_i\, \sigma_{kj})_{,k}\, dv = \int_{\partial v} x_i\, t_j\, ds\,,$$

Theorem I is immediately obtained.

Let now N_G be the number of granules per unit volume of v, and denote the volume fraction of the α'th granule of volume Δv^α by $c^\alpha = \Delta v^\alpha / v$. Let $\bar{\boldsymbol{\sigma}}^\alpha$ be the volume average of the Cauchy stress tensor, taken over the volume of the α'th granule. Then expression (7.2.3a) may also be written as

$$\bar{\boldsymbol{\sigma}} = \sum_{\alpha=1}^{N_G} c^\alpha\, \bar{\boldsymbol{\sigma}}^\alpha\,, \tag{7.2.3c}$$

provided that only the granules carry the applied loads. This, of course, is the case when v contains only granules and voids.

Consider now a large body of a granular mass of volume v^∞, and let v define a corresponding *representative volume element* (RVE), *i.e.*, a suitably large sample which is statistically representative of the granular mass viewed as a continuum; see Hill (1956), Nemat-Nasser and Hori (1993, 1999), and Section 8.1. The tractions, $\mathbf{t}(\mathbf{x})$, on the boundary, ∂v, of this RVE, therefore, represent the action of the granules outside of this RVE upon those within it. These tractions act on points where the interior granules which are situated at the boundary, ∂v, of v, are in contact with those immediately outside of v. The overall stress tensor, $\bar{\boldsymbol{\sigma}}$, is again defined by the unweighted volume average (7.2.3a).

Contact Forces and Branch Vectors: Consider two granules, labeled A and B, with centroids at $\bar{\mathbf{x}}^A$ and $\bar{\mathbf{x}}^B$, in contact at AB; see Figure 7.2.2. The contact force exerted by granule B on A is denoted by $\mathbf{f}^{AB}$, and that by A on B is $\mathbf{f}^{BA}$. It thus follows that $\mathbf{f}^{AB} + \mathbf{f}^{BA} = \mathbf{0}$. The balance of forces acting on granule A requires

$$\sum_{\alpha=1}^{n_c} \mathbf{f}^{A\alpha} = \mathbf{0}\,, \tag{7.2.4a}$$

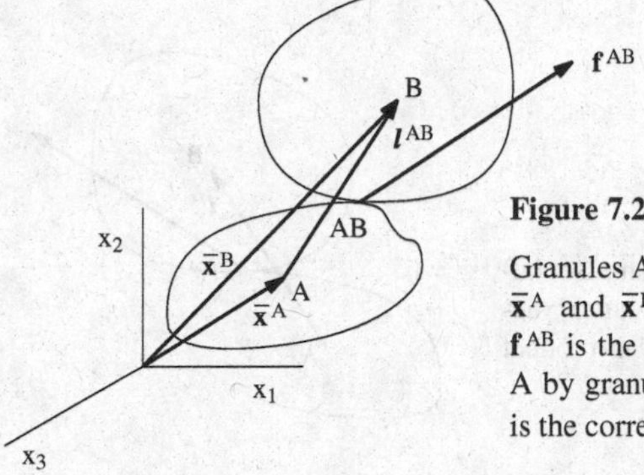

Figure 7.2.2

Granules A and B, with centroids at $\bar{\mathbf{x}}^A$ and $\bar{\mathbf{x}}^B$, are in contact at AB; $\mathbf{f}^{AB}$ is the force exerted on granule A by granule B, and $\boldsymbol{l}^{AB} = \bar{\mathbf{x}}^B - \bar{\mathbf{x}}^A$ is the corresponding *branch vector*

where n_c is the *coordination number* for granule A, *i.e.*, the number of granules contacting A. The balance of moments leads to

$$\sum_{\alpha=1}^{n_c} \mathbf{f}^{A\alpha} \otimes (\mathbf{x}^{A\alpha} - \bar{\mathbf{x}}^A) = \sum_{\alpha=1}^{n_c} (\mathbf{x}^{A\alpha} - \bar{\mathbf{x}}^A) \otimes \mathbf{f}^{A\alpha}\,. \tag{7.2.4b}$$

Sum over all granules interior to v, and noting that each contact contributes, *e.g.*, to the left-hand side, the following term:

$$\mathbf{f}^{AB} \otimes (\mathbf{x}^{AB} - \bar{\mathbf{x}}^A) + \mathbf{f}^{BA} \otimes (\mathbf{x}^{AB} - \bar{\mathbf{x}}^B)\,, \tag{7.2.4c}$$

obtain the following symmetry condition:

$$\sum_{\alpha=1}^{N} \mathbf{f}^{\alpha} \otimes \boldsymbol{l}^{\alpha} = \sum_{\alpha=1}^{N} \boldsymbol{l}^{\alpha} \otimes \mathbf{f}^{\alpha}\,, \tag{7.2.5}$$

where the sums are over all contacts in a unit volume of v, N denotes the *total* number of contacts per unit volume,[8] and $\boldsymbol{l}^{\alpha}$ is the *branch vector* associated with contact α and contact force $\mathbf{f}^{\alpha}$; $\boldsymbol{l}^{\alpha}$ being defined by

[8] In Section 7.1, N is the total number of *active* contacts per unit volume, whereas here it is the total (active and inactive) number of all contacts in a unit volume.

$$\boldsymbol{l}^{\alpha} = \bar{\mathbf{x}}^{B} - \bar{\mathbf{x}}^{A}. \tag{7.2.6}$$

Note that, in (7.2.5), from each pair of contact forces at a given contact, only one enters the summation, according to the choice of the branch vector, $\boldsymbol{l}^{\alpha}$, *i.e.*, $\mathbf{f}^{\alpha} = \mathbf{f}^{AB}$ when $\boldsymbol{l}^{\alpha} = \boldsymbol{l}^{AB}$. Thus, to each contact α, of a typical pair of contacting granules, say, A and B, a pair of vectors, $\boldsymbol{l}^{\alpha}$ and $\mathbf{f}^{\alpha}$, with a *unique tensorial product*, $\boldsymbol{l}^{\alpha} \otimes \mathbf{f}^{\alpha}$, is assigned. If the role of granules A and B is changed, the sign, but not the magnitude, of both vectors changes while their tensorial product, $\boldsymbol{l}^{\alpha} \otimes \mathbf{f}^{\alpha}$, which enters (7.2.5), remains the same. Similar comments apply to $\mathbf{f}^{\alpha} \otimes \boldsymbol{l}^{\alpha}$. Hence, it is not necessary to identify branch vectors and the corresponding contact forces with the associated granule; rather, it is sufficient to assign tensors $\boldsymbol{l}^{\alpha} \otimes \mathbf{f}^{\alpha}$ and $\mathbf{f}^{\alpha} \otimes \boldsymbol{l}^{\alpha}$ to the corresponding contact, α; see Figure 7.2.3.

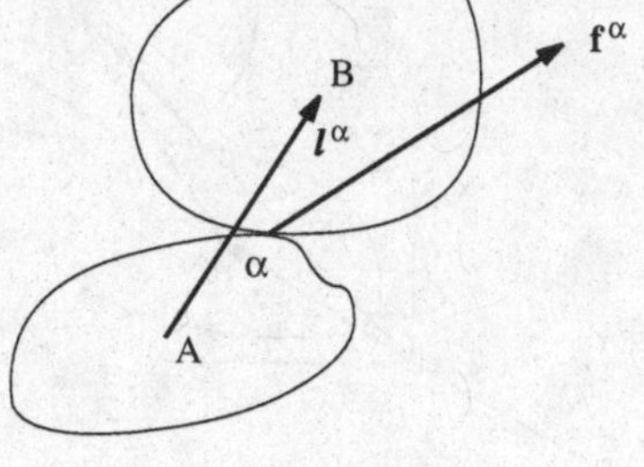

Figure 7.2.3

A typical contact α with an associated contact force $\mathbf{f}^{\alpha}$ and branch vector $\boldsymbol{l}^{\alpha}$ of unique tensorial product $\mathbf{f}^{\alpha} \otimes \boldsymbol{l}^{\alpha}$

Theorem II: *Whatever the nature and composition of the granules and the properties of their contact interfaces, the average Cauchy stress,* $\bar{\boldsymbol{\sigma}}$, *is given by*

$$\bar{\boldsymbol{\sigma}} = \sum_{\alpha=1}^{N} \frac{1}{2} (\boldsymbol{l}^{\alpha} \otimes \mathbf{f}^{\alpha} + \mathbf{f}^{\alpha} \otimes \boldsymbol{l}^{\alpha}) \quad \text{or} \quad \bar{\sigma}_{ij} = \sum_{\alpha=1}^{N} \frac{1}{2} (l_i^{\alpha} f_j^{\alpha} + f_i^{\alpha} l_j^{\alpha}). \tag{7.2.7a,b}$$

Theorem II can be proved directly, by considering the equilibrium of each granule (Nemat-Nasser, 1980), or indirectly, by using the virtual-work method (Christoffersen *et al.*, 1981).

Direct Proof: Consider a typical granule α, in equilibrium under the action of the contact forces applied to it at its, say, n_c contacts with its neighboring granules. Since granule α is in equilibrium, (7.2.3b) and (7.2.4a) apply, yielding

$$\bar{\boldsymbol{\sigma}}^{\alpha} = \frac{1}{v^{\alpha}} \sum_{\beta=1}^{n_c} \mathbf{x}^{\beta} \otimes \mathbf{f}^{\beta} = \frac{1}{v^{\alpha}} \sum_{\beta=1}^{n_c} (\mathbf{x}^{\beta} - \bar{\mathbf{x}}^{\alpha}) \otimes \mathbf{f}^{\beta}. \tag{7.2.7c,d}$$

Summing over all granules in a unit volume, arrive at

$$\bar{\boldsymbol{\sigma}} = \sum_{\alpha=1}^{N_G} \sum_{\beta=1}^{n_c} (\mathbf{x}^{\beta} - \bar{\mathbf{x}}^{\alpha}) \otimes \mathbf{f}^{\beta}. \tag{7.2.8a}$$

Now, noting (7.2.4c), obtain

$$\bar{\boldsymbol{\sigma}} = \sum_{\alpha=1}^{N} \boldsymbol{l}^{\alpha} \otimes \mathbf{f}^{\alpha}. \tag{7.2.8b}$$

Then, in view of the symmetry condition (7.2.5), (7.2.7a,b) results.

Proof by Virtual-work Method: Consider a kinematically admissible (see Section 3.7, Chapter 3) virtual velocity field $\mathbf{v}(\mathbf{x})$ which produces at a typical contact α, a relative *virtual* separation rate $\dot{\boldsymbol{\Delta}}^\alpha$. Then the virtual work principle yields

$$\sum_{\alpha=1}^{N} \dot{\boldsymbol{\Delta}}^\alpha \cdot \mathbf{f}^\alpha = \frac{1}{V} \int_V \mathrm{tr}\{\boldsymbol{\sigma}(\boldsymbol{\nabla} \otimes \mathbf{v})\}\, dV. \tag{7.2.9a}$$

Let the virtual velocity field be linear,

$$\mathbf{v} = \boldsymbol{\phi} \cdot \mathbf{x} + \mathbf{c} \quad \text{or} \quad v_i = \phi_{ij}\, x_j + c_i, \tag{7.2.9b,c}$$

where $\boldsymbol{\phi}$ with components ϕ_{ij} and $\mathbf{c}$ with components c_i, (i, j = 1, 2, 3), are constants. Then, it follows that the virtual separation rate at the α'th contact is given by

$$\dot{\boldsymbol{\Delta}}^\alpha = \boldsymbol{\phi} \cdot \boldsymbol{l}^\alpha \quad \text{or} \quad \dot{\Delta}_i^\alpha = \phi_{ij}\, l_j^\alpha. \tag{7.2.9d,e}$$

Substitution into (7.2.9a) now yields

$$\boldsymbol{\phi} \cdot \left\{ \bar{\boldsymbol{\sigma}} - \sum_{\alpha=1}^{N} \boldsymbol{l}^\alpha \otimes \mathbf{f}^\alpha \right\} = 0 \quad \text{or} \quad \phi_{ij} \left\{ \bar{\sigma}_{ij} - \sum_{\alpha=1}^{N} l_i^\alpha\, f_j^\alpha \right\}. \tag{7.2.10a,b}$$

Since $\boldsymbol{\phi}$ is arbitrary, and in view of the symmetry condition (7.2.5), Theorem II follows.

7.2.2. Other Overall Stress Measures in Granular Media

Various exact averaging theorems are presented in Chapter 8 for finitely deformed heterogeneous continua. These theorems also apply to granular masses when they are viewed as continua. As pointed out before, when the granules are very small relative to the relevant macroscale, then their assembly in an RVE can be considered as a continuum. Then, other overall stress measures can be defined, once the overall deformation gradient and nominal stress tensor are defined. Similar comments apply to the deformation rate and the rate of change of the nominal stress tensor.

On the other hand, the deformation of a granular mass which consists of hard or essentially rigid granules under the action of an overall applied load, occurs through sliding and rolling of grains over each other. In this process, some contacts are lost, and new contacts are continually formed. Thus, it is not possible to relate *individual* contacts and the associated contact forces and branches, from one configuration, say, the reference undeformed one, to another configuration, say, the current finitely deformed one.

Since the overall Cauchy stress tensor, $\bar{\boldsymbol{\sigma}}$, refers to the current configuration of the granular assembly, identification of contact forces and branch vectors, as well as the contact unit normals, presents no fundamental difficulties. Similar comments apply to the overall Kirchhoff stress tensor which will be denoted by $\bar{\boldsymbol{\tau}}$, and which relates to $\bar{\boldsymbol{\sigma}}$ by $\bar{\rho}\,\bar{\boldsymbol{\tau}} = \bar{\rho}_0 \bar{\boldsymbol{\sigma}}$, where $\bar{\rho}_0$ and $\bar{\rho}$ are the overall mass densities of the RVE in its reference and current configurations, respectively. They relate to the Jacobian, $\bar{J}$, of the overall deformation tensor, $\bar{\mathbf{F}}$, by $\bar{J} = \det|\bar{\mathbf{F}}| = \bar{\rho}_0/\bar{\rho}$.

When the current configuration is used as the reference one, then all stress measures (but not their rates) coincide. *It thus appears that the current state of a granular mass is its most natural reference configuration, where a clear relation between the overall stress measure and the microstructure of the granular mass can be established;* this relation is given by (7.2.7a,b). All other stress measures may then be defined in terms of the Cauchy stress tensor, $\bar{\boldsymbol{\sigma}}$, and the overall deformation measures, using the usual continuum relations; see Table 3.1.1, page 99, of Section 3.1 of Chapter 3. For example, if $\bar{\mathbf{F}}$ is the overall deformation gradient, measured from some arbitrary reference configuration, then

$$\bar{\mathbf{F}}\,\bar{\mathbf{S}}^{\mathrm{N}} \equiv \bar{J}\,\bar{\boldsymbol{\sigma}} = \bar{\boldsymbol{\tau}}, \qquad \bar{J} = \det|\bar{\mathbf{F}}| = \bar{\rho}_0/\bar{\rho}\,, \tag{7.2.11a,b}$$

define the overall nominal stress tensor $\bar{\mathbf{S}}^{\mathrm{N}}$ and the overall Kirchhoff stress $\bar{\boldsymbol{\tau}}$, in terms of the Cauchy stress. Other relations are obtained in a similar manner, as discussed in Chapter 8.

7.2.3. Classification of Contacts

The concept of contact classes can be used to relate in an average manner, the microstructural quantities from one configuration to another. Assume there are currently N *contacts per unit volume* within a typical RVE of volume v of a granular mass. Divide these contacts into, say, Q classes, according to the distribution of the orientation of the associated *branch vectors*. To this end, let $\mathbf{m}^{\alpha}$ denote the α'th unit branch vector, defined by

$$\boldsymbol{l}^{\alpha} = l^{\alpha}\,\mathbf{m}^{\alpha}, \qquad \mathbf{m}^{\alpha} \cdot \mathbf{m}^{\alpha} = 1 \quad (\alpha \text{ not summed})\,, \tag{7.2.12a,b}$$

where l^{α} is the length of the branch vector $\boldsymbol{l}^{\alpha}$. Choose an elementary solid angle, $d\Omega$, which contains $\mathbf{m}^{\alpha}$, and identify all contacts with unit branch vectors within this solid angle, as class α contacts. Let there be N^{α} contacts in class α per unit volume of this RVE, and denote the average length of their branch vectors by $\hat{l}^{\alpha}$ and the corresponding average contact force by $\hat{\mathbf{f}}^{\alpha}$,

$$\hat{l}^{\alpha} = \frac{1}{N^{\alpha}} \sum_{\beta=1}^{N^{\alpha}} l^{\beta}, \qquad \hat{\mathbf{f}}^{\alpha} = \frac{1}{N^{\alpha}} \sum_{\beta=1}^{N^{\alpha}} \mathbf{f}^{\beta}. \tag{7.2.13a,b}$$

Assume that in each class of contacts, the branch lengths are not correlated with the contact forces. Then, the expression for the overall Cauchy stress tensor, (7.2.7a), can be written as[9]

[9] Equation (7.2.14) is based on the assumption that $\sum_{\beta=1}^{N^{\alpha}} l^{\beta}\mathbf{f}^{\beta} = \hat{l}^{\alpha} \sum_{\beta=1}^{N^{\alpha}} \mathbf{f}^{\beta}$.

$$\bar{\boldsymbol{\sigma}} = \mathrm{N} \sum_{\alpha=1}^{Q} \mathrm{f}^{\alpha} \frac{1}{2} (\boldsymbol{l}^{\alpha} \otimes \hat{\mathbf{f}}^{\alpha} + \hat{\mathbf{f}}^{\alpha} \otimes \boldsymbol{l}^{\alpha}), \tag{7.2.14}$$

where $\mathrm{f}^{\alpha} = \mathrm{N}^{\alpha}/\mathrm{N}$ is the fractional number of contacts belonging to class α in the current state, and the summation is over the classes of contacts.

Distribution Density Function of Branch Orientations: Assume that there are a large number of elements in each class, per unit volume of the RVE, and the number of classes is very large. *Introduce the distribution density function* $\mathrm{E}(\mathbf{m})$, *to describe the distribution of classes which are then characterized by unit branches with continuously*[10] *varying orientations.* The distribution density function satisfies

$$\int_{\Omega} \mathrm{E}(\mathbf{m})\, d\Omega = \int_{\Omega_{1/2}} 2\,\mathrm{E}(\mathbf{m})\, d\Omega = 1, \tag{7.2.15a,b}$$

where $\mathrm{E}(\mathbf{m}) = \mathrm{E}(-\mathbf{m})$, and $d\Omega = \sin\psi\, d\theta\, d\psi$ is an elementary solid angle of the unit sphere Ω defined by $0 \le \theta < 2\pi$ and $0 \le \psi < \pi$; see Figure 7.2.4. Since the distribution density function $\mathrm{E}(\mathbf{m})$ accounts for both $\mathbf{m}$ amd $-\mathbf{m}$, it is often clearer to use $2\,\mathrm{E}(\mathbf{m})$ over half of the unit sphere, $\Omega_{1/2}$, defined by $0 \le \theta < 2\pi$ and $0 \le \psi < \pi/2$. Then only one, either $\mathbf{m}$ or $-\mathbf{m}$, orientation corresponds to each branch.

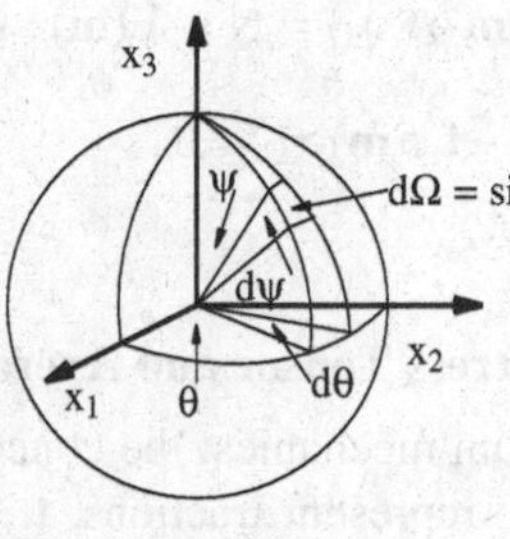

Figure 7.2.4

A unit sphere and an elementary solid angle, $d\Omega = \sin\psi\, d\theta\, d\psi$

The average of any physical quantity, ϕ, which is defined for discrete values of the branch orientation, say, $\phi = \phi(\mathbf{m}^{\alpha})$, $\alpha = 1, 2, ..., Q$, may now be expressed as[11]

$$< \phi > \equiv \sum_{\alpha=1}^{Q} \mathrm{f}^{\alpha} \phi(\mathbf{m}^{\alpha}), \tag{7.2.16a}$$

where f^{α} is the fractional number of the occurrence of $\mathbf{m}^{\alpha}$ in the data set, *i.e.,* if there are N^{α} branches of orientation $\mathbf{m}^{\alpha}$ in the total set of N orientations, then $\mathrm{f}^{\alpha} = \mathrm{N}^{\alpha}/\mathrm{N}$. Upon introduction of the continuous distribution density function,

[10] The transition from discrete to continuously varying branch orientations is considered in Section 7.3.

[11] When the average is over the initial discrete set of data points, the notation $< ... >$ is used. On the other hand, when the average is over a domain, *e.g.,* the surface of a unit sphere, Ω, then the notation $< ... >_{\Omega}$ is used. The subscript A in the expression $< ... >_{A}$ always defines the integration domain.

E(**m**), the quantity ϕ may be viewed as a function of the continuously varying branch orientation, **m**, with its average value defined by

$$< \phi >_{\Omega} = \int_{\Omega} \mathrm{E}(\mathbf{m})\, \phi(\mathbf{m}) d\Omega \,. \tag{7.2.16b}$$

With this notation, and when the classes of contacts are so numerous that their branch orientations vary continuously, then the average Cauchy stress becomes

$$\bar{\boldsymbol{\sigma}} = \mathrm{N} \int_{\Omega} \mathrm{E}(\mathbf{m}) \frac{1}{2} (\boldsymbol{l} \otimes \mathbf{f} + \mathbf{f} \otimes \boldsymbol{l})\, d\Omega \,, \tag{7.2.17}$$

where the caret, ˆ, on $\boldsymbol{l}$ and **f** is dropped for notational simplicity, even though the integration is over (continuously varying) classes of granules, and the integrand corresponds to a representative class. For example, $\mathrm{E}(\mathbf{m})\, d\Omega$ is the fractional number of contacts whose corresponding branch orientations fall in the solid angle $d\Omega$ of orientation **m**, but may have different lengths and contact forces. For these contacts, l and **f** are the average branch length and contact force, taken over this class of contacts. They are assumed to be uncorrelated within each class.

In view of the notation (7.2.16), expression (7.2.8b) and the symmetry condition (7.2.5) become

$$\bar{\boldsymbol{\sigma}} = \mathrm{N} < l\, \mathbf{m} \otimes \mathbf{f} >_{\Omega} = \mathrm{N} \int_{\Omega} \mathrm{E}(\mathbf{m})\, l\, \mathbf{m} \otimes \mathbf{f}\, d\Omega \,, \tag{7.2.18}$$

$$< l\, (\mathbf{m} \otimes \mathbf{f} - \mathbf{f} \otimes \mathbf{m}) >_{\Omega} = \mathbf{0} \,. \tag{7.2.19}$$

7.2.4. Overall Stress Tensor and Average Traction Vectors

In continuum mechanics, the concept of the *Cauchy stress tensor* is introduced in order to represent tractions, **t**, transmitted across *any* plane with arbitrary orientation, $\boldsymbol{\nu}$, passing through a point within a continuum, in terms of a symmetric second-order tensor, $\boldsymbol{\sigma}$, by the simple relation, $\mathbf{t} = \boldsymbol{\nu} \cdot \boldsymbol{\sigma}$. This stress tensor is completely defined by the tractions acting on three, say, orthogonal planes passing through that point. It is thus, of some fundamental importance to examine the relation between the average tractions defined for granular masses, and the overall stress measure, $\bar{\boldsymbol{\sigma}}$; Mehrabadi *et al.* (1982).

Consider a plane with unit normal $\boldsymbol{\nu}$, passing through an assembly of rigid granules. Identify as positive the face of the plane in the direction of $\boldsymbol{\nu}$, the other face being the negative side. Assume that the granules are so small that there are a large number of them located in the vicinity of every unit area of the $\boldsymbol{\nu}$-plane. Divide the granules located adjacent to the $\boldsymbol{\nu}$-plane into two groups, A and B, according to the position of their centroids with respect to this plane; see Figure 7.2.5. Granules in group A (hatched in Figure 7.2.5) have their centroids on the negative side of the $\boldsymbol{\nu}$-plane, whereas those in group B (unhatched) have their centroids on the positive side of this plane. A vector joining the centroids of two contacting granules, one in group A and the other in group B, forms a branch which intersects the $\boldsymbol{\nu}$-plane.

Figure 7.2.5

Branches intersect the **ν**-plane when the centroids of the contacting granules are on opposite sides of the **ν**-plane

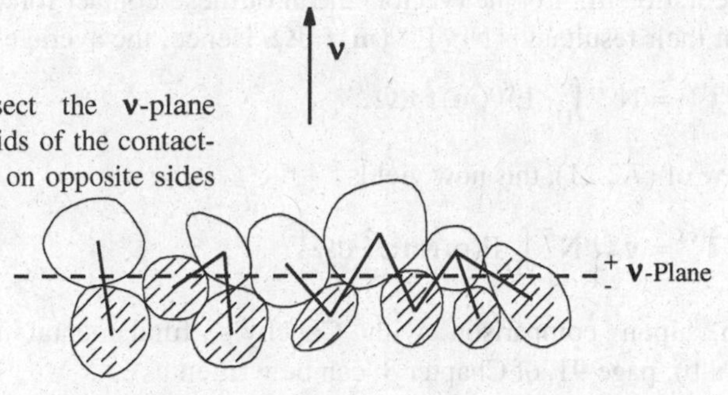

Distribution Density Function of Branch Intersections with a Plane: Let $N^{(\nu)}$ be *the density of branch intersections with the* **ν***-plane, i.e.*, the number (per unit area) of branches which point toward the positive side of the **ν**-plane and which intersect this plane. Similarly to the volume distribution density function of branch orientations, $E(\mathbf{m})$, introduce *the surface distribution density function*, $E^{(\nu)}(\mathbf{m})$, in such a manner that $E^{(\nu)}(\mathbf{m})\, d\Omega$ represents *the number of branches which intersect a unit area of the* **ν***-plane and which have orientations within the elementary solid angle* $d\Omega$ *of orientation* **m**. Then it follows that

$$\int_{\Omega_{1/2}} E^{(\nu)}(\mathbf{m})\, d\Omega = 1\,. \tag{7.2.20}$$

Note that the surface distribution density function $E^{(\nu)}(\mathbf{m})$ depends on the orientation of the considered plane, and this is stressed by the superscript **ν**. Note also that the surface density function is defined over half of the unit sphere, $\Omega_{1/2}$, while $E(\mathbf{m})$, the volume distribution density function of **m**'s, is defined over the entire unit sphere, Ω.

The surface quantities $N^{(\nu)}$ and $E^{(\nu)}(\mathbf{m})$ are related to the spatial quantities N and $E(\mathbf{m})$. This relation has been developed in Oda *et al.* (1982b) for spherical granules, and the required steps are outlined in Subsection 7.3.2, where it is shown that

$$N^{(\nu)} E^{(\nu)}(\mathbf{m}) = 2 N \bar{l}\, E(\mathbf{m})\, \mathbf{m} \cdot \boldsymbol{\nu}\,, \tag{7.2.21}$$

where $\bar{l}$ is the average branch length; see (7.3.9). This relation is used to establish the correspondence between the stress tensor defined in Subsection 7.2.3, expressions (7.2.18) and (7.2.19), and that which will be developed in the sequel, in terms of average tractions transmitted across the **ν**-plane.

The average traction vector, $\bar{\mathbf{t}}^{(\nu)}$, acting on the **ν**-plane, is the sum of the contact forces, $\mathbf{f}^{\alpha}$, transmitted (per unit area) from, say, the negative side of the **ν**-plane to the positive side, as shown in Figure 7.2.6,

$$\bar{\mathbf{t}}^{(\nu)} = \sum_{\alpha=1}^{N^{(\nu)}} \mathbf{f}^{\alpha}\,. \tag{7.2.22a}$$

Next, consider all contact forces associated with branches that intersect a unit area of the **ν**-plane and their unit branches fall in the elementary solid angle $d\Omega$

of orientation **m**. Let the (vector) mean of these contact forces be denoted by $\hat{\mathbf{f}}$, so that their resultant is $N^{(\nu)} E^{(\nu)}(\mathbf{m})\, \hat{\mathbf{f}}\, d\Omega$. Hence, the average traction becomes

$$\bar{\mathbf{t}}^{(\nu)} = N^{(\nu)} \int_{\Omega_{1/2}} E^{(\nu)}(\mathbf{m})\, \hat{\mathbf{f}}\, d\Omega\,. \tag{7.2.22b}$$

In view of (7.2.21), this now yields

$$\bar{\mathbf{t}}^{(\nu)} = \boldsymbol{\nu} \cdot \left\{ N \bar{l} \int_{\Omega} E(\mathbf{m})\, \mathbf{m} \otimes \hat{\mathbf{f}}\, d\Omega \right\} \tag{7.2.23a}$$

which, upon comparison with Cauchy's fundamental theorem, equation (3.1.1a,b), page 91, of Chapter 3, can be written as

$$\bar{\mathbf{t}}^{(\nu)} = \boldsymbol{\nu} \cdot \bar{\boldsymbol{\sigma}}\,,$$
$$\bar{\boldsymbol{\sigma}} = N \bar{l} \int_{\Omega} E(\mathbf{m})\, \mathbf{m} \otimes \hat{\mathbf{f}}\, d\Omega\,. \tag{7.2.23b,c}$$

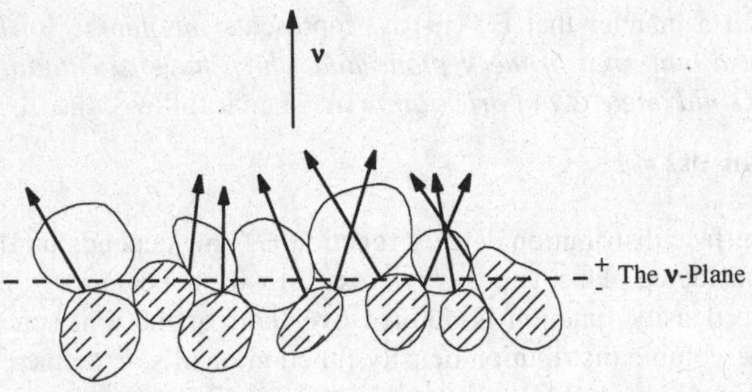

Figure 7.2.6

Average traction, $\bar{\mathbf{t}}^{(\nu)}$, on the **ν**-plane, is the sum of the contact forces, $\mathbf{f}^{\alpha}$, transmitted per unit area of this plane

Observe that, in this equation, the assumption that the distribution of branch lengths is correlated with neither their orientations nor the contact forces is incorporated, since this assumption is used to arrive at equation (7.2.21); see Subsection 7.3.2.

The stress tensor defined by (7.2.23c) has the same form as that defined by (7.2.18). The mean contact forces, $\hat{\mathbf{f}}$ in (7.2.23c) and **f** in (7.2.18), do not, however, have the same definition, but, if statistical homogeneity exists and a suitably large sample is used, then it follows that they are essentially the same. Thus, the two definitions of the overall macroscopic stress tensor do indeed coincide, in the presence of statistical homogeneity.

7.2.5. Symmetry of Overall Cauchy Stress Tensor

At this point, it may be of interest to point out that the symmetry of the overall Cauchy stress tensor, $\boldsymbol{\sigma}$, is the result of assumption (7.2.4b) which ensures the moment equilibrium for each granule. This symmetry fails if the rotation of the individual grains does not coincide with the value of the overall smooth rotation field calculated at the contacting points. Since, in general, the rotation of individual grains may differ from the overall macroscopic rotation field, a nonsymmetric stress tensor is perhaps more appropriate for the description of granular masses. In such a case, the skewsymmetric part of the stress tensor becomes

$$\bar{\boldsymbol{\sigma}}^{(\text{skew})} \equiv \mathrm{N}\,\bar{l} \int_{\Omega} \mathrm{E}(\mathbf{m})\,(\mathbf{f} \otimes \mathbf{m} - \mathbf{m} \otimes \mathbf{f})\, \mathrm{d}\Omega$$

$$= \mathrm{N}\,\bar{l} < \mathbf{f} \otimes \mathbf{m} - \mathbf{m} \otimes \mathbf{f} >_{\Omega}. \qquad (7.2.23\text{d,e})$$

However, a nonsymmetric stress tensor leads to more complex constitutive relations, while the importance of the additional nonsymmetric part has not as yet been experimentally established.[12] On the contrary, most experimental observations on an assembly of photoelastic rods suggest that the stress tensor is essentially symmetric; see Subhash *et al.* (1991) and Section 7.4. Hence, the skewsymmetric part may be insignificant. In fact, the analytical derivation of Mehrabadi *et al.* (1982) leads naturally to a symmetric stress tensor for two-dimensional deformation of granular media, as is discussed in Subsection 7.2.8; see (7.2.34f).

7.2.6. Stress-fabric Relations

The concept of fabric tensor is discussed in Section 7.3, where several index measures for granular materials are presented. The tractions acting on the $\boldsymbol{\nu}$-plane (Figure 7.2.6) are the resultant of the contact forces transmitted by granules on one side, to those on the other side. It is thus natural to expect a close relation with a *fabric vector* defined as the sum of the branches intersecting a unit area of this plane; see Figure 7.2.5. In the same manner, it is expected that the stress tensor should relate to the fabric tensors which are defined by suitable combinations of the branch vectors, contact unit normal vectors, and other parameters which characterize the microstructure of the granular mass. Some of these relations are discussed below. Further comments are made in Section 7.3, where fabric measures are detailed. Some relevant experimental results are presented in Section 7.4.

Consider the $\boldsymbol{\nu}$-plane of Figure 7.2.5, and observe that the resultant of all unit branch vectors, $\mathbf{m}$'s, which intersect a unit area of the $\boldsymbol{\nu}$-plane, and which point in the positive side of this plane, is given by

---

[12] In Subsection 6.5.17 of Chapter 6, a brief account of a simple couple-stress theory is given; see page 483.

$$\mathbf{F}^{(v)} = \sum_{\alpha=1}^{N^{(v)}} \mathbf{m}^{\alpha} = N^{(v)} \int_{\Omega_{1/2}} E^{(v)}(\mathbf{m})\, \mathbf{m}\, d\Omega, \tag{7.2.24a}$$

where a transition to continuously distributed unit branch vectors is made, as discussed before; see also Section 7.3. In view of the identity (7.2.21), expression (7.2.24a) can be written as

$$\mathbf{F}^{(v)} = \mathbf{v} \cdot \left\{ N\bar{l} \int_{\Omega} E(\mathbf{m})\, \mathbf{m} \otimes \mathbf{m}\, d\Omega \right\} \equiv \mathbf{v} \cdot \boldsymbol{\Phi}, \tag{7.2.24b}$$

where $\boldsymbol{\Phi}$ with components Φ_{ij} is a *fabric* tensor (second-order and symmetric),

$$\boldsymbol{\Phi} = N\bar{l} \int_{\Omega} E(\mathbf{m})\, \mathbf{m} \otimes \mathbf{m}\, d\Omega = N\bar{l} < \mathbf{m} \otimes \mathbf{m} >_{\Omega}. \tag{7.2.25a,b}$$

In Section 7.3 various properties of this and other fabric tensors are discussed and compared with other index measures. Here, relations between the fabric and stress tensors are examined.

Consider (7.2.13b) and (7.2.14), and observe that, without a loss in generality, the mean contact force, $\hat{\mathbf{f}}^{\alpha}$, for class α contacts, can be expressed as

$$\hat{\mathbf{f}}^{\alpha} = \mathbf{m}^{\alpha} \cdot \mathbf{K}^{\alpha} \quad \text{or} \quad \hat{f}_i^{\alpha} = m_j K_{ji}^{\alpha}, \quad i, j = 1, 2, 3, \tag{7.2.26a,b}$$

where $\mathbf{K}^{\alpha}$ is a second-order tensor associated with class α contacts. The tensor $\mathbf{K}^{\alpha}$ may be viewed as a *concentration tensor*, defining the local mean contact force, $\hat{\mathbf{f}}^{\alpha}$, of class α contacts as a linear and homogeneous function of the branch orientation $\mathbf{m}^{\alpha}$. From (7.2.17) and (7.2.26), obtain

$$\bar{\boldsymbol{\sigma}} = \frac{1}{2} N\bar{l} < (\mathbf{m} \cdot \mathbf{K}) \otimes \mathbf{m} + \mathbf{m} \otimes (\mathbf{m} \cdot \mathbf{K}) >_{\Omega} \tag{7.2.27a}$$

or, in components,

$$\bar{\sigma}_{ij} = \frac{1}{2} N\bar{l} < K_{ki}\, m_k\, m_j + K_{kj}\, m_k\, m_i >_{\Omega}. \tag{7.2.27b}$$

In general, it cannot be assumed that $\mathbf{K}^{\alpha}$ is uncorrelated with the unit branch vectors, $\mathbf{m}^{\alpha}$'s. This correlation must be defined before additional explicit results can be obtained. Thus, only some special cases are considered in what follows. Additional results are given in Sections 7.3 and 7.4.

7.2.7. Stress-fabric Relations for Spherical Granules

Restrict attention to spherical granules. Then, the unit branch vector, $\mathbf{m}$, coincides with the corresponding contact unit normal, $\mathbf{n}$. For typical class α contacts of a common contact unit normal, $\mathbf{n}^{\alpha}$, decompose the representative contact force, $\hat{\mathbf{f}}^{\alpha}$, into normal and tangential components, as follows:

$$\hat{\mathbf{f}}^{\alpha} = \hat{\sigma}^{\alpha}\, \mathbf{n}^{\alpha} + \hat{\tau}^{\alpha}\, \mathbf{s}^{\alpha}, \tag{7.2.28a}$$

where $\mathbf{s}^{\alpha}$ is a unit vector in the direction of the maximum shear force; $\hat{\sigma}^{\alpha}$ and $\hat{\tau}^{\alpha}$ are the normal and shear forces associated with the representative contact force $\hat{\mathbf{f}}^{\alpha}$. Introduce another unit vector, $\mathbf{r}^{\alpha}$, in such a manner that $\mathbf{s}^{\alpha}$, $\mathbf{n}^{\alpha}$, and $\mathbf{r}^{\alpha}$ form a right-handed unit triad,

$$s_i^\alpha = e_{ijk}\, n_j^\alpha\, r_k^\alpha, \tag{7.2.28b}$$

where e_{ijk} is the permutation symbol. Then, substitution from (7.2.28b) into (7.2.28a), and then into (7.2.26) leads to the conclusion that, for spherical granules,

$$K_{ij}^\alpha = \hat{\sigma}^\alpha \delta_{ij} + e_{ijk}\, t_k^\alpha, \quad t_k^\alpha = \hat{\tau}^\alpha r_k^\alpha, \tag{7.2.29a,b}$$

where δ_{ij} is the Kronecker delta. It follows from (7.2.29a) that the symmetric part of the matrix K_{ij}^α is spherical, and therefore, if the average shear stress associated with the class α contacts is zero, then, by necessity,

$$\hat{\mathbf{f}}^\alpha = \hat{\sigma}^\alpha \mathbf{n}^\alpha. \tag{7.2.29c}$$

For this particular case, the symmetric stress is given by

$$\bar{\boldsymbol{\sigma}} = N\bar{l} < \hat{\sigma}\, \mathbf{n} \otimes \mathbf{n} >_\Omega, \tag{7.2.29d}$$

where $\hat{\sigma}^\alpha$, in general, is correlated with the contact normals $\mathbf{n}^\alpha$. If there is no such correlation, then it follows from (7.2.29d) that

$$\bar{\boldsymbol{\sigma}} = \bar{\sigma}\hat{\boldsymbol{\Phi}}, \quad \hat{\boldsymbol{\Phi}} = N\bar{l} < \mathbf{n} \otimes \mathbf{n} >_\Omega \quad \text{or} \quad \hat{\Phi}_{ij} = N\bar{l} < n_i\, n_j >_\Omega, \tag{7.2.29e-g}$$

i.e., the stress tensor is proportional to the fabric tensor,[13] $\hat{\boldsymbol{\Phi}}$.

Equation (7.2.29e) is in some qualitative agreement with experimental results of Oda and Konishi (1974) in the sense that it predicts coaxiality between the stress and fabric tensors. Quantitatively, however, the agreement is rather poor. The equation predicts the ratio of the principal stresses to be equal to the ratio of the corresponding principal values of the fabric tensor. Experimental results, however, do not support this conclusion; see Oda *et al.* (1982b), Mehrabadi *et al.* (1982), and the following subsections.

A more general description of the stress in terms of the fabric results if $\hat{\sigma}^\alpha$ is regarded to depend on the overall macroscopic stress, $\bar{\boldsymbol{\sigma}}$, and on an *anisotropy tensor* $\hat{\mathcal{J}}$ which is defined by (Satake, 1978)

$$\hat{\mathcal{J}} = < \mathbf{n} \otimes \mathbf{n} >_\Omega, \tag{7.2.30a}$$

as well as on the contact normal, $\mathbf{n}^\alpha$, which represents the corresponding class of contacts, *i.e.*,

$$\hat{\sigma}^\alpha = \hat{\sigma}^\alpha(\bar{\boldsymbol{\sigma}}, \hat{\mathcal{J}}, \mathbf{n}^\alpha). \tag{7.2.30b}$$

---

[13] In Section 4.7, Chapter 4, the backstress, $\boldsymbol{\beta}$, is assumed to be linear in the fabric tensor $\hat{\mathcal{J}}$ which, in turn is proportional to $\hat{\boldsymbol{\Phi}}$; compare equations (4.7.1a-c), page 198, and (7.2.29e,f) and (7.2.37). It is thus seen that the continuum model of Section 4.7 corresponds to the simplest stress-fabric relation. A better representation is given by (7.2.37a) which, however, is a nonlinear relation. Note that, for granular media with a vanishingly small elastic range (discussed in Subsection 4.7.10, page 210), the backstress coincides with the Kirchhoff stress, $\boldsymbol{\tau}$, which is proportional to the Cauchy stress, $\boldsymbol{\tau} = (\rho_0/\rho)\,\boldsymbol{\sigma}$, where ρ_0 and ρ are the initial and current mass densities.

Additional special assumptions are required to obtain specific results. As an illustration, consider

$$\hat{\sigma}^\alpha = \hat{A}_{ij}\, n_i^\alpha\, n_j^\alpha\,, \tag{7.2.30c}$$

where $\hat{\mathbf{A}}$, a symmetric second-order tensor with components $\hat{A}_{ij}$, in general is a function of the overall stress, $\bar{\boldsymbol{\sigma}}$, as well as the anisotropy tensor, $\hat{\mathcal{J}}$; equation (7.2.30c) is in line with the notion of normal tractions in continuous bodies. Assuming furthermore that $\hat{\mathbf{A}}$ depends on the overall stress only through the stress invariants, *i.e.*,

$$\hat{\mathbf{A}} = \hat{\mathbf{A}}(\hat{\mathcal{J}}, \mathrm{I}, \mathrm{II}, \mathrm{III})\,, \tag{7.2.30d}$$

and that $\hat{\mathbf{A}}$ is an isotropic function of $\hat{\mathcal{J}}$, it follows that

$$\hat{\mathbf{A}} = \alpha_0\, \mathbf{1} + \alpha_1\, \hat{\mathcal{J}} + \alpha_2\, \hat{\mathcal{J}} \cdot \hat{\mathcal{J}} \tag{7.2.30e}$$

or, in components,

$$\hat{A}_{ij} = \alpha_0\, \delta_{ij} + \alpha_1\, \hat{\mathcal{J}}_{ij} + \alpha_2\, \hat{\mathcal{J}}_{ik}\, \hat{\mathcal{J}}_{kj}\,, \tag{7.2.30f}$$

where, in (7.2.30d), I, II, and III are the stress invariants which may be taken as

$$\mathrm{I} = \mathrm{tr}\bar{\boldsymbol{\sigma}}\,, \quad \mathrm{II} = \frac{1}{2}\,\mathrm{tr}(\bar{\boldsymbol{\sigma}} \cdot \bar{\boldsymbol{\sigma}})\,, \quad \mathrm{III} = \frac{1}{3!}\,\mathrm{tr}(\bar{\boldsymbol{\sigma}} \cdot \bar{\boldsymbol{\sigma}} \cdot \bar{\boldsymbol{\sigma}})\,.$$

In (7.2.30e), α_0, α_1, and α_2 are, in general, functions of the above stress invariants, as well as of the basic invariants of the symmetric anisotropy tensor $\hat{\mathcal{J}}$.[14]

Now, substitute from (7.2.30e,f) into (7.2.30c), and then the result into (7.2.29d), to arrive at

$$\begin{aligned} \bar{\sigma}_{ij} = N\bar{l}\,\{\, &\alpha_0 < n_i\, n_j >_\Omega + \alpha_1 < n_i\, n_j\, n_k\, n_l >_\Omega < n_k\, n_l >_\Omega + \\ &+ \alpha_2 < n_i\, n_j\, n_k\, n_l >_\Omega < n_k\, n_r >_\Omega < n_r\, n_l >_\Omega \}\,. \end{aligned} \tag{7.2.31a}$$

Define the fourth-order fabric tensor,

$$\hat{\mathcal{A}} = \int_\Omega E(\mathbf{n})\, \mathbf{n} \otimes \mathbf{n} \otimes \mathbf{n} \otimes \mathbf{n}\, d\Omega = \, < \mathbf{n} \otimes \mathbf{n} \otimes \mathbf{n} \otimes \mathbf{n} >_\Omega\,, \tag{7.2.32a}$$

or, in components,

$$\hat{\mathcal{A}}_{ijkl} = \, < n_i\, n_j\, n_k\, n_l >_\Omega\,. \tag{7.2.32b}$$

Equation (7.2.31a) now becomes

$$\bar{\boldsymbol{\sigma}} = \{\alpha_0\, \mathbf{1}^{(4s)} + \alpha_1\, \hat{\mathcal{A}}\} : \hat{\boldsymbol{\Phi}} + \alpha_3\, \hat{\mathcal{A}} : (\hat{\boldsymbol{\Phi}} \cdot \hat{\boldsymbol{\Phi}})\,, \quad \alpha_3 = \frac{\alpha_2}{N\bar{l}}\,, \tag{7.2.31b,c}$$

or, in components,

$$\bar{\sigma}_{ij} = \{\alpha_0\, 1_{ijkl}^{(4s)} + \alpha_1\, \hat{\mathcal{A}}_{ijkl}\}\hat{\Phi}_{kl} + \alpha_3\, \hat{\mathcal{A}}_{ijkl}\, \hat{\Phi}_{kr}\, \hat{\Phi}_{rl}\,, \tag{7.2.31d}$$

where $1_{ijkl}^{(4s)}$ are the components of the fourth-order symmetric identity tensor

[14] Note that $\mathrm{tr}\hat{\mathcal{J}} = 1$.

$\mathbf{1}^{(4s)}$. In Section 7.3, some of the properties of the fabric tensors $\hat{\mathcal{J}}$, $\hat{\mathbf{\Phi}}$, and $\hat{\mathcal{A}}$ are further examined.

7.2.8. Stress-fabric Relations for Nonspherical Granules

Consider now a general case where the granules may not be spherical. Also, the symmetry of the Cauchy stress is not assumed, but, rather, its possibility is explored. Let $\mathbf{s}^\alpha$ be a unit vector normal to the unit branch vector $\mathbf{m}^\alpha$ which represents the α class of contacts, and choose $\mathbf{s}^\alpha$ in the plane formed by $\mathbf{m}^\alpha$ and the representative contact force $\hat{\mathbf{f}}^\alpha$; see Figure 7.2.7. Decompose this representative contact force $\hat{\mathbf{f}}^\alpha$, as follows:

$$\hat{\mathbf{f}}^\alpha = \sigma^\alpha \mathbf{m}^\alpha + \tau^\alpha \mathbf{s}^\alpha . \tag{7.2.33a}$$

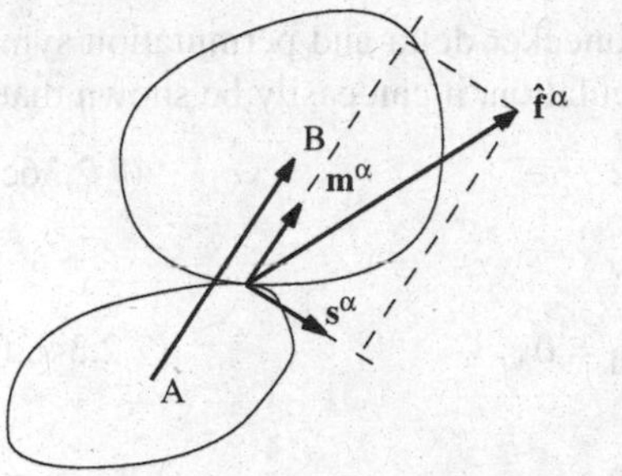

Figure 7.2.7
Decomposition of the representative contact force $\hat{\mathbf{f}}^\alpha$ of the class α contacts, into a component in the $\mathbf{m}^\alpha$-direction and another (coplanar) normal to this direction; $\mathbf{s}^\alpha$ is a unit vector normal to $\mathbf{m}^\alpha$

In view of this, the symmetric, $\bar{\boldsymbol{\sigma}}^{(\mathrm{sym})}$, and skewsymmetric, $\bar{\boldsymbol{\sigma}}^{(\mathrm{skew})}$, parts of the overall Cauchy stress tensor become

$$\bar{\boldsymbol{\sigma}}^{(\mathrm{sym})} = \mathrm{N}\,\bar{l} < \sigma\,\mathbf{m}\otimes\mathbf{m} + \tfrac{1}{2}\,\tau\,(\mathbf{m}\otimes\mathbf{s}+\mathbf{s}\otimes\mathbf{m}) >_\Omega ,$$

$$\bar{\boldsymbol{\sigma}}^{(\mathrm{skew})} = \tfrac{1}{2}\,\mathrm{N}\,\bar{l} < \tau\,(\mathbf{s}\otimes\mathbf{m} - \mathbf{m}\otimes\mathbf{s}) >_\Omega . \tag{7.2.34a,b}$$

Consider now a symmetric second-order tensor $\mathbf{A}$ and assume that

$$\sigma^\alpha = \mathrm{A}_{ij}\, m_i^\alpha\, m_j^\alpha , \qquad \tau^\alpha = \mathrm{A}_{ij} m_i^\alpha\, s_j^\alpha . \tag{7.2.33b,c}$$

Assume furthermore that tensor $\mathbf{A}$ can be expressed in terms of the fabric tensor

$$\mathcal{J} = < \mathbf{m}\otimes\mathbf{m} >_\Omega \quad \text{or} \quad \mathcal{J}_{ij} = < m_i\, m_j >_\Omega , \tag{7.2.33d,e}$$

as follows:

$$\mathrm{A}_{ij} = \beta_0\, \delta_{ij} + \beta_1\, \mathcal{J}_{ij} , \tag{7.2.33f}$$

with coefficients depending on the basic invariants of the overall Cauchy stress.

Use the component representation, and upon substitution from (7.2.33e) into (7.2.33f), then the result into (7.2.33b,c), and finally into (7.2.34a,b), obtain

$$\bar{\sigma}_{(ij)} = \{\, \beta_0\, 1_{ijkl}^{(4s)} + \beta_1\, \mathcal{A}_{ijkl} + \tfrac{1}{4} < (m_i\, s_j + m_j\, s_i)\,(m_k\, s_l + m_l\, s_k) >_\Omega \}\, \Phi_{kl} ,$$

$$\bar{\sigma}_{[ij]} = \tfrac{1}{4} < (s_i\, m_j - s_j\, m_i)\,(m_k\, s_l + m_l\, s_k) >_\Omega \Phi_{kl} , \tag{7.2.34c,d}$$

where $\bar{\sigma}_{(ij)}$ and $\bar{\sigma}_{[ij]}$ are the components of $\bar{\boldsymbol{\sigma}}^{(sym)}$ and $\bar{\boldsymbol{\sigma}}^{(skew)}$, respectively, and the second- and fourth-order fabric tensors, $\boldsymbol{\Phi}$ with components Φ_{ij} and $\boldsymbol{\mathcal{A}}$ with components $\mathcal{A}_{ijkl}$, are defined by

$$\Phi_{ij} = N\bar{l} < m_i\, m_j >_\Omega ,$$

$$\mathcal{A}_{ijkl} = < m_i\, m_j\, m_k\, m_l >_\Omega . \qquad (7.2.35a,b)$$

For spherical granules, these fabric tensors reduce to the corresponding (denoted by superscript caret, ^) quantities of the preceding subsection.

For two-dimensional cases, which apply to the photoelastic granular rods of any cross section, *e.g.,* those shown in Figure 7.1.1, page 505, expressions (7.2.34c,d) take on very simple forms. To see this, first note that, in two dimensions, the following identities hold for any pair of orthogonal unit vectors:

$$s_i\, s_j + m_i\, m_j = \delta_{ij} , \quad s_i\, m_j - s_j\, m_i = e_{ij} , \quad i, j = 1, 2 , \qquad (7.2.36a,b)$$

where δ_{ij} and e_{ij} are the two-dimensional Kronecker delta and permutation symbol, respectively.[15] In addition, by direct calculation, it can easily be shown that,

$$< m_k\, s_l + m_l\, s_k >_\Omega \Phi_{kl} = 0 , \quad i, j = 1, 2 . \qquad (7.2.36c)$$

From these, (7.2.34c,d) now become

$$\bar{\sigma}_{(ij)} = (\beta_0 + \beta_1)\, \Phi_{ij} - \frac{\beta_1}{N\bar{l}}\, II_{\boldsymbol{\Phi}}\, \delta_{ij} , \quad \bar{\sigma}_{[ij]} = 0 , \qquad (7.2.34e,f)$$

where

$$II_{\boldsymbol{\Phi}} = \tfrac{1}{2} \{ (\mathrm{tr}\boldsymbol{\Phi})^2 - \mathrm{tr}(\boldsymbol{\Phi} \cdot \boldsymbol{\Phi}) \} . \qquad (7.2.34g)$$

Equations (7.2.34e,f) have emerged naturally from a systematic development. They seem to be supported by the existing experimental observations on the behavior of rigid rods (Konishi, 1978) and photoelastic rods (Oda *et al.*, 1982a) in plane strain simple shearing. Konishi concludes from his experiments that the corresponding overall stress tensor in such two-dimensional flow is symmetric and that the stress and fabric tensors are coaxial. Reporting some of the experimental results performed in the present author's laboratories using photoelastic rods of oval cross section, Oda *et al.* (1982a) compare the stress ratio given by (7.2.34e) for biaxial loadings, with the corresponding experimental data. Since, using the Cayley-Hamilton theorem, (7.2.34e) can also be expressed as

$$\bar{\sigma}_{(ij)} = \beta_0\, \Phi_{ij} + \beta_3\, \Phi_{ik}\, \Phi_{kj} , \quad \beta_3 = \frac{\beta_1}{N\bar{l}} , \qquad (7.2.37a,b)$$

it follows that the principal values, $\bar{\sigma}_i$ of $\bar{\sigma}_{ij}$ and Φ_i of Φ_{ij}, also satisfy the same equation,

$$\bar{\sigma}_i = \beta_0\, \Phi_i + \beta_3\, (\Phi_i)^2 , \quad i = 1, 2 . \qquad (7.2.37c)$$

---

15 *I.e.*, $e_{12} = -e_{21} = 1$, $e_{11} = e_{22} = 0$.

Let ξ be the angle between the $\bar{\sigma}_1$-axis and the major principal direction of Φ_{ij}. Figure 7.2.8 shows a frequency histogram for ξ on sixteen photoelastic images of oval I particles[16] with contact friction angle of $\phi_\mu = 26°$. The analysis is performed on the assemblies after they have been fully sheared up to the peak stress state. Prior to this, the value of angle ϕ_μ was somewhere between 0 and 90°, depending on the initial packing. However, as seen from Figure 7.2.8, ξ has changed in the course of deformation, and has attained values within 15° of zero. Hence, it is reasonable to assume that the stress and fabric tensors become coaxial, as suggested by (7.2.37c).

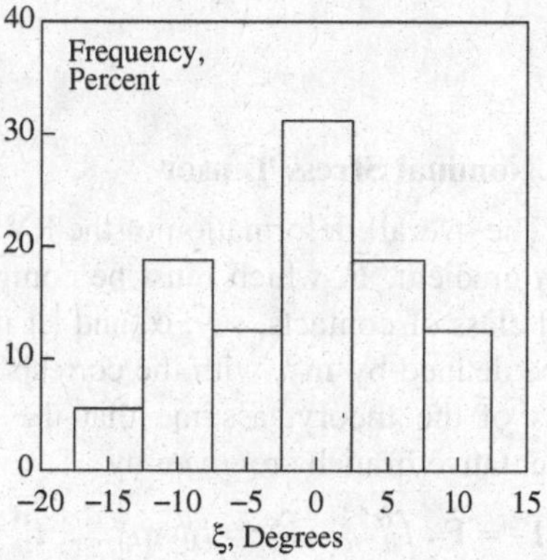

Figure 7.2.8
Relation between principal directions of stress and fabric

Since in the experiment, $\bar{\sigma}_2$ has been kept nearly constant, if it is assumed that β_0 and β_3 are constants, then (7.2.37c) with i = 2, shows that Φ_2 must also be constant. Therefore, (7.2.37c) can be written as

$$\frac{\bar{\sigma}_1}{\bar{\sigma}_2} = A\,\frac{\Phi_1}{\Phi_2} + B\left[\frac{\Phi_1}{\Phi_2}\right]^2, \tag{7.2.37d}$$

where A and B are new constants. Figure 7.2.9 shows the experimentally observed relation between the principal *stress ratio* and the principal *fabric ratio*. If it is assumed that, for $\bar{\sigma}_1/\bar{\sigma}_2 = 1$, the fabric ratio is $\Phi_1/\Phi_2 = 1$, and that for $\bar{\sigma}_1/\bar{\sigma}_2 = 6$, the fabric ratio is $\Phi_1/\Phi_2 = 2$, then (7.2.37d) becomes

$$\frac{\bar{\sigma}_1}{\bar{\sigma}_2} = -\frac{\Phi_1}{\Phi_2} + 2\left[\frac{\Phi_1}{\Phi_2}\right]^2. \tag{7.2.37e}$$

This equation is plotted in Figure 7.2.9 by a solid curve. The correlation with experimental data is rather good.

---

[16] The experimental details are given in Section 7.4.

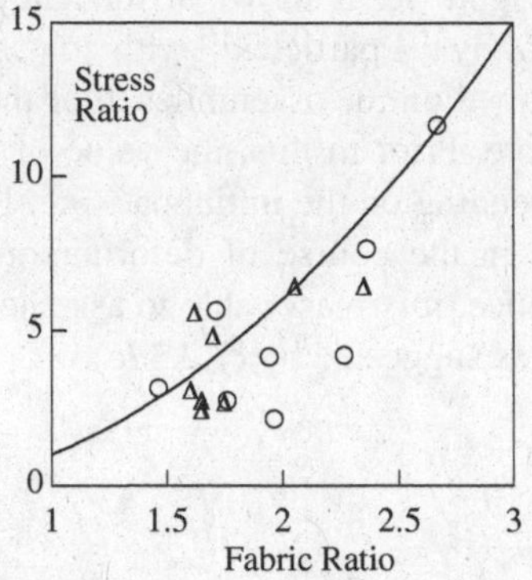

Figure 7.2.9
Comparison between experimental observations and theoretical prediction of relation between stress ratio and fabric ratio in biaxial loading of photoelastic granules of oval cross section; see Section 7.4 for the description of the experiments

7.2.9. Nominal Stress Tensor

The overall deformation of the RVE is characterized by the overall deformation gradient, $\bar{\mathbf{F}}$, which must be computed incrementally.[17] Consider now a typical class of contacts, say, α, and let its branch orientation in some reference state be defined by $\mathbf{m}_0^\alpha$, with the corresponding (average) branch length $\hat{l}_0^\alpha$. As a basis of the theory, assume that the current orientation and length of this representative branch are given by

$$\hat{\boldsymbol{l}}^\alpha = \bar{\mathbf{F}} \cdot \hat{\boldsymbol{l}}_0^\alpha, \quad \hat{\boldsymbol{l}}_0^\alpha = \hat{l}_0^\alpha \, \mathbf{m}_0^\alpha, \quad \hat{\boldsymbol{l}}^\alpha = \hat{l}^\alpha \, \mathbf{m}^\alpha. \tag{7.2.38a-c}$$

Hence, while the number of members of each class may change, as contacts are lost and new ones are formed in the course of the deformation of the RVE, their representative branch of average initial length, $\hat{l}_0^\alpha$, is assumed to transform as an elementary continuum length, according to (7.2.38a).

It is reasonable to expect that the average density of an RVE is proportional to the density, N, of its contacts, since the denser the packing the greater the density. From (7.2.11b), it then follows that

$$\bar{J} = \frac{\bar{\rho}_0}{\bar{\rho}} = \frac{N_0}{N}. \tag{7.2.39a,b}$$

With this and relation (7.2.38a), it is now possible to express the overall nominal stress tensor, $\bar{\mathbf{S}}^N$, in terms of the representative branches which are measured in the reference state, and contact forces which are measured in the current state of the RVE, as follows (Mehrabadi *et al.*, 1993):

$$\begin{aligned} \bar{\mathbf{S}}^N &= \bar{J}\,\bar{\mathbf{F}}^{-1}\,\bar{\boldsymbol{\sigma}} \\ &= N_0 \sum_{\alpha=1}^{\Omega} \hat{\boldsymbol{l}}_0^\alpha \otimes \hat{\mathbf{f}}^\alpha = N_0 < \boldsymbol{l}_0 \otimes \mathbf{f} >_\Omega . \end{aligned} \tag{7.2.40a-c}$$

---

[17] Indeed, evaluation of $\bar{\mathbf{F}}$ and the corresponding stress field is one of the main objectives of the analysis.

When the current configuration is the reference one, then this representation reduces to (7.2.18). Note that the superscript carets are shown in (7.2.40b) to emphasize the averaged nature of the corresponding quantities, but not in expression (7.2.40c), in which transition to a continuously varying branch orientation is also incorporated.

7.2.10. Stress-rate Measures

With the current configuration of a granular mass as the reference one, all overall stress measures relative to this configuration, coincide with the Cauchy stress tensor, $\bar{\boldsymbol{\sigma}}$. Their rates, however, are different, being related by expressions (3.5.1a,b), page 109, and other relevant equations developed in Section 3.5 of Chapter 3. It is thus necessary to establish the relation between only one stress-rate measure, and relevant microstructural rate-parameters. The corresponding relations for other stress-rate measures may then be obtained, using the expressions given in Section 3.5.

A simple way of approaching this problem for granular masses, is to directly calculate the nominal stress rate from (7.2.40c),

$$\dot{\bar{\mathbf{S}}}^{\mathrm{N}} = \mathrm{N}_0 < \boldsymbol{l}_0 \otimes \dot{\mathbf{f}} >_{\Omega} . \tag{7.2.41a}$$

Here, again, $\dot{\mathbf{f}}$ is the time-rate-of-change of the contact force representing the class of contacts which are associated with branch orientation $\mathbf{m}$. While $\mathbf{f}$ varies continuously, and, hence admits time differentiation, the same is not true for individual contact forces.

When the current state of the RVE is used as the reference one, then $\boldsymbol{l}_0 = \boldsymbol{l}$, and (7.2.41a) becomes

$$\dot{\bar{\mathbf{S}}}^{\mathrm{N}} = \mathrm{N} < \boldsymbol{l} \otimes \dot{\mathbf{f}} >_{\Omega} . \tag{7.2.41b}$$

7.3. FABRIC MEASURES IN AN ASSEMBLY OF RIGID GRANULES

As is shown in the preceding sections, the mechanical behavior of granular masses is strongly affected by their microstructure, namely the relative arrangement of voids and particles. *The microstructure is referred to as the granular fabric.* The parameters which characterize the granular fabric enter in a natural manner into the description of the overall macroscopic stress and deformation measures. In this section a number of scalar, vector, and tensor measures of granular fabric are examined for a random assembly of rigid spherical (Oda *et al.*, 1982b) and nonspherical granules. In Section 7.4, fabric tensors for general granular media are reexamined in light of experimental data.

7.3.1. Scalar Measures

A number of scalar measures are commonly used to characterize the microstructure of granular masses. These measures include:

- void ratio and porosity, e and n;
- density of grains, *i.e.,* the number of grains per unit volume, N_G;
- average coordination number, *i.e.,* the average number of contact points per particle, $\bar{n}_c$;
- density of contacts, *i.e.,* the number of contacts per unit volume, N; and
- average branch length, $\bar{l}$.

These are briefly discussed below.

Void Ratio and Porosity: Consider an RVE of an assembly of granules (not necessarily spherical) of total volume v, consisting of rigid particles of total volume v_s, and the associated voids of total volume v_v. The void ratio, e, and porosity, n, are defined by

$$e = \frac{v_v}{v_s}, \qquad n = \frac{v_v}{v_v + v_s} = \frac{e}{1+e}. \tag{7.3.1a-c}$$

The average mass density, $\bar{\rho}$, then is given by

$$\bar{\rho} = \frac{\rho_s\, v_s}{v_v + v_s} = \frac{\rho_s}{1+e}, \tag{7.3.1d,e}$$

where ρ_s is the mass density of the solid particles.

Density of Grains: Consider a random assembly of *rigid spheres*, and an RVE containing a very large number of particles. Denote by r_M and r_m the maximum and the minimum radii of the particles, $r_m \le r \le r_M$. Let f(r) define the distribution of radii, r, in the assembly, so that the fraction of the particles with radii ranging from r to r + dr is given by f(r) dr. Hence,

$$\int_{r_m}^{r_M} f(r)\, dr = 1. \tag{7.3.2a}$$

For any physical quantity, $\phi = \phi(r)$, define its average value by

$$< \phi >_R \equiv \int_{r_m}^{r_M} f(r)\, \phi(r)\, dr, \qquad R = [r_m, r_M]. \tag{7.3.2b,c}$$

Here, $R = [r_m, r_M]$ is the domain of integration.

Let N_G be the number of grains (spheres) per unit volume. Then the number of spheres of radii ranging from r to r + dr, per unit volume, is $N_G f(r)\, dr$, and the total solid volume per unit volume becomes

$$v_s = \frac{4}{3}\pi N_G \int_{r_m}^{r_M} f(r)\, r^3\, dr. \tag{7.3.3a}$$

Since for a unit volume, $v = 1 = (1+e)\,v_s$, it follows that

$$N_G = \frac{3}{4\pi(1+e)\,<r^3>_R}\,. \tag{7.3.3b}$$

This expression connects the number of grains per unit volume to the void ratio, e, and the grain-size distribution measured by $<r^3>_R$.

Average Coordination Number: Coordination number, n_c, refers to the number of contacts a grain has with its neighboring granules. The average value of the coordination number, taken over an RVE, is denoted by $\bar{n}_c$. Attempts have been made to relate $\bar{n}_c$ to the void ratio, e; see, *e.g.*, Smith *et al.* (1929) and Gray (1968). Figure 7.3.1 shows this relation, obtained experimentally by Oda (1977) for a homogeneous, two-sized mixture, and a four-sized mixture of spheres. The results suggest a unique relation between the average coordination number, $\bar{n}_c$, and the void ratio, e,

$$\bar{n}_c = G(e)\,, \tag{7.3.4}$$

independent of the grain-size distribution. This result has been obtained based on only limited experimental data for special distributions of spheres. It may not have general validity, although the average number of contacting points per particle should be related to the void ratio.

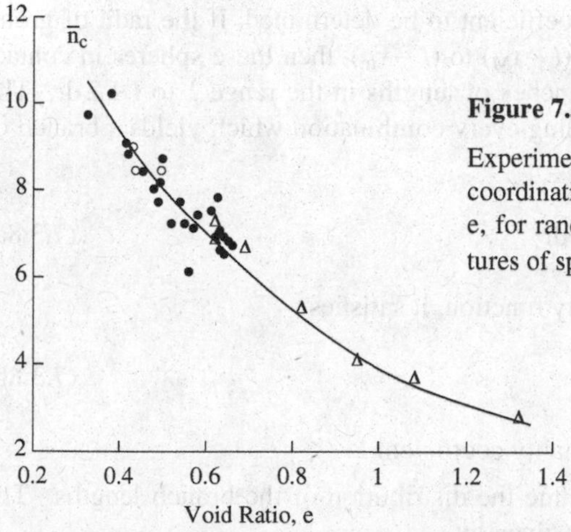

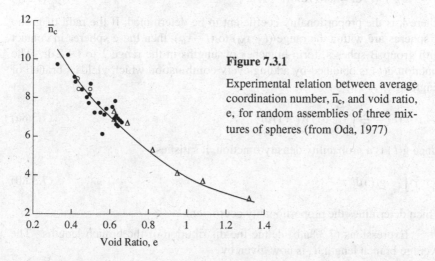

Figure 7.3.1
Experimental relation between average coordination number, $\bar{n}_c$, and void ratio, e, for random assemblies of three mixtures of spheres (from Oda, 1977)

Density of Contacts: A contact involves two contacting grains, each contributing a contacting point. Hence, the number of contacting points in an RVE, is twice the number of contacts. Let the density of contacts be denoted by N. With $\bar{n}_c$, the average coordination number, and N_G the number of grains per unit volume of the RVE, the contact density becomes, $N = \frac{1}{2}\bar{n}_c N_G$. Hence, for spherical granules,

$$N = \frac{3\,G(e)}{8\,\pi\,(1+e)\,<r^3>_R}. \tag{7.3.5}$$

Average Branch Length: Consider now the distribution of branch lengths, l, and seek to calculate their average, $\bar{l}$, for an RVE. For spheres, a typical branch length, l, is given by $l = r_1 + r_2$, where r_1 and r_2 are the radii of the corresponding contacting granules. In a mixture of spheres with maximum and minimum radii, r_M and r_m, the branch lengths fall in the range,

$$2\,r_m \leq l \leq 2\,r_M. \tag{7.3.6}$$

Let $g(l)$ be the density function defining the distribution of the branch lengths, l. It is reasonable to expect that $g(l)$ is related to the density function $f(r)$ which characterizes the distribution of the particle radii. Consider two groups (A and B) of spheres, group A with radii ranging from r to r + dr, and group B with radii ranging from $(l - r)$ to $(l - r) + dr$. For the A-spheres in contact with the B-spheres, the branch lengths fall in the range l to $l + 2\,dr$. In randomly distributed spheres, each branch is considered to be selected at random. The probability, $2\,g^{AB}(l)\,dr$, of selecting a branch with length in the interval l to $l + 2\,dr$ is proportional to the fractional number of particles belonging to the two groups, and hence it may be assumed that[18]

$$2\,g^{AB}(l)\,dr = k\,f(r)\,dr\,f(l - r)\,dr, \tag{7.3.7}$$

where k is the proportionality coefficient to be determined. If the radii of group A spheres are within the range $(l - r_M)$ to $(l - r_m)$, then these spheres in contact with group B spheres, form branches of lengths in the range l to $l + 2\,dr$. The function $g(l)$ is obtained by adding every combination which yields a branch of length l,

$$g(l) = \frac{k}{2}\int_{l - r_M}^{l - r_m} f(r)\,f(l - r)\,dr. \tag{7.3.8a}$$

Since $g(l)$ is a probability density function, it satisfies

$$\int_{2r_m}^{2r_M} g(l)\,dl = 1 \tag{7.3.8b}$$

which determines the proportionality coefficient k.

Expressions (7.3.8a,b) define the distribution of the branch lengths. The average branch length, $\bar{l}$, is now given by

$$\bar{l} = \int_{2r_m}^{2r_M} l\,g(l)\,dl. \tag{7.3.9}$$

---

18 Equation (7.3.7) assumes that the conditional probability of contacts is constant; this, in general, is not the case.

7.3.2. Vector Measures

The density and angular distribution of branches which intersect a given plane passing through a granular mass, can be expressed in terms of a *vectorial fabric measure* defined by $\mathbf{N} = 2\,N\,\bar{l} < \mathbf{m} >_{\Omega_{1/2}}$.

Density of Branch Intersections with a Plane: Consider an arbitrary plane with unit normal $\mathbf{v}$, intersecting an RVE of spherical granules. Identify as positive the face of this plane which is in the direction of $\mathbf{v}$, and the other face, as negative. Examine all branches which are intersected by this plane and which are formed by spheres with centers immediately on the negative side of the plane, contacting the spheres with centers immediately on the positive side of the plane; see Figure 7.3.2. Let the density, *i.e., the number per unit area,* of these intersections be $N^{(\mathbf{v})}$, and seek to express this in terms of the contact density, N, the distribution density function, $E(\mathbf{m})$, of the branch orientations, other relevant microstructural parameters, and the unit vector $\mathbf{v}$ which defines the orientation of the intersecting plane.

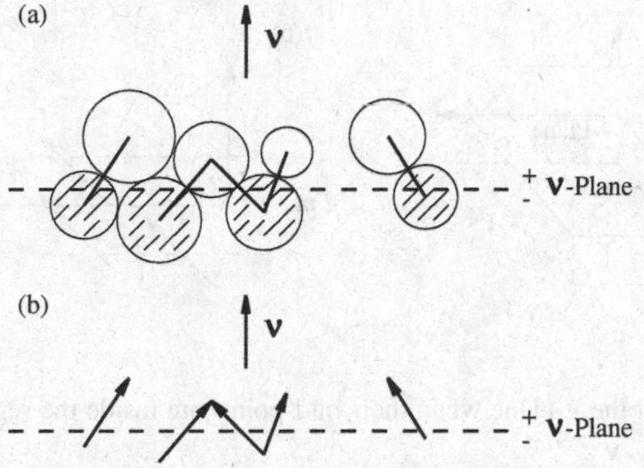

Figure 7.3.2

(a) An arbitrary plane of unit normal $\mathbf{v}$ intersecting an RVE of contacting spherical granules; (b) the $\mathbf{v}$-plane intersects branches of granules which are located in its immediate neighborhood, and $N^{(\mathbf{v})}$ is the density of the intersection points

It is shown in the sequel (Oda *et al.*, 1982b) that

$$N^{(\mathbf{v})} = N\,\bar{l} < |\mathbf{m} \cdot \mathbf{v}| >_{\Omega},$$

$$N^{(\mathbf{v})}\,E^{(\mathbf{v})}(\mathbf{m}) = 2\,N\,\bar{l}\,E(\mathbf{m})\,\mathbf{m} \cdot \mathbf{v}, \tag{7.3.10a,b}$$

where the notation (7.2.16b) is used, and $E^{(\mathbf{v})}(\mathbf{m})$ is the *surface* distribution density function, defined such that $E^{(\mathbf{v})}(\mathbf{m})\,d\Omega$ is the fractional number of branches with orientations in $d\Omega$, which intersect the $\mathbf{v}$-plane.[19]

[19] Note that only the line segments (branches) which connect centers of the spheres immediately next to the plane, can intersect this plane.

To show the validity of (7.3.10a), divide all particles located in the immediate vicinity of the **ν**-plane into two groups (see Figure 7.3.2): Group A members (hatched) have their centers on the negative side of the plane, and Group B members (unhatched) have their centers on the positive side of the plane. Solid lines in Figure 7.3.2 are branches connecting the centers of the hatched particles with the contacting unhatched ones.[20] *Assume that the branch lengths and orientations are uncorrelated.* Then the density of the mid-branch points which: (1) belong to either group; (2) have lengths in the range l to $l + \mathrm{d}l$; and (3) their orientations fall in the solid angle $\mathrm{d}\Omega$ of orientation **m**, is given by $2\,g(l)\,E(\mathbf{m})\,\mathrm{d}l\,\mathrm{d}\Omega$. If a branch of this kind has its mid-branch point at a distance less than or equal to $\frac{1}{2}\,l\,\mathbf{m}\cdot\boldsymbol{\nu}$ from the **ν**-plane, then this branch intersects the **ν**-plane; see Figure 7.3.3. To find the density of such intersections, multiply the above expression by the volume $1 \times 1 \times l\,\mathbf{m}\cdot\boldsymbol{\nu}$, obtaining $2\,N\,g(l)\,l\,E(\mathbf{m})\;\mathbf{m}\cdot\boldsymbol{\nu}\,\mathrm{d}l\,\mathrm{d}\Omega$. Upon integrating over l and Ω, and assuming the branch lengths and branch orientations are uncorrelated, arrive at the average number per unit area, $N^{(\nu)}$, of branches which intersect the **ν**-plane,

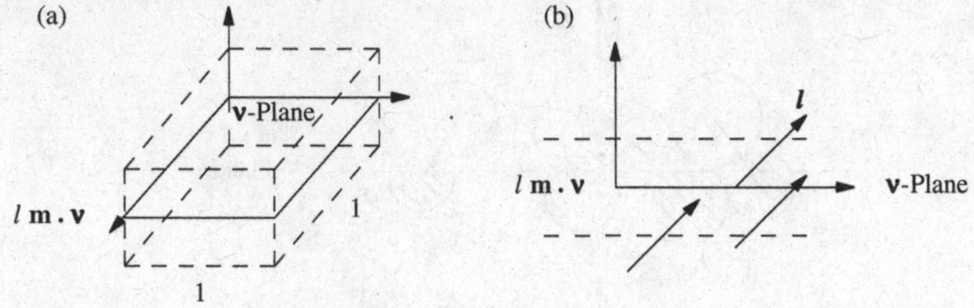

Figure 7.3.3

Branches intersect the **ν**-plane when their mid-points are inside the region with volume $1 \times 1 \times l\,\mathbf{m}\cdot\boldsymbol{\nu}$

$$N^{(\nu)} = 2\,N \int_{2r_m}^{2r_M} l\,g(l)\,\mathrm{d}l \int_{\Omega_{1/2}} E(\mathbf{m})\,\mathbf{m}\cdot\boldsymbol{\nu}\,\mathrm{d}\Omega$$

$$= 2\,N\bar{l}\,<\mathbf{m}>_{\Omega_{1/2}}\cdot\,\boldsymbol{\nu}\,, \qquad (7.3.11a,b)$$

where $\bar{l}$ is the average branch length defined by (7.3.9). This result may also be expressed in the following alternative way:

$$N^{(\nu)} = N \int_{2r_m}^{2r_M} l\,g(l)\,\mathrm{d}l \int_{\Omega} E(\mathbf{m})\;|\,\mathbf{m}\cdot\boldsymbol{\nu}\,|\,\mathrm{d}\Omega$$

---

[20] When an attempt is made to relate the (discrete) microscopic contact forces to the overall macroscopic (continuum) tractions transmitted across an interior imagined plane, then the density, $N^{(\nu)}$, of the branches connecting the two groups across this plane, and the angular distribution, $E^{(\nu)}(\mathbf{m})$, of these branches, become of paramount importance; see Subsection 7.2.4, page 522. Note that $E^{(\nu)}(\mathbf{m})$ is a surface density function, different from $E(\mathbf{m})$ which defines the angular distribution of all branches within an RVE.

$$= \mathrm{N}\bar{l} < |\mathbf{m} \cdot \mathbf{v}| >_{\Omega}. \tag{7.3.11c,d}$$

Introduce the notation

$$\mathbf{N} = 2\mathrm{N}\bar{l} \int_{\Omega_{1/2}} \mathrm{E}(\mathbf{m})\, \mathbf{m} = 2\mathrm{N}\bar{l} < \mathbf{m} >_{\Omega_{1/2}}, \tag{7.3.12a,b}$$

and note that

$$\mathrm{N}^{(\mathbf{v})} = \mathbf{N} \cdot \mathbf{v}. \tag{7.3.13}$$

In particular, the density of intersections with coordinate planes, x_1, x_2, and x_3, becomes

$$\mathrm{N}_i = \mathbf{N} \cdot \mathbf{e}_i, \quad i = 1, 2, 3, \tag{7.3.14}$$

where $\mathbf{e}_i$'s are the coordinate unit base vectors. The quantity $\mathbf{N} = 2\mathrm{N}\bar{l} < \mathbf{m} >_{\Omega_{1/2}}$ is a *vectorial fabric measure.*

Angular Distribution of Branches Intersecting the $\mathbf{v}$-plane: Let $d\mathrm{N}^{(\mathbf{v})}(\mathbf{m})$ be the density (per unit area of the $\mathbf{v}$-plane) of branches which intersect the $\mathbf{v}$-plane and lie within $d\Omega$ of orientation $\mathbf{m}$. It is given by

$$d\mathrm{N}^{(\mathbf{v})}(\mathbf{m}) = 2\mathrm{N}\bar{l}\, \mathbf{m} \cdot \mathbf{v}\, \mathrm{E}(\mathbf{m})\, d\Omega. \tag{7.3.15a}$$

On the $\mathbf{v}$-plane, introduce a probability density function, $\mathrm{E}^{(\mathbf{v})}(\mathbf{m})$, to describe the angular distribution of the branches intersecting this plane. Then, $d\mathrm{N}^{(\mathbf{v})}(\mathbf{m})$ is also given by

$$d\mathrm{N}^{(\mathbf{v})}(\mathbf{m}) = \mathrm{N}^{(\mathbf{v})}\mathrm{E}^{(\mathbf{v})}(\mathbf{m})\, d\Omega. \tag{7.3.15b}$$

From (7.3.15a,b), expression (7.3.10b) now follows. In view of (7.3.11d), this expression may also be written as

$$< \mathbf{m} >_{\Omega_{1/2}} \cdot\, \mathbf{v}\, \mathrm{E}^{(\mathbf{v})}(\mathbf{m}) = \mathbf{m} \cdot \mathbf{v}\, \mathrm{E}(\mathbf{m}). \tag{7.3.16a}$$

Since $< \mathbf{m} >_{\Omega_{1/2}} \cdot\, \mathbf{v}$ is defined in terms of $\mathrm{E}(\mathbf{m})$, it is seen that $\mathrm{E}^{(\mathbf{v})}(\mathbf{m})$ is completely characterized by this distribution density function,

$$\mathrm{E}^{(\mathbf{v})}(\mathbf{m}) = \frac{\mathbf{m} \cdot \mathbf{v}}{< \mathbf{m} >_{\Omega_{1/2}} \cdot\, \mathbf{v}}\, \mathrm{E}(\mathbf{m}). \tag{7.3.16b}$$

Solid Paths: The photoelastic images of Figure 7.1.1, page 505, show that the load-carrying contacts in frictional granular media, line up more or less in the general directions of the principal stresses. The notion of *solid paths* was introduced by Horne (1965) as a quantitative measure of this characteristic.

Consider a general direction defined by a unit vector, say, $\mathbf{v}$, and planes normal to this direction, which pass through the centers of the spherical granules; see Figure 7.3.4. Each particle is divided into two portions, one on the positive and the other on the negative side of the intersecting plane, as shown in the figure. For each particle, select at random one contact on each side of the intersecting plane, say, particle g_i and contact c_{i-j}. The contact c_{i-j} is on the positive side of particle g_i, but at the same time, it is on the negative side of particle g_j. A contact is then chosen at random among those on the positive portion of particle g_j, and it is connected by a straight line to c_{i-j}. This process is continued, connecting consecutively the selected contacts to obtain a zigzag path which is called a *solid path* by Horne (1965).

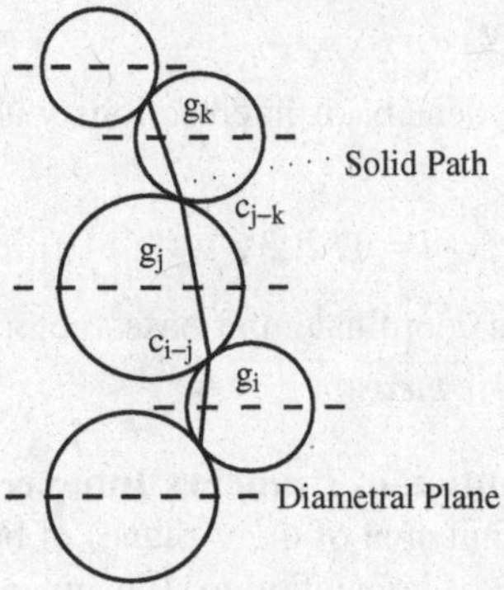

Figure 7.3.4

A solid path in spherical granules

If $\mathbf{v}$ defines a symmetry axis of $E(\mathbf{m})$, the general trend of the corresponding solid paths would correspond to this direction; otherwise the solid paths would deviate from the $\mathbf{v}$-direction. Let $\mathbf{v}$ define a symmetry axis of $E(\mathbf{m})$. The average number $N^{(SP)}(\mathbf{v})$ of contacts per unit distance measured in the $\mathbf{v}$-direction, which fall on the corresponding solid path, is given by

$$N^{(SP)}(\mathbf{v}) = \frac{1}{4\bar{r} < \mathbf{m} >_{\Omega_{1/2}} \cdot \mathbf{v}}, \tag{7.3.17a}$$

where $\bar{r}$ is the average radius of the spheres. This is obtained by noting that $\mathbf{m} \cdot \mathbf{v}$ is the projection length of the orientation $\mathbf{m}$ in the $\mathbf{v}$-direction, so that the average length of the segments of a solid path is given by $2\bar{r} < |\mathbf{m} \cdot \mathbf{v}| >_{\Omega}$.

From (7.3.17a), (7.3.11), and (7.3.13), it follows that

$$\mathbf{N} \cdot \mathbf{v} = \frac{\bar{l}}{2\bar{r}} \frac{N}{N^{(SP)}(\mathbf{v})}. \tag{7.3.17b}$$

For the assembly of uniform spheres, every branch length is 2r. In this case, $\bar{l} = 2r$, and it follows that

$$\mathbf{N} \cdot \mathbf{v} = \frac{N}{N^{(SP)}(\mathbf{v})}. \tag{7.3.17c}$$

Choosing $\mathbf{v}$ along the coordinate directions, expressions for N_i, i = 1, 2, 3, are obtained.

7.3.3. Fabric Tensors

Several fabric tensors appeared in Section 7.2 in connection with the description of stress tensors in terms of the contact forces and parameters which characterize the microstructure of the granular mass. For an assembly of non-spherical contacting granules, tensorial quantities which describe the distribution of contact unit normal vectors, $\mathbf{n}^\alpha$, and unit branch vectors, $\mathbf{m}^\alpha$, are suitable measures of the fabric or the microstructure of the material. Some of these measures are examined in the sequel.

Angular Distribution of Branches and Contact Normals: When there are a large number of contact classes, then the distribution of unit branches, $\mathbf{m}^\alpha$, can be described by a continuous function $E(\mathbf{m}) = E(-\mathbf{m})$ which satisfies (7.2.15). Similar comments apply to the distribution of contact unit normal vectors, $\mathbf{n}^\alpha$. Since the mathematical description of these two sets of unit vectors, is essentially the same, in what follows, the representation of the distribution of unit branch vectors is examined in detail and then the results are directly applied to the distribution of contact normals by a simple notational change.

Transition from discrete to continuous representation of the distribution of unit vectors, requires a *model* and involves certain assumptions; see Mardia (1972) and Kanatani (1984). Usually, a functional form with a number of free parameters is first assumed, and then these parameters are evaluated by seeking to minimize a measure of the difference or the *distance* between the corresponding original data points and the model results. Perhaps the most common measure of distance is *the mean-square error,* but other equally appealing measures have been used.

Consider a given set of unit vectors, $\mathbf{m}^a$, a = 1, 2, ..., N, defining, *e.g., experimentally measured* orientations of the branch vectors; see Section 7.4, page 544, for actual data. This set can be described in terms of its tensorial products, averaged over the data set. As before, denote the mean value of a physical quantity, say, ϕ^a, taken over their discrete data points, by $<\phi>$,

$$<\phi> \equiv \frac{1}{N}\sum_{a}^{N}\phi^a. \tag{7.3.18}$$

Now, using the *measured* branch orientations, form the following n'th-order symmetric tensor:

$$<m_{k_1} m_{k_2} \dots m_{k_n}> = \frac{1}{N}\sum_{a}^{N} m_{k_1}^a m_{k_2}^a \dots m_{k_n}^a\,. \tag{7.3.19a}$$

Since for each unit branch vector $\mathbf{m}^a$, the data point includes $-\mathbf{m}^a$, these tensors are of even order, *i.e.,* n = 0, 2, ... Furthermore, since the $\mathbf{m}^a$'s are unit vectors, the contraction of any two indices reduces the order of the tensor by two, *i.e.,* setting, say, $k_n = k_{n-1}$ and summing over $k_{n-1} = 1, 2, 3$ (in three dimensions), or $k_n = 1, 2$ (in two dimensions), a tensor of order $n-2$ results,

$$<m_{k_1} m_{k_2} \dots m_{k_{n-1}} m_{k_{n-1}}> = <m_{k_1} m_{k_2} \dots m_{k_{n-2}}>,$$

$$m_{k_{n-1}} m_{k_{n-1}} = 1\,. \tag{7.3.19b,c}$$

For n = 2, the fabric tensor $\boldsymbol{\mathcal{J}}$ with components $\mathcal{J}_{ij} = \langle m_i m_j \rangle$, and for n = 4, the fabric tensor $\boldsymbol{\mathcal{A}}$ with components $\mathcal{A}_{ijkl}$, is obtained; see (7.2.33d,e) and (7.2.35b), pages 529, 530.

To unify the notation, introduce a 2n'th-order fabric tensor $\boldsymbol{\mathcal{J}}^{(2n)}$ with components

$$\mathcal{J}_{k_1 k_2 \ldots k_{2n}} = \langle m_{k_1} m_{k_2} \ldots m_{k_{2n}} \rangle\,, \tag{7.3.20a}$$

and then specialize the results to specific cases. Hence, for n = 0, 1, and 2,

$$1 = \boldsymbol{\mathcal{J}}^{(0)}, \quad \boldsymbol{\mathcal{J}} = \boldsymbol{\mathcal{J}}^{(2)}, \quad \text{and} \quad \boldsymbol{\mathcal{A}} = \boldsymbol{\mathcal{J}}^{(4)} \tag{7.3.20b-d}$$

are obtained.

For a given discrete data set, a continuous distribution density function, E(**m**), is now defined in terms of the fabric tensors $\boldsymbol{\mathcal{J}}^{(2n)}$'s, as follows:

$$E(\mathbf{m}) = \frac{1}{2\pi(r-1)}\,(1 + \mathcal{E}_{ij} m_i m_j + \mathcal{E}_{ijkl} m_i m_j m_k m_l + \ldots)\,, \tag{7.3.21a}$$

where r is given by

$$r = \begin{cases} 2 & \text{for two dimensions} \\ 3 & \text{for three dimensions}\,. \end{cases} \tag{7.3.21b}$$

The coefficients in this representation of E(**m**) are obtained by minimizing the sum of the squared deviations from the actual data points; see Kanatani (1984) for details. This procedure leads to a set of symmetric even-order tensors, $\mathcal{E}_{ij}$, $\mathcal{E}_{ijkl}$, ..., which are all deviatoric, and, more importantly, they form an orthogonal set, so that each term can be computed independently of the others. They can be expressed in terms of $\mathcal{J}_{k_1 k_2 \ldots k_{2n}}$'s. For application to granular media, attention is confined only to second- and fourth-order fabric tensors of this kind, and hence only these are listed below. Thus, for two dimensions,

$$\mathcal{E}_{ij} = 4\,(\mathcal{J}_{ij} - \frac{1}{2}\,\delta_{ij})\,,$$

$$\mathcal{E}_{ijkl} = 16\,\{\mathcal{J}_{ijkl} - \delta_{(ij}\,\mathcal{J}_{kl)} + \frac{1}{8}\,\delta_{(ij}\,\delta_{kl)}\}\,, \tag{7.3.22a,b}$$

where parentheses around subscripts designate symmetrization of the corresponding indices.[21] For three dimensions, on the other hand, obtain

$$\mathcal{E}_{ij} = \frac{15}{2}\,(\mathcal{J}_{ij} - \frac{1}{3}\,\delta_{ij})\,,$$

$$\mathcal{E}_{ijkl} = \frac{315}{8}\,\{\mathcal{J}_{ijkl} - \frac{6}{7}\,\delta_{(ij}\,\mathcal{J}_{kl)} + \frac{3}{35}\,\delta_{(ij}\,\delta_{kl)}\}\,. \tag{7.3.23a,b}$$

---

[21] For example, $\delta_{(ij}\,\mathcal{J}_{kl)} = (\delta_{ij}\,\mathcal{J}_{kl} + \delta_{li}\,\mathcal{J}_{jk} + \delta_{kl}\,\mathcal{J}_{ij} + \delta_{jk}\,\mathcal{J}_{li})/4$.

For the contact unit normals, a similar procedure leads to a continuous distribution density function, $\hat{E}(\mathbf{n})$, which can be expressed as follows:

$$\hat{E}(\mathbf{n}) = \frac{1}{2\pi(r-1)}\,(1 + \hat{\mathcal{E}}_{ij}\,n_i\,n_j + \hat{\mathcal{E}}_{ijkl}\,n_i\,n_j\,n_k\,n_l + ...)\,, \tag{7.3.24}$$

where the $\hat{\boldsymbol{\mathcal{E}}}$-tensors are given by the same expressions as in (7.3.22) and (7.2.23) with the fabric tensors $\boldsymbol{\mathcal{J}}^{(2n)}$'s now being identified with $\hat{\mathcal{J}}^{(2n)}$'s which are defined by

$$\hat{\mathcal{J}}_{k_1 k_2 \dots k_{2n}} = \langle n_{k_1}\,n_{k_2} \dots n_{k_{2n}} \rangle\,. \tag{7.3.25}$$

7.3.4. Other Fabric Measures

The distribution of the orientations of a given set of data points defined by, say, the measured unit branch vectors, $\mathbf{m}^{\alpha}$, $\alpha = 1, 2, ..., 2N$, may also be represented directly in terms of the tensor $\boldsymbol{\mathcal{J}}^{(2n)}$, inasmuch as an approximation of the (continuous) distribution function, $E(\mathbf{m})$, up to the 2n'th order, is actually provided by

$$E(\mathbf{m}) \approx A\,\mathcal{F}_{k_1 k_2 \dots k_{2n}}\,m_{k_1}\,m_{k_2} \dots m_{k_{2n}}\,. \tag{7.3.26}$$

This follows from the fact that the members of the set $(1, m_i\,m_j, m_i\,m_j\,m_k\,m_l, ...)$ are not linearly independent, and that each member can produce all other preceding members as special cases, *i.e.*, by simple contraction of the relevant indices. For n = 0, 1, and 2, the coefficients are

$$\mathcal{F} = 1\,, \quad \mathcal{F}_{ij} = \frac{15}{2}\,\{\mathcal{J}_{ij} - \frac{1}{5}\,\delta_{ij}\}\,,$$

$$\mathcal{F}_{ijkl} = \frac{315}{8}\,\{\mathcal{J}_{ijkl} - \frac{2}{3}\,\delta_{(ij}\,\mathcal{J}_{kl)} + \frac{1}{21}\,\delta_{(ij}\,\delta_{kl)}\}\,, \tag{7.3.27a-c}$$

for three dimensions, and, for two dimensions they are given by

$$\mathcal{F} = 1\,, \quad \mathcal{F}_{ij} = 4\,\{\mathcal{J}_{ij} - \frac{1}{4}\,\delta_{ij}\}\,,$$

$$\mathcal{F}_{ijkl} = 16\,\{\mathcal{J}_{ijkl} - \frac{3}{4}\,\delta_{(ij}\,\mathcal{J}_{kl)} + \frac{1}{16}\,\delta_{(ij}\,\delta_{kl)}\}\,. \tag{7.3.28a-c}$$

It should be noted that, for the same degree of approximation, representations (7.3.21a) and (7.3.26) produce the same expressions for $E(\mathbf{m})$, as they should.

In Section 7.2, two other fabric tensors appear in the description of the stress tensor, *i.e.*, $\boldsymbol{\Phi} = N\bar{l} \langle \mathbf{m} \otimes \mathbf{m} \rangle_{\Omega}$ and $\hat{\boldsymbol{\Phi}} = N\bar{l} \langle \mathbf{n} \otimes \mathbf{n} \rangle_{\Omega}$, where N is the density of contacts and $\bar{l}$ is the average branch length. These fabric tensors include additional information on the microstructure, inasmuch as the density of the contacts and the average length of the branches (*i.e.*, the average grain size) are also incorporated in their definitions.

Since the overall Cauchy stress tensor may be expressed as $\bar{\boldsymbol{\sigma}} = N\bar{l} \langle \mathbf{m} \otimes \mathbf{f} \rangle_{\Omega}$, or, for spherical granules, as $\bar{\boldsymbol{\sigma}} = N\bar{l} \langle \mathbf{n} \otimes \mathbf{f} \rangle_{\Omega}$, other fabric tensors which include some measure of the intensity of the contact forces, are also useful in describing the load-carrying mechanisms in granular masses; see Nemat-Nasser and Mehrabadi (1983). In the next section, the distribution of a measure called the *fringe bias* is considered for this purpose. *The fringe bias measures the orientation and the intensity of the contact force, based on the structure of the fringes which occur in rod-shaped photoelastic granules.*

7.4. EXPERIMENTAL EVALUATION OF FABRIC-STRESS RELATIONS

This section is concerned with the experimental observation of the microstructure (fabric) and its relation to the overall stress and deformation measures in granular materials. Experimental procedures used to make direct observation of the microstructural changes during biaxial loading and simple shearing of photoelastic granules are briefly examined. The results of such experiments are used to provide an understanding of the overall mechanical response of this class of materials, and their microstructure-properties relations. The main objectives are:

(1) to measure and compare the components of various fabric tensors;

(2) to verify experimentally the relation between the overall stress and fabric measures;

(3) to observe how the orientations of the principal axes of each tensorial fabric measure change over several cycles of deformation;

(4) to examine the representation of the distribution density functions of the unit contact normals and the unit branch vectors, and to establish the required accuracy in the order of their harmonic expansion; and finally

(5) to study the relation between the overall macroscopic stress tensor and the local quantities such as the distribution of contact forces.

The biaxial compression tests were performed on assemblies of oval cross-sectional rods by the author and coworkers.[22] The objectives of these experiments have been to evaluate the effects of interparticle friction, particle shape, and initial fabric on the overall strength of granular materials. The variation in the spatial arrangement of the particles and particle rolling and sliding are monitored by recording photoelastic images of the microstructure at various stages during the course of deformation. Based on these observations, it is

[22] These experiments and the analyses have been performed at Northwestern University in collaboration with Professors Konishi and Oda, and have been reported in Oda, Konishi, and Nemat-Nasser (1982a,1985) and Konishi, Oda, and Nemat-Nasser (1982). The results on biaxial experiments reported in this section are based on these articles.

concluded that:

(1) particle rolling is a major microscopic deformation mechanism, especially when interparticle friction is large;

(2) there are relatively few contacts at which relative sliding is dominant, and this seems to be true even when the assembly reaches the overall failure state; this observation is in contradiction to the common assumption that particle sliding is the major microscopic deformation mode;

(3) during the course of deformation and up to the peak stress, new contacts are continually formed in such a manner that the contact unit normals tend to concentrate more in a direction parallel to the maximum principal compression. This concentration of unit normals is closely related to the formation of new column-like load paths which carry the load as the axial stress is increased. After the peak stress, such a column-like microstructure disappears and considerable rearrangement of the load paths takes place, leading to a more diffused (homogeneous) microstructure in the critical state; and

(4) the overall stress tensor, $\bar{\boldsymbol{\sigma}}$, tends to become coaxial with the fabric tensor $\mathcal{J} = <\mathbf{m} \otimes \mathbf{m}>$, as the deformation continues (here, $\mathbf{m}$ is the unit branch vector).

The cyclic simple shearing[23] experiments were performed on the same photoelastic granules. Results of these experiments also show a close relation between the fabric and stress tensors. In particular, it is found that:

(1) the off-diagonal terms in the fabric tensors, $\mathcal{J} = <\mathbf{m} \otimes \mathbf{m}>$ and $\hat{\mathcal{J}} = <\mathbf{n} \otimes \mathbf{n}>$, closely follow the overall stress-strain relation (here, $\mathbf{n}$ is the unit contact normal vector);

(2) the second-order representation of the distribution density function of contact normals and unit branches, *i.e.*, $\hat{\mathrm{E}}(\mathbf{n})$ and $\mathrm{E}(\mathbf{m})$, may not adequately capture the involved anisotropy; hence, the fourth-order terms, *i.e.*, $\mathcal{J}^{(4)} = <\mathbf{m} \otimes \mathbf{m} \otimes \mathbf{m} \otimes \mathbf{m}>$ and $\hat{\mathcal{J}}^{(4)} = <\mathbf{n} \otimes \mathbf{n} \otimes \mathbf{n} \otimes \mathbf{n}>$, must be included;

(3) the tensor, $\frac{1}{2} <\mathbf{m} \otimes \mathbf{f} + \mathbf{f} \otimes \mathbf{m}>$, is indeed proportional to the macroscopic stress; and

(4) the diagonal terms, $<\mathrm{m}_1 \mathrm{f}_1>$ and $<\mathrm{m}_2 \mathrm{f}_2>$, remain constant, representing the constant confining pressure, whereas the off-diagonal terms, $\frac{1}{2} <\mathrm{m}_1 \mathrm{f}_2 + \mathrm{m}_2 \mathrm{f}_1>$, follow the variation of the applied shear stress; this is also the case for the fabric tensors $\hat{\mathcal{J}}$ and $\mathcal{J}$.

---

[23] These experiments and the analyses were performed at University of California, San Diego (UCSD) in collaboration with Professor Mehrabadi and former graduate students, Dr. Shodja and Professor Subhash. The results have been reported in Mehrabadi, Nemat-Nasser, Shodja, and Subhash (1988) and Subhash, Nemat-Nasser, Mehrabadi, and Shodja (1991), and are used in this section.

These and related issues are discussed in the present section.

7.4.1. Photoelastic Granules

Rod-like particles of common length, 19mm, with oval cross sections are cast from polyurethane rubber with a photoelastic constant of 82.5mm/kg. Two kinds of cross-sectional shapes, referred to as oval I and oval II in the sequel, are employed. For each shape, three different size particles, *i.e.*, large, medium, and small, are prepared. The particle dimensions are given in Table 7.4.1. The ratio of the maximum, r_1, to the minimum, r_2, of the principal radii of oval I is 1.1, whereas that of oval II is 1.4.

Table 7.4.1

Dimensions of the Photoelastic Oval Particles

Shape	Aspect Ratio	Diameter, r_1/r_2, mm
Oval II	1.4	Large : 16.0/11.3 Medium : 10.7/7.4 Smal : 7.1/4.9
Oval I	1.1	Large : 14.8/13.4 Medium : 9.9/8.9 Small : 6.3/5.7

To assess the influence of intergranular friction on the overall response of the granular mass, two sets of experiments are performed, one with nonlubricated particles of overall friction angle 52°, the other with particles which are lubricated with talcum powder, resulting in an overall friction angle of 26°. Figure 7.4.1 gives the measured values of the required forces used to evaluate the coefficient of friction for each case. Here, T and N, respectively, are the tangential and normal loads necessary to overcome the frictional resistance in each case. The solid circles are for the lubricated, and the open circles are for the nonlubricated particles.

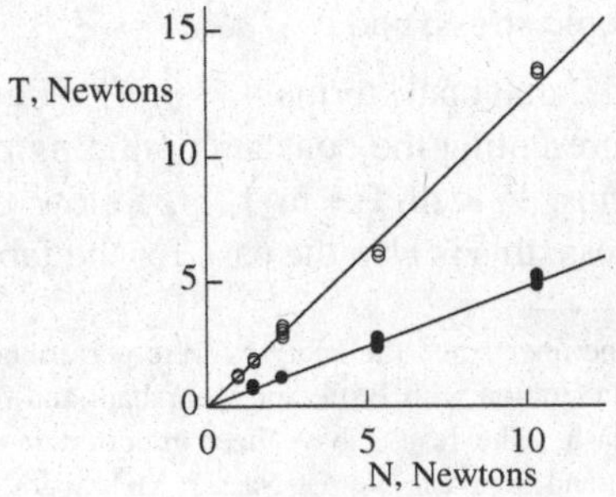

Figure 7.4.1

Experimental evaluation of the intergranular friction (from Oda *et al.*, 1982a)

7.4.2. Biaxial Experiments

Apparatus: The testing apparatus for the biaxial experiments, is shown in Figure 7.4.2. It consists of an *overall frame* within which a biaxial *loading frame* is mounted. The overall frame can be tilted in its plane and held at a desired angle; it can also be tilted and held out of its plane. This flexibility permits sample preparation with various *bedding angles* θ, Figure 7.4.3, and hence, various initial (inherent) anisotropies.

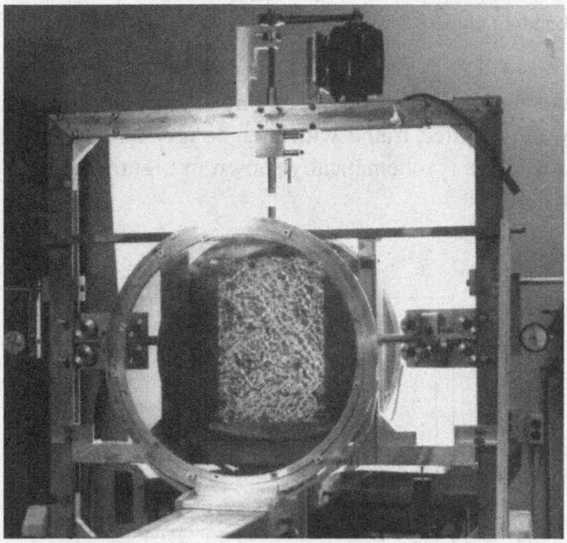

Figure 7.4.2

Experimental setup, showing the overall frame within which the biaxial loading frame is mounted (author's laboratory, Northwestern University, 1981)

The biaxial loading frame is designed[24] such that its opposite boundaries (bars) remain parallel, as the sample deforms. When the lateral boundaries (bars) are subjected to a constant load, then they can expand or contract, while remaining parallel and vertical, during the deformation of the granular assembly. Hence, the sample may dilate or densify, depending on the state of deformation imposed by the relative movements of the upper and lower bars. Figure 7.4.4 schematically shows the test apparatus and identifies its various components.

Sample Preparation: The sample is formed within the loading frame of initial dimensions of about 330mm in height and 207mm in width. Each sample

[24] The apparatus was designed and constructed at Northwestern University by the author and his technical assistant, Mr. John Schmidt, 1980.

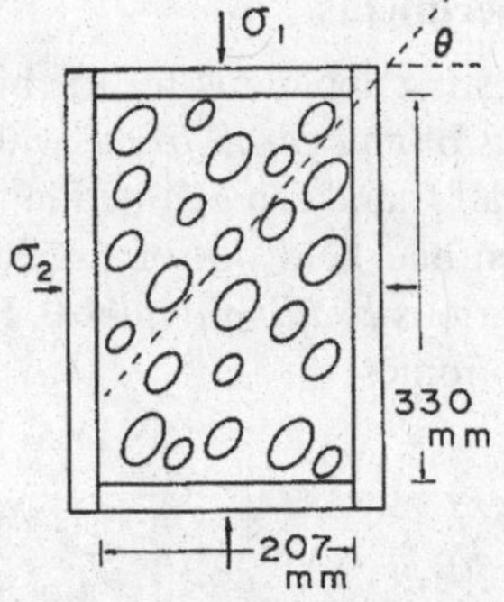

Figure 7.4.3

Sample dimensions and definition of the bedding angle θ

consists of equal numbers of particles of three sizes. Particles are stacked by hand randomly in the tilted frame with their longer axis essentially parallel to the bedding plane. This is schematically shown in Figure 7.4.3.

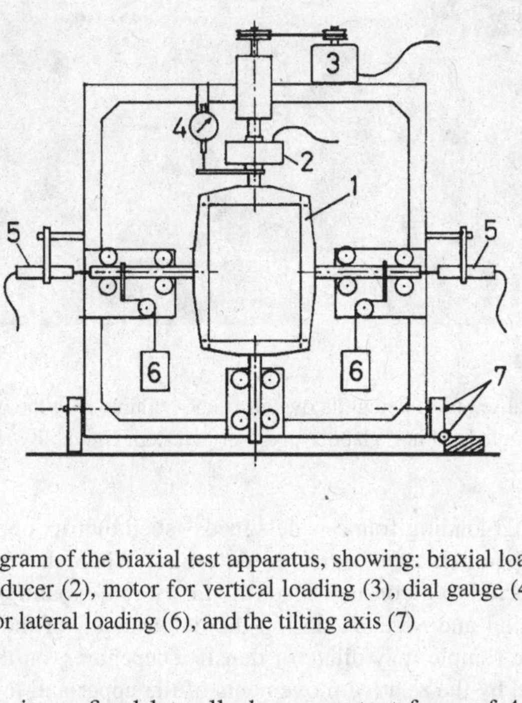

Figure 7.4.4

Schematic diagram of the biaxial test apparatus, showing: biaxial loading frame (1), load transducer (2), motor for vertical loading (3), dial gauge (4), LVDT's (5), weights for lateral loading (6), and the tilting axis (7)

The sample is confined laterally by a constant force of 4.41N (0.45kg), and it is compressed vertically downward by incremental displacement of the upper bar of the frame. The vertical load is measured by a load transducer, and the vertical and lateral displacements by a dial gauge and two laterally mounted LVDT's, respectively. The sample is placed in the field of a circular polariscope consisting of a monochromatic light source, a polarizer, two quarter-wave plates, and an analyzer. At each stage of loading, photographs of the isochromatic fringe pattern in the stressed assembly are taken. Each stage is identified by a *stage number* which then appears in the corresponding stress-strain graph. These images were later analyzed manually.[25]

Test Results: In what follows, the axial stress and strain are denoted by σ_1 and ε_1, respectively,[26] and the lateral ones by σ_2 and ε_2. The interparticle friction angle is denoted by ϕ_μ, and is assumed to be constant. Results of sixteen tests are used in the analysis. These tests are listed in Table 7.4.2, according to the particle shape (whether oval I or oval II); the interparticle friction angle (ϕ_μ = 26° or 52°); the value of the bedding angle, θ, which defines the initial anisotropy of the sample; the value of the initial void ratio, e; the number of particles used in each sample; the value of the observed peak stress ratio, $(\sigma_1/\sigma_2)_{max}$; and finally, the value of the effective internal friction for the overall sample, ϕ_{eff}.

Table 7.4.2

Properties of samples used in experiments

Shape of Particles	Friction Angle ϕ_μ	Bedding Angle θ	Void Ratio e	Number of Particles	Peak Stress Ratio $(\sigma_1/\sigma_2)_{max}$	Effective Friction ϕ_{eff}
		0°	0.190	674	7.0	49°
Oval I	52°	30°	0.183	679	4.7	40°
		60°	0.185	676	4.1	37°
		90°	0.192	676	5.9	45°
		0°	0.177	719	14.4	60°
Oval II	52°	30°	0.154	724	11.4	57°
		60°	0.159	721	8.3	52°
		90°	0.155	729	10.4	55°
		0°			4.9	42°
Oval I	26°	30°	0.169	689	6.0	46°
		60°	0.174	686	5.4	43°
		90°	0.168	693	4.8	41°
		0°	0.158	726	11.8	58°
Oval II	26°	30°	0.160	720	7.5	50°
		60°	0.155	726	4.3	39°
		90°		727	5.5	44°

The results of biaxial compression tests are given in Figures 7.4.5a,b, as relations between the stress ratio σ_1/σ_2 and the axial strain ε_1, and between the

[25] This is in contrast to the analyses of the simple shearing test results, reported in Subsection 7.4.3, which were performed using powerful computer programs which reduced the required overall effort by a factor of 20.

[26] In general, the principal directions of the stress and strain tensors do not coincide for frictional granules, as has been discussed in Chapter 4 and the present chapter; see Rudnicki and Rice (1975), Nemat-Nasser *et al.* (1981), and Nemat-Nasser (1983).

volumetric strain $\varepsilon_v = \varepsilon_1 + \varepsilon_2$ and ε_1. These results show that all assemblies tend to densify first, followed by extensive dilatancy, as is the case for all frictional granular masses.

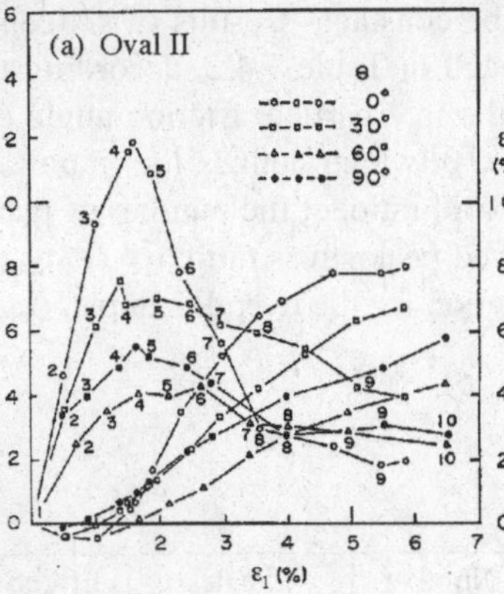

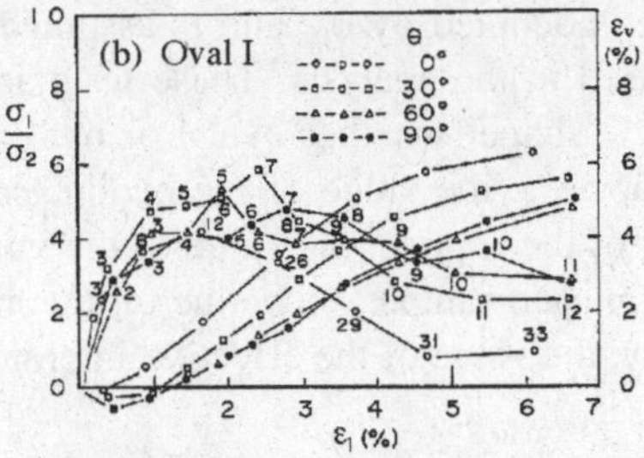

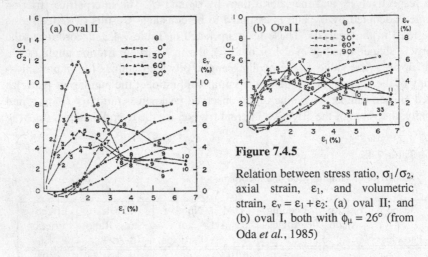

Figure 7.4.5

Relation between stress ratio, σ_1/σ_2, axial strain, ε_1, and volumetric strain, $\varepsilon_v = \varepsilon_1 + \varepsilon_2$: (a) oval II; and (b) oval I, both with $\phi_\mu = 26°$ (from Oda *et al.*, 1985)

In Figure 7.4.6, the stress ratio at failure, $(\sigma_1/\sigma_2)_f$, is plotted against the bedding angle, θ. As is seen from these data, for oval II (flat) particles, the value of the stress ratio, σ_1/σ_2, at failure depends strongly on the bedding angle, θ. For both 26° and 52° interparticle friction angles, the peak value of the stress ratio, $(\sigma_1/\sigma_2)_{max}$, is greatest for $\theta = 0°$, becomes a minimum at $\theta = 60°$, and then again increases. For the oval I (more round) particles, on the other hand, no such distinct variation of the peak stress ratio with the bedding angle, is observed. This suggests that, to a large extent, the inherent anisotropy in natural deposits of granular masses may stem from the parallel alignment of flat grains. For the assembly of oval II particles, the peak stress ratio exhibits strong dependence on the magnitude of the internal friction. No such dependence is observed for oval I particles.[27]

It is thus concluded that, a realistic micromechanically-based macroscopic description of the overall mechanical properties of cohesionless granular masses should include at a minimum the influence of interparticle friction, particle shape, and initial fabric (inherent anisotropy). These effects are, in general, coupled.

[27] Similar results have been reported by Skinner (1969) who sheared wet and dry assemblies of 1mm glass ballotini in a shear box. Although the wet glass ballotini interparticle friction angle (27° to 38°) is considerably higher than that of dry ones (2° to 5°), both exhibit essentially the same overall shear resistance, *i.e.*, the same overall effective friction coefficient. Skinner attributes this to the dominance of particle rolling during shear deformation. The experimental results of Oda *et al.* (1982a), reported here, seem to support Skinner's conclusion.

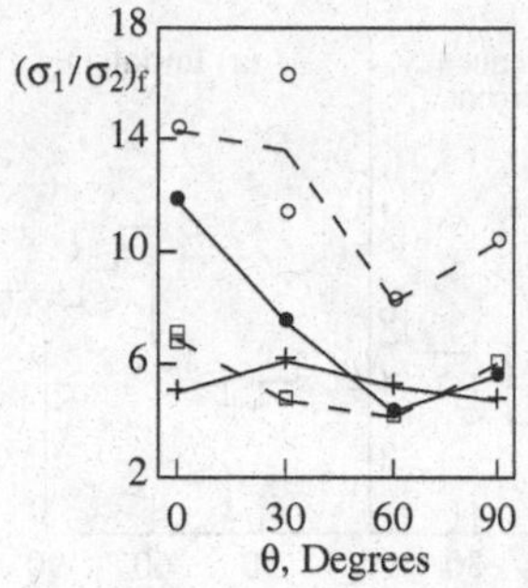

Figure 7.4.6

Variation of stress ratio at failure, $(\sigma_1/\sigma_2)_f$, with the bedding angle, θ
Oval I: ○, 52° ●, 26°
Oval II: □, 52° +, 26°
(from Oda *et al.*, 1982a)

Evolution of Distribution of Contact Normals: The experiments allow direct observation of the evolution of the fabric during the deformation of the sample. As noted before, new contacts are continually formed as some existing ones disappear, leading to continual change in the microstructure. The experiments show that the evolution of the microstructure is closely related to the variation in the distribution of contact normals, and that this distribution has a close relation to the overall applied stress.

The distribution of contact normals changes in such a manner as to produce a greater concentration of contact normals along an orientation which parallels the direction of maximum principal compression. To quantify this, photoelastic pictures are taken at various stages during the tests. These stages are marked on the corresponding curve in Figure 7.4.5. Figures 7.4.7a,b,c show the frequency distribution at positions (1, initial), (5, peak), and (9, residual) for oval II with 90° bedding angle (initial anisotropic packing) and 26° internal friction angle. These frequency distributions are plotted against the angle ν which measures the orientation of the contact unit normal vectors relative to the vertical x_2-axis; see Figure 7.4.8. In producing these frequency diagrams, a contact normal is included when it is associated with visible photoelastic fringes. Since the bedding angle is 90°, the initial concentration of contact normals favors $\nu = \pm 90°$, as is seen in Figure 7.4.7(a). This, however, changes during the course of deformation, and the concentration of unit normals tends to favor the orientation $\nu = 0$ at the peak stress, Figure 7.4.7(b), and the residual state, Figure 7.4.7(c). This orientation defines the direction of the maximum principal compression.

Figures 7.4.9a,b display the change in the frequency diagrams of the contact normals from the initial (1) to the peak (5) stress states. Figure 7.4.9(a) is the frequency diagram for the new contact normals which have appeared, whereas Figure 7.4.9(b) is that for contacts which have disappeared. As is seen, most new contacts are generated close to the σ_2-direction, while among the contacts that disappeared, most had orientations close to ±90°. Similar observations have been made by Dantu (1957), Wakabayashi (1957), and Drescher (1976), in assemblies of glass granules.

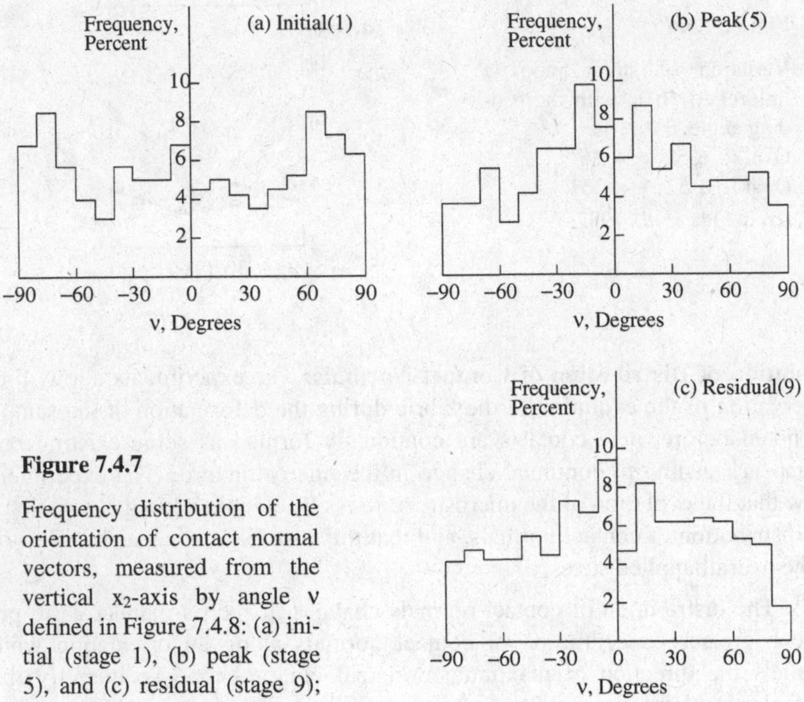

Figure 7.4.7

Frequency distribution of the orientation of contact normal vectors, measured from the vertical x_2-axis by angle ν defined in Figure 7.4.8: (a) initial (stage 1), (b) peak (stage 5), and (c) residual (stage 9); oval II with $\phi_\mu = 26°$ (from Oda *et al.*, 1982a)

Figures 7.4.10a,b show the change in the frequency diagram of contact normals from the peak (5) to the residual (9) stress states. Because of extensive dilatation, the column-like load paths lose stability during this process, leading to a fabric which should closely correspond to the critical state of Roscoe *et al.* (1958). Figures 7.4.11a,b are the photoelastic images of the microstructure at these two stages.

As pointed out before, the experimental results show a considerable amount of rolling being involved in the motion of particles relative to each other, especially for granules with the larger internal friction, *i.e.*, $\phi_\mu = 52°$. In Figure 7.4.8, angle δ defines the inclination of the contact force $\mathbf{f}^\alpha$ relative to the orientation of the contact normal, **n**. It is reasonable to assume that sliding cannot occur at a contact for which δ is less than the interparticle friction angle, ϕ_μ. The angle δ may therefore be called the *mobilized angle of friction.* This angle can be measured from the fringe patterns in the photoelastic pictures. Figures 7.4.12 and 7.4.13 are the histograms for oval I and II assemblies with interparticle friction angles of 26° and 52°, respectively, and the indicated initial bedding angles, θ. These histograms are obtained from the analysis of each assembly sheared up to the peak stress state. They show that:

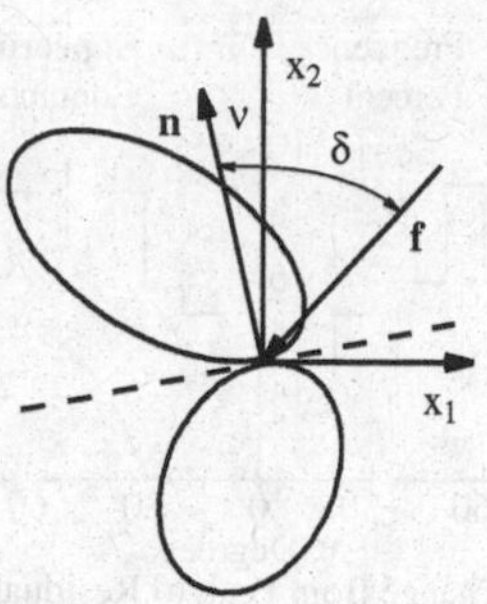

Figure 7.4.8

The angle ν measures the orientation of the contact normal vector from the vertical x_2-axis

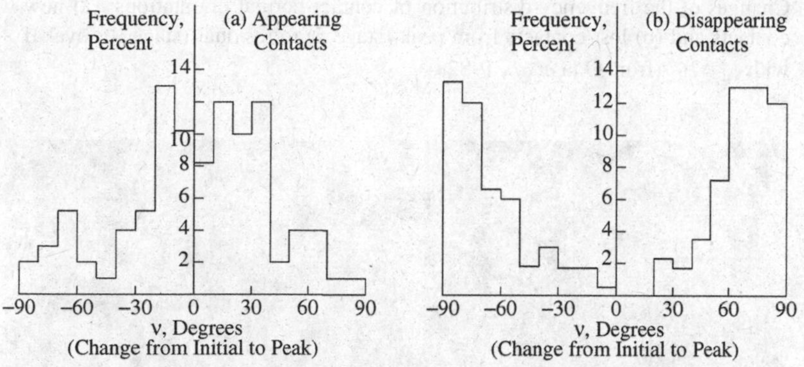

Figure 7.4.9

Changes of the frequency distribution of the orientation of contact normal vectors, measured from the vertical x_2-axis by angle ν: (a) contacts which appeared from initial (stage 1) to the peak (stage 5), and (b) contacts which disappeared from initial (stage 1) to the peak (stage 5); oval II with $\phi_\mu = 26°$ (from Oda *et al.*, 1982a)

(1) the mobilized angle of friction, δ, falls in the range $\pm 40°$ when $\phi_\mu = 26°$, and in the range $\pm 50°$ when $\phi_\mu = 52°$;

(2) the distribution range is wider for the higher interparticle friction $\phi_\mu = 52°$, than for interparticle friction $\phi_\mu = 26°$; and

(3) there are few contacts at which $|\delta| = \phi_\mu$, even in assemblies that are just at failure.

Close examination of Figures 7.4.12 and 7.4.13 shows that condition (1) is satisfied at more contacts in assemblies with the lower interparticle friction than with the higher one. This observation is in harmony with the result that rolling is more dominant at the higher interparticle friction.[28]

[28] Similar results have been reported by Oda and Konishi (1974) and Konishi (1978) for simple

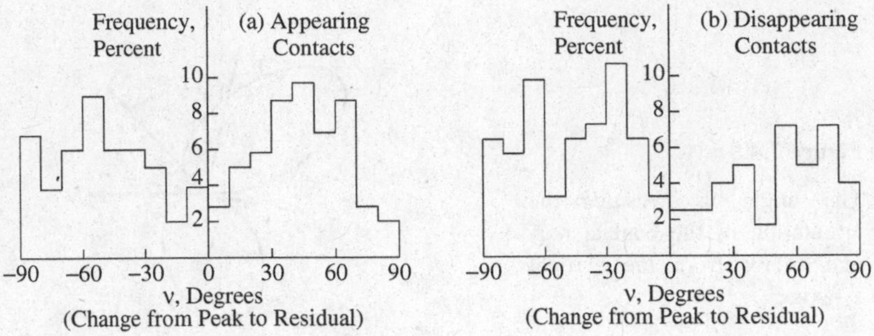

Figure 7.4.10

Changes of the frequency distribution of contact normal orientations: (a) new contacts and (b) lost contacts from peak (stage 5) to residual (stage 9); oval II with $\phi_\mu = 26°$ (from Oda *et al.*, 1982a)

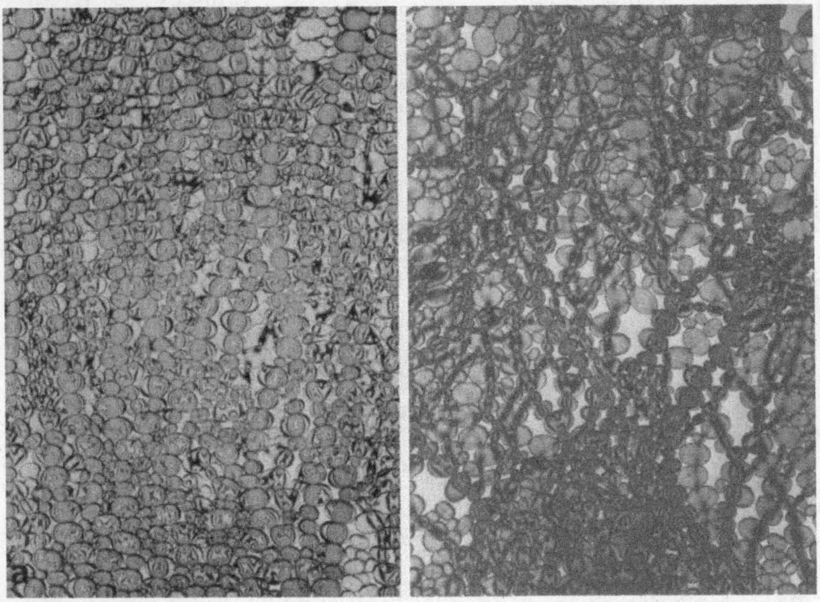

Figure 7.4.11

Photoelastic images of the microstructural change from stage (5), peak stress with well-defined column-like structure, to residual stage (9) with a diffused fabric structure

7.4.3. Simple Shearing Experiments

Apparatus: The testing equipment is shown in Figure 7.4.14. It consists of a rigid outer frame and an internal frame which can be deformed in shear while allowing for volumetric straining.[29] On this internal frame, two horizontal and

---

shearing of two-dimensional granules with circular cross sections. These authors, however, explain their results in terms of a hypothesis proposed by Horne (1965) who suggested that the overall de-

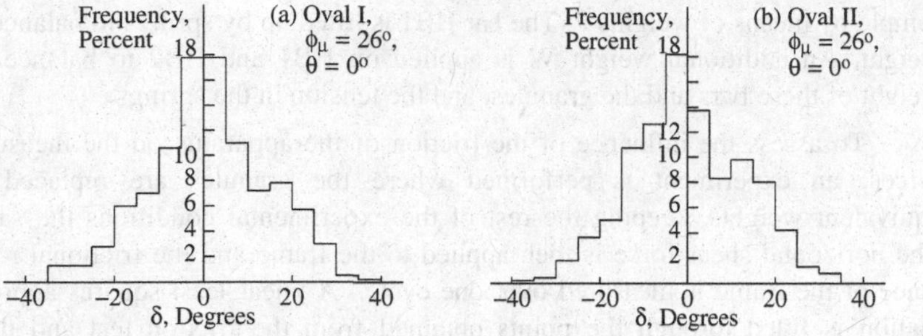

Figure 7.4.12

Histograms of mobilized friction angle for oval I and II assemblies with interparticle friction angle of 26° and initial bedding angle, $\theta = 0°$ (from Oda *et al.*, 1982a)

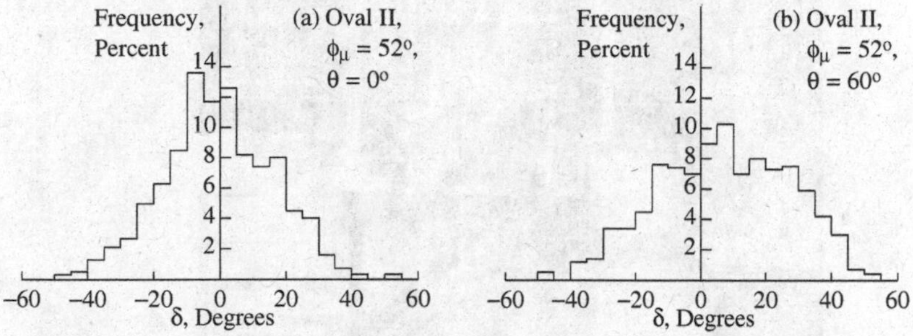

Figure 7.4.13

Histograms of mobilized friction angle for oval II assembly with interparticle friction angle of 52° and initial bedding angles, $\theta = 0, 60°$ (from Oda *et al.*, 1982a)

two vertical bars are mounted which can be moved in parallel, on the internal frame. These are denoted by HB1, HB2, VB1, and VB2, respectively, in Figure

formation takes place by relative sliding motion between instantaneously rigid groups of particles. These groups continually reform by the division and coalescence of contacting points. The experimental observations reported above, do not support such an explanation which places too much emphasis on the role of interparticle sliding as a major microdeformation mechanism.

[29] The apparatus was designed and constructed at Northwestern University by the author and his technical assistant, Mr. John Schmidt, 1983-1985, and moved to University of California, San Diego (UCSD), in 1985. All the experiments were performed by the author and coworkers in the author's laboratory at UCSD.

7.4.15 which provides a sketch of the apparatus. The granules can be packed inside the frame formed by these bars. The confining pressure is applied on the sample by means of weights P. The bar HB1 is lifted up by springs to balance its weight. An additional weight W is applied on HB1 and HB2 to balance the weight of these bars and the granules, and the tension in the springs.

To assess the influence of the friction of the apparatus on the measured forces, an experiment is performed where the granules are replaced by equivalent weights, keeping the rest of the experimental conditions the same. The horizontal shear force is then applied to the frame and the frictional resistance of the frame is measured over one cycle. A linear least-squares approximation is fitted through the points obtained from the friction test and these values are subtracted from the corresponding horizontal shear forces applied to the granules. Note that the friction test and the actual test on the granules are performed in the same direction and the same sequence over one cycle.

Figure 7.4.14

Experimental setup for simple shearing of oval-shaped photoelastic granules (author's laboratory at University of California, San Diego, 1986)

Sample Preparation and Test Procedure: The granules (1.9cm long) are packed in the internal frame of 20cm by 20cm, formed by bars HB1, HB1, VB1, and VB2. To minimize the effect of the boundaries, analyses are performed on granules within a central part which constitutes the sample; see Figure 7.4.16. The granular mass is subjected to a confining force of 1,200g (3.1kPa pressure) and is sheared horizontally by incremental displacements. The applied load on the granules is measured by a load cell, Lc. The horizontal and vertical movements of the bars are measured by the transducers Pl to P8. The shear strain is

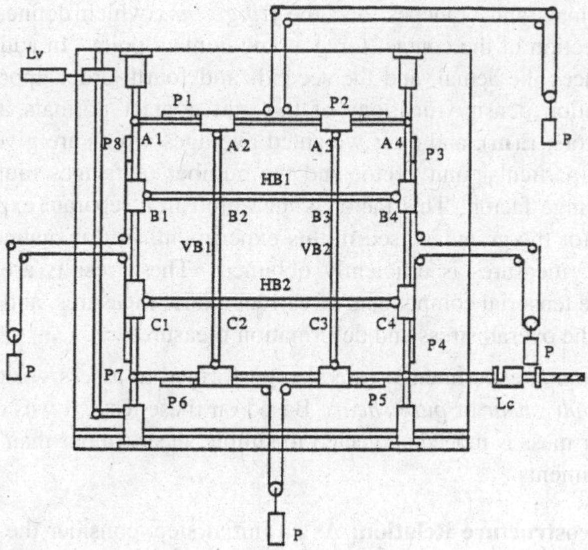

Figure 7.4.15

Schematic of the experimental apparatus for simple shearing of photoelastic granules

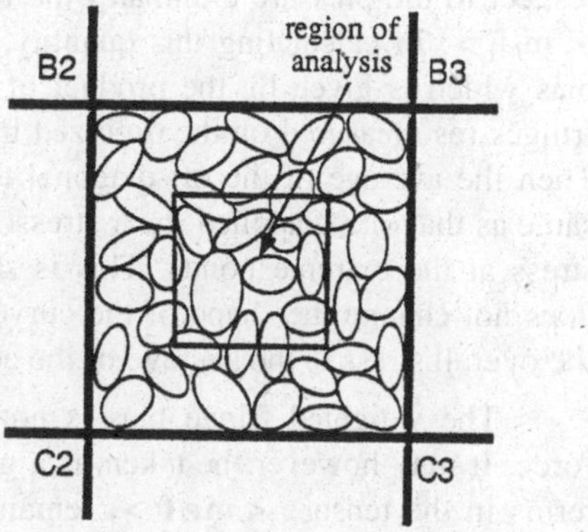

Figure 7.4.16

Schematic of the central part of the sample which is used for the analysis of the simple shearing of oval-shaped photoelastic granules

recorded by an LVDT, Lv.

The loading frame is placed in the field of a circular polariscope consisting of a monochromatic light source, a polarizer, two quarter-wave plates, and an analyzer. After the packing is complete, under a confining load, the shearing is applied incrementally. At each stage of loading, photographs of the isochromatic fringe pattern in the stressed assembly are taken. These photographs are later analyzed using a digital image analysis system. With the aid of a program specifically developed for automatic analysis of the relevant data in photoelastic images, a number of microscopic quantities are calculated for further consideration. These quantities include the number of fringes, orientation of contact normals and branches, and the *fringe bias* which defines the intensity and the direction of the contact force at any contact point. In addition, the program produces the actual, and the second- and fourth-order approximations of the distribution density functions of the unit contact normals, E(**n**), the unit branch vectors, E(**m**), and their weighted averages which are given by the product of the particular unit vector and the number of fringes multiplied by the force-per-fringe factor. This factor is measured in a separate experiment to be 27g/fringe for the granules used in this experiment. In this manner a wealth of microscopic measures is efficiently obtained. These results are then used to compute the tensorial components of various fabric measures, and to check their relation to the overall stress and deformation measures.

Additional experiments were also performed to assess whether or not the test is a *simple shear* or *pure shear*. Based on these tests, it was concluded that the granular mass is indeed subjected to simple shear (rather than pure shear) in these experiments.

Stress-microstructure Relation: As an initial step, consider the results of one cycle of simple shearing of a relatively dense sample. In Figure 7.4.17, the measured stress ratio, τ/p, is compared with the expression $< m_1 f_2 + m_2 f_1 >$ whose values are calculated from the photoelastic images which have been taken at each indicated stage of the loading shown on the stress-strain curve. Since the stress tensor is given by $\frac{1}{2} N\bar{l} < \mathbf{m} \otimes \mathbf{f} + \mathbf{f} \otimes \mathbf{m} >$, normalization with respect to the pressure eliminates the unknown factor $N\bar{l}$. Consider the tensor $< m_i f_j >$. In calculating this quantity, f_j is interpreted as the weighted fringe bias which is given by the product of the unit fringe bias and the number of fringes (as measured on the digitized image) multiplied by the force per fringe. Then the average of the off-diagonal terms is scaled so that their range is the same as that of the applied shear stress. The graph is then translated to match the stress at the extreme points. This is shown in Figure 7.4.17. This procedure does not change the shape of the curve, but determines a scale factor to relate the overall stress to the average of the corresponding microstructural quantities.

The weighted fringe bias is not an exact representation of the contact force. It may however, be taken as a good measure of this force. The diagonal terms in the tensor, $< \mathbf{m} \otimes \mathbf{f} >$, remain almost constant, representing the constant confining pressure; see Figure 7.4.18.

The off-diagonal terms of the tensors $< \mathbf{m} \otimes \mathbf{m} >$ and $< \mathbf{n} \otimes \mathbf{n} >$ are plotted in Figure 7.4.19, and show a very good agreement with the overall stress ratio.

From these results, it is concluded that the diagonal terms in each tensor, $< \mathbf{m} \otimes \mathbf{f} >$, $< \mathbf{m} \otimes \mathbf{m} >$, and $< \mathbf{n} \otimes \mathbf{n} >$, are proportional to the confining pressure, and that the off-diagonal terms are proportional to the applied shear stress. The corresponding components of the tensors $< \mathbf{m} \otimes \mathbf{m} >$ and $< \mathbf{n} \otimes \mathbf{n} >$ are almost identical.

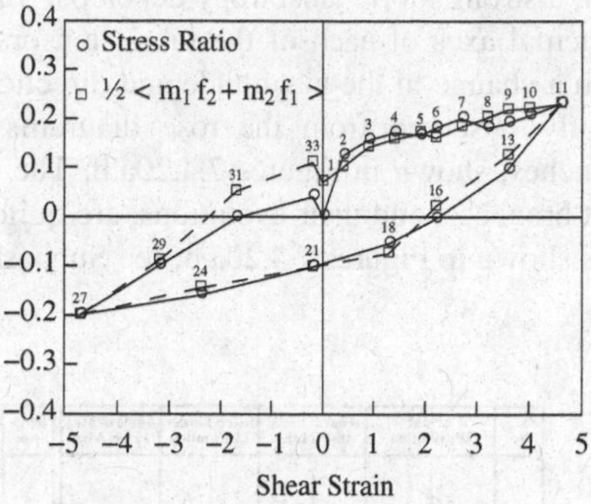

Figure 7.4.17

Stress ratio, τ/p, and the off-diagonal terms of the tensor $\frac{1}{2}<\mathbf{m}\otimes\mathbf{f} + \mathbf{f}\otimes\mathbf{m}>$, vs. shear strain, showing good agreement

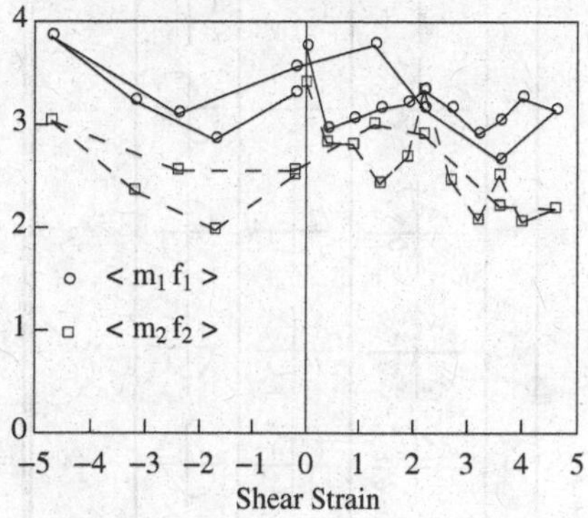

Figure 7.4.18

The diagonal terms of the tensor $<\mathbf{m}\otimes\mathbf{f}>$ remain essentially constant throughout the deformation process

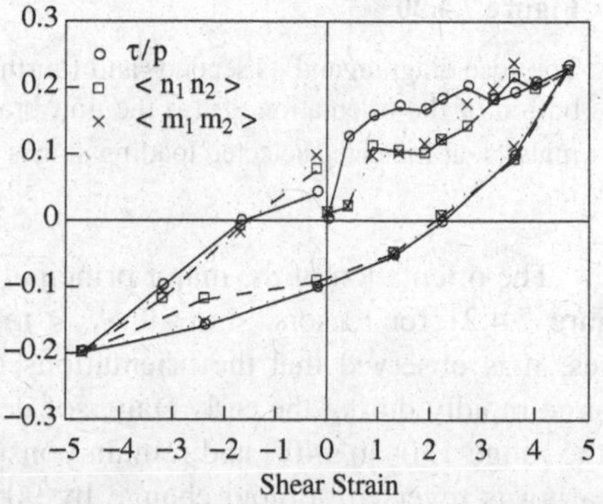

Figure 7.4.19

The off-diagonal terms of the tensors $<\mathbf{m}\otimes\mathbf{m}>$ and $<\mathbf{n}\otimes\mathbf{n}>$, correspond well to the overall stress ratio

Mechanism of Strain Hardening: In a monotonically increasing stress ratio, τ/p, particle rearrangement occurs to withstand the increasing load. Similarly to the biaxial experiment, contact normals tend to concentrate in the direction of the maximum (compressive) principal stress axis. This is clearly observed in the form of chains of heavily stressed particles in the photoelastic images; see, *e.g.*, Figure 7.1.1. The average direction of these "chains" tends in the direction of the major principal stress (greatest compression) axis. As a result of this concentration, a strong fabric anisotropy develops. In the simple shearing experiment, the principal axes of each of the fabric tensors, $<\mathbf{m}\otimes\mathbf{m}>$ and $<\mathbf{n}\otimes\mathbf{n}>$, rotate with a change in the magnitude and direction of the applied shear stress. This is clearly observed from the rose diagrams of unit contact normals and unit branches, shown in Figures 7.4.20a,b. The distributions of contact normals and unit branches, and their evolutions, are quite similar, as is seen from a few samples shown in Figures 7.4.20a,b; see Subhash *et al.* (1991) for more details.

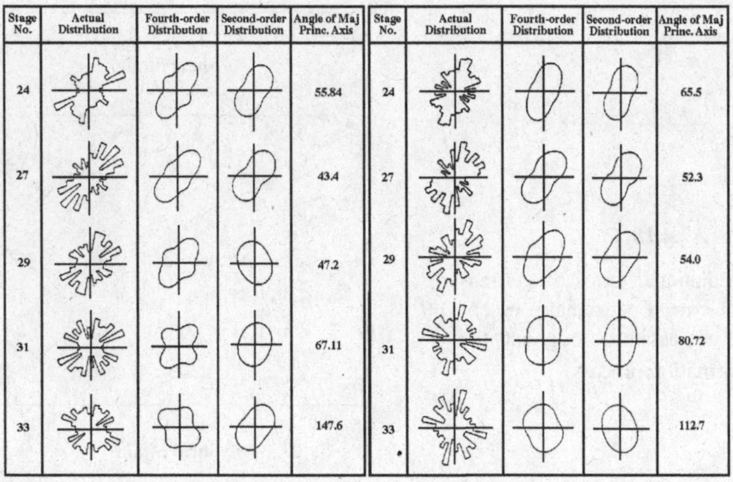

(a) (b)

Figure 7.4.20

The rose diagram and its second- and fourth-order representations, of the distribution of the orientation of: (a) the unit branches, **m**, and (b) unit contact normals, **n**, at indicated selected loading stages

The orientation of the major principal axis vs. the shear strain is plotted in Figure 7.4.21 for tensors $<\mathbf{m}\otimes\mathbf{f}>$, $<\mathbf{m}\otimes\mathbf{m}>$, and $<\mathbf{n}\otimes\mathbf{n}>$. In all three cases, it is observed that the orientations of the corresponding principal axes change rapidly during the early stages of deformation, approach constant values in the range 130° to 140°, and remain constant thereafter. Once the direction of shearing is reversed, a rapid change by 90° in the orientations of the principal axes of the fabric occurs. Their orientations then remain constant until a further change in the direction of shearing occurs. This process repeats in each cycle.

Dilatancy: Figure 7.4.22 shows the variation of the volumetric strain with the shear strain. An initial densification is seen to be followed by dilatation, typical of almost all frictional granular materials. As discussed in Section 7.1, this phenomenon can be explained in terms of the distribution of the dilatancy angle,

Figure 7.4.21

Variation of the orientation of the principal axes of tensors $< \mathbf{m} \otimes \mathbf{f} >$, $< \mathbf{m} \otimes \mathbf{m} >$, and $< \mathbf{n} \otimes \mathbf{n} >$ with shear strain; the orientations of these principal axes change rapidly and approach constant values in the range 130° to 140°, until the direction of shearing is changed

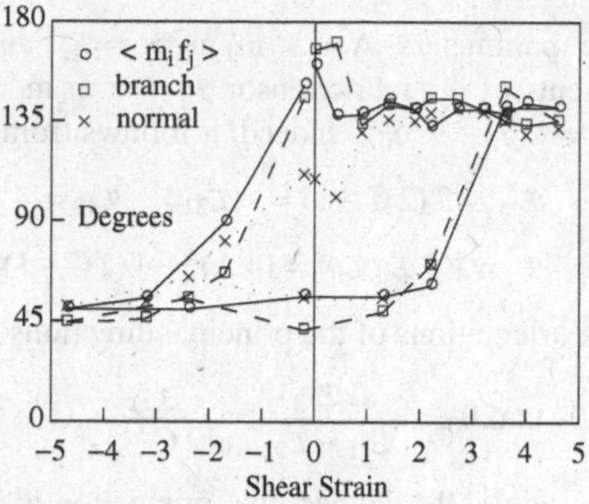

ν, and its evolution in the course of shearing under a confining pressure; see expressions (7.1.11a), page 510, (7.1.12b,e) and (7.13a,b), pages 513-14, and the corresponding discussion.

Figure 7.4.22

Variation of the volumetric strain with shear strain: initial densification is followed by dilatancy, and extensive densification upon unloading and load reversal

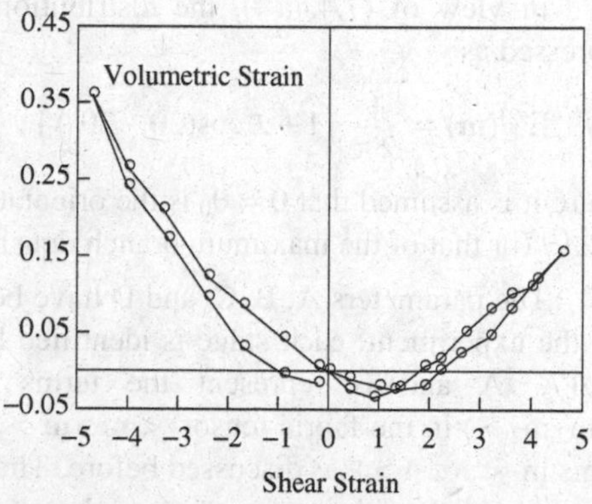

Representation of Distribution Density Functions: Consider the second- and fourth-order representations of the distribution density functions E(**m**) and E(**n**), discussed in Subsection 7.3.3; see (7.3.21) to (7.3.24), pages 542-3. Denote the second- and fourth-order representations of E by $E^{(2)}$ and $E^{(4)}$, respectively.

Consider the distribution density function of the unit branches. For two dimensions, $m_1 = \cos\theta$ and $m_2 = \sin\theta$, where θ measures the angle from the x_1-axis to **m**. Introduce the notation

$$A = < \sin\theta \cos\theta > , \quad B = < \sin\theta \cos^3\theta > ,$$
$$C = < \cos^2\theta > , \quad D = < \cos^4\theta > , \tag{7.4.1a-d}$$

and obtain

$$E^{(2)}(\mathbf{m}) = \frac{1}{2\pi} + \frac{1}{\pi} \{ 2A \sin 2\theta + (2C - 1) \cos 2\theta \} , \tag{7.4.2a}$$

and

$$E^{(4)}(\mathbf{m}) = E^{(2)}(\mathbf{m}) + \frac{1}{\pi}\{(1-8\,C+8\,D)\cos 4\theta + 4\,(2\,B-A)\sin 4\theta\}\,. \quad (7.4.3)$$

The parameters $A = \langle m_1 m_2 \rangle = \mathcal{J}_{12}$ and $C = \langle m_1^2 \rangle = \mathcal{J}_{11}$ are the components of the fabric tensor $\mathcal{J}_{ij} = \langle m_i m_j \rangle$. They also define the fabric tensor $\mathcal{E}_{ij} = 4\,(\mathcal{J}_{ij} - \frac{1}{2}\,\delta_{ij})$. Indeed, it follows from their definition that

$$\mathcal{E}_{11} = 2\,(2\,C-1) = -\,\mathcal{E}_{22}, \qquad \mathcal{E}_{12} = 4\,A\,,$$
$$\mathcal{E}^2 = (\tfrac{1}{2}\,\mathcal{E}_{ij}\,\mathcal{E}_{ij}) = (4\,A)^2 + 4\,(2\,C-1)^2\,. \quad (7.4.4a\text{-}d)$$

The orientations of the principal directions of $\mathcal{E}_{ij}$ are defined by

$$\tan 2\theta_0 = \frac{2\,\mathcal{E}_{12}}{\mathcal{E}_{11} - \mathcal{E}_{22}} = \frac{2\,A}{2\,C-1}\,. \quad (7.4.4e,f)$$

The angle θ_0 defines the orientation of the extreme values (maximum or minimum) of the distribution function $E^{(2)}$. The maximum (minimum) is the same as the angle of the greatest (least) principal value of tensor $\boldsymbol{\mathcal{E}}^{(2)}$. These angles identify the orientations of the greatest and least concentration of the unit branches (or contact normals for $E^{(2)}(\mathbf{n})$). These directions are orthogonal.[30]

In view of (7.4.4a-f), the distribution density function $E^{(2)}(\mathbf{m})$, can be expressed as

$$E^{(2)}(\mathbf{m}) = \frac{1}{2\,\pi}\{1 + \mathcal{E}\cos(2\theta - 2\theta_0)\}\,, \quad (7.4.2b)$$

where it is assumed that $\theta = \theta_0$ is the orientation of the major principal direction of $\boldsymbol{\mathcal{E}}^{(2)}$ (or that of the maximum branch density).

The parameters A, B, C, and D have been measured at various stages during the experiment; each stage is identified by a *stage number*, given in Figure 7.4.17. A and C represent the terms $\langle m_1 m_2 \rangle$ and $\langle m_1 m_1 \rangle$ (or $\langle m_2 m_2 \rangle$) in the fabric tensor $\langle \mathbf{m} \otimes \mathbf{m} \rangle$, respectively (or the corresponding terms in $\langle \mathbf{n} \otimes \mathbf{n} \rangle$), as discussed before. The parameters C and D are the diagonal terms in the fabric tensors $\mathcal{E}_{ij}$ and $\mathcal{E}_{ijkl}$, respectively, see (7.3.22a,b), page 542. These are shown in Figure 7.4.23. From this figure, it appears that they remain almost constant throughout the deformation process, representing the constant confining pressure. The off-diagonal terms, *e.g.*, A and B, behave similarly to the applied shear stress. The parameter A is plotted against the shear strain in Figure 7.4.19 for both unit normals and unit branches, and B vs. shear strain is shown in Figure 7.4.24. The parameters B and D behave similarly to A and C, respectively. Therefore, in the distribution density function, even terms like $\langle \cos^2\theta \rangle$ and $\langle \cos^4\theta \rangle$, or $\langle \sin^2\theta \rangle$ and $\langle \sin^4\theta \rangle$, relate to the applied confining pressure, and odd terms involving both sine and cosine, like A and B, relate to the overall applied shear stress.

---

30 Note that, while $0 \le C \le 1$, A is sign-indefinite. The deviatoric tensor $\boldsymbol{\mathcal{E}}$ may be represented by a Mohr circle of radius $\mathcal{E}$.

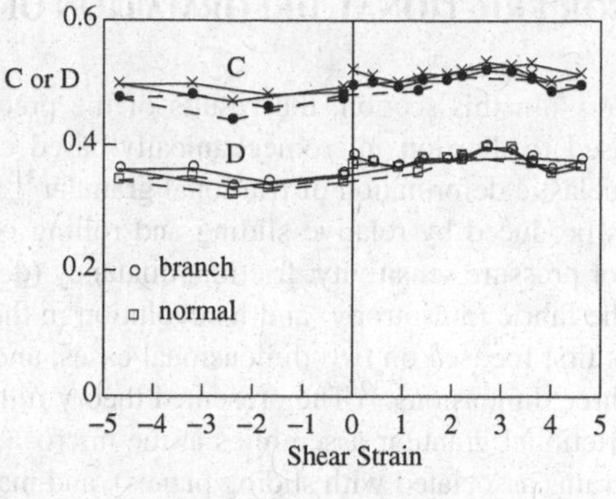

Figure 7.4.23

The parameters C and D which are the diagonal terms in the fabric tensors $\mathcal{E}_{ij}$ and $\mathcal{E}_{ijkl}$, remain essentially constant throughout the deformation process

Note that *the values of* A, B, C, *and* D, *for unit branches and unit contact normals, are almost the same. Furthermore, the distributions of the unit contact normals and the unit branches which are shown in Figures 7.4.20a,b, and the orientations of the principal axes which are shown in Figure 7.4.21 for both of these fabric tensors, are similar.* Therefore, for all practical purposes, the consideration of either unit branches or contact normals seems sufficient for the analysis of fabric. This has also been observed in earlier experiments by Mehrabadi *et al.* (1988). This is an important result which suggests that, even for the oval II particles with 1.4 aspect ratio, the contact normals can be used to express the granular fabric. This observation is used in Section 7.5 in connection with developing micromechanically-based constitutive models for frictional granular assemblies.

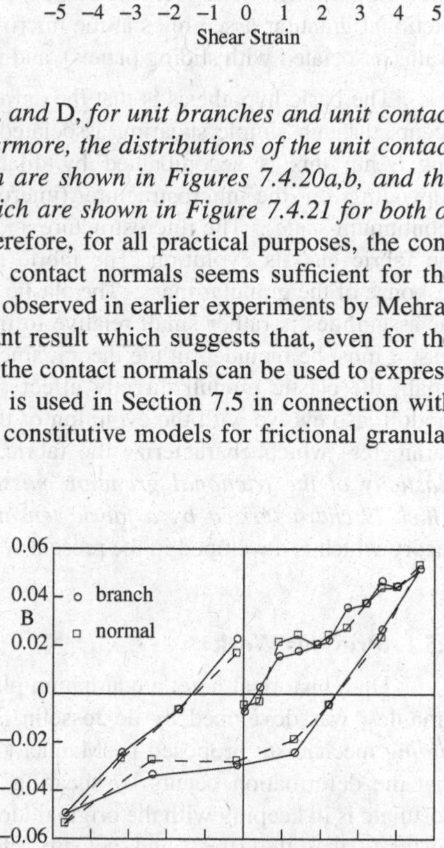

Figure 7.4.24

The variation of the parameter B with the shear strain is similar to that of the overall shear stress

Figures 7.4.20a,b show the actual, fourth-order and second-order, distributions of the contact normals and unit branches at various stages throughout the experiment. It is clear that the fourth-order approximation reveals the inherent anisotropy more accurately than does the second-order approximation.

7.5. MICROMECHANICALLY-BASED CONSTITUTIVE MODELS FOR FRICTIONAL DEFORMATION OF GRANULAR MASSES

In this section, the results of the preceding sections of this chapter are used to develop micromechanically-based constitutive models for the overall inelastic deformation of frictional granular[31] assemblies, where this deformation is produced by relative sliding and rolling of granules. These models account for pressure sensitivity, friction, dilatancy (densification), and, most importantly, the fabric (anisotropy) and its evolution in the course of deformation. Attention is first focused on two-dimensional cases, and then the results are generalized to three dimensions.[32] The presented theory fully integrates the micromechanics of frictional granular assemblies at the micro- (grains), meso- (large collections of grains associated with sliding planes), and macro- (continuum) scales.

The basic hypothesis is that the deformation of frictional granular masses occurs through simple shearing associated with grain-on-grain sliding and rolling. And, this is accompanied by dilatation or densification (meso-scale), depending on the microstructure (micro-scale) and the loading conditions (continuum-scale). The microstructure and its evolution are defined in terms of the fabric and its evolution. The fabric affects both the inelastic and elastic response of the granular mass. The elastic deformation of most frictional granular assemblies is rather small relative to their inelastic deformation. Nevertheless, it must be included in the theory, since even when the elastic range is very small, the elastic moduli directly affect the overall inelastic response. These moduli also evolve with the evolution of the fabric. They are dependent on the parameters which characterize the fabric. *Hence, the overall inelasticity and elasticity of the frictional granular masses are coupled through their fabric which is characterized by a fabric tensor.* These effects are included in the theory which is developed in the present section.

7.5.1. Previous Work

On a historical note, a continuum planar deformation model for frictional granules, was developed by de Josselin De Jong (1959), based on the *double sliding* mechanism proposed by Mandel (1947). In this model, it is postulated that the deformation occurs by shearing along the stress characteristics. This postulate is in keeping with the original idea of Coulomb and is equivalent to the postulate that the stress and velocity characteristics coincide. Based on de

[31] As before, granular masses which support the applied loads through contact friction, are called *frictional granules.*

[32] Only the rate-independent models are considered, since the examination of the rate effects associated with rapid shearing deformations is beyond the scope of the present work, and indeed requires a treatise of its own. Under great pressures, for example, frictional granules can undergo melting due to plastic heating. This kind of issue is not considered in the present work.

Josselin De Jong's postulate, Spencer (1964) constructed a general theory for the planar deformation of non-dilatant granular materials. In addition to De Josselin De Jong's postulate, Spencer assumes that the material spin relative to the stress characteristics is due to the inelastic shearing along the stress characteristics. This theory predicts that the principal axes of stress- and strain-rate tensors do not, in general, coincide, *i.e., they are noncoaxial.* De Josselin De Jong (1971, 1977) has disagreed with this last assumption. He considers that the inelastic sliding along the stress characteristics is only in the direction of the shear stress and never against it.[33] Mehrabadi and Cowin (1978) extend Spencer's theory to dilatant granular materials by including the possibility of volumetric expansion (contraction) in the direction perpendicular to the plane of shearing. A generalized version of the double-sliding model is given by Nemat-Nasser *et al.* (1981), and the results are reviewed and compared with those of crystal plasticity by Nemat-Nasser (1983); see also Anand (1983) and Lance and Nemat-Nasser (1986).

Although these models provide a good basis for understanding the stress path-dependent behavior of granular materials, they are fundamentally applicable only to granular masses which can be viewed as isotropic materials. As has been demonstrated in the preceding sections, *the shearing behavior of granular materials is highly anisotropic.* This necessitates the consideration of induced anisotropy (fabric), even when the initial state of the material is isotropic. Such anisotropic behavior can be predicted only by models which directly include the material anisotropy that develops during initial consolidation (inherent fabric) and subsequent loading (induced fabric). Indeed, even the elasticity of frictional granular masses depends on their stress-state and hence on their fabric, and changes with the fabric evolution.

In Chapter 4, several continuum models of granular materials are examined in considerable detail. The fabric effects are incorporated in these models through a backstress and its evolution with deformation; see Section 4.7, page 192. Other continuum models of granular materials include the application of the *bounding surface* model of Dafalias (1975) and Dafalias and Popov (1976), (which was originally proposed for metals, and uses certain geometric features to account for the history effects), to simulate cyclic behavior of soils, as discussed by Mróz (1980). Other continuum models with similar features have been proposed, using the concepts from classical plasticity, particularly kinematic hardening and nonassociative continuum plasticity together with the Mohr-Coulomb friction criterion (*e.g.*, Laude *et al.*, 1988), or other less conventional considerations (*e.g.*, Vardoulakis, 1996). Models of this kind are designed by relating the macroscopic variables through geometric and other phenomenological arguments.

---

[33] It is of interest to note that inelastic shearing against an applied shear stress can and does in fact occur, as discussed in Section 7.1. This, however, is accompanied by densification, under the action of the confining pressure which provides the necessary energy. In fact, the balance of energy can be used to obtain an expression for dilatancy; see expressions (7.5.44a,b), page 585.

Micromechanically-based models aim to give the continuum relations by applying the concepts of mechanics at the level of contacting granules. This requires the description of the overall stress, the characterization of fabric, the representation of kinematics, and then development of rate-constitutive relations in terms of the local (micro) quantities. Many of these issues have been examined in this chapter, and will now be reconsidered and simplified for application to constitutive modeling.[34] As has been pointed out before, the great difficulty in this kind of formulation is that the contacts between granules are not permanent. During the course of deformation, new contacts are formed while some existing contacts are lost. This precludes analytic identification of the history of individual contacts.

A direct approach which follows in detail the evolution of contacts and other interaction effects in granular materials, is large-scale computer simulation. A large-scale computer model for assemblies of contiguous solid particles was developed by Cundall (1971) and later further expanded and improved by Cundall and Strack (1979), to simulate the quasi–static shear deformation of dense assemblies of discs. Results of such simulations have been reported by Cundall *et al.* (1982), Cundall and Strack (1983), Cundall and Hart (1992), Kishino (1988), Thornton and Barnes (1986), Thornton (1987, 1998, 2000) and others; for an overview of recent similar large-scale numerical modeling and references, see Kishino (2001).

An alternative approach is to statistically represent the fabric microstructure, and seek to develop the rate constitutive relations analytically. Based on the earlier work of Christoffersen *et al.* (1981)[35], and making use of Hill's self-consistent method, Nemat-Nasser and Mehrabadi (1984) develop a rate-constitutive model of this kind for the overall response of frictional granules. They use the Coulomb friction criterion at the local (contact) level. Mehrabadi *et al.* (1993) have further improved this model. Though this model is physically-based, it fails to predict the initial densification that is generally observed in shearing under confining pressures, and it overestimates dilatancy in monotonic shearing.

A physically-based two-dimensional constitutive model is developed by Balendran and Nemat-Nasser (1993a, 1993b),[36] by considering the frictional anisotropic deformation of cylindrical granular masses, *e.g.,* the photoelastic granules discussed in Section 7.4. This model predicts rather well the observed

[34] The simplification is generally necessary to render the resulting expressions manageable. Even as the computational power increases, it still may be necessary to maintain a certain level of simplicity in the model in order to retain a clear understanding of its various essential features, and to identify potential computational errors.

[35] This work was completed in the fall of 1979, and was subsequently distributed as a report dated January, 1980. The results were presented by this writer at the 1980 Congress of Theoretical and Applied Mechanics, in Toronto, Canada.

[36] Many aspects of the mechanics of inelastic deformation of granular materials have been examined by Balendran (1993), including two- and three-dimensional flow of frictional granules, the micromechanics of deformation, the effect of fabric on overall response, and effective numerical al-

dilatancy and densification effects in monotonic and cyclic loading. An improved version of this model has been given recently by Nemat-Nasser (2000), and its generalization to a three-dimensional continuum model by Nemat-Nasser and Zhang (2002). The material presented in the remaining part of the present chapter, is based on these more recent results.

In what follows, first the basic assumptions which underlie the theory are discussed in relation to the micromechanics of the frictional deformation of a granular mass, and physics-based expressions are obtained which define the resistance to deformation due to friction and the fabric anisotropy; Subsections 7.5.2 and 7.5.3. For the two-dimensional case, explicit expressions for this resistance are given in Subsection 7.5.4. Then, these results are used in Subsections 7.5.5 and 7.5.6, to develop rate-independent yield conditions for the two-dimensional deformations, and it is shown that the model encompasses many observed attributes of frictional granules in both loading and unloading. The double-sliding results of Subsections 7.5.5 and 7.5.6, are then used to formulate rate-constitutive relations in Subsections 7.5.7 to 7.5.9, including elastic anisotropy and its relation to the fabric tensor. The fabric evolution is considered in Subsection 7.5.10, and a two-dimensional continuum model is presented in Subsections 7.5.11 to 7.5.15, with an illustrative example given in Subsection 7.5.16. In Subsection 7.5.17, the results are generalized to three dimensions, and a full set of equations is presented.

7.5.2. Model Assumptions

As discussed in Section 7.4, *rolling and sliding* at the individual grain level (at the micro-scale) are the basic micromechanisms underlying the inelastic deformation of a frictional granular mass which consists of hard grains that are under relatively small pressures. The rolling becomes more dominant as the internal grain-to-grain frictional resistance increases. At the meso-scale, it may be assumed that *dilatant simple shearing over several interacting sliding planes* produces the resulting macroscopic deformation. Each of these active sliding planes involves a shearing deformation similar to the simple shearing model of Section 7.1, which is corroborated experimentally in the photoelastic tests reported in Section 7.4. The dilatant simple shearing is accompanied by a strong induced anisotropy, whereby the distribution of contact normals, **n**'s (or the unit branch vectors, **m**'s), becomes strongly biased, with its maximum density occurring in the direction of the maximum compression. The model seeks to integrate these features into its basic structure.

In developing a fundamentally-based model, it is necessary to be guided by the basic experimental results, especially since for analytical formulations, simplifying assumptions are inevitable. As pointed out before, the results of photoelastic experiments suggest that:

---

gorithms that address coupled shearing, dilatancy, and noncoaxiality.

(1) the distributions of the unit contact normals, **n**'s, and the unit branch vectors, **m**'s, are essentially the same and may be used interchangeably;

(2) the fabric tensors $<\mathbf{n}\otimes\mathbf{n}>$ and $<\mathbf{m}\otimes\mathbf{m}>$ are essentially the same;

(3) the diagonal elements of these fabric tensors are almost constant in simple shearing under a constant confining pressure;

(4) the off-diagonal elements of these fabric tensors behave similarly to the applied shear stress; and

(5) the *second-order* distribution density function of the unit contact normals, $E^{(2)}(\mathbf{n})$, which is essentially the same as that of the unit branch vectors, $E^{(2)}(\mathbf{m})$, is represented by (in two dimensions)

$$E(\mathbf{n}) = \frac{1}{2\pi}\{1 + \mathcal{E}\cos(2\theta - 2\theta_0)\}, \tag{7.5.1a}$$

where $\mathcal{E} = (\frac{1}{2}\,\mathcal{E}_{ij}\,\mathcal{E}_{ij})^{1/2}$ is the second invariant of the fabric tensor $\boldsymbol{\mathcal{E}}^{(2)} \equiv \boldsymbol{\mathcal{E}}$, of components

$$\mathcal{E}_{ij} = 4\,(\mathcal{J}_{ij} - \frac{1}{2}\,\delta_{ij}), \tag{7.5.1b}$$

where $\mathcal{J}_{ij} = <n_i\,n_j>$, and the superscript (2) is dropped to simplify the notation; see Section 7.3 for more detail.[37] The quantity $\mathcal{E}$ defines the degree of anisotropy of distribution (7.5.1a), and θ_0 gives the orientation of the greatest density of the contact normals; $\theta_0 + \pi/2$ then gives the orientation of the corresponding least density (see equations (7.4.1) and (7.4.2) of Section 7.4).

For illustration, several shapes of the distribution $E(\mathbf{n})$ are shown in Figure 7.5.1, where θ and θ_0 are measured from the horizontal axis. The angle θ_0 also defines the orientation of the principal directions of the second-order fabric tensor $\boldsymbol{\mathcal{E}}$.

7.5.3. Resistance to Sliding

The overall deformation of a granular mass consists of a number of dilatant simple shearing deformations, along active shearing planes. At the microscale, this dilatant simple shear flow occurs on an active shearing plane through sliding and rolling of grains over each other at active contacts. In a granular sample with a large number of contacting granules, a mesoscopic shearing plane passes through a large number of contacting granules of various orientations of contact normals. Figure 7.5.2(a) schematically shows a mesoscopic shearing plane with the unit normal vector $\boldsymbol{\nu}$, a typical set of contacting granules, and the

[37] In Section 7.3, the fabric tensors associated with the unit contact normals are denoted by superimposed carets. The carets are dropped in the present section to simplify the notation. The context should eliminate any possible confusion.

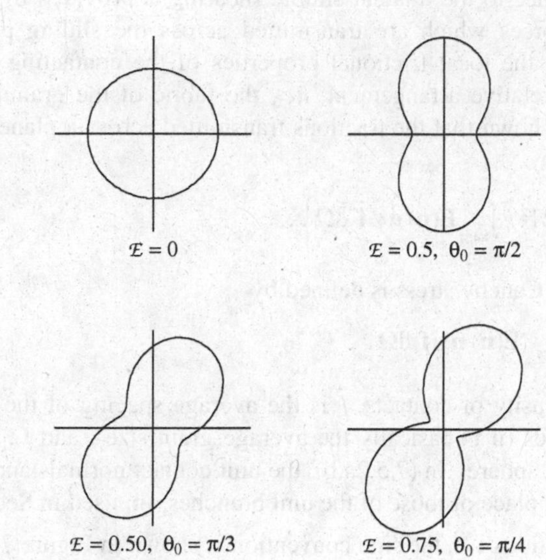

Figure 7.5.1

The second-order distribution density function (7.5.1) for indicated values of the anisotropy parameter, $\mathcal{E}$, and orientation angle, θ_0

shearing direction defined by the unit vector **s**. For future referencing, this plane will also be called the *sliding plane*, even though both particle rolling and sliding are involved at the micro-scale. The term, sliding plane, has been used in the literature by many authors for this purpose.

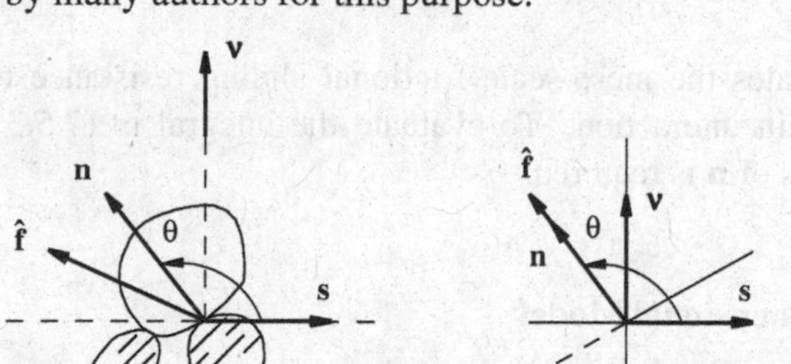

Figure 7.5.2

Schematic representation of a mesoscopic dilatant shearing plane with unit normal $\boldsymbol{\nu}$: (a) a typical set of contacting granules, and the shearing direction defined by the unit vector **s**; and (b) a simple model for estimating the resistance to sliding, $\bar{\tau}_{\text{fab}}$, due to fabric

The resistance to the dilatant simple shearing is provided by the average of the contact forces which are transmitted across the sliding plane. These forces depend on the local frictional properties of the contacting granules, as well as on their relative arrangement, *i.e.,* the fabric of the granular mass. In Section 7.2, it is shown that the tractions transmitted across a plane of unit normal $\mathbf{v}$, are given by

$$\bar{\mathbf{t}}^{(\mathbf{v})} = \mathbf{v} \cdot \left\{ 2\mathrm{N}\bar{l} \int_{\Omega_{1/2}} \mathrm{E}(\mathbf{n})\, \mathbf{n} \otimes \hat{\mathbf{f}}\, d\Omega \right\}, \tag{7.5.2a}$$

where the overall Cauchy stress is defined by

$$\bar{\boldsymbol{\sigma}} = 2\mathrm{N}\bar{l} \int_{\Omega_{1/2}} \mathrm{E}(\mathbf{n})\, \mathbf{n} \otimes \hat{\mathbf{f}}\, d\Omega. \tag{7.5.2b}$$

Here, N is the density of contacts, $\bar{l}$ is the average spacing of the centroids of contacting granules (it is basically the average grain size), and $\Omega_{1/2}$ is the surface of a half-unit sphere. In (7.5.2a,b), the unit contact normals and their distribution are used in place of those of the unit branches, $\mathbf{m}$, used in Section 7.2.

For the tractions $\bar{\mathbf{t}}^{(\mathbf{v})}$, the sign convention is shown in Figure 7.5.2(a). At a typical contact, there are two contact forces, $\hat{\mathbf{f}}$ and $-\hat{\mathbf{f}}$, and two contact normals, $\mathbf{n}$ and $-\mathbf{n}$. Choose the pair which points in the positive $\mathbf{v}$-direction. The normal tractions are then positive in *tension*. As pointed out before, each such unit normal vector represents a class of contacts, having an associated average contact force, $\hat{\mathbf{f}}$. The distribution of these contact normals and forces, is thus continuous. Since the contact forces are never tensile, any chosen pair of $\mathbf{n}$ and $\hat{\mathbf{f}}$ corresponds to negative normal tractions acting on the $\mathbf{v}$-plane.

The resistance to shearing on the sliding $\mathbf{v}$-plane is given by

$$\bar{\tau}_r = \bar{\mathbf{t}}^{(\mathbf{v})} \cdot \mathbf{s} = 2\mathrm{N}\bar{l} \int_{\Omega_{1/2}} \mathrm{E}(\mathbf{n})\, (\mathbf{v} \cdot \mathbf{n})\, (\hat{\mathbf{f}} \cdot \mathbf{s})\, d\Omega. \tag{7.5.3}$$

This equation relates the meso-scale frictional sliding resistance to the micro-scale grain-to-grain interaction. To evaluate the integral in (7.5.3), an expression for $\hat{\mathbf{f}}$ in terms of $\mathbf{n}$ is required.

7.5.4. A Two-dimensional Model

Consider two-dimensional deformation of a granular mass, *e.g.,* simple shearing or biaxial loading of a sample of rod-shaped photoelastic granules. Set $\hat{\mathrm{f}} = \mathbf{n} \cdot \hat{\mathbf{f}}$. From (7.5.2b), the hydrostatic *tension* becomes

$$\begin{aligned} \tfrac{1}{2}\,\mathrm{tr}\bar{\boldsymbol{\sigma}} &= \frac{\mathrm{N}\bar{l}}{\pi} \int_0^{\pi} \{1 + \mathcal{E}\cos(2\theta - 2\theta_0)\}\, \hat{\mathrm{f}}\, d\theta \\ &= \frac{\mathrm{N}\bar{l}\hat{\mathrm{f}}^*}{\pi} \int_0^{\pi} \{1 + \mathcal{E}\cos(2\theta - 2\theta_0)\}\, d\theta \\ &= \tfrac{1}{2}\mathrm{N}\bar{l}\hat{\mathrm{f}}^*, \end{aligned} \tag{7.5.4a-c}$$

where[38] $\hat{\mathrm{f}}^*$ is some suitable intermediate value of $\hat{\mathrm{f}}(\theta)$ for $0 < \theta < \pi$. The pressure p is therefore given by

$$p = -\tfrac{1}{2} N \bar{l}\, \hat{f}^{*}. \tag{7.5.5}$$

From (7.3.5), page 536, it is seen that $N = O(\bar{l}^{-3})$. Thus, $p = O(\hat{f}^{*}/\bar{l}^{2})$. This observation has a bearing on the phenomenon of pulverization of brittle materials, such as ceramics, under high-velocity impact.

It is difficult to obtain an explicit expression for the contact force $\hat{\mathbf{f}}$ in (7.5.2b), in terms of the corresponding contact normals, **n**, which can produce simple and effective results. Hence, consider the following alternative approach. Assume that the resistance to shearing, $\bar{\tau}_r$, can be divided into two parts, one due to a Coulomb-type isotropic frictional resistance, denoted by $\bar{\tau}_{iso}$, and the other due to the fabric anisotropy, denoted by $\bar{\tau}_{fab}$,

$$\bar{\tau}_r = \bar{\tau}_{iso} + \bar{\tau}_{fab}. \tag{7.5.6a}$$

From Figure 7.1.3e, page 507, the Coulomb frictional resistance is given by

$$\bar{\tau}_{iso} = \tfrac{1}{2} p \sin 2\phi_\mu, \tag{7.5.6b}$$

where ϕ_μ is an *effective* friction angle associated with the sliding and rolling of granules relative to one another. For cohesive granules, an additional cohesion stress must also be included.

To obtain an expression for $\bar{\tau}_{fab}$, the resistance due to the fabric anisotropy, use the second-order distribution density function, E(**n**), given by (7.5.1a), and, as a simplest model, let $\hat{\mathbf{f}}$ be defined by[39] (7.2.29c), *i.e.*, $\hat{\mathbf{f}} \approx \hat{\sigma}\mathbf{n}$. Hence, reduce (7.5.2b) to (see Figure 7.5.2(b))

$$\bar{\boldsymbol{\sigma}}_A = \frac{N\bar{l}}{\pi} \int_0^{\pi} \{1 + \mathcal{E}\cos(2\theta - 2\theta_0)\}\, \mathbf{n} \otimes \mathbf{n}\, \hat{\sigma}\, d\theta, \tag{7.5.6c}$$

where the subscript A emphasizes the approximate nature of the expression for the overall stress tensor. To estimate the resisting shear stress due to fabric anisotropy, now set

$$\begin{aligned} \bar{\tau}_{fab} &\approx \mathbf{v} \cdot \bar{\boldsymbol{\sigma}}_A \cdot \mathbf{s} \\ &= \frac{N\bar{l}}{\pi} \int_0^{\pi} \{1 + \mathcal{E}\cos(2\theta - 2\theta_0)\} \cos\theta \sin\theta\, \hat{\sigma}\, d\theta \\ &= -\tfrac{1}{2} p\, \mathcal{E} \hat{\mu} \sin 2\theta_0, \qquad \hat{\mu} = \frac{\hat{\sigma}^{*}}{\hat{f}^{*}}, \end{aligned} \tag{7.5.6d-g}$$

where $\hat{\sigma}^{*}$ is some suitable intermediate value of $\hat{\sigma}(\theta)$ for $0 < \theta < \pi$, and (7.5.5) is also used. Since, in the absence of interparticle friction, a granular mass can only support hydrostatic compression, but no shear, the parameter $\hat{\mu}$ directly relates to the coefficient of contact friction; see comments following expressions

[38] Since contact forces are always compressive, $\hat{f}^{*} < 0$.

[39] Since this is an approximation for the actual $\hat{\mathbf{f}}$ which need not be coaxial with **n**, the notation $\hat{\mathbf{f}} = \hat{\sigma}\mathbf{n}$, rather than $\hat{\mathbf{f}} = \hat{f}\mathbf{n}$, is used.

(4.7.1a-c), page 198.

Referring to Figure 7.5.1, note that the resistance to sliding due to fabric is *negative* when the angle θ_0 is between 0 and $\pi/2$, which corresponds to negative dilatancy angles in Figure 7.1.6, page 513; see also Figure 7.5.3. In this case, the orientation of the maximum density of contact normals, *i.e.,* the principal direction of the fabric tensor $\boldsymbol{\mathcal{E}}$, makes an acute angle with the sliding direction **s**. For $\pi/2 < \theta_0 < \pi$, on the other hand, both the overall isotropic friction and the fabric anisotropy contribute to the resistance to sliding.

Figure 7.5.3
The angle θ_0 measures the orientation of the major principal axis of the fabric tensor $\boldsymbol{\mathcal{E}}$ from the sliding direction, **s**; the greatest density of contact normals is along this principal orientation, associated with the principal value $E_I \geq 0$

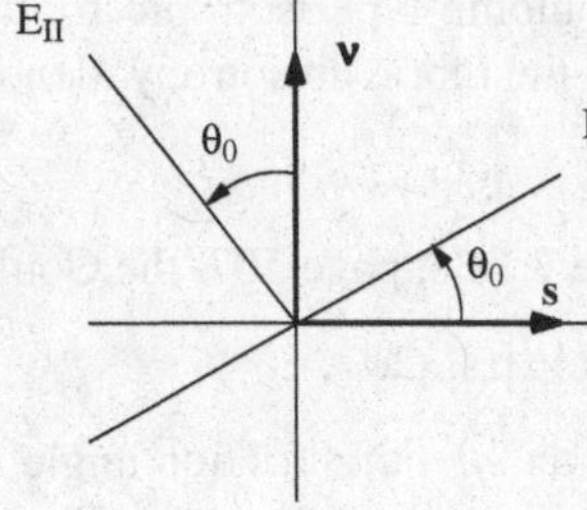

The total resistance to sliding now is

$$\bar{\tau}_r = \tfrac{1}{2}\,p\sin 2\phi_\mu - \tfrac{1}{2}\,p\,\hat{\mu}\,\mathcal{E}\sin 2\theta_0\,. \qquad (7.5.7a)$$

This expression relates to concepts from metal plasticity. The first term on the right defines the radius of the yield surface and the second term the location of the center of the yield surface in the stress space; see Chapter 4. With this in mind, introduce a deviatoric *anisotropy tensor (backstress)*,

$$\boldsymbol{\beta} \equiv -\tfrac{1}{2}\,p\,\hat{\mu}\,\boldsymbol{\mathcal{E}} \qquad (7.5.7b)$$

and rewrite (7.5.7a), as

$$\bar{\tau}_r = \tfrac{1}{2}\,p\sin 2\phi_\mu + \beta\sin 2\nu_0\,, \qquad (7.5.7c)$$

where the following relations are introduced:

$$(\tfrac{1}{2}\beta_{ij}\beta_{ij})^{1/2} = \beta\,, \quad \tan 2\nu_0 = \frac{2\beta_{12}}{\beta_{11}-\beta_{22}}\,, \quad \beta_{22} = -\beta_{11}\,.$$

$$\beta_{11} = \beta\cos 2\nu_0\,, \quad \beta_{12} = \beta\sin 2\nu_0\,, \quad \nu_0 = \theta_0 \pm \pi/2\,. \qquad (7.5.8a\text{-}f)$$

Note that, for frictional granular masses, the backstress $\boldsymbol{\beta}$ is proportional to pressure p and the friction coefficient $\hat{\mu}$. This is a fundamental feature of this class of materials, arrived at by a micromechanical consideration.

Expressions (7.5.7a-c) and (7.5.8a-f) provide a micromechanical basis for the application of an isotropic-kinematic hardening plasticity model to characterize the deformation of frictional granular masses. The kinematic hardening defined by the backstress $\boldsymbol{\beta}$, is related to the fabric tensor $\boldsymbol{\mathcal{E}}$ through (7.5.7b). The major principal axis of $\boldsymbol{\beta}$ coincides with the minor principal axis of $\boldsymbol{\mathcal{E}}$. The major principal value β_I of $\boldsymbol{\beta}$, therefore, corresponds to the least density of contact normals, or the maximum dilatancy direction; see Section 7.1. This direction makes an angle ν_0 with the sliding direction, **s**. For $0 < \nu_0 < \pi/2$, the resistance to sliding due to fabric, $\bar{\tau}_{fab}$, is *positive*, whereas for $\pi/2 < \nu_0 < \pi$, it is *negative*. In the first case, sliding in the positive **s**-direction is accompanied by *dilatancy*, while in the second case, it is accompanied by *densification*.

7.5.5. Meso-scale Yield Condition

Consider a typical sliding plane at the meso-scale. The resistance to shearing of the granules over this plane is due to interparticle friction and fabric anisotropy, as is expressed by (7.5.7a) or (7.5.7c). The micromechanical formulation of the preceding subsections provides expressions for the parameters which define this resistance. The resulting quantities, ϕ_μ, $\boldsymbol{\beta}$, ν_0, have physical meanings, are related to the microstructure of the granular mass, and can be associated with the continuum field variables. In (7.5.7a), p is viewed as the pressure, externally applied to the granular mass, and ϕ_μ is the overall effective friction angle which can be measured and experimentally related to the void ratio and the interparticular properties.[40] The quantities $\beta = (\frac{1}{2}\boldsymbol{\beta} : \boldsymbol{\beta})^{1/2}$ and ν_0 characterize the fabric anisotropy, and their evolution may be defined by rate-constitutive equations similar to the continuum model of Section 4.7, page 192.

Based on expressions (7.5.7a,c), consider the following sliding rule, a variant of Coulomb's criterion, for the sliding in the **s**-direction, on a plane with unit normal $\boldsymbol{\nu}$:

$$f \equiv \tau_\nu - \tfrac{1}{2}\,p\sin 2\phi_\mu - \beta\sin 2\nu_0 \le 0\,. \tag{7.5.9a}$$

The shear, τ_ν, and normal, σ_ν, stresses acting on the **s**-plane, as well as the pressure, p, are given by

$$\tau_\nu = \boldsymbol{\tau} : (\boldsymbol{\nu} \otimes \mathbf{s})\,, \qquad \sigma_\nu = \boldsymbol{\tau} : (\boldsymbol{\nu} \otimes \boldsymbol{\nu}) < 0\,, \qquad p = -\tfrac{1}{2}\,\mathrm{tr}(\boldsymbol{\tau})\,, \tag{7.5.9b-d}$$

where $\boldsymbol{\tau}$ is the Kirchhoff stress.[41] Here and in the sequel, the usual continuum mechanics sign convention is used, so that *tension is regarded positive*.

In (7.5.9a), the first term is the driving shear stress. The other two terms denote the resistance due to the interparticle (isotropic) friction, and the fabric anisotropy, respectively. The shear resistance due to the fabric anisotropy is not a constant but is a function of the angle between the sliding plane and the principal axes of the fabric tensor, as well as a function of the anisotropy parameter β.

---

[40] High strain-rate deformations of confined frictional granules produce considerable heat at interparticle contact points, which may lead to melting of the interface material. The friction angle ϕ_μ then depends on the interface temperature which in turn is a function of the deformation history. Effects of this kind can be included in the model, but will not be pursued in the present work.

[41] As in Chapters 4 to 6, the Kirchhoff stress is used throughout the rest of this chapter. When the current configuration is used as the reference one, the Cauchy and Kirchhoff stresses coincide (but not their rates).

In (7.5.9c), the normal stress σ_v is assumed to be compressive, since frictional granules cannot sustain tension.

In view of (7.5.9a), consider a decomposition of the Kirchhoff stress tensor[42] $\boldsymbol{\tau}$ into three parts, as follows:

$$\boldsymbol{\tau} = -p\,\mathbf{1} + \mathbf{S} + \boldsymbol{\beta}\,, \tag{7.5.10a}$$

where the backstress, $\boldsymbol{\beta}$, defined by (7.5.7b) and (7.5.8a-f), and *the stress-difference*, $\mathbf{S}$, are *deviatoric* and symmetric tensors. Let the principal values of these tensors be denoted by β_i and S_i, i = I, II, and assume that S_I and β_I make angles θ and ν_0 with the sliding direction, $\mathbf{s}$. The resolved shear stress on the sliding direction is now expressed as

$$\tau_v = S\sin 2\theta + \beta\sin 2\nu_0\,, \qquad S = (\tfrac{1}{2}\mathbf{S}:\mathbf{S})^{1/2}\,. \tag{7.5.10b,c}$$

The sliding condition (7.5.9a), now becomes

$$f \equiv S\sin 2\theta - \tfrac{1}{2}p\sin 2\phi_\mu \leq 0\,. \tag{7.5.11a}$$

The sliding occurs on planes for which (7.5.11a) attains its maximum value of zero,

$$S = p\sin\phi_\mu\,. \tag{7.5.11b}$$

There are two planes for which (7.5.11b) is satisfied. These are given by $\theta = \pm\,(\pi/4 + \phi_\mu/2)$. These planes are situated symmetrically about the greater principal stress, S_I (see Figure 7.5.4), making an angle $\pi/4 + \phi_\mu/2$ with this direction.

The decomposition of the stress tensor $\boldsymbol{\tau}$, according to (7.5.10a), may be implemented as follows. At a given deformation state, the fabric is known, being characterized by the fabric tensor $\boldsymbol{\mathcal{E}}$ which is calculated incrementally in the course of a prescribed loading history. In the x_1, x_2-plane, identify the angle θ_E of the orientation of the *minor* principal value, E_{II}, of the fabric tensor $\boldsymbol{\mathcal{E}}$, relative to the x_1-direction; see Figure 7.5.4. This corresponds to the *minimum* density of the contact unit normals, and coincides with the *major* principal direction of $\boldsymbol{\beta}$. Denote by the unit vectors $\hat{\mathbf{e}}_i$, i = 1, 2, the principal directions of $\boldsymbol{\beta}$. Then, this tensor is given by

$$\boldsymbol{\beta} = \beta\,(\hat{\mathbf{e}}_1 \otimes \hat{\mathbf{e}}_1 - \hat{\mathbf{e}}_2 \otimes \hat{\mathbf{e}}_2)\,, \qquad \beta_I = -\beta_{II} = \beta = \tfrac{1}{2}p\,\hat{\mu}\,\mathcal{E} \geq 0\,. \tag{7.5.12a,b}$$

The tensor $\mathbf{S}$ is thus obtained from

$$\mathbf{S} = \boldsymbol{\tau}' - \boldsymbol{\beta}\,, \tag{7.5.12c}$$

where prime denotes the deviatoric part. Figure 7.5.4 shows the S_I-axis and the corresponding two sliding planes, $\mathbf{s}^\alpha$, $\alpha = 1, 2$, together with the corresponding unit normals, $\mathbf{n}^\alpha$. The S_I-axis makes an angle ψ with the x_1-axis. Because of the

---

42 In Chapter 4, *the stress-difference tensor*, $\mathbf{S}$ in (7.5.10a), is denoted by the lower case letter, s, and its magnitude, S in (7.5.10c), by τ, respectively.

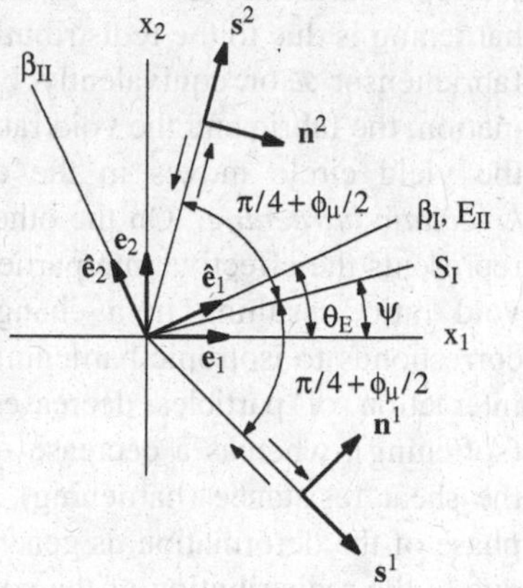

Figure 7.5.4

To decompose the stress deviator $\boldsymbol{\tau}'$: (1) identify the angle θ_E of the direction of the *minor* principal axis, E_{II}, of the fabric tensor, $\mathcal{E}$, relative to the x_1-axis; (2) measure from the x_1-axis an angle θ_E to the direction $\hat{\mathbf{e}}_1$ of β_I, as shown; (3) then $\mathbf{S} = \boldsymbol{\tau}' - \boldsymbol{\beta}$, and sliding directions with unit vectors $\mathbf{s}^1$ and $\mathbf{s}^2$, form angles $\pm(\pi/4+\phi_\mu/2)$ with the major principal axis, S_I, which makes angle ψ with the x_1-axis; the pair of unit vectors $\mathbf{s}^\alpha$ and $\mathbf{n}^\alpha$, $\alpha = 1, 2$, form a *sliding system*

sign convention, $S_I > 0$ (tension) and $S_{II} < 0$ (compression). The sliding directions are shown by arrows in this figure.

The decomposition (7.5.10a) can be interpreted in terms of the continuum plasticity models of Chapter 4. The tensor $\boldsymbol{\beta}$, which is proportional to the pressure p, represents the kinematic hardening. The tensor $\mathbf{S}$, with $S \le p\sin\phi_\mu$, is the yield circle in the deviatoric stress space; see Figure 7.5.5, where

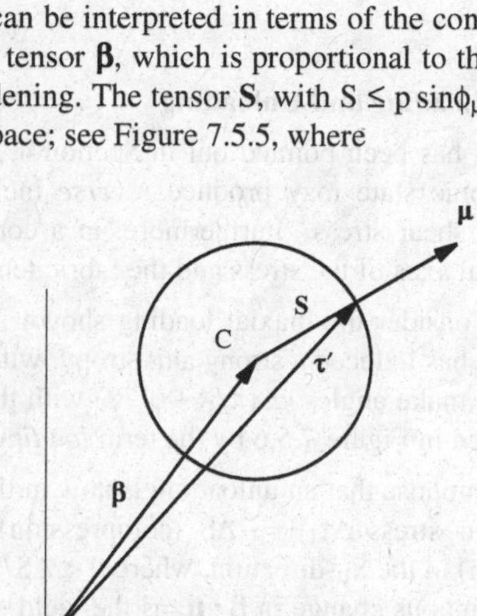

Figure 7.5.5

Yield surface in deviatoric stress space; $\boldsymbol{\mu} = \mathbf{S}/(\sqrt{2}\,S)$ is a unit tensor normal to the yield circle

$$\boldsymbol{\mu} = \mathbf{S}/(\sqrt{2}\,S) \tag{7.5.12d}$$

is a unit tensor normal to the yield circle. *Unlike for metals, the origin of the coordinates in this space, can fall outside the yield circle.* Indeed, under small to moderate pressures, the radius of the yield circle is generally very small, as most frictional granular masses have a negligibly small elastic range.

For each sliding plane, both isotropic and kinematic hardening may occur. Isotropic hardening (softening) is due to densification (dilatancy), and kinematic hardening is due to the redistribution of contact normals, being measured by the fabric tensor $\mathcal{E}$ or, equivalently, by the backstress $\boldsymbol{\beta}$. During the course of deformation, the fabric and the void ratio change. As the fabric changes, the center of the yield circle moves in the deviatoric stress space. This corresponds to *kinematic hardening*. On the other hand, the effective friction angle, ϕ_μ, which represents the effective interparticle friction at the meso-scale, changes with the void ratio, resulting in a change in the radius of the yield surface. This corresponds to isotropic hardening or softening. As the void ratio increases, the interaction of particles decreases and hence the shear resistance decreases (softening), whereas a decrease in the void ratio corresponds to an increase in the shear resistance (hardening). The isotropic softening during the dilatancy phase of the deformation, is generally accompanied by an anisotropic hardening due to the redistribution of the contact normals, *i.e.*, an increase in their density in the direction of maximum compression. At small pressures, common in soil mechanics applications, the elastic range is generally very small. The resistance due to the fabric stress, hence dominates and defines the stress state. For moderate pressures, the isotropic resistance to deformation, **S**, and the backstress, $\boldsymbol{\beta}$, are equally significant, while the influence of the fabric is expected to diminish under great pressures.

7.5.6. Loading and Unloading

It has been pointed out in Section 4.7, page 192, that *unloading* from an anisotropic state may produce *reverse* inelastic deformation, even against an applied shear stress. Furthermore, in a continued monotonic deformation, the principal axes of the stress and the fabric tensors tend to coincide.

Consider the biaxial loading shown in Figure 7.5.6, and assume that the loading has induced a strong anisotropy, with[43] $\beta >> S$. The sliding directions in loading make angles $\pm(\pi/4+\phi_\mu/2)$ with the S_I-direction. These directions are identified in Figure 7.5.6 by the term *loading*.

Suppose that an unloading is now initiated by the addition of an incremental shear stress $\Delta\tau_{11} = -\Delta S$ (compression) in the S_I-direction and $\Delta\tau_{22} = \Delta S$ (tension) in the S_{II}-direction, where $0 < \Delta S/p << 1$. This results in a sudden and discontinuous change in $\boldsymbol{\beta} : \boldsymbol{\mu}$, as the yield surface is reached by the stress point in unloading. Reverse plastic deformation then begins on the sliding planes denoted by the term *unloading* in Figure 7.5.6. These planes make angles $\pm(\pi/4-\phi_\mu/2)$ with the S_I-direction. It is important to note that, at the inception of unloading, the granular mass is in equilibrium, having a highly biased fabric and generally very small elastic range. The unloading from this state is equivalent to the reverse incremental loading with the fabric now assisting the

---

[43] Note from (7.5.11b) that $S/p < 1$.

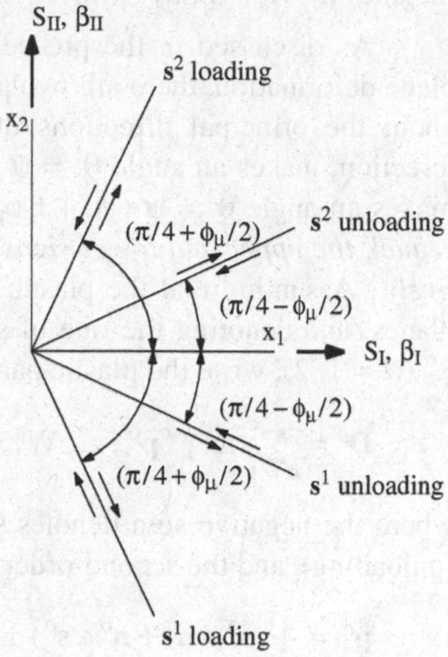

Figure 7.5.6

Sliding planes in loading and unloading: in biaxial loading, the stress state is given by $\tau_{11} = -p + S + \beta$ and $\tau_{22} = -p - S - \beta$; in unloading, the driving shear stress is defined by $\Delta\tau_{11} = -\Delta S - \beta$ and $\Delta\tau_{22} = \Delta S + \beta$ which may produce reverse plastic deformation, where $\Delta S > 0$ is the unloading shear stress increment

corresponding reverse deformation. Densification accompanies this reverse loading.[44]

7.5.7. A Rate-independent Double-sliding Model

The meso-scale double-sliding formulation of the preceding subsection is now used to develop a model for planar deformation of frictional granules. This model includes as special cases various other models which are based on the double-sliding idea.

Assume that the total plastic deformation at the continuum level, is produced by two superimposed shearing deformations along the active sliding planes. This sliding is accompanied by volumetric changes and induced anisotropy. Based on these concepts, a complete set of constitutive relations is produced in what follows.

Let, at the continuum level, the kinematics of instantaneous granular deformation be expressed by the velocity gradient $\mathbf{L} = (\partial\mathbf{v}/\partial\mathbf{x})^T$, consisting of a symmetric part, $\mathbf{D}$, the deformation rate tensor, and a skewsymmetric part, $\mathbf{W}$, the spin tensor. Each of these rates is separated into elastic and plastic parts, as follows:

$$\mathbf{D} = \mathbf{D}^* + \mathbf{D}^p, \qquad \mathbf{W} = \mathbf{W}^* + \mathbf{W}^p, \tag{7.5.13a,b}$$

where superscript p denotes the plastic part which is due to shearing along the

---

[44] Examination of experimental data reported by Okada (1992) supports this observation; see Nemat-Nasser and Zhang (2002).

sliding directions, and superscript * denotes the elastic part; note that $\mathbf{W}^*$ also includes the rigid-body spin.

As discussed in the preceding subsection and shown in Figure 7.5.4, in plane deformation, there are two preferred sliding planes, symmetrically situated about the principal directions of the stress-difference, $\mathbf{S}$. The first, the $\mathbf{s}^1$-direction, makes an angle $\theta^1 = \psi - \pi/4 \mp \phi_\mu/2$, and the second, the $\mathbf{s}^2$-direction, makes an angle $\theta^2 = \psi + \pi/4 \pm \phi_\mu/2$, with the positive x_1-axis; *here and in the sequel, the upper and lower signs correspond to loading and unloading, respectively.* Assuming that the plastic deformation is due to shearing on the sliding planes, and denoting the rate of shearing in the *positive* $\mathbf{s}^\alpha$-sliding direction by $\dot{\gamma}^\alpha$ ($\alpha = 1, 2$), write the plastic part of the velocity gradient as

$$\mathbf{D}^p = \sum_{\alpha=1}^{2} \pm\, \dot{\gamma}^\alpha \mathbf{p}^\alpha, \qquad \mathbf{W}^p = \sum_{\alpha=1}^{2} \pm\, \dot{\gamma}^\alpha \mathbf{r}^\alpha, \tag{7.5.14a,b}$$

where the negative sign denotes sliding in the *negative* direction of the $\mathbf{s}^\alpha$-axis (unloading), and the second-order tensors $\mathbf{p}^\alpha$ and $\mathbf{r}^\alpha$ are defined by

$$\mathbf{p}^\alpha = \frac{1}{2}(\mathbf{s}^\alpha \otimes \mathbf{n}^\alpha + \mathbf{n}^\alpha \otimes \mathbf{s}^\alpha) \pm \mathbf{n}^\alpha \otimes \mathbf{n}^\alpha \tan\delta^\alpha,$$

$$\mathbf{r}^\alpha = \frac{1}{2}(\mathbf{s}^\alpha \otimes \mathbf{n}^\alpha - \mathbf{n}^\alpha \otimes \mathbf{s}^\alpha). \tag{7.5.15a,b}$$

Here δ^α, $\alpha = 1,2$, are the dilatancy angles associated with the sliding planes. They represent the orientation of the *effective* microscopic planes of motion relative to the corresponding mesoscopic sliding planes that are shown in Figure 7.5.4.

The vectors $\mathbf{s}^\alpha$ and $\mathbf{n}^\alpha$ are unit vectors in the positive direction of the sliding and normal to the plane of sliding, respectively. The α'th sliding plane and the sliding direction on this plane form a *sliding system*. In the double-sliding model, there are two such systems which can be activated simultaneously. On each sliding direction, the sliding rate is positive for loading and negative for unloading. These sliding systems with their unit vectors, are shown in Figure 7.5.4. The unit vectors are defined by

$$\mathbf{s}^1 = \{\cos\theta^1, \sin\theta^1\}, \qquad \mathbf{n}^1 = \{-\sin\theta^1, \cos\theta^1\},$$

$$\mathbf{s}^2 = \{\cos\theta^2, \sin\theta^2\}, \qquad \mathbf{n}^2 = \{\sin\theta^2, -\cos\theta^2\}, \tag{7.5.16a-d}$$

where

$$\theta^1 + \theta^2 = 2\psi, \quad \theta^2 - \theta^1 = \pi/2 \pm \phi_\mu, \quad \tan 2\psi = \frac{2\,S_{12}}{S_{11} - S_{22}},$$

$$S_{22} = -S_{11}, \quad S = (\tfrac{1}{2} S_{ij} S_{ij})^{1/2}. \tag{7.5.17a-e}$$

7.5.8. Deformation Rate and Spin, Based on Double Sliding

Start with definition (7.5.14a,b), and noting (7.5.17a-e), obtain the following relations for the components of the plastic part of the velocity gradient $\mathbf{L}^p = \mathbf{D}^p + \mathbf{W}^p$:

$$D^p_{kk} = (\dot\gamma^1 \tan\delta^1 + \dot\gamma^2 \tan\delta^2), \qquad W^p_{12} = \frac{1}{2}(\dot\gamma^1 - \dot\gamma^2),$$

$$D^p_{11}{}' = \frac{1}{2}\dot\gamma^1 \frac{\cos(2\psi - (\phi_\mu - \delta^1))}{\cos\delta^1} + \frac{1}{2}\dot\gamma^2 \frac{\cos(2\psi + (\phi_\mu - \delta^2))}{\cos\delta^2} = -D^p_{22}{}',$$

$$D^p_{12} = \frac{1}{2}\dot\gamma^1 \frac{\sin(2\psi - (\phi_\mu - \delta^1))}{\cos\delta^1} + \frac{1}{2}\dot\gamma^2 \frac{\sin(2\psi + (\phi_\mu - \delta^2))}{\cos\delta^2}. \qquad (7.5.18\text{a-e})$$

In terms of the components of the stress-difference tensor **S**, the deviatoric components of the deformation rate tensor can be rewritten as

$$D^p_{11}{}' = \frac{S_{11}}{2S}\left\{\dot\gamma^1 \frac{\cos(\phi_\mu - \delta^1)}{\cos\delta^1} + \dot\gamma^2 \frac{\cos(\phi_\mu - \delta^2)}{\cos\delta^2}\right\}$$

$$+ \frac{S_{12}}{2S}\left\{\dot\gamma^1 \frac{\sin(\phi_\mu - \delta^1)}{\cos\delta^1} - \dot\gamma^2 \frac{\sin(\phi_\mu - \delta^2)}{\cos\delta^2}\right\},$$

$$D^p_{12} = \frac{S_{12}}{2S}\left\{\dot\gamma^1 \frac{\cos(\phi_\mu - \delta^1)}{\cos\delta^1} + \dot\gamma^2 \frac{\cos(\phi_\mu - \delta^2)}{\cos\delta^2}\right\}$$

$$- \frac{S_{11}}{2S}\left\{\dot\gamma^1 \frac{\sin(\phi_\mu - \delta^1)}{\cos\delta^1} - \dot\gamma^2 \frac{\sin(\phi_\mu - \delta^2)}{\cos\delta^2}\right\}. \qquad (7.5.19\text{a,b})$$

It is seen that the deviatoric plastic deformation rate tensor, $\mathbf{D}^{p\prime}$, is *not* coaxial with the stress-difference tensor, **S**, unless $\delta^1 = \delta^2 = \phi_\mu$, or in special cases, when $\dot\gamma^1 = \dot\gamma^2$ and $\delta^1 = \delta^2$.

To relate the double-sliding model to the continuum model of Section 4.7, consider equations (7.5.19a,b), and introduce the notation

$$A^\alpha = \frac{\cos(\phi_\mu - \delta^\alpha)}{\cos\delta^\alpha}, \qquad B^\alpha = \frac{\sin(\phi_\mu - \delta^\alpha)}{\cos\delta^\alpha}, \qquad \alpha = 1, 2. \qquad (7.5.20\text{a,b})$$

Then set

$$\dot\gamma = \dot\gamma^1 A^1 + \dot\gamma^2 A^2, \qquad \dot\omega = \dot\gamma^1 B^1 - \dot\gamma^2 B^2, \qquad (7.5.21\text{a,b})$$

and note that $\dot\gamma^1$ and $\dot\gamma^2$ are given by

$$\dot\gamma^1 = \frac{\dot\gamma B^2 + \dot\omega A^2}{A^1 B^2 + A^2 B^1}, \qquad \dot\gamma^2 = \frac{\dot\gamma B^1 - \dot\omega A^1}{A^1 B^2 + A^2 B^1}. \qquad (7.5.21\text{c,d})$$

It is convenient to use $\dot\gamma$ and $\dot\omega$ instead of the sliding rates, $\dot\gamma^1$ and $\dot\gamma^2$. When $\delta^1 = \delta^2$, then $\dot\gamma$ is proportional to $\dot\gamma^1 + \dot\gamma^2$ and $\dot\omega$ is proportional to the plastically-induced spin, ½ $(\dot\gamma^1 - \dot\gamma^2)$. All quantities which describe the plastic deformation rate can be expressed in terms of $\dot\gamma$ and $\dot\omega$. In particular, (7.5.19a,b) reduce to

$$D^p_{11}{}' = \dot\gamma \frac{S_{11}}{2S} + \dot\omega \frac{S_{12}}{2S}, \qquad D^p_{12} = \dot\gamma \frac{S_{12}}{2S} - \dot\omega \frac{S_{11}}{2S}, \qquad (7.5.22\text{a,b})$$

and (7.5.18a,b) become

$$D^p_{kk} = \dot{\gamma}\,\Gamma + \dot{\omega}\,\Omega\,, \qquad W^p_{12} = \dot{\gamma}\,\frac{B^2 - B^1}{2\,\Delta} + \dot{\omega}\,\frac{A^2 + A^1}{2\,\Delta}\,, \tag{7.5.22c,d}$$

where

$$\Delta = A^1 B^2 + A^2 B^1 = \frac{\sin(2\phi_\mu - (\delta^1 + \delta^2))}{\cos\delta^1 \cos\delta^2}\,,$$
$$\Gamma = (B^2 \tan\delta^1 + B^1 \tan\delta^2)/\Delta\,, \qquad \Omega = (A^2 \tan\delta^1 - A^1 \tan\delta^2)/\Delta\,. \tag{7.5.22e-g}$$

Solving (7.5.22a,b) for $\dot{\gamma}$ and $\dot{\omega}$, obtain

$$\frac{1}{2}\,\dot{\gamma} = D^{p\prime}_{11}\,\frac{S_{11}}{S} + D^{p\prime}_{12}\,\frac{S_{12}}{S}\,, \qquad \frac{1}{2}\,\dot{\omega} = D^{p\prime}_{11}\,\frac{S_{12}}{S} - D^{p\prime}_{12}\,\frac{S_{11}}{S}\,. \tag{7.5.23a,b}$$

Hence, $S\,\dot{\gamma}$ is the rate of the distortional plastic work associated with the stress-difference, **S**, measured per unit reference volume.

The plastic deformation rates (7.5.22a,b) can be expressed in the following coordinate-independent form:

$$\mathbf{D}^{p\prime} = \dot{\gamma}\,\frac{\boldsymbol{\mu}}{\sqrt{2}} + \dot{\omega}\,\frac{\tilde{\boldsymbol{\mu}}}{\sqrt{2}}\,, \tag{7.5.24a}$$

where, as before, $\boldsymbol{\mu} = \mathbf{S}/(\sqrt{2}\,S)$, and the unit tensor $\tilde{\boldsymbol{\mu}}$, orthogonal to $\boldsymbol{\mu}$, is defined by

$$\tilde{\mu}_{ij} = e_{ik}\,\mu_{kj}\,, \qquad \tilde{\boldsymbol{\mu}} : \tilde{\boldsymbol{\mu}} = 1\,, \qquad \boldsymbol{\mu} : \tilde{\boldsymbol{\mu}} = \mathbf{0}\,; \tag{7.5.24b-d}$$

here e_{ij} is the two-dimensional permutation symbol, *i.e.*, $e_{12} = -e_{21} = 1$, $e_{11} = e_{22} = 0$. The tensor $\tilde{\boldsymbol{\mu}}$ is proportional to the tensor $\mathring{\boldsymbol{\mu}}$ of Section 4.7, page 192.

It is convenient to express $\tilde{\boldsymbol{\mu}}$ directly in terms of the deviatoric part of the *total* deformation rate tensor $\mathbf{D}'$, as follows:

$$\tilde{\boldsymbol{\mu}} = a\,(\mathbf{1}^{(4s)} - \boldsymbol{\mu} \otimes \boldsymbol{\mu}) : \mathbf{D}'\,, \qquad a = \{\,|\mathbf{D}'|^2 - (\boldsymbol{\mu} : \mathbf{D})^2\}^{-1/2}\,. \tag{7.5.24e,f}$$

From (7.5.22) and (7.5.24), now obtain

$$\mathbf{D}^p = \dot{\gamma}\left\{\frac{\boldsymbol{\mu}}{\sqrt{2}} + \tfrac{1}{2}\Gamma\,\mathbf{1}\right\} + \dot{\omega}\left\{\frac{\tilde{\boldsymbol{\mu}}}{\sqrt{2}} + \tfrac{1}{2}\Omega\,\mathbf{1}\right\}. \tag{7.5.25}$$

The effective plastic deformation rate, $\dot{\gamma}_{\text{eff}}$, is

$$\dot{\gamma}_{\text{eff}} \equiv (2\,\mathbf{D}^{p\prime} : \mathbf{D}^{p\prime})^{1/2} = (\dot{\gamma}^2 + \dot{\omega}^2)^{1/2}\,,$$
$$\dot{\gamma} = \sqrt{2}\,\boldsymbol{\mu} : \mathbf{D}^p\,, \qquad \dot{\omega} = \sqrt{2}\,\tilde{\boldsymbol{\mu}} : \mathbf{D}^p\,. \tag{7.5.26a-d}$$

The final constitutive equation is therefore given by

$$\mathring{\boldsymbol{\tau}} = \boldsymbol{\mathcal{L}} : (\mathbf{D} - \mathbf{D}^p) - \mathbf{W}^p\,\boldsymbol{\tau} + \boldsymbol{\tau}\,\mathbf{W}^p\,, \tag{7.5.27}$$

where $\boldsymbol{\mathcal{L}}$ is the elasticity tensor yet to be defined, and $\mathbf{D}^p$ and $\mathbf{W}^p$ are given by (7.5.25) and (7.5.22d), respectively.

For the special case when $\delta^1 = \delta^2 = \delta$, the results reported by Balendran and Nemat-Nasser (1993a) are obtained[45], as follows:

$$B^1 = B^2 \equiv B = \frac{\sin(\phi_\mu - \delta)}{\cos\delta},$$

$$A^1 = A^2 \equiv A = \frac{\cos(\phi_\mu - \delta)}{\cos\delta}, \quad \Delta = 2AB, \tag{7.5.28a-c}$$

leading to

$$\dot{\omega} = B(\dot{\gamma}^1 - \dot{\gamma}^2) = 2B W^p_{12}, \quad \dot{\gamma} = A(\dot{\gamma}^1 + \dot{\gamma}^2),$$

$$\Omega = 0, \quad \Gamma = \tan\delta / A, \quad D^p_{kk} = \dot{\gamma}\,\Gamma. \tag{7.5.29a-f}$$

Equation (7.5.25) gives the plastic deformation rate for a general continuum model of deformation of frictional granules, based on the double-sliding concept. It includes various other models as special cases. For example, when the fabric effects are neglected, $\boldsymbol{\beta} = \mathbf{0}$, and there are no plastically-induced volumetric strains, the model of Spencer (1964, 1982) is obtained. With $\boldsymbol{\beta} = \mathbf{0}$ and $\delta^1 = \delta^2$, the model of Mehrabadi and Cowin (1978) is retrieved. When only $\delta^1 = \delta^2$, the model of Balendran and Nemat-Nasser (1993a) results. Note that $\delta^1 = \delta^2$ only when the fabric tensor is either coaxial or 90 degrees out of phase with the stress-difference tensor **S**, as shown in Figure 7.5.6. In either case, $\sin 2\nu_0 = \cos\phi_\mu$.

7.5.9. Elasticity Relations

The general elasticity relations of Subsection 4.7.7, equations (4.7.23), can be used as the starting, most general relations which can then be simplified by invoking relevant symmetries. To this end, consider the instantaneous elastic modulus tensor $\mathcal{L}$ and the decomposition (4.7.23a,b). In two dimensions, the elasticity tensor can be expressed as,

$$\mathcal{L} = \mathcal{L}' + \frac{1}{2}(\mathbf{N} \otimes \mathbf{1} + \mathbf{1} \otimes \mathbf{N}) - \frac{\kappa}{4}\mathbf{1} \otimes \mathbf{1}$$

$$= \mathcal{L}' + \frac{1}{2}(\mathbf{N}' \otimes \mathbf{1} + \mathbf{1} \otimes \mathbf{N}') + \frac{\kappa}{4}\mathbf{1} \otimes \mathbf{1}, \tag{7.5.30a,b}$$

where

$$N_{ij} = \mathcal{L}_{ijkk} = \mathcal{L}_{kkij}$$

$$\mathbf{N} = \mathbf{N}' + \frac{\kappa}{2}\mathbf{1}, \quad \kappa = \mathrm{tr}(\mathbf{N}) = \mathcal{L}_{iijj} \quad i, j = 1, 2. \tag{7.5.30c-e}$$

---

[45] Note that, in such a case, at a miminum it must be required that a principal direction of $\boldsymbol{\beta}$ coincides with a principal direction of **S** as shown in Figure 7.5.6. In (7.5.29f), Γ is the dilatancy parameter, which is denoted by B in Section 4.7; see (4.7.3b), page 199.

The elastic part of the deformation-rate tensor is now expressed in terms of the Jaumann rate of change of the Kirchhoff stress, corotational with the sliding systems, $\overset{\circ}{\boldsymbol{\tau}}{}^*$, as follows:

$$\overset{\circ}{\boldsymbol{\tau}}{}^* = \boldsymbol{\mathcal{L}} : \mathbf{D}^* = \boldsymbol{\mathcal{L}} : (\mathbf{D} - \mathbf{D}^p)\,. \tag{7.5.31a,b}$$

For frictional granular materials, the elasticity tensor $\boldsymbol{\mathcal{L}}$, in general, is not constant, but rather, it depends on the fabric tensor. When the distribution of the contact unit normals is represented by a second-order tensor, then it is reasonable to assume an orthotropic elasticity tensor with the axes of orthotropy defined by the principal axes of the fabric tensor, $\boldsymbol{\beta}$. Hence, in the $\hat{\mathbf{e}}_i$-coordinate system of Figure 7.5.4, the elastic response is expressed as

$$\hat{\overset{\circ}{\tau}}{}^*_{11} = C_{11}\,\hat{D}^*_{11} + C_{12}\,\hat{D}^*_{22}\,, \qquad \hat{\overset{\circ}{\tau}}{}^*_{22} = C_{12}\,\hat{D}^*_{11} + C_{22}\,\hat{D}^*_{22}\,,$$
$$\hat{\overset{\circ}{\tau}}{}^*_{12} = C_{33}\,\hat{D}^*_{12}\,, \tag{7.5.32a-c}$$

where superposed ^ is used to denote the tensor components in the principal axes of the fabric tensor. These equations can be rewritten in tensor form as,

$$\overset{\circ}{\tau}{}^*_{kk} = 2\,K\,D^*_{kk} + 2\sqrt{2}\,\overline{K}\,(\boldsymbol{\mu}_\beta : \mathbf{D}^*)\,,$$
$$\overset{\circ}{\boldsymbol{\tau}}{}^{*\prime} = \sqrt{2}\,\overline{K}\,D^*_{kk}\,\boldsymbol{\mu}_\beta + 2\,G\,\mathbf{D}^{*\prime} + 2\,\overline{G}\,(\boldsymbol{\mu}_\beta : \mathbf{D}^*)\,\boldsymbol{\mu}_\beta\,, \tag{7.5.33a,b}$$

where $\boldsymbol{\mu}_\beta$ is the unit fabric tensor, defined by

$$\boldsymbol{\mu}_\beta = (\hat{\mathbf{e}}_1 \otimes \hat{\mathbf{e}}_1 - \hat{\mathbf{e}}_2 \otimes \hat{\mathbf{e}}_2)/\sqrt{2} = \boldsymbol{\beta}/(\sqrt{2}\,\beta)\,, \tag{7.5.33c}$$

and the other parameters are given in terms of C_{ij}, by

$$K = (C_{11} + C_{22} + 2\,C_{12})/4\,, \qquad \overline{K} = (C_{11} - C_{22})/4$$
$$G = C_{33}/2\,, \qquad \overline{G} = (C_{11} + C_{22} - 2\,C_{12})/4\,. \tag{7.5.33d-g}$$

In view of (7.5.30a,b), it follows that

$$\boldsymbol{\mathcal{L}}' = 2\,G\,(\mathbf{1}^{(4s)} - \tfrac{1}{2}\,\mathbf{1} \otimes \mathbf{1}) + 2\,\overline{G}\,\boldsymbol{\mu}_\beta \otimes \boldsymbol{\mu}_\beta\,, \qquad \mathbf{N}' = 2\sqrt{2}\,\overline{K}\,\boldsymbol{\mu}_\beta\,, \qquad \kappa = 4\,K\,,$$
$$\boldsymbol{\mathcal{L}} = 2\,G\,(\mathbf{1}^{(4s)} - \tfrac{1}{2}\,\mathbf{1} \otimes \mathbf{1}) + 2\,\overline{G}\,\boldsymbol{\mu}_\beta \otimes \boldsymbol{\mu}_\beta + \sqrt{2}\,\overline{K}\,(\boldsymbol{\mu}_\beta \otimes \mathbf{1} + \mathbf{1} \otimes \boldsymbol{\mu}_\beta) + K\,\mathbf{1} \otimes \mathbf{1}\,. \tag{7.5.34a,b}$$

The elasticity parameters G, $\overline{G}$, K, and $\overline{K}$, in general depend on the pressure and the void-volume fraction.

7.5.10. Fabric and its Evolution

The backstress $\boldsymbol{\beta}$ representing the fabric of the granular mass, is defined by (7.5.7b) in terms of the fabric tensor $\boldsymbol{\mathcal{E}}$ which characterizes the distribution of the unit contact normals. The evolution of $\boldsymbol{\beta}$ is therefore directly related to the evolution of the distribution of the contact normals. To relate the micromechanically obtained results of the present chapter to those reported in Chapter 4, and particularly to those in Section 4.7, introduce the following representation for the backstress:

$$\boldsymbol{\beta} = \sqrt{2}\,p\,\mathbf{J}\,, \tag{7.5.35a}$$

and note that

$$\beta = (\tfrac{1}{2}\,\boldsymbol{\beta} : \boldsymbol{\beta})^{1/2} = p\,J\,, \qquad J = (\mathbf{J} : \mathbf{J})^{1/2}. \tag{7.5.35b,c}$$

The principal directions of the fabric tensor are given by the unit fabric tensor $\boldsymbol{\mu}_\beta$, defined by (7.5.33c).

The fabric changes with the continued plastic flow of the granular mass. This change must be quantified in terms of the deformation or stress measures. Experimental observations of the photoelastic granules suggest that the fabric changes with the stress, tending to become coaxial with it. Since the pressure p is expected to affect the magnitude of the contact forces, rather than the distribution of the contact unit normals, it is reasonable to assume that

$$\mathring{\mathbf{J}} = \Lambda\,\dot{\gamma}\,\boldsymbol{\mu}\,, \qquad \mathring{\boldsymbol{\beta}} = \sqrt{2}\,p\,\mathring{\mathbf{J}}\,, \tag{7.5.36a,b}$$

where Λ is a material function, depending on the fabric measure, say, **J**, and possibly its history (*e.g.*, the last extreme state of of the fabric, in cyclic loading). In monotonic loading, the induced fabric (anisotropy) will eventually saturate. Hence, Λ must then tend to zero. Upon reverse loading, a jump in the value of Λ is expected.

7.5.11. Continuum Approximation

In continuum plasticity, the terms associated with the plastically-induced spin, $\mathbf{W}^p$, are generally neglected in defining the elasticity relations. Using this, consider the continuum constitutive relation

$$\mathring{\boldsymbol{\tau}} \approx \boldsymbol{\mathcal{L}} : (\mathbf{D} - \mathbf{D}^p)\,, \tag{7.5.37}$$

together with relations (7.5.36a,b) which define the evolution of the backstress. In (7.5.37), the elasticity tensor $\boldsymbol{\mathcal{L}}$ is defined by (7.5.34b), but the plastically-induced deformation rate is now *approximated* by

$$\mathbf{D}^p = \dot{\gamma}\,\frac{\boldsymbol{\mu}}{\sqrt{2}} + \alpha\,(\mathbf{1}^{(4s)} - \boldsymbol{\mu} \otimes \boldsymbol{\mu}) : \mathbf{D}' + \tfrac{1}{2}\,\dot{\gamma}\,\Gamma\,\mathbf{1}\,, \tag{7.5.38}$$

where $\alpha = \dot{\omega}\,a/\sqrt{2}$ is the (dimensionless) noncoaxiality factor, with a, given by (7.5.24f).

7.5.12. Void Ratio

The ratio of the void volume to the solid volume is called the void ratio and is denoted by e; see (4.3.14a), page 170. For small pressures, the volume of the solid may be assumed to remain essentially the same, as compared with the changes in the void volume. Hence, it follows that

$$D^p_{kk} = \dot{\gamma}\,\Gamma \approx \frac{\dot{e}}{1+e}\,, \tag{7.5.39a,b}$$

so that the void ratio e is given by

$$e = (1 + e_0)\exp\left\{\int_0^{\gamma} \Gamma\, d\gamma\right\} - 1, \tag{7.5.39c}$$

where e_0 is the initial void ratio. As is seen, the exponential dependency of the void ratio e on the dilatancy parameter Γ, makes the result sensitive to the variation of the parameter Γ.

7.5.13. Consistency Condition

From the yield condition (7.5.11b), consider the consistency relation

$$\dot{S} - \dot{p}M = p\dot{M}, \qquad M = \sin\phi_\mu . \tag{7.5.40a,b}$$

Since pM defines the size of the yield circle, its rate of change gives the workhardening parameter. In general, M is a function of the void ratio and other parameters which characterize the microstructure of the granular mass, *e.g.*, the fabric measure J. As an illustration, let

$$M = M(e, J) . \tag{7.5.41a}$$

Then, it follows that

$$\dot{M} = H\dot{\gamma}, \qquad H = (1+e)\Gamma\frac{\partial M}{\partial e} + \mu_\beta \Lambda \frac{\partial M}{\partial J}, \qquad \mu_\beta = \boldsymbol{\mu}_\beta : \boldsymbol{\mu} . \tag{7.5.41b-d}$$

Hence, H is the workhardening parameter associated with the size of the yield surface, and μ_β gives the orientation of the principal directions of the stress-difference tensor, relative to those of the fabric tensor. The workhardening associated with the backstress is included in the $\dot{S}$-term.

Both sides of (7.5.40a) can be computed explicitly, using (7.5.34b), (7.5.36a,b), (7.5.37), and (7.5.41b). One obtains

$$\begin{aligned}
&\dot{S} - \dot{p}M = A_1 - \sqrt{2}\,\alpha A_2 - \dot{\gamma} A_3 = \dot{\gamma}\, p\,(H + \Lambda),\\
&A_1 = \sqrt{2}\,(G D_\mu + \overline{G} D_\beta \mu_\beta) + (\overline{K}\mu_\beta + KM)D_{kk} + \sqrt{2}\,\overline{K} M D_\beta,\\
&A_2 = (D_\beta - D_\mu \mu_\beta)(\overline{G}\mu_\beta + \overline{K}M),\\
&A_3 = G + \overline{G}\mu_\beta^2 + \overline{K}\mu_\beta(M + \Gamma) + KM\Gamma,
\end{aligned} \tag{7.5.42a-d}$$

where the following notation is used:

$$D_\mu = \boldsymbol{\mu} : \mathbf{D}, \qquad D_\beta = \boldsymbol{\mu}_\beta : \mathbf{D} . \tag{7.5.42e,f}$$

Here, D_μ and D_β give the orientation of the principal axes of the stress, **S**, and the fabric, **β**, tensors, relative to those of the total deformation rate tensor, **D**. These parameters characterize the current stress state and the anisotropic microstructure of the material relative to the loading parameter. Solving (7.5.42a) for $\dot{\gamma}$, obtain

$$\dot{\gamma} = \frac{A_1 - \sqrt{2}\,\alpha A_2}{p\,(H + \Lambda) + A_3} . \tag{7.5.43}$$

7.5.14. Dilatancy

To obtain an expression for the dilatancy parameter Γ, consider the rate of stress work, $\boldsymbol{\tau} : \mathbf{D}^{p}$, and equate this to the rate of dissipation. Estimate the rate of dissipation by $p\,M_f\dot{\gamma}$, where M_f is an effective overall friction coefficient. Now obtain

$$\Gamma \approx -M_f + \frac{S}{p} + \frac{\beta}{p}\mu_\beta ,$$

$$= -M_f + M + \mu_f , \qquad \mu_f = J\,\mu_\beta , \qquad (7.5.44a\text{-}c)$$

where, as in Section 4.7, the term proportional to $D_\beta - D_\mu \mu_\beta$ is excluded. This term is associated with the part of the deformation-rate tensor which is tangent to the yield surface. Since $\boldsymbol{\mu}$ and $\boldsymbol{\mu}_\beta$ are unit deviatoric tensors, $\mu_\beta = \cos 2\theta$, where θ is the angle between the major principal directions of the stress-difference, $\mathbf{S}$, and the fabric, $\boldsymbol{\beta}$, tensors. Hence, $-1 \le \mu_\beta \le 1$. The maximum fabric-induced dilatancy occurs for $\theta = 0$, *i.e.*, when S_I and β_I are coincident, while there is a maximum densification due to the fabric when these directions are 90 degrees out of phase. The latter occurs in unloading; see Figure 7.5.6.

7.5.15. Material Parameters

In the continuum approximation of the theory, there are four basic material functions which characterize the inelastic response of the granular mass. These are:

(1) M which defines the elastic range of the material. At small pressures, M is very small for most frictional ganules.

(2) M_f which defines the overall-frictional resistance of the material, and, in general, is a function of the void ratio e and the fabric measure μ_f. The following form for this function has been suggested by Nemat-Nasser and Zhang (2002), based on experimental results of Okada (1992):

$$M_f = (a + b(e - e_m)^{n_1})(\mu_0 - \mu_f)^{n_2} , \qquad (7.5.45a)$$

where e_m is the minimum value of the void ratio e, and μ_0 is a reference value of the fabric parameter μ_f. Expression (7.5.45a) is suitable for application to cohesionless sands at small pressures.

(3) Λ which defines the evolution of the fabric tensor through the backstress. In monotonic shearing, a saturation state for induced anisotropy (fabric) is eventually attained. Hence, Λ must tend to zero, in such loadings. Upon unloading, Λ undergoes a finite jump; see Nemat-Nasser and Zhang (2002).

(4) α which is the noncoaxiality parameter. From (7.5.21b) and (7.5.29a), α depends on the friction and dilatancy parameters. It is zero when the fabric and stress-difference tensors, $\boldsymbol{\beta}$ and $\mathbf{S}$, are coaxial, and the plastically-induced spin vanishes.

In addition, the elasticity is defined in terms of four elastic moduli, as well as the fabric unit tensor μ_β. The elastic moduli are G, $\bar{G}$, K, and $\bar{K}$. These moduli depend on the void ratio and pressure. They have the general form of

$$A = A_0(1 + c(e_M - e)^{n_3}), \tag{7.5.45b}$$

where A stands for any of the four elastic moduli, and e_M is the maximum value of the void ratio. The parameter A_0 is proportional to pressure, p.

In the continuum model, all the free parameters must be established based on experimental results. The procedure has been illustrated by Nemat-Nasser and Zhang (2002). An example, given by these authors, is used below for illustration.

7.5.16. Example

Okada (1992) has performed experiments on large hollow cylindrical samples of Silica 60, for which $e_m = 0.631$ and $e_M = 1.095$; see Chapter 9. Some of the results are examined and modeled by Nemat-Nasser and Zhang (2002), where details of the modeling are given. The samples have 25cm outside and 20cm inside diameters, and 25cm height. They are cyclically sheared under a constant confining pressure, in a drained condition. A pure shearing with $\tau_{rr} = \tau_{\theta\theta} = \tau_{zz} = -p$ and $\tau_{r\theta} = \tau_{rz} = 0$, may be assumed, where a polar coordinate system is used. With the subscripts $\theta = 1$ and $z = 2$, and assuming a negligible elastic range, now obtain,

$$\dot{\tau}_{12} \approx \dot{\beta}_{12} = \pm p \Lambda \dot{\gamma}, \quad \dot{\gamma} = \pm \frac{2D_{12}}{1 + \Lambda/\kappa} > 0, \quad \kappa = \frac{(G + \bar{G})}{p}. \tag{7.5.46a-c}$$

The combined elastic modulus κ has the functional form given by (7.5.45b), and for Λ, Nemat-Nasser and Zhang use

$$\Lambda(t) = \frac{\kappa}{1 + d\,|\varepsilon(t) - \varepsilon_e|^{n_4}}, \qquad \varepsilon(t) = \int_0^t D_{12}(t')\,dt', \tag{7.5.46d,e}$$

where ε_e is the value of $\varepsilon(t)$ attained just before unloading. From (7.5.46a-e), the shear stress $\tau_{12} = \tau_{\theta z}$ is obtained as a function of the strain ε. A typical example is given in Figure 7.5.7, where the solid curve represents the experimental and the dashed curve the theoretical results. The material constants used are: $\kappa = 130$, $c = 1.42$, $d = 1.40 \times 10^3$, and $n_4 = 1.1$. The pressure is 195Pa. While the material constants are selected to fit the experimental results, the general features are inherent in the structure of the model.

The void ratio is obtained from (7.5.39), with

$$\Gamma = -M_f + \zeta\mu_f, \qquad \mu_f = \pm \frac{\beta_{12}}{p}. \tag{7.5.47a,b}$$

The results for the same example are shown in Figure 7.5.7b, where the experimental data are presented by open circles. The material constants are: $a = 1/4$, $b = 0.1$, $n_1 = 0.40$, $n_2 = n_3 = 1$, and $\zeta = 3/4$. In this calculation, μ_0 in (7.5.45a),

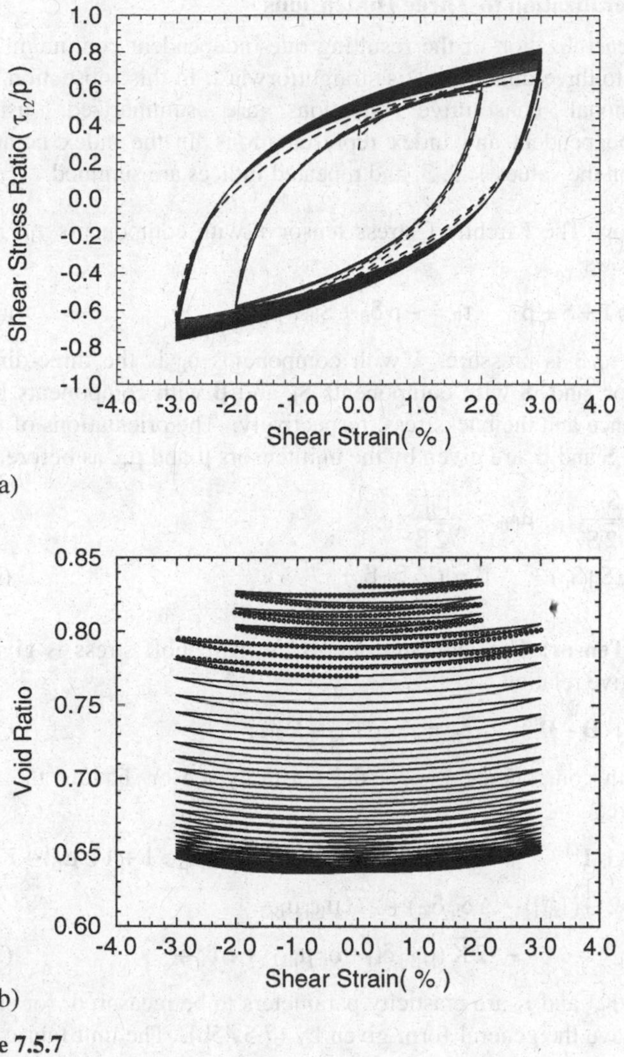

Figure 7.5.7

(a) Normalized shear stress and (b) void ratio versus shear strain in multi-cycle shearing of a hollow cylindrical sample of Silica 60 with 0.831 initial void ratio (from Nemat-Nasser and Zhang, 2002)

is the last maximum value of the fabric parameter μ_f, attained just before unloading. Note that the value of μ_0 changes twice during each cycle of shearing.

7.5.17. Generalization to Three Dimensions

The generalization of the resulting rate-independent continuum model of this section, to three dimensions is straightforward. In this subsection, the basic three-dimensional constitutive equations are summarized, using both coordinate-independent and index representations. In the index notation, subscripts take on the values 1, 2, 3, and repeated indices are summed.

Stress Tensor: The Kirchhoff stress tensor $\boldsymbol{\tau}$ with components τ_{ij}, is decomposed as follows:

$$\boldsymbol{\tau} = -p\,\mathbf{1} + \mathbf{S} + \boldsymbol{\beta}, \quad \tau_{ij} = -p\,\delta_{ij} + S_{ij} + \beta_{ij}, \tag{7.5.48a,b}$$

where $p = -\tau_{ii}/3$ is pressure, $\mathbf{1}$ with components δ_{ij} is the three-dimensional identity tensor, and $\mathbf{S}$ with components S_{ij} and $\boldsymbol{\beta}$ with components β_{ij} are the stress difference and the backstress, respectively. The orientations of the deviatoric tensors, $\mathbf{S}$ and $\boldsymbol{\beta}$, are given by the unit tensors $\boldsymbol{\mu}$ and $\boldsymbol{\mu}_\beta$, as before,

$$\mu_{ij} = \frac{S_{ij}}{\sqrt{2}\,S}, \quad \mu_{\beta ij} = \frac{\beta_{ij}}{\sqrt{2}\,\beta},$$
$$S = (\tfrac{1}{2} S_{ij} S_{ij})^{1/2}, \quad \beta = (\tfrac{1}{2} \beta_{ij} \beta_{ij})^{1/2}. \tag{7.5.48c-f}$$

Stress-rate Tensor: The Jaumann rate of the Kirchhoff stress is given by the rate-constitutive relation

$$\overset{\circ}{\boldsymbol{\tau}} = \boldsymbol{\mathcal{L}} : (\mathbf{D} - \mathbf{D}^p), \quad \overset{\circ}{\tau}_{ij} = \mathcal{L}_{ijkl}\,(D_{kl} - D^p_{kl}), \tag{7.5.49a,b}$$

where $\boldsymbol{\mathcal{L}}$ with components $\mathcal{L}_{ijkl}$ is the elasticity tensor, having the following basic structure:

$$\boldsymbol{\mathcal{L}} = 2G\,(\mathbf{1}^{(4s)} - \tfrac{1}{3}\,\mathbf{1}\otimes\mathbf{1}) + 2\overline{G}\,\boldsymbol{\mu}_\beta\otimes\boldsymbol{\mu}_\beta + \sqrt{2}\,\overline{K}\,(\boldsymbol{\mu}_\beta\otimes\mathbf{1} + \mathbf{1}\otimes\boldsymbol{\mu}_\beta) + K\,\mathbf{1}\otimes\mathbf{1},$$
$$\mathcal{L}_{ijkl} = 2G\,(1^{(4s)}_{ijkl} - \tfrac{1}{3}\,\delta_{ij}\,\delta_{kl}) + 2\overline{G}\,\mu_{\beta ij}\,\mu_{\beta kl}$$
$$+ \sqrt{2}\,\overline{K}\,(\mu_{\beta ij}\,\delta_{kl} + \delta_{ij}\,\mu_{\beta kl}) + K\delta_{ij}\,\delta_{kl}, \tag{7.5.50a,b}$$

where G, $\overline{G}$, K, and $\overline{K}$ are elasticity parameters to be measured; for small pressures, they have the general form given by (7.5.45b). The unit fabric tensor $\boldsymbol{\mu}_\beta$ is defined by (7.5.48e). The fourth-order symmetric identity tensor $\mathbf{1}^{(4s)}$ is defined by (4.2.17c), page 157, *i.e.*,

$$1^{(4s)}_{ijkl} = \tfrac{1}{2}\,(\delta_{ik}\,\delta_{jl} + \delta_{il}\,\delta_{jk}). \tag{7.5.50c}$$

In (7.5.49a,b), $\mathbf{D}^p$ with components D^p_{ij}, is the plastic part of the deformation rate tensor, given by (7.5.38), *i.e.*,

$$\mathbf{D}^p = \dot{\gamma}\,\frac{\boldsymbol{\mu}}{\sqrt{2}} + \alpha\,(\mathbf{1}^{(4s)} - \boldsymbol{\mu}\otimes\boldsymbol{\mu}) : \mathbf{D}' + \frac{1}{3}\,\dot{\gamma}\,\Gamma\,\mathbf{1},$$
$$D^p_{ij} = \dot{\gamma}\,\frac{\mu_{ij}}{\sqrt{2}} + \alpha\,(1^{(4s)}_{ijkl} - \mu_{ij}\,\mu_{kl})\,D'_{kl} + \frac{1}{3}\,\dot{\gamma}\,\Gamma\,\delta_{ij}, \tag{7.5.51a,b}$$

where α is the noncoaxiality parameter, and Γ is the dilatancy, defined below in (7.5.56).

Fabric and its Evolution: The backstress $\boldsymbol{\beta}$ is related to the fabric tensor $\boldsymbol{\mathcal{E}}$ of components $\mathcal{E}_{ij}$, by

$$\beta_{ij} = \sqrt{2}\, p\, J_{ij} = -\sqrt{2}\, p\, \hat{\mu}\, \mathcal{E}_{ij}\,, \qquad (7.5.52a,b)$$

where $\hat{\mu}$ is an effective particle-to-particle friction coefficient.

The distribution of the contact unit normals, **n**'s, is defined in terms of the fabric tensor $\boldsymbol{\mathcal{E}}$ and hence $\boldsymbol{\beta}$, as follows:

$$\mathcal{E}_{ij} = \frac{15}{2}\,(< n_i\, n_j > -\frac{1}{3}\delta_{ij})\,,$$

$$E(\mathbf{n}) = \frac{1}{4\pi}\,(1 + \mathcal{E}_{ij}\, n_i\, n_j)\,, \qquad (7.5.52c,d)$$

where < ... > stands for volume average.

The evolution of the fabric tensor β_{ij} is defined by

$$\overset{\circ}{\beta}_{ij} = \sqrt{2}\, p\, \Lambda\, \mu_{ij}\,, \qquad (7.5.53)$$

where Λ is a material function which must tend to zero in a monotonic shearing, as the induced fabric anisotropy saturates.

Yield and Consistency Conditions: The yield condition is given by

$$f \equiv S - p\,M = 0\,, \qquad (7.5.54)$$

where $p\,M$ gives the elastic range (radius of the yield surface), and generally is very small at small pressures. The parameter M is a function of the void ratio e and the fabric measure $J = (J_{ij}\, J_{ij})^{1/2}$, as given by (7.5.41a). For continued inelastic deformation, the consistency condition requires that $f = 0$. This yields an expression for $\dot{\gamma}$ which is the same as (7.5.43), *i.e.*,

$$\dot{\gamma} = \frac{A_1 - \sqrt{2}\,\alpha\, A_2}{p\,(H + \Lambda) + A_3}\,, \qquad (7.5.55)$$

with coefficients A_i, i = 1, 2, 3 given by (7.5.42b-g).

Dilatancy: The expression for dilatancy Γ remains the same as in the two-dimensional case, *i.e.*,

$$\Gamma = -M_f + M + \zeta\,\mu_f\,. \qquad (7.5.56)$$

Once the functional form of the free material parameters, M, M_f, Λ, Γ, and the elastic coeficients are established experimentally, a complete set of constitutive relations results. For small pressures, these parameters will have forms similar to those given for the two-dimensional case.

7.6. REFERENCES

Anand, L. (1983), Plane deformations of ideal granular materials, *J. Mech. Phys. Solids*, Vol. 31, 105-122.

Balendran, B. (1993), Physically-based constitutive modeling of granular materials, Ph.D. thesis, University of California, San Diego.

Balendran, B. and Nemat-Nasser, S. (1993a), Double sliding model cyclic deformation of granular materials, including dilatancy effects,. *J. Mech. Phys. Solids*, Vol. 41, No. 3, 573-612.

Balendran, B. and Nemat-Nasser, S. (1993b), Viscoplastic flow of planar granular materials, *Mech. Mat.*, Vol 16, 1-12.

Christoffersen, J., Mehrabadi, M. M., and Nemat-Nasser, S. (1981), A micromechanical description of granular material behavior, *J. Appl. Mech.*, Vol. 48, 339-344.

Cundall, P. A. (1971), A computer model for simulating progressive large scale movements in blocky rock systems, *Proc. Symp. ISRM II*, Nancy, France, 129-136.

Cundall, P. A. and Hart, R. D. (1992), Numerical modelling of discontinua, *Engrg. Compt.*, Vol. 9, 101-113.

Cundall, P. A. and Strack, O. D. L. (1979), A discrete numerical model for granular assemblies, *Géotechnique*, Vol. 29, 47-65.

Cundall, P. A. and Strack, O. D. L. (1983), Modeling of microscopic mechanisms in granular material, in *Mechanics of granular materials: new models and constitutive relations*, Proc. US-Japan Sem. Mech. Granular Mat. (Jenkins, J. T. and Satake, M. eds.), Elsevier, Amsterdam, 137-149.

Cundall, P. A., Drescher, A., and Strack, O. D. L. (1982), Numerical experiments on granular assemblies: measurements and observations, in *Deformation and failure of granular materials* (Vermeer P. A. and Luger, H. J. eds.), Balkema, Rotterdam, 355-370.

Dafalias, Y. F. (1975), On cyclic and anisotropic plasticity, *PhD Thesis*, University of California, Berkeley.

Dafalias, Y. F. and Popov, E. P. (1976), Plastic internal variables formalism of cyclic plasticity, *J. Appl. Mech.*, Vol. 43, 645-651.

Dantu, P. (1957), Contribution à l'étude mécanique et géométrique des milieux pulvérulents, *Proc. 4th Int. Conf. Soil Mech. Found. Eng.,* Vol. 1, 144.

de Josselin de Jong, G. (1959), Statics and kinematics in the failable zone of granular materials, *Thesis*, University of Delft.

de Josselin de Jong, G. (1971), The double sliding, free rotating model for granular assemblies, *Géotechnique*, Vol. 21, 155-163.

de Josselin de Jong, G. (1977), Mathematical elaboration of the double sliding free rotating model, *Arch. Mech.*, Vol. 29, 561-591.

Drescher, A. (1976), An experimental investigation of flow rules for granular materials using optically sensitive glass particles, *Géotechnique*, Vol. 26, 591-601.

Gray, W. A. (1968), *The packing of solid particles*, Chapman and Hall, London.

Hill, R. (1956), New horizons in the mechanics of solids, *J. Mech. Phys. Solids*, Vol. 5, 66-74.

Horne, M. R. (1965), The behaviour of an assembly of rotund, rigid cohesionless particles - I, II, *Proc. R. Soc. Lond. A*, Vol. 286, 62-97.

Kanatani, K. (1984), Distribution of directional data and fabric tensors, *Int. J. Eng. Sci.*, Vol. 22, No. 2, 149-164.

Kishino, Y. (1988), Disc model analysis of granular media, in *Micromechanics of granular materials* Proc. 2nd US-Japan Sem. Mech. Granular Mat. (Satake, M. and Jenkins, J. T. eds.), Elsevier, Amsterdam, 143-152.

Kishino, Y. (editor) (2001), *Powders and grains 2001*, A. A. Bulkema Publishers, Tokyo.

Konishi, J. (1978), Microscopic model studies on the mechanical behaviour of granular materials, in *Proc. US-Japan Sem. Continuum-Mech. Stat. Appr. Mech. Granular Mat.* (Cowin, S.C. and Satake, M. eds.), Gakujutsu Bunken Fukyukai, Tokyo, 27-45.

Konishi, J., Oda, M., and Nemat-Nasser, S. (1982), Inherent anisotropy and shear strength of assembly of oval cross-sectional rods, in *Deformation and failure of granular materials* (Vermeer, P. A. and Luger, H. J. eds.), Balkema, Rotterdam, 403-412.

Lance, G. and Nemat-Nasser, S. (1986), Slip-induced plastic flow of geomaterials and crystals, *Mech. Mat.*, Vol. 5, 1-11.

Laude, P. V., Nelson, R. B., and Ito, Y. M. (1988), Instability of granular materials with nonassociated flow, *J. Engng. Mech. Am. Soc. Civ. Engrs.* Vol. 114, 2173-2191.

Mandel, J. (1947), Sur les lignes de glissement et le calcul des déplacements dans la déformation plastique, *Comptes rendus de l'Académie des Sciences*, Vol. 225, 1272-1273.

Mardia, K. V. (1972), *Statistics of directional data*, Academic Press, New York.

Mehrabadi, M. M. and Cowin, S. C. (1978), Initial planar deformation of dilatant granular materials, *J. Mech. Phys. Solids*, Vol. 26, 269-284.

Mehrabadi, M. M., Loret, B., and Nemat-Nasser, S. (1993), Incremental constitutive relations for granular materials based on micromechanics, *Proc. Roy. Soc. Lond. A*, Vol. 441, 433-463.

Mehrabadi, M. M., Nemat-Nasser, S., and Oda, M. (1982), On Statistical description of stress and fabric in granular materials, *Int. J. Num. Anal. Meth. Geomech.*, Vol. 6, 95-108.

Mehrabadi, M. M., Nemat-Nasser, S., Shodja, H. M., and Subhash, G. (1988), Some basic theoretical and experimental results on micromechanics of granular flow, in *Micromechanics of granular materials*, Proc. 2nd US-Japan Sem. Mech. Granular Mat., Sendai-Zao, Japan (Satake, M. and Jenkins, J.T. eds.), Elsevier, Amsterdam, 253-262.

Mróz, Z. (1980), Deformation and flow of granular materials, in *Theoretical and applied mechanics* (Rimrott, F. P. J. and Tabarrok, B. eds.), North-Holland, Amsterdam, 119-132.

Nemat-Nasser, S. (1980), On behavior of granular materials in simple shear,

Soils and Foundations, Vol. 20, 59-73.

Nemat-Nasser, S. (1983), Fabric and its influence on mechanical behavior of granular materials, *IUTAM Conf. Deform. Failure Granular Mat.*, 37-42.

Nemat-Nasser, S. (2000), A micromechanically-based constitutive model for frictional deformation of granular materials, *J. Mech. Phys. Solids*, Vol. 38, No. 6&7, 1541-1563.

Nemat-Nasser, S. and Hori, M. (1993), *Micromechanics: overall properties of heterogeneous solids*, Elsevier, Amsterdam; second revised edition (1999), Elsevier, Amsterdam.

Nemat-Nasser, S. and Mehrabadi, M. M. (1983), Stress and fabric in granular masses, *Mechanics of Granular Materials: New Models and Constitutive Relations*, (Jenkins, J. T. and Satake, M. eds.), Elsevier Science Publishers, 1-8.

Nemat-Nasser, S. and Mehrabadi, M. M. (1984), Micromechanically based rate constitutive descriptions for granular materials, in *Mechanics of engineering materials*, Proc. Int. Conf. Constitutive Law for Eng. Mat., Theory and Application (Desai, C. S., and Gallagher, R. H. eds.), John Wiley & Sons, New York.

Nemat-Nasser, S. and Shokooh, A. (1980), On finite plastic flow of compressible materials with internal friction, *Int. J. Solids Struct.*, Vol. 16, No. 6, 495-514.

Nemat-Nasser, S. and Zhang, J. (2002), Constitutive relations for cohesionless frictional granular materials, *J. of Plasticity*, Special Issue for Tim Wright, Vol. 18, 531-547.

Nemat-Nasser, S., Mehrabadi, M. M., and Iwakuma, T. (1981), On certain macroscopic and microscopic aspects of plastic flow of ductile materials, in *Three-dimensional constitutive relations and ductile fracture* (Nemat-Nasser, S. ed.), North-Holland, Amsterdam, 157-172.

Oda, M. (1977), Co-ordination number and its relation to shear strength of granular materials, *Soils and Foundations*, Vol. 17, No. 2, 29-42.

Oda, M. and Konishi, J. (1974), Microscopic deformation mechanism of granular materials in simple shear, *Soils and Foundations*, Vol. 14, No. 4, 25-38.

Oda, M., Konishi, J., and Nemat-Nasser, S. (1980), Some experimentally based fundamental results on the mechanical behavior of granular materials, *Géotechnique*, Vol. 30, 479-495.

Oda, M., Konishi, J., and Nemat-Nasser, S. (1982a), Experimental micromechanical evaluation of strength of granular materials: effects of particle rolling, *Mech. Mat.*, Vol. 1, No. 4, 269-283.

Oda, M., Nemat-Nasser, S., and Konishi, J. (1985), Stress induced anisotropy in granular masses, *Soils and Foundations*, Vol. 25, 85-97.

Oda, M., Nemat-Nasser, S., and Mehrabadi, M. M. (1982b), A statistical study of fabric in a random assembly of spherical granules, *Int. J. Num. Anal. Methods Geomech.*, Vol. 6, 77-94.

Okada, N. (1992), *Energy dissipation in inelastic flow of cohesionless granular media,* Ph.D. thesis, University of California, San Diego.

Okada, N. and Nemat-Nasser, S. (1994), Energy dissipation in inelastic flow of saturated cohesionless granular media, *Géotechnique*, Vol. 44, No. 1, 1-19.

Roscoe, K. H., Schofield, A. N., and Wroth, C. P. (1958), On the yielding of soils, *Géotechnique*, Vol. 8, 22-53.

Rowe, P. W. (1962), The stress-dilatancy relation for static equilibrium of an assembly of particles in contact, *Proc. Roy. Soc. Lond. A*, Vol. 269, 500-527.

Rudnicki, J. W. and Rice, J. R. (1975), Conditions for the localization of deformation in pressure-sensitive dilatant materials, *J. Mech. Phys. Solids*, Vol. 23, 371-394.

Satake, M. (1978), Constitution of mechanics of granular materials through graph representation, in *Theoretical and applied mechanics*, Vol. 26, University of Tokyo Press, 257-266.

Skinner, A. E. (1969), A note on the influence of interparticle friction on the shearing strength of a random assembly of spherical particles, *Géotechnique*, Vol. 19, 150-157.

Smith, W. O, Foote, P. D., and Busang, P. F. (1929), Packing of homogeneous spheres, *J. Phys. Rev. Ser. 2*, Vol. 34, 1271-1274.

Spencer, A. J. M. (1964), A theory of the kinematics of ideal soils under plane strain conditions, *J. Mech. Phys. Solids*, Vol. 12, 337-351.

Spencer, A. J. M. (1982), Deformation of ideal granular materials, in *Mechanics of solids* (Hopkins, H. G. and Sewell, M. J. eds.), Pergamon, Oxford, 607-652.

Subhash, G., Nemat-Nasser, S., Mehrabadi, M. M., and Shodja, H. M. (1991), Experimental investigation of fabric-stress relations in granular materials, *Mech. Mat.*, Vol. 11, No. 2, 87-106.

Taylor, G. I. (1934), The mechanism of plastic deformation of crystals - I , II, *Proc. R. Soc. Lond. A*, Vol. 145, 362-387, 388-404.

Taylor, G. I. (1938), Plastic strain in metals, *J. Inst. Metals*, Vol. 62, 307-324.

Thornton, C. (1987), Computer simulated experiments on particulate material, in *Tribology in particulate technology* (Briscoe, B. J. and Adams, M. J. eds.), Adam Hilger, Bristol, UK, 292-302.

Thornton, C. (1998), Quasi-static deformation of particulate media, *Phil. Trans. Roy. Soc. Lond.*, A356, 2763-2782.

Thornton, C. (2000), Numerical simulations of deviatoric shear deformation of granular media, *Géotechnique*, Vol. 50, 43-53.

Thornton, C. and Barnes, D. J. (1986), Computer simulated deformation of compact granular assemblies, *Acta Mech.*, Vol. 64, 45-61.

Vardoulakis, I. (1996), Deformation of water-saturated sand: I. uniform undrained deformation and shear banding, II. effect of pore water flow and shear banding, *Géotechnique*, Vol. 46, 441-472.

Wakabayashi, T. (1957), Photoelastic method for determination of stress in powdered mass, in *Proc. 7th Japan Nat. Cong. Appl. Mech.*, 133-140.

8

AVERAGE QUANTITIES AND HOMOGENIZATION MODELS

The transition from micro- to macro-variables of a representative volume element of a finitely deformed aggregate (e.g., a composite or a polycrystal) is discussed in this chapter. A number of exact fundamental results on averaging techniques, valid at finite deformations and rotations of any arbitrary heterogeneous continuum, are outlined. These results depend on the choice of suitable kinematical and dynamical variables. For finite deformations, the deformation gradient and its rate, and the nominal stress and its rate, are optimally suited for the averaging purposes, as discussed in Section 8.1, where a set of exact identities is presented in terms of these variables. These identities are valid for both rate-independent and rate-dependent materials, whereas only the rate-independent materials are considered in Sections 8.2 to 8.6 and most of Section 8.7.

An exact method for homogenization of an ellipsoidal inclusion in an unbounded finitely deformed homogeneous solid is presented, generalizing Eshelby's method for application to finite deformation problems; Section 8.2. In terms of the nominal stress rate and the rate of change of the deformation gradient, a general phase-transformation problem is considered, and the concepts of eigenvelocity gradient and eigenstress rate are introduced. The generalized Eshelby's tensor and its conjugate are defined and used to obtain the field quantities in an ellipsoidal inclusion, leading to exact expressions for the concentration tensors and a set of identities relating these quantities.

The Green functions for the rate quantities measured from a finitely deformed state, are formulated in Section 8.3, and their properties are discussed, arriving at explicit formulae for the generalized Eshelby's and the concentration tensors.

In Section 8.4, exact values of the average nominal stress rate and the average velocity gradient, taken over an ellipsoidal region in a finitely deformed unbounded homogeneous solid, are obtained when arbitrary (variable) eigenvelocity gradients or eigenstress rates are prescribed in the ellipsoid. The double- and multi-inclusion problems are then examined in this section. First, two nested ellipsoidal domains in an infinitely extended, finitely deformed solid of a rate-independent material are considered. Exact expressions are obtained for the average field quantities, taken over the region between the two, as well as within each of these ellipsoidal domains, when arbitrary eigenvelocity gradients are prescribed within an arbitrary region contained in the inner ellipsoid.

Then, these results are generalized to cases where a nested set of ellipsoidal regions within an infinitely extended, homogeneously deformed, homogeneous domain of a rate-independent material, undergoes phase transformations which result in various constant eigenvelocity gradients being generated (prescribed) within each annulus, with the innermost ellipsoid having an eigenvelocity gradient of arbitrary variation. Exact expressions are obtained for the average field quantitities within each subdomain of this nested set of ellipsoids.

The details for calculating the Green function, the generalized Eshelby tensor and its conjugate, and the concentration tensors, are presented in Section 8.5. In this section, it is proved that the velocity gradient (and hence the nominal stress rate, for rate-independent models) in an ellipsoidal region within an unbounded uniform and uniformly deformed rate-independent solid, remains uniform when this region undergoes a phase transformation corresponding to a constant eigenvelocity gradient. The relation between strong ellipticity and Green's function is examined in connection with the problem of a possible loss of stability of a uniformly stressed homogeneous solid.

Selected homogenization models are considered in Section 8.6 for general composites, and in Section 8.7 for polycrystals, where the application of the exact results to the problem of estimating the overall mechanical response of a finitely deformed heterogeneous representative volume element (RVE) is outlined, and the overall effective pseudo-modulus tensor (or pseudo-compliance tensor) of the RVE is calculated for rate-independent elastoplastic materials.

In addition, other averaging methods, such as the Taylor and Sack models, are briefly reviewed for both rate-independent and rate-dependent polycrystals, in Section 8.7.

Finally, in Appendix 8.A, explicit formulae are given for calculating the generalized Eshelby tensor in two dimensions.

8.1. AVERAGING THEOREMS

There are a number of general averaging theorems, valid at finite deformations and rotations of continua of any arbitrary heterogeneity and material composition. These theorems are discussed in this section in terms of the kinematical and dynamical quantities which are presented in Chapters 2 and 3.

8.1.1. Comments on Continuum Length Scale

The continuum theory of materials deals with objects which on a certain local length scale can be regarded as homogeneous, although globally (*i.e.*, at a greater length scale) they are generally heterogeneous. In this approach, at a typical material point, an infinitesimal material element is considered, and it is assumed that within this material element the strains and stresses are homogeneous. As a prerequisite, the *continuum* itself must be locally homogeneous in a certain sense.

Since, in reality most materials are heterogeneous (microscopically, as well as macroscopically), whether or not a continuum approach can provide useful information depends on the relative values of three basic length scales. These are: (1) the greatest dimension of the essential microconstituents; (2) the smallest dimension of the elementary continuum material neighborhood; and (3) the smallest dimension of the overall continuum. For example, in a polycrystalline solid consisting of crystals of tens of microns, each crystal may be viewed as a microconstituent. The elementary continuum material neighborhood must then be, at smallest, of the order of fractions of a millimeter, and the overall solid, *i.e.*, the structural component, may have dimensions of the order of centimeters or greater. On the other hand, an earth dam may be viewed as a continuum with aggregates as microconstituents and continuum elements of the order of meters. In either case, *the objective is to define an equivalent continuum material element which, in a certain sense, has the same average mechanical response as the actual heterogeneous material element*; for a more detailed discussion and references, see Hashin (1983), Nemat-Nasser and Hori (1993, or 1999), and Torquato (2002).

For the purpose of constructing continuum models of microscopically heterogeneous materials, the concept of a *representative volume element* (RVE) is introduced, and it is assumed that the average response of an RVE under uniform boundary data, represents the corresponding response of the associated continuum material element. In this approach, the actual dimensions of an RVE are of no concern. It is only its dimension relative to the dimension of its *essential* microstructure that is of importance. If the smallest dimension of the RVE is D and the greatest dimension of its microconstituent is d, Hill (1956) suggests that an effective continuum may be produced for d/D less than 10^{-3}.

8.1.2. Choice of Deformation and Stress Measures

At finite strains and rotations, relations among kinematical and dynamical variables are *nonlinear*. These relations hold within each grain in an RVE. It is desirable to define the *overall average* kinematical and dynamical quantities for the RVE such that similar relations remain valid at the *macrolevel*. Because of this constraint, the nonlinearity precludes representation of all deformation and stress measures and their rates, as unweighted volume averages over the RVE, of the corresponding measure at the microscale.

Notation: *In this chapter, the overall macroscopic quantities, whether obtained by simple unweighted volume averaging or defined in terms of other averaged measures, will be denoted by superposed bars.*[1] *When such an overall quantity is an unweighted volume average over the RVE, then it is denoted by* $< ... >$. For example, $\bar{\mathbf{F}}$, $\bar{\mathbf{L}}$, $\bar{\mathbf{S}}^{\mathrm{N}}$, and $\bar{\boldsymbol{\sigma}}$ are, respectively, the overall measure of the deformation gradient, the velocity gradient, the nominal stress, and the Cauchy stress. If the overall deformation gradient is *defined* by

$$\bar{\mathbf{F}} = \frac{1}{V} \int_V \mathbf{F}(\mathbf{X}, t)\, dV, \tag{8.1.1a}$$

then,

$$\bar{\mathbf{F}} \equiv < \mathbf{F} >. \tag{8.1.1b}$$

On the other hand, the overall, say, Lagrangian strain, (2.4.5), is *defined* by

$$\bar{\mathbf{E}}^{\mathrm{L}} \equiv \frac{1}{2}(\bar{\mathbf{F}}^{\mathrm{T}}\bar{\mathbf{F}} - \mathbf{1}) = \frac{1}{2}(< \mathbf{F} >^{\mathrm{T}} < \mathbf{F} > - \mathbf{1}). \tag{8.1.2}$$

Clearly, in general, $\bar{\mathbf{E}}^{\mathrm{L}} \neq < \mathbf{E}^{\mathrm{L}} > = < \tfrac{1}{2}(\mathbf{F}^{\mathrm{T}}\mathbf{F} - \mathbf{1}) >$.

In view of the above observation, it is seen that, for finite deformation, *there is an inherent arbitrariness in the selection of suitable kinematical and dynamical quantities whose overall measures are defined in terms of unweighted volume averages of the corresponding micromeasures, which are then employed to define other overall quantities, using the usual continuum mechanics relations, i.e.,* relations delineated in Chapters 2 and 3. Therefore, the selection must be made with care and with due regard for the physical basis of the basic phenomenon and the experimental procedures and results.

In an experimental setting, attempts are made to subject the specimen over its gauge length to measurable *uniform* deformation and stresses. The uniformity here is in the sense of continuum mechanics. At the microlevel, neither the deformation nor the stress is uniform. Thus, it appears natural to consider an RVE subjected to either *uniform surface tractions* or *linear surface displacements, i.e., to uniform boundary data.* This will then suggest certain macroscopically uniform deformation and stress measures which are completely

[1] This notation should not be confused with the superposed bar used in other chapters, *e.g.*, Chapter 6, to denote quantities referred to an intermediate configuration. In the present chapter, *all barred quantities are the overall macroscopic quantities.*

described by the boundary data, and which are volume averages of the corresponding *nonuniform* micromeasures. It turns out that, for finite deformation, the deformation gradient, $\mathbf{F}$, its rate, $\dot{\mathbf{F}}$, and the nominal stress, $\mathbf{S}^N$, and its rate, $\dot{\mathbf{S}}^N$, are suitable deformation and stress measures for the purpose of averaging (Hill, 1972; and Havner, 1982, 1992). *Their unweighted volume averages are completely defined in terms of the surface data (whether uniform or not), and for either the uniform traction or the linear displacement boundary data, they lead to many useful relations which can be employed to effectively characterize the overall aggregate response.* These and related issues are discussed in this section.

8.1.3. Average Deformation and Deformation-rate Measures

Let the deformation of an RVE be defined by the general mapping $\mathbf{x} = \mathbf{x}(\mathbf{X}, t)$. *The surface displacements are arbitrary and need not be uniform.* The volume average deformation gradient is given in terms of the surface data, by

$$\bar{\mathbf{F}} = <\mathbf{F}> = \frac{1}{V}\int_{\partial V} \mathbf{x} \otimes \mathbf{N}\, dA \tag{8.1.3a}$$

or

$$\bar{F}_{iB} \equiv <F_{iB}> = \frac{1}{V}\int_{\partial V} x_i N_B\, dA, \quad i, B = 1, 2, 3. \tag{8.1.3b}$$

This follows from the Gauss theorem. Here, V is the initial volume with surface ∂V of exterior unit normal $\mathbf{N}$. Similarly, for an *arbitrary* but compatible surface velocity, $\dot{\mathbf{x}} = \dot{\mathbf{x}}(\mathbf{X}, t)$,

$$\dot{\bar{\mathbf{F}}} \equiv <\dot{\mathbf{F}}> = \frac{1}{V}\int_{\partial V} \dot{\mathbf{x}} \otimes \mathbf{N}\, dA \equiv \dot{\bar{\mathbf{F}}},$$

$$\dot{\bar{F}}_{iB} \equiv <\dot{F}_{iB}> = \frac{1}{V}\int_{\partial V} \dot{x}_i N_B\, dA \equiv \dot{\bar{F}}_{iB}. \tag{8.1.4a,b}$$

The overall velocity gradient $\bar{\mathbf{L}}$ is then *defined* by

$$\bar{\mathbf{L}} \equiv \dot{\bar{\mathbf{F}}}\bar{\mathbf{F}}^{-1}. \tag{8.1.5}$$

In general, $<\mathbf{L}> \neq \bar{\mathbf{L}}$ when the reference configuration does not coincide with the current one, since $\mathbf{L} = \dot{\mathbf{F}}\,\mathbf{F}^{-1}$. On the other hand, when the current configuration is chosen as reference, it follows that $\mathbf{F} = \mathbf{1}$, and hence, $\bar{\mathbf{L}} = <\mathbf{L}>$. This is an important case, often used for incremental calculation of aggregate properties. Note also that, in this case, both the overall average deformation rate and the spin tensors equal the corresponding unweighted volume average, $\bar{\mathbf{D}} = <\mathbf{D}>$ and $\bar{\mathbf{W}} = <\mathbf{W}>$. In the general case, however, these relations are *not* valid.

8.1.4. Average Stress Measures

Let an RVE be subjected to arbitrary but self-equilibrating (zero body forces) surface tractions $\mathbf{T}$, referred to and measured per unit area of the reference configuration. *These tractions need not be uniform.* Let $\mathbf{S}^N = \mathbf{S}^N(\mathbf{X}, t)$ be the corresponding variable nominal stress field in the RVE, in equilibrium with these boundary tractions. Equilibrium requires[2],

$$\nabla \cdot \mathbf{S}^N = \mathbf{0} \quad \text{in V},$$

$$\mathbf{N} \cdot \mathbf{S}^N = \mathbf{T} \quad \text{on } \partial\text{V}, \tag{8.1.6a,b}$$

where $\mathbf{N}$ is the exterior unit normal of ∂V. From these conditions and the Gauss theorem, the overall average nominal stress $\overline{\mathbf{S}}^N$ is uniquely defined in terms of the boundary tractions, by

$$\overline{\mathbf{S}}^N \equiv < \mathbf{S}^N > = \frac{1}{V} \int_{\partial V} \mathbf{X} \otimes \mathbf{T} \, dA. \tag{8.1.7}$$

If the applied self-equilibrating tractions, referred to and measured per unit area of the *current* configuration, are denoted by $\mathbf{t}$, and if $\boldsymbol{\sigma} = \boldsymbol{\sigma}(\mathbf{x}, t)$ is the variable Cauchy stress in equilibrium with these tractions, then

$$\nabla \cdot \boldsymbol{\sigma} = \mathbf{0} \quad \text{in v}$$

$$\mathbf{n} \cdot \boldsymbol{\sigma} = \mathbf{t} \quad \text{on } \partial\text{v}, \tag{8.1.8a,b}$$

where v is the current volume with boundary ∂v. Application of the Gauss theorem again shows

$$\overline{\boldsymbol{\sigma}} \equiv < \boldsymbol{\sigma} > = \frac{1}{v} \int_{\partial v} \mathbf{x} \otimes \mathbf{t} \, da. \tag{8.1.9}$$

At the *local level*, the Cauchy and the nominal stresses are related by $\boldsymbol{\sigma} = \mathbf{F}\,\mathbf{S}^N/J$; see Table 3.1.1, page 99. Therefore, if surface data are given with reference to the *initial* configuration, and the overall nominal stress is defined by (8.1.7), then, in general, $\overline{\mathbf{F}}\,\overline{\mathbf{S}}^N / \overline{J} \neq < \boldsymbol{\sigma} >$. Hence, once (8.1.7) is chosen for the definition of the overall nominal stress, then (8.1.9) must be abandoned in favor of

$$\overline{\boldsymbol{\sigma}} \equiv \frac{1}{\overline{J}} \overline{\mathbf{F}}\,\overline{\mathbf{S}}^N,$$

$$\overline{J} \equiv \det \overline{\mathbf{F}}. \tag{8.1.10a,b}$$

On the other hand, *when the current configuration is the reference one, all stress measures become identical, and (8.1.7) reduces to (8.1.9).*

---

[2] Even though the stress field may vary in time, in the description of the average properties of an RVE, this variation is viewed as being *quasi-static*. The inertia forces are then included in the equations of motion of the homogenized RVE.

8.1.5. Average Stress-rate Measures

Let $\dot{\mathbf{T}}$ be the self-equilibrating traction rates (*not necessarily uniform*) prescribed on ∂V. Let $\dot{\mathbf{S}}^N = \dot{\mathbf{S}}^N(\mathbf{X}, t)$ be the variable nominal stress rate in the RVE. Equilibrium requires,

$$\nabla \cdot \dot{\mathbf{S}}^N = \mathbf{0} \quad \text{in V},$$
$$\mathbf{N} \cdot \dot{\mathbf{S}}^N = \dot{\mathbf{T}} \quad \text{on } \partial V. \qquad (8.1.11a,b)$$

From these conditions and the Gauss theorem, the overall average nominal stress rate $\bar{\dot{\mathbf{S}}}^N$ is uniquely defined in terms of the boundary data, by

$$\bar{\dot{\mathbf{S}}}^N \equiv \langle \dot{\mathbf{S}}^N \rangle = \frac{1}{V} \int_{\partial V} \mathbf{X} \otimes \dot{\mathbf{T}} \, dA \equiv \dot{\bar{\mathbf{S}}}^N. \qquad (8.1.12a)$$

This equation is valid whether or not the reference configuration coincides with the current configuration. The overall Cauchy stress rate $\dot{\bar{\boldsymbol{\sigma}}}$, however, will not enjoy a representation similar to (8.1.9). That is, while

$$\dot{\bar{\mathbf{S}}}^N \equiv \frac{\partial}{\partial t} \langle \mathbf{S}^N \rangle = \frac{1}{V} \int_{\partial V} \mathbf{X} \otimes \dot{\mathbf{T}} \, dA$$
$$= \langle \dot{\mathbf{S}}^N \rangle \equiv \bar{\dot{\mathbf{S}}}^N, \qquad (8.1.12b)$$

similar relations cannot be written for other stress measures, including the Cauchy stress, even if the current configuration is the reference one. Hence, once (8.1.7) and (8.1.12) are accepted as the overall stress and stress rate measures, all other stress and stress rate measures must then be *defined* in terms of these quantities and the overall deformation gradient $\bar{\mathbf{F}} = \langle \mathbf{F} \rangle$ and its rate $\dot{\bar{\mathbf{F}}} \equiv \langle \dot{\mathbf{F}} \rangle = \bar{\dot{\mathbf{F}}}$, using the relations discussed in Chapters 2 and 3.

8.1.6. General Identities

Whatever the material composition, and whatever the deformation and rotation history of an arbitrary heterogeneous material of initial volume V, there is a set of fundamental averaging identities that follows from equilibrium, the Gauss theorem, and the definition of the deformation gradient and its rate, for any consistent boundary data; Hill (1984). Consider an arbitrary self-equilibrating nominal stress field, $\mathbf{S}^N$, which satisfies the equilibrium conditions (8.1.6), for some self-equilibrating boundary tractions $\mathbf{T}$. Let $\dot{\mathbf{F}}$ in V, be the time-rate-of-change of a deformation gradient associated with an arbitrary self-compatible boundary velocity field, $\dot{\mathbf{x}} = \dot{\mathbf{U}}$ on ∂V. *The nominal stress,* $\mathbf{S}^N$, *and the deformation rate,* $\dot{\mathbf{F}}$, *need not be related*[3]. Now, through the application of the Gauss theorem, it follows that[4]

---

[3] Similar comments apply to $\mathbf{F}$ and $\dot{\mathbf{S}}^N$ in identities (8.1.13b-d).

[4] The trace of the first term in (8.1.13a) is the rate of stress work per unit initial volume; see (3.1.34a,b), page 98, Subsection 3.1.10.

$$<\dot{\mathbf{F}}\,\mathbf{S}^{N}> - <\dot{\mathbf{F}}><\mathbf{S}^{N}>$$
$$= \frac{1}{V}\int_{\partial V} (\dot{\mathbf{x}} - <\dot{\mathbf{F}}>\mathbf{X}) \otimes \{\mathbf{N} \cdot (\mathbf{S}^{N} - <\mathbf{S}^{N}>)\}\, dA\,, \tag{8.1.13a}$$

where $\dot{\mathbf{x}} = \dot{\mathbf{U}}$ and $\mathbf{N} \cdot \mathbf{S}^{N} = \mathbf{T}$ on ∂V.

Similarly, if $\dot{\mathbf{S}}^{N}$ is a self-equilibrating stress-rate in V, then identity (8.1.13a) remains valid when $\mathbf{S}^{N}$ is replaced by $\dot{\mathbf{S}}^{N}$ or if $\dot{\mathbf{F}}$ is replaced by $\mathbf{F}$ (self-compatible) and $\mathbf{S}^{N}$ is replaced by $\dot{\mathbf{S}}^{N}$, or when only $\dot{\mathbf{F}}$ is replaced by $\mathbf{F}$. In this manner, the following identities are obtained:

$$<\mathbf{F}\,\dot{\mathbf{S}}^{N}> - <\mathbf{F}><\dot{\mathbf{S}}^{N}>$$
$$= \frac{1}{V}\int_{\partial V} (\mathbf{x} - <\mathbf{F}>\mathbf{X}) \otimes \{\mathbf{N} \cdot (\dot{\mathbf{S}}^{N} - <\dot{\mathbf{S}}^{N}>)\}\, dA\,,$$

$$<\mathbf{F}\,\mathbf{S}^{N}> - <\mathbf{F}><\mathbf{S}^{N}>$$
$$= \frac{1}{V}\int_{\partial V} (\mathbf{x} - <\mathbf{F}>\mathbf{X}) \otimes \{\mathbf{N} \cdot (\mathbf{S}^{N} - <\mathbf{S}^{N}>)\}\, dA\,,$$

$$<\dot{\mathbf{F}}\,\dot{\mathbf{S}}^{N}> - <\dot{\mathbf{F}}><\dot{\mathbf{S}}^{N}>$$
$$= \frac{1}{V}\int_{\partial V} (\dot{\mathbf{x}} - <\dot{\mathbf{F}}>\mathbf{X}) \otimes \{\mathbf{N} \cdot (\dot{\mathbf{S}}^{N} - <\dot{\mathbf{S}}^{N}>)\}\, dA\,. \tag{8.1.13b-d}$$

In the same manner, it is shown that,

$$<\dot{\mathbf{F}}\,\mathbf{S}^{N}> - <\dot{\mathbf{F}}><\mathbf{S}^{N}> = \frac{1}{V}\int_{\partial V} (\dot{\mathbf{x}} - <\dot{\mathbf{F}}>\mathbf{X}) \otimes (\mathbf{N} \cdot \mathbf{S}^{N})\, dA\,,$$

$$<\mathbf{F}\,\dot{\mathbf{S}}^{N}> - <\mathbf{F}><\dot{\mathbf{S}}^{N}> = \frac{1}{V}\int_{\partial V} (\mathbf{x} - <\mathbf{F}>\mathbf{X}) \otimes (\mathbf{N} \cdot \dot{\mathbf{S}}^{N})\, dA\,,$$

$$<\mathbf{F}\,\mathbf{S}^{N}> - <\mathbf{F}><\mathbf{S}^{N}> = \frac{1}{V}\int_{\partial V} (\mathbf{x} - <\mathbf{F}>\mathbf{X}) \otimes (\mathbf{N} \cdot \mathbf{S}^{N})\, dA\,,$$

$$<\dot{\mathbf{F}}\,\dot{\mathbf{S}}^{N}> - <\dot{\mathbf{F}}><\dot{\mathbf{S}}^{N}> = \frac{1}{V}\int_{\partial V} (\dot{\mathbf{x}} - <\dot{\mathbf{F}}>\mathbf{X}) \otimes (\mathbf{N} \cdot \dot{\mathbf{S}}^{N})\, dA\,,$$

(8.1.14a-d)

and that,

$$<\dot{\mathbf{F}}\,\mathbf{S}^{N}> - <\dot{\mathbf{F}}><\mathbf{S}^{N}> \frac{1}{V}\int_{\partial V} \dot{\mathbf{x}} \otimes \{\mathbf{N} \cdot (\mathbf{S}^{N} - <\mathbf{S}^{N}>)\}\, dA\,,$$

$$<\mathbf{F}\,\dot{\mathbf{S}}^{N}> - <\mathbf{F}><\dot{\mathbf{S}}^{N}> = \frac{1}{V}\int_{\partial V} \mathbf{x} \otimes \{\mathbf{N} \cdot (\dot{\mathbf{S}}^{N} - <\dot{\mathbf{S}}^{N}>)\}\, dA\,,$$

$$<\mathbf{F}\,\mathbf{S}^{N}> - <\mathbf{F}><\mathbf{S}^{N}> = \frac{1}{V}\int_{\partial V} \mathbf{x} \otimes \{\mathbf{N} \cdot (\mathbf{S}^{N} - <\mathbf{S}^{N}>)\}\, dA\,,$$

$$<\dot{\mathbf{F}}\,\dot{\mathbf{S}}^{N}> - <\dot{\mathbf{F}}><\dot{\mathbf{S}}^{N}> = \frac{1}{V}\int_{\partial V} \dot{\mathbf{x}} \otimes \{\mathbf{N} \cdot (\dot{\mathbf{S}}^{N} - <\dot{\mathbf{S}}^{N}>)\}\, dA\,.$$

(8.1.15a-d)

When the boundary data are such that the integrals on the right-hand side of, say, (8.1.13) vanish, then it follows that the average of the corresponding product equals the product of the averages. This leads to the following useful results:

$$<\dot{\mathbf{F}}\,\mathbf{S}^{\mathrm{N}}> = <\dot{\mathbf{F}}><\mathbf{S}^{\mathrm{N}}>, \quad <\mathbf{F}\,\dot{\mathbf{S}}^{\mathrm{N}}> = <\mathbf{F}><\dot{\mathbf{S}}^{\mathrm{N}}>,$$

$$<\mathbf{F}\,\mathbf{S}^{\mathrm{N}}> = <\mathbf{F}><\mathbf{S}^{\mathrm{N}}>, \quad <\dot{\mathbf{F}}\,\dot{\mathbf{S}}^{\mathrm{N}}> = <\dot{\mathbf{F}}><\dot{\mathbf{S}}^{\mathrm{N}}>. \tag{8.1.16a-d}$$

8.1.7. Uniform Boundary Tractions and Traction Rates

Suppose now the prescribed surface tractions are *uniform*, so that

$$\mathbf{T} = \mathbf{N}\cdot\mathbf{S}^{\mathrm{N0}} \quad \text{or} \quad \mathrm{T_i} = \mathrm{S^{N0}_{Ai}\,N_A} \quad \text{on } \partial\mathrm{V}, \tag{8.1.17a}$$

and

$$\dot{\mathbf{T}} = \mathbf{N}\cdot\dot{\mathbf{S}}^{\mathrm{N0}} \quad \text{or} \quad \dot{\mathrm{T}}_{\mathrm{i}} = \dot{\mathrm{S}}^{\mathrm{N0}}_{\mathrm{Ai}}\,\mathrm{N_A} \quad \text{on } \partial\mathrm{V}, \tag{8.1.17b}$$

where $\mathbf{S}^{\mathrm{N0}}$ and $\dot{\mathbf{S}}^{\mathrm{N0}}$ are spatially uniform, possibly time-dependent tensors. It then follows from (8.1.7) and (8.1.12) that,

$$\overline{\mathbf{S}}^{\mathrm{N}} = \mathbf{S}^{\mathrm{N0}} \quad \text{and} \quad \dot{\overline{\mathbf{S}}}^{\mathrm{N}} = \overline{\dot{\mathbf{S}}}^{\mathrm{N}} = \dot{\mathbf{S}}^{\mathrm{N0}}. \tag{8.1.18a,b}$$

The Kirchhoff stress, $\boldsymbol{\tau}$, relates to the deformation gradient, $\mathbf{F}$, by $\boldsymbol{\tau} = \mathbf{F}\,\mathbf{S}^{\mathrm{N}}$; see Table 3.1.1, page 99. As pointed out before, in general, the overall average Kirchhoff stress must be *defined* in terms of the average deformation gradient, $\overline{\mathbf{F}}$, and nominal stress, $\overline{\mathbf{S}}^{\mathrm{N}}$, because, in general, $<\mathbf{F}\,\mathbf{S}^{\mathrm{N}}> \neq\, <\mathbf{F}><\mathbf{S}^{\mathrm{N}}>$. For uniform boundary tractions and traction rates, on the other hand, it is seen that the overall average Kirchhoff stress and its rate enjoy the usual relations,

$$\overline{\boldsymbol{\tau}} \equiv\, <\boldsymbol{\tau}> =\, <\mathbf{F}\,\mathbf{S}^{\mathrm{N}}> = \overline{\mathbf{F}}\,\overline{\mathbf{S}}^{\mathrm{N}} = \overline{\mathbf{F}}\,\mathbf{S}^{\mathrm{N0}},$$

$$\overline{\dot{\boldsymbol{\tau}}} \equiv\, <\dot{\boldsymbol{\tau}}> =\, <\dot{\mathbf{F}}\,\mathbf{S}^{\mathrm{N}} + \mathbf{F}\,\dot{\mathbf{S}}^{\mathrm{N}}>$$

$$= \overline{\dot{\mathbf{F}}}\,\overline{\mathbf{S}}^{\mathrm{N}} + \overline{\mathbf{F}}\,\overline{\dot{\mathbf{S}}}^{\mathrm{N}} = \overline{\dot{\mathbf{F}}}\,\mathbf{S}^{\mathrm{N0}} + \overline{\mathbf{F}}\,\dot{\mathbf{S}}^{\mathrm{N0}} \equiv \dot{\overline{\boldsymbol{\tau}}}. \tag{8.1.19a,b}$$

These relations follow from (8.1.16).

8.1.8. Uniform Boundary Displacements and Displacement Rates

Suppose an RVE is subjected to linear boundary displacements and velocities, such that

$$\mathbf{x} = \mathbf{F}^0\,\mathbf{X} \quad \text{and} \quad \dot{\mathbf{x}} = \dot{\mathbf{F}}^0\,\mathbf{X} \quad \text{on } \partial\mathrm{V}, \tag{8.1.20a,b}$$

where $\mathbf{F}^0$ and $\dot{\mathbf{F}}^0$ are time-dependent, spatially *uniform* deformation and velocity gradients, measured with respect to the initial configuration. Direct substitution into (8.1.3) and (8.1.4) reveals,

$$\overline{\mathbf{F}} = \mathbf{F}^0 \quad \text{and} \quad \dot{\overline{\mathbf{F}}} = \dot{\mathbf{F}}^0. \tag{8.1.21a,b}$$

Because of microheterogeneities of the RVE, $\mathbf{F}$ and $\dot{\mathbf{F}}$ are both, in general, spatially nonuniform, $\mathbf{F} = \mathbf{F}(\mathbf{X}, t)$ and $\dot{\mathbf{F}} = \dot{\mathbf{F}}(\mathbf{X}, t)$. In this case, again, it is easy to show that the overall average Kirchhoff stress and its rate, satisfy the relations,

$$\begin{aligned} \overline{\boldsymbol{\tau}} &= \mathbf{F}^0\, \overline{\mathbf{S}}^{\mathrm{N}}, \\ \dot{\overline{\boldsymbol{\tau}}} &= \dot{\mathbf{F}}^0\, \overline{\mathbf{S}}^{\mathrm{N}} + \mathbf{F}^0\, \dot{\overline{\mathbf{S}}}^{\mathrm{N}}. \end{aligned} \tag{8.1.22a,b}$$

These follow from (8.1.16).

8.1.9. General Identities for Uniform Boundary Data

When the boundary data are uniform, as in Subsections 8.1.7 and 8.1.8, then the integrals in the right-hand side of (8.1.13) to (8.1.15) vanish identically. This leads to identities (8.1.16), with the right-hand side appropriately specialized. Hence, for uniform boundary tractions and traction rates, it follows that,

$$\begin{aligned} &< \dot{\mathbf{F}}\, \mathbf{S}^{\mathrm{N}} > = < \dot{\mathbf{F}} > \mathbf{S}^{\mathrm{N0}}, \quad < \mathbf{F}\, \dot{\mathbf{S}}^{\mathrm{N}} > = < \mathbf{F} > \dot{\mathbf{S}}^{\mathrm{N0}}, \\ &< \mathbf{F}\, \mathbf{S}^{\mathrm{N}} > = < \mathbf{F} > \mathbf{S}^{\mathrm{N0}}, \quad < \dot{\mathbf{F}}\, \dot{\mathbf{S}}^{\mathrm{N}} > = < \dot{\mathbf{F}} > \dot{\mathbf{S}}^{\mathrm{N0}}. \end{aligned} \tag{8.1.23a-d}$$

Similarly, for uniform surface displacements and their rates, identities (8.1.16) reduce to

$$\begin{aligned} &< \dot{\mathbf{F}}\, \mathbf{S}^{\mathrm{N}} > = \dot{\mathbf{F}}^0 < \mathbf{S}^{\mathrm{N}} >, \quad < \mathbf{F}\, \dot{\mathbf{S}}^{\mathrm{N}} > = \mathbf{F}^0 < \dot{\mathbf{S}}^{\mathrm{N}} >, \\ &< \mathbf{F}\, \mathbf{S}^{\mathrm{N}} > = \mathbf{F}^0 < \mathbf{S}^{\mathrm{N}} >, \quad < \dot{\mathbf{F}}\, \dot{\mathbf{S}}^{\mathrm{N}} > = \dot{\mathbf{F}}^0 < \dot{\mathbf{S}}^{\mathrm{N}} >. \end{aligned} \tag{8.1.24a-d}$$

8.2. HOMOGENIZATION AND CONCENTRATION TENSORS

It is common to assume that the stress and deformation fields are uniform within each grain of a suitably large aggregate representing an RVE, since, by necessity, the grains must be very small relative to the size of the RVE. Such an assumption is in accord with the usual continuum formulation of the flow and deformation of matter. It is also the basis of the general theory of single crystals presented in Chapter 6.

Consider an RVE subjected to suitable uniform boundary data. The stresses and strains are, in general, nonuniform within the RVE. Thus, even though they are assumed to be uniform over each grain, they change from grain to grain. A fundamental issue is to calculate the local quantities, *i.e.*, the quantities within a typical grain, in terms of suitable uniform data prescribed on the boundary of the RVE. This issue is addressed in the present section.

8.2.1. Eshelby's Tensor

For infinitesimal deformations, Eshelby's tensor is often used to estimate the local quantities in terms of their overall values. Eshelby (1957) considered an ellipsoidal inclusion within an infinitely extended linearly elastic homogeneous solid. He then showed that, if the inclusion undergoes a transformation resulting in a *uniform transformation strain*, $\boldsymbol{\varepsilon}^*$, when the inclusion is free from the constraints imposed by the surrounding matrix, then, in the presence of such a constraint, the strain in the inclusion is still *uniform* and is given by[5] $\boldsymbol{\varepsilon} = \boldsymbol{\mathcal{S}} : \boldsymbol{\varepsilon}^*$ or, in rectangular Cartesian component form, by $\varepsilon_{ij} = \mathcal{S}_{ijkl}\,\varepsilon^*_{kl}$. Here $\boldsymbol{\varepsilon}$ is the linearized strain, *i.e.,* the symmetric part of the displacement gradient, and $\boldsymbol{\mathcal{S}}$ is Eshelby's tensor, symmetric only with respect to the exchange of i and j, and k and l, but not in general, of ij and kl. *This tensor depends on the elastic properties of the matrix and the aspect ratios of the ellipsoid, but not on the properties of the ellipsoid nor the origin of the transformation strain.* It has been calculated for isotropic and certain anisotropic cases, and has been used extensively to estimate aggregate properties and to solve various problems in infinitesimal deformation theories; Kröner (1958, 1961), Budiansky and Wu (1962), Hutchinson (1970, 1976), Berveiller and Zaoui (1979), Weng (1981, 1982, 1984, 1990), Dvorak (1992), and others, see, *e.g.,* Mura (1987) and Nemat-Nasser and Hori (1993 and 1999) for discussion and other references. Again, for infinitesimal deformations, the *idea* has been used to obtain average properties of composites with periodic microstructures, for linearly elastic, power-law creep, and elastoplastic solids; see Nemat-Nasser *et al.* (1982), Iwakuma and Nemat-Nasser (1983), and Nemat-Nasser (1986). A comprehensive account is given by Nemat-Nasser and Hori (1993, 1999).

In general, an estimate of the overall properties of an aggregate at finite deformations requires an incremental formulation. It is expedient and theoretically precise, to cast this formulation in terms of the rate of change of the field quantities measured in terms of the actual time for rate-dependent deformations, or some other monotonically changing parameter, when appropriate (e.g., for rate-independent models of plasticity). In view of the results presented in the preceding subsection of the present chapter, large deformation elastoplasticity problems are most effectively formulated in terms of the rate of change of the *nominal stress* and the *velocity gradient.* In this context, Eshelby's technique has been exploited by Iwakuma and Nemat-Nasser (1984), Nemat-Nasser and Obata (1986), and Harren (1991), for finite strains and rotations, to construct an explicit procedure for estimating the local rate quantities in terms of their overall average values. In particular, using the *current* configuration as the reference one, Iwakuma and Nemat-Nasser show that, for an ellipsoidal grain, Ω, embedded in a uniform matrix (both fully nonlinear, having undergone finite deformations), the velocity gradient, L_{ij}, is uniform and is given by $L_{ij} = \mathcal{A}^{\Omega}_{ijkl}\,L^0_{lk}$, where

---

[5] The phenomenon had already been discovered by Hardiman (1954) in the context of infinite elastic plates containing elliptical inhomogeneities.

$\mathbf{L}^0$ is the uniform velocity gradient prescribed at infinity, and $\boldsymbol{\mathcal{A}}^{\Omega}$ is the concentration tensor, for whose calculation explicit formulae are given by Iwakuma and Nemat-Nasser (1984). When the average nominal stress rate and the average velocity gradient are calculated using the current aggregate configuration as the reference one, then an estimate of the *instantaneous* overall modulus tensor is obtained directly.

In the following subsections, an arbitrary reference configuration is used to obtain general *exact* results. Then, these results are specialized in Section 8.6, wherever necessary, and applied to develop several averaging *models*.

8.2.2. General Phase-transformation Problem

Consider an arbitrary reference configuration. The nominal stress rate and the velocity gradient, denoted by $\dot{\mathbf{S}}^{\mathrm{N}}$ and $\dot{\mathbf{F}}$, with components $\dot{S}^{\mathrm{N}}_{\mathrm{Ai}}$ and $\dot{F}_{\mathrm{iA}}$, are used as the basic stress rate and deformation rate measures. These quantities are assumed to be connected through certain instantaneous pseudo-moduli. More specifically, consider an extended homogeneous solid of reference volume V, with instantaneous modulus tensor $\boldsymbol{\mathcal{F}}$, of components $\mathcal{F}_{\mathrm{AiBj}}$. Assume that the nominal stress rate, $\dot{\mathbf{S}}^{\mathrm{N}}$, is a linear and homogeneous function of the velocity gradient, $\dot{\mathbf{F}}$, *both being referred to the same arbitrary reference configuration, i.e.,* set[6]

$$\dot{\mathbf{S}}^{\mathrm{N}} = \boldsymbol{\mathcal{F}} : \dot{\mathbf{F}} \quad \text{or} \quad \dot{S}^{\mathrm{N}}_{\mathrm{Ai}} = \mathcal{F}_{\mathrm{AiBj}}\, \dot{F}_{\mathrm{jB}}, \tag{8.2.1a,b}$$

where i, j, A, B = 1, 2, 3, and the summation convention is used. In many applications, the modulus tensor $\boldsymbol{\mathcal{F}}$ is symmetric with respect to the exchange of Ai and Bj, but not with respect to the exchange of A and i or B and j. The former symmetry property will be used in the sequel wherever appropriate.

Suppose that the pseudo-modulus $\boldsymbol{\mathcal{F}}$ admits an inverse, $\boldsymbol{\mathcal{G}}$, so that

$$\boldsymbol{\mathcal{F}} : \boldsymbol{\mathcal{G}} = \mathbf{1}^{(4)}, \qquad \text{or} \qquad \mathcal{F}_{\mathrm{AiBj}}\, \mathcal{G}_{\mathrm{jBkC}} = \delta_{\mathrm{AC}}\, \delta_{\mathrm{ik}}\,. \tag{8.2.1c,d}$$

Then, it follows that

$$\dot{\mathbf{F}} = \boldsymbol{\mathcal{G}} : \dot{\mathbf{S}}^{\mathrm{N}}, \qquad \text{or} \qquad \dot{F}_{\mathrm{iA}} = \mathcal{G}_{\mathrm{iAjB}}\, \dot{S}^{\mathrm{N}}_{\mathrm{Bj}}\,. \tag{8.2.1e,f}$$

The tensor $\boldsymbol{\mathcal{G}}$ with components $\mathcal{G}_{\mathrm{iAjB}}$ will be referred to as the *pseudo-compliance*.

Let now a region Ω in V undergo a phase transformation which, if Ω were free from the constraint imposed by the surrounding material,[7] it would attain a *constant* transformation (inelastic) velocity gradient $\dot{\mathbf{F}}^*$ with components $\dot{F}^*_{\mathrm{iA}}$.

---

[6] Since neither $\dot{F}_{\mathrm{iA}}$ nor $\dot{S}^{\mathrm{N}}_{\mathrm{Ai}}$ is symmetric, care must be taken in using coordinate-invariant contractions. Here, the double contraction denoted by **:** , is defined in (8.2.1b) and will be followed throughout this chapter. Note that, $\boldsymbol{\mathcal{F}} : \boldsymbol{\mathcal{G}} = \boldsymbol{\mathcal{G}} : \boldsymbol{\mathcal{F}} = \mathbf{1}^{(4)}$.

[7] *I.e.*, if Ω is cut out and allowed to change without any constraints imposed on its boundary ∂Ω.

Let the resulting velocity gradient of Ω in the presence of the constraint from the surrounding matrix, be $\dot{\mathbf{F}}$ with components $\dot{F}_{iA}$. In general, $\dot{\mathbf{F}}$ is spatially nonuniform. However, when V is homogeneous and unbounded, and Ω is ellipsoidal, then, following Eshelby's procedure, it is shown by Iwakuma and Nemat-Nasser (1984) that the resulting final velocity gradient $\dot{\mathbf{F}}$ in Ω (for any constant transformation velocity gradient $\dot{\mathbf{F}}^*$) is constant. In this case, $\dot{\mathbf{F}}$ is related linearly to the transformation velocity gradient $\dot{\mathbf{F}}^*$, by

$$\dot{\mathbf{F}} = \boldsymbol{\mathcal{S}} : \dot{\mathbf{F}}^* \quad \text{or} \quad \dot{F}_{iA} = \mathcal{S}_{iABj}\,\dot{F}^*_{jB}\,. \tag{8.2.2a,b}$$

The fourth-order tensor $\boldsymbol{\mathcal{S}}$, in general, has *no* symmetries. It generalizes Eshelby's tensor, which is mentioned in Subsection 8.2.1.

Iwakuma and Nemat-Nasser (1984) outline a method for computing $\boldsymbol{\mathcal{S}}$ for a general ellipsoidal Ω. When the operator $\mathcal{F}_{AiBj}\,\partial^2(...)/(\partial X_A\,\partial X_B)$ is elliptic, a real-valued tensor $\boldsymbol{\mathcal{S}}$ exists and can be computed in terms of the aspect ratios of the ellipsoid Ω and the modulus tensor $\mathcal{F}_{AiBj}$. For the sake of referencing, in what follows, the tensor $\boldsymbol{\mathcal{S}}$ with components $\mathcal{S}_{iABj}$ will be called the *generalized Eshelby tensor*.

Since $\dot{\mathbf{F}}^*$ is the *stress-free inelastic* velocity gradient in Ω, the nominal stress rate in this region is produced by the differential velocity gradient $(\dot{\mathbf{F}} - \dot{\mathbf{F}}^*)$, and it follows that

$$\begin{aligned}\dot{\mathbf{S}}^{N} &= \boldsymbol{\mathcal{F}} : (\dot{\mathbf{F}} - \dot{\mathbf{F}}^*)\\ &= \boldsymbol{\mathcal{F}} : (\boldsymbol{\mathcal{S}} - \mathbf{1}^{(4)}) : \dot{\mathbf{F}}^* \quad \text{in } \Omega\end{aligned} \tag{8.2.3a,b}$$

or, in component form,

$$\begin{aligned}\dot{S}^{N}_{Ai} &= \mathcal{F}_{AiBj}\,(\dot{F}_{jB} - \dot{F}^*_{jB})\\ &= \mathcal{F}_{AiBj}\,(\mathcal{S}_{jBCk} - \delta_{BC}\,\delta_{jk})\,\dot{F}^*_{kC} \quad \text{in } \Omega\,,\end{aligned} \tag{8.2.3c,d}$$

where $\mathbf{1}^{(4)}$ with components $\delta_{ij}\,\delta_{AB}$, is the general, fourth-order identity tensor.[8]

It is convenient to define a *transformation nominal stress rate* by

$$\dot{\mathbf{S}}^{N*} \equiv -\,\boldsymbol{\mathcal{F}} : \dot{\mathbf{F}}^* \quad \text{or} \quad \dot{S}^{N*}_{Ai} \equiv -\,\mathcal{F}_{AiBj}\,\dot{F}^*_{jB}\,. \tag{8.2.4a,b}$$

Then, (8.2.3a) becomes

$$\dot{\mathbf{S}}^{N} = \boldsymbol{\mathcal{F}} : \dot{\mathbf{F}} + \dot{\mathbf{S}}^{N*}\,. \tag{8.2.3e}$$

8.2.3. Homogenization and Eigenvelocity Gradient

The transformation velocity gradient, $\dot{\mathbf{F}}^*$, and the generalized Eshelby tensor, $\boldsymbol{\mathcal{S}}$, can be used to homogenize an unbounded V of modulus tensor $\boldsymbol{\mathcal{F}}$, which contains an ellipsoidal inhomogeneity Ω of pseudo-modulus tensor $\boldsymbol{\mathcal{F}}^{\Omega}$, and which is subjected to farfield uniform nominal stress rate $\dot{\mathbf{S}}^{N0}$. Denote the

---

[8] See Nemat-Nasser and Hori (1993, 1999), Section 15.5.

corresponding farfield uniform velocity gradient by $\dot{\mathbf{F}}^0$, and note that

$$\dot{\mathbf{S}}^{N0} = \boldsymbol{\mathcal{F}} : \dot{\mathbf{F}}^0 \quad \text{or} \quad \dot{S}^{N0}_{Ai} = \mathcal{F}_{AiBj}\, \dot{F}^0_{jB}\,. \tag{8.2.5a,b}$$

If the solid were uniform throughout its entire volume, then the nominal stress-rate field and hence the corresponding velocity gradient field would be uniform when the farfield data are uniform. These fields would be given by $\dot{\mathbf{F}}^0$ and $\dot{\mathbf{S}}^{N0} \equiv \boldsymbol{\mathcal{F}} : \dot{\mathbf{F}}^0$, respectively. The presence of region Ω with a different modulus tensor, *i.e.,* the existence of a material mismatch, disturbs the uniform nominal stress rate and the velocity gradient fields. Denote the resulting variable velocity gradient and nominal stress rate fields, respectively, by $\dot{\mathbf{F}}(\mathbf{X})$ and $\dot{\mathbf{S}}^N(\mathbf{X})$, and set

$$\dot{\mathbf{F}}(\mathbf{X}) = \dot{\mathbf{F}}^0 + \dot{\mathbf{F}}^d(\mathbf{X}), \qquad \dot{\mathbf{S}}^N(\mathbf{X}) = \dot{\mathbf{S}}^{N0} + \dot{\mathbf{S}}^{Nd}(\mathbf{X}). \tag{8.2.6a,b}$$

Here, $\dot{\mathbf{F}}^d(\mathbf{X})$ and $\dot{\mathbf{S}}^{Nd}(\mathbf{X})$ are the disturbance velocity gradient and the nominal stress rate fields caused by the presence of the inclusion Ω, with mismatched moduli. The fields $\dot{\mathbf{S}}^N(\mathbf{X})$ and $\dot{\mathbf{F}}(\mathbf{X})$ are related through (Figure 8.2.1a)

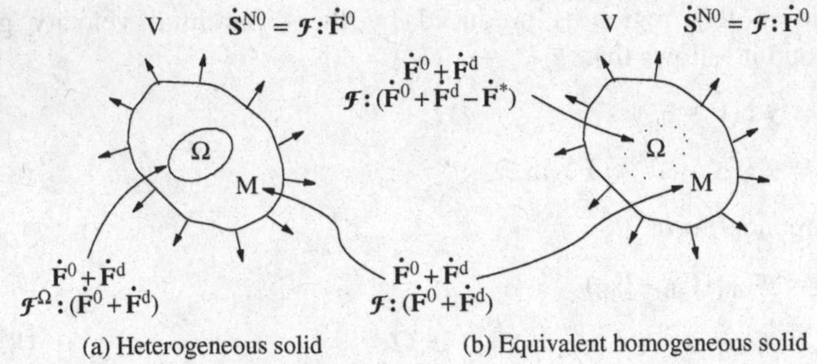

Figure 8.2.1

Equivalent homogeneous solid and eigenvelocity gradient

$$\dot{\mathbf{S}}^N(\mathbf{X}) = \begin{cases} \boldsymbol{\mathcal{F}}^{\Omega} : \dot{\mathbf{F}} & \text{in } \Omega \\ \boldsymbol{\mathcal{F}} : \dot{\mathbf{F}} & \text{in } V - \Omega \end{cases} \tag{8.2.7a}$$

or, in component form,

$$\dot{S}^N_{Ai}(\mathbf{X}) = \begin{cases} \mathcal{F}^{\Omega}_{AiBj}\, \dot{F}_{jB}(\mathbf{X}) & \text{in } \Omega \\ \mathcal{F}_{AiBj}\, \dot{F}_{jB}(\mathbf{X}) & \text{in } V - \Omega\,. \end{cases} \tag{8.2.7b}$$

Equilibrium requires

$$\dot{S}^N_{Ai,A} = 0 \ \text{ in } V,$$

$$\dot{\mathbf{S}}^N(\mathbf{X}) \rightarrow \dot{\mathbf{S}}^{N0} \quad \text{as } |\mathbf{X}| \rightarrow \infty\,. \tag{8.2.8a,b}$$

From (8.2.8b) and (8.2.1), it also follows that

$$\dot{\mathbf{F}}(\mathbf{X}) \to \dot{\mathbf{F}}^0 \quad \text{as } |\mathbf{X}| \to \infty. \tag{8.2.8c}$$

To solve this boundary-value problem, (8.2.8a) must be integrated in conjunction with (8.2.7), subject to the farfield conditions (8.2.8b) or (8.2.8c) and the continuity of the traction rates and the velocity field across the boundary $\partial\Omega$. Instead of dealing with this rather complex problem, it is convenient and effective to consider an *equivalent homogeneous* solid which has the *uniform* modulus tensor $\mathcal{F}$ of the matrix material *everywhere, including in* Ω. *Then, in order to account for the mismatch of the material properties of the inclusion and the matrix, a suitable additional velocity gradient field*[9] $\dot{\mathbf{F}}^*(\mathbf{X})$ *is introduced in* Ω, *such that the equivalent homogeneous solid has the same velocity gradient and nominal stress rate fields as the actual heterogeneous solid under the prescribed farfield data.* The velocity gradient field $\dot{\mathbf{F}}^*$ necessary for this *homogenization* will be called the *eigenvelocity gradient*. Figure (8.2.1b) illustrates this procedure for the case when the boundary traction rates, corresponding to the constant farfield nominal stress rate $\dot{\mathbf{S}}^{N0}$, are prescribed on ∂V. In this figure, the eigenvelocity gradient field is given by

$$\dot{\mathbf{F}}^*(\mathbf{X}) = \begin{cases} \mathbf{0} & \text{in M} \\ \dot{\mathbf{F}}^* & \text{in } \Omega, \end{cases} \tag{8.2.9a}$$

where $\mathrm{M} = \mathrm{V} - \Omega$. For this *equivalent* problem, the modulus tensor is *uniform everywhere, including in* Ω. It is given by $\mathcal{F}$. Therefore, the corresponding nominal stress-rate field becomes, in view of (8.2.6),

$$\dot{\mathbf{S}}^{\mathrm{N}}(\mathbf{X}) = \mathcal{F} : (\dot{\mathbf{F}}(\mathbf{X}) - \dot{\mathbf{F}}^*(\mathbf{X})) = \begin{cases} \mathcal{F} : (\dot{\mathbf{F}}^0 + \dot{\mathbf{F}}^{\mathrm{d}}(\mathbf{X})) & \text{in M} \\ \mathcal{F} : (\dot{\mathbf{F}}^0 + \dot{\mathbf{F}}^{\mathrm{d}}(\mathbf{X}) - \dot{\mathbf{F}}^*(\mathbf{X})) & \text{in } \Omega. \end{cases} \tag{8.2.9b,c}$$

Noting (8.2.6a) and (8.2.9a), observe from (8.2.9b) that the disturbance fields are related through

$$\dot{\mathbf{S}}^{\mathrm{Nd}}(\mathbf{X}) = \mathcal{F} : (\dot{\mathbf{F}}^{\mathrm{d}}(\mathbf{X}) - \dot{\mathbf{F}}^*(\mathbf{X})). \tag{8.2.10}$$

Furthermore, in the *homogenized* solid, the velocity gradient $\dot{\mathbf{F}}^{\mathrm{d}}(\mathbf{X})$ is produced by the eigenvelocity gradient $\dot{\mathbf{F}}^*$. From (8.2.2) it now follows that, for an ellipsoidal Ω,

$$\dot{\mathbf{F}}^{\mathrm{d}} = \mathcal{S} : \dot{\mathbf{F}}^* \quad \text{or} \quad \dot{F}^{\mathrm{d}}_{\mathrm{iA}} = \mathcal{S}_{\mathrm{iABj}} \dot{F}^*_{\mathrm{jB}} \quad \text{in } \Omega. \tag{8.2.11a,b}$$

The resulting nominal stress rate in Ω then is

$$\dot{\mathbf{S}}^{\mathrm{Nd}} = \mathcal{F} : (\mathcal{S} - \mathbf{1}^{(4)}) : \dot{\mathbf{F}}^* \tag{8.2.11c}$$

or, in component form,

---

[9] *I.e.*, inelastic, transformation velocity gradient.

$$\dot{S}^{Nd}_{Ai} = \mathcal{F}_{AiBj}\,(\mathcal{S}_{jBCk} - \delta_{BC}\,\delta_{jk})\,\dot{F}^*_{kC}\,. \tag{8.2.11d}$$

8.2.4. Eigenstress Rate

In the preceding subsection, the heterogeneous solid consisting of a uniform matrix M and a single inclusion Ω with different moduli, defined in (8.2.7), is homogenized by the introduction of the eigenvelocity gradient $\dot{\mathbf{F}}^*(\mathbf{X})$. The homogenization can be performed by the introduction of an *eigenstress rate* $\dot{\mathbf{S}}^{N*}(\mathbf{X})$, instead. To this end, set

$$\dot{\mathbf{S}}^{N*}(\mathbf{X}) = \begin{cases} \mathbf{0} & \text{in M} \\ \dot{\mathbf{S}}^{N*} & \text{in } \Omega. \end{cases} \tag{8.2.12a}$$

For this *alternative equivalent* problem, the modulus tensor is again *uniform everywhere, including in* Ω. The corresponding velocity gradient and nominal stress rate fields are

$$\dot{\mathbf{F}}(\mathbf{X}) = \dot{\mathbf{F}}^0 + \dot{\mathbf{F}}^d(\mathbf{X}),$$

$$\dot{\mathbf{S}}^N(\mathbf{X}) = \boldsymbol{\mathcal{F}} : \dot{\mathbf{F}}(\mathbf{X}) + \dot{\mathbf{S}}^{N*}(\mathbf{X}) = \begin{cases} \boldsymbol{\mathcal{F}} : (\dot{\mathbf{F}}^0 + \dot{\mathbf{F}}^d(\mathbf{X})) & \text{in M} \\ \boldsymbol{\mathcal{F}} : (\dot{\mathbf{F}}^0 + \dot{\mathbf{F}}^d(\mathbf{X})) + \dot{\mathbf{S}}^{N*}(\mathbf{X}) & \text{in } \Omega. \end{cases}$$

(8.2.12b,c)

From (8.2.3e), the disturbance velocity gradient and the nominal stress rate must satisfy

$$\dot{\mathbf{S}}^{Nd}(\mathbf{X}) = \boldsymbol{\mathcal{F}} : \dot{\mathbf{F}}^d(\mathbf{X}) + \dot{\mathbf{S}}^{N*}(\mathbf{X}) \quad \text{in V}, \tag{8.2.13}$$

for the required eigenstress rate, $\dot{\mathbf{S}}^{N*}(\mathbf{X})$. As is discussed in Subsection 8.3.4, in general, the disturbance nominal stress rate field $\dot{\mathbf{S}}^{Nd}(\mathbf{X})$ can be expressed in terms of an *integral operator* acting on the corresponding eigenstress rate $\dot{\mathbf{S}}^{N*}(\mathbf{X})$. However, when Ω is ellipsoidal, and when the eigenvelocity gradient $\dot{\mathbf{F}}^*$ and hence the resulting disturbance velocity gradient $\dot{\mathbf{F}}^d$ are uniform in Ω, then the corresponding eigenstress rate $\dot{\mathbf{S}}^{N*}$ and the associated disturbance nominal stress rate $\dot{\mathbf{S}}^{Nd}$ are also uniform in Ω. Therefore, a fourth-order tensor $\boldsymbol{\mathcal{T}}^{\Omega}$, may be introduced such that

$$\dot{\mathbf{S}}^{Nd} = \boldsymbol{\mathcal{T}}^{\Omega} : \dot{\mathbf{S}}^{N*} \quad \text{or} \quad \dot{S}^{Nd}_{Ai} = \mathcal{T}^{\Omega}_{AijB}\,\dot{S}^{N*}_{Bj} \quad \text{in } \Omega\,. \tag{8.2.14a,b}$$

In general, tensor $\boldsymbol{\mathcal{T}}^{\Omega}$ has no symmetries. It is conjugate to the generalized Eshelby tensor, $\boldsymbol{\mathcal{S}}^{\Omega}$. Note from (8.2.9) and (8.2.12) that the eigenvelocity gradient and the eigenstress rate are related by

$$\dot{\mathbf{S}}^{N*} + \boldsymbol{\mathcal{F}} : \dot{\mathbf{F}}^* = \mathbf{0}, \qquad \dot{\mathbf{F}}^* + \boldsymbol{\mathcal{G}} : \dot{\mathbf{S}}^{N*} = \mathbf{0} \tag{8.2.15a,b}$$

or, in component form,

$$\dot{S}^{N*}_{Ai} + \mathcal{F}_{AiBj}\,\dot{F}^*_{jB} = 0\,, \qquad \dot{F}^*_{iA} + \mathcal{G}_{iAjB}\,\dot{S}^{N*}_{Bj} = 0\,. \tag{8.2.15c,d}$$

In view of (8.2.1c,d) and (8.2.15a,b), expression (8.2.13) can be rewritten as

$$\dot{\mathbf{F}}^{d} = \boldsymbol{\mathcal{G}} : \dot{\mathbf{S}}^{Nd} - \dot{\mathbf{F}}^{*}$$
$$= \boldsymbol{\mathcal{G}} : (\boldsymbol{\mathcal{T}}^{\Omega} - \mathbf{1}^{(4)}) : \dot{\mathbf{S}}^{N*} \quad \text{in } \Omega \qquad (8.2.16a,b)$$

or, in component form,

$$\dot{F}^{d}_{iA} = \mathcal{G}_{iAjB}\,\dot{S}_{Bj} - \dot{F}^{*}_{iA}$$
$$= \mathcal{G}_{iAjB}\,(\mathcal{T}^{\Omega}_{BjkC} - \delta_{jk}\delta_{BC})\,\dot{S}^{N*}_{Ck} \quad \text{in } \Omega. \qquad (8.2.16c,d)$$

Hence,

$$\boldsymbol{\mathcal{S}}^{\Omega} : \dot{\mathbf{F}}^{*} = \boldsymbol{\mathcal{G}} : (\boldsymbol{\mathcal{T}}^{\Omega} - \mathbf{1}^{(4)}) : (-\boldsymbol{\mathcal{F}} : \dot{\mathbf{F}}^{*}),$$
$$\boldsymbol{\mathcal{T}}^{\Omega} : \dot{\mathbf{S}}^{N*} = \boldsymbol{\mathcal{F}} : (\boldsymbol{\mathcal{S}}^{\Omega} - \mathbf{1}^{(4)}) : (-\boldsymbol{\mathcal{G}} : \dot{\mathbf{S}}^{N*}). \qquad (8.2.17a,b)$$

Therefore, the tensors $\boldsymbol{\mathcal{S}}^{\Omega}$ and $\boldsymbol{\mathcal{T}}^{\Omega}$ satisfy

$$\boldsymbol{\mathcal{S}}^{\Omega} + \boldsymbol{\mathcal{G}} : \boldsymbol{\mathcal{T}}^{\Omega} : \boldsymbol{\mathcal{F}} = \mathbf{1}^{(4)}, \qquad \boldsymbol{\mathcal{T}}^{\Omega} + \boldsymbol{\mathcal{F}} : \boldsymbol{\mathcal{S}}^{\Omega} : \boldsymbol{\mathcal{G}} = \mathbf{1}^{(4)}. \qquad (8.2.17c,d)$$

In component form, these are

$$\mathcal{S}^{\Omega}_{iABj} + \mathcal{G}_{iAkC}\,\mathcal{T}^{\Omega}_{CklD}\,\mathcal{F}_{DlBj} = \delta_{ij}\,\delta_{AB},$$
$$\mathcal{T}^{\Omega}_{AijB} + \mathcal{F}_{AiCk}\,\mathcal{S}^{\Omega}_{kCDl}\,\mathcal{G}_{lDjB} = \delta_{AB}\,\delta_{ij}, \qquad (8.2.17e,f)$$

where all indices take on the values 1, 2, and 3, and repeated indices are summed.

8.2.5. Consistency Conditions

For finite V, the eigenvelocity gradients or eigenstress rates necessary for homogenization are, in general, nonuniform in Ω, even if Ω is ellipsoidal. Also, for a nonellipsoidal Ω, the required eigenvelocity gradients or eigenstress rates are in general, variable in Ω (they are zero outside of Ω), even if V is unbounded. For the general case, the eigenvelocity gradient, $\dot{\mathbf{F}}^{*}(\mathbf{X})$, or the eigenstress rate, $\dot{\mathbf{S}}^{N*}(\mathbf{X})$, is defined by a set of *consistency conditions* which ensures that the resulting nominal stress rate field $\dot{\mathbf{S}}^{N}(\mathbf{X})$, and the velocity gradient field $\dot{\mathbf{F}}(\mathbf{X})$, are the same under the applied overall loads, whether they are calculated through homogenization or directly by solving the corresponding boundary-value problem for the rate quantities in the original heterogeneous solid. Hence, the resulting nominal stress rate field in Ω is expressed as follows:

$$\dot{\mathbf{S}}^{N}(\mathbf{X}) = \boldsymbol{\mathcal{F}}^{\Omega} : \{\dot{\mathbf{F}}^{0} + \dot{\mathbf{F}}^{d}(\mathbf{X})\}$$
$$= \boldsymbol{\mathcal{F}} : \{\dot{\mathbf{F}}^{0} + \dot{\mathbf{F}}^{d}(\mathbf{X}) - \dot{\mathbf{F}}^{*}(\mathbf{X})\} \quad \text{in } \Omega, \qquad (8.2.18a)$$

and the resulting velocity gradient field in Ω, as

$$\dot{\mathbf{F}}(\mathbf{X}) = \boldsymbol{\mathcal{G}}^{\Omega} : \{\dot{\mathbf{S}}^{N0} + \dot{\mathbf{S}}^{Nd}(\mathbf{X})\}$$

$$= \mathcal{G} : \{\dot{\mathbf{S}}^{N0} + \dot{\mathbf{S}}^{Nd}(\mathbf{X}) - \dot{\mathbf{S}}^{N*}(\mathbf{X})\} \quad \text{in } \Omega. \tag{8.2.18b}$$

It is noted that both (8.2.18a) and (8.2.18b) are valid, whether uniform tractions produced by $\dot{\mathbf{S}}^{N0}$ or linear velocities produced by $\dot{\mathbf{F}}^0$ are prescribed on ∂V. If the overall nominal stress rate $\dot{\mathbf{S}}^{N0}$ is *given*, then $\dot{\mathbf{F}}^0$ is *defined* by $\mathcal{G} : \dot{\mathbf{S}}^{N0}$, whereas if the overall velocity gradient $\dot{\mathbf{F}}^0$ is *given*, then $\dot{\mathbf{S}}^{N0}$ is *defined* by $\mathcal{F} : \dot{\mathbf{F}}^0$.

Whether V is bounded or not, and for any homogeneous inclusion Ω in a homogeneous matrix M which satisfies (8.2.7), consistency conditions (8.2.18a,b) yield

$$\dot{\mathbf{F}}^0 + \dot{\mathbf{F}}^d(\mathbf{X}) = \mathcal{C}^{\Omega} : \dot{\mathbf{F}}^*(\mathbf{X}),$$
$$\dot{\mathbf{S}}^{N0} + \dot{\mathbf{S}}^{Nd}(\mathbf{X}) = \mathcal{D}^{\Omega} : \dot{\mathbf{S}}^{N*}(\mathbf{X}) \quad \text{in } \Omega, \tag{8.2.19a,b}$$

where

$$\mathcal{C}^{\Omega} \equiv (\mathcal{F} - \mathcal{F}^{\Omega})^{-1} : \mathcal{F}, \qquad \mathcal{D}^{\Omega} \equiv (\mathcal{G} - \mathcal{G}^{\Omega})^{-1} : \mathcal{G}, \tag{8.2.20a,b}$$

provided that the indicated inverses do exist; see Section 8.5 for comments. By definition, constant tensors $\mathcal{C}^{\Omega}$ and $\mathcal{D}^{\Omega}$ satisfy

$$\mathcal{G} : \mathcal{F}^{\Omega} = \mathbf{1}^{(4)} - (\mathcal{C}^{\Omega})^{-1} = (\mathbf{1}^{(4)} - (\mathcal{D}^{\Omega})^{-1})^{-T} \tag{8.2.20c}$$

or

$$\mathcal{F} : \mathcal{G}^{\Omega} = \mathbf{1}^{(4)} - (\mathcal{D}^{\Omega})^{-1} = (\mathbf{1}^{(4)} - (\mathcal{C}^{\Omega})^{-1})^{-T}, \tag{8.2.20d}$$

where the superscript -T stands for the inverse of the transpose or the transpose of the inverse.

Similarly to the generalized Eshelby tensors, *in general,* $\mathcal{C}^{\Omega}$ *and* $\mathcal{D}^{\Omega}$ *do not have any symmetries,* as can be seen from their components,

$$\mathcal{C}^{\Omega}_{jBAi} \equiv (\delta_{ij}\delta_{AB} - \mathcal{G}_{iAkC}\,\mathcal{F}^{\Omega}_{CkBj})^{-1},$$
$$\mathcal{D}^{\Omega}_{BjiA} \equiv (\delta_{AB}\delta_{ij} - \mathcal{F}_{AiCk}\,\mathcal{G}^{\Omega}_{kCjB})^{-1}. \tag{8.2.20e,f}$$

In the sequel, attention is confined to the case when V is unbounded and Ω is ellipsoidal, so that the homogenization eigenvelocity gradient and eigenstress rate are both uniform in Ω. When V is *unbounded*, there is no distinction between the cases when the velocity gradient $\dot{\mathbf{F}}^0$ or the nominal stress rate $\dot{\mathbf{S}}^{N0}$ is prescribed; this is, however, not true when V is *bounded.* Thus $\dot{\mathbf{F}}^0 = \mathcal{G} : \dot{\mathbf{S}}^{N0}$ or $\dot{\mathbf{S}}^{N0} = \mathcal{F} : \dot{\mathbf{F}}^0$. Also, when, in addition, Ω is *ellipsoidal*, then $\dot{\mathbf{F}}^d$, $\dot{\mathbf{F}}^*$, $\dot{\mathbf{S}}^{Nd}$, and $\dot{\mathbf{S}}^{N*}$ are all *constant* tensors in Ω. Hence, for unbounded V and ellipsoidal Ω, substitution for $\dot{\mathbf{F}}^d$ in (8.2.19a) or for $\dot{\mathbf{S}}^{Nd}$ in (8.2.19b) provides explicit expressions for the eigenvelocity gradient $\dot{\mathbf{F}}^*$ and eigenstress rate $\dot{\mathbf{S}}^{N*}$ which are necessary for homogenization,

$$\dot{\mathbf{F}}^* = (\mathcal{C}^{\Omega} - \mathcal{S}^{\Omega})^{-1} : \dot{\mathbf{F}}^0, \qquad \dot{\mathbf{S}}^{N*} = (\mathcal{D}^{\Omega} - \mathcal{T}^{\Omega})^{-1} : \dot{\mathbf{S}}^{N0} \quad \text{in } \Omega. \tag{8.2.21a,b}$$

These and (8.2.19a,b), now lead to

$$\dot{\mathbf{F}} = \dot{\mathbf{F}}^0 + \dot{\mathbf{F}}^d = \mathcal{C}^{\Omega} : (\mathcal{C}^{\Omega} - \mathcal{S}^{\Omega})^{-1} : \dot{\mathbf{F}}^0,$$
$$\dot{\mathbf{S}}^N = \dot{\mathbf{S}}^{N0} + \dot{\mathbf{S}}^{Nd} = \mathcal{D}^{\Omega} : (\mathcal{D}^{\Omega} - \mathcal{T}^{\Omega})^{-1} : \dot{\mathbf{S}}^{N0} \quad \text{in } \Omega. \qquad (8.2.22a,b)$$

Note that the velocity gradient $\dot{\mathbf{F}}$ and the nominal stress rate $\dot{\mathbf{S}}^N$ in Ω, given by (8.2.22a) and (8.2.22b), are equivalent. In view of constitutive relations (8.2.1), substitution of (8.2.20a,b) into (8.2.22a,b) yields

$$\dot{\mathbf{S}}^N = \mathcal{F}^{\Omega} : \mathcal{C}^{\Omega} : (\mathcal{C}^{\Omega} - \mathcal{S}^{\Omega})^{-1} : \dot{\mathbf{F}}^0$$
$$= \{\mathcal{F}^{\Omega} : \{1^{(4)} - \mathcal{S}^{\Omega} : (1^{(4)} - \mathcal{G} : \mathcal{F}^{\Omega})\}^{-1} : \mathcal{G}\} : \dot{\mathbf{S}}^{N0},$$
$$\dot{\mathbf{F}} = \mathcal{G}^{\Omega} : \mathcal{D}^{\Omega} : (\mathcal{D}^{\Omega} - \mathcal{T}^{\Omega})^{-1} : \dot{\mathbf{S}}^{N0}$$
$$= \{\mathcal{G}^{\Omega} : \{1^{(4)} - \mathcal{T}^{\Omega} : (1^{(4)} - \mathcal{F} : \mathcal{G}^{\Omega})\}^{-1} : \mathcal{F}\} : \dot{\mathbf{F}}^0 \quad \text{in } \Omega. \qquad (8.2.23a,b)$$

Noting identities (8.2.17c,d), observe that the fourth-order tensors in the right-hand sides of (8.2.23a,b), become

$$\mathcal{F}^{\Omega} : \{1^{(4)} - \mathcal{S}^{\Omega} : (1^{(4)} - \mathcal{G} : \mathcal{F}^{\Omega})\}^{-1} : \mathcal{G} = \{1^{(4)} - \mathcal{T}^{\Omega} : (1^{(4)} - \mathcal{F} : \mathcal{G}^{\Omega})\}^{-1},$$
$$\mathcal{G}^{\Omega} : \{1^{(4)} - \mathcal{T}^{\Omega} : (1^{(4)} - \mathcal{F} : \mathcal{G}^{\Omega})\}^{-1} : \mathcal{F} = \{1^{(4)} - \mathcal{S}^{\Omega} : (1^{(4)} - \mathcal{G} : \mathcal{F}^{\Omega})\}^{-1}.$$
$$(8.2.23c,d)$$

Therefore, (8.2.23c,d) compared with (8.2.22a,b), yield the following equivalence relations[10] between $(\mathcal{C}^{\Omega}, \mathcal{S}^{\Omega})$ and $(\mathcal{D}^{\Omega}, \mathcal{T}^{\Omega})$:

$$\mathcal{F}^{\Omega} : \mathcal{C}^{\Omega} : (\mathcal{C}^{\Omega} - \mathcal{S}^{\Omega})^{-1} : \mathcal{G} = \mathcal{D}^{\Omega} : (\mathcal{D}^{\Omega} - \mathcal{T}^{\Omega})^{-1},$$
$$\mathcal{G}^{\Omega} : \mathcal{D}^{\Omega} : (\mathcal{D}^{\Omega} - \mathcal{T}^{\Omega})^{-1} : \mathcal{F} = \mathcal{C}^{\Omega} : (\mathcal{C}^{\Omega} - \mathcal{S}^{\Omega})^{-1}. \qquad (8.2.23e,f)$$

8.2.6. Concentration Tensors

Equations (8.2.22a,b) yield the velocity gradient, $\dot{\mathbf{F}}$, and the nominal stress rate, $\dot{\mathbf{S}}^N$, in an ellipsoidal inclusion, Ω, of pseudo-modulus tensor $\mathcal{F}^{\Omega}$, which is embedded in an unbounded homogeneous solid of pseudo-modulus tensor $\mathcal{F}$, under farfield uniform data, $\dot{\mathbf{S}}^{N0} = \mathcal{F} : \dot{\mathbf{F}}^0$. These, thus, define the *concentration tensors* which relate the local fields in Ω to the farfield data, as follows:

$$\dot{\mathbf{F}} = \mathcal{A}^{\Omega} : \dot{\mathbf{F}}^0, \qquad \dot{\mathbf{S}}^N = \mathcal{B}^{\Omega} : \dot{\mathbf{S}}^{N0}, \qquad (8.2.24a,b)$$

where

$$\mathcal{A}^{\Omega} \equiv \mathcal{C}^{\Omega} : (\mathcal{C}^{\Omega} - \mathcal{S}^{\Omega})^{-1} \quad \text{and} \quad \mathcal{B}^{\Omega} \equiv \mathcal{D}^{\Omega} : (\mathcal{D}^{\Omega} - \mathcal{T}^{\Omega})^{-1} \qquad (8.2.24c,d)$$

are the concentration tensors; here, $\mathcal{C}^{\Omega}$ and $\mathcal{D}^{\Omega}$ are defined by (8.2.20a,b). The

[10] Except for the symmetries that exist in linear elasticity, but not in the present case, the above *exact* relations are identical (only in form) to the corresponding relations in linear elasticity; see Nemat-Nasser and Hori (1993, 1999).

concentration tensors may be rewritten as

$$\boldsymbol{\mathcal{A}}^{\Omega} = \{\mathbf{1}^{(4)} - \boldsymbol{\mathcal{P}}^{\Omega} : (\boldsymbol{\mathcal{F}} - \boldsymbol{\mathcal{F}}^{\Omega})\}^{-1},$$

$$\boldsymbol{\mathcal{B}}^{\Omega} = \{\mathbf{1}^{(4)} - \boldsymbol{\mathcal{Q}}^{\Omega} : (\boldsymbol{\mathcal{G}} - \boldsymbol{\mathcal{G}}^{\Omega})\}^{-1}. \quad (8.2.25a,b)$$

Here, in addition to (8.2.20a,b), the following notation is used:

$$\boldsymbol{\mathcal{P}}^{\Omega} \equiv \boldsymbol{\mathcal{S}}^{\Omega} : \boldsymbol{\mathcal{G}}, \qquad \boldsymbol{\mathcal{Q}}^{\Omega} \equiv \boldsymbol{\mathcal{T}}^{\Omega} : \boldsymbol{\mathcal{F}}. \quad (8.2.25c,d)$$

In component form, the concentration tensors become,

$$\mathcal{A}^{\Omega}_{jBAi} \equiv \{\delta_{ij}\delta_{AB} - \mathcal{P}^{\Omega}_{iAkC}(\mathcal{F}_{CkBj} - \mathcal{F}^{\Omega}_{CkBj})\}^{-1},$$

$$\mathcal{B}^{\Omega}_{BjiA} \equiv \{\delta_{AB}\delta_{ij} - \mathcal{Q}^{\Omega}_{AiCk}(\mathcal{G}_{kCjB} - \mathcal{G}^{\Omega}_{kCjB})\}^{-1}, \quad (8.2.25e,f)$$

where

$$\mathcal{P}^{\Omega}_{iAjB} = \mathcal{S}^{\Omega}_{iACk}\,\mathcal{G}^{\Omega}_{kCjB}, \qquad \mathcal{Q}^{\Omega}_{AiBj} = \mathcal{T}^{\Omega}_{AikC}\,\mathcal{F}^{\Omega}_{CkBj}. \quad (8.2.25g,h)$$

Tensors $\boldsymbol{\mathcal{P}}^{\Omega}$ and $\boldsymbol{\mathcal{Q}}^{\Omega}$ are obtained in Subsection 8.3.5 in terms of the Green function of the unbounded V of pseudo-modulus tensor $\boldsymbol{\mathcal{F}}$ (or, pseudo-compliance $\boldsymbol{\mathcal{G}}$) and the aspect ratios of the ellipsoid Ω. These tensors do *not* depend on the properties of Ω. They are real-valued and can be computed, as long as the operator $\mathcal{L}_{ij} \equiv \mathcal{F}_{AiBj}\,\partial^2(...)\,/\,(\partial X_A\,\partial X_B)$ is elliptic[11].

8.3. THE GREEN FUNCTION AND CONCENTRATION TENSORS

In Subsection 8.2.2, the generalized Eshelby tensor, $\boldsymbol{\mathcal{S}}^{\Omega}$, and its conjugate, $\boldsymbol{\mathcal{T}}^{\Omega}$, are introduced for an infinitely extended homogeneous solid of pseudo-modulus and pseudo-compliance tensors, $\boldsymbol{\mathcal{F}}$ and $\boldsymbol{\mathcal{G}}$. These are then used to determine the constant velocity-gradient and the nominal stress rate fields, $\dot{\mathbf{F}}$ and $\dot{\mathbf{S}}^{N}$, in an ellipsoidal subdomain Ω, in which a uniform eigenvelocity gradient (or eigenstress rate) is distributed. As pointed out in Subsection 8.2.3, the generalized Eshelby tensor and its conjugate can be used to estimate the average velocity gradients and the nominal stress rates in an inclusion of pseudo-modulus and pseudo-compliance tensors, $\boldsymbol{\mathcal{F}}^{\Omega}$ and $\boldsymbol{\mathcal{G}}^{\Omega}$, which is embedded in an unbounded uniform and uniformly deformed matrix. Under suitable settings, these tensors can be computed. Therefore, they provide an effective means for estimating the overall pseudo-moduli of a finitely deformed heterogeneous solid, using a homogenization model, as discussed in Section 8.6.

In this section, an infinitely extended, homogeneously deformed, and homogeneous solid is considered, within a portion of which either eigenvelocity gradients or eigenstress rates are distributed (not necessarily uniformly). The Green function for the unbounded solid is then used to formulate the resulting

[11] For a detailed discussion, see Subsection 8.3.5.

velocity field in terms of two integral operators, $\mathcal{S}^{\infty}(\mathbf{X}; \dot{\mathbf{F}}^*)$ and $\mathcal{T}^{\infty}(\mathbf{X}; \dot{\mathbf{S}}^{N*})$, where $\dot{\mathbf{F}}^* = \dot{\mathbf{F}}^*(\mathbf{X})$ and $\dot{\mathbf{S}}^{N*} = \dot{\mathbf{S}}^{N*}(\mathbf{X})$ are the corresponding prescribed eigenvelocity gradient and eigenstress rate.[12] The integral operators reduce to tensor operators, when the distribution of eigenvelocity gradients or eigenstress rates is uniform and the domain Ω in which they are distributed is an isolated ellipsoid in an unbounded uniform medium, resulting in the generalized Eshelby tensor, $\mathcal{S}^{\Omega}$, and its conjugate, $\mathcal{T}^{\Omega}$. Equations are derived for these tensors in the general case of a matrix with an instantaneous pseudo-modulus of elliptic (*i.e.*, with complex-valued eigenvalues) characteristics. Then a number of interesting properties of the generalized Eshelby tensor are examined,[13] showing how certain results from linear elasticity can be generalized for application to *finite deformation plasticity* which involves *full geometric and material nonlinearities.* For example, it is shown how the double-inclusion problem is solved for the velocity gradient and the nominal stress rate imposed on the *finite deformation and rotation of fully nonlinear elastoplastic heterogeneous solids.* From this, the disturbances in the average velocity gradient and the nominal stress rate (or in other related quantities), produced by a prescribed eigenvelocity gradient or eigenstress rate, are obtained. As before, throughout this section, quasi-static deformation is considered. All field quantities are measured with respect to a monotonically changing parameter. Thus, the rate of change of any field represents its instantaneous quasi-static variation, from which its incremental change is calculated. As in Sections 8.1 and 8.2, the results presented in the present section are all *exact.*

8.3.1. Reciprocal Relations

As pointed out in Subsection 8.2.2, for a broad class of problems, the pseudo-modulus tensor $\mathcal{F}$, with components $\mathcal{F}_{AiBj}$, is symmetric with respect to the exchange of Ai and Bj, *i.e.*, the nine by nine matrix of $\mathcal{F}_{AiBj}$ has diagonal symmetry. For this class of problems, the reciprocal theorem of linear elasticity remains valid.

To show this, consider a finite solid with volume V and boundary ∂V, having a modulus tensor such that $\mathcal{F}_{AiBj} = \mathcal{F}_{BjAi}$, and let it be in equilibrium under two different sets of boundary data, as follows:

$$\nabla \cdot \dot{\mathbf{S}}^{N(\alpha)}(\mathbf{X}) + \dot{\mathbf{f}}^{(\alpha)}(\mathbf{X}) = \mathbf{0} \quad \text{in V},$$
$$\mathbf{N} \cdot \dot{\mathbf{S}}^{N(\alpha)} = \dot{\mathbf{T}}^{(\alpha)} \quad \text{on } \partial\text{V}, \quad \alpha = 1, 2. \tag{8.3.1a,b}$$

These are *two different problems*, associated with the *same solid*, but for two

[12] Only the eigenvelocity gradient or the eigenstress rate, but not both, can be prescribed *arbitrarily* in a given region. Hence, when both eigenvelocity gradient and eigenstress rate are mentioned, either two separate problems are considered, or the two fields are mutually dependent. The intended alternative should be clear from the context.

[13] These results have been recently reported by the author; see Nemat-Nasser (1999, 2000).

different sets of loading conditions. Now take the dot product of (8.3.1a) with $\dot{\mathbf{x}}^{(\beta)}$ and integrate the resulting expression over V to obtain,

$$\int_{\partial V} \dot{\mathbf{x}}^{(\beta)} \cdot \dot{\mathbf{T}}^{(\alpha)}\,dV + \int_V \dot{\mathbf{x}}^{(\beta)} \cdot \dot{\mathbf{f}}^{(\alpha)}\,dV = \int_V \dot{\mathbf{F}}^{(\beta)} : \dot{\mathbf{S}}^{N(\alpha)}\,dV = \int_V \dot{\mathbf{F}}^{(\beta)} : \boldsymbol{\mathcal{F}} : \dot{\mathbf{F}}^{(\alpha)}\,dV, \tag{8.3.1c}$$

where the Gauss theorem and the boundary data (8.3.1b) are used. Since the nine by nine matrix of the components of $\boldsymbol{\mathcal{F}}$ is diagonally symmetric,

$$\int_V \dot{\mathbf{F}}^{(\beta)} : \boldsymbol{\mathcal{F}} : \dot{\mathbf{F}}^{(\alpha)}\,dV = \int_V \dot{\mathbf{F}}^{(\alpha)} : \boldsymbol{\mathcal{F}} : \dot{\mathbf{F}}^{(\beta)}\,dV, \tag{8.3.1d}$$

and, hence,

$$\int_{\partial V} \dot{\mathbf{x}}^{(\beta)} \cdot \dot{\mathbf{T}}^{(\alpha)}\,dV + \int_V \dot{\mathbf{x}}^{(\beta)} \cdot \dot{\mathbf{f}}^{(\alpha)}\,dV = \int_{\partial V} \dot{\mathbf{x}}^{(\alpha)} \cdot \dot{\mathbf{T}}^{(\beta)}\,dV + \int_V \dot{\mathbf{x}}^{(\alpha)} \cdot \dot{\mathbf{f}}^{(\beta)}\,dV. \tag{8.3.1e}$$

From this result it follows that, in an unbounded solid, the displacement rate $\dot{\mathbf{x}}^{(1)}(\mathbf{X}^{(2)})$ at point $\mathbf{X}^{(2)}$ due to a unit $\dot{\mathbf{f}}$ applied at point $\mathbf{X}^{(1)}$, is equal to the displacement rate $\dot{\mathbf{x}}^{(2)}(\mathbf{X}^{(1)})$ at point $\mathbf{X}^{(1)}$ due to a unit $\dot{\mathbf{f}}$ applied at point $\mathbf{X}^{(2)}$, *i.e.*,

$$\dot{\mathbf{x}}^{(1)}(\mathbf{X}^{(2)}) = \dot{\mathbf{x}}^{(2)}(\mathbf{X}^{(1)}). \tag{8.3.1f}$$

Expressions (8.3.1e) and (8.3.1f) are reciprocal relations, valid when the pseudo-modulus tensor $\boldsymbol{\mathcal{F}}$, with components $\mathcal{F}_{AiBj}$, is symmetric with respect to the exchange of Ai and Bj.

8.3.2. Green's Function

Consider an infinitely extended domain, denoted by V^∞, consisting of a homogeneous solid of pseudo-modulus $\boldsymbol{\mathcal{F}}$ and pseudo-compliance $\boldsymbol{\mathcal{G}}$. Formally, the Green function for this solid, $G^\infty_{jm}(\mathbf{X}, \mathbf{Y})$, is the vector-valued *fundamental*[14] *solution of the operator* $\mathcal{L}_{ij} \equiv \mathcal{F}_{AiBj}\,\partial^2(\ldots) / (\partial X_A\,\partial X_B)$, defined by the following boundary-value problem:

$$\mathcal{F}_{AiBj}\,G^\infty_{jm,AB}(\mathbf{X}, \mathbf{Y}) + \delta(\mathbf{X} - \mathbf{Y})\,\delta_{im} = 0 \qquad \mathbf{X} \text{ in } V^\infty, \tag{8.3.2a}$$

where $\delta(\mathbf{X} - \mathbf{Y})$ is the delta function, and

$$G^\infty_{jm}(\mathbf{X}, \mathbf{Y}) \to 0 \qquad \text{as } |\mathbf{X}| \to \infty. \tag{8.3.2b}$$

Note that, in the expression of the Green function, the first subscript denotes the corresponding component, and the second subscript is the direction of an applied unit "force-rate". Hence, G^∞_{jm} is the $\mathbf{e}_j$-component of the vector-valued function $\mathbf{G}^\infty(\mathbf{X}, \mathbf{Y})$, measured at point $\mathbf{X}$, due to the unit "force-rate" applied at point $\mathbf{Y}$ in the direction[15] $\mathbf{e}_m$.

---

[14] See, *e.g.*, Morse and Feshbach (1953).

[15] See Chapter 1, Section 1.1, for definition of terms.

Since the domain V^∞ is homogeneous and unbounded, only the difference between the **X**-point (where the "displacement rate" is measured) and the **Y**-point (where the unit "force-rate" is applied) determines the Green function,

$$\mathbf{G}^\infty(\mathbf{X}, \mathbf{Y}) = \mathbf{G}^\infty(\mathbf{X} - \mathbf{Y}). \tag{8.3.3}$$

The farfield condition (8.3.2b) now becomes

$$\mathbf{G}^\infty(\mathbf{Z}) \to \mathbf{0} \quad \text{as } |\mathbf{Z}| \to \infty. \tag{8.3.4a}$$

The gradient of $\mathbf{G}^\infty$ also vanishes for large values of **Z**, and hence the farfield velocity gradients and the nominal stress rates are zero. For an arbitrary finite domain W within V^∞ to be in equilibrium, the resultant traction rates on the surface ∂W of W must satisfy

$$\int_{\partial W} \mathbf{N} \cdot \{\mathcal{F} : (\nabla_\mathbf{X} \otimes \mathbf{G}^\infty(\mathbf{X} - \mathbf{Y})\} \, dS_\mathbf{X} = \begin{cases} -\mathbf{1}^{(2)} & \text{if } \mathbf{Y} \text{ in W} \\ \mathbf{0} & \text{otherwise,} \end{cases} \tag{8.3.4b}$$

where **N** is the outer unit normal of ∂W, and subscript **X** stands for differentiation (integration) with respect to **X**. Indeed, with the aid of the Gauss theorem, (8.3.4b) is obtained by integrating the governing equation (8.3.2a) over W.

The Green function $\mathbf{G}^\infty$ has the following symmetry property, derived from the reciprocal theorem (8.3.1f):

$$\mathbf{G}^\infty(\mathbf{Z}) = \mathbf{G}^{\infty T}(-\mathbf{Z}) \quad \text{or} \quad G^\infty_{ij}(\mathbf{Z}) = G^\infty_{ji}(-\mathbf{Z}). \tag{8.3.5a,b}$$

From the governing equations (8.3.2), **G** is an even function of **Z**. Thus, in view of (8.3.5),

$$\mathbf{G}^\infty(\mathbf{Z}) = \mathbf{G}^{\infty T}(\mathbf{Z}) \quad \text{or} \quad G^\infty_{ij}(\mathbf{Z}) = G^\infty_{ji}(\mathbf{Z}). \tag{8.3.6a,b}$$

8.3.3. The Body-force-rate Problem

Consider the velocity field $\dot{\mathbf{x}}(\mathbf{X})$ produced by distributed body force-rate, $\dot{\mathbf{f}}(\mathbf{X})$, which vanishes toward infinity. The boundary-value problem for $\dot{\mathbf{x}}$ is,

$$\nabla \cdot \{\mathcal{F} : (\nabla \otimes \dot{\mathbf{x}}(\mathbf{X}))^T\} + \dot{\mathbf{f}}(\mathbf{X}) = \mathbf{0} \qquad \mathbf{X} \text{ in } V^\infty, \tag{8.3.7a}$$

with

$$\dot{\mathbf{x}}(\mathbf{X}) \to \mathbf{0} \quad \text{as } |\mathbf{X}| \to \infty. \tag{8.3.7b}$$

In component form, (8.3.7a) becomes,

$$\mathcal{F}_{AiBj}\, \dot{x}_{j,AB} + \dot{f}_i = 0 \qquad \mathbf{X} \text{ in } V^\infty. \tag{8.3.7c}$$

From linearity, the solution is given by integrating the Green function $\mathbf{G}^\infty$, weighted by the prescribed body force-rate $\dot{\mathbf{f}}$. Assuming that $\dot{\mathbf{f}}$ vanishes sufficiently quickly toward infinity, this gives,

$$\dot{\mathbf{x}}(\mathbf{X}) = \int_{V^\infty} \mathbf{G}^\infty(\mathbf{X} - \mathbf{Y}) \cdot \dot{\mathbf{f}}(\mathbf{Y}) \, dV_\mathbf{Y}, \tag{8.3.8a}$$

or, in component form,

$$\dot{x}_i(\mathbf{X}) = \int_{V^\infty} G^\infty_{im}(\mathbf{X}-\mathbf{Y})\,\dot{f}_m(\mathbf{Y})\,dV_\mathbf{Y}. \tag{8.3.8b}$$

This is the *unique* solution of the boundary-value problem[16] (8.3.7), valid for any $\dot{\mathbf{f}}$ (discontinuous or not) which renders the integral finite.

Suppose the body force-rate, $\dot{\mathbf{f}}$, is given by the divergence of a tensor field $\dot{\mathbf{S}}(\mathbf{X})$ which is sufficiently smooth in W but suffers a jump to **0** across ∂W, *i.e.*,

$$\dot{\mathbf{S}}(\mathbf{X}) = \begin{cases} \dot{\mathbf{S}}(\mathbf{X}) \neq \mathbf{0} & \mathbf{X} \text{ in W} \\ \mathbf{0} & \text{otherwise,} \end{cases} \tag{8.3.9a}$$

and

$$\dot{\mathbf{f}}(\mathbf{X}) = \boldsymbol{\nabla} \cdot \dot{\mathbf{S}}(\mathbf{X}) \qquad \mathbf{X} \text{ in W}. \tag{8.3.9b}$$

While $\dot{\mathbf{f}}(\mathbf{X})$ is finite within W, it behaves as a delta function across ∂W. This behavior is represented by concentrated forces distributed within a thin layer about ∂W, representing the jump in $\dot{\mathbf{S}}(\mathbf{X})$ across this boundary; the overall effect, therefore, is represented by additional tractions acting on W over its boundary ∂W. Denote these tractions by $[\dot{\mathbf{T}}](\mathbf{X})$ for **X** on ∂W, and obtain

$$\begin{aligned}[\dot{\mathbf{T}}](\mathbf{X}) &\equiv \mathbf{N}(\mathbf{X}) \cdot [\dot{\mathbf{S}}](\mathbf{X}) \\ &= -\mathbf{N}(\mathbf{X}) \cdot \{\lim_{\mathbf{X}^+ \to \mathbf{X}} \dot{\mathbf{S}}(\mathbf{X}^+)\} \qquad \mathbf{X} \text{ on } \partial\text{W},\end{aligned} \tag{8.3.9c}$$

where the minus sign is due to the fact that the unit outer normal **N** points from the inside toward the outside of W, and $\mathbf{X}^+$ is a point inside W. The resulting velocity field produced by body force-rate $\dot{\mathbf{f}}(\mathbf{X})$ distributed within W, and tractions $[\dot{\mathbf{T}}](\mathbf{X})$ acting on ∂W, *i.e.*, the velocity field corresponding to (8.3.9b) and (8.3.9c), is given by

$$\dot{\mathbf{x}}(\mathbf{X}) = \int_W \mathbf{G}^{\infty T}(\mathbf{Y}-\mathbf{X}) \cdot \dot{\mathbf{f}}(\mathbf{Y})\,dV_\mathbf{Y} + \int_{\partial W} \mathbf{G}^{\infty T}(\mathbf{Y}-\mathbf{X}) \cdot [\dot{\mathbf{T}}](\mathbf{Y})\,dS_\mathbf{Y}, \tag{8.3.10}$$

where the symmetry of $\mathbf{G}^\infty$, (8.3.5), is used. In view of (8.3.9b,c), use the Gauss theorem to rewrite (8.3.10), as

$$\dot{\mathbf{x}}(\mathbf{X}) = -\int_W \{\boldsymbol{\nabla}_\mathbf{Y} \otimes \mathbf{G}^{\infty T}(\mathbf{Y}-\mathbf{X})\} : \dot{\mathbf{S}}(\mathbf{Y})\,dV_\mathbf{Y} \tag{8.3.11a}$$

or, in component form,[17]

$$\dot{x}_i(\mathbf{X}) = -\int_W G^\infty_{ki,A}(\mathbf{Y}-\mathbf{X})\,\dot{S}_{Ak}(\mathbf{Y})\,dV_\mathbf{Y}, \tag{8.3.11b}$$

where subscript A following a comma denotes derivative with respect to Y_A.

---

[16] Uniqueness depends on $\mathcal{L}_{ij} \equiv \mathcal{F}_{AiBj}\,\partial^2(...)/(\partial X_A\,\partial X_B)$ being an elliptic operator; see Section 8.5.

[17] In view of the limited symmetries, the correct order of contracions must be preserved in the coordinate-independent notation.

8.3.4. The Eigenvelocity Gradient or Eigenstress Rate Problem

Consider an eigenvelocity gradient field $\dot{\mathbf{F}}^*$ (or eigenstress rate field $\dot{\mathbf{S}}^{N*}$) prescribed[18] in W and vanishing identically outside of W,

$$\dot{\mathbf{F}}^*(\mathbf{X}) = \begin{cases} \dot{\mathbf{F}}^*(\mathbf{X}) & \mathbf{X} \text{ in W} \\ \mathbf{0} & \text{otherwise} \end{cases} \tag{8.3.12a}$$

or

$$\dot{\mathbf{S}}^{N*}(\mathbf{X}) = \begin{cases} \dot{\mathbf{S}}^{N*}(\mathbf{X}) & \mathbf{X} \text{ in W} \\ \mathbf{0} & \text{otherwise.} \end{cases} \tag{8.3.12b}$$

Let $\dot{\mathbf{x}}^v$ (let $\dot{\mathbf{x}}^s$) be the velocity field produced by $\dot{\mathbf{F}}^*$ (by $\dot{\mathbf{S}}^{N*}$) and denote by $\dot{\mathbf{F}}^v$ and $\dot{\mathbf{S}}^{Nv}$ (by $\dot{\mathbf{F}}^s$ and $\dot{\mathbf{S}}^{Ns}$) the corresponding velocity-gradient and nominal stress-rate fields. These velocity gradient and nominal stress-rate fields satisfy

$$\dot{\mathbf{F}}^v = (\boldsymbol{\nabla} \otimes \dot{\mathbf{x}}^v)^T, \qquad \dot{\mathbf{S}}^{Nv} = \boldsymbol{\mathcal{F}} : \dot{\mathbf{F}}^v - \boldsymbol{\mathcal{F}} : \dot{\mathbf{F}}^*, \tag{8.3.13a,b}$$

and

$$\dot{\mathbf{F}}^s = (\boldsymbol{\nabla} \otimes \dot{\mathbf{x}}^s)^T, \qquad \dot{\mathbf{S}}^{Ns} = \boldsymbol{\mathcal{F}} : \dot{\mathbf{F}}^s + \dot{\mathbf{S}}^{N*}; \tag{8.3.13c,d}$$

see Section 8.2.

The divergence of the eigenvelocity gradient $\dot{\mathbf{F}}^*$ and eigenstress rate $\dot{\mathbf{S}}^{N*}$ can be regarded as *equivalent body force-rates.* Indeed, direct substitution of the constitutive relations (8.3.13b) and (8.3.13d) into the equations of equilibrium yields

$$\boldsymbol{\nabla} \cdot \dot{\mathbf{S}}^{Nv} = \boldsymbol{\nabla} \cdot \{\boldsymbol{\mathcal{F}} : (\boldsymbol{\nabla} \otimes \dot{\mathbf{x}}^v)^T\} + \boldsymbol{\nabla} \cdot (-\boldsymbol{\mathcal{F}} : \dot{\mathbf{F}}^*) = \mathbf{0} \tag{8.3.13e,f}$$

and

$$\boldsymbol{\nabla} \cdot \dot{\mathbf{S}}^{Ns} = \boldsymbol{\nabla} \cdot \{\boldsymbol{\mathcal{F}} : (\boldsymbol{\nabla} \otimes \dot{\mathbf{x}}^s)^T\} + \boldsymbol{\nabla} \cdot \dot{\mathbf{S}}^{N*} = 0. \tag{8.3.13g,h}$$

The divergence of $-\boldsymbol{\mathcal{F}} : \dot{\mathbf{F}}^*$ (of $\dot{\mathbf{S}}^{N*}$) appears like a distribution of body force-rates in the governing equations for the velocity field $\dot{\mathbf{x}}^v$ (field $\dot{\mathbf{x}}^s$); see (8.3.7a). Hence, the velocity field $\dot{\mathbf{x}}^v$ (field $\dot{\mathbf{x}}^s$) may be expressed in terms of the Green function $\mathbf{G}^\infty$, as

$$\dot{\mathbf{x}}^v(\mathbf{X}) = \int_W \{\boldsymbol{\nabla}_Y \otimes \mathbf{G}^{\infty T}(\mathbf{Y} - \mathbf{X})\} : \{\boldsymbol{\mathcal{F}} : \dot{\mathbf{F}}^*(\mathbf{Y})\}\, dV_Y \tag{8.3.14a}$$

and

$$\dot{\mathbf{x}}^s(\mathbf{X}) = -\int_W \{\boldsymbol{\nabla}_Y \otimes \mathbf{G}^{\infty T}(\mathbf{Y} - \mathbf{X})\} : \dot{\mathbf{S}}^{N*}(\mathbf{Y})\, dV_Y \tag{8.3.14b}$$

or, in component form,

$$\dot{x}_i^v(\mathbf{X}) = \int_W G^\infty_{ki,A}(\mathbf{Y} - \mathbf{X})\, \mathcal{F}_{AkBj}\, \dot{F}^*_{jB}(\mathbf{Y})\, dV_Y \tag{8.3.14c}$$

and

---

[18] These are two separate problems which are being examined simultaneously.

$$\dot{x}_i^s(\mathbf{X}) = -\int_W G^{\infty}_{ki,A}(\mathbf{Y}-\mathbf{X})\,\dot{S}^{N*}_{Ak}(\mathbf{Y})\,dV_Y, \tag{8.3.14d}$$

where comma followed by the subscript A denotes differentiation with respect to Y_A. The velocity field $\dot{\mathbf{x}}^v$ (field $\dot{\mathbf{x}}^s$) satisfies $\dot{\mathbf{x}}^v(\mathbf{Z}) \rightarrow \mathbf{0}$, ($\dot{\mathbf{x}}^s(\mathbf{Z}) \rightarrow \mathbf{0}$), as $|\mathbf{Z}| \rightarrow \infty$, and is continuous across ∂W.

In general, the velocity-gradient and nominal stress-rate fields, $\dot{\mathbf{F}}^v$ and $\dot{\mathbf{S}}^{Nv}$, (fields $\dot{\mathbf{F}}^s$ and $\dot{\mathbf{S}}^{Ns}$) are discontinuous across surfaces where the eigenvelocity gradient (eigenstress rate) admits finite discontinuities; the velocity field $\dot{\mathbf{x}}^v$ (field $\dot{\mathbf{x}}^s$) is, of course, continuous, as are the traction-rates $\mathbf{N} \cdot \dot{\mathbf{S}}^{Nv}$ (traction-rates $\mathbf{N} \cdot \dot{\mathbf{S}}^{Ns}$).

Using the velocity gradient and the constitutive relations given by (8.3.13c,d), define the *integral operator* $\mathcal{S}^{\infty}$ (operator $\mathcal{T}^{\infty}$) which determines $\dot{\mathbf{F}}^v$ (determines $\dot{\mathbf{S}}^{Ns}$), in terms of $\dot{\mathbf{F}}^*$ (of $\dot{\mathbf{S}}^{N*}$), as follows:

$$\mathcal{S}^{\infty}(\mathbf{X}; \dot{\mathbf{F}}^*) \equiv \int_W \mathbf{\Gamma}^{\infty}(\mathbf{Y}-\mathbf{X}) : \mathcal{F} : \dot{\mathbf{F}}^*(\mathbf{Y})\,dV_Y \tag{8.3.15a}$$

and

$$\mathcal{T}^{\infty}(\mathbf{X}; \dot{\mathbf{S}}^{N*}) \equiv -\mathcal{F} : \{\int_W \mathbf{\Gamma}^{\infty}(\mathbf{Y}-\mathbf{X}) : \dot{\mathbf{S}}^{N*}(\mathbf{Y})\,dV_Y\} + \dot{\mathbf{S}}^{N*}(\mathbf{X}), \tag{8.3.15b}$$

or, in component form,

$$\mathcal{S}^{\infty}_{iA}(\mathbf{X}; \dot{\mathbf{F}}^*) \equiv \int_W \Gamma^{\infty}_{iAmC}(\mathbf{Y}-\mathbf{X})\,\mathcal{F}_{CmBl}\,\dot{F}^*_{lB}(\mathbf{Y})\,dV_Y \tag{8.3.15c}$$

and

$$\mathcal{T}^{\infty}_{Ai}(\mathbf{X}; \dot{\mathbf{S}}^{N*}) \equiv -\mathcal{F}_{AiBj}\{\int_W \Gamma^{\infty}_{jBkC}(\mathbf{Y}-\mathbf{X})\,\dot{S}^{N*}_{Ck}(\mathbf{Y})\,dV_Y\} + \dot{S}^{N*}_{Ai}(\mathbf{X}). \tag{8.3.15d}$$

In (8.3.15), the tensor field $\mathbf{\Gamma}^{\infty}(\mathbf{Y}-\mathbf{X})$ is defined by

$$\Gamma^{\infty}_{iAjB}(\mathbf{Y}-\mathbf{X}) \equiv -\frac{1}{2}\{G^{\infty}_{ij,AB}(\mathbf{Y}-\mathbf{X}) + G^{\infty}_{ji,BA}(\mathbf{Y}-\mathbf{X})\}, \tag{8.3.15e}$$

where comma followed by the subscripts A and B denotes differentiation with respect to Y_A and Y_B or X_A and X_B. In view of the symmetries of the Green function, (8.3.6), the tensor field Γ^{∞}_{iAjB} has the following symmetries:

$$\Gamma^{\infty}_{iAjB} = \Gamma^{\infty}_{jAiB} = \Gamma^{\infty}_{jBiA} = \Gamma^{\infty}_{iBjA}. \tag{8.3.15f}$$

Hence, while $\mathcal{F}_{AiBj}$, in general, may have, at most, 45 distinct components, Γ^{∞}_{iAjB} has, at most, only 36 distinct[19] components. In (8.3.2a), which defines the Green function, only the following combination of the components of the pseudo-modulus tensor is involved: $(\mathcal{F}_{AiBj} + \mathcal{F}_{BiAj}) / 2$. In light of the symmetry with respect to Ai and Bj, it is seen that only 36 distinct combinations of the components of $\mathcal{F}_{AiBj}$ enter the solution of the Green function for an unbounded domain. This combination is, $(\mathcal{F}_{AiBj} + \mathcal{F}_{BiAj} + \mathcal{F}_{AjBi} + \mathcal{F}_{BjAi}) / 4$.

---

[19] There are six distinct components for each fixed pair of A and B, or for each fixed pair of i and j. Hence, there are only 36 possible distinct components for Γ^{∞}_{iAjB}, i, j, A, B = 1, 2, 3.

When the eigenfields $\dot{\mathbf{F}}^*$ and $\dot{\mathbf{S}}^{N*}$ *correspond to each other* in the sense that

$$\dot{\mathbf{S}}^{N*}(\mathbf{X}) = -\mathcal{F} : \dot{\mathbf{F}}^*(\mathbf{X}) \tag{8.3.16a}$$

or

$$\dot{\mathbf{F}}^*(\mathbf{X}) = -\mathcal{G} : \dot{\mathbf{S}}^{N*}(\mathbf{X}) \qquad \mathbf{X} \text{ in W}, \tag{8.3.16b}$$

then the resulting velocity fields, $\dot{\mathbf{x}}^v$ and $\dot{\mathbf{x}}^s$, are the same, as are the resulting velocity-gradient and nominal stress-rate fields. Then (8.3.13a) and (8.3.13c) agree, and

$$\dot{\mathbf{S}}^{Nv} = \mathcal{F} : (\dot{\mathbf{F}}^v - \dot{\mathbf{F}}^*) = \mathcal{F} : \dot{\mathbf{F}}^s + \dot{\mathbf{S}}^{N*} = \dot{\mathbf{S}}^{Ns}. \tag{8.3.17}$$

Therefore, the integral operators $\mathcal{S}^\infty$ and $\mathcal{T}^\infty$ satisfy

$$\mathcal{T}^\infty(\mathbf{X}; -\mathcal{F} : \dot{\mathbf{F}}^*) = \mathcal{F} : \{\mathcal{S}^\infty(\mathbf{X}; \dot{\mathbf{F}}^*) - \dot{\mathbf{F}}^*(\mathbf{X})\} \tag{8.3.18a}$$

and

$$\mathcal{S}^\infty(\mathbf{X}; -\mathcal{G} : \dot{\mathbf{S}}^{N*}) = \mathcal{G} : \{\mathcal{T}^\infty(\mathbf{X}; \dot{\mathbf{S}}^{N*}) - \dot{\mathbf{S}}^{N*}(\mathbf{X})\}, \tag{8.3.18b}$$

provided that $\dot{\mathbf{S}}^{N*}$ and $\dot{\mathbf{F}}^*$ are related by (8.3.16a,b). Equations (8.3.18a,b) are the *equivalence relations* between the integral operators $\mathcal{S}^\infty$ and $\mathcal{T}^\infty$.

8.3.5. Generalized Eshelby Tensor and its Conjugate

In the preceding subsection, the integral operator $\mathcal{S}^\infty_{iA}(\mathbf{X}; \dot{\mathbf{F}}^*)$ and its conjugate $\mathcal{T}^\infty_{Ai}(\mathbf{X}; \dot{\mathbf{S}}^{N*})$, are expressed in terms of the tensor field $\Gamma^\infty_{jBkC}(\mathbf{Y}-\mathbf{X})$ which is given by the gradient of the Green function $G^\infty_{ij}(\mathbf{Z})$ of the unbounded domain V^∞; see (8.3.15a,c) and (8.3.15b,d). The integral of $\Gamma^\infty_{jBkC}(\mathbf{Y}-\mathbf{X})$ over an ellipsoidal domain Ω turns out to be constant, as shown in Subsection 8.5.3. From this, the generalized Eshelby tensor $\mathcal{S}^\Omega$ and its conjugate $\mathcal{T}^\Omega$ are obtained.

Define the tensor field

$$\mathcal{P}_{iAjB}(\mathbf{X}; \mathrm{W}) \equiv \int_\mathrm{W} \Gamma^\infty_{iAjB}(\mathbf{Y}-\mathbf{X})\, dV_\mathbf{Y}. \tag{8.3.19a}$$

In terms of $\mathcal{P}_{iAjB}(\mathbf{X}; \mathrm{W})$, the generalized Eshelby tensor $\mathcal{S}^\Omega$ for an ellipsoidal domain is given by,

$$\mathcal{S}^\Omega_{iABj} = \mathcal{P}^\Omega_{iAkC}\, \mathcal{F}_{CkBj}, \tag{8.3.19b}$$

where

$$\mathcal{P}_{iAjB}(\mathbf{X}; \Omega) \equiv \mathcal{P}^\Omega_{iAjB} = \text{constant} \qquad \text{for } \mathbf{X} \text{ in } \Omega; \tag{8.3.19c}$$

see (8.2.25c,d,g,h) and (8.3.15). There are, at most, 36 distinct components for $\mathcal{P}^\Omega_{iAjB}$; see (8.3.15f). From the symmetry, $\mathcal{P}^\Omega_{iAjB} = \mathcal{P}^\Omega_{jBiA}$, it follows that the generalized Eshelby tensor satisfies the following two symmetry properties:

$$\mathcal{S}^\Omega_{iABj}\, \mathcal{G}_{jBkC} = \mathcal{S}^\Omega_{kCBj}\, \mathcal{G}_{jBiA}\, (= \mathcal{P}^\Omega_{iAkC}) \tag{8.3.19d}$$

and

$$\mathcal{F}_{AiCk}\,\mathcal{S}^{\Omega}_{kCBj} = \mathcal{F}_{BjCk}\,\mathcal{S}^{\Omega}_{kCAi}\,(= \mathcal{F}_{BjCk}\,\mathcal{P}^{\Omega}_{kCID}\,\mathcal{F}_{DIAi})\,. \tag{8.3.19e}$$

Consider now the integral operator $\boldsymbol{\mathcal{T}}^{\infty}(\mathbf{X};\,\dot{\mathbf{S}}^{N*})$ given by (8.3.15b), and using the equivalence relations (8.3.18a), define the tensor field $\mathcal{T}^{\infty}_{AijB}(\mathbf{X};\,W)$ for a finite domain W embedded in an unbounded domain V^{∞} of the uniform modulus tensor $\boldsymbol{\mathcal{F}}$, as follows:

$$\mathcal{T}^{\infty}_{AijB}(\mathbf{X};\,W) \equiv -\,\mathcal{F}_{AiCk}\int_{W}\Gamma^{\infty}_{kCjB}(\mathbf{Y}-\mathbf{X})\,dV_{\mathbf{Y}} + H(\mathbf{X};\,W)\,\delta_{AB}\,\delta_{ij} \tag{8.3.20a}$$

which, in terms of $\mathcal{P}_{AiBj}(\mathbf{X};\,W)$, becomes

$$\mathcal{T}^{\infty}_{AijB}(\mathbf{X};\,W) = -\,\mathcal{F}_{AiCk}\,\mathcal{P}_{kCjB}(\mathbf{X};\,W) + H(\mathbf{X};\,W)\,\delta_{AB}\,\delta_{ij}\,, \tag{8.3.20b}$$

where $H(\mathbf{X};\,W)$ is the Heaviside step function[20].

When W is an ellipsoid, Ω, $\boldsymbol{\mathcal{T}}^{\infty}(\mathbf{X};\,\Omega)$ with components $\mathcal{T}^{\infty}_{AijB}(\mathbf{X};\,\Omega)$, is constant for $\mathbf{X}$ in Ω. It is then denoted by $\boldsymbol{\mathcal{T}}^{\Omega}$, or in component form, by $\mathcal{T}^{\Omega}_{AijB}$. Indeed, from (8.2.17b),

$$\mathcal{T}^{\Omega}_{AijB} = -\,\mathcal{F}_{AiCk}\,\mathcal{P}^{\Omega}_{kCjB} + \delta_{AB}\,\delta_{ij}\,. \tag{8.3.20c}$$

The tensor $\mathcal{T}^{\Omega}_{AijB}$ is *conjugate* to the generalized Eshelby tensor $\mathcal{S}^{\Omega}_{iABj}$. The equivalence relations between $\mathcal{S}^{\Omega}_{iABj}$ and $\mathcal{T}^{\Omega}_{AijB}$ are given by (8.2.17c-f). From the symmetry, $\mathcal{P}^{\Omega}_{iAjB} = \mathcal{P}^{\Omega}_{jBiA}$, it follows that the generalized conjugate Eshelby tensor satisfies the following two symmetry properties:

$$\mathcal{G}_{iAkC}\,\mathcal{T}^{\Omega}_{CkjB} = \mathcal{G}_{jBkC}\,\mathcal{T}^{\Omega}_{CkiA} \tag{8.3.20d}$$

and

$$\mathcal{T}^{\Omega}_{AikC}\,\mathcal{F}_{CkBj} = \mathcal{T}^{\Omega}_{BjkC}\,\mathcal{F}_{CkAi}\,. \tag{8.3.20e}$$

8.4. AVERAGE QUANTITIES

For an ellipsoidal region, Ω, the tensor field $\boldsymbol{\mathcal{P}}(\mathbf{X};\,\Omega)$ is constant when $\mathbf{X}$ is in Ω. This leads to a number of interesting *exact* results, similar to those for linearly elastic solids; see Nemat-Nasser and Hori (1993, 1999), Section 11. The similarity, however, is only in the form of the expressions. Therefore, the differences must be carefully noted. In the present work, the focus is on *finite inelastic deformations of solids, involving large strains, rotations, strain rates, and spins.* The instantaneous modulus and compliance tensors, *i.e.*, $\boldsymbol{\mathcal{F}}$ and $\boldsymbol{\mathcal{G}}$, display minimal symmetries, and do not possess the properties that generally exist in linear elasticity. Moreover, *in general, the results for the finite*

[20] Function $H(\mathbf{X};\,W)$ is also called the characteristic function of W.

deformation case are valid only when appropriate kinematical and dynamical variables are used. As is evident from the results of the preceding sections, *the velocity gradient and the nominal stress rate,* $\dot{\mathbf{F}}(\mathbf{X})$ *and* $\dot{\mathbf{S}}^N$, *are indeed suitable quantities for this purpose.* The reference configuration can be arbitrary, but it must be fixed. Other strain rate and stress rate tensors *do not*, in general, have similar characteristics, and, hence, may not lead to similar simple and powerful results. On the other hand, no such distinction is necessary in linear elasticity.

8.4.1. Double-inclusion Problem

Consider an arbitrary finite region W in an unbounded domain V^∞ of pseudo-modulus tensor $\mathcal{F}$. The region W may consist of several disconnected subregions, say, W_α, $\alpha = 1, 2, ..., n$, of arbitrary shape.

Suppose that *an arbitrary eigenvelocity-gradient field*, $\dot{\mathbf{F}}^*(\mathbf{X})$, is distributed in W. Let Ω be an arbitrary ellipsoidal domain in V^∞, such that W is totally contained within Ω; see Figure 8.4.1. Then, the *average* velocity gradient and the corresponding *average* nominal stress rate, taken over domain Ω, are completely determined by the generalized Eshelby tensor for Ω.

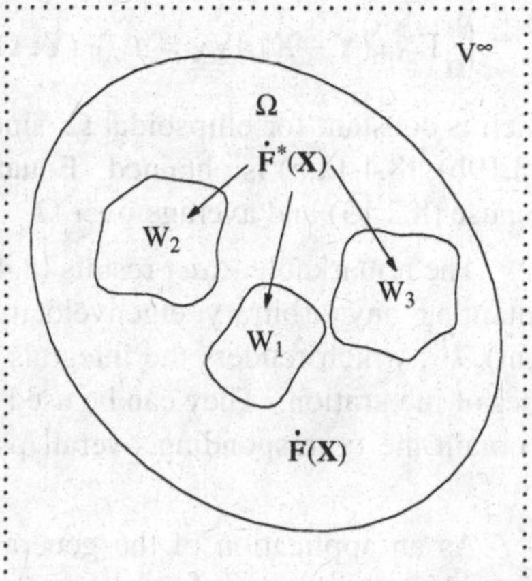

Figure 8.4.1

An ellipsoidal Ω in an unbounded uniform V^∞, contains $W = W_1 + W_2 + W_3$; arbitrary eigenvelocity gradients $\dot{\mathbf{F}}^*(\mathbf{X})$ are distributed within W

Indeed, denoting the average value of $\dot{\mathbf{F}}^*(\mathbf{X})$ over W by $< \dot{\mathbf{F}}^* >_W$, and those of $\dot{\mathbf{F}}(\mathbf{X})$ and $\dot{\mathbf{S}}^N(\mathbf{X})$ over Ω by $< \dot{\mathbf{F}} >_\Omega$ and $< \dot{\mathbf{S}}^N >_\Omega$, respectively, obtain

$$< \dot{\mathbf{F}} >_\Omega = \frac{W}{\Omega}\, \mathcal{S}^\Omega : < \dot{\mathbf{F}}^* >_W, \tag{8.4.1a}$$

$$< \dot{\mathbf{S}}^N >_\Omega = \frac{W}{\Omega}\, \mathcal{F} : (\mathcal{S}^\Omega - \mathbf{1}^{(4)}) : < \dot{\mathbf{F}}^* >_W \tag{8.4.2a}$$

or, in component form,

$$< \dot{F}_{iA} >_\Omega = \frac{W}{\Omega}\, \mathcal{S}^\Omega_{iABj} < \dot{F}^*_{jB} >_W, \tag{8.4.1b}$$

$$\dot{S}^{N}_{Ai} = \frac{W}{\Omega}\, \mathcal{F}_{AiBj}\,(\mathcal{S}_{jBCk} - \delta_{jk}\,\delta_{BC}) < \dot{F}^{*}_{kC} >_{W}, \qquad (8.4.2b)$$

where W and Ω stand for their volumes.

Consider the proof of this result, using the tensor field $\mathcal{P}_{AiBj}(\mathbf{X}; W)$. For any eigenvelocity gradient $\dot{\mathbf{F}}^{*}(\mathbf{X})$ distributed in the finite domain W within an unbounded uniform region V^{∞}, the resulting velocity-gradient field is

$$\dot{F}_{iA}(\mathbf{X}) = \int_{W} \Gamma^{\infty}_{iAjB}(\mathbf{Y}-\mathbf{X})\, \mathcal{F}_{BjCk}\, \dot{F}^{*}_{kC}(\mathbf{Y})\, dV_{Y}. \qquad (8.4.3)$$

Therefore, the average velocity gradient in the ellipsoidal domain Ω, becomes

$$< \dot{F}_{iA} >_{\Omega} = \frac{1}{\Omega} \int_{\Omega} \{ \int_{W} \Gamma^{\infty}_{iAjB}(\mathbf{Y}-\mathbf{X})\, \mathcal{F}_{BjCk}\, \dot{F}^{*}_{kC}(\mathbf{Y})\, dV_{Y} \}\, dV_{\mathbf{X}}, \qquad (8.4.4a)$$

where $\mathbf{X}$ and $\mathbf{Y}$ are in Ω and W, respectively. When the integrand $\Gamma^{\infty}_{iAjB}(\mathbf{X}-\mathbf{Y})$ in (8.4.4a) is not singular, then with the minimal assumption of integrability, the order of integration can be changed,

$$\int_{\Omega} \{ \int_{W} \Gamma^{\infty}_{iAjB}(\mathbf{Y}-\mathbf{X})\, \mathcal{F}_{BjCk}\, \dot{F}^{*}_{kC}(\mathbf{Y})\, dV_{Y} \}\, dV_{\mathbf{X}} =$$

$$\int_{W} \{ \int_{\Omega} \Gamma^{\infty}_{iAjB}(\mathbf{Y}-\mathbf{X})\, dV_{\mathbf{X}} \}\, \mathcal{F}_{BjCk}\, \dot{F}^{*}_{kC}(\mathbf{Y})\, dV_{Y}. \qquad (8.4.4b)$$

The inner integral in the right-hand side is given by

$$\int_{\Omega} \Gamma^{\infty}_{iAjB}(\mathbf{Y}-\mathbf{X})\, dV_{\mathbf{X}} = \mathcal{P}_{iAjB}(\mathbf{Y}; \Omega) \equiv \mathcal{P}^{\Omega}_{iAjB} \qquad (8.4.4c)$$

which is constant for ellipsoidal Ω, since $\mathbf{Y}$ is in $W \subset \Omega$. Therefore, in view of (8.3.19b), (8.4.1a,b) is obtained. Equations (8.4.2a,b) then follow by definition; *e.g.*, use (8.2.13) and average over Ω.

The remarkable *exact* results (8.4.1) and (8.4.2) are valid for any finite W, containing any arbitrary eigenvelocity gradient (transformation velocity gradient), $\dot{\mathbf{F}}^{*}$, which renders the integrals finite and permits the interchange of the order of integration. They can be used to homogenize a heterogeneous solid and to obtain the corresponding overall pseudo-moduli, as is discussed in Section 8.6.

As an application of the general results (8.4.1) and (8.4.2), consider an ellipsoid Ω_1 within another ellipsoid Ω_2, embedded in a uniform unbounded domain V^{∞} of pseudo-modulus tensor $\boldsymbol{\mathcal{F}}$; see Figure 8.4.2. Let Ω_1 contain an arbitrary region W, in which an arbitrary eigenvelocity gradient $\dot{\mathbf{F}}^{*}(\mathbf{X})$ is distributed. Under similar minimal assumptions, and using a similar procedure which produced (8.4.1), it can easily be shown that the average value of the velocity gradient over the annulus $\Omega_2 - \Omega_1$, is given by

$$< \dot{\mathbf{F}} >_{\Omega_2 - \Omega_1} = \frac{W}{\Omega_2 - \Omega_1} \{ \boldsymbol{\mathcal{S}}(\Omega_2) - \boldsymbol{\mathcal{S}}(\Omega_1) \} : < \dot{\mathbf{F}}^{*} >_{W} \qquad (8.4.5a)$$

or, in component form, by

$$< \dot{F}_{iA} >_{\Omega_2 - \Omega_1} = \frac{W}{\Omega_2 - \Omega_1} \{ \mathcal{S}_{iABj}(\Omega_2) - \mathcal{S}_{iABj}(\Omega_1) \} < \dot{F}^{*}_{jB} >_{W}, \qquad (8.4.5b)$$

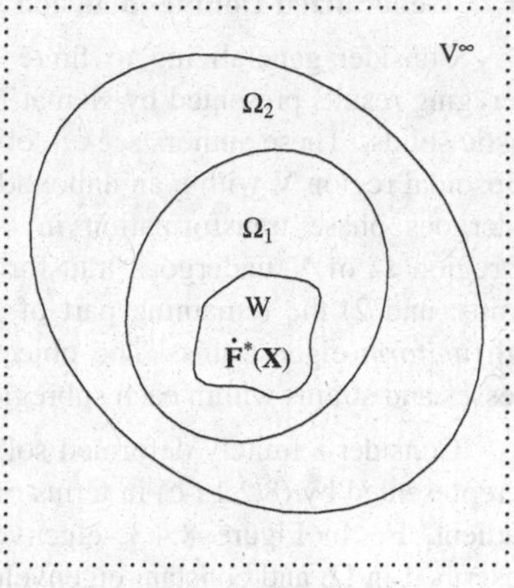

Figure 8.4.2

An ellipsoidal Ω_2 in an unbounded uniform V^∞, contains an ellipsoidal Ω_1 which contains an arbitrary region W; arbitrary eigenvelocity gradients $\dot{\mathbf{F}}^*(\mathbf{X})$ are distributed within W

where $\boldsymbol{\mathcal{S}}(\Omega_1)$ and $\boldsymbol{\mathcal{S}}(\Omega_2)$ are the generalized Eshelby tensors corresponding to the ellipsoidal domains Ω_1 and Ω_2, respectively. The average stress rate in $\Omega_2 - \Omega_1$ is given by

$$< \dot{\mathbf{S}}^N >_{\Omega_2 - \Omega_1} = \frac{W}{\Omega_2 - \Omega_1} \boldsymbol{\mathcal{F}} : \{\boldsymbol{\mathcal{S}}(\Omega_2) - \boldsymbol{\mathcal{S}}(\Omega_1)\} : < \dot{\mathbf{F}}^* >_W . \tag{8.4.5c}$$

Equations (8.4.5a,b,c) are the finite-deformation version of the so-called Tanaka-Mori result (Tanaka and Mori, 1972; see also, Nemat-Nasser and Hori, 1993 and 1999, Section 11).

The generalized Eshelby tensor depends on both the shape and the orientation of the ellipsoid. However, only the ratios of the axes enter the components of this tensor in a coordinate system coincident with the directions of the principal axes of the ellipsoid. Therefore, if the two ellipsoidal regions Ω_1 and Ω_2 have the same shape and orientation, *i.e.*, if the corresponding principal axes have common ratios and directions, and since both are embedded in the same unbounded domain, then

$$\boldsymbol{\mathcal{S}}(\Omega_2) = \boldsymbol{\mathcal{S}}(\Omega_1) . \tag{8.4.6a}$$

Hence, the average velocity gradient (and, hence, the average nominal stress rate) in $\Omega_2 - \Omega_1$ vanishes,

$$< \dot{\mathbf{F}} >_{\Omega_2 - \Omega_1} = \mathbf{0}, \qquad < \dot{\mathbf{S}}^N >_{\Omega_2 - \Omega_1} = \mathbf{0} . \tag{8.4.6b,c}$$

These *exact* results hold for any W of any arbitrary shape, containing any arbitrary eigenvelocity gradient, $\dot{\mathbf{F}}^*(\mathbf{X})$.

8.4.2. Generalized Double-inclusion Problem

Consider generalizing to finite deformations and rotations, some exact averaging results presented by Nemat-Nasser and Hori (1993, 1995) for linearly elastic solids. These authors seek to obtain the average strains and stresses in an ellipsoidal region V within an unbounded uniform linearly elastic solid, when V undergoes phase transformation in the following manner: 1) an ellipsoidal subregion Ω of V undergoes transformation corresponding to *variable* eigenstrains; and 2) the remaining part of V, $\Gamma \equiv V - \Omega$, undergoes transformation with *uniform* eigenstrains. The objective is to calculate exactly, the average stresses and strains within each subregion.

Consider a finitely deformed solid of any constitutive property which can be represented by (8.2.1a-c) in terms of the nominal stress rate, $\dot{\mathbf{S}}^{\mathrm{N}}$, and velocity gradient, $\dot{\mathbf{F}}$. In Figure 8.4.3, eigenvelocity gradients (variable), $\dot{\mathbf{F}}^{*1}(\mathbf{X})$, are prescribed in Ω, and constant eigenvelocity gradients, $\dot{\mathbf{F}}^{*2}$, are prescribed in the remaining part, $V - \Omega$. Ellipsoids Ω and V need not be coaxial or similar. The objective is to calculate exactly, the average nominal stress rate and the average velocity gradient within each subregion.

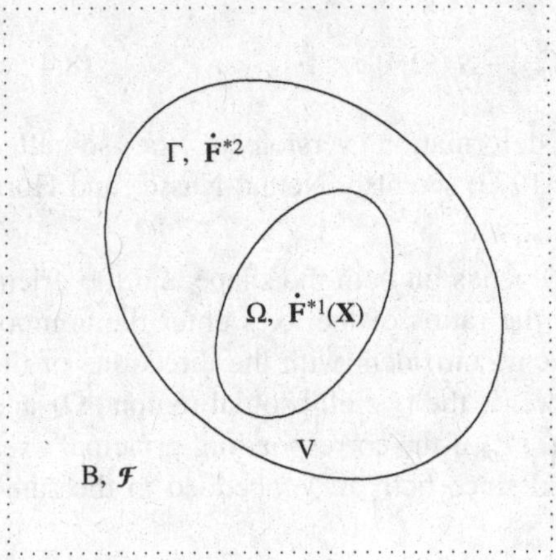

Figure 8.4.3

A finitely (uniformly) deformed unbounded uniform solid B contains two ellipsoidal regions, V and Ω with $\Omega \subset$ V; variable eigenvelocity gradients $\dot{\mathbf{F}}^{*1}(\mathbf{X})$ and uniform eigenvelocity gradients $\dot{\mathbf{F}}^{*2}$ are distributed in Ω and $\Gamma = V - \Omega$, respectively

Although the resulting velocity gradient and nominal stress-rate fields, $\dot{\mathbf{F}}(\mathbf{X})$ and $\dot{\mathbf{S}}^{\mathrm{N}}(\mathbf{X})$, in general are *not constant in* V and Ω, the average velocity gradient and nominal stress rate in Ω are *exactly* given by

$$< \dot{\mathbf{F}} >_\Omega = \mathcal{S}^\Omega : < \dot{\mathbf{F}}^{*1} >_\Omega + (\mathcal{S}^{\mathrm{V}} - \mathcal{S}^\Omega) : \dot{\mathbf{F}}^{*2},$$

$$< \dot{\mathbf{S}}^{\mathrm{N}} >_\Omega = \mathcal{F} : (\mathcal{S}^\Omega - \mathbf{1}^{(4)}) : < \dot{\mathbf{F}}^{*1} >_\Omega + \mathcal{F} : (\mathcal{S}^{\mathrm{V}} - \mathcal{S}^\Omega) : \dot{\mathbf{F}}^{*2}, \qquad (8.4.7a,b)$$

where $\mathcal{S}^{\mathrm{V}}$ and $\mathcal{S}^\Omega$ are the generalized Eshelby tensors for ellipsoids V and Ω, respectively. Similarly, the average velocity gradient and nominal stress rate over V are *exactly* given by

$$<\dot{\mathbf{F}}>_V = \mathcal{S}^V : \{f<\dot{\mathbf{F}}^{*1}>_\Omega + (1-f)\,\dot{\mathbf{F}}^{*2}\},$$

$$<\dot{\mathbf{S}}^N>_V = \mathcal{F} : (\mathcal{S}^V - \mathbf{1}^{(4)}) : \{f<\dot{\mathbf{F}}^{*1}>_\Omega + (1-f)\,\dot{\mathbf{F}}^{*2}\}, \qquad (8.4.7c,d)$$

where f is the volume fraction of Ω in V, $f = \Omega / V$. Finally, in view of (8.4.7a-d), the average velocity gradient and nominal stress rate over the annulus Γ, are given by

$$<\dot{\mathbf{F}}>_\Gamma = \mathcal{S}^V : \dot{\mathbf{F}}^{*2} + \frac{f}{1-f}\,(\mathcal{S}^V - \mathcal{S}^\Omega) : (<\dot{\mathbf{F}}^{*1}>_\Omega - \dot{\mathbf{F}}^{*2}),$$

$$<\dot{\mathbf{S}}^N>_\Gamma = \mathcal{F} : (\mathcal{S}^V - \mathbf{1}^{(4)}) : \dot{\mathbf{F}}^{*2} + \frac{f}{1-f}\,\mathcal{F} : (\mathcal{S}^V - \mathcal{S}^\Omega) : (<\dot{\mathbf{F}}^{*1}>_\Omega - \dot{\mathbf{F}}^{*2}). \qquad (8.4.7e,f)$$

Again, these equations are *exact*.

The proof of (8.4.7a-f) directly follows from the result presented in the preceding subsection. The volume average of the velocity gradient produced by $\dot{\mathbf{F}}^{*1}(\mathbf{X})$ in Ω, is

$$<\text{(velocity gradient due to } \dot{\mathbf{F}}^{*1})>_D = \begin{cases} \mathcal{S}^\Omega : <\dot{\mathbf{F}}^{*1}>_\Omega & D = \Omega \\ f\,\mathcal{S}^V : <\dot{\mathbf{F}}^{*1}>_\Omega & D = V \end{cases} \qquad (8.4.8a)$$

and the volume average of the corresponding nominal stress-rate field is

$$<\text{(nominal stress rate due to } \dot{\mathbf{F}}^{*1})>_D = \begin{cases} \mathcal{F} : (\mathcal{S}^\Omega - \mathbf{1}^{(4)}) : <\dot{\mathbf{F}}^{*1}>_\Omega & D = \Omega \\ f\,\mathcal{F} : (\mathcal{S}^V - \mathbf{1}^{(4)}) : <\dot{\mathbf{F}}^{*1}>_\Omega & D = V. \end{cases} \qquad (8.4.8b)$$

Fields due to the constant $\dot{\mathbf{F}}^{*2}$ in Γ can be obtained by superposing the fields due to the constant $-\dot{\mathbf{F}}^{*2}$ distributed over the entire Ω and the fields due to the constant $\dot{\mathbf{F}}^{*2}$ distributed over the entire V. Hence, the volume average of the velocity gradient and nominal stress-rate fields produced by $\dot{\mathbf{F}}^{*2}$ in Γ are computed by applying expressions (8.4.1,2) separately to the fields due to $-\dot{\mathbf{F}}^{*2}$ in Ω and $\dot{\mathbf{F}}^{*2}$ in V, *i.e.*,

$$<\text{(velocity gradient due to } \dot{\mathbf{F}}^{*2})>_D = \begin{cases} (\mathcal{S}^V - \mathcal{S}^\Omega) : \dot{\mathbf{F}}^{*2} & D = \Omega \\ (1-f)\,\mathcal{S}^V : \dot{\mathbf{F}}^{*2} & D = V, \end{cases}$$

$$<\text{(nominal stress rate due to } \dot{\mathbf{F}}^{*2})>_D = \begin{cases} \mathcal{F} : (\mathcal{S}^V - \mathcal{S}^\Omega) : \dot{\mathbf{F}}^{*2} & D = \Omega \\ (1-f)\,\mathcal{F} : (\mathcal{S}^V - \mathbf{1}^{(4)}) : \dot{\mathbf{F}}^{*2} & D = V. \end{cases} \qquad (8.4.8c,d)$$

These expressions are the generalized version of equations (10.4.17c,d) of Nemat-Nasser and Hori (1993, 1999).[21] The volume averages of $\dot{\mathbf{F}}$ and $\dot{\mathbf{S}}^N$ taken over Ω and V, (8.4.7a-d), are obtained directly from (8.4.8a-d).

---

[21] In Nemat-Nasser and Hori (1993), there are typographical errors (superscripts) in their equations (10.4.17c,d), page 349. Their (10.4.17d) should read:

8.4.3. A Nested Sequence of Transforming Ellipsoidal Regions

The above results can be generalized to the case when V consists of a series of annulus subregions, *i.e.,* a nested sequence of ellipsoidal regions, where, in each of the annuli, a distinct but constant eigenvelocity gradient is prescribed. As in the case of linear elasticity (Nemat-Nasser and Hori, 1993, 1995), the resulting average field quantities can be computed exactly, using the procedure outlined in the preceding subsection.

To show this, consider a nested series of ellipsoidal regions, Ω_α ($\alpha = 1, 2, ..., m$), with $\Omega_m \equiv V$, which satisfy $\Omega_1 \subset \Omega_2 \subset ... \subset \Omega_m$, and denote the annulus between Ω_α and $\Omega_{\alpha-1}$ by $\Gamma_\alpha \equiv \Omega_\alpha - \Omega_{\alpha-1}$ ($\alpha = 2, 3, ..., m$); see Figure 8.4.4. These ellipsoids need not be coaxial or similar. Now, consider the following distribution of eigenvelocity gradients:

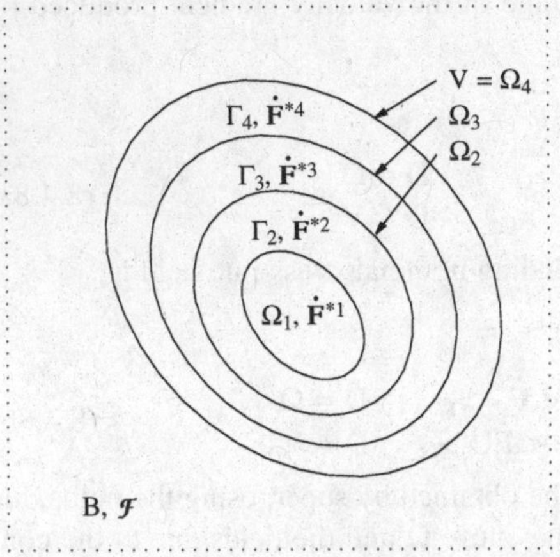

Figure 8.4.4

A nested sequence of 4 inclusions embedded in infinite domain B; Ω_α with eigenvelocity gradients $\dot{\mathbf{F}}^{*\alpha}$ ($\alpha = 1, 2, 3, 4$); B with pseudo-modulus tensor $\mathcal{F}$

$$\dot{\mathbf{F}}^{*}(\mathbf{X}) = \mathrm{H}(\mathbf{X};\,\Omega_1)\,\dot{\mathbf{F}}^{*1}(\mathbf{X}) + \sum_{\alpha=2}^{m} \mathrm{H}(\mathbf{X};\,\Gamma_\alpha)\,\dot{\mathbf{F}}^{*\alpha}, \tag{8.4.9}$$

where each $\dot{\mathbf{F}}^{*\alpha}$ ($\alpha = 2, 3, ..., m$) is constant, and $\mathrm{H}(\mathbf{X};\,\mathrm{D})$ is the Heaviside step function, taking on the value 1 when $\mathbf{X}$ is in D, and 0 otherwise; note that $\dot{\mathbf{F}}^{*1}(\mathbf{X})$ need not be constant, and that each annulus has a different but constant eigenvelocity gradient. The resulting velocity gradient and nominal stress-rate fields are denoted by $\dot{\mathbf{F}}$ and $\dot{\mathbf{S}}^{\mathrm{N}}$, respectively.

Apply expressions (8.4.1,2) to the fields produced by $\dot{\mathbf{F}}^{*1}(\mathbf{X})$ in Ω_1. The volume average of the resulting velocity-gradient field over any $\Omega_\alpha \supseteq \Omega_1$, then

---

$$<(\text{stress due to } \boldsymbol{\varepsilon}^{*2})>_{\mathrm{D}} = \begin{cases} \mathbf{C} : (\mathbf{S}^{\mathrm{V}} - \mathbf{S}^{\Omega}) : \boldsymbol{\varepsilon}^{*2} & \mathrm{D} = \Omega \\ (1-\mathrm{f})\,\mathbf{C} : (\mathbf{S}^{\mathrm{V}} - \mathbf{1}^{(4s)}) : \boldsymbol{\varepsilon}^{*2} & \mathrm{D} = \mathrm{V}. \end{cases}$$

Corrected expressions are given in the second edition of their book, Nemat-Nasser and Hori (1999).

is

$$< \text{velocity gradient due to } \dot{\mathbf{F}}^{*1} >_\alpha = \frac{\Omega_1}{\Omega_\alpha} \boldsymbol{\mathcal{S}}^{\alpha} : < \dot{\mathbf{F}}^{*1} >_1 \quad , \tag{8.4.10a}$$

where α is not summed, $\boldsymbol{\mathcal{S}}^{\alpha}$ is the generalized Eshelby tensor for Ω_α, and subscript α or 1 on $< >$ emphasizes that the volume average is taken over Ω_α or Ω_1. The velocity-gradient field due to $\dot{\mathbf{F}}^{*\beta}$ in Γ_β ($\beta = 2, 3, ..., m$) is obtained by superposing the velocity-gradient field due to $-\dot{\mathbf{F}}^{*\beta}$ distributed in $\Omega_{\beta-1}$, and the velocity-gradient field due to $\dot{\mathbf{F}}^{*\beta}$ distributed in Ω_β. Hence, the average velocity gradient over Ω_α due to $\dot{\mathbf{F}}^{*\beta}$, is

$$< \text{velocity gradient due to } \dot{\mathbf{F}}^{*\beta} >_\alpha =$$

$$\begin{cases} (\boldsymbol{\mathcal{S}}^{\beta} - \boldsymbol{\mathcal{S}}^{\beta-1}) : \dot{\mathbf{F}}^{*\beta} & \Omega_\alpha \subset \Omega_\beta \\ \{(\Omega_\beta - \Omega_{\beta-1})/\Omega_\alpha\} \, \boldsymbol{\mathcal{S}}^{\alpha} : \dot{\mathbf{F}}^{*\beta} & \Omega_\alpha \supseteq \Omega_\beta \end{cases}$$
$$(\beta \text{ not summed}). \tag{8.4.10b}$$

Since all subregions have the same pseudo-modulus tensor, $\boldsymbol{\mathcal{F}}$, the resulting average nominal stress rates are obtained directly from the corresponding average (velocity gradient - eigenvelocity gradient).

The volume average of the velocity-gradient field $\dot{\mathbf{F}}$, over Ω_α, is obtained by superposition of (8.4.10a) and (8.4.10b) for $\beta = 2, 3, ..., m$, and then the volume average of $\dot{\mathbf{F}}$ over annular region Γ_α is computed. This leads to

$$< \dot{\mathbf{F}} >_1 = \boldsymbol{\mathcal{S}}^{1} : < \dot{\mathbf{F}}^{*1} >_1 + \sum_{\beta=2}^{m} (\boldsymbol{\mathcal{S}}^{\beta} - \boldsymbol{\mathcal{S}}^{\beta-1}) : \dot{\mathbf{F}}^{*\beta} \tag{8.4.11a}$$

and

$$\begin{aligned} < \dot{\mathbf{F}} >'_\alpha = \; & \frac{1}{F_\alpha - F_{\alpha-1}} (\boldsymbol{\mathcal{S}}^{\alpha} - \boldsymbol{\mathcal{S}}^{\alpha-1}) : \{F_1 < \dot{\mathbf{F}}^{*1} >_1 + \sum_{\beta=2}^{\alpha-1} (F_\beta - F_{\beta-1}) \dot{\mathbf{F}}^{*\beta}\} \\ & + \frac{1}{F_\alpha - F_{\alpha-1}} \{(F_\alpha - 2F_{\alpha-1}) \boldsymbol{\mathcal{S}}^{\alpha} + F_{\alpha-1} \boldsymbol{\mathcal{S}}^{\alpha-1}\} : \dot{\mathbf{F}}^{*\alpha} \\ & + \sum_{\beta=\alpha+1}^{m} (\boldsymbol{\mathcal{S}}^{\beta} - \boldsymbol{\mathcal{S}}^{\beta-1}) : \dot{\mathbf{F}}^{*\beta}, \end{aligned} \tag{8.4.11b}$$

for [22]$\alpha = 2, 3, ..., m$, where $< >'_\alpha$ denotes the volume average taken over the annular region Γ_α, and F_α is the volume fraction of Ω_α relative to $\Omega_m \equiv V$, *i.e.*,

$$< >'_\alpha \equiv \frac{1}{F_\alpha - F_{\alpha-1}} (F_\alpha < >_\alpha - F_{\alpha-1} < >_{\alpha-1}),$$

$$F_\alpha \equiv \frac{\Omega_\alpha}{V}. \tag{8.4.12a,b}$$

The corresponding average nominal stress rates are computed in the same manner, and are expressed in terms of $< \dot{\mathbf{F}} >$, as

---

[22] Note that in (8.4.11b), the first summation is omitted for $\alpha = 2$, and the second summation is omitted for $\alpha = m$.

$$< \dot{\mathbf{S}}^N >_1 = \boldsymbol{\mathcal{F}} : (< \dot{\mathbf{F}} >_1 - < \dot{\mathbf{F}}^{*1} >_1),$$
$$< \dot{\mathbf{S}}^N >_\alpha' = \boldsymbol{\mathcal{F}} : (< \dot{\mathbf{F}} >_\alpha - \dot{\mathbf{F}}^{*\alpha}), \tag{8.4.11c,d}$$

for $\alpha = 2, 3, ..., m$. In particular, if all Ω_α's are similar and coaxial, then, (8.4.11a) and (8.4.11b) become

$$< \dot{\mathbf{F}} >_1 = \boldsymbol{\mathcal{S}} : < \dot{\mathbf{F}}^{*1} >_1,$$
$$< \dot{\mathbf{F}} >_\alpha' = \boldsymbol{\mathcal{S}} : \dot{\mathbf{F}}^{*\alpha}, \tag{8.4.13a,b}$$

for $\alpha = 2, 3, ..., m$, where $\boldsymbol{\mathcal{S}}$ is the generalized Eshelby tensor common to all Ω_α's.

8.5. CALCULATION OF GREEN'S FUNCTION

As pointed out in Subsection 8.3.2, the Green function, $G^\infty_{jm}(\mathbf{X})$, for an infinitely extended homogeneous domain, V^∞, is the vector-valued *fundamental solution of the operator*

$$\mathcal{L}_{ij} \equiv \mathcal{F}_{AiBj} \frac{\partial^2}{\partial X_A \, \partial X_B}, \tag{8.5.1a}$$

defined by the following boundary-value problem:[23]

$$\mathcal{F}_{AiBj} \, G^\infty_{jm,AB}(\mathbf{X}) + \delta(\mathbf{X}) \, \delta_{im} = 0 \qquad \mathbf{X} \text{ in } V^\infty, \tag{8.5.2a}$$

where $\delta(\mathbf{X})$ is the delta function, and

$$G^\infty_{jm}(\mathbf{X}) \to 0 \qquad \text{as } |\mathbf{X}| \to \infty. \tag{8.5.2b}$$

Since the matrix of the pseudo-modulus tensor $\boldsymbol{\mathcal{F}}$, has diagonal symmetry, *i.e.*, since $\mathcal{F}_{AiBj} = \mathcal{F}_{BjAi}$, it follows that the operator $\mathcal{L}_{ij}$ is symmetric,

$$\mathcal{L}_{ij} \equiv \mathcal{L}_{ji} \qquad i, j, = 1, 2, 3. \tag{8.5.1b}$$

As shown in Subsection 8.3.2, this symmetry leads to the symmetry of the Green function, $G^\infty_{jm}(\mathbf{X}) = G^\infty_{mj}(\mathbf{X})$. It thus follows that only the following combination of the components of $\boldsymbol{\mathcal{F}}$ enters the expression of the Green function:

$$\hat{\mathcal{F}}_{AiBj} = \frac{1}{4} (\mathcal{F}_{AiBj} + \mathcal{F}_{BiAj} + \mathcal{F}_{AjBi} + \mathcal{F}_{BjAi}). \tag{8.5.2c}$$

The tensor $\hat{\mathcal{F}}_{AiBj}$, in general, has a total of 36 distinct components.

---

[23] Throughout this section, all repeated indices are summed over 1, 2, and 3, unless explicitly stated otherwise.

8.5.1. Strong Ellipticity and Green's Function

The operator $\mathcal{L}_{ij}$ is called *strongly elliptic*, if

$$\mathcal{F}_{AiBj}\,\xi_A\,\xi_B\,n_i\,n_j > 0 \qquad A, B, i, j = 1, 2, 3, \tag{8.5.3}$$

for any nonzero pair of vectors, $\boldsymbol{\xi}$ and $\mathbf{n}$, with rectangular Cartesian components, ξ_A and n_i, A, i = 1, 2, 3. The existence of strong ellipticity precludes the inception of strain localization in the form of discontinuity surfaces, across which certain components of the velocity gradient undergo finite jumps.[24]

Indeed, suppose $\mathbf{N}$ is a unit vector normal to such a surface, say, Σ, at point $\mathbf{X}$; see Figure 8.5.1.

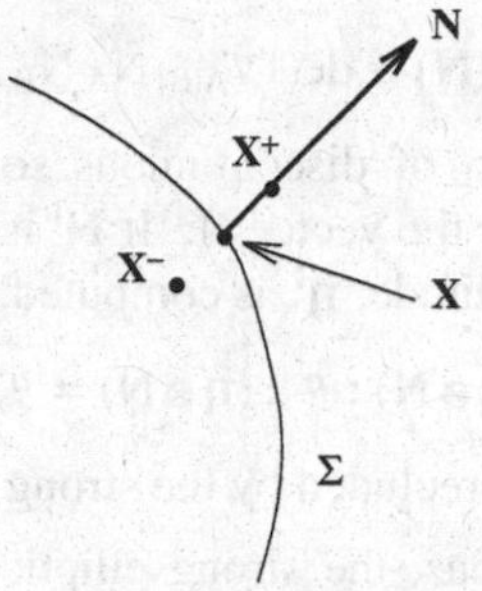

Figure 8.5.1

A surface of discontinuity Σ, across which certain components of $\dot{\mathbf{F}}$ may be discontinuous

Since the traction rates must remain continuous across Σ, it follows that the jump in the tracion rates, defined by $[\dot{\mathbf{T}}](\mathbf{X})$, must be zero. This jump is given in terms of the nominal stress rate and the velocity gradient, by

$$[\dot{\mathbf{T}}](\mathbf{X}) = \lim_{\mathbf{X}^-,\,\mathbf{X}^+ \to \mathbf{X}} \mathbf{N} \cdot \{\dot{\mathbf{S}}^N(\mathbf{X}^+) - \dot{\mathbf{S}}^N(\mathbf{X}^-)\}$$
$$= \mathbf{N} \cdot \boldsymbol{\mathcal{F}} : [\dot{\mathbf{F}}](\mathbf{X}), \tag{8.5.4a,b}$$

where

$$[\dot{\mathbf{F}}](\mathbf{X}) = \lim_{\mathbf{X}^-,\,\mathbf{X}^+ \to \mathbf{X}} \{\dot{\mathbf{F}}(\mathbf{X}^+) - \dot{\mathbf{F}}(\mathbf{X}^-)\} = \boldsymbol{\eta} \otimes \mathbf{N}. \tag{8.5.5a,b}$$

Here, $\boldsymbol{\eta}$ is the vector-valued amplitude of the jump. In component form, these expressions become

$$[\dot{T}_i](\mathbf{X}) = \lim_{\mathbf{X}^-,\,\mathbf{X}^+ \to \mathbf{X}} N_A\,\{\dot{S}^N_{Ai}(\mathbf{X}^+) - \dot{S}^N_{Ai}(\mathbf{X}^-)\}$$
$$= N_A\,\mathcal{F}_{AiBj}\,[\dot{F}_{jB}](\mathbf{X}), \tag{8.5.4c,d}$$

where

[24] See Hill (1961, 1962) for a comprehensive account; see also Hadamard (1903, 1949), Thomas (1956, 1958a,b, 1961), and Hill (1970).

$$[\dot{F}_{iA}](\mathbf{X}) = \lim_{\mathbf{X}^-, \mathbf{X}^+ \to \mathbf{X}} \{\dot{F}_{iA}(\mathbf{X}^+) - \dot{F}_{iA}(\mathbf{X}^-)\} = \eta_i N_A . \tag{8.5.5c,d}$$

From (8.5.4) and (8.5.5), the condition for the continuity of the tracion rates across Σ, becomes

$$[\dot{\mathbf{T}}](\mathbf{X}) = \mathbf{N} \cdot \boldsymbol{\mathcal{F}} : (\boldsymbol{\eta} \otimes \mathbf{N}) = \mathbf{0} \tag{8.5.6a}$$

or, in components,

$$[\dot{T}_i](\mathbf{X}) = (\mathcal{F}_{AiBj} N_A N_B)\, \eta_j = 0 . \tag{8.5.6b}$$

For nontrivial solutions, the determinant of the coefficients of $\boldsymbol{\eta}$ in the set of three homogeneous linear equations (8.5.6) must vanish, leading to the following *characteristic equation* for the operator $\mathcal{L}_{ij}$:

$$\mathcal{D}(\mathbf{N}) \equiv \det |\, \mathcal{F}_{AiBj} N_A N_B \,| = 0 . \tag{8.5.7}$$

Inception of discontinuous solutions is possible if (8.5.7) admits real-valued roots for the vector $\mathbf{N}$. If $\mathbf{N}^I$ is one such root, then from (8.5.6), the corresponding amplitude, $\boldsymbol{\eta}^I$, is computed, leading to

$$(\boldsymbol{\eta} \otimes \mathbf{N}) : \boldsymbol{\mathcal{F}} : (\boldsymbol{\eta} \otimes \mathbf{N}) = \mathcal{F}_{AiBj} N_A^I N_B^I \eta_i^I \eta_j^I = 0 . \tag{8.5.8a,b}$$

This is precluded by the strong ellipticity condition (8.5.3).

Thus, the strong ellipticity guarantees complex-valued solutions, $\mathbf{N}$, for the characteristic equation (8.5.7). Since the coefficients of this sextic equation are real, the roots occur as pairs of complex-conjugates. In such a case, (8.5.2) yields a real-valued Green function.

8.5.2. Green's Function

The Green function can be obtained by applying the Fourier transformation to (8.5.2). Denote the Fourier transform of the Green function $\mathbf{G}^\infty(\mathbf{X})$ by $\hat{\mathbf{G}}^\infty(\boldsymbol{\xi})$, *i.e.*,

$$G_{ij}^\infty(\mathbf{X}) = \frac{1}{(2\pi)^3} \int_{-\infty}^{\infty} \hat{G}_{ij}^\infty(\boldsymbol{\xi}) \exp(i\,\boldsymbol{\xi} \cdot \mathbf{X})\, dV_\xi , \tag{8.5.9a}$$

where dV_ξ is the elementry volume element in the $\boldsymbol{\xi}$-space. Hence, the Fourier transform of $G_{ij,AB}^\infty(\mathbf{X})$ is $-\hat{G}_{ij}^\infty(\boldsymbol{\xi})\, \xi_A \xi_B$. Since

$$\delta(\mathbf{X}) = \frac{1}{(2\pi)^3} \int_{-\infty}^{\infty} \exp(i\,\boldsymbol{\xi} \cdot \mathbf{X})\, dV_\xi , \tag{8.5.9b}$$

the Fourier transform of (8.5.2a) results in

$$K_{ij}(\boldsymbol{\xi})\, \hat{G}_{jm}^\infty(\boldsymbol{\xi}) = \delta_{im} , \tag{8.5.10a}$$

where

$$K_{ij}(\boldsymbol{\xi}) = \mathcal{F}_{AiBj}\, \xi_A \xi_B . \tag{8.5.10b}$$

With strong ellipticity, matrix $\mathbf{K}(\xi)$ admits an inverse for all nonzero values of

$\boldsymbol{\xi}$. Hence, the Fourier transform of the Green function becomes

$$\hat{G}^{\infty}_{ij}(\boldsymbol{\xi}) = K^{-1}_{ij}(\boldsymbol{\xi}) = \frac{N_{ij}(\boldsymbol{\xi})}{\mathcal{D}(\boldsymbol{\xi})}, \tag{8.5.11a}$$

where

$$N_{ij}(\boldsymbol{\xi}) = \frac{1}{2} e_{ikm} e_{jln} K_{kl}(\boldsymbol{\xi}) K_{mn}(\boldsymbol{\xi}) = \text{cofactor}\,(K_{ij}(\boldsymbol{\xi})),$$

$$\mathcal{D}(\boldsymbol{\xi}) = \frac{1}{6} e_{ikm} e_{jln} K_{ij}(\boldsymbol{\xi}) K_{kl}(\boldsymbol{\xi}) K_{mn}(\boldsymbol{\xi}) = \det |K_{ij}(\boldsymbol{\xi})|, \tag{8.5.11b,c}$$

and e_{ijk} is the permutation symbol.

Equations (8.5.9a) and (8.5.11a) show the relation between the structure of the Green function and the possibility of a loss of stability of the uniformly stressed homogeneous solid, because the nature of the roots of $\mathcal{D}(\boldsymbol{\xi}) = 0$ defines the ellipticity, parabolicity, or hyperbolicity of the operator (8.5.1a). As shown in the preceding subsection, the necessary conditions for the possibility of jumps of certain components of the velocity gradient across a discontinuity surface, coincide with the existence of the *real* roots of $\mathcal{D}(\boldsymbol{\xi}) = 0$. The existence of real roots precludes the existence of a real-valued Green function, since, in that case, the operator (8.5.1a) will no longer be elliptic; Iwakuma and Nemat-Nasser (1984).

To evaluate (8.5.9a), set[25]

$$\xi = (\boldsymbol{\xi} \cdot \boldsymbol{\xi})^{1/2}, \quad \bar{\boldsymbol{\xi}} = \boldsymbol{\xi} / \xi, \tag{8.5.12a,b}$$

and rewrite it as follows:

$$G^{\infty}_{ij}(\mathbf{X}) = \frac{1}{(2\pi)^3} \int_0^{\infty} d\xi \int_{S(\bar{\boldsymbol{\xi}})} \frac{N_{ij}(\bar{\boldsymbol{\xi}})}{\mathcal{D}(\bar{\boldsymbol{\xi}})} \exp\{i\xi\bar{\boldsymbol{\xi}} \cdot \mathbf{X}\}\, dS(\bar{\boldsymbol{\xi}}), \tag{8.5.13a}$$

where $dS(\bar{\boldsymbol{\xi}})$ is an elementary surface of a unit sphere, $S(\bar{\boldsymbol{\xi}})$, in the $\bar{\boldsymbol{\xi}}$ – space. In deriving (8.5.13a), the property that $N_{ij}(\boldsymbol{\xi})$ and $\mathcal{D}(\boldsymbol{\xi})$ are fourth- and sixth-degree polynomials in $\boldsymbol{\xi}$, is used.

Since the delta function satisfies

$$\delta(x) = \frac{1}{2\pi} \int_{-\infty}^{\infty} \exp\{i\xi x\}\, d\xi, \tag{8.5.14}$$

(8.5.13a) can be written as

$$G^{\infty}_{ij}(\mathbf{X}) = \frac{1}{8\pi^2} \int_{S(\bar{\boldsymbol{\xi}})} \frac{N_{ij}(\bar{\boldsymbol{\xi}})}{\mathcal{D}(\bar{\boldsymbol{\xi}})} \delta(\bar{\boldsymbol{\xi}} \cdot \mathbf{X})\, dS(\bar{\boldsymbol{\xi}}). \tag{8.5.13b}$$

---

[25] See Kröner (1953), Kneer (1965), Willis (1964), Indenbom and Orlov (1968), Kinoshita and Mura (1971), and Mura (1987).

8.5.3. Generalized Eshelby Tensor and its Conjugate

For the present application, substitute (8.3.15e,f) into (8.3.19a) and use (8.5.9a), to arrive at

$$\mathcal{P}_{iAjB}(\mathbf{X};\Omega) = \frac{-1}{(2\pi)^3}\,\frac{\partial^2}{\partial X_A\,\partial X_B}\int_W dV_Y \int_{-\infty}^{\infty} \hat{G}^{\infty}_{ij}(\boldsymbol{\xi})\exp\{i\,\boldsymbol{\xi}\cdot(\mathbf{X}-\mathbf{Y})\}\,dV_\xi. \tag{8.5.15a}$$

Now, introduce the following coordinate transformation:

$$\zeta_A = a_A\,\xi_A, \quad \overline{X}_A = X_A / a_A, \quad \overline{Y}_A = Y_A / a_A,$$
$$A = 1,\ 2,\ 3, \quad (\text{A not summed}), \tag{8.5.16}$$

where a_A, A = 1, 2, 3, are constants. An ellipsoidal Ω whose principal axes are parallel to the global coordinate axes X_A, with length $2\,a_i$, $i = 1, 2, 3$, is then transformed into a *unit sphere* $\overline{\Omega}$ in the $\overline{\mathbf{X}}$-space.[26] In terms of the coordinates $\overline{\mathbf{X}}$, $\overline{\mathbf{Y}}$, and $\boldsymbol{\zeta}$, (8.5.15a) reduces to

$$\mathcal{P}_{iAjB}(\overline{\mathbf{X}};\overline{\Omega})$$
$$= \frac{-1}{(2\pi)^3\,a_A\,a_B}\,\frac{\partial^2}{\partial \overline{X}_A\,\partial \overline{X}_B}\int_{\overline{\Omega}} dV_{\overline{Y}} \int_{-\infty}^{\infty} \hat{G}^{\infty *}_{ij}(\boldsymbol{\zeta})\exp\{i\,\boldsymbol{\zeta}\cdot(\overline{\mathbf{X}}-\overline{\mathbf{Y}})\}\,dV_\zeta,$$
$$(\text{A, B not summed}), \tag{8.5.15b}$$

where

$$\hat{G}^{\infty *}_{ij}(\boldsymbol{\zeta}) = \frac{N^*_{ij}(\boldsymbol{\zeta})}{\mathcal{D}^*(\boldsymbol{\zeta})},$$
$$N^*_{ij}(\boldsymbol{\zeta}) = \frac{1}{2}\,e_{ikm}\,e_{jln}\,K^*_{kl}(\boldsymbol{\zeta})\,K^*_{mn}(\boldsymbol{\zeta}) = \text{cofactor}\,(K^*_{ij}(\boldsymbol{\zeta})),$$
$$\mathcal{D}^*(\boldsymbol{\zeta}) = \frac{1}{6}\,e_{ikm}\,e_{jln}\,K^*_{ij}(\boldsymbol{\zeta})\,K^*_{kl}(\boldsymbol{\zeta})\,K^*_{mn}(\boldsymbol{\zeta}) = \det|\,K^*_{ij}(\boldsymbol{\zeta})\,|,$$
$$K^*_{ij}(\boldsymbol{\zeta}) = \mathcal{F}^*_{AiBj}\,\zeta_A\,\zeta_B \quad (\text{all repeated indices are summed}),$$
$$\mathcal{F}^*_{AiBj} = \frac{1}{a_A\,a_B}\,\mathcal{F}_{AiBj} \quad (\text{A, B not summed}), \tag{8.5.17a-e}$$

and $dV_{\overline{Y}}$ and dV_ζ are elementary areas in the $\overline{\mathbf{Y}}$- and $\boldsymbol{\zeta}$-space, respectively.

To evaluate (8.5.15b), set

$$\zeta = (\boldsymbol{\zeta}\cdot\boldsymbol{\zeta})^{1/2}, \quad \overline{\boldsymbol{\zeta}} = \boldsymbol{\zeta}/\zeta, \tag{8.5.18a,b}$$

and rewrite it as follows:

---

[26] As is seen, the actual calculation of the Green function for an ellipsoidal domain and a unit sphere, entails similar complexity. The differences are due to the normalization of the coordinates with respect to the principal axes of the ellipsoid, in expressions (8.5.15-17). The real complexity, however, stems from the fact that the pseudo-modulus tensor, $\boldsymbol{\mathcal{F}}$, in general has limited symmetry and varies with the deformation and the stress state.

$$\mathcal{P}_{iAjB}(\overline{\mathbf{X}};\,\overline{\Omega}) = \frac{-1}{(2\pi)^3\, a_A\, a_B}\,\frac{\partial^2}{\partial \overline{X}_A\, \partial \overline{X}_B} \int_{\overline{\Omega}} dV_{\overline{Y}} \int_0^\infty d\zeta \int_{S(\overline{\zeta})} \hat{G}^{\infty *}_{ij}(\overline{\boldsymbol{\zeta}})\, \exp(i\,\zeta\,\overline{\boldsymbol{\zeta}} \cdot \overline{\mathbf{Z}})\, dS(\overline{\boldsymbol{\zeta}}),$$

(A, B not summed), (8.5.15c)

where $\overline{\mathbf{Z}} = \overline{\mathbf{X}} - \overline{\mathbf{Y}}$, and $dS(\overline{\boldsymbol{\zeta}})$ is an elementary surface of a unit sphere, $S(\overline{\boldsymbol{\zeta}})$, in the $\overline{\zeta}$-space. In view of (8.5.14), this is further reduced to

$$\mathcal{P}_{iAjB}(\overline{\mathbf{X}};\,\overline{\Omega}) = \frac{-1}{2\,\pi^2\, a_A\, a_B}\,\frac{\partial^2}{\partial \overline{X}_A\, \partial \overline{X}_B} \int_{S(\overline{\zeta})} \hat{G}^{\infty *}_{ij}(\overline{\boldsymbol{\zeta}})\, \phi(\overline{\boldsymbol{\zeta}},\, \overline{\mathbf{X}};\, \overline{\Omega})\, dS(\overline{\boldsymbol{\zeta}}),$$

(A, B not summed), (8.5.15d)

where

$$\phi(\overline{\boldsymbol{\zeta}},\, \overline{\mathbf{X}};\, \overline{\Omega}) = \int_{\overline{\Omega}} \delta(\overline{\boldsymbol{\zeta}} \cdot (\overline{\mathbf{X}} - \overline{\mathbf{Y}}))\, dV_{\overline{Y}}. \qquad (8.5.19a)$$

The argument of the delta function in the integrand of (8.5.19a) is zero only if the vector $\overline{\mathbf{Z}} = \overline{\mathbf{X}} - \overline{\mathbf{Y}}$ is normal to $\overline{\boldsymbol{\zeta}}$. Hence, $\phi(\overline{\boldsymbol{\zeta}},\, \overline{\mathbf{X}};\, \overline{\Omega})$ is the area of a section cut by $\overline{\Omega}$ on the plane which contains the point $\overline{\mathbf{X}}$ and is normal to $\overline{\boldsymbol{\zeta}}$. For any point $\overline{\mathbf{Y}}$ on this plane, $\overline{\boldsymbol{\zeta}} \cdot (\overline{\mathbf{X}} - \overline{\mathbf{Y}}) = 0$. Since $\overline{\Omega}$ is a unit sphere, the argument of the delta function in (8.5.19a) vanishes for all $\overline{\mathbf{Y}}$ on a circle with radius $\sqrt{1 - (\overline{\boldsymbol{\zeta}} \cdot \overline{\mathbf{X}})^2}$; see Figure 8.5.2.

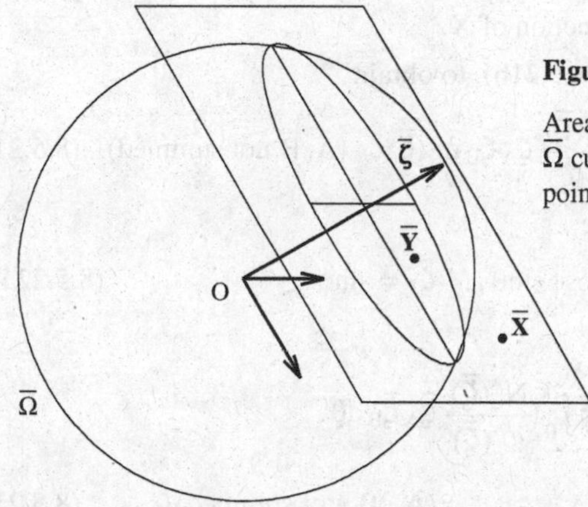

Figure 8.5.2

Area of the section in the unit sphere $\overline{\Omega}$ cut by the plane which contains the point $\overline{\mathbf{X}}$ and is normal to $\overline{\boldsymbol{\zeta}}$

To evaluate (8.5.19a), define

$$z \equiv \overline{\boldsymbol{\zeta}} \cdot \overline{\mathbf{Y}}, \quad z_0 \equiv \overline{\boldsymbol{\zeta}} \cdot \overline{\mathbf{X}}, \quad R(z) \equiv \sqrt{1 - z^2}. \qquad (8.5.20a\text{-}c)$$

Then, with respect to a cylindrical coordinate system whose z-axis coincides with $\overline{\boldsymbol{\zeta}}$, (8.5.19a) becomes

$$\phi^{\overline{\Omega}}(\overline{\boldsymbol{\zeta}},\ \overline{\mathbf{X}}) \equiv \phi(\overline{\boldsymbol{\zeta}},\ \overline{\mathbf{X}};\ \overline{\Omega}) = \int_{-1}^{1} \delta(z - z_0)\, dz \int_0^{R(z)} r\, dr \int_0^{2\pi} d\theta$$

$$= \begin{cases} \pi R^2(z_0) & \text{for } z_0 \le 1 \\ 0 & \text{for } z_0 > 1\,. \end{cases} \qquad (8.5.19b)$$

Hence,

$$\phi(\overline{\boldsymbol{\zeta}},\ \overline{\mathbf{X}};\ \overline{\Omega}) = \begin{cases} \pi\{1 - (\overline{\boldsymbol{\zeta}} \cdot \overline{\mathbf{X}})^2\} & \text{for } \overline{\boldsymbol{\zeta}} \cdot \overline{\mathbf{X}} \le 1 \\ 0 & \text{for } \overline{\boldsymbol{\zeta}} \cdot \overline{\mathbf{X}} > 1\,. \end{cases} \qquad (8.5.19c)$$

Note that, $\overline{\boldsymbol{\zeta}} \cdot \overline{\mathbf{X}} \le 1$ for all $\overline{\mathbf{X}} \in \overline{\Omega}$.

Substitute (8.5.19b) into (8.5.15d), to obtain

$$\mathcal{P}^{\Omega}_{iAjB} \equiv \mathcal{P}_{iAjB}(\overline{\mathbf{X}} \in \overline{\Omega};\ \overline{\Omega}) = \frac{1}{4\pi\, a_A\, a_B} \int_{S(\overline{\zeta})} \hat{G}^{\infty *}_{ij}(\overline{\boldsymbol{\zeta}})\, \overline{\zeta}_A\, \overline{\zeta}_B\, dS(\overline{\boldsymbol{\zeta}}),$$

$$(\text{A, B not summed})\,. \qquad (8.5.21a)$$

Since ϕ is a second-order polynomial in $\overline{\mathbf{X}}$, for $\overline{\mathbf{X}} \in \overline{\Omega}$, $\boldsymbol{\mathcal{P}}^{\Omega}$ is independent of $\overline{\mathbf{X}}$. For $\overline{\mathbf{X}}$ outside of the unit sphere $\overline{\Omega}$, the quantity $\boldsymbol{\mathcal{P}}(\overline{\mathbf{X}} \notin \overline{\Omega};\ \overline{\Omega})$ is

$$\mathcal{P}_{iAjB}(\overline{\mathbf{X}} \notin \overline{\Omega};\ \overline{\Omega}) = \frac{1}{4\pi\, a_A\, a_B} \int_{S^*(\overline{\zeta})} \hat{G}^{\infty *}_{ij}(\overline{\boldsymbol{\zeta}})\, \overline{\zeta}_A\, \overline{\zeta}_B\, dS(\overline{\boldsymbol{\zeta}}),$$

$$(\text{A, B not summed})\,, \qquad (8.5.21b)$$

where $S^*(\overline{\boldsymbol{\zeta}})$ is the subset of the surface area of the unit sphere where $\overline{\boldsymbol{\zeta}} \cdot \overline{\mathbf{X}} \le 1$. Therefore, $\boldsymbol{\mathcal{P}}(\overline{\mathbf{X}} \notin \overline{\Omega},\ \overline{\Omega})$ is a function of $\overline{\mathbf{X}}$.

Substitute (8.5.17a) into (8.5.21b), to obtain

$$\mathcal{P}^{\Omega}_{iAjB} = \frac{1}{4\pi\, a_A\, a_B} \int_{S(\overline{\zeta})} \frac{N^*_{ij}(\overline{\boldsymbol{\zeta}})}{\mathcal{D}^*(\overline{\boldsymbol{\zeta}})}\, \overline{\zeta}_A\, \overline{\zeta}_B\, dS(\overline{\boldsymbol{\zeta}}),\quad (\text{A, B not summed}). \qquad (8.5.21c)$$

To evaluate (8.5.21c), substitute

$$\overline{\zeta}_1 = \cos\phi \cos\theta, \qquad \overline{\zeta}_2 = \cos\phi \sin\theta, \qquad \overline{\zeta}_3 = \sin\phi, \qquad (8.5.22a\text{-}c)$$

into it, and obtain

$$\mathcal{P}^{\Omega}_{iAjB} = \frac{1}{4\pi\, a_A\, a_B} \int_{-\pi/2}^{\pi/2} d\phi \int_0^{2\pi} \frac{N^*_{ij}(\overline{\boldsymbol{\zeta}})}{\mathcal{D}^*(\overline{\boldsymbol{\zeta}})}\, \overline{\zeta}_A\, \overline{\zeta}_B\, d\theta,$$

$$(\text{A, B not summed})\,. \qquad (8.5.21d)$$

Now, define

$$\mathcal{H}_{iAjB}(\phi) \equiv \frac{1}{2\pi\, a_A\, a_B} \int_0^{2\pi} \frac{N^*_{ij}(\overline{\boldsymbol{\zeta}})}{\mathcal{D}^*(\overline{\boldsymbol{\zeta}})}\, \overline{\zeta}_A\, \overline{\zeta}_B\, d\theta, \quad (\text{A, B not summed}),$$

$$(8.5.23a)$$

and, with $z = e^{i\theta}$, obtain

$$\mathcal{H}_{iAjB}(\phi) = \frac{1}{2\pi i \, a_A \, a_B} \oint_\gamma \frac{N^*_{ij}(\overline{\zeta})}{z \, \mathcal{D}^*(\overline{\zeta})} \overline{\zeta}_A \overline{\zeta}_B \, dz \,, \quad \text{(A, B not summed)},$$

(8.5.23b)

where γ is the unit circle $|z| = 1$ in the complex z-plane. Then, (8.5.21d) becomes

$$\mathcal{P}^{\Omega}_{iAjB} = \frac{1}{2} \int_{-\pi/2}^{\pi/2} \mathcal{H}_{iAjB}(\phi) \, d\phi \,. \tag{8.5.21e}$$

Explicit results emerge for two-dimensional problems whose corresponding Green functions are deduced by the limiting process of $a_3 \to \infty$ which renders the integral (8.5.20b) independent of ϕ. Thus, in two dimensions,

$$\mathcal{P}^{\Omega}_{iAjB} = \mathcal{H}_{iAjB}(\phi \to 0) \,, \tag{8.5.19d}$$

where the components of $\overline{\zeta}$ are identified with (cosθ, sinθ, 0). The details are summarized in Appendix 8.A, taken from Iwakuma and Nemat-Nasser (1984); the subscripts are suitably modified to comply with the present subscript notation defined in Section 1.1.

8.6. AVERAGING MODELS

The results presented in the preceding sections of this chapter are all *mathematically exact*. Approximations are made when these exact results are used to develop *averaging models*. To discuss these models in a general setting, consider a finitely deformed heterogeneous RVE, consisting of a matrix M, and n micro-elements Ω_α, $\alpha = 1, 2, ..., n$. Assume rate-independent materials, and let the instantaneous pseudo-modulus tensors of the constituents be given by,

$$\mathcal{F}(\mathbf{X}) = \begin{cases} \mathcal{F}^\alpha & \text{if } \mathbf{X} \text{ in } \Omega_\alpha \quad (\alpha = 1, 2, ..., n-1) \\ \mathcal{F}^M . & \text{if } \mathbf{X} \text{ in M} . \end{cases} \tag{8.6.1a}$$

The corresponding pseudo-compliance tensors are

$$\mathcal{F}(\mathbf{X}) = \begin{cases} \mathcal{G}^\alpha & \text{if } \mathbf{X} \text{ in } \Omega_\alpha \quad (\alpha = 1, 2, ..., n-1) \\ \mathcal{G}^M . & \text{if } \mathbf{X} \text{ in M} . \end{cases} \tag{8.6.1b}$$

Denote the volume fractions of Ω_α and M, respectively by $f_\alpha = \Omega_\alpha / V$ and $f_M = M/V$, where Ω_α, M, and V are volumes of Ω_α, the matrix, and the RVE, respectively. Note that, $f_M + \sum_{\alpha=1}^{n} f_\alpha = 1$.

The aim of an *averaging model* is to obtain the overall average pseudo-modulus, $\overline{\mathcal{F}}$, and pseudo-compliance, $\overline{\mathcal{G}}$, tensors of the RVE. For a prescribed constant overall velocity gradient, $\dot{\mathbf{F}}^0$, it follows that $< \dot{\mathbf{F}} >_V = \dot{\mathbf{F}}^0$. The overall average pseudo-modulus tensor, $\overline{\mathcal{F}}$, is then *defined* such that

$$< \dot{\mathbf{S}}^N >_V = \overline{\mathcal{F}} : \dot{\mathbf{F}}^0 \,. \tag{8.6.2a}$$

Direct calculation now reveals that

$$(\overline{\mathcal{F}} - \mathcal{F}^{M}) : \dot{\mathbf{F}}^0 = \sum_{\alpha=1}^{n} f_\alpha (\mathcal{F}^\alpha - \mathcal{F}^{M}) : < \dot{\mathbf{F}} >_\alpha , \tag{8.6.2b}$$

where $< ... >_\alpha$ denotes volume average over Ω_α.

When, on the other hand, a constant overall nominal stress rate, $\dot{\mathbf{S}}^{N0}$, is prescribed, it follows that $< \dot{\mathbf{S}}^N >_V = \dot{\mathbf{S}}^{N0}$. Then, the overall average pseudo-compliance tensor, $\overline{\mathcal{G}}$, is *defined* such that

$$< \dot{\mathbf{F}} >_V = \overline{\mathcal{G}} : \dot{\mathbf{S}}^{N0} . \tag{8.6.3a}$$

Again, simple calculation yields,

$$(\overline{\mathcal{G}} - \mathcal{G}^{M}) : \dot{\mathbf{S}}^{N0} = \sum_{\alpha=1}^{n} f_\alpha (\mathcal{G}^\alpha - \mathcal{G}^{M}) : < \dot{\mathbf{S}}^N >_\alpha . \tag{8.6.3b}$$

The above *exact* results are the starting expressions for developing *averaging models*, such as the self-consistent model of Hill (1965b) for finite deformation elastoplasticity, as applied to polycrystalline solids by Iwakuma and Nemat-Nasser (1984).

In general, the local quantities, $< \dot{\mathbf{F}} >_\alpha$ and $< \dot{\mathbf{S}}^N >_\alpha$, in (8.6.2b) and (8.6.3b), are related to the overal quantities, $\dot{\mathbf{F}}^0$ and $\dot{\mathbf{S}}^{N0}$, through the *concentration tensors*, $\mathcal{A}^\alpha$ and $\mathcal{B}^\alpha$, as follows:

$$< \dot{\mathbf{F}} >_\alpha = \mathcal{A}^\alpha : \dot{\mathbf{F}}^0 , \tag{8.6.2c}$$

$$< \dot{\mathbf{S}}^N >_\alpha = \mathcal{B}^\alpha : \dot{\mathbf{S}}^{N0} . \tag{8.6.3c}$$

Expressions (8.6.2b) and (8.6.3b) now respectively, yield

$$\overline{\mathcal{F}} = \mathcal{F}^{M} + \sum_{\alpha=1}^{n} f_\alpha (\mathcal{F}^\alpha - \mathcal{F}^{M}) : \mathcal{A}^\alpha , \tag{8.6.4a}$$

$$\overline{\mathcal{G}} = \mathcal{G}^{M} + \sum_{\alpha=1}^{n} f_\alpha (\mathcal{G}^\alpha - \mathcal{G}^{M}) : \mathcal{B}^\alpha . \tag{8.6.5a}$$

The averaging models vary depending on the manner by which the concentration tensors, $\mathcal{A}^\alpha$ and $\mathcal{B}^\alpha$, are defined. In this section, several models which apply to finite deformation elastoplasticity are presented.

8.6.1. Consistency Restrictions for Concentration Tensors

For a given constant $\dot{\mathbf{F}}^0$, it follows that

$$< \dot{\mathbf{F}} >_V = f_M < \dot{\mathbf{F}} >_M + \sum_{\alpha=1}^{n} f_\alpha < \dot{\mathbf{F}} >_\alpha = \dot{\mathbf{F}}^0 . \tag{8.6.6a}$$

Consider a concentration tensor, $\mathcal{A}^M$, for the matrix material such that

$$< \dot{\mathbf{F}} >_M = \mathcal{A}^M : \dot{\mathbf{F}}^0 . \tag{8.6.6b}$$

Now, introduce a concentration function, $\mathcal{A}(\mathbf{X})$, such that

$$\mathcal{A}(\mathbf{X}) = \begin{cases} \mathcal{A}^{\alpha} & \text{if } \mathbf{X} \text{ in } \Omega_{\alpha} \quad (\alpha = 1, 2, ..., n-1) \\ \mathcal{A}^{M} & \text{if } \mathbf{X} \text{ in M}. \end{cases} \tag{8.6.6c}$$

In light of (8.6.6a), the volume average of this variable concentration tensor must satisfy the following consistency condition:

$$< \mathcal{A} >_V = f_M \mathcal{A}^M + \sum_{\alpha=1}^{n} f_{\alpha} \mathcal{A}^{\alpha} = \mathbf{1}^{(4)}. \tag{8.6.6d}$$

In a similar manner, when a constant overall nominal stress rate $\dot{\mathbf{S}}^{N0}$ is given, it follows that,

$$< \dot{\mathbf{S}}^N >_V = f_M < \dot{\mathbf{S}}^N >_M + \sum_{\alpha=1}^{n} f_{\alpha} < \dot{\mathbf{S}}^N >_{\alpha} = \dot{\mathbf{S}}^{N0}. \tag{8.6.7a}$$

Define a concentration tensor, $\mathcal{B}^M$, for the matrix material through

$$< \dot{\mathbf{S}}^N >_M = \mathcal{B}^M : \dot{\mathbf{S}}^{N0}, \tag{8.6.7b}$$

and introduce the spatially variable tensor $\mathcal{B}(\mathbf{X})$, by

$$\mathcal{B}(\mathbf{X}) = \begin{cases} \mathcal{B}^{\alpha} & \text{if } \mathbf{X} \text{ in } \Omega_{\alpha} \quad (\alpha = 1, 2, ..., n-1) \\ \mathcal{B}^{M} & \text{if } \mathbf{X} \text{ in M}, \end{cases} \tag{8.6.7c}$$

whose volume average yields the following consistency condition:

$$< \mathcal{B} >_V = f_M \mathcal{B}^M + \sum_{\alpha=1}^{n} f_{\alpha} \mathcal{B}^{\alpha} = \mathbf{1}^{(4)}. \tag{8.6.7d}$$

When the modeling procedure is such that the concentration tensors do not satisfy consistency conditions (8.6.6d) and (8.6.7d), then they must be suitably *normalized* in order to comply with these conditions. Normalization methods have been suggested by Walpole (1969) for linearly elastic RVE's, and by Iwakuma and Nemat-Nasser (1984) and Nemat-Nasser and Obata (1986) for finitely deformed polycrystals.

When conditions (8.6.6d) and (8.6.7d) are satisfied, then the general results (8.6.4a) and (8.6.5a) take on the following forms:

$$\overline{\mathcal{F}} = f_M \mathcal{F}^M : \mathcal{A}^M + \sum_{\alpha=1}^{n} f_{\alpha} \mathcal{F}^{\alpha} : \mathcal{A}^{\alpha}, \tag{8.6.4b}$$

$$\overline{\mathcal{G}} = f_M \mathcal{G}^M : \mathcal{B}^M + \sum_{\alpha=1}^{n} f_{\alpha} \mathcal{G}^{\alpha} : \mathcal{B}^{\alpha}. \tag{8.6.5b}$$

For a composite with a set of inclusions within a matrix, conditions (8.6.6d) and (8.6.7d) are viewed as the definition of the matrix concentration tensors, $\mathcal{A}^M$ and $\mathcal{B}^M$, respectively.

8.6.2. Voigt and Reuss Models

The simplest models are obtained when the concentration tensors are the fourth-order unit tensor $\mathbf{1}^{(4)}$. Then, (8.6.4b) and (8.6.5b) respectively, become

$$\overline{\mathcal{F}} = f_M\,\mathcal{F}^M + \sum_{\alpha=1}^{n} f_\alpha\,\mathcal{F}^\alpha, \tag{8.6.8}$$

$$\overline{\mathcal{G}} = f_M\,\mathcal{G}^M + \sum_{\alpha=1}^{n} f_\alpha\,\mathcal{G}^\alpha. \tag{8.6.9}$$

Estimates (8.6.8) and (8.6.9) are the finite-deformation versions of the Voigt (1889) and Reuss (1929) models.

8.6.3. Dilute-distribution Model

For a dilute distribution of inhomogeneities, the interaction among the inhomogeneities may be neglected. This yields reasonable estimates when the volume fraction of micro-elements is relatively small and the micro-elements are far apart. With the assumption of a dilute distribution of inhomogeneities, consider a fictitious unbounded homogeneous solid, whose pseudo-moduli are those of the *matrix* material, and in which an *isolated* micro-element, Ω_α, is embedded. The corresponding average velocity gradient and nominal stress rate in Ω_α due to farfield uniform velocity gradient $\dot{\mathbf{F}}^0$ and nominal stress rate $\dot{\mathbf{S}}^{N0}$, are respectively given by (8.2.24a) and (8.2.24b); replace superscript Ω in these equations by α. Using (8.2.24c,d), obtain the following final equations:

$$\overline{\mathcal{F}} = \mathcal{F}^M + \sum_{\alpha=1}^{n} f_\alpha\,(\mathcal{F}^\alpha - \mathcal{F}^M) : \mathcal{C}^\alpha : (\mathcal{C}^\alpha - \mathcal{S}^\alpha)^{-1}, \tag{8.6.10}$$

$$\overline{\mathcal{G}} = \mathcal{G}^M + \sum_{\alpha=1}^{n} f_\alpha\,(\mathcal{G}^\alpha - \mathcal{G}^M) : \mathcal{D}^\alpha : (\mathcal{D}^\alpha - \mathcal{T}^\alpha)^{-1}, \tag{8.6.11}$$

where $\mathcal{S}^\alpha$ and $\mathcal{T}^\alpha$ are the generalized Eshelby tensor and its conjugate, for an ellipsoidal inhomogeneity Ω_α embedded in an infinite homogeneous elstoplastic solid of the uniform instantaneous pseudo-modulus and pseudo-compliance tensors, $\mathcal{F}$ and $\mathcal{G}$. Note that, in this case, the concentration tensors for the matrix are defined by the consistency conditions.

8.6.4. Self-consistent Model

In the self-consistent method, one considers a typical micro-inclusion embedded in an unbounded homogeneous elastoplastic solid which has the *yet-unknown overall pseudo-moduli* of the RVE, and then calculates the average nominal stress rate or velocity gradient in the embedded inclusion. Since an unbounded solid is used, computation of the average quantities does not depend on whether the overall nominal stress rate or velocity gradient is regarded prescribed. Moreover, these overall quantities are related through the *overall pseudo-moduli*, i.e., by $<\dot{\mathbf{S}}^N>_V = \overline{\mathcal{F}} : \dot{\mathbf{F}}^0$ or $<\dot{\mathbf{F}}>_V = \overline{\mathcal{G}} : \dot{\mathbf{S}}^{N0}$. The results obtained in Subsection 8.2.6 now give the average nominal stress rate or velocity gradient in Ω_α; replace $\mathcal{F}$ and $\mathcal{G}$ in (8.2.25a-d) by $\overline{\mathcal{F}}$ and $\overline{\mathcal{G}}$. The resulting concentration tensors respectively, are

$$\overline{\mathcal{A}}^\alpha = \overline{\mathcal{C}}^\alpha : (\overline{\mathcal{C}}^\alpha - \overline{\mathcal{S}}^\alpha)^{-1}, \qquad \overline{\mathcal{B}}^\alpha = \overline{\mathcal{D}}^\alpha : (\overline{\mathcal{D}}^\alpha - \overline{\mathcal{T}}^\alpha)^{-1}, \tag{8.6.12a,b}$$

where $\overline{\mathcal{S}}^\alpha$ and $\overline{\mathcal{T}}^\alpha$ are the generalized Eshelby tensor and its conjugate for an isolated ellipsoidal region Ω_α, in an infinite region of uniform pseudo-modulus and pseudo-compliance tensors $\overline{\mathcal{F}}$ and $\overline{\mathcal{G}}$. In (8.6.12a,b), $\overline{\mathcal{C}}^\alpha$ and $\overline{\mathcal{D}}^\alpha$ are defined by

$$\overline{\mathcal{C}}^\alpha = (\overline{\mathcal{F}} - \mathcal{F}^\alpha)^{-1} : \overline{\mathcal{F}}, \qquad \overline{\mathcal{D}}^\alpha = (\overline{\mathcal{G}} - \mathcal{G}^\alpha)^{-1} : \overline{\mathcal{G}}. \tag{8.6.12c,d}$$

The overall pseudo-moduli then are

$$\overline{\mathcal{F}} = \mathcal{F}^M + \sum_{\alpha=1}^{n} f_\alpha (\mathcal{F}^\alpha - \mathcal{F}^M) : \overline{\mathcal{C}}^\alpha : (\overline{\mathcal{C}}^\alpha - \overline{\mathcal{S}}^\alpha)^{-1}, \tag{8.6.13}$$

$$\overline{\mathcal{G}} = \mathcal{G}^M + \sum_{\alpha=1}^{n} f_\alpha (\mathcal{G}^\alpha - \mathcal{G}^M) : \overline{\mathcal{D}}^\alpha : (\overline{\mathcal{D}}^\alpha - \overline{\mathcal{T}}^\alpha)^{-1}. \tag{8.6.14}$$

Again, the concentration tensors for the matrix are automatically given by the consistency conditions.

8.6.5. Double-inclusion Model

The self-consistent model may be viewed as a special case of the double-inclusion model, introduced by Nemat-Nasser and Hori (1993, 1999) and Hori and Nemat-Nasser (1994). This latter model includes also the Mori-Tanaka (1973) model as a special case, and hence reduces to the Hashin-Shtrikman bounds for linearly elastic composites; see Benveniste (1987)and Weng (1990).

Consider generalizing the double-inclusion model of Nemat-Nasser and Hori (1993, 1999) to finite-deformation problems, and then specializing the results for application to other cases, following Nemat-Nasser (2000). Consider two ellipsoids, one, Ω, contained in the other, V, in an unbounded region, with the following pseudo-moduli (see Figure 8.6.1):

$$\mathcal{F} = \mathcal{F}(\mathbf{X}) = \begin{cases} \mathcal{F}^\Omega & \mathbf{X} \text{ in } \Omega \\ \mathcal{F}^M & \mathbf{X} \text{ in M} \\ \mathcal{F} & \mathbf{X} \text{ outside of V}. \end{cases} \tag{8.6.15}$$

The composite inclusion V can be homogenized by the introduction of a spatially variable eigenvelocity gradient, $\dot{\mathbf{F}}^*$ in V. Since this homogenizing eigenvelocity gradient is not constant, apply the *consistency conditions* to the *average* field quantities. To this end, consider (8.4.7a-d), and identify $\dot{\mathbf{F}}^{*1}$ with $\dot{\mathbf{F}}^{*\Omega} = <\dot{\mathbf{F}}^*>_\Omega$, and $\dot{\mathbf{F}}^{*2}$ with $\dot{\mathbf{F}}^{*M} = <\dot{\mathbf{F}}^*>_M$, respectively; here $\dot{\mathbf{F}}^{*\Omega}$ and $\dot{\mathbf{F}}^{*M}$ are constant tensors. Let the farfield velocity gradient, $\dot{\mathbf{F}}^\infty$, be prescribed. Write the consistency conditions for the average field quantities, as follows:

$$\mathcal{F}^\Omega : \{\dot{\mathbf{F}}^\infty + \mathcal{S}^\Omega : \dot{\mathbf{F}}^{*\Omega} + (\mathcal{S}^V - \mathcal{S}^\Omega) : \dot{\mathbf{F}}^{*M}\}$$

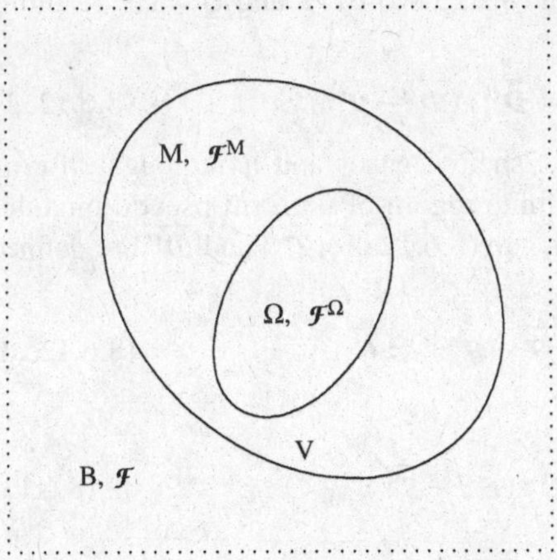

Figure 8.6.1

An unbounded uniform solid B of elasticity $\mathcal{F}$, contains two ellipsoidal regions, V and Ω (Ω ⊂ V), of elasticity tensors $\mathcal{F}^\Omega$ in Ω and $\mathcal{F}^M$ in Γ = V − Ω

$$= \mathcal{F} : \{\dot{\mathbf{F}}^\infty + (\mathcal{S}^\Omega - \mathbf{1}^{(4)}) : \dot{\mathbf{F}}^{*\Omega} + (\mathcal{S}^V - \mathcal{S}^\Omega) : \dot{\mathbf{F}}^{*M}\}$$

$$\mathcal{F}^M : \{\dot{\mathbf{F}}^\infty + \mathcal{S}^V : \dot{\mathbf{F}}^{*M} + \frac{f}{1-f}(\mathcal{S}^V - \mathcal{S}^\Omega) : (\dot{\mathbf{F}}^{*\Omega} - \dot{\mathbf{F}}^{*M})\}$$

$$= \mathcal{F} : \{\dot{\mathbf{F}}^\infty + (\mathcal{S}^V - \mathbf{1}^{(4)}) : \dot{\mathbf{F}}^{*M} + \frac{f}{1-f}(\mathcal{S}^V - \mathcal{S}^\Omega) : (\dot{\mathbf{F}}^{*\Omega} - \dot{\mathbf{F}}^{*M})\}\,, \tag{8.6.16a,b}$$

where f = Ω/V. Solve these equations for $\dot{\mathbf{F}}^{*\Omega}$ and $\dot{\mathbf{F}}^{*M}$, and substitute the results into (8.4.7c,d). The overall pseudo-modulus tensor, $\overline{\mathcal{F}}$, may be defined through the relation $< \dot{\mathbf{S}}^N >_V = \overline{\mathcal{F}} : < \dot{\mathbf{F}} >_V$ which gives

$$\overline{\mathcal{F}} = \mathcal{F} : \{\mathbf{1}^{(4)} + (\mathcal{S}^V - \mathbf{1}^{(4)}) : \mathcal{A}\} : \{\mathbf{1}^{(4)} + \mathcal{S}^V : \mathcal{A}\}^{-1}, \tag{8.6.17a}$$

where $\mathcal{A}$ is defined by

$$f\,\dot{\mathbf{F}}^{*\Omega} + (1-f)\,\dot{\mathbf{F}}^{*M} \equiv \mathcal{A} : \dot{\mathbf{F}}^\infty. \tag{8.6.17b}$$

This last expression defines the average of the (variable) homogenizing eigenvelocity gradient over V. When V and Ω are similar and coaxial, then $\mathcal{A}$ is given exactly by

$$\mathcal{A} = f\,(\mathcal{C}^\Omega - \mathcal{S})^{-1} + (1-f)\,(\mathcal{C}^M - \mathcal{S})^{-1}, \tag{8.6.17c}$$

where

$$\mathcal{C}^\Omega \equiv (\mathcal{F} - \mathcal{F}^\Omega)^{-1} : \mathcal{F}, \qquad \mathcal{C}^M \equiv (\mathcal{F} - \mathcal{F}^M)^{-1} : \mathcal{F} \tag{8.6.17d,e}$$

are the concentration tensors. In this case, the average homogenizing eigenvelocity gradient within each subregion is *exactly* defined by

$$\dot{\mathbf{F}}^{*\Omega} = (\mathcal{C}^\Omega - \mathcal{S})^{-1} : \dot{\mathbf{F}}^\infty, \qquad \dot{\mathbf{F}}^{*M} = (\mathcal{C}^M - \mathcal{S})^{-1} : \dot{\mathbf{F}}^\infty. \tag{8.6.17f,g}$$

For this case, the overall pseudo-modulus, $\overline{\boldsymbol{\mathcal{F}}}$, is obtained by substituting $\boldsymbol{\mathcal{A}}$ from (8.6.17c) into (8.6.17a). The final expression encompasses the results of several commonly used models, now generalized to finite deformations.

By specializing the model of Figure 8.6.1, and hence the expression for the concentration tensor $\boldsymbol{\mathcal{A}}$, the results of the double-inclusion model can be reduced to those of other models, generalized to finite deformations.

8.6.6. Reduction to Self-consistent Model

To obtain the self-consistent model as a special case of the double-inclusion model, let in Figure 8.6.1, $\boldsymbol{\mathcal{F}}$ be equal to $\overline{\boldsymbol{\mathcal{F}}}$, the yet unknown overall modulus tensor of the RVE, and from (8.6.17a), obtain

$$\mathrm{f}\,(\overline{\boldsymbol{C}}^{\Omega}-\overline{\boldsymbol{S}})^{-1}+(1-\mathrm{f})\,(\overline{\boldsymbol{C}}^{\mathrm{M}}-\overline{\boldsymbol{S}})^{-1}=\mathbf{0}, \tag{8.6.18a}$$

where $\overline{\boldsymbol{S}}$ is the generalized Eshely tensor for Ω embedded in an infinite homogeneous matrix of the pseudo-modulus $\overline{\boldsymbol{\mathcal{F}}}$, $\overline{\boldsymbol{C}}^{\mathrm{M}} = (\overline{\boldsymbol{\mathcal{F}}}-\boldsymbol{\mathcal{F}}^{\mathrm{M}})^{-1} : \overline{\boldsymbol{\mathcal{F}}}$, and $\overline{\boldsymbol{C}}^{\Omega} = (\overline{\boldsymbol{\mathcal{F}}}-\boldsymbol{\mathcal{F}}^{\Omega})^{-1} : \overline{\boldsymbol{\mathcal{F}}}$. As in the linear case discussed by Nemat-Nasser and Hori (1993, pages 352, 353), it is most remarkable that (8.6.18a) yields the self-consistent overall pseudo-modulus tensor. The proof is essentially the same as that originally given by Nemat-Nasser and Hori (1993) for the linear elasticity case. Starting with (8.6.18a), after some manipulation, obtain

$$\overline{\boldsymbol{\mathcal{F}}} = \boldsymbol{\mathcal{F}}^{\mathrm{M}}+\mathrm{f}\,(\boldsymbol{\mathcal{F}}^{\Omega}-\boldsymbol{\mathcal{F}}^{\mathrm{M}}) : \{\mathbf{1}^{(4)}+\overline{\boldsymbol{S}} : [(\overline{\boldsymbol{\mathcal{F}}}-\boldsymbol{\mathcal{F}}^{\Omega})^{-1} : \overline{\boldsymbol{\mathcal{F}}}-\overline{\boldsymbol{S}}\,]^{-1}\} \tag{8.6.18b}$$

which is recognized as the self-consistent result,

$$\overline{\boldsymbol{\mathcal{F}}}^{SC} = \boldsymbol{\mathcal{F}}^{\mathrm{M}}+\mathrm{f}\,(\boldsymbol{\mathcal{F}}^{\Omega}-\boldsymbol{\mathcal{F}}^{\mathrm{M}}) : \{\mathbf{1}^{(4)}+\overline{\boldsymbol{S}} : (\overline{\boldsymbol{C}}^{\Omega}-\overline{\boldsymbol{S}})^{-1}\}, \tag{8.6.18c}$$

since $\mathbf{1}^{(4)}+\overline{\boldsymbol{S}} : (\overline{\boldsymbol{C}}^{\Omega}-\overline{\boldsymbol{S}})^{-1} = \overline{\boldsymbol{C}}^{\Omega} : (\overline{\boldsymbol{C}}^{\Omega}-\overline{\boldsymbol{S}})^{-1}$; see Iwakuma and Nemat-Nasser (1984). Note that only when it is assumed that V and Ω are coaxial and similar, that the general double-inclusion model (8.6.17a,b) reduces to the self-consistent model for $\boldsymbol{\mathcal{F}} = \overline{\boldsymbol{\mathcal{F}}}$. Therefore, the model defined by (8.6.17a,b) is more general than the self-consistent one.

8.6.7. Two-phase Model

Assume that V and Ω are coaxial and similar, and that $\boldsymbol{\mathcal{F}}^{\mathrm{M}} \equiv \boldsymbol{\mathcal{F}}$. This model is shown in Figure 8.6.2. The material properties are defined by

$$\boldsymbol{\mathcal{F}} = \boldsymbol{\mathcal{F}}(\mathbf{X}) = \begin{cases} \boldsymbol{\mathcal{F}}^{\Omega} & \mathbf{X} \text{ in } \Omega \\ \boldsymbol{\mathcal{F}} & \mathbf{X} \text{ in M} \\ \boldsymbol{\mathcal{F}} & \mathbf{X} \text{ outside of V}; \end{cases} \tag{8.6.19}$$

that is, V is a subdomain of B, within which a coaxial and similar ellipsoid, Ω, is embedded. In this case, $(\boldsymbol{C}^{\mathrm{M}}-\boldsymbol{S})^{-1}$ vanishes, and (8.6.17a,c) yield

$$\overline{\boldsymbol{\mathcal{F}}}^{TP} = \boldsymbol{\mathcal{F}} : \{\,\mathbf{1}^{(4)}+\mathrm{f}\,(\boldsymbol{S}^{\Omega}-\mathbf{1}^{(4)}) : (\boldsymbol{C}^{\Omega}-\boldsymbol{S}^{\Omega})^{-1}\,\} : \{\,\mathbf{1}^{(4)}$$

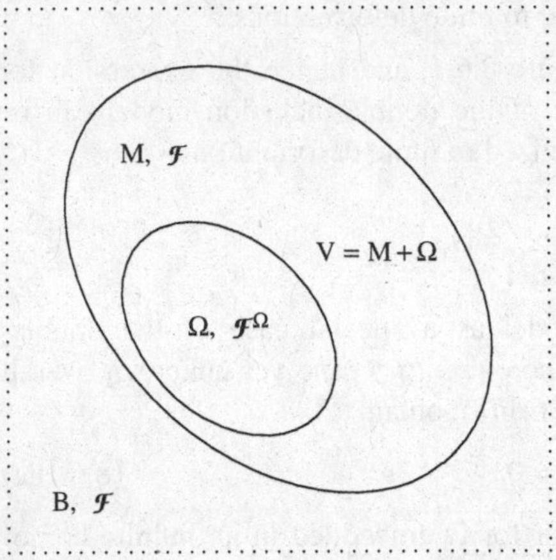

Figure 8.6.2

A double inclusion $V = M + \Omega$ is embedded in a uniform infinite domain B; the elastic-plastic mudulus tensor of Ω is $\mathcal{F}^{\Omega}$, that of $M = V - \Omega$, and the infinite domain is $\mathcal{F}$; this is a two-phase model

$$+ f\, \boldsymbol{\mathcal{S}}^{\Omega} : (\boldsymbol{\mathcal{C}}^{\Omega} - \boldsymbol{\mathcal{S}}^{\Omega})^{-1} \}^{-1}. \tag{8.6.20a}$$

For the linearly elastic case, this reduces to the result of the Mori-Tanaka model which is now generalized and applied to finite deformations; see also Nemat-Nasser (2000). Note that the mathematical derivation is *exact,* in the sense that

$$< \dot{\mathbf{S}}^{\mathrm{N}} >_{\mathrm{V}} = \overline{\mathcal{F}}^{TP} : < \dot{\mathbf{F}} >_{\mathrm{V}}, \tag{8.6.20b}$$

exactly.

8.6.8. Overall Pseudo-compliance Tensor

With a similar approach which led to the effective pseudo-modulus tensor defined by (8.6.17a) for the double-inclusion model, the overall pseudo-compliance tensor, $\overline{\mathcal{G}}$, can be formulated. *The resulting* $\overline{\mathcal{G}}$ *simply is the inverse of* $\overline{\mathcal{F}}$. As in the case of linear elasticity, the resulting overall pseudo-modulus and pseudo-compliance tensors:

1) are each other's inverse;
2) do not depend on the surface data on ∂V; and
3) do not depend on the location of Ω relative to V.

These comments apply to the double-inclusion model and its special cases discussed above.

8.6.9. Multi-inclusion Model

Similarly to the linear case developed by Nemat-Nasser and Hori (1993, page 353), the results of the double-inclusion model can be generalized and applied to an ellipsoid, V, which contains a nested series of ellipsoids, Ω_α

($\alpha = 1, 2, ..., n$), such that $\Omega_1 \subset \Omega_2 \subset ... \subset \Omega_n \equiv V$. The ellipsoid Ω_1 and each annular region $\Gamma_\alpha = \Omega_\alpha - \Omega_{\alpha-1}$ have uniform elastoplastic properties, defined by the corresponding pseudo-modulus tensor, $\mathcal{F}^\alpha$, $\alpha = 1, 2, ..., n$. The multi-inclusion V is embedded in an infinite domain of uniform modulus tensor $\mathcal{F}$. The properties of the composite, thus are defined by

$$\mathcal{F} = \mathcal{F}(\mathbf{X}) = \begin{cases} \mathcal{F}^1 & \text{if } \mathbf{X} \text{ in } \Omega_1 \\ \mathcal{F}^\alpha & \text{if } \mathbf{X} \text{ in } \Gamma_\alpha \quad (\alpha = 2, 3, ..., n) \\ \mathcal{F} & \text{otherwise}, \end{cases} \tag{8.6.21}$$

where $\mathbf{X}$ is the reference particle position.

Applying the average consistency conditions, as before, it is easy to show that the volume average of the nominal stress rate and the velocity gradient, $<\dot{\mathbf{S}}^N>_V$ and $<\dot{\mathbf{F}}>_V$, are related by the modulus tensor $\overline{\mathcal{F}}$ which is given by equation (8.6.17a) in which the tensor $\mathcal{A}$ is now defined by

$$\sum_{\alpha=1}^{n} f_\alpha \dot{\mathbf{F}}^{*\alpha} \equiv \mathcal{A} : \dot{\mathbf{F}}^\infty. \tag{8.6.22a}$$

The same result is obtained for an ellipsoidal V which contains $n-1$ ellipsoidal heterogeneities Ω_α ($\alpha = 1, 2, ..., n-1$). The remaining part of V defines the region Γ. The properties of this multi-inclusion system are given by

$$\mathcal{F}(\mathbf{X}) = \begin{cases} \mathcal{F}^\alpha & \text{if } \mathbf{X} \text{ in } \Omega_\alpha \quad (\alpha = 1, 2, ..., n-1) \\ \mathcal{F}^n & \text{if } \mathbf{X} \text{ in } \Gamma \\ \mathcal{F} & \text{otherwise}. \end{cases} \tag{8.6.23}$$

The volume fractions of Ω_α and Γ are defined by $f_\alpha = \Omega_\alpha/V$ ($\alpha = 1, 2, ..., n-1$) and $f_n = \Gamma/V$. For each Ω_α, it is assumed that $V - \Omega_\alpha$ is a region where the average eigenvelocity gradients, $\sum_{\beta \neq \alpha}^{n} \{f_\beta/(1-f_\alpha)\} \dot{\mathbf{F}}^{*\beta}$, are uniformly distributed, and the corresponding average velocity gradient is estimated. From these the required homogenizing eigenvelocity gradients are obtained, leading eventually to the overall pseudo-modulus tensor given by (8.6.17a) and (8.6.22a).

When in either multi-inclusion model, all ellipsoids are similar and coaxial, then (8.6.22a) reduces to

$$\mathcal{A} = \sum_{\alpha=1}^{n} f_\alpha (\mathcal{C}^\alpha - \mathcal{S})^{-1}. \tag{8.6.22b}$$

Hence, the overall effective pseudo-modulus tensor becomes,

$$\overline{\mathcal{F}} = \mathcal{F} : \{1^{(4)} + \sum_{\alpha=1}^{n} f_\alpha (\mathcal{S} - 1^{(4)}) : (\mathcal{C}^\alpha - \mathcal{S})^{-1}\} : \{1^{(4)} + \sum_{\alpha=1}^{n} f_\alpha \mathcal{S} : (\mathcal{C}^\alpha - \mathcal{S})^{-1}\}^{-1}. \tag{8.6.24}$$

8.7. POLYCRYSTALS

In a polycrystal RVE, there is no distinction among the grains. Each crystal may be treated as an inclusion embedded in the remaining crystals. Therefore, the concept of a matrix with embedded inclusions is not relevant. The averaging models provide a means for calculating the local velocity gradient and stress rate for a typical crystal of known geometry and lattice orientation, in terms of the corresponding overall quantities. This is done through the introduction of suitable concentration tensors. On physical grounds, some models, *e.g.*, the dilute distribution of inclusions, no longer apply. Models have been developed specifically for polycrystals. For example, the self-consistent scheme was originally proposed by Hershey (1954), Kröner (1958, 1967), and Kneer (1965), to estimate the overall moduli of polycrystals.[27]

The simplest averaging model was proposed by Taylor (1938), who neglected the elastic strain and assumed a uniform plastic strain, equal to the overall applied strain, for all the crystals within a polycrystal. Later works based on the Taylor model have shown relatively good agreement with experimental results, as has been illustrated in Section 6.5. The model has also produced reasonable results on anisotropy and texture development in polycrysalline metals subjected to large plastic deformations; see *e.g.*, Honneff and Mecking (1978), and Kocks and Canova (1981). Since many unfavorably oriented grains within a crystalline aggregate may actually have much smaller strains than others with favorably oriented slip systems, the assumption of constant strain for all grains, has been relaxed by some authors, leading to the so-called *relaxed constraints theory*; see, Honneff and Mecking (1978), and Kocks and Canova (1981).

In the self-consistent estimate, a single crystal is embedded in an unbounded uniform matrix which has the effective overall pseudo-modulus or pseudo-compliance tensor of the polycrystal. The local average nominal stress rate and velocity gradient in the embedded crystal are then calculated and used to obtain the overall effective pseudo-moduli. The self-consistency then guarantees that the effective overall pseudo-modulus and pseudo-compliance tensors are each other's inverse. A modified version of the self-consistent model has been proposed by Molinari *et al.*, (1987, 1997), following the work of Zeller and Dederichs (1973). In this approach, both equilibrium and compatibility are used to develop an integral-equation representation of the local velocity gradient. This integral equation must then be solved, using some approximate method.

---

[27] Earlier work on the elastic properties of polycrystals is due to Voigt (1910) and Reuss (1929). The self-consistent method was later applied to composites by Budiansky (1965) and Hill (1965a,b). Other related work in this area is by Hill (1952), Hashin and Shtrikman (1962), Peselnick and Meister (1965), Morris (1970, 1971), Korringa (1973), Zeller and Dederichs (1973), and Gubernatis and Krumhansl (1975); see also reviews by Hashin (1964, 1983), Watt *el al.* (1976), Kröner (1980), and Mura (1987). A comprehensive account can be found in Nemat-Nasser and Hori (1993, 1999).

For a single crystal embedded in a homogeneous equivalent medium, the method leads to a nonlinear relation between the stress and the strain rate that must be solved numerically. The method also applies to rate-dependent materials. In what follows, first rate-independent cases are considered, and then some comments on rate-dependent models are made in Subsection 8.7.5.

8.7.1. Overall Moduli for Polycrystals

Consider an RVE consisting of an aggregate of finitely deformed single crystals. Let f_n be the volume fraction of the n'th crystal of the pseudo-modulus and pseudo-compliance tensors $\mathcal{F}^n$ and $\mathcal{G}^n$, respectively. Assume that these tensors are each other's inverse. Equations (8.6.2a) and (8.6.3a) now define the overall average pseudo-modulus and pseudo-compliance tensors of the aggregate. Furtheremore, equations (8.6.2c) and (8.6.3c) give the average velocity gradient and the nominal stress tensors in the n'th crystal. In view of these expressions, it follows that

$$\overline{\mathcal{F}} = \sum_{n=1}^{N} f_n \mathcal{F}^n : \mathcal{A}^n \equiv < \mathcal{F} : \mathcal{A} >_V ,$$

$$\overline{\mathcal{G}} = \sum_{n=1}^{N} f_n \mathcal{G}^n : \mathcal{B}^n \equiv < \mathcal{G} : \mathcal{B} >_V , \qquad (8.7.1a,b)$$

where there are a total of N crystals in the RVE, and $\mathcal{A}^n$ and $\mathcal{B}^n$ are the concentration tensors for the n'th crystal. In the above expressions, the quantities $\mathcal{F} = \mathcal{F}(\mathbf{X})$ and $\mathcal{G} = \mathcal{G}(\mathbf{X})$ take on the values $\mathcal{F}^n$ and $\mathcal{G}^n$ when $\mathbf{X}$ is in the n'th crystal. Similar remarks appy to $\mathcal{A}$ and $\mathcal{B}$.

The concentration tensors, $\mathcal{A}$ and $\mathcal{B}$, must satisfy the consistency restrictions (8.6.6d) and (8.6.7d), respectively, *i.e.*,

$$< \mathcal{A} >_V = \sum_{n=1}^{N} f_n \mathcal{A}^n = \mathbf{1}^{(4)} , \qquad (8.7.2a)$$

$$< \mathcal{B} >_V = \sum_{n=1}^{N} f_n \mathcal{B}^n = \mathbf{1}^{(4)} . \qquad (8.7.2b)$$

Because of the modeling approximations, these conditions are most often violated. Hence, they must be explicitly imposed by suitably normalizing the concentration tensors. For linear elasticity, Walpole (1969) has suggested a normalization technique which also emerges from a variational consideration; see Willis (1981). Iwakuma and Nemat-Nasser (1984) have applied Walpole's method to the finite-deformation case. In this method, the concentration tensors are replaced by

$$\hat{\mathcal{A}} = \mathcal{A} (< \mathcal{A} >_V)^{-1} , \quad \hat{\mathcal{B}} = \mathcal{B} (< \mathcal{B} >_V)^{-1} . \qquad (8.7.3a,b)$$

Then, (8.7.2a,b) are automatically satisfied.[28]

[28] Other normalization methods have been examined by Iwakuma and Nemat-Nasser (1984) and Nemat-Nasser and Obata (1986).

8.7.2. Self-consistent Model

For the self-consistent method, the concentration tensors are defined by (8.6.12a,b). Hence, (8.7.1a,b) become

$$\overline{\mathcal{F}} = \sum_{n=1}^{N} f_n \, \mathcal{F}^n : \overline{\mathcal{A}}^n, \qquad \overline{\mathcal{G}} = \sum_{n=1}^{N} f_n \, \mathcal{G}^n : \overline{\mathcal{B}}^n. \tag{8.7.4a,b}$$

In view of (8.6.12c,d), it is seen that consistency restrictions (8.7.2a,b) are, in general, violated and hence must additionally be enforced.

Suppose that, for the purposes of computing the concentration tensors, initially spherical grains are used. Then, the generalized Eshelby tensors, $\overline{\mathcal{S}}^{\,n}$, would have a common value for all grains, independently of their initial crystallographic orientations, since, for spheres, these tensors depend only on the overall pseudo-modulus tensor $\overline{\mathcal{F}}$. Notwithstanding this, the pseudo-modulus tensor $\mathcal{F}^n$ of the n'th grain changes with the deformation and stress-state, and will be different for different initial crystal orientations. Hence, $\overline{\mathcal{C}}^{\,n} = (\overline{\mathcal{F}} - \mathcal{F}^n)^{-1} : \overline{\mathcal{F}}$ which corresponds to $\overline{\mathcal{C}}^{\,\alpha}$ in (8.6.12c), will depend on the initial orientation of the corresponding crystal. This means that (8.7.2a,b) do not, in general, automatically hold, whether or not the initial state is used as the reference one, and whether or not spherical grains are assumed in the calculation of the concentration tensors.

8.7.3. Choice of Reference State

As pointed out before, the reference configuration is arbitrary and can be chosen to simplify the analysis while corresponding to the physics of the process. Since the plastic response of most materials is history-dependent, it is often necessary to calculate the instantaneous elastic-plastic moduli, measured with respect to the current deformed state of the material. Once the overall pseudo-modulus tensor, $\overline{\mathcal{F}}$, of the RVE is obtained, the expressions developed in Section 8.1 can be used to calculate the overall instantaneous elastic-plastic modulus tensor with respect to the current configuration.

As an illustration, consider a case where the overall velocity gradient $\dot{\mathbf{F}}^0$ is prescribed with respect to the *initial undeformed state*, so that

$$\dot{\mathbf{x}} = \dot{\mathbf{F}}^0 \mathbf{X}, \qquad \mathbf{x} = \mathbf{F}^0 \mathbf{X}, \tag{8.7.5a,b}$$

where $\mathbf{F}^0$ is obtained by integrating $\dot{\mathbf{F}}^0$ with respect to time. From these it follows that

$$< \dot{\mathbf{F}} >_V = \dot{\mathbf{F}}^0, \qquad < \mathbf{F} >_V = \mathbf{F}^0. \tag{8.7.5c,d}$$

The overall nominal stress rate, measured with respect to the *initial undeformed state*, now is

$$< \dot{\mathbf{S}}^N >_V = \overline{\mathcal{F}} : \dot{\mathbf{F}}^0. \tag{8.7.5e}$$

From (8.7.5a,b), the velocity gradient, $\mathbf{L}^0$, measured relative to the *current deformed state*, is given by

$$\mathbf{L}^0 = \dot{\mathbf{F}}^0\,\mathbf{F}^{0-1}\,, \tag{8.7.6a}$$

from which the deformation rate and spin tensors can be computed,

$$\mathbf{L}^0 = \mathbf{D}^0 + \mathbf{W}^0\,. \tag{8.7.6b}$$

Since for elastic-plastic deformations, the current state of the material is of special importance, consider the following *new notation for the stress rates measured with respect to the current deformed state:*

$\dot{\boldsymbol{N}}$, $\dot{N}_{ij} \equiv$ the overall nominal stress rate and its components;

$\dot{\boldsymbol{T}}$, $\dot{T}_{ij} \equiv$ the overall Kirchhoff stress rate and its components;

$\overset{\circ}{\boldsymbol{T}}$, $\overset{\circ}{T}_{ij} \equiv$ the overall Jaumann rate of the Kirchhoff stress and its components. (8.7.7)

Note that, with reference to the current configuration, all the overall stress measures are equal to the overall Cauchy stress, *i.e.*, $\boldsymbol{N} = \boldsymbol{T} = \boldsymbol{\Sigma} = \bar{\boldsymbol{\sigma}}$; see Table 3.1.1, page 99. Their rates, however, are not the same, *e.g.*,

$$\dot{\boldsymbol{T}} = \dot{\boldsymbol{N}} + \mathbf{L}^0\boldsymbol{\Sigma}\,, \qquad \dot{T}_{ij} = \dot{N}_{ij} + L^0_{ik}\,\Sigma_{kj}\,. \tag{8.7.8a,b}$$

The overall nominal stress rate, $\dot{N}_{ij}$, is related to the velocity gradient, L^0_{ij}, by

$$\dot{N}_{ij} = \mathcal{F}^t_{ijkl}\,L^0_{lk}\,, \qquad \mathcal{F}^t_{ijkl} = F^0_{iA}\,\bar{\mathcal{F}}_{AjBk}\,F^0_{lB}\,. \tag{8.7.9a,b}$$

The modulus tensor $\bar{\boldsymbol{\mathcal{F}}}$ with components $\bar{\mathcal{F}}_{AjBk}$, and the Cauchy stress $\boldsymbol{\Sigma}$ with components Σ_{kj}, are calculated incrementally, starting from, and using the *initial undeformed configuration* of the polycrystalline RVE. In general, because of the modeling approximations, there is no guarantee that the resulting Cauchy stress tensor, Σ_{ij}, and the Kirchhoff stress rate, $\dot{T}_{ij}$, in (8.7.8a,b), would have the required symmetries. Furthermore, the Jaumann rate of the Kirchhoff stress,

$$\overset{\circ}{\boldsymbol{T}} = \dot{\boldsymbol{T}} - \mathbf{W}^0\boldsymbol{T} + \boldsymbol{T}\,\mathbf{W}^0\,,$$

$$\overset{\circ}{T}_{ij} = \dot{T}_{ij} - W^0_{ik}\,T_{kj} - W^0_{jk}\,T_{ki}\,, \tag{8.7.9c,d}$$

calculated using (8.7.8b) and the fact that $\boldsymbol{T} = \boldsymbol{\Sigma}$, may not, in general, be objective and hence may not be independent of the material spin tensor, $\mathbf{W}^0$. Therefore, these restrictions must be imposed at each calculation increment.

To this end, first define the instantneous elastic-plastic modulus tensor, $\boldsymbol{\mathcal{L}}$, with components $\mathcal{L}_{ijkl}$, in terms of the Jaumann rate of the Kirchhoff stress, as follows:

$$\overset{\circ}{\boldsymbol{T}} = \boldsymbol{\mathcal{L}} : \mathbf{D}^0\,. \tag{8.7.10a}$$

Then require that $\dot{T}_{ij}$ given by (8.7.8a,b) be symmetric, *i.e.*, set

$$\dot{T}_{ij} = \tfrac{1}{2}\,(\dot{N}_{ij} + \dot{N}_{ji} + \Sigma_{ik}\,L^0_{kj} + \Sigma_{jk}\,L^0_{ki})\,. \tag{8.7.10b}$$

Finally note that, in view of (8.7.10a,b),

$$\overset{\circ}{T}_{ij} = \tfrac{1}{2}(\dot{N}_{ij} + \dot{N}_{ji} + L^0_{ik}\Sigma_{kj} + L^0_{jk}\Sigma_{ki}) = \mathcal{L}_{ijkl} D^0_{kl}, \tag{8.7.10c,d}$$

which requires that $(\dot{N}_{ij} + \dot{N}_{ji} + L^0_{ik}\Sigma_{kj} + L^0_{jk}\Sigma_{ki})$ be independent of the spin tensor. This is satisfied by requiring that

$$\mathcal{F}^t_{ijkl} + \mathcal{F}^t_{jikl} + \delta_{il}\Sigma_{kj} + \delta_{jl}\Sigma_{ki} = \mathcal{F}^t_{ijlk} + \mathcal{F}^t_{jilk} + \delta_{ik}\Sigma_{lj} + \delta_{jk}\Sigma_{li}\,. \tag{8.7.10e}$$

With this in mind, then $\boldsymbol{\mathcal{L}}$ is given by

$$\mathcal{L}_{ijkl} = \tfrac{1}{2}(\mathcal{F}^t_{ijkl} + \mathcal{F}^t_{jikl} + \delta_{il}\Sigma_{kj} + \delta_{jl}\Sigma_{ki})\,. \tag{8.7.11a}$$

8.7.4. Updated Lagrangian Calculation

In most applications involving large elastic-plastic deformations, it is often more appropriate to use the current configuration as the reference one. This is usually referred to as the *updated Lagrangian method.* The method has many practical implications. For example, the evolution of texture in a polycrystal is generally examined within the deformed configuration of the material.

The calculations are performed incrementally, where for each increment, the deformed configuration is used for the reference state; this has been illustrated by the example presented in Section 6.5. As pointed out before, all stress tensors are then equal to the Cauchy stress, $\bar{\boldsymbol{\sigma}}$, and it also follows that $\mathbf{F}^0 = \mathbf{1}$ and $\dot{\mathbf{F}}^0 = \mathbf{L}^0$. The instantaneous elastic-plastic modulus tensor, $\boldsymbol{\mathcal{L}}$, then becomes

$$\mathcal{L}_{ijkl} = \tfrac{1}{2}(\bar{\mathcal{F}}_{ijkl} + \bar{\mathcal{F}}_{jikl} + \delta_{il} < \sigma_{kj} >_V + \delta_{jl} < \sigma_{ki} >_V)\,, \tag{8.7.11b}$$

where $\bar{\mathcal{F}}_{ijkl}$ is calculated incrementally, and at each increment the current deformed state is used as the reference one; see Iwakuma and Nemat-Nasser (1984) for details and illustrative examples. Note that, $\boldsymbol{T} = \boldsymbol{N} = \boldsymbol{\Sigma} \equiv < \boldsymbol{\sigma} >_V \equiv \bar{\boldsymbol{\sigma}}$.

8.7.5. Rate-dependent Models

It has been emphasized in Section 4.8 and in Chapter 6 that the response of essentially all materials is rate- and temperature-dependent. This is particularly true for most polycrystals, as can be seen from the experimental data presented in Sections 4.8 and 6.5; see also Chapter 9. Moreover, the relations among temperature, strain rate, and stress are highly nonlinear and history-dependent; see, *e.g.,* equations (6.5.13a,c), page 449, and (6.5.15d), page 450. In these cases, the single-crystal calculations are complex, requiring special algorithms, as has been illustrated in Section 6.5 for both fcc and bcc metals. The strain-rate sensitivity of crystalline metals is explained in terms of two possible mechanisms: one associated with the finite velocity of dislocations, and the other associated with the rate-sensitivity of the evolution of the dislocation substructure. As discussed in Sections 4.8 and 6.5, the first of these constituents, *i.e., the instantaneous rate-sensitivity,* is related to the wait-times associated with thermally activated dislocation motion. The second component pertains to the relative importance of dislocation generation and annihilation at different strain rates. Rashid, *et al.* (1992) refer to this constituent as the *strain-rate-history effect,* and provide experimental support for its occurrence.

In relation to the averaging techniques for aggregates of rate- and temperature-dependent single crystals, two models can be used directly. These are the Voigt and Reuss models which are examined in Subsection 8.6.2. In the first model, the local deformation rate, say, $\dot{\mathbf{F}}^n$, of the n'th crystal, is assumed to be the same as the overall prescribed deformation rate, $\dot{\mathbf{F}}^0$. In the second model, the local nominal stress rate, $\dot{\mathbf{S}}^{Nn}$, is assumed to be the same as the overall nominal stress rate, $\dot{\mathbf{S}}^{N0}$. In the crystal-plasticity literature, these models are named after Taylor (1934) and Sachs (1928). For the rate-independent cases, these assumptions lead to equations which are equivalent to (8.6.8) and (8.6.9), respectively. For the rate-dependent case, the Taylor model has been used in Section 6.5 for both fcc and bcc metals.

Suppose an updated Lagrangian method of calculation is used. Let the overall velocity gradient, $\mathbf{L}^0$, be prescribed as a function of time. Starting from the undeformed state of the polycrystal, assume that the Kirchhoff stress, $\boldsymbol{\tau}^n$, the temperature, T, the crystal orientation, and all internal parameters which characterize the microstructure (*e.g.*, dislocation density) of the n'th crystal are known at time t. For the sake of illustration, let $\mathbf{L}^0$ remain constant over the time interval from t to $t+\Delta t$; see Chapter 5 for comments on numerical integration methods. The following steps are used to obtain the overall Kirchhoff stress, $\boldsymbol{T}(t+\Delta t) = \,<\boldsymbol{\tau}(t+\Delta t)>_V$:

(1) Use a suitable model to calculate the local velocity gradient, $\mathbf{L}^n$, for the typical n'th crystal. Let the corresponding (normalized) concentration tensor be $\boldsymbol{\mathcal{A}}^n$. Then $\mathbf{L}^n = \boldsymbol{\mathcal{A}}^n : \mathbf{L}^0$. For the Taylor model, $\boldsymbol{\mathcal{A}}^n = \mathbf{1}^{(4)}$.

(2) Use a numerical algorithm similar to that in Section 6.5, to calculate the Kirchhoff (or Cauchy) stress tensor at time $t+\Delta t$.

(3) Use the unweighted volume average to obtain the overall stress tensor, measured relative to the state at time $t+\Delta t$.

For step (2), the forward-gradient method, or its modified and improved versions, have been used; see Peirce *et al.* (1984), Nemat-Nasser and Obata (1986), Zikry and Nemat-Nasser (1990), and Rashid and Nemat-Nasser (1992). However, the method of *plastic-predictor/elastic-corrector* is superior to all of these methods and should be used when large plastic deformations are involved. As illustrated in Subsections 6.5.10 to 6.5.15, this method is efficient and accurate. In this method, the resolved shear stresses of all active slip systems are calculated directly at the *end* of the timestep, $t+\Delta t$. From these, then, the corresponding stress tensor referred to the configuration at time $t+\Delta t$, is obtained and averaged over all crystals. Using a physics-based constitutive model for the slip rates, the strain-rate and temperature effects can readily be included, as has been shown in Section 6.5; see also Nemat-Nasser and Okinaka (1996) and Nemat-Nasser *et al.* (1998a,b). Note that the averaging process may be based on the

Taylor model, the self-consistent model, or the more appropriate and also more elaborate method suggested by Molinari *et al.* (1987, 1997). It is also possible to use a statistical approach by seeking to calculate various correlation functions, using an appropriate Green's function, as has been proposed by Garmestani *et al.* (2001), following the work of Adams *et al.* (1987, 1989).

8.8. REFERENCES

Adams, B. L., Morris, P. R., Wang, T. T., Willden, K. S., and Wright, S. I. (1987), Description of orientation coherence in polycrystalline materials, *Acta Metall.*, Vol. 35, 2935-2946.

Adams, B. L., Canova, G. R., and Molinari, A. (1989), A statistical formulation of viscoplastic behavior in heterogeneous polycrystals, *Textures Microstruct.*, Vol. 11, 57-71.

Benveniste, Y. (1987), A new approach to the application of Mori-Tanaka's theory in composite materials, *Mech. Matr.*, Vol. 6, 147-157.

Berveiller, M. and Zaoui, A. (1979), An extension of the self-consistent scheme to plastically-flowing polycrystals, *J. Mech. Phys. Solids*, Vol. 26, 325-344.

Budiansky, B. (1965), On the elastic moduli of some heterogeneous materials, *J. Mech. Phys. Solids*, Vol. 13, 223-227.

Budiansky, B. and Wu, T. T. (1962), Theoretical prediction of plastic strains of polycrystals, *Proc. 4th U.S. Natl. Congr. Appl. Mech., ASME*, 1,175-1,185.

Dvorak, G. J. (1992), Transformation field analysis of inelastic composite materials, *Proc. Royal Soc. Lond.*, Vol. A437, 311-327.

Eshelby, J.D. (1957), The determination of the elastic field of an ellipsoidal inclusion, and related problems, *Proc. R. Soc. Lond.*, Vol. A241, 376-396.

Garmestani, H., Lin, S., Adams, B. L., and Ahzi. S. (2001), Statistical continuum theory of large plastic deformation of polycrystalline materials, *J. Mech. Phys. Solids*, Vol. 49, 589-607.

Gubernatis, J. E. and Krumhansl, J. A. (1975), Macroscopic engineering properties of polycrystalline materials: Elastic properties, *J. Appl. Phys.*, Vol. 46, 1,875-1,883.

Hadamard, J. (1903, 1949), *Lecons sur la Propagations des Ondes et les Équations de l'Hydrodynamique,* Chapter 6 (Paris, 1903; Reprinted in New York, 1949) 241-262.

Hardiman, N. J. (1954), Elliptic elastic inclusion in an infinite elastic plate, *Q. J. Mech. Appl. Math.*, Vol. 52, 226-230.

Harren, S. V. (1991), The finite deformation of rate dependent polycrystals; I. A self-consistent frame work, *J. Mech. Phys. Solids,* Vol. 39, 345-360.

Hashin, Z. (1964), Theory of mechanical behaviour of heterogeneous media, *Appl. Mech. Rev.*, Vol. 17, 1-9.

Hashin, Z. (1983), Analysis of composite materials: a survey, *J. Appl. Mech.*,

Vol. 50, 481-505.

Hashin, Z. and Shtrikman, S. (1962), A variational approach to the theory of the elastic behaviour of polycrystals, *J. Mech. Phys. Solids*, Vol. 10, 343-352.

Havner, K. S. (1982), The theory of finite plastic deformation of crystalline solids, *Mechanics of Solids, The Rodney Hill 60th Anniversary Volume* (ed. H.G. Hopkins and M.J. Sewell) Pergamon Press, 265-302.

Havner, K. S. (1992), *Finite plastic deformation of crystalline solids,* Cambridge University Press, 235 pages.

Hershey, A. V. (1954), The elasticity of an isotropic aggregate of anisotropic cubic crystals, *J. Appl. Mech.*, Vol. 21, 236-241.

Hill, R. (1952), The elastic behaviour of a crystalline aggregate, *Proc. Phys. Soc. London, Sect. A*, Vol. 65, 349-354.

Hill, R. (1956), The mechanics of quasi-static plastic deformation in metals, *Surveys in Mechanics - The G.I. Taylor 70th Anniversary Volume* (ed. G.K. Batchelor and R.M. Davies), Cambridge University Press, 7-31.

Hill, R. (1961), Bifurcation and uniqueness in non-linear mechanics of continua, *Problems of Continuum Mechanics* (ed. M.A. Lavrent'ev, *et al.*) SIAM, 155-164.

Hill, R. (1962), Acceleration waves in solids, *J. Mech. Phys. Solids*, Vol. 10, 1-16.

Hill, R. (1965a), Continuum micro-mechanics of elastoplastic polycrystals, *J. Mech. Phys. Solids*, Vol. 13, 89-101.

Hill, R. (1965b), A self-consistent mechanics of composite materials, *J. Mech. Phys. Solids*, Vol. 13, 213-222.

Hill, R. (1970), Constitutive inequalities for isotropic elastic solids under finite strain, *Proc. R. Soc. Lond.*, Vol. A314, 457-472.

Hill, R. (1972), On constitutive macro-variables for heterogeneous solids at finite strains, *Proc. R. Soc. Lond.*, Vol. A326, 131-147.

Hill, R. (1984), On macroscopic effects of heterogeneity in elastoplastic media at finite strains, *Math. Proc. Camb. Phil. Soc.*, Vol. 95, 481-494.

Honneff, H. and Mecking, H. (1978), *Textures of materials*, Springer Berlin, 265.

Hori, M. and Nemat-Nasser, S. (1994), Double-inclusion model and overall moduli of multi-phase composites, *J. Eng. Mater. Tech.*, Vol. 116, 305-309.

Hutchinson, J. W. (1970), Elastic-plastic behavior of polycrystalline metals and composites, *Proc. R. Soc. Lond.*, Vol. A319, 247-272.

Hutchinson J. W. (1976), Bounds and self-consistent estimates for creep of polycrystalline materials, *Proc. R. Soc. Lond.*, Vol. A348, 101-127.

Indenbom, V. L. and Orlov, S. S. (1968), Construction of Green's function in terms of Green's function of lower dimension, *J. Appl. Math. Mech.*, Vol. 32, 414-420.

Iwakuma, T. and Nemat-Nasser, S. (1983), Composites with periodic microstructure, *Comput. Struct.*, Vol. 16, 13-19.

Iwakuma, T. and Nemat-Nasser, S. (1984), Finite elastic plastic deformation of polycrystalline metals, *Proc. R. Soc. Lond.*, Vol. A394, 87-119.

Kinoshita, N. and Mura, T. (1971), Elastic fields of inclusions in anisotropic media, *Phys. Stat. Solidi. (a)*, Vol. 5, 759-768

Kneer, G. (1965), Über die berechnung der elastizitätsmodulen vielkristalliner aggregate mit textur, *Phys. Stat. Solidi*, Vol. 9, 825-838.

Kocks, U. F. and Canova, G. R. (1981), *Deformation of polycrystals*, Riso National Laboratory, Roskilde, Denmark.

Korringa, J. (1973), Theory of elastic constants of heterogeneous media, *J. Math. Phys.,* Vol. 14, 509-513.

Kröner, E. (1953), Das fundamentalintegral der anisotropen elastischen differentialgleichungen, *Z. Phys.*, Vol. 136, 402-410.

Kröner, E. (1958), Berechnung der elastischen konstanten des vielkristalls aus den konstanten des einkristalls, *Z. Phys.*, Vol. 151, 504-518.

Kröner, E. (1961), Zur plastischen verformung des vielkristalls, *Acta Metall.*, Vol. 9, 155-161.

Kröner, E. (1967), Elastic moduli of perfectly disordered composite materials, *J. Mech. Phys. Solids*, Vol. 15, No. 319.

Kröner, E. (1980), Graded and perfect disorder in random media elasticity, *J. Eng. Mech. Division*, Vol. 106, No. EM5, 889-914.

Molinari, A., Canova, G. R., and Ahzi, S. (1987), A self-consistent approach of the large deformation polycrystal viscoplasticity, *Acta Metall.*, Vol. 35, 2,983-2,994.

Molinari, A., Ahzi, S., and Kouddane, R. (1997), A self-consistent modeling of elastic-plastic behavior of polycrystals, *Mech. Mat.*, Vol. 26, 43-62.

Mori, T. and Tanaka, K. (1973), Average stress in matrix and average elastic energy of materials with misfitting inclusions, *Acta Mèt.*, Vol. 21, 571-574.

Morris, P. R. (1970), Elastic constants of polycrystals, *Int. J. Eng. Sci.*, Vol. 8, 49-61.

Morris, P. R. (1971), Iterative scheme for calculating polycrystal elastic constants, *Int. J. Eng. Sci.*, Vol. 9, 917-920.

Morse, P. M. and Feshbach, H. (1953), *Methods of theoretical physics*, Parts I and II, McGraw-Hill.

Mura, T. (1987), *Micromechanics of defects in solids* (2nd Edition), Martinus Nijhoff Publishers, 587 pages.

Nemat-Nasser, S. (1986), Overall stresses and strains in solids with microstructure, in *Modeling small deformations of polycrystals*, (Gittus, J. and Zarka, J. eds.), Elsevier Publishers, Netherlands, 41-64.

Nemat-Nasser, S. (1999), Averaging theorems in finite deformation plasticity, *Mech. Mat.,* Vol. 31, 8493-523.

Nemat-Nasser, S. (2000), Multi-inclusion method for finite deformations: exact results and applications, *Mat. Sci. Engrg A*, Vol 285, Nos. 1-2, 239-245.

Nemat-Nasser, S. and Hori, M. (1993), *Micromechanics: overall properties of heterogeneous solids,* Elsevier Science Publishers; second, revised and extended edition, (1999).

Nemat-Nasser, S. and Hori, M. (1995), Universal bounds for overall properties of linear and nonlinear heterogeneous solids, *J. Eng. Mat. Tech.*Vol. 117,

412-432.

Nemat-Nasser, S. and Obata, M. (1986), Rate-dependent finite elasto-plastic deformation of polycrystals, *Proc. R. Soc. Lond.*, Vol. A407, 343-375.

Nemat-Nasser, S. and Okinaka, T. (1996), A new computational approach to crystal plasticity: fcc single crystal, *Mech. Mat.*, Vol. 24, 43-58.

Nemat-Nasser, S., Iwakuma, T., and Hejazi, M. (1982), On composites with periodic structure, *Mech. Mat.*, Vol. 1, 239-267.

Nemat-Nasser, S., Okinaka, T., and Ni, L. (1998a), A physically-based constitutive model for bcc crystals with application to polycrystalline tantalum, *J. Mech. Phys. Solids*, Vol. 46, 1009-1038.

Nemat-Nasser, S., Okinaka, T., and Ni, L. (1998b), A constitutive model for fcc crystals with application to polycrystalline OFHC copper, *Mech. Mat.*, Vol. 30, 325-341.

Peirce, D., Shih, C. F., and Needleman, A. (1984), A tangent modulus method for rate dependent solids, *Comput. Struct.*, Vol. 18, 875-887.

Peselnick, L. and Meister, R. (1965), Variational method for determining effective moduli of polycrystals: (A) hexagonal symmetry, (B) trigonal symmetry, *J. Appl. Phys.*, Vol. 36, 2,879-2,884.

Rashid, M. and Nemat-Nasser, S. (1992), A constitutive algorithm for rate-dependent crystal plasticity, *Comput. Meth. Appl. Mech. Engnrng*, Vol. 94, 201-228.

Rashid, M. M., Gray, G. T., and Nemat-Nasser, S. (1992), Heterogeneous deformations in copper single crystals at high and low strain rates, *Phil. Mag.*, Vol. 65, No. 3, 707-735.

Reuss, A. (1929), Berechnung der Fliessgrenze von Mischkristallen auf Grund der Plastizitätsbedingung für Einkristalle, *Z. Angew. Math. Mech.*, Vol. 9, 49-58.

Sachs, G. (1928), Zur Ableitung Fliessbedingungen, *Z. d. Ver. deut. Ing.*, Vol. 72, 734-736.

Tanaka, K., and Mori, T. (1972), Note on volume integrals of the elastic field around an ellipsoidal inclusion, *J. Elasticity*, Vol. 2, 199-200.

Taylor, G. I. (1934), The mechanism of plastic deformation of crystals- I , II, *Proc. R. Soc. Lond.*, Vol. A145, 362-387, 388-404.

Taylor, G. I. (1938), Plastic strain in metals, *J. Inst. Metals*, Vol. 62, 307-324.

Thomas, T. Y. (1956), Characteristic surfaces in the Prandtl-Reuss plasticity theory, *J. Rat. Mech. Anal.*, Vol. 5, 251-262.

Thomas, T. Y. (1958a), On the velocity of formation of Lüders bands, *J. Math. Mech.*, Vol. 7, 141-148.

Thomas, T. Y. (1958b), Plastic disturbances whose speed of propagation is less than the velocity of a shear wave, *J. Rat. Mech. Anal.*, Vol. 7, 893-900.

Thomas, T. Y. (1961), *Plastic flow and fracture in solids*, Academic Press.

Torquato, S. (2002), *Random heterogeneous materials*, Springer.

Voigt, W. (1889), Über die Beziehung zwischen den beiden Elastizitätskonstanten isotroper Körper, *Wied. Ann. Physik*, Vol. 38, 573-587.

Voigt, W. (1910), *Lehrbuch der Kristallphysik*, Teubner, Leipzig.

Walpole, L. J. (1969), On the overall elastic moduli of composite materials, *J. Mech. Phys. Solids*, Vol. 17, 235-251.

Watt, J. P., Davies, G. F., and O'Connell, R. J. (1976), The elastic properties of composite materials, *Rev. Geophys. Space Phys.*, Vol. 14, 541-563.

Weng, G. J. (1981), Self-consistent determination of time-dependent behavior of metals, *J. Appl. Mech.*, Vol. 48, 41-46.

Weng, G. J. (1982), A unified, self-consistent theory for the plastic-creep deformation of metals, *J. Appl. Mech.*, Vol. 104,728-734.

Weng, G. J. (1984), Some elastic properties of reinforced solids, with special reference to isotropic ones containing spherical inclusions, *Int. J. Eng. Sci.*, Vol. 22, 845-856.

Weng, G. J. (1990), The theoretical connection between Mori-Tanaka's theory and the Hashin-Shtrikman-Walpole variational bounds, *Int. J. Engng. Sci.*, Vol. 28, 1111-1120.

Willis, J. R. (1964), Anisotropic elastic inclusion problems, *Q. J. Mech. Appl. Math.*, Vol. 17, 157-174.

Willis, J. R. (1981), Variational and related methods for the overall properties of composites, *Adv. Appl. Mech.*, Vol. 21, 1-78.

Zeller, R. and Dederichs, P. H. (1973), Elastic constant of polycrystals, *Phys. Status Solidi B*, Vol. 55, 831-842.

Zikry, M. A. and Nemat-Nasser, S. (1990), High strain-rate localization and failure of crystalline materials, *Mech. Mat.*, Vol. 10, 215-237.

8.A CALCULATION OF THE ESHELBY TENSOR IN TWO DIMENSIONS

For simplicity in notation, let the principal axes of a typical elliptical grain coincide with the global coordinate directions. The results for other cases are then obtained by the usual tensor transformation. Define

$$\alpha = \alpha_1 / \alpha_2 , \tag{8.A.1}$$

where α_1 and α_2 are the principal radii of the elliptical grain, and set

$$\begin{aligned}
&F_1 = \mathcal{F}_{1111}, \quad F_2 = \mathcal{F}_{2111}, \quad F_3 = \mathcal{F}_{1211}, \quad F_4 = \mathcal{F}_{2211},\\
&F_5 = \mathcal{F}_{1112}, \quad F_6 = \mathcal{F}_{2112}, \quad F_7 = \mathcal{F}_{1212}, \quad F_8 = \mathcal{F}_{2212},\\
&F_9 = \mathcal{F}_{1121}, \quad F_{10} = \mathcal{F}_{2121}, \quad F_{11} = \mathcal{F}_{1221}, \quad F_{12} = \mathcal{F}_{2221},\\
&F_{13} = \mathcal{F}_{1122}, \quad F_{14} = \mathcal{F}_{2122}, \quad F_{15} = \mathcal{F}_{1222}, \quad F_{16} = \mathcal{F}_{2222}.
\end{aligned} \tag{8.A.2}$$

From (8.A.2), define the following additional non-dimensional coefficients:

$$c_1 = F_1 F_7 - F_3 F_5 ,$$

$$c_2 = F_{10} F_{16} - F_{12} F_{14} ,$$

$$c_3 = F_7 F_{13} - F_5 F_{15} ,$$

$$c_4 = F_6 F_{16} - F_8 F_{14},$$

$$c_5 = F_4 F_{10} - F_2 F_{12},$$

$$c_6 = F_1 F_{11} - F_3 F_9,$$

$$c_7 = F_5 F_{16} + F_6 F_{15} - F_7 F_{14} - F_8 F_{12},$$

$$c_8 = F_4 F_9 + F_3 F_{10} - F_2 F_{11} - F_1 F_{12},$$

$$c_9 = F_2 F_{16} - F_4 F_{14},$$

$$c_{10} = F_3 F_{13} - F_1 F_{15},$$

$$c_{11} = F_5 F_{11} - F_7 F_9,$$

$$c_{12} = F_8 F_{10} - F_6 F_{12},$$

$$c_{13} = F_5 F_{12} + F_{11}(F_6 + F_{13}) - F_7 F_{10} - F_9 (F_8 + F_{15}),$$

$$c_{14} = F_9 F_{16} + F_{10}(F_8 + F_{15}) - F_{11} F_{14} - F_{12}(F_6 + F_{13}),$$

$$c_{15} = F_5 F_{12} + F_6 (F_4 + F_{11}) - F_7 F_{10} - F_8 (F_2 + F_9),$$

$$c_{16} = F_3 F_6 + F_5 (F_4 + F_{11}) - F_1 F_8 - F_7 (F_2 + F_9),$$

$$c_{17} = F_3 F_{14} + F_4 (F_6 + F_{13}) - F_1 F_{16} - F_2 (F_8 + F_{15}),$$

$$c_{18} = F_4 F_5 + F_3 (F_6 + F_{13}) - F_2 F_7 - F_1 (F_8 + F_{15}),$$

$$c_{19} = F_3 F_{14} + F_{13}(F_4 + F_{11}) - F_1 F_{16} - F_{15}(F_2 + F_9),$$

$$c_{20} = F_{10} F_{15} + F_{16}(F_2 + F_9) - F_{12} F_{13} - F_{14}(F_4 + F_{11}) \,. \tag{8.A.3}$$

The calculation of the generalized Eshelby tensor requires $\mathcal{P}^{\Omega}_{iAjB}$ in (8.3.19b), where $\mathcal{P}_{iAjB}$ in two dimensions can be calculated directly by complex integration (8.5.22b) which involves the following line integrals on the unit circle γ:

$$R_1 = \frac{2}{2\pi i} \oint_\gamma \frac{(z^2+1)^4}{z D_0}\, dz,$$

$$R_2 = \frac{\alpha^4}{2\pi i} \oint_\gamma \frac{(z^2-1)^4}{z D_0}\, dz,$$

$$R_3 = \frac{\alpha^2}{2\pi i} \oint_\gamma \frac{(z^2+1)^2 (z^2-1)^2}{z D_0}\, dz,$$

$$R_4 = \frac{i\alpha}{2\pi i} \oint_\gamma \frac{(z^2+1)^3 (z^2-1)}{z D_0}\, dz,$$

$$R_5 = \frac{i\alpha^3}{2\pi i} \oint_\gamma \frac{(z^2+1)(z^2-1)^3}{z D_0}\, dz, \tag{8.A.4}$$

where

$$D_0 = D_0(z) = D_1 z^8 + D_2 z^6 + D_3 z^4 + \overline{D}_2 z^2 + \overline{D}_1 \,. \tag{8.A.5}$$

Here the superimposed bar stands for the complex conjugate, and

$$
\begin{aligned}
D_1 &= \{c_1 + \alpha^2(c_{13} + c_{17}) + \alpha^4 c_2\} + i\alpha\,\{c_{11} + c_{18} + \alpha^2(c_9 + c_{14})\}\,, \\
D_2 &= 4(c_1 - \alpha^4 c_2) + 2i\alpha\,\{c_{11} + c_{18} - \alpha^2(c_9 + c_{14})\}\,, \\
D_3 &= 6c_1 - 2\alpha^2(c_{13} + c_{17}) + 6\alpha^4 c_2\,.
\end{aligned}
\tag{8.A.6}
$$

Using (8.A.2) and (8.A.4), express $\mathcal{P}^{\Omega}_{\mathrm{iAjB}}$ explicitly as

$$
\begin{aligned}
\mathcal{P}^{\Omega}_{\hat{1}111} &= F_7\,R_1 + F_{16}\,R_3 + (F_8 + F_{15})\,R_4\,, \\
\mathcal{P}^{\Omega}_{\hat{2}111} &= F_3\,R_1 - F_{12}\,R_3 - (F_4 + F_{11})\,R_4\,, \\
\mathcal{P}^{\Omega}_{\hat{1}211} &= (F_8 + F_{15})\,R_3 + R_3 + F_7\,R_4 - F_{16}\,R_5\,, \\
\mathcal{P}^{\Omega}_{\hat{2}211} &= -(F_4 + F_{11})\,R_3 - F_3\,R_4 + F_{12}\,R_5\,, \\
\mathcal{P}^{\Omega}_{\hat{1}112} &= \mathcal{P}^{\Omega}_{\hat{1}211}\,, \quad \mathcal{P}^{\Omega}_{\hat{2}112} = \mathcal{P}^{\Omega}_{\hat{2}211}\,, \\
\mathcal{P}^{\Omega}_{\hat{1}212} &= -F_{16}\,R_2 + F_7\,R_3 - (F_8 + F_{15})\,R_5\,, \\
\mathcal{P}^{\Omega}_{\hat{2}212} &= F_{12}\,R_2 - F_3\,R_3 + (F_4 + F_{11})\,R_5\,, \\
\mathcal{P}^{\Omega}_{\hat{1}112} &= F_5\,R_1 - F_{14}\,R_3 - (F_6 + F_{13})\,R_4\,, \\
\mathcal{P}^{\Omega}_{\hat{2}121} &= -F_1\,R_1 + F_{10}\,R_3 + (F_2 + F_9)\,R_4\,, \\
\mathcal{P}^{\Omega}_{\hat{1}221} &= -(F_6 + F_{13})\,R_3 - F_5\,R_4 + F_{14}\,R_5\,, \\
\mathcal{P}^{\Omega}_{\hat{2}221} &= (F_2 + F_9)\,R_3 + F_1\,R_4 - F_{10}\,R_5\,, \\
\mathcal{P}^{\Omega}_{\hat{1}122} &= \mathcal{P}^{\Omega}_{\hat{1}221}\,, \quad \mathcal{P}^{\Omega}_{\hat{2}122} = \mathcal{P}^{\Omega}_{\hat{2}221}\,, \\
\mathcal{P}^{\Omega}_{\hat{1}222} &= F_{14}\,R_2 - F_5\,R_3 + (F_6 + F_{13})\,R_5\,, \\
\mathcal{P}^{\Omega}_{\hat{2}222} &= -F_{10}\,R_2 + F_1\,R_3 - (F_2 + F_9)\,R_5\;.
\end{aligned}
\tag{8.A.7}
$$

Therefore, from (8.3.19b), the two-dimensional components of the generalized Eshelby-like tensor, $\boldsymbol{\mathcal{S}}^{\Omega}$, are:

$$
\begin{aligned}
\mathcal{S}^{\Omega}_{\hat{1}111} &= c_1\,R_1 + c_{17}\,R_3 + c_{18}\,R_4 + c_9\,R_5\,, \\
\mathcal{S}^{\Omega}_{\hat{1}211} &= c_9\,R_2 + c_{18}\,R_3 - c_1\,R_4 - c_{17}\,R_5\,, \\
\mathcal{S}^{\Omega}_{\hat{1}112} &= -c_7\,R_3 + c_3\,R_4 + c_4\,R_5\,, \\
\mathcal{S}^{\Omega}_{\hat{1}212} &= c_4\,R_2 + c_3\,R_3 + c_7\,R_5\,, \\
\mathcal{S}^{\Omega}_{\hat{1}121} &= -c_{11}\,R_1 - c_{14}\,R_3 + c_{13}\,R_4 + c_2\,R_5\,, \\
\mathcal{S}^{\Omega}_{\hat{1}221} &= c_2\,R_2 + c_{13}\,R_3 + c_{11}\,R_4 + c_{14}\,R_5\,, \\
\mathcal{S}^{\Omega}_{\hat{1}122} &= c_3\,R_1 + c_4\,R_3 + c_7\,R_4\,, \\
\mathcal{S}^{\Omega}_{\hat{1}222} &= c_7\,R_3 - c_3\,R_4 - c_4\,R_5\,, \\
\mathcal{S}^{\Omega}_{\hat{2}111} &= -c_8\,R_3 + c_6\,R_4 + c_5\,R_5\,, \\
\mathcal{S}^{\Omega}_{\hat{2}11} &= c_5\,R_2 + c_6\,R_3 + c_8\,R_5\,, \\
\mathcal{S}^{\Omega}_{\hat{2}112} &= -c_1\,R_1 + c_{15}\,R_3 + c_{16}\,R_4 + c_{12}\,R_5\,,
\end{aligned}
$$

$$
\begin{aligned}
\mathcal{S}^{\Omega}_{2212} &= c_{12}\,R_2 + c_{16}\,R_3 - c_1\,R_4 - c_{15}\,R_5\,, \\
\mathcal{S}^{\Omega}_{2121} &= c_6\,R_1 + c_5\,R_3 + c_8\,R_4\,, \\
\mathcal{S}^{\Omega}_{2221} &= c_8\,R_3 - c_6\,R_4 - c_5\,R_5\,, \\
\mathcal{S}^{\Omega}_{2122} &= -c_{10}\,R_1 - c_{20}\,R_3 + c_{19}\,R_4 + c_2\,R_5\,, \\
\mathcal{S}^{\Omega}_{2222} &= c_2\,R_2 + c_{19}\,R_3 + c_{10}\,R_4 + c_{20}\,R_5\,.
\end{aligned} \tag{8.A.8}
$$

9

SPECIAL EXPERIMENTAL TECHNIQUES

Two classes of special experimental techniques and some related experimental results are presented in this chapter. The first class is for the characterization of the response of metals over a broad range of strains, strain rates, and temperatures, and the second is for the characterization of quasi-static deformation of frictional granules, supplementing the experimental techniques and results on the biaxial loading and simple shearing of photoelastic granules, presented in Section 7.5.

More specifically, after a brief historical account of the origin of the Hopkinson technique, the classical Kolsky bar method is examined in Section 9.1. Then, in Sections 9.2 and 9.3, some recent novel Hopkinson techniques that allow for recovery experiments at various strain rates and temperatures, are discussed in detail. This includes in Section 9.2, methods to trap both tensile and compressive elastic stress pulses at the free end of a long elastic bar; methods to generate a compression pulse that is trailed by a tension pulse of similar (but reversed) time-profile in such a bar; methods to subject a sample to a single compression pulse, or a single tension pulse, or to single combined compression that is followed by a tension pulse, and then recover the sample without it having been subjected to any other stress pulses; and methods to produce a strain-rate jump (an increase or a decrease in the imposed strain rate) during a high strain-rate deformation of a sample. Then, in Section 9.3, more recent novel techniques are discussed, by means of which recovery tests similar to those described in Section 9.2, can be performed over a broad range of temperatures, from 77 to 1,300K. By these techniques it is possible to obtain the isothermal, as well as adiabatic, flow stress of many metals at large strains, over a broad range of temperatures and strain rates, as has been mentioned and illustrated in Sections 4.8 and 6.5. In Section 9.4, a recently developed dynamic triaxial Hopkinson device is briefly reviewed.

Section 9.5 of this chapter includes a detailed discussion of the triaxial deformation of cohesionless saturated granular materials, using large hollow cylindrical samples, 25cm high, with inner and outer diameters of 20cm and 25cm, respectively. Special techniques used for sample preparation and testing are reviewed, and typical results are given for Silica 60 as the granular material. In particular, it is shown that a unique relation between the pore water pressure and the energy dissipation exists in undrained cyclic shearing of saturated samples. Some experimental results on densification of drained samples and liquefaction of saturated undrained samples are also presented.

Section 9.6 summarizes novel X-ray techniques recently developed to monitor the deformation of frictional granules in both monotonic and cyclic shearing under confining pressure. To capture the internal microstructural changes during the shearing deformation of the sample, thin columns and rows of lead silicate granules of distinct color are embedded inside the wall of the specimen of granular materials, and a series of X-ray photographs is taken, in both drained and undrained experiments. In drained tests, the microstructure of shearbands that are produced by monotonic deformation, is documented. In undrained tests, the local deformation of the granular mass during liquefaction, induced by cyclic shearing, is captured, and it is directly observed that shear localization does actually occur in liquefied specimens.

9.1. HOPKINSON BAR, KOLSKY BAR, AND RECOVERY SYSTEMS

To obtain the stress-strain relations for metals in the range of 10^2 to 10^4/s strain rates, the split Hopkinson bar is widely used. In this technique, uniaxial compression and tension, as well as torsional states of stress are produced in a sample, at various deformation rates and various stress-pulse durations that control the total plastic deformation. In this technique, if the sample does not fail during the passage of the first stress pulse, it is then loaded repeatedly by pulses which reflect off the free ends of the bars. Thus, the technique is suitable for obtaining the dynamic stress-strain relations to failure. It does not easily allow the recovery of the specimen at various levels of loading, for the microscopic and related analyses which are necessary for the understanding of the microstructural evolution associated with loading histories.

To relate the microstructure to the mechanical properties of materials, *recovery experiments with time-resolved data* at various strain rates are required. In a dynamic recovery experiment, the sample is subjected to a single preassigned stress pulse, and then recovered without it having been subjected to any additional loading. The recovery Hopkinson techniques are discussed in this and the next two sections of this chapter, and in Section 9.4, a new dynamic triaxial Hopkinson technique is briefly outlined.

9.1.1. Historical Origin of Hopkinson Techniques

Historically, the Hopkinson bar experimental technique finds its origin in the work of John Hopkinson (1901) that dates back to the 1870's, and later his son, Bertram Hopkinson (1905, 1914), who sought to establish the dynamic strength of metal wires, experimentally. Bertram Hopkinson is credited with developing the first time-resolved measurement technique to produce a pressure-pulse profile. For this contribution and for the pioneering effort of his father, the technique has been named after them.

The next important contribution was made by Davies (1948). Davies, in a detailed critical study of the Hopkinson technique, describes a method in which the displacement of the free end of the bar is measured using a parallel-plate condenser, with the end of the bar serving as the grounded plate. The amplified output of this microphone is then recorded and used to obtain the displacement history of the bar's end. Thus, this provided a viable technique for direct measurement of the stress (or strain) profile in a long elastic bar.

Davies' work was followed by a major contribution by Kolsky (1949), who used *two* elastic bars[1] with the specimen sandwiched in between, and, hence, created a technique which now is known as *the split Hopkinson bar*, or *the Kolsky bar method*, and is widely used to study the dynamic properties of many materials over a range of strain rates in essentially uniform uniaxial stress

[1] This is in contrast to the Hopkinsons and to Davies, who focused on a single bar.

states. Kolsky used Davies' technique to measure the stress profiles in both bars, and related these to the stress-strain history of the sample, through a one-dimensional elastic-wave analysis.

First Harding *et al.*(1960), and later Lindholm and Yeakley[2] (1968), introduced novel modifications to this technique which allowed experiments in uniaxial *tension*. The next innovation was by Baker and Yew (1966), who modified the split Hopkinson bar for dynamic *torsion* tests. For a review of these classical techniques and additional references, the reader is referred to Follansbee (1985), Hartley *et al.* (1985), Nicholas and Bless (1985), and Gray (2000).

9.1.2. Classical Kolsky Method

In the classical compression or tension split Hopkinson bar technique, *i.e.*, in the Kolsky bar method, there are two *elastic* bars, called the *incident* and the *transmission* bar, respectively, and a dynamic loading device, usually a gas gun with an elastic *striker* bar or tube. Figure 9.1.1 is a sketch of a compression Kolsky bar system.

Figure 9.1.1

The classical compression split Hopkinson bar, or the Kolsky bar, consisting of: a gas gun with an elastic striker bar, an elastic incident bar, and an elastic transmission bar

In a compression test, the gas gun propels a striker bar that imparts an elastic stress pulse into the incident bar. This pulse travels along the incident bar towards the sample (specimen) that is sandwiched between the incident and the transmission bars. Once the pulse reaches the specimen, it is partly transmitted to the transmission bar through the specimen (loading), and is partly reflected as tension back into the incident bar. From the transmitted pulse, the stress in the sample is calculated using the strain record obtained through a strain gauge placed appropriately (usually at midpoint) on the transmission bar. From the reflected pulse, the strain in the sample is estimated by integrating the strain measured by a strain gauge suitably attached (also, usually at midpoint) to the incident bar.[3]

---

[2] See also Lindholm (1964).

[3] The collected data generally requires corrections for the specimen's size effect, the end friction between the specimen and the bars, the lateral inertia constraint in the sample, and most impor-

The tension Kolsky system is essentially the same as the compression one, except that the loading mechanism imparts a tensile elastic pulse into the incident bar. This is accomplished by, for example, using a tubular striker that is propelled by a gas gun through which the incident bar passes, as further detailed in Section 9.2; see Figure 9.2.2, page 668.

9.1.3. Limitation of the Kolsky Method

In the Kolsky technique, if the sample does not fail in the course of the first loading, it will be subjected to repeated loading by the elastic waves that reflect off the free ends of the two bars and travel back and forth along the bars. This makes the recovery tests difficult by this classical technique. There are many applications, where it is necessary to subject the sample to a single loading and then *recover* it, without the sample having been subjected to any additional loads. Such a capability can be used, for example, to relate the microstructural changes to the loading history, or, as is discussed in Sections 4.8 and 6.5, to produce isothermal stress-strain relations for metals, at high strain rates.[4] For tension tests, and in the absence of wave trapping, both the transmitted and reflected pulses return and reload the sample repeatedly.

For both compression and tension tests, it is possible to trap both the reflected and transmitted pulses at the far-end of the bars, once the sample is subjected to the initial loading. This ensures that the sample is subjected to a single stress pulse. The time-variation of this pulse can also be controlled using a *pulse shaper*, as is discussed in Subsection 9.2.8.

In addition to the single compression or tension loading, it is also possible to subject a sample to a single compression pulse which is then followed by a single tension pulse, and then recover the sample without it having been subjected to any other loading pulses. This technique permits the study of the Bauschinger effect and the strain-rate history effects on the mechanical properties and microstructural evolution of materials under various dynamic loading

---

tantly, the elastic-wave dispersion to account for the fact that the interaction between the sample and the bars involves three-dimensional elastic deformation of the bars, whereas the actual measurements are made far away at the middle of the bars. Choosing a proper length to diameter ratio and properly lubricating the contact surfaces, some of these effects are minimized and rendered insignificant. In addition, dispersion corrections to data are generally performed. For these and related issues, see Davies and Hunter (1963), Lindholm (1964), and Follansbee (1985).

[4] For compression tests one may use special fixtures, such as "stopper rings", to limit the total axial strain of the sample and to transmit the remaining compression pulse through the stopper ring, once the sample length equals that of the ring; see, for example, Hartman *et al.* (1986) and Nemat-Nasser *et al.* (1988a). A similar technique does not exist for tension or torsion tests. Moreover, even in compression, such a technique works only if large axial permanent straining of the sample is involved, and, even then, the sample with the stopper ring is repeatedly loaded elastically. In addition, for hard brittle materials, such as ceramics and their composites, the total axial strain to failure is very small, and, therefore, a "stopper-ring approach" is difficult to implement. For this class of materials, therefore, once the initial compressive pulse has produced microcracks in the brittle sample, the subsequent reflected compression pulses may shatter the specimen, making recovery essentially impossible by the classical technique.

conditions.[5]

9.2. MOMENTUM TRAPPING FOR TENSION AND COMPRESSION HOPKINSON BARS

In a recovery tension Hopkinson bar, both the compression pulse that reflects off the sample, and the tension pulse that is transmitted through the sample, must be trapped, once the sample has been loaded. In a compression bar, on the other hand, only the tension pulse that reflects off the sample into the incident bar, must be trapped, since the transmitted compression returns to the sample as tension and cannot reload the sample.

Figure 9.2.1a shows the momentum trapping scheme for a tension pulse. The free end of the bar terminates in a *transfer flange* F, in contact with a *momentum-trap tube* MT which has the same impedance as the bar; *e.g.*, the tube has the same cross-sectional area and is made of the same material as the bar. The tensile pulse T in Figure 9.2.1a, reflects off the free end of the transfer flange F, as compression. This compressive pulse is then transferred into the momentum-trap tube MT in contact with the flange, because of their matched impedances. The compression then reflects off the free end of the tube as tension, and is trapped in the tube which begins to move away from the transfer flange, once this reflected tension pulse reaches the tube's end in contact with the flange. This process is similar to transferring *compression* pulses across contacting bars with matched impedances. In Figure 9.2.1b, the compression C is completely transmitted from bar A to bar B across the contact surface S_c, when the bars have the same impedance. This transmitted compressive pulse then reflects from the free end of the bar B as tension, and is trapped in B.

9.2.1. Tension Hopkinson Bar with Momentum Traps

In a tension Hopkinson bar, the momentum trapping scheme of Figure 9.2.1a,b is used to trap both:

(1) the compression pulse which reflects off the sample, back into the incident bar, and

(2) the tension pulse which is transmitted through the sample into the transmission bar.

A recovery tension Hopkinson system therefore, includes special momentum traps at the loading end of the incident bar and at the free end of the transmission bar. A complete system consists of the following basic elements:

[5] Note that, using a tension recovery test, the sample can be subjected to a single tensile pulse and recovered. This recovered sample may then be used to make new compression samples to be tested in a compression recovery system; see Thakur (1994) and Thakur *et al.* (1996a,b).

Figure 9.2.1

Momentum trapping procedure in a recovery split Hopkinson bar technique: (a) momentum-trap tube MT, in contact with transfer flange F, traps the tension pulse T, once the pulse reflects off the free end of the bar as compression; and (b) bar B in contact with bar A traps the compression pulse C

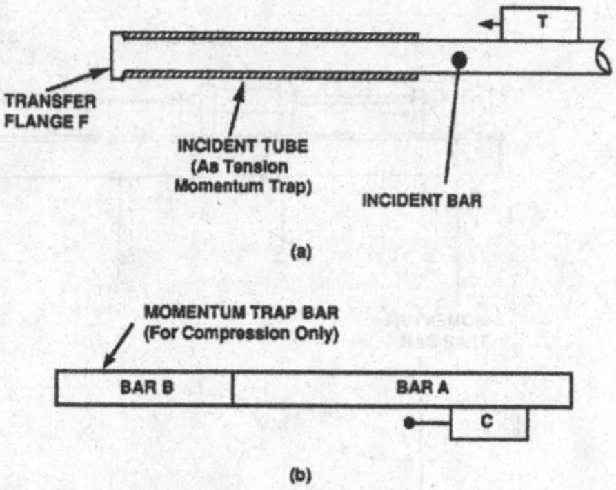

(a) a striker tube at the loading end,

(b) an incident bar with a transfer flange at its loading end,

(c) a gas gun through which the incident bar passes,

(d) a momentum-trap bar at the loading end of the incident bar,

(f) a transmission bar with a transfer flange at its free end, and

(g) a momentum-trap tube on the transmission bar, contacting the transfer flange at the free end of this bar.

Figure 9.2.2a is a sketch of a complete system, and Figure 9.2.2b is the photograph of the loading end of the first system, designed and constructed at the author's laboratory in 1988.[6] A complete system is shown in photograph 9.2.2c. The transmission bar terminates in a momentum-trap tube which is the mirror image of the one sketched in Figure 9.2.1a.

A pre-set precision gap separates the momentum-trap bar and the transfer flange of the incident bar. The precision gap is set such that the end of the momentum-trap bar and the face of the transfer flange are brought into contact, once the entire tensile pulse, produced by a tubular striker, is transferred into the incident bar through the transfer flange. This tensile pulse then travels towards the sample, where it is partly transmitted into the transmission bar, and is partly reflected as compression back into the incident bar. This reflected compression (often only a few percent less than the incident pulse) is then transmitted into the momentum-trap bar, and reflects off the other (free) end of this bar as a tensile pulse. This pulse is then trapped in the momentum-trap bar, since the contact interface with the transfer flange cannot support tension. The momentum of the trapped pulse causes the bar to fly off. The incident bar then is everywhere at

[6] This and other novel experimental devices for high deformation-rate tests and the corresponding data acquisition systems were developed by Jon Isaacs in collaboration with Dr. John E. (Skip) Starrett and the author, and later reported in Nemat-Nasser *et al.* (1988b, 1991) and elsewhere. Dr. Starrett passed away suddenly on May 13, 1990, at the young age of 47.

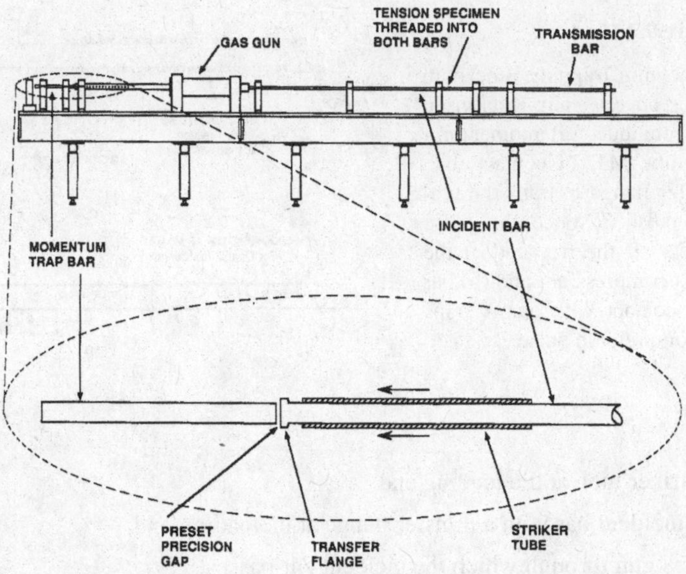

(a)

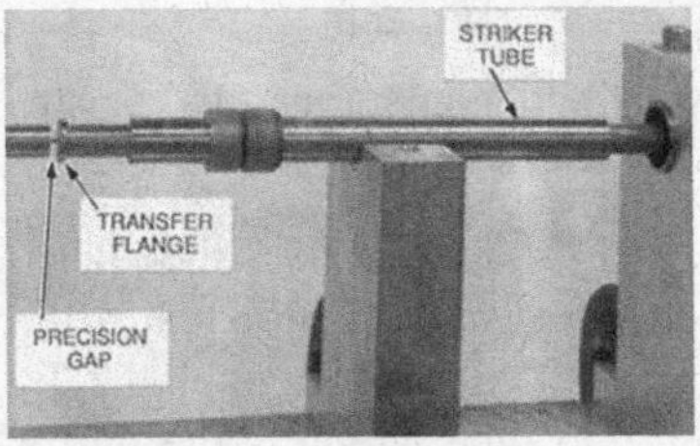

(b)

(c)

Figure 9.2.2

(a) Sketch of a complete recovery tension Hopkinson bar system; (b) photograph of the first recovery tension Hopkinson system (photo from Nemat-Nasser *et al.*, 1991); and (c) photograph of a complete tension Hopkinson system

rest. The pulse which is transmitted through the sample into the transmission bar, is also trapped at the free end of this bar by the momentum-trap tube. This free end of the transmission bar includes a transfer flange and a momentum-trap tube in contact with the flange.

At the striker end of the incident bar, the precision gap between the transfer flange and the momentum-trap bar must be calculated and set with some care to obtain satisfactory results. If the gap is too large, the reflected compression pulse in the incident bar will not be transmitted to the momentum-trap bar. On the other hand, if the gap is too small, it will be closed during the loading of the incident bar. This affects the stress pulse, moves the momentum-trap bar into an improper position, and, hence, produces undesirable results. The momentum-trap bar is correctly positioned when the gap is set such that, once the initial collision between the striker and the incident bar is complete, the incident bar is fully in contact with the momentum-trap bar. While the necessary separation can be calculated, in practice this provides a starting point which then is optimized by trial and error.

9.2.2. Stress Reversal Technique

Using a flanged incident bar that carries a momentum-trap tube and a *reaction mass*, it is possible to produce a compression pulse followed by a tension pulse in the incident bar, *i.e.*, to create a *stress reversal Hopkinson bar* system. Figure 9.2.3 is the photograph of the system's loading end. The incident tube, the striker bar, and the incident bar in this photograph have the same impedance. The striker bar and the incident tube are the same length. The incident tube rests against the transfer flange at one end and against the reaction mass at the other. The reaction mass is a large steel cylinder through which the incident bar passes.

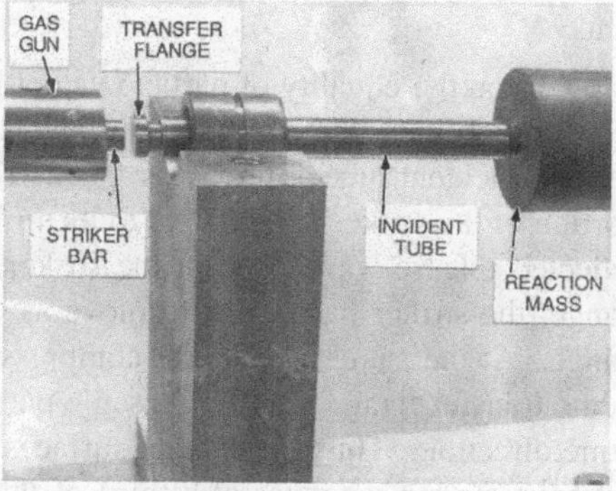

Figure 9.2.3

Photograph of the first stress reversal Hopkinson system, showing the gas gun barrel, the striker bar, the incident bar with a transfer flange, the momentum-trap tube, and the reaction mass (photo from Nemat-Nasser *et al.*, 1991)

When the striker bar impacts the transfer flange at velocity V_0, it imparts a common axial strain to the incident tube and incident bar. The compression pulse in the incident bar travels along this bar towards the specimen. The compression pulse in the incident tube reflects from the reaction mass *as compression*, and reaches the transfer flange at the same instant that the tension release pulse which is reflected from the free end of the striker, reaches the striker bar's end in contact with the transfer flange. Since the combined cross section of the incident bar and tube is twice that of the striker bar, having the same material properties, the striker bar begins to bounce back, away from the transfer flange, as the transfer flange is loaded by the compression pulse traveling along the incident tube. This compression pulse then imparts a tensile pulse to the incident bar, which follows the then existing compression pulse in the incident bar, both pulses traveling towards the sample. Figure 9.2.4 is a typical result.

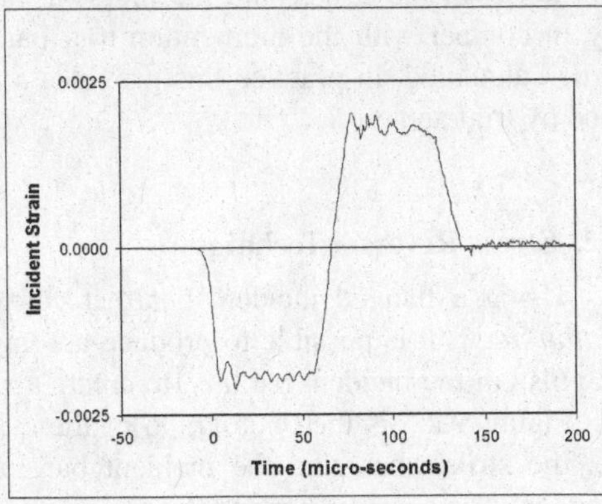

Figure 9.2.4

The time-resolved record of compression pulse followed by a tension pulse, produced using the system shown in Figure 9.2.3

In this technique, it is important to ensure a perfect impedance match of the striker bar, the incident bar, and the incident tube. Also, the configuration of the transfer flange plays an important role in the proper functioning of the system.

From the equality of particle velocities at contacting surfaces and the conservation of linear momentum, it follows that the particle velocity in the incident bar and incident tube at impact is $V_0/3$. The particle velocity of the compressed portion of the striker bar, relative to its unstressed part, then is $2\,V_0/3$. When the reflected release-wave front in the striker reaches the contact with the transfer flange, the striker bar begins to move at velocity $V_0/3$, away from the transfer flange. At the same instant, the compression wave in the incident tube imparts to the transfer flange exactly the same particle velocity, *i.e.*, $V_0/3$, in exactly the same direction. Thus the contact surfaces remain in contact for a short period,[7] until the striker separates at a third of the initial impact velocity, as the flange comes to rest.

9.2.3. Compression Hopkinson Bar with Momentum Trap

To render the classical compression Kolsky bar as a recovery system, it is only necessary to trap the tension pulse that reflects off the sample into the incident bar and travels back towards the loading end of this bar. This is accomplished by simply adding a momentum trap system to the loading end of the incident bar, as is shown in Figure 9.2.3. A complete recovery compression

[7] This time period is given by $t_0 = 2\,l_0/C_0$, where l_0 is the common length of the striker bar and the incident tube, and C_0 is their common longitudinal elastic-wave speed.

system is sketched in Figure 9.2.5a.

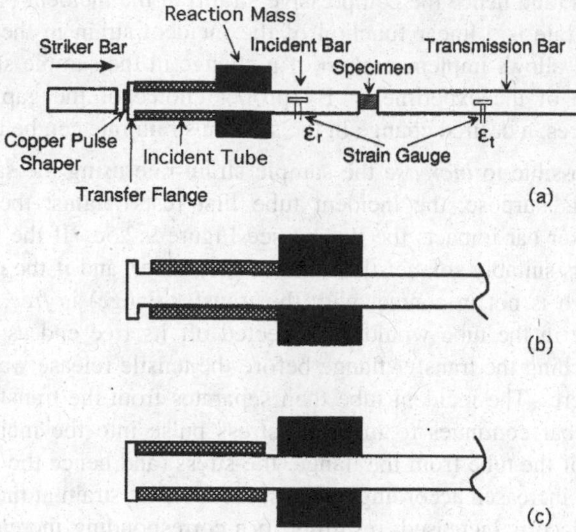

Figure 9.2.5

(a) Recovery compression Hopkinson system; (b) a gap placed between the incident tube and the transfer flange produces a *decrease in the strain rate*, once the gap is closed; and (c) a gap placed between the incident tube and the reaction mass allows the incident tube to separate from the transfer flange, producing an *increase in the strain rate* (Nemat-Nasser *et al.*, 1994)

9.2.4. Strain-rate Jump

The stress-reversal technique can be used to change the strain rate during the course of Hopkinson bar compression or tension experiments. Consider a *recovery compression system* first. The loading (striker) end of the incident bar includes a transfer flange, an incident tube, and a reaction mass, as sketched in Figure 9.2.5a.' Suppose the transfer tube is in contact with the transfer flange, prior to loading. Then, as the striker bar impacts the incident bar over the transfer flange, elastic compressive stresses are induced in both the incident bar and the incident tube, as well as in the striker bar, in accordance with their impedances. To change (decrease) the stress (and hence the strain) in the incident bar in the course of the experiment, *a predetermined gap is introduced between the transfer flange and the incident tube*, as is sketched in Figure 9.2.5b. As the striker bar now impacts the incident bar, a compressive stress pulse is imparted into this bar, resulting in the shortening of the bar. Once the incident bar is sufficiently shortened to come in contact with the incident tube over the transfer flange, a stress pulse is also imparted into the incident tube, reducing the compressive stress (and hence the compressive strain) in the incident bar. Since the sample strain rate is a linear function of the incident strain in the incident bar, the procedure allows implementation of a change in the sample strain rate during the course of the experiment. By proper choice of the gap and the involved impedances, a desired change in the sample strain rate can be achieved.

It is also possible to *increase* the sample strain rate using the same construction. For this purpose, the incident tube first rests against the transfer flange, as the striker bar impacts the flange; see Figure 9.2.5c. If the length of the incident tube is suitably smaller than that of the striker, and if the other end of this tube (which is not in contact with the transfer flange) is *free*, then the *compressive* pulse in the tube would be reflected off its free end as a *tensile* release pulse, reaching the transfer flange before the tensile release wave in the striker reaches there. The incident tube then separates from the transfer flange while the striker bar continues to impart a stress pulse into the incident bar. Upon separation of the tube from the flange, the stress (and hence the strain) in the incident bar is increased accordingly. Thus, the incident strain at the sample-incident bar interface is increased, resulting in a corresponding increase in the sample strain rate.

For numerical estimates, let ρ, C, E, A, u, ε, and σ, respectively denote the mass density, longitudinal wave speed, Young's modulus, area, particle velocity, axial strain, and axial stress, and use subscripts st, i, and tb for the striker bar, incident bar, and the incident tube, respectively. Set $k = \rho C A$, and note that $C^2 = E/\rho$.

Since the bar and the tube remain elastic throughout the experiment, in general,

$$\sigma = \rho C u = (\rho E)^{\frac{1}{2}} u = E \varepsilon, \tag{9.2.1a-c}$$

$$\varepsilon = \frac{u}{C} = \frac{\sigma}{E}. \tag{9.2.2a,b}$$

While a gap separates the incident tube and the transfer flange (see Figure 9.2.2b), the following relations hold:

$$k_i u_i = k_{st} u_{st}, \qquad V_0 - u_{st} = u_i, \tag{9.2.3a,b}$$

where V_0 is the (constant) initial velocity of the striker bar. From (9.2.1) to (9.2.3), it follows that

$$\varepsilon_i = \frac{k_{st}}{k_{st} + k_i} \frac{V_0}{C_i}. \tag{9.2.4a}$$

This strain in the incident bar is the incident strain at the sample-end of the incident bar. If the strain in the transmission bar is denoted by ε_t, then the strain rate in the sample, $\dot{\varepsilon}_s$, is given by

$$\dot{\varepsilon}_s = 2\,\frac{C_i}{l_s}(\varepsilon_i - \varepsilon_t)$$

$$= 2\,\frac{C_i}{l_s}\left[\frac{k_{st}}{k_{st}+k_i}\,\frac{V_0}{C_i} - \varepsilon_t\right], \tag{9.2.5a,b}$$

where l_s is the sample length.[8]

When the gap between the transfer flange and the incident tube is closed (see Figure 9.2.5c), the incident strain is reduced to

$$\varepsilon_i = \frac{k_{st}}{k_{st}+k_i+k_{tb}}\,\frac{V_0}{C_i}, \tag{9.2.4b}$$

resulting in the sample strain rate,

$$\dot{\varepsilon}_s = 2\,\frac{C_i}{l_s}\left[\frac{k_{st}}{k_{st}+k_i+k_{tb}}\,\frac{V_0}{C_i} - \varepsilon_t\right]. \tag{9.2.5c}$$

If the sample material is rate-independent, then ε_t will be the same in (9.2.5b) and (9.2.5c), and the change in the sample strain rate becomes

$$\Delta\dot{\varepsilon}_s = 2\,\frac{V_0}{l_s}\left[\frac{k_{st}}{k_{st}+k_i+k_{tb}} - \frac{k_{st}}{k_{st}+k_i}\right]$$

$$= -2\,\frac{V_0}{l_s}\left[\frac{k_{st}k_{tb}}{(k_{st}+k_i)(k_{st}+k_i+k_{tb})}\right]. \tag{9.2.6a,b}$$

When $k_{st} = k_i = k_{tb}$, this yields

$$\Delta\dot{\varepsilon}_s = -\frac{1}{3}\,\frac{V_0}{l_s}. \tag{9.2.6c}$$

Expressions (9.2.6b,c) also apply to the case when the strain rate is *increased*; Figure 9.2.5c. In this case, the minus sign on the right-hand side changes to a plus.

As an example, consider a striker velocity of 30 m/s and sample length of 5 mm. Then, $\Delta\dot{\varepsilon}_s = \mp 2\times10^3/\mathrm{s}$, which applies to both a *reduction* and an *increase* in the sample strain rate, depending on whether or not the incident tube is in contact with the transfer flange prior to the striker impact.

For rate-dependent materials, the flow stress changes with the strain rate. Hence the transmitted strain ε_t in the transmission bar also changes accordingly. Then,

$$\varepsilon_t = \frac{\sigma_s A_s}{E_t A_t}, \tag{9.2.7}$$

where subscripts s and t refer to the sample and transmission bar, respectively.

---

[8] Note that, in general, the incident and transmission bars have common area, wave speed, etc.

Hence, if the change in the sample flow stress is $\Delta\sigma_s$, we obtain

$$\Delta\dot{\varepsilon}_s = -2\frac{V_0}{l_s}\left[\frac{k_{st}k_{tb}}{(k_{st}+k_i)(k_{st}+k_i+k_{tb})}\right] - 2\frac{C_i\,A_s}{l_s\,E_t\,A_t}\Delta\sigma_s\,, \tag{9.2.6d}$$

where $\Delta\dot{\varepsilon}_s$ is negative when the strain rate (and hence the flow stress) is decreased.

Figure 9.2.6 illustrates the results of this method. Point A on the lower curve in this figure marks the strain in the sample, at which the strain rate is changed from about 1,600/s to about 850/s; the strain-rate difference is about 750/s. The impact velocity of the striker bar is 8 m/s; the Young modulus of the incident bar is 200 GPa. The diameter of the transmission bar is 1.27 cm. The length of the sample is 0.35 cm. The stress difference in Figure 9.2.6 due to the change of the strain rate is 50 ksi (350 MPa). From (9.2.6b) and these data, a strain-rate difference of 757/s is obtained; see Nemat-Nasser *et al.* (1994). This calculated result is very close to the observed one.

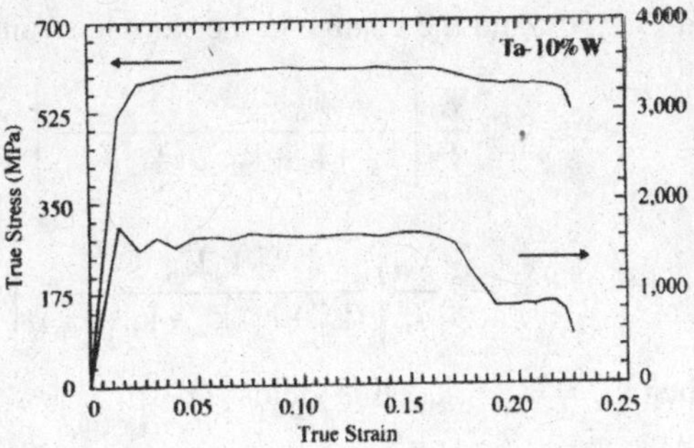

Figure 9.2.6

Example of the stress-strain relation with a sudden change in strain rate: the strain rate (lower curve) is changed from about 1,600/s to about 850/s around 18% true strain

9.2.5. Strain-rate Jump in Tension

The tension bar with a momentum trap can be used to change the strain rate during the course of tension Hopkinson tests. First consider a case where it is required to *decrease* the strain rate during a test. To accomplish this, provide a gap between the momentum-trap bar and the transfer flange of the incident bar in Figure 9.2.2b such that this gap is closed while the striker tube is transferring the load to the transfer flange. By suitable choice of the impedances of the incident bar, momentum-trap bar, and the striker tube, it is then possible to suddenly decrease the strain rate by a desired amount. The basic equations follow from those given in Subsection 9.2.4.

To *increase* the strain rate in a tension Hopkinson test, use a suitably short momentum-trap bar in contact with the flange, *before* the striker tube impacts the transfer flange. Then, the striker tube imparts a tensile pulse into the incident bar and, simultaneously, a compression pulse into the momentum-trap bar.

This compression pulse reflects off the other end of the momentum-trap bar as a tensile release wave, separating the bar from the flange, once it reaches the contact between the bar and the flange. For a suitably long striker tube, this can occur while the striker is continuing to load the transfer flange. Again, the necessary equations are similar to those given in Subsection 9.2.4. Note that the impedances must be properly selected so that a desired increase in the strain rate is achieved.

9.2.6. Recovery After Combined Compression-tension Loading

In many high strain-rate applications, materials may be subjected to dynamic compressive loads, followed by tension. An example is the examination of the dynamic Bauschinger effect and its relation to the material's microstructure. Experimental investigations of this kind require techniques to recover the sample after it has been subjected to a cycle of compression followed by tension.

The stress reversal technique of Subsection 9.2.2 can be used to subject a specimen to a combined compression-tension pulse *without interruption*, and then to recover it for post-test characterization. The loading fixture is the same as in the case of stress reversal, Figure 9.2.3. The sample is subjected to a combined stress pulse similar to that shown in Figure 9.2.4. The sample configuration and the manner by which it is attached to the incident and transmission bars are designed such that, after the specimen is subjected to the desired compression-tension pulse, it is detached from the incident bar without additional loading of its *central gauge portion*. This is accomplished as follows.

As shown in Figure 9.2.7a (see also photograph in Figure 9.2.7b), in addition to the reaction mass R_0, there are two other reaction masses, denoted by R_1 and R_2. The sample construction (see photograph in Figure 9.2.7c) includes two threaded ends and a reaction flange F_S which, at the start of the experiment, rests against the reaction mass R_1, with the sample end fully threaded into the incident bar.

The sample is threaded into the transmission bar through the reaction mass R_2. The transmission bar rests against a thin flange F_2, constructed in R_2 by reducing its opening diameter around the sample, as shown in Figure 9.2.7a. The other (output) end of the transmission bar is in contact with a bar of the same impedance, which can trap compression pulses transmitted through the sample into the transmission bar. When the striker hits the transfer flange of the incident bar, a compression pulse trailed by a tension pulse is imparted into the incident bar, as discussed in Subsection 9.2.2. The compression-tension pulse travels towards the sample, and subjects the sample to compression first (which shortens the sample), followed by tension (which almost restores the initial length of the sample). A part of the compression pulse is transmitted through the sample into the transmission bar and is trapped at the far end of this bar by the momentum-trap bar (hatched in Figure 9.2.7a). The remaining part of the compression pulse is reflected off the specimen into the incident bar, as tension. This tension then reflects back from the striker end of the incident bar and returns to the sample as *tension*, because of the presence of the transfer flange, incident tube, and the reaction mass R_0, as discussed in Subsection 9.2.2.

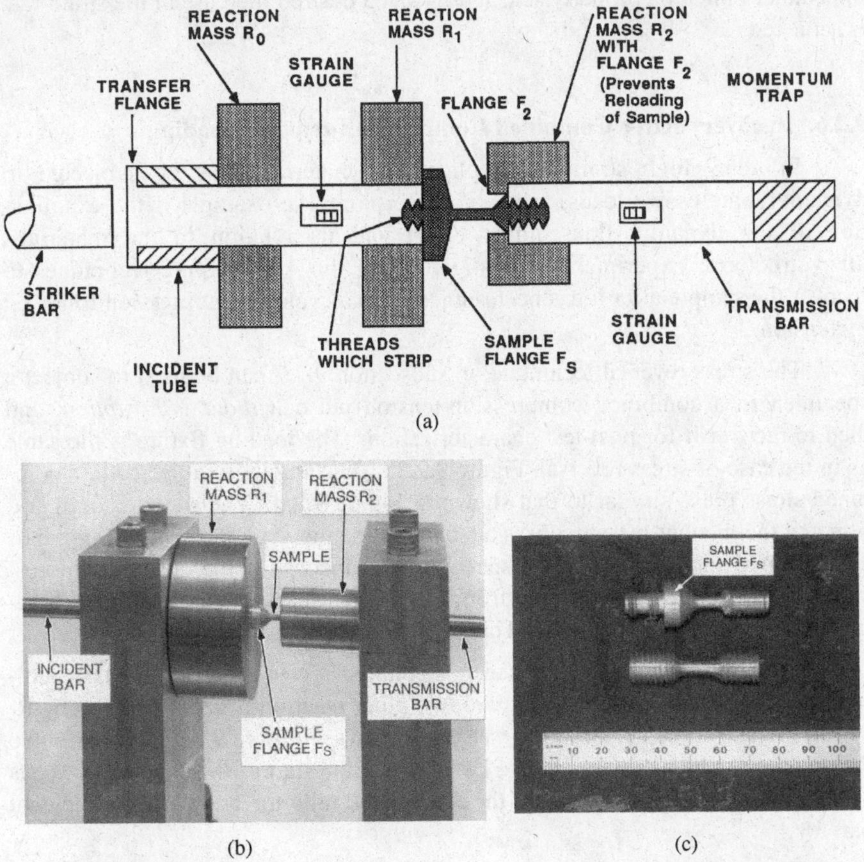

Figure 9.2.7

(a) Sketch of a complete setup for combined compression-tension dynamic *recovery* experiment; (b) photograph showing how the sample is connected to the incident and transmission bars; and (c) two samples of the same material (Al-Cu), one flanged for recovery test, and the other unflanged for data acquisition (from Nemat-Nasser *et al.*, 1991)

When the tensile segment of the initial compression-tension pulse reaches the specimen, it is partly transmitted through the sample into the transmission bar, and is partly reflected off the sample as compression, back into the incident bar. As the sample shortens during the compressive loading, prior to the arrival of the tension, its flange, F_S, separates from the reaction mass R_1. The subsequent tensile loading (almost) restores the sample's initial length, bringing the sample's flange (almost) in contact with the reaction mass R_1. The tensile pulse which reflects as compression from the far end of the transmission bar, returns to the sample, loading it (just slightly) until the transmission bar comes in contact with the flange F_2 of the reaction mass R_2. At this instant, the sample flange F_S is in complete contact with the reaction mass R_1. Shortly after, the tensile pulses which are reflected from the loading end of the incident bar reach the sample, pulling the incident bar away from the sample, while the sample flange F_S is engaged against the reaction mass R_1.[9] The threads which connect the sample to the incident bar are designed to strip off during this final tensile loading process. Note that the threads are pulled against the flange F_S resting on R_1. This prevents loading the sample over its gauge length, during this recovery stage. The sample is then recovered, having been subjected to a cycle of compression-tension loading only over its gauge length.

The flange F_S in combination with the threaded ends of the sample, makes the impedance matching difficult, and hence complicates the interpretation of the resulting data. This problem can be overcome to a great extent by performing two tests, one to recover the sample (which has a flange) for post-test analysis, and the other (using a sample which does not have a flange) to collect the required data; Figure 9.2.8. Even though the data obtained with the flanged sample are noisy, they agree quite well with those obtained with the unflanged sample. The unflanged sample, on the other hand, is subjected to additional tensile pulses, which may cause necking, or even tensile failure of the sample.

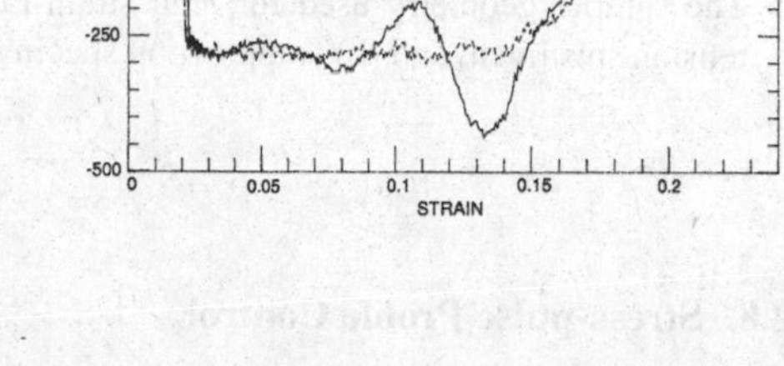

Figure 9.2.8

Data obtained at 3,200/s strain rate, using flanged (solid curve) and unflanged (dashed curve) specimens (from Nemat-Nasser *et al.*, 1991)

[9] As discussed in Subsection 9.2.3, both compression and tension pulses are reflected as *tension* from the loading end back into the incident bar.

9.2.7. Recovery After Tension and Compression Loading

The recovery tension and compression Hopkinson techniques can be used to subject a sample to a *tensile* pulse, recover the sample, produce from it compression samples (see Figure 9.2.9), and then subject these samples to compression and recover them. This allows one to study the response and microstructural changes when the material is subjected to a tension, recovered, and then subjected to a compression (Thakur *et al.*, 1996a,b). In this approach, the tensile loading is not followed immediately by a compression pulse. Hence, if there are any microstructural changes during the recovery and preparation of the compression samples from the original tension sample, then these changes may affect the subsequent compression test results. Indeed, the effects of such changes on the material's response can be studied by this method.

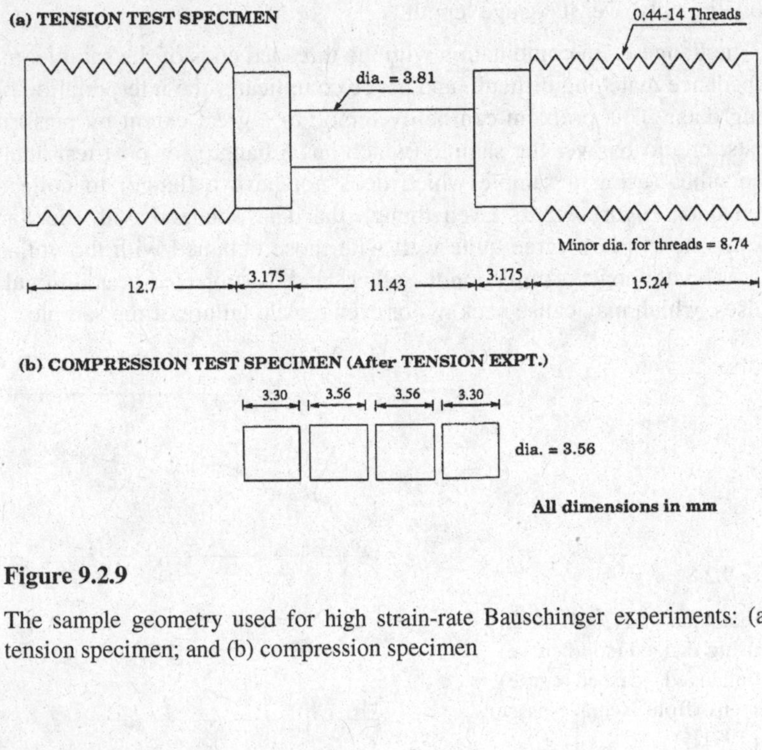

Figure 9.2.9

The sample geometry used for high strain-rate Bauschinger experiments: (a) tension specimen; and (b) compression specimen

9.2.8. Stress-pulse Profile Control

There are inherent difficulties with testing extremely hard, brittle materials, using the Hopkinson techniques. In the compression split Hopkinson bar, the reflected pulse, in general, no longer accurately measures the strain rate in the sample. Since it is often required that the diameter of a very hard sample be considerably smaller than that of the bars, then the hard sample indents the bars' ends, as is sketched in Figure 9.2.10. This indentation in general introduces a significant error in the measured results, if the reflected pulse is used to calculate the strain in the sample. The transmitted pulse, however, is a good indication of the stress in the sample. Modifications have been introduced to remedy this problem; see, Nemat-Nasser *et al.* (1988b), and Ravichandran and Nemat-Nasser (1989). These involve: (1) direct measurement of the axial and lateral strains of the sample; and (2) *pulse shaping* through a suitable material (*e.g.*, metal cushion) placed at the loading end of the incident bar. When a small metallic cushion is used, its response can be modeled through simple assumptions and using its stress-strain relation, obtained directly at a desired strain rate by the same Hopkinson system that will be used with this pulse shaper.[10] It is thus possible to control the pulse profile that loads the sample to a great measure, using a suitable pulse shaper.

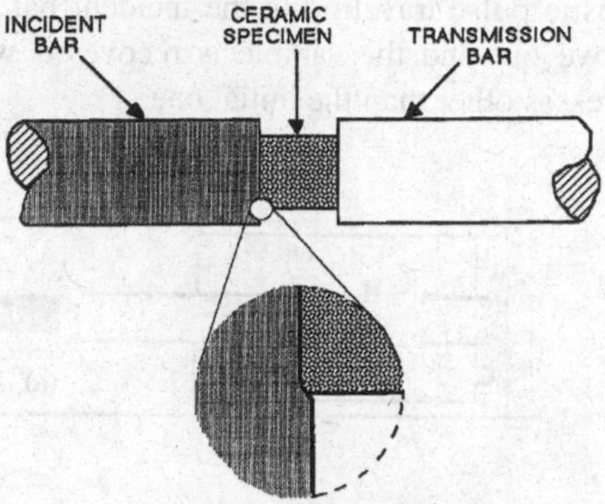

Figure 9.2.10

Because of the indentation of the bar by a hard (*e.g.*, ceramic) specimen, the reflected pulse in this case is no longer a measure of the strain in the sample

9.3. HIGH-TEMPERATURE DYNAMIC RECOVERY EXPERIMENTS

For high strain-rate tests at elevated temperatures, it is necessary to heat the sample to the required temperature, while keeping the incident and transmission bars at suitably low temperatures. If the bars are in contact with the specimen within the furnace, their temperature will increase, having a variable distribution along the bars. This affects the bars' elastic properties and hence, the stress pulses. Moreover, the bars being good heat conductors (usually of maraging steel), the considerable heat loss that occurs, makes controlling the experiment difficult.

[10] This has been detailed and illustrated for copper cushions by Nemat-Nasser *et al.* (1991); see their Appendix A.

9.3.1. High-temperature, High Strain-rate Compression Testing of Metals

In this case, the sample usually is very small. The bars are kept outside the range of the heating unit in the furnace, while keeping the specimen at the center of the furnace. The bars are then brought into contact with the specimen, milliseconds before the stress pulse reaches the end of the incident bar. This is accomplished by two *bar movers* which are activated by the same gas gun that propels the striker bar towards the incident bar. Figure 9.3.1a illustrates the experimental setup just before the gun is fired, and Figure 9.3.1b shows the configuration after the bar movers have brought the incident and transmission bars in contact with the sample. The motion of the bar movers is controlled by the area of its piston and the gas pressure in the breech of the gas gun. In a properly designed system of this kind (Nemat-Nasser and Isaacs, 1997), the bars are brought into contact with the specimen just before the sample is loaded by the elastic pulse traveling in the incident bar. Once the sample is loaded, the bars move out and the sample is recovered without having been subjected to any stresses other than the initial one.

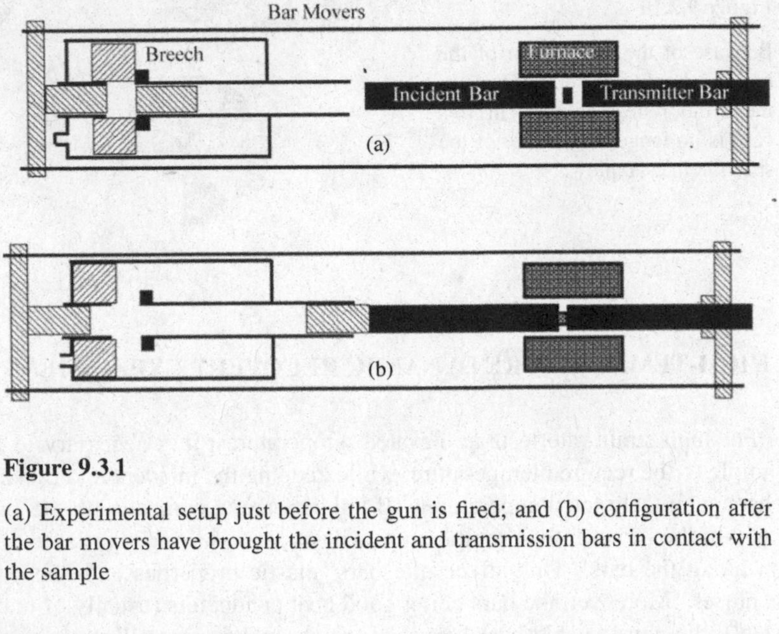

Figure 9.3.1

(a) Experimental setup just before the gun is fired; and (b) configuration after the bar movers have brought the incident and transmission bars in contact with the sample

Figure 9.3.2a shows the furnace and the ends of the incident and transmission bars. Figure 9.3.2b shows the relative position of the sample held by thermocouples, and the ends of the incident and transmission bars; this is the configuration maintained in the furnace just before firing the gas gun. The sample is attached by suitable wires to a sleeve, which is a thin tube. The bar movers then move the transmission bar, bring it in contact with the sample, and then the sample, the sleeve, and the transmission bar are brought in contact with the incident bar. Figure 9.3.2c shows the furnace, and the sample attached to the sleeve, outside of the furnace. Figure 9.3.2d shows the bar movers attached to the breech of the gas gun of the Hopkinson construction.

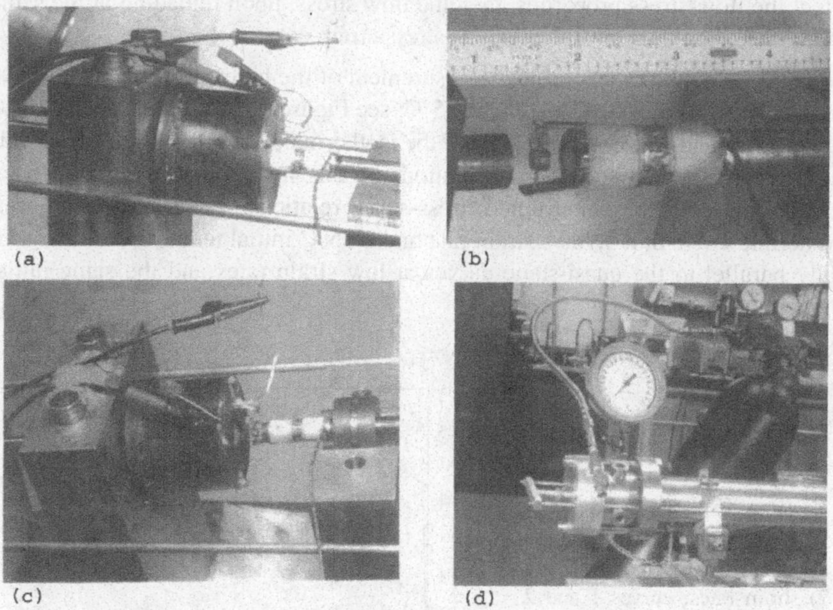

Figure 9.3.2

(a) Furnace and ends of incident and transmission bars; (b) relative position of sample held by thermocouples, and ends of incident and transmission bars; (c) furnace and sample attached to the sleeve outside of the furnace; and (d) bar movers attached to gas-gun breech

9.3.2. Isothermal Flow Stress of Metals at High Temperatures and High Strain Rates

To produce an isothermal flow stress at a high strain rate and high temperature, the sample is heated to the required temperature in the furnace attached to the recovery Hopkinson bar, and then loaded incrementally. After the application of each load increment, the sample is unloaded without being subjected to any additional stress pulses. The sample is then allowed to return to room temperature, its dimensions are measured, and then it is reheated in the furnace to its initial temperature before the application of the next strain increment.

Since the unloading, the cooling of the sample, the reheating to its initial temperature, and the reloading, may affect the microstructure and hence the thermomechanical properties of the material, it is necessary to check this in each case. To this end, a sample which has been loaded, unloaded, and cooled to its initial temperature, may then be reheated to the temperature that it had just prior to its unloading. If there are no substantial changes in the microstructure that affect the flow-stress properties, then the flow stress, upon reloading at the same strain rate, should follow the previous stress-strain curve.

As an application, consider measurement of the high strain-rate isothermal flow stress of the Ta-10%W alloy, at 25°C; see Figure 9.3.3. In this figure, curve 1 (at 10^{-3}/s) and curve 2 (at 10^{-1}/s) are the isothermal, and curve 7 (at 5,700/s) is the quasi-isothermal stress-strain relations of this material. Curves 4 to 6 (at 5,700/s) are adiabatic, interrupted stress-strain relations. As is seen, the quasi-isothermal curve at a 5,700/s strain rate and a 25°C initial temperature, is essentially parallel to the quasi-static curves at low strain rates and the same initial temperature.

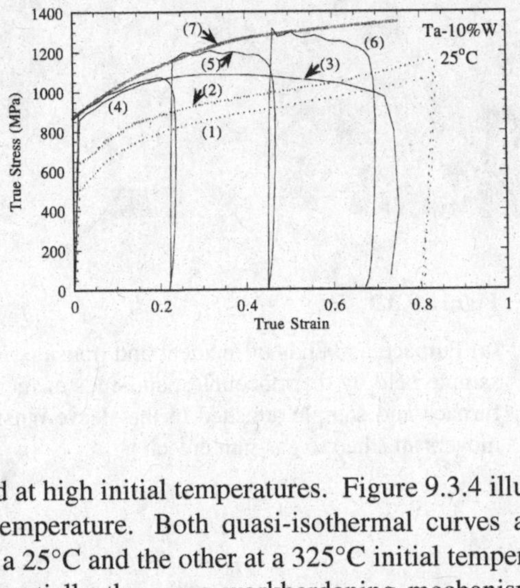

Figure 9.3.3

True stress-true strain relations for Ta-10%W at 25°C and at 10^{-3}/s (curves 1), 10^{-1}/s (curve 2), and at 5,700/s (curves 3 and 7) strain rates; curves 1 and 2 are isothermal and curve 7 is quasi-isothermal; curves 3 to 6 are adiabatic at a 5,700/s strain rate

Similar results are obtained at high initial temperatures. Figure 9.3.4 illustrates this for a 325°C initial temperature. Both quasi-isothermal curves are obtained at 5,700/s strain, one at a 25°C and the other at a 325°C initial temperature. This figure shows that essentially the same workhardening mechanisms must exist at these two different temperatures for the Ta-10%W alloy.

Thus, the isothermal workhardening seems to be independent of the temperature for this material, within the considered temperature and strain-rate ranges, suggesting that the workhardening due to plastic straining is not coupled with either temperature or strain rate. This is an important property, since, then, the flow stress can be divided into an athermal part and a thermally-activated part, where the athermal part may be assumed to depend solely on the microstructure. The thermal one is due to the short-range barriers to the motion of dislocations. Note that, here again it is necessary to establish whether or not a uniform stress state is attained in the sample, prior to yielding in each incremental straining, and whether or not the structure of the material is affected by the cycle of unloading, cooling, reheating, and reloading.

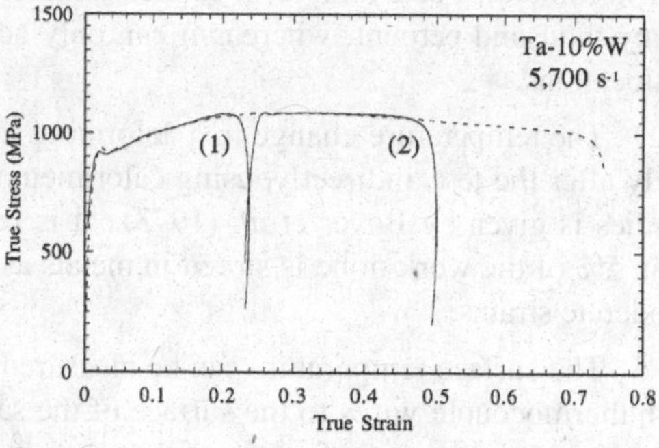

Figure 9.3.4

Quasi-isothermal stress-strain relations for Ta-10%W alloy, for 25 and 325°C initial temperature, showing essentially the same workhardening; curves 1, 2, 3, and 4 are for 25°C, and 5, 6, 7, and 8 are for 325°C initial temperatures; strain rate is 5,700/s

9.3.3. Temperature Rise During High Strain-rate Plastic Deformation

Plastic deformations generate heat. In uniaxial deformations, the applied work per unit volume is equal to the area under the true stress-strain curve.[11] This work is partly used in generating heat, and is partly stored in the material as the internal elastic energy of defects such as vacancies, interstitials, and dislocations.

The stored elastic energy of the defects depends on the nature, amount, and distribution of these defects. Randomly distributed dislocations and dislocations that are piled up against barriers store different amounts of elastic energy. Stroh (1953) and Moore and Kuhlmann-Wilsdorf (1970) calculated the internal energy of a metal sample assuming dislocations are locked in pileups. Zehnder (1991) calculated internal energy as a function of strain, assuming randomly distributed dislocations.

The heat generated in the material is either dissipated to the surroundings or is used to increase the temperature of the material. When the rate of heat generation is greater than the rate of heat loss, the temperature of the material increases.

For materials with temperature-dependent flow stress, a continuous rise of temperature during deformation results in simultaneous lowering of the flow stress. When the deformation occurs with complete heat loss to the surroundings (isothermal), the flow stress generally is higher than when there is no heat loss (adiabatic). It is thus important to determine the fraction of work which goes to increase the temperature of the material during its stress-strain characterization.

Since many metals can be deformed plastically to very large strains, the amount of the associated plastic work can correspondingly be very large, monotonically increasing with the total plastic strain. On the other hand, the stored

[11] In general, the energy of the overall elastic deformation of materials, is negligibly small, compared with the plastic work, for the large overall strains considered here.

elastic energy of the defects is limited, since the density of the defects in a material is limited. For example, a copper sample may be deformed plastically by many thousand percent, whereas it can only accommodate a limited density of dislocations.

The temperature change in a deforming sample may be measured immediately after the test, indirectly, using calorimetry. A review of such experimental studies is given by Bever *et al.* (1973). It is reported that, in general, no more than 5% of the work done is stored in metals as the elastic energy of defects, for moderate strains.

The surface temperature can be measured during deformation by attaching thin thermocouple wires to the surface of the sample. How thin the wire should be depends on the response time of the wires.[12] Another technique for measuring the surface temperature, is to use infrared radiation of the sample. Moss and Pond (1975) used pneumatic loading devices to deform copper samples at about a 100/s strain rate. For a strain of 0.2, a temperature change of 20 to 30°C was observed. Hartley *et al.* (1985) and Marchand and Duffy (1988) have carried out Hopkinson bar tests on steel and observed a 500°C temperature rise within shearbands. Mason *et al.* (1994) measured the surface temperature of materials deformed uniaxially and uniformly in the Hopkinson bar. They report for Al and steel, that the fraction of work converted to heat varies from 80 to 90%. In the case of Ti-6Al-4V, this fraction is reported to decrease to as low as 60% at a strain of about 0.2, suggesting that 40% of the work done is stored within Ti-6Al-4V as elastic energy of the defects, which seems difficult to justify. No explanation for the balance of energy is given.

Another indirect method to estimate the fraction of work that converts into heat, is to use the recovery Hopkinson method, as is briefly reviewed in the sequel. With this approach, Nemat-Nasser and Isaacs (1997) conclude that essentially all of the plastic work is converted to heat, for a Ta-10%W alloy, deformed at a strain rate of 5,700/s to large strains. More recently, Kapoor and Nemat-Nasser (1998) have performed a systematic study of this phenomenon, using both the infrared radiation measurement and the recovery Hopkinson method, on commercially pure Ti, 1018 steel, 6061 Al, and OFHC Cu, as well as on a Ta-10%W alloy. They report that the infrared measurement yields a 70% conversion of work to heat for Ta-2.5%W, generally underestimating this factor for all tested materials. The final temperature at a given strain can be determined indirectly, based on the calculated plastic work.

[12] Some preliminary experiments of this kind were carried out in the author's laboratory at UCSD by coworker Jon Isaacs. Comparison of the results with those obtained indirectly using the recovery Hopkinson method suggested that such surface temperature measurements may underestimate the actual temperature rise by at least 10%.

9.3.4. Interrupted Hopkinson Experiments and Typical Results

Several identical samples of a given metal are prepared. One sample is deformed adiabatically at a given high strain rate to very large strains, say 60 to 100%, starting with a desired initial temperature. Figure 9.3.5 illustrates this for Ta-10%W. Then, using the recovery compression Hopkinson technique, a second sample is loaded at the same strain rate and the same starting temperature, but now only to, say, about 30% strain, and then unloaded. The resulting stress-strain curve should follow that of the first sample, as it does in the data reported in Figure 9.3.5. The work done is then calculated and the associated temperature rise is estimated using an assumed percentage of the plastic work that has been converted to heat. The sample is then heated to the corresponding temperature, and is deformed at the same strain rate for another strain increment, say, 30%. If the assumed conversion factor is correct, and provided that there are no essential microstructural changes occurring during this unloading, cooling, reheating, and reloading of the sample, then the resulting stress-strain curve should follow the first adiabatic curve.

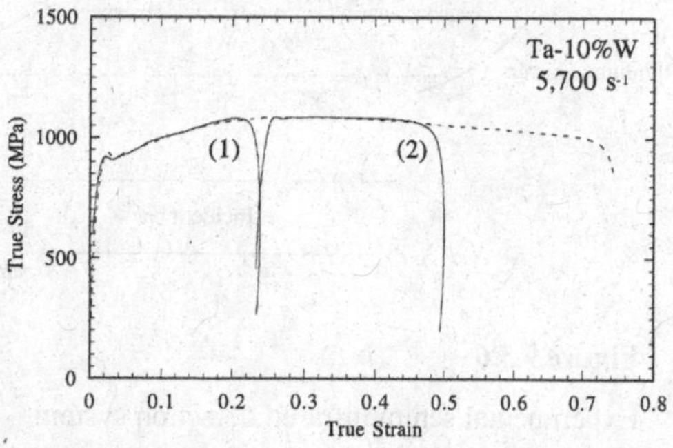

Figure 9.3.5

The true stress-true strain relations for Ta-10%W at 5,700/s strain rate; all three curves are adiabatic, with an initial temperature of 25°C for the dashed curve and for curve 1, and an initial temperature of 135°C for curve 3 (from Nemat-Nasser and Isaacs, 1997)

For the data shown in Figure 9.3.5, it was assumed that all of the work has been converted to heat. Since the resulting stress-strain curve does follow the adiabatic one, it is concluded that, for this material: (1) essentially all the plastic work has been converted to heat, with a negligible[13] amount being stored as the elastic energy of the dislocations; and (2) little or essentially no microstructural changes have occurred during the process of unloading, cooling to room temperature, reheating, and reloading the sample, which could affect the corresponding stress-strain relation at strain rates and temperatures close to those used in the experiment.

Kapoor and Nemat-Nasser (1998) have also used the infrared measurement with the interrupted Hopkinson technique. A 4-channel, EG&G Judson, Indium Antimonide, photovoltaic detector was used. The InSb photodiodes provide excellent performance in the 1 to 5.5μm wavelength region. The detector is

[13] Negligible to within the accuracy of the measurement.

mounted in a liquid nitrogen cooled dewar with a sapphire window. Since the detector is at 77K, the noise is significantly reduced. The detector area is a 2mm diameter InSb semiconductor etched with a cross to yield four quadrants. The signals pass through 1MHz amplifiers and are recorded on a Nicolet 4094 oscilloscope. Figure 9.3.6 is a schematic diagram of the infrared imaging system. The imaging system consists of two, 7.6mm diameter, off axis paraboloid rhodium mirrors. The first mirror is positioned so that the sample is located at the focus of the paraboloid mirror. This results in a 7.6mm diameter collimated infrared beam from radiation emitted by 1mm of the sample surface. The second mirror positioned approximately a meter away, reflects and focuses the radiation into the infrared detector. The infrared detector is positioned at the focal point of this second mirror.

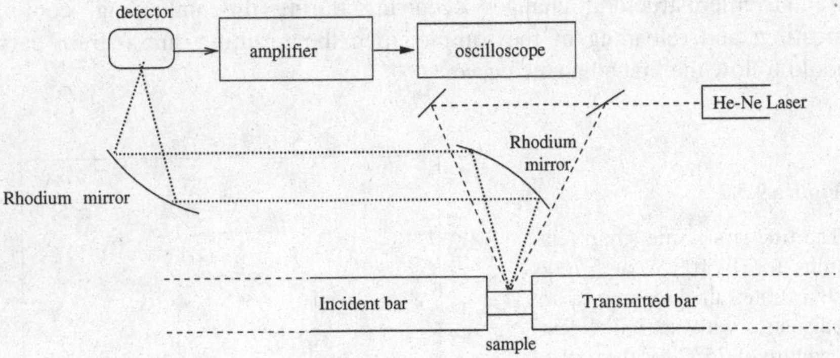

Figure 9.3.6

Experimental setup: infrared detection system

An aperture is installed both in front of the sample and in front of the detector. The aperture in front of the sample ensures that radiation emitted from bars does not interfere with radiation emitted from the sample. The whole experimental setup is covered by a black cloth to prevent external light from affecting the detector signal. Alignment of the system is performed by illuminating the sample with the He-Ne laser and imaging the spot into the detector. The use of mirrors, instead of lenses, allows visible light for alignment of the infrared optics. This optical system is mounted on a sturdy, movable and adjustable optical bench.

Infrared Detector Calibration: Cylindrical samples of Ta-2.5%W were used for calibrating the infrared (IR) detector signal. The sample was internally heated and the IR signal was recorded as a function of temperature, as measured by a thermocouple. A hole was drilled on the lateral surface of the calibration sample such that the sample could be heated internally using a soldering iron. Wires of constantum and chromel were spot welded on either side of the sample.

Constantum and chromel produce a thermocouple junction with a sensitivity of about 6.3mV per 100°C. Figures 9.3.7a,b schematically depict the sample between the bars with the thermocouple and soldering iron. As shown in Figure 9.3.7b, the sample is placed between the incident and the transmitted bars with the infrared detection system aligned. The sample is heated with a soldering rod placed in its hole. After reaching a stable temperature of about 150°C, the heat source is withdrawn. The signal, as measured by the IR detector and the thermocouple is recorded simultaneously. A comparison between the two, results in the calibration of the IR detector signal. A similar calibration was carried out using a sample deformed to 30% strain. This was done in order to check if the detector signal changes because of a change in surface conditions (as a result of straining). No appreciable change in calibration was detected. Other aspects of this IR measurement are found in Kapoor and Nemat-Nasser (1998).

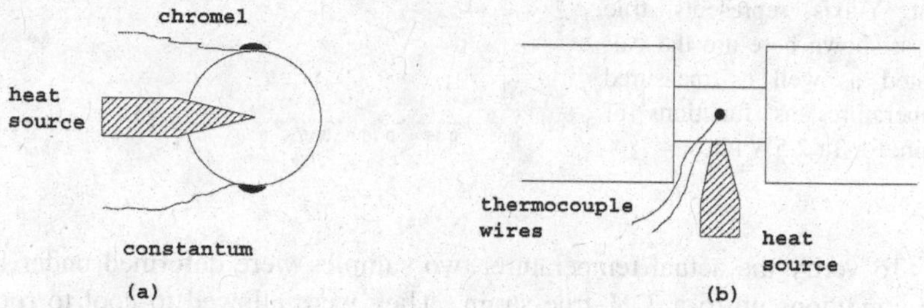

Figure 9.3.7

Configuration of sample with heat source and thermocouple

Results and Discussion: From the true stress-true strain relation in a uniaxial experiment, the temperature increase is estimated by the following equation:

$$\Delta T = \frac{\eta}{\rho C_v} \int_0^{\varepsilon} \sigma \, d\varepsilon , \qquad (9.3.1)$$

were η is the fraction of the energy used to heat the sample, ρ is the mass density, C_v is the heat capacity at constant volume, and σ and ε are the measured uniaxial true stress and strain, respectively. In the range of the considered temperatures, both ρ and C_v can be assumed to remain constant. For $\eta = 1$, all the work is assumed to be used to heat the sample, resulting in a maximum increase in the sample temperature. Thus, in general, $\eta \leq 1$.

From the true stress-true strain curve of Ta-2.5%W deformed at 3,000/s and 22°C (initial temperature), the total work done on the sample and the resulting temperature of the sample are calculated as functions of strain and η. Figure 9.3.8 is a double Y-axis plot. On the left Y-axis is the temperature of the sample as a function of the strain, whereas on the right Y-axis is the true stress as a function of the strain. There are two curves showing the temperature. The lower one is that recorded by the IR detectors, while the upper curve is calculated from equation (9.3.1), with $\eta = 1$. The IR detectors show a temperature rise of 63°C (85°C - 22°C), while the calculated curve shows a temperature rise of 93°C (115°C - 22°C). The results from the IR detector suggest that about 68% of input energy is converted to heat. A 68% conversion of work to heat implies a 32% storage of work as internal energy in the form of defects.[14]

Figure 9.3.8

Double Y-axis plot: Left Y-axis represents temperature, and right Y-axis represents true stress; shown here are the calculated as well as measured temperatures as functions of strain, for Ta-2.5W alloy

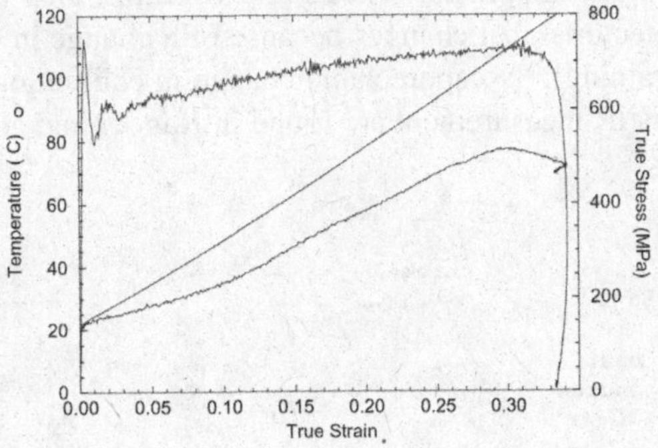

To verify the actual temperature, two samples were deformed under the same conditions up to a 32% true strain. They were allowed to cool to room temperature and were then heated to different temperatures before reloading. One sample was heated to 85°C and reloaded, whereas the other sample was heated to 115°C and reloaded. Another undeformed sample was strained (starting at room temperature) up to a strain of 54%. Figure 9.3.9 is the stress-strain plot showing the results of these three experiments. It is observed that reloading the sample at 115°C ($\approx$ 100% conversion of work to heat) results in a yield stress which is essentially the same as that just before unloading. Furthermore, the flow stress of this sample is essentially identical to the flow stress of the sample deformed to a strain of 54%. Reloading the sample at 85°C (infrared detector measurement) results in a flow stress higher than that before unloading.

The question is for which other materials in adiabatic deformations can $\eta \approx 1$ be assumed at large strains. Similar tests (loading-unloading-reloading at a calculated temperature) were carried out on commercially pure Ti, OFHC Cu, 6061 Al in T6 condition, and 1018 Steel, by Kapoor and Nemat-Nasser (1998). Mason *et al.* (1994) suggest only a 60% conversion factor for Ti-6Al-4V, 80% for 4340 steel, and 85% for 2024 Al. These values were used by Kapoor and Nemat-Nasser as guidelines in performing the interrupted Hopkinson tests on similar materials. In each case, only for $\eta \approx 1$, the stress-strain curve of the second interrupted-test sample followed that of the original uninterrupted-test

[14] This is with the assumption of no heat loss, which is justified at the considered high strain rate; see Kapoor and Nemat-Nasser (1998, Appendix).

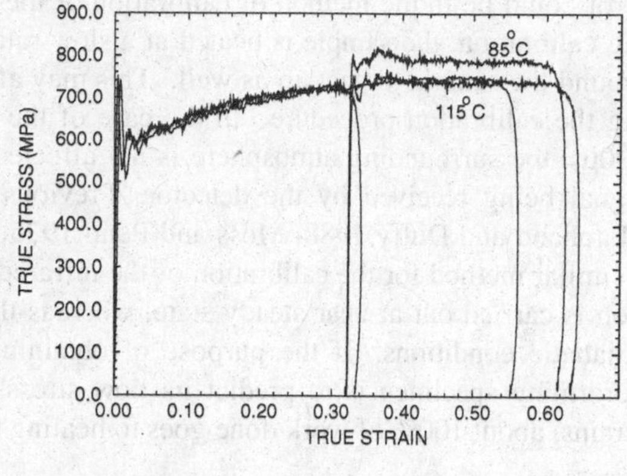

Figure 9.3.9

True stress-true strain plot of Ta-2.5%W alloy, loaded at 3,000/s strain rate and 22°C initial temperature; the samples are then unloaded after a strain of 32%, and then reloaded at 85°C ($\eta = 0.74$) and 115°C ($\eta = 1$)

sample. The results for Ti-6Al-4V are reproduced in Figure 9.3.10. From these tests, it was concluded that, at high strain rates and for large plastic strains, a good estimate of the flow stress can be made with the assumption of full conversion of work to heat during adiabatic deformation.

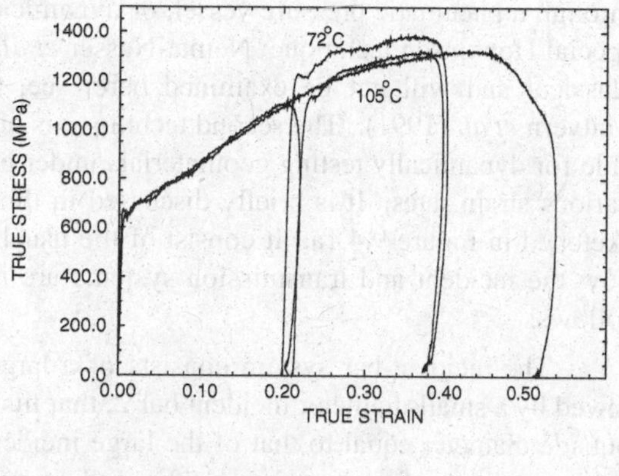

Figure 9.3.10

True stress-true strain plot of Ti-6Al-4V alloy, loaded at 3,000/s strain rate and 22°C initial temperature; the samples are then unloaded after a strain of 20%, and then reloaded at 72°C ($\eta = 0.6$) and 95°C ($\eta = 1$)

Elastic Energy of Dislocations: The elastic energy of screw and edge dislocations may be estimated, based on equations (6.1.10) and (6.1.20), pages 395 and 397. For the rather high dislocation density of $10^{15}/\mathrm{m}^2$, these expressions suggest the elastic energy to be about $0.6 \times 10^6 \mathrm{J/m}^3$. With the inclusion of the interaction energy of the dislocation, a value of the order of $10^6 \mathrm{J/m}^3$ results. The input energy depends on the total plastic strain. For the strains considered here, it is of the order of $200 \times 10^6/\mathrm{m}^3$. Even with the inclusion of the pileup effects, only a very small percentage of the total plastic work is stored in the material as internal elastic energy of the defects, for large plastic deformations. Most of the applied energy produces heat within the sample. It thus appears that the infrared measurement system underestimates the sample temperature, which in turn results in a higher estimate of the stored energy within the material. A possible error could be in the method of calibration of the infrared detector signal. During calibration, the sample is heated at a slow rate, allowing for the atmosphere around the sample to heat up as well. This may affect the infrared radiation during the calibration procedure. In the case of the actual test, lasting only about 100μs, the surrounding atmosphere is not affected, resulting in a lower infrared signal being received by the detector. Previous studies (Hartley *et al.* 1985, Marchand and Duffy 1988, Moss and Pond 1975, Mason *et al.* 1994) have used a similar method for the calibration of the infrared detector system. The calibration is carried out at near steady state, whereas the actual testing is done under adiabatic conditions. If the purpose of obtaining a temperature rise within a deforming specimen is to predict its flow stress, the assumption that, at large strains, about 100% of work done goes to heating the sample is reasonable.

9.4. HOPKINSON TECHNIQUES FOR DYNAMIC TRIAXIAL COMPRESSION TESTS

Triaxial Hopkinson techniques can be used to subject a sample simultaneously to axial and lateral compression. The lateral compression may be applied through a pneumatic pressure vessel, or dynamically, using a recently developed special Hopkinson technique; Nemat-Nasser *et al.* (2000). The first technique is classical and will not be examined here; see, Christensen *et al.* (1972) and Malvern *et al.* (1991). The second technique is of a more recent origin, and suitable for dynamically testing geomaterials under biaxial states of compression, at various strain rates. It is briefly discussed in this section. The entire system is sketched in Figure 9.4.1a. It consist of the usual gas gun and a striker bar, but, now the incident and transmission systems are modified, as explained in what follows.

The incident-bar system consists of a large-diameter incident bar 1, followed by a small-diameter incident bar 2, that fits inside an incident tube with an outside diameter equal to that of the large incident bar 1. The transmission-bar system consists of a transmission tube, and a transmission bar that fits inside the tube. The sample is placed inside a sleeve of suitable material, *e.g.*, Teflon, and the sleeve is placed inside another suitable sleeve, *e.g.*, aluminum or steel. Figure 9.4.1b is a photograph of the confining system and a sample, and Figure 9.4.1c shows a Teflon sleeve as an illustration of the confining material; see comments below.

The confining sleeve is loaded during the test by the stress pulse traveling in the incident tube, producing a radial stress on the sample. The stress pulse traveling in the incident bar loads the sample that is sandwiched between the small-diameter incident and transmission bars. The radial stress is calculated from the hoop strain measured by a strain gauge placed on the outer surface of the metal sleeve. Before the test, the entire Hopkinson bar apparatus is pressed together to ensure that there are no gaps between the contacting bars, the sample or the confining sleeve. This is vital to ensure that the radial load is applied at the same time as the axial load, and to ensure that the data collected during the experiment are an accurate measure of the sample response. Incident bar 1 must be in direct contact with both incident bar 2 and the incident tube. The incident tube must be in tight contact with the confining sleeve, which in turn must be in contact with the transmission tube. A tight contact is ensured when the incident bar 2 is slightly longer than the incident tube.

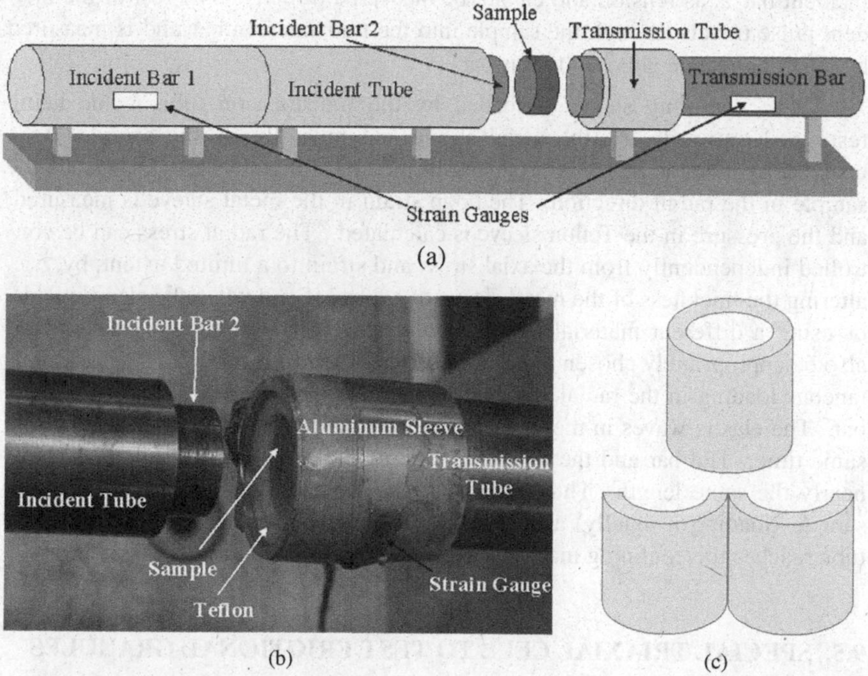

Figure 9.4.1

(a) Schematic of triaxial Hopkinson-bar system; (b) photograph of the first dynamic triaxial Hopkinson system (photo from Nemat-Nasser *et al.*, 2000); and (c) a Teflon sleeve before and after being sliced in half

A radial expansion accompanies the sample deformation. It is thus necessary to use a pre-split confining sleeve to facilitate sample recovery without any damage; see Figure 9.4.1c. Since a tight fit is important, the inside diameter of the confining sleeve should be slightly less than the sample diameter. In the case of a Teflon sleeve, slicing is done with a knife instead of machining, to produce a tight fit with the sample and the metal sleeve.

9.4.1. Operation of Triaxial Hopkinson System

The gas gun propels the striker bar at a prescribed velocity, impacting incident bar 1 and generating the incident pulse in that bar. The incident pulse travels through incident bar 1 and is measured by the strain gauge that is placed on this bar. The incident pulse then loads incident bar 2 and the incident tube. The incident pulse in incident bar 2 then loads the sample in the axial direction, while the stress pulse in the incident tube is loading the confining sleeve. Part of the incident pulse, proportional to the sample strain rate, is reflected back into incident bar 2, as tension and cannot be measured directly. The rest of the incident pulse travels through the sample into the transmission bar and is measured by the strain gauge attached to this bar.

The confining sleeve is loaded by the transmission tube, while being restrained outwardly by the metal sleeve and inwardly by the sample. As a result, a hydrostatic stress is produced in the confining sleeve, which loads the sample in the radial direction. The hoop strain in the metal sleeve is measured, and the pressure in the Teflon sleeve is calculated. The radial stress can be controlled independently from the axial stress and strain to a limited extent, by, *e.g.*, altering the thickness of the metal sleeve to control if and when the sleeve yields, or using a different material for the outer sleeve. The confining sleeve could also be appropriately chosen to produce different loading conditions. The simultaneous loading in the radial and axial directions is assured by the design of the bar. The elastic waves in the incident bar and incident tube are generated at the same time. The bar and the tube are made of the same material and they have nearly the same length. Thus, the elastic wave in the incident bar reaches the sample (loading it axially) at the same time as the stress wave in the incident tube reaches the confining material (loading the sample in the radial direction).

9.5. SPECIAL TRIAXIAL CELL TO TEST FRICTIONAL GRANULES

In this and the next section, some recent experimental techniques to characterize the deformation of cohesionless granular materials, are discussed. As has been explained in Section 4.7 and in Chapter 7, materials of this kind support externally imposed shearing through contact friction which develops at the contact points of the granules, under an applied overall compression. In the absence of pressure, frictional granules cannot resist shearing. An experimental system must therefore allow for the application of compression and shearing to reproducible samples in a controlled manner with reliable data acquisition. This requires a closed-loop feedback system to control the experiment and to monitor the specimen deformation. The author and coworkers have developed such a system over the past two decades, as is briefly discussed in the sequel.

The specimen is a large hollow cylinder, 25cm high, with inner and outer diameters of 20cm and 25cm, respectively. This geometry is such that in torsion, the shear stress remains (approximately) homogeneous throughout the thickness of the specimen. The specimen is supported by a triaxial load frame which, together with a specimen, is shown in Figure 9.5.1. When connected to an MTS servohydraulic loading system, the axial and torsional deformations are controlled independently. In addition, the specimen is subjected to lateral hydrostatic pressure, on both its inside and outside cylindrical surfaces, while being compressed axially. In this manner, triaxial states of stress can be imposed on the material under fully computer-controlled conditions, with data acquisition capability.

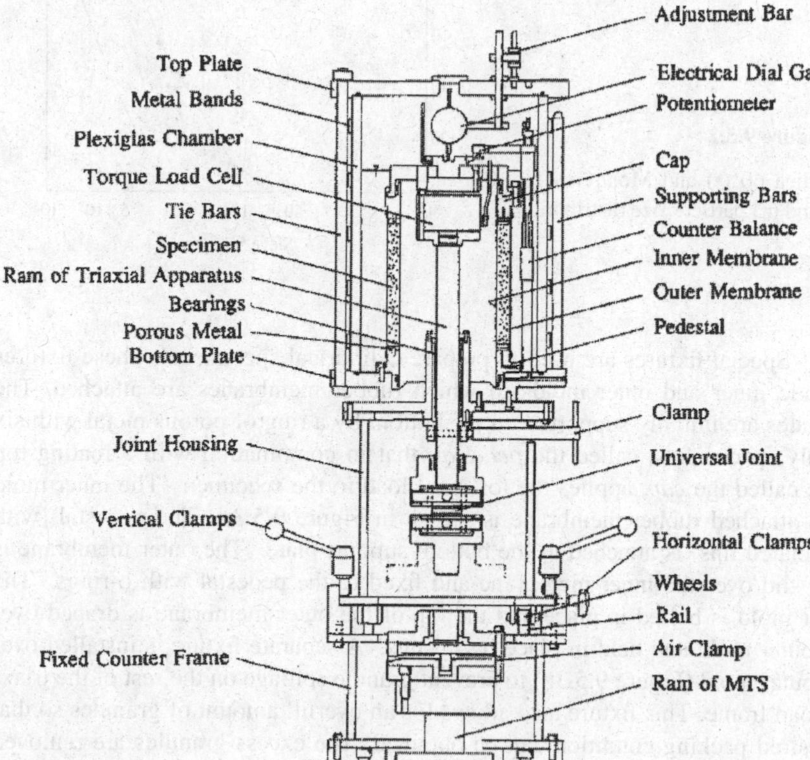

Figure 9.5.1

Triaxial load frame

9.5.1. Specimen Preparation and Installation

Silica 60, manufactured by U.S. Silica, may be used for the material, since it consists of fine particles, minimizing the membrane penetration phenomenon that otherwise may invalidate the test results. The particle size distribution is shown in Figure 9.5.2. The mean particle diameter is 220μm and the specific gravity of the granules is 2.645. Depending on the packing conditions, different void ratios are obtained. For Silica 60, the minimum and maximum void ratios are about 0.631 and 1.095, respectively.

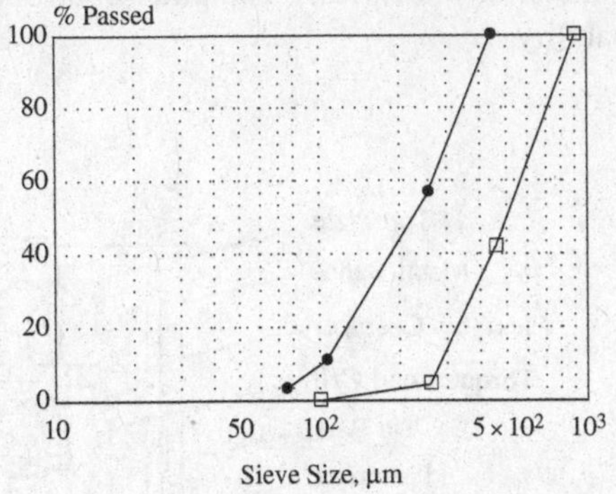

Figure 9.5.2

Silica 60 (•) and Montrerey No. 0 sand (□) particle size distributions

Special fixtures are used to prepare cylindrical specimens. These fixtures include inner and outer molds to which rubber membranes are attached. The granules are initially supported on the bottom by a ring of porous metal with six evenly spaced fins, called the *pedestal*, that in combination with a mating top ring, called the *cap*, applies the torsional load to the specimen. The inner mold with attached rubber membrane is shown in Figure 9.5.3a. The pedestal (with associated fins) is attached to the bottom support plate. The outer membrane is then slid over the inner membrane and fixed to the pedestal with o-rings. The outer mold is bolted in place and the top of the outer membrane is draped over the outer mold and held in place by o-rings. A separate fixture is installed over the outer mold (Figure 9.5.3b) to prevent granule spillage on the rest of the triaxial load frame. This fixture also allows for an overfill amount of granules so that a desired packing condition can be obtained. The excess granules are removed later. The tube extending from the outer mold in Figure 9.5.3b is used to create a vacuum between the outer mold and outer membrane, thereby holding the latter securely in place.

The initial packing condition of the granules has a significant effect on the overall response of the specimen, as has been discussed by Arthur and Menzies (1972), Oda (1972), Miura and Toki (1982), and others. The specimen preparation method must therefore achieve a consistent initial packing condition to allow reproducibility. For this, a technique has been adopted in the geotechnical engineering community, called the rodding method. It consists of pouring an approximately 2cm deep layer of granules into the mold and then inserting a rod into the last layer approximately 1-1.5cm deep. The rod is moved around the circumference of the granules in an up-and-down motion for 2-3 revolutions. This

(a) (b)

Figure 9.5.3

(a) Inner mold with attached rubber membrane and pedestal; and (b) outer fixture and overall sample construction molds

procedure is continued until the mold is filled. Experience with Silica 60 shows that 10 layers yield a loosely packed, and 14 layers a densely packed sample. In the first case, granules in wet form are used, where the air-dried materials are mixed with 8 weight percent water before pouring in the mold. This prevents inhomogeneous initial packing conditions. The resulting void ratio for this condition varies between 0.865 and 0.871. To produce densely packed samples, only air-dried granules are used, with the resulting void ratio varying between 0.708 and 0.725.

The fixture attached to the top of the outer mold is then removed and the amount of overfilled granules is 'cut' away. The cap is installed next; it consists of the same porous metal as the pedestal and also has six fins. A second vacuum system is connected to the cap and pedestal. The purpose of this vacuum system is to make the specimen rigid under atmospheric pressure. The vacuum level is maintained at 29.4kN. The first vacuum system that keeps the outer membrane affixed to the outer mold is then released. The outer and then the inner molds are removed. A torque load-cell unit is first bolted onto the ram of the triaxial load frame (Figure 9.5.1), and then bolted onto the cap. Next, a potentiometer is attached to the load frame (Figure 9.5.4). The potentiometer measures the twist angle during the experiment. A plexiglass chamber with steel bands is installed over the entire specimen, and a top plate is installed. The top plate is affixed to the bottom plate via stainless steel tie bars. The bars hold the chamber firmly in place. The purpose of the chamber is manyfold. First, it provides confinement of the experiment if the granular mold loses integrity. Second, it holds the water that is used to apply hydrostatic pressure to the specimen. Finally, it is used as a viewport to observe the progress of the experiment.

Figure 9.5.4

Specimen under vacuum with torque load cell and potentiometer

The specimen assembly is now complete. The assembly is then raised to the level of the MTS load frame via a forklift. A special work frame has been built onto the MTS load frame. This work frame allows one to attach all connections to the specimen assembly as well as providing a railway for installation and removal of the specimen into and from the MTS load frame. The overall assembly is shown in Figure 9.5.5.

The MTS load frame used for this experiment has an axial capability of 89kN (20,000lb), and a torsional capability of 565N-m (5,000lb-in) which can be used independently. The system is closed-loop so that feedback from any selected transducer can be used to control the test.

Once the triaxial load frame has been rolled into place over the ram of the MTS load frame, it is secured in place by both vertical and horizontal clamps (Figure 9.5.1). The hydraulics for the MTS system are turned on and the MTS ram is raised to the level of the universal joint, using displacement control. An air clamp that is fixed to the top of the MTS ram is then actuated and grips the universal joint on the bottom of the ram of the triaxial load frame. The universal joint is required to accommodate any misalignment between the ram of the MTS load frame and the ram of the triaxial load frame.

9.5.2. Experimental Procedure

The first step in the experimental procedure is to fill the plexiglass chamber with water until the specimen is completely submerged. The remaining space above the specimen is pressurized with air to 29.4kN, which is the same value as the vacuum inside the specimen. During this operation the vacuum in

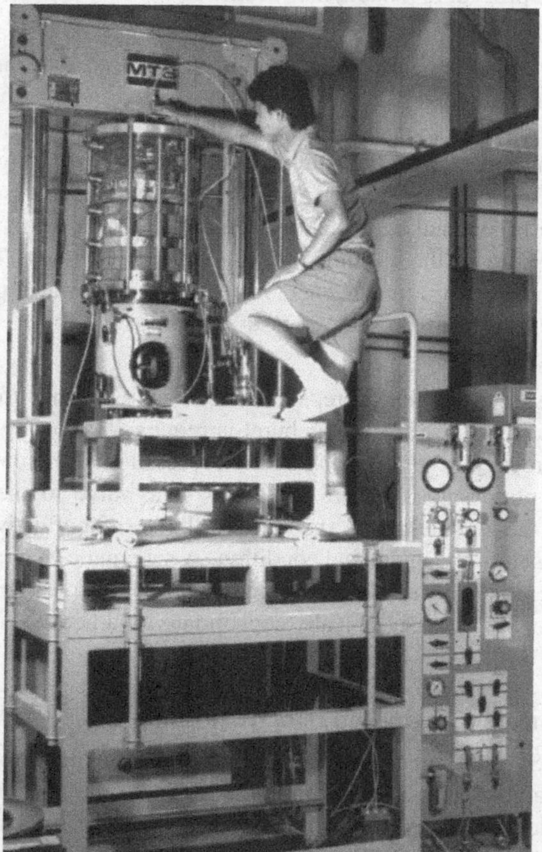

Figure 9.5.5

Overall assembly of MTS and triaxial load frame

the specimen is released and water pressurized in such a manner as to keep the effective pressure in the specimen constant, 29.4kN. The specimen is then water saturated in the following manner. To attain full saturation, first the specimen is saturated with CO_2 gas through the porous metal in the pedestal and cap. The flow of gas is continued until all air is removed from the specimen. CO_2 gas is used because of its high solubility in water. A fixed amount, 4 liters, of de-aired water is used to saturate the specimen. The small amounts of air and CO_2 gas remaining in the specimen must then be removed as much as possible. The pore water pressure is then increased to 196kN as back pressure, using a buret system, while at the same time the external hydrostatic pressure is increased to

225.4kN, in order to keep the effective pressure constant, 29.4kN, during this procedure. This procedure reduces the volume of the excess gas in the specimen due to the relatively high pore water pressure.

To perform experiments of this type, it is required that the specimen be highly saturated. The degree of saturation is measured by the b-value. To measure the b-value the specimen must be in the undrained condition. This condition is met by closing the valve to the buret, ensuring that the specimen remains at a fixed volume. The specimen is said to be perfectly saturated (b = 1) if an incremental increase in external hydrostatic pressure has the effect of increasing the pore water pressure in the specimen by an identical amount. The b-value is defined as the ratio of the incremental increase of pore water pressure to the incremental increase of hydrostatic pressure. The values for all of our experiments are in excess of 0.99.

The last step of specimen preparation is to increase the effective pressure to 196kN. To do this, the valve to the buret is reopened, allowing water to drain from the specimen. The external hydrostatic pressure is thereby increased to 392kN, where pore water pressure is 196kN. The valve to the buret is then closed, leaving the granules in the specimen at the desired effective pressure and the specimen itself in the undrained condition. A schematic diagram of the pressurization system is shown in Figure 9.5.6. Finally, the specimen is left undisturbed in this condition to isotropically consolidate for a period of 3 hours.

9.5.3. Experimental Control and Data Acquisition

The MTS load frame used in the author's laboratory for this class of experiments, had a computer operating three independent controllers, each with three independent feedback channels. Controller 1 was used for the vertical movement of the MTS/triaxial load frame ram assembly, with one channel to monitor the load from the torque load cell, and the other to monitor the vertical displacement of the specimen. Controller 2 was used to monitor the chamber pressure and pore water pressure. Controller 3 was dedicated to the twist of the ram assembly, with one channel to monitor the torque from the torque load cell and another to monitor the angle of twist from the potentiometer.

The experiment was conducted using two closed-loop feedback systems. System one used channel one of controller 1 in load control to keep the specimen in a state of hydrostatic compression in accord with the external pressure. The second feedback system used channel two of controller 2 in displacement control to cyclically twist the specimen to desired shear strain amplitudes and at desired shear strain rates. All experiments were conducted at a shear strain rate of 0.667%/minute. The loading wave form was triangular. Shear strain amplitudes were 0.2%, 0.5%, and 1.0% for both loose and dense specimens. Tests at 0.4% and 2.0% shear strain amplitudes were performed on dense specimens. All tests were continued until the excess pore water pressure reached 95% of the initial effective pressure, *i.e.*, 186.2kN. Operationally, the tests were stopped at

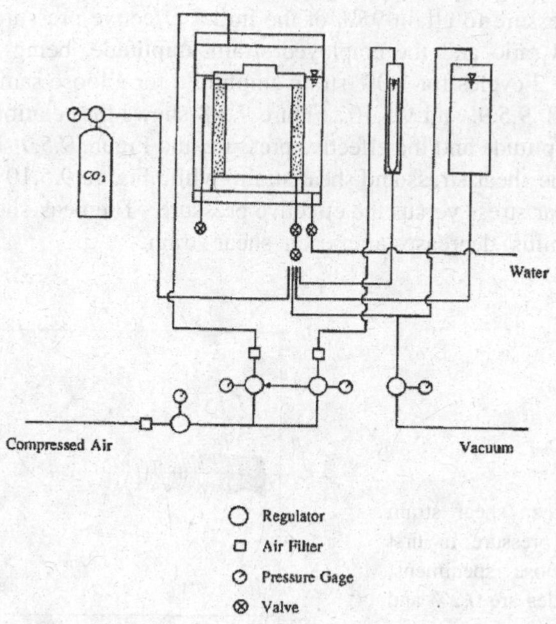

Figure 9.5.6

Schematic diagram of air pressure and water supply system

the end of the cycle after which transducer two of controller 2 (pore pressure transducer) reached a value of 382.2kN. This process was called *the first loading*. The specimens were then reconsolidated under the same initial effective pressure, 196kN, as in the first loading, and a second loading was performed.

9.5.4. Typical Experimental Results

For illustration, typical results of the first and second loadings, for both loose and dense samples, are presented in this subsection,[15] focusing on the cyclic shear deformation characteristics of water-saturated specimens in strain controlled conditions.[16] Figure 9.5.7 shows the relation between the shear strain and the effective pressure for loose samples, for two strain amplitudes, 0.2% and 1.0%. The effective pressure decreases after each cycle, with the reduction after

[15] See Okada (1992) and Okada and Nemat-Nasser (1994) for extensive studies of both densification and liquefaction of cohesionless granules, including a quantitative examination of the relation between the external work, measured per unit volume, and the corresponding excess pore water pressure.

[16] This is in contrast to most other researchers who have conducted the undrained cyclic shear tests under stress controlled conditions; see, for example, Ishihara and Yasuda (1975), Seed (1979), and Tatsuoka *et al.* (1982).

the first cycle being especially large. The number of cycles required for the excess pore pressure to attain 95% of the initial effective pressure, depends on the initial void ratio and the employed strain amplitude, being 27 cycles for 0.2%, and only 2 cycles for 1.0% strain amplitude for a loose sample, as shown in Figures 9.5.8, 9.5.9, and 9.5.10. Figure 9.5.8 shows the relation between the shear strain amplitude and the effective pressure and Figure 9.5.9 shows the relation between the shear stress and shear strain, while Figure 9.5.10 is for the corresponding shear stress versus the effective pressure. The peak shear stress (and the secant modulus) decreases after each shear strain.

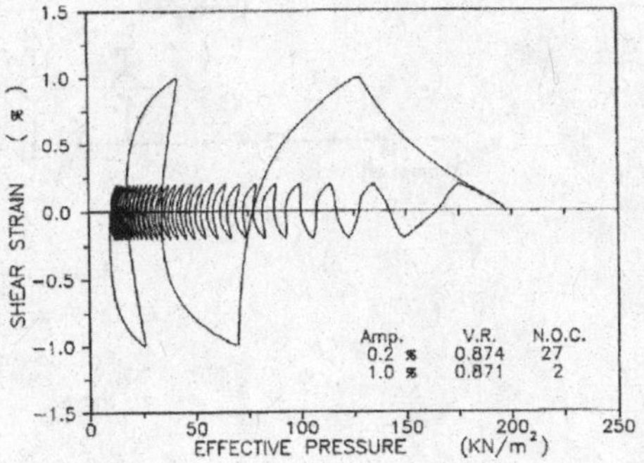

Figure 9.5.7

Relation between shear strain and effective pressure in first loading of loose specimens; strain amplitudes are 0.2% and 1.0%

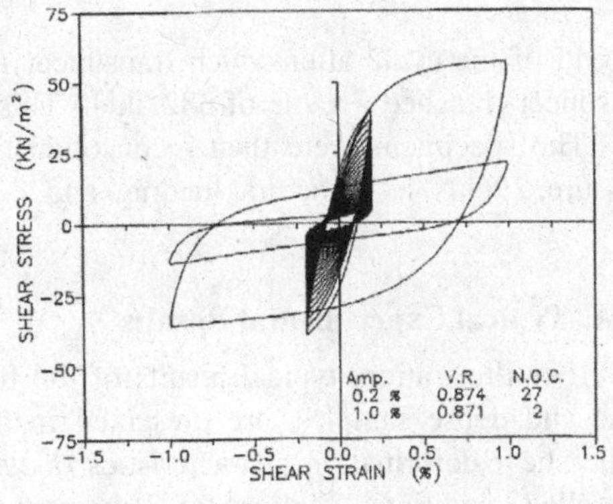

Figure 9.5.8

Relation between shear stress and shear strain in first loading of loose specimens; strain amplitudes are 0.2% and 1.0%

After a first loading, the specimen was reconsolidated as mentioned above, and a second loading was conducted. Various tests were performed, with the strain amplitude in the second loading being the same as, or different from, that in the first loading. The shear deformation characteristics in the second loading were compared with the results of the first loading. Figure 9.5.10 shows the first two cycles of the relation between the shear strain and the effective pressure in both first (solid curve) and second (dotted curve) loading for a dense specimen deformed at a strain amplitude of 0.5%. The excess pore water pressure

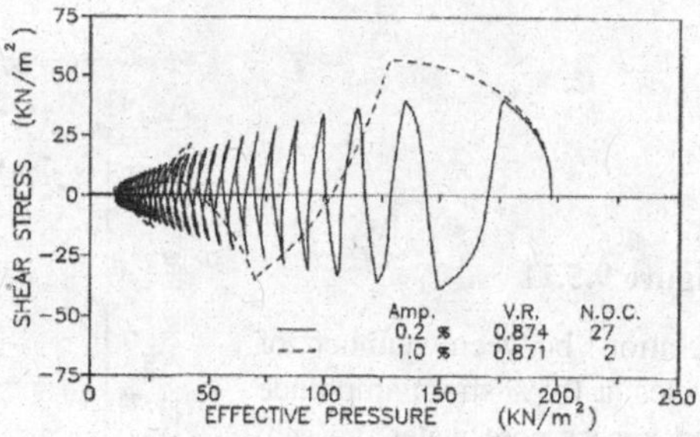

Figure 9.5.9

Relation between shear stress and effective pressure in first loading of loose specimens; strain amplitudes are 0.2% and 1.0%

accumulated during the second loading is much less than that in the first loading, and the number of cycles required to reach 95% of the initial effective pressure in the second loading is much greater than that in the first loading.

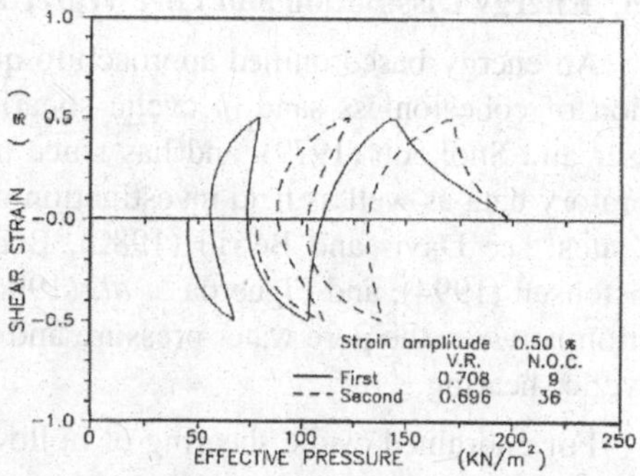

Figure 9.5.10

Relation between shear strain and effective pressure in first and second loading of dense specimen whose strain amplitude is 0.5%

Figure 9.5.11 is a direct comparison of the pore water pressure variation in loose and dense specimens, deformed at a strain amplitude of 0.5%. It takes a greater number of cycles for the pore water pressure to reach a specified level in the second loading than it does in the first loading, for both cases. This is due to the ordered arrangement of the granules, attained upon the completion of the first loading. Specifically, the contact normals tend to be oriented after the first loading such that the specimen is better able to resist a similar shearing. Another trend to notice is the faster pore water pressure build-up during the first loading of the dense specimen than that in the second loading of the loose specimen. This is also explained by the oriented contact normals, despite the large difference in densities. These trends were seen in all experiments when comparison was made between the behavior of loose and dense specimens deformed at a constant strain amplitude. It should be noted, however, that, in a strain-controlled test, the deformation of the specimen is limited by the prescribed strain amplitude. This prevents extensive particle rearrangement which often occurs in stress-controlled tests, once sufficiently high pore pressures are attained, as discussed by Nemat-Nasser and Tobita (1982).

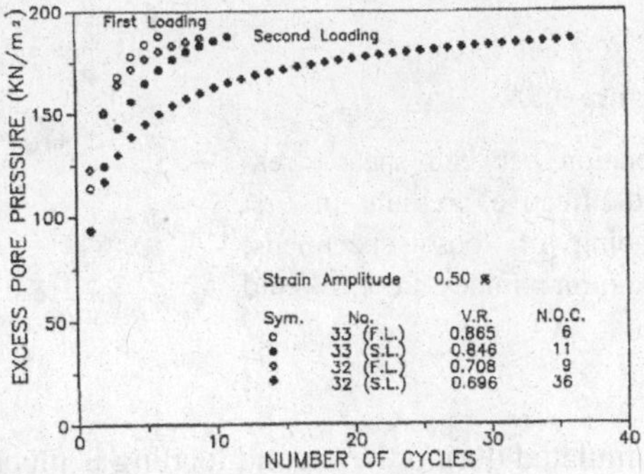

Figure 9.5.11

Relation between number of cycles at 0.5% strain amplitude and excess pore water pressure in first and second loading of both loose and dense specimens

9.5.5. Energy Dissipation and Pore Water Pressure

An energy-based unified approach to quantify the densification and liquefaction of cohesionless sand in cyclic shearing, has been proposed by Nemat-Nasser and Shokooh (1979), and has since been extended and used to analyze laboratory data as well as field investigation of the liquefaction potential of various sites; see Davis and Berrill (1982), Berrill and Davis (1985), Berrill and Christensen (1994), and Figueroa *et al.* (1994). Here, the existence of a unique relation between the pore water pressure and the energy dissipation is illustrated in cyclic shearing.[17]

For undrained cyclic shearing of hollow cylindrical specimens under constant confining pressure, the volumetric strain is essentially zero for the small pressures that are involved in the present case. Hence, the rate of dissipation can easily be estimated. Denote by $\dot{\mathrm{w}}$ the energy dissipation per unit volume, and note that

$$\dot{\mathrm{w}} = \frac{1}{\mathrm{v}} \int_{\partial \mathrm{v}} \mathbf{t} \cdot \dot{\mathbf{u}} \, \mathrm{da}\,, \tag{9.5.1a}$$

where v is the volume of the sample, and **t** denotes the applied boundary tractions. Assuming uniform shear stresses applied on the top and bottom of the cross section of the cylinder, (9.5.1a) is approximated by

$$\dot{\mathrm{w}} = \tau \dot{\gamma}\,, \tag{9.5.1b}$$

where τ is the shear stress and $\dot{\gamma}$ is the torsional shear strain rate. Integration of (9.5.1b) over time now yields the corresponding energy dissipation,

---

[17] For an extensive study with many examples, see Okada (1992) and Okada and Nemat-Nasser (1994).

$$\Delta w = \int_{0}^{\Delta t} \tau \dot{\gamma}\, dt. \tag{9.5.2}$$

The external work is calculated from the experimental results, and correlated with the accumulated pore water pressure for both loose and dense specimens. The relation between the external work per unit volume and the excess pore water pressure for both the loose and dense specimens is shown in Figure 9.5.12. Three strain amplitudes, 0.2%, 0.5% and 1.0%, for the loose samples, and five different strain amplitudes, 0.2%, 0.4%, 0.5%, 1.0% and 2.0%, for the dense samples are employed. The data for the dense case also include results obtained by a random-amplitude (with zero crossing on each cycle) loading that produced liquefaction in 39 cycles. The pore pressure at the end of each cycle in each experiment is plotted in Figure 9.5.12. Clearly, there seems to exist a unique nonlinear relation between the external work and the accumulated pore water pressure. This relation is independent of the employed shear-strain amplitude, but the number of cycles to the 95% initial effective pressure does depend on the strain amplitude.

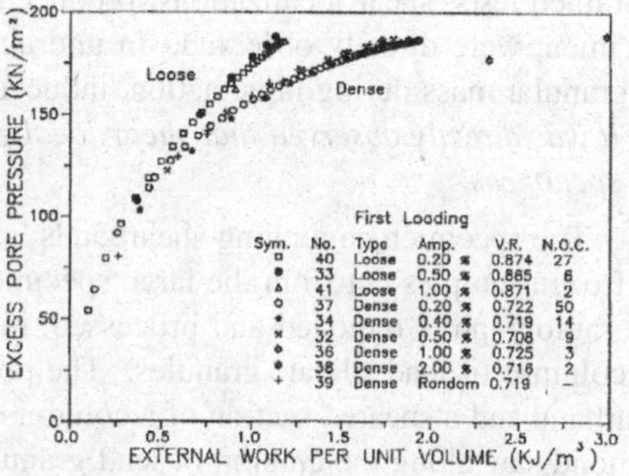

Figure 9.5.12

Relation between external work per unit volume and excess pore water pressure in first and second loading of both loose and dense specimens at various strain amplitudes

9.6. RADIOGRAPHIC OBSERVATION OF SHEARBANDS IN LIQUEFACTION

Monotone shearing of a specimen of frictional granules under a confining pressure, invariably leads to shear strain localization, or shearband formation, where the overall sample deformation is accommodated through localized shearing within narrow bands. Pressure sensitivity, dilatancy, and shear strain localization are integral parts of the mechanical response of masses of frictional granules. The relations between the shearband orientation and thickness, and the internal friction and dilatancy have been the subject of many investigations, providing considerable insight into this phenomenon. In addition, radiography has been applied by a number of investigators to study various aspects of the response of particulates.[18]

Shear localization in liquefied granular materials has been observed through radiographic investigations, using a modified version of the triaxial cell discussed in Section 9.5; Nemat-Nasser and Okada (1995, 1998, 2001). The relatively thin wall of the specimen, with the X-ray films suitably placed inside the cylinder, reduces the film-to-object distance to a minimum, improving the sharpness of the radiograph, without influencing the deformation of the specimen. Several stages of shear deformation are then captured on a single film through a narrow *window* which is provided in order to expose only a limited portion of the film to the X-ray. The position of the film relative to the narrow window is controlled by a computer-controlled stepping motor placed in the chamber, again without affecting the sample deformation.

To capture the internal microstructural changes during the shearing deformation of the sample, thin columns and rows of lead silicate granules of distinct color are embedded inside the wall of the specimen of granular materials, and a series of X-ray photographs is taken, in both drained and undrained experiments. In drained tests, shear localizations, produced by monotonic deformation of the specimen, were directly observed. In undrained tests, the local deformation of the granular mass during liquefaction, induced by cyclic shearing, was captured, and *it was directly observed that shear localization does actually occur in liquefied specimens.*

The specimen containing shearbands is frozen *in situ* and, later, the ice in the frozen samples cut from the large specimen, is replaced by a polymer resin. The radiograph is removed and processed, recording on film the movements of the columns of lead silicate granules. The portion of the specimen containing a shearband and a sheared section of a column of lead silicate granules, is cut by a diamond saw along the column of lead granules, and the cut surface is polished in order to directly observe the arrangement of the particles within the shearband. The lead silicate granules trace the localized zone, and provide details of the deformation pattern within and close to the localized zone. The shear strain at the center of the shearband zone exceeds 500%, although the overall nominal shear strain is only 10%. The localization is accompanied by extensive necking. In the rest of this section, the special experimental techniques used in this class of novel experiments, are summarized and some experimental results are presented.

---

[18] See, for example, Bergfelt (1956), Roscoe *et al.* (1963), Arthur *et al.* (1964), Kirkpatric and Belshaw (1968), Krinitzsky (1970), Roscoe (1970), Balasubramaniam (1976), Vardoulakis *et al.* (1978), Scarpelli and Wood (1982), Vardoulakis (1980), Vardoulakis and Graf (1982), Molenkamp (1985), Mühlhaus and Vardoulakis (1987), Kolymbas and Rombach (1989), and Tatsuoka *et al.* (1990).

9.6.1. Experimental Setup

The geometry of the hollow cylindrical specimen provides enough inside space to house the X-ray film and related equipment, without any interference with the specimen deformation. Since the specimen is only 2.5cm thick, clear radiographs are obtained. A plexiglass tube is placed within the hollow specimen. It carries the X-ray film and provides a narrow window for radiography. The window is formed by two lead foils attached to the plexiglass ahead of the film, and two lead plates placed in front of the plexiglass tube, as shown in Figure 9.6.1 which is a horizontal cross section of the test cell. It shows from inside out, the supporting rod, the plexiglass tube with attached film, two 1mm lead foils, the hollow cylindrical specimen containing lead silicate granules, two 5mm lead plates, and a plastic box containing air. This construction allows for obtaining a clear image on the X-ray film, as it minimizes the film-to-object distance by placing the X-ray film inside the hollow cylindrical specimen, with the X-ray source only 45cm away from the film. The two 1mm lead foils in front of the film expose only a limited area of the film; see Figure 9.6.2. The spacing of the foils can be adjusted to suit the experimental requirements, since a wider space is needed for larger deformations. The two thick lead plates with adjustable spacing, in front of the specimen, allow X-ray penetration only through the window provided by the two lead foils. A lead foil is also placed behind the film in order to absorb the X-rays passing through the film and the plexiglass tube. The chamber[19] in Figure 9.6.1 is filled with water during the experiment. The hollow plastic box between the specimen and the chamber wall improves the sharpness of the X-ray photos, without interfering with the water pressure on the specimen.

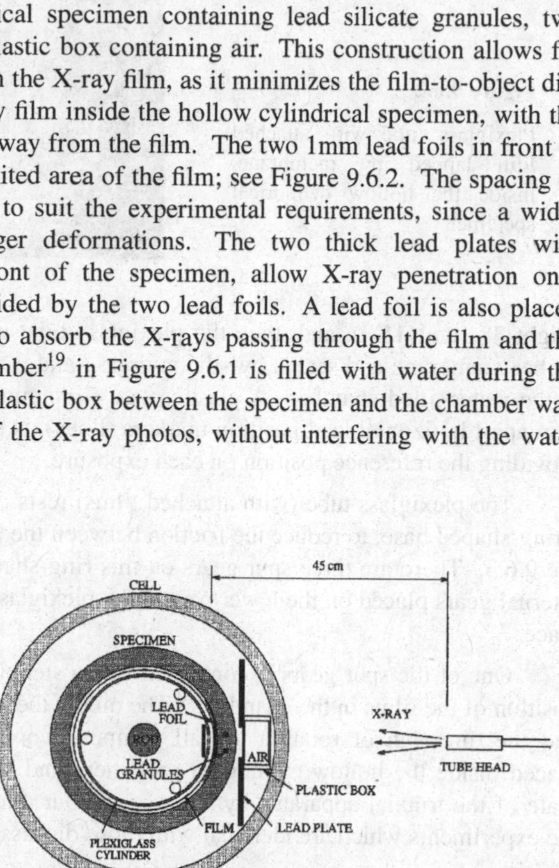

Figure 9.6.1

Horizontal cross section of the test cell containing the specimen and other indicated attachments

The films (the negatives attached to fluorescent screens) are attached on the outer surface of the plexiglass tube (outer diameter 18.4cm (7.25") and

[19] See also Figures 9.5.1 and 9.5.5, pages 693 and 697.

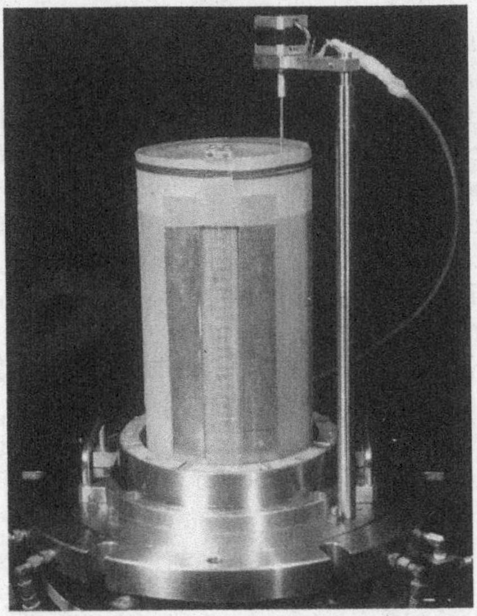

Figure 9.6.2
Plexiglass tube with attached film lapped by membrane, inside the hollow cylindrical specimen

height 38.1cm (15")), and, then the plexiglass tube with film is lapped with a rubber membrane and sealed by silicon grease and o-rings to keep the films dry in the water-filled chamber. A plastic plate containing one vertical and several 1cm apart horizontal lead position markers (lines), is placed in front of the film, providing the reference position on each exposure.

The plexiglass tube (with attached films) rests on a teflon sheet placed on a ring-shaped base, to reduce the friction between the tube and the base; see Figure 9.6.3. There are three spur gears on this ring-shaped base. They match the internal gears placed on the lower part of the plexiglass tube, holding the tube in place.

One of the spur gears is connected to the stepping motor that controls the position of the films in the chamber.[20] The mode, the number of steps, the speed, and the direction of rotation are all computer-controlled. The entire unit is placed inside the hollow cylindrical specimen, and it is attached to the bottom plate of the triaxial apparatus by screws, without affecting the other aspects of the experiments which are identical with those discussed in Section 9.5.

The number of exposures depends on the film size and the window width provided by the lead foils. For a 4cm (1.6") window, 5 exposures are possible on an 8×10 film, and with 2 films attached to the plexiglass tube, a total of 10 exposures are obtained.[21] The initial positions of the lead silicate granules in the

[20] The motor used by Nemat-Nasser and Okada (1995, 1998, 2001) was controlled by a CY545 Stepper System Controller (made by Cybernetic Micro System, Inc.) and can be operated either at 200 steps (1.8 degree per step), or at 400 steps (0.9 degree per step) per revolution.

[21] The Flash X-ray System, the radiographic film used, and other details are described in Nemat-

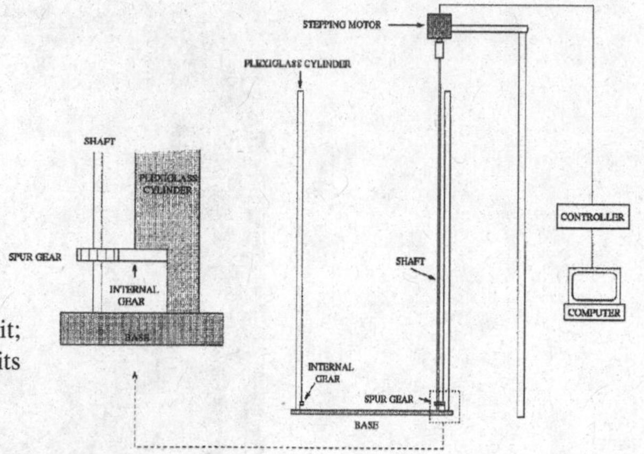

Figure 9.6.3

Film-location control unit; stepping motor and its control system

specimen define the reference configuration prior to loading. First, the plexiglass tube is rotated by the computer-controlled stepping motor to bring the unexposed film in line with the X-ray tubehead. Second, the specimen is deformed to a desired state, and the radiograph is taken. This procedure is then repeated.

9.6.2. Experimental Observation

Some experimental observations on Monterey No. 0 sand are reported here for illustration. A typical sample containing shearbands is shown in Figure 9.6.4. Shearbands appear at overall strains exceeding 5%. In general, several shearbands occur within the same sample. Their configuration varies with the initial void ratio.

For an initial void ratio of 0.673, a radiograph of six different deformation stages in a drained experiment, is given in Figure 9.6.5, and Figure 9.6.6 shows the polished surface of a cross section taken through the shearband. A deformed column of lead silicate granules with distinct color, is seen within the shearband zone. Extensive stretching along the center line of the shearband zone is accompanied by lateral contraction and extensive necking of the specimen over the shearband zone, as shown in Figure 9.6.7 which is a typical profile of the surface indentation of the sample over this zone. The greatest thickness reduction at the center of the shearband exceeds 2mm. The width of the necked zone is 15-20mm. Even though the final void ratio within the strain-localized region may not be the same as its void ratio prior to localization, the differences are small, compared to the large strains that occur within the shearband zone, and can be neglected in preliminary estimates.

Nasser and Okada (2001).

Figure 9.6.4

A sample containing shear-bands

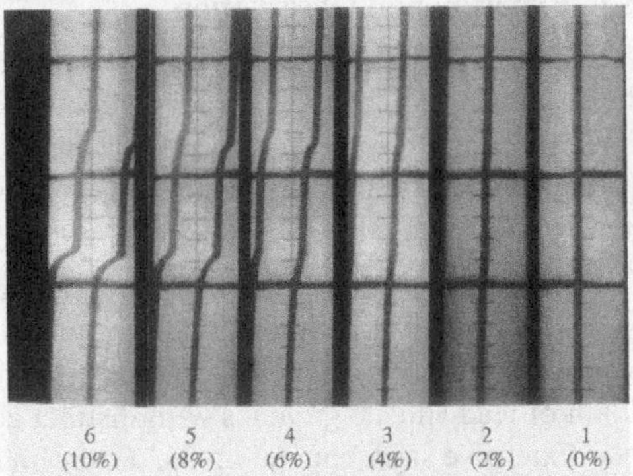

Figure 9.6.5

Radiograph of six different deformation stages in a drained experiment

From the measured dimensional variation within the shearband, and using some reasonable assumptions, it is possible to estimate the radial and shear strains over the shearband. These are given in Figures 9.6.8a and b, respectively. With ε standing for the overall strain (about 10%), and λ_{rr} for the radial stretch, the experimental data suggest the following form for the deformation gradient, **F**, over the shearband zone:

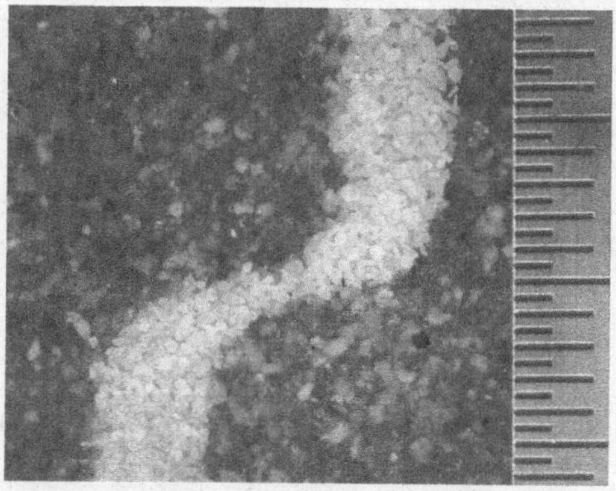

Figure 9.6.6

Polished surface of a section through shearband, showing deformed configuration of lead granules

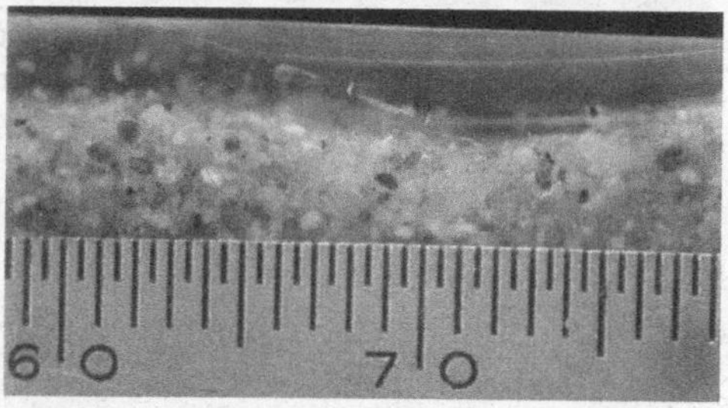

Figure 9.6.7

Surface indentation over the necked region

$$\mathbf{F} \approx \begin{bmatrix} 1+O(\varepsilon) & \tan\gamma & O(\varepsilon) \\ O(\varepsilon) & 1/\lambda_{rr}+O(\varepsilon) & O(\varepsilon) \\ O(\varepsilon) & O(\varepsilon) & \lambda_{rr}+O(\varepsilon) \end{bmatrix}, \tag{9.6.1}$$

where $O(\varepsilon)$ stands for a quantity of the order of ε. Then, the distribution of the stretches in the shearband zone is estimated using the deformation gradient tensor and the distribution of the shear strain given in Figure 9.6.8. The results are presented in Figure 9.6.9a. The maximum stretch is estimated to exceed 5 at the center of the shearband. The minimum stretch is estimated graphically.[22]

The change of the width (normal to the center line) of the deformed lead silicate granular column in the shearband zone is measured using a 50-times-magnified image. Straight lines are drawn every 0.4mm parallel to the shearband direction, and, then, the width, W, of the lead silicate material is measured. The results are presented in Figure 9.6.9b. If the particles move parallel to the shearband direction, the width of the lead silicate granules must be nearly the same along the shearband. However, the width is minimum at the center of the

[22] The distance between two corresponding points, one at the right and the other at the left of the edges of the lead silicate granules, is measured directly and divided by the original length. This process is continued until, finally, the minimum stretch is found.

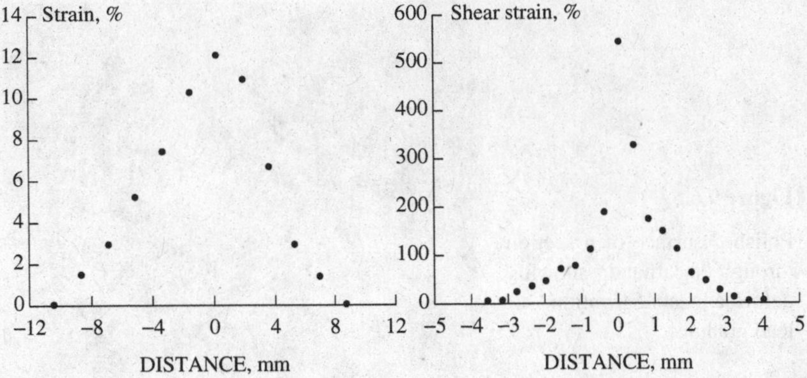

Figure 9.6.8

(a) Radial strain due to shearband necking; and (b) shear strain distribution in shearband zone; DISTANCE = 0 is at the center of shearband

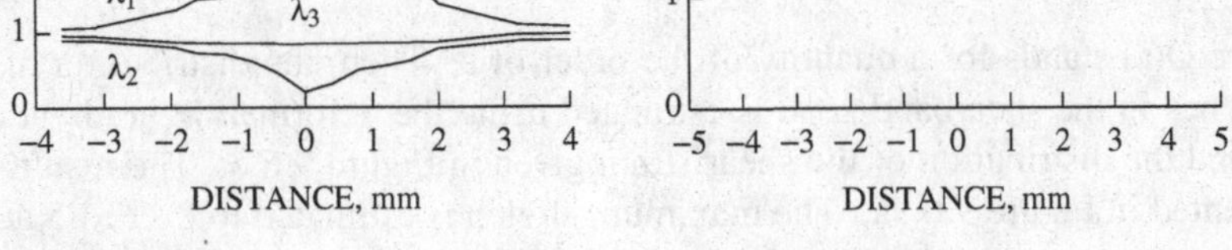

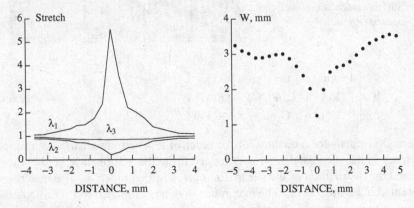

Figure 9.6.9

(a) Distribution of estimated stretches within shearband zone; and (b) deformed width, W, of lead silicate granules within shearband zone

shearband, indicating that extensional deformations take place along the center line of the shearband zone. The total number of granules within an initially identified material volume in the shearband zone remains the same. The reduction in the width of the lead silicate granules together with the observed necking in the radial direction, suggest extensive extensional deformations along the shearband center line. This is accompanied by contractional deformations in planes normal to the center line of the shearband.

9.7. REFERENCES

Arthur, J. R. F. and Menzies, B. K. (1972), Inherent anisotropy in a sand, *Géotechnique,* Vol. 22, No. 1, 115-128.

Arthur, J. R. F., James, R. G., and Roscoe, K. H. (1964), The determination of stress fields during plane strain of a sand mass, *Géotechnique,* Vol. 14, 283-308.

Baker, W. and Yew, C. H. (1966), Strain rate effects in the propagation of torsional plastic waves, *J. Appl. Mech.*, Vol. 33, 917-923.

Balasubramaniam, A. S. (1976), Local strains and displacement patterns in triaxial specimens of a saturated clay, *Soils and Foundations*, Vol. 16, No. 1, 101-114.

Bergfelt, A. (1956), Loading tests on clay, *Géotechnique,* Vol.6, 15-31.

Berrill, J. B. and Davis, R. O. (1985), Energy dissipation and seismic liquefaction of sands: revised model, *Soils and Foundations*, Vol. 25, 106-118.

Berrill, J. B. and Christensen, S. A. (1994), Some liquefaction results from New Zealand, *Proc. Workshop on Bridge Design*, held in Queenstown, New Zealand, 164-182.

Bever, M. B., Holt, D. L., and Titchener, A. L. (1973), The stored energy of cold work, *Progress in Materials Science,* B. Chalmers, (eds. J. W. Christian, T. B. Massalski), Vol. 17, 5-88.

Christensen, R. J., Swanson, S. R., and Brown, W. S. (1972), Split Hopkinson bar test on rock under confining pressure, *Exp. Mech.*, Vol. 12, 508-541.

Davies, E. D. H. and Hunter, S. C. (1963), The dynamic compression testing of solids by the method of the split Hopkinson pressure bar, *J. Mech. Phys. Solids,* Vol. 11, 155-179.

Davies, R. M. (1948), A critical study of the Hopkinson pressure bar, *Phil. Trans. A.*, Vol. 240, 375-457.

Davis, R. O. and Berrill, J. B. (1982), Energy dissipation and seismic liquefaction, *Earthquake Engrg. Struct. Dynamics-ASCE*, Vol. 10, 59-68.

Figueroa, J. L., Saada, A. S., Liang, L., and Dahisaria, N. M. (1994), Evaluation of soil liquefaction by energy principles, *J. Geotech. Engrg. Div. ASCE*, Vol. 120, 1554-1569.

Follansbee, P. S. (1985), *Metals Handbook, American Society of Metals,* Vol. 8, 198-203.

Gray, G. T. (2000), Classic split-Hopkinson pressure bar testing, *Mechanical Testing and Evaluation Handbook, Amer. Soc. Metals*, Vol. 8, 462-476.

Harding, J., Wood, E. O., and Campbell, J. D. (1960), Tensile testing of materials at impact rates of strain, *J. Mech. Eng. Sci.*, Vol. 2, No. 2, 88-96.

Hartley, K. A., Duffy, J., and Hawley, R. H. (1985), The torsional Kolsky (split-Hopkinson) bar, *Mechanical Testing, ASM Handbook*, Vol. 8, 218-228.

Hartman, K. H., Kunze, H. D., and Meyer, L. W. (1986), Metallurgical Effects in Impact Loaded Materials, in *Shock Waves and High-Strain-Rate Phenomena in Metals*, (eds. M. A. Meyers & L. E. Murr), Plenum, NY, Vol. 21, 325.

Hopkinson, B. (1905), The effects of momentary stresses in metals, *Proc. R. Soc. A.*, Vol. 74, 498-506.

Hopkinson, B. (1914), A method of measuring the pressure produced in the detonation of high explosives or by the impact of bullets, *Phil. Trans. A.*, Vol.213, 437-456.

Hopkinson, J. (1901), in *Original papers by J. Hopkinson*, Vol. II, (ed. B. Hopkinson), Cambridge University Press, 316-324.

Ishihara, K. and Yasuda, S. (1975), Sand liquefaction in hollow cylinder torsion under irregular excitation, *Soils and Foundations*, Vol. 15, No. 1, 45-59.

Kapoor, R. and Nemat-Nasser, S (1998), Determination of temperature rise during high strain rate deformation, *Mech. Mat.*, Vol 27, 1-12.

Kirkpatrick, W. M. and Belshaw, D. J. (1968), On the interpretation of the triaxial test, *Géotechnique*, Vol. 18, 336-350.

Kolsky, H. (1949), An investigation of the mechanical properties of materials at very high rates of loading, *Proc. R. Soc. B.*, Vol. 62, 676-700.

Kolymbas, D. and Rombach, G. (1989), Shear band formation in generalized hypoelasticity, *Ingenjeur-Archiv* Vol. 59, 177-186.

Krinitzsky, E. L. (1970), *Radiography in the Earth Sciences and Soil Mechanics,* Plenum Press, New York.

Lindholm, U. S. (1964), Some experiments with split Hopkinson pressure bar, *J. Mech. Phys. Solids*, Vol. 12, 317-335.

Lindholm, U. S. and Yeakley, L. M. (1968), High strain-rate testing: tension and compression, *Expr. Mech.*, Vol. 8, 1-9.

Malvern, L, E., Jenkins, D. A., Tang, T., and McClure, S. (1991), Dynamic testing of laterally confined concrete, *Micromechanics of failure of quasi brittle materials*, Elsevier Applied Science, 343-352.

Marchand, J. J. and Duffy, J. (1988), An experimental study of the formation process of adiabatic shear bands in a structural steel, *J. Mech. Phys. Solids*, Vol. 36, No. 3, 251-283.

Mason, J. J., Rosakis, A. J., and Ravichandran, G. (1994), On the strain and strain rate dependence of the fraction of plastic work converted to heat: An experimental study using high speed infrared detectors and Kolsky bar, *Mech. Mat.*, Vol. 17, 135-145.

Miura, S. and Toki, S. (1982), A sample preparation method and its effect on static and cyclic deformation-strength properties of sand, *Soils and Foundations,* Vol. 22, No. 1, 61-77.

Molenkamp, F. (1985), Comparison of frictional material models with respect to shear band initiation, *Géotechnique*, Vol. 35, No. 2, 127-143.

Moore, J. T. and Kuhlmann-Wilsdorf, D. (1970), The rate of energy storage in workhardened metals, In *2nd International Conference on the strength of metals and alloys*, Vol. 2, 484-488.

Moss, G. L. and Pond, R. B. (1975), Inhomogeneous thermal changes in copper during plastic elongation, *Metall. Trans.*, Vol. 6A, 1223-1235.

Mühlhaus, H.-B. and Vardoulakis, I. (1987), The thickness of shear bands in granular materials, *Géotechnique,* Vol. 37, No. 3, 271-283.

Nemat-Nasser, S. and Isaacs, J. B. (1997), Direct measurement of isothermal flow stress of metals at elevated temperatures and high strain rates with application to Ta and Ta-W alloys, *Acta Metall.*, Vol 45, 907-919.

Nemat-Nasser, S. and Okada, N. (1995), Direct observation of deformation of granualar materials through x-ray photography, *Proceedings of 10th ASCE Engineering Mechanics Speciality Conference*, Vol. 1, 605-608.

Nemat-Nasser, S. and Okada, N. (1998), Strain localization in particulate media, *Proc. of ASCE 12th Engineering Mechanics: A Force for the 21st Century*, La Jolla, May 17-20, ASCE CD-ROM, 1009-1012.

Nemat-Nasser, S. and Okada, N. (2001), Radiographic and microscopic observation of shear bands in granular materials, *Géotechnique*, Vol. 51, No. 9, 753-765.

Nemat-Nasser, S. and Shokooh, A. (1979), A unified approach to densification and liquefaction of cohesionless sand in cyclic shearing, *Can. Geotech. J.*, Vol. 16, 659-678.

Nemat-Nasser, S. and Tobita, Y. (1982), Influence of fabric on liquefaction and densification potential of cohesionless sand, *Mech. Mat.*, Vol. 1, No. 1, 43-62.

Nemat-Nasser, S., Isaacs, J., and Rome, J. (2000), Triaxial Hopkinson techniques, *Mechanical Testing and Evaluation Handbook, Amer. Soc. Metals*, Vol. 8, 516-518.

Nemat-Nasser, S., Isaacs, J. B., and Starrett, J. E. (1991), Hopkinson techniques for dynamic recovery experiments, *Proc. R. Soc.,* Vol. 435A, 371-391.

Nemat-Nasser, S., Li, Y. F, and Isaacs, J. B. (1994) Experimental/computational evaluation of flow stress at high strain rates with application to adiabatic shearbanding, *Mech. Mat.*, Vol. 17, 111-134.

Nemat-Nasser, S., Isaacs, J. B., Ravichandran, G., and Starrett, J. E. (1988b), High strain rate testing in the U.S., Proc. of TTCP TTP-1 Workshop on New Techniques of Small Scale High Strain Rate Studies, Australia.

Nemat-Nasser, S., Hori, M., Starrett, J. E., Altman, B., Chang, S., and Isaacs, J. (1988a), *Impact Loading and Dynamic Behaviour of Materials* (ed. C. Y. Chiem, H.-D. Kunze, L. W. Meyer), Vol. 1, 343. Verlag: DGM Informationsgesellschaft.

Nicholas, T. and Bless, S. J. (1985), High strain rate tension testing, *Mechanical Testing Handbook, Amer. Soc. Metals,* Vol. 8, 208-214.

Oda, M. (1972), The mechanism of fabric changes during compressional deformation of sand, *Soils and Foundations*, Vol. 12, No. 2, 1-18.

Okada, N. (1992), Energy Dissipation in Inelastic Flow of Cohensionless Granular Media, Ph.D. Thesis, University of California, San Diego.

Okada, N. and Nemat-Nasser, S. (1994), Energy dissipation in inelastic flow of saturated cohensionless granular media, *Géotechnique*, Vol. 44 , No. 1, 1-19.

Ravichandran, G. and Nemat-Nasser, S. (1989), Micromechanics of dynamic fracturing of ceramic composites, *Proc. ICF-7, Advances in Fracture Research*, Vol.1, No.4., 41-48.

Roscoe, K. H. (1970), The influence of strains in soil mechanics, 10th Rankine Lecture, *Géotechnique*, Vol. 20, No. 2, 129-179.

Roscoe, K. H., Arthur, J. R. F., and James, R. G. (1963), The determination of strains in soils by an X-ray method, *Civil Engineering and Public Works Review*, Vol. 58, 873-876, 1009-1012.

Scarpelli, G. and Wood, D. M. (1982), Experimental observations of shear band patterns in direct shear tests, *IUTAM Conference on Deformation and Failure of Granular Materials*, Delft, 473-484.

Seed, H. B. (1979), Soil liquefaction and cyclic mobility evaluation for level ground during earthquakes, *J. Geotech. Engr. Div, ASCE*, Vol. 105, No. GT2, 201-255.

Stroh, A. N. (1953), A theoretical calculation of the stored energy in a work-hardened material, *Proc. R. Soc.*, Vol. 218A, 391-400.

Tatsuoka, F., Muramatsu, M., and Sasaki, T. (1982), Cyclic undrained stress-strain behavior of dense sands and torsional simple shear test, *Soils and Foundations*, Vol. 22, No. 2, 55-70.

Tatsuoka, F, Nakamura, S., Huang, C.-C. and Tani, K. (1990), Strength anisotropy and shear band direction in plane strain tests of sand, *Soils and Foundations*, Vol. 30, No. 1, 35-54.

Thakur, A. (1994), Deformation behavior and Bauschinger effect in super alloy, Ph.D. Thesis, University of California, San Diego.

Thakur, A., Nemat-Nasser, S., and Vecchio, K. S. (1996a), Bauschinger effect in Haynes 230 alloy: influences of strain rate and temperature, *Metall. & Mat. Trans.*, Vol 27A, 1739-1748.

Thakur, A., Nemat-Nasser, S. and Vecchio, K. S. (1996b), Dynamic Bauschinger effect, *Acta Mat.*, Vol 44, 2797-2807.

Vardoulakis, I. (1980), Shear band inclination and shear modulus on sand in biaxial tests, *Int. J. Num. Anal. Meth. Geomech.*, Vol. 4, 103-119.

Vardoulakis, I. and Graf, B. (1982), Imperfection sensitivity of the biaxial test on dry sand, *IUTAM Conference on Deformation and Failure of Granular Materials*, Delft, 485-491.

Vardoulakis, I., Goldscheider, M., and Gudehus, G. (1978), Formation of shear bands in sand bodies as a bifurcation problem, *Int. J. Numer. Anal. Methods Geomech.*, Vol. 2, 99-128.

Zehnder, A. T. (1991), A model for the heating due to plastic work, *Mech. Res. Commun*, Vol. 18, 23-28.

CITED AUTHORS[1]

[1] Numbers refer to the page where a complete reference to the citation is given.

L

M

N

O

SUBJECT INDEX

D

E